TABLE II (cont.)
Areas under the standard normal curve

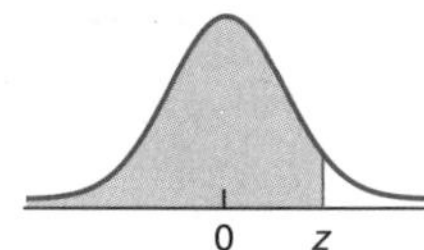

z	0.00	0.01	0.02	0.03	0.04	0.05	0.06	0.07	0.08	0.09
	Second decimal place in z									
0.0	0.5000	0.5040	0.5080	0.5120	0.5160	0.5199	0.5239	0.5279	0.5319	0.5359
0.1	0.5398	0.5438	0.5478	0.5517	0.5557	0.5596	0.5636	0.5675	0.5714	0.5753
0.2	0.5793	0.5832	0.5871	0.5910	0.5948	0.5987	0.6026	0.6064	0.6103	0.6141
0.3	0.6179	0.6217	0.6255	0.6293	0.6331	0.6368	0.6406	0.6443	0.6480	0.6517
0.4	0.6554	0.6591	0.6628	0.6664	0.6700	0.6736	0.6772	0.6808	0.6844	0.6879
0.5	0.6915	0.6950	0.6985	0.7019	0.7054	0.7088	0.7123	0.7157	0.7190	0.7224
0.6	0.7257	0.7291	0.7324	0.7357	0.7389	0.7422	0.7454	0.7486	0.7517	0.7549
0.7	0.7580	0.7611	0.7642	0.7673	0.7704	0.7734	0.7764	0.7794	0.7823	0.7852
0.8	0.7881	0.7910	0.7939	0.7967	0.7995	0.8023	0.8051	0.8078	0.8106	0.8133
0.9	0.8159	0.8186	0.8212	0.8238	0.8264	0.8289	0.8315	0.8340	0.8365	0.8389
1.0	0.8413	0.8438	0.8461	0.8485	0.8508	0.8531	0.8554	0.8577	0.8599	0.8621
1.1	0.8643	0.8665	0.8686	0.8708	0.8729	0.8749	0.8770	0.8790	0.8810	0.8830
1.2	0.8849	0.8869	0.8888	0.8907	0.8925	0.8944	0.8962	0.8980	0.8997	0.9015
1.3	0.9032	0.9049	0.9066	0.9082	0.9099	0.9115	0.9131	0.9147	0.9162	0.9177
1.4	0.9192	0.9207	0.9222	0.9236	0.9251	0.9265	0.9279	0.9292	0.9306	0.9319
1.5	0.9332	0.9345	0.9357	0.9370	0.9382	0.9394	0.9406	0.9418	0.9429	0.9441
1.6	0.9452	0.9463	0.9474	0.9484	0.9495	0.9505	0.9515	0.9525	0.9535	0.9545
1.7	0.9554	0.9564	0.9573	0.9582	0.9591	0.9599	0.9608	0.9616	0.9625	0.9633
1.8	0.9641	0.9649	0.9656	0.9664	0.9671	0.9678	0.9686	0.9693	0.9699	0.9706
1.9	0.9713	0.9719	0.9726	0.9732	0.9738	0.9744	0.9750	0.9756	0.9761	0.9767
2.0	0.9772	0.9778	0.9783	0.9788	0.9793	0.9798	0.9803	0.9808	0.9812	0.9817
2.1	0.9821	0.9826	0.9830	0.9834	0.9838	0.9842	0.9846	0.9850	0.9854	0.9857
2.2	0.9861	0.9864	0.9868	0.9871	0.9875	0.9878	0.9881	0.9884	0.9887	0.9890
2.3	0.9893	0.9896	0.9898	0.9901	0.9904	0.9906	0.9909	0.9911	0.9913	0.9916
2.4	0.9918	0.9920	0.9922	0.9925	0.9927	0.9929	0.9931	0.9932	0.9934	0.9936
2.5	0.9938	0.9940	0.9941	0.9943	0.9945	0.9946	0.9948	0.9949	0.9951	0.9952
2.6	0.9953	0.9955	0.9956	0.9957	0.9959	0.9960	0.9961	0.9962	0.9963	0.9964
2.7	0.9965	0.9966	0.9967	0.9968	0.9969	0.9970	0.9971	0.9972	0.9973	0.9974
2.8	0.9974	0.9975	0.9976	0.9977	0.9977	0.9978	0.9979	0.9979	0.9980	0.9981
2.9	0.9981	0.9982	0.9982	0.9983	0.9984	0.9984	0.9985	0.9985	0.9986	0.9986
3.0	0.9987	0.9987	0.9987	0.9988	0.9988	0.9989	0.9989	0.9989	0.9990	0.9990
3.1	0.9990	0.9991	0.9991	0.9991	0.9992	0.9992	0.9992	0.9992	0.9993	0.9993
3.2	0.9993	0.9993	0.9994	0.9994	0.9994	0.9994	0.9994	0.9995	0.9995	0.9995
3.3	0.9995	0.9995	0.9995	0.9996	0.9996	0.9996	0.9996	0.9996	0.9996	0.9997
3.4	0.9997	0.9997	0.9997	0.9997	0.9997	0.9997	0.9997	0.9997	0.9997	0.9998
3.5	0.9998	0.9998	0.9998	0.9998	0.9998	0.9998	0.9998	0.9998	0.9998	0.9998
3.6	0.9998	0.9998	0.9999	0.9999	0.9999	0.9999	0.9999	0.9999	0.9999	0.9999
3.7	0.9999	0.9999	0.9999	0.9999	0.9999	0.9999	0.9999	0.9999	0.9999	0.9999
3.8	0.9999	0.9999	0.9999	0.9999	0.9999	0.9999	0.9999	0.9999	0.9999	0.9999
3.9	1.0000†									

† For $z \geq 3.90$, the areas are 1.0000 to four decimal places.

Minitab Quick Reference

Following is a quick reference to most of the Minitab procedures and menu instructions discussed in the book, alphabetically by procedure. See the page number references for more detail. Also see Section 1.4 (pages 11–21) for Minitab basics and elements of input/output.

Procedure	Menu instructions
Bar graph	Graph ➤ Chart... *90*
Binomial probabilities	Calc ➤ Probability Distributions ➤ Binomial... *323, 324, 325*
Boxplot	Graph ➤ Boxplot... *168*
Contingency tables	Stat ➤ Tables ➤ Cross Tabulation... *789*
Correlation test for normality	Stat ➤ Basic Statistics ➤ Normality Test... *922*
Dotplot	Graph ➤ Dotplot... *88*
Estimation and prediction	Stat ➤ Regression ➤ Regression... *907*
Exit Minitab	File ➤ Exit *21*
Five-number summary	Stat ➤ Basic Statistics ➤ Display Descriptive Statistics... *169*
Histogram	Graph ➤ Histogram... *87*
Independence test	Stat ➤ Tables ➤ Chi-Square Test... *803*
Kruskal–Wallis test	Stat ➤ Nonparametrics ➤ Kruskal-Wallis... *977*
Linear correlation coefficient	Stat ➤ Basic Statistics ➤ Correlation... *860*
Linear regression	Stat ➤ Regression ➤ Regression... *838*
Mann–Whitney test	Stat ➤ Nonparametrics ➤ Mann-Whitney... *634*
Mean	Calc ➤ Column Statistics... *132*
Median	Calc ➤ Column Statistics... *133*
Mode	Stat ➤ Tables ➤ Tally... *133*
Nonpooled *t*-procedures	Stat ➤ Basic Statistics ➤ 2-Sample t... *620*
Normal probabilities	Calc ➤ Probability Distributions ➤ Normal... *381*
Normal probability plot	Graph ➤ Plot... *389*
Normal scores	Calc ➤ Calculator... *389*
One-way ANOVA	Stat ➤ ANOVA ➤ Oneway (Unstacked)... *952*
Paired *t*-procedures	Stat ➤ Basic Statistics ➤ Paired t... *650*
Paired Wilcoxon signed-rank test	Stat ➤ Nonparametrics ➤ 1-Sample Wilcoxon... *661*
Pie Chart	Graph ➤ Pie Chart... *89*
Poisson probabilities	Calc ➤ Probability Distributions ➤ Poisson... *339, 340*
Pooled *t*-procedures	Stat ➤ Basic Statistics ➤ 2-Sample t... *608*
Proportion inferences (one)	Stat ➤ Basic Statistics ➤ 1 Proportion... *733, 743*
Proportion inferences (two)	Stat ➤ Basic Statistics ➤ 2 Proportions... *755*
Range	Calc ➤ Column Statistics... *153*
Residual analysis	Stat ➤ Regression ➤ Regression... *882*
Sample standard deviation	Calc ➤ Column Statistics... *153*
Scatter diagram	Graph ➤ Plot... *837*
Simple random sample	Calc ➤ Random Data ➤ Sample From Columns... *30*
Stem-and-leaf diagram	Graph ➤ Stem-and-Leaf... *98*
Store patterned data	Calc ➤ Make Patterned Data ➤ Simple Set of Numbers... *30*
Tally	Stat ➤ Tables ➤ Tally... *76*
***t*-interval procedure**	Stat ➤ Basic Statistics ➤ 1-Sample t... *476*
***t*-test**	Stat ➤ Basic Statistics ➤ 1-Sample t... *556*
Tukey multiple comparison	Stat ➤ ANOVA ➤ Oneway... *964*
Wilcoxon signed-rank test	Stat ➤ Nonparametrics ➤ 1-Sample Wilcoxon... *569*
***z*-interval procedure**	Stat ➤ Basic Statistics ➤ 1-Sample Z... *456*
***z*-test**	Stat ➤ Basic Statistics ➤ 1-Sample Z... *544*

Introductory Statistics

FIFTH EDITION

Introductory Statistics

FIFTH EDITION

Neil A. Weiss
Arizona State University

Biographies by Carol A. Weiss

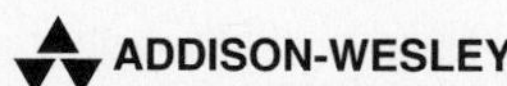

An imprint of Addison Wesley Longman, Inc.

Reading, Massachusetts • Menlo Park, California • New York • Harlow, England
Don Mills, Ontario • Sydney • Mexico City • Madrid • Amsterdam

Publisher: Greg Tobin
Sponsoring Editor: Jennifer Albanese
Editorial Assistant: Laura Potter
Project Manager: Jennifer Bagdigian
Production Supervisor: Karen Guardino
Text Designer: Cynthia Crampton
Design Supervisor: Barbara T. Atkinson
Cover Designer: Karen Rappaport
Marketing Manager: Brenda Bravener
Marketing Coordinator: Melissa Perrelli
Prepress Buyer: Caroline Fell
Technical Art Consultant: Joseph Vetere
Print Buyer: Evelyn Beaton
Illustrations: Scientific Illustrators, Inc.
Compositors: Carol and Neil Weiss

Photo Credits: Florence Nightingale, page 56, courtesy of The Bettmann Archive. Adolphe Quetelet, page 123, Copyright Bibliothèque royale Albert Ier Cabinet des Estampes, Bruxelles, J. Odevaère. John Tukey, page 193, courtesy of John W. Tukey. Andrei Kolmogorov, page 286, courtesy of The Bettmann Archive. James Bernoulli, page 350, courtesy of The Bettmann Archive. Carl Friedrich Gauss, page 405, courtesy of The Bettmann Archive. Pierre-Simon Laplace, page 440, courtesy of The Bettmann Archive. William Gosset, page 486, courtesy of The Granger Collection, New York. Jerzy Neyman, page 589, courtesy of the Department of Statistics, University of California, Berkeley. Gertrude Cox, page 680, courtesy of Research Triangle Institute. W. Edwards Deming, page 721, Steve Barth/Corbis-Bettmann. Abraham de Moivre, page 766, courtesy of The Granger Collection, New York. Karl Pearson, page 815, courtesy of Brown Brothers. Adrien Legendre, page 868, courtesy of The Bettmann Archive. Sir Francis Galton, page 931, courtesy of Stock Montage. Sir Ronald Fisher, page 988, courtesy of The Bettmann Archive.

Library of Congress Cataloging-in-Publication Data

Weiss, N. A. (Neil A.)
Introductory statistics. — 5th ed. / Neil Weiss; biographies by Carol Weiss.
p. cm.
Includes index.
ISBN 0-201-88330-9 (book)
ISBN 0-201-59877-9 (book/disk pkg.)
1. Statistics. I. Title.
QA276.12.W45 1999
519.5—dc21 98-20513
CIP

Reprinted with corrections, March 1999

Many of the designations used by manufacturers and sellers to distinguish their products are claimed as trademarks. Where those designations appear in this book, and Addison-Wesley was aware of a trademark claim, the designations have been printed in initial caps or all caps.

2 3 4 5 6 7 8 9 10 DOC 010099

Preface

Using and understanding statistics or statistical procedures have become required skills in virtually every profession and every academic discipline. The purpose of this book is to help students grasp basic statistical concepts and techniques, and to present real-life opportunities for applying them.

The text is intended for a one- or two-semester course and for quarter-system courses as well. Instructors can easily fit the text to the pace and depth they prefer. Introductory high-school algebra is a sufficient prerequisite.

New advances in technology and new insights into the practice of teaching statistics have inspired many of the changes in the Fifth Edition of *Introductory Statistics,* leading to more emphasis on conceptual understanding and less emphasis on computation. We have also worked hard to refine and improve the exercise sets, the clarity of the examples and explanations, and the real-world applications.

Highlights of the Approach

Several aspects of the approach used in the Fifth Edition give the text distinction and prominence among basic statistics books. Here are a few of those aspects:

ASA/MAA-guidelines compliant. We have followed ASA/MAA guidelines to stress the interpretation of statistical results, the contemporary applications of statistics, and the importance of critical thinking.

Data analysis and exploration. We agree wholeheartedly with the trend of including more exploratory and confirmatory data analysis in statistics courses and have incorporated an extensive amount into the text and exercises. Recognizing that not all readers will have access to computers, we have provided ample opportunity to analyze and explore data without the use of a computer.

Detailed and careful explanations. We have included every step of explanation we think a typical reader might need. Our guiding principle is to avoid cognitive jumps, making the learning process smooth and enjoyable. We believe detailed and careful explanations result in better understanding.

Emphasis on application. We have concentrated on the application of statistical techniques to the analysis of data. Although statistical theory has been kept to a minimum, we have provided a thorough explanation of the rationale for using each statistical procedure.

Real-world examples. Because we believe that the majority of students learn by example, every concept discussed in the book is illustrated by at least one detailed example. The examples are, for the most part, based on real-life situations and have been chosen for their interest as well as for their illustrative value.

Real-world exercises. Most exercises in the book are based on information found in newspapers, magazines, statistical abstracts, and journal articles; sources are explicitly cited. The exercises are designed not only to help the reader learn the material but also to show that statistics is a lively and relevant discipline. Answers to selected exercises are included in Appendix B.

Technology. We have chosen Release 12 of Minitab® to illustrate the use of statistical software, but the text has been written so that the instructor is free to select other packages, such as SAS® or SPSS®. Additionally, Excel® and TI-83® manuals have been prepared to accompany the book. The examples and exercises have been carefully correlated to these technologies by using icons to represent

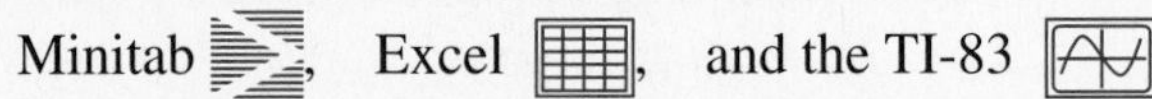

All computer and calculator material is *optional,* but recommended.

Minitab sections are integrated as optional subsections occurring immediately following the particular statistical concept under consideration. In each subsection we explain how Minitab can be used to solve problems that were solved by hand earlier in the section. Each solution consists of introducing the required menu instructions, displaying the computer output, and interpreting the results.

Most exercise sets contain a group of exercises using technology. Three types of technology exercises have been included.

- *Interpretation.* These exercises ask the reader to interpret computer printouts; no knowledge of or access to statistical software is necessary for these exercises.
- *Data Analysis and Inference.* These exercises ask the reader to use Minitab or some other statistical software or calculator to solve exercises that were presented previously for hand solution. All basic Minitab instructions required for these technology exercises will have been covered in the text, with more detailed instructions contained in the *Minitab Manual.* Excel and TI-83 instructions are discussed in the *Excel Manual* and *TI-83 Manual,* respectively.
- *Simulation.* These exercises ask the reader to use statistical software or a calculator to perform a simulation. They are designed to provide concrete illustrations of some of the more complex concepts (e.g., sampling distributions) and to show the reader how a computer or statistical calculator can be used to reveal statistical facts.

New Content and Technology Support

Content changes have been made throughout the Fifth Edition and new ancillaries provide even greater technology support. Here are some highlights:

Option for brief probability coverage. The probability required for statistical inference, presented in Chapter 4, can now be covered in two or three class periods. Further probability is available at the instructor's option. (See pages 197 and 289 and note that sections marked with an asterisk are optional.)

Optional linear models modules. Three new optional chapter-length modules that contain additional material on linear models are available for customizing your course. Contact your Addison Wesley Longman representative for details.

- *Multiple Regression Analysis*
- *Model Building in Regression*
- *Design of Experiments and Analysis of Variance*

Minitab for Windows® 95/NT™. We feature the Professional version of Minitab for Windows 95/NT, the latest version of Minitab for PCs. Aside from worksheet size, only minor differences exist between the Professional version and *The Student Edition of MINITAB for Windows 95/NT.* The Fifth Edition of *Introductory Statistics* and *The Student Edition of MINITAB for Windows 95/NT* are available packaged together at a special discount to students. Contact your Addison Wesley Longman representative for details.

TI-83® and Excel® support. The integration of technology has been expanded in the Fifth Edition to include support for using the TI-83 graphing calculator and Excel spreadsheet. Exercises and examples in the book are keyed to the *TI-83 Manual* and *Excel Manual.* We also continue to offer a dedicated *Minitab Manual.* The three icons

 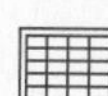 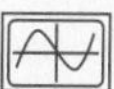

have been used throughout the book to denote technology integration supplemented by the manuals.

New Features

We've also made changes to improve how the book supports effective teaching and learning. Some of the most significant new features include:

All-new design. We have redesigned the text for improved readability.

Case studies. Each chapter begins with a classic or contemporary case study that highlights the real-world relevance of the material under consideration. At the end of the chapter, the case study is reviewed and discussed in light of the chapter's major points and then problems are presented for the students to solve.

New concept and technology exercises. The exercise sets have been extensively updated and revised, including hundreds of new concept exercises and many new technology exercises. Each section exercise set is now divided into the following three categories:

- *Statistical Concepts and Skills.* These exercises help the students master the skills and concepts explicitly discussed in the section. (See page 134.)
- *Extending the Concepts and Skills.* These exercises invite the students to extend their skills by examining material not necessarily covered in the text. (See page 135.)
- *Using Technology.* These exercises provide the students with an opportunity to interpret and apply the computing and statistical capabilities of Minitab, Excel, and the TI-83 calculator to conduct data analyses and simulations. Three types of technology exercises have been included in most exercise sets: interpretation of computer output, use of Minitab or some other statistical software to solve exercises that were presented previously for hand solution, and use of statistical software to perform a simulation. (See page 137.)

Comprehensive end-of-chapter exercises. We have expanded the review tests at the end of each chapter to include concept questions, basic-skill problems, and exercises using technology. These provide a comprehensive collection of problems for reviewing the chapter. (See pages 760–764.) Answers to the review tests are given in Appendix B.

Internet projects with dedicated Web site. Each chapter now presents an Internet project. These projects, which are keyed to the text,

- engage the students in active and collaborative learning through simulations, demonstrations, and other activities as a supplement to the book,
- guide the students through applications using Internet links to access real data and other information provided by the vast resources of the World Wide Web.

The Internet projects (see page 54) can be completed individually or in a collaborative learning environment and are featured on our dedicated Web site. The URL for the Web site can be found on the back cover of the book.

DataDisk CD. This CD-ROM contains all data sets used in examples, exercises, and case studies. It also includes the *Focus Database,* a database that was obtained by randomly selecting 500 Arizona State University sophomores. Seven variables are considered for each student: sex, high-school GPA, SAT math score, cumulative GPA, SAT verbal score, age, and total hours completed. Compatible with PC and Macintosh, DataDisk CD is packaged with every copy of the book.

Minitab Quick Reference. For quick access and reference to the Minitab menu instructions discussed in this book, we have provided a Minitab Quick Reference (MQR) for Minitab for Windows 95/NT. Located inside the front cover of the book, the MQR includes the procedures, their menu instructions, and page-number references for more details.

Continuing Features

Following are some features of the Fifth Edition retained from previous editions.

Data sets. In most examples and exercises, we have presented raw data in addition to summary statistics. This gives a more realistic view of statistics and provides an opportunity for the problems to be solved by computer or statistical calculator, if so desired. Hundreds of new data sets have been included and most of those from previous editions have been updated.

Procedure boxes. To help the reader learn statistical procedures, we have developed easy-to-follow, step-by-step methods for carrying out those procedures. For ease in locating, each procedure is displayed with a color background. A unique feature of this book is that each step in the procedure is presented again within the example that illustrates the procedure. This serves a twofold purpose: it shows how the procedure is applied and helps the reader master the steps in the procedure.

Procedure index. Given the numerous statistical procedures, it is sometimes difficult to find a specific one, especially when the book is being used for reference purposes. Consequently, we have included a procedure index. Located inside the back cover of the book, the procedure index provides a quick and easy way to find the required procedure for performing any particular statistical analysis.

Computer simulations. Computer simulations appear in both the text and the exercises. The simulations serve as pedagogical aids for understanding complex concepts such as sampling distributions.

General objectives and chapter outlines. Included at the beginning of each chapter is a general description of the chapter, an explanation of how the chapter relates to the text as a whole, and an outline that lists the sections in the chapter.

Biographical sketches. Each chapter ends with a brief biography of a famous statistician. Besides being of general interest, these biographies help the reader obtain a perspective on how the science of statistics developed.

Chapter reviews. The end-of-chapter material begins with a chapter review. The review includes (1) chapter objectives, (2) a list of key terms with page references, and (3) a review test. These pedagogical aids provide the students with an organized method for reviewing and studying.

Database exercises. Following each chapter review, a section entitled "Using the Focus Database" asks the students to conduct various statistical analyses on the Focus Database, which is contained in the `Focus` folder of DataDisk CD. These exercises are optional and are to be done by computer.

Formula/Table card. A detachable formula/table card (FTC) is provided with the book. This card contains all of the formulas and many of the tables that appear in the text. The FTC is helpful for quick-reference purposes; many instructors also find it convenient for use with examinations.

Organization and Chapter-by-Chapter Changes

The text offers a great deal of flexibility in choosing material to cover.

- Chapter 1 presents the nature of statistics, sampling designs, and—new to this edition—an introduction to experimental design. (The new optional modular chapter *Design of Experiments and Analysis of Variance* provides a more comprehensive treatment.) Also new to Chapter 1 is a presentation of the basics of Minitab for Windows 95/NT needed for subsequent use of the software.
- Chapters 2 and 3 provide the fundamentals of descriptive statistics. The discussion of descriptive measures is now more concise for quicker coverage.
- Chapters 4 and 5 examine probability and discrete random variables. In this edition only the first three sections of Chapter 4 are prerequisite to coverage of inferential statistics; the remaining five sections of Chapter 4 and all four sections of Chapter 5 are optional.
- In Chapter 6 we provide a concise discussion of the normal distribution, including an optional section on the normal approximation to the binomial distribution.
- Chapter 7 introduces the concept of sampling distributions and presents an improved and simplified introduction to the sampling distribution of the mean.
- Chapters 8 and 9 now give an easily accessible introduction to confidence intervals and hypothesis tests for one population mean by using the terminology of variables and avoiding formal probability. Both chapters employ the σ-known versus σ-unknown criterion for deciding which parametric procedure to use; this approach makes confidence intervals and hypothesis tests easier to understand and apply, and provides a method consistent with Minitab. We consider Chapters 1–9 the core of an introductory statistics course.
- Chapter 10 discusses inferences for two population means and now contains a detailed discussion of the meaning of independent samples, including graphics for quick assimilation. The two-sample z-procedures have been relegated to the exercises so that the presentation can focus on the more practical two-sample t-procedures. Now included in Chapter 10 is a separate optional section devoted to the Wilcoxon signed-rank test for paired samples.
- Chapter 11 is new to this edition and is optional. It presents material on inferences for one and two population standard deviations (or variances).
- In Chapter 12 we discuss inferences for one and two population proportions, including Minitab's new dedicated procedures for proportion inferences.
- New to Chapter 13, which examines the chi-square goodness-of-fit test and the chi-square independence test, is a section on grouping bivariate data into contingency tables and an improved presentation of association.
- Chapter 14 now gives a more informal treatment of regression and correlation, relying on intuitive and graphical presentation of important concepts. The placement is flexible—the chapter can be covered any time after Chapter 3.

- Chapter 15 examines inferential methods in regression and correlation. Multiple regression and model building are now covered in the new optional modular chapters *Multiple Regression Analysis* and *Model Building in Regression,* which also include topics such as transformations, polynomial models, qualitative predictors, and model selection.
- In Chapter 16 we introduce analysis of variance, including one-way ANOVA, multiple comparisons, and the Kruskal–Wallis test. Other types of ANOVA, including two-way ANOVA and randomized block design, are discussed in the new optional module *Design of Experiments and Analysis of Variance.*

The following flowchart summarizes the preceding discussion and shows the interdependence among chapters. In the flowchart, the prerequisites for a given chapter consist of all chapters having a path leading to that chapter.

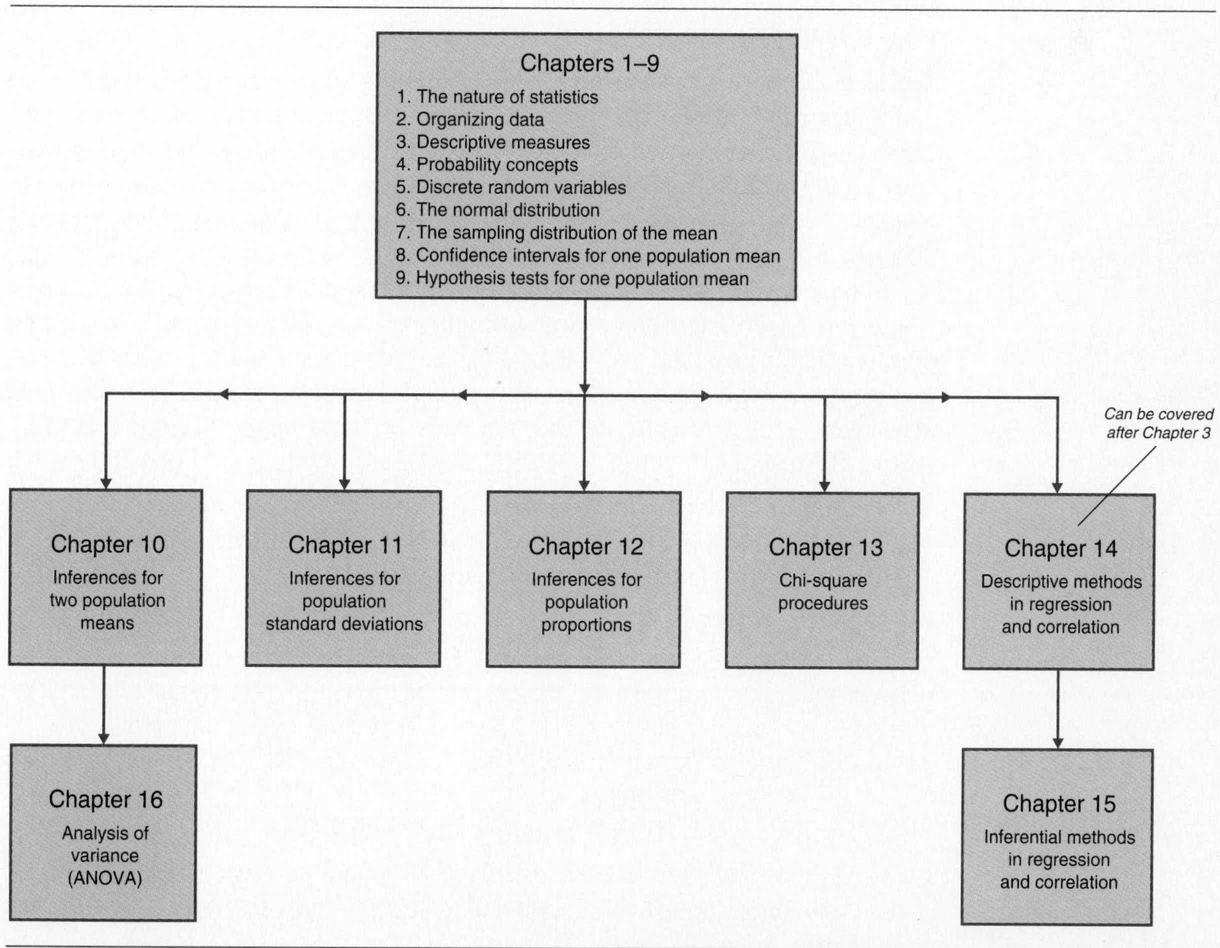

Supplements and Other Support

The following supplements have been prepared to accompany the Fifth Edition of *Introductory Statistics.*

For the Instructor

Instructor's Solutions Manual (ISBN 0-201-88321-X). This supplement, by David Lund of the University of Wisconsin at Eau Claire, contains detailed, worked-out solutions to all section exercises, review-test problems, Focus Database exercises, and case studies in the text.

PowerPoint Presentation Slides CD (ISBN 0-201-88323-6). New to this edition, PowerPoint Presentation Slides CD contains most of the text's figures and tables, and all of its definitions, formulas, key facts, and procedure boxes.

Printed Test Bank (ISBN 0-201-88325-2). This supplement provides several printed examinations for each chapter of the text.

TestGen-EQ with QuizMaster. Available for Windows (ISBN 0-201-39864-8) and Macintosh (ISBN 0-201-39865-5) operating systems, TestGen-EQ is a computerized test generator with algorithmically defined problems organized specifically for the Fifth Edition. The user-friendly graphical interface enables instructors to select, view, edit, and add test items, and then print tests in a variety of fonts and forms.

Seven question types are available; search and sort features allow the instructor to quickly locate questions and arrange them in a preferred order. The built-in question editor gives the user the power to create graphs, import graphics, insert mathematical symbols, templates, and variable numbers or text. Instructors can also create practice tests for posting to a Web site by using the "Export to HTML" feature. Tests created with TestGen-EQ can be used with QuizMaster-EQ, which enables students to take exams on a computer network.

For the Student

Minitab Manual (ISBN 0-201-88324-4). Written by Peter W. Zehna of the Naval Postgraduate School, the *Minitab Manual* is keyed to the book and provides detailed Minitab instructions to complement those given in the book.

TI-83 Manual (ISBN 0-201-39863-X). This new step-by-step guide, written by Ellen Fischer of Georgia Southern University, presents instructions for using the TI-83 calculator to solve selected examples and exercises found in the book.

Excel Manual (ISBN 0-201-43449-0). Correlated to exercises and examples in the book, this new manual by Peter W. Zehna provides step-by-step instruction in Excel.

Student's Solutions Manual (ISBN 0-201-88322-8). Also prepared by David Lund, this manual includes detailed solutions to all odd-numbered section exercises and all review-test problems in the text.

DataDisk CD. Included with every copy of the book, DataDisk is now on CD-ROM, compatible for the PC and Macintosh. This disk contains text (ASCII) files for the Focus Database and data sets appearing in all the exercises, examples, and case studies. It also contains the Minitab macros discussed and applied in the text. The convenience of the CD-ROM allows for easy storage, analysis, and access to these data sets without having to enter them manually.

The Student Edition of MINITAB® for Windows® 95/NT™. This release of the *Student Edition,* Release 12 for Windows 95/NT (ISBN 0-201-39715-3), allows for 5000 data points. It has improved graphing capabilities and has increased its functionality to include multiple regression and more. Also provided with the software are data sets drawn from business, the social sciences, and the physical sciences. In addition to the software, case studies and 16 tutorial sessions are featured in a tutorial manual written by John McKenzie of Babson College with Robert Goldman of Simmons College.

Modular Chapters. Three optional modular chapters written by Dennis Young of Arizona State University can be custom bound into the text or purchased separately. Modules include *Multiple Regression Analysis* (ISBN 0-201-43710-4), *Model Building in Regression* (ISBN 0-201-43711-2), and *Design of Experiments and Analysis of Variance* (ISBN 0-201-43715-5).

Web site and Internet Projects. The Web site supporting the Fifth Edition of *Introductory Statistics* includes downloadable practice tests, data sets, and formula/table card. It also provides access to the Internet projects, prepared by Tim Arnold of SAS Institute, Inc. The URL can be found on the back cover of the book.

ActivStats™ 2.0 (ISBN 0-201-61478-2). Developed by Paul Velleman and Data Description, Inc., *ActivStats 2.0* presents a complete introductory statistics course on CD-ROM using a full range of multimedia. Integrating video, simulation, animation, narration, text, interactive experiments, World Wide Web access, and Data Desk, a fully functioning statistics package, this product brings each student into a rich learning environment. Also included are exercises for reinforcement of key concepts, an index, and a glossary. The list of topics and a number of homework problems taken directly from *Introductory Statistics* make this program a strong complement to the text. *ActivStats 2.0* is PC and Macintosh compatible.

Acknowledgments

It is our pleasure to thank the following reviewers, whose comments and suggestions resulted in significant improvements to this edition.

Beth Chance
University of the Pacific

Carol DeVille
Louisiana Tech University

Dennis M. O'Brien
University of Wisconsin, La Crosse

Dwight M. Olson
John Carroll University

Steven E. Rigdon
Southern Illinois University, Edwardsville

George W. Schultz
St. Petersburg Jr. College

Cid Srinivasan
University of Kentucky, Lexington

W. Ed Stephens
McNeese State University

Kathy Taylor
Clackamas Community College

We would also like to thank the following reviewers who provided comments and suggestions for earlier editions of *Introductory Statistics:*

Jasper Adams
Stephen F. Austin State University

Larry Ammann
University of Texas, Dallas

George Anderson
Central Piedmont Community College

Gwen Applebaugh
University of Wisconsin, Eau Claire

Jerald T. Ball
Las Positas College

Mary Sue Beersman
Northeast Missouri State University

Gary B. Beus
Brigham Young University

William Beyer
University of Akron

Jerry Bloomberg
Essex Community College

Patricia Buchanan
Penn State University

Toni Carroll
Siena Heights College

Curtis Church
Middle Tennessee State University

Gabie Church
Louisiana State University

Patti Collings
Brigham Young University

Constance Cutchins
University of Central Florida

Rickie J. Domangue
James Madison University

Elizabeth Eltinge
Texas A&M University

Eugene Enneking
Portland State University

Ruby Evans
Sante Fe Community College

Dale O. Everson
University of Idaho

Charles M. Farmer
James Madison University

Chris Franklin
University of Georgia

Jeff Frost
Johnson County Community College

Joe Fred Gonzalez, Jr.
University of Maryland

Larry Griffey
Florida Community College, Jacksonville

Shu-Ping Hodgson
Central Michigan University

Mark E. Johnson
University of Central Florida

Debra Landre
San Joaquin Delta College

Benny Lo
Ohlone College

David R. Lund
University of Wisconsin, Eau Claire

Rhonda Magel
North Dakota State University

Linda Malone
University of Central Florida

Jacinta Mann
Seton Hill College

Joseph McWilliams
Stephen F. Austin State University

Bernard J. Morzuch
University of Massachusetts, Amherst

Charles B. Peters
University of Houston

Tom Ribley
Valencia Community College

Larry Ringer
Texas A&M University

Gaspard Rizzuto
University of Southwestern Louisiana

C. Bradley Russell
Clemson University

Leroy Sathre
Valencia Community College

Robert L. Schaefer
Miami University

James Schott
University of Central Florida

Franklin Sheehan
San Francisco State University

Rana P. Singh
Virginia State University

Donald Sisson
Utah State University

Duane Steffey
San Diego State University

Larry Stephens
University of Nebraska, Omaha

Ram C. Tiwari
University of North Carolina

Joseph J. Walker
Georgia State University

Lyndon C. Weberg
University of Wisconsin, River Falls

Calvin L. Williams
Clemson University

Our thanks as well to Professor Michael Driscoll for his help in selecting the statisticians for the biographical sketches; and Professors Charles Kaufman, Sharon Lohr, Kathy Prewitt, Walter Reid, and Bill Steed with whom we have had several illuminating discussions. Thanks also to Professors Matthew Hassett and Ronald Jacobowitz for their many helpful comments and suggestions.

To Professor Larry Griffey, we express our appreciation for his formula/table card. We are grateful to Professor Peter Zehna for preparing the *Minitab Manual* and *Excel Manual* and to Professor Ellen Fischer for the *TI-83 Manual.* Our thanks go as well to Dean David Lund for writing the *Instructor's Solutions Manual* and *Student's Solutions Manual.* Many thanks to Professor Dennis Young for his linear models modules and to Tim Arnold for his Internet projects. In addition, we extend our appreciation to Phyllis Barnidge and Joe Fred Gonzalez, Jr., for providing the answers to the exercises that appear in the back of the book.

We are grateful to Dr. William Feldman and Mr. Frank Crosswhite for supplying the data from their study on the Golden Torch Cactus; to Professor Thomas A. Ryan, Jr., for his correspondence concerning the correlation test for normality; and to Dr. George McManus and Mr. Gregory Weiss for supplying the data from their study of zooplankton nutrition in the Gulf of Mexico.

We also extend our appreciation to Professor John Tukey for taking the time to provide us with autobiographical information and a photo of himself; to Alison Stern-Dunyak of the American Statistical Association for supplying biographical information on Gertrude Cox; to Maureen Quinn of The Gallup Organization for providing the information used in the case study of Chapter 13; and to all the people at Minitab for their excellent Author Assistance Program and technical support. Our appreciation also goes to Howard Blaut and Rick Hanna for providing data on real estate; to Jeffrey Jirele for supplying data on automobile insurance; and to Mary Neary for furnishing the Focus database.

Many thanks to our text designer Cynthia Crampton for an outstanding design. We are also grateful to our proofreaders Jenny Bagdigian, Elka Block, Trent Buskirk, Frank Purcell, and Gregory Weiss. Our appreciation goes as well to our macro writer Don DeLand, to Berthold Horn and Louis Vosloo at Y&Y, Inc., and to George and Brian Morris of Scientific Illustrators.

Without the help of many people at Addison Wesley Longman, this book and its numerous ancillaries would not have been possible; to all of them go our heartfelt thanks. We would, however, like to give special thanks to the following people at Addison Wesley Longman: Greg Tobin, Jennifer Albanese, Brenda Bravener, Karen Guardino, Jenny Bagdigian, Kim Ellwood, Laura Potter, Melissa Perrelli, Caroline Fell, Joseph Vetere, Barbara Atkinson, Karen Scott, and Alex Levering.

Finally, we would like to express our appreciation to Carol Weiss. Apart from writing the text, she was involved in every aspect of development and production. Moreover, Carol researched and wrote the biographies and took on the task of typesetter using the TEX typesetting system.

Tempe, Arizona *N.A.W.*

Contents

P A R T

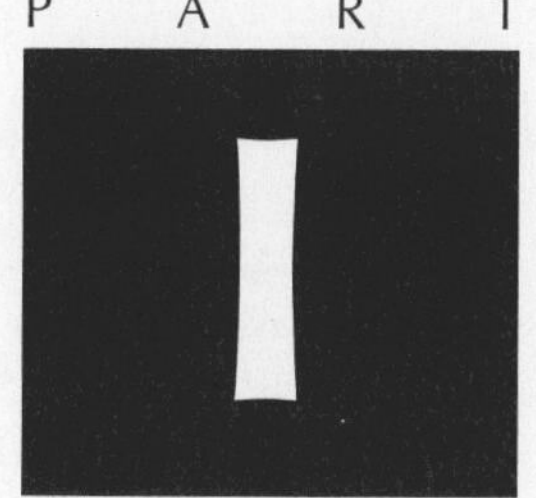

Introduction

CASE STUDY

TOP FILMS OF ALL TIME

The American Film Institute (AFI) conducted a survey as part of a celebration of the 100th anniversary of cinema. AFI polled 1500 filmmakers, actors, critics, politicians, and film historians, asking them to pick their 100 favorite films from a list of 400. The films on the list were made between 1896 and 1996.

After tallying the responses, AFI compiled a list representing the top 100 films. *Citizen Kane,* made in 1941, finished in first place, followed by *Casablanca,* which was made in 1942. Only one movie filmed in the 1990s ranked in the top 10—*Schindler's List* (1993)—and, in fact, the next highest ranking film of the 90s was *The Silence of the Lambs* (1991), which ranked 65. The following table gives the top 40 finishers in the poll.

Rank	Film	Year	Rank	Film	Year
1	Citizen Kane	1941	21	The Grapes of Wrath	1940
2	Casablanca	1942	22	2001: A Space Odyssey	1968
3	The Godfather	1972	23	The Maltese Falcon	1941
4	Gone With the Wind	1939	24	Raging Bull	1980
5	Lawrence of Arabia	1962	25	E.T. The Extra-Terrestrial	1982
6	The Wizard of Oz	1939	26	Dr. Strangelove	1964
7	The Graduate	1967	27	Bonnie & Clyde	1967
8	On the Waterfront	1954	28	Apocalypse Now	1979
9	Schindler's List	1993	29	Mr. Smith Goes to Washington	1939
10	Singin' in the Rain	1952	30	The Treasure of the Sierra Madre	1948
11	It's a Wonderful Life	1946	31	Annie Hall	1977
12	Sunset Blvd.	1950	32	The Godfather, Part II	1974
13	The Bridge on the River Kwai	1957	33	High Noon	1952
14	Some Like It Hot	1959	34	To Kill a Mockingbird	1962
15	Star Wars	1977	35	It Happened One Night	1934
16	All About Eve	1950	36	Midnight Cowboy	1969
17	The African Queen	1951	37	The Best Years of Our Lives	1946
18	Psycho	1960	38	Double Indemnity	1944
19	Chinatown	1974	39	Doctor Zhivago	1965
20	One Flew Over the Cuckoo's Nest	1975	40	North by Northwest	1959

Armed with the knowledge gained in this chapter, we will return to further analyze the AFI poll at the end of the chapter.

The Nature of Statistics

1

GENERAL OBJECTIVES

What does the word *statistics* bring to mind? Most people immediately think of numerical facts or data, such as unemployment figures, farm prices, or the number of marriages and divorces. *Webster's New World Dictionary* gives two definitions of the word *statistics:*

> 1. facts or data of a numerical kind, assembled, classified, and tabulated so as to present significant information about a given subject. 2. [construed as sing.], the science of assembling, classifying, and tabulating such facts or data.

But statistics encompasses much more than these definitions convey. Not only do statisticians assemble, classify, and tabulate data, but they also analyze data in order to make generalizations and decisions. For example, a political analyst can use data from a portion of the voting population to predict the political preferences of the entire voting population. And a city council can decide where to build a new airport runway based on environmental impact statements and demographic reports that include a variety of statistical data. In this chapter we introduce some basic terminology so that the various meanings of the word *statistics* will become clear to you.

Note: As you read the chapter outline in the box to the right you will observe that an asterisk appears next to the section on Using the Computer. Here and throughout this book, an asterisk denotes optional material.

CHAPTER OUTLINE

1.1 TWO KINDS OF STATISTICS

You probably already know something about statistics. If you read newspapers, watch the news on television, or follow sports, then you see and hear the word *statistics* frequently. In this section we will use familiar examples such as baseball statistics and voter polls to introduce the two major types of statistics: *descriptive statistics* and *inferential statistics.*

Each spring in the late 1940s, the major-league baseball season was officially opened when President Harry S Truman threw out the "first ball" of the season at the opening game of the Washington Senators. Both President Truman and the Washington Senators had reason to be interested in statistics. Consider, for instance, the year 1948.

EXAMPLE 1.1 DESCRIPTIVE STATISTICS

In 1948 the Washington Senators played 153 games, winning 56 and losing 97. They finished seventh in the American League and were led in hitting by Bud Stewart, whose batting average was .279. These and many other statistics were compiled by baseball statisticians who took the complete records for each game of the season and organized that large mass of information effectively and efficiently.

Although baseball fans take baseball statistics for granted, a great deal of time and effort is required to gather and organize them. Moreover, without such statistics, baseball would be much harder to understand. For instance, picture yourself trying to select the best hitter in the American League with only the official score sheets for each game. (More than 600 games were played in 1948; the best hitter was Ted Williams, who led the league with a batting average of .369.) ◨

The work of baseball statisticians provides an excellent illustration of descriptive statistics. A formal definition of the term *descriptive statistics* is presented in Definition 1.1.

DEFINITION 1.1 DESCRIPTIVE STATISTICS

Descriptive statistics consists of methods for organizing and summarizing information.

Descriptive statistics includes the construction of graphs, charts, and tables, and the calculation of various descriptive measures such as averages, measures of variation, and percentiles. We will discuss descriptive statistics in detail in Chapters 2 and 3.

As we said, descriptive statistics is one of the two major types of statistics. The other major type, inferential statistics, is illustrated in Example 1.2.

EXAMPLE 1.2 INFERENTIAL STATISTICS

In the fall of 1948, President Truman was also concerned about statistics. The Gallup Poll taken just prior to the election predicted that he would win only 44.5% of the vote and be defeated by the Republican nominee, Thomas E. Dewey. But this time the statisticians had predicted incorrectly. Truman won more than 49% of the vote and with it the presidency. The Gallup Organization modified some of its procedures and has correctly predicted the winner ever since. □

Political polling provides an example of inferential statistics. It would be expensive and unrealistic to interview all Americans on their voting preferences. Statisticians who wish to gauge the sentiment of the entire *population* of U.S. voters can afford to interview only a carefully chosen group of a few thousand voters. This group is referred to as a *sample* of the population. Statisticians analyze the information obtained from a sample of the voting population to make inferences (draw conclusions) about the preferences of the entire voting population. Inferential statistics provides methods for making such inferences.

The terminology introduced in the context of political polling is used in general in statistics. Specifically, we have the following definitions. See also Fig. 1.1 at the top of the next page.

DEFINITION 1.2 POPULATION AND SAMPLE

Population: The collection of all individuals or items under consideration in a statistical study.

Sample: That part of the population from which information is collected.

Now that we have discussed the terms population and sample, we can present the definition of inferential statistics.

DEFINITION 1.3 INFERENTIAL STATISTICS

Inferential statistics consists of methods for drawing and measuring the reliability of conclusions about a population based on information obtained from a sample of the population.

FIGURE 1.1
Relationship between population and sample

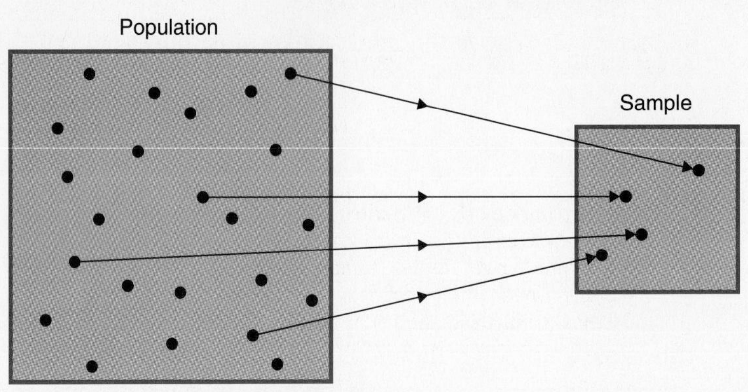

Descriptive statistics and inferential statistics are interrelated. It is almost always necessary to invoke techniques of descriptive statistics to organize and summarize the information obtained from a sample before carrying out an inferential analysis. Furthermore, the preliminary descriptive analysis of a sample often reveals features that lead to the choice of (or to a reconsideration of the choice of) the appropriate inferential method.

EXERCISES 1.1

STATISTICAL CONCEPTS AND SKILLS

1.1 Define the following terms.
a. population **b.** sample

1.2 What are the two major types of statistics? Describe them in detail.

1.3 Identify some of the methods that are used in descriptive statistics.

1.4 Explain two ways in which descriptive statistics and inferential statistics are interrelated.

1.2 CLASSIFYING STATISTICAL STUDIES

As you proceed through this book, you will obtain a thorough understanding of the principles of descriptive and inferential statistics and of related topics, such as exploratory data analysis. At this point, you should be able to classify statistical studies as either descriptive or inferential.

Examples 1.3 and 1.4 will give you some practice in making the distinction between descriptive and inferential studies. In each example we have presented the result of a statistical study and have classified the study as either descriptive or inferential. Try to classify each study yourself before reading our explanation.

EXAMPLE 1.3

CLASSIFYING STATISTICAL STUDIES

The study Table 1.1 displays the voting results for the 1948 presidential election.

TABLE 1.1
Final results of the 1948 presidential election

Ticket	Votes	Percentage
Truman-Barkley (Democratic)	24,179,345	49.7
Dewey-Warren (Republican)	21,991,291	45.2
Thurmond-Wright (States Rights)	1,176,125	2.4
Wallace-Taylor (Progressive)	1,157,326	2.4
Thomas-Smith (Socialist)	139,572	0.3

Classification This study is descriptive. It is a summary of the votes cast by U.S. voters in the 1948 presidential election. No inferences were made. ◨

EXAMPLE 1.4

CLASSIFYING STATISTICAL STUDIES

The study For the 101 years preceding 1977, baseballs used by the major leagues were purchased from the Spalding Company. In 1977 that company stopped manufacturing major-league baseballs, and the major leagues arranged to buy their baseballs from the Rawlings Company.

Early in the 1977 season, pitchers began to complain that the Rawlings ball was "livelier" than the Spalding ball. They claimed it was harder, bounced farther and faster, and gave hitters an unfair advantage. There was some evidence for this. In the first 616 games of 1977, 1033 home runs were hit, compared to only 762 home runs in the first 616 games of 1976.

Sports Illustrated magazine sponsored a careful study of the liveliness question, and the results appeared in the June 13, 1977, issue. In this study an independent testing company randomly selected 85 baseballs from the current (1977) supplies of various major-league teams. The bounce, weight, and hardness of the baseballs chosen were carefully measured. Those measurements were then compared with measurements obtained from similar tests on baseballs used in the years 1952, 1953, 1961, 1963, 1970, and 1973. The conclusion, presented on page 24 of the *Sports Illustrated* article, was that "... the 1977 Rawlings ball is livelier than the 1976 Spalding, but not as lively as it could be under big league rules, or as the ball has been in the past."

Classification This is an inferential study. The independent testing company used a sample of 85 baseballs from the 1977 supplies of major-league teams to make an inference about the population of all such baseballs. (It has been estimated that approximately 360,000 baseballs were used by the major leagues in 1977.) ◨

The *Sports Illustrated* study also provides an excellent illustration of a situation in which it is not feasible to obtain information for the entire population. Indeed, after the bounce and hardness tests, all of the baseballs sampled were taken to a butcher in Plainfield, New Jersey, to be sliced in half so that researchers could look inside them. Clearly, it would not have been practical to test every baseball in this way.

In closing, we emphasize that it is possible to perform a descriptive study on a sample as well as on a population. Only when an inference is made about the population based on information obtained from the sample does the study become inferential.

EXERCISES 1.2

STATISTICAL CONCEPTS AND SKILLS

In Exercises 1.5–1.12, classify each of the studies as either descriptive or inferential.

1.5 The A. C. Nielsen Company collects and publishes information on the television-viewing habits of Americans. Data from a sample of Americans yielded the following estimates of average TV viewing time per week for all Americans. The times are in hours and minutes. [SOURCE: Nielsen Media Research, *Nielsen Report on Television.*]

Group (by age)		Time
Average all persons		*30:14*
Women	Total 18+	34:47
	18–24	28:54
	25–54	31:05
	55+	44:11
Men	Total 18+	30.41
	18–24	23:31
	25–54	28:44
	55+	38:47
Teens	12–17	21:50
Children	2–11	23:01

1.6 In 1936 the voters of North Carolina cast their presidential votes as follows.

Candidate and party	Number of votes
Roosevelt (Democratic)	616,414
Landon (Republican)	223,283
Thomas (Socialist)	21
Browder (Communist)	11
Lemke (Union)	2

1.7 The U.S. National Center for Health Statistics published the following rate estimates in *Vital Statistics of the United States* for the leading causes of death in 1994. The estimates are based on a 10% sampling of all 1994 U.S. death certificates. Rates are per 100,000 population.

Cause	Rate
Major cardiovascular diseases	362.6
Malignancies (cancers)	205.2
Chronic obstructive pulmonary diseases	39.1
Accidents	34.6
Pneumonia and influenza	31.5

1.8 The U.S. Substance Abuse and Mental Health Services Administration collects and publishes data on drug use, by type of drug and age group, in *National Household Survey on Drug Abuse.* The following table provides information for the years 1985 and 1995. The percentages shown are estimates obtained from national samples.

FORMULA/TABLE CARD FOR WEISS'S *INTRODUCTORY STATISTICS, FIFTH EDITION*

Larry R. Griffey

NOTATION In the formulas below, unless stated otherwise, we employ the following notation which may or may not appear with subscripts:

n = sample size	σ = population stdev
$\bar{x}$ = sample mean	d = paired difference
s = sample stdev	$\hat{p}$ = sample proportion
Q_j = jth quartile	p = population proportion
N = population size	O = observed frequency
μ = population mean	E = expected frequency

CHAPTER 3 Descriptive Measures

- Sample mean: $\bar{x} = \dfrac{\Sigma x}{n}$
- Range: Range = Max − Min
- Sample standard deviation:

$$s = \sqrt{\frac{\Sigma(x-\bar{x})^2}{n-1}} \quad \text{or} \quad s = \sqrt{\frac{\Sigma x^2 - (\Sigma x)^2/n}{n-1}}$$

- Quartile positions: $(n+1)/4$, $(n+1)/2$, $3(n+1)/4$
- Interquartile range: $\text{IQR} = Q_3 - Q_1$
- Lower limit $= Q_1 - 1.5 \cdot \text{IQR}$, Upper limit $= Q_3 + 1.5 \cdot \text{IQR}$
- Population mean (mean of a variable): $\mu = \dfrac{\Sigma x}{N}$
- Population standard deviation (standard deviation of a variable):

$$\sigma = \sqrt{\frac{\Sigma(x-\mu)^2}{N}} \quad \text{or} \quad \sigma = \sqrt{\frac{\Sigma x^2}{N} - \mu^2}$$

- Standardized variable: $z = \dfrac{x-\mu}{\sigma}$

CHAPTER 4 Probability Concepts

- Probability for equally likely outcomes:

$$P(E) = \frac{f}{N},$$

where f denotes the number of ways event E can occur and N denotes the total number of outcomes possible.

- Special addition rule:

$$P(A \text{ or } B \text{ or } C \text{ or } \cdots) = P(A) + P(B) + P(C) + \cdots$$

(A, B, C, ... mutually exclusive)

- Complementation rule: $P(E) = 1 - P(\text{not } E)$
- General addition rule: $P(A \text{ or } B) = P(A) + P(B) - P(A \,\&\, B)$
- Conditional-probability rule: $P(B \mid A) = \dfrac{P(A \,\&\, B)}{P(A)}$
- General multiplication rule: $P(A \,\&\, B) = P(A) \cdot P(B \mid A)$
- Special multiplication rule:

$$P(A \,\&\, B \,\&\, C \,\&\, \cdots) = P(A) \cdot P(B) \cdot P(C) \cdots$$

(A, B, C, ... independent) $P(B|A) = P(B)$

- Rule of total probability:

$$P(B) = \sum_{j=1}^{k} P(A_j) \cdot P(B \mid A_j)$$

($A_1, A_2, \ldots, A_k$ mutually exclusive and exhaustive)

- Bayes's rule:

$$P(A_i \mid B) = \frac{P(A_i) \cdot P(B \mid A_i)}{\sum_{j=1}^{k} P(A_j) \cdot P(B \mid A_j)}$$

($A_1, A_2, \ldots, A_k$ mutually exclusive and exhaustive)

- Factorial: $k! = k(k-1)\cdots 2 \cdot 1$
- Permutations rule: $(m)_r = \dfrac{m!}{(m-r)!}$
- Special permutations rule: $(m)_m = m!$
- Combinations rule: $\dbinom{m}{r} = \dfrac{m!}{r!\,(m-r)!}$
- Number of possible samples: $\dbinom{N}{n} = \dfrac{N!}{n!\,(N-n)!}$

CHAPTER 5 Discrete Random Variables

- Mean of a discrete random variable X: $\mu = \Sigma x P(X = x)$
- Standard deviation of a discrete random variable X:

$$\sigma = \sqrt{\Sigma(x-\mu)^2 P(X=x)} \quad \text{or} \quad \sigma = \sqrt{\Sigma x^2 P(X=x) - \mu^2}$$

- Factorial: $k! = k(k-1)\cdots 2 \cdot 1$
- Binomial coefficient: $\dbinom{n}{x} = \dfrac{n!}{x!\,(n-x)!}$
- Binomial probability formula:

$$P(X = x) = \binom{n}{x} p^x (1-p)^{n-x},$$

where n denotes the number of trials and p denotes the success probability.

- Mean of a binomial random variable: $\mu = np$
- Standard deviation of a binomial random variable: $\sigma = \sqrt{np(1-p)}$
- Poisson probability formula: $P(X = x) = e^{-\lambda} \dfrac{\lambda^x}{x!}$
- Mean of a Poisson random variable: $\mu = \lambda$
- Standard deviation of a Poisson random variable: $\sigma = \sqrt{\lambda}$

- Total sum of squares: $SST = \Sigma(y - \bar{y})^2 = S_{yy}$
- Regression sum of squares: $SSR = \Sigma(\hat{y} - \bar{y})^2 = S_{xy}^2/S_{xx}$
- Error sum of squares: $SSE = \Sigma(y - \hat{y})^2 = S_{yy} - S_{xy}^2/S_{xx}$
- Regression identity: $SST = SSR + SSE$
- Coefficient of determination: $r^2 = \dfrac{SSR}{SST}$
- Linear correlation coefficient:

$$r = \frac{\frac{1}{n-1}\,\Sigma(x - \bar{x})(y - \bar{y})}{s_x s_y} \quad \text{or} \quad r = \frac{S_{xy}}{\sqrt{S_{xx}S_{yy}}}$$

CHAPTER 15 Inferential Methods in Regression and Correlation

- Population regression equation: $y = \beta_0 + \beta_1 x$
- Standard error of the estimate: $s_e = \sqrt{\dfrac{SSE}{n-2}}$
- Test statistic for H_0: $\beta_1 = 0$:

$$t = \frac{b_1}{s_e/\sqrt{S_{xx}}}$$

with df $= n - 2$.

- Confidence interval for β_1:

$$b_1 \pm t_{\alpha/2} \cdot \frac{s_e}{\sqrt{S_{xx}}}$$

with df $= n - 2$.

- Confidence interval for the conditional mean of the response variable corresponding to x_p:

$$\hat{y}_p \pm t_{\alpha/2} \cdot s_e\sqrt{\frac{1}{n} + \frac{(x_p - \Sigma x/n)^2}{S_{xx}}}$$

with df $= n - 2$.

- Prediction interval for an observed value of the response variable corresponding to x_p:

$$\hat{y}_p \pm t_{\alpha/2} \cdot s_e\sqrt{1 + \frac{1}{n} + \frac{(x_p - \Sigma x/n)^2}{S_{xx}}}$$

with df $= n - 2$.

- Test statistic for H_0: $\rho = 0$:

$$t = \frac{r}{\sqrt{\dfrac{1 - r^2}{n-2}}}$$

with df $= n - 2$.

- Test statistic for a correlation test for normality:

$$R_p = \frac{\Sigma xw}{\sqrt{S_{xx}\,\Sigma w^2}}$$

where x and w denote, respectively, observations of the variable and the corresponding normal scores.

CHAPTER 16 Analysis of Variance (ANOVA)

- Notation in one-way ANOVA:

k = number of populations
n = total number of observations
$\bar{x}$ = mean of all n observations
n_j = size of sample from Population j
$\bar{x}_j$ = mean of sample from Population j
s_j^2 = variance of sample from Population j
T_j = sum of sample data from Population j

- Defining formulas for sums of squares in one-way ANOVA:

$$SST = \Sigma(x - \bar{x})^2$$
$$SSTR = \Sigma n_j(\bar{x}_j - \bar{x})^2$$
$$SSE = \Sigma(n_j - 1)s_j^2$$

- One-way ANOVA identity: $SST = SSTR + SSE$
- Computing formulas for sums of squares in one-way ANOVA:

$$SST = \Sigma x^2 - (\Sigma x)^2/n$$
$$SSTR = \Sigma(T_j^2/n_j) - (\Sigma x)^2/n$$
$$SSE = SST - SSTR$$

- Mean squares in one-way ANOVA:

$$MSTR = \frac{SSTR}{k-1}, \qquad MSE = \frac{SSE}{n-k}$$

- Test statistic for one-way ANOVA (independent samples, normal populations, and equal population standard deviations):

$$F = \frac{MSTR}{MSE}$$

with df $= (k - 1, n - k)$.

- Confidence interval for $\mu_i - \mu_j$ in the Tukey multiple-comparison method (independent samples, normal populations, and equal population standard deviations):

$$(\bar{x}_i - \bar{x}_j) \pm \frac{q_\alpha}{\sqrt{2}} \cdot s\sqrt{(1/n_i) + (1/n_j)},$$

where $s = \sqrt{MSE}$ and q_α is obtained for a q-curve with parameters k and $n - k$.

- Test statistic for a Kruskal–Wallis test (independent samples, same shape populations, all sample sizes 5 or greater):

$$H = \frac{SSTR}{SST/(n-1)} \quad \text{or} \quad H = \frac{12}{n(n+1)}\sum \frac{R_j^2}{n_j} - 3(n+1),$$

where $SSTR$ and SST are computed for the ranks of the data, and R_j denotes the sum of the ranks for the sample data from Population j. H is approximately chi-square with df $= k - 1$.

Type of drug	Percentage, 12 years old and over			
	Ever used		*Current user*	
	1985	1995	1985	1995
Marijuana	29.4	31.0	9.7	4.7
Cocaine	11.2	10.3	3.0	0.7
Inhalants	7.9	5.7	0.6	0.4
Hallucinogens	6.9	9.5	1.2	0.7
Heroin	0.9	1.2	0.1	0.1
Stimulants[1]	7.3	4.9	1.8	0.4
Sedatives[1]	4.8	2.7	0.5	0.2
Tranquilizers[1]	7.6	3.9	2.2	0.4
Analgesics[1]	7.6	6.1	1.4	0.6
Alcohol	84.9	82.3	60.2	52.2

1 = Nonmedical use.

1.9 The following table, obtained from the *Statistical Abstract of the United States, 1997,* displays 1994 attendance figures for selected spectator sports. Data are in thousands, rounded to the nearest thousand.

Sport	Attendance (thousands)
Baseball, major leagues	50,010
Basketball	
NCAA Men's college	28,390
NCAA Women's college	4,557
Professional	19,350
Football	
NCAA College	36,460
Professional	(NA)
Horseracing	42,065
Jai alai	3,684

1.10 Newspapers publish weather data for cities all over the world. The following table gives high and low temperatures and sky conditions for some selected cities on March 18, 1992. (nr = no report)

City	High	Low	Sky condition
Athens	63	45	Rain
Bogota	68	45	Pcy
Cairo	72	54	Clear
Hong Kong	79	72	Cloudy
Moscow	33	27	Sunny
Perth	85	66	Cloudy
Rome	61	41	Clear
Stockholm	43	25	Cloudy
Tel Aviv	nr	52	nr

1.11 The U.S. Council of Economic Advisers compiles data on the the annual averages of daily figures for the Dow Jones Industrial Averages and publishes its findings in *Economic Report of the President.* Here are the data for the years 1990 through 1996.

Year	Annual average
1990	2678.9
1991	2929.3
1992	3284.3
1993	3522.1
1994	3793.8
1995	4493.8
1996	5742.9

1.12 A study by the Gallup Organization concluded that in 1995, an estimated 72% of U.S. households were involved in at least one form of gardening. [SOURCE: National Gardening Association, *National Gardening Survey.*]

1.13 In each part below, decide whether the specified study would be descriptive or inferential. Provide a reason for each of your answers.

a. A tire manufacturer wants to estimate the average life of a new type of steel-belted radial.

b. A sports writer plans to list the winning times for all swimming events in the 2000 Olympics.

c. A politician obtains the exact number of votes that were cast for her opponent in 1998.

d. A medical researcher tests an anticancer drug that may have harmful side effects.

e. A candidate for governor estimates the percentage of voters that will vote for him in the upcoming gubernatorial election.

f. An economist estimates the average income of all California residents.

g. The owner of a small business determines the average salary of her 20 employees.

1.14 The chairperson of the mathematics department at a large state university wanted to estimate the average final exam score for the 2476 students in basic algebra. She randomly selected 50 exams from the 2476 and found the average score on the 50 exams to be 78.3%. From this she estimated that the average score for all 2476 students was roughly 78.3%.

a. What kind of study did the chairperson do?
b. What kind of study would she have done had she averaged all 2476 exam scores?

EXTENDING THE CONCEPTS AND SKILLS

1.15 A *Newsweek* poll of a sample of Americans revealed that "84% of those surveyed would choose organically grown produce over produce grown using chemical fertilizers, pesticides, and herbicides."
a. Is the statement in quotes an inferential statement, or is it simply descriptive?
b. What if, based on the same information, the statement had been "84% of Americans would choose organically grown produce over produce grown using chemical fertilizers, pesticides, and herbicides"?

1.16 In a press release dated September 24, 1997, dateline Washington, DC, the Mellman Group for the Coalition to Reduce Nuclear Dangers reported that "a new nationwide poll shows that 70.3% of Americans think the U.S. Senate should approve a treaty with 140 countries that would prohibit underground nuclear weapons explosions worldwide."
a. Do you think the statement in the press release is inferential or descriptive? Can you be sure?
b. Actually, the Mellman Group conducted an opinion survey of 800 adults and determined that 70.3% of them thought the U.S. Senate should approve a treaty with 140 countries that would prohibit underground nuclear weapons explosions worldwide. How would you rephrase the statement in the press release to make it clear that it is a descriptive statement?

1.3 THE DEVELOPMENT OF STATISTICS

According to the *Dictionary of Scientific Biography,* "The word *'Statistik,'* first printed in 1672, meant *Staatswissenschaft,* or, rather, a science concerning the states. It was cultivated at the German universities, where it consisted of more or less systematically collecting 'state curiosities' rather than quantitative material."

As we know, the modern science of statistics is much broader than just collecting "state curiosities" and includes both descriptive statistics and inferential statistics. Historically, descriptive statistics appeared first. Censuses were taken as long ago as Roman times. Over the years, records of such things as births, deaths, marriages, and taxes have led naturally to the development of descriptive statistics.

Inferential statistics is a newer arrival. Major developments began to occur with the research of Karl Pearson (1857–1936) and Ronald Fisher (1890–1962), who published their results in the early years of the twentieth century. Since the work of Pearson and Fisher, inferential statistics has evolved rapidly and is now applied in many fields. In fact, an understanding of the basic concepts of inferential statistics has become mandatory for virtually every professional.

Familiarity with statistics will also help you make more sense of many things you read in newspapers and magazines and on the Internet. For instance, in the description of the *Sports Illustrated* baseball test (Example 1.4), it may have struck you as unreasonable that a sample of only 85 baseballs could be used to draw a conclusion about a population of roughly 360,000 baseballs. By the time you have completed Chapter 9, you will understand why such inferences are not unreasonable.

1.4 USING THE COMPUTER*

Computers and statistical calculators are ideal for doing descriptive and inferential statistics. A person rarely needs to write his or her own programs, since they already exist for almost all aspects of statistics.

Today, programs for conducting statistical and data analyses are available in general-use spreadsheet software, graphing calculators, and dedicated statistical software. In this book, we will explain the use of dedicated statistical software. We refer readers interested in using spreadsheet software to the *Excel Manual* by Peter Zehna (Reading, MA: Addison-Wesley, 1999) and those interested in using graphing calculators to the *TI-83 Manual* by Ellen Fischer (Reading, MA: Addison-Wesley, 1999).

Many high-quality statistical software packages are available. We have chosen Release 12 of the professional version of Minitab® for Windows® to illustrate the basic ideas of statistical software. Only minor differences exist between the professional versions of Minitab for Windows and Macintosh® and, aside from worksheet size, the same is true for *The Student Edition of MINITAB for Windows* (abbreviated as "Student Edition").[†]

At the end of most sections, we will discuss how to use Minitab to solve problems that were solved by hand earlier in the section. Each solution first introduces the appropriate procedure and then displays and interprets the computer output. For best results you should be at a computer and perform the steps taken in the example under consideration.

To use Minitab you must first gain access to it on the computer. Because the procedure for accessing Minitab depends on the type of computer system being used, the precise method should be obtained from your instructor or an appropriate staff member at the computer center.

Minitab Basics

To start Minitab, we double-click the Minitab icon, , located in the Minitab program folder or on the Desktop; that is, we first move the mouse cursor on the Minitab icon and then press and release the mouse button twice in rapid succession.

When Minitab is started, we will see something similar to what is shown in Fig. 1.2. Pictured in Fig. 1.2 is the default configuration of the windows at start-up for Release 12 of the professional version of Minitab; the configuration can be changed.

---

* This section begins the optional coverage of Minitab statistical software and can be omitted by those not wishing to learn how a computer can be used to conduct statistical and data analyses.

[†] Minitab is a registered trademark of Minitab Inc. Windows is a registered trademark of Microsoft Corporation. Macintosh is a registered trademark of Apple Computers, Inc.

FIGURE 1.2
Minitab's windows at start-up

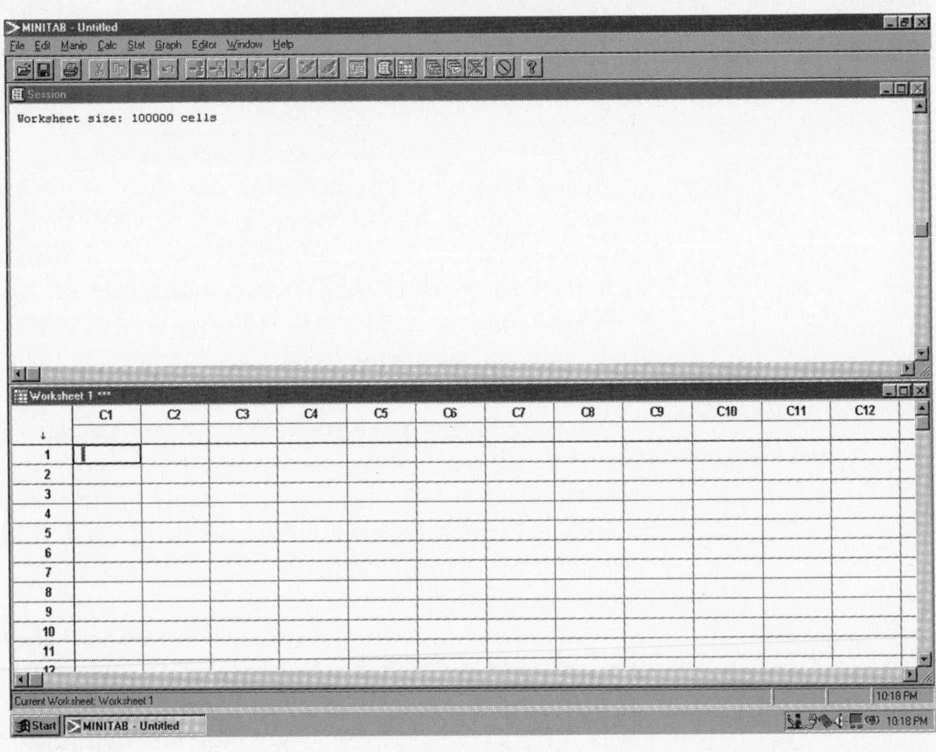

Figure 1.2 shows two open windows, the **Session window** and a **worksheet**. The Session window displays non-graphical output such as tables of statistics and character graphs. A worksheet is where we enter, name, view, and edit data.

From Fig. 1.2 we observe that the menu bar is across the top. This bar contains the main menus: File, Edit, Manip, Calc, Stat, Graph, Editor, Window, and Help. It is from one of these menus that we begin executing Minitab commands. Beneath the Menu bar is the Toolbar which provides shortcuts for several important actions.

Table 1.2 presents some mouse basics. Study this table carefully, as we will be using the terminology presented there throughout the book.

Two typographical conventions that we will often use involve text to be typed and menu instructions to be performed.

- Text to be typed is set in typewriter type and is underlined. For example, `HOURS` means that HOURS should be typed in the place indicated.
- Menu instructions are set in boldface type with the entries separated by pointers. For example, **Stat ➤ Tables ➤ Chi-Square Test...** means first click **Stat** in the menu bar, next click **Tables**, and then click **Chi-Square Test...**. See Fig. 1.3.

TABLE 1.2
Mouse basics

Term	Meaning
Point	Move the mouse so that the mouse cursor is on the desired item.
Click	Point to the desired item and then press and release the mouse button.
Double-click	Point to the desired item and then rapidly press and release the mouse button twice.
Drag	Press and hold down the mouse button at the desired start location, move the mouse so that the mouse cursor is at the desired end location, and then release the mouse button.

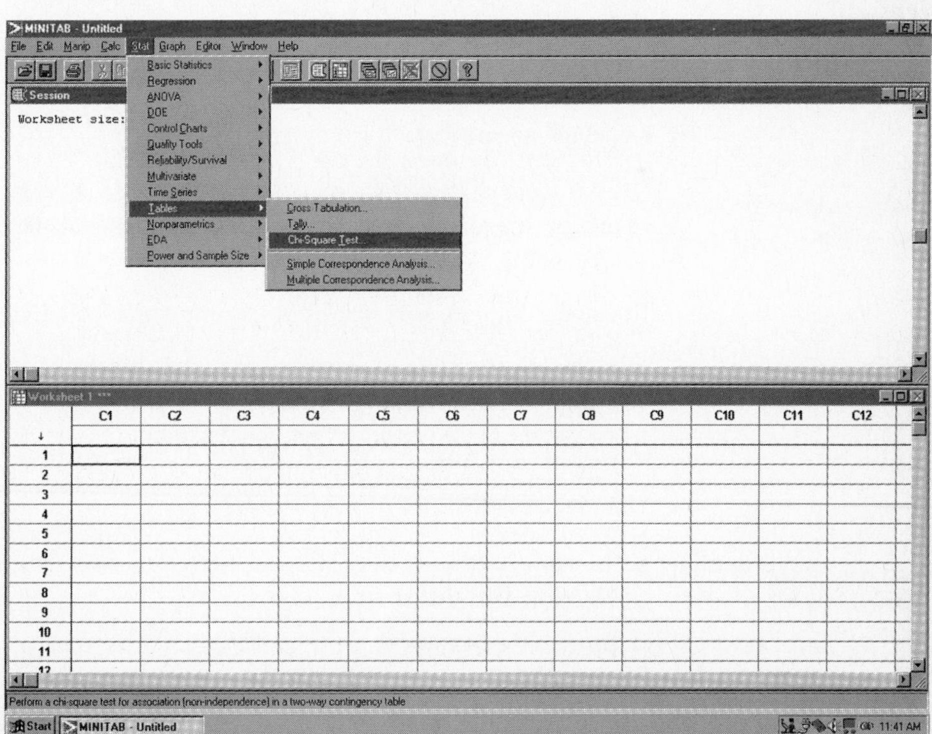

FIGURE 1.3
Display of menu instructions

Storing Data

To prepare for the Minitab sections in this book, we will explain how to store data. Roughly speaking, *data* are the information collected in a statistical study and a *data set* consists of a particular collection of data. Each data set is stored in a **column.** A column is designated by a "C" followed by a number; thus C1 stands for Column 1, C2 for Column 2, and so forth.

From Fig. 1.2 we see that the column designations are displayed along the top of the worksheet. The numbers at the left of the worksheet represent positions within a column and are referred to as **rows.** Each rectangle occurring at the intersection of a column and a row is called a **cell.** A cell can hold one observation (piece of data) and is specified by giving its column number and row number.

The **active cell** has the worksheet cursor inside it and a dark rectangle around it. In Fig. 1.2 the active cell is the first row of Column 1 (C1). An easy way to change the location of the active cell is to click in the cell to be made active. To enter an observation in a cell, we first make the cell active and then type the value. Changing the value in a cell is done in the same way.

Directly below each column label in the worksheet is a cell optionally used for naming the column. Naming a column is often handy because we can then refer to the column by its name instead of trying to remember its number. There are restrictions on naming columns. The name of a column *cannot:*

- be longer than 31 characters,
- begin or end with a blank,
- include an apostrophe (') or a number sign (#),
- start with or consist entirely of asterisks (*), or
- be repeated, that is, two columns in the same worksheet cannot have the same name.

To name a column, we click in its column-name cell and type the desired name.

Storing Data: Entering Manually

One of several ways a data set can be stored in a Minitab column is to manually enter the data into a worksheet. Example 1.5 explains how that is done.

EXAMPLE 1.5 MANUALLY STORING DATA

In 1908, W. S. Gosset published a pioneering paper entitled "The Probable Error of a Mean" (*Biometrika,* Vol. 6, pp. 1–25). Table 1.3 displays one of several data sets discussed in the paper. The data give the additional sleep, in hours, obtained by 10 patients using laevohysocyamine hydrobromide.

TABLE 1.3 Additional sleep, in hours, for 10 patients

1.9	0.8	1.1	0.1
−0.1	4.4	5.5	1.6
4.6	3.4		

Store these data in a column named HOURS by manually entering them into a worksheet.

SOLUTION To begin, be sure that the data-entry arrow in the upper-left corner of the worksheet is pointing down. This arrow controls the direction of data entry. The direction can be changed by clicking on the arrow.

Suppose we decide to store the data in C4. To name that column HOURS, we click in the column-name cell for C4 (the cell directly below "C4"), type <u>HOURS</u>, and then press the Enter key.

Now we are ready to store the data from Table 1.3 in C4. We type <u>1.9</u> (the first observation in Table 1.3) in row 1 of C4 and press Enter. Next we type <u>0.8</u> in row 2 of C4 and press Enter. We continue in this manner until all 10 observations have been entered. Figure 1.4 shows the results.

FIGURE 1.4 Worksheet after naming C4 HOURS and entering the data from Table 1.3

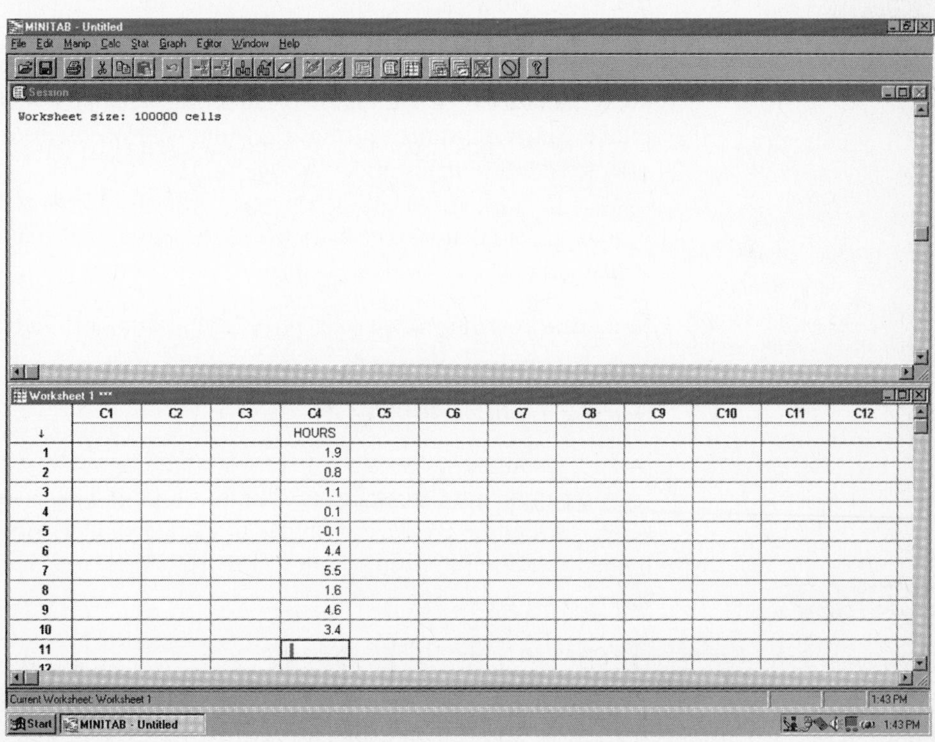

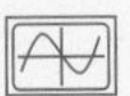

The data from Table 1.3 are now stored in a column named HOURS and are available for analysis. By the way, if an observation is entered incorrectly, we can correct the error by making its cell active and typing the correct observation. ■

Storing Data: Importing From a File

Storing one or more data sets manually works well for small data sets. But for large data sets, manual data entry is tedious and invites errors. If the data are available in an electronic file, we can easily import the data into Minitab. In fact, Minitab supports importing for several file types.

Included with this book is DataDisk CD, a compact disk that contains, among other things, ASCII files for the data sets appearing in the exercises. We will illustrate a method for importing data into Minitab from ASCII files by showing how that is done with DataDisk CD. In doing so, we assume that Minitab has been accessed and that DataDisk CD is in drive `E:`. If DataDisk CD is in a different drive, simply replace `E:` by that drive's designation.

EXAMPLE 1.6 IMPORTING DATA FROM A FILE

The Food and Nutrition Board of the National Academy of Sciences states that the RDA of iron for adult females under the age of 51 is 18 mg. The following iron intakes, in milligrams, during a 24-hour period were obtained for 45 randomly selected adult females under the age of 51.

TABLE 1.4 Iron intakes, in milligrams, for a sample of 45 adult females under the age of 51

15.0	18.1	14.4	14.6	10.9	18.1	18.2	18.3	15.0
16.0	12.6	16.6	20.7	19.8	11.6	12.8	15.6	11.0
15.3	9.4	19.5	18.3	14.5	16.6	11.5	16.4	12.5
14.6	11.9	12.5	18.6	13.1	12.1	10.7	17.3	12.4
17.0	6.3	16.8	12.5	16.3	14.7	12.7	16.3	11.5

These data are from Exercise 9.45 on page 522 and can be found in the file `9-45.dat` in the `Exercise` folder of DataDisk CD. Store these data in a column named IRON by importing from DataDisk CD.

SOLUTION We proceed in the following manner.

1 Choose **File ➤ Other Files ➤ Import Special Text...**

2 Type `IRON` in the **Store data in column(s)** text box

3 Click **OK**

4 Type E:\IS5\Exercise\9-45 in the **File name** text box

5 Click **Open**

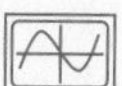

As a result of the above procedure, the iron-intake data in Table 1.4 are now stored in a column named IRON, as you can see by looking in the worksheet. ◨

EXAMPLE 1.7 IMPORTING DATA FROM A FILE

Current Population Reports, a publication of the U.S. Bureau of the Census, provides information on the ages of married people. The ages of 10 married couples are displayed in Table 1.5.

TABLE 1.5 Ages of 10 married couples

Husband	Wife
54	53
21	22
32	33
78	74
70	64
33	35
68	67
32	28
54	41
52	44

These data are from Exercise 10.93 on page 652 and can be found in the file 10-93.dat in the Exercise folder of DataDisk CD. Simultaneously store these data in columns named HUSBAND and WIFE, respectively, by importing from DataDisk CD.

SOLUTION We proceed in the following manner.

1 Choose **File ➤ Other Files ➤ Import Special Text...**

2 Type HUSBAND WIFE in the **Store data in column(s)** text box

3 Click **OK**

4 Type E:\IS5\Exercise\10-93 in the **File name** text box

5 Click **Open**

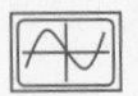

As a result of the above procedure, the age-data for husbands and wives in Table 1.5 are now stored, respectively, in columns named HUSBAND and WIFE, as you can see by looking in the worksheet. ◨

If the data sets for an exercise are contained in more than one file, then each file will be named by the exercise number followed by a letter. For instance, as we see from DataDisk CD, the two data sets for Exercise 10.21 on page 610 are contained, respectively, in the files `10-21a.dat` and `10-21b.dat`. To import these data sets into Minitab, we apply the procedure explained in Example 1.6 twice.

Session, Worksheet, and Graph Windows

We discovered earlier that when we start Minitab, the default is two open windows: the Session window and a worksheet. As we said, the Session window displays non-graphical output such as tables of statistics and character graphs; a worksheet is where we enter, name, view, and edit data.

Another important type of window is a **graph window.** Graph windows are where high-resolution graphs (also called professional graphs) are displayed. Many graph windows can be open simultaneously.

To illustrate different types of windows and how output appears in them, we will apply Minitab procedures to the data sets we stored in Examples 1.6 and 1.7. Don't be concerned right now with the details of the Minitab procedures or the resulting output. Be aware, however, that the output you get may vary slightly in appearance from what is presented here and elsewhere in this book if you are not using Release 12 of the professional version of Minitab for Windows.

EXAMPLE 1.8 SESSION-WINDOW OUTPUT

Use Minitab to obtain several descriptive statistics for the iron-intake data considered in Example 1.6 on page 16.

SOLUTION We have already stored the iron-intake data from Table 1.4 in a column named IRON. To obtain several descriptive statistics for that data, we proceed as follows.

1 Choose **Stat ➤ Basic Statistics ➤ Display Descriptive Statistics...**

2 Specify IRON in the **Variables** text box

3 Click **OK**

As you can see by looking in Minitab's Session window, the output resulting is as shown in Printout 1.1. ◨

PRINTOUT 1.1
Minitab descriptive statistics for the iron-intake data

```
Variable         N       Mean     Median     TrMean      StDev   SE Mean
IRON            45     14.680     14.700     14.741      3.083     0.460

Variable   Minimum    Maximum         Q1         Q3
IRON         6.300     20.700     12.450     16.900
```

EXAMPLE 1.9 GRAPH-WINDOW OUTPUT

Use Minitab to obtain a plot of the wives' ages versus the husbands' ages for the data considered in Example 1.7 on page 17.

SOLUTION We have already stored the age data from Table 1.5 in columns named HUSBAND and WIFE. To obtain a plot of the wives' ages versus the husbands' ages, we proceed as follows.

1 Choose **Graph ➤ Plot...**
2 Specify WIFE in the **Y** text box for **Graph 1**
3 Click in the **X** text box for **Graph 1** and specify HUSBAND
4 Click **OK**

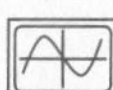

The resulting plot will appear in a Minitab graph window and will look like the plot shown in Printout 1.2. ◨

PRINTOUT 1.2
Minitab plot of the husband-and-wife age data

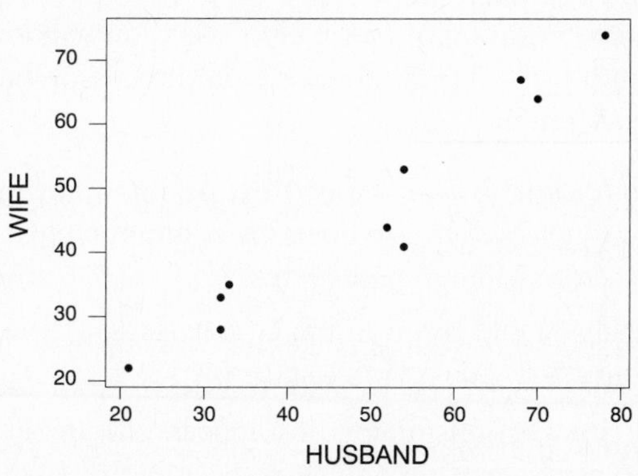

Printing, Saving, and Retrieving in Minitab

There are various ways that we can retain work that we have done in Minitab. The exact procedure depends on what and how we want to retain. Sometimes we want to retain everything and other times only separate pieces such as portions of the Session window, worksheets, or graphs. Generally, retaining can be done either by printing (hard copy) or saving to a file (electronic copy).

Retaining and retrieving a project: Beginning with Release 12 of Minitab, we can save all the work we have done since starting Minitab as a **project.** Saving a project retains the Session window, all graphs, and all worksheets.

- To save a project to a file, we choose **File ➤ Save Project As...**, type the file name (including path, if required) in the **File name** text box, and then click **Save**.
- To retrieve a project from a file, we choose **File ➤ Open Project...**, specify the file name (including path, if required) in the **File name** text box, and then click **Open**. When we retrieve a project, the current project is lost unless we save it (or have already saved it).

Retaining and retrieving a portion of the Session window: To retain a portion of the Session window, we first use a mouse to select (highlight) the required material; we do this by dragging the mouse over the material to be retained. If we want to save the entire Session window, we omit the selection step, but click in the Session window to make that window active.

- To print the selection (or the entire Session window), we choose **File ➤ Print Session Window...** and then click **OK**.
- To save the selection (or the entire Session window) to a file, we choose **File ➤ Save Session Window As...**, type the file name (including path, if required) in the **File name** text box, and then click **Save**.
- When we save a portion of or the entire Session window to a file, we can view, edit, or print it using a text editor or word processor, but we cannot open it from within Minitab.

Retaining and retrieving a worksheet: As we know, a worksheet contains the data that we enter or retrieve. It also contains column names, constants, matrices, and data generated from Minitab procedures.

- To print a worksheet, we first make that worksheet active by clicking any cell of the worksheet. Next we choose **File ➤ Print Worksheet...**. A **Data Window Print Options** dialog box appears; for the default, click **OK** twice. We can also print a portion of a worksheet by first using a mouse to select the required material.
- To save a worksheet to a file, we first make that worksheet active by clicking anywhere in the worksheet. Then we choose **File ➤ Save Worksheet As...**, type the file name (including path, if required) in the **File name** text box, and then click **Save**.

- To retrieve a worksheet from a file, we choose **File ➤ Open Worksheet...**, type the file name (including path, if required) in the **File name** text box, and then click **Open**.

Retaining and retrieving a graph: To retain a graph, we first ensure it is active by clicking with the mouse anywhere on the graph window.

- To print a graph, we choose **File ➤ Print Graph...** and click **OK**.
- To save a graph to a file, we choose **File ➤ Save Graph As...**, type the file name (including path, if required) in the **File name** text box, and then click **Save**.
- To retrieve a graph from a file, we choose **File ➤ Open Graph...**, type the file name (including path, if required) in the **File name** text box, and then click **Open**.

Exiting Minitab

When we are finished using Minitab, we must exit the software. This is accomplished in the following manner.

1 Choose **File ➤ Exit**

2 Decide whether and what to save

It may also be necessary to log off the computer. This will involve a special local procedure that you can obtain from your instructor or an appropriate staff member at the computer center.

EXERCISES 1.4

STATISTICAL CONCEPTS AND SKILLS

1.17 Name three technologies that can be used to conduct statistical and data analyses.

1.18 Explain a function of each of the following windows in Minitab.
a. Session window **b.** worksheet
c. graph window

1.19 Describe the meaning of each of the following mouse terms.
a. point **b.** click
c. double-click **d.** drag

1.20 Regarding Minitab menu instructions, what does underlined text set in typewriter-type refer to?

1.21 Regarding Minitab menu instructions, what does **Calc ➤ Random Data ➤ Normal...** ask the user to do?

1.22 In Minitab, we store a data set in a _____.

1.23 How is the fifth column in Minitab designated?

1.24 Identify the meaning of each of the following in a Minitab worksheet.
a. column **b.** row
c. cell **d.** active cell

USING TECHNOLOGY  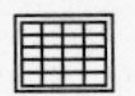 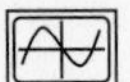

In Exercises 1.25–1.30, store the following data in a Minitab worksheet or in a corresponding place if you are using a different technology.

1.25 The Energy Information Administration collects data on residential energy consumption and expenditures and publishes its findings in *Residential Energy Consumption Survey: Consumption and Expenditures.* The following table shows one year's energy consumptions, in millions of BTU, for a sample of 50 households in the South.

130	55	45	64	155	66	60	80	102	62
58	101	75	111	151	139	81	55	66	90
97	77	51	67	125	50	136	55	83	91
54	86	100	78	93	113	111	104	96	113
96	87	129	109	69	94	99	97	83	97

a. Manually enter the data on energy consumptions in a column named ENERGY1.

b. The energy-consumption data are stored in the file `1-25.dat` which is found in the `Exercise` folder of DataDisk CD. Import the data from DataDisk CD into a column named ENERGY2.

1.26 The Bureau of Economic Analysis publishes data on the length of stay in Europe and the Mediterranean by U.S. travelers in *Survey of Current Business.* A sample of 36 U.S. residents who traveled to Europe and the Mediterranean one year yielded the following data on length of stay, in days.

41	16	6	21	1	21
5	31	20	27	17	10
3	32	2	48	8	12
21	44	1	56	5	12
3	13	15	10	18	3
1	11	14	12	64	10

a. Manually enter the length-of-stay data in a column named STAY1.

b. The length-of-stay data are in the file `1-26.dat`, found in the `Exercise` folder of DataDisk CD. Store the data in a column named STAY2 by importing from DataDisk CD.

1.27 The U.S. National Center for Health Statistics publishes data on heights and weights by age and sex in *Vital and Health Statistics.* A sample of 11 males age 18–24 years yielded the following data on height, in inches, and weight, in pounds.

Height	Weight
65	175
67	133
71	185
71	163
66	126
75	198
67	153
70	163
71	159
69	151
69	155

a. Manually enter the data on heights and weights in columns named HEIGHT1 and WEIGHT1.

b. The height and weight data are in the file `1-27.dat`, found in the `Exercise` folder of DataDisk CD. Store the data in columns named HEIGHT2 and WEIGHT2 by importing from DataDisk CD.

1.28 Hanna Properties specializes in custom-home resales in the Equestrian Estates, an exclusive subdivision in Phoenix, Arizona. A sample of nine custom homes currently listed for sale provided the following information on size and price. The size data are in hundreds of square feet; the price data are in thousands of dollars.

Size	26	27	33	29	29	34	30	40	22
Price	235	249	267	269	295	345	415	475	195

a. Manually enter the size and price data in columns named SIZE1 and PRICE1, respectively.

b. The size and price data are in the file `1-28.dat`, found in the `Exercise` folder of DataDisk CD. Store the data in columns named SIZE2 and PRICE2 by importing from DataDisk CD.

1.29 The National Center for Education Statistics surveys college libraries to obtain information on the number of volumes held. Results of the surveys are published in *Digest of Education Statistics* and *Academic Libraries.* Samples of public and private colleges yield the following data on number of volumes held, in thousands rounded to the nearest thousand.

Public	Private
79	139
41	603
516	113
15	27
24	67
411	500
265	

a. Manually enter the data on number of volumes in columns named PUBLIC1 and PRIVATE1.

b. The data on number of volumes are in the files `1-29a.dat` and `1-29b.dat`, found in the `Exercise` folder of DataDisk CD. Store the data in columns named PUBLIC2 and PRIVATE2 by importing from DataDisk CD.

1.30 Data on household vehicle miles of travel (VMT) are compiled annually by the Federal Highway Administration and are published in *National Personal Transportation Survey, Summary of Travel Trends.* Samples of 15 midwestern households and 14 southern households provide the following data on last year's VMT, in thousands of miles.

Midwest			South		
16.2	12.9	17.3	22.2	19.2	9.3
14.6	18.6	10.8	24.6	20.2	15.8
11.2	16.6	16.6	18.0	12.2	20.1
24.4	20.3	20.9	16.0	17.5	18.2
9.6	15.1	18.3	22.8	11.5	

a. Manually enter the VMT data in columns named MIDWEST1 and SOUTH1.

b. The VMT data are in the files `1-30a.dat` and `1-30b.dat` which are found in the `Exercise` folder of DataDisk CD. Store these two data sets in columns named MIDWEST2 and SOUTH2 by importing from DataDisk CD.

1.5 IS A STUDY NECESSARY?

Throughout this book we will see examples of organizations or people conducting studies: a consumer group wants information about the gas mileage of a particular make of car, so it performs mileage tests on a sample of such cars and statistically analyzes the resulting data; or a teacher wants to know about the comparative merits of two teaching methods, so he tests those methods on two groups of students.

This reflects a healthy attitude—to obtain information about a subject of interest, plan and conduct a study. However, the possibility always exists that a study being considered has already been done. Repeating it would be a waste of time, energy, and money. Therefore before a study is planned and conducted, a literature search should be made. This does not require going through all the books in the library. Many information-collection agencies specialize in finding studies on specific topics in specific areas.

For instance, the Educational Resources Information Center assembles educational studies; publications entitled *Psychological Abstracts* gather the results of studies in psychology; the National Library of Medicine compiles lists of medical studies and makes them accessible to research centers and universities. A considerable amount of information can also be obtained from publications by government agencies such as the Bureau of the Census and the Environmental Protection Agency. Data are published on income, age, energy consumption, and hundreds of other variables. Many of the examples and exercises in this book are based on information obtained from the Census Bureau's *Statistical Abstract of the United States.*

It is not our purpose to explain how to search through journal articles, abstracts, or census data. The important point here is this: *It is often possible to avoid the effort and expense of a study if someone else has already done that study and published the results.*

1.6 SIMPLE RANDOM SAMPLING

If it has been determined that the information required is not already available from a previous study, a new study can be planned to obtain the information. One method for acquiring information is to conduct a **census,** meaning that information is obtained for the entire population of interest. However, conducting a census is generally time consuming and costly, frequently impractical, and sometimes impossible.

Two methods other than a census for obtaining information are sampling and experimentation. In most of this book, we will concentrate on sampling. Experimentation is introduced in Section 1.8 and is discussed sporadically throughout the text. Readers interested in a detailed treatment of experimentation are referred to the module *Design of Experiments and Analysis of Variance* by Dennis L. Young (Reading, MA: Addison-Wesley, 1999).

If sampling has been deemed appropriate, we must then decide how to select the sample, that is, we must choose the method for obtaining a sample from the population. In making that choice, we should keep in mind that the sample will be used to draw conclusions about the entire population. Consequently, it is essential that the sample be a **representative sample**—it should reflect as closely as possible the relevant characteristics of the population under consideration.

For instance, it would not make much sense to use the average weight of a sample of football players to make an inference about the average weight of all adult males. Nor would it be reasonable to try to estimate the median income of California residents by sampling the incomes of Beverly Hills residents.

To see what can happen when a sample is not representative, let's consider the presidential election of 1936. Before the election, the *Literary Digest* magazine conducted an opinion poll of the voting population. Its survey team asked a sample of the voting population whether they would vote for Franklin D. Roosevelt, the Democratic candidate, or for Alfred Landon, the Republican candidate.

Based on the results of the survey, the magazine predicted an easy win for Landon. The actual election results, of course, were that Roosevelt won by the greatest landslide in the history of presidential elections! What happened? Here are two reasons given for why the poll failed:

- The sample was obtained from among people who owned a car or had a telephone. In 1936 that group included only the more well-to-do people, and historically such people tend to vote Republican.
- The response rate was low (less than 25% of those polled responded) and there was a nonresponse bias (a disproportionate number of those responding to the poll were Landon supporters).

Whatever the reason for the poll's failure, the sample obtained by the *Literary Digest* was obviously not representative.

Most modern sampling procedures employ **probability sampling.** In probability sampling, a random device, such as tossing a coin or consulting a table of random numbers, is used to decide which members of the population will constitute the sample instead of leaving such decisions to human judgment.

It is still possible to obtain a nonrepresentative sample when probability sampling is used. However, probability sampling eliminates unintentional selection bias and permits the researcher to control the chance of obtaining a nonrepresentative sample. Furthermore, the use of probability sampling guarantees that the techniques of inferential statistics can be applied. In this section and the next, we will examine the most important probability-sampling methods.

Simple Random Sampling

The inferential techniques considered in this book are intended for use with only one particular sampling procedure: **simple random sampling,** or more briefly, **random sampling.** Simple random sampling is the simplest type of probability sampling and is also the basis for the more complex types of probability sampling.

DEFINITION 1.4 SIMPLE RANDOM SAMPLING; SIMPLE RANDOM SAMPLES

Simple random sampling: A sampling procedure for which each possible sample of a given size is equally likely to be the one obtained.

Simple random sample: A sample obtained by simple random sampling.

There are two types of simple random sampling. One is **simple random sampling with replacement,** where a member of the population can be selected more than once; the other is **simple random sampling without replacement,** where a member of the population can be selected at most once. *Unless otherwise specified, we will assume that simple random sampling is done without replacement.*

In Example 1.10 we have chosen a very small population—the five top Oklahoma state officials—to illustrate simple random sampling. In practice we would not sample from such a small population but would instead take a census. We are using a small population here to make it easier to understand the concept of simple random sampling.

EXAMPLE 1.10 ILLUSTRATES DEFINITION 1.4

As reported by the *World Almanac, 1998,* the top five state officials of Oklahoma are as shown in Table 1.6. Consider these five officials a population of interest.

TABLE 1.6 Five top Oklahoma state officials

Official
Governor (G)
Lieutenant Governor (L)
Secretary of State (S)
Attorney General (A)
Treasurer (T)

a. List the possible samples (without replacement) of two officials from this population of five officials.

b. Describe a method for obtaining a simple random sample of two officials from this population of five officials.

c. For the sampling method described in part (b), what are the chances that any particular sample of two officials will be the one selected?

d. Repeat parts (a)–(c) for samples of size four.

SOLUTION For convenience we will use the letters placed parenthetically after the officials in Table 1.6 to represent the officials.

a. There are 10 possible samples of two officials from the population of five officials. They are listed in Table 1.7.

TABLE 1.7 Possible samples of two officials from the population of five officials

G, L	G, S
G, A	G, T
L, S	L, A
L, T	S, A
S, T	A, T

b. Here is one method we could use to obtain a simple random sample of size two: First write each of the letters corresponding to the five officials, G, L, S, A, and T, on separate pieces of paper. Next place the five slips of paper into a box and shake the box. Then, while blindfolded, pick two of the slips of paper.

c. The sampling procedure described in part (b) ensures that we are taking a simple random sample. Consequently, each of the possible samples of two officials is equally likely to be the one selected. Since there are 10 possible samples, the chances are $\frac{1}{10}$ (1 in 10) that any particular sample will be the one selected.

d. For samples of size four, there are five possibilities, as indicated in Table 1.8.

TABLE 1.8 Possible samples of four officials from the population of five officials

G, L, S, A
G, L, S, T
G, L, A, T
G, S, A, T
L, S, A, T

In this case a simple-random-sampling procedure, such as picking four slips of paper out of a box, gives each of the five possible samples in Table 1.8 a 1 in 5 chance of being the one selected. ◨

Random-Number Tables

Obtaining a simple random sample by picking slips of paper out of a box is usually not practical, especially when the population being sampled is large. But there are several practical procedures for getting simple random samples. One common method uses a **table of random numbers,** a table of randomly chosen digits. Example 1.11 explains how a table of random numbers can be used to obtain a simple random sample.

EXAMPLE 1.11 USING RANDOM-NUMBER TABLES

Student questionnaires, known as "teacher evaluations," gained widespread use about 30 years ago. Generally, student evaluations of teachers are not done at final exam time. It is more common for professors to hand out evaluation forms a week or so before the final.

That practice, however, poses several problems. On some days less than 60% of the students registered for a class may actually attend. Moreover, because many of those who are present have other classes to prepare for, they often fill out their teacher-evaluation forms in a hurry so that they can leave class early. It may well be better, therefore, to select a sample of students from the class and interview them individually. This is the kind of situation in which a simple random sample should be obtained.

During one semester, Professor Hassett wanted to sample the attitudes of the students taking college algebra at his school. He decided to interview 15 of the 728 students enrolled in the course. Since Professor Hassett had a registration list on which the 728 students were numbered 1–728, he could obtain a simple random sample of 15 students by randomly selecting 15 numbers between 1 and 728. To do this he used a table of random numbers. The random-number table employed by Professor Hassett is presented as Table I in Appendix A. For ease of reference, we repeat it here as Table 1.9.

TABLE 1.9
Random numbers

Line number	Column number									
	00–09		*10–19*		*20–29*		*30–39*		*40–49*	
00	15544	80712	97742	21500	97081	42451	50623	56071	28882	28739
01	01011	21285	04729	39986	73150	31548	30168	76189	56996	19210
02	47435	53308	40718	29050	74858	64517	93573	51058	68501	42723
03	91312	75137	86274	59834	69844	19853	06917	17413	44474	86530
04	12775	08768	80791	16298	22934	09630	98862	39746	64623	32768
05	31466	43761	94872	92230	52367	13205	38634	55882	77518	36252
06	09300	43847	40881	51243	97810	18903	53914	31688	06220	40422
07	73582	13810	57784	72454	68997	72229	30340	08844	53924	89630
08	11092	81392	58189	22697	41063	09451	09789	00637	06450	85990
09	93322	98567	00116	35605	66790	52965	62877	21740	56476	49296
10	80134	12484	67089	08674	70753	90959	45842	59844	45214	36505
11	97888	31797	95037	84400	76041	96668	75920	68482	56855	97417
12	92612	27082	59459	69380	98654	20407	88151	56263	27126	63797
13	72744	45586	43279	44218	83638	05422	00995	70217	78925	39097
14	96256	70653	45285	26293	78305	80252	03625	40159	68760	84716
15	07851	47452	66742	83331	54701	06573	98169	37499	67756	68301
16	25594	41552	96475	56151	02089	33748	65289	89956	89559	33687
17	65358	15155	59374	80940	03411	94656	69440	47156	77115	99463
18	09402	31008	53424	21928	02198	61201	02457	87214	59750	51330
19	97424	90765	01634	37328	41243	33564	17884	94747	93650	77668
							↓	↑		

To select 15 random numbers between 1 and 728, we first pick a random starting point, say, by closing our eyes and putting our finger down on Table 1.9. Then beginning with the three digits under our finger, we go down the table and record the numbers as we go. Since we want numbers between 1 and 728 only, we discard the number 000 and numbers between 729 and 999. To avoid repetition we also eliminate numbers that have occurred previously. If not enough numbers have been found by the time we reach the bottom of the table, we move over to the next column of three-digit numbers and go up.

Using this procedure, Professor Hassett obtained 069, circled in Table 1.9, as a starting point. Reading down from 069 to the bottom of Table 1.9 and then up the next column of three-digit numbers, he found the 15 random numbers displayed in Fig. 1.5 and in Table 1.10.

FIGURE 1.5
Procedure used by Professor Hassett to obtain 15 random numbers between 1 and 728 from Table 1.9

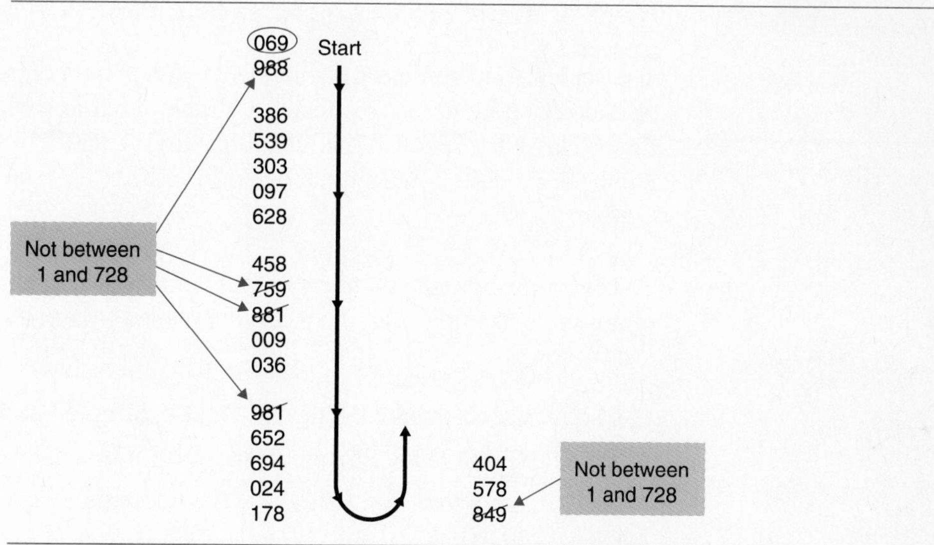

TABLE 1.10
Registration numbers of students interviewed

69	303	458	652	178
386	97	9	694	578
539	628	36	24	404

  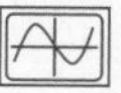

Thus Professor Hassett interviewed the 15 students whose numbers on the registration list are the ones shown in Table 1.10. ◨

Many calculators and most computers have a **random-number generator,** which makes it possible to automatically obtain a list of random numbers within any specified range. Tables of random numbers, such as Table 1.9, were used much more frequently before inexpensive machines with random-number generators became available. If you have a calculator or computer with a random-number generator, you may find it easier to use than a table of random numbers. Using random-number generators is discussed in the exercises at the end of this section.

Using the Computer (Optional)

Minitab can be employed to obtain a simple random sample from a population. To illustrate how this is accomplished, we return to the situation of Example 1.11 where Professor Hassett wanted to obtain a sample of the students taking college algebra at his school.

EXAMPLE 1.12 USING MINITAB TO OBTAIN A SIMPLE RANDOM SAMPLE

The students in Professor Hassett's college-algebra course were numbered 1–728 on the registration list. To obtain a simple random sample of 15 students, Professor Hassett selected 15 numbers between 1 and 728 from a table of random numbers. Explain how Minitab could have been used to obtain a simple random sample.

SOLUTION To begin, we store the numbers 1–728, corresponding to the registration numbers of the students in the class, in a column named NUMBERS. An easy way to do that is as follows.

1 Choose **Calc ➤ Make Patterned Data ➤ Simple Set of Numbers...**
2 Type NUMBERS in the **Store patterned data in** text box
3 Click in the **From first value** text box and type 1
4 Click in the **To last value** text box and type 728
5 Click **OK**

Note: In listing the steps above, we assume the **Simple Set of Numbers** dialog box contains the default settings so that, for example, the **In steps of** text box contains the number 1. Minitab retains settings in dialog boxes until we create a new project or exit and restart Minitab. This is convenient for a number of reasons, as we will see. Unless otherwise specified, however, we will presume that dialog boxes contain the default settings. To ensure this, press the F3 key immediately after entering a dialog box.

As you can see by looking in the worksheet, the numbers 1–728 are now stored in a column named NUMBERS. To obtain 15 randomly selected numbers between 1 and 728, we proceed in the following manner.

1 Choose **Calc ➤ Random Data ➤ Sample From Columns...**
2 Type 15 in the small text box after **Sample**
3 Click in the **Sample** [ ] **rows from column(s)** text box and specify NUMBERS
4 Click in the **Store samples in** text box and type SRS
5 Click **OK**

The 15 randomly selected numbers are now stored in a column named SRS, as you can see by looking in the worksheet. Professor Hassett could have used these 15 numbers as the registration numbers of the students to be interviewed. ◨

We can obtain a hard copy of the 15 numbers in SRS in various forms. One form is obtained by printing the appropriate portion of the worksheet, as explained on page 20. When we print in this way, the 15 numbers appear in one long column, just as they do in the worksheet.

Another form is obtained by first displaying the 15 numbers in SRS (and only those numbers) in the Session window. To do that, we proceed as follows.

1 Choose **Manip ➤ Display Data...**
2 Specify SRS in the **Columns, constants, and matrices to display** text box
3 Click **OK**

As you can see by looking in the Session window, the 15 numbers in SRS are displayed on two lines, each line containing several numbers. If we want a hard copy of the 15 numbers displayed in this form, we can print the appropriate portion of the Session window, as explained on page 20.

EXERCISES 1.6

STATISTICAL CONCEPTS AND SKILLS

1.31 Explain why a census is often not the best way to obtain information about a population.

1.32 Identify two methods other than a census for obtaining information.

1.33 In sampling, why is it important to obtain a representative sample?

1.34 The political opinions of 150 voters in the retirement community of Sun City, Arizona, is used as a sample of the political opinions of all Arizona voters. Is the sample representative? Explain your answer.

1.35 Explain why the following sample is not representative: A sample of 30 dentists from Seattle is taken in order to estimate the median income of all Seattle residents.

1.36 Provide a scenario of your own in which a sample is not representative.

1.37 Regarding probability sampling:
a. What is it?
b. Answer true or false to the following statement and explain your answer: Probability sampling always yields a representative sample.
c. Identify some advantages of probability sampling.

1.38 Regarding simple random sampling:
a. What is simple random sampling?
b. What is a simple random sample?
c. Identify two forms of simple random sampling and explain the difference between the two.

1.39 The inferential procedures discussed in this book are intended for use with only one particular sampling procedure. What sampling procedure is that?

1.40 Identify two methods for obtaining a simple random sample.

1.41 The five top Oklahoma state officials are displayed in Table 1.6 on page 26. Use that table to solve the following problems.

a. List the 10 possible samples (without replacement) of size three that can be obtained from the population of five officials.

b. If a simple-random-sampling procedure is used to obtain a sample of three officials, what are the chances that it is the first sample on your list in part (a)? the second sample? the tenth sample?

1.42 According to the *World Almanac,* the members of the U.S. House of Representatives from South Carolina as of October 15, 1997, are Mark Sanford (MS), Floyd Spence (FS), Lindsey Graham (LG), Bob Inglis (BI), John Spratt, (JS), and James Clyburn (JC).

a. List the 15 possible samples (without replacement) of two representatives that can be selected from the six. For brevity use the initials provided.

b. Describe a procedure for taking a simple random sample of two representatives from the six.

c. If a simple-random-sampling procedure is used to obtain two representatives, what are the chances of selecting MS and FS? BI and JC?

1.43 Refer to Exercise 1.42.

a. List the 15 possible samples (without replacement) of four representatives that can be selected from the six.

b. Describe a procedure for taking a simple random sample of four representatives from the six.

c. If a simple-random-sampling procedure is used to obtain four representatives, what are the chances of selecting MS, FS, LG, and BI? FS, JS, BI, and JC?

1.44 Refer to Exercise 1.42.

a. List the 20 possible samples (without replacement) of three representatives that can be selected from the six.

b. Describe a procedure for taking a simple random sample of three representatives from the six.

c. If a simple-random-sampling procedure is used to obtain three representatives, what are the chances of selecting MS, FS, and LG? BI, JS, and MS?

In each of Exercises 1.45–1.48, use Table I in Appendix A to obtain the required list of random numbers.

1.45 The owner of a business that employs 685 people wants to select 25 of them at random for extensive interviewing. Construct a list of 25 random numbers between 1 and 685 that can be used in obtaining the required simple random sample.

1.46 A university committee on parking has been formed to gauge the sentiment of the people using the university's parking facilities. Each person that uses the facilities has a parking sticker with a number. This year the numbers range from 1 to 8493. Make a list of 30 numbers that can be employed to obtain a simple random sample of 30 people who use the parking facilities.

1.47 Each year *Fortune Magazine* publishes an article entitled "The International 500" that provides a ranking by sales of the top 500 firms outside the United States. Suppose you want to examine various characteristics of successful firms. Further suppose that for your study you decide to take a simple random sample of 10 firms from *Fortune Magazine*'s list of "The International 500." Determine 10 numbers you can use to obtain your sample.

1.48 In the game of keno, there are 80 balls, numbered 1–80, and 20 of the 80 balls are selected at random. Simulate one game of keno by obtaining 20 random numbers between 1 and 80.

EXTENDING THE CONCEPTS AND SKILLS

1.49 Refer to Exercise 1.41.

a. Repeat part (a) for samples of size one.

b. What is the difference between obtaining a sample of size one and selecting one official at random?

1.50 Refer to Exercise 1.41.

a. Repeat part (a) for samples of size five.

b. What is the difference between obtaining a sample of size five and taking a census?

Random-number generators. As we mentioned earlier, a random-number generator makes it possible to automatically obtain a list of random numbers within any specified range. Usually a random-number generator returns a number, r, between 0 and 1. To obtain random integers in an arbitrary range, A to B, use the conversion $A + (B - A + 1)r$ and round down to the nearest integer. For example, to obtain random integers in the range from 1 to 728, use the conversion $1 + (728 - 1 + 1)r = 1 + 728r$ and round down to the nearest integer. We will discuss the use of a random-number generator in Exercises 1.51 and 1.52.

1.51 Refer to Exercise 1.45.

a. Explain how a random-number generator can be used to obtain 25 random numbers between 1 and 685, and thereby identify 25 employees to be interviewed.

b. If you have access to a random-number generator, implement part (a).

1.52 Refer to Exercise 1.46.

a. Explain how a random-number generator can be used to obtain 30 random numbers between 1 and 8493, and thereby identify 30 people who use the parking facilities.

b. If you have access to a random-number generator, implement part (a).

USING TECHNOLOGY  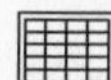 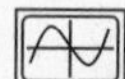

1.53 Refer to Exercise 1.47. Use Minitab or some other statistical software to determine 10 random numbers that can be used to obtain a simple random sample of 10 firms from *Fortune Magazine*'s list of "The International 500."

1.54 Refer to Exercise 1.48. Use Minitab or some other statistical software to simulate one game of keno, that is, to obtain a simple random sample of 20 numbers between 1 and 80.

1.7 OTHER SAMPLING PROCEDURES

Simple random sampling is the most natural and easiest to understand method of probability sampling—it corresponds to our intuitive notion of random selection by lot. However, simple random sampling does have drawbacks. For instance, it may fail to provide sufficient coverage when information about subpopulations is required and may be impractical when the members of the population are widely scattered geographically. In this section we will examine some other commonly used sampling procedures that are often more appropriate than simple random sampling.

Systematic Random Sampling

One method that takes less effort to implement than simple random sampling is **systematic random sampling.** Consider Example 1.13.

EXAMPLE 1.13 SYSTEMATIC RANDOM SAMPLING

Let's return to the situation in Example 1.11 where Professor Hassett wanted to obtain a sample of 15 of the 728 students enrolled in college algebra at his school. Use systematic random sampling to obtain the sample.

SOLUTION To begin, we divide the population size by the sample size and round the answer down to the nearest whole number: $\frac{728}{15} = 48$ (rounded down). Next we select a number at random between 1 and 48 using, say, a table of random numbers. We did this and obtained the number 22. Then we list every 48th number, starting at 22, until we have 15 numbers. This yields the 15 numbers displayed in Table 1.11.

TABLE 1.11 Numbers obtained by systematic random sampling

22	166	310	454	598
70	214	358	502	646
118	262	406	550	694

Had Professor Hassett used systematic random sampling to obtain his sample of students and had he obtained the number 22 as his starting point, he would have interviewed the 15 students whose numbers on the registration list are the ones shown in Table 1.11.

 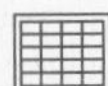

As illustrated in Example 1.13, we implement systematic random sampling using the following three steps:

1. Divide the population size by the sample size and round the result down to the nearest whole number, m.
2. Use a random number table (or a similar device) to obtain a number, k, between 1 and m.
3. Select for the sample those members of the population that are numbered $k, k + m, k + 2m, \ldots$.

Systematic random sampling is not only easier to execute than simple random sampling, but it also provides results comparable to simple random sampling unless there is some kind of cyclical pattern in the listing of the members of the population (e.g., male, female, male, female, . . .), a phenomenon that is relatively rare.

Cluster Sampling

Another alternative to simple random sampling is **cluster sampling.** This method is particularly useful when the members of the population under consideration are widely scattered geographically.

EXAMPLE 1.14 CLUSTER SAMPLING

At one time the city council of Tempe, Arizona, was being pressured by citizens' groups to install bike paths in the city. The members of the council wanted to be sure they had the support of a majority of the taxpayers, so they decided to poll the homeowners in the city.

Their first attempt at surveying public opinion was a questionnaire mailed out with the city's 18,000 homeowner water bills. Unfortunately, this method did not work very well. Only 19.4% of the questionnaires were returned, and a large number of those had comments written on them indicating that they came from avid bicyclists or people strongly resenting bicyclists. The questionnaire generally had not been returned by the average voter, and the city council realized that.

The city had an employee in the planning department with sample-survey experience. The council called her in and asked her to do a survey. She was given two assistants to help interview a representative sample of voters and was instructed to report back in 10 days.

The planner thought about taking a simple random sample of 300 voters, 100 interviews for herself and for each of her two assistants. However, using a simple random sample created some time problems. The city was so spread out that an interviewer with a list of 100 voters randomly scattered around the city would have to drive an average of 18 minutes from one interview to the next. This would require approximately 30 hours of driving time for each interviewer and could delay completion of the report. Obviously, simple random sampling would not do.

To save time the planner decided to use cluster sampling. The residential portion of the city was divided into 947 blocks, each containing approximately 20 houses, as seen in Fig. 1.6.

FIGURE 1.6
A typical block of homes

The planner numbered the blocks (clusters) on the city map from 1 to 947 and then used a table of random numbers to obtain a simple random sample of 15 of the 947 blocks. Each of the three interviewers was then assigned 5 of the 15 blocks obtained. This method gave each interviewer roughly 100 homes to visit but saved a great deal of travel time; an interviewer could work on a block for nearly a full day without having to drive to another neighborhood. The report was finished on time. ◨

In the simplest case, as illustrated by Example 1.14, cluster sampling is implemented using the following three steps:

1. Divide the population into groups (clusters).
2. Obtain a simple random sample of the clusters.
3. Use all of the members of the clusters obtained in step 2 as the sample.

Although cluster sampling can save time and money, it does have disadvantages. Ideally, each cluster should mirror the entire population. However, that is often not the case, as members of a cluster are frequently more homogeneous than the members of the population as a whole. This situation can cause problems.

For instance, let's look at a simplified small town, as depicted in Fig. 1.7. The town council is thinking about building a town swimming pool. A planner for the town needs to sample voter sentiment on using public funds to build the pool. Many upper-income and middle-income homeowners will probably say "No" because they own pools or can use a neighbor's. Many low-income voters will probably say "Yes" because they generally do not have access to pools.

FIGURE 1.7
Clusters for a small town

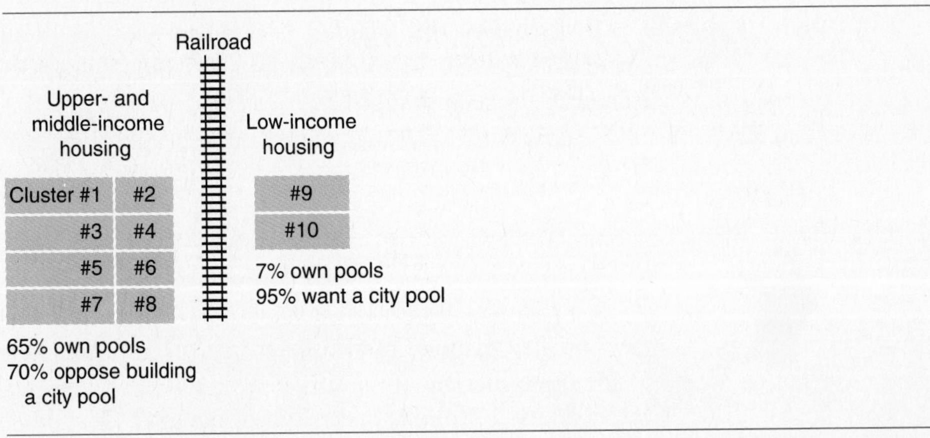

If the planner uses cluster sampling and interviews the voters of, say, three randomly selected clusters, then there is a good chance that no low-income voters will be interviewed.† And if no low-income voters are interviewed, the results of the survey will be misleading. Suppose, for instance, the planner obtained clusters #3, #5, and #8. Then his survey would show that only about 30% of the voters

† There are 120 possible three-cluster samples and 56 of those contain neither of the low-income clusters, #9 and #10. In other words, 46.7% of the possible three-cluster samples contain neither of the low-income clusters.

want a pool. But that is not true. More than 40% of the voters want a pool. The clusters that most strongly support the town swimming pool would not have been included in the survey.

In this hypothetical example, the town is so small that common sense indicates that a cluster sample may not be representative. However, in situations where there are hundreds of clusters, such problems may be more difficult to detect.

Stratified Sampling

Another sampling method, known as **stratified sampling,** is often more reliable than cluster sampling. In stratified sampling the population is first divided into subpopulations, called **strata,** and then sampling is done from each stratum. Ideally, the members of each stratum should be homogeneous relative to the characteristic under consideration.

EXAMPLE 1.15 STRATIFIED SAMPLING

Consider again the town-swimming-pool illustration. In stratified sampling, voters could be divided into three strata: upper-income, middle-income, and low-income. A simple random sample could then be taken from each of the three strata.

This stratified-sampling procedure would ensure that no income group is missed. It would also improve the precision of the statistical estimates (since the voters within each income group tend to be somewhat homogeneous) and would make it possible to estimate the separate opinions of each of the three strata. □

In stratified sampling the strata are often sampled in proportion to their size; this is called **proportional allocation.** For instance, suppose the strata consisting of the three income groups (upper, middle, and low) in Example 1.15 comprise, respectively, 10%, 70%, and 20% of the town. Then for a sample size of 50, say, the number of upper-income, middle-income, and low-income individuals sampled would be, respectively, 5 (10% of 50), 35 (70% of 50), and 10 (20% of 50).

The simplest type of stratified sampling, called **stratified random sampling with proportional allocation,** is implemented using the following three steps:

1. Divide the population into subpopulations (strata).
2. From each stratum obtain a simple random sample of size proportional to the size of the stratum; that is, the sample size for a stratum equals the total sample size times the stratum size divided by the population size.
3. Use all of the members obtained in step 2 as the sample.

Multistage Sampling

Most large-scale surveys combine one or more of simple random sampling, systematic random sampling, cluster sampling, and stratified sampling in ways that can be quite complex. Such **multistage sampling** is used frequently by pollsters and government agencies.

For instance, the U.S. National Center for Health Statistics conducts surveys of the civilian noninstitutional U.S. population to obtain information on illnesses, injuries, and other health issues. Data collection is by a multistage probability sample of approximately 42,000 households. Information obtained from the surveys is published in *National Health Interview Survey.*

EXERCISES 1.7

Statistical Concepts and Skills

1.55 In Exercise 1.45 on page 32, we used simple random sampling to obtain 25 numbers between 1 and 685, and thereby identified 25 employees to be interviewed.

a. Employ systematic random sampling to accomplish that same task.

b. Which method is easier: simple random sampling or systematic random sampling?

c. Does it seem reasonable to use systematic random sampling to obtain a representative sample? Explain your answer.

1.56 In Exercise 1.46 on page 32, we used simple random sampling to obtain 30 numbers between 1 and 8493, and thereby identified 30 people who use a university's parking facilities.

a. Employ systematic random sampling to accomplish the same thing.

b. Which method is easier: simple random sampling or systematic random sampling?

c. Does it seem reasonable to use systematic random sampling to obtain a representative sample? Why?

1.57 Refer to Exercise 1.47 on page 32. Would it be reasonable to use systematic random sampling? Explain your answer.

1.58 Refer to Exercise 1.48 on page 32. Would it be reasonable to use systematic random sampling? Why?

1.59 Students in the dormitories of a university in the state of New York live in clusters of four double rooms, called *suites*. There are 48 suites, with eight students per suite.

a. Describe a cluster-sampling procedure for obtaining a sample of 24 dormitory residents.

b. Students typically choose friends from their classes as suitemates. With that in mind, do you think cluster sampling is a good procedure for obtaining a representative sample of dormitory residents? Explain your answer.

c. The university housing office has separate lists of dormitory residents by class level. Using those lists, the following table was obtained showing the number of dormitory residents at each class level.

Class level	Number of dorm residents
Freshman	128
Sophomore	112
Junior	96
Senior	48

Employ the table to design a procedure for obtaining a stratified sample of 24 dormitory residents. Use stratified random sampling with proportional allocation.

EXTENDING THE CONCEPTS AND SKILLS

1.60 In simple random sampling, all samples of a given size are equally likely. Is that true in systematic random sampling? Explain your answer.

1.61 On June 27, 1996, an article appeared in *The Wall Street Journal* presenting the results of a nationwide poll regarding the White House procurement of FBI files on prominent Republicans and related ethical controversies. The article was entitled "White House Assertions on FBI Files Are Widely Rejected, Survey Shows." At the end of the article, the explanation of the sampling procedure found in the box at the right was given. Discuss the different aspects of sampling that appear in this explanation.

> The Wall Street Journal/NBC News poll was based on nationwide telephone interviews of 2,010 adults, including 1,637 registered voters, conducted Thursday to Tuesday by the polling organizations of Peter Hart and Robert Teeter. Questions related to politics were asked only of registered voters; questions related to economics and health were asked of all adults.
>
> The sample was drawn from 520 randomly selected geographic points in the continental U.S. Each region was represented in proportion to its population. Households were selected by a method that gave all telephone numbers, listed and unlisted, an equal chance of being included.
>
> One adult, 18 years or older, was selected from each household by a procedure to provide the correct number of male and female respondents.
>
> Chances are 19 of 20 that if all adults with telephones in the U.S. had been surveyed, the finding would differ from these poll results by no more than 2.2 percentage points in either direction among all adults and 2.5 among registered voters. Sample tolerances for subgroups are larger.

1.8 EXPERIMENTAL DESIGN

As we mentioned earlier, two methods for obtaining information other than a census are sampling and experimentation. In Sections 1.6 and 1.7, we discussed some of the basic principles and techniques of sampling. Now we will do the same for experimentation. To begin, we introduce some terminology that further helps us differentiate among types of studies.

Observational Studies and Designed Experiments

Often the purpose of a statistical study is to investigate whether a relationship exists between two characteristics, such as smoking and lung cancer, height and weight, or educational attainment and annual income. For these kinds of studies it is important to distinguish between two types of procedures: observational studies and designed experiments.

In an **observational study,** researchers simply observe characteristics and take measurements, as in a sample survey. On the other hand, in a **designed experiment,** researchers impose treatments and controls and then observe characteristics and take measurements. Observational studies can reveal only *association,* whereas designed experiments can help establish *causation.* Examples 1.16 and 1.17 illustrate differences between observational studies and designed experiments.

EXAMPLE 1.16 AN OBSERVATIONAL STUDY

Approximately 450,000 vasectomies are performed each year in the United States. In this surgical procedure for contraception, the tube carrying sperm from the testicles is cut.

Several studies have been conducted to analyze the relationship between vasectomies and prostate cancer. One such study appeared in a February 1993 issue of *The Journal of the American Medical Association.* Dr. Edward Giovannucci, leader of the study and epidemiologist at Harvard-affiliated Brigham and Women's Hospital, said that "... we found 113 cases of prostate cancer among 22,000 men who had a vasectomy. This compares to a rate of 70 cases per 22,000 among men who didn't have a vasectomy."

Dr. Giovannucci's study shows about a 60% elevated risk of prostate cancer for men who have had a vasectomy, thereby revealing an association between vasectomy and prostate cancer. But does it establish causation: that having a vasectomy causes an increased risk of prostate cancer?

The answer is no, because the study is observational. Dr. Giovannucci simply observed two groups of men, one having vasectomies and the other not. Thus, although an association was established between vasectomy and prostate cancer, the association might be due to other factors (e.g., temperament) that make some men more likely to have vasectomies and also put them at greater risk of prostate cancer. In the words of Dr. Stuart Howards, a urology professor at the University of Virginia Medical School who did not participate in the study, "... [these results] have to be considered seriously but do not prove that vasectomy causes prostate cancer." □

EXAMPLE 1.17 A DESIGNED EXPERIMENT

For several years, evidence has been mounting that folic acid reduces major birth defects. An issue of *The Arizona Republic* reported on a Hungarian study that provides the strongest evidence yet. The results of the study, directed by Dr. Andrew E. Czeizel and Dr. Istvan Dudas of the National Institute of Hygiene in Budapest, were published in the *New England Journal of Medicine.*

For the study, the doctors enrolled 4753 women prior to conception. The women were divided randomly into two groups. One group took daily multivitamins containing 0.8 mg of folic acid, whereas the other group received only trace elements. A drastic reduction in the rate of major birth defects occurred among the women who took folic acid: 13 per 1000 as compared to 23 per 1000 for those women who did not take folic acid.

In contrast to the observational study considered in Example 1.16, this is a designed experiment and does help establish causation. The researchers did not simply observe two groups of women, but instead randomly assigned one group to take daily doses of folic acid and the other group to take only trace elements. ◨

The study in Example 1.17 illustrates three basic principles of experimental design: control, randomization, and replication.

- *Control:* The doctors compared the rate of major birth defects for the women who took folic acid to that for the women who took only trace elements. This comparison controlled for such things as the *placebo effect,* where subjects respond to the idea of a specific treatment rather than to the treatment itself.
- *Randomization:* The women were divided randomly into two groups to avoid unintentional selection bias in constituting the groups and thereby help eliminate the problem of potential confounding factors such as life-style and emotional state.
- *Replication:* A large number of women were recruited for the study to make it likely that the two groups created by randomization would be similar and also to increase the chances of detecting an effect due to the folic acid if such an effect exists.

In the folic-acid study, both dosages of folic acid (0.8 mg and essentially none) are referred to as *treatments* in the context of experimental design. Generally, each experimental condition is called a **treatment,** of which there may be several. Key Fact 1.1 summarizes our discussion about the principles of experimental design.

KEY FACT 1.1 **PRINCIPLES OF EXPERIMENTAL DESIGN**

The following principles of experimental design enable a researcher to conclude that differences in the results of an experiment not reasonably attributable to chance are likely caused by the treatments.

- ***Control:*** Some method should be used to control for effects due to factors other than the ones of primary interest.
- ***Randomization:*** Subjects should be randomly divided into groups to avoid unintentional selection bias in constituting the groups, that is, to make the groups as similar as possible.
- ***Replication:*** A sufficient number of subjects should be used to ensure that randomization creates groups that resemble each other closely and to increase the chances of detecting differences among the treatments when such differences actually exist.

An important method of control is to compare several treatments. In fact, one of the most common experimental situations involves a specified treatment and a *placebo,* an inert or innocuous medical substance. Technically, both the specified treatment and the placebo are treatments. The group receiving the specified treatment is called the **treatment group** and the one receiving the placebo is called the **control group.** In the folic-acid study, the women who took folic acid constitute the treatment group and the ones who took only trace elements constitute the control group.

Terminology of Experimental Design

Having discussed the principles of experimental design, we will now introduce some additional terminology used in experimental design. Each woman in the folic-acid study is, in the language of experimental design, a *subject.* More generally, we have the following definition.

DEFINITION 1.5 EXPERIMENTAL UNITS; SUBJECTS

In a designed experiment, the individuals or items on which the experiment is performed are called ***experimental units.*** When the experimental units are humans, the term ***subject*** is often used in place of experimental unit.

Referring again to the folic-acid study, we see that the researchers were interested in the effect of folic acid on major birth defects. Whether or not the baby of a woman in the study had a major birth defect is the **response variable** for this study. The daily dosage of folic acid is called the **factor.** In this case, the factor has two **levels,** namely, 0.8 mg and essentially none.

When there is only one factor, as in the folic-acid study, the treatments are the same as the levels of the factor. On the other hand, if there is more than one factor, each treatment results as a combination of levels of the various factors. Example 1.18 presents an experiment in which there are two factors.

EXAMPLE 1.18 TERMINOLOGY OF EXPERIMENTAL DESIGN

Golden Torch Cactus (botanical name, *Trichocereus spachianus*), a columnar cactus native to Argentina, is regarded as having excellent landscape potential. Feldman and Crosswhite, two researchers at the Boyce Thompson Southwestern Arboretum, conducted a thorough investigation of the optimal method for producing these cacti.

The researchers examined, among other things, the effects of a hydrophilic polymer and irrigation regime on weight gain. Hydrophilic polymers are used as soil additives to keep moisture in the root zone. For this study the researchers chose Broadleaf P-4 polyacrylamide, abbreviated P4. The hydrophilic polymer was either used or not used and five irrigation regimes were employed: none, light, medium, heavy, and very heavy. Identify the

a. experimental units.

b. response variable.

c. factors.

d. levels of each factor.

e. treatments.

SOLUTION **a.** The experimental units are the cacti used in the study.

b. The response variable is weight gain.

c. The factors are hydrophilic polymer and irrigation regime.

d. Hydrophilic polymer has two levels: with and without. Irrigation regime has five levels: none, light, medium, heavy, and very heavy.

e. Each treatment is a combination of a level of hydrophilic polymer and a level of irrigation regime. Table 1.12 depicts the treatments. In the table, we have abbreviated "very heavy" by "Xheavy."

TABLE 1.12 Schematic for the 10 treatments in the cactus study

Polymer	Irrigation regime: None	Light	Medium	Heavy	Xheavy
No P4	No water No P4 (Treatment 1)	Light water No P4 (Treatment 2)	Medium water No P4 (Treatment 3)	Heavy water No P4 (Treatment 4)	Xheavy water No P4 (Treatment 5)
With P4	No water With P4 (Treatment 6)	Light water With P4 (Treatment 7)	Medium water With P4 (Treatment 8)	Heavy water With P4 (Treatment 9)	Xheavy water With P4 (Treatment 10)

As we can observe from Table 1.12, there are 10 different treatments for this experiment. □

Keeping in mind our discussion in Example 1.18, we now present formal definitions of several important terms used in experimental design.

DEFINITION 1.6 RESPONSE VARIABLE, FACTORS, LEVELS, AND TREATMENTS

Response variable: The characteristic of the experimental outcome that is to be measured or observed.

Factor: A variable whose effect on the response variable is of interest in the experiment.

Levels: The possible values of a factor.

Treatment: Each experimental condition. For one-factor experiments, the treatments are the levels of the single factor. For multi-factor experiments, each treatment is a combination of levels of the factors.

Statistical Designs

Once we have chosen the treatments, we must decide how the experimental units are to be assigned to the treatments (or vice versa). The women in the folic-acid study were randomly divided into two groups; one group received folic acid and the other only trace elements. In the cactus study, 40 cacti were divided randomly into 10 groups of four cacti each and then each group was assigned a different treatment from among the 10 depicted in Table 1.12. Both of these experiments employed a **completely randomized design.**

DEFINITION 1.7 COMPLETELY RANDOMIZED DESIGN

In a ***completely randomized design,*** all the experimental units are assigned randomly among all the treatments.

The completely randomized design is one of the most commonly used and simplest designs, but is not always the best design. There are several alternatives to the completely randomized design.

For instance, in a **randomized block design,** experimental units that are similar in ways that are expected to affect the response variable are grouped in **blocks.** Then the random assignment of experimental units to the treatments is done on a block-by-block basis.

DEFINITION 1.8 RANDOMIZED BLOCK DESIGN

In a ***randomized block design,*** the experimental units are assigned randomly among all the treatments separately within each block.

In Example 1.19 we contrast completely randomized designs and randomized block designs.

EXAMPLE 1.19 COMPLETELY RANDOMIZED DESIGNS AND RANDOMIZED BLOCK DESIGNS

Suppose we want to compare the driving distances for five different brands of golf ball. With 40 golfers, discuss a method of comparison using a

a. completely randomized design.

b. randomized block design.

SOLUTION Here the experimental units are the golfers, the response variable is driving distance, the factor is brand of golf ball, and the levels (and treatments) are the five brands.

a. For a completely randomized design, we would randomly divide the 40 golfers into five groups of eight golfers each and then randomly assign each group to drive a different brand of ball. See Fig. 1.8.

FIGURE 1.8 Completely randomized design for golf-ball experiment

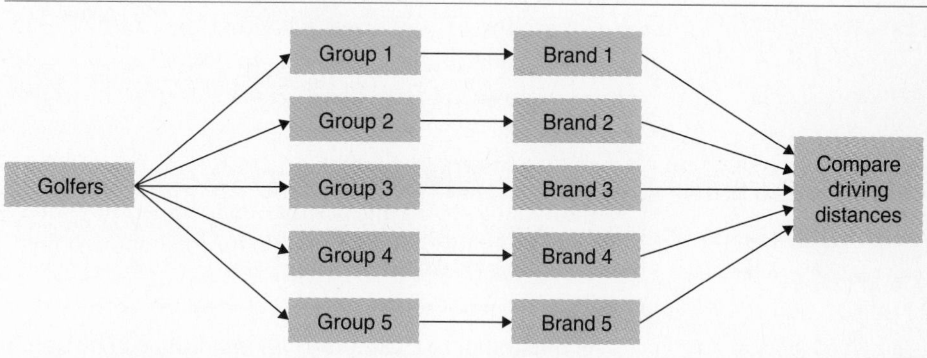

b. Since driving distance is affected by gender, it is probably better to use a randomized block design, with blocking by gender. This could be done with 40 golfers, say, 20 men and 20 women. We would randomly divide the 20 men into five groups of four men each and then randomly assign each group of men to drive a different brand of ball. Likewise, we would randomly divide the 20 women into five groups of four women each and then randomly assign each group of women to drive a different brand of ball. See Fig. 1.9 at the top of the following page.

By blocking we can isolate and remove the variation in driving distances between men and women and thereby make it easier to detect differences in driving distances among the five brands of golf ball, if such differences exist.

FIGURE 1.9 Randomized block design for golf-ball experiment

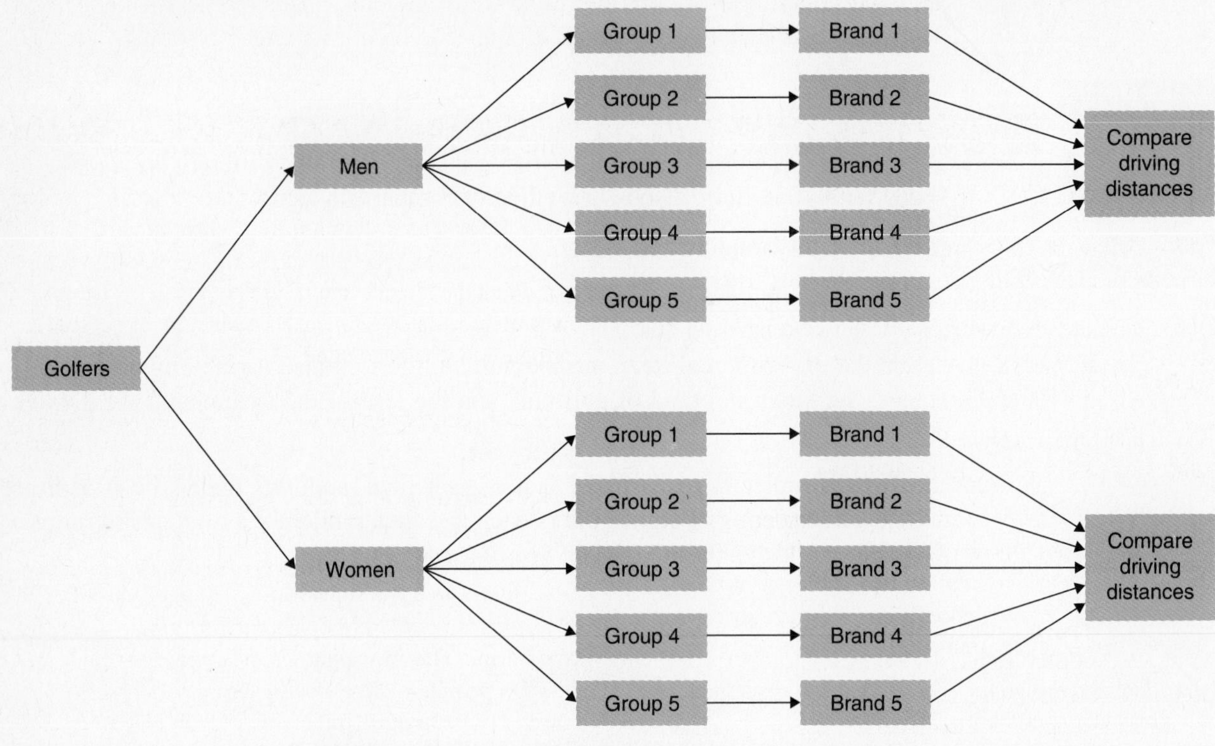

Additionally, blocking permits us to separately analyze differences in driving distances among the five brands for men and women. ◨

As illustrated in Example 1.19, blocking can be used to isolate and remove systematic differences among blocks and thereby make it easier to detect differences among treatments when such differences exist. Blocking also makes it possible to separately analyze the effects of the treatments on each block.

The simplest and one of the most common randomized block designs occurs when

- there is only one factor,
- the number of experimental units in each block equals the number of treatments, and
- within each block, each experimental unit is assigned to a different treatment.

In this section we have introduced some of the basic terminology and principles of experimental design. However, we have just scratched the surface of this vast and important topic to which entire courses and books are devoted. Further discussion of experimental design can be found in the module *Design of Experiments and Analysis of Variance* by Dennis L. Young (Reading, MA: Addison-Wesley, 1999).

EXERCISES 1.8

Statistical Concepts and Skills

1.62 Define the following terms:
a. observational study **b.** designed experiment

1.63 Fill in the following blank: Observational studies can reveal only association, whereas designed experiments can help establish _____.

1.64 State and explain the significance of the three basic principles of experimental design.

1.65 In the folic-acid study, 4753 women were enrolled by the doctors. Explain how a table of random numbers or a random-number generator could be used to divide the women randomly into two groups, one of size 2376 and the other of size 2377.

1.66 Refer to the vasectomy/prostate-cancer study discussed in Example 1.16 on page 40.
a. How could the study be modified to make it a designed experiment?
b. Comment on the feasibility of the designed experiment that you described in part (a).

In Exercises 1.67–1.70, identify each study as an observational study or a designed experiment. Justify your answers.

1.67 In the 1940s and early 1950s, there was great public concern over epidemics of polio. In an attempt to alleviate this serious problem, Jonas Salk of the University of Pittsburgh developed a vaccine for polio. Various preliminary experiments indicated that the vaccine was safe and potentially effective. Nonetheless, it was deemed necessary to conduct a large-scale study to determine whether the vaccine would truly work. A test involving nearly 2 million grade-school children was devised. All of the children were inoculated, but only half received the Salk vaccine; the other half were given a placebo, in this case an injection of salt dissolved in water. Neither the children nor the doctors performing the diagnoses knew which children belonged to which group. Instead an evaluation center kept records of who received the Salk vaccine and who did not. The center found that the incidence of polio was far less among the children inoculated with the Salk vaccine. From that information it was concluded that the Salk vaccine would be effective in preventing polio for all U.S. schoolchildren and, consequently, was then made available for general use.

1.68 According to a study published in the February 1993 issue of the *Journal of the American Public Health Association,* left-handed people do not die at an earlier age than right-handed people, contrary to the conclusion of a highly publicized report done two years prior. The investigation involved a 6-year study of 3800 people in East Boston older than age 65. Researchers at Harvard University and the National Institute of Aging found that the "lefties" and "righties" died at exactly the same rate. "There was no difference, period," said Dr. Jack Guralnik, an epidemiologist at the institute and one of the co-authors of the report.

1.69 Catherine E. Ross, a sociologist at the University of Illinois, presented the results of her research on mental depression in families at the 1993 annual meeting of the American Association for the Advancement of Science. Her results were based on a study of mental depression in 1000 families and showed in part that, on the average, the most depressed women are those that remain at home with their children, whereas the least depressed are those that have no children and a job. Ross emphasized that her conclusions concern averages, meaning, for example, that some mothers are quite happy staying home with their children. But she added that the data show that "... any plea to return to the 'traditional' family of the 1950s is a plea to return wives and mothers to a psychologically disadvantaged position in which husbands have much better mental health than wives."

1.70 For environmental purposes, efforts are being made to use industrial wastes on agricultural soils, since many such wastes contain nutrients that enhance crop growth. In a 1984 issue of *Environmental Pollution (Series A)*, Mohammad Ajmal and Ahsan Ullah Khan reported their findings on experiments with brewery wastes used for agricultural purposes. The researchers studied the physico-chemical properties of effluent from Mohan Meakin Breweries Ltd, Ghazibad, UP, India, and "... its effects on the physico-chemical characteristics of agricultural soil, seed germination pattern, and the growth of two common crop plants." They assessed the impact using different concentrations of the effluent: 25%, 50%, 75%, and 100%. Various chemical properties of the treated soil were measured, in particular, available nutrients.

1.71 In a designed experiment:

a. What are the experimental units?

b. If the experimental units are human beings, what term is often used in place of experimental unit?

1.72 Prozac® (fluoxetine hydrochloride), a product of Eli Lilly and Company, is used for the treatment of depression, obsessive-compulsive disorder (OCD), and bulimia nervosa. A 1998 issue of the magazine *Arthritis Today* contained an advertisement reporting on the "... treatment-emergent adverse events that occurred in 2% or more patients treated with Prozac and with incidence greater than placebo in the treatment of depression, OCD, or bulimia." In the study, 2444 patients took Prozac and 1331 patients were given a placebo. Identify the

a. treatment group. **b.** control group.

c. treatments.

In each of Exercises 1.73–1.78, we have presented a description of a designed experiment. In each case, identify the

a. experimental units,

b. response variable,

c. factor(s),

d. levels of each factor,

e. treatments.

1.73 To compare the lifetimes of four brands of flashlight battery, 20 flashlights were used. The 20 flashlights were randomly divided into four groups of five flashlights each. Then each group of flashlights was equipped with a different brand of battery.

1.74 A chain of convenience stores wanted to compare three different advertising policies for their effect on dollar amount of sales:

- Policy 1: No advertising.
- Policy 2: Advertise in neighborhoods with circulars.
- Policy 3: Use circulars and advertise in newspapers.

Eighteen stores were randomly selected and divided randomly into three groups of six stores. Each group used (a different) one of the three policies. Following the implementation of the policies, sales figures were obtained for each of the stores during a 1-month period.

1.75 Refer to Example 1.18 on page 42. Another variable Feldman and Crosswhite investigated in their study of Golden Torch cacti was total length of cuttings at the end of 16 months.

1.76 Storage of perishable items is an important concern for many companies. One study examined the effect of storage time and storage temperature on the deterioration of a particular item. Three different storage temperatures and five different storage times were used.

1.77 Supermarkets are interested in strategies to temporarily increase unit sales of a product. In one study, researchers compared the effect of display type and price on unit sales for a particular product. The following display types and pricing schemes were employed:

- Display types: normal display space interior to an aisle, normal display space at the end of an aisle, and enlarged display space.
- Pricing schemes: regular price, reduced price, and cost.

1.78 In a classic study, described by F. Yates in *The Design and Analysis of Factorial Experiments,* the effect on oat yield was compared for three different varieties of oats and four different concentrations of manure (0, 0.2, 0.4, and 0.6 cwt per acre).

1.79 Refer to Exercise 1.73. Is this a completely randomized design or a randomized block design? Explain your answer.

1.80 Refer to Exercise 1.73. Suppose we compare the lifetimes of the four brands of flashlight battery by employing the following method. We use 20 flashlights, five different brands of four flashlights each. Each of the four flashlights of a given brand, uses a different brand of battery. Is this a completely randomized design or a randomized block design? Explain your answer.

EXTENDING THE CONCEPTS AND SKILLS

1.81 In Exercise 1.67, we discussed the Salk-vaccine experiment. As we mentioned in that exercise, neither the children nor the doctors performing the diagnoses knew which children had been given the Salk vaccine and which had been given the placebo. This technique is called **double-blinding.** Explain the advantages of using double-blinding in the Salk-vaccine experiment.

1.82 In sampling from a population, state which type of sampling design corresponds to each of the following experimental designs:
a. completely randomized design
b. randomized block design

1.83 Comment on the statement in quotes given at the end of Exercise 1.69. *(Hint:* Recall that the study is observational.*)*

USING TECHNOLOGY  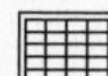 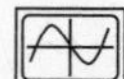

1.84 Use Minitab or some other statistical software to carry out the process described in Exercise 1.65.

CHAPTER REVIEW

You Should Be Able To

1. classify statistical studies as either descriptive or inferential.
2. identify the population and the sample in an inferential study.
3. explain what is meant by a representative sample.
4. describe simple random sampling, systematic random sampling, cluster sampling, and stratified sampling.
5. use a table of random numbers to obtain a simple random sample.
6. explain the difference between an observational study and a designed experiment.
7. classify a study concerning the relationship between two variables or characteristics as either an observational study or a designed experiment.
8. state the three basic principles of experimental design.
9. identify the treatment group and control group in a study.
10. identify the experimental units, response variable, factor(s), levels of each factor, and treatments in a designed experiment.
11. distinguish between a completely randomized design and a randomized block design.

*12. start and exit Minitab.
*13. store and name a data set in Minitab, both manually and by importing from a file.
*14. identify the different types of windows in Minitab.
*15. print, save, and retrieve in Minitab.
*16. use the Minitab procedures covered in this chapter.

* Recall that an asterisk indicates material that is optionally covered.

Key Terms

active cell,* *14*
blocks, *44*
cell,* *14*
census, *24*
cluster sampling, *34*
column,* *14*
completely randomized design, *44*
control, *41*
control group, *42*
descriptive statistics, *4*
designed experiment, *39*
experimental units, *42*
factor, *44*
graph window,* *18*
inferential statistics, *5*
levels, *44*
multistage sampling, *38*
observational study, *39*
population, *5*
probability sampling, *25*
project,* *20*
proportional allocation, *37*
random-number generator, *29*
random sampling, *25*
randomization, *41*
randomized block design, *44*
replication, *41*
representative sample, *24*
response variable, *44*
rows,* *14*
sample, *5*
Session window,* *12*
simple random sample, *25*
simple random sampling, *25*
simple random sampling with replacement, *25*
simple random sampling without replacement, *25*
strata, *37*
stratified random sampling with proportional allocation, *37*
stratified sampling, *37*
subject, *42*
systematic random sampling, *33*
table of random numbers, *27*
treatment, *44*
treatment group, *42*
worksheet,* *12*

REVIEW TEST

STATISTICAL CONCEPTS AND SKILLS

1. In a newspaper, magazine, or on the Internet, find an example of
a. a descriptive study.
b. an inferential study.

2. Almost any inferential study involves aspects of descriptive statistics. Explain why this is so.

In Problems 3–7, classify each of the studies as either descriptive or inferential.

3. On May 14, 1998, the baseball scores shown in the table at the top of the next column were posted on *USA TODAY*'s web site.

NATIONAL LEAGUE BASEBALL

Expos	9	Pirates	0
Giants	5	Astros	1
Marlins	4	Cubs	9
Reds	10	Rockies	3
Braves	10	Mets	4
Cardinals	2	Padres	3
Phillies	4	Brewers	8
Dodgers	9	Diamondbacks	3

4. A National Institute of Mental Health survey concluded that "about 20% of adult Americans suffer from at least one psychiatric disorder." This and other estimates were obtained from results of interviews with thousands of Americans in St. Louis, Baltimore, and New Haven.

5. On February 14, 1985, the Census Bureau released a survey stating that "fewer Americans have health insurance coverage than previously thought." The survey, which was based on a multistage probability sample of 20,000 households, concluded that about 85% of the population is covered by health insurance—a far cry from the 97.3% figure found in a 1978 survey by the Department of Health and Human Services. By the way, the 1996 estimate given by the Census Bureau is that roughly 84.4% of the U.S. population is covered by health insurance.

6. The International Civil Aviation Organization compiles data on worldwide airline fatalities and publishes its findings in *Civil Aviation Statistics of the World.* The following table provides statistics on the number of passenger deaths for selected years.

Year	1970	1975	1980	1985	1990	1995
Deaths	700	467	814	1066	495	710

7. The Bureau of Justice Statistics (BJS) conducts monthly surveys of approximately 60,000 U.S. households. These monthly surveys are then used to determine annual estimates of criminal victimization which are subsequently published in *Criminal Victimization.* For example, the following table shows the victimization-rate estimates reported by the BJS for property crimes against households in 1995, by residence. Rates are per 1000 households. *(Note:* MV = motor vehicle.*)*

Residence	Burglary	MV theft	Theft
Urban	59.9	25.7	262.3
Suburban	39.0	15.0	213.0
Rural	46.8	6.9	164.8

8. Before planning and conducting a study to obtain information, what should be done?

9. Explain the meaning of each of the following terms.
a. representative sample
b. probability sampling
c. simple random sampling

10. A researcher wants to estimate the average income of parents of college students. To accomplish that, he surveys a sample of 250 students at Yale. Is this a representative sample? Explain your answer.

11. Which of the following sampling procedures employ probability sampling?
a. A college student is hired to interview a sample of voters in her town. She stays on campus and interviews 100 students in the cafeteria.
b. A pollster wants to interview 20 gasoline-station managers in Baltimore. He posts a list of all such managers on his wall, closes his eyes, and tosses a dart at the list 20 times. He interviews the people whose names the dart hits.

12. The Pacific region of the United States consists of five states: Washington (WA), Oregon (OR), California (CA), Alaska (AK), and Hawaii (HI).
a. List the 10 possible samples (without replacement) of size three that can be obtained from the population of five Pacific states.
b. If a simple-random-sampling procedure is used to obtain a sample of three Pacific states, what are the chances that it is the first sample on your list in part (a)? the second? the tenth?

13. The mayor of a small town with 7246 registered voters wants to send a detailed questionnaire to a simple random sample of 50 voters.
a. Explain how Table I in Appendix A can be employed to obtain the sample.
b. Starting at the four-digit number in line number 14 and column numbers 16–19 of Table I, read down the column, up the next, etc., to find 50 numbers that can be used to identify the voters who will receive the questionnaire.

14. Describe each of the following sampling methods and indicate conditions under which each is appropriate.
a. systematic random sampling
b. cluster sampling
c. stratified random sampling with proportional allocation

15. Refer to Problem 13.

a. Use the technique of systematic random sampling to obtain 50 numbers from the numbers 1–7246, thereby identifying 50 registered voters to receive the questionnaire.

b. In this case do you think systematic random sampling is an appropriate alternative to simple random sampling? Explain your answer.

16. The faculty of a college consists of 820 members. A new president has just been appointed. The president wants to get an idea of what the faculty considers the most important issues currently facing the school. She does not have time to interview all the faculty members and so decides to stratify the faculty by rank and use stratified random sampling with proportional allocation to obtain a sample of 40 faculty members. There are 205 full professors, 328 associate professors, 246 assistant professors, and 41 instructors.

a. How many of each rank should be selected for the interviewing?

b. Use Table I in Appendix A to obtain the required sample. Explain your procedure in detail.

17. Regarding observational studies and designed experiments:

a. Describe each type of study.

b. With respect to possible conclusions, what important difference exists between the two kinds of studies?

18. An article appearing in an issue of *The Arizona Republic* reported on a study conducted by Greg Duncan of the University of Michigan. According to the report, "Persistent poverty during the first 5 years of life leaves children with IQs 9.1 points lower at age 5 than children who suffer no poverty during that period ...." Is this an observational study or is it a designed experiment?

19. The Spring 1998 issue of *Inside MS* contained an article describing AVONEX™ (Interferon beta-1a), a drug used in the treatment of relapsing forms of multiple sclerosis. Included in the article was a report on "... adverse events and selected laboratory abnormalities that occurred at an incidence of 2% or more among the 158 multiple sclerosis patients treated with 30 mcg of AVONEX™ once weekly by IM injection. In the study, 158 patients took AVONEX™ and 143 patients were given a placebo.

a. Is this an observational study or is it a designed experiment?

b. Identify the treatment group, control group, and treatments.

20. Identify and explain the significance of the three basic principles of experimental design.

21. A classic study, conducted in 1935 by B. Lowe, analyzed differences in the amount of fat absorbed by doughnuts in cooking using four different fats. For the experiment, 24 batches of donuts were randomly divided into four groups of six batches each. The four groups were then randomly assigned to the four fats. Identify the

a. experimental units,

b. response variable,

c. factor(s),

d. levels of each factor,

e. treatments.

22. In the paper "Effects of Plant Density on Tomato Yields in Nigeria," appearing in an issue of *Experimental Agriculture,* researchers reported on the effect of tomato variety and planting density on yield. Identify the

a. experimental units,

b. response variable,

c. factor(s),

d. levels of each factor,

e. treatments.

23. An experiment is to be conducted to compare four different brands of gasoline for gas mileage.

a. Suppose we randomly divide 24 cars into four groups of six cars each and then randomly assign the four groups to the four brands of gasolines, one group per brand. Is this a completely randomized design or a randomized block design? If it is the latter, what are the blocks?

b. Suppose, instead, we use six different models of car whose varying characteristics (e.g., weight, horsepower) affect gas mileage. Four cars of each model are randomly assigned to the four different brands of gasoline. Is this a completely randomized design or a randomized block design? If it is the latter, what are the blocks?

c. Which design is better, the one in part (a) or the one in part (b)? Explain your answer.

USING TECHNOLOGY  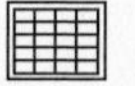 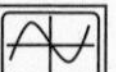

In Problems 24–26, store the following data in a Minitab worksheet or in a corresponding place if you are using a different technology.

24. A research physician conducted a study on the ages of people with diabetes. The following data were obtained for the ages of a sample of 35 diabetics.

48	41	57	83	41	55	59
61	38	48	79	75	77	7
54	23	47	56	79	68	61
64	45	53	82	68	38	70
10	60	83	76	21	65	47

a. Manually enter the data on ages in a column named AGE1.

b. The age data are in the file `1r-24.dat`, found in the `Exercise` folder of DataDisk CD. Store the data in a column named AGE2 by importing from DataDisk CD.

25. A *Consumer Reports* article revealed the data on horsepower and gas mileage for a sample of 12 automobiles found at the top of the next column.

a. Manually enter the horsepower and gas-mileage data in columns named HP1 and MPG1.

b. The data on horsepower and gas mileage are in the file `1r-25.dat` which is found in the `Exercise` folder of DataDisk CD. Store the data in columns named HP2 and MPG2 by importing from DataDisk CD.

Horsepower	Mileage (mpg)
155	16.9
68	30.0
95	27.5
97	27.4
125	17.0
115	21.6
110	18.6
120	18.1
68	34.1
80	27.4
70	34.2
78	30.5

26. The U.S. Bureau of Labor Statistics publishes data on weekly earnings of full-time wage and salary workers in *Employment and Earnings.* Random samples of male and female workers gave the following data on weekly earnings, in dollars.

Men		Women	
826	2523	1994	2109
1790	288	510	291
477	317	426	274
307	718	290	1097
		1361	328

a. Manually enter the data on wages in columns named MEN1 and WOMEN1.

b. The data on wages can be found in the files `1r-26a.dat` and `1r-26b.dat` in the `Exercise` folder of DataDisk CD. Import the data from DataDisk CD into columns named MEN2 and WOMEN2.

27. Refer to Problem 13. Use Minitab or some other statistical software to obtain a simple random sample of 50 numbers from the numbers 1–7246, thus identifying 50 registered voters to receive the questionnaire.

INTERNET PROJECT

The Titanic Disaster

The Internet Projects Page for your book, *Introductory Statistics, fifth edition,* is a location on the Internet designed to help you understand statistics. The starting Internet address (URL) for the page is

`http://hepg.awl.com` *keyword:* Weiss.

From this Web page, you can reach the Internet Projects Page. We suggest that you bookmark the Internet Projects Page for easy access in the future.

Each chapter in the book includes an Internet project that provides a set of simulations, demonstrations, and other activities as a supplement to those found in the text. The material comes from universities, individuals, and companies from all over the world.

In this first Internet project, you will become acquainted with the controversy and data behind the Titanic disaster. You will visit web sites at several locations, viewing and thinking about the data and facts presented there. You will also see statistical demonstrations and animations that illustrate the concepts you have covered in class.

In the year 1912, the Titanic was the largest, most luxurious, and technologically advanced liner in the world. At 11:40 P.M. on Sunday, April 14 of that year, the Titanic struck an iceberg. In two and one half hours, the liner that many thought unsinkable went down. Only 705 of her 2224 passengers and crew were saved.

Some believe that the rescue procedures used that night unfairly favored the wealthier passengers. In this project, you will use statistical thinking to explore the data behind the disaster to arrive at your own (statistically informed) conclusion.

USING THE FOCUS DATABASE

The file `focus.dat`, residing in the `Focus` folder of DataDisk CD, contains information on 500 randomly selected Arizona State University sophomores. Seven variables are considered for each student: sex, high-school GPA, SAT math score, cumulative GPA, SAT verbal score, age, and total hours completed. For reference purposes we will call this database the **Focus database.**

The data in the Focus database will be used in the "Using the Focus Database" exercises that appear at the end of each chapter. Large data sets such as these are almost always analyzed by computer, and that is how you should handle these exercises.

In preparation for the "Using the Focus Database" exercises, you should store the seven data sets contained in `focus.dat` in a Minitab worksheet or in a corresponding place if you are using a different technology. For those running Release 12 of Minitab for Windows, access and start Minitab and place DataDisk CD in the `E:` drive (or whatever drive name your computer uses for your CD-ROM). Then proceed as follows.

1 Choose **File ➤ Other Files ➤ Import Special Text...**

2 Type `C1-C7` in the **Store data in column(s)** text box

3 Click **OK**
4 Type `E:\IS5\Focus\focus.dat` in the **File name** text box
5 Click **Open**

The seven data sets in `focus.dat` are now stored in Columns C1–C7. Notice that for the sex data, which are in C1, we have employed the coding 1 for female and 2 for male.

For ease of reference, it is useful to name the columns containing the data sets in the Focus database. The naming can be accomplished as explained on page 14. Using that or any other method, you should name Columns C1–C7 as suggested in Table 1.13.

TABLE 1.13
Names for data sets in the Focus database

Column	Name
C1	SEX
C2	HS GPA
C3	SAT MATH
C4	CUM GPA
C5	SAT VERB
C6	AGE
C7	HOURS

Now that you have the data sets in the Focus database stored and named in Minitab, save the worksheet to a file named `focus.mtw`, as explained on page 20. Then any time you want to analyze the Focus database, retrieve the worksheet `focus.mtw` as explained on page 21.

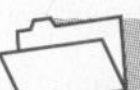

CASE STUDY DISCUSSION

Top Films of All Time

At the beginning of this chapter, we discussed the results of a survey by the American Film Institute (AFI). Now that we have learned some of the basic terminology of statistics, we can examine that survey in more detail.

Answer each of the following questions regarding the survey. In doing so, you may want to reread the description of the survey given on page 2.

a. Identify the population.
b. Identify the sample.
c. Is the sample representative relative to the population of all American moviegoers? Explain your answer.
d. Consider the following statement: "Among the 1500 filmmakers, actors, critics, politicians, and film historians polled by AFI, the top ranking film was *Citizen Kane*." Is this statement descriptive or inferential? Why?

e. Suppose the statement in part (d) is changed to the following statement: "Based on the AFI poll, *Citizen Kane* is the top ranking film among all filmmakers, actors, critics, politicians, and film historians." Is this statement descriptive or inferential? Explain your answer.

BIOGRAPHY FLORENCE NIGHTINGALE

Florence Nightingale (1820–1910), the founder of modern nursing, was born in Florence, Italy, into a wealthy English family. In 1849, over the objections of her parents, she entered the Institution of Protestant Deaconesses at Kaiserswerth, Germany, which "... trained country girls of good character to nurse the sick."

The Crimean War began in March 1854 when England and France declared war on Russia. Nightingale, after serving as superintendent of the Institution for the Care of Sick Gentlewomen in London, was appointed by the English secretary of state at war, Sidney Herbert, to be in charge of 38 nurses who were to be stationed at military hospitals in Turkey.

Nightingale found the conditions in the hospitals appalling—overcrowded, filthy, and without sufficient facilities. In addition to the administrative duties she undertook to alleviate those conditions, she spent many hours tending patients; after 8:00 P.M. she allowed none of her nurses in the wards, but made rounds herself every night, a deed that earned her the epithet Lady of the Lamp.

Nightingale was an ardent believer in the power of statistics and used statistics extensively to gain an understanding of social and health issues. She lobbied to introduce statistics into the curriculum at Oxford and invented the coxcomb chart, a type of pie chart. Nightingale felt that charts and diagrams were a means of making statistical information understandable to people who would otherwise be unwilling to digest the dry numbers.

In May 1857, as a result of Nightingale's interviews with officials ranging from the secretary of state to Queen Victoria herself, the Royal Commission on the Health of the Army was established. Under the auspices of the commission, the Army Medical School was founded. In 1860 Nightingale used a fund set up by the public to honor her work in the Crimea to create the Nightingale School for Nurses at St. Thomas's Hospital. During that same year, at the International Statistical Congress in London, she authored one of the three papers discussed in the Sanitary Section and also met Adolphe Quetelet (see Chapter 2 biography) who had greatly influenced her work.

After 1857, Nightingale lived as an invalid, although it has never been determined that she had any specific illness. In fact, it has been speculated that her invalidism was a stratagem she employed in order to devote herself to her work.

Nightingale was elected an Honorary Member of the American Statistical Association in 1874. In 1907 she was presented the Order of Merit, an order for meritorious service established by King Edward VII; she was the first woman to receive that award.

Florence Nightingale died in 1910. An offer of a national funeral and burial at Westminster Abbey was declined, and, according to her wishes, Nightingale was buried in the family plot in East Mellow, Hampshire, England.

PART

Descriptive Statistics

CASE STUDY

INFANT MORTALITY

Infant mortality is concerned with infant deaths during the first year of life. Generally, the infant mortality rate provides the number of such deaths per 1000 live births during a calendar year. In 1987, the U.S. Congress established the National Commission to Prevent Infant Mortality, whose charge is to create a national strategy for reducing the infant mortality rate of the United States.

From the *Statistical Abstract of the United States,* we obtained information on 1997 infant mortality rates for nations having a 1997 population of 6 million people or more. If we rank those nations according to their infant mortality rates, the top 24 are as shown in the following table.

Rank	Country	Infant mortality	Rank	Country	Infant mortality
1	Japan	4.4	13	Belgium	6.3
2	Sweden	4.5	**14**	**United States**	**6.6**
3	Netherlands	4.8	15	Italy	6.8
4	Hong Kong	5.0	16	Greece	7.2
5	Switzerland	5.4	17	Portugal	7.5
6	Australia	5.4	18	South Korea	8.0
7	Germany	5.9	19	Czech Republic	8.3
8	Canada	6.0	20	Cuba	8.9
9	France	6.0	21	Hungary	12.2
10	Spain	6.1	22	Poland	12.3
11	Austria	6.1	23	Belarus	12.6
12	United Kingdom	6.3	24	Chile	13.2

As we see from the table, even among nations with the lowest rates, infant mortality varies considerably, from a low of 4.4 (Japan) to a high of 13.2 (Chile). Although the United States has a relatively poor showing, its infant mortality rate has dropped by almost one-third from what it was in 1989.

At the end of this chapter, we will revisit the information presented in the table and apply some of our newly learned statistical skills to identify and analyze various aspects of these infant mortality rates.

Organizing Data

2

GENERAL OBJECTIVES

As we discovered in Chapter 1, descriptive statistics consists of methods for organizing and summarizing information clearly and effectively. In this chapter we will begin our study of descriptive statistics. Specifically, we will learn how to classify data by type, organize data into tables, and summarize data with graphical displays. And because graphical displays can often be misleading, we will examine ways of analyzing and interpreting them carefully.

CHAPTER OUTLINE

2.1 VARIABLES AND DATA

A characteristic that varies from one person or thing to another is called a **variable.** Examples of variables for humans are height, weight, number of siblings, sex, marital status, and eye color. The first three of these variables yield numerical information and are examples of **quantitative variables;** the last three yield nonnumerical information and are examples of **qualitative variables,** also referred to as **categorical variables.**

Quantitative variables can be classified as either discrete or continuous. A **discrete variable** is one whose possible values form a finite (or countably infinite[†]) set of numbers, usually some collection of whole numbers. The number of siblings a person has is an example of a discrete variable. A discrete variable usually involves a count of something, like the number of siblings a person has, the number of cars owned by a family, or the number of students in an introductory statistics class.

A **continuous variable** is a variable whose possible values form some interval of numbers. The height of a person is an example of a continuous variable. Typically, a continuous variable involves a measurement of something, like the height of a person, the weight of a newborn baby, or the length of time a car battery lasts.

The preceding discussion is summarized graphically in the diagram below and verbally in Definition 2.1.

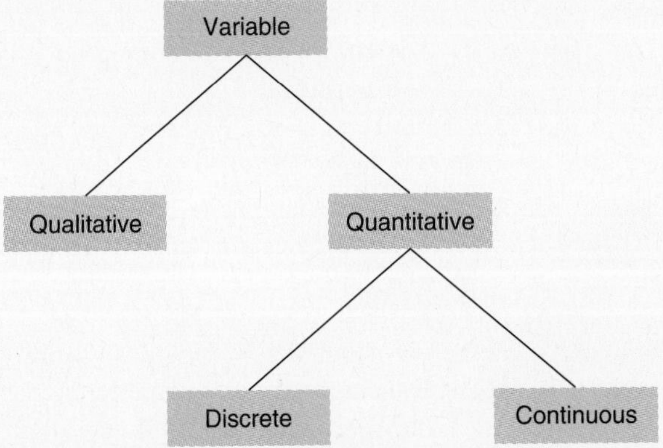

DEFINITION 2.1 VARIABLES

Variable: A characteristic that varies from one person or thing to another.

Qualitative variable: A nonnumerically valued variable.

Quantitative variable: A numerically valued variable.

[†] A countably infinite set is an infinite set whose members can be arranged in a sequence. Mathematically, this means that the members of the set can be put in one-to-one correspondence with the positive integers.

Discrete variable: A quantitative variable whose possible values form a finite (or countably infinite) set of numbers.

Continuous variable: A quantitative variable whose possible values form some interval of numbers.

Observing the values of a variable for one or more people or things yields **data.** Thus the information collected, organized, and analyzed by statisticians is data. The terms *qualitative, quantitative, discrete,* and *continuous* are used to describe data as well as variables: qualitative data are data obtained by observing values of a qualitative variable; quantitative data are data obtained by observing values of a quantitative variable; and so forth.

DEFINITION 2.2 DATA

Data: Information obtained by observing values of a variable.

Qualitative data: Data obtained by observing values of a qualitative variable.

Quantitative data: Data obtained by observing values of a quantitative variable.

Discrete data: Data obtained by observing values of a discrete variable.

Continuous data: Data obtained by observing values of a continuous variable.

Each individual piece of data is called an **observation** and the collection of all observations for a particular variable is called a **data set.**†

EXAMPLE 2.1 VARIABLES AND DATA

At noon on April 20, 1998, almost 11,500 men and women set out to run from Hopkinton Center to the John Hancock Building in Boston. Their run, covering 26 miles and 385 yards, would be watched by thousands of people lining the streets leading into Boston and by millions more on television. It was the 102nd running of the Boston Marathon.

A great deal of information was accumulated and recorded that afternoon by the Boston Athletic Association. The men's competition was won by Moses Tanui of Kenya with a time of 2 hours, 7 minutes, and 34 seconds. The winner of the women's competition was Fatuma Roba of Ethiopa; her time was 2 hours, 23 minutes, and 21 seconds. There were 6989 men and 2982 women who finished the marathon before the official cutoff time of 5 hours.

---

† Sometimes *data set* is used to refer to all of the data for all of the variables under consideration.

The Boston Marathon provides examples of different types of variables and data. The simplest type is illustrated by the classification of each entrant as either male or female. "Sex" is a qualitative variable because its possible values (male or female) are nonnumerical. Thus, for instance, the information that Moses Tanui is a male is qualitative data—data obtained by observing the value of the variable "sex" for Moses Tanui.

Most racing fans are interested in the places of the finishers. "Place of finish" is a quantitative variable, which is also a discrete variable because there are only a finite number of possible finishing places. The information that among the women, Fatuma Roba and Renata Paradowska finished first and second, respectively, and Cindy Barber-Keeler and Yumiko Furuya finished 14th and 15th, respectively, is discrete, quantitative data—data obtained by observing the values of the variable "place of finish" for the four runners.

More can be learned about what happened in a race by looking at the times of the finishers. For instance, Fatuma Roba finished 3 minutes and 56 seconds ahead of Renata Paradowska, but Cindy Barber-Keeler beat Yumiko Furuya by only 51 seconds. Differences between times indicate exactly how far apart two runners finished, whereas differences in places do not. "Finishing time" is a quantitative variable, which is also a continuous variable because the finishing time of a runner can conceptually be any positive number. The information that Moses Tanui ran his race in 2:07:34 and Fatuma Roba ran hers in 2:23:21 is continuous, quantitative data—data obtained by observing the values of the variable "finishing time" for Moses Tanui and Fatuma Roba. ◨

EXAMPLE 2.2 **VARIABLES AND DATA**

Humans are classified as having one of four blood types: A, B, AB, or O. What kind of data do you receive when you are told your blood type?

SOLUTION Your blood type is qualitative data, data obtained by observing your value of the variable "blood type." ◨

EXAMPLE 2.3 **VARIABLES AND DATA**

The U.S. Bureau of the Census collects data on household size and publishes the information in *Current Population Reports.* What kind of data is the number of people in your household?

SOLUTION The number of people in your household is discrete, quantitative data—data obtained by observing the value of the variable "household size" for your particular household. ◨

EXAMPLE 2.4 VARIABLES AND DATA

The *Information Please Almanac* lists the world's highest waterfalls. The list shows that Angel Falls in Venezuela is 3281 feet high, more than twice as high as Ribbon Falls in Yosemite, California, which is 1612 feet high. What kind of data are these heights?

SOLUTION The waterfall heights are continuous, quantitative data—data obtained by observing the values of the variable "height" for the two waterfalls. ◨

Classification and the Choice of a Statistical Method

Some of the descriptive and inferential procedures that we will study are valid for only certain types of data; that is one reason why it is important to be able to correctly classify data. Statisticians use other classifications besides the ones presented here. But the types we have discussed are sufficient for the majority of applications.

Data classification is sometimes difficult; statisticians themselves occasionally disagree over data type. For example, some classify data involving amounts of money as discrete data, whereas others classify such data as continuous data. In most cases, however, the appropriate classification of data is fairly clear and may aid in the choice of the correct statistical method.

EXERCISES 2.1

STATISTICAL CONCEPTS AND SKILLS

2.1 Give an example, other than the ones in this section, of each of the following variables.
a. qualitative variable
b. discrete, quantitative variable
c. continuous, quantitative variable

2.2 Explain in your own words the meaning of each of the following terms.
a. qualitative variable
b. discrete, quantitative variable
c. continuous, quantitative variable

2.3 Explain in your own words the meaning of each of the following terms.
a. qualitative data
b. discrete, quantitative data
c. continuous, quantitative data

2.4 Give a reason why the classification of data is important.

2.5 Of the variables we have studied so far, which type yields nonnumerical data?

For each part of Exercises 2.6–2.13, classify the data as either qualitative or quantitative; if quantitative, further classify it as discrete or continuous. Also identify the variable under consideration.

2.6 According to Professor Sidney S. Culbert of the University of Washington, the principal languages of the world in 1997 were as follows.

Rank	Language	Speakers (millions)
1	Mandarin	1025
2	English	497
3	Hindi	476
4	Spanish	409
5	Russian	279
6	Arabic	235
7	Bengali	207

a. What type of data is in the first column of the table?
b. What type of data is provided by the information that Sally Ride speaks English?
c. What type of data is provided by the information in the third column of the table?

2.7 As reported by the U.S. Department of Agriculture in *Agricultural Statistics,* tobacco production in the United States for the years 1991–1996 is as displayed in the following table. What type of data is provided by the second column of the table?

Year	Pounds (millions)
1991	1664
1992	1722
1993	1613
1994	1583
1995	1269
1996	1565

2.8 On May 4, 1961, Commander Malcolm Ross, U.S. Naval Reserves, ascended 113,739.9 ft in a free balloon. What kind of data is the height given here?

2.9 The U.S. Bureau of Labor Statistics publishes information on employment in *Employment and Earnings.* Following is a table providing 1996 employment figures for selected industries in the United States. What kind of data is given by the employee numbers?

Industry	Employees (1000s)
Agriculture	3,443
Mining	569
Construction	7,943
Manufacturing	20,518
Trade	26,497
Services	45,043

2.10 What type of data is provided by the numbers in the following table?

DO YOUR WORRIES MATCH THOSE OF THE EXPERTS?

Experts and laypeople were asked to rank the risk of dying in any year from various activities and technologies. The experts' ranking closely matches known fatality statistics.

Public		Experts
1	Nuclear power	20
2	Motor vehicles	1
3	Handguns	4
4	Smoking	2
5	Motorcycles	6
6	Alcoholic beverages	3
7	General (private) aviation	12
8	Police work	17
9	Pesticides	8
10	Surgery	5
11	Fire fighting	18
12	Large construction	13
13	Hunting	23
14	Spray cans	26
15	Mountain climbing	29
16	Bicycles	15
17	Commercial aviation	16
18	Electric power (nonnuclear)	9
19	Swimming	10
20	Contraceptives	11

Note: The table above was reprinted by permission from an issue of *Science.* Copyright by the American Association for the Advancement of Science.

2.11 The following table of continental statistics provides approximate land areas and 1997 population estimates. [SOURCE: Bureau of the Census, U.S. Department of Commerce.]

Continent	Land area (1000 sq mi)	Population (millions)
Africa	11,707	750
Asia	10,644	3477
Europe	3,800	508
North America	9,360	464
Oceania	3,284	29
South America	6,883	329

a. What type of data is provided by the land area figures displayed in the second column of the table?
b. What type of data is contained in the statement, "Africa is largest in land area and second largest in population"?
c. What type of data is provided by the population figures shown in the third column of the table?
d. What type of data do we obtain from the fact that Marie Curie was born in Europe?

2.12 The following table displays the rank by population and the population in thousands of the 10 largest cities in the United States in 1980 and 1990. [SOURCE: U.S. Bureau of the Census, *Census of Population.*]

City	1980		1990	
	Rank	Population (thous.)	Rank	Population (thous.)
New York	1	7,072	1	7,323
Los Angeles	3	2,967	2	3,485
Chicago	2	3,005	3	2,784
Houston	5	1,595	4	1,631
Philadelphia	4	1,688	5	1,586
San Diego	8	876	6	1,111
Detroit	6	1,203	7	1,028
Dallas	7	904	8	1,007
Phoenix	9	790	9	983
San Antonio	11	786	10	936

a. What type of data is provided by the statement, "In 1980 Houston was the fifth largest city in the United States"?
b. What type of data is given in the 1990 "Population" column of the table?
c. What type of data is provided by the information that Theodore Roosevelt was born in New York?

2.13 The *American Banker* reports that the five largest U.S. commercial banks, by deposits, as of December 31, 1996, were as follows.

Bank	Rank	Deposits ($millions)
Chase Manhattan Bank, New York, NY	1	272,429
Citibank, New York, NY	2	241,006
Bank of America, San Francisco, CA	3	180,480
Morgan Guaranty Trust Co., New York, NY	4	172,563
Wells Fargo Bank, San Francisco, CA	5	99,165

a. What kind of data is displayed in the second column of the table?
b. What kind of data is presented in the third column of the table?

2.14 What kinds of data would be collected in each of the following situations?
a. A quality-control engineer measures the lifetimes of electric light bulbs.
b. A businessperson wants to know the number of families with preteen children in Pueblo, CO.
c. A manufacturer of sporting goods classifies each major-league baseball player as either right-handed or left-handed and counts the number of players in each category.
d. A sociologist needs to estimate the average annual income of the residents of Ossining, New York.
e. A pollster plans to classify each individual in a sample of voters as Democrat or Republican and count the total number in each group.
f. An administrator at a college needs to know how many men and women participated in varsity sports

during the spring semester and how much money was spent on men's sports and on women's sports.

Extending the Concepts and Skills

Ordinal data. Another important type of data is **ordinal data,** data about order or rank given on a scale such as 1, 2, 3, ... or A, B, C, ...

2.15 In each of Exercises 2.6–2.13, identify ordinal data, if any.

2.16 Following are several variables. Which, if any, yield ordinal data? Explain your answer.

a. height
b. weight
c. age
d. sex
e. number of siblings
f. religion
g. place of birth
h. high-school class rank

2.2 GROUPING DATA

When we studied data types in Section 2.1, we used examples that did not present large quantities of data. But the amount of data collected in real-world situations can sometimes be overwhelming. For example, a list of U.S. colleges and universities with information on enrollment, number of teachers, highest degree offered, and governing official can be found in the *World Almanac.* These data occupy 28 pages of small type!

By suitably organizing data, we can often make a large and complicated set of data more compact and easier to understand. In this section we will discuss **grouping,** which involves, as the term implies, putting data into groups rather than treating each observation individually. Grouping is one of the most common methods for organizing data.

EXAMPLE 2.5 Grouping Data

Table 2.1 displays the number of days to maturity for 40 short-term investments. The data are from *Barron's National Business and Financial Weekly.*

TABLE 2.1 Days to maturity for 40 short-term investments

70	64	99	55	64	89	87	65
62	38	67	70	60	69	78	39
75	56	71	51	99	68	95	86
57	53	47	50	55	81	80	98
51	36	63	66	85	79	83	70

It is difficult to get a clear picture of the data in Table 2.1. By grouping the data into categories, or **classes,** we can make it much simpler to comprehend. The first step is to decide on the classes. One convenient way to group these data is by 10s.

Since the shortest maturity period is 36 days, our first class will be for maturity periods from 30 days up to, but not including, 40 days. We will use the symbol $<$ as a shorthand for "up to, but not including." So, our first class is depicted as 30 ⩤ 40.[†] Noticing that the longest maturity period is 99 days, we see that grouping by 10s results in the seven classes given in the first column of Table 2.2.

TABLE 2.2
Classes and counts for the days-to-maturity data in Table 2.1

Days to maturity	Tally	Number of investments
30 ⩤ 40	|||	3
40 ⩤ 50	|	1
50 ⩤ 60	𝍸 |||	8
60 ⩤ 70	𝍸 𝍸	10
70 ⩤ 80	𝍸 ||	7
80 ⩤ 90	𝍸 ||	7
90 ⩤ 100	||||	4
		40

The final step for grouping the data is to determine how many investments are in each class. We do this by going through the data in Table 2.1 and placing a tally mark for each investment in the appropriate line of Table 2.2. For instance, the first investment in Table 2.1 has a 70-day maturity period. This calls for a tally mark on the line for the class 70 ⩤ 80. The results of the tallying procedure are shown in the second column of Table 2.2. Now we count the tallies for each class and record the totals in the third column of Table 2.2.

By simply glancing at Table 2.2, we can easily obtain various pieces of useful information. For instance, we observe that more investments are in the 60s range than in any other. Comparing Tables 2.1 and 2.2, we see that grouping the data makes it much simpler to read and understand. ■

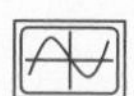

In Example 2.5 we used a common-sense approach to grouping data into classes. Some of that common sense can be used as guidelines for grouping. Three of the most important guidelines follow.

1. *The number of classes should be small enough to provide an effective summary but large enough to display the relevant characteristics of the data.*

In Example 2.5, seven classes are used. Usually the number of classes should be between 5 and 20; but that is only a rule of thumb.

2. *Each observation must belong to one, and only one, class.*

Careless planning in Example 2.5 could have led to classes like 30–40, 40–50, 50–60, and so on. Then, for instance, to which class would the investment with a 50-day maturity period belong? The classes in Table 2.2 do not cause such confusion; they cover all maturity periods and do not overlap.

[†] The symbol ⩤ is obtained by superimposing a less-than sign over a dash. It seems preferable to use the notation 30 ⩤ 40 instead of the more awkward 30–< 40 or 30–under 40.

3. *Whenever feasible, all classes should have the same width.*

The classes in Table 2.2 all have a width of 10 days. Among other things, choosing classes of equal width facilitates the graphical display of the data.

The list could go on, but for our purposes these three guidelines provide a solid basis for grouping data. And we should always keep in mind that the reason for grouping is to organize the data into a sensible number of classes in order to make the data more accessible and understandable.

Frequency and Relative-Frequency Distributions

The number of observations that fall into a particular class is called the **frequency** (or **count**) of that class. For example, as we see from Table 2.2, the frequency of the class 50 < 60 is eight, since eight investments are in the 50s days-to-maturity range. A table listing all classes and their frequencies is called a **frequency distribution.** The first and third columns of Table 2.2 constitute a frequency distribution for the days-to-maturity data.

In addition to the frequency of a class, we are often interested in the **percentage** of a class. We find the percentage by dividing the frequency of the class by the total number of observations and multiplying the result by 100. Referring again to Table 2.2, we see that the percentage of investments in the class 50 < 59 is

$$\frac{8}{40} = 0.20 \text{ or } 20\%.$$

Thus 20% of the investments have a number of days to maturity in the 50s.

The percentage of a class, expressed as a decimal, is usually referred to as the **relative frequency** of the class. For the class 50 < 60, the relative frequency is 0.20. A table listing all classes and their relative frequencies is called a **relative-frequency distribution.** Table 2.3 displays a relative-frequency distribution for the days-to-maturity data. Notice that the relative frequencies sum to 1 (100%).

TABLE 2.3
Relative-frequency distribution for the days-to-maturity data in Table 2.1

Days to maturity	Relative frequency		
30 < 40	0.075	←	*3/40*
40 < 50	0.025	←	*1/40*
50 < 60	0.200	←	*8/40*
60 < 70	0.250	←	*10/40*
70 < 80	0.175	←	*7/40*
80 < 90	0.175	←	*7/40*
90 < 100	0.100	←	*4/40*
	1.000		

When comparing two data sets, relative-frequency distributions are better than frequency distributions. This is because relative frequencies are always between 0 and 1 and hence provide a standard for comparison. Two data sets having identical frequency distributions will, of course, have identical relative-frequency distributions, but two data sets having identical relative-frequency distributions will have identical frequency distributions only if both data sets have the same number of observations.

Grouping Terminology

To become adept at grouping, you must first become familiar with and understand the various terms associated with it. We have already discussed several of these terms. To introduce some additional ones, let's return to the days-to-maturity data.

Consider, for example, the class $50 < 60$. The smallest maturity period that can go in this class is 50; this value is called the **lower cutpoint** of the class. The smallest maturity period that can go in the next higher class is 60; this value is called the **upper cutpoint** of the class and, of course, is also the lower cutpoint of the next higher class, $60 < 70$.

The number in the middle of the class $50 < 60$ is $(50 + 60)/2 = 55$, and this is called the **midpoint** (or **mark**) of the class. Midpoints provide single numbers for representing classes and are often used in graphical displays and for computing descriptive measures. The **width** of the class $50 < 60$, obtained by subtracting its lower cutpoint from its upper cutpoint, is $60 - 50 = 10$. Definition 2.3 summarizes the terminology of grouping.

DEFINITION 2.3 TERMS USED IN GROUPING

Classes: Categories for grouping data.

Frequency: The number of observations that fall in a class.

Frequency distribution: A listing of all classes along with their frequencies.

Relative frequency: The ratio of the frequency of a class to the total number of observations.

Relative-frequency distribution: A listing of all classes along with their relative frequencies.

Lower cutpoint: The smallest value that can go in a class.

Upper cutpoint: The smallest value that can go in the next higher class. The upper cutpoint of a class is the same as the lower cutpoint of the next higher class.

Midpoint: The middle of a class, obtained by taking the average of its lower and upper cutpoints.

Width: The difference between the upper and lower cutpoints of a class.

A table giving the classes, frequencies, relative frequencies, and midpoints of a data set is called a **grouped-data table.** A grouped-data table for the data on the number of days to maturity is presented in Table 2.4.

TABLE 2.4
Grouped-data table for the days-to-maturity data

Days to maturity	Frequency (no. of investments)	Relative frequency	Midpoint
30 ⩤ 40	3	0.075	35
40 ⩤ 50	1	0.025	45
50 ⩤ 60	8	0.200	55
60 ⩤ 70	10	0.250	65
70 ⩤ 80	7	0.175	75
80 ⩤ 90	7	0.175	85
90 ⩤ 100	4	0.100	95
	40	1.000	

EXAMPLE 2.6 GROUPED-DATA TABLES

The U.S. National Center for Health Statistics publishes data on weights and heights by age and sex in *Vital and Health Statistics.* The weights in Table 2.5, given to the nearest tenth of a pound, were obtained from a sample of 18–24-year-old males. Construct a grouped-data table for these weights. Use a class width of 20 and a first cutpoint of 120.

TABLE 2.5
Weights of 37 males, age 18–24 years

129.2	185.3	218.1	182.5	142.8	155.2	170.0	151.3	158.6	149.9
187.5	145.6	167.3	161.0	178.7	165.0	172.5	191.1	173.6	182.0
150.7	187.0	173.7	178.2	161.7	170.1	165.8	175.4	209.1	146.4
214.6	136.7	278.8	175.6	188.7	132.1	158.5			

SOLUTION Because we are to use a class width of 20 and a first cutpoint of 120, the first class will be 120 ⩤ 140. From Table 2.5 we see that the largest weight is 278.8 lb. Thus we choose the classes displayed in the first column of Table 2.6.

Applying the tallying procedure to the data in Table 2.5, we obtain the frequencies in the second column of Table 2.6. To illustrate some typical computations for the third and fourth columns of Table 2.6, let's consider the class 160 ⩤ 180:

$$\text{relative frequency} = \frac{14}{37} = 0.378$$

$$\text{midpoint} = \frac{160 + 180}{2} = 170.$$

 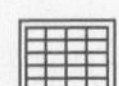

The other entries in the third and fourth columns are computed similarly. ◨

TABLE 2.6
Grouped-data table for the weights of 37 males, age 18–24 years

Weight (lb)	Frequency	Relative frequency	Midpoint
120 < 140	3	0.081	130
140 < 160	9	0.243	150
160 < 180	14	0.378	170
180 < 200	7	0.189	190
200 < 220	3	0.081	210
220 < 240	0	0.000	230
240 < 260	0	0.000	250
260 < 280	1	0.027	270
	37	0.999	

As we know, relative frequencies must always sum to 1. However, the sum of the relative frequencies in the third column of Table 2.6 is given as 0.999. The reason for this discrepancy is that each relative frequency is rounded to three decimal places; and when those rounded relative frequencies are added, the resulting sum differs from 1 by a little. This phenomenon is usually referred to as **rounding error** or **roundoff error.**

An Alternate Method for Depicting Classes

Often, especially when the data are expressed as whole numbers, classes may be depicted using an alternate method. For instance, the first class for the days-to-maturity data is for maturity periods from 30 days up to, but not including, 40 days. Since, in this case, the maturity periods are expressed to the nearest whole day, we see that the first class can also be characterized as being for maturity periods from 30 days up to, and including, 39 days.

So, we can depict the first class by 30–39 as well as by 30 < 40. Using this method for depicting classes, we obtain in Table 2.7 an alternate grouped-data table for the days-to-maturity data. Compare Tables 2.4 and 2.7.

TABLE 2.7
Grouped-data table for the days-to-maturity data using alternative method for depicting classes

Days to maturity	Frequency (no. of investments)	Relative frequency	Midpoint
30–39	3	0.075	35
40–49	1	0.025	45
50–59	8	0.200	55
60–69	10	0.250	65
70–79	7	0.175	75
80–89	7	0.175	85
90–99	4	0.100	95
	40	1.000	

The weight data in Table 2.5 are presented to one decimal place. Consequently, the alternate method for depicting the classes shown in the first column of Table 2.6 would be 120–139.9, 140–159.9, 160–179.9, and so forth.

Each of the two methods we have discussed thus far for depicting classes are in common use and each has its advantages and disadvantages. We will freely use both methods throughout this book. So be sure you have a clear understanding of the two different approaches.

Single-Value Grouping

Up to this point, each class we have used for grouping data represents a range of possible values. For instance, in Table 2.6, the first class, 120 < 140, is for weights from 120 lb up to, but not including, 140 lb. In some cases, however, it is more appropriate to use classes that each represent a single possible value. This is particularly true for discrete data in which there are only relatively few distinct observations. Consider, for instance, Example 2.7.

EXAMPLE 2.7 SINGLE-VALUE GROUPING

A city planner collected data on the number of school-age children in each of 30 families. The data are displayed in Table 2.8. Construct a grouped-data table for these data using classes based on a single value.

TABLE 2.8
Number of school-age children in each of 30 families

0	3	0	0	3	0
2	2	0	1	2	1
0	0	1	2	4	0
4	2	1	0	1	0
0	2	0	1	3	2

SOLUTION We first note that since each class is to represent a single numerical value, the classes are 0, 1, 2, 3, and 4. These classes are displayed in the first column of Table 2.9.

TABLE 2.9
Grouped-data table for number of school-age children

Number of school-age children	Frequency	Relative frequency
0	12	0.400
1	6	0.200
2	7	0.233
3	3	0.100
4	2	0.067
	30	1.000

Tallying the data in Table 2.8, we obtain the frequencies in the second column of Table 2.9. Dividing each frequency by the total number of observations, 30, yields the relative frequencies shown in the third column of Table 2.9. The table indicates, for example, that 7 of the 30 families, or 0.233 (23.3%), have two school-age children.

Since each class is based on a single value, the midpoint of each class is the same as the class; for instance, the midpoint of the class 3 is 3. It is therefore unnecessary to include a midpoint column in Table 2.9 since such a column would be identical to the first column. In other words, Table 2.9 can serve as a grouped-data table.[†] ◨

Frequency and Relative-Frequency Distributions for Qualitative Data

Although the concepts of cutpoints and midpoints apply to quantitative data, they are not appropriate for qualitative data. For instance, with data that categorize people as male or female, the classes are "male" and "female," and it makes no sense to look for cutpoints or midpoints. We can, of course, still compute frequencies and relative frequencies for qualitative data, as illustrated in Example 2.8.

EXAMPLE 2.8 FREQUENCY AND RELATIVE-FREQUENCY DISTRIBUTIONS FOR QUALITATIVE DATA

Professor Weiss asked his introductory statistics students to state their political party affiliations as Democratic (D), Republican (R), or Other (O). The responses are given in Table 2.10. Determine the frequency and relative-frequency distributions for these data.

TABLE 2.10 Political party affiliations of the students in introductory statistics

D	R	O	R	R	R	R	R
D	O	R	D	O	O	R	D
D	R	O	D	R	R	O	R
D	O	D	D	D	R	O	D
O	R	D	R	R	R	R	D

SOLUTION The classes for grouping the data are "Democratic," "Republican," and "Other." Tallying the data in Table 2.10, we obtain the frequency distribution displayed in the first two columns of Table 2.11.

[†] Although it is generally unnecessary, we can also obtain the cutpoints for single-value grouped data such as the grouped data in Table 2.9. For that grouped data, the cutpoints are −0.5, 0.5, 1.5, 2.5, 3.5, and 4.5.

TABLE 2.11
Frequency and relative-frequency distributions for political party affiliations

Party	Frequency	Relative frequency
Democratic	13	0.325
Republican	18	0.450
Other	9	0.225
	40	1.000

Dividing each frequency in the second column of Table 2.11 by the total number of students, 40, we get the relative frequencies in the third column. The first and third columns of Table 2.11 provide the relative-frequency distribution for the data. □

Using the Computer (Optional)

We can use Minitab to group both quantitative and qualitative data. We will illustrate the methods for doing that using data sets considered earlier in this section. Our first illustration (Example 2.9) is optional and requires DataDisk CD, the compact disk included in the back of your book.†

EXAMPLE 2.9 USING MINITAB TO OBTAIN A GROUPED-DATA TABLE

Table 2.1 on page 66 gives the number of days to maturity for 40 short-term investments. Apply Minitab to obtain a grouped-data table for those weights. Use a class width of 10 and a first cutpoint of 30.

SOLUTION First we store the data from Table 2.1 in a column named MATURITY. Although Minitab does not have a dedicated procedure for obtaining grouped-data tables, we can still construct them by applying several existing Minitab procedures. More simply, a **Minitab macro**—a computer program written in Minitab's programming language—can be used to automate those procedures. Such a macro, `group.mac`, can be found in the `Macro` folder of DataDisk CD.

To invoke a Minitab macro, we must first enable the command language. This is done as follows:

1 Click in the Session window

2 Choose **Editor ➤ Enable Command Language**

† DataDisk CD is a compact disk that contains text files for the data sets appearing in the exercises, computer sections, Focus-database sections, and case studies. DataDisk CD makes it possible to store those data sets in a computer or statistical calculator without having to enter them manually. Also included on DataDisk CD are several useful Minitab macros written by the author of this book.

Now we can run `group.mac` by typing, in the Session window, a percent sign (%) followed by the path for `group.mac` and the location of the data enclosed by apostrophes. So, for instance, if DataDisk CD is in drive `E:`, we type

`%E:\IS5\Macro\group.mac 'MATURITY'`

and press the [Enter] key.

After invoking the `group` macro, we see from the Session window that we are given three alternatives for specifying the classes. Since we want a class width of 10 and a first cutpoint of 30, we select the third option (3), press the [Enter] key, and then type 30 10 when prompted to enter the lower cutpoint of the first class and the class width. The resulting output is shown in Printout 2.1.

PRINTOUT 2.1
Minitab output for the group macro

```
Grouped-data table for MATURITY     N = 40


 Row  LowerCut  UpperCut   Freq  RelFreq  Midpoint

   1        30        40      3    0.075        35
   2        40        50      1    0.025        45
   3        50        60      8    0.200        55
   4        60        70     10    0.250        65
   5        70        80      7    0.175        75
   6        80        90      7    0.175        85
   7        90       100      4    0.100        95
```

Compare the output in Printout 2.1 to the grouped-data table we obtained by hand in Table 2.4 on page 70. ◨

We can use the `group` macro discussed in Example 2.9 to obtain a grouped-data table for any quantitative data set, including ones in which each class is based on a single value. However, for single-value grouping, it is quicker to use a built-in Minitab procedure. We illustrate the use of that procedure in Example 2.10.

EXAMPLE 2.10 USING MINITAB TO OBTAIN A SINGLE-VALUE GROUPED-DATA TABLE

Table 2.8 on page 72 displays the number of school-age children in each of 30 families. Use Minitab to obtain a grouped-data table for these data using classes based on a single value.

SOLUTION First we store the data from Table 2.8 in a column named CHILDREN. Then we proceed as follows.

1 Choose **Stat ➤ Tables ➤ Tally...**
2 Specify CHILDREN in the **Variables** text box
3 Select the **Counts** and **Percents** check boxes
4 Click **OK**

The output that results is shown in Printout 2.2.

PRINTOUT 2.2
Minitab output for the tally procedure

```
CHILDREN  Count Percent
       0     12   40.00
       1      6   20.00
       2      7   23.33
       3      3   10.00
       4      2    6.67
      N=     30
```

Compare the output in Printout 2.2 to the grouped-data table we obtained by hand in Table 2.9 on page 72. Notice, in particular, that Minitab uses percents instead of relative frequencies. □

Minitab's tally procedure, discussed in Example 2.10, can also be used to obtain frequency and relative-frequency distributions for qualitative data. For instance, suppose we want to obtain frequency and relative-frequency distributions for the data on political party affiliations in Example 2.8. First we store the data from Table 2.10 on page 73 in a column named POLITIC. Then we proceed exactly as in the four steps above Printout 2.2 except that in the second step, we replace CHILDREN by POLITIC.

EXERCISES 2.2

STATISTICAL CONCEPTS AND SKILLS

2.17 Identify an important reason for grouping data.

2.18 Do the concepts of cutpoints and midpoints make sense for qualitative-data classes? Explain your answer.

2.19 State three of the most important guidelines in choosing the classes for grouping a data set.

2.20 Explain the difference between each of the following pairs of terms.
a. frequency and relative frequency
b. percentage and relative frequency

2.21 When comparing two data sets, is it better to use frequency distributions or relative-frequency distributions? Explain your answer.

2.22 What are the four elements of a grouped-data table? Explain the meaning of each element.

2.23 When grouping data using classes that each represent a range of possible values, we discussed two methods for depicting the classes. Identify the two methods and explain the relative advantages and disadvantages of each.

2.24 When grouping quantitative data, we have examined three types of classes: (1) $a < b$, (2) a–b, and (3) single-value grouping. In each part, decide which of these three types is usually best. Explain your answers.

a. Continuous data displayed to one or more decimal places.

b. Discrete data in which there are relatively few distinct observations.

2.25 When grouping data using classes that each represent a single possible numerical value, why is it unnecessary to include a midpoint column in the grouped-data table?

2.26 The Bureau of Economic Analysis gathers information on the length of stay in Europe and the Mediterranean by U.S. travelers. Data are published in *Survey of Current Business*. A sample of 36 U.S. residents who traveled to Europe and the Mediterranean one year yielded the following data, in days, on length of stay.

41	16	6	21	1	21
5	31	20	27	17	10
3	32	2	48	8	12
21	44	1	56	5	12
3	13	15	10	18	3
1	11	14	12	64	10

Use classes of equal width starting with the class $1 < 8$ to construct a grouped-data table for these data on length of stay.

2.27 The U.S. Energy Information Administration collects data on residential energy consumption and expenditures. Results are published in the document *Residential Energy Consumption Survey: Consumption and Expenditures*. The following table gives one year's energy consumptions for a sample of 50 households in the South. Data are in millions of BTUs.

130	55	45	64	155	66	60	80	102	62
58	101	75	111	151	139	81	55	66	90
97	77	51	67	125	50	136	55	83	91
54	86	100	78	93	113	111	104	96	113
96	87	129	109	69	94	99	97	83	97

Use classes of equal width beginning with $40 < 50$ to construct a grouped-data table for these data on energy consumption.

2.28 The Food and Nutrition Board of the National Academy of Sciences states that the recommended daily allowance of iron is 18 mg for adult females under the age of 51. The amounts of iron intake, in milligrams, during a 24-hour period for a sample of 45 such females follows.

15.0	18.1	14.4	14.6	10.9	18.1	18.2	18.3	15.0
16.0	12.6	16.6	20.7	19.8	11.6	12.8	15.6	11.0
15.3	9.4	19.5	18.3	14.5	16.6	11.5	16.4	12.5
14.6	11.9	12.5	18.6	13.1	12.1	10.7	17.3	12.4
17.0	6.3	16.8	12.5	16.3	14.7	12.7	16.3	11.5

Construct a grouped-data table for these iron intakes. Use a first cutpoint of 6 and classes of equal width 2.

2.29 Data on starting salaries for college graduates are provided by *The Northwestern Endicott-Lindquist Report*. A sample of 35 liberal-arts graduates yielded the following starting annual salaries. Data are in thousands of dollars, rounded to the nearest hundred dollars.

29.0	25.8	30.3	29.6	30.0	27.7	32.8
27.3	26.7	27.0	28.1	30.1	28.6	28.0
27.7	29.8	29.4	26.1	28.5	28.9	28.2
28.1	26.2	27.3	31.7	29.0	28.2	29.9
28.1	29.8	29.5	30.4	25.3	25.3	29.5

Using 25 as the first cutpoint and classes of equal width 1, construct a grouped-data table for these starting annual salaries.

In each of Exercises 2.30–2.33, redo the specified exercise using the alternative method for depicting classes as discussed on page 71.

2.30 Exercise 2.26.

2.31 Exercise 2.27.

2.32 Exercise 2.28.

2.33 Exercise 2.29.

2.34 The U.S. Bureau of the Census conducts nationwide surveys on characteristics of U.S. households. Following are data on the number of people per household for a sample of 40 households.

2	5	2	1	1	2	3	4
1	4	4	2	1	4	3	3
7	1	2	2	3	4	2	2
6	5	2	5	1	3	2	5
2	1	3	3	2	2	3	3

Construct a grouped-data table for these household sizes using classes based on a single value.

2.35 A car salesperson keeps track of the number of cars she sells per week. The number of cars she sold per week last year are as follows.

1	0	3	3	1	0	2	1	4	0	4	1	2
3	6	4	3	0	2	2	1	1	2	2	2	3
5	1	0	2	5	3	1	3	1	1	1	1	2
2	3	0	4	4	1	0	1	1	3	2	5	2

Construct a grouped-data table for the number of sales per week. Use classes based on a single value.

2.36 A research physician conducted a study on the ages of people with diabetes. The following data were obtained for the ages of a sample of 35 diabetics. Construct an appropriate grouped-data table.

48	41	57	83	41	55	59
61	38	48	79	75	77	7
54	23	47	56	79	68	61
64	45	53	82	68	38	70
10	60	83	76	21	65	47

2.37 Cudahey Masonry employs 80 bricklayers. The number of days each employee misses is recorded. Absentee records for the past year are as follows. Construct a grouped-data table for these data

2	3	6	2	6	5	5	2	4	7	5	3	6	4	4	4
2	2	4	5	4	0	2	1	6	3	5	3	6	6	4	7
5	2	5	0	5	6	5	2	4	2	6	2	4	3	5	4
2	4	4	3	3	4	0	5	6	3	5	5	2	4	4	2
0	7	5	5	7	6	1	5	3	3	4	7	7	2	5	5

2.38 According to the *World Almanac, 1998,* the all-time top television programs by rating (percentage of TV-owning households tuned in to the program) are as shown in the following table. [SOURCE: Nielsen Media Research, Jan. 1961–Jan. 1997.]

Program	Telecast date	Network	Rating (%)	Audience (millions)
M*A*S*H (last episode)	02/28/83	CBS	60.2	50.2
Dallas (Who shot J.R.?)	11/21/80	CBS	53.3	41.5
Roots-Pt. 8	01/30/77	ABC	51.1	36.4
Super Bowl XVI	01/24/82	CBS	49.1	40.0
Super Bowl XVII	01/30/83	NBC	48.6	40.5
XVII Winter Olympics-2d Wed.	02/23/94	CBS	48.5	45.7
Super Bowl XX	01/26/86	NBC	48.3	41.5
Gone With the Wind-Pt.1	11/07/76	NBC	47.7	34.0
Gone With the Wind-Pt.2	11/08/76	NBC	47.4	33.8
Super Bowl XII	01/15/78	CBS	47.2	34.4
Super Bowl XIII	01/21/79	NBC	47.1	35.1
Bob Hope Christmas Show	01/15/70	NBC	46.6	27.3
Super Bowl XVIII	01/22/84	CBS	46.4	38.8
Super Bowl XIX	01/20/85	ABC	46.4	39.4
Super Bowl XIV	01/20/80	CBS	46.3	35.3
Super Bowl XXX	01/28/96	NBC	46.0	44.2
ABC Theatre (The Day After)	11/20/83	ABC	46.0	38.6
Roots-Pt. 6	01/28/77	ABC	45.9	32.7
The Fugitive	08/29/67	ABC	45.9	25.7
Super Bowl XXI	01/25/87	CBS	45.8	40.0

Construct frequency and relative-frequency distributions for the network data. (*Hint:* The classes are "ABC," "CBS," and "NBC.")

2.39 According to the *World Almanac, 1998,* the National Collegiate Athletic Association wrestling champions for the years 1968–1997 are as shown in the table at the top of the following page. Construct both a frequency distribution and a relative-frequency distribution for the champions.

Year	Champion	Year	Champion
1968	Oklahoma State	1983	Iowa
1969	Iowa State	1984	Iowa
1970	Iowa State	1985	Iowa
1971	Oklahoma State	1986	Iowa
1972	Iowa State	1987	Iowa State
1973	Iowa State	1988	Arizona State
1974	Oklahoma	1989	Oklahoma State
1975	Iowa	1990	Oklahoma State
1976	Iowa	1991	Iowa
1977	Iowa State	1992	Iowa
1978	Iowa	1993	Iowa
1979	Iowa	1994	Oklahoma State
1980	Iowa	1995	Iowa
1981	Iowa	1996	Iowa
1982	Iowa	1997	Iowa

EXTENDING THE CONCEPTS AND SKILLS

The table at the top of the next column gives the closing Dow Jones Industrial Averages (DJIA) on November 20, 1997. We will use the data in Exercises 2.40–2.42.

2.40 The column headed "Volume" in the DJIA table shows the number of shares sold, in hundreds, for each of the DJIA stocks. Construct a grouped-data table for these sales volumes. Begin with the class $0 < 1$, where the values are in millions of sales.

2.41 The column headed "Last" in the DJIA table gives the closing price per share, in dollars, for each of the DJIA stocks.

a. Construct a grouped-data table for these prices using the classes $30 < 40, 40 < 50, \ldots, 90 < 100$, and "100 & over."

b. Why is there no midpoint for the last class?

2.42 The column headed "Change" in the DJIA table gives the difference, in dollars, between the closing price per share given in the second column of the table and the closing price per share on the previous trading day. Construct a grouped-data table for the changes using classes of your choice. Explain your choice of classes.

Symbol	Last	Change	Volume (hds)
AA	70 5/16	+1 3/16	6462
ALD	37 7/8	+1/2	14593
AXP	79 1/2	+1/2	12499
BA	49 15/16	-1/16	31773
CAT	46 3/4	+1/8	20404
CHV	84 3/16	+1 7/8	8782
DD	61 1/4	+1 1/8	23919
DIS	90 11/16	-11/16	23516
EK	60 9/16	+1/2	11261
GE	71 15/16	+1 1/4	54604
GM	61 9/16	+3/8	18546
GT	61 1/16	-7/16	3339
HWP	62 7/8	+1 3/4	41387
IBM	104 3/4	+1 11/16	33541
IP	46 3/4	-1/4	30109
JNJ	64 1/4	+1 1/8	25896
JPM	116 1/8	+3/4	7569
KO	63 1/2	+1 3/8	50483
MCD	48 5/8	+5/8	19344
MMM	96 7/8	+2 9/16	10013
MO	42 11/16	+3/8	34733
MRK	92 9/16	-1/4	39102
PG	76 1/16	+1 1/2	30548
S	47 3/8	-3/16	13787
T	54 9/16	+1 3/4	62822
TRV	51	+2 1/16	39918
UK	45 13/16	+1	4748
UTX	74 11/16	+1 7/16	6704
WMT	39 15/16	+1/2	25511
XON	62 11/16	+5/8	31897

2.43 The exam scores for the students in an introductory statistics class are as follows.

88	82	89	70	85
63	100	86	67	39
90	96	76	34	81
64	75	84	89	96

a. Group these exam scores using the classes 30–39, 40–49, 50–59, 60–69, 70–79, 80–89, and 90–100.

b. What are the widths of the classes?

c. If you wanted all the classes to have the same width, what classes would you use?

Contingency tables. The methods we have examined in this section apply to grouping data obtained from observing values of one variable. Such data are called **univariate.** For instance, in Example 2.6 we considered data obtained from observing the values of the variable "weight" for a sample of 18–24-year-old males. Consequently, those data are univariate. We could have considered not only the weights of the males, but also their heights. Then we would have data on two variables, height and weight. Data obtained from observing values of two variables are called **bivariate.** Bivariate data can be grouped using tables called **contingency tables.** Exercises 2.44 and 2.45 consider the grouping of bivariate data using contingency tables.

2.44 The following bivariate data on age (in years) and sex were obtained from the students in a freshman calculus course. The data show, for example, that the first student on the list is 21 years old and is a male.

Age	Sex	Age	Sex	Age	Sex	Age	Sex	Age	Sex
21	M	29	F	22	M	23	F	21	F
20	M	20	M	23	M	44	M	28	F
42	F	18	F	19	F	19	M	21	F
21	M	21	M	21	M	21	F	21	F
19	F	26	M	21	F	19	M	24	F
21	F	24	F	21	F	25	M	24	F
19	F	19	M	20	F	21	M	24	F
19	M	25	M	20	F	19	M	23	M
23	M	19	F	20	F	18	F	20	F
20	F	23	M	22	F	18	F	19	M

We will discuss the grouping of these data into the following contingency table.

		Age (yrs)			
		Under 21	21–25	Over 25	Total
Sex	Male		|		
	Female				
	Total				

a. To tally the data for the first student, place a tally mark in the box labeled by the "21–25" column and the "Male" row, as indicated. Tally the data for the other 49 students.

b. Construct a table like the one in part (a) but with frequencies replacing the tally marks. Add the frequencies in each row and column of your table and record the sums in the proper "Total" boxes.

c. What do the row and column totals represent?

d. Add the row totals and add the column totals. Why are those two sums equal, and what does their common value represent?

e. Construct a table giving the relative frequencies for the data. *(Hint:* Divide each frequency obtained in part (b) by the grand total of 50 students.*)*

f. Interpret the entries in your table from part (e) in terms of percentages.

2.45 The heights (in inches) and weights (in pounds) of the students in Exercise 2.44 are as follows.

Height	Weight	Height	Weight	Height	Weight
68	140	67	155	74	215
67	140	67	130	67	129
72	145	68	160	72	275
69	145	64	127	68	135
66	115	74	170	60	95
72	185	73	180	75	175
64	130	63	142	61	120
65	145	69	170	73	180
62	127	62	103	64	125
69	135	68	160	66	130
66	110	75	185	63	105
69	178	64	122	69	155
63	130	64	130	70	170
72	185	70	215	64	105
67	120	63	105	65	132
68	135	76	200	65	115
64	130	71	169		

Repeat parts (a)–(f) of Exercise 2.44 for the bivariate data on heights and weights. Use a contingency table with classes for weight of equal width 40 starting with 90–129 and classes for height of equal width 6 starting with 60–65.

USING TECHNOLOGY   

2.46 Use Minitab or some other statistical software to obtain the grouped-data table required in Exercise 2.26 on page 77.

2.47 Use Minitab or some other statistical software to obtain the grouped-data table required in Exercise 2.27 on page 77.

2.48 Use Minitab or some other statistical software to obtain the grouped-data table required in Exercise 2.34 on page 78.

2.49 Use Minitab or some other statistical software to obtain the grouped-data table required in Exercise 2.35 on page 78.

2.50 Use Minitab or some other statistical software to obtain the frequency and relative-frequency distributions required in Exercise 2.38 on page 78.

2.51 Use Minitab or some other statistical software to obtain the frequency and relative-frequency distributions required in Exercise 2.39 on page 78.

2.3 GRAPHS AND CHARTS

Besides grouping, another method for organizing and summarizing data is to draw a picture of some kind. The old saying "a picture is worth a thousand words" has particular relevance in statistics—a graph or chart of a data set often provides the simplest and most efficient display. In this section we will examine various techniques for organizing and summarizing data using graphs and charts. We begin by discussing histograms.

EXAMPLE 2.11 HISTOGRAMS

Table 2.4 shows a grouped-data table for the number of days to maturity for 40 short-term investments. The first three columns of that table are repeated below in Table 2.12. Obtain graphical displays for these grouped data.

TABLE 2.12 Frequency and relative-frequency distributions for the days-to-maturity data

Days to maturity	Frequency (no. of investments)	Relative frequency
30 < 40	3	0.075
40 < 50	1	0.025
50 < 60	8	0.200
60 < 70	10	0.250
70 < 80	7	0.175
80 < 90	7	0.175
90 < 100	4	0.100
	40	1.000

SOLUTION One way to display these grouped data pictorially is to construct a graph with the classes depicted on the horizontal axis and the frequencies depicted on the vertical axis. This can be done using a **frequency histogram.** A frequency histogram for the days-to-maturity data is shown in Fig. 2.1(a).

FIGURE 2.1
Days-to-maturity:
(a) frequency histogram
(b) relative-frequency histogram

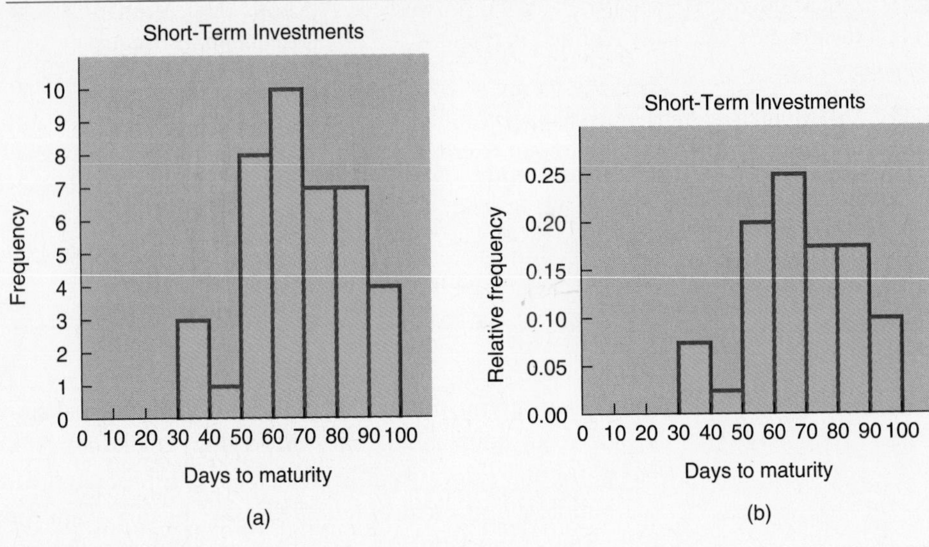

Here are some important observations about Fig. 2.1(a).

- The height of each bar is equal to the frequency of the class it represents.
- The bar for each class extends from the lower cutpoint of the class to the upper cutpoint of the class.[†]
- Each axis of the frequency histogram has a label, and the frequency histogram as a whole has a title.

A frequency histogram displays the frequencies of the classes. To display the relative frequencies (or percentages), we can use a **relative-frequency histogram,** which is similar to a frequency histogram. The only difference is that the height of each bar in a relative-frequency histogram is equal to the relative frequency of the class instead of the frequency of the class. A relative-frequency histogram for the days-to-maturity data is shown in Fig. 2.1(b).

 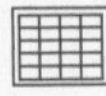 

Notice that the shapes of the relative-frequency histogram in Fig. 2.1(b) and the frequency histogram in Fig. 2.1(a) are identical. This is because the frequencies and relative frequencies are proportional. □

[†] This is only one of several methods for depicting the classes on the horizontal axis. Another common method is to use midpoints instead of cutpoints to label the horizontal axis. In that case each bar is centered over the midpoint of the class it represents.

DEFINITION 2.4 FREQUENCY AND RELATIVE-FREQUENCY HISTOGRAMS

Frequency histogram: A graph that displays the classes on the horizontal axis and the frequencies of the classes on the vertical axis. The frequency of each class is represented by a vertical bar whose height is equal to the frequency of the class.

Relative-frequency histogram: A graph that displays the classes on the horizontal axis and the relative frequencies of the classes on the vertical axis. The relative frequency of each class is represented by a vertical bar whose height is equal to the relative frequency of the class.

For purposes of visually comparing the distributions of two data sets, it is better to use relative-frequency histograms than frequency histograms. This is because the same vertical scale is used for all relative-frequency histograms—a minimum of 0 and a maximum of 1. On the other hand, the vertical scale of a frequency histogram depends on the number of observations.

Histograms for Single-Value Grouping

For the days-to-maturity data, each class represents a range of possible days to maturity, and in Figs. 2.1(a) and 2.1(b) the histogram bar for each class extends over that range. When data are grouped using classes based on a single value, we proceed somewhat differently. In that case each bar is centered over the only possible value in the class, as illustrated in Example 2.12.

EXAMPLE 2.12 HISTOGRAMS FOR SINGLE-VALUE GROUPED DATA

In Example 2.7 we considered data on the number of school-age children in each of 30 families. We grouped those data using classes based on a single value. The frequency and relative-frequency distributions are given in Table 2.9 and are repeated in Table 2.13. Construct a frequency histogram and a relative-frequency histogram for these grouped data.

TABLE 2.13 Frequency and relative-frequency distributions for number of school-age children

Number of school-age children	Frequency	Relative frequency
0	12	0.400
1	6	0.200
2	7	0.233
3	3	0.100
4	2	0.067
	30	1.000

SOLUTION As we just mentioned, for single-value grouping, we place the middle of each histogram bar directly over the single value represented by the class. Hence the frequency and relative-frequency histograms for the grouped data in Table 2.13 are as pictured in Figs. 2.2(a) and 2.2(b).

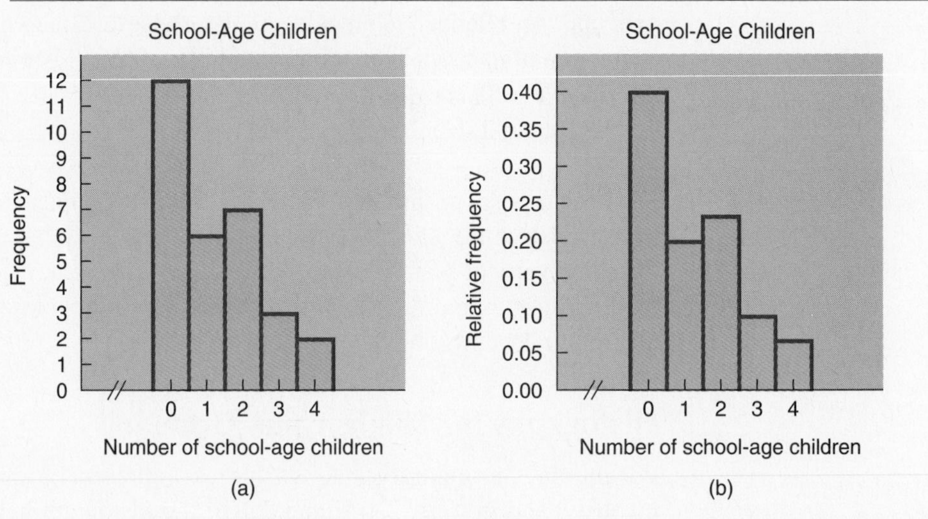

FIGURE 2.2 School-age children: (a) frequency histogram (b) relative-frequency histogram

Notice the symbol // on the horizontal axes in Figs. 2.2(a) and 2.2(b). This symbol indicates that the zero point on that axis is not in its usual position at the intersection of the horizontal and vertical axes. Whenever any such modification is made, whether on the horizontal or vertical axis, the symbol // or some similar symbol should be used to indicate that fact.

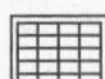
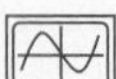

Dotplots

Another type of graphical display for quantitative data is the **dotplot.** Dotplots are particularly useful for showing the relative positions of the data in a data set or for comparing two or more data sets. We introduce dotplots in Example 2.13.

EXAMPLE 2.13 DOTPLOTS

A farmer is interested in estimating his yield of oats if he farms organically. He uses the method on a sample of 15 one-acre plots. The yields, in bushels, are depicted in Table 2.14. Construct a dotplot for the data.

TABLE 2.14
Oat yields

67	65	55	57	58
61	61	61	64	62
62	60	62	60	67

SOLUTION To construct a dotplot for the data in Table 2.14, we begin by drawing a horizontal axis that displays the possible oat yields. Then we record each yield by placing a dot over the appropriate value on the horizontal axis. For instance, the first yield is 67 bushels. This calls for a dot over the "67" on the horizontal axis. The dotplot for the data in Table 2.14 is pictured in Fig. 2.3.

 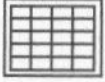 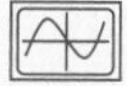

FIGURE 2.3
Dotplot for oat yields

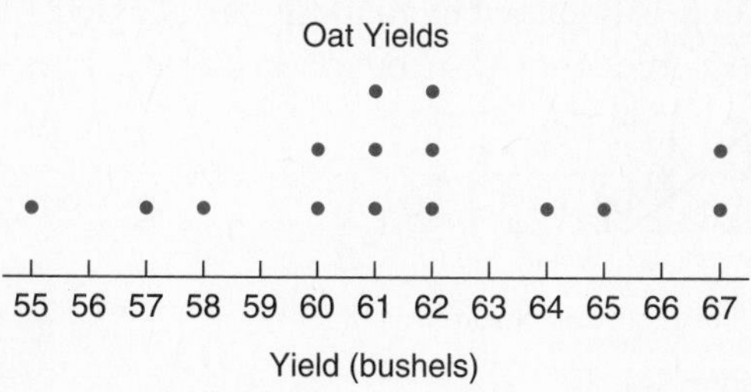

As we see, dotplots are similar to histograms. In fact, when data are grouped in classes based on a single value, a dotplot and a frequency histogram are essentially identical. However, for single-value grouped data that involve decimals, dotplots are generally preferable to histograms because they are easier to construct and use.

Graphical Displays for Qualitative Data

Histograms and dotplots are designed for use with quantitative data. Qualitative data are portrayed using different techniques. Two common methods for displaying qualitative data graphically are *pie charts* and *bar graphs.*

EXAMPLE 2.14 PIE CHARTS AND BAR GRAPHS

In Example 2.8 we obtained frequency and relative-frequency distributions for the political party affiliations of the students in Professor Weiss's introductory statistics class. We repeat those distributions in Table 2.15. Display the relative-frequency distribution of these qualitative data using

a. a pie chart. **b.** a bar graph.

TABLE 2.15 Frequency and relative-frequency distributions for political party affiliations

Party	Frequency	Relative frequency
Democratic	13	0.325
Republican	18	0.450
Other	9	0.225
	40	1.000

SOLUTION **a.** A **pie chart** is a disk divided into pie-shaped pieces proportional to the relative frequencies. In this case we need to divide a disk into three pie-shaped pieces comprising 32.5%, 45.0%, and 22.5% of the disk. We can do this by using a protractor and the fact that there are 360° in a circle. Thus, for instance, the first piece of the disk is obtained by marking off 117° (32.5% of 360°). The pie chart for the relative-frequency distribution in Table 2.15 is shown in Fig. 2.4(a).

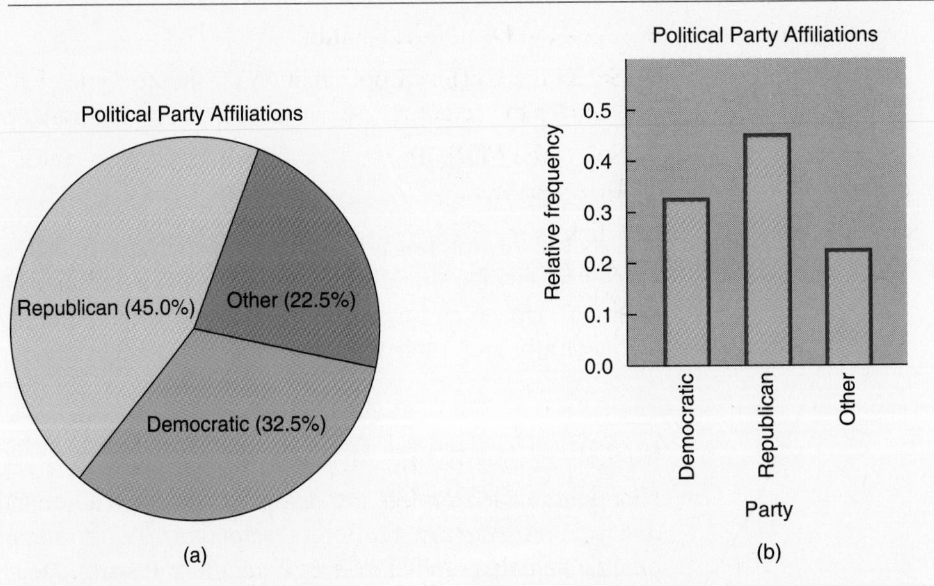

FIGURE 2.4 Political party affiliations: (a) pie chart (b) bar graph

b. A **bar graph** is like a histogram except that its bars do not touch each other. The bar graph for the relative-frequency distribution in Table 2.15 is pictured in Fig. 2.4(b). ◨

Histograms, dotplots, pie charts, and bar graphs are only a few of the countless ways that data can be portrayed pictorially. We will consider some additional graphical displays in the exercises for this section.

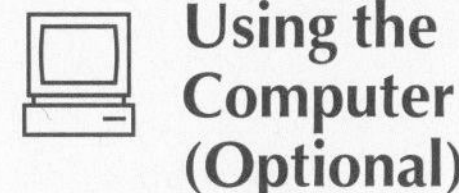

Using the Computer (Optional)

Minitab can be applied to obtain a frequency histogram for a data set. To illustrate we return once again to the number of days to maturity for 40 short-term investments first discussed in Example 2.5.

EXAMPLE 2.15 USING MINITAB TO OBTAIN A HISTOGRAM

Table 2.1 on page 66 gives the number of days to maturity for 40 short-term investments. Use Minitab to construct a frequency histogram for these data based on the classes 30 < 40, 40 < 50, . . . , 90 < 100.

SOLUTION To begin we store the data from Table 2.1 in a column named MATURITY. Then we proceed as follows.

1 Choose **Graph ➤ Histogram...**
2 Specify MATURITY in the **X** text box for **Graph 1**
3 Click the **Options...** button[†]
4 Select the **CutPoint** option button from the **Type of Intervals** field (This tells Minitab to use cutpoints, not midpoints, on the horizontal axis.)
5 Select the **Midpoint/cutpoint positions** option button from the **Definition of Intervals** field
6 Click in the **Midpoint/cutpoint positions** text box and type `30:100/10` (This tells Minitab to use cutpoints from 30 to 100 (`30:100`) at increments of 10 (`/10`).)
7 Click **OK**
8 Click **OK**

Printout 2.3 at the top of the next page shows the output that results. Compare Minitab's frequency histogram in Printout 2.3 to the one we drew by hand in Fig. 2.1(a) on page 82.

We can also use Minitab to obtain a relative-frequency histogram (actually, a percent histogram). To do so, we apply the same steps as we just did for obtaining a frequency histogram except that after the third step, we insert the following step: Select the **Percent** option button from the **Type of Histogram** field. Do that and compare the resulting percent histogram to the relative-frequency histogram that we drew by hand in Fig. 2.1(b) on page 82. ◨

[†] The steps from clicking the **Options...** button through the first "Click **OK**" are optional. If we do not perform these steps, Minitab will choose its own classes and use midpoints under the histogram bars to identify the classes.

PRINTOUT 2.3
Minitab histogram for the days-to-maturity data

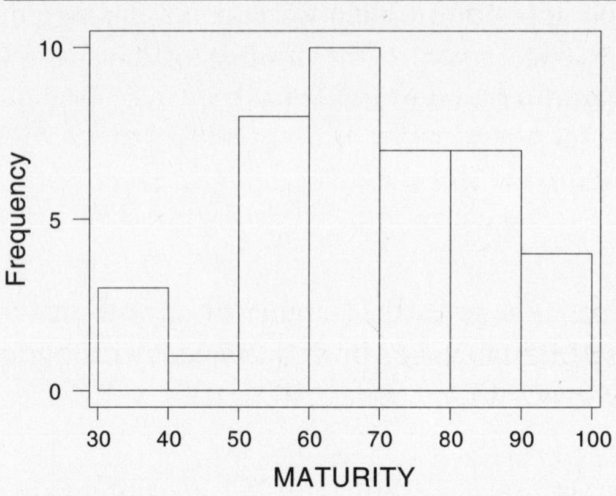

Earlier in this section we discussed dotplots. As with histograms, we can use Minitab to construct dotplots. To illustrate we will apply Minitab to obtain a dotplot of the data on oat yields considered in Example 2.13.

EXAMPLE 2.16 USING MINITAB TO OBTAIN A DOTPLOT

Table 2.14 on page 85 gives the oat yields, in bushels, obtained by a farmer growing oats organically on a sample of 15 one-acre plots. Use Minitab to construct a dotplot of the data.

SOLUTION To begin we store the data from Table 2.14 in a column named OATS. The way we proceed next depends on which release of Minitab we are using. We will indicate the procedure used for Release 12 and higher and then, following that, explain how to proceed when using other releases.

1 Choose **Graph ➤ Dotplot...**
2 Specify OATS in the **Variables** text box
3 Click **OK**

The output that results is shown in Printout 2.4.

PRINTOUT 2.4
Minitab dotplot for the oat-yield data

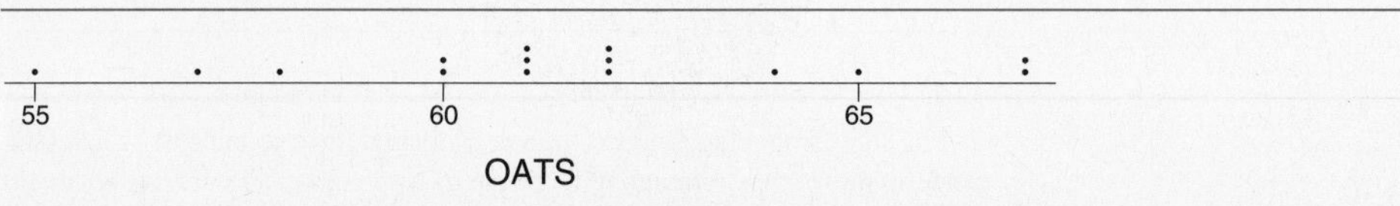

Versions of Minitab earlier than Release 12 do not offer a high-resolution dotplot but instead provide a character dotplot that appears in the Session window. To obtain a character dotplot in earlier versions of Minitab, proceed as we did above for the high-resolution dotplot except replace the first step by: Choose **Graph ➤ Character Graphs ➤ Dotplot...**. ◨

Minitab can also be used to obtain pie charts and bar graphs for qualitative data. We illustrate the methods by returning to the data on political party affiliations discussed previously in Example 2.14.

EXAMPLE 2.17

USING MINITAB TO OBTAIN PIE CHARTS AND BAR GRAPHS

Table 2.10 on page 73 gives the political party affiliations of the students in Professor Weiss's introductory statistics class. Use Minitab to obtain a pie chart and bar graph for these data.

SOLUTION To begin we store the data from Table 2.10 in a column named POLITIC. For the pie chart, we proceed as follows.

1 Choose **Graph ➤ Pie Chart...**
2 Specify POLITIC in the **Chart data in** text box
3 Click **OK**

The resulting pie chart is displayed in Printout 2.5.

PRINTOUT 2.5
Minitab pie chart for the data on political party affiliations

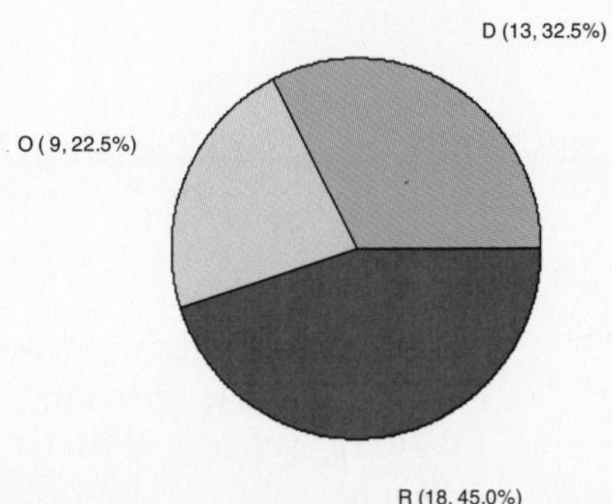

Compare Minitab's pie chart with the one we drew by hand in Fig. 2.4(a) on page 86. Notice, in particular, that Minitab's pie chart displays both frequencies and relative frequencies.

To obtain a bar graph of the data, we do the following.

1 Choose **Graph ➤ Chart...**
2 Specify POLITIC in the **X** text box for **Graph 1**
3 Click **OK**

The resulting bar graph is displayed in Printout 2.6.

PRINTOUT 2.6
Minitab bar graph for the data on political party affiliations

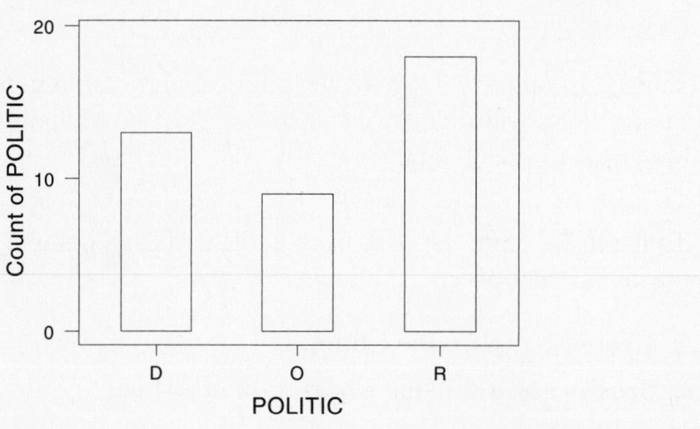

Compare Minitab's bar graph with the one we drew by hand in Fig. 2.4(b) on page 86. Notice, in particular, that Minitab's bar graph uses frequencies (counts) instead of relative frequencies on the vertical axis of the bar graph. ◨

EXERCISES 2.3

STATISTICAL CONCEPTS AND SKILLS

2.52 Explain in your own words advantages and disadvantages of histograms relative to grouped-data tables.

2.53 Explain the difference between a frequency histogram and a relative-frequency histogram.

2.54 For data that are grouped not using classes based on a single value, we have used cutpoints on the horizontal axis of a histogram for depicting the classes. But we can also use midpoints, in which case each bar is centered over the midpoint of the class it represents. Explain in your own words advantages and disadvantages of each method.

2.55 In a bar graph, unlike in a histogram, the bars do not abut each other. Give a reason why you think that is done.

2.56 Some users of statistics prefer pie charts to bar graphs because people are accustomed to having the horizontal axis of a graph show order. For example, someone might infer from Fig. 2.4(b) on page 86 that "Republican" is less than "Other" because "Republican" is shown to the left of "Other" on the horizontal axis. Pie charts do not lead to such inferences. Give other advantages and disadvantages of each method.

2.57 Refer to Example 2.13 on page 84.

a. Explain why a frequency histogram of the oat yields using classes based on a single value would be essentially identical to the dotplot in Fig. 2.3.

b. Would the dotplot and a frequency histogram be essentially identical if the classes for the histogram are not each based on a single value? Explain your answer.

In Exercises 2.58–2.63, we have presented the frequency and relative-frequency distributions obtained for selected exercises considered in Section 2.2. For each exercise,

a. construct a frequency histogram.

b. construct a relative-frequency histogram.

2.58 The lengths of stay in Europe and the Mediterranean obtained from a sample of 36 U.S. residents who traveled there one year:

Length of stay (days)	Frequency	Relative frequency
1 < 8	10	0.278
8 < 15	10	0.278
15 < 22	8	0.222
22 < 29	1	0.028
29 < 36	2	0.056
36 < 43	1	0.028
43 < 50	2	0.056
50 < 57	1	0.028
57 < 64	0	0.000
64 < 71	1	0.028

2.59 One year's energy consumptions for a sample of 50 households in the South:

Energy consumption (millions of BTU)	Frequency	Relative frequency
40 < 50	1	0.02
50 < 60	7	0.14
60 < 70	7	0.14
70 < 80	3	0.06
80 < 90	6	0.12
90 < 100	10	0.20
100 < 110	5	0.10
110 < 120	4	0.08
120 < 130	2	0.04
130 < 140	3	0.06
140 < 150	0	0.00
150 < 160	2	0.04

2.60 The 24-hour iron intakes, in milligrams, for a sample of 45 women under the age of 51:

Iron intake (mg)	Frequency	Relative frequency
6 < 8	1	0.022
8 < 10	1	0.022
10 < 12	7	0.156
12 < 14	9	0.200
14 < 16	9	0.200
16 < 18	9	0.200
18 < 20	8	0.178
20 < 22	1	0.022

2.61 The starting annual salaries for a sample of 35 liberal-arts graduates:

Starting salary ($thousands)	Frequency	Relative frequency
25 < 26	3	0.086
26 < 27	3	0.086
27 < 28	5	0.143
28 < 29	9	0.257
29 < 30	9	0.257
30 < 31	4	0.114
31 < 32	1	0.029
32 < 33	1	0.029

2.62 The number of people per household for a sample of 40 U.S. households:

Number of people	Frequency	Relative frequency
1	7	0.175
2	13	0.325
3	9	0.225
4	5	0.125
5	4	0.100
6	1	0.025
7	1	0.025

2.63 The number of cars sold per week for last year by a car salesperson:

Number of cars sold	Frequency	Relative frequency
0	7	0.135
1	15	0.288
2	12	0.231
3	9	0.173
4	5	0.096
5	3	0.058
6	1	0.019

2.64 Construct a dotplot for the following exam scores of the students in an introductory statistics class.

88	82	89	70	85
63	100	86	67	39
90	96	76	34	81
64	75	84	89	96

2.65 The Motor Vehicle Manufacturers Association of the United States publishes information on the ages of cars and trucks currently in use in *Motor Vehicle Facts and Figures.* A sample of 37 trucks provided the ages, in years, displayed in the following table. Construct a dotplot for the ages.

8	12	14	16	15	5	11	13
4	12	12	15	12	3	10	9
11	3	18	4	9	11	17	
7	4	12	12	8	9	10	
9	9	1	7	6	9	7	

In Exercises 2.66 and 2.67, we have displayed the frequency and relative-frequency distributions obtained in Exercises 2.38 and 2.39, respectively. For each exercise,

a. draw a pie chart for the relative frequencies.

b. construct a bar graph for the relative frequencies.

2.66 The network data for the all-time top TV programs by rating as of January 1997:

Network	Frequency	Relative frequency
CBS	8	0.40
ABC	5	0.25
NBC	7	0.35

2.67 The winners of the NCAA wrestling championships for the years 1968–1997:

Champion	Frequency	Relative frequency
Oklahoma State	5	0.167
Iowa State	6	0.200
Oklahoma	1	0.033
Iowa	17	0.567
Arizona State	1	0.033

2.68 The Internal Revenue Service publishes data on adjusted gross incomes in *Statistics of Income, Individual Income Tax Returns.* Below is a relative-frequency histogram for one year's individual income tax returns showing an adjusted gross income less than $50,000.

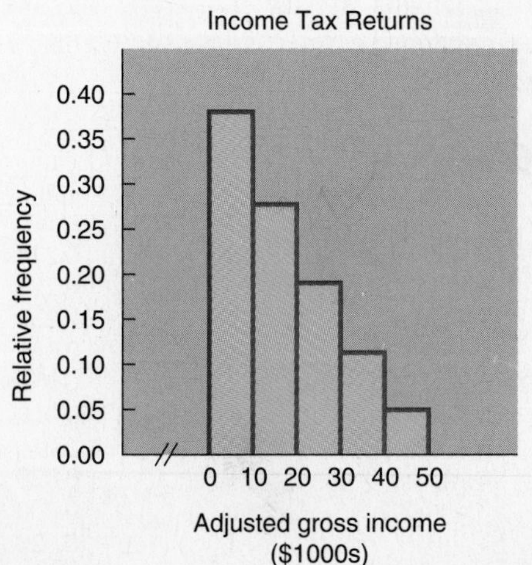

Using the histogram and noting that adjusted gross incomes are expressed to the nearest whole dollar, answer each of the following questions.

a. Approximately what percentage of the individual income tax returns had an adjusted gross income between \$10,000 and \$19,999, inclusive?

b. Approximately what percentage had an adjusted gross income less than \$30,000?

c. Given that 89,928,000 individual income tax returns had an adjusted gross income of less than \$50,000, roughly how many had an adjusted gross income between \$30,000 and \$49,999, inclusive?

2.69 A pediatrician who tested the cholesterol levels of several young patients was alarmed to find that many had levels over 200 mg per 100 mL. Following is a relative-frequency histogram for the readings of some patients having high cholesterol levels.

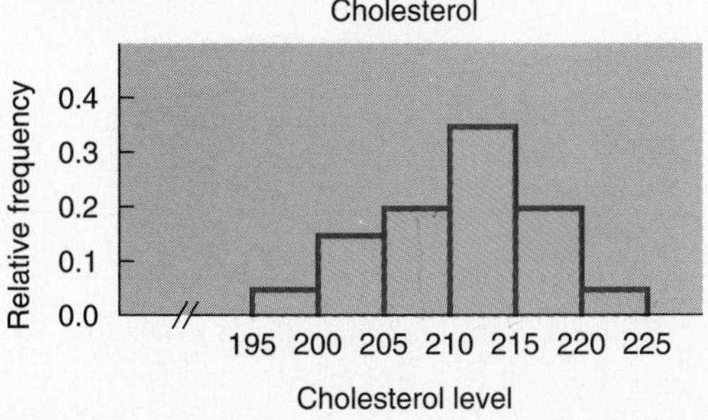

Answer the following questions using the graph. Note that cholesterol levels are always expressed as whole numbers.

a. What percentage of the patients have cholesterol levels between 205 and 209, inclusive?

b. What percentage have levels of 215 or higher?

c. Given that the number of patients is 20, how many have levels between 210 and 214, inclusive?

Extending the Concepts and Skills

Relative-frequency polygons. Another graphical display in common use is the relative-frequency polygon. In a **relative-frequency polygon,** a point is plotted above each class midpoint at a height equal to the relative frequency of the class. Then the points are joined with connecting lines. For instance, referring to Table 2.4 on page 70, we see that the relative-frequency polygon for the days-to-maturity data is as follows.

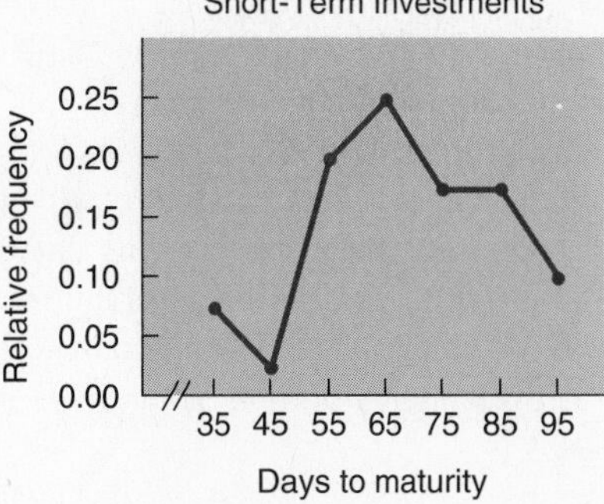

2.70 Construct a relative-frequency polygon for the length-of-stay data in Exercise 2.58.

2.71 Construct a relative-frequency polygon for the energy-consumption data in Exercise 2.59.

Ogives. Cumulative information can be portrayed using a graph called an ogive (ō′jīv). To construct an ogive, we first make a table displaying cumulative frequencies and cumulative relative frequencies. Table 2.16 provides such a table for the days-to-maturity data.

TABLE 2.16 Cumulative information for days-to-maturity data

Less than	Cumulative frequency	Cumulative relative frequency
30	0	0.000
40	3	0.075
50	4	0.100
60	12	0.300
70	22	0.550
80	29	0.725
90	36	0.900
100	40	1.000

The first column of Table 2.16 gives the cutpoints of the classes and the second column gives the cumulative frequencies. A **cumulative frequency** is obtained by summing the frequencies of all classes representing values less than the specified cutpoint. For instance, by referring to Table 2.12 on page 81, we can find the cumulative frequency of investments with a maturity period of less than 50 days. We see that the

$$\text{cumulative frequency} = 3 + 1 = 4.$$

This means that four of the investments have a maturity period of less than 50 days.

The third column of Table 2.16 gives the cumulative relative frequencies. A **cumulative relative frequency** is found by dividing the corresponding cumulative frequency by the total number of observations, which in this case is 40. For instance, the cumulative relative frequency of investments with a maturity period of less than 50 days is

$$\text{cumulative relative frequency} = \frac{4}{40} = 0.100.$$

This means that 10% of the investments have a maturity period of less than 50 days.

Using Table 2.16, we can now construct an ogive for the days-to-maturity data. In an **ogive** a point is plotted above each cutpoint at a height equal to the cumulative relative frequency. Then the points are joined with connecting lines. Consequently, the ogive for the days-to-maturity data is as follows.

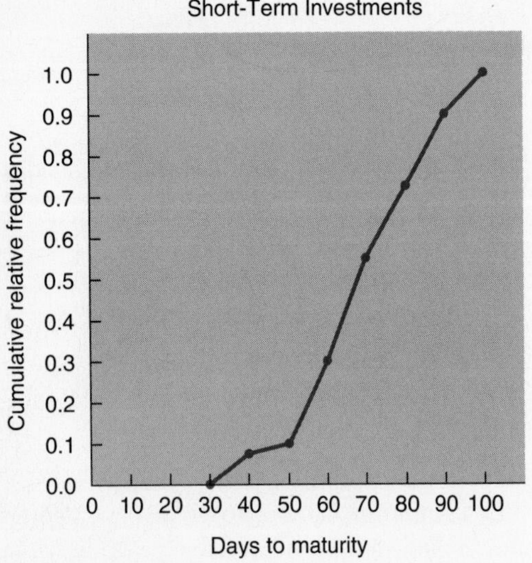

2.72 Refer to Exercise 2.58.

a. Construct a table similar to Table 2.16 for the length-of-stay data. Interpret your results.

b. Draw an ogive for the data.

2.73 Refer to Exercise 2.59.

a. Construct a table similar to Table 2.16 for the energy-consumption data. Interpret your results.

b. Draw an ogive for the data.

USING TECHNOLOGY  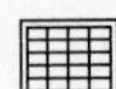 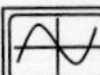

2.74 Use Minitab or some other statistical software to obtain a frequency histogram and relative-frequency (or percent) histogram as required in Exercise 2.58. The raw (ungrouped) data are given in Exercise 2.26 on page 77.

2.75 Use Minitab or some other statistical software to obtain a frequency histogram and relative-frequency (or percent) histogram as required in Exercise 2.59. The raw (ungrouped) data are given in Exercise 2.27 on page 77.

2.76 Use Minitab or some other statistical software to obtain the frequency histogram required in Exercise 2.62. The raw data are given in Exercise 2.34 on page 78.

2.77 Use Minitab or some other statistical software to obtain the frequency histogram required in Exercise 2.63. The raw data are given in Exercise 2.35 on page 78.

2.78 We used Minitab's histogram procedure to obtain the following frequency histogram for the lengths of stay of a sample of U.S. residents who traveled to Europe and the Mediterranean one year. Each length of stay is given to the nearest day.

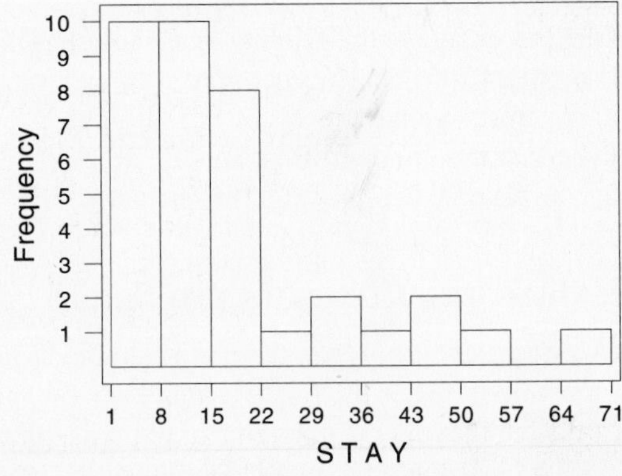

a. How many U.S. residents were sampled?

b. Identify the midpoint of the sixth class.

c. What common class width was used to construct the frequency distribution?

d. How many stayed between 15 and 21 days?

2.79 We used Minitab's histogram procedure to obtain the following frequency histogram for one year's energy consumptions for a sample of households in the South. The data are in millions of BTU.

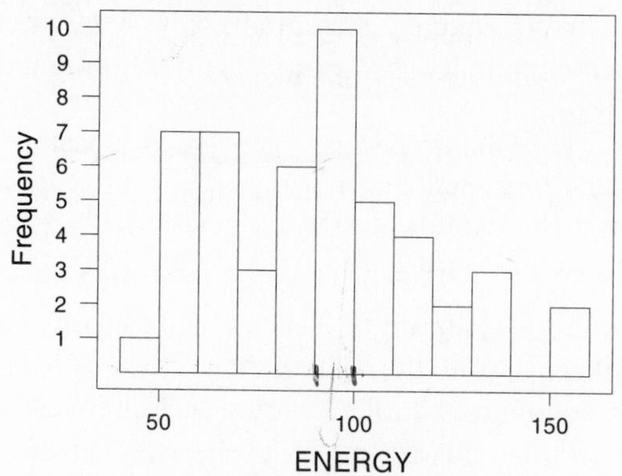

a. How many households were sampled?

b. Identify the midpoint of the sixth class.

c. What common class width was used to construct the frequency distribution?

d. How many of the households sampled had an energy consumption of at least 100 million BTU but less than 110 million BTU?

2.80 Use Minitab or some other statistical software to obtain a dotplot for the data from Exercise 2.64.

2.81 Use Minitab or some other statistical software to obtain a dotplot for the data from Exercise 2.65.

2.82 We used Minitab's character-graph dotplot procedure to obtain the following dotplot for the exam scores from Exercise 2.64.

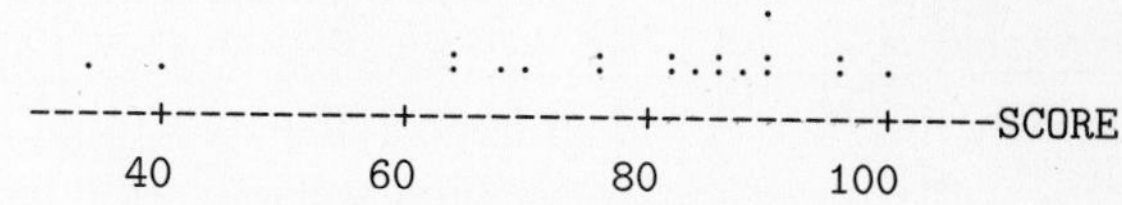

a. How many exam scores are there?

b. How many students scored between 80 and 92, inclusive?

2.83 The following dotplot was obtained by applying Minitab's character-graph dotplot procedure to the age data from Exercise 2.65.

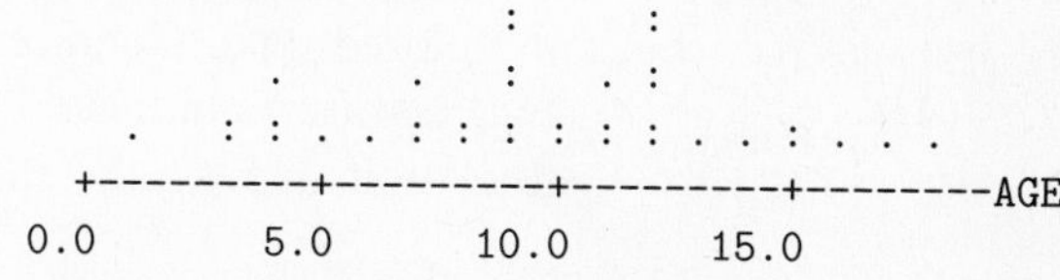

a. How many trucks were sampled?

b. How many of the trucks were between 5 and 8 years old, inclusive?

2.4 STEM-AND-LEAF DIAGRAMS

New ways of displaying data are constantly being invented. One method, developed in the late 1960s by Professor John Tukey of Princeton University, is called a **stem-and-leaf diagram,** or **stemplot.** This ingenious diagram is often easier to construct than either a frequency distribution or a histogram and generally displays more information. To illustrate stem-and-leaf diagrams, we return once more to the days-to-maturity data.

EXAMPLE 2.18 STEM-AND-LEAF DIAGRAMS

The data on the number of days to maturity for 40 short-term investments are repeated in Table 2.17.

TABLE 2.17
Days to maturity for 40 short-term investments

70	64	99	55	64	89	87	65
62	38	67	70	60	69	78	39
75	56	71	51	99	68	95	86
57	53	47	50	55	81	80	98
51	36	63	66	85	79	83	70

In Table 2.2 (page 67) we grouped these data by 10s and in Fig. 2.1(a) (page 82) we portrayed the data graphically with a frequency histogram. Now we will construct a stem-and-leaf diagram which simultaneously groups the data and yields a graphical display similar to a histogram.

First we select the leading digits from the data in Table 2.17. This yields the numbers 3, 4, . . . , 9. Next we list those leading digits in a column, as shown by the colored numbers in Fig. 2.5(a).

Then we go through the data in Table 2.17 and write the final digit of each number to the right of the appropriate leading digit: The first investment has a maturity period of 70 days. This calls for a "0" to the right of the "7" in colored type in Fig. 2.5(a). Reading down the first column of Table 2.17, the second investment has a maturity period of 62 days, so this calls for a "2" to the right of the "6" in colored type. Continuing in this manner, we obtain the diagram displayed in Fig. 2.5(a). As indicated in the figure, the leading digits are called **stems** and the final digits **leaves;** the entire diagram is called a **stem-and-leaf diagram.**

FIGURE 2.5
Diagrams for days-to-maturity data:
(a) stem-and-leaf
(b) shaded stem-and-leaf
(c) ordered stem-and-leaf

(a)

Stems	Leaves
3	8 6 9
4	7
5	7 1 6 3 5 1 0 5
6	2 4 7 3 6 4 0 9 8 5
7	0 5 1 0 9 8 0
8	5 9 1 7 0 3 6
9	9 9 5 8

(b)

3	8 6 9
4	7
5	7 1 6 3 5 1 0 5
6	2 4 7 3 6 4 0 9 8 5
7	0 5 1 0 9 8 0
8	5 9 1 7 0 3 6
9	9 9 5 8

(c)

3	6 8 9
4	7
5	0 1 1 3 5 5 6 7
6	0 2 3 4 4 5 6 7 8 9
7	0 0 0 1 5 8 9
8	0 1 3 5 6 7 9
9	5 8 9 9

The stem-and-leaf diagram for the days-to-maturity data is similar to a frequency histogram for those data because the length of the row of leaves for a class equals the frequency of the class. [Turn the stem-and-leaf diagram 90° counterclockwise and compare it to the frequency histogram in Fig. 2.1(a) on page 82.]

By shading each row of leaves, as in Fig. 2.5(b), we get a diagram that looks even more like a frequency histogram of the data. The diagram in Fig. 2.5(b) is called a **shaded stem-and-leaf diagram.** Because the numbers in a shaded stem-and-leaf diagram are still visible under the shading, a shaded stem-and-leaf diagram exhibits the raw (ungrouped) data in addition to providing a graphical display of a frequency distribution. On the other hand, although a frequency histogram provides a graphical display of a frequency distribution, it is generally not possible to recover the raw data from a frequency histogram.

Another form of stem-and-leaf diagram is called an **ordered stem-and-leaf diagram.** For this type of stem-and-leaf diagram, the leaves in each row are ordered from smallest to largest. This makes it easier to comprehend the data and also facilitates the computation of descriptive measures such as the median (to be discussed in Chapter 3). The ordered stem-and-leaf diagram for the days-to-maturity data is presented in Fig. 2.5(c). ◨

EXAMPLE 2.19 STEM-AND-LEAF DIAGRAMS

A pediatrician who tested the cholesterol levels of several young patients was alarmed to find that many had levels over 200 mg per 100 mL. The readings of 20 patients with high levels are presented in Table 2.18. Construct a stem-and-leaf diagram for these cholesterol-level data.

TABLE 2.18 Cholesterol levels for 20 high-level patients

210	209	212	208	202	218	200	214	218	210
217	207	210	203	215	221	213	210	199	208

SOLUTION Because these data are three-digit numbers, we use the first two digits as the stems and the third digit as the leaves. A stem-and-leaf diagram for the cholesterol levels is displayed in Fig. 2.6(a).

FIGURE 2.6 Stem-and-leaf diagram for cholesterol levels: (a) using one line per stem (b) using two lines per stem

(a)

```
19 | 9
20 | 8 2 9 7 0 8 3
21 | 0 7 5 0 8 2 0 0 3 8 4
22 | 1
```

(b)

```
19 |
19 | 9
20 | 2 0 3
20 | 8 9 7 8
21 | 0 0 2 0 0 3 4
21 | 7 5 8 8
22 | 1
22 |
```

The stem-and-leaf diagram in Fig. 2.6(a) is only moderately helpful because there are so few stems. We can construct a better stem-and-leaf diagram by using two lines for each stem, with the first line for the leaf digits 0–4 and the second line for the leaf digits 5–9. This stem-and-leaf diagram is shown in Fig. 2.6(b).

As we have seen, stem-and-leaf diagrams have several advantages over the more classical techniques for grouping and graphing. However, they do have some drawbacks. For instance, they are generally not useful with large data sets and can be awkward with data containing many digits; histograms are usually preferable to stem-and-leaf diagrams in such cases.

Minitab can be applied to obtain a stem-and-leaf diagram for a set of data. To illustrate we return to the data on cholesterol levels which we introduced in the previous example.

EXAMPLE 2.20 USING MINITAB TO OBTAIN A STEM-AND-LEAF DIAGRAM

Table 2.18 gives the cholesterol levels for 20 high-level patients. Apply Minitab to obtain a stem-and-leaf diagram for those data with two lines per stem.

SOLUTION We begin by storing the cholesterol-level data in a column named CHOLLEV. Then we proceed as follows.

1 Choose **Graph ➤ Stem-and-Leaf...** (In Release 11, choose **Graph ➤ Character Graphs ➤ Stem-and-Leaf...**)
2 Specify CHOLLEV in the **Variables** text box
3 Click in the **Increment** text box and type 5 (This tells Minitab that the distance between the smallest possible number on one line and the smallest possible number on the next line should be 5, thus providing for a stem-and-leaf diagram with two lines per stem)
4 Click **OK**

The output resulting is depicted in Printout 2.7.

The first line in Printout 2.7 describes what follows (`Stem-and-leaf of CHOLLEV`) and also displays the number of observations (`N = 20`). The next line (`Leaf Unit = 1.0`) indicates where the decimal point goes, in this case directly after each leaf digit. The second and third columns of numbers in Printout 2.7 give the stems and leaves, respectively. Since the leaves in each row are ordered, we see that the output in Printout 2.7 is actually an ordered stem-and-leaf diagram.

PRINTOUT 2.7
Minitab stem-and-leaf diagram for the cholesterol-level data

```
Stem-and-leaf of CHOLLEV   N  = 20
Leaf Unit = 1.0


     1    19 9
     4    20 023
     8    20 7889
    (7)   21 0000234
     5    21 5788
     1    22 1
```

The numbers in the first column of the stem-and-leaf diagram, called **depths,** are used to display cumulative frequencies. Starting from the top, the depths indicate the number of observations (leaves) that lie in a given row or before. For instance, the 8 in the third row shows that there are eight observations in the first three rows.

When the row containing the middle observation(s) is reached, the cumulative frequency is replaced by the number of observations in that row. Thus in this case we see that the middle observations lie somewhere between 210 and 214, inclusive, and that seven observations are in that range. The depths following the row that contains the middle observation(s) indicate the number of observations that lie in a given row or after. For instance, the 5 in the fifth row shows that there are five observations in the last two rows.

Compare the stem-and-leaf diagram in Printout 2.7 to the one obtained by hand in Fig. 2.6(b) on page 97. ◨

EXERCISES 2.4

Statistical Concepts and Skills

2.84 Discuss relative advantages and disadvantages of stem-and-leaf diagrams and frequency histograms.

2.85 Suppose a data set contains a large number of observations. Which graphical display is generally preferable: a histogram or a stem-and-leaf diagram? Why?

2.86 Explain why it is generally not possible to recover the raw data from a frequency histogram. Under what circumstances is it possible?

2.87 A stem-and-leaf diagram has been discovered to be only moderately useful because there are too few stems. How can this problem be remedied?

In each of Exercises 2.88–2.91,

a. construct a stem-and-leaf diagram.

b. construct an ordered stem-and-leaf diagram.

2.88 A research physician conducted a study on the ages of people with diabetes. The following data were obtained for the ages of a sample of 35 diabetics.

48	41	57	83	41	55	59
61	38	48	79	75	77	7
54	23	47	56	79	68	61
64	45	53	82	68	38	70
10	60	83	76	21	65	47

2.89 A soft-drink bottler sells "one-liter" bottles of soda. A consumer group is concerned that the bottler may be shortchanging customers. Thirty bottles of soda are randomly selected. The contents, in milliliters, of the bottles chosen are shown below. *(Hint:* Use the stems 91, 92, ..., 106.*)*

1025	977	1018	975	977
990	959	957	1031	964
986	914	1010	988	1028
989	1001	984	974	1017
1060	1030	991	999	997
996	1014	946	995	987

2.90 The Bureau of Economic Analysis gathers information on the length of stay in Europe and the Mediterranean by U.S. travelers. Data are published in *Survey of Current Business.* A sample of 36 U.S. residents who traveled to Europe and the Mediterranean one year yielded the following data, in days, on length of stay.

41	16	6	21	1	21
5	31	20	27	17	10
3	32	2	48	8	12
21	44	1	56	5	12
3	13	15	10	18	3
1	11	14	12	64	10

2.91 The U.S. Energy Information Administration collects data on residential energy consumption and expenditures. Results are published in *Residential Energy Consumption Survey: Consumption and Expenditures.* The following table provides the data on one year's energy consumptions for a sample of 50 households in the South. Data are given in millions of BTU.

130	55	45	64	155	66	60	80	102	62
58	101	75	111	151	139	81	55	66	90
97	77	51	67	125	50	136	55	83	91
54	86	100	78	93	113	111	104	96	113
96	87	129	109	69	94	99	97	83	97

2.92 As reported by the U.S. Bureau of the Census in *Current Population Reports,* the percentage of the adult population completing high school in each state is as follows.

State	Percent	State	Percent	State	Percent
AL	67	LA	68	OH	76
AK	87	ME	79	OK	75
AZ	79	MD	78	OR	81
AR	66	MA	80	PA	75
CA	76	MI	77	RI	72
CO	84	MN	82	SC	68
CT	79	MS	64	SD	77
DE	77	MO	74	TN	67
FL	74	MT	81	TX	72
GA	71	NE	82	UT	85
HI	80	NV	79	VT	81
ID	80	NH	82	VA	75
IL	76	NJ	77	WA	84
IN	76	NM	75	WV	66
IA	80	NY	75	WI	79
KS	81	NC	70	WY	83
KY	65	ND	77		

Construct a stem-and-leaf diagram for the percentages

a. using one line per stem.
b. using two lines per stem.
c. using five lines per stem.

2.93 The U.S. Federal Bureau of Investigation published the following annual crime rates in *Crime in the United States.* Rates are per 1000 population.

State	Rate	State	Rate	State	Rate
AL	49	LA	67	OH	45
AK	57	ME	33	OK	56
AZ	79	MD	61	OR	63
AR	48	MA	44	PA	33
CA	62	MI	54	RI	41
CO	53	MN	43	SC	60
CT	45	MS	48	SD	31
DE	41	MO	53	TN	51
FL	83	MT	50	TX	59
GA	60	NE	44	UT	53
HI	67	NV	67	VT	33
ID	41	NH	27	VA	40
IL	56	NJ	47	WA	60
IN	46	NM	62	WV	25
IA	37	NY	51	WI	39
KS	49	NC	56	WY	43
KY	35	ND	27		

Construct a stem-and-leaf diagram for the crime rates

a. using one line per stem.
b. using two lines per stem.

2.94 The U.S. National Oceanic and Atmospheric Administration publishes temperature data in *Climatography of the United States.* According to that document, the annual average maximum temperatures for selected cities in the United States are as follows.

City	Annual average max. temp.	City	Annual average max. temp.
Mobile, AL	77	Reno, NV	67
Juneau, AK	47	Concord, NH	57
Phoenix, AZ	85	Atlantic City, NJ	63
Little Rock, AR	73	Albuquerque, NM	70
Los Angeles, CA	70	Albany, NY	58
Sacramento, CA	73	Buffalo, NY	56
San Francisco, CA	65	New York, NY	62
Denver, CO	64	Charlotte, NC	71
Hartford, CT	60	Raleigh, NC	70
Wilmington, DE	64	Bismarck, ND	54
Washington, DC	67	Cincinnati, OH	64
Jacksonville, FL	79	Cleveland, OH	59
Miami, FL	83	Columbus, OH	62
Atlanta, GA	71	Oklahoma City, OK	71
Honolulu, HI	84	Portland, OR	62
Boise, ID	63	Philadelphia, PA	63
Chicago, IL	59	Pittsburgh, PA	60
Peoria, IL	60	Providence, RI	59
Indianapolis, IN	62	Columbia, SC	75
Des Moines, IA	59	Sioux Falls, SD	57
Wichita, KS	68	Memphis, TN	72
Louisville, KY	66	Nashville, TN	70
New Orleans, LA	78	Dallas-Ft. Worth, TX	77
Portland, ME	55	El Paso, TX	78
Baltimore, MD	65	Houston, TX	79
Boston, MA	59	Salt Lake City, UT	64
Detroit, MI	58	Burlington, VT	54
Sault Ste. Marie, MI	49	Norfolk, VA	68
Duluth, MN	48	Richmond, VA	69
Mnpls-St. Paul, MN	54	Seattle-Tacoma, WA	59
Jackson, MS	76	Spokane, WA	57
Kansas City, MO	64	Charleston, WV	66
St. Louis, MO	66	Milwaukee, WI	55
Great Falls, MT	56	Cheyenne, WY	58
Omaha, NE	62	San Juan, PR	86

Construct a stem-and-leaf diagram for these data

a. using two lines per stem.

b. using five lines per stem.

2.95 The U.S. National Oceanic and Atmospheric Administration publishes temperature data in *Climatography of the United States.* According to that document, the annual average minimum temperatures for selected cities in the United States are as follows.

City	Annual average min. temp.	City	Annual average min. temp.
Mobile, AL	58	Reno, NV	32
Juneau, AK	33	Concord, NH	33
Phoenix, AZ	57	Atlantic City, NJ	43
Little Rock, AR	51	Albuquerque, NM	42
Los Angeles, CA	55	Albany, NY	37
Sacramento, CA	48	Buffalo, NY	39
San Francisco, CA	48	New York, NY	47
Denver, CO	36	Charlotte, NC	49
Hartford, CT	40	Raleigh, NC	48
Wilmington, DE	45	Bismarck, ND	29
Washington, DC	49	Cincinnati, OH	45
Jacksonville, FL	57	Cleveland, OH	41
Miami, FL	69	Columbus, OH	42
Atlanta, GA	51	Oklahoma City, OK	49
Honolulu, HI	70	Portland, OR	44
Boise, ID	39	Philadelphia, PA	45
Chicago, IL	40	Pittsburgh, PA	41
Peoria, IL	41	Providence, RI	41
Indianapolis, IN	42	Columbia, SC	51
Des Moines, IA	40	Sioux Falls, SD	34
Wichita, KS	45	Memphis, TN	52
Louisville, KY	46	Nashville, TN	49
New Orleans, LA	59	Dallas-Ft. Worth, TX	55
Portland, ME	35	El Paso, TX	49
Baltimore, MD	45	Houston, TX	57
Boston, MA	44	Salt Lake City, UT	39
Detroit, MI	39	Burlington, VT	35
Sault Ste. Marie, MI	31	Norfolk, VA	51
Duluth, MN	29	Richmond, VA	47
Mnpls-St. Paul, MN	35	Seattle-Tacoma, WA	44
Jackson, MS	53	Spokane, WA	37
Kansas City, MO	44	Charleston, WV	44
St. Louis, MO	45	Milwaukee, WI	38
Great Falls, MT	33	Cheyenne, WY	33
Omaha, NE	40	San Juan, PR	73

Construct a stem-and-leaf diagram for these data

a. using two lines per stem.

b. using five lines per stem.

Extending the Concepts and Skills

Further stem-and-leaf techniques. We mentioned earlier that stem-and-leaf diagrams can be awkward with data containing many digits. In such cases we can either **round** or **truncate** each observation to a suitable number of digits. Exercises 2.96 and 2.97 discuss rounding and truncating in relation to stem-and-leaf diagrams.

2.96 The U.S. National Center for Health Statistics publishes data on weights and heights by age and sex in *Vital and Health Statistics.* The following weights, given to the nearest tenth of a pound, were obtained from a sample of 18–24-year-old males.

129.2	185.3	218.1	182.5	142.8	155.2
170.0	151.3	187.5	145.6	167.3	161.0
178.7	165.0	172.5	191.1	150.7	
187.0	173.7	178.2	161.7	170.1	
165.8	214.6	136.7	278.8	175.6	
188.7	132.1	158.5	146.4	209.1	
175.4	182.0	173.6	149.9	158.6	

a. First round each observation to the nearest pound and then construct a stem-and-leaf diagram of the rounded data.
b. Truncate each observation by dropping the decimal part and then construct a stem-and-leaf diagram of the truncated data.
c. Compare the stem-and-leaf diagrams obtained in parts (a) and (b).

2.97 Refer to Exercise 2.89.
a. Round each observation to the nearest 10 ml, drop the terminal 0s, and then obtain a stem-and-leaf diagram of the resulting data.
b. Truncate each observation by dropping the units digit and then construct a stem-and-leaf diagram of the truncated data.
c. Compare the stem-and-leaf diagrams obtained in parts (a) and (b) with each other and with the one obtained in Exercise 2.89.

Using Technology

 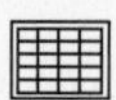 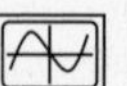

2.98 Use Minitab or some other statistical software to construct a stem-and-leaf diagram for the data in Exercise 2.90 with
a. one line per stem.
b. two lines per stem.
c. five lines per stem.

2.99 Use Minitab or some other statistical software to construct a stem-and-leaf diagram for the data in Exercise 2.91 with
a. one line per stem.
b. two lines per stem.

2.100 A hardware manufacturer produces 10-mm-diameter bolts. The manufacturer knows that the diameters of the bolts produced vary somewhat from 10 mm and also from each other. To examine the variation, the manufacturer takes a sample of bolts and obtains the following stem-and-leaf diagram for the diameters.

```
Stem-and-leaf of BOLTDIAM  N  = 20
Leaf Unit = 0.010

   1    98 9
   1    99
   1    99
   3    99 45
   5    99 67
   9    99 8899
 (2)   100 01
   9   100 2333
   5   100 55
   3   100
   3   100 88
   1   101 0
```

a. How many bolts were sampled?
b. What is the smallest diameter, in millimeters, of the bolts sampled? (*Hint:* Note that the leaf unit is 0.010.)
c. How many lines per stem are used in this stem-and-leaf diagram?
d. Use the depths to determine the number of bolts sampled having diameters of 10.04 mm or greater.
e. Use the depths to determine the number of bolts sampled having diameters of 9.97 mm or smaller.
f. List the diameters that are less than 10 mm.

2.101 The U.S. Bureau of the Census reports the percentage of the adult population completing high school in each state in *Current Population Reports.* We applied Minitab's stem-and-leaf procedure to obtain the following stem-and-leaf diagram for those data.

```
Stem-and-leaf of HSCOMP    N  = 50
Leaf Unit = 1.0


   1    6 4
   8    6 5667788
  14    7 012244
 (20)   7 55555666677777899999
  16    8 00001111222344
   2    8 57
```

a. How many observations are there?
b. How many lines per stem are used?
c. Use the depths to determine how many of the percentages are 80% or greater.
d. Use the depths to determine how many of the percentages are 69% or less.
e. Identify the largest percentage.
f. List the percentages that are in the 60s.

2.102 Following is a Minitab stem-and-leaf diagram for the weights from Exercise 2.96.

```
Stem-and-leaf of WEIGHT    N  = 37
Leaf Unit = 1.0


   1    12 9
   3    13 26
   7    14 2569
  12    15 01588
  17    16 11557
  (9)   17 002335588
  11    18 225778
   5    19 1
   4    20 9
   3    21 48
   1    22
   1    23
   1    24
   1    25
   1    26
   1    27 8
```

Did Minitab use rounding or truncation to obtain this stem-and-leaf diagram? Explain your answer.

2.5 DISTRIBUTION SHAPES; SYMMETRY AND SKEWNESS

The **distribution** of a data set is a table, graph, or formula that tells us the values of the observations and how often they occur. Up to now, we have seen distributions of data sets portrayed by frequency distributions, relative-frequency distributions, frequency histograms, relative-frequency histograms, dotplots, stem-and-leaf diagrams, pie charts, and bar graphs.

An important aspect of the distribution of a quantitative data set is its shape. For, as we will discover in later chapters, the shape of a distribution frequently plays a role in determining the appropriate method of statistical analysis. In discussing distribution shapes, it is better to use smooth curves that approximate the overall shape.

For instance, Fig. 2.7 displays a relative-frequency histogram for the heights of the 3264 female students attending a midwestern college. Also included in Fig. 2.7 is a smooth curve that approximates the overall shape of the distribution. Both the histogram and the smooth curve show that this distribution of heights is bell-shaped (or mound-shaped), but the smooth curve makes it a little easier to see the shape.

FIGURE 2.7 Relative-frequency histogram and approximating smooth curve for the distribution of heights

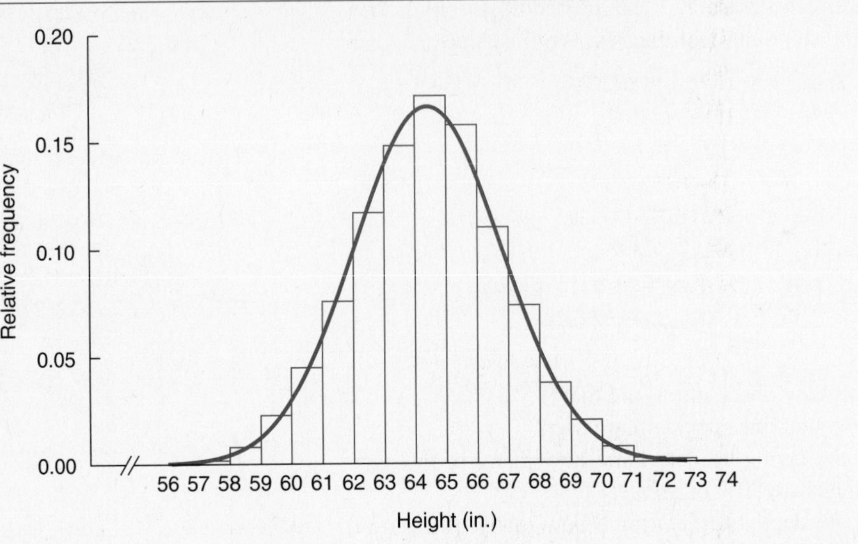

Another advantage of using smooth curves to describe distribution shapes is that we need not worry about minor differences in shape but can instead concentrate on overall patterns. This, in turn, allows us to designate relatively few shapes in order to classify most distributions.

Distribution Shapes

Figure 2.8 displays some common distribution shapes: **bell-shaped, triangular, uniform, reverse J-shaped, J-shaped, right skewed, left skewed, bimodal,** and **multimodal.** These shapes are idealized forms; in practice, distributions rarely have these exact shapes. Consequently, in identifying the shape of a distribution, we do not require exact conformance, especially when considering small data sets. So, for example, we describe the distribution of heights displayed in Fig. 2.7 as bell-shaped even though the histogram does not form a perfect bell.

EXAMPLE 2.21 IDENTIFYING DISTRIBUTION SHAPES

A relative-frequency histogram for household size in the United States, shown in Fig. 2.9(a), is based on data found in *Current Population Reports,* a publication of the U.S. Census Bureau.[†] Describe the distribution shape for sizes of U.S. households.

[†] Actually, the class "7" portrayed in Fig. 2.9 is for seven or more people.

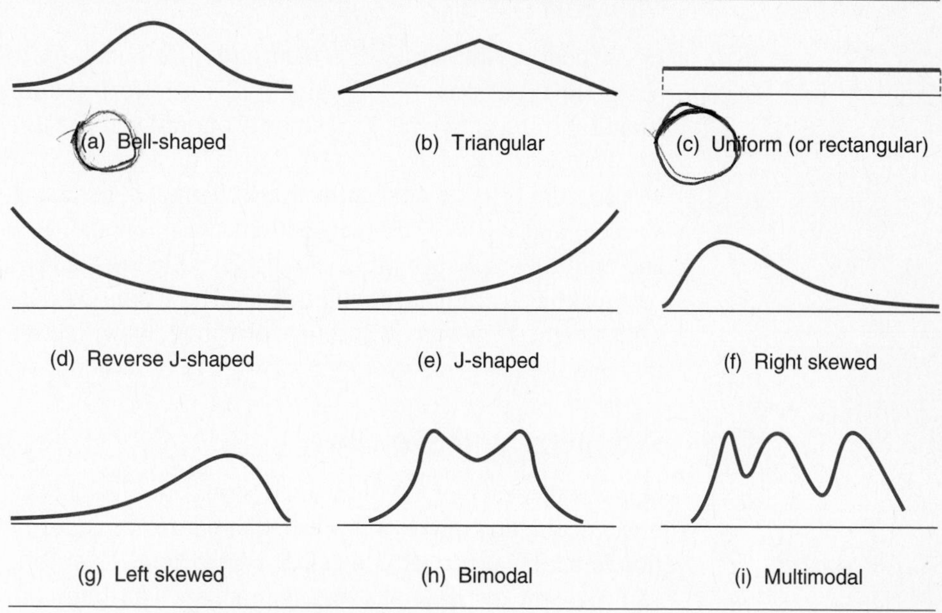

FIGURE 2.8 Common distribution shapes

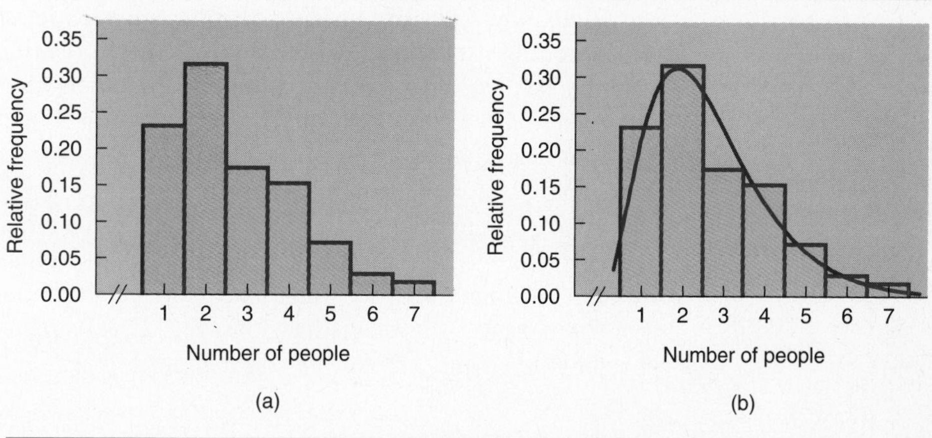

FIGURE 2.9 Relative-frequency histogram for household size

SOLUTION First we draw a smooth curve through the histogram in Fig. 2.9(a), as seen in Fig. 2.9(b). Then, by referring to Fig. 2.8, we find that the distribution of household sizes is right skewed. ◨

There are other distribution shapes besides those represented in Fig. 2.8. However, the types in Fig. 2.8 comprise the most commonly encountered distribution shapes and will suffice for our purposes.

Modality

In considering the shape of a distribution, it is helpful to observe the number of peaks (highest points). A distribution is said to be **unimodal** if it has one peak; **bimodal** if it has two peaks; and **multimodal** if it has three or more peaks.

The distribution of heights in Fig. 2.7 on page 104 is unimodal. More generally, we see from Fig. 2.8 that bell-shaped, triangular, reverse-J-shaped, J-shaped, right-skewed, and left-skewed distributions are unimodal. Representations of bimodal and multimodal distributions are displayed in Figs. 2.8(h) and (i), respectively.†

Technically, a distribution is bimodal or multimodal only if the peaks are the same height. However, in practice, distributions with pronounced but not necessarily equal-height peaks are often referred to as bimodal or multimodal.

Symmetry and Skewness

Observe that each of the three distributions in Figs. 2.8(a)–(c) has the property that it can be divided into two pieces that are mirror images of one another. A distribution having that property is called **symmetric.** Therefore, bell-shaped, triangular, and uniform distributions are symmetric. The bimodal distribution pictured in Fig. 2.8(h) is also symmetric, but that is not always true for bimodal or multimodal distributions. Figure 2.8(i) shows an asymmetric multimodal distribution.

Again, when classifying distributions, we must be somewhat flexible. Thus we do not insist on exact symmetry to classify a distribution as symmetric. For example, the distribution of heights in Fig. 2.7 is considered symmetric.

A unimodal distribution that is not symmetric is either right skewed, as in Fig. 2.8(f), or left skewed, as in Fig. 2.8(g). A right-skewed distribution rises to its peak rapidly and comes back toward the horizontal axis more slowly—its "right tail" is longer than its "left tail." On the other hand, a left-skewed distribution rises to its peak slowly and comes back toward the horizontal axis more rapidly—its "left tail" is longer than its "right tail." Notice that reverse-J-shaped distributions (Fig. 2.8(d)) and J-shaped distributions (Fig. 2.8(e)) are, respectively, special types of right-skewed and left-skewed distributions.

Population and Sample Distributions

Recall that a variable is a characteristic that varies from one person or thing to another and that observing one or more values of a variable yields data. The data set obtained by observing the values of a variable for an entire population is called **population data** or **census data;** a data set obtained by observing the values of a variable for a sample of the population is called **sample data.** To distinguish the distributions of population data and sample data, we use the following terminology.

† A uniform distribution has either no peaks or infinitely many peaks, depending on how one looks at it. In any case, we do not classify a uniform distribution according to modality.

DEFINITION 2.5 POPULATION AND SAMPLE DISTRIBUTIONS; DISTRIBUTION OF A VARIABLE

The distribution of population data is called the ***population distribution*** or the ***distribution of the variable.***

The distribution of sample data is called a ***sample distribution.***

It is important to note that for a particular population and variable, sample distributions vary from sample to sample, but there is only one population distribution, namely, the distribution of the variable under consideration on the population under consideration. Example 2.22 illustrates this point and some others as well.

EXAMPLE 2.22 COMPARES POPULATION AND SAMPLE DISTRIBUTIONS

In Example 2.21 we considered household size for U.S. households. Here the variable is household size and the population consists of all U.S. households. Figure 2.9(a) on page 105 provides a relative-frequency histogram for the sizes of all U.S. households, that is, for the distribution of the variable "household size."

We obtained six (random) samples of 100 households each from the population of all U.S. households. Figure 2.10 on the next page shows relative-frequency histograms of household size for all six samples. Compare the distributions of the six samples to each other and to the population distribution.

SOLUTION As we see from Fig. 2.10, the distributions of the six samples, although similar, do have definite differences. This is not surprising since we would expect variation from one sample to another. Nonetheless, the overall shapes of the six sample distributions are roughly the same and are also similar in shape to the population distribution—all are right skewed. □

In practice, we usually do not know the population distribution. As Example 2.22 suggests, we can, under those circumstances, use the distribution of a sample from the population to get a rough idea of the population distribution.

KEY FACT 2.1 POPULATION AND SAMPLE DISTRIBUTIONS

The distribution of a random sample from a population approximates the population distribution. In other words, if a random sample is taken from a population, then the distribution of the observed values of the variable under consideration will approximate the distribution of the variable. The larger the sample, the better the approximation tends to be.

FIGURE 2.10
Relative-frequency histograms for household size for six random samples of size 100 from the population of U.S. households

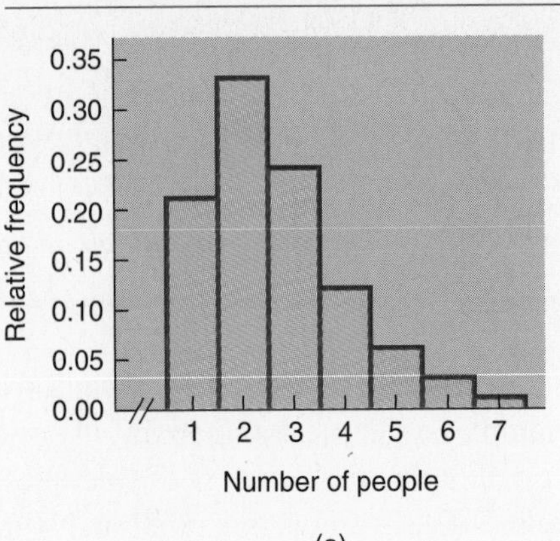

(a)

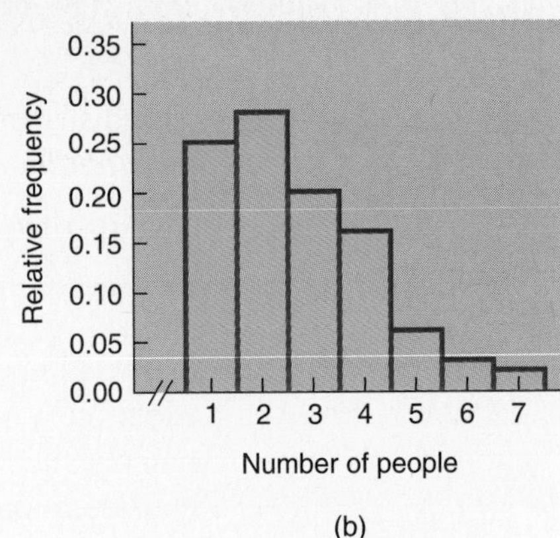

(b)

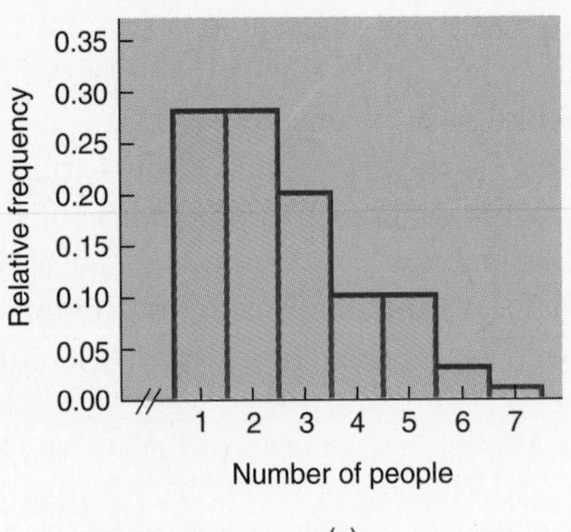

(c)

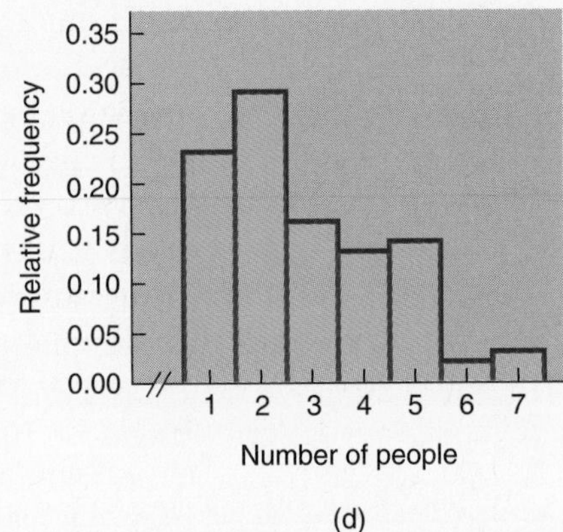

(d)

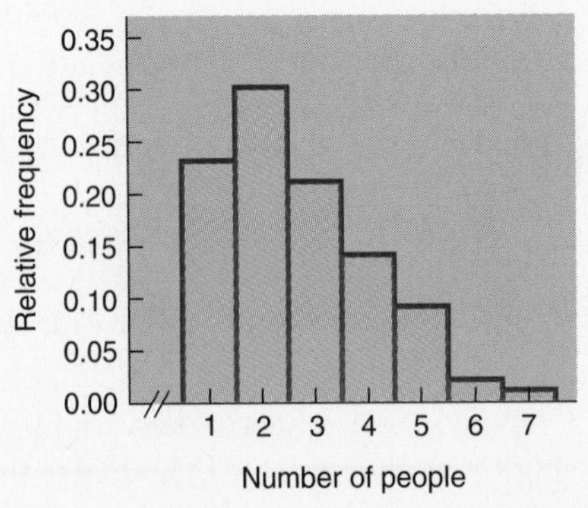

(e)

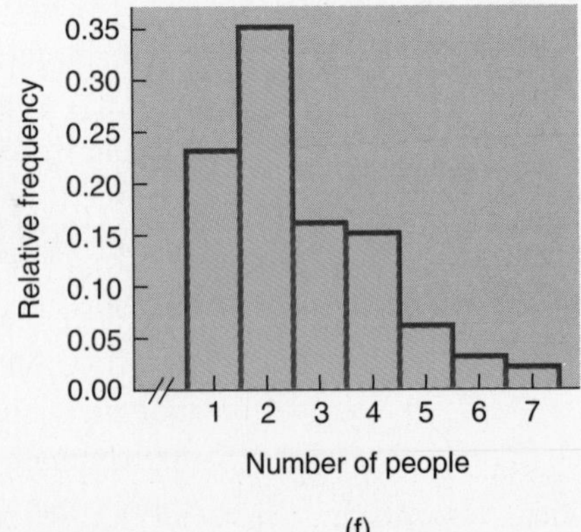

(f)

EXERCISES 2.5

Statistical Concepts and Skills

2.103 Explain the meaning of each of the following terms.
a. distribution of a data set
b. sample data
c. population data
d. census data
e. sample distribution
f. population distribution
g. distribution of a variable

2.104 Give two reasons why it is useful to use smooth curves in describing shapes of distributions.

2.105 Suppose a variable of a population has a bell-shaped distribution. If you take a large sample from the population, roughly what shape would you expect for the distribution of the sample?

2.106 Suppose that a variable of a population has a reverse J-shaped distribution and that two samples are taken from the population.
a. Would you expect the distributions of the two samples to have roughly the same shape? If so, what shape?
b. Would you expect some variation in shape for the distributions of the two samples? Explain your answer.

2.107 Identify and sketch three distribution shapes that are symmetric.

In each of Exercises 2.108–2.117, we have provided a graphical display of a data set. For each exercise,
a. identify the overall shape of the distribution by referring to Fig. 2.8 on page 105.
b. state whether the distribution is (roughly) symmetric, right skewed, or left skewed.

2.108 A frequency histogram for the number of questions answered incorrectly on an eight-question fraction quiz by each of the 25 students in a fifth-grade class:

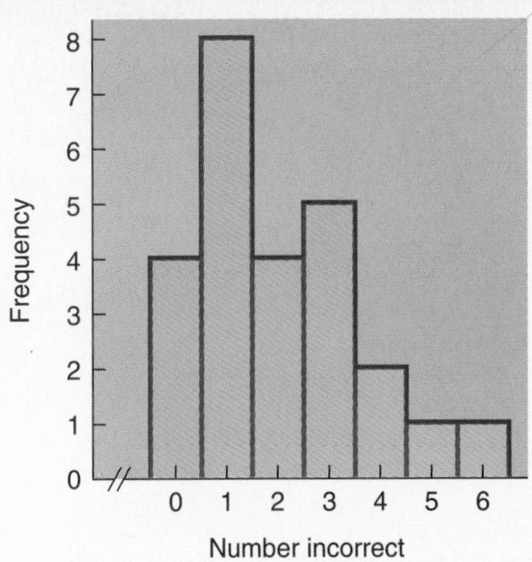

2.109 A frequency histogram for the number of cars sold per week by a car salesperson one year:

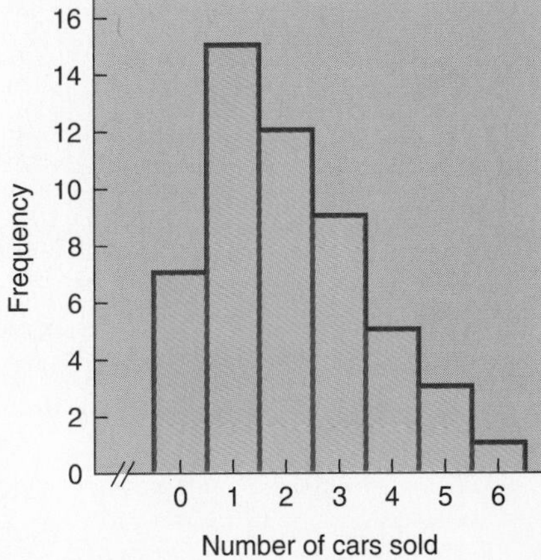

2.110 A relative-frequency histogram for the ages of a sample of 35 diabetics obtained by a research physician:

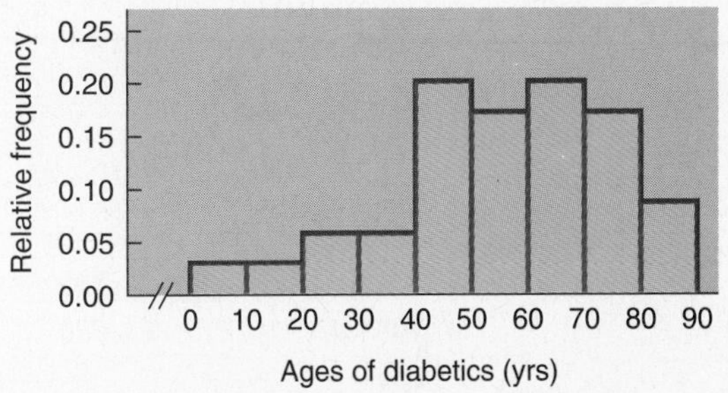

2.111 A relative-frequency histogram for the starting salaries of a sample of 35 liberal-arts graduates:

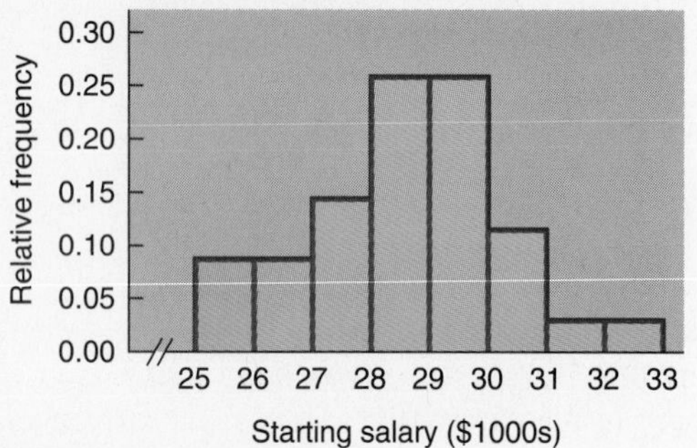

2.112 A relative-frequency histogram for one year's federal individual income tax returns showing an adjusted gross income less than $50,000:

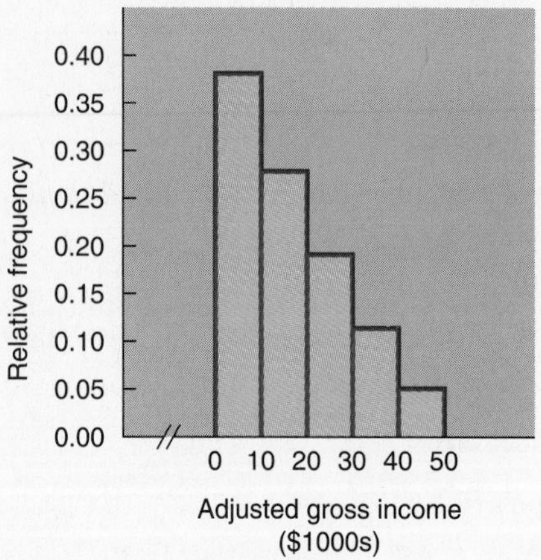

2.113 A relative-frequency histogram for the cholesterol levels of 20 high-level patients:

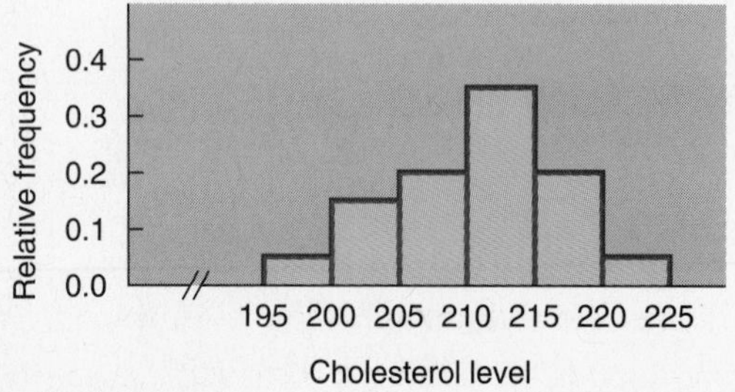

2.114 A stem-and-leaf diagram for the lengths of stay in Europe and the Mediterranean obtained from a sample of 36 U.S. residents who traveled there one year:

0	5 3 3 1 6 2 1 1 8 5 3
1	6 3 1 5 4 0 2 7 8 0 2 2 0
2	1 0 1 7 1
3	1 2
4	1 4 8
5	6
6	4

2.115 A stem-and-leaf diagram for the contents, in milliliters, of a sample of 30 "one-liter" bottles obtained by a consumer group from a soft-drink bottler:

91	4
92	
93	
94	6
95	9 7
96	4
97	7 5 4 7
98	6 9 4 8 7
99	0 6 1 9 5 7
100	1
101	4 8 0 7
102	5 8
103	0 1
104	
105	
106	0

2.116 A stem-and-leaf diagram for the percentage of adults in each state that have completed high school:

6	4 5
6	6 6 7 7
6	8 8
7	0 1
7	2 2
7	4 4 5 5 5 5 5
7	6 6 6 6 7 7 7 7 7
7	8 9 9 9 9 9
8	0 0 0 0 1 1 1 1
8	2 2 2 3
8	4 4 5
8	7

2.117 A stem-and-leaf diagram for the annual crime rates in the United States by state:

2	5 7 7
3	1 3 3 3
3	5 7 9
4	0 1 1 1 3 3 4 4
4	5 5 6 7 8 8 9 9
5	0 1 1 3 3 3 4
5	6 6 6 7 9
6	0 0 0 1 2 2 3
6	7 7 7
7	
7	9
8	3

Extending the Concepts and Skills

2.118 This exercise is a class project and works best in relatively large classes.

a. Obtain the number of siblings for each student in the class.

b. Obtain a relative-frequency histogram of the number of siblings using single-value grouping.

c. Obtain a simple random sample of roughly one-third of the students in the class.

d. Determine the number of siblings for each student in the sample.

e. Obtain a relative-frequency histogram of the number of siblings for the sample using single-value grouping.

f. Repeat parts (c)–(e) three more times.

g. Compare the histograms for the samples to each other and to that for the entire population. Relate your observations to Key Fact 2.1.

2.119 This exercise can be done individually or, better yet, as a class project.

a. Use a table of random numbers or a random-number generator to obtain 50 random integers between 0 and 9.

b. Without graphing the distribution of the 50 numbers you obtained, guess its shape. Explain your reasoning.

c. Construct a relative-frequency histogram for the 50 numbers you obtained using classes each based on a single value. Is its shape roughly what you expected?

d. If your answer to part (c) was "no," provide an explanation.

e. What would you do to make it more plausible to get a "yes" answer to part (c)?

f. If you are doing this exercise as a class project, repeat parts (a)–(c) for 1000 random integers.

Using Technology  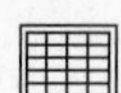 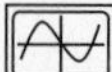

It is often useful, both for purposes of understanding and research, to simulate variables. Simulating a variable means that we use a computer or statistical calculator to generate observations of the variable. In Exercises 2.120–2.122, we will use simulation to enhance our discussion of distribution shapes and the relation between population and sample distributions.

2.120 In this exercise we will use technology to work Exercise 2.119.

a. Use Minitab or some other statistical software to obtain 50 random integers between 0 and 9. If you are using Minitab, proceed as follows.

1 Choose **Calc ➤ Random Data ➤ Integer...**

2 Type 50 in the **Generate rows of data** text box

3 Click in the **Store in column(s)** text box and type DIGIT

4 Click in the **Minimum value** text box and type 0

5 Click in the **Maximum value** text box and type 9

6 Click **OK**

b. Obtain a histogram of the numbers that you obtained in part (a) based on single-value grouping.

c. Repeat parts (a) and (b) five more times.

d. Are the shapes of the distributions that you obtained in parts (a)–(c) roughly what you expected?

e. Repeat parts (a)–(d), but generate 1000 random integers each time instead of 50.

2.121 One of the most important distributions is called the *standard normal distribution*. Using Minitab, a sample of 3000 observations from a variable having the standard normal distribution can be obtained in the following way.

1 Choose **Calc ➤ Random Data ➤ Normal...**

2 Type 3000 in the **Generate rows of data** text box

3 Click in the **Store in column(s)** text box and type STDNORM

4 Click **OK**

a. Use Minitab or some other statistical software to generate a sample of 3000 observations from a variable having the standard normal distribution.
b. Obtain a histogram of the 3000 observations.
c. Based on your histogram in part (b), what shape does the standard normal distribution have? Explain your reasoning.

2.122 Another important distribution in statistics is called a *t-distribution.* If you are using Minitab, a sample of 2500 observations from a variable having a t-distribution with six degrees of freedom can be obtained as follows.

1 Choose **Calc ➤ Random Data ➤ T...**

2 Type 2500 in the **Generate rows of data** text box

3 Click in the **Store in column(s)** text box and type TDIST

4 Click in the **Degrees of freedom** text box and type 6

5 Click **OK**

a. Use Minitab or some other statistical software to generate a sample of 2500 observations from a variable having a t-distribution with six degrees of freedom.
b. Obtain a histogram of the 2500 observations from part (a).
c. Based on your histogram in part (b), what shape does a t-distribution with six degrees of freedom have? Explain your reasoning.

2.123 Another important distribution is called a *chi-square distribution.* If you are using Minitab, a sample of 500 observations from a variable having a chi-square distribution with five degrees of freedom can be obtained as follows.

1 Choose **Calc ➤ Random Data ➤ Chisquare...**

2 Type 500 in the **Generate rows of data** text box

3 Click in the **Store in column(s)** text box and type CHISQ

4 Click in the **Degrees of freedom** text box and type 5

5 Click **OK**

a. Use Minitab or some other statistical software to generate a sample of 500 observations from a variable having a chi-square distribution with five degrees of freedom.
b. Obtain a histogram of the 500 observations from part (a).
c. Based on your histogram in part (b), what shape does a chi-square distribution with five degrees of freedom have? Explain your reasoning.

2.6 MISLEADING GRAPHS

Graphs and charts are frequently constructed in a manner that causes them to be misleading. Sometimes this is intentional and sometimes it is not. Regardless of the intent, it is important to read and interpret graphs and charts with a great deal of care. In this section we will examine some misleading graphs and charts.

EXAMPLE 2.23 TRUNCATED GRAPHS

Figure 2.11(a) shows a bar graph from an article in a major metropolitan newspaper. The graph displays the unemployment rates in the United States from September of one year to March of the next year.

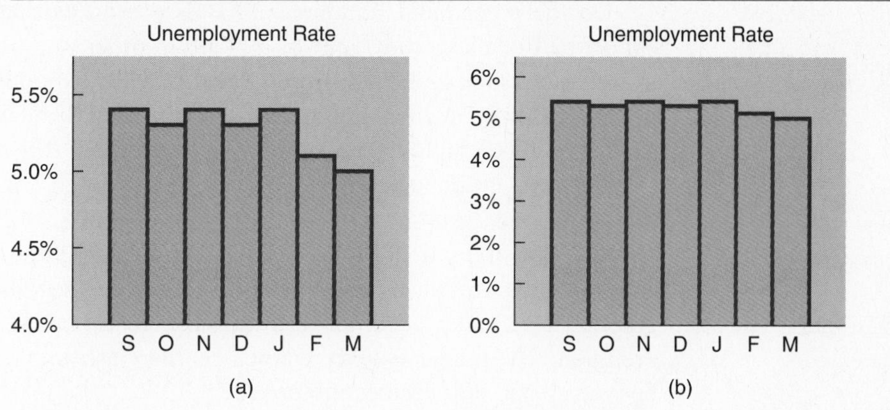

FIGURE 2.11
Unemployment rates: (a) truncated graph, (b) nontruncated graph

A quick look at Fig. 2.11(a) might lead you to conclude that the unemployment rate dropped by roughly one-fourth between January and March, since the bar for March is about one-fourth smaller than the bar for January. In reality, however, the unemployment rate dropped less than one-thirteenth, from 5.4% to 5.0%. Consequently, we see that we must analyze the graph more carefully to figure out what it truly represents.

The unemployment-rate graph in Fig. 2.11(a) is an example of a **truncated graph** because the vertical axis, which should start at 0%, starts at 4% instead. Thus the part of the graph from 0% to 4% has been cut off, or truncated. This truncation causes the bars to be out of correct proportion and hence creates a misleading impression. The graph would be even more deceptive if it started at 4.5%. To see this, slide a piece of paper over the bottom of Fig. 2.11(a) so that the bars begin at 4.5%. By how much does it now appear that the unemployment rate dropped between January and March?

As we have observed, the truncated graph in Fig. 2.11(a) is potentially misleading. However, it is probably safe to say that the truncation was done to present a picture of the "ups" and "downs" in the unemployment-rate pattern rather than to intentionally mislead the reader.

A nontruncated version of Fig. 2.11(a) is shown in Fig. 2.11(b). Although Fig. 2.11(b) provides a correct graphical display of the unemployment-rate data, the "ups" and "downs" are not as easy to spot as they are in the truncated graph in Fig. 2.11(a). ◨

Truncated graphs have long been a target of statisticians. Many statistics books warn against their use. Nonetheless, as we saw in Example 2.23, truncated graphs are still used today, even in reputable publications.

On the other hand, Example 2.23 also suggests that it may be desirable to cut off part of the vertical axis of a graph in order to more easily convey relevant information, such as the "ups" and "downs" of the monthly unemployment rates. In these cases a truncated graph should not be used. Instead, a special symbol, such as //, should be employed to signify that the vertical axis has been modified.

The two graphs shown in Fig. 2.12 provide an excellent illustration. Both graphs portray the number of new single-family homes sold per month over several months. The first graph, Fig. 2.12(a), is a truncated graph, truncated most likely in an attempt to present a clear visual display of the variation in sales. The second graph, Fig. 2.12(b), accomplishes the same result but is less subject to misinterpretation. The reader is aptly warned by the slashes that part of the vertical axis between 0 and 500 has been removed.

FIGURE 2.12
New single-family home sales

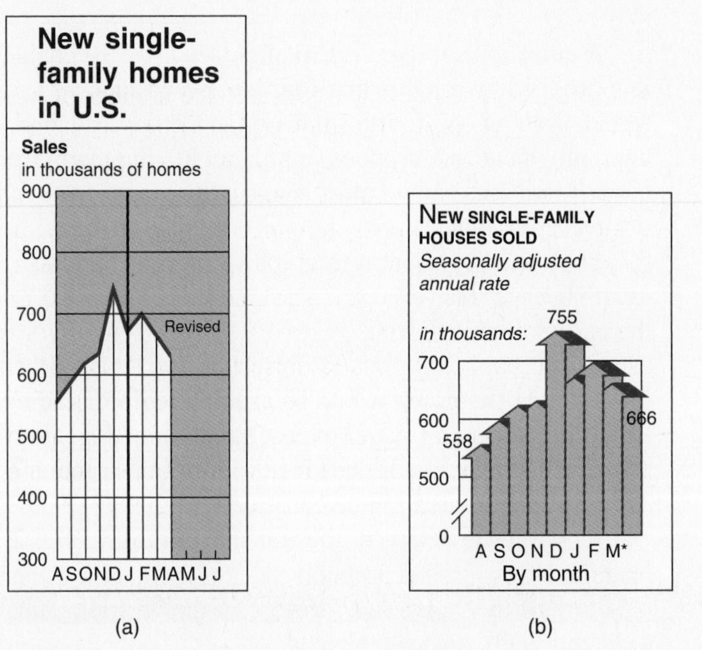

(a) (b)

SOURCES: Figure 2.12(a) reprinted by permission of Tribune Media Services. Figure 2.12(b) data from U.S. Department of Commerce and U.S. Department of Housing and Urban Development.

Improper Scaling

Misleading graphs and charts can also result from **improper scaling.** Example 2.24 shows how this can happen.

EXAMPLE 2.24 IMPROPER SCALING

A developer is preparing a brochure to attract investors for a new shopping center that is to be built in an area of Denver, Colorado. The area is growing rapidly; this year twice as many homes will be built there as last year. To illustrate that fact, the developer draws a **pictogram,** as in Fig. 2.13.

FIGURE 2.13
Pictogram for home building

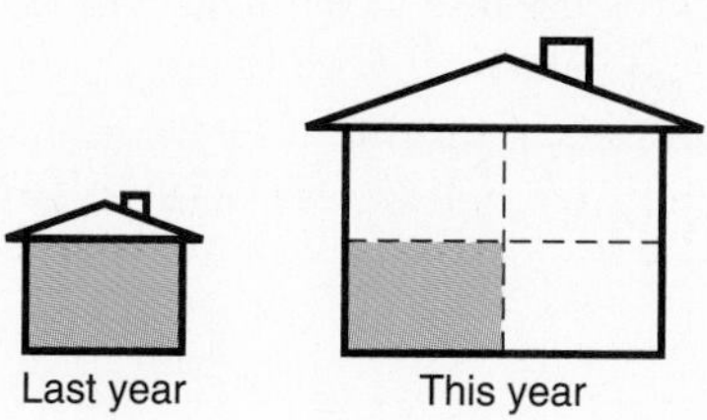

The house on the left represents the number of homes built last year. Since the number of homes that will be built this year is double the number built last year, the developer makes the house on the right twice as tall and twice as wide as the house on the left. However, this scaling is improper because it gives the visual impression that four times as many homes will be built this year as last. So the developer's brochure may mislead the unwary investor. □

Graphs and charts can be misleading in countless ways besides the two ways we have discussed. Many more examples of misleading graphs can be found in the entertaining and classic book *How to Lie with Statistics* by Darrell Huff (New York: Norton, 1955). The main purpose of this section has been to show you that graphs and charts should be constructed and read carefully.

EXERCISES 2.6

STATISTICAL CONCEPTS AND SKILLS

2.124 Give one reason why it is important to construct and read graphs and charts carefully.

2.125 This exercise discusses truncated graphs.

a. What is a truncated graph?

b. Give a legitimate motivation for truncating the axis of a graph.

c. If we have a legitimate motivation for truncating the axis of a graph, how can we correctly obtain that objective without creating the possibility of misinterpretation?

2.126 Find two examples of graphs in a current newspaper or magazine that might be misleading. Explain why you think the graphs are potentially misleading.

2.127 Each year the director of the reading program in a school district administers a standard test of reading skills. Then the director compares the average score for his district with the national average. Figure 2.14 was presented to the school board in 1998.

FIGURE 2.14 Average reading scores

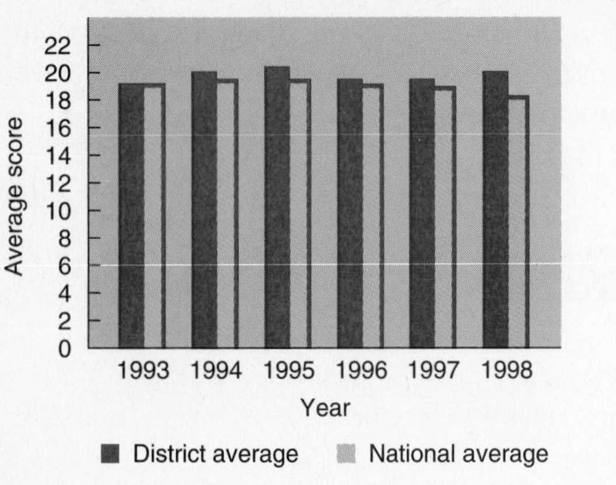

a. Obtain a truncated version of Fig. 2.14 by sliding a piece of paper over the bottom of the graph so that the bars start at 16.

b. Repeat part (a) but have the bars start at 18.

c. What misleading impression about the 1998 scores is given by the truncated graphs you observed in parts (a) and (b)?

2.128 The following bar graph is based on a newspaper article entitled "Immigrants add seasoning to America's melting pot." [Used with permission from American Demographics, Ithaca, NY.]

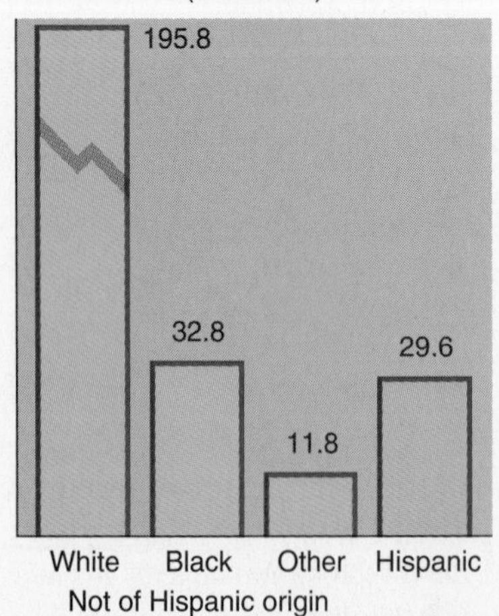

a. Explain why a break is shown in the first bar.

b. Why was the graph constructed with a broken bar?

c. Do you think this graph is potentially misleading?

2.129 The following bar graph, taken from *The Arizona Republic,* provides data on the M2 money supply over several months. M2 consists of cash in circulation, deposits in checking accounts, nonbank traveler's checks, accounts such as savings deposits, and money-market mutual funds.

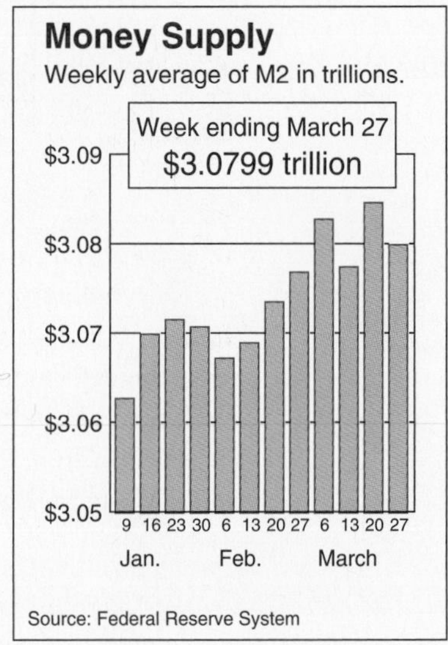

a. What is wrong with the bar graph?

b. Construct a version of the bar graph with a nontruncated and unmodified vertical axis.

c. Construct a version of the bar graph in which the vertical axis is modified in an acceptable manner.

Extending the Concepts and Skills

2.130 Refer to Example 2.24 on page 115. Indicate a way in which the developer can accurately illustrate that twice as many homes will be built in the area this year as last.

2.131 A manufacturer of golf balls has determined that a newly developed process results in a ball that lasts roughly twice as long as a ball produced using the current process. To illustrate this graphically, she designs a brochure showing a "new" ball having twice the radius of the "old" ball.

a. What is wrong with this? (*Hint:* The volume of a sphere is proportional to the cube of its radius.)

b. How can the manufacturer accurately illustrate the fact that the "new" ball lasts twice as long as the "old" ball?

CHAPTER REVIEW

You Should Be Able To

1. classify variables and data as either qualitative or quantitative.
2. distinguish between discrete and continuous variables and data.
3. identify terms associated with the grouping of data.
4. group data into a frequency distribution and a relative-frequency distribution.
5. construct a grouped-data table.
6. draw a frequency histogram and a relative-frequency histogram.
7. construct a dotplot.
8. draw a pie chart and a bar graph.
9. construct stem-and-leaf, shaded stem-and-leaf, and ordered stem-and-leaf diagrams.
10. identify the shape and modality of the distribution of a data set.
11. specify whether a unimodal distribution is symmetric, right skewed, or left skewed.
12. understand the relationship between sample distributions and the population distribution (distribution of the variable under consideration).
13. identify and correct misleading graphs.

*14. use the Minitab procedures covered in this chapter.

*15. interpret the output obtained from the application of the Minitab procedures discussed in this chapter.

Key Terms

bar graph, *86*
bell-shaped, *104*
bimodal, *104, 106*
categorical variable, *60*
census data, *106*
classes, *69*
continuous data, *61*
continuous variable, *61*
count, *68*
data, *61*
data set, *61*
depths,* *99*
discrete data, *61*
discrete variable, *61*
distribution, *103*
distribution of a variable, *107*
dotplot, *84*
frequency, *69*
frequency distribution, *69*
frequency histogram, *83*
grouped-data table, *70*
grouping, *66*
improper scaling, *114*
J-shaped, *104*
leaves, *96*
left skewed, *104*
lower cutpoint, *69*
mark, *69*
midpoint, *69*
multimodal, *104, 106*
observation, *61*
ordered stem-and-leaf diagram, *97*

REVIEW TEST

STATISTICAL CONCEPTS AND SKILLS

1. This problem is about variables and data.
a. What is a variable?
b. Identify two main types of variables.
c. Identify the two kinds of quantitative variables.
d. What is data?
e. How is data type determined?

2. Explain why it is important to group data.

3. To which type of data do the concepts of cutpoints and midpoints not apply? Why?

4. A quantitative data set has been grouped into a grouped-data table using equal-width classes of width 8.
a. If the midpoint of the first class is 10, what are its lower and upper cutpoints?
b. What is the midpoint of the second class?
c. What are the lower and upper cutpoints of the third class?
d. Into which class would an observation of 22 go?

5. A quantitative data set has been grouped into a grouped-data table using equal-width classes.
a. If the lower and upper cutpoints of the first class are 5 and 15, respectively, what is the common class width?
b. What is the midpoint of the second class?
c. What are the lower and upper cutpoints of the third class?

6. When is it particularly appropriate to use single-value grouping?

7. Explain the relative positioning of the bars in a histogram to the numbers labeling the horizontal axis in each of the following cases.
a. Cutpoints are used to label the horizontal axis.
b. Midpoints are used to label the horizontal axis.

8. Identify two main types of graphical displays used for qualitative data.

9. Which is preferable as a graphical display for a large, quantitative data set, a histogram or a stem-and-leaf diagram? Explain your answer.

10. Sketch the curve corresponding to each of the following distribution shapes.
a. bell-shaped **b.** right skewed
c. reverse J-shaped **d.** uniform

11. Make an educated guess as to the distribution shape of each of the following variables. Explain your answers.
a. Height of American adult males.
b. Annual income of U.S. households.
c. Age of full-time college students.
d. Cumulative GPA of college seniors.

12. A variable of a population has a left-skewed distribution.

a. If a large random sample is taken from the population, roughly what shape will the distribution of the sample have? Explain your answer.

b. If two random samples are taken from the population, would you expect the two sample distributions to have identical shapes? Explain your answer.

c. If two random samples are taken from the population, would you expect the two sample distributions to have similar shapes? If so, what shape would that be? Explain your answers.

13. The world's five largest hydroelectric plants, based on ultimate capacity, are as shown in the following table. Capacities are in megawatts. [SOURCE: T. W. Mermel, *Intl. Waterpower & Dam Construction Handbook.*]

Rank	Name	Country	Capacity
1	Turukhansk	Russia	20,000
2	Itaipu	Brazil/Para.	13,320
3	Grand Coulee	U.S.A.	10,830
4	Guri	Venezuela	10,300
5	Tucurui	Brazil	7,260

a. What type of data is given in the first column of the table?

b. What type of data is given in the fourth column?

c. What type of data is given in the third column?

14. The ages at inauguration for the first 42 presidents of the United States are as shown in the table at the top of the next column.

a. Construct a grouped-data table for these inauguration ages using equal-width classes and beginning with the class 40–44.

b. Identify the lower and upper cutpoints of the first class. *(Hint:* Be careful!*)*

c. Identify the common class width.

d. Draw a frequency histogram for the inauguration ages based on your grouping in part (a).

15. Refer to Problem 14. Construct a dotplot for the ages at inauguration of the first 42 presidents of the United States.

President	Age at inaug.	President	Age at inaug.
G. Washington	57	G. Cleveland	47
J. Adams	61	B. Harrison	55
T. Jefferson	57	G. Cleveland	55
J. Madison	57	W. McKinley	54
J. Monroe	58	T. Roosevelt	42
J. Q. Adams	57	W. Taft	51
A. Jackson	61	W. Wilson	56
M. Van Buren	54	W. Harding	55
W. Harrison	68	C. Coolidge	51
J. Tyler	51	H. Hoover	54
J. Polk	49	F. Roosevelt	51
Z. Taylor	64	H. Truman	60
M. Fillmore	50	D. Eisenhower	62
F. Pierce	48	J. Kennedy	43
J. Buchanan	65	L. Johnson	55
A. Lincoln	52	R. Nixon	56
A. Johnson	56	G. Ford	61
U. Grant	46	J. Carter	52
R. Hayes	54	R. Reagan	69
J. Garfield	49	G. Bush	64
C. Arthur	50	W. Clinton	46

16. Refer to Problem 14. Construct an ordered stem-and-leaf diagram for the inauguration ages of the first 42 presidents of the United States using

a. one line per stem.

b. two lines per stem.

c. Which of the two stem-and-leaf diagrams you just constructed corresponds to the frequency distribution of Problem 14(a)?

17. The Prescott National Bank has six tellers available to serve customers. The data in the following table provide the number of busy tellers observed at 25 spot checks.

6	5	4	1	5
6	1	5	5	5
3	5	2	4	3
4	5	0	6	4
3	4	2	3	6

a. Construct a grouped-data table for these data using single-value grouping.

b. Draw a relative-frequency histogram for the data based on the grouping in part (a).

18. The class levels of the students in Professor Weiss's introductory statistics course are shown in the following table. We have used the abbreviations Fr, So, Ju, and Se, respectively, for Freshman, Sophomore, Junior, and Senior.

Fr	So	Ju	So	Ju	Ju	Se	Ju
Se	So	Fr	Ju	So	Ju	So	Se
So	So	Se	So	So	Se	So	Fr
Ju	So	Ju	Fr	Fr	Ju	Ju	Fr
So	Se	Ju	Ju	So	So	So	Se

a. Obtain frequency and relative-frequency distributions for these data.
b. Draw a pie chart of the data that displays the percentage of students at each class level.
c. Draw a bar graph of the data that displays the relative frequency of students at each class level.

19. According to the *World Almanac,* the highs for the Dow Jones Industrial Averages between 1961–1996 are as follows.

Year	High	Year	High
1961	734.91	1979	897.61
1962	726.01	1980	1000.17
1963	767.21	1981	1024.05
1964	891.71	1982	1071.55
1965	969.26	1983	1287.20
1966	995.15	1984	1286.64
1967	943.08	1985	1553.10
1968	985.21	1986	1955.57
1969	968.85	1987	2722.42
1970	842.00	1988	2183.50
1971	950.82	1989	2791.41
1972	1036.27	1990	2999.75
1973	1051.70	1991	3168.83
1974	891.66	1992	3413.21
1975	881.81	1993	3794.33
1976	1014.79	1994	3978.36
1977	999.75	1995	5216.47
1978	907.74	1996	6560.91

a. Construct a grouped-data table for the highs using classes of equal width starting with the class 400 < 1000.
b. Draw a relative-frequency histogram for the highs based on your result from part (a).

20. Identify the distribution shapes of
a. the inauguration ages of the first 42 presidents of the United States (from Problem 14).
b. the number of tellers busy with customers at Prescott National Bank during 25 spot checks (from Problem 17).

21. Draw a smooth curve that represents a symmetric trimodal (three-peak) distribution.

22. Following is a graph based on one that appeared in a newspaper article entitled "Hand that rocked cradle turns to work as women reshape U.S. labor force." The graph depicts the labor force participation rates for the years 1960 and 1980 and projected figures for the year 2000.

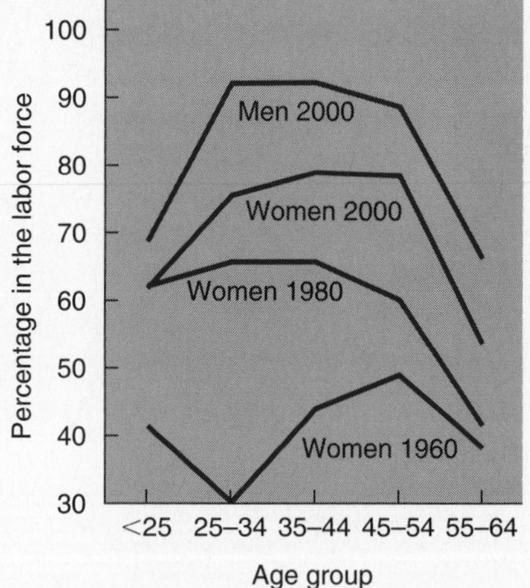

a. Cover up the numbers on the vertical axis of the graph with a piece of paper.
b. Look at the 1960 and 2000 graphs for women. Focus on the 35–44-year-old age group. What impression does the graph convey regarding the ratio of the percentages of women in the labor force for 1960 and 2000?
c. Now remove the piece of paper from the graph. Using the vertical scale, find the actual ratio of the percentages of 35–44-year-old women in the labor force for 1960 and 2000.
d. Why is the graph potentially misleading?
e. What can be done to make the graph less potentially misleading?

USING TECHNOLOGY 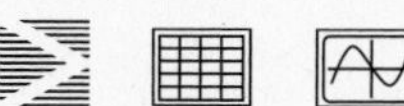

23. Refer to the age data in Problem 14. Use Minitab or some other statistical software to obtain a

a. frequency histogram of the data similar to the one found in Problem 14(d).

b. dotplot of the data.

c. stem-and-leaf diagram similar to the one constructed in Problem 16(b).

24. A city planner working on bikeways needs information about local bicycle commuters. She designs a questionnaire. One question asks how long it takes the rider to pedal from home to his or her destination. We used Minitab to obtain the following histogram for the times, rounded to the nearest minute, of a sample of local bicycle commuters.

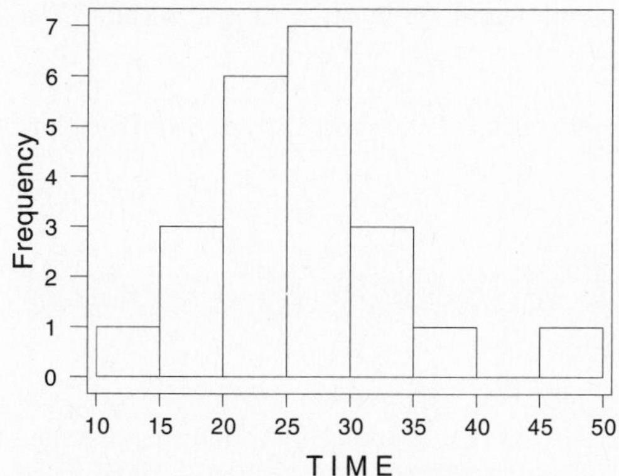

a. How many times are in the sample?

b. Identify the midpoint of the fourth class.

c. How many times are between 30 and 34 minutes, inclusive?

d. What column name was chosen for the times?

25. The Minitab output shown at the top of the next column gives a stem-and-leaf diagram for the sample of bicycle-commuter times in Problem 24.

a. How many lines per stem are used in this stem-and-leaf diagram?

b. Use the depths to determine how many of the times are 35 minutes or more.

c. Use the depths to determine how many of the times are less than 20 minutes.

d. What is the longest time in the sample?

e. List the times that are in the 30s.

```
Stem-and-leaf of TIME     N  = 22
Leaf Unit = 1.0

    1    1 2
    4    1 569
   10    2 122334
  (7)    2 6678999
    5    3 011
    2    3 7
    1    4
    1    4 8
```

26. Refer to Problem 18. Use Minitab or some other statistical software to obtain a

a. pie chart of the data.

b. bar graph of the data.

27. *F-distributions* are important in statistics. If you are using Minitab, a sample of 1000 observations from a variable having an F-distribution with degrees of freedom (3, 16) can be obtained as follows.

1 Choose **Calc ➤ Random Data ➤ F...**

2 Type 1000 in the **Generate rows of data** text box

3 Click in the **Store in column(s)** text box and type FDIST

4 Click in the **Numerator degrees of freedom** text box and type 3

5 Click in the **Denominator degrees of freedom** text box and type 16

6 Click **OK**

a. Use Minitab or some other statistical software to generate a sample of 1000 observations from a variable having an F-distribution with degrees of freedom (3, 16).

b. Obtain a histogram of the 1000 observations from part (a).

c. Based on your histogram in part (b), what shape does an F-distribution with degrees of freedom (3, 16) have? Explain your reasoning.

INTERNET PROJECT

Infant Mortality

In this Internet project, we will broaden our investigation of infant mortality. The data shown in the table on page 58 provide a window into a complex human issue. The global picture of infant mortality furnishes us with one of the best examples of how statistics can help us understand questions of human life, society, and progress. The problem of infant mortality can be seen as a set of physiological, cultural, economic, racial, and political issues.

Around 1900, the rate of infant mortality in the United States was 150. Knowledge has helped to bring down that number. And knowledge (much of it gained through statistical techniques) will help to bring the rate down even further.

The last part of this Internet project provides you with a hands-on demonstration of a dynamic histogram. You will see how the choices you make when creating a histogram can change the way others perceive your data. You will also see some examples of "dangerous" graph types—graphs that you must create with care in order to give a truthful and valid representation of the data.

URL for access to Internet Projects Page: `http://hepg.awl.com` *keyword:* Weiss

USING THE FOCUS DATABASE

In Chapter 1 we explained how to store the Focus database in a Minitab worksheet named `focus.mtw`. If you haven't already created that worksheet, follow the instructions on pages 54–55 to create it now.

The Focus database contains information on 500 randomly selected Arizona State University sophomores. Seven variables are considered for each student: sex, high-school GPA, SAT math score, cumulative GPA, SAT verbal score, age, and total hours. Use Minitab or some other statistical software to solve the following problems.

a. Obtain a histogram for the ages of the sophomores in the sample.

b. Obtain individual histograms for the ages of the female sophomores and the male sophomores in the sample. Compare the two histograms and discuss the differences you observe.

c. Construct a stem-and-leaf diagram for the cumulative GPAs of the sophomores in the sample. Use five lines per stem.

d. Construct individual stem-and-leaf diagrams for the cumulative GPAs of the female sophomores and the male sophomores in the sample. Use five lines per stem for both diagrams. Compare the two diagrams and discuss any differences you observe.

e. Obtain dotplots for both the SAT math and SAT verbal scores of the sophomores in the sample. Compare the two dotplots.

f. Identify the shapes of the distributions in parts (a)–(e). Which distributions are symmetric?

CASE STUDY DISCUSSION

Infant Mortality

Recall that the infant mortality rate of a nation represents the number of deaths of children under 1 year old per 1000 live births in a calendar year. At the beginning of this chapter on page 58, we presented data on infant mortality for nations with 6 million or more population having the lowest rates. Referring to the table given there, solve each of the following problems.

a. What type of data is displayed in the second column of the table?
b. What type of data is given by the statement that the United States ranks 14th in infant mortality rate among nations with 6 million or more population?
c. Construct a grouped-data table for the infant mortality rates using classes of equal width and starting with the class $4 < 5$.
d. Construct a frequency histogram for the infant mortality rates based on your grouping in part (c).
e. Construct an ordered stem-and-leaf diagram for the infant mortality rates using one line per stem.
f. Use Minitab or some other statistical software to solve parts (c)–(e).

BIOGRAPHY ADOLPHE QUETELET

Lambert Adolphe Jacques Quetelet was born in Ghent, Belgium, on February 22, 1796. He attended school locally and in 1819 received the first doctorate of science degree granted at the newly established University of Ghent. In that same year, he obtained a position as a professor of mathematics at the Brussels Athenaeum.

Quetelet was elected to the Belgian Royal Academy in 1820 and served as its secretary from 1834 until his death in 1874. He was founder and director of the Royal Observatory in Brussels, founder and a major contributor to the journal *Correspondance mathématique et physique*, and, according to Stephen M. Stigler in *The History of Statistics,* "... active in the founding of more statistical organizations than any other individual in the nineteenth century." Among the organizations he established was the International Statistical Congress, initiated in 1853.

In 1835 Quetelet wrote a two-volume set entitled *A Treatise on Man and the Development of His Faculties,* the publication in which he introduced his concept of the "average man" and that firmly established his international reputation as a statistician and sociologist. A review in the *Athenaeum* stated, "We consider the appearance of these volumes as forming an epoch in the literary history of civilization."

In 1855 Quetelet suffered a stroke that limited his work but not his popularity. He died on February 17, 1874. His funeral was attended by royalty and famous scientists from around the world. A monument to his memory was erected in Brussels in 1880.

CASE STUDY

PER CAPITA INCOME BY STATE

Per capita income is defined as the annual total personal income of the residents of a region divided by the resident population of that region as of July 1 of that year. How does per capita income in the United States vary from state to state? Where does your state rank?

A report published in *Survey of Current Business* by the U.S. Bureau of Economic Analysis revealed the following data on per capita income on a state-by-state basis for the year 1996.

Rank	State	Per capita income	Rank	State	Per capita income	Rank	State	Per capita income
1	D.C.	$34,932	18	R.I.	$24,765	35	S.D.	$21,516
2	Conn.	$33,189	19	Pa.	$24,668	36	Wyo.	$21,245
3	N.J.	$31,053	20	Alaska	$24,558	37	Ariz.	$20,989
4	Mass.	$29,439	21	Fla.	$24,104	38	Maine	$20,826
5	N.Y.	$28,782	22	Ohio	$23,537	39	N.D.	$20,710
6	Del.	$27,622	23	Kan.	$23,281	40	Ala.	$20,055
7	Md.	$27,221	24	Wis.	$23,269	41	La.	$19,824
8	Ill.	$26,598	25	Neb.	$23,047	42	S.C.	$19,755
9	N.H.	$26,520	26	Mo.	$22,864	43	Ky.	$19,687
10	Minn.	$25,580	27	Ga.	$22,709	44	Idaho	$19,539
11	Nev.	$25,451	28	Ore.	$22,668	45	Okla.	$19,350
12	Hawaii	$25,159	29	Iowa	$22,560	46	Utah	$19,156
13	Calif.	$25,144	30	Ind.	$22,440	47	Mont.	$19,047
14	Colo.	$25,084	31	Vt.	$22,124	48	Ark.	$18,928
15	Va.	$24,925	32	Texas	$22,045	49	N.M.	$18,770
16	Wash.	$24,838	33	N.C.	$22,010	50	W.Va.	$18,444
17	Mich.	$24,810	34	Tenn.	$21,764	51	Miss.	$17,471

In this chapter we will learn several additional techniques to help us analyze data. At the end of the chapter, after you have mastered those techniques, we will ask you to apply them to conduct data analyses of the data on per capita income by state and to answer the questions posed at the beginning of this discussion.

Descriptive Measures

3

GENERAL OBJECTIVES

In Chapter 2 we began our study of descriptive statistics. We learned how to organize data into tables and summarize data with graphical displays. Another method of summarizing data is to compute numbers, such as an average, that describe the data set. Numbers that are used to describe data sets are called **descriptive measures.** In this chapter we will continue our study of descriptive statistics by examining some of the most important descriptive measures.

CHAPTER OUTLINE

3.1 MEASURES OF CENTER

Descriptive measures that indicate where the center or most typical value of a data set lies are called **measures of central tendency** or, more simply, **measures of center.** Measures of center are often referred to as *averages.*

In this section we will discuss the three most important measures of center: the *mean, median,* and *mode.* The mean and median apply only to quantitative data, whereas the mode can be used with either quantitative or qualitative data.

The Mean

The most commonly used measure of center is the **mean.** When people speak of taking an average, it is the mean that they are most often referring to.

DEFINITION 3.1 MEAN OF A DATA SET

The ***mean*** of a data set is the sum of the observations divided by the number of observations.

EXAMPLE 3.1 ILLUSTRATES DEFINITION 3.1

A mathematician spent one summer working for a small mathematical consulting firm. The firm employed a few senior consultants, who made between $800 and $1050 per week; a few junior consultants, who made between $400 and $450 per week; and several clerical workers, who made $300 per week.

Because the first half of the summer was busier than the second half, more employees were required during the first half. Tables 3.1 and 3.2 display typical lists of weekly earnings for the two halves of the summer. Find the mean of each of the two data sets.

TABLE 3.1 Data Set I

$300	300	300	940	300	300	400
300	400	450	800	450	1050	

TABLE 3.2 Data Set II

$300	300	940	450	400
400	300	300	1050	300

SOLUTION According to Definition 3.1, the mean of a data set is obtained by summing all the observations and then dividing that sum by the total number of observations.

As we see from Table 3.1, Data Set I has 13 observations. The sum of those 13 observations is \$6290. Consequently,

$$\text{Mean of Data Set I} = \frac{\$6290}{13} = \$483.85 \text{ (rounded to the nearest cent).}$$

Similarly,

$$\text{Mean of Data Set II} = \frac{\$4740}{10} = \$474.00.$$

 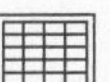 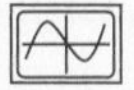

Thus the mean salary of the 13 employees in Data Set I is \$483.85 and that of the 10 employees in Data Set II is \$474.00. ◨

The Median

Another frequently used measure of center is the median. Essentially, the **median** of a data set is the number that divides the bottom 50% of the data from the top 50%. To obtain the median of a data set, we arrange the data in increasing order and then determine the middle value in the ordered list. A more precise definition of the median is found in Definition 3.2.

DEFINITION 3.2 MEDIAN OF A DATA SET

Arrange the data in increasing order.

- If the number of observations is odd, then the ***median*** is the observation exactly in the middle of the ordered list.
- If the number of observations is even, then the ***median*** is the mean of the two middle observations in the ordered list.

In both cases, if we let n denote the number of observations, then the median is at position $(n + 1)/2$ in the ordered list.

EXAMPLE 3.2 ILLUSTRATES DEFINITION 3.2

Consider again the two sets of salary data shown in Tables 3.1 and 3.2. Determine the median of each of the two data sets.

SOLUTION To find the median of Data Set I, we apply Definition 3.2. First we arrange the data in increasing order:

300 300 300 300 300 300 **400** 400 450 450 800 940 1050

The number of observations in Data Set I is 13, which is an odd number. Since $n = 13$, $(n+1)/2 = (13+1)/2 = 7$. Consequently, the median is the seventh observation in the ordered list, which is 400 (shown in boldface type). The median salary of the 13 employees in Data Set I is \$400.

To find the median of Data Set II, we again apply Definition 3.2. First we arrange the data in increasing order:

300 300 300 300 **300** **400** 400 450 940 1050

The number of observations in Data Set II is 10, which is an even number. Since $n = 10$, $(n+1)/2 = (10+1)/2 = 5.5$. Consequently, the median is halfway between the fifth and sixth observations (shown in boldface type) in the ordered list. In other words, the median salary of the 10 employees in Data Set II is $(300 + 400)/2 = \$350$. □

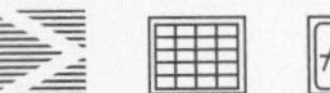

As we have just seen, to determine the median of a data set we must first arrange the data in increasing order. It is often helpful to construct a stem-and-leaf diagram as a preliminary step to ordering the data.

The Mode

The final measure of center we will discuss is the mode. Basically, the **mode** is the value that occurs most frequently in a data set. A more exact definition of the mode is provided in Definition 3.3.

DEFINITION 3.3 MODE OF A DATA SET

Obtain the frequency of occurrence of each value and note the greatest frequency.

- If the greatest frequency is 1 (i.e., no value occurs more than once), then the data set has no mode.
- If the greatest frequency is 2 or greater, then any value that occurs with that greatest frequency is called a ***mode*** of the data set.

To obtain the mode(s) of a data set, we first construct a frequency distribution for the data using classes based on a single value. The mode(s) can then be determined easily from the frequency distribution, as explained in Example 3.3.

EXAMPLE 3.3 ILLUSTRATES DEFINITION 3.3

Determine the mode(s) of each of the two sets of salary data given in Tables 3.1 and 3.2 on page 126.

SOLUTION First we consider the salary data in Data Set I. Referring to Table 3.1, we obtain the frequency distribution of the data using classes based on a single value, as shown in Table 3.3.

TABLE 3.3
Frequency distribution for Data Set I using single-value grouping

Salary	300	400	450	800	940	1050
Frequency	6	2	2	1	1	1

We see from Table 3.3 that the greatest frequency of occurrence is 6 and that 300 is the only value occurring with that frequency. So the mode of the 13 salaries in Data Set I is $300.

Proceeding in the same way, we find that for Data Set II, the greatest frequency of occurrence is 5 and that 300 is the only value occurring with that frequency. So the mode of the 10 salaries in Data Set II is $300. ◨

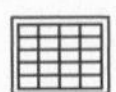

A data set can have more than one mode if there is more than one value that occurs with the greatest frequency. For instance, suppose two of the clerical workers in Data Set I, who make $300 per week, were promoted to $400-per-week jobs. Then both the value 300 and the value 400 would occur with greatest frequency 4. This new data set would thus have two modes, $300 and $400.

Comparison of the Mean, Median, and Mode

The mean, median, and mode of a data set are often different. Table 3.4 summarizes the definitions of these three measures of center and gives their values for Data Set I and Data Set II, which we computed in Examples 3.1–3.3.

TABLE 3.4
Means, medians, and modes of salaries in Data Set I and Data Set II

Measure of center	Definition	Data Set I	Data Set II
Mean	$\dfrac{\text{Sum of observations}}{\text{Number of observations}}$	$483.85	$474.00
Median	Middle value in ordered list	$400.00	$350.00
Mode	Most frequent value	$300.00	$300.00

In both Data Sets I and II, the mean is larger than the median. This is because the mean is strongly affected by the few large salaries in each data set. Generally speaking, the mean is sensitive to extreme (very large or very small) observations, whereas the median is not. Consequently, when the choice for the measure of center is between the mean and the median, the median is usually preferred for data sets that have extreme observations.

Figure 3.1 shows the relative positions of the mean and median for right-skewed, symmetric, and left-skewed distributions. As we see from the figure, the mean is pulled in the direction of skewness, that is, in the direction of the extreme observations. For a right-skewed distribution, the mean is greater than the median; for a symmetric distribution, the mean and median are equal; and for a left-skewed distribution, the mean is less than the median.

FIGURE 3.1 Relative positions of the mean and median for (a) right-skewed, (b) symmetric, and (c) left-skewed distributions

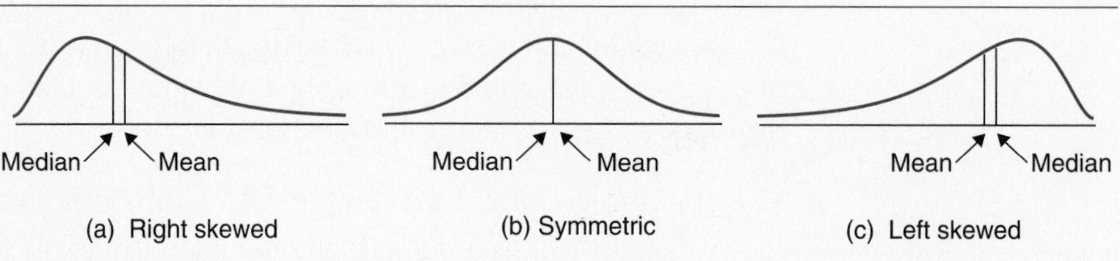

A descriptive measure is called **resistant** if it is not sensitive to the influence of a few extreme observations. Thus the median is a resistant measure of center, whereas the mean is not. The resistance of the mean can be improved by using **trimmed means,** where a specified percentage of the smallest and largest observations are removed before computing the mean. Exercise 3.17 discusses trimmed means in more detail.

The mode for each of Data Sets I and II differs from both the mean and the median. Whereas the mean and the median are aimed at finding the center of a data set, the mode is really not—the value that occurs most frequently may not be near the center.

It should now be clear that the mean, median, and mode generally provide different information. There is no simple rule for deciding which measure of center to use in a given situation. Although skill in making such decisions is attained through practice, even experts may disagree on the most suitable measure of center for a particular data set. Example 3.4 discusses three data sets and suggests the most appropriate measure of center for each.

EXAMPLE 3.4 SELECTING AN APPROPRIATE MEASURE OF CENTER

a. A student takes four exams in a biology class. His grades are 88, 75, 95, and 100. If asked for his average, which measure of center is the student likely to report?

b. The National Association of REALTORS® publishes data on resale prices of U.S. homes. Which measure of center is most appropriate for such resale prices?

c. In the 1998 Boston Marathon, there were two categories of official finishers: male and female. The following table provides a frequency distribution for those data. Which measure of center should be used here?

Sex	Frequency
Male	6989
Female	2982

SOLUTION **a.** Chances are that the student would report the mean of his four exam scores, which is 89.5. The mean is probably the most suitable measure of center for the student to use since it takes into account the numerical value of each score and therefore indicates total overall performance.

b. The most appropriate measure of center for resale home prices is the median, because it is aimed at finding the center of the data on resale home prices and because it is not strongly affected by the relatively few homes with extremely high or low resale prices. Thus the median provides a better indication of the "typical" resale price than either the mean or the mode does.

c. The only suitable measure of center for these data is the mode, which is "male." Each observation in this data set is either "male" or "female." There is no way to compute a mean or median for such data. Of the mean, median, and mode, the mode is the only measure of center that can be used for qualitative data. ◨

Many measures of center appearing in newspapers or reported by government agencies are medians, as is the case for household income and years of school completed. Furthermore, in an attempt to provide a clearer picture, some reports include both the mean and the median. For instance, the National Center for Health Statistics does this for daily intake of nutrients in the publication *Vital and Health Statistics.*

Population and Sample Mean

Recall that a variable is a characteristic that varies from one person or thing to another and that observing one or more values of a variable yields data. The data set obtained by observing the values of a variable for an entire population is called *population data;* a data set obtained by observing the values of a variable for a sample of the population is called *sample data.*

The mean of population data is called the **population mean** or the **mean of the variable;** the mean of sample data is called a **sample mean.** And likewise for the median and mode and, for that matter, any descriptive measure. Figure 3.2 shows the two ways in which the mean of a data set can be interpreted.

FIGURE 3.2
Possible interpretations for the mean of a data set

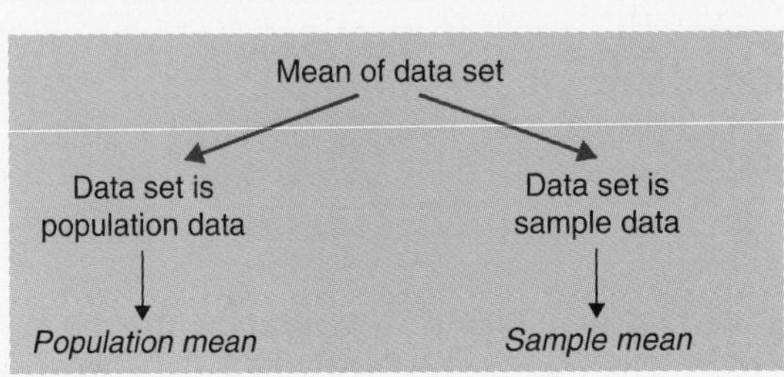

In the next three sections, we will concentrate on descriptive measures of samples. Then, in Section 3.5, we will discuss descriptive measures of populations and their relationship to descriptive measures of samples.

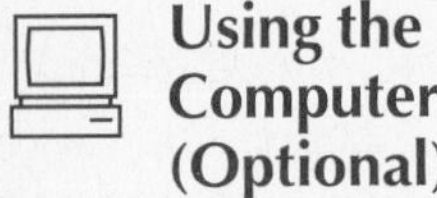

Using the Computer (Optional)

For small data sets such as the two sets of salary data in Tables 3.1 and 3.2 on page 126, it is easy to obtain the mean, median, and mode by hand. However, for even moderately large data sets, obtaining these descriptive measures is tedious and prone to error and, so, it is best to use a statistical calculator (i.e., a calculator with statistical functions) or a computer.

We can apply Minitab in several ways to obtain the mean, median, and mode of a data set. One method is illustrated here and another in Section 3.4.

EXAMPLE 3.5 USING MINITAB TO OBTAIN THE MEAN, MEDIAN, AND MODE

Use Minitab to obtain the mean, median, and mode of the salary data for Data Set I, displayed in Table 3.1 on page 126.

SOLUTION First we store the salary data from Table 3.1 in a column named DATASETI. Then we proceed in the following manner to obtain the mean of Data Set I.

1 Choose **Calc ➤ Column Statistics...**
2 Select the **Mean** option button from the **Statistic** field
3 Click in the **Input variable** text box and specify DATASETI
4 Click **OK**

Using the same steps except selecting the **Median** option button instead of the **Mean** option button, we obtain the median of Data Set I.† The resulting output is displayed in Printout 3.1 from which we see that for Data Set I, the mean salary is $483.85 and the median salary is $400.00.

PRINTOUT 3.1
Minitab output for the mean and median of Data Set I

```
Mean of DATASETI = 483.85

Median of DATASETI = 400.00
```

To obtain the mode of Data Set I, we first apply Minitab's tally procedure as follows.

1 Choose **Stat ➤ Tables ➤ Tally...**
2 Specify DATASETI in the **Variables** text box
3 Select the **Counts** check box
4 Click **OK**

The output that results is shown in Printout 3.2.

PRINTOUT 3.2
Minitab output for the tally procedure

```
DATASETI  Count
     300      6
     400      2
     450      2
     800      1
     940      1
    1050      1
      N=     13
```

Referring to the `Count` column in Printout 3.2, we see that the greatest frequency of occurrence is 6 and that 300 is the only value occurring with that frequency. So the mode of the 13 salaries in Data Set I is $300. ◨

† Once we have obtained the mean, it is easier to obtain the median by editing the last command dialogue: Choose **Edit ➤ Edit Last Dialog**, select the **Median** option button from the **Statistic** field, and then click **OK**.

EXERCISES 3.1

STATISTICAL CONCEPTS AND SKILLS

3.1 Explain in detail the purpose of a measure of center.

3.2 Name and describe the three most important measures of center.

3.3 Of the mean, median, and mode, which is the only one appropriate for use with qualitative data?

3.4 True or false: The mean, median, and the mode can all be used with quantitative data.

3.5 Consider the data set 1, 2, 3, 4, 5, 6, 7, 8, 9.

a. Obtain the mean and median of the data.

b. Replace the 9 in the data set by 99 and again compute the mean and median. Decide which measure of center works better here and explain your answer.

c. For the data set in part (b), the mean is neither central nor typical for the data. The lack of what property of the mean accounts for this?

3.6 Complete the following statement: A descriptive measure is resistant if . . . .

3.7 The U.S. Department of Housing and Urban Development and the U.S. Bureau of the Census compile information on new privately owned one-family houses. According to the document *Characteristics of New Housing,* in 1996 the mean floor space of such homes was 2120 sq ft and the median was 1950 sq ft. Which measure of center do you think is more appropriate? Justify your answer.

3.8 The Board of Governors of the Federal Reserve System publishes information on family net worth in *Federal Reserve Bulletin.* In 1995, the mean net worth of families in the United States was \$205.9 thousand and the median net worth was \$56.4 thousand. Which measure of center do you think is more appropriate? Explain your answer.

In Exercises 3.9–3.12, determine the mean, median, and mode(s) for each of the data sets by hand, that is, using only basic calculator functions. For the mean and the median, round each answer to one more decimal place than that used for the observations.

3.9 The National Center for Education Statistics surveys college and university libraries to obtain information on the number of volumes held. The number of volumes, in thousands, for a sample of seven public colleges and universities are as follows.

79	516	24	265	41	15	411

3.10 The American Hospital Association publishes figures on the costs to community hospitals per patient per day in *Hospital Statistics.* A sample of 10 such costs in New York yields the data, in dollars, below.

669	1134	816	1023	582
1023	786	1053	849	1163

3.11 The U.S. National Science Foundation, Division of Science Resources Studies, collects data on the ages of recipients of science and engineering doctoral degrees. Results are published in *Survey of Earned Doctorates.* A sample of one year's recipients yields the following ages.

37	28	36	33	37	43
41	28	24	44	27	24

3.12 The U.S. National Center for Health Statistics compiles data on the length of stay by patients in short-term hospitals and publishes its findings in *Vital and Health Statistics.* A random sample of nine patients yielded the following data on length of stay, in days.

4	12	18	9	12	6	7	3	55

3.13 According to the *World Almanac, 1998,* the National Collegiate Athletic Association wrestling champions for the years 1968–1997 are as follows.

Year	Champion	Year	Champion
1968	Oklahoma State	1983	Iowa
1969	Iowa State	1984	Iowa
1970	Iowa State	1985	Iowa
1971	Oklahoma State	1986	Iowa
1972	Iowa State	1987	Iowa State
1973	Iowa State	1988	Arizona State
1974	Oklahoma	1989	Oklahoma State
1975	Iowa	1990	Oklahoma State
1976	Iowa	1991	Iowa
1977	Iowa State	1992	Iowa
1978	Iowa	1993	Iowa
1979	Iowa	1994	Oklahoma State
1980	Iowa	1995	Iowa
1981	Iowa	1996	Iowa
1982	Iowa	1997	Iowa

a. Determine the mode of the champion data.

b. Would it be appropriate to use either the mean or the median here? Explain your answer.

3.14 According to the *World Almanac, 1998,* the all-time top television programs by rating (percentage of TV-owning households tuned in to the program) are as shown in the table at the top of the next column. [SOURCE: Nielsen Media Research.]

a. Determine the mode of the network data.

b. Would it be appropriate to use either the mean or the median here? Explain your answer.

Program	Telecast date	Network	Rating (%)	Audience (millions)
M*A*S*H (last episode)	02/28/83	CBS	60.2	50.2
Dallas (Who shot J.R.?)	11/21/80	CBS	53.3	41.5
Roots-Pt. 8	01/30/77	ABC	51.1	36.4
Super Bowl XVI	01/24/82	CBS	49.1	40.0
Super Bowl XVII	01/30/83	NBC	48.6	40.5
XVII Winter Olympics-2d Wed.	02/23/94	CBS	48.5	45.7
Super Bowl XX	01/26/86	NBC	48.3	41.5
Gone With the Wind-Pt.1	11/07/76	NBC	47.7	34.0
Gone With the Wind-Pt.2	11/08/76	NBC	47.4	33.8
Super Bowl XII	01/15/78	CBS	47.2	34.4
Super Bowl XIII	01/21/79	NBC	47.1	35.1
Bob Hope Christmas Show	01/15/70	NBC	46.6	27.3
Super Bowl XVIII	01/22/84	CBS	46.4	38.8
Super Bowl XIX	01/20/85	ABC	46.4	39.4
Super Bowl XIV	01/20/80	CBS	46.3	35.3
Super Bowl XXX	01/28/96	NBC	46.0	44.2
ABC Theatre (The Day After)	11/20/83	ABC	46.0	38.6
Roots-Pt. 6	01/28/77	ABC	45.9	32.7
The Fugitive	08/29/67	ABC	45.9	25.7
Super Bowl XXI	01/25/87	CBS	45.8	40.0

Extending the Concepts and Skills

Ordinal data are data about order or rank. Most statisticians recommend using the median for indicating the center of an ordinal data set, but some researchers also use the mean. In Exercises 3.15 and 3.16, we have presented ordinal data sets. For each exercise,

a. compute the mean of the data.

b. compute the median of the data.

c. decide which of the two measures of center is best.

3.15 A distance runner entered seven marathons. His finishing places in the first six races were 4, 5, 3, 2, 7, and 4. In the seventh race, he decided to go all out to win and ran in first place for 20 miles. This tired him out so badly that he ended up walking parts of the last 6 miles. He did finish, but in 72nd place. The runner's places provide the following ordinal data set: 4, 5, 3, 2, 7, 4, 72.

3.16 Twenty-one algebra students were asked to rate the change in "test anxiety" produced by their algebra course. Negative ratings meant that they worried more about tests at the end of the course than at the beginning, whereas positive ratings meant they worried less at the end of the course than at the beginning. Their ratings were as follows.

0	0	0	−1	−1	1	0
−2	1	0	−2	0	−2	2
−2	1	0	−1	−3	−3	−1

3.17 Some data sets contain **outliers,** observations that fall well outside the overall pattern of the data. (Outliers are discussed in more detail in Section 3.4.) Suppose, for instance, you are interested in the ability of high-school algebra students to compute square roots. You decide to give a square-root exam to 10 of these students. Unfortunately, one of the students had a fight with his girlfriend and cannot concentrate—he gets a 0. The 10 scores are displayed in order in the following table. The score of 0 is an outlier.

0	58	61	63	67	69	70	71	78	80

Statisticians have a systematic method for avoiding extreme observations and outliers when they calculate means. They compute **trimmed means,** in which high and low observations are deleted or "trimmed off" before the mean is calculated. For instance, to compute the 10% trimmed mean of the test-score data, we first delete both the bottom 10% and the top 10% of the ordered data, that is, 0 and 80. Then we calculate the mean of the remaining data. Thus the 10% trimmed mean of the test-score data is

$$\frac{58 + 61 + 63 + 67 + 69 + 70 + 71 + 78}{8} = 67.1.$$

The following table displays a set of algebra final-exam scores for a 40-question test.

2	15	16	16	19	21	21	25	26	27
4	15	16	17	20	21	24	25	27	28

a. Do any of the scores look like outliers?
b. Compute the usual mean of the data.
c. Compute the 5% trimmed mean of the data.
d. Compute the 10% trimmed mean of the data.
e. Compare the means you obtained in parts (b)–(d). Which of the three means do you think provides the best measure of center for the data?

The midrange. Another measure of center is the midrange. The **midrange** of a data set is defined as the mean of the minimum (smallest) and maximum (largest) observations in the data set:

$$\text{Midrange} = \frac{\text{Min} + \text{Max}}{2}.$$

For instance, the midrange of the four exam scores, 88, 75, 95, and 100, in Example 3.4(a) on page 130 is

$$\text{Midrange} = \frac{75 + 100}{2} = 87.5.$$

We will apply the midrange in Exercises 3.18 and 3.19.

3.18 Determine the midrange of each of the following data sets.
a. The ages in Exercise 3.11.
b. The lengths of stay in Exercise 3.12.

3.19 Identify advantages and disadvantages of the midrange as a measure of center.

The modal class. Suppose we are given a data set that is already grouped into a frequency distribution and we do not have access to the raw data. Then, in general, it is not possible to determine the mode(s). In such cases we can find the modal class instead. The **modal class** is defined as the class(es) having the largest frequency. Note that the mode of a data set may or may not be contained in the modal class. We will apply the concept of the modal class in Exercises 3.20–3.22.

3.20 A frequency distribution for the number of days to maturity for 40 short-term investments is displayed in the following table.

Days to maturity	Frequency
30 < 40	3
40 < 50	1
50 < 60	8
60 < 70	10
70 < 80	7
80 < 90	7
90 < 100	4

a. Determine the modal class.
b. Now refer to the raw data presented in Table 2.1 on page 66. Obtain the mode of the data.
c. Is the mode contained in the modal class?

3.21 The following table gives a frequency distribution for the cholesterol levels of 20 patients with high levels.

Cholesterol level	Frequency
195–199	1
200–204	3
205–209	4
210–214	7
215–219	4
220–224	1

a. Determine the modal class.
b. Now refer to the raw data presented in Table 2.18 on page 97. Find the mode of the data.
c. Is the mode contained in the modal class?

3.22 Regarding data that are grouped in a frequency distribution:
a. There is a case in which it is always possible to obtain the mode of a data set from the frequency distribution. When is that?
b. True or false: Suppose all the classes in the frequency distribution are based on a single value. Then the modal class and the mode are identical.

USING TECHNOLOGY  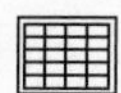 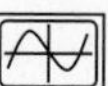

3.23 Use Minitab or some other statistical software to determine the mean, median, and mode of the age data in Exercise 3.11.

3.24 Use Minitab or some other statistical software to determine the mean, median, and mode of the length-of-stay data in Exercise 3.12.

3.25 Use Minitab or some other statistical software to determine the mode of the champion data in Exercise 3.13.

3.26 Use Minitab or some other statistical software to determine the mode of the network data in Exercise 3.14.

Obtain the mean, median, and mode(s) for each of the data sets in Exercises 3.27–3.30 using any technology that you have available.

3.27 The average retail price for bananas in 1994 was 46.0 cents per pound, as reported by the U.S. Department of Agriculture in *Food Cost Review.* Recently, a random sample of 15 markets gave the following prices for bananas in cents per pound.

51	48	50	48	45
52	53	49	43	42
45	52	52	46	50

3.28 The Bureau of Labor Statistics collects data on employment and hourly earnings in private industry groups and publishes its findings in *Employment and Earnings.* Twenty people working in the manufacturing industry are selected at random; their hourly earnings, in dollars, are as follows.

16.70	7.44	13.78	16.49	7.49
17.92	17.21	5.51	10.40	10.75
15.27	19.72	10.68	13.10	14.70
15.55	16.67	15.07	14.02	15.99

3.29 The *Physician's Handbook* provides statistics on heights and weights of children by age. The heights, in inches, of 20 randomly selected 6-year-old girls follow.

44	44	47	46	38
42	46	41	50	43
40	51	47	43	47
48	48	45	41	46

3.30 As reported by the College Entrance Examination Board in *National College-Bound Senior,* the mean (non-recentered) verbal score on the Scholastic Assessment Test (SAT) in 1995 was 428 points out of a possible 800. A random sample of 25 verbal scores for last year yielded the following data.

344	494	350	376	313
489	358	383	498	556
379	301	432	560	494
418	483	444	477	420
492	287	434	514	613

3.2 THE SAMPLE MEAN

In this section we will discuss the sample mean in more detail. We will also introduce some mathematical notation that is useful for expressing the formula for the sample mean and many other descriptive measures.

To begin, we note that in statistics, as in algebra, we use letters such as x, y, and z to denote variables. So, for instance, if we are studying heights and weights of college students, we might let x denote the variable "height" and y denote the variable "weight."

In Definition 3.1 we expressed the definition of the mean of a data set in words: The mean of a data set is the sum of the observations divided by the number of observations. Using mathematical notation, we can express such definitions much more concisely. First we introduce the mathematical notation for "sum of the observations."

EXAMPLE 3.6 SUMMATION NOTATION

The exam scores for the student in Example 3.4(a) are 88, 75, 95, and 100. If we let x denote the variable "exam score," then the symbol x_i denotes the ith observation in the data set. For the exam scores, we have

$$\begin{aligned} x_1 &= \text{score on Exam } 1 = 88, \\ x_2 &= \text{score on Exam } 2 = 75, \\ x_3 &= \text{score on Exam } 3 = 95, \\ x_4 &= \text{score on Exam } 4 = 100. \end{aligned}$$

More simply, we can just write $x_1 = 88$, $x_2 = 75$, $x_3 = 95$, and $x_4 = 100$. The numbers 1, 2, 3, and 4 written below the xs are called **subscripts.** Using this notation, the sum of the exam-score data can be expressed symbolically as

$$x_1 + x_2 + x_3 + x_4.$$

We can use **summation notation** to obtain a shorthand description for this sum. The notation uses the uppercase Greek letter Σ (sigma). That letter, which corresponds to the English letter S, stands for the phrase "the sum of." Thus in place of the lengthy expression $x_1 + x_2 + x_3 + x_4$, we can use Σx, read as "summation x" or "the sum of the observations of the variable x."

For the exam-score data,

$$\Sigma x = x_1 + x_2 + x_3 + x_4 = 88 + 75 + 95 + 100 = 358.$$

 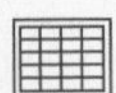 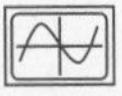

In words, the sum of the four exam scores is 358. □

For clarity it is sometimes useful to incorporate subscripts into the summation notation. This can be done by writing Σx_i instead of Σx. The subscript i is a generic subscript. To be even more precise, we can use *indices* and write $\sum_{i=1}^{n} x_i$, where n denotes the number of observations.

Notation for a Sample Mean

Recall that a data set obtained by observing the values of a variable for a sample of a population is called *sample data* and that the mean of sample data is called a *sample mean.* The symbol used for a sample mean is a bar over the letter representing the variable. So, for a variable x, a sample mean is denoted by $\bar{x}$, read as "x bar." If we also use the letter n to denote the **sample size** or, equivalently, the number of observations, then we can express the definition of a sample mean very concisely as in Definition 3.4.

DEFINITION 3.4 SAMPLE MEAN

For a variable x, the mean of the observations for a sample is called a ***sample mean*** and is denoted $\bar{x}$. Symbolically, we have

$$\bar{x} = \frac{\Sigma x}{n},$$

where n is the sample size.

EXAMPLE 3.7 ILLUSTRATES DEFINITION 3.4

Each year, automobile manufacturers perform mileage tests on their new car models and submit the results of their analyses to the Environmental Protection Agency (EPA). The EPA then tests the vehicles to find whether the manufacturers' results are correct.

In 1998, one company reported that a particular model, equipped with a four-speed manual transmission, averaged 29 miles per gallon (mpg) on the highway. Let's suppose the EPA tested 15 of the cars and obtained the gas mileages shown in Table 3.5. Determine the sample mean of these gas mileages.

TABLE 3.5 Gas mileages (mpg)

27.3	31.2	29.4	31.6	28.6
30.9	29.7	28.5	27.8	27.3
25.9	28.8	28.9	27.8	27.6

SOLUTION Summing the gas mileages in Table 3.5, we obtain $\Sigma x = 431.3$. Since the sample size (number of observations) is 15, we have $n = 15$. Thus

$$\bar{x} = \frac{\Sigma x}{n} = \frac{431.3}{15} = 28.75.$$

The mean gas mileage of the sample of 15 cars is 28.75 mpg. ◨

Other Important Sums

We must often find sums other than the sum of the observations, Σx. One such sum is the sum of the squares of the observations, Σx^2. In Section 3.3, we will need to obtain Σx, Σx^2, and various other sums. So that we can concentrate on the concepts to be presented there instead of the computations, we will discuss computing those sums now.

EXAMPLE 3.8 OTHER IMPORTANT SUMS

The exam-score data from Example 3.6 are repeated in the first column of Table 3.6. The remaining columns of the table contain some related quantities whose significance will become apparent in Section 3.3.

TABLE 3.6 Exam-score data and related quantities

x	x^2	$x - \bar{x}$	$(x - \bar{x})^2$
88	7,744	−1.5	2.25
75	5,625	−14.5	210.25
95	9,025	5.5	30.25
100	10,000	10.5	110.25
358	32,394	0	353.00

In Example 3.6 we found that the sum of the exam-score data is 358, a fact which we record at the bottom of the first column of Table 3.6. The second column of Table 3.6 displays the squares, x^2, of the exam scores. The sum of those squares is 32,394; that is, $\Sigma x^2 = 32{,}394$.

To obtain the third column of Table 3.6, we must first compute the mean, $\bar{x}$, of the four exam scores. Since $n = 4$ and $\Sigma x = 358$,

$$\bar{x} = \frac{\Sigma x}{n} = \frac{358}{4} = 89.5.$$

Subtracting 89.5 from each of the four exam scores in the first column of Table 3.6, we get the $x - \bar{x}$ values shown in the third column. The sum of those values is 0; that is, $\Sigma(x - \overline{x}) = 0$. The fourth column of Table 3.6 gives the squares, $(x - \overline{x})^2$, of the $x - \overline{x}$ values. The sum of those squares is 353; that is, $\Sigma(x - \overline{x})^2 = 353$. □

EXERCISES 3.2

STATISTICAL CONCEPTS AND SKILLS

3.31 Explain in your own words why it is useful to have mathematical notation.

3.32 Explain what each of the following symbols represents.
a. Σ **b.** n **c.** $\overline{x}$

3.33 For a given population, is the population mean a variable? What about a sample mean?

3.34 Let $x_1 = 1, x_2 = 7, x_3 = 4, x_4 = 5$, and $x_5 = 10$.
a. Compute Σx. **b.** Find n. **c.** Determine $\overline{x}$.

3.35 Let $x_1 = 12, x_2 = 8, x_3 = 9$, and $x_4 = 17$.
a. Compute Σx. **b.** Find n. **c.** Determine $\overline{x}$.

For each data set in Exercises 3.36–3.39,
a. compute Σx.
b. find n.
c. determine the sample mean. Round your answer to one more decimal place than that used for the observations.

3.36 As reported by the R. R. Bowker Company of New York in *Library Journal,* the mean annual subscription rate to law periodicals was \$97.33 in 1995. A sample of this year's law periodicals yields the following subscription rates to the nearest dollar.

106	122	120	123
118	114	138	131
128	124	119	130

3.37 In 1995 the average annual gasoline and motor-oil expenditure per U.S. consumer unit (families and single consumers) was \$1006, as reported by the U.S. Bureau of Labor Statistics in *Consumer Expenditure Survey.* That same year a sample of eight high-income consumer units yielded the following gasoline and motor-oil expenditures, in dollars.

744	1688	2107	1866
1496	2045	1792	1233

population

3.38 A team of medical researchers has developed an exercise program to help reduce hypertension. To ascertain whether the program is effective, the team selects a sample of 10 hypertensive individuals and places them on the exercise program for 1 month. The following table displays the diastolic blood pressures of the 10 hypertensive individuals before they began the exercise program.

106	118	118	99	109
94	109	95	97	106

3.39 In 1908 W. S. Gosset published the article "The Probable Error of a Mean" (*Biometrika,* Vol. 6, pp. 1–25). It is in this pioneering paper, published under the pseudonym "Student," that he introduced what later became known as Student's t-distribution, which we will discuss in Chapter 8. As an example, Gosset used the following data set, which gives the additional sleep in hours obtained by a sample of 10 patients using laevohysocyamine hydrobromide.

1.9	0.8	1.1	0.1	−0.1
4.4	5.5	1.6	4.6	3.4

In Exercises 3.40 and 3.41,
a. compute $\overline{x}$.
b. compute Σx^2, $\Sigma(x - \overline{x})$, and $\Sigma(x - \overline{x})^2$ by constructing a table similar to Table 3.6.

3.40 Five families have the following number of children: 2, 3, 4, 4, 3.

3.41 The amount of money, in dollars, a salesperson earned on six randomly selected days yielded the following data set: 75, 98, 130, 63, 115, 107.

Extending the Concepts and Skills

3.42 Explain the difference between the quantities $(\Sigma x)^2$ and Σx^2. Construct an example to show that, in general, those two quantities are unequal.

3.43 Explain the difference between the quantities Σxy and $\Sigma x \Sigma y$. Provide an example to show that, in general, those two quantities are unequal.

3.44 For the exam-score data in Example 3.8, we found that $\Sigma(x - \bar{x}) = 0$. Explain why this is true for any data set. *(Hint:* Write out the sum and use the fact that $\Sigma x = n\bar{x}$.)

3.3 MEASURES OF VARIATION; THE SAMPLE STANDARD DEVIATION

Up to this point, we have discussed only descriptive measures of center, specifically, the mean, median, and mode. However, two data sets can have the same mean, median, or mode and yet still be quite different in other respects. For example, consider the heights of the five starting players on each of two men's college basketball teams, as shown in Fig. 3.3.

FIGURE 3.3
Five starting players on each of two men's college basketball teams and their heights

The two teams have the same mean heights, 75 inches (6′ 3″); the same median heights, 76 inches (6′ 4″); and the same modes, 76 inches (6′ 4″). Nonetheless, it is clear that the two data sets differ. In particular, the heights of the players on Team II vary much more than those on Team I. To describe that difference quantitatively, we use a descriptive measure that indicates the amount of variation or spread in a data set. Such descriptive measures are referred to as **measures of variation** or **measures of spread.**

Just as there are several different measures of center, there are also several different measures of variation. In this section we will examine two of the most frequently used measures of variation—the *range* and *sample standard deviation.* We will begin with the range since it is the simplest to understand and compute.

The Range

The contrast between the heights of the two teams shown in Fig. 3.3 becomes clear if we place the shortest player on each team next to the tallest, as shown in Fig. 3.4.

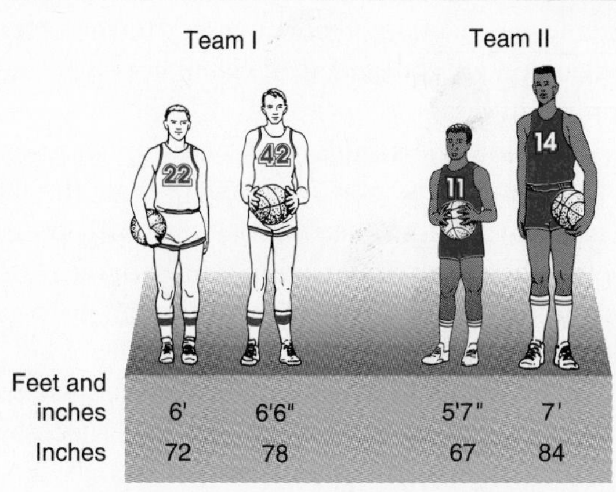

FIGURE 3.4 Shortest and tallest starting players on each of two men's college basketball teams and their heights

The **range** of a data set is obtained by computing the difference between the maximum (largest) and minimum (smallest) observations. Hence, as we see from Fig. 3.4,

$$\text{Team I: Range} = 78 - 72 = 6 \text{ inches},$$
$$\text{Team II: Range} = 84 - 67 = 17 \text{ inches}.$$

DEFINITION 3.5 RANGE OF A DATA SET

The ***range*** of a data set is the difference between its maximum and minimum observations: Range = Max − Min.

 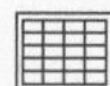

The range of a data set is quite easy to compute. However, in using the range, a great deal of information is ignored—only the largest and smallest observations are considered; the other observations are disregarded.

For that reason, two other measures of variation, the *standard deviation* and the *interquartile range,* are generally favored over the range. The standard deviation is the preferred measure of variation when the mean is used as the measure of center; the interquartile range is preferred when the median is used as the measure of center. As we said, the standard deviation will be discussed in this section; the interquartile range will be considered in Section 3.4.

The Sample Standard Deviation

In contrast to the range, the standard deviation takes into account all of the observations. The calculations required to determine a standard deviation are more involved than those needed to obtain a range. However, this problem is not serious because most computers and statistical calculators have built-in functions to do the necessary computations.

Roughly speaking, the **standard deviation** measures variation by indicating how far, on the average, the observations are from the mean. For a data set with a large amount of variation, the observations will, on the average, be far from the mean; hence the standard deviation will be large. For a data set with a small amount of variation, the observations will, on the average, be close to the mean; consequently, the standard deviation will be small.

To compute the standard deviation of a data set, we need to know whether it is population data or sample data. This information is necessary because the formulas for the standard deviations of sample data and population data differ slightly. In this section we will concentrate on the sample standard deviation. The population standard deviation will be discussed in Section 3.5.

The first step in computing a sample standard deviation is to find how far each observation is from the mean, the **deviations from the mean.** We show how to calculate the deviations from the mean in Example 3.9.

EXAMPLE 3.9 DEVIATIONS FROM THE MEAN

The heights, in inches, of the five starting players on Team I are 72, 73, 76, 76, and 78, as we see from Fig. 3.3 on page 142. Find the deviations from the mean.

SOLUTION The mean height of the starting players on Team I is

$$\bar{x} = \frac{\Sigma x}{n} = \frac{72 + 73 + 76 + 76 + 78}{5} = \frac{375}{5} = 75 \text{ inches.}$$

To obtain the deviation from the mean for a particular observation, we subtract the mean from it; that is, we compute $x - \bar{x}$. For instance, the deviation from the mean for the height of 72 inches is $x - \overline{x} = 72 - 75 = -3$. The deviations from the mean for all five observations are given in the second column of Table 3.7 and are displayed graphically in Fig. 3.5. ◨

TABLE 3.7 Deviations from the mean

Height x	Deviation from the mean $x - \overline{x}$
72	−3
73	−2
76	1
76	1
78	3

FIGURE 3.5 Graphical display of the deviations from the mean (dots represent observations)

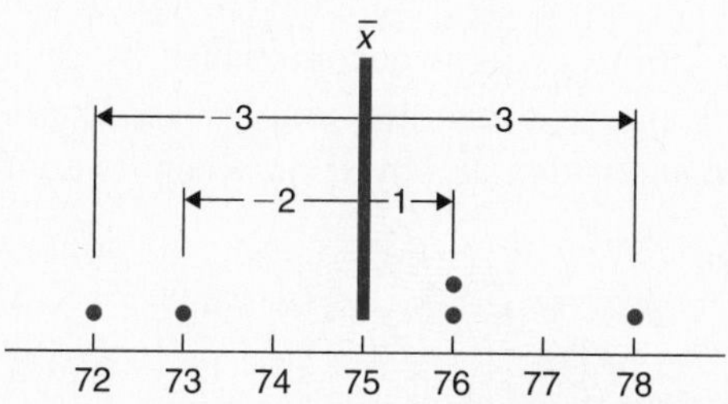

The second step in computing a sample standard deviation is to obtain a measure of the total deviation from the mean for all the observations. Although the quantities $x - \overline{x}$ represent deviations from the mean, adding them to get a total deviation from the mean is of no value because their sum, $\Sigma(x - \overline{x})$, always equals zero. Summing the data in the second column of Table 3.7 shows this to be true for the height data of Team I but, in fact, it is true in general.

In computing a sample standard deviation, the deviations from the mean, $x - \overline{x}$, are squared to obtain quantities that do not sum to zero. The sum of the squared deviations from the mean, $\Sigma(x - \overline{x})^2$, is called the **sum of squared deviations** and provides a measure of total deviation from the mean for all the observations.

EXAMPLE 3.10 SUM OF SQUARED DEVIATIONS

Compute the sum of squared deviations for the heights of the starting players on Team I.

SOLUTION In Table 3.8 we have appended a column for $(x - \overline{x})^2$ to Table 3.7.

TABLE 3.8
Table for computing the sum of squared deviations for the heights of Team I

Height x	Deviation from mean $x - \bar{x}$	Squared deviation $(x - \bar{x})^2$
72	−3	9
73	−2	4
76	1	1
76	1	1
78	3	9
		24

From the third column of Table 3.8, we find that $\Sigma(x - \bar{x})^2 = 24$. The sum of squared deviations is 24 inches^2. ◨

The third step in computing a sample standard deviation is to take an average of the squared deviations. This is accomplished by dividing the sum of squared deviations by $n - 1$, one less than the sample size. The resulting quantity is called a **sample variance** and is denoted by s_x^2 or, when no confusion can arise, by s^2. In symbols,

$$s^2 = \frac{\Sigma(x - \bar{x})^2}{n - 1}.$$

Note: If we divided by n instead of by $n - 1$, then the sample variance would be the mean of the squared deviations. Although dividing by n seems more natural, we divide by $n - 1$ for the following reason: One of the main uses of the sample variance is to estimate the population variance (to be defined in Section 3.5). Division by n tends to underestimate the population variance, whereas division by $n - 1$ gives, on the average, the correct value.

EXAMPLE 3.11 SAMPLE VARIANCE

Obtain the sample variance of the heights of the starting players on Team I.

SOLUTION From Example 3.10 the sum of squared deviations is 24 inches^2. Noting that $n = 5$, we find that

$$s^2 = \frac{\Sigma(x - \bar{x})^2}{n - 1} = \frac{24}{5 - 1} = 6.$$

The sample variance is 6 inches^2. ◨

It is important to realize that a sample variance is in units that are the square of the original units. This results from squaring the deviations from the mean. For instance, as we know from Example 3.11, the sample variance of the heights of the players on Team I is 6 inches^2. Since it is desirable to have descriptive measures in the original units, the final step in computing a sample standard deviation is to take the square root of the sample variance. In other words, the **sample standard deviation,** denoted by s_x or s, is

$$s = \sqrt{\frac{\Sigma(x - \overline{x})^2}{n - 1}}.$$

EXAMPLE 3.12 SAMPLE STANDARD DEVIATION

Determine the sample standard deviation of the heights of the starting players on Team I.

SOLUTION From Example 3.11 the sample variance is 6 inches^2. Thus the sample standard deviation equals

$$s = \sqrt{\frac{\Sigma(x - \overline{x})^2}{n - 1}} = \sqrt{6} = 2.4 \text{ inches},$$

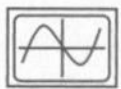

rounded to the nearest tenth of an inch.

Definition 3.6 summarizes our discussion of the sample standard deviation.

DEFINITION 3.6 SAMPLE STANDARD DEVIATION

For a variable x, the standard deviation of the observations for a sample is called a ***sample standard deviation.*** It is denoted by s_x or, when no confusion will arise, simply by s. We have

$$s = \sqrt{\frac{\Sigma(x - \overline{x})^2}{n - 1}},$$

where n is the sample size.

The steps required to obtain a sample standard deviation were illustrated in Examples 3.9–3.12. The computations were performed in four separate examples to explain the interpretation of the sample standard deviation as well as the calculations involved. Now that we have done that, we can present a simple procedure for computing a sample standard deviation.

Step 1 *Calculate the sample mean,* $\overline{x}$.

Step 2 *Construct a table to obtain the sum of squared deviations,* $\Sigma(x - \overline{x})^2$.

Step 3 *Apply Definition 3.6 to determine the sample standard deviation,* s.

We apply this three-step procedure in the next example.

EXAMPLE 3.13 ILLUSTRATES DEFINITION 3.6

The heights, in inches, of the five starting players on Team II are 67, 72, 76, 76, and 84. Obtain the sample standard deviation of these heights.

SOLUTION We apply the three-step procedure just described.

Step 1 *Calculate the sample mean,* $\overline{x}$.

We have

$$\overline{x} = \frac{\Sigma x}{n} = \frac{67 + 72 + 76 + 76 + 84}{5} = \frac{375}{5} = 75 \text{ inches.}$$

Step 2 *Construct a table to obtain the sum of squared deviations,* $\Sigma(x - \overline{x})^2$.

Table 3.9 provides a table with columns for x, $x - \overline{x}$, and $(x - \overline{x})^2$. From the third column, we see that $\Sigma(x - \overline{x})^2 = 156$ inches^2.

TABLE 3.9
Table for computing the sum of squared deviations for the heights of Team II

x	$x - \overline{x}$	$(x - \overline{x})^2$
67	−8	64
72	−3	9
76	1	1
76	1	1
84	9	81
		156

Step 3 *Apply Definition 3.6 to determine the sample standard deviation,* s.

We have $n = 5$ and $\Sigma(x - \overline{x})^2 = 156$. Consequently, the sample standard deviation of the heights for Team II equals

$$s = \sqrt{\frac{\Sigma(x - \overline{x})^2}{n - 1}} = \sqrt{\frac{156}{5 - 1}} = \sqrt{39} = 6.2 \text{ inches,}$$

rounded to the nearest tenth of an inch. □

In Examples 3.12 and 3.13, we found that the sample standard deviations of the heights of the starting players on Teams I and II are, respectively, 2.4 inches and 6.2 inches. Hence we see that Team II, which has more variation in height than Team I, also has a larger standard deviation. That is the way a measure of variation is supposed to work.

KEY FACT 3.1 VARIATION AND THE STANDARD DEVIATION

The more variation there is in a data set, the larger its standard deviation.

Key Fact 3.1 shows that the standard deviation satisfies the basic criterion for a measure of variation and, in fact, it is the most commonly used measure of variation. However, the standard deviation does have its drawbacks. For instance, it is not resistant—its value can be strongly affected by a few extreme observations.

A Computing Formula for s

We are going to present an alternative formula for obtaining a sample standard deviation. Thus it will be convenient to have a name for the original formula given in Definition 3.6. Since that is the formula used to define the sample standard deviation, we will call it the *defining formula* for s. The alternative formula for obtaining a sample standard deviation is given in Formula 3.1. We will call that formula the *computing formula* for s.

FORMULA 3.1 COMPUTING FORMULA FOR A SAMPLE STANDARD DEVIATION

A sample standard deviation can be computed using the formula

$$s = \sqrt{\frac{\Sigma x^2 - (\Sigma x)^2/n}{n - 1}},$$

where n is the sample size.

The computing formula for s is equivalent to the defining formula—both formulas give the same answer, although differences due to roundoff error are possible. However, the computing formula is usually faster and easier for doing calculations by hand and also reduces the chance for roundoff error.

Before illustrating the computing formula for s, we comment on the similar-looking expressions, Σx^2 and $(\Sigma x)^2$, that occur in that formula. The expression Σx^2 represents the sum of the squares of the data; it is obtained by first squaring each observation and then summing those squared values. The expression $(\Sigma x)^2$ represents the square of the sum of the data; it is obtained by first summing the observations and then squaring that sum.

We also emphasize that in the numerator of the computing formula, the division of $(\Sigma x)^2$ by n should be performed before the subtraction from Σx^2. In other words, first compute $(\Sigma x)^2/n$ and then subtract the result from Σx^2.

EXAMPLE 3.14 ILLUSTRATES FORMULA 3.1

In Example 3.13 we obtained the sample standard deviation of the heights for the five starting players on Team II using the defining formula for s. Obtain that sample standard deviation using the computing formula.

SOLUTION To apply the computing formula for s, we need the sums Σx and Σx^2. These are determined in Table 3.10.

TABLE 3.10 Table for computation of s using the computing formula

x	x^2
67	4,489
72	5,184
76	5,776
76	5,776
84	7,056
375	28,281

We have $n = 5$, and from the bottom row of Table 3.10 we see that $\Sigma x = 375$ and $\Sigma x^2 = 28{,}281$. Thus by Formula 3.1,

$$s = \sqrt{\frac{\Sigma x^2 - (\Sigma x)^2/n}{n-1}} = \sqrt{\frac{28{,}281 - (375)^2/5}{5-1}}$$

$$= \sqrt{\frac{28{,}281 - 28{,}125}{4}} = \sqrt{\frac{156}{4}} = \sqrt{39} = 6.2 \text{ inches,}$$

 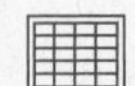

rounded to the nearest tenth of an inch.

We have now obtained the sample standard deviation of the heights of the players on Team II in two ways—using the defining formula (Example 3.13) and using the computing formula (Example 3.14). As we see, both formulas give the same value of 6.2 inches for the sample standard deviation. For these height data, either formula is relatively easy to apply. However, for most data sets, and especially for those in which the mean is not a whole number, the computing formula is preferable.

Here is an important rule to remember when obtaining a sample standard deviation or, for that matter, any other descriptive measure using only basic calculator functions: *Do not perform any rounding until the computation is complete; otherwise, significant roundoff error can result.*

Further Interpretation of the Standard Deviation

As we know, the standard deviation is a measure of variation—the more variation there is in a data set, the larger its standard deviation. Table 3.11 displays two data sets, each having 10 observations. A brief inspection of the table reveals that Data Set II has more variation than Data Set I.

TABLE 3.11
Data sets having different variation

Data Set I	51	44	41	58	48	47	53	47	45	66
Data Set II	37	61	49	20	70	53	48	48	50	64

We computed the sample mean and sample standard deviation of each data set and summarized the results in Table 3.12. As expected, the standard deviation of Data Set II is larger than that of Data Set I.

TABLE 3.12
Means and standard deviations of the data sets in Table 3.11

Data Set I	Data Set II
$\bar{x} = 50.0$	$\bar{x} = 50.0$
$s = 7.4$	$s = 14.2$

To enable us to visually compare the variations in the two data sets, we have drawn the graphs in Figs. 3.6 and 3.7. On each graph we have marked the observations with dots. In addition, we have located the sample mean, $\bar{x} = 50$, and have measured off intervals equal in length to the standard deviation: 7.4 for Data Set I and 14.2 for Data Set II.

FIGURE 3.6 Data Set I; $\bar{x} = 50$, $s = 7.4$

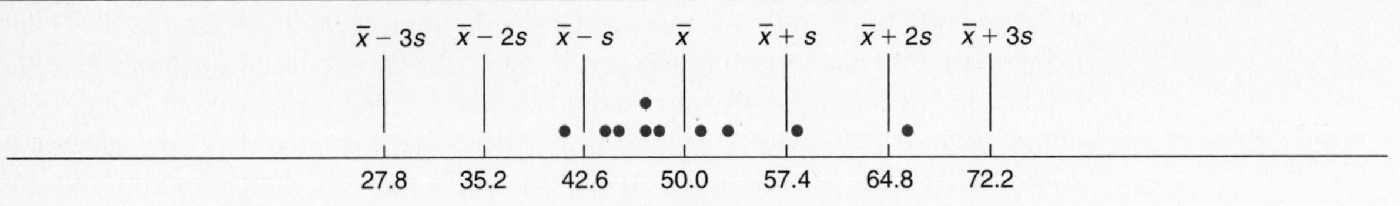

FIGURE 3.7 Data Set II; $\bar{x} = 50$, $s = 14.2$

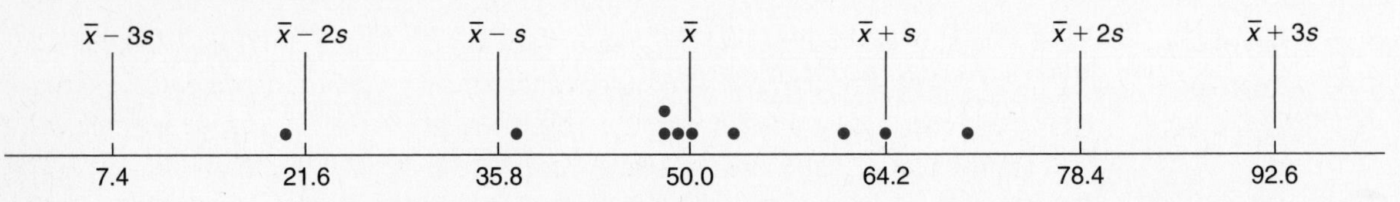

Let's examine Fig. 3.6. Notice that the horizontal position labeled $\bar{x} + 2s$ represents the number that is two standard deviations to the right of the mean, which in this case is

$$\bar{x} + 2s = 50.0 + 2 \cdot 7.4 = 50.0 + 14.8 = 64.8.^{\dagger}$$

Likewise, the horizontal position labeled $\bar{x} - 3s$ represents the number that is three standard deviations to the left of the mean, which in this case is

$$\bar{x} - 3s = 50.0 - 3 \cdot 7.4 = 50.0 - 22.2 = 27.8.$$

Figure 3.7 is interpreted in a similar manner.

The graphs in Figs. 3.6 and 3.7 vividly illustrate that Data Set II has more variation than Data Set I. They also show that for each data set, all observations lie within a few standard deviations to either side of the mean. This is no accident.

KEY FACT 3.2 **THREE-STANDARD-DEVIATIONS RULE**

Almost all of the observations in any data set lie within three standard deviations to either side of the mean.

A data set with a great deal of variation will have a large standard deviation and, consequently, three standard deviations to either side of its mean will be extensive, as in Fig. 3.7. A data set with little variation will have a small standard deviation and, hence, three standard deviations to either side of its mean will be narrow, as in Fig. 3.6.

† Recall that for an expression of the form $a \pm b \cdot c$, the multiplication should be done before the addition or the subtraction. Thus $50.0 + 2 \cdot 7.4 = 50.0 + 14.8 = 64.8$.

The three-standard-deviations rule is somewhat vague—what does "almost all" mean? It can be made more precise in several ways. We can apply **Chebychev's rule** which implies, in particular, that at least 89% of the observations in any data set lie within three standard deviations to either side of the mean. And, if the distribution of the data is approximately bell-shaped, we can apply the **empirical rule** which implies, in particular, that roughly 99.7% of the observations lie within three standard deviations to either side of the mean. These applications will be considered in the exercises for this section.

Using the Computer (Optional)

In this section we discussed two measures of variation: the range and sample standard deviation. For purposes of understanding, it is essential to calculate a few of these and other descriptive measures by hand. However, in practice, since such calculations are usually tedious, we generally employ a computer or statistical calculator to obtain descriptive measures whenever possible. Our next example shows how Minitab can be used to obtain a range and sample standard deviation.

EXAMPLE 3.15 USING MINITAB TO OBTAIN THE RANGE AND SAMPLE STANDARD DEVIATION

The heights of the five starting players on Team II are 67, 72, 76, 76, and 84 inches. Use Minitab to obtain the range and sample standard deviation of these heights.

SOLUTION First we store the height data in a column named TEAMII. Then we proceed in the following manner to obtain the range.

1 Choose **Calc ➤ Column Statistics...**
2 Select the **Range** option button from the **Statistic** field
3 Click in the **Input variable** text box and specify TEAMII
4 Click **OK**

Using the same steps except selecting the **Standard deviation** option button instead of the **Range** option button, we obtain the sample standard deviation. The resulting output is displayed in Printout 3.3.

PRINTOUT 3.3 Minitab output for the range and standard deviation

```
Range of TEAMII = 17.000

Standard deviation of TEAMII = 6.2450
```

From Printout 3.3 we see that the range of the heights for the starting players on Team II is 17 inches and the sample standard deviation of the heights is 6.2450 inches. ❑

EXERCISES 3.3

STATISTICAL CONCEPTS AND SKILLS

3.45 Explain the purpose of a measure of variation.

3.46 Why is the standard deviation preferable to the range as a measure of variation?

3.47 When we use the standard deviation as a measure of variation, what is the reference point?

3.48 Discuss one major drawback to the standard deviation as a measure of variation.

3.49 Consider the data set 1, 2, 3, 4, 5, 6, 7, 8, 9.

a. Obtain the sample standard deviation of the data using the defining formula.

b. Replace the 9 in the data set by 99 and again compute the sample standard deviation using the defining formula.

c. Comparing your answers in parts (a) and (b), the lack of what property of the standard deviation accounts for its extreme sensitivity to the change of 9 to 99?

3.50 Consider the following four data sets.

Data Set I		Data Set II		Data Set III		Data Set IV	
1	5	1	9	5	5	2	4
1	8	1	9	5	5	4	4
2	8	1	9	5	5	4	4
2	9	1	9	5	5	4	10
5	9	1	9	5	5	4	10

a. Compute the mean of each data set.

b. Although the four data sets have the same means, they are quite different in another respect. How are they different?

c. Which data set appears to have the least variation? the greatest variation?

d. Compute the range of each data set.

e. Compute the sample standard deviation of each data set using the defining formula.

f. From your answers to parts (d) and (e), which measure of variation better distinguishes the spread in the four data sets—the range or the standard deviation? Explain your answer.

g. Are your answers from parts (c) and (e) consistent?

3.51 Below are 10 IQ scores.

110	122	132	107	101
97	115	91	125	142

Time each of the following calculations.

a. Obtain the sample standard deviation of the 10 IQs using the defining formula.

b. Obtain the sample standard deviation of the 10 IQs using the computing formula.

c. Did the computing formula save time? Explain why it did or did not.

3.52 Consider the data set 3, 3, 3, 3, 3, 3.

a. Guess the value of the sample standard deviation without calculating it? Explain your reasoning.

b. Calculate the sample standard deviation using the defining formula.

c. Complete the following statement and explain your reasoning: If all observations in a data set are equal, then the sample standard deviation is ____.

d. Complete the following statement and explain your reasoning: If the sample standard deviation of a data set is 0, then . . . .

In Exercises 3.53–3.56, we have repeated the data from Exercises 3.9–3.12. For each exercise obtain, using only basic calculator functions, the

a. range of the data.

b. sample standard deviation of the data using the defining formula.

c. sample standard deviation of the data using the computing formula.

d. State which formula you found easier to use in obtaining s.

Note: In parts (b) and (c), round your final answers to one more decimal place than that used for the data.

3.53 The National Center for Education Statistics surveys college and university libraries to obtain information on the number of volumes held. The number of volumes, in thousands, for a sample of seven public colleges and universities are as follows.

79	516	24	265	41	15	411

3.54 The American Hospital Association publishes figures on the costs to community hospitals per patient per day in *Hospital Statistics.* A sample of 10 such costs in New York yields the data, in dollars, below.

669	1134	816	1023	582
1023	786	1053	849	1163

3.55 The U.S. National Science Foundation, Division of Science Resources Studies, collects data on the ages of recipients of science and engineering doctoral degrees. Results are published in *Survey of Earned Doctorates.* A sample of one year's recipients yields the following ages.

37	28	36	33	37	43
41	28	24	44	27	24

3.56 The U.S. National Center for Health Statistics compiles data on the length of stay by patients in short-term hospitals and publishes its findings in *Vital and Health Statistics.* A random sample of nine patients yielded the following data on length of stay, in days.

4	12	18	9	12	6	7	3	55

EXTENDING THE CONCEPTS AND SKILLS

3.57 In Exercise 3.17 on page 135, we discussed *outliers,* observations that fall well outside the overall pattern of the data. The following table contains two data sets. Data Set II is obtained by removing the outliers from Data Set I.

Data Set I	Data Set II
0 12 14 15 23	10 14 15 17
0 14 15 16 24	12 14 15
10 14 15 17	14 15 16

a. Compute the sample standard deviation of each of the two data sets.
b. Compute the range of each of the two data sets.
c. What effect do outliers have on variation? Explain your answer.

3.58 Another measure of variation is the **mean absolute deviation (MAD).** The MAD of a data set is the mean of the absolute values of the deviations from the mean:

$$\text{MAD} = \frac{\Sigma|x - \bar{x}|}{n}.$$

a. Obtain the MAD of the ages in Exercise 3.55.
b. Obtain the MAD of the stays in Exercise 3.56.

Chebychev's rule. A more precise version of the three-standard-deviations rule (Key Fact 3.2 on page 152) can be obtained from *Chebychev's rule,* which can be stated as follows:

> **CHEBYCHEV'S RULE** For any data set and any number $k > 1$, at least $100(1 - 1/k^2)\%$ of the observations lie within k standard deviations to either side of the mean.

Two special cases of Chebychev's rule are applied frequently, namely, when $k = 2$ and $k = 3$. These state, respectively, that:

- At least 75% of the observations in any data set lie within two standard deviations to either side of the mean.
- At least 89% of the observations in any data set lie within three standard deviations to either side of the mean.

We will discuss and apply Chebychev's rule in Exercises 3.59–3.62.

3.59 Verify that the two statements in the above bulleted list are indeed special cases of Chebychev's rule with $k = 2$ and $k = 3$, respectively.

3.60 Consider the data sets portrayed in Figs. 3.6 and 3.7 on page 152.

a. Chebychev's rule says that at least 75% of the observations lie within two standard deviations of the mean. What percentage of the observations portrayed in Fig. 3.6 actually lie within two standard deviations of the mean?
b. Chebychev's rule says that at least 89% of the observations lie within three standard deviations of the mean. What percentage of the observations portrayed in Fig. 3.6 actually lie within three standard deviations of the mean?

c. Repeat parts (a) and (b) for the data portrayed in Fig. 3.7.

d. From parts (a)–(c), we see that Chebychev's rule does not necessarily provide precise estimates for the percentage of observations that lie within a specified number of standard deviations to either side of the mean. Nonetheless, Chebychev's rule is quite important for several reasons. Can you think of some?

3.61 Consider the following sample of exam scores, arranged in increasing order.

28	57	58	64	69	74
79	80	83	85	85	87
87	89	89	90	92	93
94	94	95	96	96	97
97	97	97	98	100	100

(*Note:* The sample mean and sample standard deviation of these exam scores are, respectively, 85 and 16.1.)

a. Use Chebychev's rule to estimate the percentage of the observations that lie within two standard deviations to either side of the mean.

b. Use the data to obtain the exact percentage of the observations that lie within two standard deviations to either side of the mean. Compare your answer here to that in part (a).

c. Use Chebychev's rule to estimate the percentage of the observations that lie within three standard deviations to either side of the mean.

d. Use the data to obtain the exact percentage of the observations that lie within three standard deviations to either side of the mean. Compare your answer here to that in part (c).

3.62 Chebychev's rule also permits us to make pertinent statements about a data set when we know only its mean and standard deviation; and frequently that is all we do know. Here is an example of this use of Chebychev's rule. The R. R. Bowker Company of New York compiles information on costs of new books and publishes its findings in *Library Journal.* A sample of 40 sociology books taken this year has a mean cost of \$53.75 and a standard deviation of \$10.42. Use this information and the two aforementioned special cases of Chebychev's rule to complete the following statements.

a. At least 30 of the 40 sociology books cost between _____ and _____.

b. At least _____ of the 40 sociology books cost between \$22.49 and \$85.01.

The empirical rule. For data sets that have approximately bell-shaped distributions, we can improve on the estimates given by Chebychev's rule by using the *empirical rule,* which is as follows:

> **EMPIRICAL RULE** For any data set having approximately a bell-shaped distribution, we have that:
>
> - Roughly 68% of the observations lie within one standard deviation to either side of the mean.
> - Roughly 95% of the observations lie within two standard deviations to either side of the mean.
> - Roughly 99.7% of the observations lie within three standard deviations to either side of the mean.

We will discuss and apply the empirical rule in Exercises 3.63–3.66.

3.63 This exercise compares Chebychev's rule and the empirical rule.

a. Compare the estimates given by the two rules for the percentage of observations that lie within two standard deviations of the mean. Comment on the differences.

b. Compare the estimates given by the two rules for the percentage of observations that lie within three standard deviations of the mean. Comment on the differences.

3.64 The following table gives the crime rates per 1000 population, arranged in increasing order, for the states in the United States.

25	27	27	31	33
33	33	35	37	39
40	41	41	41	43
43	44	44	45	45
46	47	48	48	49
49	50	51	51	53
53	53	54	56	56
56	57	59	60	60
60	61	62	62	63
67	67	67	79	83

(*Note:* The sample mean and sample standard deviation of these crime rates are, respectively, 49.48 and 12.75.)

a. Use the empirical rule to estimate the percentages of the observations that lie within one, two, and three standard deviations to either side of the mean.

b. Use the data to obtain the exact percentages of the observations that lie within one, two, and three standard deviations to either side of the mean.

c. Compare your answers in parts (a) and (b).

d. A stem-and-leaf diagram of the crime rates is presented in Exercise 2.117 on page 111. Based on that diagram, comment on your comparisons in part (c).

e. Is it appropriate to use the empirical rule for these data? Explain your answer.

3.65 Refer to the exam scores displayed in Exercise 3.61.

a. Use the empirical rule to estimate the percentages of the observations that lie within one, two, and three standard deviations to either side of the mean.

b. Use the data to obtain the exact percentages of the observations that lie within one, two, and three standard deviations to either side of the mean.

c. Compare your answers in parts (a) and (b).

d. Construct a histogram or stem-and-leaf diagram of the data. Based on your graph, comment on your comparisons in part (c).

e. Is it appropriate to use the empirical rule for these data? Explain your answer.

3.66 Refer to Exercise 3.62. Assuming the distribution of costs for the 40 sociology books is approximately bell-shaped, apply the empirical rule to complete the following statements and compare your answers to those obtained in Exercise 3.62 where Chebychev's rule was used.

a. Roughly 38 of the 40 sociology books cost between _____ and _____.

b. Roughly _____ of the 40 sociology books cost between \$22.49 and \$85.01.

Grouped-data formulas. When data are grouped in a frequency distribution, we use formulas different from the ones we have previously discussed to obtain a sample mean and sample standard deviation.

GROUPED-DATA FORMULAS

$$\bar{x} = \frac{\Sigma xf}{n} \quad \text{and} \quad s = \sqrt{\frac{\Sigma(x - \bar{x})^2 f}{n - 1}}$$

where x denotes class midpoint, f class frequency, and n $(= \Sigma f)$ the sample size.

In general, these formulas yield only approximations to the actual sample mean and sample standard deviation. We will discuss the grouped-data formulas in Exercises 3.67 and 3.68.

3.67 In the following table, we repeat the salary data in Data Set II which we discussed in Example 3.1.

300	300	940	450	400
400	300	300	1050	300

a. Obtain the sample mean and sample standard deviation of this (ungrouped) data set using Definitions 3.4 and 3.6 on pages 139 and 147, respectively.

b. A frequency distribution for Data Set II using single-value grouping is given in the first two columns of the following table. The third column of the table is for the xf-values, that is, class midpoint (which here is the same as the class) times class frequency.

Salary x	Frequency f	Salary · Frequency xf
300	5	1500
400	2	
450	1	
940	1	
1050	1	
	10	

Complete the missing entries in the table and then use the grouped-data formula given above to obtain the sample mean.

c. Compare the answers that you obtained for the sample mean in parts (a) and (b). Explain why the grouped-data formula always yields the actual sample mean when the data are grouped in classes each based on a single value. (*Hint:* What does xf represent for any particular class?)

d. Construct a table similar to the one in part (b) but with columns for x, f, $x - \overline{x}$, $(x - \overline{x})^2$, and $(x - \overline{x})^2 f$. Use the table and the grouped-data formula given above to obtain the sample standard deviation.

e. Compare your answers for the sample standard deviation in parts (a) and (d). Explain why the grouped-data formula always yields the actual sample standard deviation when the data are grouped in classes each based on a single value.

3.68 Following is a grouped-data table for the days to maturity for 40 short-term investments.

Days to maturity	Frequency f	Relative frequency	Midpoint x
30 < 40	3	0.075	35
40 < 50	1	0.025	45
50 < 60	8	0.200	55
60 < 70	10	0.250	65
70 < 80	7	0.175	75
80 < 90	7	0.175	85
90 < 100	4	0.100	95
	40	1.000	

a. Use the grouped-data formulas to estimate the sample mean and sample standard deviation of the days-to-maturity data. Round your final answers to one decimal place.

b. The table below gives the raw days-to-maturity data.

70	64	99	55	64	89	87	65
62	38	67	70	60	69	78	39
75	56	71	51	99	68	95	86
57	53	47	50	55	81	80	98
51	36	63	66	85	79	83	70

Using Definitions 3.4 and 3.6 on pages 139 and 147, respectively, we find that the true sample mean and sample standard deviation of the days-to-maturity data are 68.3 and 16.7, rounded to one decimal place. Compare these actual values of $\overline{x}$ and s to the estimates from part (a). Explain why the grouped-data formulas generally yield only approximations to the sample mean and sample standard deviation for non–single-value grouping.

Using Technology   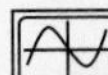

3.69 Use Minitab or some other statistical software to determine the range and sample standard deviation of the age data in Exercise 3.55.

3.70 Use Minitab or some other statistical software to determine the range and sample standard deviation of the length-of-stay data in Exercise 3.56.

Obtain the range and sample standard deviation of each of the data sets in Exercises 3.71–3.74 using any technology that you have available.

3.71 The average retail price for bananas in 1994 was 46.0 cents per pound, as reported by the U.S. Department of Agriculture in *Food Cost Review.* Recently, a random sample of 15 markets gave the following prices for bananas in cents per pound.

51	48	50	48	45
52	53	49	43	42
45	52	52	46	50

3.72 The Bureau of Labor Statistics collects data on employment and hourly earnings in private industry groups and publishes its findings in *Employment and Earnings.* Twenty people working in the manufacturing industry are selected at random; their hourly earnings, in dollars, are as follows.

16.70	7.44	13.78	16.49	7.49
17.92	17.21	5.51	10.40	10.75
15.27	19.72	10.68	13.10	14.70
15.55	16.67	15.07	14.02	15.99

3.73 The *Physician's Handbook* provides statistics on heights and weights of children by age. The heights, in inches, of 20 randomly selected 6-year-old girls follow.

44	44	47	46	38
42	46	41	50	43
40	51	47	43	47
48	48	45	41	46

3.74 As reported by the College Entrance Examination Board in *National College-Bound Senior,* the mean (non-recentered) verbal score on the Scholastic Assessment Test (SAT) in 1995 was 428 points out of a possible 800. A random sample of 25 verbal scores for last year yielded the following data.

344	494	350	376	313
489	358	383	498	556
379	301	432	560	494
418	483	444	477	420
492	287	434	514	613

3.4 THE FIVE-NUMBER SUMMARY; BOXPLOTS

Up to this point, we have concentrated mostly on the mean and standard deviation to measure center and variation. Now we will examine several descriptive measures based on percentiles. Unlike the mean and standard deviation, descriptive measures based on percentiles are resistant—they are not sensitive to the influence of a few extreme observations. For this reason, descriptive measures based on percentiles are often preferred over those based on the mean and standard deviation.

Percentiles, Deciles, and Quartiles

As we learned in Section 3.1, the median of a data set divides the data into two equal parts—the bottom 50% and the top 50%. The **percentiles** of a data set divide it into hundredths, or 100 equal parts. A data set has 99 percentiles, denoted by $P_1, P_2, \ldots, P_{99}$. Roughly speaking, the first percentile, P_1, is the number that divides the bottom 1% of the data from the top 99%; the second percentile, P_2, is the number that divides the bottom 2% of the data from the top 98%; and so forth. Note that the median is the 50th percentile.

Deciles are also useful. The **deciles** of a data set divide it into tenths, or 10 equal parts. A data set has nine deciles, which we denote by $D_1, D_2, \ldots, D_9$. Basically, the first decile, D_1, is the number that divides the bottom 10% of the data from the top 90%; the second decile, D_2, is the number that divides the bottom 20% of the data from the top 80%; and so on. Note that the first decile is the 10th percentile, the second decile is the 20th percentile, and so forth.

The most commonly used percentiles are quartiles. The **quartiles** of a data set divide it into quarters, or four equal parts. A data set has three quartiles, which we denote by Q_1, Q_2, and Q_3. Roughly speaking, the first quartile, Q_1, is the number that divides the bottom 25% of the data from the top 75%; the second quartile, Q_2, is the median, which, as we know, is the number that divides the bottom 50% of the data from the top 50%; and the third quartile, Q_3, is the number that divides the bottom 75% of the data from the top 25%. Note that the first and third quartiles are the 25th and 75th percentiles, respectively.

Figure 3.8 depicts the quartiles for uniform, bell-shaped, right-skewed, and left-skewed distributions.

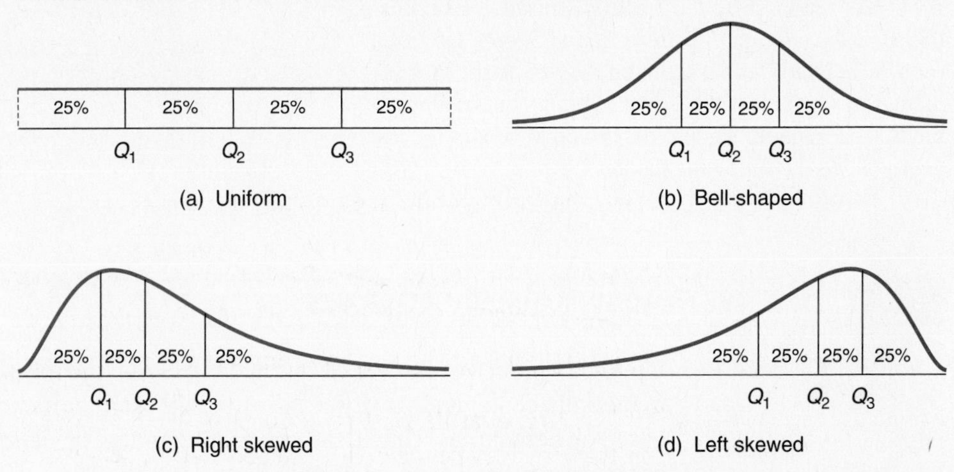

FIGURE 3.8 Quartiles for (a) uniform, (b) bell-shaped, (c) right-skewed, and (d) left-skewed distributions

At this point our intuitive definitions of percentiles and deciles will suffice. However, quartiles need to be defined more precisely, which is done in Definition 3.7.

DEFINITION 3.7 QUARTILES

Let n denote the number of observations. Arrange the data in increasing order.

- The ***first quartile*** is at position $(n + 1)/4$.
- The ***second quartile*** is the median, which is at position $(n + 1)/2$.
- The ***third quartile*** is at position $3(n + 1)/4$.

If a position is not a whole number, linear interpolation is used.

Note: Not all statisticians define quartiles in exactly the same way. We have defined quartiles as Minitab does for consistency in our computer discussion. Other definitions may lead to slightly different values. Be especially aware of this when using statistical software other than Minitab.

EXAMPLE 3.16 ILLUSTRATES DEFINITION 3.7

The A. C. Nielsen Company publishes data on the TV viewing habits of Americans by various characteristics in *Nielsen Report on Television.* A sample of 20 people yields the weekly viewing times, in hours, displayed in Table 3.13.

TABLE 3.13 Weekly TV viewing times

25	41	27	32	43
66	35	31	15	5
34	26	32	38	16
30	38	30	20	21

Determine and interpret the quartiles for these data.

SOLUTION To find the quartiles, we apply Definition 3.7. The number of observations is 20, so $n = 20$. Arranging the data in increasing order, we get the following ordered list:

5 15 16 20 **21** **25** 26 27 30 **30** **31** 32 32 34 **35** **38** 38 41 43 66

The first quartile is at position $(n+1)/4 = (20+1)/4 = 5.25$, that is, one-fourth of the way between the fifth and sixth observations (the first pair of numbers shown in boldface type). Thus the first quartile is obtained by linear interpolation as follows:

$$21 + 0.25 \cdot (25 - 21) = 22.$$

In symbols, $Q_1 = 22$.

The second quartile is at position $(n+1)/2 = (20+1)/2 = 10.5$, that is, halfway between the tenth and eleventh observations (the second pair of numbers shown in boldface type). Thus the second quartile is

$$30 + 0.5 \cdot (31 - 30) = 30.5.$$

In symbols, $Q_2 = 30.5$.

The third quartile is at position $3(n+1)/4 = 3 \cdot (20+1)/4 = 15.75$, that is, three-fourths of the way between the fifteenth and sixteenth observations (the third pair of numbers shown in boldface type). Thus the third quartile is

$$35 + 0.75 \cdot (38 - 35) = 37.25.$$

In symbols, $Q_3 = 37.25$.

Interpreting our results, we conclude that 25% of the viewing times are less than 22 hours, 25% are between 22 and 30.5 hours, 25% are between 30.5 and 37.25 hours, and 25% are greater than 37.25 hours. □

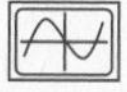

The Interquartile Range

Next we discuss the **interquartile range.** Since the interquartile range is defined using quartiles, it is the preferred measure of variation when the median is used as the measure of center. Like the median, the interquartile range is a resistant measure.

DEFINITION 3.8 INTERQUARTILE RANGE

The ***interquartile range,*** denoted **IQR,** is the difference between the first and third quartiles; that is,

$$\text{IQR} = Q_3 - Q_1.$$

Roughly speaking, the IQR gives the range of the middle 50% of the observations.

EXAMPLE 3.17 ILLUSTRATES DEFINITION 3.8

Obtain the IQR for the TV-viewing-time data displayed in Table 3.13 on page 161.

SOLUTION As we discovered in Example 3.16, the first and third quartiles are 22 and 37.25, respectively. Therefore the interquartile range is

$$Q_3 - Q_1 = 37.25 - 22 = 15.25.$$

In symbols, IQR = 15.25. ◨

The Five-Number Summary

From the three quartiles, we can obtain a measure of center (the median, Q_2) and measures of variation of the two middle quarters of the data, $Q_2 - Q_1$ for the second quarter and $Q_3 - Q_2$ for the third quarter. But the three quartiles don't tell us anything about the variation of the first and fourth quarters.

To gain that information, we need only include the minimum and maximum observations also. Then the variation of the first quarter can be measured as the difference between the minimum and the first quartile, $Q_1 - \text{Min}$, and the variation of the fourth quarter can be measured as the difference between the third quartile and the maximum, $\text{Max} - Q_3$.

For the TV-viewing-time data in Table 3.13, we have Min = 5, $Q_1 = 22$, $Q_2 = 30.5$, $Q_3 = 37.25$, and Max = 66. Consequently, the variations of the four quarters are, respectively, 17, 8.5, 6.75, and 28.75. There is less variation in the middle two quarters than in the first and fourth quarters, and the fourth quarter has the greatest variation of all.

Thus we see that the minimum, maximum, and quartiles together provide, among other things, information on center and variation. Written in increasing order, they comprise what is called the **five-number summary** of a data set.

DEFINITION 3.9 FIVE-NUMBER SUMMARY

The ***five-number summary*** of a data set consists of the minimum, maximum, and quartiles written in increasing order: Min, Q_1, Q_2, Q_3, Max.

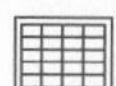

So, for example, the five-number summary of the data on TV-viewing times is given by 5, 22, 30.5, 37.25, and 66.

Outliers

In data analysis it is important to identify **outliers,** observations that fall well outside the overall pattern of the data. An outlier requires special attention: It may be the result of a measurement or recording error, an observation from a different population, or an unusual extreme observation. Note that an extreme observation need not be an outlier; it may instead be an indication of skewness.

As an example of an outlier, consider the data set consisting of the individual wealths (in dollars) of all U.S. residents. For this data set, the wealth of Bill Gates is an outlier, in this case an unusual extreme observation.

When an outlier is observed, we should always try to determine its cause. If it is discovered that an outlier is due to a measurement or recording error or that for some other reason it clearly does not belong in the data set, then the outlier can be removed without further ado. However, if no explanation for the outlier is apparent, then the decision whether to retain it in the data set can often be difficult and calls for a judgment by the researcher.

We can use quartiles and the IQR to identify potential outliers, that is, as a diagnostic tool for spotting observations that may be outliers. To accomplish that we define the lower and upper limits, the numbers that lie, respectively, 1.5 IQRs below the first quartile and 1.5 IQRs above the third quartile.

DEFINITION 3.10 LOWER AND UPPER LIMITS

The ***lower limit*** and ***upper limit*** are defined as follows:

$$\text{Lower limit} = Q_1 - 1.5 \cdot \text{IQR}$$
$$\text{Upper limit} = Q_3 + 1.5 \cdot \text{IQR}$$

Observations that lie outside the lower and upper limits are potential outliers. Further data analysis should be done (e.g., histograms, stem-and-leaf diagrams) to ascertain whether such observations are truly outliers.

For the TV-viewing-time data in Table 3.13, we have $Q_1 = 22$, $Q_3 = 37.25$, and IQR = 15.25. Therefore,

$$\text{Lower limit} = Q_1 - 1.5 \cdot \text{IQR} = 22 - 1.5 \cdot 15.25 = -0.875$$

and

$$\text{Upper limit} = Q_3 + 1.5 \cdot \text{IQR} = 37.25 + 1.5 \cdot 15.25 = 60.125.$$

These limits are portrayed graphically in Fig. 3.9.

FIGURE 3.9
Lower and upper limits for TV-viewing times

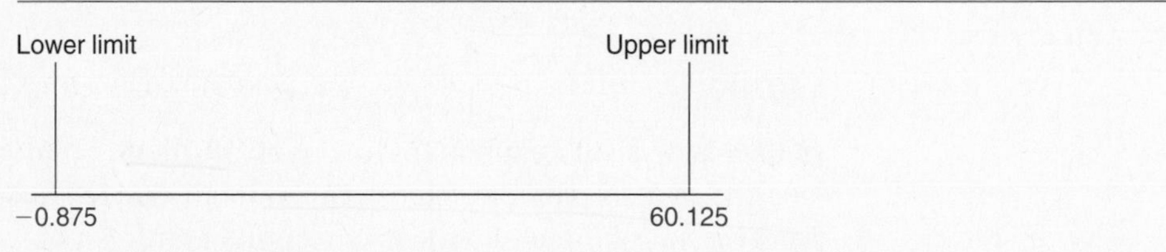

Referring now to Table 3.13, or more simply to the ordered list of the data on page 161, we see that there is one observation, 66, that lies outside the lower and upper limits. Consequently, 66 is a potential outlier. A histogram and a stem-and-leaf diagram both indicate that 66 is indeed an outlier.

Boxplots

A **boxplot,** also called a **box-and-whisker diagram,** is based on the five-number summary and can be used to provide a graphical display of the center and variation of a data set. These diagrams were also invented by Professor John Tukey.

Actually, two types of boxplots are in common use. One is simply called a boxplot; the other is called a **modified boxplot.** The main difference between the two types of boxplots is that potential outliers are plotted individually in a modified boxplot, but not in a boxplot. Thus when outliers are of concern, modified boxplots are the preferred type of boxplot. Procedures 3.1 and 3.2 provide, respectively, step-by-step methods for constructing boxplots and modified boxplots.

PROCEDURE 3.1 TO CONSTRUCT A BOXPLOT

Step 1 Determine the five number summary.

Step 2 Draw a horizontal axis on which the numbers obtained in Step 1 can be located. Above this axis, mark the quartiles and the minimum and maximum with vertical lines.

Step 3 Connect the quartiles to each other to make a box, and then connect the box to the minimum and maximum with lines.

To construct a modified boxplot, we need to introduce the concept of adjacent values. The **adjacent values** of a data set are the most extreme observations still lying within the lower and upper limits, that is, the most extreme observations that are not potential outliers. Note that if a data set has no potential outliers, then the adjacent values are just the minimum and maximum observations.

PROCEDURE 3.2 **TO CONSTRUCT A MODIFIED BOXPLOT**

Step 1 Determine the quartiles.

Step 2 Determine potential outliers and the adjacent values.

Step 3 Draw a horizontal axis on which the numbers obtained in Steps 1 and 2 can be located. Above this axis, mark the quartiles and the adjacent values with vertical lines.

Step 4 Connect the quartiles to each other to make a box, and then connect the box to the adjacent values with lines.

Step 5 Plot each potential outlier with an asterisk.†

If a data set has no potential outliers, then its boxplot and modified boxplot are identical. Also we should point out that in both types of boxplots, the two lines emanating from the box are called **whiskers.**

EXAMPLE 3.18 ILLUSTRATES PROCEDURES 3.1 AND 3.2

The weekly TV-viewing times for a sample of 20 people are displayed in Table 3.13 on page 161. Construct a boxplot and, if appropriate, a modified boxplot.

SOLUTION To obtain a boxplot for the TV-viewing times, we apply the step-by-step method presented in Procedure 3.1.

Step 1 *Determine the five-number summary.*

We have already obtained the five-number summary for the TV-viewing times. It is: Min = 5, $Q_1 = 22$, $Q_2 = 30.5$, $Q_3 = 37.25$, and Max = 66.

Step 2 *Draw a horizontal axis on which the numbers obtained in Step 1 can be located. Above this axis, mark the quartiles and the minimum and maximum with vertical lines.*

This is done in Fig. 3.10(a).

† The symbol used for plotting potential outliers varies.

Step 3 *Connect the quartiles to each other to make a box, and then connect the box to the minimum and maximum with lines.*

This is also done in Fig. 3.10(a).

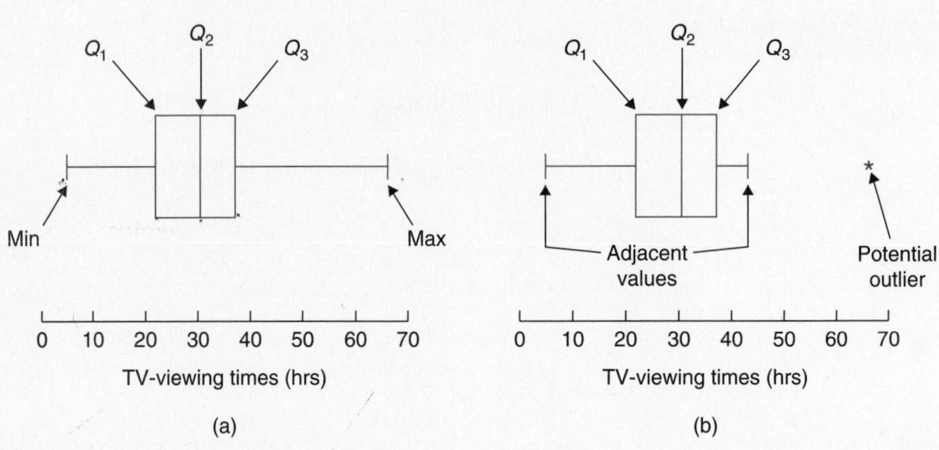

FIGURE 3.10
(a) Boxplot for TV-viewing times
(b) Modified boxplot for TV-viewing times

Figure 3.10(a) is a boxplot for the data on TV-viewing times. Observe that the two boxes in the boxplot indicate the spread of the second and third quarters of the data and that since the ends of the combined box are at the quartiles, the width of the combined box equals the interquartile range, IQR. Also notice that the two whiskers indicate the spread of the first and fourth quarters. Thus we see easily from the boxplot that there is less variation in the middle two quarters of the TV-viewing times than in the first and fourth quarters, and that the fourth quarter has the greatest variation of all.

We discovered on page 164 that the TV-viewing times contain a potential outlier. Therefore it is appropriate to construct a modified boxplot since it will be different from the (ordinary) boxplot we just obtained. We apply Procedure 3.2.

Step 1 *Determine the quartiles for the data.*

The quartiles for the TV-viewing times were obtained in Example 3.16: We have $Q_1 = 22$, $Q_2 = 30.5$, and $Q_3 = 37.25$.

Step 2 *Determine the potential outliers and the adjacent values.*

We found earlier that the TV-viewing-times contain one potential outlier, 66. Referring now to the ordered list of the data on page 161, we see that the adjacent values are 5 and 43.

Step 3 *Draw a horizontal axis on which the numbers obtained in Steps 1 and 2 can be located. Above this axis, mark the quartiles and the adjacent values with vertical lines.*

This is done in Fig. 3.10(b).

Step 4 *Connect the quartiles to each other to make a box, and then connect the box to the adjacent values with lines.*

This is also done in Fig. 3.10(b).

Step 5 *Plot each potential outlier with an asterisk.*

 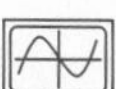

As we noted in Step 2, the data set contains one potential outlier, namely 66. It is plotted with an asterisk in Fig. 3.10(b). ◨

Other Uses of Boxplots

Boxplots are especially suited for comparing two or more data sets. In doing so, it is important to plot all of the boxplots using the same scale. Exercises 3.93 and 3.94 examine this use of boxplots.

We can also use a boxplot to identify the approximate shape of the distribution of a data set. Figure 3.11 at the top of the next page displays uniform, bell-shaped, right-skewed, and left-skewed distributions and their corresponding boxplots. Study Fig. 3.11 carefully, noticing especially how box width and whisker length relate to skewness and symmetry.

Employing boxplots to identify the shape of a distribution is most useful with large data sets. For small data sets, boxplots can be unreliable in identifying distribution shape; it is generally better to use a histogram, stem-and-leaf diagram, or dotplot to ascertain distribution shape for a small data set.

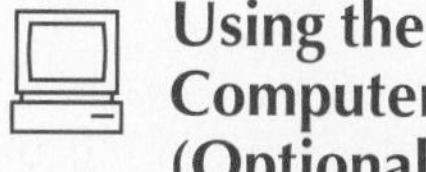

Using the Computer (Optional)

Minitab can be used to obtain a boxplot for one or more data sets. We will illustrate the use of Minitab's boxplot procedure by applying it to the TV-viewing-times data set considered in this section.

EXAMPLE 3.19 USING MINITAB TO OBTAIN A BOXPLOT

Use Minitab to obtain a modified boxplot for the TV-viewing times displayed in Table 3.13 on page 161.

FIGURE 3.11 Distribution shapes and boxplots for (a) uniform, (b) bell-shaped, (c) right-skewed, and (d) left-skewed distributions

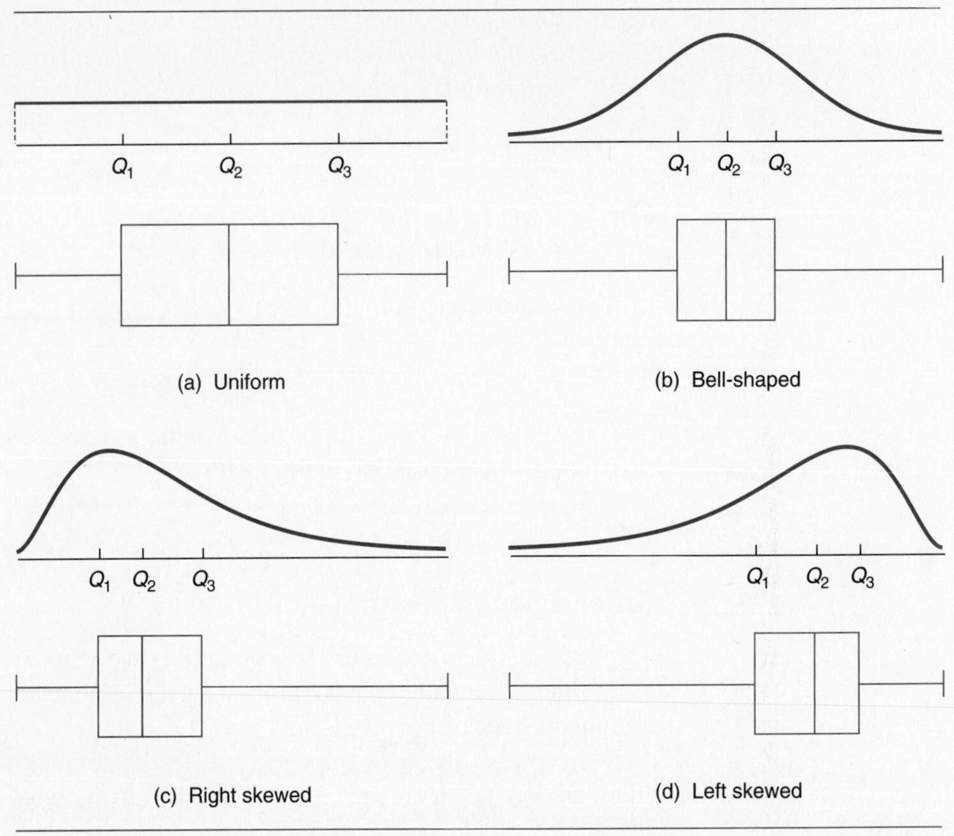

SOLUTION First we store the TV-viewing times in a column named TIMES. Then we proceed in the following way.

1 Choose **Graph ➤ Boxplot...**

2 Specify TIMES in the **Y** text box for **Graph 1**

3 Click **OK**

The resulting output is shown in Printout 3.4.†

Notice that in Minitab's boxplot, the axis scaling the boxplot runs vertically instead of horizontally. This is the default; but it can be changed by doing the following while in the **Boxplot** dialog box: Click the **Options...** button, select the **Transpose X and Y** check box, and click **OK**.

† To obtain modified boxplots like the ones we have drawn by hand, we must have **IQRange Box** and **Outlier Symbol** in the **Display** text boxes for **Item 1** and **Item 2**, respectively, and **Graph** in the **For each** text boxes for **Item 1** and **Item 2**. These are the Minitab defaults; if they have been changed by you or another user, you should reset the defaults by pressing the F3 key before beginning your interaction with the **Boxplot** dialog box.

PRINTOUT 3.4
Minitab boxplot for the TV-viewing-times data

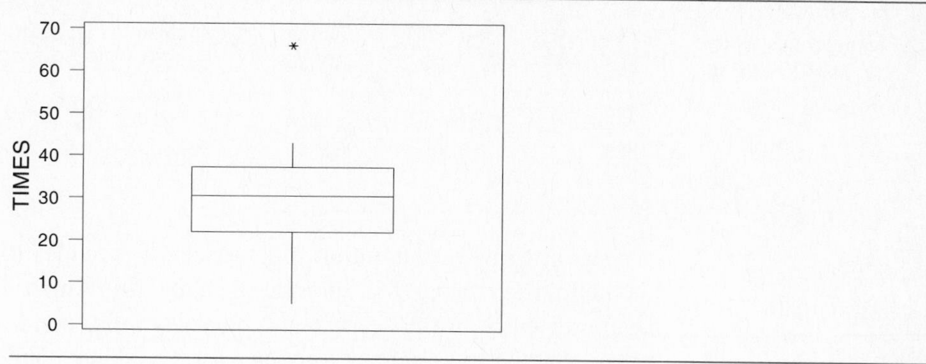

Compare the boxplot in Printout 3.4 to the one we obtained by hand in Fig. 3.10(b) on page 166. ◨

We can use a boxplot generated by Minitab to estimate the five-number summary (Min, Q_1, Q_2, Q_3, Max), since the five numbers are all represented in the boxplot. But we don't have to be content with estimating those numbers because Minitab can be used to simultaneously obtain the five-number summary (and several other descriptive measures as well, including the mean and sample standard deviation). Example 3.20 shows how this is done for the TV-viewing-times data.

EXAMPLE 3.20 USING MINITAB TO OBTAIN THE FIVE-NUMBER SUMMARY

Consider once again the data on TV-viewing times displayed in Table 3.13 on page 161. Use Minitab to determine the five-number summary for the data.

SOLUTION In Example 3.19 we stored the TV-viewing times in a column named TIMES. To obtain the five-number summary for that data, we proceed as follows.

1 Choose **Stat ➤ Basic Statistics ➤ Display Descriptive Statistics...** (In Release 11, choose **Stat ➤ Basic Statistics ➤ Descriptive Statistics...**)
2 Specify TIMES in the **Variables** text box
3 Click **OK**

Printout 3.5 shows the output obtained.

PRINTOUT 3.5
Minitab descriptive statistics for the TV-viewing-times data

```
Variable        N      Mean    Median    TrMean     StDev   SE Mean
TIMES          20     30.25     30.50     29.67     12.65      2.83

Variable  Minimum   Maximum        Q1        Q3
TIMES        5.00     66.00     22.00     37.25
```

The last row of the output in Printout 3.5 provides the minimum (`Minimum`), maximum (`Maximum`), first quartile (`Q1`), and third quartile (`Q3`), respectively; and the fourth entry in the second row provides the second quartile (`Median`). Thus we see that the five-number summary for the TV-viewing-time data is 5, 22, 30.5, 37.25, and 66. □

As we observe from Printout 3.5, in addition to supplying the five-number summary, the output also provides the sample size (`N`), sample mean (`Mean`), a 5% trimmed mean (`TrMean`), the sample standard deviation (`StDev`), and the estimated standard error of the mean (`SE Mean`), which is $s/\sqrt{n}$.

EXERCISES 3.4

Statistical Concepts and Skills

3.75 Identify an advantage that the median and interquartile range have, respectively, over the mean and standard deviation.

3.76 Explain why we add the minimum and maximum observations to the three quartiles in order to better describe the variation in a data set.

3.77 Is an extreme observation necessarily an outlier? Explain your answer.

3.78 Under what conditions are boxplots useful for identifying the shape of a distribution?

3.79 Regarding the interquartile range:
a. What type of descriptive measure is it?
b. What is it measuring?

3.80 What are the lower and upper limits used for? Explain your answer.

3.81 When is a modified boxplot the same as an ordinary boxplot? Explain your answer.

3.82 Which measure of variation is preferred when
a. the mean is used as a measure of center?
b. the median is used as a measure of center?

In each of Exercises 3.83–3.88, determine the quartiles using only basic calculator functions.

3.83 The exam scores for the students in an introductory statistics course are as follows.

88	67	64	76	86
85	82	39	75	34
90	63	89	90	84
81	96	100	70	96

3.84 The table below gives the hourly temperature readings (in degrees Fahrenheit) for Colorado Springs, Colorado, on Tuesday, August 22, 1978.

69	63	61	74	88	87	74	68
66	61	64	79	87	85	71	65
65	60	70	84	85	73	70	62

3.85 The U.S. National Center for Health Statistics compiles data on the length of stay by patients in short-term hospitals and publishes its findings in *Vital and Health Statistics.* A random sample of 21 patients yielded the following data on length of stay, in days.

4	4	12	18	9	6	12
3	6	15	7	3	55	1
10	13	5	7	1	23	9

3.86 Intelligence quotients (IQs) measured on the Stanford Revision of the Binet-Simon Intelligence Scale have a mean of 100 points and a standard deviation of 16 points. Suppose that 25 randomly selected people are given that IQ test and that the results are as follows.

91	96	106	116	97
102	96	124	115	121
95	111	105	101	86
88	129	112	82	98
104	118	127	66	102

3.87 The U.S. Federal Highway Administration conducts studies on motor vehicle travel by type of vehicle. Results are published annually in *Highway Statistics*. A sample of 15 cars yields the following data on number of miles driven, in thousands, for last year.

10.2	10.3	8.9	12.7	8.3
9.2	13.7	7.7	3.3	10.6
11.8	6.6	8.6	5.7	12.0

3.88 The U.S. Energy Information Administration reports figures on residential energy consumption and expenditures in *Residential Energy Consumption Survey: Consumption and Expenditures.* A sample of 18 households using electricity as their primary energy source yields the following data on one year's energy expenditures, in dollars.

1376	1452	1235	1480	1185	1327
1059	1400	1227	1102	1168	1070
949	1351	1259	1179	1393	1456

In each of Exercises 3.89–3.92,

a. *determine the interquartile range.*
b. *obtain the five-number summary.*
c. *identify potential outliers, if any.*
d. *construct and interpret a boxplot and, if appropriate, a modified boxplot.*

3.89 The exam scores for the 20 students in an introductory statistics class given in Exercise 3.83.

3.90 The data on hourly temperatures for Colorado Springs, Colorado, on Tuesday, August 22, 1978, given in Exercise 3.84.

3.91 The lengths of stay in short-term hospitals by 21 randomly selected patients given in Exercise 3.85.

3.92 The IQs on the Stanford Revision of the Binet-Simon Intelligence Scale for a sample of 25 people given in Exercise 3.86.

Extending the Concepts and Skills

3.93 Surveys are conducted by the Northwestern University Placement Center, Evanston, Illinois, on starting salaries for college graduates. Results of the surveys can be found in *The Northwestern Lindquist-Endicott Report.* The following diagram shows boxplots for the starting annual salaries obtained from samples of 32 computer-science graduates (top boxplot) and 35 liberal-arts graduates (bottom boxplot). The data are in thousands of dollars.

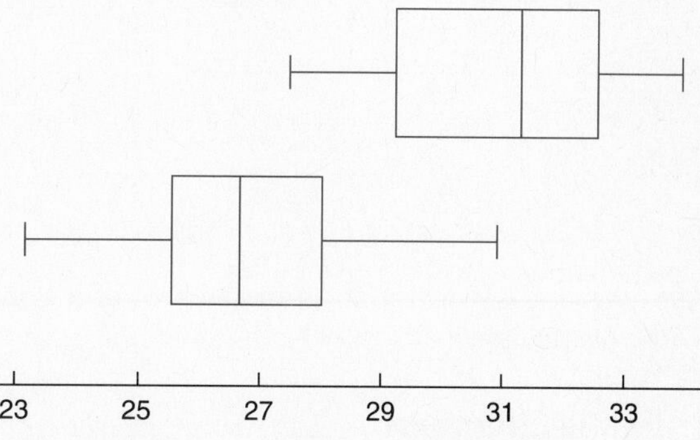

Use the boxplots to compare the starting salaries of the computer-science and liberal-arts graduates sampled.

3.94 Researchers in obesity wanted to compare the effectiveness of dieting with exercise against dieting without exercise. Seventy-three patients were randomly divided into two groups. Group 1, composed of 37 patients, was put on a program of dieting with exercise. Group 2, composed of 36 patients, dieted only. The results for weight loss, in pounds, after 2 months are summarized in the following boxplots. The top boxplot is for Group 1 and the bottom boxplot is for Group 2.

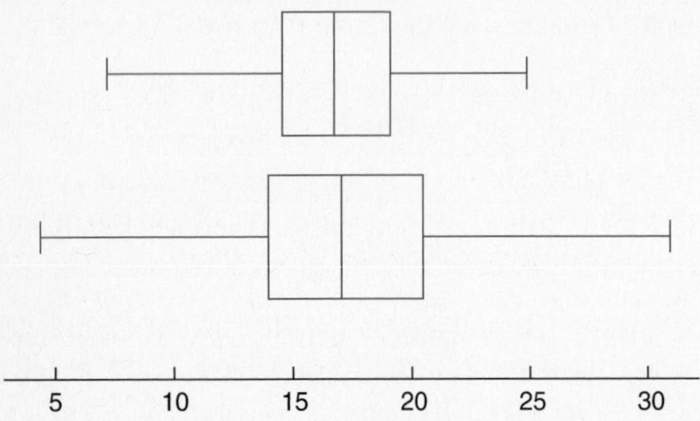

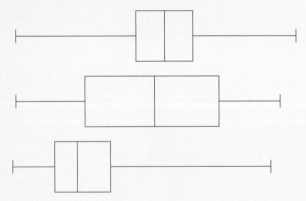

Employ the boxplots to compare the weight losses for the two groups.

3.95 Each boxplot below was obtained from a very large data set. Use the boxplots to identify the approximate shape of the distribution of each data set.

3.96 What can you say about the boxplot of a symmetric distribution?

USING TECHNOLOGY   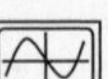

3.97 Refer to the exam-score data in Exercise 3.83. Use Minitab or some other statistical software to

a. obtain a boxplot for the data.

b. determine the five-number summary of the data.

3.98 Refer to the data on hourly temperatures in Exercise 3.84. Use Minitab or some other statistical software to

a. obtain a boxplot for the data.

b. determine the five-number summary of the data.

Obtain a boxplot and the five-number summary for each of the data sets in Exercises 3.99–3.102 using any technology that you have available.

3.99 The average retail price for bananas in 1994 was 46.0 cents per pound, as reported by the U.S. Department of Agriculture in *Food Cost Review.* Recently, a random sample of 15 markets gave the following prices for bananas in cents per pound.

51	48	50	48	45
52	53	49	43	42
45	52	52	46	50

3.100 The Bureau of Labor Statistics collects data on employment and hourly earnings in private industry groups and publishes its findings in *Employment and Earnings.* Twenty people working in the manufacturing industry are selected at random; their hourly earnings, in dollars, are as follows.

16.70	7.44	13.78	16.49	7.49
17.92	17.21	5.51	10.40	10.75
15.27	19.72	10.68	13.10	14.70
15.55	16.67	15.07	14.02	15.99

3.101 *Physician's Handbook* provides statistics on heights and weights of children by age. The heights, in inches, of 20 randomly selected 6-year-old girls follow.

44	44	47	46	38
42	46	41	50	43
40	51	47	43	47
48	48	45	41	46

3.102 As reported by the College Entrance Examination Board in *National College-Bound Senior,* the mean (non-recentered) verbal score on the Scholastic Assessment Test (SAT) in 1995 was 428 points out of a possible 800. A random sample of 25 verbal scores for last year yielded the following data.

344	494	350	376	313
489	358	383	498	556
379	301	432	560	494
418	483	444	477	420
492	287	434	514	613

3.103 The U.S. Department of Agriculture collects data on annual per capita beef consumption and publishes the results in *Food Consumption, Prices, and Expenditures*. Each person in a sample is asked to estimate his or her beef consumption, in pounds, for last year.

a. Printout 3.6 depicts a boxplot, generated by Minitab, for the beef-consumption data. Use the printout to discuss the variation in the data set and to determine approximately the five-number summary.

b. Are there any potential outliers in the data? If so, identify them approximately and provide an explanation for their possible cause.

c. In Printout 3.7 we have displayed the output obtained by applying Minitab's descriptive-statistics procedure to the beef-consumption data. Use the output to determine exactly the five-number summary of the data. Compare your answers to the approximate values you obtained in part (a).

3.104 A prospective investor in a limited partnership wants information on the monthly rental charges for three-bedroom apartments in the area. She obtains the monthly rental charges for a sample of three-bedroom apartments.

a. Printout 3.8 on the next page provides the output obtained by applying Minitab's boxplot procedure to the sample of monthly rental charges. Use the printout to discuss the variation in the data set and to find approximately the five-number summary.

b. Are there any potential outliers in the data? If so, identify them approximately and provide an explanation for their possible cause.

c. Printout 3.9 on the next page presents the output resulting from applying Minitab's descriptive-statistics procedure to the monthly rental-charge data. Use the output to determine exactly the five-number summary of the data. Compare your answers to the approximate values determined in part (a).

PRINTOUT 3.6 Minitab output for Exercise 3.103(a)

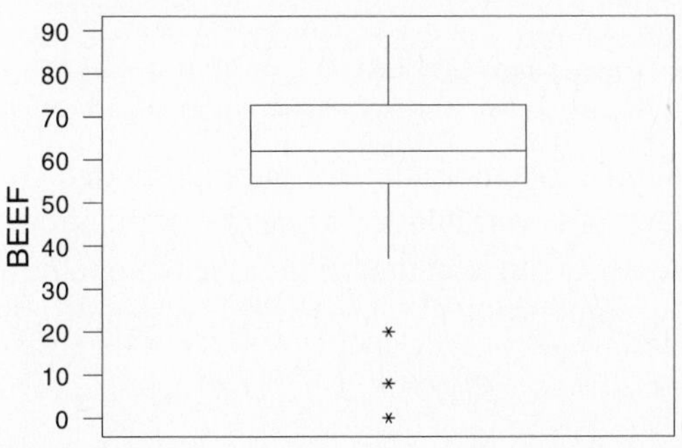

PRINTOUT 3.7 Minitab output for Exercise 3.103(c)

```
Variable          N       Mean     Median     TrMean      StDev    SE Mean
BEEF             40      58.40      62.00      60.22      20.42       3.23

Variable    Minimum    Maximum         Q1         Q3
BEEF           0.00      89.00      54.50      72.75
```

PRINTOUT 3.8 Minitab output for Exercise 3.104(a)

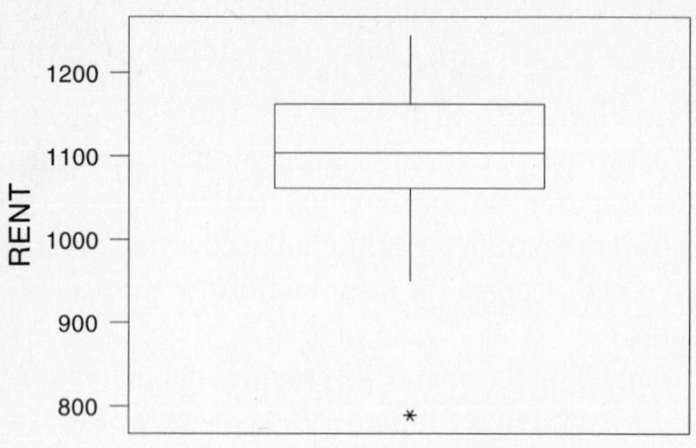

PRINTOUT 3.9 Minitab output for Exercise 3.104(c)

```
Variable        N      Mean    Median    TrMean     StDev   SE Mean
RENT           32    1099.2    1103.5    1105.8      91.2      16.1

Variable  Minimum   Maximum        Q1        Q3
RENT        789.0    1245.0    1061.2    1162.5
```

3.5 DESCRIPTIVE MEASURES FOR POPULATIONS; USE OF SAMPLES

In this section we will discuss several descriptive measures for population data—the data obtained by observing the values of a variable for an entire population. Although in reality we often don't have access to population data, it is nonetheless helpful to see the notation and formulas used for descriptive measures of population data.

The Population Mean

Recall that for a sample of size n, the sample mean is

$$\bar{x} = \frac{\Sigma x}{n}.$$

The mean of a finite population is obtained in the same way: first sum all possible observations of the variable and then divide by the size of the population.

However, to distinguish the population mean from a sample mean, we use the Greek letter μ (pronounced "mew") to denote the mean of the population. We also use the uppercase English letter N to represent the size of the population. Table 3.14 summarizes the notation employed for both a sample and the population.

TABLE 3.14 Notation used for a sample and for a population

	Size	Mean
Sample	n	$\bar{x}$
Population	N	μ

DEFINITION 3.11 **POPULATION MEAN (MEAN OF A VARIABLE)**

For a variable x, the mean of all possible observations for the entire population is called the ***population mean*** or ***mean of the variable*** x. It is denoted by μ_x or, when no confusion will arise, simply by μ. For a finite population, we have

$$\mu = \frac{\Sigma x}{N},$$

where N is the population size.

EXAMPLE 3.21 ILLUSTRATES DEFINITION 3.11

Table 3.15 on the next page presents data for the (active) players on the 1997–98 Dallas Cowboys football team. Obtain the population mean weight of these players.

SOLUTION The sum of the weights in Table 3.15 is 12,859 lb. Since there are 53 players, $N = 53$. Thus

$$\mu = \frac{\Sigma x}{N} = \frac{12{,}859}{53} = 242.6,$$

  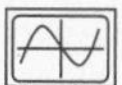

rounded to one decimal place. The population mean weight of the players on the Dallas Cowboys is 242.6 lb. □

TABLE 3.15
Dallas Cowboys, 1997–98

Name	Position	Height	Weight (lb)	College
Richie Cunningham	K	5′10″	167	SW Louisiana
Toby Gowin	P	5′10″	167	North Texas
Troy Aikman	QB	6′4″	219	UCLA
Jason Garrett	QB	6′2″	195	Princeton
Wade Wilson	QB	6′3″	208	East Texas State
Sherman Williams	RB	5′8″	202	Alabama
Emmitt Smith	RB	5′9″	209	Florida
Kevin Mathis	CB	5′9″	172	East Texas State
Omar Stoutmire	S	5′11″	198	Fresno State
Kevin Smith	CB	5′11″	190	Texas A&M
Singor Mobley	S	5′11″	195	Washington State
Darren Woodson	S	6′1″	219	Arizona State
Kenny Wheaton	CB	5′10″	190	Oregon
Brock Marlon	S	5′11″	197	Nevada-Reno
Herschel Walker	FB	6′1″	225	Georgia
Wendell Davis	CB	5′10″	183	Oklahoma
Bill Bates	S	6′1″	213	Tennessee
Charlie Williams	S	6′0″	189	Bowling Green
Nicky Sualua	FB	5′11″	257	Ohio State
Daryl Johnston	FB	6′2″	242	Syracuse
Clay Shiver	C	6′2″	294	Florida State
Broderick Thomas	LB	6′4″	254	Nebraska
Dexter Coakley	LB	5′10″	215	Appalachian State
Darryl Hardy	LB	6′2″	230	Tennessee
Fred Strickland	LB	6′2″	251	Purdue
Randall Godfrey	LB	6′2″	237	Georgia
Vinson Smith	LB	6′2″	248	East Carolina
Alan Campos	LB	6′3″	236	Louisville
Nate Newton	G	6′3″	320	Florida A&M
Leonard Renfro	DT	6′3″	308	Colorado
John Flannery	G/C	6′3″	304	Syracuse
George Hegamin	T	6′7″	331	North Carolina State
Dale Hellestrae	C-G	6′5″	291	Southern Methodist
Mark Tuinei	T	6′5″	314	Hawaii
Larry Allen	G-T	6′3″	326	Sonoma State
Tony Casillas	DT	6′3″	278	Oklahoma
Steve Scifres	G/T	6′4″	300	Wyoming
Erik Williams	T	6′6″	328	Central State (Ohio)
Stepfret Williams	WR	6′0″	170	Northeast Louisiana
Scott Galbraith	TE	6′2″	255	Southern California
Macey Brooks	WR	6′5″	220	James Madison
Anthony Miller	WR	5′11″	190	Tennessee
Eric Bjornson	TE	6′4″	236	Washington
Billy Davis	WR	6′1″	205	Pittsburgh
Michael Irvin	WR	6′2″	207	Miami (Florida)
David LaFleur	TE	6′7″	280	Louisiana State
Darren Benson	DT	6′7″	308	Trinity Valley C.C.
Tony Tolbert	DE	6′6″	263	Texas El-Paso
Chad Hennings	DT	6′6″	291	Air Force
Antonio Anderson	DT	6′6″	318	Syracuse
Kavika Pittman	DE	6′6″	267	McNeese State
Shante Carver	DE	6′5″	263	Arizona State
Hurvin McCormack	DL	6′5″	284	Indiana

Using a Sample Mean to Estimate a Population Mean

Although we usually deal with sample data in inferential studies, the objective is to describe the entire population. The reason for resorting to a sample is that it is generally more practical. We illustrate this point in Example 3.22.

EXAMPLE 3.22 A USE OF A SAMPLE MEAN

The U.S. Bureau of the Census reports the mean (annual) income of U.S. households in its annual publication *Current Population Reports.* To obtain the complete population data—the incomes of all U.S. households—would be extremely expensive and time-consuming. It is also unnecessary, because accurate estimates for the mean income of all U.S. households can be obtained from the mean income of a sample of such households. In reality the Census Bureau samples 60,000 households out of a total of more than 97 million U.S. households.

The variable under consideration here is income and the population of interest is all U.S. households; the mean income of all U.S. households is the population mean, μ. The sample consists of the 60,000 households obtained by the Census Bureau and the mean income of those household is a sample mean, $\bar{x}$. Figure 3.12 summarizes this discussion graphically.

FIGURE 3.12 Population and sample for incomes of U.S. households

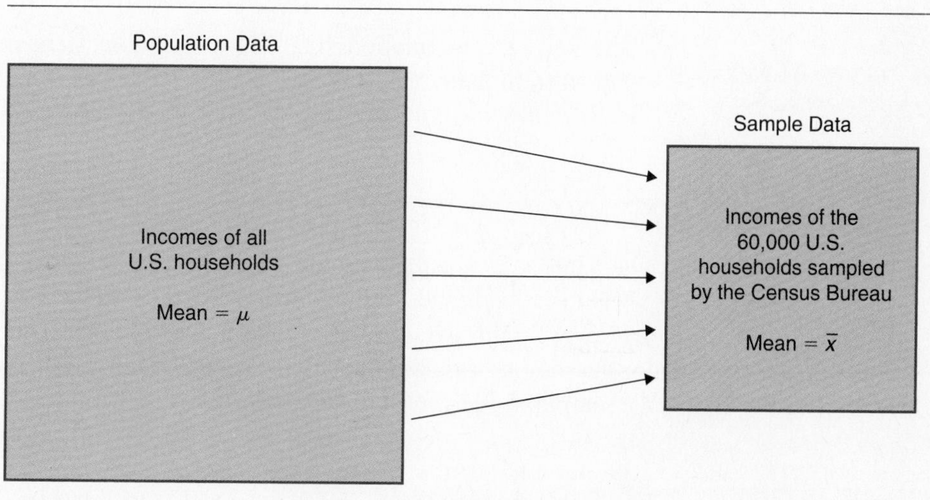

After the sample has been taken, the Census Bureau can compute the sample mean income, $\bar{x}$, of the 60,000 households obtained. Using the value of $\bar{x}$, the

Census Bureau can then estimate the population mean income, μ, of all U.S. households. We will study these kinds of inferences in Chapter 8. ◨

The Population Standard Deviation

Recall that for a sample of size n, the sample standard deviation is

$$s = \sqrt{\frac{\Sigma(x - \bar{x})^2}{n - 1}}.$$

As we will see in a moment, the standard deviation of a finite population is obtained in a similar but slightly different way. To distinguish the population standard deviation from a sample standard deviation, we use the Greek letter σ (pronounced "sigma") to denote the standard deviation of the population.

DEFINITION 3.12 **POPULATION STANDARD DEVIATION (STANDARD DEVIATION OF A VARIABLE)**

For a variable x, the standard deviation of all possible observations for the entire population is called the ***population standard deviation*** or ***standard deviation of the variable*** x. It is denoted by $\boldsymbol{\sigma_x}$ or, when no confusion will arise, simply by $\boldsymbol{\sigma}$. For a finite population, we have

$$\sigma = \sqrt{\frac{\Sigma(x - \mu)^2}{N}},$$

where N is the population size. The population standard deviation can also be obtained from the computing formula

$$\sigma = \sqrt{\frac{\Sigma x^2}{N} - \mu^2}.$$

Note: Just as s^2 is called a sample variance, $\boldsymbol{\sigma^2}$ is called the **population variance.**

Notice that in the defining formula for the population standard deviation, σ, we divide by N, the size of the population. On the other hand, in the defining formula for the sample standard deviation, s, we divide by $n - 1$, one less than the size of the sample. A reason for this was discussed in the note preceding Example 3.11 on page 146.

EXAMPLE 3.23 ILLUSTRATES DEFINITION 3.12

Obtain the population standard deviation of the weights of the players on the Dallas Cowboys football team.

SOLUTION We will apply the computing formula given in Definition 3.12. To implement that formula, we need the sum of the squares of the weights and the population mean weight, μ. From Example 3.21 on page 175, we know that $\mu = 242.6$ lb. Squaring each weight in Table 3.15 and adding the results, we find that $\Sigma x^2 = 3{,}243{,}569$. Recalling that there are 53 players, we have

$$\sigma = \sqrt{\frac{\Sigma x^2}{N} - \mu^2} = \sqrt{\frac{3{,}243{,}569}{53} - (242.6)^2} = 48.4,$$

rounded to one decimal place. The population standard deviation of the weights of the players on the Dallas Cowboys is 48.4 lb. ◨

Note: For purposes of illustration, we used a rounded value for μ to compute the population standard deviation in Example 3.23. However, in practice, so as to avoid roundoff error, no rounding should be done until the computation is complete. Waiting until the end to round, we find that the correct value of the population standard deviation is 48.3 lb (to one decimal place).

Using a Sample Standard Deviation to Estimate a Population Standard Deviation

In Example 3.24, we discuss how a sample standard deviation can be used to estimate a population standard deviation.

EXAMPLE 3.24 A USE OF A SAMPLE STANDARD DEVIATION

A hardware manufacturer produces "10-mm" bolts. The manufacturer knows that the diameters of the bolts produced vary somewhat from 10 mm and also from each other. But even if he is willing to accept some variation in bolt diameters, he cannot tolerate too much variation—if the variation is too large, too many of the bolts will be unusable. In other words, the manufacturer must ensure that the standard deviation of bolt diameters is not unduly large.

As a first step, the manufacturer needs to know the population standard deviation, σ, of bolt diameters. Since in this case it is not possible to determine σ exactly (do you know why?), the manufacturer must use the standard deviation of the diameters of a sample of bolts to estimate σ. To that end, the manufacturer decides to take a sample of 20 bolts.

The variable under consideration here is diameter and the population of interest consists of all "10-mm" bolts that have been or ever will be produced by the manufacturer; the standard deviation of the diameters of all such bolts is the population standard deviation, σ. The sample consists of the 20 bolts obtained by the manufacturer and the standard deviation of the diameters of those bolts is a sample standard deviation, s. Figure 3.13 summarizes this discussion graphically.

FIGURE 3.13 Population and sample for bolt diameters

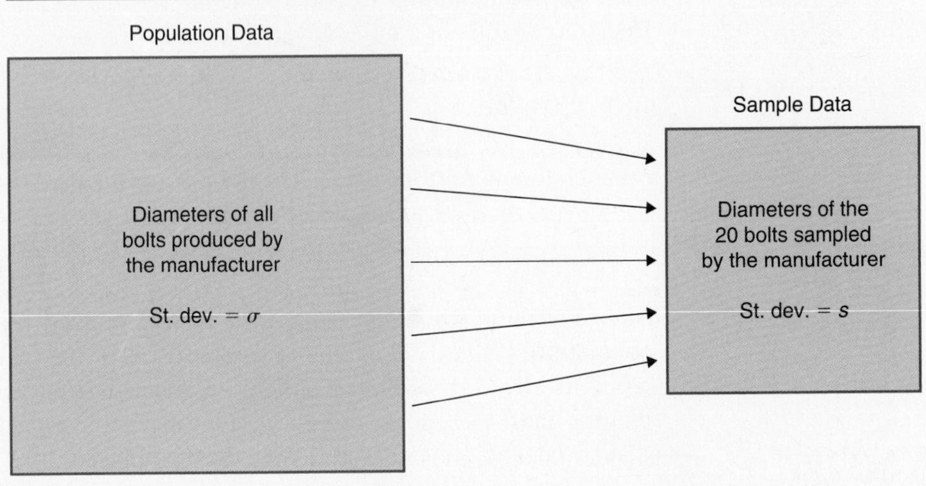

After the sample has been taken, the manufacturer can compute the sample standard deviation, s, of the diameters of the 20 bolts obtained. Using the value of s, he can then estimate the population standard deviation, σ, of the diameters of all bolts being produced. Such inferences will be examined in Chapter 11. ◨

Parameter and Statistic

Some specific statistical terminology helps us distinguish between descriptive measures for populations and descriptive measures for samples.

DEFINITION 3.13 PARAMETER AND STATISTIC

Parameter: A descriptive measure for a population.

Statistic: A descriptive measure for a sample.

Thus, for example, μ and σ are parameters, whereas $\bar{x}$ and s are statistics.

Standardized Variables and z-Scores

We can associate with any variable x, a new variable obtained by subtracting from x its mean and then dividing by its standard deviation. More formally, we have the following definition.

DEFINITION 3.14 STANDARDIZED VARIABLE

For a variable x, the variable

$$z = \frac{x - \mu}{\sigma}$$

is called the ***standardized version*** of x or the ***standardized variable*** corresponding to the variable x.

A standardized variable always has mean 0 and standard deviation 1. Because of this and other reasons, standardized variables play an important role in many aspects of statistical practice and theory. A few applications of standardized variables are presented in this section; several others will appear throughout the remainder of the book.

EXAMPLE 3.25 ILLUSTRATES DEFINITION 3.14

Let us consider a very simple variable x, namely, one with possible observations shown in the first row of Table 3.16.

TABLE 3.16
Possible observations of x and z

x	−1	3	3	3	5	5
z	−2	0	0	0	1	1

a. Determine the standardized version of x.

b. Determine the observed value of z corresponding to an observed value of x of 5.

c. Obtain all possible observations of z.

d. Find the mean and standard deviation of z using Definitions 3.11 and 3.12. Was it necessary to do the calculations in order to obtain the mean and standard deviation?

e. Obtain dotplots of the distributions of both x and z.

SOLUTION **a.** Using Definitions 3.11 and 3.12, we find that the mean and standard deviation of x are 3 and 2, respectively: $\mu = 3$ and $\sigma = 2$. Consequently, the standardized version of x is

$$z = \frac{x - 3}{2}.$$

b. The observed value of z corresponding to an observed value of x of 5 is

$$z = \frac{x-3}{2} = \frac{5-3}{2} = 1.$$

c. Applying the formula $z = (x-3)/2$ to each of the possible observations of the variable x shown in the first row of Table 3.16, we obtain the possible observations of the standardized variable z shown in the second row of Table 3.16.

d. Referring to the second row of Table 3.16, we find that

$$\mu_z = \frac{\Sigma z}{N} = \frac{0}{6} = 0$$

and

$$\sigma_z = \sqrt{\frac{\Sigma(z-\mu_z)^2}{N}} = \sqrt{\frac{6}{6}} = 1.$$

The results of these two computations illustrate something that we mentioned earlier, namely, that the mean of a standardized variable is always 0 and its standard deviation is always 1. Thus it really wasn't necessary to perform the calculations in order to obtain the mean and standard deviation of z.

e. Figures 3.14(a) and 3.14(b) show, respectively, dotplots of the distributions of x and z.

FIGURE 3.14 Dotplots of the distributions of x and its standardized version z

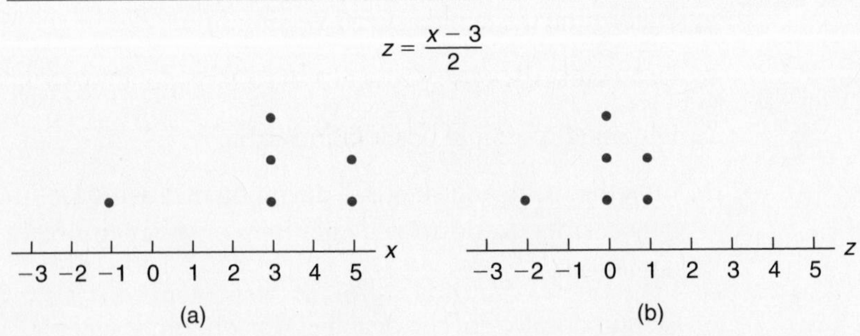

The two graphs illustrate visually something that we already know: standardizing shifts a distribution so that the new mean is 0 and changes the scale so that the new standard deviation is 1. ◨

When we observe a value of a variable x, the corresponding value of the standardized variable z is called the **standard score** or **z-score.** The z-score of an observation tells us how many standard deviations the observation is from the mean; that is, how far the observation is from the mean in units of standard deviation. A negative z-score indicates that the observation is below (smaller than) the mean, whereas a positive z-score indicates that the observation is above (greater than) the mean.

EXAMPLE 3.26 z-SCORES

Consider again the weight data for the Dallas Cowboys given in the fourth column of Table 3.15 on page 176. We determined earlier that the mean and standard deviation of the weights are, respectively, 242.6 lb and 48.3 lb.[†] So, in this case, the standardized variable is

$$z = \frac{x - 242.6}{48.3}.$$

a. Determine and interpret the z-score of Richie Cunningham's weight of 167 lb.

b. Determine and interpret the z-score of Larry Allen's weight of 326 lb.

c. Construct a graph showing the results obtained in parts (a) and (b).

SOLUTION **a.** The z-score for Richie Cunningham's weight of 167 lb is

$$z = \frac{x - 242.6}{48.3} = \frac{167 - 242.6}{48.3} = -1.57.$$

Thus Richie Cunningham's weight is 1.57 standard deviations below the mean.

b. The z-score for Larry Allen's weight of 326 lb is

$$z = \frac{x - 242.6}{48.3} = \frac{326 - 242.6}{48.3} = 1.73.$$

Thus Larry Allen's weight is 1.73 standard deviations above the mean.

c. In Fig. 3.15, we have marked Richie Cunningham's weight of 167 lb with a color dot and Larry Allen's weight of 326 lb with a black dot. Additionally, we have located the mean, $\mu = 242.6$ lb, and have measured off intervals equal in length to the standard deviation, $\sigma = 48.3$ lb.

[†] See the note on page 179 following Example 3.23.

FIGURE 3.15
Graph showing Richie Cunningham's weight (color dot) and Larry Allen's weight (black dot)

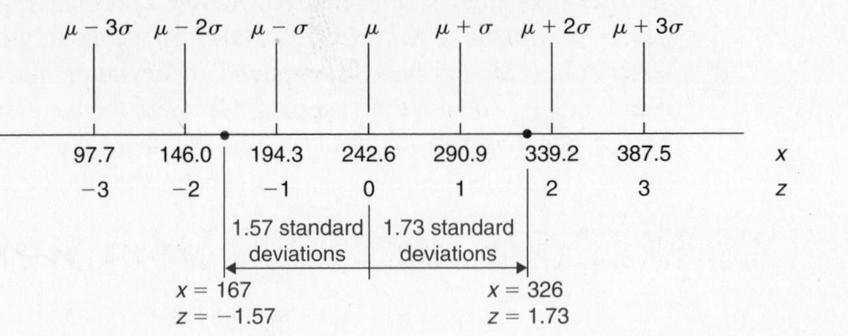

Notice that in Fig. 3.15, the numbers in the row labeled x represent weights in pounds and the numbers in the row labeled z represent z-scores (i.e., number of standard deviations from the mean). □

The three-standard-deviations rule states that almost all of the observations in any data set lie within three standard deviations to either side of the mean. In terms of variables and z-scores, this means that for any variable, almost all possible observations have z-scores between -3 and 3.

Consequently, the z-score of an observation can be used as a measure of its relative standing. A large positive z-score (i.e., a z-score roughly 3 or bigger) indicates that the observation is larger than most of the other possible observations; a large negative z-score (i.e., a z-score roughly -3 or smaller) indicates that the observation is smaller than most of the other possible observations; and a z-score near 0 indicates that the observation is located near the mean.

The statements in the preceding paragraph can be refined and made more precise by applying Chebychev's rule, as we will see in the exercises of this section. And, if the distribution of the variable is roughly bell-shaped, then we can do even better by applying the empirical rule, as we will discover in Chapter 6.

We can also use z-scores to compare the relative standings of two observations from different populations. For example, we could use z-scores to compare the exam scores of two students in different sections of a beginning English course.

Other Descriptive Measures for Populations

Up to this point, we have concentrated our discussion of descriptive measures for populations on the mean and standard deviation. This is because many of the classical inference procedures for center and variation concern those two parameters.

However, modern statistical analyses also rely heavily on descriptive measures based on percentiles, such as the median. Percentiles, deciles, quartiles, the IQR, and other descriptive measures based on percentiles are defined in the same way for (finite) populations as they are for samples. For simplicity we will use the same notation for percentiles, deciles, and quartiles whether we are considering a sample or a population. But we will employ different notations for a sample median and a population median—we'll use M to denote a sample median and η (eta) to denote a population median.

EXERCISES 3.5

STATISTICAL CONCEPTS AND SKILLS

3.105 Although in practice we generally deal with sample data in inferential studies, what is the ultimate objective of such studies?

3.106 In each part below, identify the quantity as a parameter or a statistic.
a. μ **b.** s **c.** $\bar{x}$ **d.** σ

3.107 Which statistic is used to estimate
a. a population mean?
b. a population standard deviation?

3.108 A standardized variable always has mean ____ and standard deviation ____.

3.109 The z-score corresponding to an observed value of a variable tells us ____.

3.110 A positive z-score indicates that the observation is ____ the mean, whereas a negative z-score indicates that the observation is ____ the mean.

3.111 We found earlier in this section that the population mean weight of the players on the 1997–98 Dallas Cowboys football team is 242.6 lb. In this context is the number 242.6 a parameter or a statistic? Why?

3.112 We determined earlier in this section that the population standard deviation of the weights of the starting offensive players on the 1997–98 Dallas Cowboys football team is 48.3 lb. In this context is the number 48.3 a parameter or a statistic? Explain your answer.

3.113 In Section 3.3 we analyzed the heights of the starting five players on each of two men's college basketball teams. The heights, in inches, of the players on Team II are 67, 72, 76, 76, and 84. Regarding the five players as a sample of all male starting college basketball players,

a. compute the sample mean height, $\bar{x}$.
b. compute the sample standard deviation, s.

Regarding the players now as a population,

c. compute the population mean height, μ.
d. compute the population standard deviation, σ.

Comparing your answers from parts (a) and (c) and from parts (b) and (d),

e. why are the values for $\bar{x}$ and μ equal?
f. why are the values for s and σ different?

In each of Exercises 3.114–3.117, regard the given data as population data. For each exercise,
a. obtain the population mean, μ.
b. obtain the population standard deviation, σ.

3.114 One year's monthly electric bills, in dollars, for a family living in the Southeast are as follows.

77	81	73	73
81	158	167	188
142	98	62	80

3.115 One year's monthly telephone bills, in dollars, for the family in Exercise 3.114.

95	88	87	93
90	93	94	76
90	144	112	103

3.116 As reported by the U.S. Department of Agriculture in *Crop Production,* the acreages (in thousands) of corn harvested in 1994 by the six leading corn-producing states are as shown in the following table.

State	IA	IL	NE	MN	IN	OH
Acreage	12.7	11.4	8.3	6.4	6.0	3.5

3.117 The U.S. Agency for International Development compiles information on U.S. foreign aid commitments for economic assistance. Data are published in *U.S. Overseas Loans and Grants and Assistance from International Organizations.* In 1993 the commitments to selected Latin American countries were as follows.

Country	Aid ($millions)	Country	Aid ($millions)
Bolivia	58	Haiti	47
Costa Rica	9	Honduras	33
Domin. Rep.	20	Jamaica	43
Ecuador	13	Nicaragua	128
El Salvador	172	Panama	6
Guatemala	33	Peru	41

3.118 According to *Statistical Report,* published by the U.S. Bureau of Prisons, the mean time served by prisoners released from federal institutions for the first time is 16.3 months. Assume the standard deviation of the times served is 17.9 months. Let x denote time served by a prisoner released for the first time from a federal institution.

a. Find the standardized version of x.

b. What are the mean and standard deviation of the standardized variable?

c. Determine the z-scores for prison times served of 64.7 months and 4.2 months. Round your answers to two decimal places.

d. Interpret your answers in part (c).

e. Construct a graph similar to Fig. 3.15 on page 184 showing your results from parts (b) and (c).

3.119 A local bottling plant fills bottles of soda for distribution in the surrounding area. Although the advertised content is 354 mL, the filling machine is actually set to a mean content of 356 mL. The standard deviation is 1.63 mL. Let y denote the content of a bottle of soda.

a. Find the standardized variable corresponding to y.

b. What are the mean and standard deviation of the standardized variable?

c. Determine the z-scores for contents of 352 mL and 361 mL. Round your answers to two decimal places.

d. Interpret your answers in part (c).

e. Construct a graph similar to Fig. 3.15 on page 184 showing your results from parts (b) and (c).

3.120 Suppose you take an exam with 400 possible points and are told that the mean score is 280 and the standard deviation is 20. You are also told that you got 350. Did you do well on the exam? Why?

EXTENDING THE CONCEPTS AND SKILLS

Measuring relative standing with z-scores. Chebychev's rule states that for any data set and any number $k > 1$, at least $100(1 - 1/k^2)\%$ of the observations lie within k standard deviations to either side of the mean.

We can use z-scores and Chebychev's rule to estimate the relative standing of an observation. To see how, let us consider the weights of the Dallas Cowboys given in Table 3.15. Earlier we found that the population mean and standard deviation of the weights are, respectively, 242.6 lb and 48.3 lb. We also found that the z-score for Larry Allen's weight of 326 lb is 1.73.

Applying Chebychev's rule to that z-score, we conclude that at least $100(1 - 1/1.73^2)\%$, or 66.6%, of the weights lie within 1.73 standard deviations to either side of the mean. Therefore, Larry Allen's weight, which is 1.73 standard deviations above the mean, is greater than at least 66.6% of the other player's weights.

3.121 Refer to the weights of the Dallas Cowboys in the fourth column of Table 3.15 on page 176.

a. Find the z-score for Stepfret Williams's weight.

b. Use part (a) and Chebychev's rule to estimate the relative standing of Stepfret Williams's weight.

3.122 A company produces cans of stewed tomatoes with an advertised weight of 14 oz. The standard deviation of the weights is known to be 0.4 oz. A quality-control engineer selects a can of stewed tomatoes at random and finds its net weight to be 17.28 oz.

a. Estimate the relative standing of that can of stewed tomatoes, assuming the true mean weight is 14 oz. Use the z-score and Chebychev's rule.

b. Does the quality-control engineer have reason to suspect that the true mean weight of all cans of stewed tomatoes being produced is not 14 oz? Why?

3.123 Suppose you buy a new car whose advertised mileage is 25 miles per gallon (mpg). After driving your car for several months, you find that its mileage is 21.4 mpg. You telephone the manufacturer and learn that the standard deviation of gas mileages for all cars of the model you bought is 1.15 mpg.

a. Find and interpret the z-score for the gas mileage of your car.

b. Estimate the relative standing of your car's gas mileage. Use the z-score and Chebychev's rule.

c. Does it appear that your car is getting unusually low gas mileage? Explain your answer.

3.124 Each year, thousands of students bound for college take the Scholastic Assessment Test, or SAT. The test measures the verbal and mathematical abilities of prospective college students. Student scores are reported on a scale that ranges from a low of 200 to a high of 800. In one high-school graduating class, the mean mathematics score on the SAT was 493 and the standard deviation was 105; the mean verbal score was 420 and the standard deviation was 98. A student in the graduating class scored 703 on the math and 665 on the verbal. Relative to the other students in the graduating class, on which test did the student do better? Why?

Relating population and sample standard deviations. In Exercises 3.125 and 3.126 we will examine the numerical relationship between the population and sample standard deviations computed from the same data. This relationship is helpful when the computer or statistical calculator being used has a function for sample standard deviation but not for population standard deviation.

3.125 Consider the following three data sets.

Data Set 1	Data Set 2	Data Set 3
2 4 7 3	7 5 5 3 9 8 6	4 7 8 9 7 4 5 3 4 5

a. Assuming that each of these data sets is sample data, compute the standard deviations. (Round your final answers to two decimal places.)

b. Assuming that each of these data sets is population data, compute the standard deviations. (Round your final answers to two decimal places.)

c. Using your results from parts (a) and (b), make an educated guess about the answer to the following question: Suppose both s and σ are computed for the same data set. Will they tend to be closer together if the data set is large or if it is small?

3.126 Consider a data set with m observations. If the data are sample data, we compute the sample standard deviation, s, whereas if the data are population data, we compute the population standard deviation, σ.

a. Derive a mathematical formula that gives σ in terms of s when both are computed for the same data set. (*Hint:* First note that, numerically, the values of $\bar{x}$ and μ are identical. Consider the ratio of the defining formula for σ to the defining formula for s.)

b. Refer to the three data sets in Exercise 3.125. Verify that your formula in part (a) works for each of the three data sets.

c. Suppose a data set consists of 15 observations. You compute the sample standard deviation of the data and obtain $s = 38.6$. Then you realize that the data are actually population data and that you should have obtained the population standard deviation instead. Use your formula from part (a) to obtain σ.

USING TECHNOLOGY  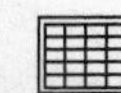 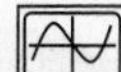

Note: Many statistical software packages have a built-in procedure for obtaining a sample standard deviation but do not have one for obtaining a population standard deviation. We can deal with this problem by using the following formula that expresses the population standard deviation in terms of the sample standard deviation when both are computed for the same data set:

$$\sigma = \sqrt{\frac{m-1}{m}} \cdot s,$$

where m is the number of observations.

3.127 The heights of the players on the 1997–98 Dallas Cowboys are displayed in Table 3.15 on page 176. Using any technology that you have available, obtain

a. the population mean height, in inches.

b. the population standard deviations of the heights.

3.128 Use Minitab or some other statistical software to obtain the population mean and population standard deviation of the acreages in Exercise 3.116.

3.129 Use Minitab or some other statistical software to obtain the population mean and population standard deviation of the aid amounts in Exercise 3.117.

CHAPTER REVIEW

You Should Be Able To

1. use and understand the formulas presented in this chapter.
2. explain the purpose of a measure of center.
3. obtain and interpret the mean, the median, and the mode(s) of a data set.
4. choose an appropriate measure of center for a data set.
5. use and understand summation notation.
6. define, compute, and interpret a sample mean.
7. explain the purpose of a measure of variation.
8. define, compute, and interpret the range of a data set.
9. define, compute, and interpret a sample standard deviation.
10. define percentiles, deciles, and quartiles.
11. obtain and interpret the quartiles, IQR, and five-number summary of a data set.
12. obtain the lower and upper limits of a data set and identify potential outliers.
13. construct and interpret a boxplot and a modified boxplot.
14. use a boxplot to identify distribution shape for large data sets.
15. define the population mean (mean of a variable).
16. define the population standard deviation (standard deviation of a variable).
17. compute the population mean and population standard deviation of a finite population.
18. distinguish between a parameter and a statistic.
19. understand how and why statistics are used to estimate parameters.
20. obtain and interpret z-scores.

***21.** use the Minitab procedures covered in this chapter.

***22.** interpret the output obtained from the application of the Minitab procedures discussed in this chapter.

Key Terms

adjacent values, *165*
box-and-whisker diagram, *164*
boxplot, *164*
Chebychev's rule, *155*
deciles, *159*
descriptive measures, *125*
deviations from the mean, *144*
empirical rule, *156*
first quartile (Q_1), *160*
five-number summary, *163*
interquartile range (IQR), *162*
lower limit, *163*
mean, *126*
mean of a variable (μ), *175*
measures of center, *126*
measures of central tendency, *126*
measures of spread, *142*
measures of variation, *142*
median, *127*
mode, *128*
modified boxplot, *164*
outliers, *163*
parameter, *180*
percentiles, *159*
population mean (μ), *175*
population standard deviation (σ), *178*
population variance (σ^2), *178*
quartiles, *160*
range, *143*
resistant, *130*
sample mean ($\bar{x}$), *139*
sample size (n), *139*
sample standard deviation (s), *147*
sample variance (s^2), *146*
second quartile (Q_2), *160*
standard deviation, *144*

REVIEW TEST

STATISTICAL CONCEPTS AND SKILLS

1. Define the following terms.
a. descriptive measures
b. measures of center
c. measures of variation

2. Identify the two most commonly used measures of center for quantitative data. Explain the relative advantages and disadvantages of each.

3. Among the major measures of center that we discussed, which is the only one appropriate for qualitative data?

4. Identify the most appropriate measure of variation corresponding to each of the following measures of center.
a. mean **b.** median

5. Specify the mathematical symbol used for each of the following descriptive measures.
a. sample mean **b.** sample standard deviation
c. population mean
d. population standard deviation

6. Data Set A has more variation than Data Set B. Decide which of the following statements are necessarily true.
a. Data Set A has a larger mean than Data Set B.
b. Data Set A has a larger standard deviation than Data Set B.

7. Complete the following statement: Almost all of the observations in any data set lie within _____ standard deviations to either side of the mean.

8. Regarding the five-number summary:
a. Identify its components.
b. How can it be employed to describe center and variation?
c. What graphical display is based on it?

9. Regarding outliers:
a. What is an outlier?
b. Explain how we can identify potential outliers knowing only the first and third quartiles.

10. Regarding z-scores:
a. How is a z-score obtained?
b. What is the interpretation of a z-score?
c. An observation has a z-score of 2.9. Roughly speaking, what is the relative standing of the observation?

11. Euromonitor Publications Limited, London, compiles data on per capita food consumption of major food commodities in various countries. Data are published in *European Marketing Data and Statistics*. Samples of 10 Germans and 15 Russians yield the following fish consumptions, in kilograms, for last year.

Germans		Russians		
10	12	16	21	12
17	12	11	5	23
14	11	19	19	22
13	8	16	23	12
9	8	18	7	17

a. Determine the mean of each data set.
b. Determine the median of each data set.
c. Determine the mode(s) of each data set.

12. The National Center for Health Statistics publishes information on the duration of marriages in *Vital Statistics of the United States.* Which measure of center is more appropriate for data on the duration of marriages: the mean or the median? Explain your answer.

13. Death certificates provide data on the causes of death. Which of the three major measures of center is appropriate here? Explain your answer.

14. Telephone companies conduct surveys to obtain information on the durations of telephone conversations. A sample of 12 phone calls yields the following durations to the nearest minute.

4	2	1	2	2	8
6	3	1	3	1	15

a. Obtain the mean duration of this sample of 12 calls.
b. Obtain the range of the durations.
c. Obtain the sample standard deviation of the durations.

15. Dr. Thomas Stanley of Georgia State University has collected information on millionaires, including their ages, since 1973. A sample of 36 millionaires has a mean age of 58.5 years and a standard deviation of 13.4 years.
a. Complete the graph below.

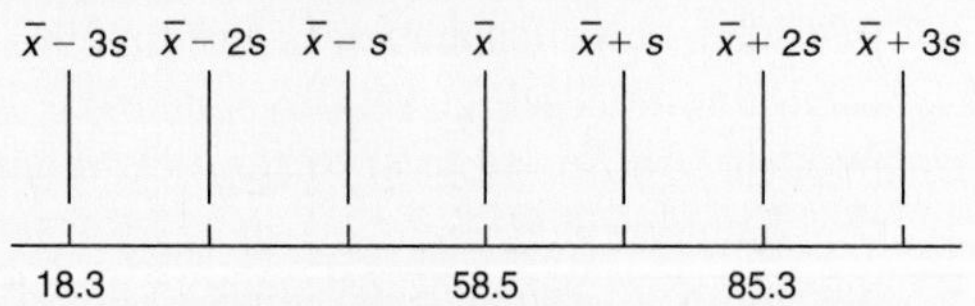

b. Fill in the following blanks: Almost all of the ages of the 36 millionaires are between ____ and ____ years old.

16. Referring to Problem 15, the actual ages of the 36 millionaires sampled, arranged in increasing order, are as follows.

31	38	39	39	42	42	45	47	48
48	48	52	52	53	54	55	57	59
60	61	64	64	66	66	67	68	68
69	71	71	74	75	77	79	79	79

a. Determine the quartiles for the data.
b. Obtain the interquartile range.
c. Find the five-number summary.
d. Calculate the lower and upper limits.
e. Identify potential outliers, if any.
f. Construct and interpret a boxplot and, if appropriate, a modified boxplot.

17. According to *Peterson's Guides,* the Fall 1996 enrollment figures for the University of California campuses were as follows.

Campus	Enrollment (1000s)
Berkeley	29.6
Davis	23.9
Irvine	17.3
Los Angeles	34.9
Riverside	9.1
San Diego	18.1
San Francisco	3.7
Santa Barbara	18.5
Santa Cruz	10.2

a. Compute the population mean enrollment, μ, of the UC campuses.
b. Compute σ.
c. Letting x denote enrollment, specify the standardized variable, z, corresponding to x.
d. Without performing any calculations, give the mean and standard deviation of z.
e. Construct dotplots for the distributions of both x and z. Interpret your graphs.
f. Obtain and interpret the z-scores for the enrollments at the Los Angeles and San Diego campuses.

18. The U.S. Energy Information Administration reports figures on retail gasoline prices in *Monthly Energy Review.* Data are obtained by sampling 10,000 gasoline service stations from a total of more than 185,000. For the 10,000 stations sampled, suppose the mean price per gallon for unleaded regular gasoline is \$1.35.
a. Is the mean price given here a sample mean or a population mean? Explain your answer.
b. What letter would you use to designate the mean of \$1.35?
c. Is the mean price given here a statistic or a parameter? Explain your answer.

USING TECHNOLOGY 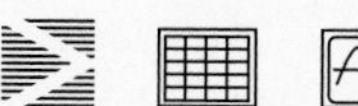

19. Use Minitab or some other statistical software to determine the following statistics for the age data in Problem 16.

a. mean
b. median
c. range
d. sample standard deviation

20. Refer to the data in Problem 16. Use Minitab or some other statistical software to obtain

a. a (modified) boxplot for the data.
b. the five-number summary of the data.

21. Refer to Problem 16.

a. Printout 3.10 shows a boxplot generated by Minitab for the age data. Use the boxplot to discuss the variation in the data set and to determine approximately the five-number summary.
b. Are there potential outliers in the data? If so, identify them approximately and provide a possible explanation for their cause.
c. Printout 3.11 displays the output obtained by applying Minitab's descriptive-statistics procedure to the age data. Use the output to determine exactly the five-number summary of the data. Compare your answers to the approximate values you obtained in part (a).

PRINTOUT 3.10 Minitab output for Problem 21(a)

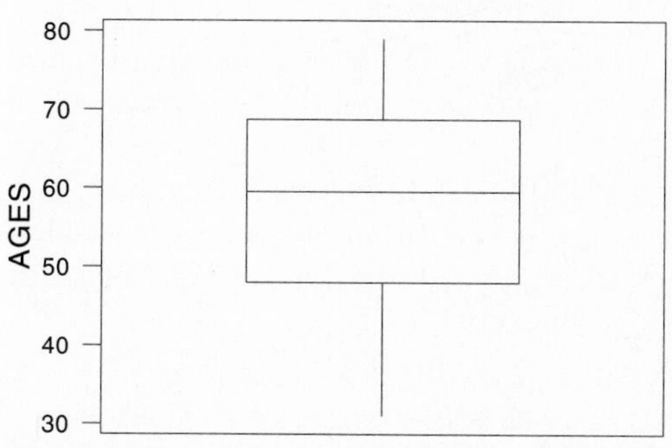

PRINTOUT 3.11 Minitab output for Problem 21(c)

```
Variable         N      Mean    Median    TrMean    StDev   SE Mean
AGES            36     58.53     59.50     58.75    13.36      2.23

Variable   Minimum   Maximum        Q1        Q3
AGES         31.00     79.00     48.00     68.75
```

INTERNET PROJECT

Old Faithful Geyser and a Survey of Wages

In this project, you will explore two topics. The first one is the Old Faithful Geyser data, which shows the time between eruptions and the duration of eruptions. In the Internet project for Chapter 2, you experimented with an interactive histogram of that data to learn more about the nature of histograms (and class sizes). In this project, you will take a deeper look into the data itself.

The second part of the project is a survey of wages. Wage and salary data provide a typical example of why it is important to use different types of descriptive measures to help explain your data. The statistic you choose can make a striking difference in the conclusions you draw: in this case, the difference is between using the mean and the median.

In each part of the project, you will see how certain measures can help describe the data and how some measures are better than others. As usual, you will look at several different views of the data to better understand the true nature of the phenomenon that the data describes.

URL for access to Internet Projects Page: `http://hepg.awl.com` *keyword:* Weiss

USING THE FOCUS DATABASE

In Chapter 1 we explained how to store the Focus database in a Minitab worksheet named `focus.mtw`. If you haven't already created that worksheet, follow the instructions on pages 54–55 to create it now.

The Focus database contains information on 500 randomly selected Arizona State University sophomores. Seven variables are considered for each student: sex, high-school GPA, SAT math score, cumulative GPA, SAT verbal score, age, and total hours. Use Minitab or some other statistical software to solve the following problems.

a. Determine the means and medians of the SAT math scores and SAT verbal scores. Use these measures of center to compare the two sets of scores.

b. Find the ranges, sample standard deviations, and interquartile ranges of the SAT math scores and SAT verbal scores. Use these measures of variation to compare the two sets of scores.

c. Construct boxplots for the SAT math scores and SAT verbal scores. Use the boxplots to compare the two sets of scores. Be sure to consider variation, distribution shape, and outliers.

CASE STUDY DISCUSSION

Per Capita Income by State

As we mentioned in the introduction to this case study at the beginning of the chapter, *per capita income* is defined as the annual total personal income of the residents of a region divided by the resident population of that region as of July 1 of that year.

The table on page 124 displays the 1996 per capita income for each state in the United States and for the District of Columbia. Referring to that table, solve each of the following problems.

a. Determine the mean and median of the per capita incomes. Explain any difference between these two measures of center.
b. Obtain the range and sample standard deviation of the per capita incomes.
c. Find and interpret the z-score for your state's per capita income.
d. Determine and interpret the quartiles of the per capita incomes.
e. Find and interpret the five-number summary of the data.
f. Find the lower and upper limits. Use them to identify potential outliers.
g. Construct a boxplot for the per capita incomes and interpret your result in terms of the variation in the data.
h. Use Minitab or some other statistical software to solve parts (a)–(g).

BIOGRAPHY JOHN TUKEY

John Wilder Tukey was born on June 16, 1915, in New Bedford, Massachusetts. After earning bachelor's and master's degrees in chemistry from Brown University in 1936 and 1937, respectively, he enrolled in the mathematics program at Princeton University, where he received a master's degree in 1938 and a doctorate in 1939.

After graduating, Tukey was appointed Henry B. Fine Instructor in Mathematics at Princeton; 10 years later he was advanced to a full professorship. In 1965, Princeton established a department of statistics, and Tukey was named its first chairperson. In addition to his position at Princeton, he was a member of the Technical Staff at AT&T Bell Laboratories from 1945 until his retirement in 1985 as Associate Executive Director–Research in the Information Sciences Division.

Tukey is among the leaders in the field of exploratory data analysis (EDA). EDA provides techniques, such as stem-and-leaf diagrams, for effectively investigating data. He has made fundamental contributions to the areas of robust estimation and time series analysis. Tukey has written numerous books and more than 350 technical papers on mathematics, statistics, and other scientific subjects. He also coined the word *bit,* a contraction of *binary digit* (a unit of information, often as processed by a computer).

Tukey's participation in educational, public, and government service is most impressive. He was appointed to serve on the President's Science Advisory Committee by President Eisenhower; was chairperson of the committee that prepared "Restoring the Quality of our Environment" in 1965; helped develop the National Assessment of Educational Progress; and was a member of the Special Advisory Panel on 1990 Census of the U.S. Department of Commerce, Bureau of the Census—to name only a few of his involvements.

Among many honors, Tukey has received the National Medal of Science, the IEEE Medal of Honor, Princeton University's James Madison Medal, and Foreign Member, The Royal Society (London). He was the first recipient of the Samuel S. Wilks Award of the American Statistical Association. Tukey remains on the faculty at Princeton as Donner Professor of Science, Emeritus; Professor of Statistics, Emeritus; and Senior Research Statistician.

P A R T

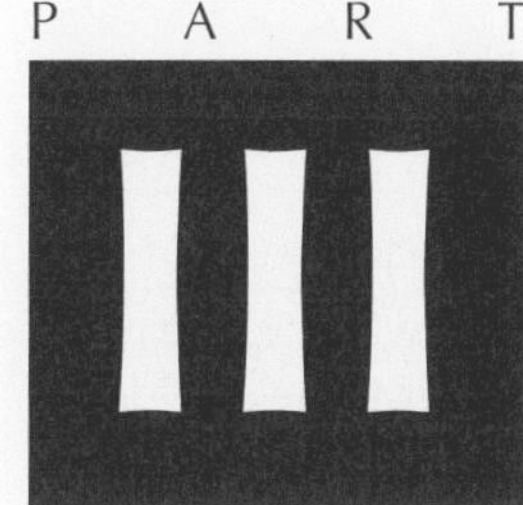

Probability, Random Variables, and Sampling Distributions

CASE STUDY

THE POWERBALL®

The Powerball is a lottery sold in Arizona, Connecticut, Delaware, District of Columbia, Idaho, Indiana, Iowa, Kansas, Kentucky, Louisiana, Minnesota, Missouri, Montana, Nebraska, New Hampshire, New Mexico, Oregon, Rhode Island, South Dakota, West Virginia, and Wisconsin. It was introduced in April of 1992 and took its present form in November of 1997.

Although the Powerball is a multi-state lottery game, it is not the first. That distinction goes to Lotto*America, which was created in 1988 when Iowa and six other states joined forces to offer a game with a large jackpot. Since the more people that play, the bigger the jackpots tend to be, multi-state lotteries offer larger prizes than those standing alone.

A Powerball jackpot starts at $10 million and grows if no one wins it. Since, as we will see, the chance of winning a jackpot is small, the jackpot often grows to huge amounts, sometimes around $300 million. In case there are multiple winners, the jackpot is divided equally among them. Drawings take place on Wednesday and Saturday evenings at 9:59 P.M. and can be watched during the 10:00 P.M. news.

Here is how the Powerball is played: A player first selects five numbers from the numbers 1–49 and then chooses a Powerball number, which can be any number between 1 and 42. A ticket costs $1. In the drawing, five white balls are drawn randomly from 49 white balls numbered 1–49; and one red Powerball is drawn randomly from 42 red balls numbered 1–42.

To win the jackpot, a ticket must match all of the balls drawn; there are other smaller prizes for matching some but not all of the balls drawn. What are the chances of winning the jackpot? What are the chances of winning any prize whatsoever? After studying probability, we will be able to answer these and other questions. We will do that when we revisit the Powerball at the end of this chapter.

Probability Concepts

4

GENERAL OBJECTIVES

Up to this point, we have been concentrating on descriptive statistics, methods for organizing and summarizing data. However, another important aspect of this text is to present the fundamentals of inferential statistics, methods of drawing conclusions about a population based on information obtained from a sample of the population.

Because inferential statistics involves using information obtained from part of a population (a sample) to draw conclusions about the entire population, we can never be certain that our conclusions are correct—uncertainty is inherent in inferential statistics. So before we can understand, develop, and apply the methods of inferential statistics, we need to become familiar with uncertainty.

The science of uncertainty is called **probability theory.** Probability theory enables us to evaluate and control the likelihood that a statistical inference is correct and, more generally, provides the mathematical basis for inferential statistics. This chapter begins our study of probability.

CHAPTER OUTLINE

4.1 PROBABILITY BASICS

Although most applications of probability theory to statistical inference involve large populations, the fundamental concepts of probability are most easily illustrated and explained using relatively small populations and games of chance. So keep in mind that many of the examples in this chapter are designed expressly to present the principles of probability in a lucid manner.

The Equal-Likelihood Model

We discussed an important aspect of probability when we examined probability sampling in Chapter 1. Let us return to the illustration of simple random sampling given in Example 1.10 on page 26.

EXAMPLE 4.1 INTRODUCES PROBABILITY

As reported by the *World Almanac, 1998,* the top five state officials of Oklahoma are as shown in Table 4.1.

TABLE 4.1 Five top Oklahoma state officials

Officials
Governor (G)
Lieutenant Governor (L)
Secretary of State (S)
Attorney General (A)
Treasurer (T)

Suppose we take a simple random sample without replacement of two officials from the five officials.

a. Determine the probability that we obtain the governor and treasurer.

b. Determine the probability that the attorney general is included in the sample.

SOLUTION For convenience we will use the letters placed parenthetically after the officials in Table 4.1 to represent the officials. As we discovered in Example 1.10, there are 10 possible samples of two officials from the population of five officials. They are listed in Table 4.2.

TABLE 4.2 Possible samples of two officials

G, L	G, S	G, A	G, T	L, S
L, A	L, T	S, A	S, T	A, T

If we take a simple random sample of size two, then each of the possible samples of two officials is equally likely to be the one selected.

a. Since there are 10 possible samples, the probability is $\frac{1}{10}$, or 0.1, of selecting the governor and treasurer (G, T). Another way of looking at this is that one out of 10, or 10% of the samples include both the governor and treasurer; so, the probability of obtaining such a sample is 10%, or 0.1. The same goes for any other two particular officials.

b. Referring to Table 4.2, we see that the attorney general (A) is included in four of the 10 possible samples of size two. Since each of the 10 possible samples is equally likely to be the one selected, the probability is $\frac{4}{10}$, or 0.4, that the attorney general is included in the sample. Again, another way of looking at this is that four out of 10, or 40% of the samples include the attorney general; so, the probability of obtaining such a sample is 40%, or 0.4. ◨

The essential idea in Example 4.1 is that when outcomes are equally likely, probabilities are nothing more than percentages (relative frequencies). In other words, probabilities are computed using the following simple formula, which we refer to as the ***f*/*N* rule.**

DEFINITION 4.1 **PROBABILITY FOR EQUALLY LIKELY OUTCOMES**

Suppose an experiment has N possible outcomes, all equally likely. Then the probability that a specified event occurs equals the number of ways, f, that the event can occur, divided by the total number of possible outcomes. In symbols,

$$\text{Probability of an event} = \frac{f}{N}.$$

(Number of ways event can occur ↙ f; Total number of possible outcomes ↖ N)

In stating Definition 4.1, we have used the terms "experiment" and "event" in their intuitive sense. Basically, by an **experiment,** we mean an action whose outcome cannot be predicted with certainty. And, by an **event,** we mean some specified result that may or may not occur when the experiment is performed.

For instance, in Example 4.1 the experiment consists of taking a random sample of size two from the five officials. It has 10 possible outcomes ($N = 10$), all equally likely. In part (b), the event is that the sample obtained includes the attorney general, which can occur in four ways ($f = 4$); so its probability equals

$$\frac{f}{N} = \frac{4}{10} = 0.4,$$

as we noted in Example 4.1(b). Examples 4.2 and 4.3 provide two additional illustrations of Definition 4.1. These examples further indicate the varied contexts under which the equal-likelihood model applies.

EXAMPLE 4.2 ILLUSTRATES DEFINITION 4.1

The U.S. Bureau of the Census compiles data on family income and publishes its findings in *Current Population Reports.* Table 4.3 gives a frequency distribution of annual income for U.S. families in 1995.

TABLE 4.3 Frequency distribution of annual income for U.S. families

Income	Frequency (1000s)
Under \$10,000	5,216
\$10,000–\$14,999	4,507
\$15,000–\$24,999	10,040
\$25,000–\$34,999	9,828
\$35,000–\$49,999	12,841
\$50,000–\$74,999	14,204
\$75,000 & over	12,961
	69,597

A 1995 U.S. family is selected **at random,** meaning that each family is equally likely to be the one obtained (simple random sample of size one). Determine the probability that the family obtained has an annual income of

a. between \$50,000 and \$74,999, inclusive.

b. between \$25,000 and \$74,999, inclusive.

c. under \$15,000.

SOLUTION The second column of Table 4.3 shows that in 1995, there were 69,597 thousand U.S. families; so $N = 69{,}597$ thousand.

a. For this part, the event in question is that the family obtained makes between \$50,000 and \$74,999. We observe from Table 4.3 that the number of such families is 14,204 thousand; so $f = 14{,}204$ thousand. We find, by applying the f/N rule, that the probability the family obtained makes between \$50,000 and \$74,999 equals

$$\frac{f}{N} = \frac{14{,}204}{69{,}597} = 0.204,$$

to three decimal places. In terms of percentages, this means that in 1995, 20.4% of U.S. families made between \$50,000 and \$74,999.

b. For this part, the event in question is that the family obtained makes between \$25,000 and \$74,999. Referring to Table 4.3, we observe that the number of such families is $9{,}828 + 12{,}841 + 14{,}204$, or 36,873 thousand; consequently, $f = 36{,}873$ thousand. So, the required probability is

$$\frac{f}{N} = \frac{36{,}873}{69{,}597} = 0.530,$$

to three decimal places. In terms of percentages, this means that in 1995, 53.0% of U.S. families made between \$25,000 and \$74,999.

c. Proceeding as in parts (a) and (b), we find that the probability that the family obtained makes under \$15,000 is

$$\frac{f}{N} = \frac{5{,}216 + 4{,}507}{69{,}597} = \frac{9{,}723}{69{,}597} = 0.140,$$

to three decimal places. In terms of percentages, this means that in 1995, 14.0% of U.S. families made under \$15,000. ◨

EXAMPLE 4.3 ILLUSTRATES DEFINITION 4.1

When a pair of balanced dice is rolled, 36 equally likely outcomes are possible, as depicted in Fig. 4.1. Find the probability that

a. the sum of the dice is 11.

b. doubles are rolled; that is, both dice come up the same number.

FIGURE 4.1
Possible outcomes for rolling a pair of dice

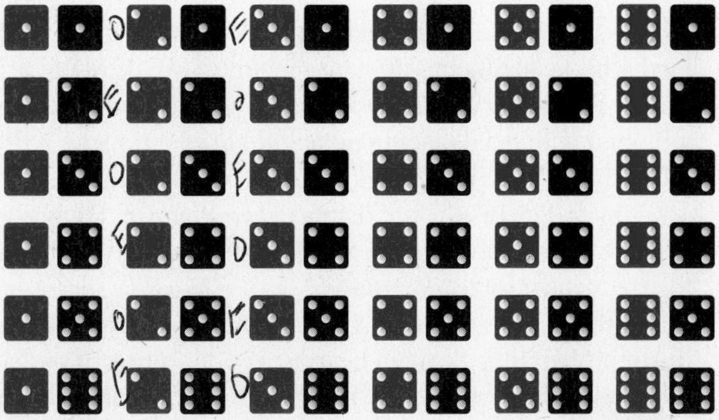

SOLUTION For this experiment, $N = 36$.

a. The sum of the dice can be 11 in two ways, as is apparent from Fig. 4.1. So the probability that the sum of the dice is 11 equals $f/N = 2/36 = 0.056$.

b. As we see from Fig. 4.1, doubles can be rolled in six ways; consequently, the probability of rolling doubles equals $f/N = 6/36 = 0.167$. □

The Meaning of Probability

Essentially, probability is a generalization of the concept of percentage. When we select a member at random from a finite population, as we did in Example 4.2, probability is nothing more than percentage. But, in general, how do we interpret probability? For instance, what do we mean by saying that

- the probability is 0.314 that the gestation period of a woman will exceed 9 months, or
- the probability is 0.667 that the favorite in a horse race finishes in the money (first, second, or third place), or
- the probability is 0.40 that a traffic fatality involves an intoxicated or alcohol-impaired driver or nonoccupant?

Some probabilities are easy to interpret: A probability near 0 indicates that the event in question is very unlikely to occur when the experiment is performed, whereas a probability near 1 (100%) suggests that the event is quite likely to occur. To gain further insight into the meaning of probability, it is useful to consider the **frequentist interpretation of probability,** which construes the probability of an event to be the proportion of times it occurs in a large number of repetitions of the experiment.

Consider, for instance, the simple experiment of tossing a balanced coin once. Because the coin is balanced, we reason that there is a 50-50 chance the coin will land with heads facing up. Consequently, we attribute a probability of 0.5 to that event. The frequentist interpretation is that in a large number of tosses, the coin will land with heads facing up about half the time.

We used a computer to perform two simulations of tossing a balanced coin 100 times. The results are displayed in Fig. 4.2. Each graph shows the number of tosses of the coin versus the proportion of heads. Both graphs seem to corroborate the frequentist interpretation.

Although the frequentist interpretation is helpful for understanding the meaning of probability, it cannot be used as a definition of probability. One common way probabilities are defined is by specifying a **probability model**—a mathematical description of the experiment based on certain primary aspects and assumptions.

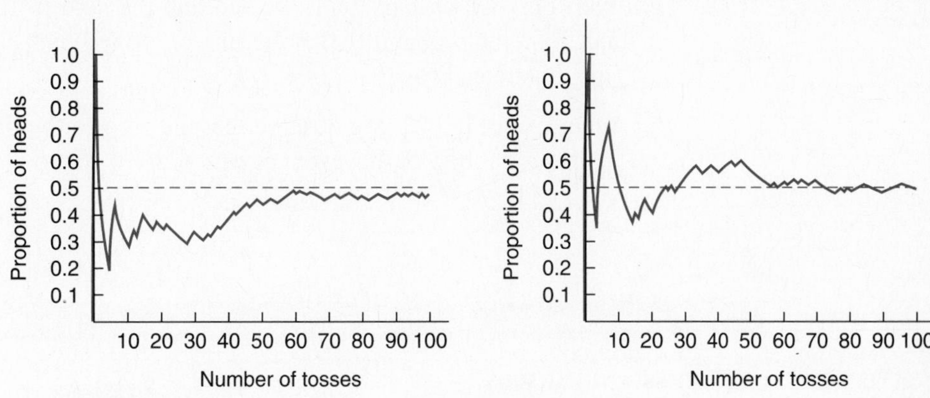

FIGURE 4.2 Two computer simulations of tossing a balanced coin 100 times

The **equal-likelihood model** discussed earlier in this section is an example of a probability model. The primary aspect and assumption is that all possible outcomes are equally likely to occur. We will discuss other probability models later in this and subsequent chapters.

Basic Properties of Probabilities

Probabilities have some simple but basic properties, as listed in Key Fact 4.1.

KEY FACT 4.1 **BASIC PROPERTIES OF PROBABILITIES**

Property 1: The probability of an event is always between 0 and 1, inclusive.

Property 2: The probability of an event that cannot occur is 0. (An event that cannot occur is called an ***impossible event.***)

Property 3: The probability of an event that must occur is 1. (An event that must occur is called a ***certain event.***)

Property 1 indicates that numbers such as 5 or −0.23 could not possibly be probabilities. Consequently, if you calculate a probability and get an answer like 5 or −0.23, then you made an error. Example 4.4 illustrates Properties 2 and 3.

EXAMPLE 4.4 ILLUSTRATES PROPERTIES 2 AND 3 OF KEY FACT 4.1

Refer to Example 4.3 on page 201, where a pair of balanced dice is rolled. Determine the probability that

a. the sum of the dice is 1. **b.** the sum of the dice is 12 or less.

SOLUTION **a.** From Fig. 4.1 on page 201, we see that the sum of the dice must be no less than 2. So the probability that the sum of the dice is 1 equals $f/N = 0/36 = 0$. This illustrates Property 2 of Key Fact 4.1.

b. Again from Fig. 4.1, we observe that the sum of the dice is always 12 or less. So the probability of that event equals $f/N = 36/36 = 1$. This illustrates Property 3 of Key Fact 4.1.

EXERCISES 4.1

STATISTICAL CONCEPTS AND SKILLS

4.1 What is an experiment? an event?

4.2 Concerning the equal-likelihood model of probability:
a. What is it?
b. How is the probability of an event found?

4.3 What is the difference between selecting a member at random from a finite population and taking a simple random sample of size one?

4.4 If a member is selected at random from a finite population, then probabilities are identical to ______.

4.5 State the frequentist interpretation of probability.

4.6 Interpret each of the following probability statements using the frequentist interpretation of probability.
a. The probability is 0.487 that a newborn baby will be a girl.
b. The probability of a single ticket winning a prize in the Powerball® lottery is 0.029.
c. If a balanced dime is tossed three times, the probability that it will come up heads all three times is 0.125.

4.7 Which of the following numbers could not possibly be probabilities? Justify your answer.
a. 0.462 **b.** −0.201 **c.** 1
d. $\frac{5}{6}$ **e.** 3.5 **f.** 0

4.8 The five top Oklahoma state officials are as shown in the following table.

Governor (G)
Lieutenant Governor (L)
Secretary of State (S)
Attorney General (A)
Treasurer (T)

a. List the possible samples without replacement of size three that can be obtained from the population of five officials. *(Hint:* There are 10 possible samples.*)*

If a simple random sample without replacement of three officials is taken from the five officials, determine the probability that
b. the governor, attorney general, and treasurer are obtained.
c. the governor and treasurer are included in the sample.
d. the governor is included in the sample.

In Exercises 4.9–4.16, express each probability as a decimal rounded to three places.

4.9 The U.S. Bureau of the Census publishes data on housing units in *American Housing Survey in the United States.* The following table provides a frequency distribution for the number of rooms in U.S. housing units. The frequencies are in thousands.

Rooms	No. of units
1	862
2	1,422
3	10,166
4	20,789
5	24,328
6	22,151
7	14,183
8+	15,555

A U.S. housing unit is selected at random. Find the probability that the housing unit obtained has
a. four rooms. **b.** more than four rooms.
c. one or two rooms. **d.** fewer than one room.
e. one or more rooms.

4.10 A *family* is defined to be a group of two or more persons related by birth, marriage, or adoption and residing together in a household. According to *Current Population Reports,* a publication of the U.S. Bureau of the Census, the size distribution of U.S. families is as follows. The frequencies are in thousands.

Size	No. of families
2	29,765
3	15,771
4	14,421
5	6,234
6	2,182
7+	1,221

A U.S. family is selected at random. Find the probability that the family obtained has
a. two persons. **b.** more than three persons.
c. between one and three persons, inclusive.
d. one person. **e.** one or more persons.

4.11 As reported by the Bureau of Labor Statistics in *Employment and Earnings,* the age distribution of employed persons 16 years old and over is as shown in the following table.

Age (years)	Frequency (1000)
16–19	6,500
20–24	12,138
25–34	32,077
35–44	35,051
45–54	25,514
55–64	11,739
65 & over	3,690

If an employed person is selected at random, find the probability that the person obtained is
a. between 25 and 34 years old, inclusive.
b. at least 45 years old, that is, 45 years old or older.
c. between 20 and 44 years old, inclusive.
d. under 20 or over 54.

4.12 According to *Survey of Graduate Science Engineering Students and Postdoctorates,* published by the National Science Foundation, the distribution of graduate science students in doctorate-granting institutions is as follows. Frequencies are in thousands.

Field	Frequency
Physical sciences	31.8
Environmental	14.0
Mathematical sciences	16.4
Computer sciences	28.1
Agricultural sciences	11.6
Biological sciences	53.9
Psychology	39.4
Social sciences	77.3

A graduate science student who is attending a doctorate-granting institution is selected at random. Determine the probability that the field of the student obtained is
a. psychology. **b.** physical or social science.
c. not computer science.

4.13 Two balanced dice are rolled. By referring to Fig. 4.1 on page 201, determine the probability that the sum of the dice is
a. 6. **b.** even. **c.** 7 or 11. **d.** 2, 3, or 12.

4.14 A balanced dime is tossed three times. The possible outcomes can be represented as follows.

HHH	HTH	THH	TTH
HHT	HTT	THT	TTT

Here, for example, HHT means that the first two tosses come up heads and the third tails. Find the probability that
a. exactly two of the three tosses are heads.
b. the last two tosses come up tails.
c. all three tosses come up the same.
d. the second toss is heads.

4.15 The Internal Revenue Service reports in *Statistics of Income* that the number of U.S. businesses by category is as shown in the table below. Frequencies are in thousands.

Proprietorships	Partnerships	Corporations
16,154	1,493	4,342

Determine the probability that a business selected at random is

a. a proprietorship. **b.** not a partnership.
c. a corporation.

4.16 According to the Census Bureau publication *Current Housing Reports,* housing in the United States is occupied as follows. Frequencies are in thousands.

Owner-occupied	Renter-occupied	Vacant year-round
63,544	34,150	8,710

Find the probability that a housing unit selected at random is

a. owner-occupied. **b.** not renter-occupied.
c. vacant year-round.

4.17 Refer to Exercise 4.9. Which, if any, of the events in parts (a)–(e) are certain? impossible?

4.18 Refer to Exercise 4.10. Which, if any, of the events in parts (a)–(e) are certain? impossible?

4.19 The following anecdote, entitled "Get your story straight," was received through an e-mail list: There were two friends taking chemistry and both had a solid A going into the final. They were so confident about their grades that the weekend before the Monday chemistry final, they decided to party with some friends at a university several miles away. They had a great time but, due to heavy Saturday partying, slept all day Sunday and didn't make it back to their school until early Monday morning.

Rather than take the final, they went to the professor after the final exam was over and explained why they missed the final. They said they had visited friends at another university and had planned to return on time to study, but on the way back they had a flat tire and, because they had no spare, were late getting back to campus. The professor thought this over and said they could make up the final on Tuesday.

The two friends were relieved and spent Monday night studying for the final. Upon arriving for the make-up final, the professor placed the two students in separate rooms and gave each a test booklet. They looked at the first problem which was a simple molarity problem worth 5 points. "Cool" they thought, "this is going to be easy." They turned the page to do the next problem but were unprepared for what they saw. It read: "(95 points) Which tire?" What is the probability that both students say the same tire? (*Hint:* List the possibilities.)

EXTENDING THE CONCEPTS AND SKILLS

4.20 Explain what is wrong with the following argument: When two balanced dice are rolled, the sum of the dice can be 2, 3, 4, 5, 6, 7, 8, 9, 10, 11, or 12. This gives 11 possibilities. Therefore the probability that the sum is 12 equals $\frac{1}{11}$.

4.21 Explain what is wrong with the following argument: When a balanced coin is tossed twice, the total number of heads obtained can be 0, 1, or 2. This gives three possibilities. So the probability of getting two heads is $\frac{1}{3}$.

4.22 Closely related to probabilities are **odds.** Newspapers, magazines, and other popular publications often express likelihood in terms of odds instead of probabilities, and odds are used much more than probabilities in gambling contexts. If the probability an event occurs is p, then the odds that the event occurs are p to $1 - p$. This is also expressed by saying that the odds are p to $1 - p$ *in favor of the event* or that the odds are $1 - p$ to p *against the event.* Conversely, if the odds in favor of an event are a to b (or, equivalently, the odds against it are b to a), then the probability the event occurs is $a/(a + b)$. For example, if an event has probability 0.75 of occurring, then the odds that the event occurs are 0.75 to 0.25 or 3 to 1; if the odds against an event are 3 to 2, then the probability that the event occurs is $2/(2 + 3)$ or 0.4.

a. When two balanced dice are rolled, the probability that the sum of the dice is 8 equals $\frac{5}{36}$. What are the odds that the sum of the dice is 8?

b. According to a study by the Alan Guttmacher Institute, reported in the *New York Times,* approximately 20% of all Americans are infected with a viral sexually transmitted disease (STD), such as herpes or hepatitis B. What are the odds against a randomly selected American being infected with a viral STD?

c. In a horse race, the tote board indicates that the odds against the #5 horse winning the race are 5 to 2. What is the probability, according to the tote board, that this horse wins?

d. A study reported in a 1994 issue of the *Journal of the American Medical Association* suggests that the odds are 1 to 4 that an American family has trouble paying medical bills. Based on this information, what is the probability that a randomly selected American family has trouble paying its medical bills?

e. Coleman & Associates, Inc., conducted a study to determine who curses at their computer. The results, which appeared in *USA TODAY,* indicated that 46% of people age 18–34 years have cursed at their computer. What are the odds against a randomly selected 18- to 34-year-old having cursed at his or her computer?

f. A roulette wheel contains 38 numbers, of which 18 are red, 18 are black, and 2 are green. When the roulette ball is spun, it is equally likely to land on any of the 38 numbers. For a bet on red, the house pays even odds (i.e., 1 to 1). What should the odds actually be to make the bet fair?

4.23 We stated earlier in this section that the frequentist interpretation cannot be used as a definition of probability. Why do you think that is so?

4.2 EVENTS

Before continuing our study of probability, we need to discuss events in greater detail. In Section 4.1 we used the word *event* intuitively. To be more precise, in probability an **event** consists of a collection of outcomes.

EXAMPLE 4.5 EVENTS

A deck of playing cards contains 52 cards, as seen in Fig. 4.3. When we perform the experiment of randomly selecting one card from the deck, exactly one of these 52 cards will be obtained. The collection of all 52 cards—the possible outcomes—is called the **sample space** for this experiment.

FIGURE 4.3
A deck of playing cards

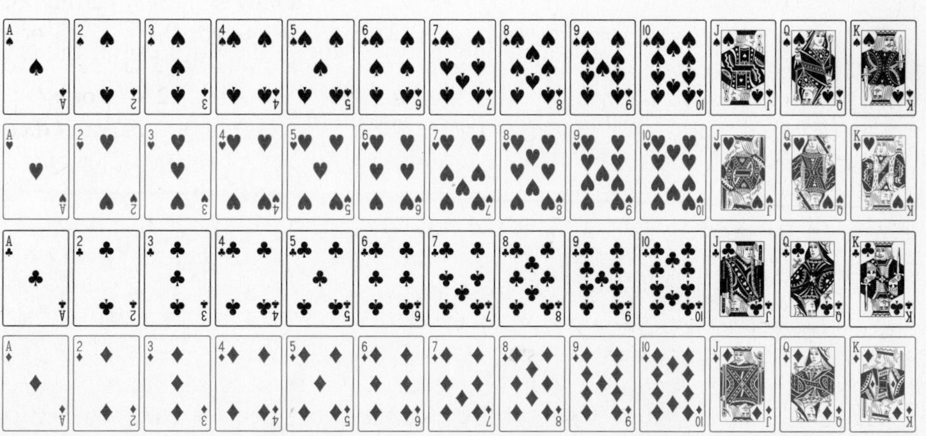

Many different events can be associated with this card-selection experiment. Let's consider the following four:

1. The event that the card selected is the king of hearts.
2. The event that the card selected is a king.
3. The event that the card selected is a heart.
4. The event that the card selected is a face card.

List the outcomes comprising each of these four events.

SOLUTION The first event, the event that the card selected is the king of hearts, consists of the single outcome, "king of hearts," and is pictured in Fig. 4.4.

FIGURE 4.4
The event the king of hearts is selected

The second event, the event that the card selected is a king, consists of the four outcomes, "king of spades," "king of hearts," "king of clubs," and "king of diamonds." This event is depicted in Fig. 4.5.

FIGURE 4.5
The event a king is selected

Thirteen outcomes comprise the third event, the event that the card selected is a heart; namely, the outcomes, "ace of hearts," "two of hearts," . . . , "king of hearts." This event is shown in Fig. 4.6.

FIGURE 4.6
The event a heart is selected

The fourth event, the event that the card selected is a face card, consists of 12 outcomes; namely, the 12 face cards shown in Fig. 4.7.

FIGURE 4.7
The event a face card is selected

When the experiment of selecting a card from the deck is performed, an event *occurs* if it includes the card selected. For instance, if the card selected turns out to be the king of spades, then the second and fourth events (Figs. 4.5 and 4.7) occur, whereas the first and third events (Figs. 4.4 and 4.6) do not. □

The term *sample space* reflects the fact that in statistics, the collection of possible outcomes often consists of the possible samples of a given size, as in Table 4.2 on page 198. Definition 4.2 summarizes the terminology discussed so far in this section.

DEFINITION 4.2 SAMPLE SPACE AND EVENTS

Sample space: The collection of all possible outcomes for an experiment.

Event: A collection of outcomes for the experiment, that is, any subset of the sample space.

Notation and Graphical Displays for Events

It is convenient and less cumbersome to employ letters such as $A, B, C, D, \ldots$ to represent events. For instance, in the card-selection experiment of Example 4.5, we might let

$$A = \text{event the card selected is the king of hearts,}$$
$$B = \text{event the card selected is a king,}$$
$$C = \text{event the card selected is a heart,}$$
$$D = \text{event the card selected is a face card.}$$

Graphical displays of events are useful for explaining and understanding probability. **Venn diagrams,** named after English logician John Venn (1834–1923), are one of the best ways to visually portray events and relationships among events. The sample space is depicted as a rectangle, and the various events are drawn as disks (or other geometric shapes) inside the rectangle. In the simplest case, only one event is displayed, as in Fig. 4.8. The colored portion of Fig. 4.8 represents event E.

FIGURE 4.8
Venn diagram for event E

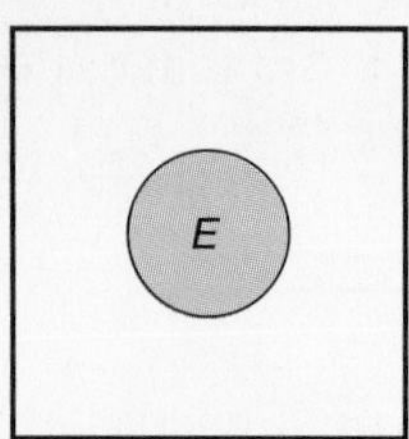

Relationships Among Events

To each event, E, there corresponds another event defined by the condition that "E does not occur." That event is called the **complement** of E and is denoted by **(not E).** Event (not E) consists of all outcomes not in E. A Venn diagram, as in Fig. 4.9(a), makes this idea clearer.

FIGURE 4.9
Venn diagrams for
(a) event (not E),
(b) event (A & B),
(c) event (A or B)

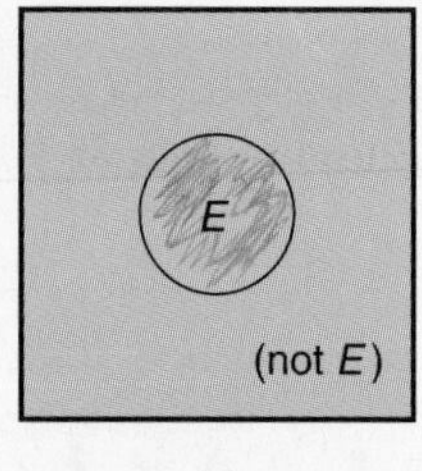

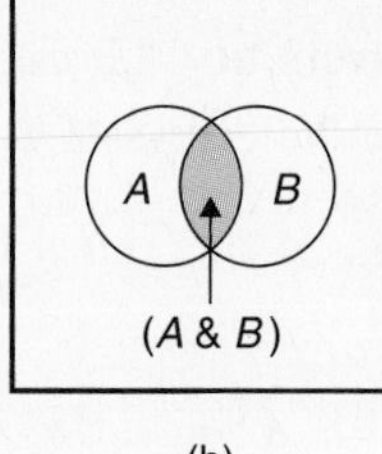

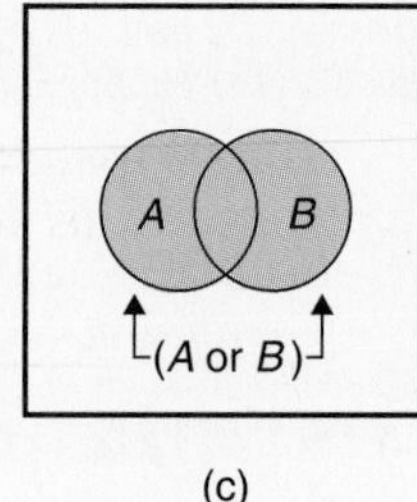

With any two events, say, A and B, we can associate two new events. One new event is defined by the condition that "both event A and event B occur" and is denoted by **(A & B).** Event (A & B) consists of all outcomes common to both event A and event B, as illustrated in Fig. 4.9(b).

The other new event associated with A and B is defined by the condition that "either event A or event B or both occur" or, equivalently, that "at least one of events A and B occurs." That event is denoted by **(A or B)** and consists of all outcomes in either event A or event B or both, as Fig. 4.9(c) shows.

DEFINITION 4.3 RELATIONSHIPS AMONG EVENTS

(not E): The event that "E does not occur."

(A & B): The event that "both A and B occur."

(A or B): The event that "either A or B or both occur."

Because the event that "both A and B occur" is the same as the event that "both B and A occur," event $(A \& B)$ is the same as event $(B \& A)$. Similarly, event $(A \text{ or } B)$ is the same as event $(B \text{ or } A)$.

EXAMPLE 4.6 ILLUSTRATES DEFINITION 4.3

For the experiment of randomly selecting one card from a deck of 52, let

A = event the card selected is the king of hearts,
B = event the card selected is a king,
C = event the card selected is a heart,
D = event the card selected is a face card.

The outcomes constituting each of those four events are shown in Figs. 4.4–4.7, respectively. Determine the following events.

a. (not D) **b.** $(B \& C)$ **c.** $(B \text{ or } C)$ **d.** $(C \& D)$

SOLUTION **a.** (not D) is the event that "D does not occur"—the event that a face card is not selected. Event (not D) consists of the 40 cards in the deck that are not face cards, as depicted in Fig. 4.10.

FIGURE 4.10
Event (not D)

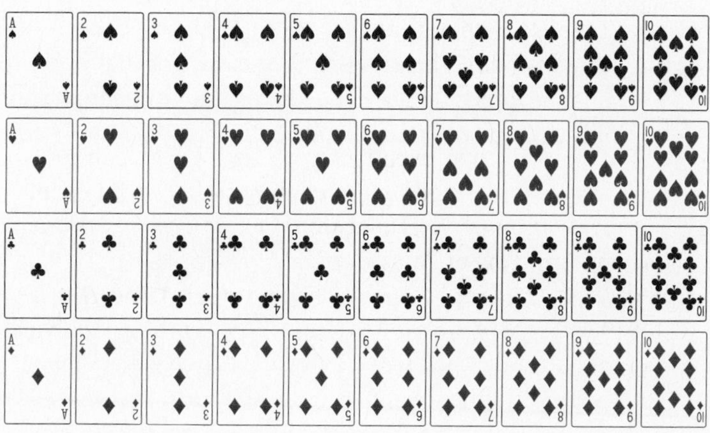

b. $(B \& C)$ is the event that "both B and C occur"—the event that the card selected is both a king and a heart. This can happen only if the card selected is the king of hearts. Consequently, $(B \& C)$ is the event that the card selected is the king of hearts and consists of the single outcome shown in Fig. 4.11. *Note:* Because event $(B \& C)$ is the same as event A, we can write $A = (B \& C)$.

FIGURE 4.11
Event (B & C)

c. (B or C) is the event that "either B or C or both occur"—the event that the card selected is either a king or a heart or both. Event (B or C) consists of 16 outcomes; namely, the 4 kings and the 12 non-king hearts, as illustrated in Fig. 4.12. *Note:* Event (B or C) can occur in 16, not 17, ways since the outcome "king of hearts" is common to both event B and event C.

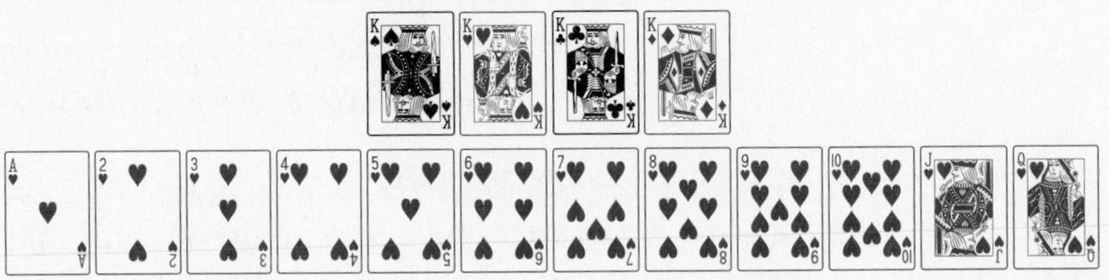

FIGURE 4.12
Event (B or C)

d. (C & D) is the event that "both C and D occur"—the event that the card selected is both a heart and a face card. For that event to occur, the card selected must be either the jack, queen, or king of hearts. Thus event (C & D) consists of the three outcomes displayed in Fig. 4.13.

FIGURE 4.13
Event (C & D)

These three outcomes are the ones common to events C and D. ❑

In Example 4.6, we described each of four events by listing their outcomes (Figs. 4.10–4.13). Sometimes it is more appropriate to describe events verbally, as in Example 4.7.

EXAMPLE 4.7 ILLUSTRATES DEFINITION 4.3

A frequency distribution for the ages of the 40 students in Professor Weiss's introductory statistics class is presented in Table 4.4.

TABLE 4.4
Frequency distribution for students' ages

Age (yrs)	17	18	19	20	21	22	23	24	26	35	36
Frequency	1	1	9	7	7	5	3	4	1	1	1

One student is selected at random. Let

A = event the student selected is under 21,
B = event the student selected is over 30,
C = event the student selected is in his or her 20s,
D = event the student selected is over 18.

Determine the following events.

a. (not D) **b.** (A & D) **c.** (A or D) **d.** (B or C)

SOLUTION

a. (not D) is the event that "D does not occur"—the event that the student selected is not over 18, that is, is 18 or under. As we see from Table 4.4, (not D) is composed of the two students in the class who are 18 or under.

b. (A & D) is the event that "both A and D occur"—the event that the student selected is both under 21 and over 18, that is, is either 19 or 20. Event (A & D) is composed of the 16 students in the class who are 19 or 20.

c. (A or D) is the event that "either A or D or both occur"—the event that the student selected is either under 21 or over 18 or both. But every student in the class is either under 21 or over 18. Consequently, event (A or D) is composed of all 40 students in the class and is certain to occur.

d. (B or C) is the event that "either B or C or both occur"—the event that the student selected is either over 30 or in his or her 20s. Table 4.4 shows that event (B or C) is composed of the 29 students in the class who are 20 or over. ◨

Mutually Exclusive Events

Next we introduce the concept of events being **mutually exclusive.** Definition 4.4 explains what that means.

DEFINITION 4.4 MUTUALLY EXCLUSIVE EVENTS

Two or more events are said to be ***mutually exclusive*** if at most one of them can occur when the experiment is performed, that is, if no two of them have outcomes in common.

The Venn diagrams in Fig. 4.14 portray the difference between two events that are mutually exclusive and two events that are not mutually exclusive. In Fig. 4.15 we show three mutually exclusive events and two cases where three events are not mutually exclusive.

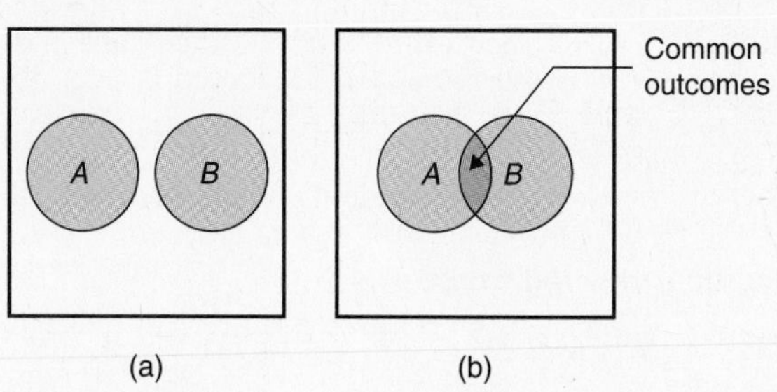

FIGURE 4.14 (a) Two mutually exclusive events (b) Two non–mutually exclusive events

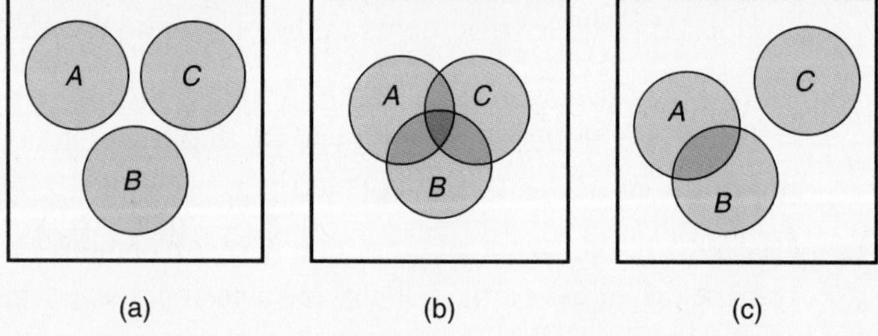

FIGURE 4.15 (a) Three mutually exclusive events (b) Three non–mutually exclusive events (c) Three non–mutually exclusive events

EXAMPLE 4.8 ILLUSTRATES DEFINITION 4.4

For the experiment of randomly selecting one card from a deck of 52, let

C = event the card selected is a heart,
D = event the card selected is a face card,
E = event the card selected is an ace,
F = event the card selected is an 8,
G = event the card selected is a 10 or a jack.

Determine which of the following collections of events are mutually exclusive.

a. C and D **b.** C and E **c.** D and E

d. D, E, and F **e.** D, E, F, and G

SOLUTION **a.** Event C and event D are not mutually exclusive because they have the common outcomes "king of hearts," "queen of hearts," and "jack of hearts." Both events occur if the card selected is the king, queen, or jack of hearts.

b. Event C and event E are not mutually exclusive but have the common outcome "ace of hearts." Both events occur if the card selected is the ace of hearts.

c. Event D and event E are mutually exclusive since they have no common outcomes. They cannot both occur when the experiment is performed because it is impossible to select a card that is both a face card and an ace.

d. Events D, E, and F are mutually exclusive because no two of them can occur simultaneously.

e. Events D, E, F, and G are not mutually exclusive since event D and event G both occur if the card selected is a jack. ◨

EXERCISES 4.2

STATISTICAL CONCEPTS AND SKILLS

4.24 Fill in the following blanks.

a. The collection of all possible outcomes for an experiment is called the ____.

b. A collection of possible outcomes is called an ____.

4.25 What type of graphical displays are useful for portraying events and relationships among events?

4.26 Construct Venn diagrams representing each of the following events.

a. (not E) **b.** (A or B) **c.** (A & B)
d. (A & B & C) **e.** (A or B or C)
f. ((not A) & B)

4.27 What does it mean for two events to be mutually exclusive? What about three events?

4.28 Answer true or false to each of the following statements and give reasons for your answers.

a. If event A and event B are mutually exclusive, then so are events A, B, and C for every event C.

b. If event A and event B are not mutually exclusive, then neither are events A, B, and C for every event C.

4.29 When one die is rolled, the following six outcomes are possible:

List the outcomes comprising each of the events below.

A = event the die comes up even,
B = event the die comes up 4 or more,
C = event the die comes up at most 2,
D = event the die comes up 3.

4.30 In a horse race, the odds against winning are given in the table below. The table shows, for example, that the odds against winning are 8 to 1 for horse #1.

Horse	#1	#2	#3	#4	#5	#6	#7	#8
Odds	8	15	2	3	30	5	10	5

List the outcomes comprising each of the events below.

A = event that one of the top two favorites wins (the top two favorites are the two horses with the lowest odds against winning),

B = event that the winning horse's number is above 5,

C = event that the winning horse's number is at most 3, that is, 3 or less,

D = event that one of the two long shots wins (the two long shots are the two horses with the highest odds against winning).

4.31 When a dime is tossed four times, 16 outcomes are possible:

HHHH	HTHH	THHH	TTHH
HHHT	HTHT	THHT	TTHT
HHTH	HTTH	THTH	TTTH
HHTT	HTTT	THTT	TTTT

Here, for example, HTTH represents the outcome that the first toss is heads, the next two tosses are tails, and the fourth toss is heads. List the outcomes that constitute each of the following four events.

A = event exactly two heads are tossed,

B = event the first two tosses are tails,

C = event the first toss is heads,

D = event all four tosses come up the same.

4.32 A committee consists of five executives, three women and two men. Their names are Maria (M), John (J), Susan (S), Bill (B), and Carol (C). The committee needs to select a chairperson and a secretary. It decides to make the selection randomly by drawing straws. The person getting the longest straw will be appointed chairperson and the one getting the shortest straw will be appointed secretary. We can represent the possible outcomes in the following manner.

MS	SM	CM	JM	BM
MC	SC	CS	JS	BS
MJ	SJ	CJ	JC	BC
MB	SB	CB	JB	BJ

Here, for example, MS represents the outcome that Maria is appointed chairperson and Susan is appointed secretary. List the outcomes comprising each of the following four events.

A = event a male is appointed chairperson,

B = event Carol is appointed chairperson,

C = event Bill is appointed secretary,

D = event only females are appointed.

4.33 Refer to Exercise 4.29. For each of the following events, list the outcomes that constitute the event and describe the event in words.
a. (not A) **b.** (A & B) **c.** (B or C)

4.34 Refer to Exercise 4.30. For each of the following events, list the outcomes that constitute the event and describe the event in words.
a. (not C) **b.** (C & D) **c.** (A or C)

4.35 Refer to Exercise 4.31. For each of the following events, list the outcomes that comprise the event and describe the event in words.
a. (not B) **b.** (A & B) **c.** (C or D)

4.36 Refer to Exercise 4.32. For each of the following events, list the outcomes that comprise the event and describe the event in words.
a. (not A) **b.** (B & D) **c.** (B or C)

4.37 The U.S. Bureau of the Census publishes data on housing units in *American Housing Survey in the United States*. The following table provides a frequency distribution for the number of rooms in U.S. housing units. The frequencies are in thousands.

Rooms	No. of units
1	862
2	1,422
3	10,166
4	20,789
5	24,328
6	22,151
7	14,183
8+	15,555

For a U.S. housing unit selected at random, let

A = event the unit has at most four rooms,

B = event the unit has at least two rooms,

C = event the unit has between five and seven rooms, inclusive,

D = event the unit has more than seven rooms.

Describe each of the following events in words and determine the number of outcomes (housing units) that comprise each event.

a. (not A) **b.** (A & B) **c.** (C or D)

4.38 A *family* is defined to be a group of two or more persons related by birth, marriage, or adoption and residing together in a household. According to *Current Population Reports,* a publication of the U.S. Bureau of the Census, the size distribution of U.S. families is as follows. The frequencies are in thousands.

Size	No. of families
2	29,765
3	15,771
4	14,421
5	6,234
6	2,182
7+	1,221

For a U.S. family selected at random, let

A = event the family has at most five members,

B = event the family has at least three members,

C = event the family has between four and six members, inclusive,

D = event the family has at least five members.

Describe each of the following events in words and determine the number of outcomes (families) that comprise each event.

a. (not B) **b.** (C & D) **c.** (A or D)

4.39 As reported by the Bureau of Labor Statistics in *Employment and Earnings,* the age distribution of employed persons 16 years old and over is as shown in the following table.

Age (years)	Frequency (1000s)
16–19	6,500
20–24	12,138
25–34	32,077
35–44	35,051
45–54	25,514
55–64	11,739
65 & over	3,690

An employed person is selected at random. Let

A = event the person is under 20,

B = event the person is between 20 and 54, inclusive,

C = event the person is under 45,

D = event the person is 55 or over.

Describe each of the following events in words and determine the number of outcomes (persons) that comprise each event.

a. (not C) **b.** (not B) **c.** (B & C)
d. (A or D)

4.40 Draw a Venn diagram portraying four mutually exclusive events.

4.41 Refer to Exercise 4.29.

a. Are events A and B mutually exclusive?
b. Are events B and C mutually exclusive?
c. Are events A, C, and D mutually exclusive?
d. Are there three mutually exclusive events among A, B, C, and D? How about four?

4.42 Each part of this exercise lists events from Exercise 4.30. In each case, decide whether the events are mutually exclusive.

a. A and B **b.** B and C **c.** A, B, and C
d. A, B, and D **e.** A, B, C, and D

4.43 For the following groups of events from Exercise 4.39, determine which are mutually exclusive.

a. C and D **b.** B and C **c.** A, B, and D
d. A, B, and C **e.** A, B, C, and D

4.44 For the following groups of events from Exercise 4.38, determine which are mutually exclusive.

a. A and B **b.** (not B) and C
c. A and D **d.** (not B), C, and D

Extending the Concepts and Skills

4.45 Construct a Venn diagram that portrays four events, A, B, C, and D, having the following properties: Events A, B, and C are mutually exclusive; events A, B, and D are mutually exclusive; no other three of the four events are mutually exclusive.

4.46 Suppose A, B, and C are three events with the property that they cannot all occur simultaneously. Does this necessarily imply that A, B, and C are mutually exclusive? Justify your answer and illustrate it with a Venn diagram.

4.3 SOME RULES OF PROBABILITY

In this section we will discuss several rules of probability. Before beginning, however, we need to introduce an additional notation used in probability.

EXAMPLE 4.9 An Additional Probability Notation

When a balanced die is rolled once, six equally likely outcomes are possible, as shown in Fig. 4.16.

FIGURE 4.16
Sample space for rolling a die once

Consider, for instance, the event that the die comes up even. This event can occur in three ways; namely, if 2, 4, or 6 is rolled. Since $f/N = 3/6 = 0.5$, we see that *the probability is 0.5 that the die comes up even.*

Employing probability notation enables us to express the italicized phrase much more concisely. Let A denote the event that the die comes up even. We use the notation $P(A)$ to represent the probability that event A occurs. So the italicized statement can be written simply as $P(A) = 0.5$, read "the probability of A is 0.5." Keep in mind that A refers to the event that the die comes up even, whereas $P(A)$ refers to the probability of that event occurring. ◨

DEFINITION 4.5 PROBABILITY NOTATION

If E is an event, then $\boldsymbol{P(E)}$ stands for the probability that event E occurs. It is read "the probability of E."

The Special Addition Rule

The first rule of probability we will study is the **special addition rule,** which states that for mutually exclusive events, the probability that one or another of the events occurs equals the sum of the individual probabilities. We can use a Venn diagram to see the validity of the special addition rule. Figure 4.17 shows two mutually exclusive events, A and B.

FIGURE 4.17
Two mutually exclusive events

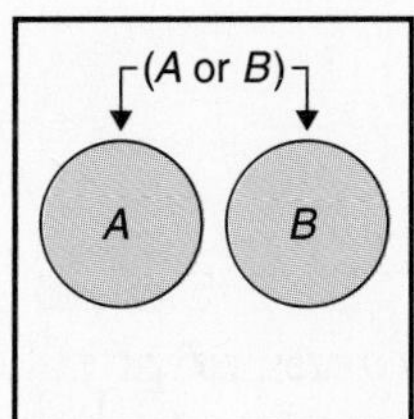

If we think of the colored regions in Fig. 4.17 as probabilities, then the colored disk on the left is $P(A)$, the colored disk on the right is $P(B)$, and the total colored region is $P(A \text{ or } B)$. Because event A and event B are mutually exclusive, the total colored region equals the sum of the two colored disks; that is, $P(A \text{ or } B) = P(A) + P(B)$.

FORMULA 4.1 THE SPECIAL ADDITION RULE

If event A and event B are mutually exclusive, then

$$P(A \text{ or } B) = P(A) + P(B).$$

More generally, if events $A, B, C, \ldots$ are mutually exclusive, then

$$P(A \text{ or } B \text{ or } C \text{ or } \cdots) = P(A) + P(B) + P(C) + \cdots.$$

In words, for mutually exclusive events, the probability that one or another of the events occurs equals the sum of the individual probabilities.

EXAMPLE 4.10 ILLUSTRATES FORMULA 4.1

The U.S. Bureau of the Census compiles information about farms and publishes its findings in *Census of Agriculture.* According to that publication, a relative-frequency distribution for the size of farms in the United States is as presented in the first two columns of Table 4.5. The table shows, for instance, that 15.6% (0.156) of U.S. farms have between 100 and 180 acres (i.e., at least 100 acres but less than 180 acres).

TABLE 4.5
Size of farms in the United States

Size (acres)	Relative frequency	Event
Under 10	0.086	A
10 < 50	0.202	B
50 < 100	0.147	C
100 < 180	0.156	D
180 < 260	0.089	E
260 < 500	0.133	F
500 < 1000	0.097	G
1000 < 2000	0.053	H
2000 & over	0.037	I

In the third column of Table 4.5, we have introduced events that correspond to each size class. For example, if a farm is selected at random, then D denotes the event that the farm obtained has between 100 and 180 acres. The probabilities of the events in the third column of Table 4.5 equal the relative frequencies displayed in the second column. Thus the probability that a randomly selected farm has between 100 and 180 acres is $P(D) = 0.156$.

Use Table 4.5 and the special addition rule to determine the probability that a randomly selected farm has between 100 and 500 acres.

SOLUTION As we see from Table 4.5, the event that the farm obtained has between 100 and 500 acres can be expressed as (D or E or F). Events D, E, and F are mutually exclusive and so by the special addition rule,

$$\begin{aligned} P(D \text{ or } E \text{ or } F) &= P(D) + P(E) + P(F) \\ &= 0.156 + 0.089 + 0.133 = 0.378. \end{aligned}$$

In terms of percentages, this means that 37.8% of U.S. farms have between 100 and 500 acres. ◨

The Complementation Rule

The second rule of probability we will study is the **complementation rule,** which states that the probability an event occurs equals 1 minus the probability it does not occur. We can use a Venn diagram to see the validity of the complementation rule. Figure 4.18 shows an event, E, and its complement, (not E).

If we think of the regions in Fig. 4.18 as probabilities, then the entire region enclosed by the rectangle is the probability of the sample space, which is 1. Furthermore, the colored region is $P(E)$ and the uncolored region is $P(\text{not } E)$. So we see that $P(E) + P(\text{not } E) = 1$ or, equivalently, $P(E) = 1 - P(\text{not } E)$.

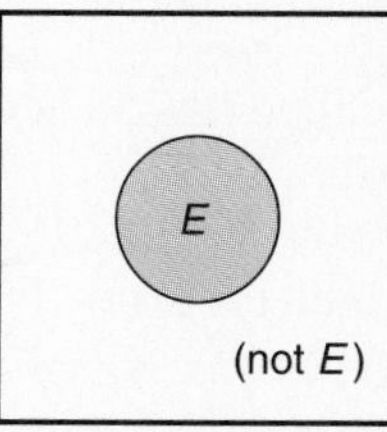

FIGURE 4.18
An event and its complement

FORMULA 4.2 **THE COMPLEMENTATION RULE**

For any event E,

$$P(E) = 1 - P(\text{not } E).$$

In words, the probability that an event occurs equals 1 minus the probability that it does not occur.

The complementation rule is useful because it is sometimes easier to compute the probability that an event does not occur than the probability that it does occur. In such cases we can, in view of the complementation rule, obtain the probability that the event occurs by first computing the probability that it does not occur and then subtracting the result from 1. Example 4.11 provides an illustration of this fact.

EXAMPLE 4.11 **ILLUSTRATES FORMULA 4.2**

The first two columns of Table 4.5 on page 220 provide a relative-frequency distribution for the size of U.S. farms. For a randomly selected farm, find the probability that the farm obtained has

a. less than 2000 acres. **b.** 50 acres or more.

SOLUTION **a.** Let

$$J = \text{event the farm obtained has less than 2000 acres.}$$

To determine $P(J)$ we will apply the complementation rule, since it is easier to compute $P(\text{not } J)$. Note that (not J) is the event that the farm obtained has 2000 or more acres, which is event I in Table 4.5. Consequently, we see that $P(\text{not } J) = P(I) = 0.037$. Applying the complementation rule, we find that

$$P(J) = 1 - P(\text{not } J) = 1 - 0.037 = 0.963.$$

The probability is 0.963 that a randomly selected farm has less than 2000 acres.

b. Let

$$K = \text{event the farm obtained has 50 acres or more.}$$

We will apply the complementation rule to find $P(K)$. Now, (not K) is the event that the farm obtained has less than 50 acres. From Table 4.5 we see that event (not K) is the same as event (A or B). Since event A and event B are mutually exclusive, the special addition rule implies that

$$P(\text{not } K) = P(A \text{ or } B) = P(A) + P(B) = 0.086 + 0.202 = 0.288.$$

Using this fact and the complementation rule, we conclude that

$$P(K) = 1 - P(\text{not } K) = 1 - 0.288 = 0.712.$$

The probability is 0.712 that a randomly selected farm has 50 acres or more. □

The General Addition Rule

The special addition rule (Formula 4.1 on page 219) gives a formula for finding the probability of event (A or B) from the probabilities of event A and event B, provided event A and event B are mutually exclusive. For events that are not mutually exclusive, we must use a different rule—the *general addition rule.* To introduce the general addition rule, we will employ the Venn diagram shown in Fig. 4.19.

FIGURE 4.19
Non–mutually exclusive events

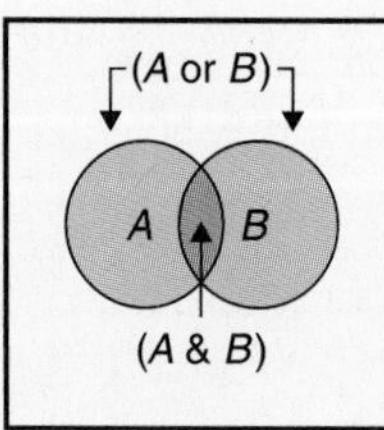

If we think of the colored regions in Fig. 4.19 as probabilities, then the colored disk on the left is $P(A)$, the colored disk on the right is $P(B)$, and the total colored region is $P(A \text{ or } B)$. To obtain the total colored region, $P(A \text{ or } B)$, we first sum the two colored disks, $P(A)$ and $P(B)$. In doing this, the common colored region, $P(A \,\&\, B)$, is counted twice. To account for that fact, we must subtract $P(A \,\&\, B)$ from the sum. So we see that $P(A \text{ or } B) = P(A) + P(B) - P(A \,\&\, B)$. This formula is the **general addition rule.**

FORMULA 4.3 **THE GENERAL ADDITION RULE**

If A and B are any two events, then

$$P(A \text{ or } B) = P(A) + P(B) - P(A \& B).$$

In words, for any two events, the probability that one or the other occurs equals the sum of the individual probabilities less the probability that both occur.

There are also general addition rules for three events, four events, five events, and so forth, but these rules are much more complicated than the one for two events. We will discuss the general addition rule for three events in the exercises.

EXAMPLE 4.12 **ILLUSTRATES FORMULA 4.3**

Consider again the experiment of selecting one card at random from a deck of 52 playing cards. Find the probability that the card selected is either a spade or a face card

a. without using the general addition rule.

b. using the general addition rule.

SOLUTION **a.** Let

$$E = \text{event the card selected is either a spade or a face card.}$$

Event E consists of 22 cards; namely, the 13 spades plus the other nine face cards that are not spades, as shown in Fig. 4.20.

FIGURE 4.20
Event E

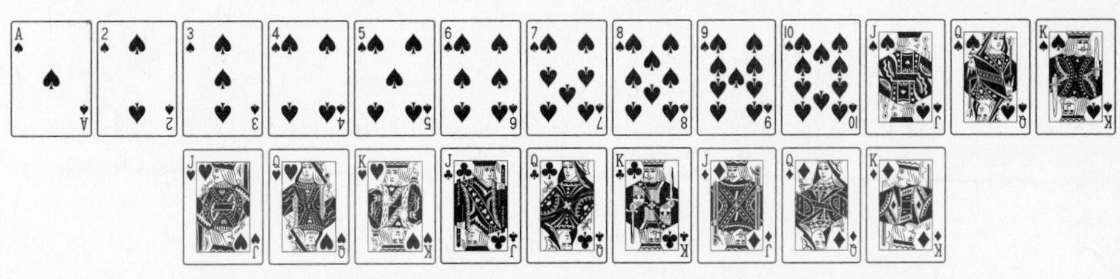

Consequently, by the f/N rule,

$$P(E) = \frac{f}{N} = \frac{22}{52} = 0.423.$$

b. To determine $P(E)$ using the general addition rule, we first note that we can write $E = (C \text{ or } D)$, where

$$C = \text{event the card selected is a spade,}$$
$$D = \text{event the card selected is a face card.}$$

Event C consists of the 13 spades and event D consists of the 12 face cards. Also, event $(C \,\&\, D)$ consists of the three spades that are face cards—the jack, queen, and king of spades. Applying the general addition rule, we get

$$P(E) = P(C \text{ or } D) = P(C) + P(D) - P(C \,\&\, D)$$
$$= \frac{13}{52} + \frac{12}{52} - \frac{3}{52} = 0.250 + 0.231 - 0.058 = 0.423.$$

This agrees with the answer obtained in part (a). ◨

In Example 4.12 we computed the probability of selecting either a spade or a face card in two ways: first without using the general addition rule and then using it. There, computing the probability was simpler without using the general addition rule. Frequently, however, the general addition rule is the easier or even the only way to compute a probability. This point is illustrated in Example 4.13.

EXAMPLE 4.13 **ILLUSTRATES FORMULA 4.3**

Data on people arrested are published by the U.S. Federal Bureau of Investigation in *Crime in the United States.* In 1995, 79.6% of people arrested were male, 18.3% were under 18 years old, and 13.5% were males under 18 years old. A person arrested in 1995 is selected at random. Find the probability that the person obtained is either male or under 18.

SOLUTION Let

$$M = \text{event the person obtained is male,}$$
$$E = \text{event the person obtained is under 18.}$$

The event that the person obtained is either male or under 18 can be expressed as $(M \text{ or } E)$. We want to determine $P(M \text{ or } E)$. From the percentage data given, we know that $P(M) = 0.796$, $P(E) = 0.183$, and $P(M \,\&\, E) = 0.135$. Applying the general addition rule, we conclude that

$$P(M \text{ or } E) = P(M) + P(E) - P(M \,\&\, E)$$
$$= 0.796 + 0.183 - 0.135 = 0.844.$$

In terms of percentages, this means that 84.4% of those arrested in 1995 were either male or under 18 years old. ◨

The general addition rule is consistent with the special addition rule—if two events are mutually exclusive, then both rules yield the same result.

EXERCISES 4.3

STATISTICAL CONCEPTS AND SKILLS

4.47 A marble is selected at random from a bowl containing 10 marbles of which 3 are red, 2 are white, and 5 are blue. Let E denote the event that the marble obtained is white. Determine the probability that the marble obtained is white. Use probability notation to express your answer.

4.48 Suppose you hold 20 out of a total of 500 tickets sold for a lottery. The grand-prize winner is determined by the random selection of one of the 500 tickets. Let G be the event that you win the grand prize. Find the probability that you win the grand prize. Express your answer in probability notation.

4.49 According to the *Congressional Directory,* the age distribution for senators in the 104th U.S. Congress is as follows.

Age (yrs)	No. of senators
Under 40	1
40–49	14
50–59	41
60–69	27
70 and over	17
	100

For a senator selected at random, let

A = event the senator is under 40,

B = event the senator is in his or her 40s,

C = event the senator is in his or her 50s,

S = event the senator is under 60.

a. Use the table and the f/N rule to find $P(S)$.
b. Express event S in terms of events A, B, and C.
c. Determine $P(A)$, $P(B)$, and $P(C)$.
d. Compute $P(S)$ using the special addition rule and your answers from parts (b) and (c). Compare your answer with the one you found in part (a).

4.50 As reported by Dun & Bradstreet in *Business Failure Record,* the numbers of commercial failures for the year 1995 by type of industry are as follows.

Industry	Failures
Agriculture, forestry, fishing	2,231
Mining	200
Construction	9,158
Manufacturing	4,383
Transportation, public utilities	2,733
Wholesale trade	4,149
Retail trade	12,952
Finance, insurance, real estate	4,293
Services	21,850
Public adminstration	24
Other	9,221
	71,194

For a 1995 failed business selected at random, let

A = event it was in wholesale trade,

B = event it was in retail trade,

T = event it was in either wholesale trade or retail trade.

a. Use the table and the f/N rule to find $P(T)$.
b. Express event T in terms of events A and B.
c. Determine $P(A)$ and $P(B)$.
d. Compute $P(T)$ using the special addition rule and your answers from parts (b) and (c). Compare your answer with the one you found in part (a).

4.51 According to *Statistics of Income,* compiled by the Internal Revenue Service, a relative-frequency distribution for U.S. partnerships by receipts received from sales and services is as follows.

Receipts	Relative frequency	Event
Under \$25,000	0.601	A
\$25,000–\$49,999	0.081	B
\$50,000–\$99,999	0.088	C
\$100,000–\$499,999	0.153	D
\$500,000–\$999,999	0.036	E
\$1,000,000 or more	0.041	F

We see, for example, that 60.1% of U.S. partnerships had receipts of under \$25,000. For a U.S. partnership selected at random, let

A = event the partnership obtained had receipts under \$25,000,

B = event the partnership obtained had receipts between \$25,000 and \$49,999, inclusive,

and so on (see the third column of the table). Find the probability that the partnership obtained had receipts of

a. under \$100,000.
b. at least \$500,000.
c. between \$25,000 and \$499,999, inclusive.
d. Interpret each of your answers in parts (a)–(c) in terms of percentages.

4.52 The U.S. Bureau of the Census compiles information on educational attainment and publishes its findings in *Current Population Reports.* Following is a percentage distribution for the number of years of school completed by U.S. adults, 25 years old and over.

Years completed	Percentage	Event
0–4	2.4	A
5–7	3.8	B
8	4.4	C
9–11	11.0	D
12	38.6	E
13–15	18.4	F
16 or more	21.4	G

The table shows, for instance, that 38.6% of adults 25 years old and over have completed exactly 12 years of school. Suppose an adult 25 years old or over is randomly selected from the population. Let

A = event the person obtained has completed between 0 and 4 years of school,

B = event the person obtained has completed between 5 and 7 years of school,

and so forth (see the third column of the table). Determine the probability that the person obtained

a. has at most an elementary-school education, that is, has completed 8 years of school or less.
b. has at most a high-school education, that is, has completed 12 years of school or less.
c. has completed at least 1 year of college, that is, 13 years of school or more.
d. Interpret each of your answers in parts (a)–(c) in terms of percentages.

In Exercises 4.53–4.56, find the designated probabilities using the complementation rule.

4.53 Refer to Exercise 4.49. Find the probability that a randomly selected senator in the 104th Congress is
a. 40 years old or older. **b.** under 60 years old.

4.54 Refer to Exercise 4.50. Find the probability that a failed business selected at random
a. was not in construction.
b. was neither in manufacturing nor services.

4.55 Refer to Exercise 4.51. Find the probability that a randomly selected partnership had receipts of
a. under \$1,000,000. **b.** \$50,000 or greater.

4.56 Refer to Exercise 4.52. Determine the probability that a randomly selected adult 25 years old or over has completed
a. less than 4 years of college.
b. 8 years or more of school.

4.57 In the game of *craps,* a player rolls two balanced dice. There are 36 equally likely outcomes possible, as shown in Fig. 4.1 on page 201. Let

A = event the sum of the dice is 7,

B = event the sum of the dice is 11,

C = event the sum of the dice is 2,

D = event the sum of the dice is 3,

E = event the sum of the dice is 12,

F = event the sum of the dice is 8,

G = event doubles are rolled.

a. Compute the probability of each of the seven events.
b. The player wins on the first roll if the sum of the dice is 7 or 11. Find the probability of that event using the special addition rule and your answers from part (a).
c. The player loses on the first roll if the sum of the dice is 2, 3, or 12. Determine the probability of that event using the special addition rule and your answers from part (a).
d. Compute the probability that either the sum of the dice is 8 or doubles are rolled without using the general addition rule.
e. Compute the probability that either the sum of the dice is 8 or doubles are rolled using the general addition rule and compare your answer to the one you obtained in part (d).

4.58 As reported by the U.S. Bureau of Justice Statistics in *Profile of Jail Inmates,* 56.5% of jail inmates are white, 94.0% are male, and 53.5% are white males. For a randomly selected jail inmate, let

W = event the inmate obtained is white,

M = event the inmate obtained is male.

a. Find $P(W)$, $P(M)$, and $P(W \& M)$.
b. Determine $P(W \text{ or } M)$ and interpret your answer in terms of percentages.
c. Obtain the probability that a randomly selected inmate is female.

4.59 According to *Current Population Reports,* published by the Census Bureau, 52.0% of U.S. adults are female, 9.5% are divorced, and 5.4% are divorced females. For a U.S. adult selected at random, let

F = event the person is female,

D = event the person is divorced.

a. Obtain $P(F)$, $P(D)$, and $P(F \& D)$.
b. Determine $P(F \text{ or } D)$ and interpret your answer in terms of percentages.
c. Find the probability that a randomly selected adult is male.

4.60 Let A and B be events such that $P(A) = \frac{1}{4}$, $P(B) = \frac{1}{3}$, and $P(A \text{ or } B) = \frac{1}{2}$.
a. Are events A and B mutually exclusive? Explain your answer.
b. Determine $P(A \& B)$.

4.61 Let A and B be events such that $P(A) = \frac{1}{3}$, $P(A \text{ or } B) = \frac{1}{2}$, and $P(A \& B) = \frac{1}{10}$. Find $P(B)$.

EXTENDING THE CONCEPTS AND SKILLS

4.62 Suppose A and B are mutually exclusive events.
a. Use the special addition rule to express $P(A \text{ or } B)$ in terms of $P(A)$ and $P(B)$.
b. Show that the general addition rule gives the same answer.

4.63 Gerald Kushel, Ed.D., was interviewed by *Bottom Line/Personal* on the secrets of successful people. To study success, Kushel questioned 1200 people, among whom were lawyers, artists, teachers, and students. He found that 15% enjoy neither their jobs nor their personal lives, 80% enjoy their jobs but not their personal lives, and 4% enjoy both their jobs and their personal lives. Determine the percentage of the 1200 people interviewed who
a. enjoy either their jobs or their personal lives.
b. enjoy their personal lives but not their jobs.

4.64 The general addition rule for three events is

$$\begin{aligned} P(A \text{ or } B \text{ or } C) = P(A) &+ P(B) + P(C) \\ &- P(A \& B) - P(A \& C) \\ &- P(B \& C) + P(A \& B \& C). \end{aligned}$$

To illustrate this rule, consider the experiment of tossing a balanced dime three times. Eight equally likely outcomes are possible:

HHH	HTH	THH	TTH
HHT	HTT	THT	TTT

Let

A = event the first toss is heads,

B = event the second toss is tails,

C = event the third toss is heads.

a. List the outcomes comprising each of those events.
b. Find $P(A)$, $P(B)$, and $P(C)$.
c. Describe each of the following events in words and list the outcomes that comprise each one: $(A \,\&\, B)$, $(A \,\&\, C)$, $(B \,\&\, C)$, and $(A \,\&\, B \,\&\, C)$.
d. Find the probability of each event in part (c).
e. Describe event (A or B or C) in words and list the outcomes that comprise it.
f. Find $P(A \text{ or } B \text{ or } C)$ using your answer from part (e) and the f/N rule.
g. Determine $P(A \text{ or } B \text{ or } C)$ using your answers from parts (b) and (d) and the general addition rule for three events.

4.65 Referring to Exercise 4.64, provide the formula for the general addition rule for four events.

4.4 CONTINGENCY TABLES; JOINT AND MARGINAL PROBABILITIES*

In Section 2.2 we discussed grouping data obtained from one variable of a population into a frequency distribution. Data obtained by observing values of one variable of a population are called **univariate data.**

We often need to group and analyze data obtained from two variables of a population. Data obtained by observing values of two variables of a population are called **bivariate data,** and a frequency distribution for bivariate data is called a **contingency table** or **two-way table.** Example 4.14 introduces contingency tables.

EXAMPLE 4.14 CONTINGENCY TABLES

The *Arizona State University Statistical Summary* provides information on various characteristics of the ASU faculty. Data on the variables age and rank of ASU faculty members yielded the contingency table shown in Table 4.6.

The small boxes inside the rectangle formed by the heavy lines of Table 4.6 are called **cells.** The number 2 in the upper left cell of Table 4.6 indicates that two faculty members are full professors under the age of 30. The number 170, diagonally below and to the right of the 2, shows that 170 faculty members are associate professors in their 30s.

The row total in the first row of Table 4.6 reveals that 68 $(2 + 3 + 57 + 6)$ of the faculty members are under 30. Similarly, the column total in the third column shows that 320 of the faculty members are assistant professors. The number 1164 in the lower right corner of the table gives the total number of faculty. That total can be found by summing either the row totals or the column totals; it can also be found by summing the frequencies in the 20 cells of the contingency table. □

* The remainder of this chapter is optional and is not needed for the presentation of inferential statistics. Sections 4.4–4.6 are recommended for those covering the binomial distribution (Section 5.3).

TABLE 4.6
Contingency table for age and rank of ASU faculty members

Age	Rank: Full professor R_1	Associate professor R_2	Assistant professor R_3	Instructor R_4	Total
Under 30 A_1	2	3	57	6	68
30–39 A_2	52	170	163	17	402
40–49 A_3	156	125	61	6	348
50–59 A_4	145	68	36	4	253
60 & over A_5	75	15	3	0	93
Total	430	381	320	33	1164

Joint and Marginal Probabilities

We will now use the age and rank data from Table 4.6 to introduce the concepts of joint probabilities and marginal probabilities.

EXAMPLE 4.15 JOINT AND MARGINAL PROBABILITIES

Suppose an ASU faculty member is selected at random. Notice that the rows and columns of Table 4.6 are labeled with subscripted letters. The subscripted letter A_1, labeling the first row, represents the event that the faculty member obtained is under 30 years old:

$$A_1 = \text{event the faculty member is under 30.}$$

Similarly,

$$R_2 = \text{event the faculty member is an associate professor,}$$

and so forth. The events A_1, A_2, A_3, A_4, and A_5 are mutually exclusive, as are the events R_1, R_2, R_3, and R_4.

In addition to considering events A_1 through A_5 and R_1 through R_4 separately, we can also consider them jointly. For example, the event that the faculty member obtained is under 30 (event A_1) *and* is also an associate professor (event R_2) can be expressed as $(A_1 \& R_2)$:

$$(A_1 \& R_2) = \text{event the faculty member is an associate professor under 30.}$$

Event $(A_1 \& R_2)$ is represented by the cell in the first row and second column of Table 4.6. Here there are 20 different joint events, one for each cell of the contingency table.

It is sometimes useful to think of a contingency table as a Venn diagram. The Venn diagram corresponding to the contingency table in Table 4.6 is shown in Fig. 4.21. That Venn diagram makes it clear that the 20 joint events, $(A_1 \& R_1)$, $(A_1 \& R_2)$, ..., $(A_5 \& R_4)$, are mutually exclusive.

FIGURE 4.21 Venn diagram corresponding to Table 4.6

	R_1	R_2	R_3	R_4
A_1	$(A_1 \& R_1)$	$(A_1 \& R_2)$	$(A_1 \& R_3)$	$(A_1 \& R_4)$
A_2	$(A_2 \& R_1)$	$(A_2 \& R_2)$	$(A_2 \& R_3)$	$(A_2 \& R_4)$
A_3	$(A_3 \& R_1)$	$(A_3 \& R_2)$	$(A_3 \& R_3)$	$(A_3 \& R_4)$
A_4	$(A_4 \& R_1)$	$(A_4 \& R_2)$	$(A_4 \& R_3)$	$(A_4 \& R_4)$
A_5	$(A_5 \& R_1)$	$(A_5 \& R_2)$	$(A_5 \& R_3)$	$(A_5 \& R_4)$

Let's now move on to an examination of probabilities. Because the total number of faculty members is 1164, we have $N = 1164$. To determine, for instance, the probability that the faculty member obtained is an associate professor (event R_2), we first note from Table 4.6 that $f = 381$ and then apply the f/N rule:

$$P(R_2) = \frac{f}{N} = \frac{381}{1164} = 0.327.$$

Similarly, the probability that the faculty member obtained is under 30 equals

$$P(A_1) = \frac{f}{N} = \frac{68}{1164} = 0.058.$$

We can also find probabilities for joint events, so-called **joint probabilities.** For instance, the probability that the faculty member obtained is an associate professor under 30 equals

$$P(A_1 \& R_2) = \frac{f}{N} = \frac{3}{1164} = 0.003.$$

frequency

In Table 4.7 we have replaced the joint frequency distribution in Table 4.6 with a **joint probability distribution.** The probabilities in Table 4.7 are determined in the same way as the three probabilities we just computed.

TABLE 4.7
Joint probability distribution corresponding to Table 4.6

Age	Rank: Full professor R_1	Associate professor R_2	Assistant professor R_3	Instructor R_4	$P(A_i)$
Under 30 A_1	0.002	0.003	0.049	0.005	0.058
30–39 A_2	0.045	0.146	0.140	0.015	0.345
40–49 A_3	0.134	0.107	0.052	0.005	0.299
50–59 A_4	0.125	0.058	0.031	0.003	0.217
60 & over A_5	0.064	0.013	0.003	0.000	0.080
$P(R_j)$	0.369	0.327	0.275	0.028	1.000

Notice that the joint probabilities are displayed in the cells of Table 4.7. Also observe that the row and column labels "Total" in Table 4.6 have been changed in Table 4.7 to $P(R_j)$ and $P(A_i)$, respectively. This is because the last row of Table 4.7 gives the probabilities of events R_1 through R_4 and the last column gives the probabilities of events A_1 through A_5. Those probabilities are often called **marginal probabilities** because they are in the margin of the joint probability distribution.

The sum of the joint probabilities in a row or column of a joint probability distribution equals the marginal probability in that row or column, with any observed discrepancy being due to roundoff error. For example, consider the A_4 row of Table 4.7. The sum of the joint probabilities in that row is

$$0.125 + 0.058 + 0.031 + 0.003 = 0.217,$$

 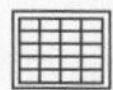 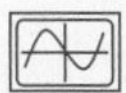

which is precisely the marginal probability at the end of the A_4 row. ◨

EXERCISES 4.4

Statistical Concepts and Skills

4.66 What is a contingency table? What are its cells?

4.67 Identify three ways in which the total number of observations of bivariate data can be obtained from the frequencies in a contingency table.

4.68 Suppose that bivariate data are to be grouped into a contingency table. Determine the number of cells that the contingency table will have if the number of possible values for the two variables are

a. two and three. **b.** four and three.
c. m and n.

4.69 Fill in the following blanks.

a. Data obtained by observing values of one variable of a population are called _____ data.
b. Data obtained by observing values of two variables of a population are called _____ data.

4.70 Give examples of univariate and bivariate data.

4.71 The U.S. National Center for Education Statistics compiles information on institutions of higher education and publishes its findings in *Digest of Education Statistics.* Following is a contingency table giving the number of institutions of higher education in the United States by region and type.

Region	Public T_1	Private T_2	Total
Northeast R_1	266	555	821
Midwest R_2	359	504	863
South R_3	533	502	1035
West R_4	313	242	555
Total	1471	1803	3274

a. How many cells does this contingency table have?
b. What is the total number of institutions of higher education in the United States?
c. How many institutions are in the Midwest?
d. How many are public?
e. How many are private schools in the South?

4.72 As reported by the Motor Vehicle Manufacturers Association of the United States in *Motor Vehicle Facts and Figures,* the numbers of cars and trucks in use by age are as shown in the following contingency table. Frequencies are in millions.

Age (yrs)	Car V_1	Truck V_2	Total
Under 6 A_1	46.2	27.8	74.0
6–8 A_2	26.9	13.1	40.0
9–11 A_3	23.3	10.7	34.0
12 & over A_4	26.8	18.6	45.4
Total	123.2	70.2	193.4

a. How many cells does this contingency table have?
b. What is the total number of cars and trucks in use?
c. How many vehicles are trucks?
d. How many vehicles are between 6 and 8 years old?
e. How many vehicles are trucks that are between 9 and 11 years old?

4.73 The American Medical Association compiles information on U.S. physicians in *Physician Characteristics and Distribution in the U.S.* Following is a contingency table for U.S. surgeons cross-classified by specialty and base of practice.

Specialty	Base of practice: Office B_1	Hospital B_2	Other B_3	Total
General surgery S_1	24,128	12,225	1,658	38,011
Obstetrics/ gynecology S_2	24,150	6,734	1,140	32,024
Orthopedics S_3	13,364	4,248	414	18,026
Ophthal- mology S_4	12,328	2,694	518	15,540
Total	73,970	25,901	3,730	103,601

a. How many surgeons are office-based?
b. How many surgeons are ophthalmologists?
c. How many are office-based ophthalmologists?
d. How many surgeons are either office-based or ophthalmologists?
e. How many general surgeons are hospital-based?
f. How many hospital-based surgeons are OB/GYNs?
g. How many surgeons are not hospital based?

4.74 The American Hospital Association publishes data about U.S. hospitals and nursing homes in *Hospital Statistics.* The contingency table at the top of the next column provides a cross classification of U.S. hospitals and nursing homes by type of facility and number of beds. In the following questions, we will use the term "hospital" to refer to either a hospital or nursing home.
a. How many hospitals have at least 75 beds?
b. How many psychiatric facilities are there?
c. How many hospitals are psychiatric facilities with at least 75 beds?
d. How many hospitals either are psychiatric facilities or have at least 75 beds?
e. How many general facilities have between 25 and 74 beds?
f. How many hospitals with between 25 and 74 beds are chronic facilities?
g. How many hospitals have more than 24 beds?

Facility	Number of beds: 24– B_1	25–74 B_2	75+ B_3	Total
General H_1	260	1586	3557	5403
Psychiatric H_2	24	242	471	737
Chronic H_3	1	3	22	26
Tuberculosis H_4	0	2	2	4
Other H_5	25	177	208	410
Total	310	2010	4260	6580

4.75 According to *Census of Agriculture,* published by the U.S. Bureau of the Census, a joint frequency distribution for the number of farms, by acreage and tenure of operator, is as shown in the following contingency table. Frequencies are in thousands.

Acreage	Tenure of operator: Full owner T_1	Part owner T_2	Tenant T_3	Total
Under 50 A_1	444	58	52	554
50 < 180 A_2	395		59	584
180 < 500 A_3	190		55	428
500 < 1000 A_4	48	111		186
1000 & over A_5	35	114	24	173
Total	1112	596	217	1925

a. Fill in the three empty cells.
b. How many cells does this contingency table have?
c. How many farms have under 50 acres?
d. How many farms are tenant operated?
e. How many farms are operated by part owners and have between 500 and 1000 acres?
f. How many farms are not full-owner operated?
g. How many tenant-operated farms have 180 acres or more?

4.76 As reported by the U.S. Bureau of the Census in *Current Population Reports,* a contingency table for annual income level of families by type of family is as follows. Frequencies are in thousands of families.

Income level	Type of family: Married couple F_1	Husband only F_2	Wife only F_3	Total
Under \$15K I_1	4,156	698	4,868	9,722
\$15K < \$35K I_2	13,942		4,617	19,869
\$35K < \$50K I_3	10,447			12,841
\$50K < \$75K I_4	12,749	518	937	14,204
\$75K & over I_5	12,276	309	377	12,962
Total	53,570	3,514	12,514	69,598

a. Fill in the three empty cells.
b. How many cells does this contingency table have?
c. Determine the number of families that make somewhere between \$15,000 and \$35,000.
d. How many families have only the husband present?
e. How many families have only the wife present and make under \$15,000?
f. How many families make at least \$35,000?
g. Determine the number of married couples that make less than \$50,000.

4.77 Refer to Exercise 4.71.
a. For a randomly selected institution of higher education, describe each of the following events in words: T_2, R_3, and $(T_1 \ \& \ R_4)$.
b. Compute the probability of each event in part (a). Interpret your answers in terms of percentages.
c. Construct a joint probability distribution similar to Table 4.7 on page 231.
d. Verify that the sum of each row and column of joint probabilities equals the marginal probability in that row or column. (*Note:* Rounding may cause slight deviations.)

4.78 Refer to Exercise 4.72.
a. For a randomly selected vehicle (car or truck), describe each of the following events in words: A_3, V_1, and $(A_3 \ \& \ V_1)$.
b. Determine the probability of each event in part (a). Interpret your answers in terms of percentages.
c. Construct a joint probability distribution similar to Table 4.7 on page 231.
d. Verify that the sum of each row and column of joint probabilities equals the marginal probability in that row or column. (*Note:* Rounding may cause slight deviations.)

4.79 Refer to Exercise 4.73.
a. For a randomly selected surgeon, describe each of the following events in words: B_1, S_3, and $(B_1 \ \& \ S_3)$.
b. Compute the probability of each event in part (a).
c. Compute $P(B_1 \text{ or } S_3)$ using the contingency table and the f/N rule.
d. Compute $P(B_1 \text{ or } S_3)$ using the general addition rule and your answers from part (b).
e. Construct a joint probability distribution.

4.80 Refer to Exercise 4.74.
a. For a U.S. hospital selected at random, describe each of the following events in words: H_2, B_2, and $(H_2 \ \& \ B_2)$.
b. Compute the probability of each event in part (a).
c. Compute $P(H_2 \text{ or } B_2)$ using the contingency table and the f/N rule.
d. Compute $P(H_2 \text{ or } B_2)$ using the general addition rule and your answers from part (b).
e. Construct a joint probability distribution.

4.81 Refer to Exercise 4.75. A U.S. farm is selected at random.

a. Use the letters in the margins of the contingency table to represent each of the following three events: The farm obtained (i) has between 180 and 500 acres, (ii) is part-owner operated, and (iii) is full-owner operated and has at least 1000 acres.

b. Compute the probability of each event in part (a).

c. Construct a **joint percentage distribution,** a table similar to a joint probability distribution except with percentages replacing probabilities.

4.82 Refer to Exercise 4.76. A U.S. family is selected at random.

a. Use the letters in the margins of the contingency table to represent each of the following three events: The family obtained (i) has only the wife present, (ii) makes at least $75,000, and (iii) has only the husband present and makes between $15,000 and $35,000.

b. Determine the probability of each event in part (a).

c. Construct a **joint percentage distribution,** a table similar to a joint probability distribution except with percentages replacing probabilities.

Extending the Concepts and Skills

4.83 Explain why the joint events in a contingency table are mutually exclusive.

4.84 In this exercise you are asked to verify that the sum of the joint probabilities in a row or column of a joint probability distribution equals the marginal probability in that row or column. Consider the following joint probability distribution.

	C_1	$\cdots$	C_n	$P(R_i)$
R_1	$P(R_1 \& C_1)$	$\cdots$	$P(R_1 \& C_n)$	$P(R_1)$
$\cdot$	$\cdot$	$\cdots$	$\cdot$	$\cdot$
$\cdot$	$\cdot$	$\cdots$	$\cdot$	$\cdot$
$\cdot$	$\cdot$	$\cdots$	$\cdot$	$\cdot$
R_m	$P(R_m \& C_1)$	$\cdots$	$P(R_m \& C_n)$	$P(R_m)$
$P(C_j)$	$P(C_1)$	$\cdots$	$P(C_n)$	1

a. Explain why we can write

$$R_1 = \big((R_1 \& C_1) \text{ or } \cdots \text{ or } (R_1 \& C_n)\big).$$

b. Why are the events $(R_1 \& C_1), \ldots, (R_1 \& C_n)$ mutually exclusive?

c. Explain why parts (a) and (b) imply that

$$P(R_1) = P(R_1 \& C_1) + \cdots + P(R_1 \& C_n).$$

This equation shows that the first row of joint probabilities sums to the marginal probability at the end of that row. A similar argument applies to any other row or column.

4.5 CONDITIONAL PROBABILITY*

In this section we introduce the concept of conditional probability. The **conditional probability** of an event is the probability that the event occurs under the assumption that another event has occurred.

DEFINITION 4.6 CONDITIONAL PROBABILITY

The probability that event B occurs given that event A has occurred is called a ***conditional probability.*** It is denoted by the symbol $P(B \mid A)$, which is read "the probability of B given A." We call A the ***given event.***

EXAMPLE 4.16 ILLUSTRATES DEFINITION 4.6

When a balanced die is rolled once, six equally likely outcomes are possible, as displayed in Fig. 4.22.

FIGURE 4.22
Sample space for rolling a die once

Let

$$F = \text{event a 5 is rolled,}$$
$$O = \text{event the die comes up odd.}$$

Determine the following probabilities:

a. $P(F)$, the probability that a 5 is rolled.

b. $P(F \mid O)$, the conditional probability that a 5 is rolled given that the die comes up odd.

c. $P\big(O \mid (\text{not } F)\big)$, the conditional probability that the die comes up odd given that a 5 is not rolled.

SOLUTION **a.** To obtain $P(F)$, the probability that a 5 is rolled, we proceed as usual. From Fig. 4.22 we see that six outcomes are possible. Also, event F can occur in only one way: if the die comes up 5. Thus the probability that a 5 is rolled equals

$$P(F) = \frac{f}{N} = \frac{1}{6} = 0.167.$$

b. Given that the die comes up odd, that is, that event O has occurred, there are no longer six possible outcomes. There are only three, as shown in Fig. 4.23.

FIGURE 4.23
Event O

Therefore the conditional probability that a 5 is rolled given that the die comes up odd equals

$$P(F \mid O) = \frac{f}{N} = \frac{1}{3} = 0.333.$$

Comparing this probability with the one that we obtained in part (a), we see that $P(F \mid O) \neq P(F)$; that is, the conditional probability that a 5 is rolled given that the die comes up odd is not the same as the (unconditional) probability that a 5 is rolled. Knowing that the die comes up odd affects the probability that a 5 is rolled.

c. Given that a 5 is not rolled, that is, that event (not F) has occurred, the possible outcomes are the five shown in Fig. 4.24.

FIGURE 4.24
Event (not F)

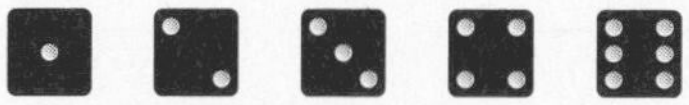

Under these circumstances, event O (odd) can occur in two ways: if a 1 or a 3 is rolled. So the conditional probability that the die comes up odd given that a 5 is not rolled equals

$$P\big(O \mid (\text{not } F)\big) = \frac{f}{N} = \frac{2}{5} = 0.4.$$

Compare this probability with the (unconditional) probability that the die comes up odd, which is 0.5. □

Conditional probability is often used to analyze bivariate data. In Section 4.4 we discussed contingency tables as a method for tabulating such data. Now we will learn how to obtain conditional probabilities for bivariate data directly from a contingency table.

EXAMPLE 4.17 **ILLUSTRATES DEFINITION 4.6**

Table 4.8, shown at the top of the next page, repeats the contingency table for age and rank of ASU faculty members. Suppose an ASU faculty member is selected at random.

a. Determine the (unconditional) probability that the faculty member selected is in his or her 50s.

b. Determine the (conditional) probability that the faculty member selected is in his or her 50s given that an assistant professor is selected.

c. Interpret the probabilities obtained in parts (a) and (b) in terms of percentages.

SOLUTION **a.** Here we are to determine the probability that the faculty member selected is in his or her 50s (event A_4). From Table 4.8 we see that $N = 1164$, since the total number of faculty members is 1164. Also, because 253 of the faculty members

TABLE 4.8
Contingency table for age and rank of ASU faculty members

Age		Rank: Full professor R_1	Associate professor R_2	Assistant professor R_3	Instructor R_4	Total
	Under 30 A_1	2	3	57	6	68
	30–39 A_2	52	170	163	17	402
	40–49 A_3	156	125	61	6	348
	50–59 A_4	145	68	36	4	253
	60 & over A_5	75	15	3	0	93
	Total	430	381	320	33	1164

are in their 50s, we have $f = 253$. So

$$P(A_4) = \frac{f}{N} = \frac{253}{1164} = 0.217.$$

b. For this part we are to find the probability that the faculty member selected is in his or her 50s (event A_4) given that an assistant professor is selected (event R_3). To obtain that probability, we restrict our attention to the assistant-professor column of Table 4.8. We have $N = 320$, since the total number of assistant professors is 320. Also, because 36 of the assistant professors are in their 50s, we have $f = 36$. Consequently,

$$P(A_4 \mid R_3) = \frac{f}{N} = \frac{36}{320} = 0.113.$$

c. In terms of percentages, $P(A_4) = 0.217$ means that 21.7% of the faculty are in their 50s; $P(A_4 \mid R_3) = 0.113$ means that 11.3% of the assistant professors are in their 50s. ◨

The Conditional-Probability Rule

In the previous two examples, we computed conditional probabilities *directly,* meaning that we first obtained the new sample space determined by the given event and then, using the new sample space, we calculated probabilities in the usual manner. For instance, in Example 4.16(b), we computed the conditional probability that a 5 is rolled given that the die comes up odd. To do that we first obtained the new sample space (in this case, 1, 3, 5) and then went on from there.

Sometimes we cannot determine conditional probabilities directly but must instead compute them in terms of unconditional probabilities. To see how this can be done, we return to the situation of Example 4.17.

EXAMPLE 4.18 INTRODUCES THE CONDITIONAL-PROBABILITY RULE

In Example 4.17(b) we determined the conditional probability that a faculty member is in his or her 50s (event A_4) given that an assistant professor is selected (event R_3). To accomplish that we restricted our attention to the R_3 column of Table 4.8 and obtained

$$P(A_4 \mid R_3) = \frac{36}{320} = 0.113.$$

This is a direct computation of the conditional probability $P(A_4 \mid R_3)$. Compute the conditional probability $P(A_4 \mid R_3)$ using unconditional probabilities.

SOLUTION First we note that the number 36 in the numerator of the above fraction is the number of assistant professors in their 50s, that is, the number of ways event $(R_3 \,\&\, A_4)$ can occur. Next we observe that the number 320 in the denominator of the above fraction is the total number of assistant professors, that is, the number of ways event R_3 can occur. So the numbers 36 and 320 are those used to compute the unconditional probabilities of events $(R_3 \,\&\, A_4)$ and R_3, respectively:

$$P(R_3 \,\&\, A_4) = \frac{36}{1164} = 0.031, \qquad P(R_3) = \frac{320}{1164} = 0.275.$$

From the previous three probabilities, we see that

$$P(A_4 \mid R_3) = \frac{36}{320} = \frac{\frac{36}{1164}}{\frac{320}{1164}} = \frac{P(R_3 \,\&\, A_4)}{P(R_3)}.$$

Consequently, the conditional probability $P(A_4 \mid R_3)$ can be obtained from the unconditional probabilities $P(R_3 \,\&\, A_4)$ and $P(R_3)$ by using the formula

$$P(A_4 \mid R_3) = \frac{P(R_3 \,\&\, A_4)}{P(R_3)}.$$

That formula holds in general and is called the **conditional-probability rule.** ◨

FORMULA 4.4 THE CONDITIONAL-PROBABILITY RULE

If A and B are any two events, then

$$P(B \mid A) = \frac{P(A \,\&\, B)}{P(A)}.$$

In words, for any two events, the conditional probability that one event occurs given that the other event has occurred equals the joint probability of the two events divided by the probability of the given event.†

For the faculty-member example, conditional probabilities can be obtained either directly or by applying the conditional-probability rule. However, as Example 4.19 illustrates, the conditional-probability rule is sometimes the only way conditional probabilities can be determined.

 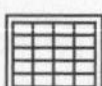

EXAMPLE 4.19 ILLUSTRATES FORMULA 4.4

Data on the marital status of U.S. adults can be found in *Current Population Reports,* a publication of the U.S. Bureau of the Census. Table 4.9 provides a joint probability distribution for the marital status of U.S. adults by sex. We have used "Single" as an abbreviation for "Never married."

TABLE 4.9 Joint probability distribution of marital status and sex

Sex	Single M_1	Married M_2	Widowed M_3	Divorced M_4	$P(S_i)$
Male S_1	0.129	0.298	0.013	0.040	0.480
Female S_2	0.104	0.305	0.057	0.054	0.520
$P(M_j)$	0.233	0.603	0.070	0.095	1.000

(Column heading: Marital status)

A U.S. adult is selected at random.

a. Determine the probability that the adult selected is divorced, given that the adult selected is a male.

b. Determine the probability that the adult selected is a male, given that the adult selected is divorced.

† To be perfectly correct, we must assume that the given event is not impossible because it is not permissible to divide by 0. In this section, we will assume that both events under consideration are not impossible.

SOLUTION Unlike our previous illustrations with contingency tables, we do not have the frequency data here, only the probability (relative-frequency) data. Because of that we cannot compute conditional probabilities directly; we must use the conditional-probability rule.

a. Here we want $P(M_4 \mid S_1)$. Using the conditional-probability rule and Table 4.9, we get

$$P(M_4 \mid S_1) = \frac{P(S_1 \;\&\; M_4)}{P(S_1)} = \frac{0.040}{0.480} = 0.083.$$

In terms of percentages, this means that 8.3% of adult males are divorced.

b. For this part we want $P(S_1 \mid M_4)$. Using the conditional-probability rule and Table 4.9, we get

$$P(S_1 \mid M_4) = \frac{P(M_4 \;\&\; S_1)}{P(M_4)} = \frac{0.040}{0.095} = 0.421.$$

In other words, 42.1% of divorced adults are males. □

EXERCISES 4.5

STATISTICAL CONCEPTS AND SKILLS

4.85 Regarding conditional probability:
a. What is it?
b. Which event is the "given event"?

4.86 Give an example where the conditional probability of an event is the same as the unconditional probability of the event. (*Hint:* Consider the experiment of tossing a coin twice.)

For Exercises 4.87–4.92, compute conditional probabilities directly; that is, do not use the conditional-probability rule.

4.87 Suppose one card is selected at random from an ordinary deck of 52 playing cards. Let

A = event a face card is selected,
B = event a king is selected,
C = event a heart is selected.

Determine the following probabilities and express your results in words.

a. $P(B)$ **b.** $P(B \mid A)$ **c.** $P(B \mid C)$
d. $P(B \mid (\text{not } A))$ **e.** $P(A)$ **f.** $P(A \mid B)$
g. $P(A \mid C)$ **h.** $P(A \mid (\text{not } B))$

4.88 A balanced dime is tossed twice. The four possible equally likely outcomes are HH, HT, TH, TT. Let

A = event the first toss is heads,
B = event the second toss is heads,
C = event at least one toss is heads.

Determine the following probabilities and express your results in words.

a. $P(B)$ **b.** $P(B \mid A)$ **c.** $P(B \mid C)$
d. $P(C)$ **e.** $P(C \mid A)$ **f.** $P(C \mid (\text{not } B))$

4.89 The U.S. Bureau of the Census publishes data on housing units in *American Housing Survey in the United States.* The following table provides a frequency distribution for the number of rooms in U.S. housing units. The frequencies are in thousands.

Rooms	No. of units
1	862
2	1,422
3	10,166
4	20,789
5	24,328
6	22,151
7	14,183
8+	15,555

A U.S. housing unit selected at random. Find the

a. probability that the unit selected has four rooms.

b. conditional probability that the unit selected has exactly four rooms given that it has at least two rooms.

c. conditional probability that the unit has at most four rooms given that it has at least two rooms.

d. Interpret your answers in parts (a)–(c) in terms of percentages.

4.90 As reported by the U.S. Bureau of the Census in *Current Population Reports,* a frequency distribution for the population of the states in the United States is as shown in the following table.

Population size (millions)	Frequency
Under 1	8
1 –< 2	8
2 –< 3	5
3 –< 5	10
5 –< 10	12
10 & over	7

For a state selected at random, find the probability that the population of the state obtained

a. is between 2 million and 3 million.

b. is between 2 million and 3 million, given that it is at least 1 million.

c. is less than 5 million, given that it is at least 1 million.

d. Interpret your answers in parts (a)–(c) in terms of percentages.

4.91 The U.S. National Center for Education Statistics compiles information on institutions of higher education and publishes its findings in *Digest of Education Statistics.* Following is a contingency table giving the number of institutions of higher education in the United States by region and type.

		Type		
		Public T_1	Private T_2	Total
Region	Northeast R_1	266	555	821
	Midwest R_2	359	504	863
	South R_3	533	502	1035
	West R_4	313	242	555
	Total	1471	1803	3274

For an institution of higher education selected at random, find the probability that the institution obtained

a. is in the Northeast.

b. is in the Northeast, given that it is a private school.

c. is a private school, given that it is in the Northeast.

d. Interpret your answers in parts (a)–(c) in terms of percentages.

4.92 As reported by the Motor Vehicle Manufacturers Association of the United States in *Motor Vehicle Facts and Figures,* the numbers of cars and trucks in use by age are as shown in the following contingency table. Frequencies are in millions.

		Type		
		Car V_1	Truck V_2	Total
Age (yrs)	Under 6 A_1	46.2	27.8	74.0
	6–8 A_2	26.9	13.1	40.0
	9–11 A_3	23.3	10.7	34.0
	12 & over A_4	26.8	18.6	45.4
	Total	123.2	70.2	193.4

For a randomly selected vehicle (car or truck), determine the probability that the vehicle obtained

a. is under 6 years old.

b. is under 6 years old, given that it is a car.

c. is a car.

d. is a car, given that it is under 6 years old.

e. Interpret your answers in parts (a)–(d) in terms of percentages.

4.93 According to *Census of Agriculture,* published by the U.S. Bureau of the Census, a joint frequency distribution for the number of farms, by acreage and tenure of operator, is as shown in the following contingency table. Frequencies are in thousands.

Tenure of operator

Acreage	Full owner T_1	Part owner T_2	Tenant T_3	Total
Under 50 A_1	444	58	52	554
50 < 180 A_2	395	130	59	584
180 < 500 A_3	190	183	55	428
500 < 1000 A_4	48	111	27	186
1000 & over A_5	35	114	24	173
Total	1112	596	217	1925

a. Find $P(T_3)$. **b.** Find $P(T_3 \& A_3)$.

c. Obtain $P(A_3 \mid T_3)$ directly from the table.

d. Obtain $P(A_3 \mid T_3)$ using the conditional-probability rule and your answers from parts (a) and (b).

e. State your results in parts (a)–(c) in words.

4.94 The American Hospital Association publishes information about U.S. hospitals and nursing homes in *Hospital Statistics.* Here is a contingency table providing a cross classification of U.S. hospitals and nursing homes by type of facility and number of beds.

Number of beds

Facility	24– B_1	25–74 B_2	75+ B_3	Total
General H_1	260	1586	3557	5403
Psychiatric H_2	24	242	471	737
Chronic H_3	1	3	22	26
Tuberculosis H_4	0	2	2	4
Other H_5	25	177	208	410
Total	310	2010	4260	6580

a. Find $P(H_1)$. **b.** Find $P(H_1 \& B_3)$.

c. Obtain $P(B_3 \mid H_1)$ directly from the table.

d. Obtain $P(B_3 \mid H_1)$ using the conditional-probability rule and your answers from parts (a) and (b).

e. State your results in parts (a)–(c) in words.

4.95 The U.S. Congress, Joint Committee on Printing, provides information on the composition of Congress in *Congressional Directory.* Here is a joint probability distribution for the members of the 105th Congress by legislative group and political party. The "other" category includes Independents and vacancies.

Group

Party	Rep C_1	Senator C_2	$P(P_i)$
Democratic P_1	0.385	0.084	0.469
Republican P_2	0.424	0.103	0.527
Other P_3	0.004	0.000	0.004
$P(C_j)$	0.813	0.187	1.000

If a member of the 105th Congress is selected at random, what is the probability that the member obtained

a. is a senator? **b.** is a Republican senator?
c. is a Republican, given that he or she is a senator?
d. is a senator, given that he or she is a Republican?
e. Interpret your answers in parts (a)–(d) in terms of percentages.

4.96 The National Center for Education Statistics publishes information on U.S. engineers and scientists in *Digest of Education Statistics.* The table below presents a joint probability distribution for engineers and scientists by highest degree obtained.

Highest degree	Type: Engineer T_1	Type: Scientist T_2	$P(D_i)$
Bachelors D_1	0.343	0.289	0.632
Masters D_2	0.098	0.146	0.244
Doctorate D_3	0.017	0.091	0.108
Other D_4	0.013	0.003	0.016
$P(T_j)$	0.471	0.529	1.000

A person is selected at random from among the engineers and scientists. Determine the probability that the person obtained

a. is an engineer. **b.** has a doctorate.
c. is an engineer with a doctorate.
d. is an engineer, given the person has a doctorate.
e. has a doctorate, given the person is an engineer.
f. Interpret your answers in parts (a)–(e) in terms of percentages.

4.97 According to the Census Bureau's *Current Population Reports,* 12.6% of U.S. residents are African-American and 6.6% are African-American women. What percentage of African-Americans are women?

4.98 As reported by the Federal Bureau of Investigation in *Crime in the United States,* 4.9% of property crimes are committed in rural areas and 1.9% of property crimes are burglaries committed in rural areas. What percentage of property crimes committed in rural areas are burglaries?

EXTENDING THE CONCEPTS AND SKILLS

4.99 Give an example of an experiment in which conditional probabilities

a. can be computed both directly and by using the conditional-probability rule.
b. cannot be computed directly but only by using the conditional-probability rule.

4.100 Refer to Exercise 4.92.

a. Construct a joint probability distribution.
b. Determine the probability distribution of age for cars in use; that is, construct a table showing the conditional probabilities that a car in use is under 6 years old, 6–8 years old, and so on.
c. Determine the probability distribution of type for vehicles 6–8 years old.
d. The probability distributions in parts (b) and (c) are examples of **conditional-probability distributions.** Determine two other conditional-probability distributions for the data on age and type of motor vehicles in use.

4.6 THE MULTIPLICATION RULE; INDEPENDENCE*

The conditional-probability rule is used to compute conditional probabilities in terms of unconditional probabilities:

$$P(B \mid A) = \frac{P(A \,\&\, B)}{P(A)}.$$

Multiplying both sides of this equation by $P(A)$, we obtain a formula for computing joint probabilities in terms of marginal and conditional probabilities.

FORMULA 4.5 **THE GENERAL MULTIPLICATION RULE**

If A and B are any two events, then

$$P(A \,\&\, B) = P(A) \cdot P(B \mid A).$$

In words, for any two events, their joint probability equals the probability that one of the events occurs times the conditional probability of the other event given that event.

The conditional-probability rule and the general multiplication rule are simply variations of each other. When the joint and marginal probabilities are known or easily determined directly, we can use the conditional-probability rule to obtain conditional probabilities. On the other hand, when the marginal and conditional probabilities are known or easily determined directly, we can use the general multiplication rule to obtain joint probabilities.

EXAMPLE 4.20 **ILLUSTRATES FORMULA 4.5**

The U.S. Congress, Joint Committee on Printing, provides information on the composition of Congress in *Congressional Directory.* For the 105th Congress, 18.7% of members are senators and 45% of senators are Democrats. What is the probability that a randomly selected member of the 105th Congress is a Democratic senator?

SOLUTION Let

$$D = \text{event the member selected is a Democrat,}$$
$$S = \text{event the member selected is a senator.}$$

The event that the member selected is a Democratic senator can be expressed as $(S \,\&\, D)$. We want to determine the probability of that event.

Because 18.7% of members are senators, $P(S) = 0.187$; and because 45% of senators are Democrats, $P(D \mid S) = 0.450$. Applying the general multiplication rule, we get

$$P(S \,\&\, D) = P(S) \cdot P(D \mid S) = 0.187 \cdot 0.450 = 0.084.$$

The probability is 0.084 that a randomly selected member of the 105th Congress is a Democratic senator. Expressed in terms of percentages, 8.4% of members of the 105th Congress are Democratic senators. ◨

Another application of the general multiplication rule relates to sampling two or more members from a population. Example 4.21 provides an illustration.

EXAMPLE 4.21 ILLUSTRATES FORMULA 4.5

In Professor Weiss's introductory statistics class, the numbers of males and females are as shown in the frequency distribution in Table 4.10.

TABLE 4.10 Frequency distribution of males and females in Professor Weiss's introductory statistics class

Sex	Frequency
Male	17
Female	23
	40

Two students are selected at random from the class. The first student obtained is not returned to the class for possible reselection; that is, the sampling is without replacement. Find the probability that the first student obtained is female and the second is male.

SOLUTION Let's use the following notation:

$$F1 = \text{event the first student obtained is female,}$$
$$M2 = \text{event the second student obtained is male.}$$

The problem is to determine $P(F1 \,\&\, M2)$. By the general multiplication rule, we can write

$$P(F1 \,\&\, M2) = P(F1) \cdot P(M2 \mid F1).$$

As we will now see, it is easy to compute the two probabilities on the right side of this equation. For $P(F1)$—the probability that the first student obtained is female—we note from Table 4.10 that 23 of the 40 students are female and, consequently,

$$P(F1) = \frac{f}{N} = \frac{23}{40}.$$

Next we find $P(M2 \mid F1)$—the conditional probability that the second student obtained is male given that the first one obtained is female. We observe that, given the first student obtained is female, there are 39 students remaining in the class, of which 17 are male. So,

$$P(M2 \mid F1) = \frac{f}{N} = \frac{17}{39}.$$

Applying the general multiplication rule, we conclude that

$$P(F1 \,\&\, M2) = P(F1) \cdot P(M2 \,|\, F1) = \frac{23}{40} \cdot \frac{17}{39} = 0.251.$$

When two students are randomly selected from the class, the probability is 0.251 that the first student obtained is female and the second is male. ◨

It is often helpful to draw a **tree diagram** when applying the general multiplication rule. An appropriate tree diagram for Example 4.21 is shown in Fig. 4.25.

FIGURE 4.25 Tree diagram for student-selection problem

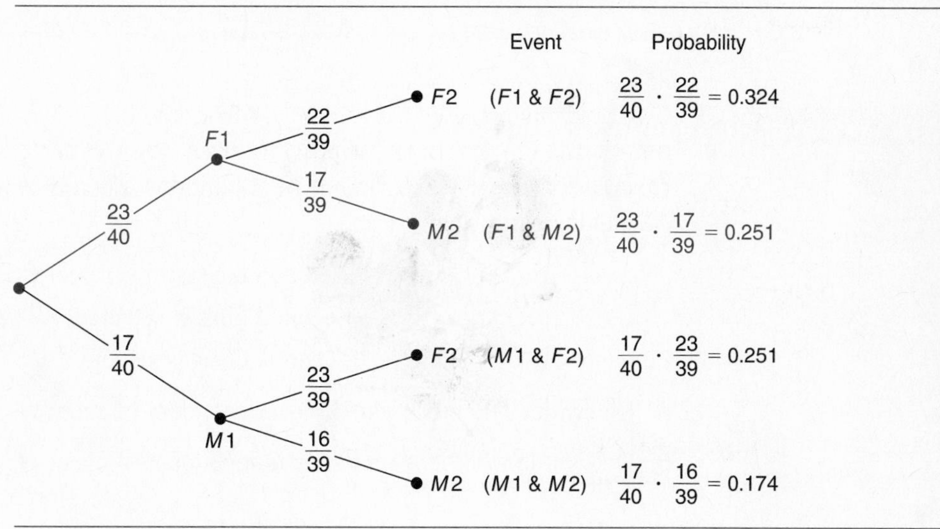

Each branch of the tree corresponds to one possibility for selecting two students at random from the class. For instance, the second branch of the tree, shown in color, corresponds to event ($F1$ & $M2$)—the event that the first student obtained is female (event $F1$) and the second is male (event $M2$).

Starting from the left on that branch, the number $\frac{23}{40}$ is the probability that the first student obtained is female, $P(F1)$; and the number $\frac{17}{39}$ is the conditional probability that the second student obtained is male given that the first student obtained is female, $P(M2 \,|\, F1)$. The product of those two probabilities is, by the general multiplication rule, the probability that the first student obtained is female and the second is male, $P(F1 \,\&\, M2)$. The second entry in the Probability column of Fig. 4.25 shows that this probability equals 0.251, as we discovered at the end of Example 4.21.

Independence

One of the most important concepts in probability is that of statistical independence of events. For two events, statistical independence or, more briefly, independence, is defined as follows.

DEFINITION 4.7 INDEPENDENT EVENTS

Event B is said to be ***independent*** of event A if the occurrence of event A does not affect the probability that event B occurs. In symbols,

$$P(B \mid A) = P(B).$$

This means that knowing whether event A has occurred provides no probabilistic information about the occurrence of event B.

EXAMPLE 4.22 ILLUSTRATES DEFINITION 4.7

Consider again the experiment of randomly selecting one card from a deck of 52 playing cards, let

$$\begin{aligned} F &= \text{event a face card is selected,} \\ K &= \text{event a king is selected,} \\ H &= \text{event a heart is selected.} \end{aligned}$$

a. Determine whether event K is independent of event F.

b. Determine whether event K is independent of event H.

SOLUTION First we note that the unconditional probability that event K occurs equals

$$P(K) = \frac{f}{N} = \frac{4}{52} = \frac{1}{13} = 0.077.$$

a. To determine whether event K is independent of event F, we must compute $P(K \mid F)$ and compare it to $P(K)$. If those two probabilities are equal, event K is independent of event F; otherwise, event K is not independent of event F. Now, given that event F has occurred, 12 outcomes are possible (four jacks, four queens, and four kings), and event K can occur in four ways out of those 12 possibilities. So

$$P(K \mid F) = \frac{f}{N} = \frac{4}{12} = 0.333.$$

This is not equal to $P(K)$, so, event K is not independent of event F. This lack of independence stems from the fact that the percentage of kings among the face cards (33.3%) is not the same as the percentage of kings among all the cards (7.7%).

b. Here we need to compute $P(K \mid H)$ and compare it to $P(K)$. Given that event H has occurred, 13 outcomes are possible (the 13 hearts), and event K can occur in one way out of those 13 possibilities. Therefore

$$P(K \mid H) = \frac{f}{N} = \frac{1}{13} = 0.077.$$

This is equal to $P(K)$ and, so, event K is independent of event H. This independence stems from the fact that the percentage of kings among the hearts is the same as the percentage of kings among all the cards; namely, 7.7%. ◨

It can be shown that if event B is independent of event A, then it is also true that event A is independent of event B. So, in such cases, we often say that event A and event B are **independent,** or that A and B are **independent events.** If two events are not independent, we say they are **dependent events.** In Example 4.22, F and K are dependent events, whereas K and H are independent events.

The Special Multiplication Rule

Recall that the general multiplication rule states that for any two events A and B,

$$P(A \,\&\, B) = P(A) \cdot P(B \mid A).$$

If A and B are independent events, then $P(B \mid A) = P(B)$. Thus for the special case of independent events, we can replace the term $P(B \mid A)$ in the general multiplication rule by the term $P(B)$. This yields the following rule.

FORMULA 4.6 **THE SPECIAL MULTIPLICATION RULE (FOR TWO INDEPENDENT EVENTS)**

If A and B are independent events, then

$$P(A \,\&\, B) = P(A) \cdot P(B),$$

and conversely, if $P(A \,\&\, B) = P(A) \cdot P(B)$, then A and B are independent events. In words, two events are independent if and only if their joint probability equals the product of their marginal probabilities.

As in Example 4.22, we can use the definition of independence (Definition 4.7) to decide whether two specified events are independent. If the two events are, say, A and B, this means determining whether $P(B \mid A) = P(B)$. Alternatively, we can decide whether event A and event B are independent by employing the special multiplication rule, that is, by determining whether $P(A \& B) = P(A) \cdot P(B)$.

The definition of independence for three or more events is more complicated than that for two events. Nevertheless, the special multiplication rule still holds.

FORMULA 4.7 **THE SPECIAL MULTIPLICATION RULE**

If events A, B, C, ... are independent, then

$$P(A \& B \& C \& \cdots) = P(A) \cdot P(B) \cdot P(C) \cdots.$$

We can use the special multiplication rule to compute joint probabilities when we know or can reasonably assume that two or more events are independent. Example 4.23 illustrates this point.

EXAMPLE 4.23 **ILLUSTRATES FORMULA 4.7**

A roulette wheel contains 38 numbers, of which 18 are red, 18 are black, and 2 are green. When the roulette ball is spun, it is equally likely to land on any of the 38 numbers.

a. In two plays at a roulette wheel, what is the probability that the ball will land on green the first time and on black the second time?

b. In five plays at a roulette wheel, what is the probability that the ball will land on red all five times?

SOLUTION First of all we note that it is reasonable to assume that outcomes on successive plays at the wheel are independent.

a. Let

$$G1 = \text{event the ball lands on green the first time,}$$
$$B2 = \text{event the ball lands on black the second time.}$$

The problem is to determine $P(G1 \& B2)$. Because outcomes on successive plays at the wheel are independent, event $G1$ and event $B2$ are independent. Applying the special multiplication rule, we conclude that

$$P(G1 \& B2) = P(G1) \cdot P(B2) = \frac{2}{38} \cdot \frac{18}{38} = 0.025.$$

In two plays at a roulette wheel, there is a 2.5% chance that the ball will land on green the first time and on black the second time.

b. Let

$$R1 = \text{event the ball lands on red the first time,}$$
$$R2 = \text{event the ball lands on red the second time,}$$

and so forth. We need to compute $P(R1 \,\&\, R2 \,\&\, R3 \,\&\, R4 \,\&\, R5)$. Since outcomes on successive plays at the wheel are independent, the special multiplication rule applies to give

$$\begin{aligned} P(R1 \,\&\, R2 \,\&\, R3 \,\&\, R4 \,\&\, R5) &= P(R1) \cdot P(R2) \cdot P(R3) \cdot P(R4) \cdot P(R5) \\ &= \frac{18}{38} \cdot \frac{18}{38} \cdot \frac{18}{38} \cdot \frac{18}{38} \cdot \frac{18}{38} = 0.024. \end{aligned}$$

In five plays at a roulette wheel, there is a 2.4% chance that the ball will land on red all five times. ◨

Mutually Exclusive Versus Independent Events

It is important to realize that the terms *mutually exclusive* and *independent* refer to different concepts. Mutually exclusive events are those that cannot occur simultaneously. Independent events are those for which the occurrence of some does not affect the probabilities of the others occurring. In fact, if two or more events are mutually exclusive, then the occurrence of one precludes the occurrence of the others. Two or more (non-impossible) events cannot be both mutually exclusive and independent.

EXERCISES 4.6

STATISTICAL CONCEPTS AND SKILLS

4.101 Regarding the general multiplication rule and the conditional-probability rule.

a. State these two rules.

b. Explain the relationship between them.

c. Why do we emphasize two different variations of essentially the same rule?

4.102 Suppose A and B are two events.

a. Explain what it means for event B to be independent of event A.

b. If event A and event B are independent, how can we obtain their joint probability in terms of the marginal probabilities?

4.103 According to the Opinion Research Corporation, 44% of U.S. women suffer from holiday depression and, from the Census Bureau's *Current Population Reports,* 52% of U.S. adults are women. Find the probability that a randomly selected U.S. adult is a woman who suffers from holiday depression. Interpret your answer in terms of percentages.

4.104 The National Center for Education Statistics states in *Digest of Education Statistics* that 43.9% of all public elementary schools have between 250 and 499 students. Moreover, 51.4% of all public schools are elementary schools. Determine the probability that a randomly selected public school is an elementary school with between 250 and 499 students. Interpret your answer in terms of percentages.

4.105 Cards numbered 1, 2, 3, ..., 10 are placed in a box. The box is shaken and a blindfolded person selects two successive cards without replacement.

a. What is the probability that the first card selected is numbered 6?

b. Given that the first card is numbered 6, what is the probability that the second is numbered 9?

c. Find the probability of selecting first a 6 and then a 9.

d. What is the probability that both cards selected are numbered over 5?

4.106 A person has agreed to participate in an ESP experiment. He is asked to randomly pick two numbers between 1 and 6. The second number must be different from the first. Let

H = event the first number picked is a 3,

K = event the second number picked exceeds 4.

Determine

a. $P(H)$. **b.** $P(K \mid H)$. **c.** $P(H \& K)$.

Find the probability that both numbers picked are

d. less than 3. **e.** greater than 3.

4.107 According to the *World Almanac, 1998,* the political-party distribution of U.S. governors, as of mid-October 1997, is as follow.

Party	Frequency
Democratic	17
Republican	32
Independent	1

Suppose that two governors are selected at random without replacement.

a. Obtain the probability that the first is a Republican and the second a Democrat.

b. Obtain the probability that both are Republicans.

c. Draw a tree diagram for this problem similar to Fig. 4.25 on page 247.

d. What is the probability that the two governors selected are both Democrats, both Republicans, or both Independents?

e. What is the probability that one of the governors selected is a Republican and the other a Democrat?

4.108 A frequency distribution for the class level of students in Professor Weiss's introductory statistics course is as follows.

Class	Frequency
Freshman	6
Sophomore	15
Junior	12
Senior	7

Two students are randomly selected without replacement. Determine the probability that

a. the first student obtained is a junior and the second a senior.

b. both students obtained are sophomores.

c. Draw a tree diagram for this problem similar to Fig. 4.25 on page 247.

d. What is the probability that one of the students obtained is a freshman and the other a sophomore?

4.109 The U.S. National Center for Health Statistics compiles data on injuries and publishes the information in *Vital and Health Statistics.* A contingency table for injuries in the United States by circumstance and sex is as follows. Frequencies are in millions.

		Circumstance			
		Work C_1	Home C_2	Other C_3	Total
Sex	Male S_1	8.0	9.8	17.8	35.6
	Female S_2	1.3	11.6	12.9	25.8
	Total	9.3	21.4	30.7	61.4

a. Find $P(C_1)$. **b.** Find $P(C_1 \mid S_2)$.

c. Are events C_1 and S_2 independent? Why?

d. Is the event that an injured person is male independent of the event that an injured person was hurt at home? Explain your answer.

4.110 A study conducted by the Census Bureau revealed the following data on the methods Americans use to get to work, by residence. The frequencies are in millions of workers.

Method		Residence: Urban R_1	Rural R_2	Total
	Automobile M_1	45.0	15.0	60.0
	Public trans. M_2	6.5	0.5	7.0
	Total	51.5	15.5	67.0

a. Find $P(M_1)$. **b.** Find $P(M_1 \mid R_2)$.
c. Are M_1 and R_2 independent events? Why?
d. Is the event that a worker resides in an urban area independent of the event that the worker uses an automobile to get to work? Justify your answer.

4.111 When a balanced dime is tossed three times, eight equally likely outcomes are possible:

HHH	HTH	THH	TTH
HHT	HTT	THT	TTT

Let

A = event the first toss is heads,
B = event the third toss is tails,
C = event the total number of heads is one.

a. Compute $P(A)$, $P(B)$, and $P(C)$.
b. Compute $P(B \mid A)$.
c. Are A and B independent events? Why?
d. Compute $P(C \mid A)$.
e. Are A and C independent events? Why?

4.112 When two balanced dice are rolled, 36 equally likely outcomes are possible. (See Fig. 4.1, p. 201.) Let

A = event the colored die comes up even,
B = event the black die comes up odd,
C = event the sum of the dice is 10,
D = event the sum of the dice is even.

a. Compute $P(A)$, $P(B)$, $P(C)$, and $P(D)$.
b. Compute $P(B \mid A)$.
c. Are events A and B independent? Why?
d. Compute $P(C \mid A)$.
e. Are events A and C independent? Why?
f. Compute $P(D \mid A)$.
g. Are events A and D independent? Why?

4.113 The U.S. Congress, Joint Committee on Printing, publishes data on the composition of Congress in *Congressional Directory*. Here is a joint probability distribution for the members of the 105th Congress by legislative group and political party. The "other" category includes Independents and vacancies.

Party		Group: Rep C_1	Senator C_2	$P(P_i)$
	Democratic P_1	0.385	0.084	0.469
	Republican P_2	0.424	0.103	0.527
	Other P_3	0.004	0.000	0.004
	$P(C_j)$	0.813	0.187	1.000

a. Determine $P(P_1)$, $P(C_2)$, and $P(P_1 \& C_2)$.
b. Use the special multiplication rule to determine whether events P_1 and C_2 are independent.

4.114 The National Center for Education Statistics publishes information on U.S. engineers and scientists in *Digest of Education Statistics*. The table below presents a joint probability distribution for engineers and scientists by highest degree obtained.

Highest degree		Type: Engineer T_1	Scientist T_2	$P(D_i)$
	Bachelors D_1	0.343	0.289	0.632
	Masters D_2	0.098	0.146	0.244
	Doctorate D_3	0.017	0.091	0.108
	Other D_4	0.013	0.003	0.016
	$P(T_j)$	0.471	0.529	1.000

a. Determine $P(T_2)$, $P(D_3)$, and $P(T_2 \,\&\, D_3)$.
b. Are T_2 and D_3 independent events? Why?

4.115 Two cards are drawn at random from an ordinary deck of 52 cards. Determine the probability that both cards are aces if
a. the first card is replaced before the second card is drawn.
b. the first card is not replaced before the second card is drawn.

4.116 In the game of *Yahtzee,* five balanced dice are rolled.
a. What is the probability of rolling all 2s? *(Hint:* Use the fact that the outcomes of different dice are independent.*)*
b. What is the probability that all the dice come up the same number? *(Hint:* Apply both the special addition rule and the special multiplication rule.*)*

4.117 Events E and F are independent, $P(E) = \frac{1}{3}$, and $P(F) = \frac{1}{4}$. Find
a. $P(E \,\&\, F)$. **b.** $P(E \text{ or } F)$.

4.118 A family has purchased two identical notebook computers. The *Reference and Troubleshooting Guide* suggests that with no power management features enabled, there is a 70% chance that the operating time will exceed 2.8 hours. Assuming no power management features are enabled, determine the probability that
a. both computers will have an operating time exceeding 2.8 hours.
b. at least one of the two computers will have an operating time exceeding 2.8 hours.

4.119 In a letter to the editor that appeared in the February 23, 1987, issue of *U.S. News and World Report,* a reader discussed the issue of space shuttle safety. Each "criticality 1" item must have 99.99% reliability, according to NASA standards, meaning that the probability of failure for such an item is 0.0001. Mission 25, the mission in which the Challenger exploded, had 748 "criticality 1" items. Determine the probability that
a. none of the "criticality 1" items would fail.
b. at least one "criticality 1" item would fail.
c. Interpret your answer in part (b) in words.

4.120 A hardware manufacturer produces nuts and bolts. Each bolt produced is attached to a nut to make a single unit. It is known that 2% of the nuts produced and 3% of the bolts produced are defective in some way. A nut-bolt unit is considered defective if either the nut or the bolt has a defect. Determine the percentage of nondefective nut-bolt units.

4.121 As reported by the Chicago Title Insurance Company in *The Guarantor,* there is a 77.3% chance that a home buyer will purchase a resale home. In the next four home purchases, find the probability that
a. the first three will be resales and the fourth will be a new home.
b. the first will be a resale, the second a new home, and the last two resales.
c. the first will be a resale, the next two new homes, and the last a resale.
d. exactly three of the four will be resales.

4.122 The U.S. Federal Bureau of Investigation compiles information on violent crimes by type and publishes its findings in *Crime in the United States.* Here is a probability distribution for violent crimes.

Violent crime	Probability
Murder	0.012
Forcible rape	0.054
Robbery	0.323
Aggravated assault	0.611

The table indicates, for example, that the probability is 0.012 that a violent crime will be a murder. Out of three violent crimes,
a. find the probability that the first two are robberies and the third is a forcible rape.
b. find the probability that the first is a murder, the second a robbery, and the third an aggravated assault.

4.123 The National Center for Health Statistics compiles information on activity limitations. Results are published in *Vital and Health Statistics.* The data show that 13.6% of males and 14.4% of females have an activity limitation. Are sex and activity limitation statistically independent? Explain your answer.

Extending the Concepts and Skills

4.124 For three events, say, A, B, and C, the general multiplication rule is

$$P(A \,\&\, B \,\&\, C) = P(A) \cdot P(B \mid A) \cdot P\big(C \mid (A \,\&\, B)\big).$$

Three cards are randomly selected without replacement from an ordinary deck of 52.

a. Find the probability that all three cards are hearts.

b. Find the probability that the first two cards are hearts and the third a spade.

c. Provide a mathematical statement of the general multiplication rule for four events.

4.125 In this exercise we will further examine the concepts of independent events and mutually exclusive events.

a. If two events are mutually exclusive, determine their joint probability.

b. If two non-impossible events are independent, explain why their joint probability is not 0.

c. Give an example of two events that are neither mutually exclusive nor independent.

4.126 Three events, say, A, B, and C, are said to be independent if

$$P(A \,\&\, B) = P(A) \cdot P(B),$$
$$P(A \,\&\, C) = P(A) \cdot P(C),$$
$$P(B \,\&\, C) = P(B) \cdot P(C),$$

and

$$P(A \,\&\, B \,\&\, C) = P(A) \cdot P(B) \cdot P(C).$$

What do you think is required for four events to be independent? Explain your definition in words.

4.127 When two balanced dice are rolled, 36 equally likely outcomes are possible, as seen in Fig. 4.1 on page 201. Let

A = event the colored die comes up even,
B = event the black die comes up even,
C = event the sum of the dice is even,
D = event the colored die comes up 1, 2, or 3,
E = event the colored die comes up 3, 4, or 5,
F = event the sum of the dice is 5.

Apply the definition of independence for three events, stated in Exercise 4.126, to solve each of the following problems.

a. Are A, B, and C independent events?

b. Show that $P(D \,\&\, E \,\&\, F) = P(D) \cdot P(E) \cdot P(F)$ but that D, E, and F are not independent events.

4.128 When a balanced coin is tossed four times, 16 equally likely outcomes are possible:

HHHH	THHH	THHT	THTT
HHHT	HHTT	THTH	TTHT
HHTH	HTHT	TTHH	TTTH
HTHH	HTTH	HTTT	TTTT

Let

A = event the first toss is heads,
B = event the second toss is tails,
C = event the last two tosses are heads.

Apply the definition of independence for three events, stated in Exercise 4.126, to show that A, B, and C are independent events.

4.7 BAYES'S RULE*

In this section we will discuss a rule of probability developed by Thomas Bayes, an eighteenth-century clergyman. This rule is aptly called Bayes's rule. One of the primary uses of Bayes's rule is to revise probabilities in accordance with newly

* This more advanced section continues the optional material on probability. Although important in its own right, it is not needed for the remainder of the book.

acquired information. Such revised probabilities are actually conditional probabilities, and so in some sense we have already examined much of the material in this section. However, as we will see, Bayes's rule involves some new concepts and techniques.

The Rule of Total Probability

In preparation for Bayes's rule, we need to study another rule of probability called the rule of total probability. First we consider the concept of exhaustive events. Events $A_1, A_2, \ldots, A_k$ are said to be **exhaustive** if one or more of them must occur.

For instance, the National Governors' Association classifies governors as Democrat, Republican, or Independent. Suppose a governor is selected at random; let E_1, E_2, and E_3 denote, respectively, the events that the governor selected is a Democrat, Republican, and Independent. Then events E_1, E_2, and E_3 are exhaustive since at least one of them must occur when a governor is selected—the governor obtained must be a Democrat, Republican, or Independent.

The events E_1, E_2, and E_3 are not only exhaustive, but are also mutually exclusive since a governor cannot have more than one political party classification at a given time. In general, if events are both exhaustive and mutually exclusive, then exactly one of them must occur. This is true because at least one of the events must occur (since the events are exhaustive) and at most one of the events can occur (since the events are mutually exclusive).

An event and its complement are always mutually exclusive and exhaustive. Figure 4.26(a) portrays three events, A_1, A_2, and A_3, that are both mutually exclusive and exhaustive. In the figure the three events do not overlap, indicating that they are mutually exclusive; furthermore, they fill out the entire region enclosed by the heavy rectangle (i.e., the sample space), indicating that they are exhaustive.

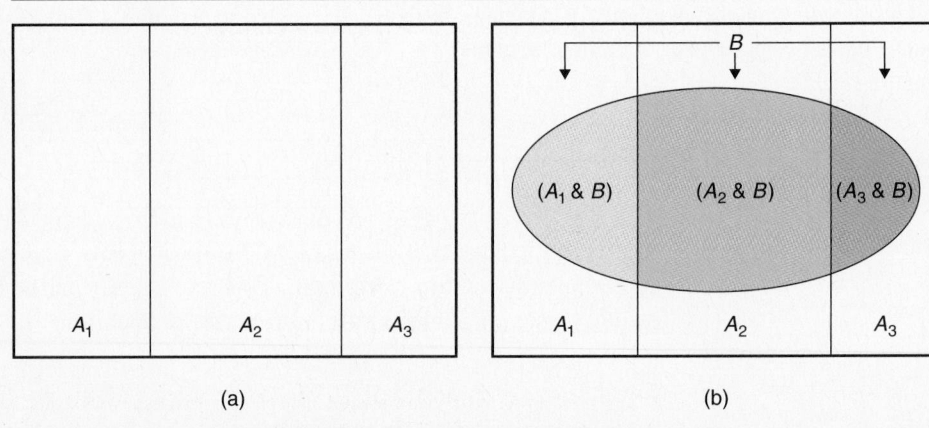

FIGURE 4.26
(a) Three mutually exclusive and exhaustive events
(b) An event B and three mutually exclusive and exhaustive events

Now consider, say, three mutually exclusive and exhaustive events, A_1, A_2, and A_3, and any event B, as portrayed in Fig. 4.26(b). As we see from Fig. 4.26(b), event B is comprised of the mutually exclusive events $(A_1 \& B)$, $(A_2 \& B)$, and $(A_3 \& B)$, shown in color. This reflects the fact that event B must occur in conjunction with exactly one of the events A_1, A_2, and A_3.

If we think of the colored regions in Fig. 4.26(b) as probabilities, then the total colored region is $P(B)$, and the three colored subregions are, from left to right, $P(A_1 \& B)$, $P(A_2 \& B)$, and $P(A_3 \& B)$. Because events $(A_1 \& B)$, $(A_2 \& B)$, and $(A_3 \& B)$ are mutually exclusive, the total colored region equals the sum of the three colored subregions; in other words,

$$P(B) = P(A_1 \& B) + P(A_2 \& B) + P(A_3 \& B).$$

Applying the general multiplication rule (page 245), to each term on the right side of this equation, we obtain

$$P(B) = P(A_1) \cdot P(B \mid A_1) + P(A_2) \cdot P(B \mid A_2) + P(A_3) \cdot P(B \mid A_3).$$

This formula holds in general and is called the **rule of total probability.** It is also referred to as the **stratified sampling theorem** because of its importance in stratified sampling.

FORMULA 4.8 THE RULE OF TOTAL PROBABILITY

Suppose events $A_1, A_2, \ldots, A_k$ are mutually exclusive and exhaustive; that is, exactly one of the events must occur. Then for any event B,

$$P(B) = \sum_{j=1}^{k} P(A_j) \cdot P(B \mid A_j).$$

EXAMPLE 4.24 ILLUSTRATES FORMULA 4.8

The U.S. Bureau of the Census collects data on the resident population, by age and region of residence, and presents its findings in *Current Population Reports.* In the first two columns of Table 4.11, we have provided a percentage distribution for region of residence; the third column displays the percentage of seniors (age 65 or over) in each region. We see, for instance, that 19.0% of U.S. residents live in the Northeast region and that 13.8% of residents living in the Northeast are seniors. Use Table 4.11 to determine the percentage of U.S. residents that are seniors.

TABLE 4.11 Percentage distribution for region of residence, and percentage of seniors in each region

Region	Percentage of U.S. population	Percentage seniors
Northeast	19.0	13.8
Midwest	23.1	13.0
South	35.5	12.8
West	22.4	11.1
	100.0	

SOLUTION To solve this problem, we first translate the information displayed in Table 4.11 into the language of probability. Suppose a U.S. resident is selected at random. Let

$$S = \text{event the resident selected is a senior,}$$

and

$$R_1 = \text{event the resident selected lives in the Northeast,}$$
$$R_2 = \text{event the resident selected lives in the Midwest,}$$
$$R_3 = \text{event the resident selected lives in the South,}$$
$$R_4 = \text{event the resident selected lives in the West.}$$

Then the percentages shown in the second and third columns of Table 4.11 translate into the probabilities displayed in Table 4.12.

TABLE 4.12 Probabilities derived from Table 4.11

$P(R_1) = 0.190$	$P(S \mid R_1) = 0.138$
$P(R_2) = 0.231$	$P(S \mid R_2) = 0.130$
$P(R_3) = 0.355$	$P(S \mid R_3) = 0.128$
$P(R_4) = 0.224$	$P(S \mid R_4) = 0.111$

The problem is to determine the percentage of U.S. residents that are seniors, or, in terms of probability, $P(S)$. Because a U.S. resident must reside in exactly one of the four regions, events R_1, R_2, R_3, and R_4 are mutually exclusive and exhaustive. Therefore, by the rule of total probability applied to the event S, we have from Table 4.12 that

$$\begin{aligned} P(S) &= \sum_{j=1}^{4} P(R_j) \cdot P(S \mid R_j) \\ &= 0.190 \cdot 0.138 + 0.231 \cdot 0.130 + 0.355 \cdot 0.128 + 0.224 \cdot 0.111 \\ &= 0.127. \end{aligned}$$

A tree diagram for this calculation is shown in Fig. 4.27. In the figure, J represents the event that the resident selected is not a senior. We obtain $P(S)$ from the tree diagram by first multiplying the two probabilities on each branch of the tree that ends with S (the colored branches) and then summing all those products.

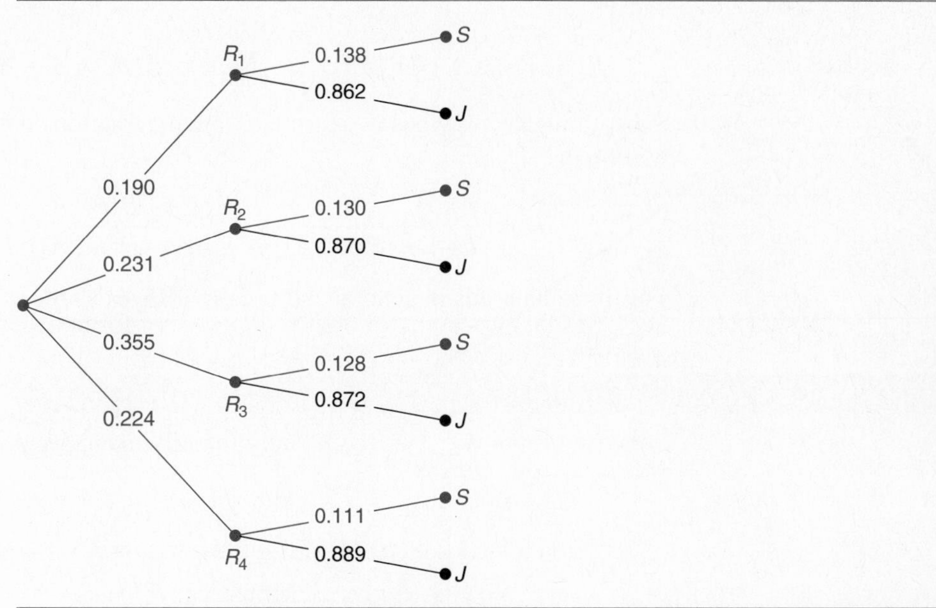

FIGURE 4.27 Tree diagram for calculating $P(S)$ using the rule of total probability

In any case, we see that the probability is 0.127 that a randomly selected U.S. resident is a senior. In other words, 12.7% of U.S. residents are seniors. □

Bayes's Rule

Using the rule of total probability, we can derive Bayes's rule. For simplicity let's consider three events, A_1, A_2, and A_3, that are mutually exclusive and exhaustive; and let B be any event. For Bayes's rule we assume the probabilities $P(A_1)$, $P(A_2)$, $P(A_3)$, $P(B \mid A_1)$, $P(B \mid A_2)$, and $P(B \mid A_3)$ are known. The problem is to use those six probabilities to determine the conditional probabilities $P(A_1 \mid B)$, $P(A_2 \mid B)$, and $P(A_3 \mid B)$.

We will show how to express $P(A_2 \mid B)$ in terms of the six known probabilities; $P(A_1 \mid B)$ and $P(A_3 \mid B)$ are handled similarly. First we apply the conditional probability rule (page 240), to write

$$(1) \qquad P(A_2 \mid B) = \frac{P(B \,\&\, A_2)}{P(B)} = \frac{P(A_2 \,\&\, B)}{P(B)}.$$

Next we apply the general multiplication rule (page 245), to the numerator of the fraction on the right, and the rule of total probability to the denominator of the fraction on the right. This gives

$$P(A_2 \,\&\, B) = P(A_2) \cdot P(B \mid A_2)$$

and

$$P(B) = P(A_1) \cdot P(B \mid A_1) + P(A_2) \cdot P(B \mid A_2) + P(A_3) \cdot P(B \mid A_3).$$

Substituting the previous two formulas into the fraction on the right of Equation (1), we obtain

$$P(A_2 \mid B) = \frac{P(A_2) \cdot P(B \mid A_2)}{P(A_1) \cdot P(B \mid A_1) + P(A_2) \cdot P(B \mid A_2) + P(A_3) \cdot P(B \mid A_3)}.$$

This formula holds in general and is called **Bayes's rule.**

FORMULA 4.9 BAYES'S RULE

Suppose events $A_1, A_2, \ldots, A_k$ are mutually exclusive and exhaustive. Then for any event B,

$$P(A_i \mid B) = \frac{P(A_i) \cdot P(B \mid A_i)}{\sum_{j=1}^{k} P(A_j) \cdot P(B \mid A_j)},$$

where A_i can be any one of the events $A_1, A_2, \ldots, A_k$.

EXAMPLE 4.25 ILLUSTRATES FORMULA 4.9

From Table 4.11 on page 258, we know that 13.8% of Northeast residents are seniors. Now we ask: what percentage of seniors are Northeast residents?

SOLUTION Referring to the notation introduced at the beginning of the solution to Example 4.24, we see that, in terms of probability, the problem is to find $P(R_1 \mid S)$—the probability that a U.S. resident lives in the Northeast given that the resident is a senior. To obtain that conditional probability, we apply Bayes's rule and Table 4.12 on page 258:

$$P(R_1 \mid S) = \frac{P(R_1) \cdot P(S \mid R_1)}{\sum_{j=1}^{4} P(R_j) \cdot P(S \mid R_j)}$$

$$= \frac{0.190 \cdot 0.138}{0.190 \cdot 0.138 + 0.231 \cdot 0.130 + 0.355 \cdot 0.128 + 0.224 \cdot 0.111}$$

$$= 0.207.$$

So we see that 20.7% of seniors are Northeast residents. ◨

EXAMPLE 4.26 **ILLUSTRATES FORMULA 4.9**

According to the Arizona Chapter of the American Lung Association, 7.0% of the population has lung disease. Of those people having lung disease, 90.0% are smokers; and of those not having lung disease, 25.3% are smokers. Determine the probability that a randomly selected smoker has lung disease.

SOLUTION Suppose a person is selected at random. Let

$$S = \text{event the person selected is a smoker,}$$

and

$$L_1 = \text{event the person selected has no lung disease,}$$
$$L_2 = \text{event the person selected has lung disease.}$$

Note that events L_1 and L_2 are complementary, which implies that they are mutually exclusive and exhaustive.

The data provided in the statement of the problem indicate that $P(L_2) = 0.070$, $P(S \mid L_2) = 0.900$, and $P(S \mid L_1) = 0.253$. Also, since $L_1 = (\text{not } L_2)$, we can conclude that $P(L_1) = P(\text{not } L_2) = 1 - P(L_2) = 1 - 0.070 = 0.930$. We summarize this information in Table 4.13.

TABLE 4.13 Known probability information

$P(L_1) = 0.930$	$P(S \mid L_1) = 0.253$
$P(L_2) = 0.070$	$P(S \mid L_2) = 0.900$

The problem is to determine the probability that a randomly selected smoker has lung disease, $P(L_2 \mid S)$. Applying Bayes's rule to the probability data in Table 4.13, we obtain

$$P(L_2 \mid S) = \frac{P(L_2) \cdot P(S \mid L_2)}{P(L_1) \cdot P(S \mid L_1) + P(L_2) \cdot P(S \mid L_2)}$$

$$= \frac{0.070 \cdot 0.900}{0.930 \cdot 0.253 + 0.070 \cdot 0.900} = 0.211.$$

So the probability is 0.211 that a randomly selected smoker has lung disease. In terms of percentages, 21.1% of smokers have lung disease. □

We observe from Example 4.26 that the rate of lung disease among smokers (21.1%) is more than three times the rate among the general population (7.0%). Using arguments similar to those in Example 4.26, we can show that the probability is 0.010 that a randomly selected nonsmoker has lung disease; in other words, 1.0% of nonsmokers have lung disease.

So we see that the rate of lung disease among smokers (21.1%) is more than 20 times that among nonsmokers (1.0%). But we should note that because this study is observational, we cannot conclude solely on the basis of this information that smoking causes lung disease; we can only infer that a strong positive association exists between smoking and lung disease.

Prior and Posterior Probabilities

Two important terms associated with Bayes's rule are *prior probability* and *posterior probability.* We will introduce these terms by referring to Example 4.26.

From the information provided, we know that the probability is 0.070 that a randomly selected person has lung disease: $P(L_2) = 0.070$. This probability does not take into consideration whether the person is a smoker. It is therefore called a **prior probability** because it represents the probability that the person selected has lung disease *before* knowing whether the person is a smoker.

Now suppose the person selected is found to be a smoker. On the basis of this additional information, we can revise the probability that the person has lung disease. This can be done by determining the conditional probability that the person selected has lung disease, given that the person selected is a smoker: $P(L_2 \mid S) = 0.211$ (from Example 4.26). This revised probability is called a **posterior probability** since it represents the probability that the person selected has lung disease *after* knowing that the person is a smoker.

EXERCISES 4.7

STATISTICAL CONCEPTS AND SKILLS

4.129 What does it mean to say that four events are exhaustive?

4.130 Explain why an event and its complement are always mutually exclusive and exhaustive.

4.131 Refer to Example 4.24 on page 257. In probability notation, we can express the percentage of Midwest residents as $P(R_2)$. Do the same for the percentage of

a. Southern residents.
b. Southern residents that are seniors.
c. seniors that are Southern residents.

4.132 An article appearing in *The Arizona Republic* reported on a study by researchers at Harvard University and the National Institute of Aging. The study compared the life spans of left-handed and right-handed people in the United States. According to the article, 9% of females and 13% of males are left-handed. Census Bureau data indicate that 51.2% of the people in the United States are females and 48.8% are males. Answer the following questions for U.S. residents.

a. What percentage of residents are left-handed?
b. What percentage of men are left-handed?
c. What percentage of left-handed residents are men?

4.133 According to an Opinion Dynamics Poll published in *USA TODAY,* roughly 54% of U.S. men and 33% of U.S. women believe in aliens. Of U.S. adults, 48% are men and 52% women. Answer the following questions for U.S. adults.

a. What percentage of adults believe in aliens?
b. What percentage of women believe in aliens?
c. What percentage of adults that believe in aliens are women?

4.134 The results of a study conducted by Interep Research appearing in *USA TODAY* indicate that, in the United States, 29% of 18–34 year olds listen to classical music, as do 44% of 35–54 year olds and 27% of those 55 years old or over. According to *Current Population Reports,* a Census Bureau publication, of U.S. adults, 33.4% are 18–34 years old, 38.5% are 35--54 years old, and 28.1% are 55 years old or over. Answer the following questions for U.S. adults.

a. What percentage of adults listen to classical music?
b. What percentage of 18–34 year olds listen to classical music?
c. What percentage of classical-music listeners are 18–34 year olds?

4.135 A survey conducted by TELENATION/Market Facts Inc., combined with information from the Census Bureau's *Current Population Reports,* yielded the following table. The first two columns give a percentage distribution of adults by age group. And the third column gives the percentage of people in each age group who go to the movies at least once a month, people whom we will call *movie goers.*

Age	Percentage of adults	Percentage movie goers
18–24	12.7	83
25–34	20.7	54
35–44	22.0	43
45–54	16.5	37
55–64	10.9	27
65 & over	17.2	20

An adult is selected at random.

a. Determine the probability that the adult obtained is a movie goer.
b. Determine the probability that the adult obtained is between 25 and 34 years old given that he or she is a movie goer.
c. Interpret your answers in parts (a) and (b) in terms of percentages.

4.136 In the first two columns of the following table, we have provided a percentage distribution for the religious affiliation of the voters in a hypothetical city. The data are based on the national distribution for religious preference found in *Emerging Trends,* published by the Princeton Religion Research Center of Princeton, New Jersey. In the third column of the table, we show the percentage of Democrats in each religious group of voters. The table shows, for instance, that 28% of the voters in the city are Catholic and that 53% of the Catholic voters in the city are Democrats.

Religion	Percentage of voters	Percentage Democrats
Catholic	28	53
Jewish	2	61
Protestant	57	42
Other	4	58
None	9	67

A voter in the city is selected at random.

a. Determine the probability that the voter obtained is a Democrat.
b. Determine the probability that the voter obtained is a Protestant given that he or she is a Democrat.
c. Interpret your answers in parts (a) and (b) in terms of percentages.

4.137 Textbook editors must estimate the sales of new (first-edition) books. The records of one major publishing company indicate that 10% of all new books sell more than projected, 30% sell close to projected, and 60% sell less than projected. Of those that sell more than projected, 70% are revised for a second edition, as are 50% of those that sell close to projected, and 20% of those that sell less than projected.

a. What percentage of books published by this publishing company go to a second edition?
b. What percentage of books published by this publishing company that go to a second edition sold less than projected in their first edition?

4.138 EDA Products, a manufacturer of customized metal fabrication items, produces forged tools for a major retailer. EDA currently uses three 50-ton forge presses to manufacture a particular model of slip joint pliers. Although each of the presses accounts for one-third of production, the presses produce defective units with varying percentages. In fact, recent tests indicate that Press 1 produces 1.1% defective units, Press 2 produces 0.8% defective units, and Press 3 produces 1.6% defective units.

a. What percentage of slip joint pliers produced by EDA are defective?
b. Of those slip joint pliers that are defective, what percentage are produced by Press 2?

4.139 The National Center for Health Statistics provides information on suicides by sex and method used. Data are published in *Vital Statistics of the United States*. In 1995, there were 31,284 suicides in the United States, of which 25,369 were males and 5,915 were females. The following table gives a relative-frequency distribution for method used by males and females who committed suicide.

Method used	Relative frequency for males	Relative frequency for females
Poisoning	0.133	0.364
Hanging/ strangulation	0.154	0.126
Firearms	0.651	0.408
Other	0.062	0.102

A suicide report is selected at random. Find the
a. probability that a firearm was used for the suicide.
b. prior probability that the person who committed suicide was a female.
c. posterior probability that the person who committed suicide was a female, given that a firearm was used.
d. Interpret the probabilities obtained in parts (a)–(c) in terms of percentages.

4.140 A Gallup poll conducted 10 years ago asked 1005 adults and 500 teenagers the question, "What is the nation's top problem?" The pie charts shown at the top of the next column were used to summarize the results of the survey. Suppose a person who participated in the survey is selected at random.
a. Determine the probability that the person selected said that drug abuse is the nation's top problem.
b. Find the prior probability that the person selected is a teenager.
c. Find the posterior probability that the person selected is a teenager, given that the person selected said that drug abuse is the nation's top problem.
d. Interpret the probabilities obtained in parts (a)–(c) in terms of percentages.

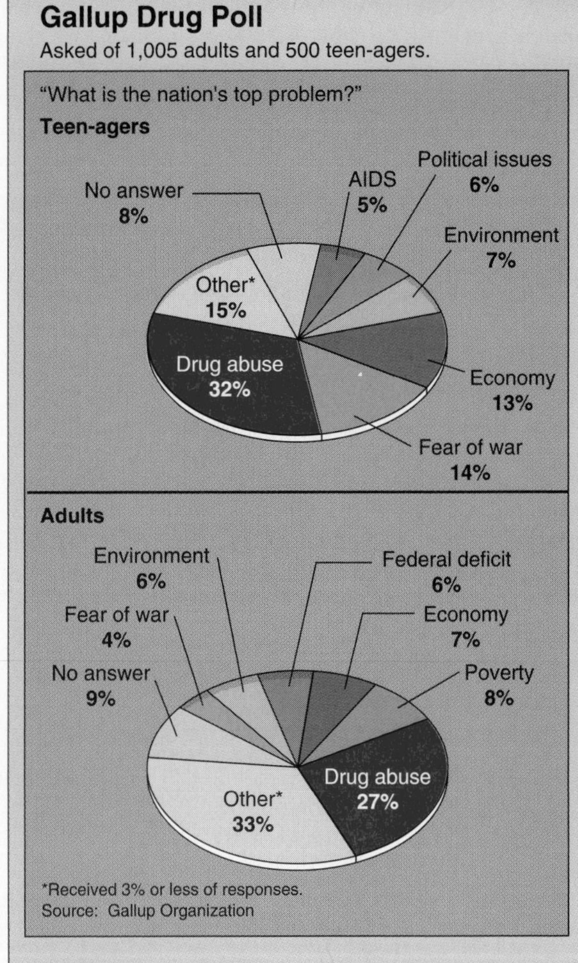

4.141 At a grocery store, eggs come in cartons that hold a dozen eggs. Experience indicates that 78.5% of the cartons have no broken eggs, 19.2% have one broken egg, 2.2% have two broken eggs, 0.1% have three broken eggs, and the percentage of cartons with four or more broken eggs is negligible. An egg selected at random from a carton is found to be broken. What is the probability that this egg is the only broken one in the carton?

Extending the Concepts and Skills

4.142 Medical tests are frequently used to decide whether a person has a particular disease. The **sensitivity** of a test is defined as the probability that a person having the disease will test positive; the **specificity** of a test is defined as the probability that a person not having the disease will test negative. A test for a certain disease has been used for many years. Experience with the test indicates that its sensitivity is 0.934 and its specificity is 0.968. Furthermore, it is known that roughly 1 in 500 people has the disease.

a. Interpret the sensitivity and specificity of this test in terms of percentages.
b. Determine the probability that a person testing positive actually has the disease.
c. Interpret your answer from part (b) in terms of percentages.

4.143 *Bottom Line/Personal* newsletter interviewed Gerald Kushel, Ed.D., on the secrets of successful people. To study success, Kushel questioned 1200 people, among whom were lawyers, artists, teachers, and students. He found that 15% enjoy neither their jobs nor their personal lives, 80% enjoy their jobs but not their personal lives, and 4% enjoy both their jobs and their personal lives.

a. Determine the percentage of the people interviewed who enjoy their jobs.
b. What percentage of the people interviewed who enjoy their jobs also enjoy their personal lives?

4.144 Refer to Example 4.26 on page 261.

a. Determine the probability that a randomly selected nonsmoker has lung disease.
b. Use part (a) and the result of Example 4.26 to compare the rates of lung disease for smokers and nonsmokers.

4.145 Regarding the concepts of exhaustive and mutually exclusive.

a. Draw a Venn diagram illustrating three events that are mutually exclusive but not exhaustive.
b. Give an example of three events that are mutually exclusive but not exhaustive.
c. Draw a Venn diagram illustrating three events that are exhaustive but not mutually exclusive.
d. Give an example of three events that are exhaustive but not mutually exclusive.

4.8 COUNTING RULES*

We often need to determine the number of ways something can happen—the number of possible outcomes for an experiment, the number of ways an event can occur, the number of ways a certain task can be performed, and so forth. Sometimes we can list the possibilities and then count them; but in most cases, the number of possibilities is so large that a direct listing is impractical.

Consequently, we need to develop techniques that do not rely on a direct listing for determining the number of ways something can happen. Such techniques are usually referred to as **counting rules.** In this section we will examine some widely used counting rules.

The Basic Counting Rule

One counting rule, called the **basic counting rule (BCR),** is fundamental to all the counting techniques we will discuss. We introduce this rule in Example 4.27.

* This more advanced section continues the optional material on probability. Although important in its own right, it is not needed for the remainder of the book.

EXAMPLE 4.27 INTRODUCES THE BASIC COUNTING RULE

Robson Communities, Inc., builds new-home communities in several parts of Arizona. In Sun Lakes, Arizona, four models are offered—the Shalimar, Palacia, Valencia, and Monterey—each in three different elevations, designated A, B, and C. How many choices are there for the selection of a home, including both model and elevation?

SOLUTION We will first use a tree diagram (Fig. 4.28) to systematically obtain a direct listing of the possibilities. In the tree diagram, we have used S for Shalimar, P for Palacia, V for Valencia, and M for Monterey.

FIGURE 4.28
Tree diagram for model and elevation possibilities

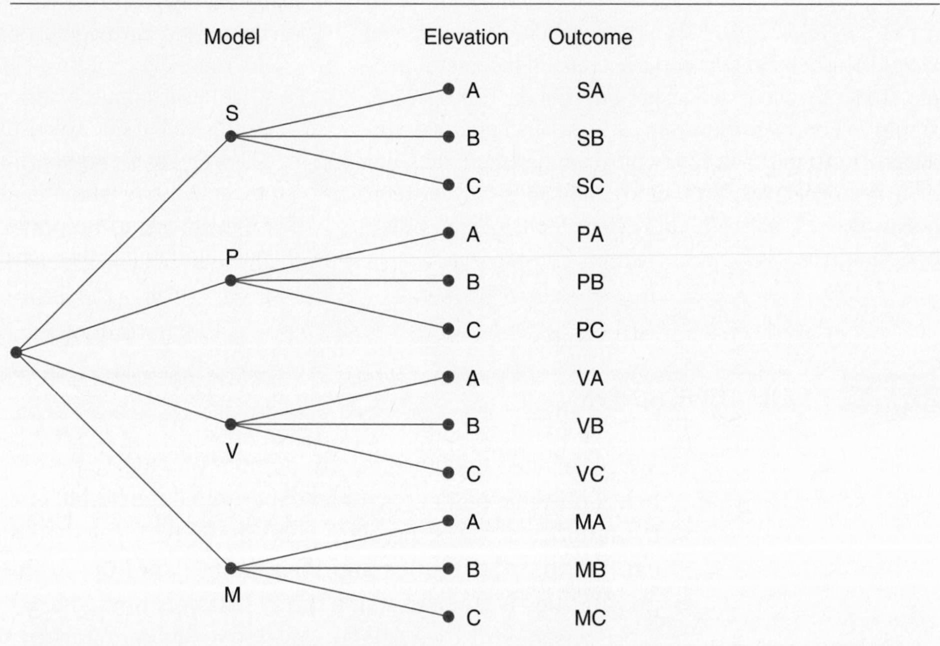

Each branch of the tree corresponds to one possibility for model and elevation. For instance, the first branch of the tree, ending in SA, corresponds to the Shalimar model with the A elevation. The total number of possibilities can be obtained by counting the number of branches at the end of the tree. So we see that there are 12 choices for the selection of a home, including both model and elevation.

Although the tree-diagram approach for determining the number of possibilities is a direct listing, it provides us with a clue for obtaining the number of possibilities without resorting to a direct listing. Specifically, there are four possibilities for model, indicated by the four sub-branches emanating from the starting point of the tree; and to each possibility for model, there correspond three possibilities for elevation, indicated by the three sub-branches emanating from the end of each model sub-branch. Consequently, there are

$$\underbrace{3+3+3+3}_{4 \text{ times}} = 4 \cdot 3 = 12$$

possibilities altogether. So we see that the total number of possibilities can be obtained by multiplying the number of possibilities for the model by the number of possibilities for the elevation. □

In Example 4.27 there are two actions (choice of model and choice of elevation) and we multiplied the number of possibilities for each action to obtain the total number of possibilities. This same multiplication principle applies regardless of the number of actions.

KEY FACT 4.2 **THE BASIC COUNTING RULE (BCR)**[†]

Suppose r actions (choices, experiments) are to be performed in a definite order. Further suppose there are m_1 possibilities for the first action, and that corresponding to each of these possibilities there are m_2 possibilities for the second action, and so on. Then there are $m_1 \cdot m_2 \cdots m_r$ possibilities altogether for the r actions.

As we noted, in Example 4.27 there are two actions ($r = 2$), selecting a model and selecting an elevation. Because there are four possibilities for model, $m_1 = 4$; and because corresponding to each model there are three possibilities for elevation, $m_2 = 3$. Therefore by the BCR, the total number of possibilities, including both model and elevation, is

$$m_1 \cdot m_2 = 4 \cdot 3 = 12,$$

as we discovered in Example 4.27.

Because the number of possibilities in the model/elevation problem is quite small, it is relatively simple to determine the number by a direct listing, as we did in the tree diagram in Fig. 4.28. It is, nonetheless, even easier to obtain the number of possibilities by applying the BCR. Moreover, in problems where the number of possibilities is large, a direct listing is not feasible and the BCR is the only practical way to proceed.

[†] The basic counting rule is also known as the **basic principle of counting,** the **fundamental counting rule,** and the **multiplication rule.**

EXAMPLE 4.28 ILLUSTRATES KEY FACT 4.2

The license plates of Arizona consist of three letters followed by three digits.

a. How many different license plates are possible?

b. How many possibilities are there for license plates in which no letter or digit is repeated?

SOLUTION For both parts (a) and (b), we will apply the BCR with six actions ($r = 6$).

a. There are 26 possibilities for the first letter, 26 for the second letter, and 26 for the third letter; and there are 10 possibilities for the first digit, 10 for the second digit, and 10 for the third digit. Applying the BCR, we see that there are

$$m_1 \cdot m_2 \cdot m_3 \cdot m_4 \cdot m_5 \cdot m_6 = 26 \cdot 26 \cdot 26 \cdot 10 \cdot 10 \cdot 10 = 17{,}576{,}000$$

possibilities altogether for different license plates. Obviously, it would not be practical to obtain the number of possibilities by a direct listing—the tree diagram would have 17,576,000 branches!

b. For this part there are again 26 possibilities for the first letter. But to each possibility for the first letter, there correspond 25 possibilities for the second letter because the second letter cannot be the same as the first. And to each possibility for the first two letters, there correspond 24 possibilities for the third letter because the third letter cannot be the same as either the first or the second. Similarly, there are 10 possibilities for the first digit, 9 for the second digit, and 8 for the third digit. So by the BCR, there are

$$m_1 \cdot m_2 \cdot m_3 \cdot m_4 \cdot m_5 \cdot m_6 = 26 \cdot 25 \cdot 24 \cdot 10 \cdot 9 \cdot 8 = 11{,}232{,}000$$

possibilities for license plates in which no letter or digit is repeated. ❑

Factorials

Before we continue our presentation of counting rules, we need to discuss factorials. Factorials are used extensively in mathematics and its applications.

DEFINITION 4.8 FACTORIALS

The product of the first k positive integers is called ***k factorial*** and is denoted $\boldsymbol{k!}$. In symbols,

$$k! = k(k-1)\cdots 2 \cdot 1.$$

We also define $0! = 1$.

EXAMPLE 4.29 ILLUSTRATES DEFINITION 4.8

Determine 3!, 4!, and 5!.

SOLUTION Applying Definition 4.8, we obtain

$$3! = 3 \cdot 2 \cdot 1 = 6,$$
$$4! = 4 \cdot 3 \cdot 2 \cdot 1 = 24,$$
$$5! = 5 \cdot 4 \cdot 3 \cdot 2 \cdot 1 = 120,$$

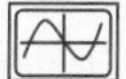

as required.

Note, for instance, that $6! = 6 \cdot 5!$, $6! = 6 \cdot 5 \cdot 4!$, $6! = 6 \cdot 5 \cdot 4 \cdot 3!$, and so on. In general, if $j \leq k$, then $k! = k(k-1) \cdots (k-j+1)(k-j)!$.

Permutations

A **permutation** of r objects from a collection of m objects is any *ordered* arrangement of r of the m objects. The number of possible permutations of r objects that can be formed from a collection of m objects is denoted by $(m)_r$ or ${}_mP_r$.† Let's look at a simple example.

EXAMPLE 4.30 INTRODUCES PERMUTATIONS

Consider the collection consisting of the five letters *a, b, c, d, e*.

a. List all possible permutations of three letters from this collection of five letters.

b. Use part (a) to determine the number of possible permutations of three letters that can be formed from the collection of five letters; that is, find $(5)_3$.

c. Use the BCR to determine the number of possible permutations of three letters that can be formed from the collection of five letters; that is, find $(5)_3$ using the BCR.

SOLUTION **a.** For this part we need to list all ordered arrangements of three letters from the first five letters of the English alphabet. This is done in Table 4.14.

† The notation ${}_mP_r$ is more common than $(m)_r$, but we prefer the latter since it is less cumbersome and has other pedagogical advantages.

TABLE 4.14
Possible permutations of three letters from the collection of five letters

abc	*abd*	*abe*	*acd*	*ace*	*ade*	*bcd*	*bce*	*bde*	*cde*
acb	*adb*	*aeb*	*adc*	*aec*	*aed*	*bdc*	*bec*	*bed*	*ced*
bac	*bad*	*bae*	*cad*	*cae*	*dae*	*cbd*	*cbe*	*dbe*	*dce*
bca	*bda*	*bea*	*cda*	*cea*	*dea*	*cdb*	*ceb*	*deb*	*dec*
cab	*dab*	*eab*	*dac*	*eac*	*ead*	*dbc*	*ebc*	*ebd*	*ecd*
cba	*dba*	*eba*	*dca*	*eca*	*eda*	*dcb*	*ecb*	*edb*	*edc*

b. From Table 4.14 we see that there are 60 possible permutations of three letters from the collection of five letters; in other words, $(5)_3 = 60$.

c. Here we want to use the BCR to determine the number of possible permutations of three letters from the collection of five letters. There are five possibilities for the first letter, four possibilities for the second letter, and three possibilities for the third letter. Hence by the BCR, there are

$$m_1 \cdot m_2 \cdot m_3 = 5 \cdot 4 \cdot 3 = 60$$

possibilities altogether. So again we see that $(5)_3 = 60$. □

We can make two relevant observations from Example 4.30. First, it is generally tedious or impractical to list all possible permutations under consideration. Second, it is not necessary to list the possible permutations in order to determine how many there are—we can use the BCR to count them.

By studying part (c) of Example 4.30, we see that the BCR can be used to obtain a general formula for $(m)_r$. The formula is $(m)_r = m(m-1)\cdots(m-r+1)$. Multiplying the numerator and denominator of this formula by $(m-r)!$, we get the equivalent expression $(m)_r = m!/(m-r)!$. We summarize this discussion in Formula 4.10.

FORMULA 4.10 **PERMUTATIONS RULE**

The number of possible permutations of r objects from a collection of m objects is given by the formula

$$(m)_r = \frac{m!}{(m-r)!}.$$

EXAMPLE 4.31 ILLUSTRATES FORMULA 4.10

In an exacta wager at the race track, the bettor picks the two horses that he or she thinks will finish first and second, in a specified order. For a race with 12 entrants, determine the number of possible exacta wagers.

SOLUTION Selecting two horses from the 12 horses for an exacta wager is equivalent to specifying a permutation of two objects from a collection of 12 objects; the first object is the horse selected to finish in first place and the second object is the horse selected to finish in second place. Thus the number of possible exacta wagers is $(12)_2$—the number of possible permutations of two objects from a collection of 12 objects. Applying the permutations rule, with $m = 12$ and $r = 2$, we obtain

$$(12)_2 = \frac{12!}{(12-2)!} = \frac{12!}{10!} = \frac{12 \cdot 11 \cdot \cancel{10!}}{\cancel{10!}} = 12 \cdot 11 = 132.$$

In a 12-horse race, there are 132 possible exacta wagers. ◨

EXAMPLE 4.32 ILLUSTRATES FORMULA 4.10

A student has 10 books to arrange on a shelf of a bookcase. In how many ways can the 10 books be arranged?

SOLUTION Any particular arrangement of the 10 books on the shelf is a permutation of 10 objects from a collection of 10 objects. So for this problem we need to determine $(10)_{10}$, the number of possible permutations of 10 objects from a collection of 10 objects, more commonly expressed as the number of possible permutations of 10 objects among themselves. Applying the permutations rule, we get

$$(10)_{10} = \frac{10!}{(10-10)!} = \frac{10!}{0!} = \frac{10!}{1} = 10! = 3{,}628{,}800.$$

There are 3,628,800 ways to arrange the 10 books on the shelf. It doesn't seem possible that there could be this many ways, but there are! ◨

Let's generalize Example 4.32 to find the number of possible permutations of m objects among themselves. Using the permutations rule, we conclude that

$$(m)_m = \frac{m!}{(m-m)!} = \frac{m!}{0!} = \frac{m!}{1} = m!.$$

Consequently, we have the following formula as a special case of the permutations rule.

FORMULA 4.11 **SPECIAL PERMUTATIONS RULE**

The number of possible permutations of m objects among themselves is $m!$.

Combinations

A **combination** of r objects from a collection of m objects is any *unordered* arrangement of r of the m objects, in other words, any subset of r objects from the collection of m objects. Note that order matters in permutations, but not in combinations. The number of possible combinations of r objects that can be formed from a collection of m objects is denoted by $\binom{m}{r}$ or ${}_mC_r$. Let's return to the situation of Example 4.30.

EXAMPLE 4.33 INTRODUCES COMBINATIONS

Consider the collection consisting of the five letters *a, b, c, d, e.*

a. List all possible combinations of three letters from this collection of five letters.

b. Use part (a) to determine the number of possible combinations of three letters that can be formed from the collection of five letters; that is, find $\binom{5}{3}$.

SOLUTION **a.** For this part we need to list all unordered arrangements of three letters from the first five letters in the English alphabet. This is done in Table 4.15.

TABLE 4.15 Combinations

$\{a, b, c\}$	$\{a, b, d\}$	$\{a, b, e\}$	$\{a, c, d\}$	$\{a, c, e\}$	$\{a, d, e\}$	$\{b, c, d\}$	$\{b, c, e\}$	$\{b, d, e\}$	$\{c, d, e\}$

b. From Table 4.15 we see that there are 10 possible combinations of three letters from the collection of five letters; in other words, $\binom{5}{3} = 10$. ◨

In Example 4.33 we obtained the number of possible combinations by a direct listing. We can avoid resorting to a direct listing by deriving a formula for determining the number of possible combinations. To see how this is done, let's return once more to the English-letters example.

Look at the first combination in Table 4.15, $\{a, b, c\}$. By the special permutations rule, there are $3! = 6$ permutations of these three letters among themselves; they are *abc, acb, bac, bca, cab,* and *cba*. These six permutations are the ones displayed in the first column of Table 4.14 on page 270. Similarly, there are $3! = 6$ permutations of the three letters in the second combination in Table 4.15, $\{a, b, d\}$. These six permutations are the ones displayed in the second column of Table 4.14. The same comments apply to the other eight combinations in Table 4.15.

So we see that to each combination of three letters from the collection of five letters, there correspond $3!$ permutations of three letters from the collection of five letters. Moreover, any such permutation is accounted for in this way. Consequently, there must be $3!$ times as many permutations as combinations; or, equivalently, the number of possible combinations of three letters from the collection of five letters must equal the number of possible permutations of three letters from the collection of five letters divided by 3!:

$$\binom{5}{3} = \frac{(5)_3}{3!} = \frac{5!/(5-3)!}{3!} = \frac{5!}{3!\,(5-3)!} = \frac{5 \cdot 4 \cdot \cancel{3!}}{\cancel{3!}\,2!} = \frac{5 \cdot 4}{2} = 10.$$

This is the number we obtained in Example 4.33 by a direct listing.

The same type of argument that we just gave holds in general. In other words, we have the following rule for determining the number of possible combinations.

FORMULA 4.12 **COMBINATIONS RULE**

The number of possible combinations of r objects from a collection of m objects is given by the formula

$$\binom{m}{r} = \frac{m!}{r!\,(m-r)!}.$$

EXAMPLE 4.34 **ILLUSTRATES FORMULA 4.12**

In order to recruit new members, a compact-disc club advertises a special introductory offer: A new member agrees to buy one compact disc at regular club prices and receives free any four compact discs of his or her choice from a collection of 69 compact discs. How many possibilities does a new member have for the selection of the four free compact discs?

SOLUTION Any particular selection of four compact discs from 69 compact discs is a combination of four objects from a collection of 69 objects. So, by the combinations rule, the number of possible selections equals

$$\binom{69}{4} = \frac{69!}{4!\,(69-4)!} = \frac{69!}{4!\,65!} = \frac{69 \cdot 68 \cdot 67 \cdot 66 \cdot \cancel{65!}}{4!\,\cancel{65!}} = 864{,}501.$$

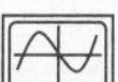

There are 864,501 possibilities for the selection of four compact discs from the collection of 69 compact discs. ◨

EXAMPLE 4.35 **ILLUSTRATES FORMULA 4.12**

An economics professor is using a new method to teach a junior-level course with an enrollment of 42 students. The professor wants to conduct in-depth interviews with the students to get feedback on the new teaching method, but does not want to interview all 42 of them. She decides to interview a sample of five students from the class. How many different samples are possible?

SOLUTION A sample of five students from the class of 42 students can be considered a combination of five objects from a collection of 42 objects. Consequently, by the combinations rule, the number of possible samples is

$$\binom{42}{5} = \frac{42!}{5!\,(42-5)!} = \frac{42!}{5!\,37!} = 850{,}668.$$

There are 850,668 different samples of five students that can be obtained from the population of 42 students in the class. ◨

Example 4.35 shows how to determine the number of possible samples of a specified size from a finite population. This is so important that we record it as Formula 4.13.

FORMULA 4.13 NUMBER OF POSSIBLE SAMPLES

The number of possible samples of size n from a population of size N is $\binom{N}{n}$.

Applications to Probability

Suppose an experiment has N equally likely possible outcomes. Then according to the f/N rule, the probability that a specified event occurs equals the number of ways, f, that the event can occur divided by the total number of possible outcomes.

Although in the probability problems we have considered up to this point it has been easy to determine f and N, that is not always the case. We must often use counting rules to obtain the number of possible outcomes and the number of ways that the specified event can occur. Example 4.36 provides an illustration of how counting rules can be applied to solve probability problems.

EXAMPLE 4.36 APPLYING COUNTING RULES TO PROBABILITY

The quality-assurance engineer of a television company inspects TVs in lots of 100. He selects five of the 100 TVs at random and inspects them thoroughly. Assuming that six of the 100 TVs in the current lot are actually defective, find the probability that exactly two of the five TVs selected by the engineer are defective.

SOLUTION Because the engineer makes his selection at random, each of the possible outcomes is equally likely. This means that we can apply the f/N rule to obtain the required probability.

First we determine the number of possible outcomes for the experiment. This is the number of ways that five TVs can be selected from the 100 TVs—the number of possible combinations of five objects from a collection of 100 objects. Applying the combinations rule, we obtain

$$\binom{100}{5} = \frac{100!}{5!\,(100-5)!} = \frac{100!}{5!\,95!} = 75{,}287{,}520.$$

We see that $N = 75{,}287{,}520$.

Next we determine the number of ways the specified event can occur, that is, the number of outcomes in which exactly two of the five TVs selected are defective. To accomplish this, it is helpful to think of the 100 TVs as partitioned into two groups, namely, the defective TVs and the nondefective TVs, as shown in the top part of Fig. 4.29.

FIGURE 4.29 Calculating the number of outcomes in which exactly two of the five TVs selected are defective

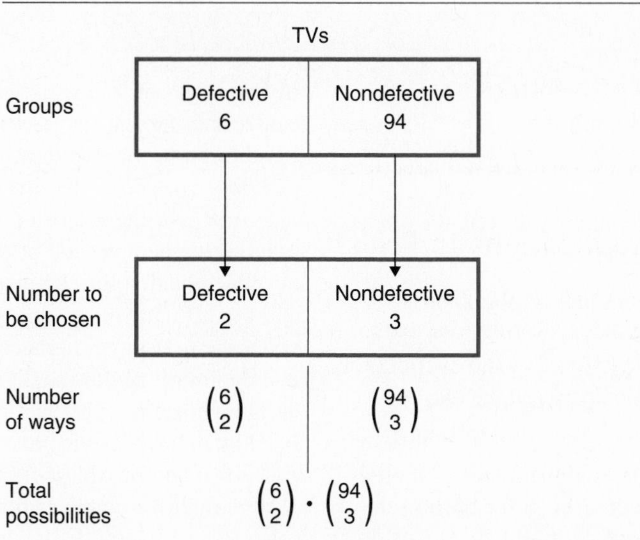

There are six TVs in the first group, of which two are to be selected. This can be done in

$$\binom{6}{2} = \frac{6!}{2!\,(6-2)!} = \frac{6!}{2!\,4!} = 15$$

ways. There are 94 TVs in the second group, of which three are to be selected. This can be done in

$$\binom{94}{3} = \frac{94!}{3!\,(94-3)!} = \frac{94!}{3!\,91!} = 134{,}044$$

ways. Consequently, by the BCR, there are a total of

$$\binom{6}{2} \cdot \binom{94}{3} = 15 \cdot 134{,}044 = 2{,}010{,}660$$

outcomes in which exactly two of the five TVs selected are defective. So we see that $f = 2{,}010{,}660$. Figure 4.29 summarizes the calculations done in this paragraph.

Applying the f/N rule, we now conclude that the probability that exactly two of the five TVs selected are defective equals

$$\frac{f}{N} = \frac{2{,}010{,}660}{75{,}287{,}520} = 0.027.$$

There is a 2.7% chance that exactly two of the five TVs selected by the engineer will be defective. □

EXERCISES 4.8

Statistical Concepts and Skills

4.146 What are counting rules? Why are these rules important?

4.147 Why is the basic counting rule (BCR) often referred to as the multiplication rule?

4.148 Regarding permutations and combinations.
a. What is a permutation?
b. What is a combination?
c. What is the major distinction between the two?

4.149 Refer to Example 4.27 on page 266. Suppose the developer discontinues the Shalimar model but provides an additional elevation choice, D, for each of the remaining three model choices.
a. Draw a tree diagram similar to Fig. 4.28 showing the possible choices for the selection of a home, including both model and elevation.
b. Use the tree diagram in part (a) to determine the total number of choices for the selection of a home, including both model and elevation.
c. Use the BCR to determine the total number of choices for the selection of a home, including both model and elevation.

4.150 The author spoke with a representative of the United States Post Office and obtained the following information about zip codes. A five-digit zip code consists of five digits of which the first three give the sectional center and the last two the post office or delivery area. In addition to the five-digit zip code, there is a trailing *plus four zip code.* The first two digits of the plus four zip code give the sector or several blocks and the last two the segment or side of the street. For the five-digit zip code, the first four digits can be any of the digits 0–9 and the fifth any of the digits 1–8. For the plus four zip code, the first three digits can be any of the digits 0–9 and the fourth any of the digits 1–9.
a. How many possible five-digit zip codes are there?
b. How many possible plus four zip codes are there?
c. How many possibilities are there including both the five-digit zip code and the plus four zip code?

4.151 Telephone numbers in the United States consist of a three-digit area code followed by a seven-digit local number. Suppose neither the first digit of an area code nor the first digit of a local number can be a zero but that all other choices are acceptable.
a. How many different area codes are possible?
b. For a given area code, how many local telephone numbers are possible?
c. How many telephone numbers are possible?

4.152 Computerized testing systems are used extensively by professors. A physics professor needs to construct a five-question quiz, one question for each of five topics. The computerized testing system she uses provides eight choices for the question on the first topic, 10 choices for the question on the second topic, seven choices for the question on the third topic, eight choices for the question on the fourth topic, and six choices for the question on the fifth topic. How many possibilities are there for the five-question quiz?

4.153 *Scientific Computing & Automation* magazine offers free subscriptions to the scientific community. The magazine does ask, however, that a person answer six questions: primary title, type of facility, area of work, brand of computer used, type of operating system in use, and type of instruments in use. There are six choices given for the first question, eight for the second, five for the third, 19 for the fourth, 16 for the fifth, and 14 for the sixth. How many possibilities are there for answering all six questions?

4.154 Determine the value of each of the following.
a. $(7)_3$ **b.** $(5)_2$ **c.** $(8)_4$
d $(6)_0$ **e.** $(9)_9$

4.155 Determine the value of each of the following.
a. $(4)_3$ **b.** $(15)_4$ **c.** $(6)_2$
d $(10)_0$ **e.** $(8)_8$

4.156 At a movie festival, a team of judges is to pick the first, second, and third place winners from the 18 films entered. How many possibilities are there?

4.157 Investment firms usually have a large selection of mutual funds from which an investor can choose. One such firm has 30 mutual funds. Suppose you plan to invest in four of these mutual funds, one during each quarter of next year. In how many different ways can you make these four investments?

4.158 The sales manager of a clothing company needs to assign seven salespeople to seven different territories. How many possibilities are there for the assignments?

4.159 An extrasensory-perception (ESP) experiment is conducted by a psychologist. For part of the experiment, the psychologist takes 10 cards, numbered 1–10, and shuffles them. Then she looks at the cards one at a time. While she looks at each card, the subject writes down the number he thinks is on the card.
a. How many possibilities are there for the order in which the subject writes down the numbers?
b. If the subject has no ESP and is just guessing each time, what is the probability that he writes down the numbers in the correct order, that is, in the order that the cards are actually arranged?

4.160 Determine the value of each of the following.
a. $\binom{7}{3}$ **b.** $\binom{5}{2}$ **c.** $\binom{8}{4}$ **d.** $\binom{6}{0}$ **e.** $\binom{9}{9}$

4.161 Determine the value of each of the following.
a. $\binom{4}{3}$ **b.** $\binom{15}{4}$ **c.** $\binom{6}{2}$ **d.** $\binom{10}{0}$ **e.** $\binom{8}{8}$

4.162 A poker hand consists of 5 cards dealt from an ordinary deck of 52 playing cards.
a. How many possible poker hands are there?
b. How many different hands are there consisting of three kings and two queens?
c. The hand in part (b) is an example of a full house: three cards of one denomination and two of another. How many different full houses are there?
d. Obtain the probability of being dealt a full house.

4.163 The U.S. Senate consists of 100 senators, two from each state. A committee consisting of five senators is to be formed.
a. How many different committees are possible?
b. How many are possible if no state may have more than one senator on the committee?
c. If the committee is selected at random from all 100 senators, what is the probability that no state will have both of its senators on the committee?

4.164 How many samples of size 5 are possible from a population of size 70?

4.165 How many samples of size 6 are possible from a population with 45 members?

4.166 Suppose you have a key ring with eight keys on it, one of which is your house key. Further suppose you get home after dark and can't see the keys on the key ring. You randomly try one key at a time, being careful not to mix the keys you have already tried with the ones you haven't. What is the probability that you get the right key
a. on the first try? **b.** on the eighth try?
c. on or before the fifth try?

4.167 Refer to Example 4.36 on page 274. Determine the probability that
a. exactly one of the TVs selected is defective.
b. at most one of the TVs selected is defective.
c. at least one of the TVs selected is defective.

4.168 *The Birthday Problem.* A biology class has 38 students. Find the probability that at least two students in the class have the same birthday. For simplicity assume there are always 365 days in a year and that birth rates are constant throughout the year. *(Hint:* First determine the probability that no two students have the same birthday and then apply the complementation rule.*)*

4.169 The Arizona state lottery, *Lotto,* is played as follows: The player selects six numbers from the numbers 1–42 and buys a ticket for $1. There are six winning numbers, which are selected at random from the numbers 1–42. To win a prize, a *Lotto* ticket must contain three or more of the winning numbers. A ticket with exactly three winning numbers is paid $2. The prize for a ticket with exactly four, five, or six winning numbers depends on sales and on how many other tickets were sold that have exactly four, five, or six winning numbers, respectively. If you buy one *Lotto* ticket, determine the probability that
a. you win the jackpot; that is, your six numbers are the same as the six winning numbers.
b. your ticket contains exactly four winning numbers.
c. you don't win a prize.

4.170 A student takes a true-false test consisting of 15 questions. Assuming the student guesses at each question, find the probability that
a. the student gets at least one question correct.
b. the student gets a 60% or better on the exam.

EXTENDING THE CONCEPTS AND SKILLS

4.171 According to the Center for Political Studies at the University of Michigan, Ann Arbor, roughly 50% of U.S. adults are Democrats. Suppose 10 U.S. adults are selected at random. Determine the approximate probability that
a. exactly five are Democrats.
b. eight or more are Democrats.

4.172 *The Birthday Problem.* Refer to Exercise 4.168, but now assume the class consists of N students. Determine the probability that at least two of the students have the same birthday.

4.173 Suppose a simple random sample of size n is to be taken without replacement from a population of size N.
a. Determine the probability that any particular sample of size n is the one selected.
b. Determine the probability that any specified member of the population is included in the sample.
c. Determine the probability that any k specified members of the population are included in the sample.

USING TECHNOLOGY  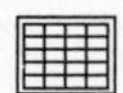 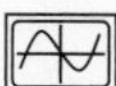

4.174 Use a computer or a programmable calculator and your answer from Exercise 4.172 to construct a table giving the probability that at least two of the students in the class have the same birthday, for $N = 2, 3, \ldots, 70$.

CHAPTER REVIEW

You Should Be Able To

1. use and understand the formulas presented in this chapter.
2. compute probabilities for experiments having equally likely outcomes.
3. interpret probabilities using the frequentist interpretation of probability.
4. state and understand the basic properties of probability.
5. construct and interpret Venn diagrams.
6. find and describe (not E), (A & B), and (A or B).
7. determine whether two or more events are mutually exclusive.
8. understand and use probability notation.

9. state and apply the special addition rule.
10. state and apply the complementation rule.
11. state and apply the general addition rule.
***12.** read and interpret contingency tables.
***13.** construct a joint probability distribution.
***14.** compute conditional probabilities both directly and by using the conditional-probability rule.
***15.** state and apply the general multiplication rule.
***16.** state and apply the special multiplication rule.
***17.** determine whether two events are independent.
***18.** understand the difference between mutually exclusive events and independent events.
***19.** determine whether two or more events are exhaustive.
***20.** state and apply the rule of total probability.
***21.** state and apply Bayes's rule.
***22.** state and apply the basic counting rule (BCR).
***23.** state and apply the permutations and combinations rules.
***24.** apply counting rules to solve probability problems when appropriate.

Key Terms

$(A \,\&\, B)$, *210*
$(A \text{ or } B)$, *210*
at random, *200*
basic counting rule (BCR),* *267*
Bayes's rule,* *260*
bivariate data,* *228*
cells,* *228*
certain event, *203*
combination,* *272*
combinations rule,* *273*
complement, *210*
complementation rule, *221*
conditional probability,* *235*
conditional-probability rule,* *240*
contingency table,* *228*
counting rules,* *265*
dependent events,* *249*
equal-likelihood model, *203*
event, *209*
exhaustive events,* *256*
experiment, *199*
f/N rule, *199*
factorials,* *268*
frequentist interpretation of probability, *202*
general addition rule, *223*
general multiplication rule,* *245*
given event,* *235*
impossible event, *203*
independent,* *248*
independent events,* *248*
joint probabilities,* *230*
joint probability distribution,* *231*
marginal probabilities,* *231*
mutually exclusive events, *214*
$(\text{not } E)$, *210*
$P(B \mid A)$,* *235*
$P(E)$, *218*
permutation,* *269*
permutations rule,* *270*
posterior probability,* *262*
prior probability,* *262*
probability model, *202*
probability theory, *197*
rule of total probability,* *257*
sample space, *209*
special addition rule, *219*
special multiplication rule*, *249, 250*
special permutations rule,* *271*
stratified sampling theorem,* *257*
tree diagram,* *247*
two-way table,* *228*
univariate data,* *228*
Venn diagrams, *209*

REVIEW TEST

STATISTICAL CONCEPTS AND SKILLS

1. Why is probability theory important to statistics?

2. Regarding the equal-likelihood model.
   a. What is it?
   b. How are probabilities computed?

3. What meaning is given to the probability of an event by the frequentist interpretation of probability?

4. Decide which of the following numbers could not possibly be probabilities. Explain your reasoning.
   a. 0.047 b. −0.047 c. 3.5 d. 1/3.5

5. Identify a commonly used graphical technique for portraying events and relationships among events.

6. What does it mean for two or more events to be mutually exclusive?

7. Suppose that E is an event. Use probability notation to represent
   a. the probability that event E occurs.
   b. the probability that event E occurs equals 0.436.

8. Answer true or false to each of the following statements and explain your answers.
   a. For any two events, the probability that one or the other occurs equals the sum of the two individual probabilities.
   b. For any event, the probability it occurs equals 1 minus the probability that it doesn't occur.

9. Identify one reason why the complementation rule is useful.

*10. Fill in the following blanks.
   a. Data obtained by observing values of one variable of a population are called ____ data.
   b. Data obtained by observing values of two variables of a population are called ____ data.
   c. A frequency distribution for bivariate data is called a ____.

*11. The sum of the joint probabilities in a row or column of a joint probability distribution equals the ____ probability in that row or column.

*12. Let A and B be events.
   a. Use probability notation to represent the conditional probability that event B occurs given that event A has occurred.
   b. In part (a), which is the given event, A or B?

*13. Identify two possible ways in which conditional probabilities can be computed.

*14. What is the relationship between the joint probability and marginal probabilities of two independent events?

*15. If two or more events have the property that at least one of them must occur when the experiment is performed, then the events are said to be ____.

*16. State the basic counting rule (BCR).

*17. For the first four letters in the English alphabet:
   a. List the possible permutations of three letters from the four.
   b. List the possible combinations of three letters from the four.
   c. Use parts (a) and (b) to obtain $(4)_3$ and $\binom{4}{3}$.

18. The U.S. Internal Revenue Service compiles data on income tax returns and summarizes its findings in *Statistics of Income.* In the first two columns of Table 4.16, we have given a frequency distribution for adjusted gross income (AGI) from 1994 federal individual income tax returns for taxable returns. Frequencies (number of returns) are in thousands.

TABLE 4.16 Adjusted gross incomes

Adjusted gross income	Frequency (1000s)	Event	Probability
Under $10K	10,547	A	
$10K < 20K	16,699	B	
$20K < 30K	17,065	C	
$30K < 40K	11,931	D	
$40K < 50K	8,992	E	
$50K < 100K	17,878	F	
$100K & over	4,508	G	
	87,620		

A 1994 federal individual income tax return is selected at random from among taxable returns.

a. Determine $P(A)$, the probability that the return selected shows an AGI under \$10K.

b. Find the probability that the return selected shows an AGI between \$30K and \$100K (i.e., at least \$30K but less than \$100K).

c. Compute the probability of each of the seven events in the third column of Table 4.16 and record those probabilities in the fourth column.

19. Refer to Problem 18. A 1994 federal individual income tax return is selected at random from among the taxable returns. Let

H = event the return shows an AGI between \$20K and \$100K,

I = event the return shows an AGI of less than \$50K,

J = event the return shows an AGI of less than \$100K,

K = event the return shows an AGI of at least \$50K.

Describe each of the following events in words and determine the number of outcomes (returns) that comprise each event.

a. (not J) **b.** (H & I) **c.** (H or K)
d. (H & K)

20. For the following groups of events from Problem 19, determine which are mutually exclusive.

a. H and I **b.** I and K **c.** H and (not J)
d. H, (not J), and K

21. Refer to Problems 18 and 19.

a. Use the second column of Table 4.16 and the f/N rule to compute the probability of each of the four events H, I, J, and K.

b. Express each of the events H, I, J, and K in terms of the mutually exclusive events in the third column of Table 4.16.

c. Compute the probability of each of the four events H, I, J, and K using your answers from part (b), the special addition rule, and the fourth column of Table 4.16, which you completed in Problem 18(c).

22. Consider the events (not J), (H & I), (H or K), and (H & K) discussed in Problem 19.

a. Find the probability of each of those four events using the f/N rule and your answers from Problem 19.

b. Compute $P(J)$ using the complementation rule and your answer for $P(\text{not } J)$ from part (a).

c. In Problem 21(a) you found that $P(H) = 0.638$ and $P(K) = 0.255$; and in part (a) of this problem you found that $P(H \,\&\, K) = 0.204$. Using those probabilities and the general addition rule, find $P(H \text{ or } K)$.

d. Compare the answers that you obtained for $P(H \text{ or } K)$ in parts (a) and (c).

***23.** The U.S. National Center for Education Statistics publishes information about school enrollment in *Digest of Education Statistics.* Table 4.17 provides a contingency table for enrollment in public and private schools by level. Frequencies are in thousands of students.

TABLE 4.17 Enrollment by level and type

Level		Type: Public T_1	Type: Private T_2	Total
	Elementary L_1	26,951	3,600	30,551
	High school L_2	12,215	1,400	13,615
	College L_3	9,612	2,562	12,174
	Total	48,778	7,562	56,340

a. How many cells are in this contingency table?
b. How many students are in high school?
c. How many students attend public schools?
d. How many students attend private colleges?

***24.** Refer to the information given in Problem 23. A student is selected at random.

a. Describe each of the following events in words: L_3, T_1, and $(T_1 \& L_3)$.

b. Find the probability of each event in part (a) and interpret your answers in terms of percentages.

c. Construct a joint probability distribution corresponding to Table 4.17.

d. Compute $P(T_1 \text{ or } L_3)$ using Table 4.17 and the f/N rule.

e. Compute $P(T_1 \text{ or } L_3)$ using the general addition rule and your answers from part (b).

***25.** Refer to the information given in Problem 23. A student is selected at random.

a. Find $P(L_3 \mid T_1)$ directly using Table 4.17 and the f/N rule. Interpret the probability you obtain in terms of percentages.

b. Use the conditional-probability rule and your answers from Problem 24(b) to find $P(L_3 \mid T_1)$.

***26.** Refer to the information given in Problem 23. A student is selected at random.

a. Using Table 4.17, find $P(T_2)$ and $P(T_2 \mid L_2)$.

b. Are events L_2 and T_2 independent? Explain your answer in terms of percentages.

c. Are events L_2 and T_2 mutually exclusive?

d. Is the event that a student is in elementary school independent of the event that a student attends public school? Justify your answer.

***27.** During one year, the College of Public Programs at Arizona State University awarded the following number of masters degrees.

Type of degree	Frequency
Master of arts	3
Master of public administration	28
Master of science	19

Two students who received such masters degrees are selected at random without replacement. Determine the probability that

a. the first student obtained received a master of arts and the second a master of science.

b. both students obtained received a master of public administration.

c. Construct a tree diagram for this problem similar to Fig. 4.25 on page 247.

d. Find the probability that the two students obtained received the same degree.

***28.** According to Maureen and Jay Neitz of the Medical College of Wisconsin Eye Institute, 9% of men are color blind. For four randomly selected men, find the probability that

a. none are color blind.

b. the first three are not color blind and the fourth is color blind.

c. exactly one of the four is color blind.

***29.** Suppose A and B are events such that $P(A) = 0.4$, $P(B) = 0.5$, and $P(A \& B) = 0.2$. Answer each of the following questions and explain your reasoning.

a. Are A and B mutually exclusive?

b. Are A and B independent?

***30.** Ten years ago a poll was taken by *The New York Times* to gauge the sentiment of Americans on the changes in the role of women over the previous 20 years. One question asked was: "Many women have better jobs and more opportunities than they did 20 years ago. Do you think women have had to give up too much in the process, or not?" The relevant data are shown in the chart displayed in the first column on the next page. Suppose a person who participated in the survey is selected at random. Find the probability that

a. the person answered "no" to the question, given that the person selected is a woman.

b. the person answered "no" to the question.

c. the person is a woman.

d. the person is a woman, given that the person selected answered "no" to the question.

e. Interpret the probabilities obtained in parts (a)–(d) in terms of percentages.

f. Of the four probabilities in parts (a)–(d), which are prior and which are posterior?

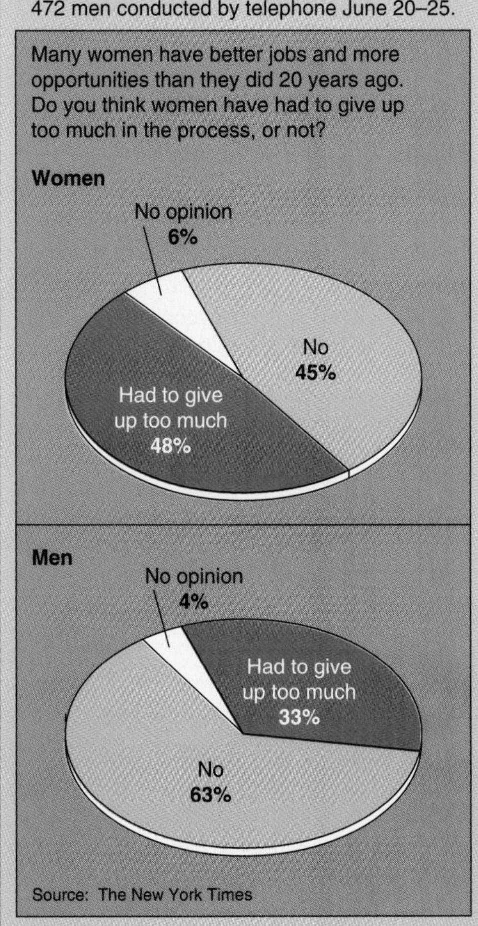
Second Thoughts
About Success
Based on interviews with 1,025 women and
472 men conducted by telephone June 20–25.
Many women have better jobs and more
opportunities than they did 20 years ago.
Do you think women have had to give up
too much in the process, or not?
Women
No opinion
6%
No
45%
Had to give
up too much
48%
Men
No opinion
4%
Had to give
up too much
33%
No
63%
Source: The New York Times

***31.** In Example 4.31 on page 270, we considered exacta wagering in horse racing. Two similar wagers are the quinella and the trifecta. In a quinella wager, the bettor picks the two horses he or she believes will finish first and second, but not in a specified order. In a trifecta wager, the bettor picks the three horses he or she thinks will finish first, second, and third in a specified order. For a 12-horse race,

a. how many different quinella wagers are there?
b. how many different trifecta wagers are there?
c. Repeat parts (a) and (b) for an eight-horse race.

***32.** A bridge hand consists of 13 cards dealt at random from an ordinary deck of 52 playing cards.

a. How many possible bridge hands are there?
b. Find the probability of being dealt a bridge hand that contains exactly two of the four aces.
c. Find the probability of being dealt an 8-4-1 distribution, that is, eight cards of one suit, four of another, and one of another.
d. Determine the probability of being dealt a 5-5-2-1 distribution.
e. Determine the probability of being dealt a hand void in a specified suit.

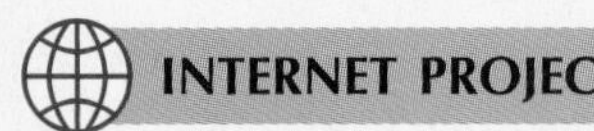

The Space Shuttle Challenger

In this Internet project, you will interactively explore the concepts of probability. The animations and demonstrations available in the Analysis section can help you understand how these concepts actually work.

Sometimes, in thinking so deeply about probability, you can overlook the connection between the concepts and their consequences in the real world. The Data Views section of the project covers an emotionally charged incident where probability concepts had the most serious consequences.

On January 20, 1986, the 25th flight of the National Aeronautics and Space Administration's (NASA) space shuttle program took off. Just after liftoff, a puff of gray smoke could be seen coming from the right solid rocket booster. Seventy-three seconds into the flight, the space shuttle Challenger had climbed 10 miles and then exploded into a fireball, killing all aboard.

The cause of the explosion was found to be an O-ring failure in the right solid rocket booster. Cold weather was a contributing factor. In this project, you will examine the data concerning temperature and its effect on the Challenger's O-rings.

When you studied the theory and concepts of probability in this chapter, you saw that everyday events follow probabilistic rules. The importance of learning how probability works is related to the significance of the events. As you will see in this project, sometimes those events have momentous consequences indeed.

URL for access to Internet Projects Page: `http://hepg.awl.com` *keyword:* Weiss

USING THE FOCUS DATABASE

In Chapter 1 we explained how to store the Focus database in a Minitab worksheet named `focus.mtw`. If you haven't already created that worksheet, follow the instructions on pages 54–55 to create it now.

The Focus database contains information on 500 randomly selected Arizona State University sophomores for seven different variables: sex, high-school GPA, SAT math score, cumulative GPA, SAT verbal score, age, and total hours. Use Minitab or some other statistical software to solve the following problems.

a. Obtain a relative-frequency distribution for the sex data.

b. Using your answer from part (a), determine the probability that a randomly selected sophomore from the sample of 500 is a female.

c. Consider the experiment of selecting a sophomore at random from the sample of 500 and observing the sex of the person obtained. Simulate that experiment 1000 times. *(Hint:* The simulation is equivalent to taking a random sample of size 1000 with replacement.*)*

d. Referring to the simulation performed in part (c), in approximately what percentage of the 1000 experiments would you expect a female to be selected? Compare that percentage to the actual percentage of the 1000 experiments in which a female was selected.

e. Let X denote the age of a randomly selected sophomore from the sample of 500. Obtain the probability distribution of the random variable X.

f. Obtain a probability histogram or similar graphic for the random variable in part (e).

g. Find the mean and standard deviation of the random variable X defined in part (e).

h. Consider the experiment of randomly selecting 10 sophomores with replacement from the sample of 500 and observing the number of those selected who are 21 years old. Simulate that experiment 2000 times.

i. Referring to the simulation from part (h), in approximately what percentage of the 2000 experiments would you expect exactly three of the 10 sophomores selected to be 21 years old? Compare that percentage to the actual percentage of the 2000 experiments in which exactly three of the 10 sophomores selected are 21 years old.

CASE STUDY DISCUSSION

The Powerball®

At the beginning of this chapter on page 196, we discussed the Powerball lottery and learned some of its rules. Recall that, for a single ticket, a player first selects five numbers from the numbers 1–49 and then chooses a Powerball number, which can be any number between 1 and 42. A ticket costs \$1. In the drawing, five white balls are drawn randomly from 49 white balls numbered 1–49; and one red Powerball is drawn randomly from 42 red balls numbered 1–42.

To win the jackpot, a ticket must match all of the balls drawn. Prizes are also given for matching some but not all of the balls drawn. Table 4.18 provides the number of matches, the prizes given, and the probabilities of winning.

TABLE 4.18 Powerball winning combinations, prizes, and probabilities

Matches	Prize	Probability
5 + 1	Jackpot	0.00000001
5	\$100,000	0.00000051
4 + 1	\$5,000	0.00000275
4	\$100	0.00011262
3 + 1	\$100	0.00011812
3	\$7	0.00484285
2 + 1	\$7	0.00165366
1 + 1	\$4	0.00847500
0 + 1	\$3	0.01355999

Here are some things to note about Table 4.18.

- In the first column, an entry of the form $w + 1$ indicates w matches out of the five plus the Powerball; one of the form w indicates w matches out of the five and no Powerball.

- The prize amount for the jackpot depends on how recently it has been won, how many people win it, and choice of jackpot payment.
- Each probability in the third column is given to eight decimal places and can be obtained using the techniques discussed in Section 4.8.

Let E be an event having probability p. It can be shown that in independent repetitions of the experiment, it takes on the average $1/p$ times until event E occurs. We will apply this fact to the Powerball.

Suppose we were to purchase one Powerball ticket per week. Let's find how long we should expect to wait before winning the jackpot. The third column of Table 4.18 shows that for the jackpot, $p = 0.00000001$ and, therefore, we have $1/0.00000001 = 100{,}000{,}000$. So, if we purchased one Powerball ticket per week, then we should expect to wait approximately 100 million weeks, or roughly 1.92 million years, before winning the jackpot.†

a. If you purchase one ticket, what is the probability that you win a prize?
b. If you purchase one ticket, what is the probability that you don't win a prize?
c. If you win a prize, what is the probability it is the \$3 prize for having only the Powerball number?
d. If you were to buy one ticket per week, approximately how long should you expect to wait before getting a ticket with exactly three winning numbers and no Powerball?
e. If you were to buy one ticket per week, approximately how long should you expect to wait before winning a prize?

BIOGRAPHY ANDREI KOLMOGOROV

Andrei Nikolaevich Kolmogorov was born on April 25, 1903, in Tambov, Russia. At the age of 17, Kolmogorov entered Moscow State University and graduated from there in 1925. His contributions to the world of mathematics, many of which appear in his numerous articles and books, encompass a formidable range of subjects.

Kolmogorov revolutionized probability theory by introducing the modern axiomatic approach to probability and by proving many of the fundamental theorems that are a consequence of that approach. He also developed two systems of partial differential equations, which bear his name. Those systems extended the development of probability theory and allowed its broader application to the fields of physics, chemistry, biology, and civil engineering.

In 1938 Kolmogorov published an extensive article entitled "Mathematics," which appeared in the first edition of the *Bolshaya Sovyetskaya Entsiklopediya* (Great Soviet Encyclopedia). This article discussed the development of mathematics from ancient to modern times and interpreted it in terms of dialectical materialism, the philosophy originated by Karl Marx and Friedrich Engels.

† The probability 0.00000001 for winning the jackpot is approximate. Using a more exact value, we find that we should expect to wait "only" about 1.54 million years before winning the jackpot.

Kolmogorov became a member of the faculty at Moscow State University in 1925, at the age of 22. In 1931 he was promoted to professor; in 1933 he was appointed a director of the Institute of Mathematics of the university; and in 1937 he became Head of the University.

In addition to his work in higher mathematics, Kolmogorov was interested in the mathematical education of schoolchildren. He was chairman of the Commission for Mathematical Education under the Presidium of the Academy of Sciences of the U.S.S.R. During his tenure as chairman, he was instrumental in the development of a new mathematics training program that was introduced into Soviet schools.

Kolmogorov remained on the faculty at Moscow State University until his death in Moscow on October 20, 1987.

CASE STUDY

ACES WILD ON THE SIXTH AT OAK HILL

A most amazing event occurred during the second round of the 1989 U.S. Open at Oak Hill in Pittsford, New York. Four golfers—Doug Weaver, Mark Wiebe, Jerry Pate, and Nick Price—made holes-in-one on the sixth hole. What are the chances of such a remarkable event occurring?

An article appeared the next day in the *Boston Globe* that discussed the event in detail. To quote the article, "... for perspective, consider this: This is the 89th U.S. Open, and through the thousands and thousands and thousands of rounds played in the previous 88, there had been only 17 holes-in-one. Yet on this dark Friday morning, there were four holes-in-one on the same hole in less than two hours. Four times into a cup 4 1/2 inches in diameter from 160 yards away."

The article also reported odds estimates obtained from several sources. These estimates varied considerably, from 1 in 10 million to 1 in 1,890,000,000,000,000 to 1 in 8.7 million to 1 in 332,000. After you have completed this chapter, you will be able to compute the odds for yourself.

Discrete Random Variables*

5

GENERAL OBJECTIVES

In Chapters 2 and 3 we examined, among other things, variables and their distributions. Most of the variables we discussed in those chapters were variables of finite populations. But there are many variables not of that type; for example, the number of people waiting in line at a bank, the lifetime of an automobile tire, and the weight of a newborn baby.

Probability theory enables us to extend concepts that apply to variables of finite populations—concepts such as relative-frequency distribution, mean, and standard deviation—to other types of variables. In doing so, we are led to the notion of a *random variable* and its *probability distribution.*

In this chapter, we will discuss the fundamentals of discrete random variables and probability distributions and examine the concepts of the mean and standard deviation of a discrete random variable. Additionally, we will study in detail two of the most important discrete random variables: the binomial and Poisson.

* This more advanced chapter continues the optional material on probability. Those covering the normal approximation to the binomial distribution (Section 6.5) should do Sections 5.1–5.3. Otherwise, although important in its own right, this chapter is not needed for the remainder of the book.

CHAPTER OUTLINE

5.1 DISCRETE RANDOM VARIABLES AND PROBABILITY DISTRIBUTIONS*

In this section we introduce discrete random variables and probability distributions. As we will discover, these concepts are natural extensions of the ideas of variables and relative-frequency distributions. Example 5.1 introduces random variables.

EXAMPLE 5.1 RANDOM VARIABLES

Professor Weiss asked his introductory statistics students to state how many siblings they have. Table 5.1 presents a grouped-data table for that information. The table shows, for instance, that 11 of the 40 students, or 27.5%, have two siblings.

TABLE 5.1 Grouped-data table for number of siblings for students in introductory statistics

Siblings x	Frequency f	Relative frequency
0	8	0.200
1	17	0.425
2	11	0.275
3	3	0.075
4	1	0.025
	40	1.000

Since the "number of siblings" varies from student to student, it is a variable. Suppose now that a student is selected at random. Then the "number of siblings" of the student obtained is called a **random variable** because its value depends on chance, namely, on which student is selected. □

DEFINITION 5.1 RANDOM VARIABLE

A ***random variable*** is a quantitative variable whose value depends on chance.

Example 5.1 shows how random variables arise naturally as quantitative variables of finite populations in the context of randomness. But random variables occur in many other ways. Here are four examples:

- The sum of the dice when a pair of fair dice is rolled.
- The number of puppies in a litter.
- The return on an investment.
- The lifetime of a flashlight battery.

As we learned in Chapter 2, a *discrete variable* is a variable whose possible values form a finite or countably infinite set of numbers. The variable "number of siblings" in Example 5.1 is a discrete variable, its possible values being 0, 1, 2, 3, and 4. We use the adjective *discrete* for random variables in the same way that we do for variables.

DEFINITION 5.2 DISCRETE RANDOM VARIABLE

A ***discrete random variable*** is a random variable whose possible values form a finite (or countably infinite) set of numbers.

Random-Variable Notation

Recall that we use letters near the end of the alphabet, like x, y, and z, to denote variables. We also use such letters to represent random variables but, in that context, we usually make the letters uppercase. For instance, we might use x to denote the variable "number of siblings;" in the context of randomness, however, we would generally use X.

By employing random-variable notation, we can develop useful shorthands for discussing and analyzing random variables. For example, suppose we use X to denote the number of siblings of a randomly selected student. Then we can represent the event that the student selected has, say, two siblings by $\{X = 2\}$, read "X equals two." And we can express the probability of that event as $P(X = 2)$, read "the probability that X equals two."

Probability Distributions and Histograms

As we know, the relative-frequency distribution of a variable gives the possible values of the variable and the proportion of times each value occurs. Using the language of probability, we can extend the notion of relative-frequency distribution—a concept applying to variables of finite populations—to any random variable.

DEFINITION 5.3 PROBABILITY DISTRIBUTION AND PROBABILITY HISTOGRAM

Probability distribution: A listing of the possible values and corresponding probabilities of a discrete random variable; or a formula for the probabilities.

Probability histogram: A graph of the probability distribution that displays the possible values of a discrete random variable on the horizontal axis and the probabilities of those values on the vertical axis. The probability of each value is represented by a vertical bar whose height is equal to the probability.

EXAMPLE 5.2 **ILLUSTRATES DEFINITION 5.3**

Refer to Example 5.1 and, as before, let X denote the number of siblings of a randomly selected student.

a. Determine the probability distribution of the random variable X.

b. Construct a probability histogram for the random variable X.

SOLUTION **a.** We want to determine the probability of each of the possible values of the random variable X. To obtain, for instance, $P(X = 2)$, the probability that the student selected has two siblings, we apply the f/N rule. Referring to Table 5.1 on page 290, we see that

$$P(X = 2) = \frac{f}{N} = \frac{11}{40} = 0.275.$$

The other probabilities are found in the same way. Table 5.2 displays those probabilities and provides the probability distribution of the random variable X.

TABLE 5.2 Probability distribution of the random variable X, the number of siblings of a randomly selected student

Siblings x	Probability $P(X = x)$
0	0.200
1	0.425
2	0.275
3	0.075
4	0.025
	1.000

b. To construct a probability histogram for X, we plot its possible values on a horizontal axis and display the corresponding probabilities using vertical bars. Referring to Table 5.2, we get the probability histogram shown in Fig. 5.1.

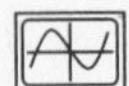

The probability histogram provides a quick and easy way to visualize how the probabilities of the random variable X are distributed. ◨

Since the variable "number of siblings" is a variable of a finite population, its probabilities are identical to relative-frequencies. As a consequence, its probability distribution, given in the first and second columns of Table 5.2, is the same as its relative-frequency distribution, shown in the first and third columns of Table 5.1 on page 290. And, apart from labeling, its probability histogram is identical to its relative-frequency histogram. These statements hold true for any variable of a finite population.

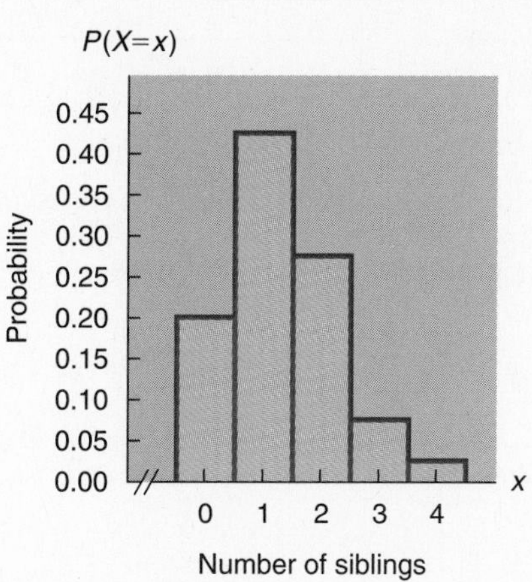

FIGURE 5.1 Probability histogram for the random variable X, the number of siblings of a randomly selected student

We also observe that the probabilities in the second column of Table 5.2 sum to 1. This is always the case for discrete random variables.

KEY FACT 5.1 SUM OF THE PROBABILITIES OF A DISCRETE RANDOM VARIABLE

For any discrete random variable, X, the sum of the probabilities of its possible values equals 1; in symbols, we have $\Sigma P(X = x) = 1$.

Examples 5.3 and 5.4 provide additional illustrations of random-variable notation and probability distributions.

EXAMPLE 5.3 RANDOM-VARIABLE NOTATION AND PROBABILITY DISTRIBUTIONS

The U.S. National Center for Education Statistics compiles enrollment data on U.S. public schools and publishes the results in *Digest of Education Statistics.* Table 5.3 on the next page displays a frequency distribution for the enrollment by grade level in public elementary schools, where $0 =$ kindergarten, $1 =$ first grade, and so on. Frequencies are in thousands of students.

For a randomly selected student in elementary school, let Y denote the grade level of the student obtained. Then Y is a discrete random variable whose possible values are $0, 1, 2, \ldots, 8$.

a. Use random-variable notation to represent the event that the student selected is in the fifth grade.

b. Determine $P(Y = 5)$ and express the result in terms of percentages.

c. Determine the probability distribution of Y.

TABLE 5.3
Frequency distribution for enrollment by grade level in U.S. public elementary schools

Grade level y	Frequency f
0	4,043
1	3,593
2	3,440
3	3,439
4	3,426
5	3,372
6	3,381
7	3,404
8	3,302
	31,400

SOLUTION **a.** The event the student selected is in the fifth grade can be represented as $\{Y = 5\}$.

b. $P(Y = 5)$ is the probability that the student selected is in the fifth grade. Using Table 5.3 and the f/N rule, we can obtain that probability:

$$P(Y = 5) = \frac{f}{N} = \frac{3{,}372}{31{,}400} = 0.107.$$

In terms of percentages, this means that 10.7% of elementary-school students are in the fifth grade.

 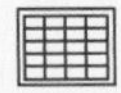

c. The probability distribution of Y is obtained by computing $P(Y = y)$ for $y = 0, 1, 2, \ldots, 8$. We have already done that for $y = 5$. The other probabilities are computed similarly and are displayed in Table 5.4. ◨

TABLE 5.4
Probability distribution of the random variable Y, the grade level of a randomly selected elementary-school student

Grade level y	Probability $P(Y = y)$
0	0.129
1	0.114
2	0.110
3	0.110
4	0.109
5	0.107
6	0.108
7	0.108
8	0.105
	1.000

Once we have the probability distribution of a discrete random variable, it is easy to determine any probability involving that random variable. The basic tool for accomplishing this is the special addition rule (page 219).

EXAMPLE 5.4 RANDOM-VARIABLE NOTATION AND PROBABILITY DISTRIBUTIONS

When a balanced dime is tossed three times, eight equally likely outcomes are possible, as shown in Table 5.5.

TABLE 5.5
Possible outcomes

HHH	HTH	THH	TTH
HHT	HTT	THT	TTT

Here, for instance, HHT means that the first two tosses are heads and the third is tails. Let X denote the total number of heads obtained in the three tosses. Then X is a discrete random variable whose possible values are 0, 1, 2, and 3.

a. Use random-variable notation to represent the event that exactly two heads are tossed.

b. Determine $P(X = 2)$.

c. Find the probability distribution of X.

d. Use random-variable notation to represent the event that at most two heads are tossed.

e. Find $P(X \leq 2)$.

SOLUTION **a.** The event that exactly two heads are tossed can be represented as $\{X = 2\}$.

b. $P(X = 2)$ is the probability that exactly two heads are tossed. We see from Table 5.5 that there are three ways to get a total of two heads and that there are eight possible outcomes altogether. So, by the f/N rule,

$$P(X = 2) = \frac{f}{N} = \frac{3}{8} = 0.375.$$

c. The remaining probabilities for X are computed as in part (b) and are shown in Table 5.6.

TABLE 5.6
Probability distribution of the random variable X, the number of heads obtained in three tosses of a balanced dime

No. of heads x	Probability $P(X = x)$
0	0.125
1	0.375
2	0.375
3	0.125
	1.000

d. The event that at most two heads are tossed can be represented as $\{X \leq 2\}$, read as "X is less than or equal to two."

e. $P(X \leq 2)$ is the probability that at most two heads are tossed. The event that at most two heads are tossed can be expressed as

$$\{X \leq 2\} = \big(\{X = 0\} \text{ or } \{X = 1\} \text{ or } \{X = 2\}\big).$$

Because the three events on the right are mutually exclusive, we can use the special addition rule and Table 5.6 to conclude that

$$\begin{aligned} P(X \leq 2) &= P(X = 0) + P(X = 1) + P(X = 2) \\ &= 0.125 + 0.375 + 0.375 = 0.875. \end{aligned}$$

The probability is 0.875 that at most two heads are tossed. □

Interpretation of Probability Distributions

Recall that the frequentist interpretation of probability construes the probability of an event to be the proportion of times it occurs in a large number of (independent) repetitions of the experiment. Using that interpretation we can clarify the meaning of probability distributions. Example 5.5 shows how.

EXAMPLE 5.5 INTERPRETING A PROBABILITY DISTRIBUTION

Consider again the random variable X discussed in Example 5.4, the number of heads obtained in three tosses of a balanced dime. Suppose we repeat the experiment of observing the number of heads obtained in three tosses of a balanced dime a large number of times. Then the proportion of those times in which, say, no heads are obtained (i.e., $X = 0$) should be approximately equal to the probability of that event [i.e., $P(X = 0)$]. The same statement holds true for the other three possible values of the random variable X.

We used a computer to simulate 1000 observations of the number of heads obtained in three tosses of a balanced dime. Table 5.7 shows the frequencies and proportions for the numbers of heads obtained in the 1000 observations. For instance, 136 of the 1000 observations resulted in no heads out of three tosses, giving a proportion of 0.136.

As expected, the proportions in the third column of Table 5.7 are fairly close to the true probabilities in the second column of Table 5.6 on page 295. We can see this more easily if we compare a histogram for the proportions to the probability histogram of the random variable X. See Fig. 5.2.

TABLE 5.7 Frequencies and proportions for the numbers of heads obtained in three tosses of a balanced dime for 1000 observations

No. of heads x	Frequency f	Proportion $f/1000$
0	136	0.136
1	377	0.377
2	368	0.368
3	119	0.119
	1000	1.000

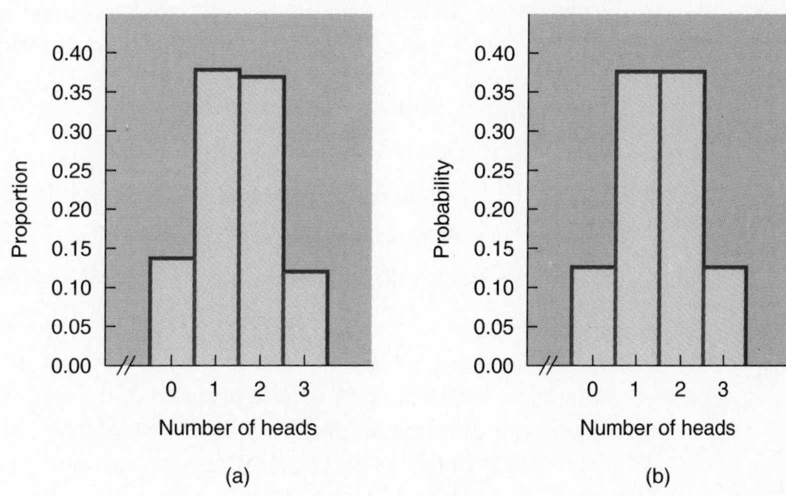

FIGURE 5.2 (a) Histogram of proportions for the numbers of heads obtained in three tosses of a balanced dime for 1000 observations (b) Probability histogram for the number of heads obtained in three tosses of a balanced dime

 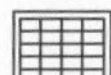

If we simulated, say, 10,000 observations instead of 1000, then the proportions that would appear in the third column of Table 5.7 would most likely be even closer to the true probabilities listed in the second column of Table 5.6. □

In view of Example 5.5, we can make the following statement concerning the interpretation of a probability distribution.

KEY FACT 5.2 **INTERPRETATION OF A PROBABILITY DISTRIBUTION**

In a large number of independent observations of a random variable X, the proportion of times each possible value occurs will approximate the probability distribution of X, or equivalently, a histogram of the proportions will approximate the probability histogram for X.

Using the Computer (Optional)

We can use Minitab to simulate observations of a specified random variable, in particular, those of a discrete random variable. To illustrate, we return to the simulation discussed in Example 5.5.

EXAMPLE 5.6 USING MINITAB TO SIMULATE DISCRETE RANDOM VARIABLES

Table 5.7 on page 297 shows the results of simulating 1000 observations of the number of heads obtained in three tosses of a balanced dime. We performed the simulation by using Minitab. The details for doing this will now be explained.

To begin, we store the probability distribution shown in Table 5.6 on page 295 in columns named x and P(X=x). Then we proceed as follows.

1 Choose **Calc ➤ Random Data ➤ Discrete...**
2 Type 1000 in the **Generate rows of data** text box
3 Click in the **Store in column(s)** text box and type NUMHEADS
4 Click in the **Values in** text box and specify x
5 Click in the **Probabilities in** text box and specify 'P(X=x)'
6 Click **OK**

As a consequence of the above procedure, the numbers of heads obtained in three tosses of a balanced dime for 1000 observations are stored in NUMHEADS. To obtain the frequencies and proportions for the numbers of heads, we proceed as follows.

1 Choose **Stat ➤ Tables ➤ Tally...**
2 Specify NUMHEADS in the **Variables** text box
3 Select **Counts** and **Percents** from the list of **Display** check boxes
4 Click **OK**

The computer output we actually got by performing the simulation and tallying is shown in Printout 5.1. Compare the table in Printout 5.1 to the frequencies and proportions in Table 5.7.

PRINTOUT 5.1 Minitab tally for the simulation of 1000 observations of the number of heads obtained in three tosses of a balanced dime

```
NUMHEADS  Count Percent
       0    136   13.60
       1    377   37.70
       2    368   36.80
       3    119   11.90
      N=   1000
```

If we performed the simulation again, it is virtually certain that we would obtain results different from (although similar to) those in Printout 5.1. □

We can also use Minitab to get a probability histogram for a discrete random variable. To illustrate, we will explain how to obtain a Minitab probability histogram for the number of heads, X, in three tosses of a balanced dime, whose probability distribution we already stored in columns named `x` and `P(X=x)`. Implement the following instructions and compare your result to the probability histogram we drew by hand in Fig. 5.2(b) on page 297.

1 Choose **Graph ➤ Chart...**
2 Click in the **Y** text box for **Graph 1** and specify `'P(X=x)'`
3 Click in the **X** text box for **Graph 1** and specify `x`
4 Click the **Edit Attributes...** button, click in the **Bar Width** text box for **Graph 1** and type `1`, and then click **OK**
5 Click the **Frame** stand-alone pop-up menu button, select **Min and Max...**, click in the **Minimum for Y** text box and type `0`, and then click **OK**
6 Click the **Frame** stand-alone pop-up menu button, select **Axis...**, click in the **Label** text box for **1** and type `Number of heads`, click in the **Label** text box for **2** and type `Probability`, and then click **OK**
7 Click **OK**

EXERCISES 5.1

STATISTICAL CONCEPTS AND SKILLS

5.1 Fill in the blanks:

a. A relative-frequency distribution is to a variable as a _____ distribution is to a random variable.

b. A relative-frequency histogram is to a variable as a _____ histogram is to a random variable.

5.2 Provide an example (other than one discussed in the text) of a random variable that does not arise from a quantitative variable of a finite population in the context of randomness.

5.3 Let X denote the number of siblings of a randomly selected student. Explain the difference between $\{X = 3\}$ and $P(X = 3)$.

5.4 Fill in the blank: For a discrete random variable, the sum of the probabilities of its possible values equals _____.

5.5 Suppose we make a large number of independent observations of a random variable and then construct a table giving the possible values of the random variable and the proportion of times each value occurs. What will this table resemble?

5.6 What rule of probability permits us to obtain any probability for a discrete random variable by simply knowing its probability distribution?

5.7 According to the Census Bureau publication *Current Population Reports,* a frequency distribution for household size (number of people per household) in the United States is as follows. Frequencies are in millions of households. (*Note:* The "7" is actually "7 or more" but we will consider it "7" for illustrative purposes.)

Size	1	2	3	4	5	6	7
Frequency	19.4	26.5	14.6	12.9	6.1	2.5	1.6

Let X denote the size of a U.S. household selected at random.

a. What are the possible values of the random variable X?

b. Employ random-variable notation to represent the event that the household selected has exactly five people.

c. Find $P(X = 5)$; interpret in terms of percentages.

d. Obtain the probability distribution of X.

e. Construct a probability histogram for X.

5.8 The U.S. National Center for Education Statistics compiles enrollment data on U.S. public schools and reports the information in *Digest of Education Statistics.* The following table displays a frequency distribution for the enrollment by grade level in public secondary schools. Frequencies are in thousands of students.

Grade	9	10	11	12
Frequency	3604	3131	2749	2488

For a randomly selected student in public secondary school, let X denote the grade level of the student obtained.

a. What are the possible values of the random variable X?

b. Employ random-variable notation to represent the event that the student obtained is in the tenth grade.

c. Find $P(X = 10)$; interpret in terms of percentages.

d. Obtain the probability distribution of X.

e. Construct a probability histogram for X.

5.9 When two balanced dice are rolled, 36 equally likely outcomes are possible, as seen in Fig. 4.1 on page 201. Let Y denote the sum of the dice.

a. What are the possible values of the random variable Y?

b. Employ random-variable notation to represent the event that the sum of the dice is 7.

c. Find $P(Y = 7)$.

d. Find the probability distribution of Y. Leave your probabilities in fraction form.

e. Construct a probability histogram for Y.

5.10 When two balanced dice are rolled, 36 equally likely outcomes are possible, as seen in Fig. 4.1 on page 201. Let X denote the larger number showing on the two dice. For example, if the outcome is

then $X = 5$. If both dice come up the same number, then X equals that common value.

a. What are the possible values of the random variable X?

b. Employ random-variable notation to represent the event that the larger number is 4.

c. Find $P(X = 4)$.

d. Find the probability distribution of X. Leave your probabilities in fraction form.

e. Construct a probability histogram for X.

5.11 Prescott National Bank has six tellers available to serve customers. The number of tellers busy with customers at, say, 1:00 P.M. varies from day to day and depends on chance; so it is a random variable, which we will call X. Past records indicate that the probability distribution of X is as shown in the following table.

x	$P(X = x)$
0	0.029
1	0.049
2	0.078
3	0.155
4	0.212
5	0.262
6	0.215

We see, for example, that the probability is 0.262 that exactly five of the tellers will be busy with customers at 1:00 P.M.; that is, about 26.2% of the time, exactly five tellers are busy with customers at 1:00 P.M. Use random-variable notation to represent each of the following events. At 1:00 P.M.,

a. exactly four tellers are busy.
b. at least two tellers are busy.
c. fewer than five tellers are busy.
d. at least two but fewer than five tellers are busy.

Use the special addition rule and the probability distribution to determine

e. $P(X = 4)$.
f. $P(X \geq 2)$.
g. $P(X < 5)$.
h. $P(2 \leq X < 5)$.

5.12 From past experience, a car salesperson knows that the number of cars she sells per week is a random variable, Y, with a probability distribution as shown here.

y	$P(Y = y)$
0	0.135
1	0.271
2	0.271
3	0.180
4	0.090
5	0.036
6	0.012
7	0.003
8	0.002

A week is selected at random. Use random-variable notation to represent the event that the salesperson sells

a. exactly three cars.
b. at least three cars.
c. fewer than seven cars.
d. at least three but fewer than seven cars.

Use the special addition rule and the probability distribution to find

e. $P(Y = 3)$.
f. $P(Y \geq 3)$.
g. $P(Y < 7)$.
h. $P(3 \leq Y < 7)$.

Extending the Concepts and Skills

5.13 Suppose that Z is a random variable and that $P(Z > 1.96) = 0.025$. Find $P(Z \leq 1.96)$. *(Hint:* Use the complementation rule.*)*

5.14 Suppose T and Z are random variables.

a. If $P(T > 2.02) = 0.05$ and $P(T < -2.02) = 0.05$, obtain $P(-2.02 \leq T \leq 2.02)$.
b. Suppose that $P(-1.64 \leq Z \leq 1.64) = 0.90$ and that $P(Z > 1.64) = P(Z < -1.64)$. Find $P(Z > 1.64)$.

5.15 Let $c > 0$ and $0 \leq \alpha \leq 1$. Also let X, Y, and T be random variables.

a. If $P(X > c) = \alpha$, determine $P(X \leq c)$ in terms of α.
b. If $P(Y > c) = \alpha/2$ and $P(Y < -c) = P(Y > c)$, find $P(-c \leq Y \leq c)$ in terms of α.
c. Suppose that $P(-c \leq T \leq c) = 1 - \alpha$ and, moreover, that $P(T < -c) = P(T > c)$. Find $P(T > c)$ in terms of α.

Using Technology  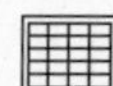 

5.16 Refer to the probability distribution displayed in Table 5.6 on page 295.

a. Use Minitab or some other statistical software to repeat the simulation done in Example 5.6 on page 298.
b. Obtain the proportions for the numbers of heads in three tosses and compare it to the probability distribution in Table 5.6.
c. Obtain a histogram of the proportions and compare it to the probability histogram in Fig. 5.2(b) on page 297.
d. What do parts (b) and (c) illustrate?

5.17 Refer to the probability distribution displayed in Table 5.2 on page 292.

a. Use Minitab or some other statistical software to simulate 2000 observations of the number of siblings of a randomly selected student.
b. Obtain the proportions for the number of siblings and compare it to the probability distribution in Table 5.2.
c. Obtain a histogram of the proportions and compare it to the probability histogram in Fig. 5.1 on page 293.
d. What do parts (b) and (c) illustrate?

5.2 THE MEAN AND STANDARD DEVIATION OF A DISCRETE RANDOM VARIABLE*

In this section we introduce the mean and standard deviation of a discrete random variable. As we will see, the mean and standard deviation of a discrete random variable are analogous to the population mean and population standard deviation.

Mean of a Discrete Random Variable

Recall that for a variable x, the mean of all possible observations for the entire population is called the *population mean* or *mean of the variable* x. In Section 3.5 we gave a formula for the mean of a variable x:

$$\mu = \frac{\Sigma x}{N}.$$

Although this formula applies only to variables of finite populations, we can use it and the language of probability to extend the concept of the mean to any discrete variable. To see how this is done, we will employ a very simple example.

EXAMPLE 5.7 MEAN OF A DISCRETE RANDOM VARIABLE

A population of eight students has the ages given in Table 5.8.

TABLE 5.8 Ages of eight students

19	20	20	19	21	27	20	21

The variable under consideration here is age and the population consists of the eight students. If we let X denote the age of a randomly selected student, then X is a random variable. In view of Table 5.8, the probability distribution of X is as shown in Table 5.9. Express the mean age of the eight students in terms of the probability distribution of the random variable X.

TABLE 5.9 Probability distribution of X, the age of a randomly selected student

Age x	Probability $P(X = x)$		
19	0.250	←	*2/8*
20	0.375	←	*3/8*
21	0.250	←	*2/8*
27	0.125	←	*1/8*

* See the note on the bottom of page 289.

SOLUTION Referring first to Table 5.8 and then to Table 5.9, we get

$$\begin{aligned}
\mu = \frac{\Sigma x}{N} &= \frac{19+20+20+19+21+27+20+21}{8} \\
&= \frac{\overbrace{19+19}^{2}+\overbrace{20+20+20}^{3}+\overbrace{21+21}^{2}+\overbrace{27}^{1}}{8} \\
&= \frac{19\cdot 2+20\cdot 3+21\cdot 2+27\cdot 1}{8} \\
&= 19\cdot\frac{2}{8}+20\cdot\frac{3}{8}+21\cdot\frac{2}{8}+27\cdot\frac{1}{8} \\
&= 19\cdot P(X=19)+20\cdot P(X=20)+21\cdot P(X=21)+27\cdot P(X=27) \\
&= \Sigma x P(X=x),
\end{aligned}$$

as required. ◨

Example 5.7 shows that we can express the mean of a variable of a finite population in terms of the probability distribution of the corresponding random variable: $\mu = \Sigma x P(X = x)$. Because the expression on the right of this equation is meaningful for any discrete random variable, we make the following definition.

DEFINITION 5.4 **MEAN OF A DISCRETE RANDOM VARIABLE**

The ***mean of a discrete random variable*** X is denoted by μ_X or, when no confusion will arise, simply by μ. It is defined by

$$\mu = \Sigma x P(X = x).$$

The terms ***expected value*** and ***expectation*** are commonly used in place of *mean.*

We now have a definition of *mean* which is consistent with that for variables of finite populations and applies to any discrete random variable. In other words, we have extended the concept of population mean to any discrete variable.[†]

[†] We can also extend the concept of population mean to any continuous variable and, using integral calculus, develop a formula analogous to the one given in Definition 5.4 for discrete variables. We will not present the formula for the mean of a continuous variable because it is not needed for this book.

As we learned in Chapter 3, constructing appropriate tables provides an efficient way to compute descriptive measures by hand. Example 5.8 applies this technique to obtaining the mean of a discrete random variable.

EXAMPLE 5.8 ILLUSTRATES DEFINITION 5.4

Prescott National Bank has six tellers available to serve customers. The number of tellers busy with customers at, say, 1:00 P.M. varies from day to day and depends on chance; so it is a random variable, which we will call X. Past records indicate that the probability distribution of X is as shown in the first two columns of Table 5.10. The table indicates, for instance, that the probability is 0.262 that exactly five tellers will be busy with customers at 1:00 P.M. Find the mean of the random variable X.

TABLE 5.10 Table for computing the mean of the random variable X, the number of tellers busy with customers

x	$P(X = x)$	$xP(X = x)$
0	0.029	0.000
1	0.049	0.049
2	0.078	0.156
3	0.155	0.465
4	0.212	0.848
5	0.262	1.310
6	0.215	1.290
		4.118

SOLUTION To obtain the mean of the random variable X, we append a column for the product of x with $P(X = x)$ to Table 5.10 and apply Definition 5.4. Summing the entries in the third column of Table 5.10, we find that $\mu = \Sigma x P(X = x) = 4.118$. The mean number of tellers busy with customers at 1:00 P.M. is 4.118.

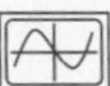

Interpretation of the Mean of a Random Variable

As we know, the mean of a variable of a finite population is the arithmetic average of all possible observations. A similar interpretation holds for the mean of a random variable.

For instance, in Example 5.8 the random variable X is the number of tellers busy with customers at 1:00 P.M. As we have seen, the mean of that random variable is 4.118. Of course, there never will be a day when 4.118 tellers are busy with customers at 1:00 P.M. The mean of 4.118 simply indicates that over the course of many days, the average number of busy tellers at 1:00 P.M. will be about 4.118.

This interpretation holds in all cases. It is known colloquially as the **law of averages** and in mathematical circles as the **law of large numbers.**

KEY FACT 5.3 **INTERPRETATION OF THE MEAN OF A RANDOM VARIABLE**

In a large number of independent observations of a random variable X, the average value of those observations will be approximately equal to the mean, μ, of X. The larger the number of observations, the closer the average tends to be to μ.

We used a computer to simulate the number of busy tellers at 1:00 P.M. on 100 randomly selected days; that is, we obtained 100 independent observations of the random variable X. The data are displayed in Table 5.11.

TABLE 5.11 One hundred observations of the random variable X, the number of tellers busy with customers

5	3	5	3	4	3	4	3	6	5	6	4	5	4	3	5	4	5	6	3
4	1	6	5	3	6	3	5	5	4	6	4	1	6	5	3	3	6	4	5
3	4	2	5	5	6	5	4	6	2	4	5	4	6	4	5	5	3	4	6
1	5	4	6	4	4	4	5	6	2	5	4	5	1	3	3	6	4	6	4
5	6	5	5	3	2	4	6	6	1	5	1	3	6	5	3	5	4	3	6

The average value of the 100 observations in Table 5.11 is 4.25. This value is quite close to the mean, $\mu = 4.118$, of the random variable X. If we made, say, 1000 observations instead of 100, then the average value of those 1000 observations would most likely be even closer to 4.118.

Figure 5.3(a) shows a plot of the average number of busy tellers versus the number of observations for the data in Table 5.11. The dashed line is at $\mu = 4.118$. In Fig. 5.3(b) is another such plot for a different simulation of the number of busy tellers at 1:00 P.M. on 100 randomly selected days. Both plots suggest that as the number of observations increases, the average number of busy tellers approaches the mean, $\mu = 4.118$, of the random variable X.

FIGURE 5.3 Graphs showing the average number of busy tellers versus the number of observations for two simulations of 100 observations each

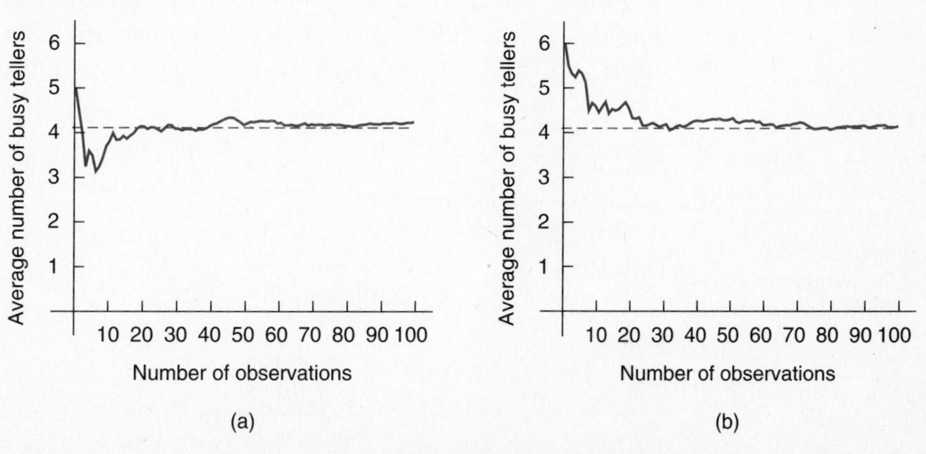

Standard Deviation of a Discrete Random Variable

By employing reasoning similar to that used for the mean, we can also extend the concept of population standard deviation (standard deviation of a variable) to any discrete variable.

DEFINITION 5.5 STANDARD DEVIATION OF A DISCRETE RANDOM VARIABLE

The ***standard deviation of a discrete random variable*** X is denoted by σ_X or, when no confusion will arise, simply by σ. It is defined by

$$\sigma = \sqrt{\Sigma(x-\mu)^2 P(X=x)}.$$

The standard deviation of a discrete random variable can also be obtained from the computing formula

$$\sigma = \sqrt{\Sigma x^2 P(X=x) - \mu^2}.$$

The square of the standard deviation, σ^2, is called the ***variance*** of the random variable X.

EXAMPLE 5.9 ILLUSTRATES DEFINITION 5.5

Refer to Example 5.8 on page 304, where X denotes the number of tellers busy with customers at 1:00 P.M. Find the standard deviation of the random variable X.

SOLUTION We will apply the computing formula given in Definition 5.5. To implement that formula, we need the mean of X, which we obtained in Example 5.8, and columns for x^2 and $x^2P(X=x)$, which are presented in the last two columns of Table 5.12.

TABLE 5.12
Table for computing the standard deviation of the random variable X, the number of tellers busy with customers

x	$P(X=x)$	x^2	$x^2P(X=x)$
0	0.029	0	0.000
1	0.049	1	0.049
2	0.078	4	0.312
3	0.155	9	1.395
4	0.212	16	3.392
5	0.262	25	6.550
6	0.215	36	7.740
			19.438

From the final column of Table 5.12, we see that $\Sigma x^2 P(X = x) = 19.438$. Recalling that $\mu = 4.118$, we can now obtain the standard deviation of X:

$$\sigma = \sqrt{\Sigma x^2 P(X = x) - \mu^2} = \sqrt{19.438 - 4.118^2} = 1.6,$$

rounded to one decimal place. The standard deviation of the number of tellers busy with customers at 1:00 P.M. is 1.6. ■

As we know, the standard deviation of a variable of a finite population indicates the variation of all possible observations by, roughly speaking, measuring how far the possible observations are from the mean, on the average. A similar interpretation can be given to the standard deviation of a random variable. In particular, the smaller the standard deviation of a random variable X, the more likely that an observed value of X will be close to the mean.

EXERCISES 5.2

STATISTICAL CONCEPTS AND SKILLS

5.18 Suppose the random variables X and Y represent the amount of return on two different investments. Further suppose the mean of X equals the mean of Y, but the standard deviation of X is greater than the standard deviation of Y.

a. On the average, is there a difference between the returns of the two investments? Explain your answer.

b. Which investment is more conservative? Why?

In Exercises 5.19–5.22, we have given the probability distributions of the random variables considered in Exercises 5.7–5.10. For each exercise,

a. find and interpret the mean of the random variable.

b. obtain the standard deviation of the random variable using one of the formulas given in Definition 5.5 on page 306.

c. draw a probability histogram for the random variable; locate the mean; and show one, two, and three standard-deviation intervals.

5.19 The random variable X is the number of people in a randomly selected U.S. household. Its probability distribution follows.

x	1	2	3	4	5	6	7
$P(X = x)$	0.232	0.317	0.175	0.154	0.073	0.030	0.019

5.20 The random variable X is the grade level of a secondary-school student selected at random. Its probability distribution is as follows.

x	9	10	11	12
$P(X = x)$	0.301	0.262	0.230	0.208

5.21 The random variable Y is the sum of the dice when two balanced dice are rolled. Its probability distribution is shown in the following table. Perform the required computations using fractions; that is, do not convert to decimals until the final answer is obtained.

y	2	3	4	5	6	7	8	9	10	11	12
$P(Y = y)$	$\frac{1}{36}$	$\frac{1}{18}$	$\frac{1}{12}$	$\frac{1}{9}$	$\frac{5}{36}$	$\frac{1}{6}$	$\frac{5}{36}$	$\frac{1}{9}$	$\frac{1}{12}$	$\frac{1}{18}$	$\frac{1}{36}$

5.22 The random variable X is the larger number showing when two balanced dice are rolled. Its probability distribution follows.

x	1	2	3	4	5	6
$P(X = x)$	$\frac{1}{36}$	$\frac{1}{12}$	$\frac{5}{36}$	$\frac{7}{36}$	$\frac{1}{4}$	$\frac{11}{36}$

Expected value. As noted in Definition 5.4, the mean of a random variable is also called its *expected value.* This terminology is especially useful in gambling and decision theory, as illustrated in Exercises 5.23 and 5.24.

5.23 A roulette wheel contains 38 numbers: 18 are red, 18 are black, and 2 are green. When the roulette ball is spun, it is equally likely to land on any of the 38 numbers. Suppose you bet $1 on red. If the ball lands on a red number, you win $1; otherwise you lose your $1. Let X be the amount you win on your $1 bet. Then X is a random variable whose probability distribution is as follows.

x	$P(X = x)$
1	0.474
−1	0.526

a. Verify that the probability distribution shown in the table is correct.
b. Find the expected value of the random variable X.
c. On the average, how much will you lose per play? *(Hint:* Refer to the interpretation of the mean of a random variable given in Key Fact 5.3 on page 305.*)*
d. Approximately how much would you expect to lose if you bet $1 on red 100 times? 1000 times?
e. Do you think roulette is a profitable game to play? Explain your answer.

5.24 An investor plans to invest $50,000 in one of four investments. The return on each investment depends on whether next year's economy is strong or weak. The following table summarizes the possible payoffs, in dollars, for the four investments.

Investment	Next year's economy: Strong	Weak
Certificate of deposit	6,000	6,000
Office complex	15,000	5,000
Land speculation	33,000	−17,000
Technical school	5,500	10,000

Let V, W, X, and Y denote, respectively, the payoffs for the certificate of deposit, office complex, land speculation, and technical school. Then V, W, X, and Y are random variables. Assume that next year's economy has a 40% chance of being strong and a 60% chance of being weak.

a. Find the probability distribution of each of the four random variables V, W, X, and Y.
b. Determine the expected value of each of the four random variables.
c. Which investment has the best expected payoff? Which has the worst?
d. Which investment would you select? Explain your answer.

5.25 A factory manager collected data on the number of equipment breakdowns per day. From those data, she derived the probability distribution shown in the table below, where W denotes the number of breakdowns on a given day.

w	$P(W = w)$
0	0.80
1	0.15
2	0.05

a. Determine μ_W and σ_W.
b. On average, how many breakdowns occur per day?
c. About how many breakdowns are expected during a one-year period, assuming 250 work days per year?

Extending the Concepts and Skills

Properties of the mean and standard deviation. Exercises 5.26 and 5.27 develop some important properties of the mean and standard deviation of a random variable. Two of these properties relate the mean and standard deviation of the sum of two random variables to the individual means and standard deviations, respectively; two others relate the mean and standard deviation of a constant times a random variable to the constant and the mean and standard deviation of the random variable, respectively.

In developing these properties, we will need the concept of independent random variables. Two discrete random variables, X and Y, are said to be **independent random variables** if

$$P\left(\{X = x\} \,\&\, \{Y = y\}\right) = P(X = x)P(Y = y)$$

for all x and y, that is, if the joint probability distribution of X and Y equals the product of their marginal probability distributions. This is equivalent to requiring that events $\{X = x\}$ and $\{Y = y\}$ are independent for all x and y. A similar definition holds for independence of more than two discrete random variables.

5.26 Refer to Exercise 5.25. Assume the number of breakdowns on different days are independent of one another. Let X and Y denote the number of breakdowns on each of two consecutive days.

a. Complete the following joint probability distribution table.

		y = 0	1	2	$P(X = x)$
x	0				
	1				
	2				
	$P(Y = y)$				

Hint: To obtain the joint probability in the first row and third column, use the definition of independence for discrete random variables and the table in Exercise 5.25:

$$P\left(\{X = 0\} \,\&\, \{Y = 2\}\right) = P(X = 0)P(Y = 2) = 0.80 \cdot 0.05 = 0.04.$$

b. Use the joint probability distribution you obtained in part (a) to determine the probability distribution of the random variable $X + Y$, the total number of breakdowns in two days; that is, complete the following table.

u	$P(X + Y = u)$
0	
1	
2	
3	
4	

c. Use part (b) to find μ_{X+Y} and σ^2_{X+Y}.

d. Use part (c) to verify that the following equations are true for this example:

$$\mu_{X+Y} = \mu_X + \mu_Y \quad \text{and} \quad \sigma^2_{X+Y} = \sigma^2_X + \sigma^2_Y.$$

(*Note:* The mean and variance of X and Y are the same as that of W in Exercise 5.25.)

e. The equations in part (d) hold in general: If X and Y are any two random variables, then

$$\mu_{X+Y} = \mu_X + \mu_Y.$$

If, in addition, X and Y are independent, then

$$\sigma^2_{X+Y} = \sigma^2_X + \sigma^2_Y.$$

Interpret these two equations in words.

5.27 The factory manager in Exercise 5.25 estimates that each breakdown costs the company \$500 in repairs and loss of production. If W is the number of breakdowns in a day, then \$500W is the cost due to breakdowns for that day.

a. Referring to the probability distribution in Exercise 5.25, determine the probability distribution of the random variable $500W$.

b. Determine the mean daily breakdown cost, μ_{500W}, using your answer from part (a).

c. What is the relationship between μ_{500W} and μ_W? (*Note:* From Exercise 5.25, $\mu_W = 0.25$.)
d. Find σ_{500W} using your answer from part (a).
e. What is the apparent relationship between σ_{500W} and σ_W? (*Note:* From Exercise 5.25, $\sigma_W = 0.536$.)
f. The results in parts (c) and (e) are true in general: If W is any random variable and c is a constant, then

$$\mu_{cW} = c\mu_W \qquad \text{and} \qquad \sigma_{cW} = |c|\sigma_W.$$

Interpret these two equations in words.

USING TECHNOLOGY  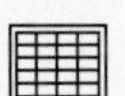 

5.28 Benny's Barber Shop in Cleveland has five chairs for waiting customers. The number of customers waiting is a random variable Y which, from previous records, is known to have the following probability distribution.

y	0	1	2	3	4	5
$P(Y = y)$	0.424	0.161	0.134	0.111	0.093	0.077

a. Compute the mean number of customers waiting, μ.
b. In a large number of independent observations, how many customers will be waiting, on the average?
c. Use Minitab or some other statistical software to simulate 100 observations of the number of customers waiting.
d. Obtain the mean of the observations in part (c), and compare it to μ.
e. What is part (d) illustrating?

5.3 THE BINOMIAL DISTRIBUTION*

Many problems in probability and statistics concern the repetition of an experiment having two possible outcomes. In such contexts, each repetition of the experiment is called a **trial.** Here are three examples.

- Testing the effectiveness of a drug: Several patients take the drug (the trials), and for each patient, the drug is either effective or not effective (the two possible outcomes).
- Weekly sales of a car salesperson: The salesperson has several customers during the week (the trials), and for each customer, the salesperson either makes a sale or does not make a sale (the two possible outcomes).
- Taste tests for colas: A number of people taste two different colas (the trials), and for each person, the preference is either for the first cola or for the second cola (the two possible outcomes).

Analyzing repeated trials of an experiment having two possible outcomes requires knowledge of factorials, binomial coefficients, Bernoulli trials, and the binomial distribution. We begin with factorials.

Factorials

Factorials are defined in Definition 5.6.

* See the note on the bottom of page 289.

DEFINITION 5.6 FACTORIALS

The product of the first k positive integers is called ***k factorial*** and is denoted $\boldsymbol{k!}$. In symbols,

$$k! = k(k-1)\cdots 2 \cdot 1.$$

We also define $0! = 1$.

EXAMPLE 5.10 ILLUSTRATES DEFINITION 5.6

Determine 3!, 4!, and 5!.

SOLUTION Applying Definition 5.6, we obtain that $3! = 3 \cdot 2 \cdot 1 = 6$, $4! = 4 \cdot 3 \cdot 2 \cdot 1 = 24$, and $5! = 5 \cdot 4 \cdot 3 \cdot 2 \cdot 1 = 120$. ◨

Note, for instance, that $6! = 6 \cdot 5!$, $6! = 6 \cdot 5 \cdot 4!$, $6! = 6 \cdot 5 \cdot 4 \cdot 3!$, and so on. In general, if $j \leq k$, then $k! = k(k-1)\cdots(k-j+1)(k-j)!$.

Binomial Coefficients

You may have already encountered binomial coefficients in algebra when you studied the binomial expansion, the expansion of $(a+b)^n$. Here is the definition of binomial coefficients.[†]

DEFINITION 5.7 BINOMIAL COEFFICIENTS

If n is a positive integer and x is a nonnegative integer less than or equal to n, then the ***binomial coefficient*** $\binom{n}{x}$ is defined as

$$\binom{n}{x} = \frac{n!}{x!\,(n-x)!}.$$

EXAMPLE 5.11 ILLUSTRATES DEFINITION 5.7

Determine the value of each of the following binomial coefficients.

a. $\binom{6}{1}$ **b.** $\binom{5}{3}$ **c.** $\binom{7}{3}$ **d.** $\binom{4}{4}$

[†] If you have read Section 4.8, you will recognize the binomial coefficient $\binom{n}{x}$ as the number of possible combinations of x objects from a collection of n objects.

SOLUTION We apply Definition 5.7.

a. $\binom{6}{1} = \frac{6!}{1!\,(6-1)!} = \frac{6!}{1!\,5!} = \frac{6 \cdot \cancel{5!}}{1!\,\cancel{5!}} = \frac{6}{1} = 6$

b. $\binom{5}{3} = \frac{5!}{3!\,(5-3)!} = \frac{5!}{3!\,2!} = \frac{5 \cdot 4 \cdot \cancel{3!}}{\cancel{3!}\,2!} = \frac{5 \cdot 4}{2} = 10$

c. $\binom{7}{3} = \frac{7!}{3!\,(7-3)!} = \frac{7!}{3!\,4!} = \frac{7 \cdot 6 \cdot 5 \cdot \cancel{4!}}{3!\,\cancel{4!}} = \frac{7 \cdot 6 \cdot 5}{6} = 35$

d. $\binom{4}{4} = \frac{4!}{4!\,(4-4)!} = \frac{4!}{4!\,0!} = \frac{\cancel{4!}}{\cancel{4!}\,0!} = \frac{1}{1} = 1$ □

Bernoulli Trials

Next we define Bernoulli trials.

DEFINITION 5.8 BERNOULLI TRIALS

Repeated identical trials are called ***Bernoulli trials*** if the following three conditions are satisfied:

1. Each trial has two possible outcomes, denoted generically by ***s***, for ***success,*** and ***f***, for ***failure.***
2. The trials are independent.
3. The probability of a success remains the same from trial to trial. We call that probability the ***success probability*** and denote it by the letter ***p***.

Introducing the Binomial Distribution

The **binomial distribution** is the probability distribution for the number of successes in a sequence of Bernoulli trials. We introduce this concept in Example 5.12.

EXAMPLE 5.12 INTRODUCES THE BINOMIAL DISTRIBUTION

Mortality tables enable actuaries to obtain the probability that a person at any particular age will live a specified number of years. This, in turn, permits the determination of life-insurance premiums, retirement pensions, annuity payments, and related items of importance to insurance companies and others.

According to tables provided by the U.S. National Center for Health Statistics in *Vital Statistics of the United States,* there is about an 80% chance that a person age 20 will be alive at age 65. Suppose three people age 20 are selected at random.

a. Formulate the process of observing which people are alive at age 65 as a sequence of three Bernoulli trials.

b. Obtain the possible outcomes of the three Bernoulli trials.

c. Determine the probability of each outcome in part (b).

d. Find the probability that exactly two of the three people will be alive at age 65.

e. Obtain the probability distribution of the number of people out of the three that are alive at age 65.

SOLUTION **a.** Each trial consists of observing whether a person currently age 20 is alive at age 65 and has two possible outcomes: alive or dead. The trials are independent. If we let a success, *s*, correspond to being alive at age 65, then the success probability is 0.8 (80%); that is, $p = 0.8$.

b. For this part we are to obtain the possible outcomes of the three Bernoulli trials, that is, the possible alive-dead results for the three people. The possible outcomes are shown in Table 5.13 (s = success = alive, f = failure = dead).

TABLE 5.13
Possible outcomes

sss	*ssf*	*sfs*	*sff*
fss	*fsf*	*ffs*	*fff*

For instance, *ssf* represents the outcome that at age 65 the first two people are alive and the third is not.

c. Here we are to determine the probability of each of the possible outcomes. As we see from Table 5.13, eight outcomes are possible. However, because these eight outcomes are not equally likely, we cannot use the f/N rule to determine their probabilities; instead we must proceed as follows. First of all, by part (a), the success probability equals 0.8:

$$P(s) = p = 0.8.$$

Therefore the failure probability is

$$P(f) = 1 - p = 1 - 0.8 = 0.2.$$

Now using the fact that the trials are independent, we can apply the special multiplication rule (page 250), to obtain the probability of each of the eight possible outcomes. For instance, the probability of the outcome *ssf* is

$$P(ssf) = P(s) \cdot P(s) \cdot P(f) = 0.8 \cdot 0.8 \cdot 0.2 = 0.128.$$

Similar computations yield the probabilities of the other seven possible outcomes. All eight possible outcomes and their probabilities are shown in Table 5.14.

TABLE 5.14 Outcomes and probabilities for observing whether each of three people is alive at age 65

Outcome	Probability
sss	$(0.8)(0.8)(0.8) = 0.512$
ssf	$(0.8)(0.8)(0.2) = 0.128$
sfs	$(0.8)(0.2)(0.8) = 0.128$
sff	$(0.8)(0.2)(0.2) = 0.032$
fss	$(0.2)(0.8)(0.8) = 0.128$
fsf	$(0.2)(0.8)(0.2) = 0.032$
ffs	$(0.2)(0.2)(0.8) = 0.032$
fff	$(0.2)(0.2)(0.2) = 0.008$

Note that outcomes containing the same number of successes have the same probability. For instance, three outcomes contain exactly two successes: *ssf, sfs,* and *fss*. Each of those three outcomes has the same probability: 0.128. This is because each probability is obtained by multiplying two success probabilities of 0.8 and one failure probability of 0.2.

A tree diagram is useful for organizing and summarizing the possible outcomes and their probabilities, as presented in Fig. 5.4.

FIGURE 5.4 Tree diagram corresponding to Table 5.14

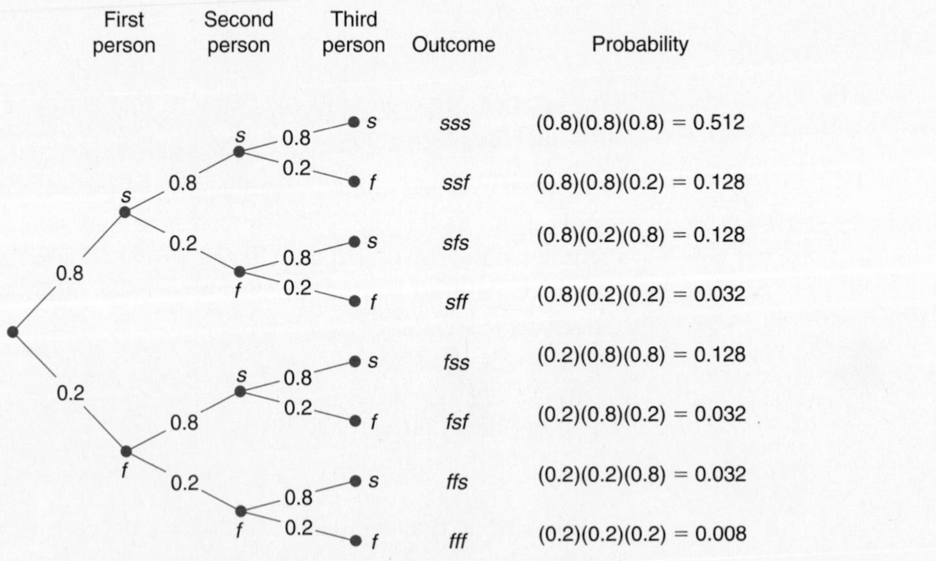

First person	Second person	Third person	Outcome	Probability
s (0.8)	s (0.8)	s (0.8)	sss	(0.8)(0.8)(0.8) = 0.512
		f (0.2)	ssf	(0.8)(0.8)(0.2) = 0.128
	f (0.2)	s (0.8)	sfs	(0.8)(0.2)(0.8) = 0.128
		f (0.2)	sff	(0.8)(0.2)(0.2) = 0.032
f (0.2)	s (0.8)	s (0.8)	fss	(0.2)(0.8)(0.8) = 0.128
		f (0.2)	fsf	(0.2)(0.8)(0.2) = 0.032
	f (0.2)	s (0.8)	ffs	(0.2)(0.2)(0.8) = 0.032
		f (0.2)	fff	(0.2)(0.2)(0.2) = 0.008

d. The problem here is to determine the probability that exactly two of the three people will be alive at age 65. As we see from Table 5.14, the event that exactly two of the three people are alive at age 65 consists of three outcomes: *ssf, sfs,* and *fss*. Each of those three outcomes has probability 0.128, as we observed in part (c). So, by the special addition rule (page 219), we have

$$P(\text{Exactly two will be alive}) = P(ssf) + P(sfs) + P(fss)$$
$$= \underbrace{0.128 + 0.128 + 0.128}_{3 \text{ times}} = 3 \cdot 0.128 = 0.384.$$

Consequently, the probability is 0.384 that exactly two of the three people will be alive at age 65.

e. For this part, we are to obtain the probability distribution of the number of people out of the three that are alive at age 65, which we denote by X. In part (d) we found $P(X = 2)$. Proceeding in the same way, we can determine the remaining three probabilities: $P(X = 0)$, $P(X = 1)$, and $P(X = 3)$. The results are displayed in Table 5.15.

TABLE 5.15 Probability distribution of the random variable X, the number of people out of three that are alive at age 65

Number alive x	Probability $P(X = x)$
0	0.008
1	0.096
2	0.384
3	0.512

A probability histogram for the distribution in Table 5.15 is given in Fig. 5.5. Note for future reference that the probability distribution is left skewed. ◨

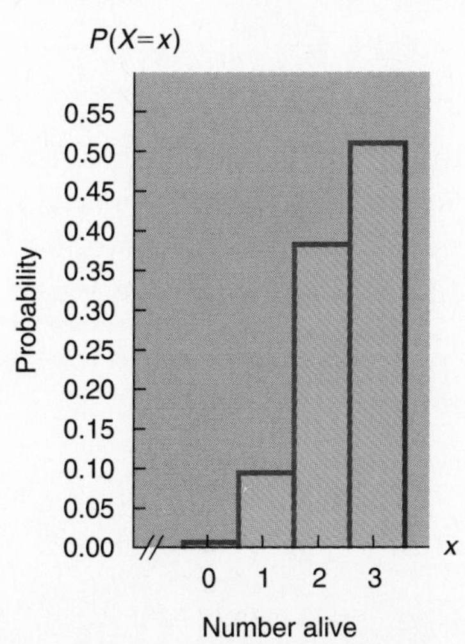

FIGURE 5.5 Probability histogram for the random variable X, the number of people out of three that are alive at age 65

The Binomial Probability Formula

Table 5.15 displays the probability distribution of the random variable "number alive at age 65" for three people currently age 20. To obtain that probability distribution, we used a tabulation method (Table 5.14), which required a significant amount of work.

But in most practical applications the work required would be considerably more and often prohibitive, because the number of trials is generally much larger than three. For instance, if twenty 20 year-olds instead of three 20 year-olds were observed, there would be over one million possible outcomes. In that case the tabulation method would certainly not be feasible.

The good news is that there is a relatively simple formula for obtaining binomial probabilities. A first step in developing that formula is the following fact.

KEY FACT 5.4 **NUMBER OF OUTCOMES CONTAINING A SPECIFIED NUMBER OF SUCCESSES**

In n Bernoulli trials, the number of outcomes containing exactly x successes is equal to the binomial coefficient $\binom{n}{x}$.

We will not stop to prove Key Fact 5.4, but let's quickly check to see that it is consistent with the results obtained in Example 5.12. For instance, the direct listing in Table 5.13 shows that there are three outcomes in which exactly two of the three people are alive at age 65, namely, *ssf, sfs,* and *fss.* Using binomial coefficients we can determine that fact without resorting to a direct listing. Applying Key Fact 5.4, we have

$$\begin{bmatrix}\text{Number of outcomes}\\ \text{comprising the event}\\ \text{exactly two alive}\end{bmatrix} = \binom{3}{2} = \frac{3!}{2!\,(3-2)!} = \frac{3!}{2!\,1!} = 3.$$

We can now develop a probability formula for the number of successes in Bernoulli trials. We indicate briefly how that formula is derived by referring to Example 5.12. For instance, to determine the probability that exactly two of the three people will be alive at age 65, $P(X = 2)$, we reason as follows.

1. Any particular outcome in which exactly two of the three people are alive at age 65 (e.g., *sfs*) has probability

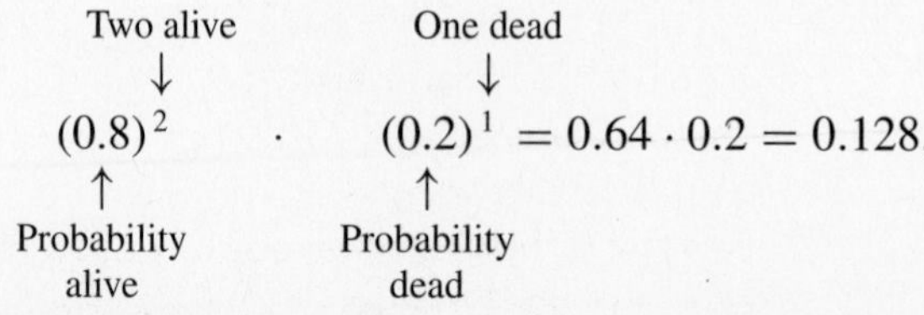

obtained by multiplying two success probabilities of 0.8 and one failure probability of 0.2.

2. By Key Fact 5.4, the number of outcomes in which exactly two of the three people are alive at age 65 is equal to

$$\underset{\text{Number alive}\ \uparrow}{\overset{\text{Number of trials}\ \downarrow}{\binom{3}{2}}} = \frac{3!}{2!\,(3-2)!} = 3.$$

3. By the special addition rule, the probability that exactly two of the three people will be alive at age 65 equals

$$P(X = 2) = \binom{3}{2} \cdot (0.8)^2(0.2)^1 = 3 \cdot 0.128 = 0.384.$$

Of course, this is the same result obtained in Example 5.12(d). However, this time we determined the probability quickly and easily—no tabulation and no listing were required. More importantly, the reasoning we used applies to any sequence of Bernoulli trials and leads to the following formula.

FORMULA 5.1 **BINOMIAL PROBABILITY FORMULA**

Let X denote the total number of successes in n Bernoulli trials with success probability p. Then the probability distribution of the random variable X is given by the formula

$$P(X = x) = \binom{n}{x} p^x (1-p)^{n-x}.$$

The random variable X is called a ***binomial random variable*** and is said to have the ***binomial distribution*** with parameters n and p.

To determine a binomial probability formula in specific problems, it is useful to have a well-organized strategy, such as the one presented in Procedure 5.1 at the top of the following page.

We will illustrate Procedure 5.1 by applying it to the random variable considered in Example 5.12.

PROCEDURE 5.1 **TO FIND A BINOMIAL PROBABILITY FORMULA**

ASSUMPTIONS

1. n identical trials are to be performed.
2. Two outcomes, success or failure, are possible for each trial.
3. The trials are independent.
4. The success probability, p, remains the same from trial to trial.

Step 1 Identify a success.

Step 2 Determine p, the success probability.

Step 3 Determine n, the number of trials.

Step 4 The binomial probability formula for the number of successes, X, is

$$P(X = x) = \binom{n}{x} p^x (1 - p)^{n-x}.$$

EXAMPLE 5.13 ILLUSTRATES PROCEDURE 5.1

According to tables provided by the U.S. National Center for Health Statistics in *Vital Statistics of the United States,* there is about an 80% chance that a person age 20 will be alive at age 65. Suppose three people age 20 are selected at random. Find the probability that the number alive at age 65 will be

a. exactly two. **b.** at most one. **c.** at least one.

d. Determine the probability distribution of the number alive at age 65.

SOLUTION Let X denote the number of people out of the three that are alive at age 65. To solve parts (a)–(d), we first apply Procedure 5.1.

Step 1 Identify a success.

In this problem a success is that a person currently age 20 will be alive at age 65.

Step 2 Determine p, the success probability.

This is the probability that a person currently age 20 will be alive at age 65, which is 80%. So $p = 0.8$.

Step 3 Determine n, the number of trials.

In this case the number of trials is the number of people in the study, which is three. So $n = 3$.

Step 4 *The binomial probability formula for the number of successes, X, is*

$$P(X = x) = \binom{n}{x} p^x (1 - p)^{n-x}.$$

Since $n = 3$ and $p = 0.8$, the formula becomes

$$P(X = x) = \binom{3}{x} (0.8)^x (0.2)^{3-x}.$$

Now that we have applied Procedure 5.1, it is relatively easy to solve the problems posed in parts (a)–(d).

a. Here we want the probability that exactly two of the three people will be alive at age 65 (i.e., two successes). Applying the binomial probability formula with $x = 2$ yields

$$P(X = 2) = \binom{3}{2} (0.8)^2 (0.2)^{3-2} = \frac{3!}{2!\,(3-2)!} (0.8)^2 (0.2)^1 = 0.384.$$

Consequently, the probability is 0.384 that exactly two of the three people will be alive at age 65.

b. The probability that at most one person will be alive at age 65 equals

$$\begin{aligned} P(X \le 1) &= P(X = 0) + P(X = 1) \\ &= \binom{3}{0} (0.8)^0 (0.2)^{3-0} + \binom{3}{1} (0.8)^1 (0.2)^{3-1} \\ &= 0.008 + 0.096 = 0.104. \end{aligned}$$

c. The probability that at least one person will be alive at age 65 is $P(X \ge 1)$. This can be obtained by first using the fact that

$$P(X \ge 1) = P(X = 1) + P(X = 2) + P(X = 3)$$

and then applying the binomial probability formula to calculate each of the three individual probabilities. However, it is easier to use the complementation rule:

$$\begin{aligned} P(X \ge 1) &= 1 - P(X < 1) = 1 - P(X = 0) \\ &= 1 - \binom{3}{0} (0.8)^0 (0.2)^{3-0} = 1 - 0.008 = 0.992. \end{aligned}$$

The chances are 99.2% of at least one of the three being alive at age 65.

d. Here we are to obtain the probability distribution of the random variable X. Thus we need to compute $P(X = x)$, for $x = 0, 1, 2,$ and 3, using the binomial probability formula. This has already been done for $x = 0, 1,$ and 2 in parts (a) and (b). For $x = 3$ we have

$$P(X = 3) = \binom{3}{3}(0.8)^3(0.2)^{3-3} = (0.8)^3 = 0.512.$$

So the probability distribution of X is as shown in Table 5.15 on page 315. But this time we computed the probabilities quickly and easily using the binomial probability formula. ◨

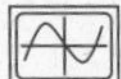

Figure 5.5 on page 315 shows that for three people currently 20 years old, the probability distribution of the number that will be alive at age 65 is left skewed; this is because the success probability, $p = 0.8$, exceeds 0.5. More generally, a binomial distribution is right skewed if $p < 0.5$, is symmetric if $p = 0.5$, and is left skewed if $p > 0.5$. Figure 5.6 illustrates these facts for binomial distributions with $n = 6$ in case $p = 0.25$, $p = 0.5$, and $p = 0.75$.

FIGURE 5.6 Probability histograms for binomial distributions with parameters $n = 6$ and (a) $p = 0.25$, (b) $p = 0.5$, (c) $p = 0.75$

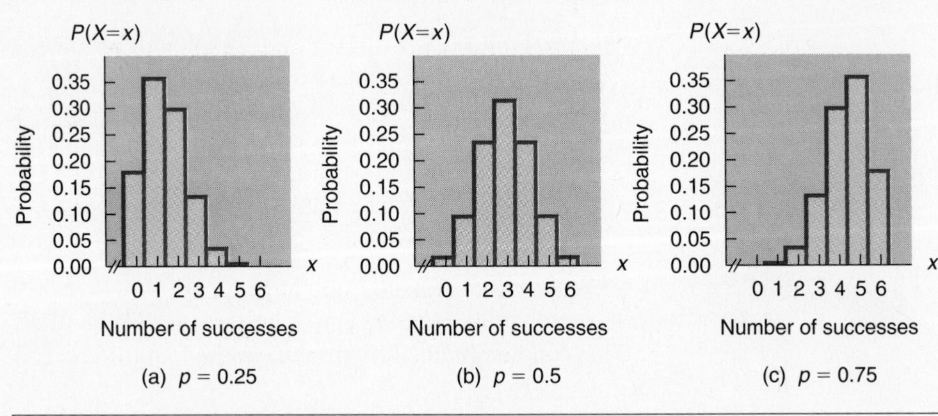

Mean and Standard Deviation of a Binomial Random Variable

In Section 5.2 we discussed the mean and standard deviation of a discrete random variable. The formulas required for computing these parameters are presented in Definition 5.4 (page 303) and Definition 5.5 (page 306).

As these formulas apply to any discrete random variable, they work in particular for a binomial random variable. So we can obtain the mean and standard deviation of a binomial random variable by first determining its probability distribution using the binomial probability formula and then applying Definitions 5.4 and 5.5.

But it turns out that there is an easier way. If we substitute the binomial probability formula into the formulas for the mean and standard deviation of a discrete random variable and then simplify mathematically, we obtain the following formulas.

FORMULA 5.2 **MEAN AND STANDARD DEVIATION OF A BINOMIAL RANDOM VARIABLE**

The mean and standard deviation of a binomial random variable with parameters n and p are

$$\mu = np \qquad \text{and} \qquad \sigma = \sqrt{np(1-p)},$$

respectively.

EXAMPLE 5.14 **ILLUSTRATES FORMULA 5.2**

For three randomly selected 20 year-olds, let X denote the number that are still alive at age 65. Obtain the mean and standard deviation of the random variable X.

SOLUTION As we have seen, X is a binomial random variable with parameters $n = 3$ and $p = 0.8$. Applying Formula 5.2 we get

$$\mu = np = 3 \cdot 0.8 = 2.4$$

and

$$\sigma = \sqrt{np(1-p)} = \sqrt{3 \cdot 0.8 \cdot 0.2} = 0.69.$$

In particular, the expected number that will be alive at age 65 equals 2.4; on the average, 2.4 of every three 20 year-olds will be alive at age 65. □

Binomial Approximation to the Hypergeometric Distribution

Many statistical studies are concerned with the proportion (percentage) of a finite population that has a specified attribute. For instance, we might be interested in the proportion of U.S. adults that have Internet access. Here the population consists of all U.S. adults, and the specified attribute is "has Internet access." Or we might want to know the proportion of U.S. businesses that are minority owned. In this case the population is composed of all U.S. businesses, and the specified attribute is "minority owned."

Generally the population under consideration is large, and it is therefore usually impractical and often impossible to determine the population proportion by taking a census; for instance, imagine trying to interview every U.S. adult in order to ascertain the proportion that have Internet access. So, in practice, we mostly rely on sampling and use the sample data to estimate the population proportion.

Suppose to that end a simple random sample of size n is taken from a population in which the proportion having a specified attribute equals p. Then a random variable of primary importance in estimating p is the number of members sampled that have the specified attribute, which we will call X. The exact probability distribution of X depends on whether the sampling is done with or without replacement.

If sampling is done with replacement, then the sampling process constitutes Bernoulli trials: Each selection of a member from the population corresponds to a trial. A success occurs on a trial if the member selected in that trial has the specified attribute; otherwise a failure occurs. The trials are independent since the sampling is done with replacement. The success probability remains the same from trial to trial—it always equals the proportion of the population having the specified attribute. As a consequence, the random variable X has the binomial distribution with parameters n (the sample size) and p (the population proportion).

In reality, however, sampling is ordinarily done without replacement. Under these circumstances the sampling process does not constitute Bernoulli trials because the trials are not independent and the success probability varies from trial to trial. This means that the random variable X does not have a binomial distribution; rather, it has what is called a **hypergeometric distribution.**

We will not present the hypergeometric probability formula here because, in practice, a hypergeometric distribution can usually be approximated by a binomial distribution. This is due to the fact that if the sample size does not exceed 5% of the population size, then there is little difference between sampling with and without replacement.

KEY FACT 5.5 SAMPLING AND THE BINOMIAL DISTRIBUTION

Suppose that a simple random sample of size n is taken from a finite population in which the proportion of members having a specified attribute is p. Then the number of members sampled that have the specified attribute

- has exactly a binomial distribution with parameters n and p if the sampling is done with replacement.
- has approximately a binomial distribution with parameters n and p if the sampling is done without replacement and the sample size does not exceed 5% of the population size.

For example, according to the Census Bureau publication *Current Population Reports,* 81.7% of U.S. adults have completed high school. Suppose eight U.S. adults are to be randomly selected without replacement, and let X denote the number of those sampled that have completed high school. Then, since the sample size does not exceed 5% of the population size, the random variable X has approximately a binomial distribution with parameters $n = 8$ and $p = 0.817$.

Other Discrete Probability Distributions

The binomial distribution is the most important and most widely used discrete probability distribution. However, many other discrete probability distributions occur frequently in practice.

We have already mentioned the hypergeometric distribution. Additional ones are the Poisson, discrete uniform, geometric, negative binomial, and multinomial distributions. We will briefly discuss the Poisson, hypergeometric, and geometric distributions in the exercises of this section. A more thorough presentation of the Poisson distribution is given in Section 5.4.

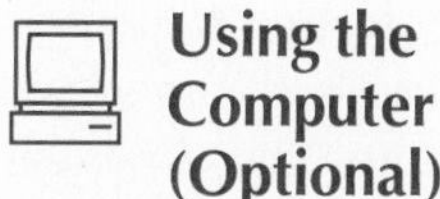

Using the Computer (Optional)

Minitab can be used to determine binomial probabilities. The appropriate Minitab procedure depends on the binomial probabilities that we need to obtain. To illustrate, we return once more to the example on survival to age 65.

EXAMPLE 5.15 USING MINITAB TO DETERMINE BINOMIAL PROBABILITIES

According to tables provided by the U.S. National Center for Health Statistics in *Vital Statistics of the United States,* there is about an 80% chance that a person age 20 will be alive at age 65. Suppose three people age 20 are selected at random. Find the probability that the number alive at age 65 will be

a. exactly two. **b.** at most one.

c. Determine the probability distribution of the number alive at age 65.

SOLUTION Let X denote the number of people out of the three that are still alive at age 65. As we have seen, X is a binomial random variable with parameters $n = 3$ and $p = 0.8$.

a. To determine the probability that exactly two of the three people will be alive at age 65, $P(X = 2)$, we use Minitab as follows.

1 Choose **Calc ➤ Probability Distributions ➤ Binomial...**

2 Select the **Probability** option button to indicate that we want to obtain an individual probability

3 Click in the **Number of trials** text box and type 3
4 Click in the **Probability of success** text box and type 0.8
5 Select the **Input constant** option button
6 Click in the **Input constant** text box and type 2
7 Click **OK**

Printout 5.2 displays the output that results.

PRINTOUT 5.2
Minitab output for $P(X = 2)$

```
Binomial with n = 3 and p = 0.800000

        x        P( X = x)
     2.00          0.3840
```

Printout 5.2 shows that the probability is 0.3840 that exactly two of the three people will be alive at age 65.

b. For this part we want the probability that at most one person will be alive at age 65, $P(X \leq 1)$. Minitab can be used in several ways to find that probability. One way is to employ Minitab's capability for obtaining cumulative probabilities. A **cumulative probability** gives the probability that a random variable is less than or equal to a specified number. Here we want the cumulative probability for the number 1. We use Minitab as follows.

1 Choose **Calc ➤ Probability Distributions ➤ Binomial...**
2 Select the **Cumulative probability** option button to indicate that we want to obtain a cumulative probability
3 Click in the **Number of trials** text box and type 3
4 Click in the **Probability of success** text box and type 0.8
5 Select the **Input constant** option button
6 Click in the **Input constant** text box and type 1
7 Click **OK**

Printout 5.3 shows the resulting output.

PRINTOUT 5.3
Minitab output for $P(X \leq 1)$

```
Binomial with n = 3 and p = 0.800000

        x       P( X <= x)
     1.00          0.1040
```

As we see from Printout 5.3, the probability is 0.1040 that at most one person will be alive at age 65.

c. Here we are to obtain the probability distribution of the random variable X. This can be accomplished by first storing the possible values for X (the integers 0–3) in a column named x and then proceeding in the following way.

1 Choose **Calc ➤ Probability Distributions ➤ Binomial...**

2 Select the **Probability** option button to indicate that we want to obtain individual probabilities

3 Click in the **Number of trials** text box and type 3

4 Click in the **Probability of success** text box and type 0.8

5 Select the **Input column** option button

6 Click in the **Input column** text box and specify x

7 Click **OK**

The resulting output is portrayed in Printout 5.4.

PRINTOUT 5.4
Minitab output for the binomial distribution with parameters $n = 3$ and $p = 0.8$

```
Binomial with n = 3 and p = 0.800000

        x        P( X = x)
     0.00          0.0080
     1.00          0.0960
     2.00          0.3840
     3.00          0.5120
```

Compare the table in Printout 5.4 to the binomial distribution displayed in Table 5.15 on page 315. ◨

EXERCISES 5.3

STATISTICAL CONCEPTS AND SKILLS

5.29 What is meant by Bernoulli trials?

5.30 Give two examples of Bernoulli trials other than ones presented in the book.

5.31 What do you think that the "bi" in "binomial" signifies?

5.32 Find 1!, 2!, and 6!.

5.33 Compute 7!, 8!, and 9!.

5.34 Evaluate the following binomial coefficients.

a. $\binom{4}{1}$ **b.** $\binom{6}{2}$ **c.** $\binom{8}{3}$ **d.** $\binom{9}{6}$

5.35 Evaluate the following binomial coefficients.

a. $\binom{5}{2}$ **b.** $\binom{7}{4}$ **c.** $\binom{10}{3}$ **d.** $\binom{12}{5}$

5.36 Determine the value of each of the following binomial coefficients.

a. $\binom{3}{2}$ **b.** $\binom{6}{0}$ **c.** $\binom{6}{6}$ **d.** $\binom{7}{3}$

5.37 Determine the value of each of the following binomial coefficients.

a. $\binom{5}{3}$ **b.** $\binom{10}{0}$ **c.** $\binom{10}{10}$ **d.** $\binom{9}{5}$

5.38 Based on data from the *Statistical Abstract of the United States,* the probability that a newborn baby will be a girl is about 0.487. Three newborns are selected at random.

a. Considering a success in a given birth to be "a girl," formulate the process of observing the sex of the three newborns as a sequence of three Bernoulli trials.

b. Construct a table similar to Table 5.14 on page 314 for the three births. Display the probabilities to three decimal places.

c. Draw a tree diagram for this problem similar to Fig. 5.4 on page 314.

d. List the outcomes in which exactly one of the three babies is a girl.

e. Find the probability of each outcome in part (d). Why are those probabilities all the same?

f. Use parts (d) and (e) to determine the probability that exactly one of the three babies is a girl.

g. Without using the binomial probability formula, obtain the probability distribution of the random variable X, the number of babies out of three that are girls.

5.39 Pinworm infestation, commonly found in children, can be treated with the drug pyrantel pamoate. According to the *Merck Manual,* the treatment is effective in 90% of cases. Three children with pinworm infestation are given pyrantel pamoate.

a. Considering a success in a given case to be "a cure," formulate the process of observing which children are cured as a sequence of three Bernoulli trials.

b. Construct a table similar to Table 5.14 on page 314 for the three cases. Display the probabilities to three decimal places.

c. Draw a tree diagram for this problem similar to Fig. 5.4 on page 314.

d. List the outcomes in which exactly two of the three children are cured.

e. Find the probability of each outcome in part (d). Why are those probabilities all the same?

f. Use parts (d) and (e) to determine the probability that exactly two of the three children will be cured.

g. Without using the binomial probability formula, obtain the probability distribution of the random variable X, the number of children out of three that are cured.

5.40 A phone-company study conducted in Phoenix, Arizona, revealed that the probability is 0.25 that a randomly selected phone call will last longer than the mean duration of 3.8 minutes. For any particular phone call, suppose we consider a success to be that the call lasts at most 3.8 minutes, that is, 3.8 minutes or less.

a. What is the success probability, p?

b. Construct a table similar to Table 5.14 on page 314 for the possible success-failure results of four randomly selected calls. Display the probabilities to three decimal places.

c. Draw a tree diagram for this problem similar to Fig. 5.4 on page 314.

d. List the outcomes in which exactly two of the four calls last at most 3.8 minutes.

e. Find the probability of each outcome in part (d). Why are those probabilities all the same?

f. Use parts (d) and (e) to determine the probability that exactly two of the four calls last at most 3.8 minutes.

g. Without using the binomial probability formula, obtain the probability distribution of the random variable Y, the number of calls out of four that last at most 3.8 minutes.

5.41 The National Institute of Mental Health reports that there is a 20% chance of an adult American suffering from a psychiatric disorder. Four randomly selected adult Americans are examined for psychiatric disorders.

a. If we let a success correspond to an adult American having a psychiatric disorder, then what is the success probability, p? *(Note:* The use of the word *success* in Bernoulli trials need not conform to the ordinarily positive connotation of the word.*)*

b. Construct a table similar to Table 5.14 on page 314 for the four people examined. Display the probabilities to four decimal places.

c. Draw a tree diagram for this problem similar to Fig. 5.4 on page 314.
d. List the outcomes in which exactly three of the four people examined have a psychiatric disorder.
e. Find the probability of each outcome in part (d). Why are those probabilities all the same?
f. Use parts (d) and (e) to determine the probability that exactly three of the four people examined have a psychiatric disorder.
g. Without using the binomial probability formula, obtain the probability distribution of the random variable Y, the number of adults out of four that have a psychiatric disorder.

5.42 Use Procedure 5.1 on page 318 to solve part (g) of Exercise 5.38.

5.43 Use Procedure 5.1 on page 318 to solve part (g) of Exercise 5.39.

5.44 Use Procedure 5.1 on page 318 to solve part (g) of Exercise 5.40.

5.45 Use Procedure 5.1 on page 318 to solve part (g) of Exercise 5.41.

5.46 Following are two probability histograms of binomial distributions. For each one, specify whether the success probability, p, is less than, equal to, or greater than 0.5. Explain your answers.

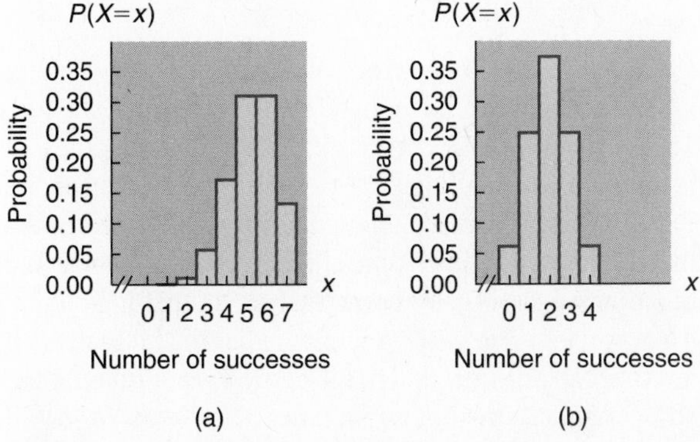

5.47 Following are two probability histograms of binomial distributions. For each one, specify whether the success probability, p, is less than, equal to, or greater than 0.5. Explain your answers.

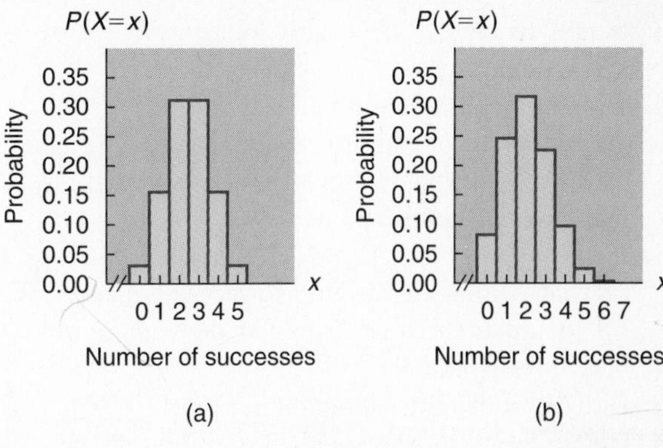

In each of Exercises 5.48–5.53, use Procedure 5.1 on page 318 to obtain the required probabilities. Express each probability answer as a decimal rounded to three places.

5.48 As reported by the Chicago Title Insurance Company in *The Guarantor,* chances are 22.7% that a home buyer will purchase a new (non-resale) home. Let a success correspond to the purchase of a new home. In the next four home purchases, find the probability that the number of new-home purchases will be
a. exactly three. **b.** exactly two.
c. at least two.
d. between two and three, inclusive.
e. Determine the probability distribution of the random variable X, the number of new-home purchases in the next four home purchases.
f. Identify the probability distribution of X as right skewed, symmetric, or left skewed without consulting its probability distribution or drawing its probability histogram.
g. Draw a probability histogram for X.
h. Obtain the mean and standard deviation of the random variable X using your answer from part (e) and Definitions 5.4 and 5.5 on pages 303 and 306, respectively.
i. Obtain the mean and standard deviation of the random variable X using Formula 5.2 on page 321.
j. Interpret your answer for the mean in words.

5.49 According to the *Daily Racing Form,* the probability is approximately 0.67 that the favorite in a horse race will finish in the money (first, second, or third place). In the next five races, what is the probability that the favorite finishes in the money

a. exactly twice? **b.** exactly four times?
c. at least four times?
d. between two and four times, inclusive?
e. Determine the probability distribution of the random variable X, the number of times the favorite finishes in the money in the next five races.
f. Identify the probability distribution of X as right skewed, symmetric, or left skewed without consulting its probability distribution or drawing its probability histogram.
g. Draw a probability histogram for X.
h. Obtain the mean and standard deviation of the random variable X using your answer from part (e) and Definitions 5.4 and 5.5 on pages 303 and 306, respectively.
i. Obtain the mean and standard deviation of the random variable X using Formula 5.2 on page 321.
j. Interpret your answer for the mean in words.

5.50 As reported by Television Bureau of Advertising, Inc., in *Trends in Television,* 81.0% of U.S. households have a VCR. If six households are randomly selected without replacement, what is the (approximate) probability that the number having a VCR is
a. exactly four? **b.** at least four?
c. at most five?
d. between two and five, inclusive?
e. Determine the (approximate) probability distribution of the random variable Y, the number of households out of six that have a VCR.
f. Determine and interpret the mean of the random variable Y.
g. Obtain the standard deviation of Y.
h. Strictly speaking, why is the probability distribution that you obtained in part (e) only approximately correct? What is the exact distribution called?

5.51 According to *Current Population Reports,* a publication of the U.S. Bureau of the Census, 25% of U.S. children are not living with both parents. If four U.S. children are selected at random without replacement, determine the (approximate) probability that the number not living with both parents is
a. exactly two. **b.** at most two.
c. between one and three, inclusive.
d. either fewer than one or more than two.
e. Determine the (approximate) probability distribution of the random variable Y, the number of children out of four that are not living with both parents.
f. Determine and interpret the mean of the random variable Y.
g. Obtain the standard deviation of Y.
h. Strictly speaking, why is the probability distribution that you obtained in part (e) only approximately correct? What is the exact distribution called?

5.52 Studies show that 60% of U.S. families use physical aggression to resolve conflict. If 10 families are selected at random, find the probability that the number that use physical aggression to resolve conflict is
a. exactly five.
b. between five and seven, inclusive.
c. over 80% of those surveyed.
d. fewer than nine.

5.53 According to the American Bankers Association, only one in 10 people are dissatisfied with their local bank. If 12 people are randomly selected, what is the probability that the number dissatisfied with their local bank is
a. exactly two? **b.** at most two?
c. at least two?
d. between one and three, inclusive?

EXTENDING THE CONCEPTS AND SKILLS

5.54 A success, *s,* in Bernoulli trials is often derived from a collection of outcomes. For example, a roulette wheel consists of 38 numbers, of which 18 are red, 18 are black, and 2 are green. When the roulette ball is spun, it is equally likely to land on any one of the 38 numbers. If we are interested in which number the ball lands on, then each play at the wheel has 38 possible outcomes. Suppose, however, that we are betting on red. Then we are interested only in whether the ball lands on a red number. From this point of view, each play at the wheel has only two possible outcomes—either the ball lands on a red number or it doesn't. Hence successive bets on red constitute a sequence of Bernoulli trials with success probability $\frac{18}{38}$. In four plays at a roulette wheel, what is the probability that the ball lands on red
a. exactly twice? **b.** at least once?

5.55 The Arizona state lottery, *Lotto,* is played as follows: The player selects six numbers from the numbers 1–42 and buys a ticket for $1. There are six winning numbers, which are selected at random from the numbers 1–42. To win a prize, a *Lotto* ticket must contain three or more of the winning numbers. Following is a probability distribution for the number of winning numbers for a single ticket.

Number of winning numbers	Probability
0	0.3713060
1	0.4311941
2	0.1684352
3	0.0272219
4	0.0018014
5	0.0000412
6	0.0000002

a. If you buy one *Lotto* ticket, determine the probability that you win a prize. Round your answer to three decimal places.

b. If you buy one *Lotto* ticket per week for a year, determine the probability that you win a prize at least once in the 52 tries.

5.56 A sales representative for a tire manufacturer claims that the company's steel-belted radials last at least 35,000 miles. A tire dealer decides to check that claim by testing eight of the tires. If 75% or more of the eight tires he tests last at least 35,000 miles, he will purchase tires from the sales representative. If, in fact, 90% of the steel-belted radials produced by the manufacturer last at least 35,000 miles, what is the probability that the tire dealer will purchase tires from the sales representative?

5.57 From past experience the owner of a restaurant knows that, on average, 4% of the parties making reservations never show. How many reservations can the owner accept and still be at least 80% sure that all parties making a reservation will show?

5.58 Sickle cell anemia is an inherited blood disease that occurs primarily in blacks. In the United States, roughly 15 of every 10,000 black children have sickle cell anemia. The red blood cells of an affected person are abnormal; the result is severe chronic anemia (inability to carry the required amount of oxygen), which causes headaches, shortness of breath, jaundice, increased risk of pneumococcal pneumonia and gallstones, and other severe problems. Sickle cell anemia arises in children inheriting an abnormal type of hemoglobin, called hemoglobin S, from both parents. If hemoglobin S is inherited from only one parent, then the person is said to have sickle cell trait and is generally free from symptoms. There is a 50% chance that a person having sickle cell trait will pass hemoglobin S to an offspring.

a. Obtain the probability that a child of two people having sickle cell trait will have sickle cell anemia.

b. If two people having sickle cell trait have five children, determine the probability that at least one of the children will have sickle cell anemia.

c. If two people having sickle cell trait have five children, find the probability distribution of the number of those children who will have sickle cell anemia.

d. Construct a probability histogram for the probability distribution in part (c).

e. If two people having sickle cell trait have five children, how many can they expect will have sickle cell anemia?

5.59 Refer to the discussion on the binomial approximation to the hypergeometric distribution beginning on page 321.

a. If sampling is with replacement, explain why the trials are independent and the success probability remains the same from trial to trial, always being the proportion of the population having the specified attribute.

b. If sampling is without replacement, explain why the trials are not independent and the success probability varies from trial to trial.

5.60 Refer to the discussion on the binomial approximation to the hypergeometric distribution beginning on page 321. Consider the following frequency distribution for students in Professor Weiss's introductory statistics class.

Sex	Frequency
Male	17
Female	23

Two students are selected at random. Find the probability that both students are male if the selection is done

a. with replacement. (*Hint:* Apply the special multiplication rule on page 249.)

b. without replacement. (*Hint:* Apply the general multiplication rule on page 245.)

c. Compare the answers obtained in parts (a) and (b).

Suppose Professor Weiss's class had 10 times the students, but in the same proportions, that is, 170 males and 230 females.

d. Repeat parts (a)–(c) using this hypothetical distribution of students.

e. In which case is there less difference between sampling without and with replacement? Explain why this is so.

5.61 In this exercise we will discuss the **hypergeometric distribution** in more detail. When sampling without replacement from a finite population, the hypergeometric distribution is the exact probability distribution for the number of members sampled that have a specified attribute. The hypergeometric probability formula is

$$P(X = x) = \frac{\binom{Np}{x}\binom{N(1-p)}{n-x}}{\binom{N}{n}},$$

where X denotes the number of members sampled having the specified attribute, N is the population size, n is the sample size, and p is the population proportion.

To illustrate, suppose a customer purchases four fuses from a shipment of 250, of which 94% are not defective. Let a success correspond to a fuse that is not defective.

a. Determine N, n, and p.

b. Use the hypergeometric probability formula to find the probability distribution of the number of nondefective fuses the customer gets.

Key Fact 5.5 shows that a hypergeometric distribution can be approximated by a binomial distribution provided the sample size does not exceed 5% of the population size. In particular, we can use the binomial probability formula

$$P(X = x) = \binom{n}{x} p^x (1-p)^{n-x},$$

with $n = 4$ and $p = 0.94$, to approximate the probability distribution of the number of nondefective fuses that the customer gets.

c. Obtain the binomial distribution with parameters $n = 4$ and $p = 0.94$.

d. Compare the hypergeometric distribution that you obtained in part (b) with the binomial distribution that you obtained in part (c).

5.62 In this exercise we will discuss the **geometric distribution,** the probability distribution for the number of trials until the first success in Bernoulli trials. The geometric probability formula is

$$P(X = x) = p(1-p)^{x-1},$$

where X denotes the number of trials until the first success and p the success probability. Using the geometric probability formula and Definition 5.4 on page 303, it can be shown that the mean of the random variable X is $1/p$.

To illustrate, let's again consider the Arizona state lottery, *Lotto,* as described in Exercise 5.55. Suppose you buy one *Lotto* ticket per week. Let X denote the number of weeks until you win a prize.

a. Find and interpret the probability formula for the random variable X. (*Note:* The appropriate success probability was obtained in Exercise 5.55(a).)

b. Compute the probability that the number of weeks until you win a prize is exactly 3; at most 3; at least 3.

c. On the average, how long will it take until you win a prize?

5.63 Another important discrete probability distribution is the **Poisson distribution,** named in honor of the French mathematician and physicist Simeon Poisson (1781–1840). This probability distribution is often used to model the frequency with which a specified event occurs during a particular period of time. The Poisson probability formula is

$$P(X = x) = e^{-\lambda}\frac{\lambda^x}{x!},$$

where X is the number of times the event occurs and λ is a parameter equal to the mean of X. The number e is the base of natural logarithms and is approximately equal to 2.7183.

To illustrate, let's consider the following problem: Desert Samaritan Hospital, located in Mesa, AZ, keeps records of emergency-room traffic. From those records we find that the number of patients arriving between 6:00 P.M. and 7:00 P.M. has a Poisson distribution with parameter $\lambda = 6.9$. Determine the probability that, on a given day, the number of patients arriving at the emergency room between 6:00 P.M. and 7:00 P.M. will be

a. exactly four. **b.** at most two.
c. between four and 10, inclusive.

USING TECHNOLOGY  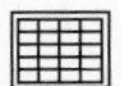 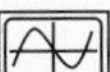

5.64 Use Minitab or some other statistical software to obtain the required probability or probabilities in parts (a)–(e) of Exercise 5.48.

5.65 Use Minitab or some other statistical software to obtain the required probability or probabilities in parts (a)–(e) of Exercise 5.49.

5.66 Following is a Minitab printout showing a binomial distribution.

```
Binomial with n = 7 and p = 0.340000

      x        P( X = x)
   0.00          0.0546
   1.00          0.1967
   2.00          0.3040
   3.00          0.2610
   4.00          0.1345
   5.00          0.0416
   6.00          0.0071
   7.00          0.0005
```

Use the printout to determine the

a. number of trials. **b.** success probability.
c. probability of exactly three successes.
d. probability of between two and five successes, inclusive.
e. probability of at most four successes.
f. probability of at least four successes.

5.67 The following printout, obtained from Minitab, gives the cumulative probabilities for a binomial distribution.

```
Binomial with n = 6 and p = 0.590000

      x        P( X <= x)
   0.00          0.0048
   1.00          0.0458
   2.00          0.1933
   3.00          0.4764
   4.00          0.7819
   5.00          0.9578
   6.00          1.0000
```

Use the printout to find the

a. number of trials. **b.** success probability.
c. probability of at most three successes.
d. probability of at least three successes. *(Hint:* Use the complementation rule and the printout.*)*
e. probability of exactly three successes. *(Hint:* Use the special addition rule to express $P(X = 3)$ as the difference of two cumulative probabilities.*)*

5.68 According to an article published in *Reader's Digest,* 10% of people are left-handed. For 10 people selected at random, let X denote the number of people out of the 10 who are left-handed. We used Minitab to obtain the probability distribution of the random variable X. The output is shown below.

```
Binomial with n = 10 and p = 0.100000

      x        P( X = x)
   0.00          0.3487
   1.00          0.3874
   2.00          0.1937
   3.00          0.0574
   4.00          0.0112
   5.00          0.0015
   6.00          0.0001
   7.00          0.0000
   8.00          0.0000
   9.00          0.0000
  10.00          0.0000
```

Employ the printout to obtain the probability that out of the 10 people chosen,

a. exactly one is left-handed.
b. at least one is left-handed.
c. at most one is left-handed.
d. more than one is left-handed.
e. between one and three, inclusive, are left-handed.

5.69 The U.S. National Center for Health Statistics reports that 27% of U.S. adults ages 20 to 74 have high cholesterol levels. Eight U.S. adults are selected at random. Let X denote the number of those chosen who have high cholesterol levels. The Minitab printout at the right gives the cumulative probabilities for the random variable X. From the printout, obtain the probability that out of the eight adults chosen, the number having high cholesterol levels is

a. at most three. **b.** at least three.
c. exactly three.
d. between two and four, inclusive.
e. either fewer than two or more than five.

```
Binomial with n = 8 and p = 0.270000

       x        P( X <= x)
    0.00          0.0806
    1.00          0.3193
    2.00          0.6282
    3.00          0.8567
    4.00          0.9623
    5.00          0.9936
    6.00          0.9994
    7.00          1.0000
    8.00          1.0000
```

5.4 THE POISSON DISTRIBUTION*

Another important discrete probability distribution is the **Poisson distribution,** named in honor of the French mathematician and physicist Simeon D. Poisson (1781–1840). The Poisson distribution is often used to model the frequency with which a specified event occurs during a particular period of time. For instance, we might employ the Poisson distribution when analyzing

- the number of patients arriving at an emergency room between 6:00 P.M. and 7:00 P.M.
- the number of telephone calls received per day at a switchboard.
- the number of alpha particles emitted per minute by a radioactive substance.

Additionally, the Poisson distribution might be used to describe the number of misprints in books, the number of defective teeth per person, or the number of bacterial colonies appearing on a petri dish smeared with a bacterial suspension.

The Poisson Probability Formula

A particular Poisson distribution is identified by one parameter, usually denoted by the Greek letter λ (lambda). As we will see, that parameter represents the mean of the distribution. Formula 5.3 provides the **Poisson probability formula,** the formula used to obtain probabilities for a random variable having a Poisson distribution.

---

* See the note on the bottom of page 289.

FORMULA 5.3 **POISSON PROBABILITY FORMULA**

Probabilities for a random variable X having a Poisson distribution are given by the formula

$$P(X = x) = e^{-\lambda} \frac{\lambda^x}{x!}, \qquad x = 0, 1, 2, \ldots,$$

where λ is a positive real number and $e \approx 2.718$. (Most calculators have an e-key.) The random variable X is called a ***Poisson random variable*** and is said to have the ***Poisson distribution*** with parameter λ.

A Poisson random variable has a (countably) infinite number of possible values, namely, all nonnegative integers. Consequently, we cannot display all the probabilities for a Poisson random variable in a probability-distribution table.

EXAMPLE 5.16 ILLUSTRATES FORMULA 5.3

Desert Samaritan Hospital, located in Mesa, Arizona, keeps records of emergency-room traffic. From those records we find that the number of patients arriving between 6:00 P.M. and 7:00 P.M. has a Poisson distribution with parameter $\lambda = 6.9$. Determine the probability that, on a given day, the number of patients arriving at the emergency room between 6:00 P.M. and 7:00 P.M. will be

a. exactly four.

b. at most two.

c. between four and 10, inclusive.

d. Obtain a table of probabilities for the random variable X, the number of patients arriving between 6:00 P.M. and 7:00 P.M. Stop when the probabilities become zero to three decimal places.

e. Use part (d) to construct a (partial) probability histogram for the random variable X.

f. Identify the shape of the probability distribution of X.

SOLUTION As we mentioned, we are letting X denote the number of patients arriving between 6:00 P.M. and 7:00 P.M. We know that the random variable X has a Poisson distribution with parameter $\lambda = 6.9$. Thus, by Formula 5.3, probabilities for X are given by the Poisson probability formula,

$$P(X = x) = e^{-6.9} \frac{(6.9)^x}{x!}.$$

Using this formula we can now solve the problems posed in parts (a)–(e).

a. Here we want the probability of exactly four arrivals. Applying the Poisson probability formula with $x = 4$ gives

$$P(X = 4) = e^{-6.9}\frac{(6.9)^4}{4!} = e^{-6.9}\cdot\frac{2266.7121}{24} = 0.095.$$

b. The probability of at most two arrivals is

$$\begin{aligned}P(X \le 2) &= P(X = 0) + P(X = 1) + P(X = 2)\\ &= e^{-6.9}\frac{(6.9)^0}{0!} + e^{-6.9}\frac{(6.9)^1}{1!} + e^{-6.9}\frac{(6.9)^2}{2!}\\ &= e^{-6.9}\left(\frac{6.9^0}{0!} + \frac{6.9^1}{1!} + \frac{6.9^2}{2!}\right)\\ &= e^{-6.9}\left(1 + 6.9 + 23.805\right) = e^{-6.9}\cdot 31.705 = 0.032.\end{aligned}$$

c. The probability of between four and 10 arrivals, inclusive, is

$$\begin{aligned}P(4 \le X \le 10) &= P(X = 4) + P(X = 5) + \cdots + P(X = 10)\\ &= e^{-6.9}\left(\frac{6.9^4}{4!} + \frac{6.9^5}{5!} + \cdots + \frac{6.9^{10}}{10!}\right) = 0.821.\end{aligned}$$

d. Proceeding as in part (a), we obtain a partial probability distribution of the random variable X, as shown in Table 5.16.

TABLE 5.16 Partial probability distribution of the random variable X, the number of patients arriving at the emergency room between 6:00 P.M. and 7:00 P.M.

Number arriving x	Probability $P(X = x)$	Number arriving x	Probability $P(X = x)$
0	0.001	10	0.068
1	0.007	11	0.043
2	0.024	12	0.025
3	0.055	13	0.013
4	0.095	14	0.006
5	0.131	15	0.003
6	0.151	16	0.001
7	0.149	17	0.001
8	0.128	18	0.000
9	0.098		

e. Using Table 5.16 we obtain a partial probability histogram for the random variable X, as depicted in Fig. 5.7.

FIGURE 5.7 Partial probability histogram for the random variable X, the number of patients arriving at the emergency room between 6:00 P.M. and 7:00 P.M.

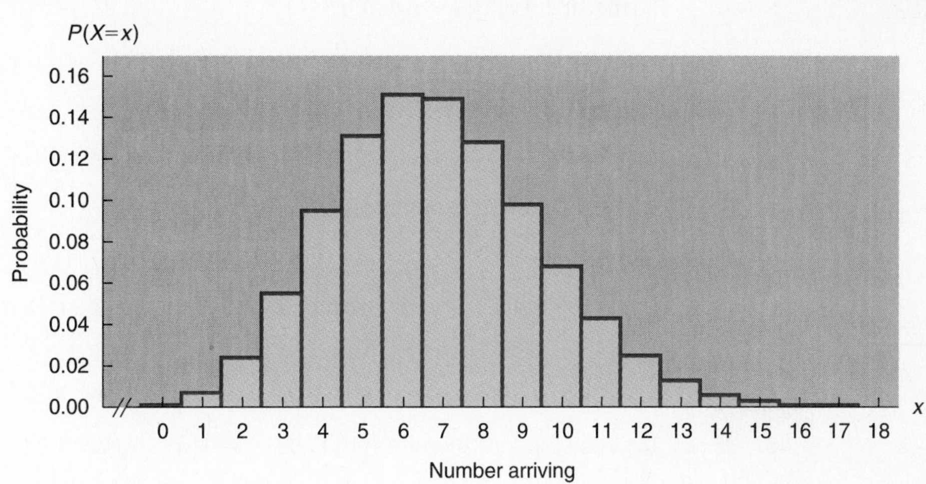

 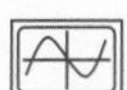

f. From Fig. 5.7 we see that the probability distribution of the random variable X is right skewed. ◨

In part (f) of Example 5.16, we discovered that the probability distribution of the number of patients arriving at the emergency room between 6:00 P.M. and 7:00 P.M. is right skewed. This is true for all Poisson distributions—the probability distribution of any Poisson random variable is right skewed.

Mean and Standard Deviation of a Poisson Random Variable

We can derive special formulas for the mean and standard deviation of a Poisson random variable. If we substitute the Poisson probability formula into the formulas for the mean and standard deviation of a discrete random variable and then simplify mathematically, we obtain the following formulas.

FORMULA 5.4 MEAN AND STANDARD DEVIATION OF A POISSON RANDOM VARIABLE

The mean and standard deviation of a Poisson random variable with parameter λ are

$$\mu = \lambda \qquad \text{and} \qquad \sigma = \sqrt{\lambda},$$

respectively.

EXAMPLE 5.17 ILLUSTRATES FORMULA 5.4

Let X denote the number of patients arriving at the emergency room of Desert Samaritan Hospital between 6:00 P.M. and 7:00 P.M.

a. Determine and interpret the mean of the random variable X.

b. Determine the standard deviation of X.

SOLUTION As we know, X has the Poisson distribution with parameter $\lambda = 6.9$. So we can apply Formula 5.4 to determine the mean and standard deviation of X.

a. The mean of X is $\mu = \lambda = 6.9$. In other words, on the average, 6.9 patients arrive at the emergency room between 6:00 P.M. and 7:00 P.M.

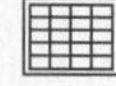

b. The standard deviation of X is $\sigma = \sqrt{\lambda} = \sqrt{6.9} = 2.6$. ◨

Poisson Approximation to the Binomial Distribution

Recall that the binomial probability formula is

$$P(X = x) = \binom{n}{x} p^x (1 - p)^{n-x}.$$

This is the formula from which we obtain probabilities for the number of successes, X, in n Bernoulli trials with success probability p.

When n is large, the binomial probability formula can be awkward or impractical to use because of computational difficulties. These difficulties persist today even with sophisticated computers and calculators. Consequently, methods have been developed that permit us to approximate binomial probabilities using formulas that are easier to work with.

One of those methods employs the Poisson probability formula and applies when n is large and p is small (as a rule of thumb, we will require that $n \geq 100$ and $np \leq 10$). In such cases we can use a Poisson distribution to approximate the binomial distribution. And, as we might expect, the appropriate Poisson distribution is the one whose mean is the same as that of the binomial distribution; that is, $\lambda = np$. Specifically, we have the procedure shown at the top of the next page.

EXAMPLE 5.18 ILLUSTRATES PROCEDURE 5.2

According to data obtained from the International Data Base and published in the *Statistical Abstract of the United States,* the infant mortality rate in Sweden is 4.5 per 1000 live births. Use the Poisson approximation to determine the probability that out of 500 randomly selected live births there are

a. no infant deaths. **b.** at most three infant deaths.

PROCEDURE 5.2

TO APPROXIMATE BINOMIAL PROBABILITIES USING A POISSON PROBABILITY FORMULA

Step 1 Determine n, the number of trials, and p, the success probability.

Step 2 Check that $n \geq 100$ and $np \leq 10$. If they are not, the Poisson approximation should not be used.

Step 3 Use the Poisson probability formula

$$P(X = x) = e^{-np} \frac{(np)^x}{x!}$$

to approximate the required binomial probabilities.

SOLUTION We apply Procedure 5.2.

Step 1 *Determine n, the number of trials, and p, the success probability.*

We have $n = 500$ and $p = \frac{4.5}{1000} = 0.0045$.

Step 2 *Check that $n \geq 100$ and $np \leq 10$.*

From Step 1 we conclude that $n = 500$ and $np = 500 \cdot 0.0045 = 2.25$. So we see that $n \geq 100$ and $np \leq 10$.

Step 3 *Use the Poisson probability formula*

$$P(X = x) = e^{-np} \frac{(np)^x}{x!}$$

to approximate the required binomial probabilities.

As we noted in Step 2, $np = 2.25$. Thus in this case the appropriate Poisson probability formula is

$$P(X = x) = e^{-2.25} \frac{(2.25)^x}{x!}.$$

a. For this part we want the probability of no infant deaths in 500 live births. That probability is

$$P(X = 0) = e^{-2.25} \frac{(2.25)^0}{0!} = 0.105,$$

approximately.

b. Here we want the probability of at most three infant deaths in 500 live births. That probability is

$$P(X \leq 3) = P(X = 0) + P(X = 1) + P(X = 2) + P(X = 3)$$

$$= e^{-2.25}\left(\frac{2.25^0}{0!} + \frac{2.25^1}{1!} + \frac{2.25^2}{2!} + \frac{2.25^3}{3!}\right) = 0.809,$$

approximately. ◨

Referring to Example 5.18, we will now illustrate the accuracy of the Poisson approximation to the binomial distribution. We used a computer to obtain both the binomial distribution with parameters $n = 500$ and $p = 0.0045$ and the Poisson distribution with parameter $\lambda = np = 500 \cdot 0.0045 = 2.25$.

Table 5.17 shows both distributions and exhibits how well the Poisson approximates the binomial. The probabilities are displayed to four decimal places in order to present a clearer picture of the differences between the two distributions. Notice that we stopped listing the probabilities once they became zero to four decimal places.

TABLE 5.17 Comparison of the binomial distribution with parameters $n = 500$ and $p = 0.0045$ to the Poisson distribution with parameter $\lambda = 2.25$

x	Binomial probability	Poisson approximation
0	0.1049	0.1054
1	0.2370	0.2371
2	0.2673	0.2668
3	0.2006	0.2001
4	0.1127	0.1126
5	0.0505	0.0506
6	0.0188	0.0190
7	0.0060	0.0061
8	0.0017	0.0017
9	0.0004	0.0004
10	0.0001	0.0001
11	0.0000	0.0000

For large n and small p, it is not always possible to use a computer instead of a Poisson approximation to determine a required binomial distribution—sometimes n is so large or p is so small that even a computer can't handle the computations to obtain a binomial distribution. Nonetheless, the Poisson approximation will still be easy to apply.

Using the Computer (Optional)

Minitab can be applied to determine Poisson probabilities. The appropriate Minitab procedure depends on the Poisson probabilities that we need to obtain. To illustrate, we return to the emergency-room example.

EXAMPLE 5.19 USING MINITAB TO DETERMINE POISSON PROBABILITIES

Use Minitab to solve the problems in parts (a)–(c) of Example 5.16 on page 333.

SOLUTION We want to determine probabilities for the number of patients arriving at the emergency room of Desert Samaritan Hospital between 6:00 P.M. and 7:00 P.M. Letting X denote the number of arrivals, we have seen that X has a Poisson distribution with parameter $\lambda = 6.9$.

a. To determine the probability of exactly four arrivals, we use Minitab as follows.

1 Choose **Calc ➤ Probability Distributions ➤ Poisson...**
2 Select the **Probability** option button to indicate that we want to obtain an individual probability
3 Click in the **Mean** text box and type `6.9`
4 Select the **Input constant** option button
5 Click in the **Input constant** text box and type 4
6 Click **OK**

Printout 5.5 displays the output that results.

PRINTOUT 5.5
Minitab output for $P(X = 4)$

```
Poisson with mu = 6.90000

          x        P( X = x)
       4.00           0.0952
```

Printout 5.5 shows that the probability is 0.0952 that exactly four patients will arrive between 6:00 P.M. and 7:00 P.M. Note that Minitab uses `mu` rather than `lambda` in specifying the parameter for the Poisson distribution.

b. For this part we want the probability of at most two arrivals, that is, two or fewer. Since that probability is a cumulative probability (see page 324), the easiest way to obtain it using Minitab is as follows.

1 Choose **Calc ➤ Probability Distributions ➤ Poisson...**
2 Select the **Cumulative probability** option button to indicate that we want to obtain a cumulative probability

3 Click in the **Mean** text box and type 6.9
4 Select the **Input constant** option button
5 Click in the **Input constant** text box and type 2
6 Click **OK**

Printout 5.6 shows the resulting output.

PRINTOUT 5.6
Minitab output for $P(X \leq 2)$

```
Poisson with mu = 6.90000

      x        P( X <= x)
   2.00          0.0320
```

From the printout we find that the probability of at most two arrivals between 6:00 P.M. and 7:00 P.M. is 0.0320.

c. Minitab can be used in several ways to obtain the probability of between four and 10 arrivals, inclusive. One way is to instruct Minitab to print the individual Poisson probabilities for the x-values 4–10. We do this by first storing those integers in a column named x and then proceeding as follows.

1 Choose **Calc ➤ Probability Distributions ➤ Poisson...**
2 Select the **Probability** option button to indicate that we want to obtain individual probabilities
3 Click in the **Mean** text box and type 6.9
4 Select the **Input column** option button
5 Click in the **Input column** text box and specify x
6 Click **OK**

The resulting output is portrayed in Printout 5.7.

PRINTOUT 5.7
Minitab output of individual Poisson probabilities for $x = 4$ to $x = 10$

```
Poisson with mu = 6.90000

      x        P( X = x)
   4.00          0.0952
   5.00          0.1314
   6.00          0.1511
   7.00          0.1489
   8.00          0.1284
   9.00          0.0985
  10.00          0.0679
```

The probability of between four and 10 arrivals is equal to the sum of the probabilities in the second column of Printout 5.7, which is 0.8214. In symbols, we have $P(4 \le X \le 10) = 0.8214$. Incidentally, more efficient ways exist of using Minitab to obtain this probability. See, for instance, Exercise 5.93. ◨

EXERCISES 5.4

STATISTICAL CONCEPTS AND SKILLS

5.70 Identify two uses of Poisson distributions.

5.71 Suppose X has a Poisson distribution with parameter $\lambda = 3$. Determine

a. $P(X = 2)$. **b.** $P(X \le 3)$.

c. $P(X > 0)$. (*Hint:* Use the complementation rule.)

d. the mean of X.

e. the standard deviation of X.

5.72 Suppose X has a Poisson distribution with parameter $\lambda = 4.7$. Find

a. $P(X = 5)$. **b.** $P(X < 2)$.

c. $P(X \ge 3)$. (*Hint:* Use the complementation rule.)

d. the mean of X.

e. the standard deviation of X.

5.73 A 1910 article, "The Probability Variations in the Distribution of α Particles," appearing in *Philosophical Magazine,* describes the results of experiments with polonium. The experiments, conducted by Ernest Rutherford and Hans Geiger, indicate that the number of α (alpha) particles reaching a small screen during an 8-minute interval has a Poisson distribution with parameter $\lambda = 3.87$. Determine the probability that, during an 8-minute interval, the number of α particles reaching the screen will be

a. exactly four. **b.** at most one.

c. between two and five, inclusive.

5.74 A paper by L. F. Richardson, published in the *Journal of the Royal Statistical Society,* analyzed the distribution of wars in time. From the data we find that the number of wars that begin during a given calendar year has approximately a Poisson distribution with parameter $\lambda = 0.7$. If a calendar year is selected at random, find the probability that the number of wars that begin during that calendar year will be

a. zero. **b.** at most two.

c. between one and three, inclusive.

5.75 M. F. Driscoll and N. A. Weiss discussed the modeling of motel reservation networks in "An Application of Queuing Theory to Reservation Networks" (*TIMS,* 22, pp. 540–546, 1976). They defined a Type 1 call to be a call from a motel's computer terminal to the national reservation center. For a certain motel, the number of Type 1 calls per hour has a Poisson distribution with parameter $\lambda = 1.7$. Determine the probability that the number of Type 1 calls made from this motel during a period of 1 hour will be

a. exactly one. **b.** at most two.

c. at least two. (*Hint:* Use the complementation rule.)

5.76 Based on past records, the owner of a fast-food restaurant knows that, on the average, 2.4 cars use the drive-through window between 3:00 P.M. and 3:15 P.M. Assuming the number of such cars has a Poisson distribution, obtain the probability that, between 3:00 P.M. and 3:15 P.M.,

a. exactly two cars will use the drive-through window.

b. at least three cars will use the drive-through window.

5.77 Refer to Exercise 5.73. Let X denote the number of α particles reaching the screen during an 8-minute interval.

a. Determine and interpret the mean of the random variable X.

b. Determine the standard deviation of X.

5.78 Refer to Exercise 5.74. Let X denote the number of wars that begin during a randomly selected calendar year.

a. Determine and interpret the mean of the random variable X.

b. Determine the standard deviation of X.

5.79 Refer to Exercise 5.75. Let X denote the number of Type 1 calls made by the motel during a 1-hour period.

a. Construct a table of probabilities for the random variable X. Compute the probabilities until they are zero to three decimal places.

b. Draw a histogram of the probabilities in part (a).

5.80 Refer to Exercise 5.76. Let X be the number of cars using the drive-through window between 3:00 P.M. and 3:15 P.M.

a. Construct a table of probabilities for the random variable X. Compute the probabilities until they are zero to three decimal places.

b. Draw a histogram of the probabilities in part (a).

In each of Exercises 5.81–5.84, determine the required probabilities by using the Poisson approximation to the binomial distribution, Procedure 5.2 on page 337.

5.81 In a letter to the editor appearing in the February 23, 1987, issue of *U.S. News and World Report,* a reader discussed the issue of space-shuttle safety. Each "criticality 1" item must have a 99.99% reliability, according to NASA standards, meaning that the probability of failure for a "criticality 1" item is only 0.0001. Mission 25, the mission in which the Challenger exploded on take-off, had 748 "criticality 1" items. Determine the probability that

a. none of the "criticality 1" items would fail.

b. at least one "criticality 1" item would fail.

5.82 An experienced and very accurate data-entry operator has a probability of 0.0002 of making an incorrect keystroke. Determine the probability that on a page containing 3680 characters the data-entry operator makes

a. no mistakes. **b.** at most one mistake.

c. at least two mistakes.

5.83 The literacy rate of a population represents the percent of the population over 15 years old that can read and write. According to the *Reader's Digest Almanac and Yearbook,* the literacy rate in Russia is 99.8%. If 1000 residents of Russia are selected at random, find the probability that the number who are illiterate is

a. exactly two. **b.** at most two. **c.** at least two.

5.84 During the second round of the 1989 U.S. Open, a strange thing happened: Four golfers made holes-in-one on the sixth hole. According to the experts, the odds against a PGA golfer making a hole-in-one are 3708-1; that is, the probability is $\frac{1}{3709}$. There were 155 golfers participating in the second round. Determine the probability that at least four of the 155 golfers would get a hole-in-one on the sixth hole.

EXTENDING THE CONCEPTS AND SKILLS

5.85 Regarding using a Poisson distribution to approximate binomial probabilities, we made the following statement on page 336: "... as we might expect, the appropriate Poisson distribution is the one whose mean is the same as that of the binomial distribution...." Explain why this makes sense.

5.86 Roughly speaking, we can use the Poisson probability formula to approximate binomial probabilities when n is large and p is small (i.e., near 0). Explain how to use the Poisson probability formula to approximate binomial probabilities when n is large and p is large (i.e., near 1).

USING TECHNOLOGY  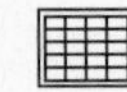 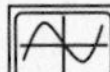

5.87 Use Minitab or some other statistical software to obtain the required probabilities in Exercise 5.73.

5.88 Use Minitab or some other statistical software to obtain the required probabilities in Exercise 5.74.

5.89 Consider the following Minitab printout showing a portion of a Poisson distribution.

```
Poisson with mu = 2.10000

        x        P( X = x)
     0.00           0.1225
     1.00           0.2572
     2.00           0.2700
     3.00           0.1890
     4.00           0.0992
     5.00           0.0417
     6.00           0.0146
     7.00           0.0044
     8.00           0.0011
     9.00           0.0003
    10.00           0.0001
```

a. What is the parameter λ for this Poisson distribution?

b. Use the printout to obtain $P(X = 3)$.
c. Use the printout to obtain $P(2 \leq X \leq 5)$.

5.90 Consider the following Minitab printout showing cumulative probabilities for a Poisson distribution.

```
Poisson with mu = 1.40000

      x        P( X <= x)
   0.00          0.2466
   1.00          0.5918
   2.00          0.8335
   3.00          0.9463
   4.00          0.9857
   5.00          0.9968
   6.00          0.9994
   7.00          0.9999
   8.00          1.0000
```

a. Identify λ for this Poisson distribution.
b. Use the printout to determine $P(x \leq 4)$, $P(x \geq 3)$, and $P(x = 2)$.

5.91 In Exercise 5.83 you used the Poisson approximation to the binomial distribution with parameters $n = 1000$ and $p = 0.002$. Use Minitab or some other statistical software to obtain both the binomial distribution and its approximating Poisson distribution. Construct a table similar to Table 5.17 on page 338 to compare the two distributions.

5.92 In Exercise 5.82 you employed the Poisson approximation to the binomial distribution with parameters $n = 3680$ and $p = 0.0002$. Use Minitab or some other statistical software to obtain both the binomial distribution and its approximating Poisson distribution. Construct a table similar to Table 5.17 on page 338 to compare the two distributions.

5.93 On page 341 we stated that there are more efficient ways to use Minitab to determine $P(4 \leq X \leq 10)$ than the method used in Example 5.19(c).
a. Explain how Minitab's cumulative probability capacity can be used to obtain $P(4 \leq X \leq 10)$ for a Poisson random variable with parameter $\lambda = 6.9$.
b. Use Minitab or some other statistical software to carry out the procedure you described in part (a).

CHAPTER REVIEW

Note: In this chapter review, we have not used asterisks for optional material because the entire chapter is optional.

You Should Be Able To

1. use and understand the formulas presented in this chapter.
2. determine the probability distribution of a discrete random variable.
3. construct a probability histogram.
4. describe events using random-variable notation, when appropriate.
5. understand the probability distribution of a random variable using the frequentist interpretation of probability.
6. find and interpret the mean and standard deviation of a discrete random variable.
7. compute factorials and binomial coefficients.
8. define and apply the concept of Bernoulli trials.
9. assign probabilities to the outcomes in a sequence of Bernoulli trials.
10. obtain binomial probabilities.
11. compute the mean and standard deviation of a binomial random variable.
12. obtain Poisson probabilities.

13. compute the mean and standard deviation of a Poisson random variable.
14. use the Poisson distribution to approximate binomial probabilities, when appropriate.
15. use the Minitab procedures covered in this chapter.
16. interpret the output obtained from the application of the Minitab procedures discussed in this chapter.

Key Terms

Bernoulli trials, *312*
binomial coefficients, *311*
binomial distribution, *312, 317*
binomial probability formula, *317*
binomial random variable, *317*
cumulative probability, *324*
discrete random variable, *291*
expectation, *303*
expected value, *303*
factorials, *311*
failure, *312*
hypergeometric distribution, *322*
law of averages, *304*
law of large numbers, *304*
mean of a discrete random variable, *303*
Poisson distribution, *333*
Poisson probability formula, *333*
Poisson random variable, *333*
probability distribution, *291*
probability histogram, *291*
random variable, *290*
standard deviation of a discrete random variable, *306*
success, *312*
success probability, *312*
trial, *310*
variance of a discrete random variable, *306*

REVIEW TEST

STATISTICAL CONCEPTS AND SKILLS

1. Fill in the blanks:
  a. A ____ is a quantitative variable whose value depends on chance.
  b. A discrete random variable is a random variable whose possible values form a ____ (or ____) set of numbers.

2. What does the probability distribution of a discrete random variable tell us?

3. How do we graphically portray the probability distribution of a discrete random variable?

4. If we sum the probabilities of the possible values of a discrete random variable, the result is always equal to ____.

5. A random variable X takes the value 2 with probability 0.386.
  a. Express that fact using probability notation.
  b. If we make repeated independent observations of the random variable X, then in approximately what percentage of those observations will we observe the value 2.
  c. Roughly how many times would we expect to observe the value 2 in 50 observations? 500 observations?

6. A random variable X has mean 3.6. If we make a large number of repeated independent observations of the random variable X, then the average value of those observations will be approximately equal to ____.

7. Two random variables, X and Y, have standard deviations 2.4 and 3.6, respectively. Which one is more likely to take a value close to its mean? Explain your answer.

8. List the three requirements for repeated trials to constitute Bernoulli trials.

9. What is the relationship between Bernoulli trials and the binomial distribution?

10. In 10 Bernoulli trials, how many outcomes contain exactly three successes?

11. Explain how the special formulas for the mean and standard deviation of a binomial or Poisson random variable are derived.

12. Suppose a simple random sample of size n is taken from a finite population in which the proportion of members having a specified attribute is p. Let X be the number of members sampled that have the specified attribute.

a. If the sampling is done with replacement, identify the probability distribution of X.

b. If the sampling is done without replacement, identify the probability distribution of X.

c. Under what conditions is it acceptable to approximate the probability distribution in part (b) by the probability distribution in part (a)? Why is it acceptable?

13. According to the *Arizona State University Main Facts Book,* a frequency distribution for the number of undergraduate students attending the main campus in fall 1997, by class level, is as shown in the following table. Here we are using the coding 1 = freshman, 2 = sophomore, 3 = junior, and 4 = senior.

Class level	No. of students
1	6,382
2	6,312
3	8,157
4	11,686

For a randomly selected ASU undergraduate, let X denote the class level of the student obtained.

a. What are the possible values of the random variable X?

b. Use random-variable notation to represent the event that the student selected is a junior (class-level 3).

c. Determine $P(X = 3)$ and interpret your answer in terms of percentages.

d. Determine the probability distribution of the random variable X.

e. Construct a probability histogram for the random variable X.

14. An accounting office has six incoming telephone lines. The probability distribution of the number of busy lines, Y, is as follows.

y	$P(Y = y)$
0	0.052
1	0.154
2	0.232
3	0.240
4	0.174
5	0.105
6	0.043

Use random-variable notation to express each of the following events. The number of busy lines is

a. exactly four. **b.** at least four.

c. between two and four, inclusive.

d. at least one.

Apply the special addition rule and the probability distribution to determine

e. $P(Y = 4)$. **f.** $P(Y \geq 4)$.

g. $P(2 \leq Y \leq 4)$. **h.** $P(Y \geq 1)$.

15. Refer to the probability distribution displayed in the table in Problem 14.

a. Find the mean of the random variable Y.

b. On the average, how many lines are busy?

c. Compute the standard deviation of Y.

d. Construct a probability histogram for Y, locate the mean, and show one, two, and three standard deviation intervals.

16. Determine 0!, 3!, 4!, and 7!.

17. Determine the value of each of the following binomial coefficients.

a. $\binom{8}{3}$ **b.** $\binom{8}{5}$ **c.** $\binom{6}{6}$ **d.** $\binom{10}{2}$

e. $\binom{40}{4}$ **f.** $\binom{100}{0}$

18. The game of craps is played by rolling two balanced dice. A first roll of a sum of 7 or 11 wins; and a first roll of a sum of 2, 3, or 12 loses. To win with any other first sum, that sum must be repeated before a sum of 7 is thrown. It can be shown that the probability is 0.493 that a player wins a game of craps. Suppose we consider a win by a player to be a success, s.

a. What is the success probability, p?

b. Construct a table showing the possible win-lose results and their probabilities for three games of craps. Round each probability to three decimal places.
c. Draw a tree diagram.
d. List the outcomes in which the player wins exactly two out of three times.
e. Determine the probability of each of the outcomes in part (d). Explain why those probabilities are equal.
f. Find the probability that the player wins exactly two out of three times.
g. Without using the binomial probability formula, obtain the probability distribution of the random variable Y, the number of times out of three that the player wins.
h. Identify the probability distribution in part (g).

19. The nation of Surinam is located on the northern coast of South America. According to the *World Almanac,* 80% of the population is literate. Use the binomial probability formula to find the probability that the number of literate people in a random sample of four is
a. exactly three. **b.** at most three.
c. at least three.
d. Determine the probability distribution of the random variable X, the number of literate Surinamese in a random sample of four.
e. Without referring to the probability distribution obtained in part (d) or constructing a probability histogram, decide whether the probability distribution is right skewed, symmetric, or left skewed. Explain your answer.
f. Draw a probability histogram for X.
g. Strictly speaking, why is the probability distribution that you obtained in part (d) only approximately correct? What is the exact distribution called?
h. Determine and interpret the mean of the random variable X.
i. Determine the standard deviation of X.

20. Following are two probability histograms of binomial distributions. For each one, specify whether the success probability is less than, equal to, or greater than 0.5.

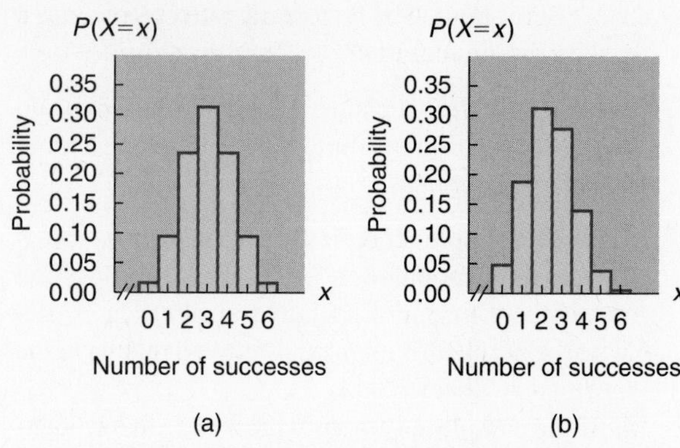

21. A classic study by F. Thorndike on the number of calls to a wrong number appeared in the paper "Applications of Poisson's Probability Summation (*Bell Systems Technical Journal,* 5, pp. 604–624, 1926). The study examined the number of calls to a wrong number from coin-box telephones in a large transportation terminal. According to the paper, the number of calls to a wrong number, X, in a 1-minute period has a Poisson distribution with parameter $\lambda = 1.75$. Find the probability that during a 1-minute period the number of calls to a wrong number is
a. exactly two.
b. between four and six, inclusive.
c. at least one.
d. Obtain a table of probabilities for X, stopping when the probabilities become zero to three decimal places.
e. Use part (d) to construct a partial probability histogram for the random variable X.
f. Identify the shape of the probability distribution of X. Is this shape typical of Poisson distributions?

22. Refer to Problem 21.
a. Determine and interpret the mean of the random variable X.
b. Determine the standard deviation of X.

23. The probability is approximately 0.00024 of being dealt four of a kind in a hand of five-card poker.
a. In 10,000 hands of five-card poker, roughly how many times would you expect to be dealt four of a kind?

b. Use the Poisson approximation to the binomial distribution to find the probability of being dealt four of a kind exactly twice in 10,000 hands of five-card poker.

c. Use the Poisson approximation to the binomial distribution to find the probability of being dealt four of a kind at least twice in 10,000 hands of five-card poker.

USING TECHNOLOGY  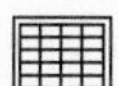 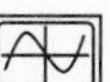

24. Refer to the probability distribution obtained in Problem 13(d).

a. Use Minitab or some other statistical software to simulate 2500 observations of the class level of a randomly selected undergraduate at ASU.

b. Obtain the proportions for the 2500 class levels simulated in part (a) and compare it to the probability distribution obtained in Problem 13(d).

c. Construct a histogram of the proportions and compare it to the probability histogram obtained in Problem 13(e).

d. What do parts (b) and (c) illustrate?

25. Refer to the probability distribution displayed in the table in Problem 14.

a. Use Minitab or some other statistical software to simulate 200 observations of the number of busy lines.

b. Obtain the mean of the observations in part (a), and compare it to μ, found in Problem 15(a).

c. What is part (b) illustrating?

26. Use Minitab or some other statistical software to obtain the required probability or probabilities in parts (a)–(d) of Problem 19.

27. The following is a Minitab printout showing a binomial distribution.

```
Binomial with n = 5 and p = 0.650000

     x         P( X = x)
  0.00           0.0053
  1.00           0.0488
  2.00           0.1811
  3.00           0.3364
  4.00           0.3124
  5.00           0.1160
```

Use the printout to determine the

a. number of trials.

b. success probability.

c. probability of exactly one success.

d. probability of between one and three successes, inclusive.

e. probability of at most one success.

f. probability of at least one success.

28. The following is a Minitab printout showing cumulative probabilities for a binomial distribution.

```
Binomial with n = 8 and p = 0.570000

     x         P( X <= x)
  0.00           0.0012
  1.00           0.0136
  2.00           0.0711
  3.00           0.2235
  4.00           0.4762
  5.00           0.7440
  6.00           0.9216
  7.00           0.9889
  8.00           1.0000
```

From the printout, obtain the

a. number of trials.

b. success probability.

c. probability of at most four successes.

d. probability of at least four successes. *(Hint:* Use the complementation rule.*)*

e. probability of exactly four successes. *(Hint:* First apply the special addition rule to express $P(X = 4)$ as the difference of two cumulative probabilities.*)*

29. The March, 1998, issue of *Consumer Reports* revealed that about 71% of chickens bought in retail markets are contaminated with some disease-causing bacteria such as campylobacter bacteria or salmonella. For seven randomly selected chickens, let X denote the number contaminated with some disease-causing bacteria. The Minitab printout presented below provides the probability distribution of the random variable X.

```
Binomial with n = 7 and p = 0.710000

      x          P( X = x)
   0.00           0.0002
   1.00           0.0030
   2.00           0.0217
   3.00           0.0886
   4.00           0.2169
   5.00           0.3186
   6.00           0.2600
   7.00           0.0910
```

From the printout, determine the probability that of the seven chickens chosen, the number contaminated with some disease-causing bacteria is

a. exactly two. **b.** at least two.

c. fewer than two.

d. between two and five, inclusive.

30. Use Minitab or some other statistical software to obtain the required probability or probabilities in parts (a)–(d) of Problem 21.

31. The following Minitab printout shows cumulative probabilities for a Poisson distribution.

```
Poisson with mu = 2.40000

      x         P( X <= x)
   0.00           0.0907
   1.00           0.3084
   2.00           0.5697
   3.00           0.7787
   4.00           0.9041
   5.00           0.9643
   6.00           0.9884
   7.00           0.9967
   8.00           0.9991
   9.00           0.9998
  10.00           1.0000
```

Use the printout to

a. identify the parameter λ for the Poisson random variable, X, under consideration.

b. obtain $P(X = 2)$. **c.** obtain $P(X \geq 2)$.

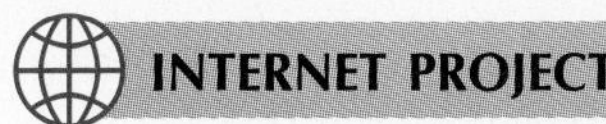

INTERNET PROJECT

Racial Bias in Jury Selection

We all have our biases—humans inevitably evaluate new circumstances based on their past experience and knowledge. These biases can effect us in systematic ways.

In this Internet project, you will examine the possibility of racial bias in jury selection by looking at two actual court cases tried in the United States. In each case, an African-American was on trial and had a jury with no African-Americans seated. One trial was in Monroe County, Alabama, which has a population of about 24,000. The proportion of whites in that county is 0.58, or 58%; the proportion of blacks is 0.40, or 40%.

For the two court cases, the possibility of racial bias in the selection of the juries was raised on their appeal and, in both instances, the higher court determined that there was insufficient evidence of racial bias. The defendant in each court case was convicted and given the death sentence.

If 40% of the residents of a community are African-Americans, what is the chance of randomly selecting 12 people from the community and obtaining no African-Americans? This question is an oversimplification of the facts underlying the cases, but it is a part of the evidence that was considered in the aftermath of both convictions. In this project you will be able to look deeper into the legal labyrinth of jury selection.

URL for access to Internet Projects Page: `http://hepg.awl.com` *keyword:* Weiss

USING THE FOCUS DATABASE

In Chapter 1 we explained how to store the Focus database in a Minitab worksheet named `focus.mtw`. If you haven't already created that worksheet, follow the instructions on pages 54–55 to create it now.

The Focus database contains information on 500 randomly selected Arizona State University sophomores for seven different variables: sex, high-school GPA, SAT math score, cumulative GPA, SAT verbal score, age, and total hours. Use Minitab or some other statistical software to solve the following problems.

a. Let X denote the age of a randomly selected sophomore from the sample of 500. Obtain the probability distribution of the random variable X.

b. Obtain a probability histogram or similar graphic for the random variable in part (a).

c. Determine the mean and standard deviation of the random variable X defined in part (a).

d. Consider the experiment of randomly selecting 10 sophomores with replacement from the sample of 500 and observing the number of those selected who are 21 years old. Simulate that experiment 2000 times.

e. Referring to the simulation from part (d), in approximately what percentage of the 2000 experiments would you expect exactly three of the 10 sophomores selected to be 21 years old? Compare that percentage to the actual percentage of the 2000 experiments in which exactly three of the 10 sophomores selected are 21 years old.

CASE STUDY DISCUSSION

Aces Wild on the Sixth at Oak Hill

As we discovered at the beginning of this chapter, on June 16, 1989, during the second round of the 1989 U.S. Open, four golfers—Doug Weaver, Mark Wiebe, Jerry Pate, and Nick Price—made holes-in-one on the sixth hole at Oak Hill in Pittsford, New York. Now that you have studied the material in this chapter, you can determine for yourself the likelihood of such an event.

Here are the relevant data: According to the experts, the odds against a professional golfer making a hole-in-one are 3708-1; in other words, the probability is $\frac{1}{3709}$ that a professional golfer will make a hole-in-one. There were 155 golfers participating in the second round.

a. Determine the probability that at least four of the 155 golfers would get a hole-in-one on the sixth hole.

b. What assumptions did you make in solving part (a)? Do those assumptions seem reasonable?

c. Use Minitab or some other statistical software to solve part (a).

BIOGRAPHY JAMES BERNOULLI

James Bernoulli was born on December 27, 1654, in Basle, Switzerland. He was the first of the Bernoulli family of mathematicians; his younger brother John and various nephews and grandnephews were also renowned mathematicians. His father, Nicolaus Bernoulli (1623–1708), planned the ministry as James's career. James rebelled, however, to him, mathematics was much more interesting.

Although Bernoulli was schooled in theology, he studied mathematics on his own. He was especially fascinated with calculus. In a 1690 issue of the journal *Acta eruditorum,* Bernoulli used the word *integral* to describe the inverse of differential. The results of his studies of calculus and the catenary (the curve formed by a cord freely suspended between two fixed points) were soon applied to the building of suspension bridges.

Some of Bernoulli's most important work was published posthumously in *Ars Conjectandi* (The Art of Conjecturing) in 1713. This book contains his theory of permutations and combinations, the Bernoulli numbers, and his writings on probability, which include the weak law of large numbers for Bernoulli trials. *Ars Conjectandi* has been regarded as the beginning of the theory of probability.

Both James and his brother John were highly accomplished mathematicians. But rather than collaborating in their work, they were most often competing. James would publish a question inviting solutions in a professional journal. John would reply in the same journal with a solution, only to find that an ensuing issue would contain another article by James, telling him that he was wrong. In their later years, they communicated only in this manner.

Bernoulli began lecturing in natural philosophy and mechanics at the University of Basle in 1682 and became a Professor of Mathematics there in 1687. He remained at the university until his death of a "slow fever" on August 10, 1705.

CASE STUDY

CHEST SIZES OF SCOTTISH MILITIAMEN

In 1817, an article entitled "Statement of the Sizes of Men in Different Counties of Scotland, Taken from the Local Militia" appeared in the *Edinburgh Medical and Surgical Journal* (13, pp. 260–264). Included in the article were data on chest circumference by height for 5732 Scottish militiamen. The data were collected by an army contractor who was responsible for providing clothing for the militia. A frequency distribution for the chest circumferences is given in the following table.

Chest size	Frequency	Chest size	Frequency
33	3	41	935
34	19	42	646
35	81	43	313
36	189	44	168
37	409	45	50
38	753	46	18
39	1062	47	3
40	1082	48	1

In his book *Lettres à S.A.R. le Duc Régnant de Saxe-Cobourg et Gotha sur la théorie des probabilités appliquée aux sciences morales et politiques* (Brussels: Hayez, 1846), Adolphe Quetelet discussed a procedure for fitting a normal curve to the data on chest circumferences. The method he used was based on the binomial distribution.

At the end of this chapter, you will be asked to fit a normal curve to the data using different techniques, ones that you will learn shortly.

The Normal Distribution

6

GENERAL OBJECTIVES

In this chapter we will discuss the most important distribution in statistics—the **normal distribution.** Its importance lies in the fact that it appears again and again in both theory and practice.

The normal distribution is used as a model for a variety of physical measurements since it has been discovered that many such measurements have distributions that are normally distributed or at least approximately so. And, as we will see throughout this book, the normal distribution is applied frequently to make statistical inferences such as estimating parameters and conducting hypothesis tests.

CHAPTER OUTLINE

6.1 INTRODUCING NORMALLY DISTRIBUTED VARIABLES

In the world around us, we observe a wide variety of variables. Many are intrinsically different. But some—such as aptitude-test scores, heights of women, and wheat yield—share an important characteristic: their distributions have roughly the shape of a **normal curve,** that is, a special type of bell-shaped curve like the one shown in Fig. 6.1.

FIGURE 6.1
A normal curve

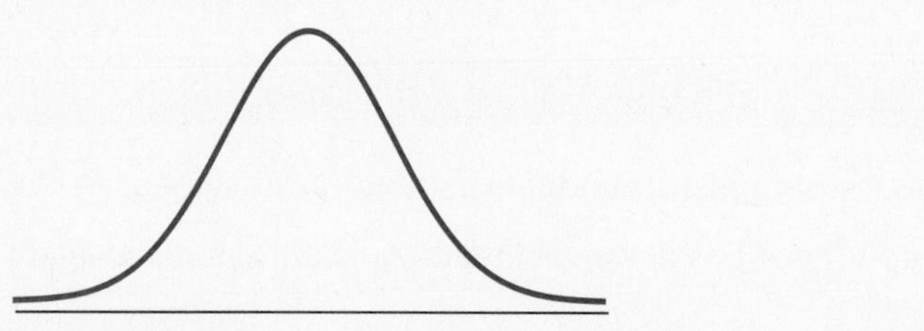

DEFINITION 6.1 NORMALLY DISTRIBUTED VARIABLE

A variable is said to be ***normally distributed*** or to have a ***normal distribution*** if its distribution has the shape of a normal curve.

Two items of importance relative to normal distributions are as follows.

- If a variable of a population is normally distributed and is the only variable under consideration, then it has become common statistical practice to say that the **population is normally distributed** or that we have a **normally distributed population.**
- In practice it is unusual for a distribution to have exactly the shape of a normal curve. If a variable's distribution is shaped roughly like a normal curve, then we say that the variable is **approximately normally distributed** or has **approximately a normal distribution.**

A normal distribution (and hence a normal curve) is completely determined by the mean and standard deviation; that is, two normally distributed variables having the same mean and standard deviation must have the same distribution. We often identify a normal curve by stating the corresponding mean and standard deviation and calling those the **parameters** of the normal curve.†

† The equation of the normal curve with parameters μ and σ is $y = e^{-(x-\mu)^2/2\sigma^2}/\sqrt{2\pi}\sigma$, where $e \approx 2.718$ and $\pi \approx 3.142$.

A normal distribution is symmetric about and centered at the mean of the variable and its spread depends on the standard deviation of the variable—the larger the standard deviation, the flatter and more spread out the distribution. Figure 6.2 displays three normal distributions.

FIGURE 6.2
Three normal distributions

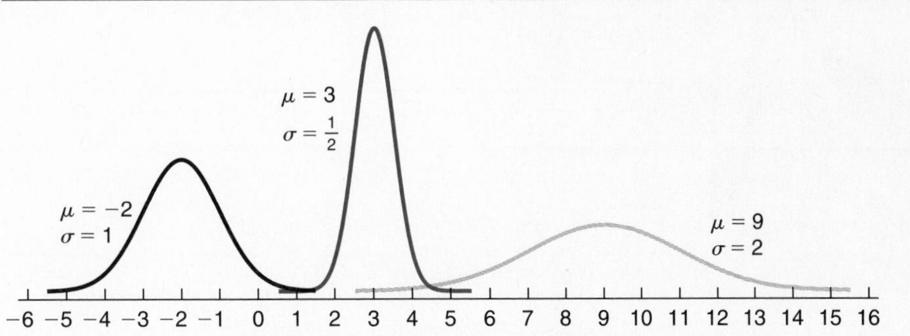

The three-standard-deviations rule, when applied to a variable, states that almost all of the possible observations of the variable lie within three standard deviations to either side of the mean. This rule is reflected by the three normal distributions in Fig. 6.2 through the fact that each normal curve is close to the horizontal axis outside of the range within three standard deviations to either side of the mean.

For instance, the third normal distribution in Fig. 6.2 has mean $\mu = 9$ and standard deviation $\sigma = 2$. Three standard deviations below (to the left of) the mean is

$$\mu - 3\sigma = 9 - 3 \cdot 2 = 3$$

and three standard deviations above (to the right of) the mean is

$$\mu + 3\sigma = 9 + 3 \cdot 2 = 15.$$

As we see from Fig. 6.2, the corresponding normal curve is close to the horizontal axis outside of the range from 3 to 15.

Conversely, keeping in mind what we have just learned helps us to sketch a normal distribution. Its associated normal curve

- is bell-shaped,
- is centered at μ, and
- is close to the horizontal axis outside of the range from $\mu - 3\sigma$ to $\mu + 3\sigma$.

See Figs. 6.2 and 6.3.

FIGURE 6.3
Graph of generic normal distribution

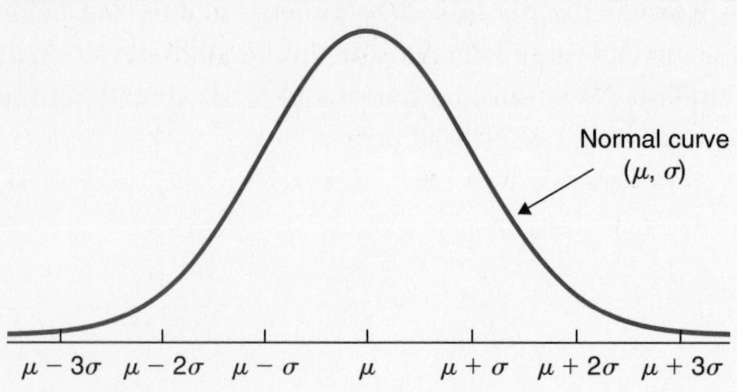

EXAMPLE 6.1 A NORMALLY DISTRIBUTED VARIABLE

A midwestern college has an enrollment of 3264 female students. Records show that the mean height of these students is 64.4 inches and that the standard deviation is 2.4 inches. Here the variable is height and the population consists of the 3264 female students attending the college.

Frequency and relative-frequency distributions for these heights are presented in Table 6.1. The table shows, for instance, that 7.35% (0.0735) of the students are between 67 and 68 inches tall.

TABLE 6.1
Frequency and relative-frequency distributions for heights

Height (in.)	Frequency f	Relative frequency
56 < 57	3	0.0009
57 < 58	6	0.0018
58 < 59	26	0.0080
59 < 60	74	0.0227
60 < 61	147	0.0450
61 < 62	247	0.0757
62 < 63	382	0.1170
63 < 64	483	0.1480
64 < 65	559	0.1713
65 < 66	514	0.1575
66 < 67	359	0.1100
67 < 68	240	0.0735
68 < 69	122	0.0374
69 < 70	65	0.0199
70 < 71	24	0.0074
71 < 72	7	0.0021
72 < 73	5	0.0015
73 < 74	1	0.0003
	3264	1.0000

A relative-frequency histogram for the heights of the 3264 female students is included in Fig. 6.4. It shows that the distribution of heights has roughly the shape of a normal curve and, consequently, that the variable height is approximately normally distributed for this population. The associated normal curve, the one with parameters $\mu = 64.4$ and $\sigma = 2.4$, is superimposed on the histogram in Fig. 6.4.

FIGURE 6.4
Relative-frequency histogram for heights with superimposed normal curve

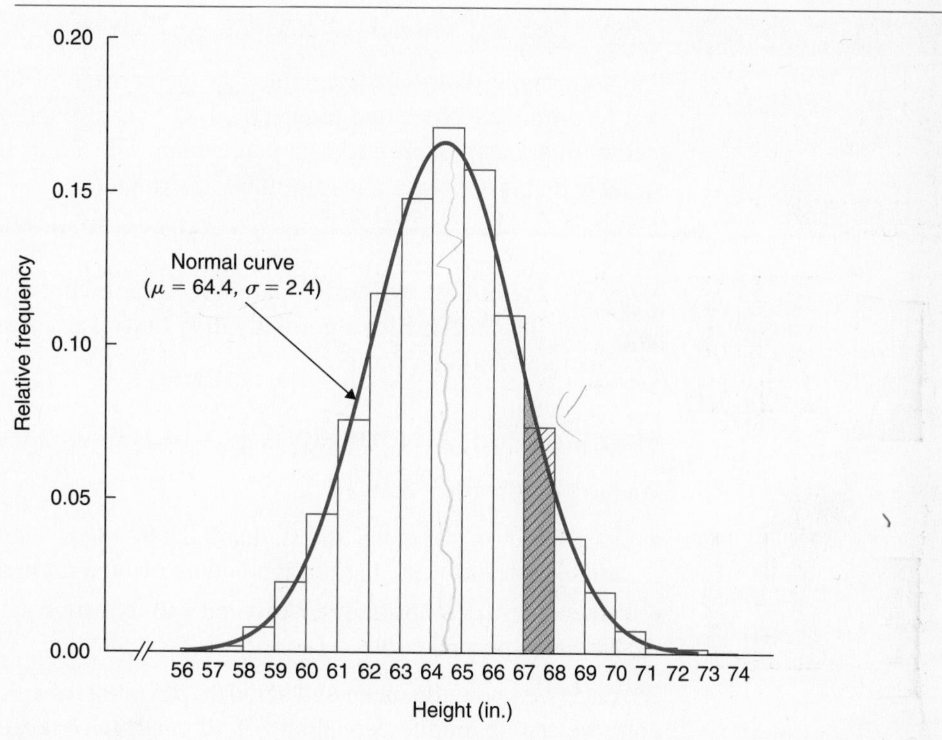

Studying Fig. 6.4, we notice another important fact, namely, that the percentage of female students whose heights lie within any specified range can be approximated by the corresponding area under the normal curve. For instance, consider the percentage of female students who are between 67 and 68 inches tall. According to Table 6.1, the exact percentage is 7.35% (0.0735). Note that 0.0735 also equals the area of the cross-hatched bar in Fig. 6.4 because the bar has height 0.0735 and width 1.

Now look at the area under the normal curve between 67 and 68, the area shaded in Fig. 6.4. Observe that this area is approximately equal to the area of the cross-hatched bar which, as we have noted, equals the percentage of students who are between 67 and 68 inches tall. Consequently, we can approximate the percentage of students who are between 67 and 68 inches tall by the area under the normal curve between 67 and 68. □

Key Fact 6.1 summarizes the important relationship that we discovered in Example 6.1 between percentages for a normally distributed variable and areas under its associated normal curve.

KEY FACT 6.1 **NORMALLY DISTRIBUTED VARIABLES AND NORMAL-CURVE AREAS**

For a normally distributed variable, the percentage of all possible observations that lie within any specified range equals the corresponding area under its associated normal curve expressed as a percentage. This holds true approximately for a variable that is approximately normally distributed.

Note: For brevity, we will often paraphrase the content of Key Fact 6.1 with the statement "percentages for a normally distributed variable are equal to areas under its associated normal curve."

Standardizing a Normally Distributed Variable

We have now learned that:

- Once we know the mean and standard deviation of a normally distributed variable, we know its distribution and associated normal curve.
- Percentages for a normally distributed variable are equal to areas under its associated normal curve.

So, once we know the mean and standard deviation of a normally distributed variable, we can obtain the percentage of all possible observations that lie within any specified range by determining the corresponding area under its associated normal curve. Now the question is: How do we find areas under a normal curve?

Conceptually, we need a table of areas for each normal curve. This, of course, is not possible because there are infinitely many different normal curves—one for each choice of μ and σ. The way out of this difficulty is standardizing which, as we will see, transforms every normal distribution into one particular normal distribution.

DEFINITION 6.2 **STANDARD NORMAL DISTRIBUTION; STANDARD NORMAL CURVE**

A normally distributed variable having mean 0 and standard deviation 1 is said to have the ***standard normal distribution.*** Its associated normal curve is called the ***standard normal curve.*** See Fig. 6.5.

FIGURE 6.5 Standard normal distribution

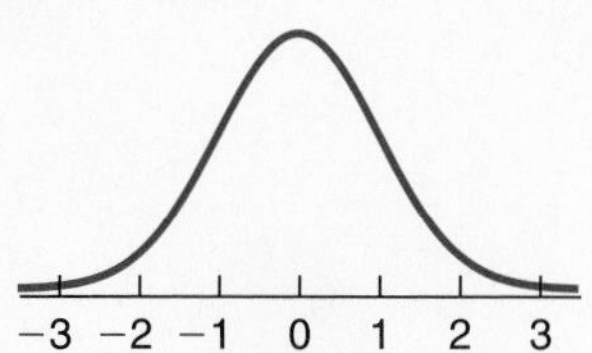

Recall that we standardize a variable x by subtracting its mean and then dividing by its standard deviation. The resulting variable, $z = (x - \mu)/\sigma$, is called the *standardized version* of x or the *standardized variable* corresponding to x and always has mean 0 and standard deviation 1. It turns out that if a variable is normally distributed, then so is its standardized version. Thus we have the following essential fact.

KEY FACT 6.2 **STANDARDIZED NORMALLY DISTRIBUTED VARIABLE**

The standardized version of a normally distributed variable x,

$$z = \frac{x - \mu}{\sigma},$$

has the standard normal distribution.

We can interpret Key Fact 6.2 in several ways. Theoretically it says that standardizing converts all normal distributions into the standard normal distribution, as depicted in Fig. 6.6 on the following page.

But here is a more practical interpretation of Key Fact 6.2. Let x be a normally distributed variable and a and b numbers. The percentage of all possible observations of x that lie between a and b is the same as the percentage of all possible observations of z that lie between $(a - \mu)/\sigma$ and $(b - \mu)/\sigma$. And, in view of Key Fact 6.2, this latter percentage equals the area under the standard normal curve between $(a - \mu)/\sigma$ and $(b - \mu)/\sigma$. Figure 6.7 on the next page summarizes this paragraph graphically.

Consequently, for a normally distributed variable, we can find the percentage of all possible observations that lie within any specified range using the following two steps:

1. Express the range in terms of z-scores.
2. Obtain the corresponding area under the standard normal curve.

We already know how to convert to z-scores. Therefore, we need only learn how to find areas under the standard normal curve, which we will do in Section 6.2.

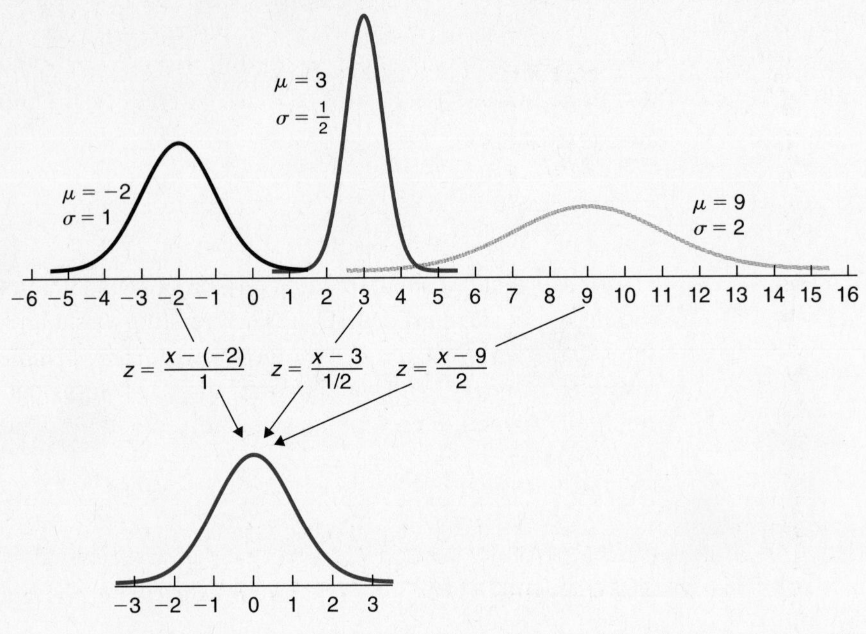

FIGURE 6.6 Standardizing normal distributions

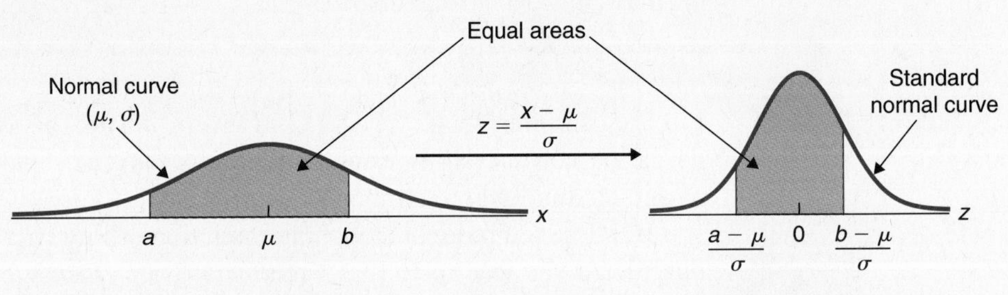

FIGURE 6.7 Finding percentages for a normally distributed variable from areas under the standard normal curve

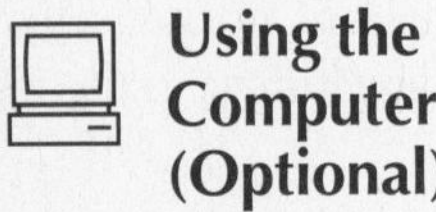

Using the Computer (Optional)

It is often useful, both for purposes of understanding and research, to simulate variables. Simulating a variable means that we use a computer or statistical calculator to generate observations of the variable. Here we will show how Minitab can be used to simulate a normally distributed variable.

EXAMPLE 6.2 USING MINITAB TO SIMULATE A NORMALLY DISTRIBUTED VARIABLE

Gestation periods of humans are normally distributed with a mean of 266 days and a standard deviation of 16 days. Use Minitab to simulate 1000 human gestation periods and then obtain a histogram of the results.

SOLUTION Here the variable x is gestation period and, for humans, it is normally distributed with mean $\mu = 266$ days and standard deviation $\sigma = 16$ days. We want to use Minitab to simulate 1000 observations of the variable x for humans. To do that and store the results in a column named DAYS, we proceed as follows.

1 Choose **Calc ➤ Random Data ➤ Normal...**
2 Type 1000 in the **Generate rows of data** text box
3 Click in the **Store in column(s)** text box and type DAYS
4 Click in the **Mean** text box and type 266
5 Click in the **Standard deviation** text box and type 16
6 Click **OK**

The procedure above produces 1000 (simulated) human gestation periods that are stored in DAYS. To obtain a histogram for the contents, we apply Minitab's histogram procedure, as explained in Section 2.3. The computer output we actually got by performing the simulation and applying Minitab's histogram procedure is shown in Printout 6.1. For purposes of comparison, we have superimposed the normal curve associated with the variable, namely, the normal curve with parameters $\mu = 266$ and $\sigma = 16$.

PRINTOUT 6.1
Minitab histogram of 1000 simulated human gestation periods

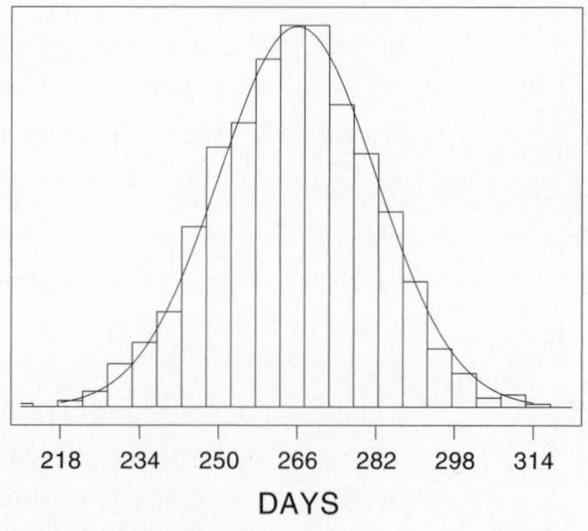

As expected, the histogram in Printout 6.1 is shaped roughly like the normal curve associated with the variable. Since we have only sample data here, we would not expect the histogram to be shaped exactly like the normal curve. But since the sample size is large, we do expect it to be close to that shape, which it is.

 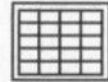 

If you do the simulation, you will (almost) certainly obtain different results than those depicted in Printout 6.1; but your results should be similar. ◨

EXERCISES 6.1

STATISTICAL CONCEPTS AND SKILLS

6.1 A variable is approximately normally distributed. If we draw a histogram of the distribution of the variable, roughly what shape will it have?

6.2 Precisely what do we mean by saying that a population is normally distributed?

6.3 Two normally distributed variables have the same means and the same standard deviations. What can you say about their distributions? Explain your answer.

6.4 Which normal distribution has a wider spread: the one with mean 1 and standard deviation 2 or the one with mean 2 and standard deviation 1? Explain your answer.

6.5 True or false: The normal distribution having mean −4 and standard deviation 3 and the normal distribution having mean 6 and standard deviation 3 have the same shape. Explain your answer.

6.6 True or false: The normal distribution having mean −4 and standard deviation 3 and the normal distribution having mean 6 and standard deviation 3 are centered at the same place. Explain your answer.

6.7 True or false: The mean of a normal distribution has no effect on its shape. Explain your answer.

6.8 What are the parameters for a normal curve?

6.9 Sketch the normal distribution with

a. $\mu = 3$ and $\sigma = 3$.
b. $\mu = 1$ and $\sigma = 3$.
c. $\mu = 3$ and $\sigma = 1$.

6.10 Sketch the normal distribution with

a. $\mu = -2$ and $\sigma = 2$.
b. $\mu = -2$ and $\sigma = \frac{1}{2}$.
c. $\mu = 0$ and $\sigma = 2$.

6.11 For a normally distributed variable, what is the relationship between the percentage of all possible observations that lie between 2 and 3 and the area under the associated normal curve between 2 and 3? What if the variable is only approximately normally distributed?

6.12 The area under a particular normal curve between 10 and 15 is 0.6874. A normally distributed variable has the same mean and standard deviation as the parameters for this normal curve. What percentage of all possible observations of the variable lie between 10 and 15? Explain your answer.

6.13 Refer to Example 6.1 on page 356.

a. Use the relative-frequency distribution in Table 6.1 to obtain the percentage of female students who are between 60 and 65 inches tall.

b. Use your answer from part (a) to estimate the area under the normal curve with parameters $\mu = 64.4$ and $\sigma = 2.4$ that lies between 60 and 65. Why do you only get an estimate of the true area?

6.14 Refer to Example 6.1 on page 356.

a. The area under the standard normal curve with parameters $\mu = 64.4$ and $\sigma = 2.4$ that lies to the left of 61 is 0.0783. Use this information to estimate the percentage of female students who are shorter than 61 inches.

b. Use the relative-frequency distribution in Table 6.1 to obtain the exact percentage of female students who are shorter than 61 inches.

c. Compare your answers to parts (a) and (b).

6.15 As reported by *Runner's World* magazine, the times of the finishers in the New York City 10-km run are normally distributed with a mean of 61 minutes and a standard deviation of 9 minutes. Let x denote finishing time for finishers in the New York City 10-km run.

a. Sketch the distribution of the variable x.

b. Obtain the standardized version, z, of x.

c. Identify and sketch the distribution of z.

d. Fill in the following blanks: The percentage of finishers in the New York City 10-km run with times between 50 and 70 minutes is equal to the area under the standard normal curve between ____ and ____.

e. Fill in the following blanks: The percentage of finishers in the New York City 10-km run with times exceeding 75 minutes is equal to the area under the standard normal curve that lies to the ____ of ____.

6.16 The length of the western rattlesnake is normally distributed with a mean of 42 inches and a standard deviation of 2 inches. Let x denote length for western rattlesnakes.

a. Sketch the distribution of the variable x.

b. Obtain the standardized version, z, of x.

c. Identify and sketch the distribution of z.

d. Fill in the following blanks: The percentage of western rattlesnakes having lengths between 35 inches and 40 inches is equal to the area under the standard normal curve between _____ and _____.

e. Fill in the following blanks: The percentage of western rattlesnakes having lengths less than 45 inches is equal to the area under the standard normal curve that lies to the _____ of _____.

Extending the Concepts and Skills

6.17 Use the footnote on page 354 to write the equation of the

a. associated normal curve of a normally distributed variable having mean 5 and standard deviation 2.

b. standard normal curve.

6.18 This exercise verifies Key Fact 6.2, namely, that the standardized version of a normally distributed variable has the standard normal distribution. The verification requires a knowledge of elementary calculus. Let x be a normally distributed variable having mean μ and standard deviation σ, and let $z = (x - \mu)/\sigma$ be the standardized version of x.

a. Show that the percentage of all possible observations of z that lie between a and b equals the percentage of all possible observations of x that lie between $\mu + a\sigma$ and $\mu + b\sigma$.

b. Explain why the percentage of all possible observations of x that lie between $\mu + a\sigma$ and $\mu + b\sigma$ equals

$$\int_{\mu+a\sigma}^{\mu+b\sigma} \frac{1}{\sqrt{2\pi}\,\sigma} e^{-(x-\mu)^2/2\sigma^2}\,dx$$

expressed as a percentage. (*Hint:* See the footnote on page 354.)

c. Making the substitution $z = (x - \mu)/\sigma$, show that the integral in part (b) equals

$$\int_a^b \frac{1}{\sqrt{2\pi}} e^{-z^2/2}\,dz.$$

d. Conclude from parts (a)–(c) that z has the standard normal distribution. (*Hint:* What area does the integral in part (c) equal?)

Using Technology   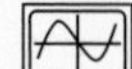

6.19 Using any technology available to you, graph the normal distribution with mean 5 and standard deviation 2.

6.20 Using any technology available to you, graph the standard normal distribution.

6.21 Refer to the simulation of human gestation periods discussed in Example 6.2 on page 360.

a. Sketch the normal curve for human gestation periods.

b. Simulate 1000 human gestation periods.

c. Approximately what values would you expect for the sample mean and sample standard deviation of the 1000 observations? Explain your answers.

d. Obtain the sample mean and sample standard deviation of the 1000 observations and compare your answers to your estimates in part (c).

e. Roughly what would you expect a histogram of the 1000 observations to look like? Explain your answer.

f. Obtain a histogram of the 1000 observations and compare your result to your expectation in part (e).

6.22 Refer to Exercise 6.15.

a. Sketch the normal curve for the finishing times.

b. Simulate 1500 finishing times.

c. Approximately what values would you expect for the sample mean and sample standard deviation of the 1500 observations? Explain your answers.

d. Obtain the sample mean and sample standard deviation of the 1500 observations and compare your answers to your estimates in part (c).

e. Roughly what would you expect a histogram of the 1500 observations to look like? Explain your answer.

f. Obtain a histogram of the 1500 observations and compare your result to your expectation in part (e).

6.2 AREAS UNDER THE STANDARD NORMAL CURVE

In Section 6.1 we introduced normally distributed variables. Among other things, we discovered the following: For a normally distributed variable, we can obtain the percentage of all possible observations that lie within any specified range by first converting to z-scores and then determining the corresponding area under the standard normal curve.

We have already learned how to convert to z-scores. Now, in this section, we will find how to implement the second step—determining areas under the standard normal curve.

Basic Properties of the Standard Normal Curve

Before we discover how to find areas under the standard normal curve, it will be worthwhile to discuss some of its basic properties. Recall that the standard normal curve is the curve associated with the standard normal distribution, that is, the normal distribution having mean 0 and standard deviation 1. Figure 6.8 shows the standard normal distribution and the standard normal curve.

FIGURE 6.8 Standard normal distribution and standard normal curve

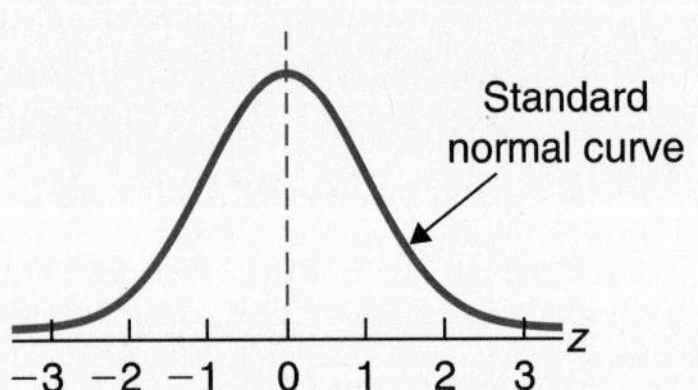

We learned in Section 6.1 that a normal curve is bell-shaped, is centered at μ, and is close to the horizontal axis outside the range from $\mu - 3\sigma$ to $\mu + 3\sigma$. Applied to the standard normal curve, this means that it is bell-shaped, is centered at 0, and is close to the horizontal axis outside the range from −3 to 3. From these facts, we also see that the standard normal curve is symmetric about 0. All of these properties are reflected in Fig. 6.8.

One property of the standard normal curve that is not obvious from Fig. 6.8 is that the total area under the curve is 1. This property is not unique to the standard normal curve; in fact, the total area under any curve that represents the distribution of a variable is equal to 1. Key Fact 6.3 summarizes our discussion about the properties of the standard normal curve.

KEY FACT 6.3 **BASIC PROPERTIES OF THE STANDARD NORMAL CURVE**

Property 1: The total area under the standard normal curve is equal to 1.

Property 2: The standard normal curve extends indefinitely in both directions, approaching, but never touching, the horizontal axis as it does so.

Property 3: The standard normal curve is symmetric about 0; that is, the part of the curve to the left of the dashed line in Fig. 6.8 is the mirror image of the part of the curve to the right of it.

Property 4: Most of the area under the standard normal curve lies between -3 and 3.

Because the standard normal curve is the associated normal curve for a standardized normally distributed variable, we have labeled the horizontal axis in Fig. 6.8 with the letter z and will refer to numbers on that axis as z-scores. For these reasons the standard normal curve is sometimes called the **z-curve.**

Using the Standard-Normal Table (Table II)

Because of the importance of areas under the standard normal curve, tables of those areas have been constructed. Such a table is Table II, which can be found inside the front cover of this book as well as in Appendix A.

A typical four-decimal-place number in the body of Table II gives the area under the standard normal curve that lies to the left of a specified z-score. The left page of Table II is for negative z-scores and the right page is for positive z-scores.

EXAMPLE 6.3 **FINDING THE AREA TO THE LEFT OF A SPECIFIED z-SCORE**

Determine the area under the standard normal curve that lies to the left of 1.23, as shown in Fig. 6.9(a).

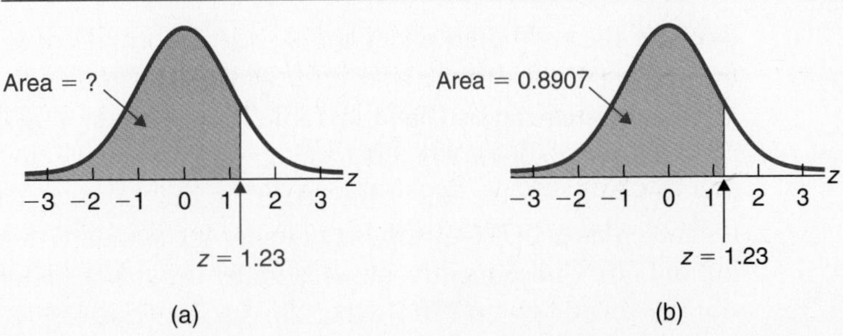

FIGURE 6.9 Finding the area under the standard normal curve to the left of 1.23

SOLUTION We use Table II, specifically the portion on the right page, since 1.23 is positive. First we go down the left-hand column, labeled z, to "1.2." Then we go across that row until we are under the "0.03" in the top row. The number in the body of the table there, 0.8907, is the area under the standard normal curve that lies to the left of 1.23. This area is shown in Fig. 6.9(b).

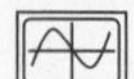

Finding the area under the standard normal curve that lies to the left of a specified z-score is one important use of Table II. Two other important uses of that table are finding the area to the right of a specified z-score and finding the area between two specified z-scores. We illustrate these two uses in Examples 6.4 and 6.5, respectively.

EXAMPLE 6.4 FINDING THE AREA TO THE RIGHT OF A SPECIFIED z-SCORE

Determine the area under the standard normal curve that lies to the right of 0.76, as seen in Fig. 6.10(a).

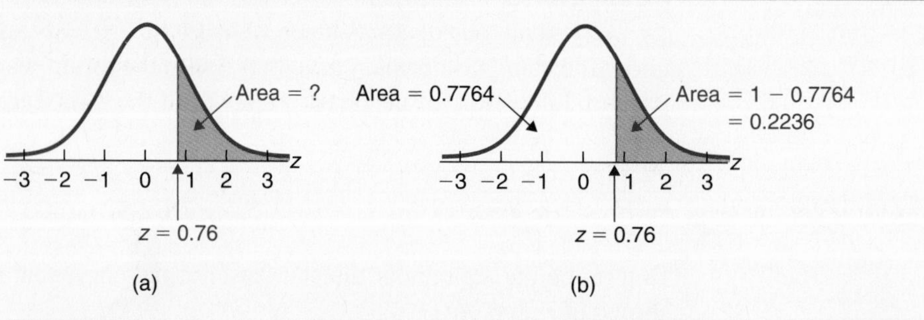

FIGURE 6.10 Finding the area under the standard normal curve to the right of 0.76

SOLUTION Because the total area under the standard normal curve is 1 (Property 1 of Key Fact 6.3), the area to the right of 0.76 equals 1 minus the area to the left of 0.76. This latter area can be found in Table II, as explained in Example 6.3.

First we go down the left-hand column, labeled z, to "0.7." Then we go across that row until we are under the "0.06" in the top row. The number in the body of the table there, 0.7764, is the area under the standard normal curve that lies to the left of 0.76. Consequently, the area under the standard normal curve that lies to the right of 0.76 is $1 - 0.7764 = 0.2236$, as shown in Fig. 6.10(b).

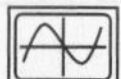

EXAMPLE 6.5 **FINDING THE AREA BETWEEN TWO SPECIFIED z-SCORES**

Determine the area under the standard normal curve that lies between −0.68 and 1.82, as seen in Fig. 6.11(a).

FIGURE 6.11 Finding the area under the standard normal curve that lies between −0.68 and 1.82

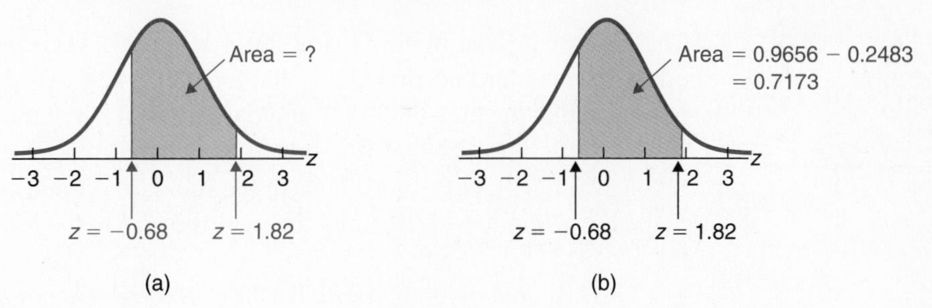

SOLUTION The area under the standard normal curve that lies between −0.68 and 1.82 equals the area to the left of 1.82 minus the area to the left of −0.68. Table II shows that the area to the left of 1.82 is 0.9656 and that the area to the left of −0.68 is 0.2483.† So, the area under the standard normal curve that lies between −0.68 and 1.82 is 0.9656 − 0.2483 = 0.7173, as depicted in Fig. 6.11(b). ◨

The discussion presented in Examples 6.3–6.5 is summarized by the three graphs in Fig. 6.12. Each graph shows how Table II, which gives areas to the left of a specified z-score, can be used to obtain a required area.

FIGURE 6.12 Using Table II to find the area under the standard normal curve that lies (a) to the left of a specified z-score, (b) to the right of a specified z-score, (c) between two specified z-scores

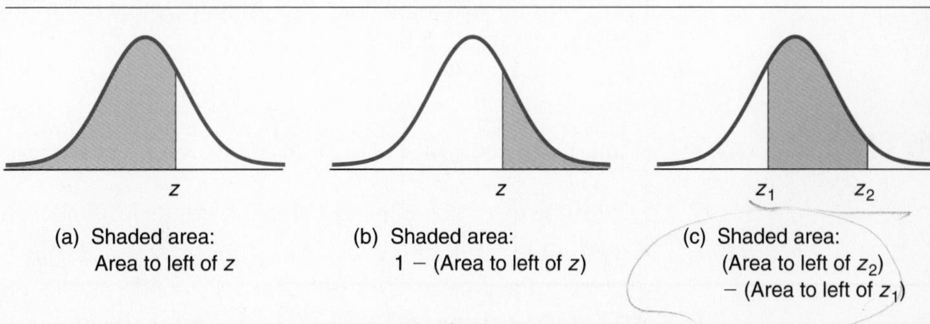

Obtaining the area to the left of a specified z-score requires one table look-up, as seen in Fig. 6.12(a); obtaining the area to the right of a specified z-score

† The area to the left of −0.68 is obtained from the left page of Table II, since −0.68 is negative. Notice that the second decimal places displayed at the top of this half of Table II go from 0.09 to 0.00, not from 0.00 to 0.09.

requires one table look-up and one subtraction (from 1), as seen in Fig. 6.12(b); and obtaining the area between two specified z-scores requires two table look-ups and one subtraction, as seen in Fig. 6.12(c).

A Note Concerning Table II

The first area given in Table II is for $z = -3.90$. According to the table, the area under the standard normal curve that lies to the left of -3.90 is 0.0000. This does not mean that the area under the standard normal curve that lies to the left of -3.90 is exactly 0, but only that it is 0 to four decimal places (the area is 0.0000481 to seven decimal places). Indeed, since the standard normal curve extends indefinitely to the left without ever touching the axis, the area to the left of any z-score is greater than 0.

The last area given in Table II is for $z = 3.90$. According to the table, the area under the standard normal curve that lies to the left of 3.90 is 1.0000. This does not mean that the area under the standard normal curve that lies to the left of 3.90 is exactly 1, but only that it is 1 to four decimal places (the area is 0.9999519 to seven decimal places). Indeed, since the total area under the standard normal curve is exactly 1 and the curve extends indefinitely to the right without ever touching the axis, the area to the left of any z-score is less than 1.

Finding the z-Score for a Specified Area

Up to this point, we have used Table II to find areas under the standard normal curve to the left of a specified z-score, to the right of a specified z-score, and between two specified z-scores. Now we will learn how to use Table II to find the z-score(s) corresponding to a specified area under the standard normal curve. Consider Example 6.6.

EXAMPLE 6.6 FINDING THE z-SCORE HAVING A SPECIFIED AREA TO ITS LEFT

Determine the z-score having area 0.04 to its left under the standard normal curve, as seen in Fig. 6.13(a).

SOLUTION We use Table II to obtain the z-score corresponding to the area 0.04. For ease of reference, we have reproduced a portion of Table II in Table 6.2.

We search the body of Table 6.2 (or Table II) for the area 0.04. Because we find no such area in the table, we use the area closest to 0.04, which is 0.0401. The z-score corresponding to that area is -1.75, as seen in Table 6.2. Figure 6.13(b) summarizes our results.

  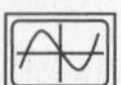

FIGURE 6.13
Finding the z-score having area 0.04 to its left

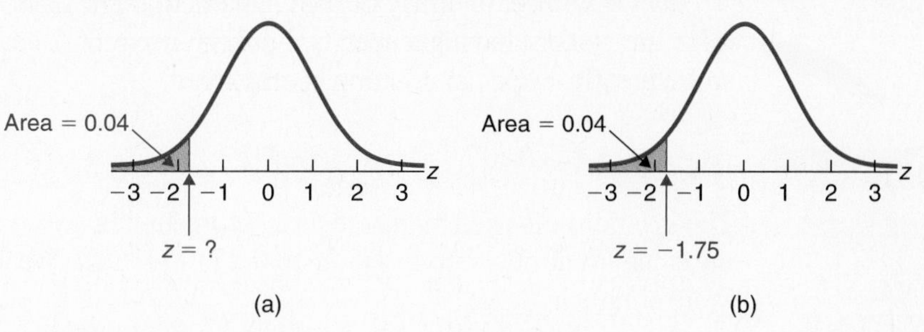

TABLE 6.2
Areas under the standard normal curve

	Second decimal place in z									
z	*0.09*	*0.08*	*0.07*	*0.06*	*0.05*	*0.04*	*0.03*	*0.02*	*0.01*	*0.00*
.	.	.	.	.	.	.	.	.	.	.
.	.	.	.	.	.	.	.	.	.	.
.	.	.	.	.	.	.	.	.	.	.
−1.9	0.0233	0.0239	0.0244	0.0250	0.0256	0.0262	0.0268	0.0274	0.0281	0.0287
−1.8	0.0294	0.0301	0.0307	0.0314	0.0322	0.0329	0.0336	0.0344	0.0351	0.0359
−1.7	0.0367	0.0375	0.0384	0.0392	0.0401	0.0409	0.0418	0.0427	0.0436	0.0446
−1.6	0.0455	0.0465	0.0475	0.0485	0.0495	0.0505	0.0516	0.0526	0.0537	0.0548
−1.5	0.0559	0.0571	0.0582	0.0594	0.0606	0.0618	0.0630	0.0643	0.0655	0.0668
.	.	.	.	.	.	.	.	.	.	.
.	.	.	.	.	.	.	.	.	.	.
.	.	.	.	.	.	.	.	.	.	.

In Example 6.6, we were to determine the z-score having area 0.04 to its left. Because we were unable to find an area-entry of 0.04 in Table II, we selected the area closest to 0.04 and took the z-score corresponding to that area as an approximation of the required z-score.

This illustrates what we do in the most typical case: When there is no area-entry in Table II exactly equal to the one desired, but there is one area-entry closest to the one desired, we take the z-score corresponding to the closest area-entry as an approximation of the required z-score.

Two other cases are possible. One case is when there is an area-entry in Table II exactly equal to the one desired; nothing more need be said in this case. The other case is when there is no area-entry in Table II exactly equal to the one desired, but there are two area-entries equally closest to the one desired; in this case, we take the mean of the two corresponding z-scores as an approximation of the required z-score. Both cases are illustrated in Example 6.7, which we will present momentarily.

As we will see in many of the chapters to come, it is often necessary to determine the z-score having a specified area to its right. Because this problem occurs so frequently, a special notation is employed.

DEFINITION 6.3 **THE z_α NOTATION**

The symbol z_α is used to denote the z-score having area α (alpha) to its right under the standard normal curve, as illustrated in Fig. 6.14. We read "z_α" as "z sub α" or more simply as "z α."

FIGURE 6.14
The z_α notation

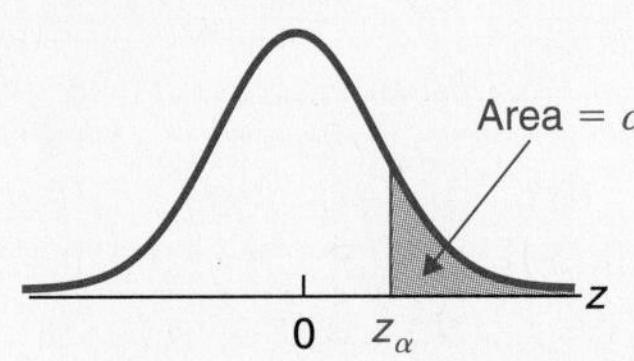

EXAMPLE 6.7 **FINDING z_α**

Use Table II to find

a. $z_{0.025}$. **b.** $z_{0.05}$.

SOLUTION **a.** $z_{0.025}$ is the z-score having area 0.025 to its right under the standard normal curve, as shown in Fig. 6.15(a).

FIGURE 6.15
Finding $z_{0.025}$

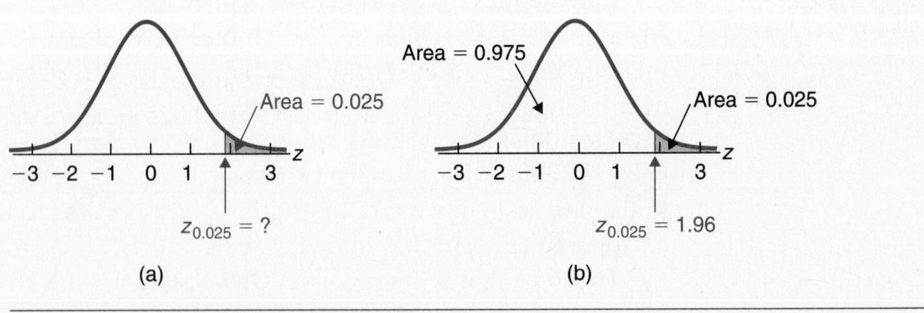

Since the area under the standard normal curve to the right of $z_{0.025}$ equals 0.025, the area to its left is $1 - 0.025 = 0.975$, as shown in Fig. 6.15(b). We search the body of Table II for the area 0.975 and find that such an area-entry does indeed exist. Its corresponding z-score is 1.96. So $z_{0.025} = 1.96$. See Fig. 6.15(b).

b. $z_{0.05}$ is the z-score having area 0.05 to its right under the standard normal curve, as shown in Fig. 6.16(a).

FIGURE 6.16
Finding $z_{0.05}$

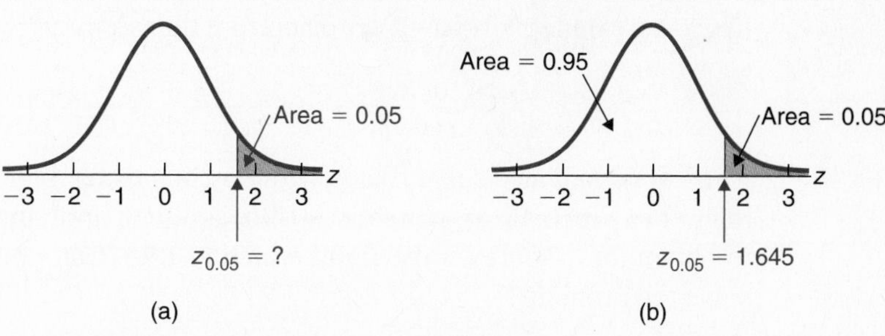

Since the area under the standard normal curve to the right of $z_{0.05}$ equals 0.05, the area to its left is $1 - 0.05 = 0.95$, as shown in Fig. 6.16(b). We search the body of Table II for the area 0.95, but find no such area. Instead we find that there are two areas closest to 0.95, namely, 0.9495 and 0.9505. The z-scores corresponding to those two areas are 1.64 and 1.65, respectively. So, we take as our approximation of $z_{0.05}$ the z-score halfway between 1.64 and 1.65; that is, $z_{0.05} = 1.645$, as shown in Fig. 6.16(b). ◨

In Example 6.8 we show how to find the two z-scores that divide the area under the standard normal curve into a specified middle area and two outside areas.

EXAMPLE 6.8 FINDING THE z-SCORES FOR A SPECIFIED AREA

Determine the two z-scores that divide the area under the standard normal curve into a middle 0.95 area and two outside 0.025 areas, as depicted in Fig. 6.17(a).

FIGURE 6.17
Finding the two z-scores dividing the area under the standard normal curve into a middle 0.95 area and two outside 0.025 areas

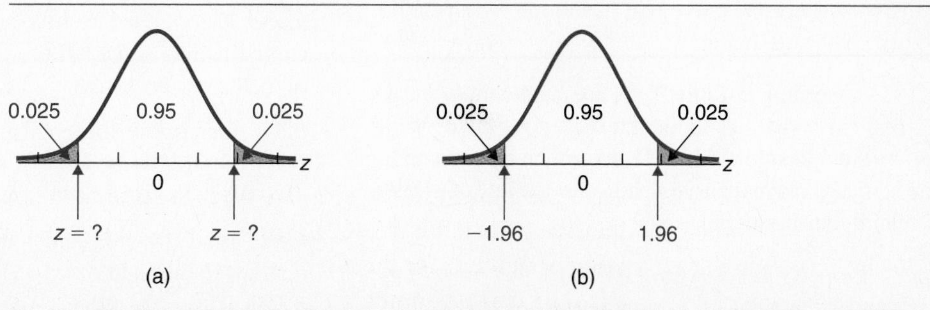

SOLUTION This problem can be solved in several ways. Here is one way. As we see from Fig. 6.17(a), the area of the shaded region on the right is 0.025. This means that the z-score on the right is $z_{0.025}$. In Example 6.7(a) we found that $z_{0.025} = 1.96$. Thus the z-score on the right is 1.96. Since the standard normal curve is symmetric about 0, the z-score on the left is -1.96. Therefore the two required z-scores are ± 1.96, as shown in Fig. 6.17(b). □

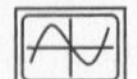

Note: We could also solve Example 6.8 by first using Table II to find the z-score on the left in Fig. 6.17(a), which is -1.96, and then applying the symmetry property to obtain the z-score on the right, which is 1.96. Can you think of a third way to solve the problem?

EXERCISES 6.2

STATISTICAL CONCEPTS AND SKILLS

6.23 Explain why it is important to be able to obtain areas under the standard normal curve.

6.24 With which normal distribution is the standard normal curve associated?

6.25 Without consulting Table II, explain why the area under the standard normal curve that lies to the right of 0 is equal to 0.5.

6.26 According to Table II, the area under the standard normal curve that lies to the left of -2.08 is 0.0188. Without further consulting Table II, determine the area under the standard normal curve that lies to the right of 2.08. Explain your reasoning.

6.27 According to Table II, the area under the standard normal curve that lies to the left of 0.43 is 0.6664. Without further consulting Table II, determine the area under the standard normal curve that lies to the right of 0.43. Explain your reasoning.

6.28 According to Table II, the area under the standard normal curve that lies to the left of 1.96 is 0.975. Without further consulting Table II, determine the area under the standard normal curve that lies to the left of -1.96. Explain your reasoning.

6.29 Property 4 of Key Fact 6.3 states that most of the area under the standard normal curve lies between -3 and 3. Use Table II to determine precisely the percentage of the area under the standard normal curve that lies between -3 and 3.

6.30 Why is the standard normal curve sometimes referred to as the z-curve?

6.31 Explain how Table II is used to determine the area under the standard normal curve that lies

a. to the left of a specified z-score.
b. to the right of a specified z-score.
c. between two specified z-scores.

6.32 Fill in the following blanks: The area under the standard normal curve that lies to the left of a z-score is always strictly between ____ and ____.

Use Table II to obtain the areas under the standard normal curve required in Exercises 6.33–6.40. Sketch a standard normal curve and shade the area of interest in each problem.

6.33 Determine the area under the standard normal curve that lies to the left of

a. 2.24. **b.** -1.56. **c.** 0. **d.** -4.

6.34 Determine the area under the standard normal curve that lies to the left of

a. -0.87. **b.** 3.56. **c.** 5.12.

6.35 Find the area under the standard normal curve that lies to the right of

a. -1.07. **b.** 0.6. **c.** 0. **d.** 4.2.

6.36 Find the area under the standard normal curve that lies to the right of
a. 2.02. **b.** −0.56. **c.** −4.

6.37 Determine the area under the standard normal curve that lies between
a. −2.18 and 1.44. **b.** −2 and −1.5.
c. 0.59 and 1.51. **d.** 1.1 and 4.2.

6.38 Determine the area under the standard normal curve that lies between
a. −0.88 and 2.24. **b.** −2.5 and −2.
c. 1.48 and 2.72. **d.** −5.1 and 1.

6.39 Find the area under the standard normal curve that lies
a. either to the left of −2.12 or to the right of 1.67.
b. either to the left of 0.63 or to the right of 1.54.

6.40 Find the area under the standard normal curve that lies
a. either to the left of −1 or to the right of 2.
b. either to the left of −2.51 or to the right of −1.

6.41 Use Table II to obtain the following shaded areas under the standard normal curve.
a.

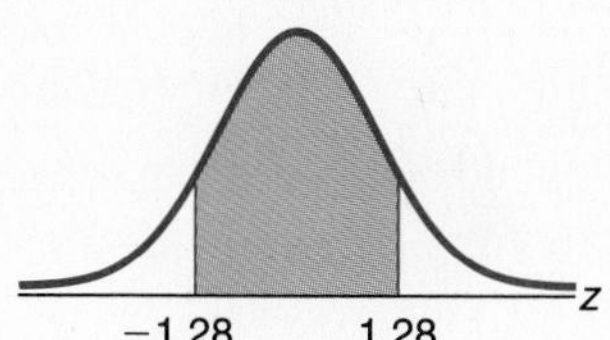

b.

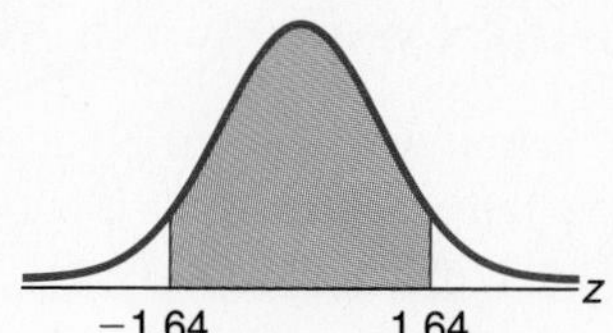

c.

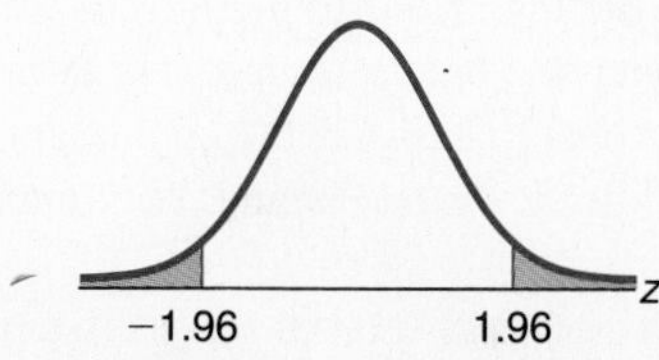

d.

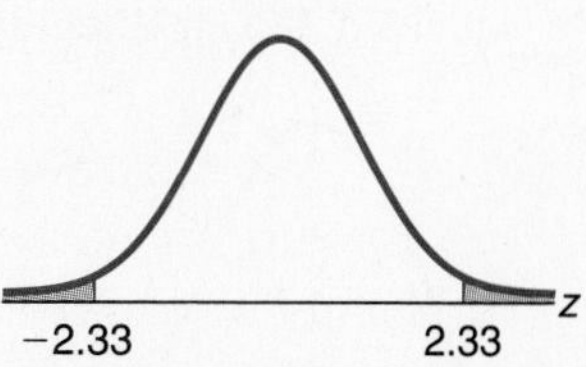

6.42 Use Table II to obtain the following shaded areas under the standard normal curve.
a.

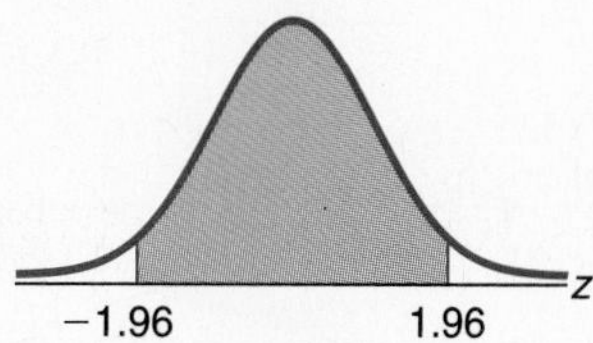

b.

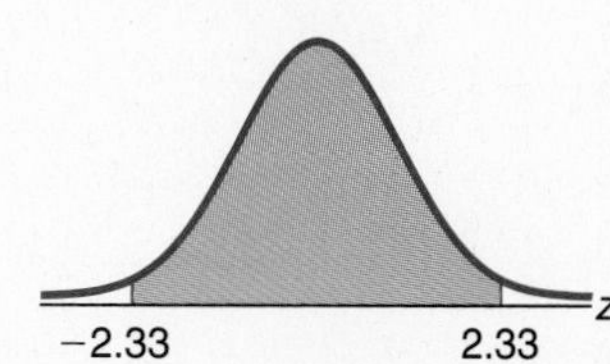

c.

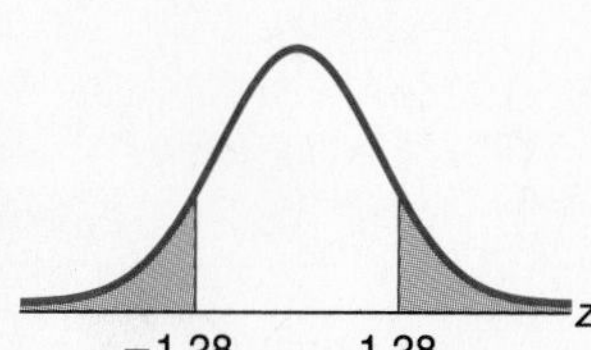

d.

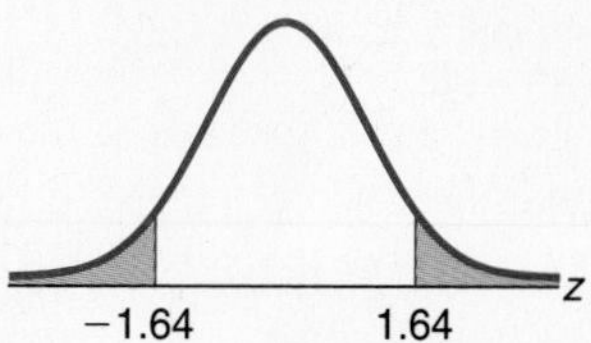

6.43 In each part, find the area under the standard normal curve that lies between the specified z-scores, sketch a standard normal curve, and shade the area of interest.
a. −1 and 1 **b.** −2 and 2 **c.** −3 and 3

6.44 Below is a standard normal curve. The total area under the curve is divided into eight regions.

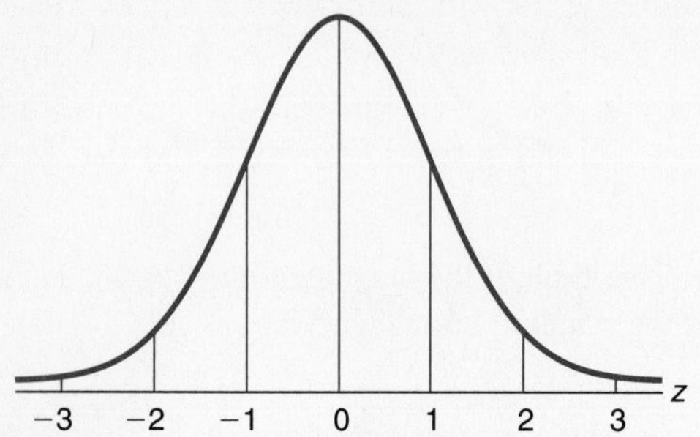

a. Determine the area of each region.
b. Complete the following table.

Region	Area	Percentage of total area
$-\infty$ to -3	0.0013	0.13
-3 to -2		
-2 to -1		
-1 to 0		
0 to 1	0.3413	34.13
1 to 2		
2 to 3		
3 to ∞		
	1.0000	100.00

In Exercises 6.45–6.56, use Table II to obtain the required z-scores. Illustrate your work with graphs.

6.45 Obtain the z-score for which the area under the standard normal curve to its left is 0.025.

6.46 Determine the z-score for which the area under the standard normal curve to its left is 0.01.

6.47 Find the z-score having area 0.75 to its left under the standard normal curve.

6.48 Obtain the z-score having area 0.80 to its left under the standard normal curve.

6.49 Obtain the z-score having area 0.95 to its right; that is, find $z_{0.95}$.

6.50 Obtain the z-score having area 0.70 to its right; that is, find $z_{0.70}$.

6.51 Determine $z_{0.33}$; that is, find the z-score having area 0.33 to its right under the standard normal curve.

6.52 Determine $z_{0.015}$; that is, find the z-score having area 0.015 to its right under the standard normal curve.

6.53 Find the following z-scores.
a. $z_{0.03}$ **b.** $z_{0.005}$

6.54 Obtain the following z-scores.
a. $z_{0.20}$ **b.** $z_{0.06}$

6.55 Determine the two z-scores that divide the area under the standard normal curve into a middle 0.90 area and two outside 0.05 areas.

6.56 Determine the two z-scores that divide the area under the standard normal curve into a middle 0.99 area and two outside 0.005 areas.

6.57 Complete the following table.

$z_{0.10}$	$z_{0.05}$	$z_{0.025}$	$z_{0.01}$	$z_{0.005}$
1.28				

EXTENDING THE CONCEPTS AND SKILLS

6.58 In this section we mentioned that the total area under any curve representing the distribution of a variable is equal to 1. Explain why this is so.

6.59 Let $0 < \alpha < 1$. Determine
a. the z-score having area α to its right in terms of z_α.
b. the z-score having area α to its left in terms of z_α.
c. the two z-scores that divide the area under the curve into a middle $1 - \alpha$ area and two outside areas of $\alpha/2$.
d. Draw graphs to illustrate your results in parts (a)–(c).

6.3 WORKING WITH NORMALLY DISTRIBUTED VARIABLES

We have now learned everything required to obtain the percentage of all possible observations of a normally distributed variable that lie within any specified range. To do so, we first convert to z-scores and then determine the corresponding area under the standard normal curve. More formally, we use the following procedure.

PROCEDURE 6.1 TO DETERMINE A PERCENTAGE OR PROBABILITY FOR A NORMALLY DISTRIBUTED VARIABLE

Step 1 Sketch the normal curve associated with the variable.

Step 2 Shade the region of interest and mark the delimiting x-values.

Step 3 Compute the z-scores for the delimiting x-values found in Step 2.

Step 4 Use Table II to obtain the area under the standard normal curve delimited by the z-scores found in Step 3.

The steps in Procedure 6.1 are portrayed graphically in Fig. 6.18. This figure represents the case where the specified range is between two numbers, a and b. The case where the specified range is to the left (or right) of a specified number is represented similarly, but there will be only one x-value and the shaded region will be the area under the normal curve that lies to the left (or right) of that x-value.

FIGURE 6.18
Graphical portrayal of Procedure 6.1

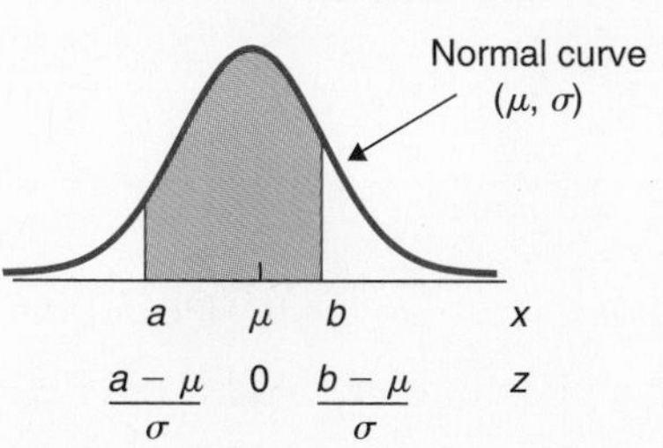

Note: When computing z-scores in Step 3 of Procedure 6.1, round to two decimal places because that is the precision provided in Table II.

Procedure 6.1 applies to probabilities as well as percentages. Remember, if a member of a population is selected at random, then the probability that the observed value of a variable will lie within any specified range equals the percentage of all possible observations of the variable lying within that range.

EXAMPLE 6.9 ILLUSTRATES PROCEDURE 6.1

Intelligence quotients (IQs) measured on the Stanford Revision of the Binet-Simon Intelligence Scale are known to be normally distributed with a mean of 100 and a standard deviation of 16. Obtain the percentage of people having IQs between 115 and 140.

SOLUTION Here the variable is IQ and the population consists of all people. Since IQs are normally distributed, we can determine the required percentage by applying Procedure 6.1.

Step 1 *Sketch the normal curve associated with the variable.*

Here $\mu = 100$ and $\sigma = 16$. We have sketched the normal curve associated with the variable in Fig. 6.19. Notice that the tick marks are 16 units apart, that is, the distance between successive tick marks is equal to the standard deviation.

FIGURE 6.19
Determination of the percentage of people having IQs between 115 and 140

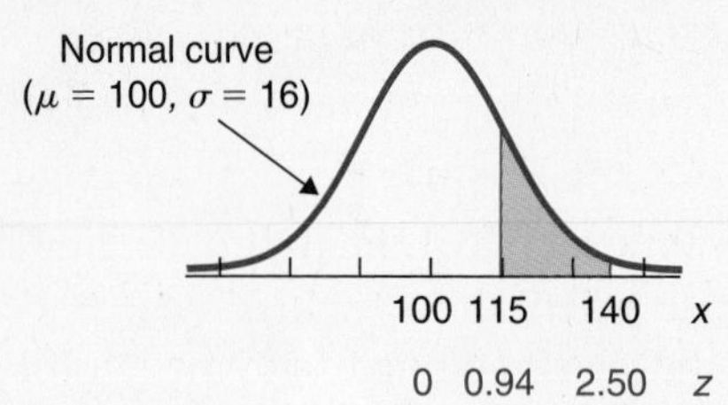

z-score computations: Area to the left of z:

$x = 115 \longrightarrow z = \dfrac{115 - 100}{16} = 0.94$ 0.8264

$x = 140 \longrightarrow z = \dfrac{140 - 100}{16} = 2.50$ 0.9938

Shaded area = 0.9938 − 0.8264 = 0.1674

Step 2 *Shade the region of interest and mark the delimiting x-values.*

See the shaded region and delimiting x-values in Fig. 6.19.

Step 3 *Compute the z-scores for the delimiting x-values found in Step 2.*

We need to obtain the z-scores for the x-values 115 and 140:

$$x = 115 \quad \longrightarrow \quad z = \frac{115 - \mu}{\sigma} = \frac{115 - 100}{16} = 0.94,$$

$$x = 140 \quad \longrightarrow \quad z = \frac{140 - \mu}{\sigma} = \frac{140 - 100}{16} = 2.50.$$

These z-scores are marked beneath the x-values in Fig. 6.19.

Step 4 *Use Table II to obtain the area under the standard normal curve delimited by the z-scores found in Step 3.*

The area under the standard normal curve to the left of 0.94 is 0.8264 and that to the left of 2.50 is 0.9938. Consequently, the required area, the area shaded in Fig. 6.19, is $0.9938 - 0.8264 = 0.1674$.

Note that Procedure 6.1 can be accomplished efficiently by performing all four steps in a "picture," as in Fig. 6.19. In any case, we see that 16.74% (0.1674) of all people have IQs between 115 and 140. We can interpret this result probabilistically by saying that the probability is 0.1674 that a randomly selected person will have an IQ between 115 and 140. ◨

  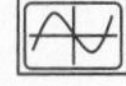

Visualizing a Normal Distribution

We will now present a rule that permits us to quickly and easily visualize and obtain useful information about a normally distributed variable. The rule gives the percentages of all possible observations that lie within one, two, and three standard deviations to either side of the mean.

Recall that the z-score of an observation tells us how many standard deviations the observation is from the mean. Thus the percentage of all possible observations that lie within one standard deviation to either side of the mean equals the percentage of all observations whose z-scores are between -1 and 1. And this we know, for a normally distributed variable, is the same as the area under the standard normal curve between -1 and 1, which is 0.6826 or 68.26%. Proceeding in the same way, we get the following rule.

KEY FACT 6.4

68.26-95.44-99.74 RULE

For any normally distributed variable,

Property 1: 68.26% of all possible observations lie within one standard deviation to either side of the mean, that is, between $\mu - \sigma$ and $\mu + \sigma$.

Property 2: 95.44% of all possible observations lie within two standard deviations to either side of the mean, that is, between $\mu - 2\sigma$ and $\mu + 2\sigma$.

Property 3: 99.74% of all possible observations lie within three standard deviations to either side of the mean, that is, between $\mu - 3\sigma$ and $\mu + 3\sigma$.

These properties are displayed graphically in Fig. 6.20.

FIGURE 6.20
68.26-95.44-99.74 rule

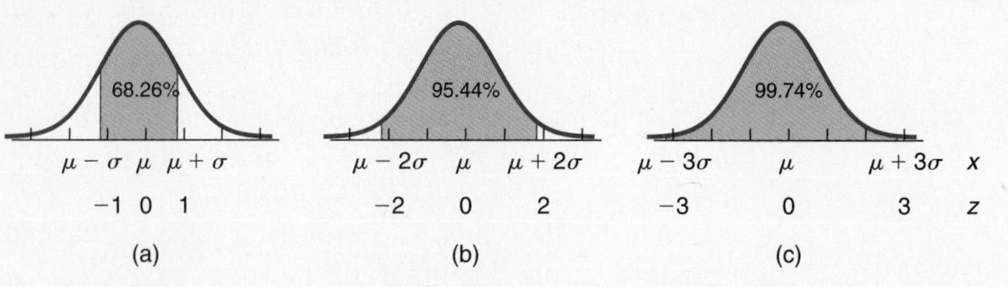

EXAMPLE 6.10 ILLUSTRATES KEY FACT 6.4

Apply the 68.26-95.44-99.74 rule to IQs.

SOLUTION Recall that $\mu = 100$ and $\sigma = 16$.

From Property 1 of the 68.26-95.44-99.74 rule, 68.26% of all people have IQs within one standard deviation to either side of the mean. One standard deviation below the mean is $\mu - \sigma = 100 - 16 = 84$; one standard deviation above the mean is $\mu + \sigma = 100 + 16 = 116$. So 68.26% of all people have IQs between 84 and 116, as seen in Fig. 6.21(a).

FIGURE 6.21
Graphical display of 68.26-95.44-99.74 rule for IQs

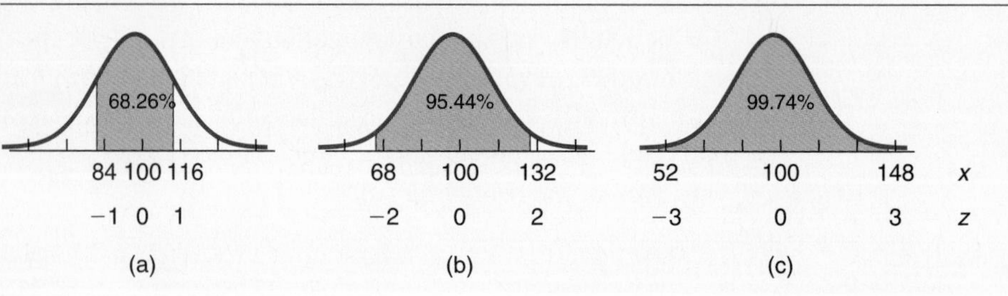

From Property 2 of the 68.26-95.44-99.74 rule, 95.44% of all people have IQs within two standard deviations to either side of the mean. Two standard deviations below the mean is $\mu - 2\sigma = 100 - 2 \cdot 16 = 100 - 32 = 68$; two standard deviations above the mean is $\mu + 2\sigma = 100 + 2 \cdot 16 = 100 + 32 = 132$. Therefore 95.44% of all people have IQs between 68 and 132, as seen in Fig. 6.21(b).

From Property 3 of the 68.26-95.44-99.74 rule, we find that 99.74% of all people have IQs between 52 ($= 100 - 3 \cdot 16$) and 148 ($= 100 + 3 \cdot 16$), as seen in Fig. 6.21(c). ◨

As we said, the 68.26-95.44-99.74 rule allows us to obtain useful information about a normally distributed variable quickly and easily, as illustrated in

Example 6.10. Note, however, that similar facts are obtainable for any number of standard deviations. For instance, from Table II, we find that for any normally distributed variable, 86.64% of all possible observations lie within 1.5 standard deviations to either side of the mean.

Experience has shown that the 68.26-95.44-99.74 rule works reasonably well for any variable having approximately a bell-shaped distribution—normal or not. This fact has become known as the **empirical rule.** See page 156 for more on the empirical rule.

Finding the Observations for a Specified Percentage

Procedure 6.1 shows how to determine the percentage of all possible observations of a normally distributed variable that lie within any specified range. Frequently, we want to carry out the reverse procedure, that is, find the observations corresponding to a specified percentage. Here is a procedure for doing that.

PROCEDURE 6.2 TO DETERMINE THE OBSERVATIONS CORRESPONDING TO A SPECIFIED PERCENTAGE OR PROBABILITY FOR A NORMALLY DISTRIBUTED VARIABLE

Step 1 Sketch the normal curve associated with the variable.

Step 2 Shade the region of interest.

Step 3 Use Table II to obtain the z-scores delimiting the region in Step 2.

Step 4 Obtain the x-values having the z-scores found in Step 3.

Among other things, we can use Procedure 6.2 to obtain quartiles, deciles, or any other percentile for a normally distributed variable. Example 6.11 shows how to find percentiles.

EXAMPLE 6.11 ILLUSTRATES PROCEDURE 6.2

Obtain the 90th percentile for IQs.

SOLUTION The 90th percentile, P_{90}, is the IQ that is higher than those of 90% of all people. Since IQs are normally distributed, we can determine the 90th percentile by applying Procedure 6.2.

Step 1 *Sketch the normal curve associated with the variable.*

Here $\mu = 100$ and $\sigma = 16$. We have sketched the normal curve associated with IQs in Fig. 6.22.

FIGURE 6.22
Finding the 90th percentile for IQs

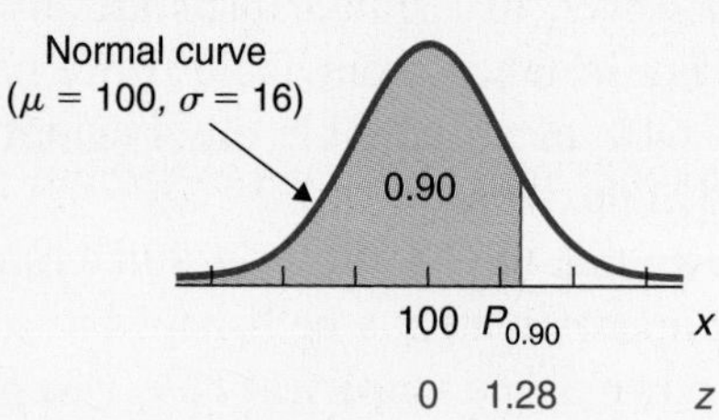

Step 2 *Shade the region of interest.*

See the shaded region in Fig. 6.22.

Step 3 *Use Table II to obtain the z-scores delimiting the region in Step 2.*

The z-score corresponding to P_{90} is the one having area 0.90 to its left under the standard normal curve. From Table II we find that z-score to be 1.28, approximately. See Fig. 6.22.

Step 4 *Obtain the x-values having the z-scores found in Step 3.*

We need to find the x-value having z-score 1.28—the IQ that is 1.28 standard deviations above the mean. It is $100 + 1.28 \cdot 16 = 100 + 20.48 = 120.48$.

  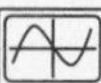

Therefore, the 90th percentile for IQs is 120.48. In other words, 90% of people have IQs below 120.48 and 10% have IQs above 120.48. ◨

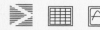

Using the Computer (Optional)

Minitab can be used to carry out the procedures that we have discussed in this section, namely, to obtain, for a normally distributed variable,

- the percentage of all possible observations that lie within any specified range.
- the observations corresponding to a specified percentage.

We will illustrate these two Minitab procedures by solving problems that we considered earlier in this section.

EXAMPLE 6.12 USING MINITAB TO OBTAIN A PERCENTAGE FOR A NORMALLY DISTRIBUTED VARIABLE

Intelligence quotients (IQs) measured on the Stanford Revision of the Binet-Simon Intelligence Scale are known to be normally distributed with a mean of 100 and a standard deviation of 16. Use Minitab to obtain the percentage of people having IQs between 115 and 140.

SOLUTION First we store the delimiting IQs, 115 and 140, in a column named IQ. Then we proceed in the following manner.

1 Choose **Calc ➤ Probability Distributions ➤ Normal...**
2 Select the **Cumulative probability** option button
3 Click in the **Mean** text box and type 100
4 Click in the **Standard deviation** text box and type 16
5 Select the **Input column** option button
6 Click in the **Input column** text box and specify IQ
7 Click **OK**

Printout 6.2 shows the resulting output.

PRINTOUT 6.2
Minitab output for determining the percentage of all people having IQs between 115 and 140

```
Normal with mean = 100.000 and standard deviation = 16.0000

        x        P( X <= x)
 115.0000          0.8257
 140.0000          0.9938
```

The IQs delimiting the range are given in the first column of Printout 6.2. In the second column, Minitab displays the cumulative probability for each delimiting IQ, that is, the percentage (expressed as a decimal) of people having IQs less than or equal to the given one. For instance, the last line of Printout 6.2 shows that 99.38% (0.9938) of people have IQs of 140 or less.

Subtracting the smaller number in the second column of Printout 6.2 from the larger number, we obtain the percentage of people having IQs between 115 and 140: $0.9938 - 0.8257 = 0.1681$ or 16.81%. Note that this differs slightly from the percentage of 16.74% that we obtained in Example 6.9. The difference is due to the fact that Minitab retains more accuracy than we can get in Table II. ◨

EXAMPLE 6.13 USING MINITAB TO FIND THE OBSERVATION FOR A SPECIFIED PERCENTAGE

Use Minitab to obtain the 90th percentile for IQs.

SOLUTION We proceed as follows.

1 Choose **Calc ➤ Probability Distributions ➤ Normal...**
2 Select the **Inverse cumulative probability** option button

3 Click in the **Mean** text box and type 100

4 Click in the **Standard deviation** text box and type 16

5 Select the **Input constant** option button

6 Click in the **Input constant** text box and type 0.90

7 Click **OK**

The output that results is shown in Printout 6.3.

PRINTOUT 6.3
Minitab printout for obtaining the 90th percentile for IQs

```
Normal with mean = 100.000 and standard deviation = 16.0000

 P( X <= x)            x
    0.9000      120.5048
```

The specified cumulative probability is given in the first column of Printout 6.3. This is the percentile-level expressed as a decimal. In the second column, Minitab displays the corresponding IQ. Thus the 90th percentile for IQs is 120.5048. Note that this differs slightly from the value of 120.48 that we obtained in Example 6.11. The difference is due to the fact that Minitab retains more accuracy. ◨

EXERCISES 6.3

STATISTICAL CONCEPTS AND SKILLS

6.60 Briefly, for a normally distributed variable, how do we obtain the percentage of all possible observations that lie within a specified range.

6.61 Explain why the percentage of all possible observations of a normally distributed variable that lie within two standard deviations to either side of the mean equals the area under the standard normal curve between −2 and 2.

6.62 What does the empirical rule say?

Apply Procedure 6.1 on page 375 to solve each of Exercises 6.63–6.68

6.63 The annual wages, excluding board, of U.S. farm laborers in 1926 were normally distributed with a mean of \$586 and a standard deviation of \$97. In 1926 what percentage of U.S. farm laborers had an annual wage of
a. between \$500 and \$700? **b.** at least \$400?

6.64 In 1905 R. Pearl published the article "Biometrical Studies on Man. I. Variation and Correlation in Brain Weight" (*Biometrika,* Vol. 4, pp. 13–104). According to the study, brain weights of Swedish men are normally distributed with a mean of 1.40 kilograms (kg) and a standard deviation of 0.11 kg. Obtain the percentage of Swedish men having brain weights
a. between 1.50 kg and 1.70 kg.
b. less than 1.6 kg.

6.65 As reported by the U.S. National Center for Health Statistics in *Vital and Health Statistics,* males who are 6 ft tall and between 18 and 24 years of age have a mean weight of 175 lb. If the weights are normally distributed with a standard deviation of 14 lb, find the percentage of such males that weigh
a. between 190 and 210 lb. **b.** less than 150 lb.

6.66 An issue of *Scientific American* reveals that the batting averages of major-league baseball players are approximately normally distributed and have a mean of 0.270 and a standard deviation of 0.031. Determine the percentage of major-league baseball players having batting averages

a. between 0.225 and 0.250. **b.** at least 0.300.

6.67 As reported by *Runner's World* magazine, the times of the finishers in the New York City 10-km run are normally distributed with a mean of 61 minutes and a standard deviation of 9 minutes. Let X be the time of a randomly selected finisher. Find

a. $P(X > 75)$.

b. $P(X < 50 \quad \text{or} \quad X > 70)$.

6.68 The length of the western rattlesnake is normally distributed with a mean of 42 inches and a standard deviation of 2 inches. Let X be the length of one of these snakes selected at random. Determine

a. $P(X > 45)$.

b. $P(35 \leq X \leq 40)$.

Apply the 68.26-95.44-99.74 rule to solve each of Exercises 6.69–6.72.

6.69 Refer to Exercise 6.65. Determine the percentage of males who are 6 ft tall and between 18 and 24 years of age that have weights within

a. 1 standard deviation to either side of the mean.

b. 2 standard deviations to either side of the mean.

c. 3 standard deviations to either side of the mean.

6.70 Refer to Exercise 6.66. Obtain the percentage of major-league baseball players who have batting averages within

a. 1 standard deviation to either side of the mean.

b. 2 standard deviations to either side of the mean.

c. 3 standard deviations to either side of the mean.

6.71 The Department of Agriculture compiles information on food costs and publishes its findings in *Human Nutrition Information Service.* According to that document, the mean weekly food cost for a couple with two children 6–11 years old is $153.10. Presuming the costs are normally distributed with a standard deviation of $17.20, fill in the following blanks.

a. 68.26% of such couples have weekly food costs between $____ and $____.

b. 95.44% of such couples have weekly food costs between $____ and $____.

c. 99.74% of such couples have weekly food costs between $____ and $____.

d. Draw graphs similar to those in Fig. 6.21 on page 378 to portray your results.

6.72 The A. C. Nielsen Company reports in *Nielsen Report on Television* that the mean weekly television viewing time for children age 2–11 years is 23.02 hours. Assuming the weekly television viewing times of such children are normally distributed with a standard deviation of 6.23 hours, fill in the following blanks.

a. 68.26% of all such children watch between ____ and ____ hours of TV per week.

b. 95.44% of all such children watch between ____ and ____ hours of TV per week.

c. 99.74% of all such children watch between ____ and ____ hours of TV per week.

d. Draw graphs similar to those in Fig. 6.21 on page 378 to portray your results.

6.73 Refer to Exercise 6.65.

a. Obtain the 15th percentile of the weights.

b. Find the 98th percentile.

c. Determine the quartiles.

6.74 Refer to Exercise 6.66.

a. Find the 82nd percentile of the batting averages.

b. Obtain the first decile (i.e., 10th percentile).

c. Determine the quartiles.

6.75 Refer to Exercise 6.71.

a. Determine the quartiles of the food costs.

b. Obtain the 3rd decile.

c. Find the 85th percentile.

6.76 Refer to Exercise 6.72.

a. Determine the quartiles of the viewing times.

b. Obtain the 45th percentile.

c. Find the 9th decile.

EXTENDING THE CONCEPTS AND SKILLS

6.77 For a normally distributed variable, fill in the following blanks.

a. ____% of all possible observations lie within 1.96 standard deviations to either side of the mean.

b. ____% of all possible observations lie within 1.64 standard deviations to either side of the mean.

6.78 For a normally distributed variable, fill in the following blanks.

a. _____% of all possible observations lie within 1.28 standard deviations to either side of the mean.

b. _____% of all possible observations lie within 2.33 standard deviations to either side of the mean.

6.79 For a normally distributed variable, fill in the following blanks.

a. 99% of all possible observations lie within _____ standard deviations to either side of the mean.

b. 80% of all possible observations lie within _____ standard deviations to either side of the mean.

6.80 For a normally distributed variable, fill in the following blanks.

a. 95% of all possible observations lie within _____ standard deviations to either side of the mean.

b. 90% of all possible observations lie within _____ standard deviations to either side of the mean.

In Example 6.1 on page 356, we considered the heights of the 3264 female students attending a midwestern college. The mean and standard deviation of the heights are, respectively, 64.4 inches and 2.4 inches. We discovered (see Fig. 6.4) that the heights are approximately normally distributed. This means that we can use the normal distribution with $\mu = 64.4$ and $\sigma = 2.4$ to approximate the percentage of these students having heights within any specified range. In each part of Exercises 6.81 and 6.82, (i) obtain the exact percentage from Table 6.1, (ii) use the normal distribution to approximate the percentage, and (iii) compare answers.

6.81 The percentage of female students with heights
a. between 62 and 63 in. **b.** between 65 and 70 in.

6.82 The percentage of female students with heights
a. between 71 and 72 in. **b.** between 61 and 65 in.

6.83 Let $0 < \alpha < 1$. For a normally distributed variable, show that $100(1 - \alpha)\%$ of all possible observations lie within $z_{\alpha/2}$ standard deviations to either side of the mean, that is, between $\mu - z_{\alpha/2} \cdot \sigma$ and $\mu + z_{\alpha/2} \cdot \sigma$. (*Hint:* Recall that $z_{\alpha/2}$ is the z-score having area $\alpha/2$ to its right under the standard normal curve.)

6.84 Let x be a normally distributed variable having mean μ and standard deviation σ.

a. Express the quartiles, Q_1, Q_2, and Q_3, in terms of μ and σ.

b. Express the kth percentile, P_k, in terms of μ, σ, and k. (*Hint:* Observe that the z-score for the kth percentile has area $1 - \frac{k}{100}$ to its right under the standard normal curve.)

USING TECHNOLOGY  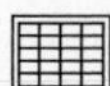 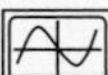

6.85 Use Minitab or some other statistical software to solve Exercise 6.65.

6.86 Use Minitab or some other statistical software to solve Exercise 6.66.

6.87 Use Minitab or some other statistical software to solve Exercise 6.73.

6.88 Use Minitab or some other statistical software to solve Exercise 6.74.

6.4 ASSESSING NORMALITY; NORMAL PROBABILITY PLOTS

We have now seen how to work with normally distributed variables. For instance, we know how to determine the percentage of all possible observations that lie within any specified range and how to obtain the observations corresponding to a specified percentage.

Another problem involves deciding whether a variable is normally distributed, or at least approximately so, based on a sample of observations. Such decisions often play a major role in subsequent analyses—from percentage or percentile calculations to statistical inferences.

As we learned in Key Fact 2.1, if a random sample is taken from a population, then the distribution of the observed values of a variable will approximate the distribution of the variable—and the larger the sample, the better the approximation tends to be. We can use this fact to help decide whether a variable is normally distributed.

If a variable is normally distributed, then, for a large sample, a histogram of the observations should be roughly bell-shaped; for a very large sample, even moderate departures from a bell shape cast doubt on the normality of the variable. On the other hand, for a relatively small sample, it is often difficult to ascertain a clear shape in a histogram and, in particular, whether it is bell-shaped. These comments hold true as well for stem-and-leaf diagrams and dotplots.

So, for relatively small samples, a more sensitive graphical technique than the ones we have learned thus far is required for assessing normality. Normal probability plots provide such a technique.

The idea behind a normal probability plot is simple: Compare the observed values of the variable to the observations we would expect to get if the variable is normally distributed. More precisely, a **normal probability plot** is a plot of the observed values of the variable versus the **normal scores**—the observations we would expect to get for a variable having the standard normal distribution. If the variable is normally distributed, then the normal probability plot should be roughly linear (i.e., fall roughly in a straight line), and vice versa.

When we use a normal probability plot to assess the normality of a variable, we must remember two things. First that the decision of whether a normal probability plot is roughly linear is a subjective one and, second, that we are using the observations of the variable for a sample to make a judgment about all possible observations of the variable (i.e., the distribution of the variable). With these considerations in mind, we present the following guidelines for assessing normality.

KEY FACT 6.5 GUIDELINES FOR ASSESSING NORMALITY USING A NORMAL PROBABILITY PLOT

To assess the normality of a variable using sample data, construct a normal probability plot.

- If the plot is roughly linear, then accept as reasonable that the variable is approximately normally distributed.
- If the plot shows systematic deviations from linearity (e.g., if it displays significant curvature), then conclude that the variable is probably not approximately normally distributed.

These guidelines should be interpreted loosely for small samples, but can be interpreted rather strictly for large samples.

In practice, normal probability plots are obtained by computer. It is helpful, however, for purposes of understanding, to construct a few by hand. To that end, we have provided in Table III of Appendix A the normal scores for sample sizes from 5 to 30. Example 6.14 explains how Table III can be used to obtain a normal probability plot.

EXAMPLE 6.14 CONSTRUCTING A NORMAL PROBABILITY PLOT

The Internal Revenue Service publishes data on federal individual income tax returns in *Statistics of Income, Individual Income Tax Returns.* A random sample of 12 returns from last year revealed the adjusted gross incomes, in thousands of dollars, shown in Table 6.3.

TABLE 6.3 Adjusted gross incomes ($1000s)

9.7	93.1	33.0	21.2
81.4	51.1	43.5	10.6
12.8	7.8	18.1	12.7

a. Construct a normal probability plot for these data.

b. Assess the normality of adjusted gross incomes.

SOLUTION Here the variable is adjusted gross income and the population consists of all last year's federal individual income tax returns.

a. To construct a normal probability plot, we first arrange the data in increasing order and obtain the normal scores from Table III. The ordered data are shown in the first column of Table 6.4 and the normal scores, from the $n = 12$ column of Table III, are shown in the second column of Table 6.4.

TABLE 6.4 Ordered data and normal scores

Adjusted gross income	Normal score
7.8	−1.64
9.7	−1.11
10.6	−0.79
12.7	−0.53
12.8	−0.31
18.1	−0.10
21.2	0.10
33.0	0.31
43.5	0.53
51.1	0.79
81.4	1.11
93.1	1.64

Next we plot the points in Table 6.4, using the horizontal axis for the adjusted gross incomes and the vertical axis for the normal scores. For instance, the first point plotted has a horizontal coordinate of 7.8 and a vertical coordinate of -1.64. Figure 6.23 shows all 12 points from Table 6.4; this is the normal probability plot for the sample of adjusted gross incomes.

FIGURE 6.23
Normal probability plot for the sample of adjusted gross incomes

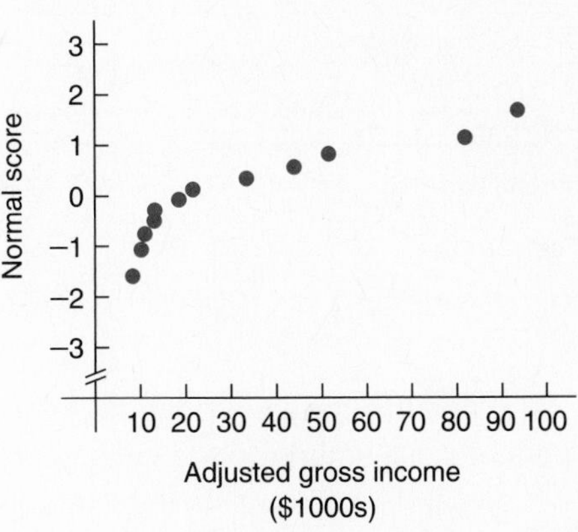

  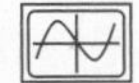

b. The normal probability plot in Fig. 6.23 displays significant curvature. Evidently, adjusted gross incomes are not approximately normally distributed. ◨

Detecting Outliers Using Normal Probability Plots

Recall that outliers are observations that fall well outside the overall pattern of the data. We can employ normal probability plots to detect outliers as explained in Example 6.15.

EXAMPLE 6.15 USING NORMAL PROBABILITY PLOTS TO DETECT OUTLIERS

The U.S. Department of Agriculture publishes data on U.S. chicken consumption in *Food Consumption, Prices, and Expenditures.* Last year's chicken consumptions, in pounds, for 17 randomly selected people are displayed in Table 6.5. A normal probability plot for these observations is presented in Fig. 6.24(a). Use the plot to discuss the distribution of chicken consumptions and to detect any outliers.

TABLE 6.5
Sample of last year's chicken consumptions (lbs)

47	39	62	49	50	70
59	53	55	0	65	63
53	51	50	72	45	

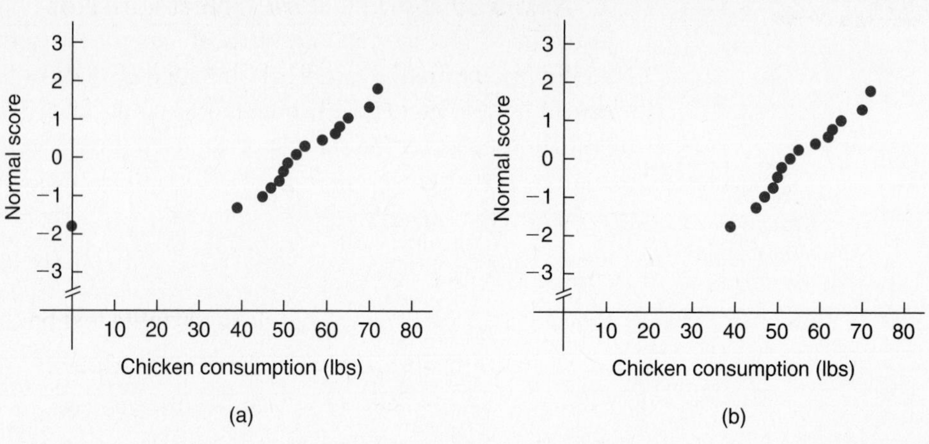

FIGURE 6.24 Normal probability plots for chicken-consumptions: (a) original data, (b) data with outlier removed

SOLUTION We see from Fig. 6.24(a) that the normal probability plot falls roughly in a straight line except for the point corresponding to the consumption of 0 lb. That point on the normal probability plot falls well outside the overall pattern of the plot; hence 0 lb is an outlier. This outlier might be a recording error or due to a person in the sample who does not eat chicken for some reason (e.g., a vegetarian).

If we remove the outlier 0 lb from the sample data and draw a normal probability plot for the abridged data, then, as we see from Fig. 6.24(b), this normal probability plot is quite linear and shows no outliers. It appears plausible that among people who eat chicken, the amounts they consume annually are approximately normally distributed.

Note: There appear to be only 15 points (instead of the expected 17) in Fig. 6.24(a). This is because an averaging process was used to assign identical normal scores to the two 50s and identical normal scores to the two 53s. So there really are 17 points in the graph, but only 15 are distinguishable because there are two sets of two identical points. An alternative procedure is to treat identical observations as being slightly different; then no averaging process is required.

In this section we have learned how a normal probability plot for a sample of observations of a variable can be used as an aid for deciding whether the variable is (approximately) normally distributed. Although this visual assessment of normality is subjective, it is sufficient for most statistical analyses.

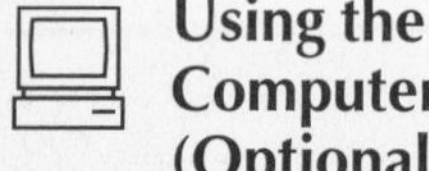

Using the Computer (Optional)

Minitab has several procedures for obtaining normal probability plots. We will discuss the one that is consistent with our earlier presentation of constructing such plots by hand.

EXAMPLE 6.16 USING MINITAB TO OBTAIN A NORMAL PROBABILITY PLOT

Use Minitab to obtain a normal probability plot for the adjusted gross incomes displayed in Table 6.3 on page 386.

SOLUTION First we store the data from Table 6.3 in a column named AGI. Then we determine the normal scores as follows.

1 Choose **Calc ➤ Calculator...**
2 Type NSCORE in the **Store result in variable** text box
3 Select **Normal scores** from the **Functions** list box
4 Specify AGI for **number** in the **Expression** text box
5 Click **OK**

The normal scores for the adjusted gross incomes are now stored in NSCORE. Next we apply Minitab to obtain the normal probability plot for the adjusted gross incomes, that is, a plot of the adjusted gross incomes versus the normal scores for that data.

1 Choose **Graph ➤ Plot...**
2 Specify NSCORE in the **Y** text box for **Graph 1**
3 Click in the **X** text box for **Graph 1** and specify AGI
4 Click **OK**

The resulting normal probability plot is shown in Printout 6.4. Compare this normal probability plot to the one we drew by hand in Fig. 6.23 on page 387. ◨

PRINTOUT 6.4 Minitab normal probability plot for the sample of adjusted gross incomes

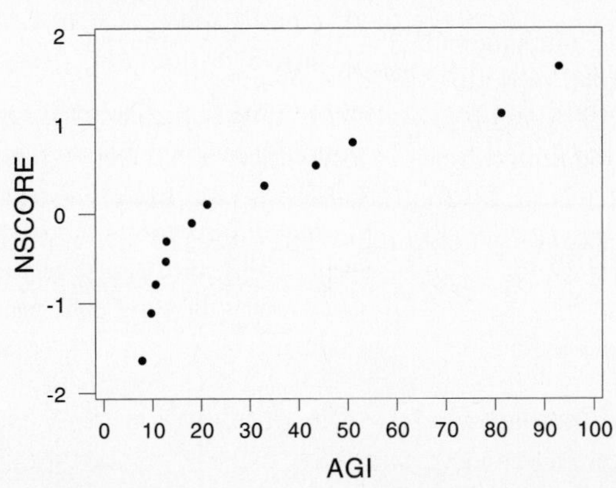

EXERCISES 6.4

STATISTICAL CONCEPTS AND SKILLS

6.89 Explain why it is often important to assess the normality of a variable.

6.90 Under what circumstances is it usually better to use a normal probability plot to assess the normality of a variable instead of using a histogram, stem-and-leaf diagram, or dotplot?

6.91 Explain in detail what a normal probability plot is and how it is used to assess the normality of a variable.

6.92 How is a normal probability plot used to detect outliers?

In each of Exercises 6.93–6.96,

a. *use Table III in Appendix A to construct a normal probability plot of the given data. For simplicity, treat equal observations as being slightly different when obtaining normal scores.*
b. *use part (a) to identify any outliers.*
c. *use part (a) to assess the normality of the variable under consideration.*

6.93 A sample of the final-exam scores in a large introductory statistics course is as follows.

88	67	64	76	86
85	82	39	75	34
90	63	89	90	84
81	96	100	70	96

6.94 As reported by the R. R. Bowker Company of New York in *Library Journal*, the mean annual subscription rate of law periodicals was \$97.33 in 1995. A sample of this year's law periodicals yields the following subscription rates to the nearest dollar.

106	122	120	123
118	114	138	131
128	124	119	130

6.95 The U.S. Federal Highway Administration conducts studies on motor vehicle travel by type of vehicle. Results are published annually in *Highway Statistics*. A sample of 15 cars yields the following data on number of miles driven, in thousands, for last year.

10.2	10.3	8.9	12.7	8.3
9.2	13.7	7.7	3.3	10.6
11.8	6.6	8.6	5.7	12.0

6.96 The Bureau of Labor Statistics publishes information on average annual expenditures by consumers in *Consumer Expenditure Survey*. In 1995 the mean amount spent by consumers on nonalcoholic beverages was \$240. A random sample of 12 consumers yielded the following data, in dollars, on last year's expenditures on nonalcoholic beverages.

361	176	184	265
259	281	240	273
259	249	194	258

USING TECHNOLOGY  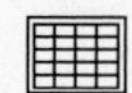 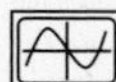

6.97 Use Minitab or some other statistical software to obtain a normal probability plot for the exam scores in Exercise 6.93.

6.98 Use Minitab or some other statistical software to obtain a normal probability plot for the subscription rates in Exercise 6.94.

In each of Exercises 6.99–6.102, use any technology that you have available to

a. *obtain a normal probability plot of the given data,*
b. *identify outliers, if any, and*
c. *assess the normality of the variable being considered.*

6.99 The U.S. Energy Information Administration reports figures on residential energy consumption and expenditures in *Residential Energy Consumption Survey: Consumption and Expenditures*. A sample of 18 households using electricity as their primary energy source yields the following data on one year's energy expenditures.

\$1376	1452	1235	1480	1185	1327
1059	1400	1227	1102	1168	1070
949	1351	1259	1179	1393	1456

6.100 In January 1994 the Department of Agriculture estimated that a typical U.S. family of four would spend \$117 per week for food. During that same year, a random sample of 10 Kansas families of four yielded the weekly food costs shown in the following table.

103	129	109	95	121
98	112	110	101	119

6.101 The U.S. National Center for Health Statistics compiles data on the length of stay by patients in short-term hospitals and publishes its findings in *Vital and Health Statistics.* A random sample of 21 patients yielded the following data on length of stay. The data are in days.

4	4	12	18	9	6	12
3	6	15	7	3	55	1
10	13	5	7	1	23	9

6.102 IQs measured on the Stanford Revision of the Binet-Simon Intelligence Scale have mean 100 points and standard deviation 16 points. Twenty-five randomly selected people are given the IQ test; the results are shown in the following table.

91	96	106	116	97
102	96	124	115	121
95	111	105	101	86
88	129	112	82	98
104	118	127	66	102

6.103 Gestation periods of humans are normally distributed with a mean of 266 days and a standard deviation of 16 days.

a. Use Minitab or some other statistical software to simulate four random samples of 50 human gestation periods each. *(Note:* If you are using Minitab, refer to Example 6.2 on page 360 for an explanation of how to simulate a normally distributed variable.*)*

b. Obtain a normal probability plot of each sample in part (a).

c. Are the normal probability plots in part (b) what you expected? Explain your answer.

6.104 Desert Samaritan Hospital, located in Mesa, Arizona, keeps records of emergency-room traffic. From those records we find that the times between arriving patients have a special type of reverse J-shaped distribution called an *exponential* distribution. We also find that the mean time between arriving patients is approximately 8.7 minutes.

a. Use Minitab or some other statistical software to simulate four random samples of 75 interarrival times each. *(Note:* If you are using Minitab, proceed in a similar way as for simulating a normal distribution, but begin by choosing **Calc ➤ Random Data ➤ Exponential. . . .***)*

b. Obtain a normal probability plot of each sample in part (a).

c. Are the normal probability plots in part (b) what you expected? Explain your answer.

6.5 NORMAL APPROXIMATION TO THE BINOMIAL DISTRIBUTION*

In this section we will discover that it is often possible to approximate binomial probabilities by areas under a suitable normal curve. The development of the mathematical theory for doing this is credited to Abraham de Moivre (1667–1754) and Pierre-Simon Laplace (1749–1827). For more information on de Moivre and Laplace, see the biographies at the end of Chapters 12 and 7, respectively.

First we need to briefly review the binomial distribution, which we discussed in detail in Section 5.3. Suppose n identical independent success-failure experiments

* This section is optional and will not be needed in subsequent sections of the book. Coverage of the binomial distribution in Chapter 5 is prerequisite to this section.

are performed, with the probability of success on any given trial being equal to p. Let X denote the total number of successes in the n trials. Then the probability distribution of the random variable X is given by the binomial probability formula,

$$P(X = x) = \binom{n}{x} p^x (1 - p)^{n-x}.$$

We say that X has the binomial distribution with parameters n and p.

You might be wondering why we would use normal-curve areas to approximate binomial probabilities when we can obtain them exactly by employing the binomial probability formula. Example 6.17 explains why.

EXAMPLE 6.17 THE NEED TO APPROXIMATE BINOMIAL PROBABILITIES

Mortality tables enable actuaries to obtain the probability that a person at any particular age will live a specified number of years. This, in turn, permits the determination of life-insurance premiums, retirement pensions, annuity payments, and related items of importance to insurance companies and others.

According to tables provided by the U.S. National Center for Health Statistics in *Vital Statistics of the United States,* there is about an 80% chance that a person age 20 will be alive at age 65. In Example 5.13 on page 318, we used the binomial probability formula to determine probabilities for the number of 20-year-olds out of three that will be alive at age 65.

For most real-world problems, the number of people under investigation is much larger than three. Although, in principle, we can use the binomial probability formula to determine probabilities regardless of the number of people being considered, the practical realities dictate otherwise. Suppose, for instance, that 500 people age 20 are selected at random. Find the probability that

a. exactly 390 of them will be alive at age 65.

b. between 375 and 425 of them, inclusive, will be alive at age 65.

SOLUTION Let X denote the number of people out of the 500 that will be alive at age 65. Then X has the binomial distribution with parameters $n = 500$ (the 500 people) and $p = 0.8$ (the probability a person age 20 will be alive at age 65). Thus, in principle, probabilities for X can be determined exactly by using the binomial probability formula,

$$P(X = x) = \binom{500}{x} (0.8)^x (0.2)^{500-x}.$$

Let's apply that formula to the problems posed in parts (a) and (b).

a. Here we want the probability that exactly 390 of the 500 people will still be alive at age 65, $P(X = 390)$. The "answer" is

$$P(X = 390) = \binom{500}{390}(0.8)^{390}(0.2)^{110}.$$

However, to actually obtain the numerical value of the expression on the right of the preceding equation is not easy, even with a calculator. In performing the computations, we must be careful to avoid such pitfalls as making roundoff errors and getting numbers so large or so small that they are outside the range of the calculator. Fortunately, as we will soon see, the computations can be sidestepped altogether by using normal-curve areas in a simple way.

b. For this part we need to determine the probability that between 375 and 425, inclusive, of the 500 people will be alive at age 65, $P(375 \le X \le 425)$. The "answer" is

$$\begin{aligned} P(375 \le X \le 425) &= P(X = 375) + P(X = 376) + \cdots + P(X = 425) \\ &= \binom{500}{375}(0.8)^{375}(0.2)^{125} + \binom{500}{376}(0.8)^{376}(0.2)^{124} \\ &\quad + \cdots + \binom{500}{425}(0.8)^{425}(0.2)^{75}. \end{aligned}$$

We have the same computational difficulties as we did in part (a), except that here we must evaluate 51 complex expressions instead of 1. Thus, although in theory we can use the binomial probability formula to determine the answer, doing so in practice is another matter.

Surprising as it might seem, there is a way to use normal-curve areas to get the (approximate) answer—and it is easy! We will return to this problem momentarily and obtain the probabilities required in parts (a) and (b). ◨

Example 6.17 makes it clear why we often need to approximate binomial probabilities, even though the binomial probability formula is available for computing them exactly: *It is not practical to use the binomial probability formula when the number of trials, n, is very large.*

Under certain conditions on n and p, the distribution of a binomial random variable is (roughly) bell-shaped. In such cases we can approximate probabilities for the random variable by areas under a suitable normal curve. To show how this is done, we present Example 6.18. For this example it is actually easy to calculate binomial probabilities exactly using the binomial probability formula. However, for purposes of illustration, we will also show how normal-curve areas approximate the binomial probabilities.

EXAMPLE 6.18 APPROXIMATING BINOMIAL PROBABILITIES BY AREAS UNDER A NORMAL CURVE

A student is taking a true-false exam with 10 questions. Assuming the student guesses at all 10 questions, use the binomial probability formula to determine the exact probability that the student gets either 7 or 8 answers correct. Then approximate that probability by an area under a suitable normal curve.

SOLUTION If we let X denote the number of correct guesses by the student, then X has a binomial distribution. There are 10 questions, so $n = 10$. Because the student guesses at each question, the success probability, p, is 0.5. Consequently, X has the binomial distribution with parameters $n = 10$ and $p = 0.5$. In other words, probabilities for X are given by the binomial probability formula,

$$P(X = x) = \binom{10}{x}(0.5)^x(1 - 0.5)^{10-x}.$$

Applying that formula, we obtain the probability distribution of X, shown below in Table 6.6.

TABLE 6.6
Probability distribution of the number of correct guesses by the student

Number correct x	Probability $P(X = x)$
0	0.0010
1	0.0098
2	0.0439
3	0.1172
4	0.2051
5	0.2461
6	0.2051
7	0.1172
8	0.0439
9	0.0098
10	0.0010

The problem is to determine the probability that the student gets either 7 or 8 questions correct, $P(X = 7 \text{ or } 8)$. From Table 6.6 we find that the exact probability is equal to

$$P(X = 7 \text{ or } 8) = P(X = 7) + P(X = 8) = 0.1172 + 0.0439 = 0.1611.$$

Let's now see how we can approximate the probability $P(X = 7 \text{ or } 8)$ by an area under a suitable normal curve. To begin, we refer to Table 6.6 in order to obtain the probability histogram of the random variable X. That histogram is shown in Fig. 6.25.

FIGURE 6.25
Probability histogram for X with superimposed normal curve

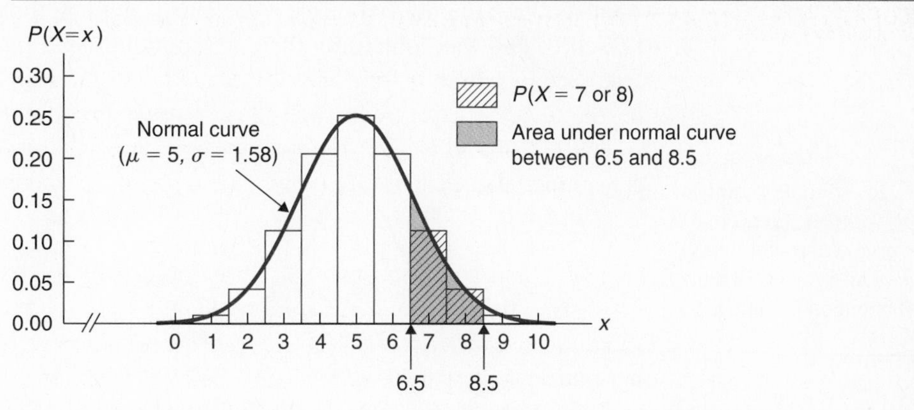

Because the probability histogram in Fig. 6.25 is bell-shaped, probabilities for X can be approximated by areas under a normal curve. As expected, the appropriate normal curve is the one whose parameters are the same as the mean and standard deviation of the random variable X. Since X has the binomial distribution with parameters $n = 10$ and $p = 0.5$, we can apply Formula 5.2 (page 321) to easily obtain the mean and standard deviation:

$$\mu = np = 10 \cdot 0.5 = 5$$

and

$$\sigma = \sqrt{np(1-p)} = \sqrt{10 \cdot 0.5 \cdot (1-0.5)} = 1.58.$$

So the normal curve used here is the one with parameters $\mu = 5$ and $\sigma = 1.58$. That normal curve is superimposed on the probability histogram in Fig. 6.25.

The probability $P(X = 7 \text{ or } 8)$ is exactly equal to the combined area of the corresponding bars of the histogram, the cross-hatched area in Fig. 6.25. By examining the figure carefully, we observe that the cross-hatched area is approximately equal to the area under the normal curve between 6.5 and 8.5, the shaded area in Fig. 6.25. It should be clear from Fig. 6.25 why we consider the area under the normal curve between 6.5 and 8.5 instead of between 7 and 8. This is called the **correction for continuity** because it is a correction required as a result of approximating the distribution of a discrete variable by that of a continuous variable.

Thus we see, at least qualitatively, that the probability $P(X = 7 \text{ or } 8)$ is approximately equal to the area under the normal curve with parameters $\mu = 5$ and $\sigma = 1.58$ that lies between 6.5 and 8.5. To compare these values quantitatively, we must obtain the normal-curve area. This is done in the usual way by converting to z-scores and then obtaining the corresponding area under the standard normal curve. See Fig. 6.26.

FIGURE 6.26 Determination of the area under the normal curve with parameters $\mu = 5$ and $\sigma = 1.58$ that lies between 6.5 and 8.5

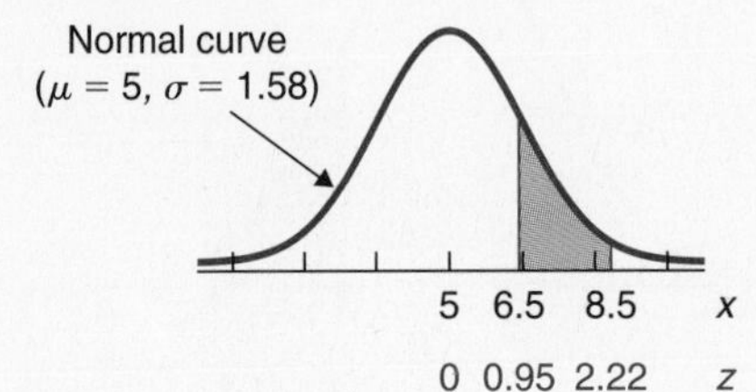

z-score computations:

Area to the left of z:

$x = 6.5 \longrightarrow z = \dfrac{6.5 - 5}{1.58} = 0.95$ 0.8289

$x = 8.5 \longrightarrow z = \dfrac{8.5 - 5}{1.58} = 2.22$ 0.9868

Shaded area $= 0.9868 - 0.8289 = 0.1579$

The last line in Fig. 6.26 shows that the area under the normal curve between 6.5 and 8.5 is 0.1579. Comparing that area to the exact value of the probability $P(X = 7 \text{ or } 8)$, which is 0.1611, we see that the normal-curve area provides an excellent approximation of the exact probability. ◨

As illustrated in Example 6.18, we can use normal-curve areas to approximate probabilities for binomial random variables that have bell-shaped distributions. Whether or not a particular binomial random variable has a bell-shaped distribution depends on its parameters, n and p. Figure 6.27 shows nine different binomial distributions.

As portrayed by Figs. 6.27(a) and 6.27(c), a binomial distribution with $p \neq 0.5$ is skewed. For small n, the skewness is enough to preclude using a normal approximation. However, as n increases, the skewness subsides and the binomial distribution becomes sufficiently bell-shaped to permit a normal approximation. On the other hand, as illustrated in Fig. 6.27(b), a binomial distribution with $p = 0.5$ is symmetric, regardless of the number of trials. Nonetheless, such a distribution will not be sufficiently bell-shaped to permit a normal approximation if n is too small.

The customary rule of thumb for using the normal approximation is that *both np and $n(1 - p)$ are 5 or greater.* This indicates, as suggested in Fig. 6.27, that the further the success probability is from 0.5 (in either direction), the larger the number of trials must be in order to employ the normal approximation.

FIGURE 6.27 Nine different binomial distributions

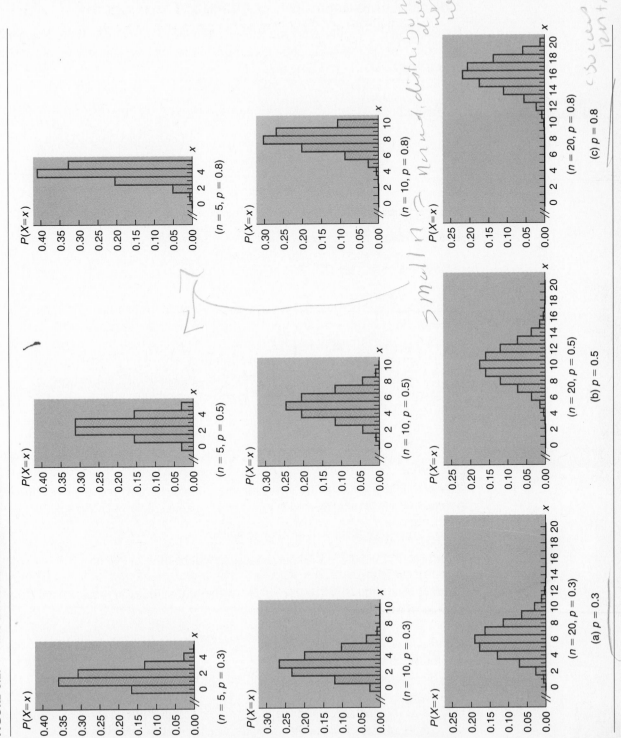

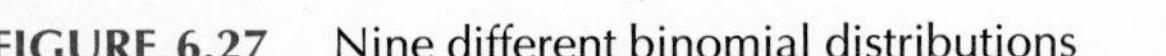

$(n = 5, p = 0.3)$ $(n = 5, p = 0.5)$ $(n = 5, p = 0.8)$

$(n = 10, p = 0.3)$ $(n = 10, p = 0.5)$ $(n = 10, p = 0.8)$

$(n = 20, p = 0.3)$ $(n = 20, p = 0.5)$ $(n = 20, p = 0.8)$

(a) $p = 0.3$ (b) $p = 0.5$ (c) $p = 0.8$

Normal Approximation to a Binomial Distribution

By examining carefully what we did in Example 6.18, we can write a general procedure for approximating binomial probabilities by areas under a normal curve. This is presented as Procedure 6.3.

PROCEDURE 6.3 TO APPROXIMATE BINOMIAL PROBABILITIES BY NORMAL-CURVE AREAS

Step 1 Determine n, the number of trials, and p, the success probability.

Step 2 Check that both np and $n(1-p)$ are 5 or greater. If they are not, the normal approximation should not be used.

Step 3 Find μ and σ using the formulas $\mu = np$ and $\sigma = \sqrt{np(1-p)}$.

Step 4 Make the correction for continuity and find the required area under the normal curve with parameters μ and σ.

Note: Step 4 of Procedure 6.3 requires the correction for continuity. As illustrated in Example 6.18, this means the following: When using normal-curve areas to approximate the probability that an observed value of a binomial random variable will be between two integers, inclusive, subtract 0.5 from the smaller integer and add 0.5 to the larger integer before finding the area under the normal curve.

We will now apply Procedure 6.3 to solve the problems posed at the beginning of this section in Example 6.17.

EXAMPLE 6.19 ILLUSTRATES PROCEDURE 6.3

The probability is 0.80 that a person age 20 will be alive at age 65. Suppose 500 people age 20 are selected at random. Determine the probability that

a. exactly 390 of them will be alive at age 65.

b. between 375 and 425 of them, inclusive, will be alive at age 65.

SOLUTION We will obtain the approximate values of the probabilities in parts (a) and (b) by applying Procedure 6.3.

Step 1 *Determine n, the number of trials, and p, the success probability.*

We have $n = 500$ and $p = 0.8$.

Step 2 *Check that both np and $n(1-p)$ are 5 or greater.*

Referring to the values for n and p obtained in Step 1, we see that

$$np = 500 \cdot 0.8 = 400 \quad \text{and} \quad n(1-p) = 500 \cdot 0.2 = 100.$$

Since both np and $n(1 - p)$ are greater than 5, we can employ the normal approximation.

Step 3 *Find μ and σ using the formulas $\mu = np$ and $\sigma = \sqrt{np(1-p)}$.*

We have $\mu = 500 \cdot 0.8 = 400$ and $\sigma = \sqrt{500 \cdot 0.8 \cdot 0.2} = 8.94$.

Step 4 *Make the correction for continuity and find the required area under the normal curve with parameters μ and σ.*

a. Here we want the probability that exactly 390 of the 500 people selected will be alive at age 65, that is, $P(X = 390)$. To make the correction for continuity, we subtract 0.5 from 390 and add 0.5 to 390. Thus we need to find the area under the normal curve with parameters $\mu = 400$ and $\sigma = 8.94$ that lies between 389.5 and 390.5. The required area is obtained in Fig. 6.28.

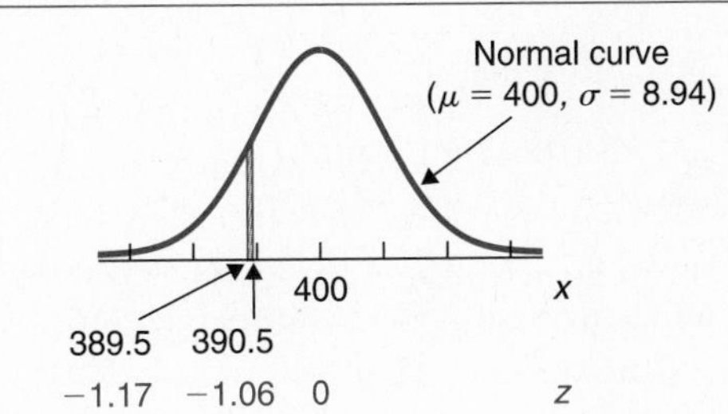

FIGURE 6.28 Determination of the area under the normal curve with parameters $\mu = 400$ and $\sigma = 8.94$ that lies between 389.5 and 390.5

z-score computations:

		Area to the left of *z*:
$x = 389.5$	$\longrightarrow z = \dfrac{389.5 - 400}{8.94} = -1.17$	0.1210
$x = 390.5$	$\longrightarrow z = \dfrac{390.5 - 400}{8.94} = -1.06$	0.1446

Shaded area $= 0.1446 - 0.1210 = 0.0236$

So $P(X = 390) = 0.0236$, approximately. The probability is roughly 0.0236 that exactly 390 of the 500 people selected will be alive at age 65.

b. For this part we want the probability that between 375 and 425, inclusive, of the 500 people selected will be alive at age 65, $P(375 \le X \le 425)$. To make the correction for continuity, we subtract 0.5 from 375 and add 0.5 to 425. Hence we need to determine the area under the normal curve with parameters $\mu = 400$ and $\sigma = 8.94$ that lies between 374.5 and 425.5. This is done as in part (a) by converting to z-scores and then obtaining the corresponding area under the standard normal curve. Doing that we find the required area to be 0.9956.

Consequently, $P(375 \leq X \leq 425) = 0.9956$, approximately. The probability that between 375 and 425 of the 500 people selected will be alive at age 65 is approximately equal to 0.9956. ◨

EXERCISES 6.5

STATISTICAL CONCEPTS AND SKILLS

6.105 Why do we sometimes use normal-curve areas to approximate binomial probabilities even though we have a formula for computing them exactly?

6.106 The rule of thumb for using the normal approximation to the binomial is that both np and $n(1 - p)$ are 5 or greater. Why is this restriction necessary?

6.107 Refer to Example 6.18 on page 394.

a. Use Table 6.6 to find the exact probability that the student guesses correctly on
  i. 4 or 5 questions, $P(X = 4 \text{ or } 5)$.
  ii. between 3 and 7 questions, $P(3 \leq X \leq 7)$.

b. Apply Procedure 6.3 on page 398 to approximate the probabilities in part (a) by areas under a normal curve. Compare your answers.

6.108 Refer to Example 6.18 on page 394.

a. Use Table 6.6 to find the exact probability that the student guesses correctly on
  i. at most 5 questions, $P(0 \leq X \leq 5)$.
  ii. at least 6 questions, $P(6 \leq X \leq 10)$.

b. Apply Procedure 6.3 on page 398 to approximate the probabilities in part (a) by areas under a normal curve. Compare your answers.

6.109 If in Example 6.18 the true-false exam had 30 questions instead of 10, which normal curve would you use to approximate probabilities for the number of correct guesses?

6.110 If in Example 6.18 the true-false exam had 25 questions instead of 10, which normal curve would you use to approximate probabilities for the number of correct guesses?

In Exercises 6.111–6.114, apply Procedure 6.3 to approximate the required binomial probabilities.

6.111 According to the *Daily Racing Form,* the probability is 0.67 that the favorite in a horse race will finish in the money (first, second, or third place). Determine the probability that in the next 200 races the favorite will finish in the money

a. exactly 140 times.

b. between 120 and 130 times, inclusive.

c. at least 150 times.

6.112 As reported by a spokesperson for Southwest Airlines, the no-show rate for reservations is 16%. In other words, the probability is 0.16 that a party making a reservation will not show up. The next flight has 42 parties with reservations. What is the probability that

a. exactly 5 parties do not show up?

b. between 9 and 12, inclusive, do not show up?

c. at least 1 does not show up?

d. at most 2 do not show up?

6.113 The U.S. National Center for Health Statistics states in *Vital and Health Statistics* that 38.3% of all injuries in the United States occur at home. Out of 500 randomly selected injuries, what is the probability that the number occurring at home will be

a. exactly 200?

b. between 180 and 210, inclusive?

c. at most 225?

6.114 *The World Almanac* reports that the infant mortality rate in India is 139 per 1000 live births. Determine the probability that out of 1000 randomly selected live births in India, there are

a. exactly 139 infant deaths.

b. between 120 and 150 infant deaths, inclusive.

c. at most 130 infant deaths.

Extending the Concepts and Skills

6.115 A roulette wheel consists of 38 numbers, of which 18 are red, 18 are black, and 2 are green. When the roulette ball is spun, it is equally likely to land on each of the 38 numbers. A gambler is playing roulette and bets \$10 on red each time. If the ball lands on a red number, the gambler wins \$10 from the house; otherwise, the gambler loses \$10. What is the probability that the gambler will be ahead after

a. 100 bets? **b.** 1000 bets? **c.** 5000 bets?
(*Hint:* The gambler will be ahead after a series of bets if and only if she has won more than half of the bets.)

6.116 A brand of flashlight battery has normally distributed lifetimes with a mean of 30 hours and a standard deviation of 5 hours. A supermarket purchases 500 of these batteries from the manufacturer. What is the probability that at least 80% of them will last longer than 25 hours?

CHAPTER REVIEW

You Should Be Able To

1. use and understand the formulas presented in this chapter.
2. explain what it means for a variable to be normally distributed or approximately normally distributed.
3. explain the meaning of the parameters for a normal curve.
4. identify the basic properties of and sketch a normal curve.
5. identify the standard normal distribution and the standard normal curve.
6. use Table II to find areas under the standard normal curve.
7. use Table II to find the z-score(s) corresponding to a specified area under the standard normal curve.
8. use and understand the z_α notation.
9. determine a percentage or probability for a normally distributed variable.
10. state and apply the 68.26-95.44-99.74 rule.
11. determine the observations corresponding to a specified percentage or probability for a normally distributed variable.
12. explain how to assess the normality of a variable using a normal probability plot.
13. construct a normal probability plot with the aid of Table III.
14. detect outliers from a normal probability plot.

*15. approximate binomial probabilities by normal-curve areas, when appropriate.
*16. use the Minitab procedures covered in this chapter.
*17. interpret the output obtained from the application of the Minitab procedures discussed in this chapter.

Key Terms

68.26-95.44-99.74 rule, *377*
approximately normally distributed, *354*
correction for continuity, *395*
empirical rule, *379*
normal curve, *354*
normal distribution, *354*
normal probability plot, *385*
normal scores, *385*
normally distributed population, *354*
normally distributed variable, *354*
parameters, *354*
standard normal curve, *358*
standard normal distribution, *358*
standardized normally distributed variable, *359*
z_α, *370*
z-curve, *365*

REVIEW TEST

STATISTICAL CONCEPTS AND SKILLS

1. Identify two primary reasons for studying the normal distribution.

2. Define the following terms.
a. normally distributed variable
b. normally distributed population
c. parameters for a normal curve

3. Answer true or false to each of the following statements. Give reasons for your answers.
a. Two variables having the same mean and standard deviation have the same distribution.
b. Two normally distributed variables having the same mean and standard deviation have the same distribution.

4. Explain the relationship between percentages for a normally distributed variable and areas under the corresponding normal curve.

5. Identify the distribution of the standardized version of a normally distributed variable.

6. Answer true or false to each of the following statements.
a. Two normal distributions having the same mean are centered at the same place, regardless of the relationship between their standard deviations.
b. Two normal distributions having the same standard deviation have the same shape, regardless of the relationship between their means.

7. Consider the normal curves having the following parameters: $\mu = 1.5$ and $\sigma = 3$; $\mu = 1.5$ and $\sigma = 6.2$; $\mu = -2.7$ and $\sigma = 3$; $\mu = 0$ and $\sigma = 1$.
a. Which curve has the largest spread?
b. Which curves are centered at the same place?
c. Which curves have the same shape?
d. Which curve is centered farthest to the left?
e. Which curve is the standard normal curve?

8. What key fact permits us to determine percentages for a normally distributed variable by first converting to z-scores and then determining the corresponding area under the standard normal curve?

9. Explain in words how Table II is used to determine the area under the standard normal curve that lies
a. to the left of a specified z-score.
b. to the right of a specified z-score.
c. between two specified z-scores.

10. Explain in words how Table II is used to determine the z-score having a specified area to its
a. left under the standard normal curve.
b. right under the standard normal curve.

11. What does the symbol z_α stand for?

12. State the 68.26-95.44-99.74 rule.

13. Roughly speaking, what are the normal scores corresponding to a sample of observations?

14. If we observe the values of a normally distributed variable for a sample, then a normal probability plot should be roughly ____.

15. Sketch the normal curve with parameters
a. $\mu = -1$ and $\sigma = 2$. **b.** $\mu = 3$ and $\sigma = 2$.
c. $\mu = -1$ and $\sigma = 0.5$.

16. According to R. R. Paul, the mean gestation period of the Morgan mare is 339.6 days ("Foaling Date," *The Morgan Horse,* 33:40, 1973). The gestation periods are normally distributed with a standard deviation of 13.3 days. Let x denote gestation period for Morgan mares.
a. Sketch the distribution of the variable x.
b. Obtain the standardized version, z, of x.
c. Identify and sketch the distribution of z.
d. Given that the area under the normal curve with parameter 339.6 and 13.3 that lies between 320 and 360 is 0.8672, determine the probability that a randomly selected Morgan mare (in foal) will have a gestation period between 320 days and 360 days.
e. The percentage of Morgan mares that have gestation periods exceeding 369 days equals the area under the standard normal curve that lies to the ____ of ____.

17. According to Table II, the area under the standard normal curve that lies to the left of 1.05 is 0.8531.

Without further reference to Table II, determine the area under the standard normal curve that lies

a. to the right of 1.05.
b. to the left of −1.05.
c. between −1.05 and 1.05.

18. Determine and sketch the area under the standard normal curve that lies

a. to the left of −3.02.
b. to the right of 0.61.
c. between 1.11 and 2.75.
d. between −2.06 and 5.02.
e. between −4.11 and −1.5.
f. either to the left of 1 or to the right of 3.

19. For the standard normal curve, find the z-score(s)

a. having area 0.30 to its left.
b. having area 0.10 to its right.
c. $z_{0.025}$, $z_{0.05}$, $z_{0.01}$, and $z_{0.005}$.
d. that divide the area under the curve into a middle 0.99 area and two outside 0.005 areas.

20. Each year, thousands of college seniors take the Graduate Record Examination (GRE). The scores are transformed so they have a mean of 500 and a standard deviation of 100. Furthermore, the scores are known to be normally distributed. Determine the percentage of students that score

a. between 350 and 625. **b.** 375 or greater.
c. below 750.

21. Use the 68.26-95.44-99.74 rule to fill in the following blanks with regard to Problem 20.

a. 68.26% of the students who take the GRE score between ____ and ____.
b. 95.44% of the students who take the GRE score between ____ and ____.
c. 99.74% of the students who take the GRE score between ____ and ____.

22. Refer to Problem 20.

a. Obtain the quartiles for GRE scores. Interpret your results in words.
b. Find the 99th percentile for GRE scores. Interpret your result in words.

23. Each year, manufacturers perform mileage tests on new car models and submit the results to the Environmental Protection Agency (EPA). The EPA then tests the vehicles to determine whether the manufacturers are correct. In 1998 one company reported that a particular model equipped with a four-speed manual transmission averaged 29 mpg on the highway. Suppose the EPA tested 15 of the cars and obtained the following gas mileages.

27.3	31.2	29.4	31.6	28.6
30.9	29.7	28.5	27.8	27.3
25.9	28.8	28.9	27.8	27.6

a. Use Table III to construct a normal probability plot for the data. For simplicity, treat equal observations as being slightly different when obtaining the normal scores.
b. Use part (a) to identify any outliers.
c. Use part (a) to assess normality.

***24.** Acute rotavirus diarrhea is the leading cause of death among children under age 5, killing an estimated 4.5 million annually in developing countries. Scientists from Finland and Belgium have claimed that a new oral vaccine is 80% effective against rotavirus diarrhea. Assuming the claim is correct, find the probability that out of 1500 cases, the vaccine will be effective in

a. exactly 1225 cases. **b.** at least 1175 cases.
c. between 1150 and 1250 cases, inclusive.

USING TECHNOLOGY

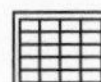

25. Refer to Problem 20. Use any technology available to you to do the following.

a. Sketch the normal curve for GRE scores.
b. Simulate 1000 GRE scores.
c. Approximately what would you expect the sample mean and sample standard deviation of the 1000 GRE scores obtained in part (b) to be? Explain your answers.
d. Determine the sample mean and sample standard deviation of the 1000 GRE scores obtained in part (b) and compare your answers to your answers in part (c).
e. Roughly what would you expect a histogram of the 1000 GRE scores obtained in part (b) to look like? Explain your answer.

f. Obtain a histogram of the 1000 GRE scores from part (b) and compare your result to your expectation in part (e).

26. Use Minitab or some other statistical software to solve Problem 20. Comment on any discrepancies.

27. Use Minitab or some other statistical software to solve Problem 22. Comment on any discrepancies.

28. Use Minitab or some other statistical software to obtain a normal probability plot for the gas mileages in Problem 23. Comment on any differences between the two plots.

INTERNET PROJECT

IQ of Boys and Girls

In this Internet project, you will interact with several simulations and animations that show why and how the normal distribution is useful. You will also explore a real data set containing the intelligence quotient (IQ) scores for boys and girls in primary school. Do you think that, in general, boys and girls are equally smart? While most people believe there is no difference in IQ between the sexes, some believe there is.

IQ is a measure constructed to be normally distributed with a mean of 100 and a standard deviation of 16. Earlier in this chapter, we calculated probabilities for IQs based on this normal distribution model. In this Internet project, you will discover how well the model fits the real data.

URL for access to Internet Projects Page: `http://hepg.awl.com` *keyword:* Weiss

USING THE FOCUS DATABASE

In Chapter 1 we explained how to store the Focus database in a Minitab worksheet named `focus.mtw`. If you haven't already created that worksheet, follow the instructions on pages 54–55 to create it now.

The Focus database contains information on 500 randomly selected Arizona State University sophomores. Seven variables are considered for each student: sex, high-school GPA, SAT math score, cumulative GPA, SAT verbal score, age, and total hours. Use Minitab or some other statistical software to solve the following problems.

a. Obtain a histogram and a dotplot for each of the data sets high-school GPA, SAT math score, cumulative GPA, SAT verbal score, age, and total hours.

b. Based on part (a), which of the six variables for ASU sophomores—high-school GPA, SAT math score, cumulative GPA, SAT verbal score, age, and total hours—appear to be approximately normally distributed?

c. Obtain a normal probability plot for each of the data sets high-school GPA, SAT math score, cumulative GPA, SAT verbal score, age, and total hours.

d. Based on part (c), which of the six variables for ASU sophomores—high-school GPA, SAT math score, cumulative GPA, SAT verbal score, age, and total hours—appear to be approximately normally distributed?

CASE STUDY DISCUSSION

Chest Sizes of Scottish Militiamen

On page 352, we presented a frequency distribution for data on chest circumference by height for 5732 Scottish militiamen. As mentioned there, Adolphe Quetelet used a procedure for fitting a normal curve to the data based on the binomial distribution. Here you will accomplish that task by using techniques that you studied earlier in this chapter.

a. Construct a relative-frequency histogram for the chest-circumference data using classes based on a single value.
b. Find the population mean and population standard deviation of the data.
c. Identify the normal curve that should be used for the chest circumferences.
d. Use the table on page 352 to find the percentage of militiamen in the survey with chest circumferences between 36 and 41 inches, inclusive. *(Note:* Since the circumferences were rounded to the nearest inch, you are actually finding the percentage of militiamen in the survey with chest circumferences between 35.5 and 41.5 inches.*)*
e. Use the normal curve you identified in part (c) to obtain an approximation to the percentage of militiamen in the survey having chest circumferences between 35.5 and 41.5 inches. Compare your answer to the exact percentage found in part (d).
f. Use Minitab or some other statistical software to solve parts (a), (b), (d), and (e).

BIOGRAPHY CARL FRIEDRICH GAUSS

Born on April 30, 1777, in Brunswick, Germany, the only son in a poor, semi-literate peasant family, Carl Friedrich Gauss taught himself to calculate before he could talk. At the age of 3, he pointed out an error in his father's calculations of wages. In addition to his arithmetic experimentation, he taught himself to read. At the age of 8, Gauss instantly solved the summing of all numbers from 1 to 100. His father was persuaded to allow him to stay in school and to study after school instead of working to help support the family.

Impressed by Gauss's brilliance, the Duke of Brunswick supported him monetarily from the ages of 14 to 30. This permitted Gauss to pursue his studies exclusively. Gauss conceived most of his mathematical discoveries by the time he was 17. He was granted his doctorate in absentia from the university at Helmstedt; his doctoral thesis developed the concept of complex numbers and proved the fundamental theorem of algebra, which had previously been only partially established. Shortly thereafter, Gauss published his theory of numbers, which is considered one of the most brilliant achievements in mathematics.

Gauss made important discoveries in many disciplines. Two of his major contributions to statistics were the development of the least-squares method and fundamental work with the normal distribution, often called the *Gaussian distribution* in his honor.

In 1807, Gauss accepted the directorship of the observatory at the University of Göttingen which ended his dependence on the Duke of Brunswick. He remained there the rest of his life. In 1833 Gauss and a colleague, Wilhelm Weber, invented a working electric telegraph, 5 years before Samuel Morse. Gauss died in Göttingen in 1855.

CASE STUDY

THE CHESAPEAKE AND OHIO FREIGHT STUDY

Can relatively small samples really provide results that are nearly as accurate as those obtained from a census? Although statisticians have shown mathematically that this is the case, a real study in which the results of a sample are compared with those of a census might make this fact even more credible.

When a freight shipment travels over several railroads, the freight charge is divided among them according to prearranged agreements. With each shipment of freight, a document called a *waybill* is issued that provides information on the goods, route, and total charges. From the waybill of any particular freight shipment, the amount due each railroad can be calculated.

For a large number of shipments, the computations required for allocating the shares properly among the railroads are time-consuming and costly. Consequently, if the division of total revenue among the railroads in question could be done accurately on the basis of a sample—as statisticians contend—then considerable savings could be realized in accounting and clerical costs.

To convince themselves of the validity of the sampling approach, officials of the Chesapeake and Ohio Railroad Company (C&O) undertook a study of freight shipments that had traveled over its Pere Marquette district and another railroad during a 6-month period. The total number of waybills for that period was known (22,984), as was the total freight revenue. The problem was to obtain an accurate estimate of the total revenue due C&O using as small a sample of waybills as possible.

Statistical theory was applied to determine the sample size required in order to obtain an estimate of the total revenue due C&O with a prescribed accuracy. In all, 2072 of the 22,984 waybills, roughly 9%, were sampled. For each waybill in the sample, the necessary calculations were performed to find the amount of freight revenue for that shipment belonging to C&O. From those amounts the total revenue due C&O for all shipments was estimated to be $64,568.

How close was the estimate of $64,568, based on a sample of only 2072 waybills, to the total revenue actually due C&O for the 22,984 waybills? Take a guess; we'll discuss the answer at the end of this chapter.

The Sampling Distribution of the Mean

7

GENERAL OBJECTIVES

In the preceding chapters, we have studied sampling, descriptive statistics, probability, random variables, and the normal distribution. Now we will learn how those seemingly diverse topics can be integrated to lay the groundwork for inferential statistics.

We will first explain why distributions must be incorporated into the design of inferential studies. Next we will obtain formulas for the mean and standard deviation of all possible sample means for samples of a given size from a population. Following that we will investigate the sampling distribution of the mean—the distribution of all possible sample means for samples of a given size. That concept sets the stage for two important statistical-inference procedures—using the mean, $\bar{x}$, of a sample from a population to estimate and draw conclusions about the mean, μ, of the entire population.

CHAPTER OUTLINE

7.1 SAMPLING ERROR; THE NEED FOR SAMPLING DISTRIBUTIONS

We have already discovered that using a sample to acquire information about a population is often preferable to conducting a census, where data for the entire population are collected. Generally, sampling is less costly and can be done more quickly than a census; it is often the only practical way to gather information.

But now we need to deal with the following problem: Because a sample from a population provides data for only a portion of the entire population, it is unreasonable to expect the sample to yield perfectly accurate information about the population. Thus we should anticipate that a certain amount of error will result simply because we are sampling. This kind of error is called **sampling error.**

DEFINITION 7.1 SAMPLING ERROR

Sampling error is the error resulting from using a sample to estimate a population characteristic.

EXAMPLE 7.1 SAMPLING ERROR AND THE NEED FOR SAMPLING DISTRIBUTIONS

The Census Bureau publishes annual figures on the mean income of U.S. households in *Current Population Reports.* Actually, the Census Bureau reports the mean income of a sample of about 60,000 households out of a total of more than 97 million households. For instance, in 1993 the mean income of U.S. households was reported to be $41,428. This amount is really the sample mean income, $\bar{x}$, of the 60,000 households surveyed, not the (population) mean income, μ, of all U.S. households.

We certainly cannot expect the mean income, $\bar{x}$, of the 60,000 households sampled by the Census Bureau to be exactly the same as the mean income, μ, of all U.S. households—some sampling error is to be anticipated. But how much sampling error should we expect; that is, how accurate are such estimates likely to be? Is it likely, for instance, that the sample mean household income reported by the Census Bureau will be within $1000 of the population mean household income?

 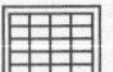

To answer such questions, we need to know the distribution of all possible sample means that could be obtained by sampling the incomes of 60,000 households. That distribution is called the *sampling distribution of the mean.* □

The distribution of a statistic (i.e., of all possible observations of the statistic for samples of a given size) is called the **sampling distribution** of the statistic. In this chapter we will concentrate on the sampling distribution of the statistic $\bar{x}$.

DEFINITION 7.2 SAMPLING DISTRIBUTION OF THE MEAN

For a variable x and a given sample size, the distribution of the variable $\bar{x}$ (i.e., of all possible sample means) is called the ***sampling distribution of the mean.***

We see that in statistics the following terms and phrases are synonymous:

- sampling distribution of the mean
- distribution of the variable $\bar{x}$
- distribution of all possible sample means of a given size

Therefore, in this book, we will use these three terms interchangeably.

Introducing the sampling distribution of the mean with an example that is both realistic and concrete is difficult because even for moderately large populations the number of possible samples is enormous, thus prohibiting an actual listing of the possibilities.† Consequently, we will use an unrealistically small population to introduce the sampling distribution of the mean. Keep in mind, however, that populations are much larger in real-life applications.

EXAMPLE 7.2 INTRODUCES THE SAMPLING DISTRIBUTION OF THE MEAN

The heights, in inches, of the five starting players, whom we will call A, B, C, D, and E, on a men's basketball team are displayed in Table 7.1. Here the population of interest consists of the five players and the variable under consideration is height.

TABLE 7.1 Heights of the five starting players

Player	A	B	C	D	E
Height	76	78	79	81	86

a. Obtain the sampling distribution of the mean for samples of size two.

b. Make some observations about sampling error when the mean height of a random sample of two players is used to estimate the population mean height.‡

c. Determine the probability that, for a random sample of size two, the sampling error made in estimating the population mean by the sample mean will be 1 inch or less; that is, determine the probability that $\bar{x}$ will be within 1 inch of μ.

† For example, the number of possible samples of size 50 from a population of size 10,000 is approximately 3×10^{135}, a 3 followed by 135 zeros.

‡ Unless otherwise specified, when we say *sample* or *random sample,* we mean *simple random sample.* Also, we assume sampling is without replacement unless stated otherwise.

SOLUTION For future reference we first compute the population mean height:

$$\mu = \frac{\Sigma x}{N} = \frac{76 + 78 + 79 + 81 + 86}{5} = 80.$$

a. For this part we need to determine the distribution of all possible sample means for samples of size two. The population under consideration here is so small that we can list the possibilities directly. There are 10 possible samples of size two. The first column of Table 7.2 displays the 10 possible samples, the second column the corresponding heights (i.e., values of the variable height), and the third column the sample means. As a visual aid, we have also drawn a dotplot in Fig. 7.1 to portray the distribution of the sample means, that is, the sampling distribution of the mean for samples of size two.

TABLE 7.2
Possible samples and sample means for samples of size two

Sample	Heights	$\bar{x}$
A, B	76, 78	77.0
A, C	76, 79	77.5
A, D	76, 81	78.5
A, E	76, 86	81.0
B, C	78, 79	78.5
B, D	78, 81	79.5
B, E	78, 86	82.0
C, D	79, 81	80.0
C, E	79, 86	82.5
D, E	81, 86	83.5

FIGURE 7.1
Dotplot for the sampling distribution of the mean for samples of size two ($n = 2$)

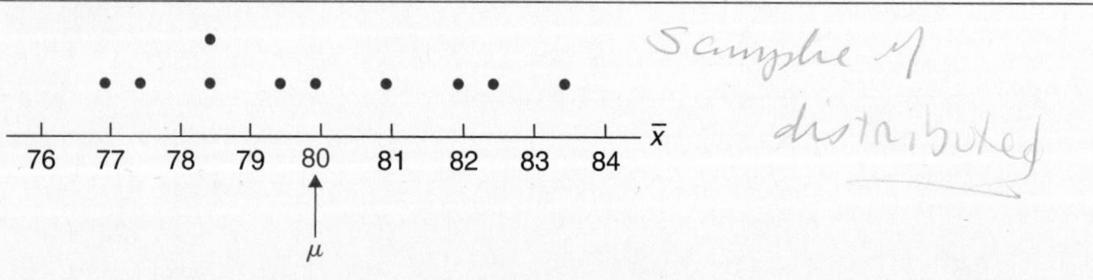

b. Referring to Table 7.2 or Fig. 7.1, we can make some simple but significant observations about sampling error when the mean height of a random sample of two players is used to estimate the population mean height. For instance, it is unlikely that the mean height of the two players selected will equal the population mean of 80 inches. In fact, only one of the 10 samples has mean 80 inches, the eighth sample in Table 7.2. Thus, in this case, the chances are only $\frac{1}{10}$, or 10%, that $\bar{x}$ will equal μ; some sampling error is likely.

c. Referring to Fig. 7.1, we see that exactly three of the 10 samples have means within 1 inch of the population mean of 80 inches. So the probability is $\frac{3}{10}$, or 0.3, that the sampling error made in estimating μ by $\bar{x}$ will be 1 inch or less. In other words, there is a 30% chance that the mean height of the two players selected will be within 1 inch of the population mean. ◨

In Example 7.2 we determined the sampling distribution of the mean for samples of size two. If we consider samples of another size, say, of size four, then we obtain a different sampling distribution of the mean.

EXAMPLE 7.3 **FURTHER ILLUSTRATES THE SAMPLING DISTRIBUTION OF THE MEAN**

The heights of the five starting players on a men's basketball team are presented in Table 7.1 on page 409.

a. Obtain the sampling distribution of the mean for samples of size four.

b. Make some observations about sampling error when the mean height of a random sample of four players is used to estimate the population mean height.

c. Determine the probability that, for a random sample of size four, the sampling error made in estimating the population mean by the sample mean will be 1 inch or less; that is, determine the probability that $\bar{x}$ will be within 1 inch of μ.

SOLUTION **a.** For this part we need to determine the distribution of all possible sample means for samples of size four. There are five possible samples of size four. The first column of Table 7.3 displays the possible samples, the second column the corresponding heights (i.e., values of the variable height), and the third column the sample means. As a visual aid, we have also drawn a dotplot in Fig. 7.2 at the top of the next page to portray the distribution of the sample means, that is, the sampling distribution of the mean for samples of size four.

TABLE 7.3 Possible samples and sample means for samples of size four

Sample	Heights	$\bar{x}$
A, B, C, D	76, 78, 79, 81	78.50
A, B, C, E	76, 78, 79, 86	79.75
A, B, D, E	76, 78, 81, 86	80.25
A, C, D, E	76, 79, 81, 86	80.50
B, C, D, E	78, 79, 81, 86	81.00

b. Referring to Table 7.3 or Fig. 7.2 on the next page, we observe that none of the samples of size four has a mean equal to the population mean of 80 inches. Thus, when the mean height of a random sample of four players is used to estimate the population mean height, some sampling error is certain.

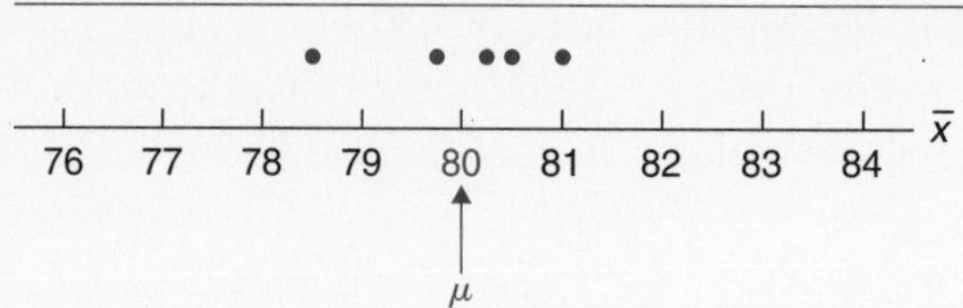

FIGURE 7.2 Dotplot for the sampling distribution of the mean for samples of size four ($n = 4$)

c. Referring to Fig. 7.2, we see that exactly four of the five samples have means within 1 inch of the population mean of 80 inches. So the probability is $\frac{4}{5}$, or 0.8, that the sampling error made in estimating μ by $\bar{x}$ will be 1 inch or less. In other words, there is an 80% chance that the mean height of the four players selected will be within 1 inch of the population mean. ◨

Sample Size and Sampling Error

In Figs. 7.1 and 7.2, we drew dotplots for the sampling distributions of the mean for samples of sizes two and four, respectively. Those two dotplots together with ones for samples of sizes one, three, and five, are displayed in Fig. 7.3.

Figure 7.3 vividly depicts that the possible sample means cluster closer around the population mean as the sample size increases. This in turn implies that sampling error tends to (although may not) be smaller for large samples than for small samples.

For example, as we see from Fig. 7.3, for samples of size one, 2 out of 5, or 40%, of the possible sample means lie within 1 inch of μ; for samples of size two, 3 out of 10, or 30%, of the possible sample means lie within 1 inch of μ; for samples of size three, 5 out of 10, or 50%, of the possible sample means lie within 1 inch of μ; for samples of size four, 4 out of 5, or 80%, of the possible sample means lie within 1 inch of μ; and for samples of size five, 1 out of 1, or 100%, of the possible sample means lie within 1 inch of μ. Table 7.4 summarizes these results and also provides another sampling-error illustration easily obtained from Fig. 7.3.

TABLE 7.4 Sample size and sampling error illustrations for the heights of the basketball players

Sample size n	No. possible samples	No. within 1″ of μ	% within 1″ of μ	No. within 0.5″ of μ	% within 0.5″ of μ
1	5	2	40%	0	0%
2	10	3	30%	2	20%
3	10	5	50%	2	20%
4	5	4	80%	3	60%
5	1	1	100%	1	100%

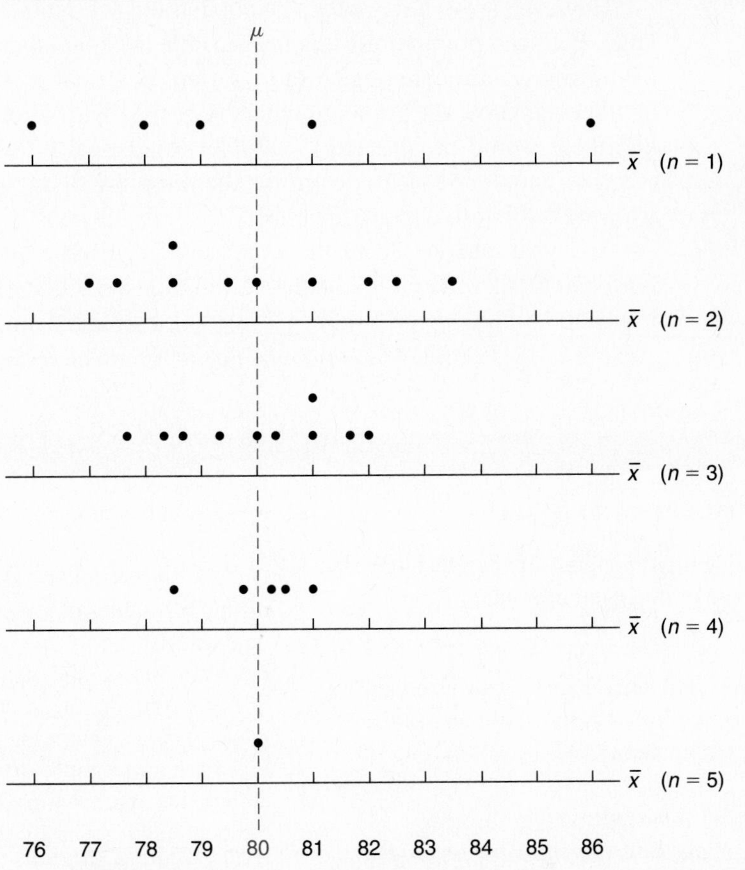

FIGURE 7.3 Dotplots for the sampling distributions of the mean for samples of sizes one, two, three, four, and five

More generally, we can make the following qualitative statement.

KEY FACT 7.1 SAMPLE SIZE AND SAMPLING ERROR

The possible sample means cluster closer around the population mean as the sample size increases. Thus the larger the sample size, the smaller the sampling error tends to be in estimating a population mean, μ, by a sample mean, $\bar{x}$.

What We Do in Practice

We have used the heights of a population of five basketball players to illustrate and explain the importance of the sampling distribution of the mean. For that small population, it is easy to obtain the sampling distribution of the mean for any sample size by listing all of the possible sample means.

However, as we have noted, the populations we deal with in practice are usually large. For such populations it is not feasible to obtain the sampling distribution of the mean by a direct listing. A more serious practical problem is that in reality we do not even know the population data (i.e., the distribution of the variable)—if we did, there would be no need to sample! Realistically then, we could not list the possible sample means to determine the sampling distribution of the mean even if we were willing to expend the effort.

So, what can we do in the usual case of a large and unknown population? Fortunately, there exist mathematical relationships that allow us to determine, at least approximately, the sampling distribution of the mean for any specified sample size. We will discuss those relationships in Sections 7.2 and 7.3.

EXERCISES 7.1

STATISTICAL CONCEPTS AND SKILLS

7.1 Why is sampling often preferable to conducting a census for the purpose of obtaining information about a population?

7.2 Why do we generally expect some error when estimating a parameter (such as a population mean) by a statistic (such as a sample mean)?

Exercises 7.3–7.15 are intended solely to provide concrete illustrations of the sampling distribution of the mean. For that reason the populations considered are unrealistically small. In each exercise assume that sampling is without replacement.

7.3 As reported by the *World Almanac, 1998,* the annual salaries, to the nearest thousand dollars, of the top five state officials of Oklahoma are as follows. Consider these officials a population of interest.

Official	Salary ($1000s)
Governor (G)	70
Lieutenant Governor (L)	63
Secretary of State (S)	44
Attorney General (A)	75
Treasurer (T)	70

a. Compute the population mean of the five salaries.
b. For samples of size two, construct a table similar to Table 7.2 on page 410. Use the letters placed parenthetically after each official to represent the officials.
c. Draw a dotplot for the sampling distribution of the mean for samples of size two like the one in Fig. 7.1 on page 410.
d. For a random sample of size two, what is the chance that the sample mean will equal the population mean? That is, determine $P(\bar{x} = \mu)$.
e. For a random sample of size two, obtain the probability that the sampling error made in estimating the population mean by the sample mean will be 4 (i.e., $4000) or less; that is, determine the probability that $\bar{x}$ will be within 4 of μ. Interpret your result in terms of percentages.

7.4 Repeat parts (b)–(e) of Exercise 7.3 for samples of size one.

7.5 Repeat parts (b)–(e) of Exercise 7.3 for samples of size three.

7.6 Repeat parts (b)–(e) of Exercise 7.3 for samples of size four.

7.7 Repeat parts (b)–(e) of Exercise 7.3 for samples of size five.

7.8 This exercise requires that you have done Exercises 7.3–7.7.
a. Draw a graph similar to Fig. 7.3 on page 413 for sample sizes of one, two, three, four, and five.

b. What does your graph in part (a) illustrate about the impact that increasing sample size has on sampling error?

c. Construct a table similar to Table 7.4 on page 412 for some values of your choice.

7.9 The lengths, in centimeters, of six bullfrogs in a small pond are as follows.

Frog	A	B	C	D	E	F
Length	19	14	15	9	16	17

Consider the six bullfrogs a population of interest.

a. Calculate the mean length, μ, of the six bullfrogs.

b. For samples of size two, construct a table similar to Table 7.2 on page 410. (There are 15 possible samples of size two.)

c. Draw a dotplot for the sampling distribution of the mean for samples of size two.

d. For a random sample of size two, what is the chance that the sample mean will equal the population mean? That is, determine $P(\bar{x} = \mu)$.

e. For a random sample of size two, determine the probability that the mean length of the sample selected will be within 1 cm of the population mean. Interpret your result in terms of percentages.

7.10 Repeat parts (b)–(e) of Exercise 7.9 for samples of size one.

7.11 Repeat parts (b)–(e) of Exercise 7.9 for samples of size three. (There are 20 possible samples.)

7.12 Repeat parts (b)–(e) of Exercise 7.9 for samples of size four. (There are 15 possible samples.)

7.13 Repeat parts (b)–(e) of Exercise 7.9 for samples of size five. (There are six possible samples.)

7.14 Repeat parts (b)–(e) of Exercise 7.9 for samples of size six. What is the relationship between the only possible sample here and the population?

7.15 Explain what the dotplots in parts (c) of Exercises 7.9–7.14 illustrate about the impact increasing sample size has on sampling error.

Extending the Concepts and Skills

7.16 Suppose a sample is to be taken without replacement from a finite population of size N. If the sample size is the same as the population size,

a. how many possible samples are there?

b. what are the possible sample means?

c. what is the relationship between the only possible sample and the population?

7.17 Suppose a random sample of size one is to be taken from a finite population of size N.

a. How many possible samples are there?

b. Identify the relation between the possible sample means and the possible observations of the variable under consideration.

c. What is the difference between taking a random sample of size one from a population and selecting a member at random from the population?

7.2 THE MEAN AND STANDARD DEVIATION OF $\bar{x}$

In Section 7.1 we discussed the sampling distribution of the mean—the distribution of all possible sample means for any specified sample size or, equivalently, the distribution of the variable $\bar{x}$. We use the sampling distribution of the mean to make inferences about a population mean based on the mean of a sample from the population.

As we said earlier, it is generally not possible to know the sampling distribution of the mean exactly. Fortunately, however, we can often approximate that sampling distribution by a normal distribution; that is, under certain conditions the variable $\bar{x}$ is approximately normally distributed.

Recall that a variable is normally distributed if its distribution has the shape of a normal curve and that a normal distribution is determined by the mean and standard deviation. Hence a first step in learning how to approximate the sampling distribution of the mean by a normal distribution is to obtain the mean and standard deviation of $\bar{x}$. That is what we will do in this section.

To begin, let's review the notation used for the mean and standard deviation of a variable. The mean of a variable is denoted by the Greek letter μ subscripted, if necessary, with the letter representing the variable. So the mean of x is written as μ_x, the mean of y as μ_y, and so forth. In particular, then, the mean of $\bar{x}$ is written as $\mu_{\bar{x}}$; similarly, the standard deviation of $\bar{x}$ is written as $\sigma_{\bar{x}}$.

The Mean of $\bar{x}$

There is a simple relationship between the mean of the variable $\bar{x}$ and the mean of the variable under consideration—they are equal: $\mu_{\bar{x}} = \mu$. In other words, for any particular sample size, the mean of all possible sample means equals the population mean. This equality holds regardless of the size of the sample. We illustrate the relationship $\mu_{\bar{x}} = \mu$ by returning in Example 7.4 to the heights of the basketball players considered in the previous section.

EXAMPLE 7.4 **ILLUSTRATES THE RELATION $\mu_{\bar{x}} = \mu$**

The heights, in inches, of the five starting players on a men's basketball team are displayed in Table 7.5. Here the population of interest consists of the five players and the variable under consideration is height.

TABLE 7.5
Heights of the five starting players

Player	A	B	C	D	E
Height	76	78	79	81	86

a. Determine the population mean, μ.

b. Obtain the mean, $\mu_{\bar{x}}$, of the variable $\bar{x}$ for samples of size two. Verify that the relation $\mu_{\bar{x}} = \mu$ holds.

c. Repeat part (b) for samples of size four.

SOLUTION **a.** To determine the population mean (i.e., the mean of the variable height), we apply Definition 3.11 on page 175 to the heights in Table 7.5:

$$\mu = \frac{\Sigma x}{N} = \frac{76 + 78 + 79 + 81 + 86}{5} = 80.$$

Thus the mean height of the five players is 80 inches.

b. To obtain the mean of the variable $\bar{x}$ for samples of size two, we again apply Definition 3.11, but this time to $\bar{x}$. Referring to the third column of Table 7.2 on page 410, we get that

$$\mu_{\bar{x}} = \frac{77.0 + 77.5 + \cdots + 83.5}{10} = 80.$$

By part (a), $\mu = 80$. So for samples of size two, we see that $\mu_{\bar{x}} = \mu$.

c. Proceeding as in part (b), but this time referring to the third column of Table 7.3 on page 411, we obtain the mean of the variable $\bar{x}$ for samples of size four:

$$\mu_{\bar{x}} = \frac{78.50 + 79.75 + 80.25 + 80.50 + 81.00}{5} = 80,$$

which again is the same as μ. ◨

FORMULA 7.1 **MEAN OF THE VARIABLE $\bar{x}$**

For samples of size n, the mean of the variable $\bar{x}$ equals the mean of the variable under consideration:

$$\mu_{\bar{x}} = \mu.$$

In other words, for each sample size, the mean of all possible sample means equals the population mean.

The Standard Deviation of $\bar{x}$

Next we will investigate the standard deviation of the variable $\bar{x}$. Our purpose is to discover any apparent relationship that the standard deviation of $\bar{x}$ has with the standard deviation of the variable under consideration. For our investigation, we once again return to the basketball players.

EXAMPLE 7.5 INVESTIGATES THE RELATION BETWEEN $\sigma_{\bar{x}}$ AND σ

Refer to Table 7.5.

a. Determine the population standard deviation, σ.

b. Obtain the standard deviation, $\sigma_{\bar{x}}$, of the variable $\bar{x}$ for samples of size two. Indicate any apparent relationship between $\sigma_{\bar{x}}$ and σ.

c. Repeat part (b) for samples of sizes one, three, four, and five.

d. Summarize and discuss the results obtained in parts (a)–(c).

SOLUTION **a.** To determine the population standard deviation (i.e., the standard deviation of the variable height), we will apply Definition 3.12 on page 178 to the heights in Table 7.5. Recalling that $\mu = 80$, we have

$$\begin{aligned}\sigma &= \sqrt{\frac{\Sigma(x-\mu)^2}{N}} \\ &= \sqrt{\frac{(76-80)^2+(78-80)^2+(79-80)^2+(81-80)^2+(86-80)^2}{5}} \\ &= \sqrt{\frac{16+4+1+1+36}{5}} = \sqrt{11.6} = 3.41.\end{aligned}$$

Thus the standard deviation of the heights of the five players is 3.41 inches.

b. To obtain the standard deviation of the variable $\bar{x}$ for samples of size two, we again apply Definition 3.12, but this time to $\bar{x}$. Referring to the third column of Table 7.2 on page 410 and recalling that $\mu_{\bar{x}} = \mu = 80$, we have

$$\begin{aligned}\sigma_{\bar{x}} &= \sqrt{\frac{(77.0-80)^2+(77.5-80)^2+\cdots+(83.5-80)^2}{10}} \\ &= \sqrt{\frac{9.00+6.25+\cdots+12.25}{10}} = \sqrt{4.35} = 2.09,\end{aligned}$$

to two decimal places. Thus for samples of size two, the standard deviation of $\bar{x}$ is $\sigma_{\bar{x}} = 2.09$. Note that this is not the same as the population standard deviation, which is $\sigma = 3.41$. Also note that $\sigma_{\bar{x}}$ is smaller than σ.

c. Using the same procedure as in part (b), we can compute $\sigma_{\bar{x}}$ for samples of sizes one, three, four, and five. We summarize the results in Table 7.6.

TABLE 7.6 The standard deviation of $\bar{x}$ for sample sizes one, two, three, four, and five

Sample size n	Standard deviation of $\bar{x}$ $\sigma_{\bar{x}}$
1	3.41
2	2.09
3	1.39
4	0.85
5	0.00

d. Table 7.6 suggests that the standard deviation of $\bar{x}$ gets smaller as the sample size gets larger. We could have predicted this phenomenon from the dotplots in Fig. 7.3 on page 413 and from the fact that the standard deviation of a variable measures the variation of its possible values. Figure 7.3 indicates graphically that the variation, and hence the standard deviation, of all possible sample means decreases with increasing sample size. Table 7.6 indicates that same thing numerically. ◨

Example 7.5 provides evidence that the standard deviation of $\bar{x}$ gets smaller as the sample size increases. Our question now is whether there is a formula that relates the standard deviation of $\bar{x}$ to the sample size and standard deviation of the population. The answer is yes! In fact, two different formulas express the precise relationship.

When sampling is done without replacement from a finite population, as in Example 7.5, the appropriate formula is

$$\sigma_{\bar{x}} = \sqrt{\frac{N-n}{N-1}} \cdot \frac{\sigma}{\sqrt{n}},$$

where, as usual, n denotes the sample size and N the population size. And when sampling is done with replacement from a finite population or when it is done from an infinite population, the appropriate formula is

$$\sigma_{\bar{x}} = \frac{\sigma}{\sqrt{n}}.$$

When the sample size is small relative to the population size, there is little difference between sampling with and without replacement.[†] So, in such cases, it is not surprising that the two formulas for $\sigma_{\bar{x}}$ yield almost the same numbers. Since in most practical applications, the sample size is in fact small relative to the population size, we will, with the understanding that the equality may be only approximate, use the second formula exclusively in this book because it is simpler than the first.

FORMULA 7.2 **STANDARD DEVIATION OF THE VARIABLE $\bar{x}$**

For samples of size n, the standard deviation of the variable $\bar{x}$ equals the standard deviation of the variable under consideration divided by the square root of the sample size:

$$\sigma_{\bar{x}} = \frac{\sigma}{\sqrt{n}}.$$

In other words, for each sample size, the standard deviation of all possible sample means equals the population standard deviation divided by the square root of the sample size.

[†] A rule of thumb is that the sample size is 5% or less of the population size, that is, $n \leq 0.05N$.

Applying the Formulas

We have seen that there are simple formulas that relate the mean and standard deviation of $\bar{x}$ to the mean and standard deviation of the population: $\mu_{\bar{x}} = \mu$ and $\sigma_{\bar{x}} = \sigma/\sqrt{n}$ (at least approximately). Those formulas are applied in Example 7.6.

EXAMPLE 7.6 ILLUSTRATES FORMULAS 7.1 AND 7.2

As reported by the U.S. Bureau of the Census in *Current Housing Reports,* the mean livable square footage for single-family detached homes is 1742 square feet. The standard deviation is 568 square feet.

a. For samples of 25 single-family detached homes, determine the mean and standard deviation of the variable $\bar{x}$.

b. Repeat part (a) for a sample of size 500.

SOLUTION Here the variable is livable square footage and the population consists of all single-family detached homes in the United States. We know that $\mu = 1742$ sq. ft and $\sigma = 568$ sq. ft.

a. For this part, the sample size is 25 and, so, $\bar{x}$ denotes the mean livable square footage of a sample of 25 single-family detached homes. We want to obtain the mean and standard deviation of all such possible sample means. Applying Formula 7.1 (page 417), we get

$$\mu_{\bar{x}} = \mu = 1742.$$

Since $n = 25$, we conclude from Formula 7.2 that

$$\sigma_{\bar{x}} = \frac{\sigma}{\sqrt{n}} = \frac{568}{\sqrt{25}} = 113.6.$$

In other words, for samples of size 25, the mean and standard deviation of all possible sample means are, respectively, 1742 square feet and 113.6 square feet.

b. Proceeding as in part (a), we find that

$$\mu_{\bar{x}} = \mu = 1742$$

and since, here, $n = 500$,

$$\sigma_{\bar{x}} = \frac{\sigma}{\sqrt{n}} = \frac{568}{\sqrt{500}} = 25.4.$$

That is, for samples of size 500, the mean and standard deviation of all possible sample means are, respectively, 1742 square feet and 25.4 square feet. ◨

Sample Size and Sampling Error (Revisited)

Key Fact 7.1 states that the possible sample means cluster closer around the population mean as the sample size increases and, therefore, the larger the sample size, the smaller the sampling error tends to be in estimating a population mean by a sample mean. Here is why that key fact is true.

- The larger the sample size, the smaller the standard deviation of $\bar{x}$.
- The smaller the standard deviation of $\bar{x}$, the more closely its possible values (the possible sample means) cluster around the mean of $\bar{x}$.
- The mean of $\bar{x}$ is the same as the population mean: $\mu_{\bar{x}} = \mu$.

Because the standard deviation of $\bar{x}$ determines the amount of sampling error to be expected when a population mean is estimated by a sample mean, it is often referred to as the **standard error of the mean.** We note that, in general, the standard deviation of a statistic that is used to estimate a parameter is called the **standard error (SE)** of the statistic.

EXERCISES 7.2

STATISTICAL CONCEPTS AND SKILLS

7.18 Although, in general, we cannot know the sampling distribution of the mean exactly, by what distribution can we often approximate it?

7.19 Why is obtaining the mean and standard deviation of $\bar{x}$ a first step in approximating the sampling distribution of the mean by a normal distribution?

7.20 Does the sample size have an effect on the mean of all possible sample means? Explain your answer.

7.21 Does the sample size have an effect on the standard deviation of all possible sample means? Explain your answer.

7.22 Explain why increasing the sample size results in a tendency for smaller sampling error when using a sample mean to estimate a population mean.

7.23 What is another name for the standard deviation of the variable $\bar{x}$? What is the reason for that name?

7.24 We stated earlier in this section that when the sample size is small relative to the population size, there is little difference between sampling with and without replacement. Explain in your own words why that statement is true.

You need to have solved Exercises 7.3–7.7 before solving Exercises 7.25–7.29.

7.25 As reported by the *World Almanac, 1998,* the annual salaries, to the nearest thousand dollars, of the top five state officials of Oklahoma are as follows. Consider these officials a population of interest.

Official	Salary ($1000s)
Governor (G)	70
Lieutenant Governor (L)	63
Secretary of State (S)	44
Attorney General (A)	75
Treasurer (T)	70

a. Find the population mean, μ, of the five salaries.

b. Consider samples of size two without replacement. Use your answer to Exercise 7.3(b) on page 414 and Definition 3.11 on page 175 to find the mean of the variable $\bar{x}$.

c. Find $\mu_{\bar{x}}$ using only the result of part (a).

7.26 Repeat parts (b) and (c) of Exercise 7.25 for samples of size one. For part (b), use your answer to Exercise 7.4(b).

7.27 Repeat parts (b) and (c) of Exercise 7.25 for samples of size three. For part (b), use your answer to Exercise 7.5(b).

7.28 Repeat parts (b) and (c) of Exercise 7.25 for samples of size four. For part (b), use your answer to Exercise 7.6(b).

7.29 Repeat parts (b) and (c) of Exercise 7.25 for samples of size five. For part (b), use your answer to Exercise 7.7(b).

7.30 According to the Census Bureau publication *Current Construction Reports,* the mean price of new mobile homes is \$38,400. The standard deviation of the prices is \$7200.

a. For samples of 50 new mobile homes, determine the mean and standard deviation of all possible sample mean prices.

b. Repeat part (a) for samples of size 100.

7.31 The U.S. National Center for Health Statistics publishes information on the length of stay by patients in short-stay hospitals in *Vital and Health Statistics.* According to that publication, the mean stay of female patients in short-stay hospitals is $\mu = 5.8$ days. The standard deviation is $\sigma = 4.3$ days.

a. For samples of 75 female patients, determine the mean and standard deviation of all possible sample mean lengths of stay.

b. Repeat part (a) for samples of size 500.

7.32 The National Council of the Churches of Christ in the United States of America reports in *Yearbook of American and Canadian Churches* that the mean number of members per local church is 458. Assume a standard deviation of 116.7. Let $\bar{x}$ denote the mean number of members for a sample of local churches.

a. For samples of size 100, find the mean and standard deviation of $\bar{x}$. Interpret your results in words.

b. Repeat part (a) with $n = 200$.

7.33 According to *Motor Vehicle Facts and Figures,* published by the Motor Vehicle Manufacturers Association of the United States, the mean age of cars in use is 8.5 years and the standard deviation is 2.6 years. Let $\bar{x}$ denote the mean age of a sample of cars.

a. For samples of size 50, determine the mean and standard deviation of $\bar{x}$. Interpret your results in words.

b. Repeat part (a) with $n = 200$.

EXTENDING THE CONCEPTS AND SKILLS

Finite population correction factor. In doing Exercises 7.34–7.41, recall the following fact: If sampling is done without replacement from a finite population of size N, then the standard deviation of $\bar{x}$ can be obtained from the formula

$$\text{(1)} \qquad \sigma_{\bar{x}} = \sqrt{\frac{N-n}{N-1}} \cdot \frac{\sigma}{\sqrt{n}},$$

where n is the sample size. The term $\sqrt{\frac{N-n}{N-1}}$ is referred to as the **finite population correction factor.** If the sample size is small relative to the population size, then we can ignore the finite population correction factor and use the simpler formula,

$$\text{(2)} \qquad \sigma_{\bar{x}} = \frac{\sigma}{\sqrt{n}}.$$

A rule of thumb is that the finite population correction factor can be ignored provided the sample size is 5% or less of the population size, that is, $n \leq 0.05N$.

7.34 In Example 7.5, we used the definition of the standard deviation of a variable to obtain the standard deviation of the heights of the five starting players on a basketball team and, also, the standard deviation of $\bar{x}$ for samples of sizes one, two, three, four, and five. The results are summarized in Table 7.6 on page 418. Since the sampling is without replacement from a finite population, we can also use (1) to obtain $\sigma_{\bar{x}}$.

a. Use (1) to compute $\sigma_{\bar{x}}$ for samples of sizes one, two, three, four, and five. Compare your answers to those in Table 7.6.

b. Use the simpler formula, (2), to compute $\sigma_{\bar{x}}$ for samples of sizes one, two, three, four, and five. Compare your answers to those in Table 7.6. Why does (2) generally yield such poor approximations to the true values?

c. What percentage of the population size is a sample of size one? two? three? four? five?

7.35 Refer to Exercise 7.25.

a. Find the population standard deviation, σ, of the five salaries.

b. Consider samples of size two without replacement. Use your answer to Exercise 7.3(b) on page 414 and Definition 3.12 on page 178 to find the standard deviation of the variable $\bar{x}$.

c. In part (b) which formula would be appropriate, (1) or (2)?

d. Use (1) to compute $\sigma_{\bar{x}}$ and compare your answer with that from part (b).

e. Use (2) to compute $\sigma_{\bar{x}}$ and compare your answer with that from part (b).

f. Why does (2) yield such a poor approximation to the true value of $\sigma_{\bar{x}}$?

7.36 Repeat parts (b)–(e) of Exercise 7.35 for samples of size one. For part (b), use your answer to Exercise 7.4(b).

7.37 Repeat parts (b)–(f) of Exercise 7.35 for samples of size three. For part (b), use your answer to Exercise 7.5(b).

7.38 Repeat parts (b)–(f) of Exercise 7.35 for samples of size four. For part (b), use your answer to Exercise 7.6(b).

7.39 Repeat parts (b)–(f) of Exercise 7.35 for samples of size five. For part (b), use your answer to Exercise 7.7(b).

7.40 Let $\bar{x}$ denote the mean of a sample of size n from a population of size N.

a. Which formula, (1) or (2), should be used to compute $\sigma_{\bar{x}}$ if sampling is done without replacement? with replacement?

b. Assume $n = 1$. Compute $\sigma_{\bar{x}}$ using both (1) and (2). Why do both formulas yield the same result? Explain in words why $\sigma_{\bar{x}} = \sigma$.

c. Assume $n = N$ and sampling is done without replacement. Compute $\sigma_{\bar{x}}$ using (1). Could you have guessed the answer without doing any computations? Explain your answer.

7.41 Consider samples of size n without replacement from a population of size N.

a. Show that if $n \leq 0.05N$, then

$$0.97 \leq \sqrt{\frac{N-n}{N-1}} \leq 1.$$

b. Use part (a) to explain why there is little difference in the values provided by (1) and (2) when the sample size is 5% or less of the population size.

Unbiased and biased estimators. A statistic is said to be an **unbiased estimator** of a parameter if the mean of all its possible values equals the parameter; otherwise, it is said to be a **biased estimator.** An unbiased estimator yields, on the average, the correct value of the parameter, whereas a biased estimator does not. We will discuss unbiased and biased estimators in Exercises 7.42 and 7.43.

7.42 Is the sample mean an unbiased estimator of the population mean? Explain your answer.

7.43 Is the sample median an unbiased estimator of the population median? (*Hint:* Refer to Example 7.2 on page 409. Consider samples of size two.)

7.44 This exercise can be done individually or, better yet, as a class project.

a. Use a random-number table or random-number generator to obtain a sample (with replacement) of four digits between 0 and 9. Do this a total of 50 times and compute the mean of each sample.

b. Theoretically, what are the mean and standard deviation of all possible sample means for samples of size four?

c. Roughly what would you expect the mean and standard deviation of the 50 sample means you obtained in part (a) to equal? Explain your answers.

d. Determine the mean and standard deviation of the 50 sample means you obtained in part (a).

e. Compare your answers in parts (c) and (d). Why are they different?

USING TECHNOLOGY  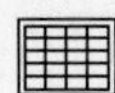 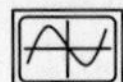

7.45 Gestation periods of humans are normally distributed with a mean of 266 days and a standard deviation of 16 days. Suppose we observe the gestation periods for a sample of nine humans.

a. Theoretically, what are the mean and standard deviation of all possible sample means?

b. Use Minitab or some other statistical software to simulate 2000 samples of nine human gestation periods each. If you are using Minitab, proceed as follows.

1 Choose **Calc ➤ Random Data ➤ Normal...**

2 Type 2000 in the **Generate rows of data** text box

3 Click in the **Store in column(s)** text box and type C1-C9

4 Click in the **Mean** text box and type 266

5 Click in the **Standard deviation** text box and type 16

6 Click **OK**

In each of the 2000 rows of C1–C9, you will have a sample of nine human gestation periods.

c. Determine the mean of each of the 2000 samples you obtained in part (b). If you are using Minitab, proceed as follows.

1 Choose **Calc ➤ Row Statistics...**

2 Select the **Mean** option button

3 Click in the **Input variables** text box and specify C1-C9

4 Click in the **Store result in** text box and type XBAR

5 Click **OK**

The means of the 2000 samples will now be stored in a column named XBAR.

d. Roughly what would you expect the mean and standard deviation of the 2000 sample means you obtained in part (c) to equal? Explain your answers.

e. Determine the mean and standard deviation of the 2000 sample means you obtained in part (c).

f. Compare your answers in parts (d) and (e). Why are they different?

7.46 Desert Samaritan Hospital, located in Mesa, Arizona, keeps records of emergency-room traffic. From those records we find that the time from the arrival of one patient to the next, called an interarrival time, has a special type of reverse J-shaped distribution called an *exponential distribution.* We also find that the mean time between arriving patients is approximately 8.7 minutes, as is the standard deviation. Suppose we observe a sample of 10 interarrival times.

a. Theoretically, what are the mean and standard deviation of all possible sample means?

b. Use Minitab or some other statistical software to simulate 1000 samples of 10 interarrival times each. If you are using Minitab, proceed in a similar way as in part (b) of Exercise 7.45 but begin with **Calc ➤ Random Data ➤ Exponential...**.

c. Obtain the mean of each sample of the 1000 samples you obtained in part (b). If you are using Minitab, proceed in a similar way as in part (c) of Exercise 7.45.

d. Roughly what would you expect the mean and standard deviation of the 1000 sample means you obtained in part (c) to equal? Explain your answers.

e. Determine the mean and standard deviation of the 1000 sample means you obtained in part (c).

f. Compare your answers in parts (d) and (e). Why are they different?

7.3 THE SAMPLING DISTRIBUTION OF THE MEAN

In Section 7.2 we took the first step in describing the sampling distribution of the mean—the distribution of the variable $\bar{x}$. There we discovered that the mean and standard deviation of $\bar{x}$ can be expressed in terms of the sample size and the population mean and standard deviation: $\mu_{\bar{x}} = \mu$ and $\sigma_{\bar{x}} = \sigma/\sqrt{n}$.

In this section we will take the final step in describing the sampling distribution of the mean. It is helpful to distinguish between the case in which the variable under consideration is normally distributed and the case in which it may not be so.

Sampling Distribution of the Mean for Normally Distributed Variables

Although it is by no means obvious, if the variable under consideration is normally distributed, then so is the variable $\bar{x}$. The proof of this fact requires advanced mathematics, but we can make it plausible by simulation. Simulating a variable means that we use a computer or statistical calculator to generate observations of the variable.

EXAMPLE 7.7 SAMPLING DISTRIBUTION OF THE MEAN FOR A NORMALLY DISTRIBUTED VARIABLE

Intelligence quotients (IQs) measured on the Stanford Revision of the Binet-Simon Intelligence Scale are normally distributed with a mean of 100 and a standard deviation of 16. Since IQ is a normally distributed variable, so is $\bar{x}$. We also know that $\mu_{\bar{x}} = \mu = 100$ and $\sigma_{\bar{x}} = \sigma/\sqrt{n} = 16/\sqrt{n}$. Thus, for instance, the possible sample means for samples of four IQs have a normal distribution with mean 100 and standard deviation $16/\sqrt{4} = 8$. Use simulation to make that fact plausible.

SOLUTION Here we will present a summary of what we did and the results. See the computer section at the end of this section for details. We simulated 1000 samples of four IQs each, determined the mean of each of the 1000 samples, and obtained a histogram of the 1000 sample means. Printout 7.1 displays the histogram. For purposes of comparison, we have superimposed on the histogram the normal curve for the sampling distribution, namely, the normal curve with parameters 100 and 8.

PRINTOUT 7.1 Minitab histogram of the sample means for 1000 samples of four IQs. The normal curve for $\bar{x}$ is superimposed.

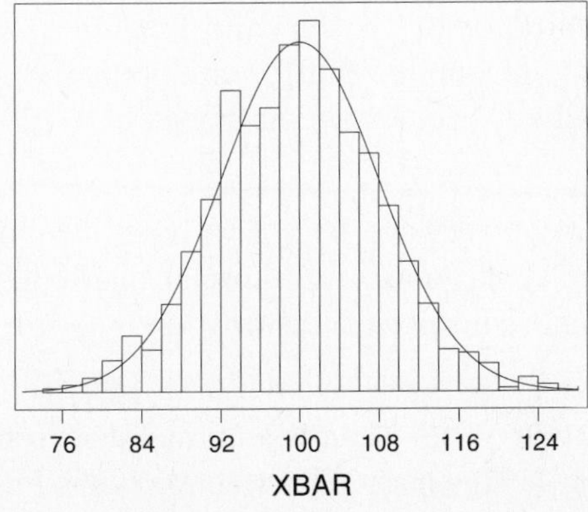

As expected, the histogram in Printout 7.1 is shaped roughly like a normal curve. Remember that we have only a sample of sample means; so we would not expect the histogram to be shaped exactly like a normal curve. On the other hand, since the number of sample means is large, we do expect it to be close to that shape, which it is. Not only does the histogram in Printout 7.1 makes it plausible that $\bar{x}$ is normally distributed, but it suggests as well that its mean is 100 and its standard deviation is 8. □

  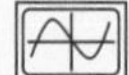

Key Fact 7.2 summarizes our discussion and simulation about the sampling distribution of the mean in case the variable under consideration is normally distributed.

KEY FACT 7.2 **SAMPLING DISTRIBUTION OF THE MEAN FOR A NORMALLY DISTRIBUTED VARIABLE**

Suppose a variable x of a population is normally distributed with mean μ and standard deviation σ. Then, for samples of size n, the variable $\bar{x}$ is also normally distributed and has mean μ and standard deviation $\sigma/\sqrt{n}$.

EXAMPLE 7.8 ILLUSTRATES KEY FACT 7.2

Consider once again the variable IQ, which is normally distributed with mean 100 and standard deviation 16. Obtain the sampling distribution of the mean for samples of size

a. 4. **b.** 16.

SOLUTION The normal distribution for IQs is shown in Fig. 7.4(a). Since IQs are normally distributed, Key Fact 7.2 implies that for any particular sample size n, the variable $\bar{x}$ is also normally distributed and has mean $\mu = 100$ and standard deviation $\sigma/\sqrt{n} = 16/\sqrt{n}$.

a. For samples of size 4, we have $16/\sqrt{n} = 16/\sqrt{4} = 8$ and, therefore, the sampling distribution of the mean is a normal distribution with mean 100 and standard deviation 8. This normal distribution is shown in Fig. 7.4(b).

b. For samples of size 16, we have $16/\sqrt{n} = 16/\sqrt{16} = 4$ and, therefore, the sampling distribution of the mean is a normal distribution with mean 100 and standard deviation 4. This normal distribution is shown in Fig. 7.4(c). □

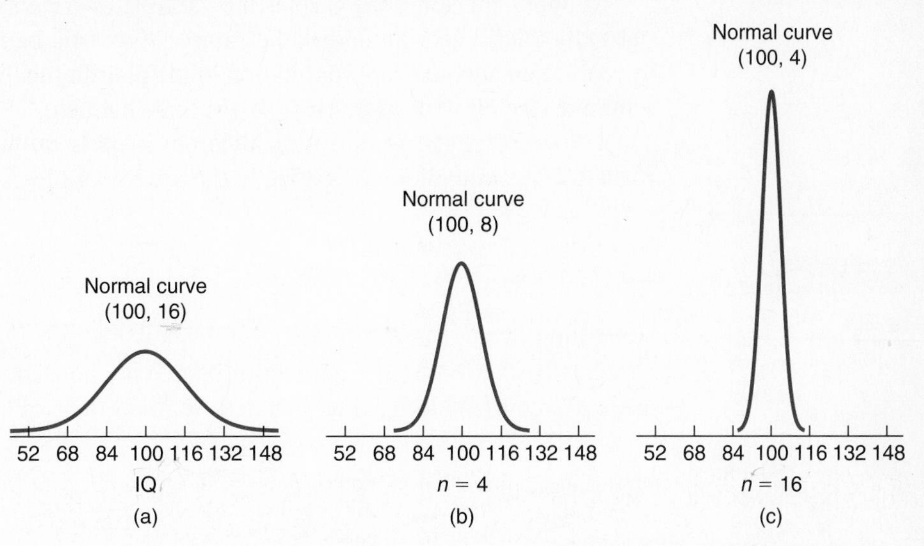

FIGURE 7.4 (a) Normal distribution for IQs (b) Sampling distribution of the mean for $n = 4$ (c) Sampling distribution of the mean for $n = 16$

The normal curves in Figs. 7.4(b) and 7.4(c) are drawn to scale so that we can compare them visually and observe two important things we already know—both curves are centered at the population mean ($\mu_{\bar{x}} = \mu$) and the spread becomes less extensive as the sample size increases ($\sigma_{\bar{x}} = \sigma/\sqrt{n}$). And, as a consequence of these two facts, we see illustrated in Fig. 7.4 something else that we already know: The possible sample means cluster closer around the population mean as the sample size increases and, therefore, the larger the sample size, the smaller the sampling error tends to be in estimating a population mean by a sample mean.

Central Limit Theorem

According to Key Fact 7.2, if the variable x under consideration is normally distributed, then so is the variable $\bar{x}$. Remarkably, that key fact holds true approximately regardless of the distribution of x, provided only that the sample size is relatively large. This extraordinary fact, called the **central limit theorem,** is one of the most important theorems in statistics.

KEY FACT 7.3 **THE CENTRAL LIMIT THEOREM**

For a relatively large sample size, the variable $\bar{x}$ is approximately normally distributed, regardless of the distribution of the variable under consideration. The approximation becomes better and better with increasing sample size.

Roughly speaking, the further the variable under consideration is from being normally distributed, the larger the sample size must be for a normal distribution to provide an adequate approximation to the distribution of $\overline{x}$. Generally, however, a sample size of 30 or more ($n \geq 30$) is large enough.

The proof of the central limit theorem is quite difficult. But we can make it plausible by simulation, as is done in the next example.

EXAMPLE 7.9 ILLUSTRATES KEY FACT 7.3

According to the Census Bureau publication *Current Population Reports,* a frequency distribution for the number of people per household in the United States is as displayed in Table 7.7. Frequencies are in millions of households.

TABLE 7.7 Frequency distribution for U.S. household size

No. of people	1	2	3	4	5	6	7
Frequency	19.4	26.5	14.6	12.9	6.1	2.5	1.6

Here the variable under consideration is household size and the population consists of all U.S. households. From Table 7.7 we find that the mean household size is $\mu = 2.685$ persons and the standard deviation is $\sigma = 1.47$ persons. Consequently, for samples of size n, we have that $\mu_{\overline{x}} = \mu = 2.685$ and $\sigma_{\overline{x}} = \sigma/\sqrt{n} = 1.47/\sqrt{n}$.

In Fig. 7.5 we have presented a relative-frequency histogram for household size, obtained by consulting Table 7.7.

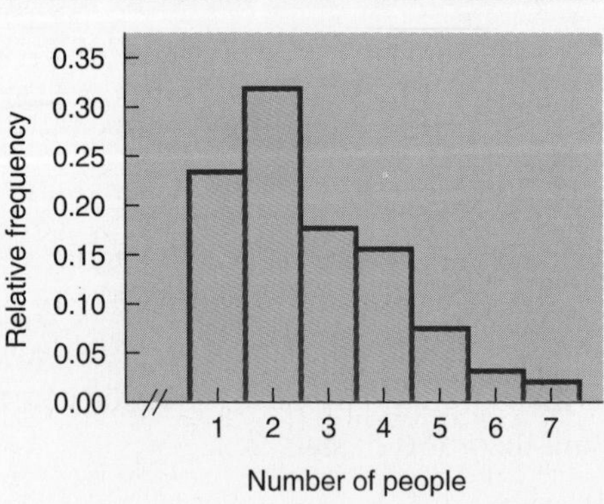

FIGURE 7.5 Relative-frequency histogram for household size

Notice that the variable household size is far from being normally distributed; it is clearly right skewed. Nonetheless, according to the central limit theorem, the sampling distribution of the mean can be approximated by a normal distribution when the sample size is large. Use simulation to make that fact plausible for a sample size of 30.

SOLUTION For samples of size 30, the variable $\bar{x}$ has mean 2.685 and standard deviation $1.47/\sqrt{30} = 0.27$. We simulated 1000 samples of 30 household sizes each, determined the mean of each of the 1000 samples, and obtained a histogram of the 1000 sample means. Printout 7.2 displays the histogram. For purposes of comparison, we have superimposed on the histogram the normal distribution with mean 2.685 and standard deviation 0.27.

PRINTOUT 7.2 Minitab histogram of the sample means for 1000 samples of 30 household sizes. The approximating normal curve for $\bar{x}$ is superimposed.

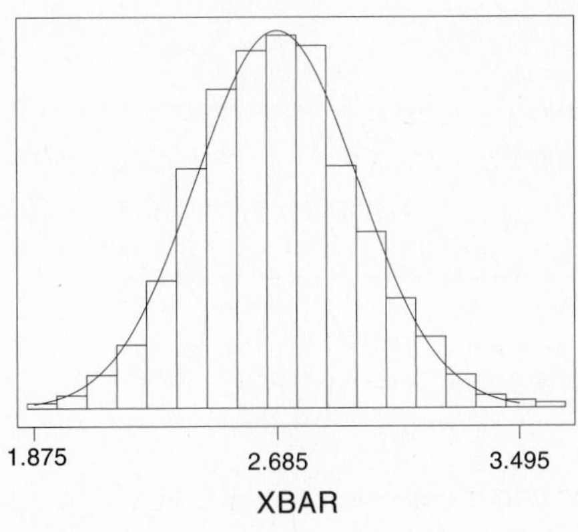

The histogram in Printout 7.2 is shaped roughly like a normal curve, specifically, like the normal curve with parameters 2.685 and 0.27. This makes it plausible that $\bar{x}$ is approximately normally distributed, as guaranteed by the central limit theorem.

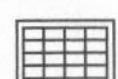

The Sampling Distribution of the Mean

We now summarize in Key Fact 7.4 the important facts that we have learned about the sampling distribution of the mean. We will use Key Fact 7.4 frequently.

KEY FACT 7.4 SAMPLING DISTRIBUTION OF THE MEAN

Suppose a variable x of a population has mean μ and standard deviation σ. Then, for samples of size n,

- The mean of $\bar{x}$ equals the population mean: $\mu_{\bar{x}} = \mu$.
- The standard deviation of $\bar{x}$ equals the population standard deviation divided by the square root of the sample size: $\sigma_{\bar{x}} = \sigma/\sqrt{n}$.
- If x is normally distributed, then so is $\bar{x}$, regardless of sample size.
- If the sample size is large, then $\bar{x}$ is approximately normally distributed, regardless of the distribution of x.

Thus if either the variable under consideration is normally distributed or the sample size is large, then the distribution of all possible sample means is, at least approximately, a normal distribution with mean μ and standard deviation $\sigma/\sqrt{n}$.

The content of Key Fact 7.4 is portrayed graphically in Fig. 7.6.

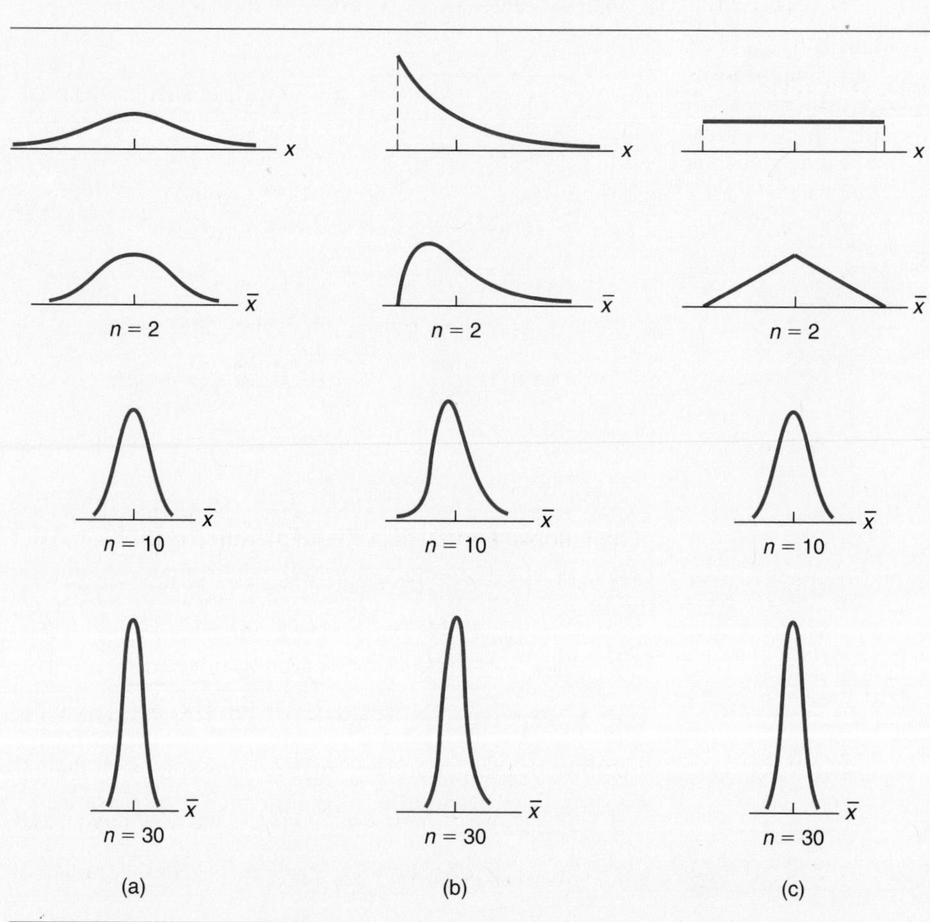

FIGURE 7.6 Sampling distributions for (a) normal, (b) reverse-J-shaped, and (c) uniform variables

We know that if the variable under consideration is normally distributed, then so is the variable $\bar{x}$, regardless of sample size. This fact is illustrated by Fig. 7.6(a).

Additionally, we know that if the sample size is large, then the variable $\bar{x}$ is approximately normally distributed, regardless of the distribution of the variable under consideration. Figs. 7.6(b) and 7.6(c) provide a visual display of this fact for two nonnormal variables, one having a reverse-J-shaped distribution and the other having a uniform distribution. In each of these two cases, we see the following:

For samples of size two, the variable $\bar{x}$ is far from being normally distributed; for samples of size 10, it is already somewhat normally distributed; and for samples of size 30, it is very close to being normally distributed.

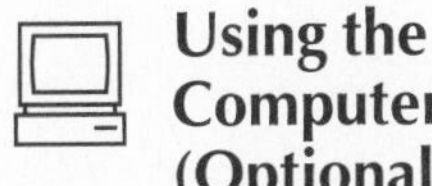

Using the Computer (Optional)

In Example 7.7 we used simulation to illustrate the fact that whenever the variable x under consideration is normally distributed, then so is the variable $\bar{x}$. We now provide the details for carrying out the simulation.

EXAMPLE 7.10 USING MINITAB TO SIMULATE THE SAMPLING DISTRIBUTION OF THE MEAN

Intelligence quotients (IQs) measured on the Stanford Revision of the Binet-Simon Intelligence Scale are known to be normally distributed with a mean of 100 and a standard deviation of 16. Use Minitab to do the following.

a. Simulate 1000 samples of four IQs each.

b. Determine the mean of each of the 1000 samples.

c. Obtain a histogram of the 1000 sample means.

SOLUTION **a.** Noting that IQ is normally distributed, we proceed as follows to simulate 1000 samples of four IQs each.

1 Choose **Calc ➤ Random Data ➤ Normal...**
2 Type `1000` in the **Generate rows of data** text box
3 Click in the **Store in column(s)** text box and type `C1-C4`
4 Click in the **Mean** text box and type `100`
5 Click in the **Standard deviation** text box and type `16`
6 Click **OK**

In each of the 1000 rows of C1–C4, we now have a sample of four IQs.

b. To determine the mean of each of the 1000 samples, we proceed in the following way.

1 Choose **Calc ➤ Row Statistics...**
2 Select the **Mean** option button
3 Click in the **Input variables** text box and specify C1-C4
4 Click in the **Store result in** text box and type `XBAR`
5 Click **OK**

The 1000 sample means are now stored in a column named XBAR.

c. To obtain a histogram of the 1000 sample means, we use Minitab's histogram procedure as discussed in Chapter 2.

1 Choose **Graph ➤ Histogram...**
2 Specify XBAR in the **X** text box for **Graph 1**
3 Click **OK**

Note: Printout 7.1 on page 425 contains a histogram obtained in this way but with some embellishments, such as the superimposed normal curve, that we don't get by using only the previous three steps. ◨

EXERCISES 7.3

STATISTICAL CONCEPTS AND SKILLS

7.47 A variable of a population has a mean of $\mu = 100$ and a standard deviation of $\sigma = 28$.

a. Identify the sampling distribution of the mean for samples of size 49.
b. In answering part (a), what assumptions did you make about the distribution of the variable under consideration?
c. Can you answer part (a) if the sample size is 16 instead of 49? Why or why not?

7.48 A variable of a population has a mean of $\mu = 35$ and a standard deviation of $\sigma = 42$.

a. If the variable is normally distributed, identify the sampling distribution of the mean for samples of size 9.
b. Can you answer part (a) if the distribution of the variable under consideration is unknown? Explain your answer.
c. Can you answer part (a) if the distribution of the variable under consideration is unknown but the sample size is 36 instead of 9? Why or why not?

7.49 A variable of a population is normally distributed with mean μ and standard deviation σ.

a. Identify the distribution of $\bar{x}$.
b. Does your answer to part (a) depend on how large the sample size is? Explain your answer.
c. What are the mean and the standard deviation of $\bar{x}$?
d. Does your answer to part (c) depend on the assumption that the variable under consideration is normally distributed?

7.50 A variable of a population has mean μ and standard deviation σ. For a large sample size n, answer the following questions.

a. Identify the distribution of $\bar{x}$.
b. Does your answer to part (a) depend on the fact that n is large? Explain your answer.
c. What are the mean and the standard deviation of $\bar{x}$?
d. Does your answer to part (c) depend on the assumption that the sample size is large?

7.51 Refer to Fig. 7.6 on page 430.

a. Why are the four graphs in Fig. 7.6(a) all centered at the same place?
b. Why does the spread of the graphs diminish with increasing sample size? How does this fact affect the sampling error when estimating a population mean, μ, by a sample mean, $\bar{x}$?
c. Why are the graphs in Fig. 7.6(a) bell-shaped?
d. Why do the graphs in Figs. 7.6(b) and 7.6(c) become bell-shaped as the sample size increases?

7.52 According to the central limit theorem, for a relatively large sample size, the variable $\bar{x}$ is approximately normally distributed.

a. What rule of thumb is used for deciding whether the sample size is relatively large?

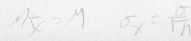

b. Roughly speaking, what property of the distribution of the variable under consideration determines how large the sample size must be for a normal distribution to provide an adequate approximation to the distribution of $\bar{x}$?

7.53 As reported by *Runner's World* magazine, the times of the finishers in the New York City 10-km run are normally distributed with a mean of 61 minutes and a standard deviation of 9 minutes.

a. Determine the sampling distribution of the mean for samples of size four. Interpret your answer in terms of the distribution of the possible sample means for samples of four finishing times.

b. Repeat part (a) for samples of size nine.

c. Construct graphs similar to those in Fig. 7.4 on page 427.

7.54 In 1905 R. Pearl published the article "Biometrical Studies on Man. I. Variation and Correlation in Brain Weight" (*Biometrika,* Vol. 4, pp. 13–104). According to the study, brain weights of Swedish men are normally distributed with a mean of 1.40 kilograms (kg) and a standard deviation of 0.11 kg.

a. Determine the sampling distribution of the mean for samples of size three. Interpret your answer in terms of the distribution of the possible sample means for samples of three brain weights.

b. Repeat part (a) for samples of size 12.

c. Construct graphs similar to those in Fig. 7.4 on page 427.

7.55 According to *Current Population Reports,* published by the U.S. Bureau of the Census, the mean annual alimony income received by women is \$8657. Assume a standard deviation of \$7500.

a. Determine the sampling distribution of the mean for samples of size 100.

b. Repeat part (a) for samples of size 1000.

c. Must you assume that annual alimony payments are normally distributed to answer parts (a) and (b)? Explain your answer.

7.56 The U.S. Energy Information Administration collects data on household vehicles and publishes the results in *Residential Transportation Energy Consumption Survey, Consumption Patterns of Household Vehicles.* According to this document, the mean monthly fuel expenditure per household vehicle is \$58.80. The standard deviation is \$30.40.

a. Determine the sampling distribution of the mean for samples of size 50.

b. Repeat part (a) for samples of size 100.

c. Must you assume that monthly fuel expenditures for household vehicles are normally distributed to answer parts (a) and (b)? Explain your answer.

7.57 Refer to Exercise 7.53.

a. Determine the percentage of samples of size four that have mean finishing times within 5 minutes of the population mean finishing time of 61 minutes. Interpret your answer in terms of sampling error.

b. Repeat part (a) for samples of size nine.

7.58 Refer to Exercise 7.54.

a. Determine the percentage of samples of size three that have mean brain weights within 0.1 kg of the population mean brain weight of 1.40 kg. Interpret your answer in terms of sampling error.

b. Repeat part (a) for samples of size 12.

7.59 Refer to Exercise 7.55.

a. What is the probability that the sampling error made in estimating the mean annual alimony income of all American women receiving alimony by that of a random sample of 100 such women will be at most \$500?

b. Repeat part (a) for samples of size 1000.

7.60 Refer to Exercise 7.56.

a. What is the probability that the sampling error made in estimating the mean monthly fuel expenditure for household vehicles by that of a random sample of 50 such vehicles will be at most \$5?

b. Repeat part (a) for samples of size 100.

7.61 An economist employed by the Department of Agriculture needs to estimate the mean weekly food cost for couples with two children 6–11 years old. The economist plans to randomly sample 500 such families and use the mean weekly food cost, $\bar{x}$, of those families as her estimate of the true mean weekly food cost, μ. If $\sigma = \$17.20$, find the probability that the economist's estimate will be within \$1 of the actual mean.

7.62 An air-conditioning contractor is preparing to offer service contracts on the brand of compressor used in all of the units her company installs. Before she can work out the details, she must estimate how long those compressors last on the average. The contractor anticipated this need and has kept detailed records on the lifetimes of a random sample of 250 compressors. She plans to use the sample mean lifetime, $\bar{x}$, of those 250 compressors as her estimate for the population mean lifetime, μ, of all such compressors. If the lifetimes of this brand of compressor have a standard deviation of 40 months, what is the probability that the contractor's estimate will be within 5 months of the true mean?

EXTENDING THE CONCEPTS AND SKILLS

Use the 68.26-95.44-99.74 rule on page 377 to answer the questions posed in parts (a)–(c) of Exercises 7.63 and 7.64.

7.63 A variable of a population is normally distributed with mean μ and standard deviation σ. For samples of size *n*, fill in the following blanks. Justify your answers.

a. 68.26% of all possible samples have means that lie within _____ of the population mean μ.

b. 95.44% of all possible samples have means that lie within _____ of the population mean μ.

c. 99.74% of all possible samples have means that lie within _____ of the population mean μ.

d. $100(1-\alpha)\%$ of all possible samples have means that lie within _____ of the population mean μ. *(Hint: Draw a graph for the distribution of $\bar{x}$ and determine the z-scores dividing the area under the normal curve into a middle $1-\alpha$ area and two outside areas of $\alpha/2$.)*

7.64 A variable of a population has mean μ and standard deviation σ. For a large sample size *n*, fill in the following blanks. Justify your answers.

a. Approximately _____% of all possible samples have means within $\sigma/\sqrt{n}$ of the population mean μ.

b. Approximately _____% of all possible samples have means within $2\sigma/\sqrt{n}$ of the population mean μ.

c. Approximately _____% of all possible samples have means within $3\sigma/\sqrt{n}$ of the population mean μ.

d. Approximately _____% of all possible samples have means within $z_{\alpha/2}$ of the population mean μ.

7.65 A brand of water-softener salt comes in packages marked "net weight 40 lb." The company that packages the salt claims that the bags contain an average of 40 lb of salt and that the standard deviation of the weights is 1.5 lb. Furthermore, it is known that the weights are normally distributed.

a. Obtain the probability that the weight of one randomly selected bag of water-softener salt will be 39 lb or less, if the company's claim is true.

b. Determine the probability that the mean weight of 10 randomly selected bags of water-softener salt will be 39 lb or less, if the company's claim is true.

c. If you bought one bag of water-softener salt and it weighed 39 lb, would you consider this evidence that the company's claim is incorrect? Explain your answer.

d. If you bought 10 bags of water-softener salt and their mean weight was 39 lb, would you consider this evidence that the company's claim is incorrect? Explain your answer.

7.66 According to a spokesperson for the Salt River Project, a company that supplies electricity to the greater Phoenix area, the mean annual electric bill in 1997 was \$1122.17. Assume that for any particular year, annual electric bills are normally distributed and have a standard deviation of \$204. At the end of 1998, an independent consumer agency wanted to determine whether the mean annual electric bill had increased over the 1997 figure of \$1122.17.

a. Determine the probability that the mean of a random sample of 25 annual electric bills for 1998 will be \$1145 or greater if the 1998 mean annual electric bill for all customers was still \$1122.17.

b. If the consumer agency takes a random sample of 25 annual electric bills for 1998 and finds that $\bar{x} = \$1145$, does this provide evidence that the 1998 mean was greater than the 1997 mean of \$1122.17?

c. Repeat parts (a) and (b) for $n = 250$.

d. To answer part (c), is it necessary to assume that annual electric bills are normally distributed? Explain your answer.

USING TECHNOLOGY 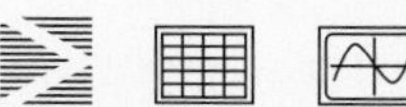

In Exercises 7.67 and 7.68, you may want to refer to Exercises 7.45 and 7.46 on page 424.

7.67 Gestation periods of humans are normally distributed with a mean of 266 days and a standard deviation of 16 days. Suppose we observe the gestation periods for a sample of nine humans.

a. Use Minitab or some other statistical software to simulate 2000 samples of nine human gestation periods each.

b. Find the sample mean of each of the 2000 samples.

c. Obtain the mean, the standard deviation, and a histogram of the 2000 sample means.

d. Theoretically, what are the mean, standard deviation, and distribution of all possible sample means for samples of size nine?

e. Compare your results from parts (c) and (d).

7.68 A variable is said to have an **exponential distribution** or to be **exponentially distributed** if its distribution has the shape of an exponential curve, that is, a curve of the form $y = e^{-x/\mu}/\mu$ for $x > 0$, where μ is the mean of the variable. The standard deviation of such a variable also equals μ. Desert Samaritan Hospital, located in Mesa, Arizona, keeps records of emergency-room traffic. From those records we find that the time from the arrival of one patient to the next, called an interarrival time, has an exponential distribution with mean $\mu = 8.7$ minutes.

a. Sketch the exponential curve for the distribution of the variable "interarrival time." Note that this variable is far from being normally distributed. What shape does its distribution have?

b. Use Minitab or some other statistical software to simulate 1000 samples of four interarrival times each.

c. Find the sample mean of each of the 1000 samples.

d. Determine the mean and standard deviation of the 1000 sample means.

e. Theoretically, what are the mean and the standard deviation of all possible sample means for samples of size four? Compare your answers to the ones you obtained in part (d).

f. Obtain a histogram of the 1000 sample means. Is the histogram bell-shaped? Would you necessarily expect it to be?

g. Repeat parts (b)–(f) for a sample size of 40.

CHAPTER REVIEW

You Should Be Able To

1. use and understand the formulas presented in this chapter.
2. define sampling error and explain the need for sampling distributions.
3. find the mean and standard deviation of the variable $\bar{x}$, given the mean and standard deviation of the population and the sample size.
4. state and apply the central limit theorem.
5. determine the sampling distribution of the mean when the variable under consideration is normally distributed.
6. determine the sampling distribution of the mean when the sample size is 30 or more.

*7. use the Minitab procedures covered in this chapter.

Key Terms

central limit theorem, *427*
sampling distribution, *408*
sampling distribution of the mean, *409*
sampling error, *408*
standard error, *421*
standard error of the mean, *421*

REVIEW TEST

Statistical Concepts and Skills

1. Define *sampling error.*

2. What is the sampling distribution of a statistic? Why is it important?

3. Provide two synonymous terms for the distribution of all possible sample means for samples of a given size.

4. Relative to the population mean, what happens to the possible sample means for samples of the same size as the sample size increases? Explain the relevance of this property in estimating a population mean by a sample mean.

5. In 1994 the Internal Revenue Service (IRS) sampled approximately 125,000 tax returns to obtain estimates of various parameters. Data were published in *Statistics of Income, Individual Income Tax Returns.* According to that document, the mean income tax per taxable return for the returns sampled was \$6104.
   a. Explain the meaning of sampling error in this context.
   b. If, in reality, the population mean income tax per taxable return in 1994 was \$6192, how much sampling error was made in estimating that parameter by the sample mean of \$6104?
   c. If the IRS had sampled 250,000 returns instead of 125,000, would the sampling error necessarily have been smaller? Explain your answer.
   d. In future surveys how can the IRS increase the likelihood of small sampling error?

6. The following table gives the monthly salaries (in \$1000s) of the six officers of a company.

Officer	A	B	C	D	E	F
Salary	8	12	16	20	24	28

   a. Calculate the population mean monthly salary μ.

There are 15 possible samples of size four from the population of six officers. They are listed in the first column of the following table.

Sample	Salaries	$\bar{x}$
A, B, C, D	8, 12, 16, 20	14
A, B, C, E	8, 12, 16, 24	15
A, B, C, F	8, 12, 16, 28	16
A, B, D, E	8, 12, 20, 24	16
A, B, D, F	8, 12, 20, 28	17
A, B, E, F	8, 12, 24, 28	18
A, C, D, E		
A, C, D, F		
A, C, E, F		
A, D, E, F		
B, C, D, E		
B, C, D, F		
B, C, E, F		
B, D, E, F		
C, D, E, F		

   b. Complete the second and third columns of the above table.
   c. Complete the following dotplot for the sampling distribution of the mean for samples of size four. Locate the population mean on the graph.

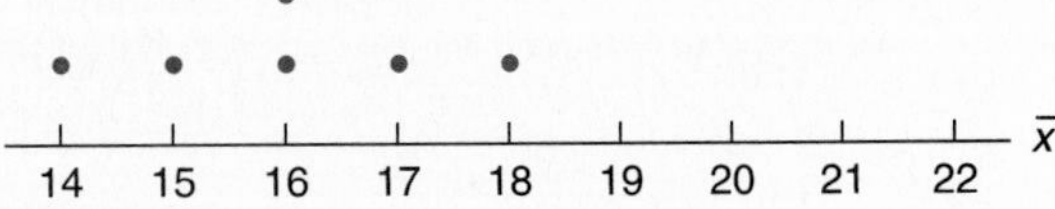

   d. Obtain the probability that the mean of a random sample of four salaries will be within 1 (i.e., \$1000) of the population mean.
   e. Use the answer you obtained in part (b) and Definition 3.11 on page 175 to find the mean of the variable $\bar{x}$. Interpret your answer.
   f. Can you obtain the mean of the variable $\bar{x}$ without doing the calculation in part (e)? Explain your answer.

7. As reported by the U.S. Department of Agriculture in *Crop Production,* the mean yield of cotton per acre for U.S. farms is 506 lb. The standard deviation is 237 lb.

a. For samples of 25 one-acre plots of cotton, determine the mean and standard deviation of all possible sample mean yields.
b. Repeat part (a) for a sample size of 200.
c. For a sample size of 500, answer the following question without doing any computations: Will the standard deviation of all possible sample mean yields be larger, smaller, or the same as that in part (b)? Explain your answer.

8. A variable x, which may or may not be normally distributed, has mean $\mu = 40$ and standard deviation $\sigma = 10$. Suppose 100 members of the population are to be selected at random. Decide whether each of the following statements is true or false or whether it is not possible to tell. Give a reason for each of your answers.
a. There is roughly a 68.26% chance that the mean of the sample will be between 30 and 50.
b. 68.26% of all possible observations of x lie between 30 and 50.
c. There is roughly a 68.26% chance that the mean of the sample will be between 39 and 41.

9. Repeat Problem 8 assuming that the variable under consideration is normally distributed.

10. The monthly rents for studio apartments in a large city are normally distributed with a mean of \$585 and a standard deviation of \$45.
a. Sketch the normal curve for the monthly rents.
b. Find the sampling distribution of the mean for samples of size three. Draw a graph of the normal curve associated with $\bar{x}$.
c. Repeat part (b) for samples of size nine.

11. Refer to Problem 10.
a. Determine the percentage of samples of size three that have mean monthly rents within \$10 of the population mean of \$585.
b. Obtain the probability that the mean monthly rent of three randomly selected studio apartments will be within \$10 of the population mean of \$585.
c. Interpret the probability you obtained in part (b) in terms of sampling error.
d. Repeat parts (a)–(c) for samples of size 75.

12. The following figure shows the curve for a normally distributed variable (in color). Superimposed are the curves for the sampling distributions of the mean for two different sample sizes.

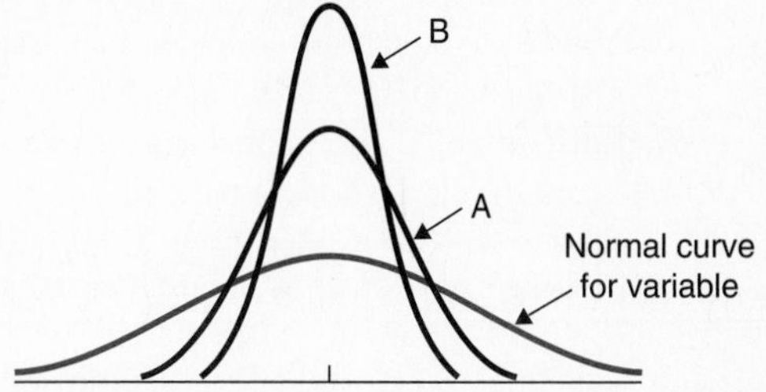

a. Explain why all three curves are centered at the same place.
b. Which curve corresponds to the larger sample size? Explain your answer.
c. Why is the spread of each curve different?
d. Which of the two sampling-distribution curves corresponds to the sample size that will tend to produce less sampling error? Explain your answer.
e. Why are the two sampling-distribution curves normal curves?

13. The American Council of Life Insurance reports the mean life insurance in force per covered family in *Life Insurance Fact Book.* Assume the standard deviation of life insurance in force is \$50,900.
a. Determine the probability that the sampling error made in estimating the (population) mean life insurance in force by that of a sample of 500 covered families will be \$2000 or less.
b. Must you assume that life-insurance amounts are normally distributed in order to answer part (a)? What if the sample size is 20 instead of 500?
c. Repeat part (a) for a sample size of 5000.

14. A paint manufacturer in Pittsburgh claims his paint will last an average of 5 years. Assuming paint life is normally distributed and has a standard deviation of 0.5 years, answer each of the following questions.
a. Suppose you paint one house with the paint and the paint lasts 4.5 years. Would you consider that evidence against the manufacturer's claim? (*Hint:* Assuming the manufacturer's claim is

correct, determine the probability that the paint life for a randomly selected house painted with the paint is 4.5 years or less.)

b. Suppose you paint 10 houses with the paint and the paint lasts an average of 4.5 years for the 10 houses. Would you consider that evidence against the manufacturer's claim?

c. Repeat part (b) if the paint lasts an average of 4.9 years for the 10 houses painted.

USING TECHNOLOGY  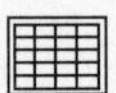 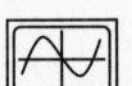

15. Each year, thousands of college seniors take the Graduate Record Examination (GRE). The scores are transformed so they have a mean of 500 and a standard deviation of 100. Furthermore, the scores are known to be normally distributed.

a. Use Minitab or some other statistical software to simulate 1000 samples of four GRE scores each.

b. Find the sample mean of each of the 1000 samples obtained in part (a).

c. Obtain the mean, the standard deviation, and a histogram of the 1000 sample means.

d. Theoretically, what are the mean, standard deviation, and distribution of all possible sample means for samples of size four?

e. Compare your answers from parts (c) and (d).

16. A variable is said to be **uniformly distributed** or to have a **uniform distribution** with parameters a and b if its distribution has the shape of the horizontal line segment $y = 1/(b - a)$, for $a < x < b$. The mean and standard deviation of such a variable are $(a + b)/2$ and $(b - a)/\sqrt{12}$, respectively. The basic random-number generator on a computer or calculator, which returns a number between 0 and 1, simulates a variable having a uniform distribution with parameters 0 and 1.

a. Sketch the distribution of a uniformly distributed variable with parameters 0 and 1. Observe from your sketch that such a variable is far from being normally distributed.

b. Use Minitab or some other statistical software to simulate 2000 samples of two random numbers between 0 and 1. If you are using Minitab, begin with **Calc ➤ Random Data ➤ Uniform....**

c. Find the sample mean of each of the 2000 samples obtained in part (b).

d. Determine the mean and standard deviation of the 2000 sample means.

e. Theoretically, what are the mean and the standard deviation of all possible sample means for samples of size two? Compare your answers to the ones you obtained in part (d).

f. Obtain a histogram of the 2000 sample means. Is the histogram bell-shaped? Would you necessarily expect it to be?

g. Repeat parts (b)–(f) for a sample size of 35.

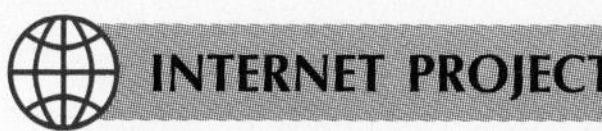

INTERNET PROJECT

Simulations

This Internet project provides you with tools, simulations, and demonstrations from around the world. With these tools, you can experiment with the ideas of this chapter for yourself.

You will simulate processes that show how random data can converge toward the predicted theory that you learned in the text. You will also investigate one of the central ideas of this chapter: how sample means tend to the population mean as the sample size increases.

URL for access to Internet Projects Page: `http://hepg.awl.com` *keyword:* Weiss

USING THE FOCUS DATABASE

In Chapter 1 we explained how to store the Focus database in a Minitab worksheet named `focus.mtw`. If you haven't already created that worksheet, follow the instructions on pages 54–55 to create it now.

The Focus database contains information on 500 randomly selected Arizona State University sophomores. Seven variables are considered for each student: sex, high-school GPA, SAT math score, cumulative GPA, SAT verbal score, age, and total hours. Suppose that in addition to studying these seven variables, we wish to conduct extensive interviews on college life with 25 of the 500 sophomores. Use Minitab or some other statistical software to obtain a simple random sample of 25 of the 500 sophomores in the Focus database. *(Hint:* Refer to the computer discussion in Example 1.12 on page 30.*)*

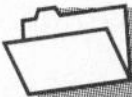

CASE STUDY DISCUSSION

The Chesapeake and Ohio Freight Study

At the beginning of this chapter on page 406, we discussed a freight study commissioned by the Chesapeake and Ohio Railroad Company (C&O). A sample of 2072 waybills from a population of 22,984 waybills was used to estimate the total revenue due C&O. The estimate arrived at was $64,568.

Since all 22,984 waybills were available, a census could be taken to determine exactly the total revenue due C&O and thereby reveal the accuracy of the estimate obtained by sampling. The exact amount due C&O was found to be $64,651.

a. What percentage of the waybills constituted the sample?
b. What percentage error was made by using the sample to estimate the total revenue due C&O?

c. At the time, the cost of a complete examination was approximately \$5000, whereas the cost of the sampling was only \$1000. Knowing this and your answers to parts (a) and (b), do you think that sampling was preferable to a census? Explain your answer.

d. In the study, the \$83 error was against C&O. Would it necessarily have to be that way?

BIOGRAPHY PIERRE-SIMON LAPLACE

Pierre-Simon Laplace was born on March 23, 1749, at Beaumount-en-Auge, Normandy, France, the son of a peasant farmer. His early schooling was at the military academy at Beaumount, where he developed his mathematical abilities. At the age of 18, he went to Paris. Within two years he was recommended for a professorship at the École Militaire by the French mathematician and philosopher Jean d'Alembert. (It is said that Laplace examined and passed Napoleon Bonaparte there in 1785.) In 1773 Laplace was granted membership in the Academy of Sciences.

Laplace held various positions in public life: he was president of the Bureau des Longitudes, professor at the École Normale, Minister of the Interior under Napoleon for six weeks (at which time he was replaced by Napoleon's brother), and Chancellor of the Senate; he was also made a marquis.

His professional interests were also varied. He published several volumes on celestial mechanics (which the Scottish geologist and mathematician John Playfair said were "the highest point to which man has yet ascended in the scale of intellectual attainment"), a book entitled *Théorie analytique des probabilités* (Analytic Theory of Probability), and other works on physics and mathematics. Laplace's primary contribution to the field of probability and statistics was the remarkable and all-important central limit theorem, which appeared in an 1809 publication and was read to the Academy of Sciences on April 9, 1810.

Astronomy was Laplace's major work; approximately half of his publications were concerned with the solar system and its gravitational interactions. These interactions were so complex that even Sir Isaac Newton had concluded "divine intervention was periodically required to preserve the system in equilibrium." Laplace, however, proved that planets' average angular velocities are invariable and periodic, and thus made the most important advance in physical astronomy since Newton.

When Laplace died in Paris on March 5, 1827, he was eulogized by the famous French mathematician and physicist Simeon Poisson as "the Newton of France."

PART

Inferential Statistics

CASE STUDY

ZOOPLANKTON NUTRITION IN THE GULF OF MEXICO

As the Mississippi River meets the Gulf of Mexico, the river's fresh water, with its high concentration of nutrient-laden sediment, mixes with gulf water. This results in a body of low salinity, nutrient-rich water (often referred to as the *plume*) that spreads westward along the Louisiana-Texas coast. The nutrients that are mixed into the gulf fuel the growth of microscopic marine algae, which in turn are eaten by tiny marine animals called zooplankton.

Dr. George McManus and Gregory Weiss, two marine scientists studying the dynamics of the lower food chain along the Gulf Coast, were interested in examining the effects of the nutrient-rich Mississippi River water on zooplankton nutrition. In early spring 1993, they collected zooplankton samples at three sites: near the mouth of the river, in the western part of the plume, and outside the plume to the east.

Lipids are biochemical compounds that serve as energy storage and cell-membrane components. The scientists used the lipid content of the zooplankton as an index of nutritional quality. Specifically, they used the ratio of two kinds of lipid compounds, triacylglycerols (energy storage compounds) and sterols (cell-membrane components), to determine the nutritional state of each zooplankton. The following table displays the data collected at the three sites. Note that the data are unitless, being the ratio of two weights.

Mouth of river	Western plume	Outside plume
1.89	1.79	1.36
2.27	2.30	0.32
1.14	2.52	0.34
1.51	1.37	0.00
	0.93	0.11

After studying confidence intervals in this chapter, you will be asked to analyze these data for the purpose of estimating the mean lipid content of zooplankton residing in each of the three regions of the Gulf Coast.

Confidence Intervals for One Population Mean

8

GENERAL OBJECTIVES

We now begin our study of inferential statistics. In this chapter we will examine methods for estimating the mean of a population. As you might suspect, the statistic used to estimate a population mean, μ, is a sample mean, $\bar{x}$. Because of sampling error, we cannot expect $\bar{x}$ to be exactly equal to μ. Thus it is important to provide information about the accuracy of the estimate. This leads to the discussion of *confidence intervals,* the main topic of this chapter.

We will examine two procedures for obtaining confidence intervals for the mean of a population. The first procedure applies when the standard deviation of the population is known, the second when the standard deviation of the population is unknown.

CHAPTER OUTLINE

8.1 ESTIMATING A POPULATION MEAN

A common problem in statistics is to obtain information about the mean, μ, of a population. For example, we might want to know

- the mean tar content of a certain brand of cigarette,
- the mean life of a newly developed steel-belted radial tire,
- the mean gas mileage of a new-model car, or
- the mean annual income of liberal-arts graduates.

If the population is small, we can ordinarily determine μ exactly by first taking a census and then computing μ from the population data. But if the population is large, as it often is in practice, then taking a census is generally impractical, extremely expensive, or impossible. Nonetheless we can usually obtain sufficiently accurate information about μ by taking a sample from the population. Let's look at an example.

EXAMPLE 8.1 USING A SAMPLE MEAN TO ESTIMATE A POPULATION MEAN

The U.S. Bureau of the Census publishes annual price figures for new mobile homes in *Current Construction Reports.* The figures are obtained from sampling, not from a census. A random sample of 36 new mobile homes yields the prices, in thousands of dollars, shown in Table 8.1. Use the data to estimate the population mean price, μ, of all new mobile homes.

TABLE 8.1 Prices ($1000s) of 36 randomly selected new mobile homes

36.5	23.1	34.1	43.6	35.7	31.2	48.5	35.7	32.4
44.8	27.9	45.1	35.3	44.2	49.4	50.3	47.2	48.6
25.8	28.4	43.3	22.2	37.6	32.9	32.4	42.3	31.0
35.9	36.4	46.9	46.5	34.4	37.9	46.8	48.8	35.1

SOLUTION As expected, we will estimate the mean price, μ, of all new mobile homes by the mean price, $\bar{x}$, of the 36 new mobile homes sampled. From Table 8.1 we find that

$$\bar{x} = \frac{\Sigma x}{n} = \frac{1378.2}{36} = 38.28.$$

Therefore, based on the sample data, we estimate the mean price, μ, of all new mobile homes to be approximately \$38.28 thousand, that is, \$38,280. An estimate of this kind is called a **point estimate** for μ because it consists of a single number, or point. ◨

As indicated in Definition 8.1, the term *point estimate* applies to the use of a statistic to estimate any parameter, not just a population mean.

DEFINITION 8.1 POINT ESTIMATE

A ***point estimate*** of a parameter is the value of a statistic that is used to estimate the parameter.

As we learned in Chapter 7, it is unreasonable to expect a sample mean to exactly equal the population mean; some sampling error is to be anticipated. Therefore, in addition to reporting a point estimate for μ, we need to provide information that indicates the accuracy of the estimate. We can do this by giving a **confidence-interval estimate** for μ. With a confidence-interval estimate for μ, we use the mean of a sample to construct an interval of numbers and state how confident we are that μ lies in that interval.

DEFINITION 8.2 CONFIDENCE-INTERVAL ESTIMATE; CONFIDENCE LEVEL

A ***confidence-interval estimate*** of a parameter consists of an interval of numbers obtained from a point estimate of the parameter together with a percentage that specifies how confident we are that the parameter lies in the interval. The confidence percentage is called the ***confidence level.***

Note: We often abbreviate *confidence interval* by CI.

In Example 8.2 we will obtain a 95.44% confidence interval for the mean price of all new mobile homes. In doing so, we will discuss in detail the logic behind determining confidence intervals. A general procedure, based on this logic, for obtaining confidence intervals at any prescribed confidence level will be presented in the next section.

EXAMPLE 8.2 INTRODUCES CONFIDENCE INTERVALS

Refer to Example 8.1. We will work in thousands of dollars. Assume that prices of new mobile homes are normally distributed and that the population standard deviation of all such prices is \$7.2 thousand, that is, \$7200.[†]

a. Identify the distribution of the variable $\bar{x}$, that is, the sampling distribution of the mean for samples of size 36.

[†] We might know the population standard deviation from previous research or from a preliminary study of prices. We will consider the more usual case, where σ is unknown, in Section 8.4.

b. Use part (a) to show that 95.44% of all samples of 36 new mobile homes have the property that the interval from $\bar{x} - 2.4$ to $\bar{x} + 2.4$ contains μ.

c. Use part (b) and the sample data in Table 8.1 to obtain a 95.44% confidence interval for the mean price of all new mobile homes.

SOLUTION **a.** Because $n = 36$, $\sigma = 7.2$, and prices of new mobile homes are normally distributed, Key Fact 7.4 on page 429 implies that

- $\mu_{\bar{x}} = \mu$ (which we don't know),
- $\sigma_{\bar{x}} = \sigma/\sqrt{n} = 7.2/\sqrt{36} = 1.2$, and
- $\bar{x}$ is normally distributed.

In other words, for samples of size 36, the variable $\bar{x}$ is normally distributed with mean μ and standard deviation 1.2.

b. The "95.44 part" of the 68.26-95.44-99.74 rule states that for a normally distributed variable, 95.44% of all possible observations lie within two standard deviations to either side of the mean. Applying this to the variable $\bar{x}$ and referring to part (a), we see that 95.44% of all samples of 36 new mobile homes have mean prices within $2 \cdot 1.2 = 2.4$ of μ. Or equivalently, 95.44% of all samples of 36 new mobile homes have the property that the interval from $\bar{x} - 2.4$ to $\bar{x} + 2.4$ contains μ.

c. As we know from part (b), 95.44% of all samples of 36 new mobile homes have the property that the interval from $\bar{x} - 2.4$ to $\bar{x} + 2.4$ contains μ. Therefore, we can be 95.44% confident that the sample of 36 new mobile homes whose prices are in Table 8.1 has that property. For that sample, $\bar{x} = 38.28$ and, so,

$$\bar{x} - 2.4 = 38.28 - 2.4 = 35.88 \quad \text{and} \quad \bar{x} + 2.4 = 38.28 + 2.4 = 40.68.$$

Consequently, our 95.44% confidence interval is from 35.88 to 40.68; we can be 95.44% confident that the mean price, μ, of all new mobile homes is somewhere between \$35,880 and \$40,680. It is essential to remember that, in reality, this or any other 95.44% confidence interval may or may not contain μ, but we can be 95.44% confident that it does. ■

A confidence interval for a population mean depends on the sample mean, $\bar{x}$, which in turn depends on the sample selected. For example, suppose the prices of the 36 new mobile homes sampled were as shown in Table 8.2 instead of as in Table 8.1.

TABLE 8.2 Prices ($1000s) of another sample of 36 randomly selected new mobile homes

48.0	47.1	36.2	28.0	39.1	35.6	41.5	39.7	47.0
50.5	38.8	31.0	50.7	37.5	36.3	37.1	43.0	43.8
40.7	28.2	41.6	40.3	54.2	44.2	43.0	35.2	39.3
43.9	33.4	44.1	40.8	47.1	29.9	41.1	39.1	52.9

Then we would have $\bar{x} = 40.83$ so that

$$\bar{x} - 2.4 = 40.83 - 2.4 = 38.43 \quad \text{and} \quad \bar{x} + 2.4 = 40.83 + 2.4 = 43.23.$$

Hence in this case the 95.44% confidence interval for μ would be from 38.43 to 43.23. We could be 95.44% confident that the mean price, μ, of all new mobile homes is somewhere between $38,430 and $43,230.

Example 8.3 stresses the importance of interpreting a confidence interval correctly. It also illustrates that the population mean, μ, may or may not lie in the confidence interval obtained.

EXAMPLE 8.3 INTERPRETING CONFIDENCE INTERVALS

Consider again the prices of new mobile homes. As we saw in part (b) of Example 8.2, 95.44% of all samples of 36 new mobile homes have the property that the interval from $\bar{x} - 2.4$ to $\bar{x} + 2.4$ contains μ. In other words, if 36 new mobile homes are selected at random and their mean price, $\bar{x}$, is computed, then the interval from

(1) $$\bar{x} - 2.4 \quad \text{to} \quad \bar{x} + 2.4$$

will be a 95.44% confidence interval for the mean price of all new mobile homes.

To illustrate that the mean price, μ, of all new mobile homes may or may not lie in the 95.44% confidence interval obtained, we used a computer to simulate 20 samples of 36 new mobile home prices each. For the simulation, we assumed that $\mu = 40$ (i.e., $40 thousand) and $\sigma = 7.2$ (i.e., $7.2 thousand). Of course, in reality, we don't know μ; we are assuming a value for μ to illustrate a point.

For each of the 20 samples of 36 new mobile home prices, we did three things: computed the sample mean price, $\bar{x}$; used (1) to obtain the 95.44% confidence interval for μ based on the sample; and noted whether the population mean, $\mu = 40$, actually lies in the confidence interval.

Figure 8.1 at the top of the next page summarizes our results. For each sample, we have drawn a graph on the right-hand side of Fig. 8.1. The dot represents the sample mean, $\bar{x}$, in thousands of dollars, and the horizontal line represents the corresponding 95.44% confidence interval. As we see, the population mean, μ, lies in the confidence interval when and only when the horizontal line crosses the dashed line.

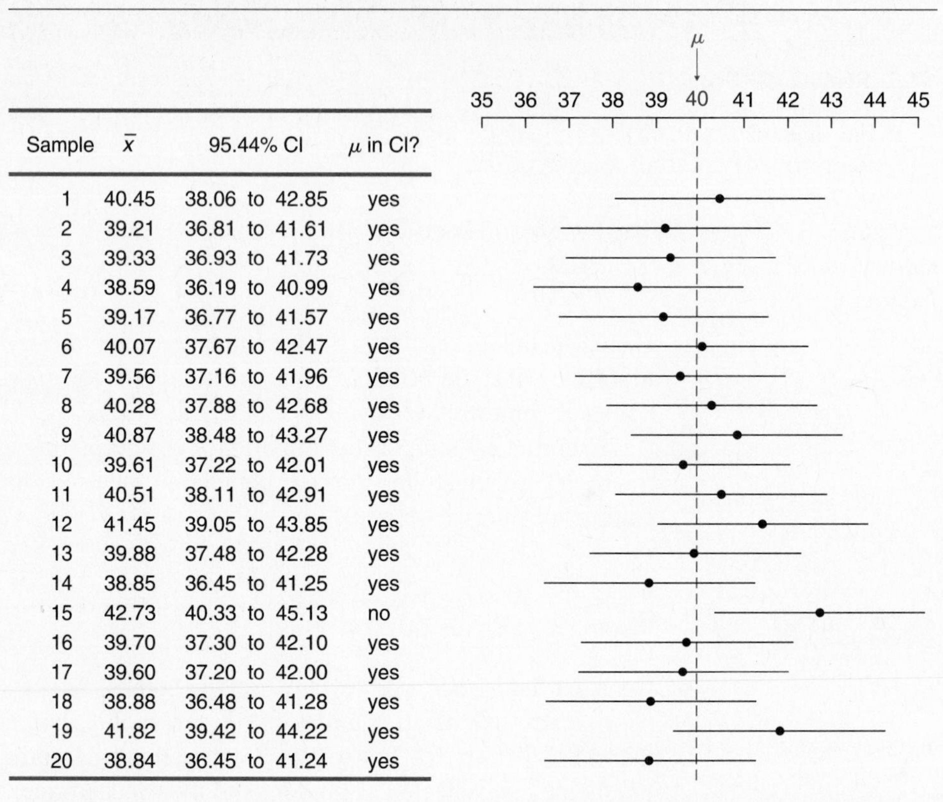

Sample	$\bar{x}$	95.44% CI	μ in CI?
1	40.45	38.06 to 42.85	yes
2	39.21	36.81 to 41.61	yes
3	39.33	36.93 to 41.73	yes
4	38.59	36.19 to 40.99	yes
5	39.17	36.77 to 41.57	yes
6	40.07	37.67 to 42.47	yes
7	39.56	37.16 to 41.96	yes
8	40.28	37.88 to 42.68	yes
9	40.87	38.48 to 43.27	yes
10	39.61	37.22 to 42.01	yes
11	40.51	38.11 to 42.91	yes
12	41.45	39.05 to 43.85	yes
13	39.88	37.48 to 42.28	yes
14	38.85	36.45 to 41.25	yes
15	42.73	40.33 to 45.13	no
16	39.70	37.30 to 42.10	yes
17	39.60	37.20 to 42.00	yes
18	38.88	36.48 to 41.28	yes
19	41.82	39.42 to 44.22	yes
20	38.84	36.45 to 41.24	yes

FIGURE 8.1 Twenty confidence intervals for the mean price of all new mobile homes, each based on a sample of 36 new mobile homes

From Fig. 8.1 we observe that μ lies in the 95.44% confidence interval in 19 of the 20 samples, that is, in 95% of the samples. If instead of 20 samples, we simulated, say, 1000 samples, then we would most likely find that the percentage of those 1000 samples for which μ lies in the 95.44% confidence interval would be even closer to 95.44%. So we can be 95.44% confident that any computed 95.44% confidence interval will contain μ. □

In Example 8.2 we obtained a 95.44% confidence interval for the mean price of all new mobile homes based on a sample of size 36. In doing so, we assumed the prices are normally distributed, which was used to conclude that $\bar{x}$ is normally distributed.

If the prices are not normally distributed, then, because of the central limit theorem, we can still say that $\bar{x}$ is approximately normally distributed. The impact on the resulting 95.44% confidence interval would be that it is only approximately correct, that is, the true confidence level would be only approximately equal to 95.44%.

EXERCISES 8.1

STATISTICAL CONCEPTS AND SKILLS

8.1 The value of a statistic that is used to estimate a parameter is called a ____ of the parameter.

8.2 What is a confidence-interval estimate of a parameter? Why is such an estimate superior to a point estimate?

8.3 The R. R. Bowker Company of New York collects data on books and periodicals. Sources for information are *Publishers Weekly, The Bowker Annual of Library and Book Trade Information,* and *Library Journal.* Twenty randomly selected science books have the prices, to the nearest dollar, shown in the table below.

91	61	99	88	93
73	100	54	100	75
85	78	70	80	87
52	97	102	80	75

a. Use the data to obtain a point estimate for the population mean price, μ, of all science books. *(Note:* The sum of the data is $1640.*)*

b. Is it likely that your point estimate in part (a) is exactly equal to μ? Explain your answer.

8.4 An educational psychologist at a large university wants to estimate the mean IQ of the students in attendance. A random sample of 15 students gives the following data on IQs.

113	120	103	118	104
88	110	126	112	120
98	110	126	101	115

a. Use the data to obtain a point estimate for the mean IQ, μ, of all students attending the university. *(Note:* The sum of the data is 1664.*)*

b. Is it likely that your point estimate in part (a) is exactly equal to μ? Explain your answer.

8.5 The U.S. National Center for Health Statistics compiles natality (birth) statistics and publishes the results in *Vital Statistics of the United States.* A random sample of 35 babies yields the following birth weights, in pounds.

7.4	6.0	8.6	4.5	2.0	7.9	4.0
2.6	5.9	7.3	7.3	7.0	6.3	8.1
7.1	7.3	6.6	5.2	9.8	8.0	10.9
6.3	3.8	5.0	8.0	10.7	9.7	6.0
6.8	10.3	7.6	6.5	7.1	5.8	6.9

Obtain a point estimate for the mean weight, μ, of all newborns. *(Note:* $\Sigma x = 240.3$ lb.*)*

8.6 *Trends in Television,* published by the Television Bureau of Advertising, contains information on the number of television sets owned by U.S. households. Fifty households are selected at random. The number of TV sets per household sampled is as follows.

1	1	1	2	6	3	3	4	2	4
3	2	1	5	2	1	3	6	2	2
3	1	1	4	3	2	2	2	2	3
0	3	1	2	1	2	3	1	1	3
3	2	1	2	1	1	3	1	5	1

Use the sample data to find a point estimate for the mean number of TV sets, μ, per U.S. household. *(Note:* $\Sigma x = 114$.*)*

For Exercises 8.7–8.10, you may want to review Example 8.2, which begins on page 445.

8.7 Refer to Exercise 8.3. Assume that science book prices are normally distributed with a standard deviation of $16.

a. Determine a 95.44% confidence interval for the mean price of all science books based on the data in Exercise 8.3.

b. Interpret your result in part (a).

c. Does the mean price of all science books lie in the confidence interval you obtained in part (a)? Explain your answer.

8.8 Refer to Exercise 8.4. Assume IQs for students attending the university are normally distributed with a standard deviation of 12.

a. Use the data in Exercise 8.4 to find a 95.44% confidence interval for the mean IQ, μ, of all students attending the university.

b. Interpret your result in part (a).

c. Does the mean IQ of all students attending the university lie in the confidence interval you obtained in part (a)? Explain your answer.

8.9 Refer to Exercise 8.5. Assume the population standard deviation of weights for newborns is 1.9 lb.

a. Use the data in Exercise 8.5 to determine an approximate 95.44% confidence interval for the mean weight, μ, of all newborns.

b. Interpret your result in part (a).

c. Why is the 95.44% confidence interval that you obtained in part (a) not necessarily exact?

8.10 Refer to Exercise 8.6. Assume $\sigma = 1.4$.

a. Obtain an approximate 95.44% confidence interval for the mean number of TV sets per U.S. household using the data from Exercise 8.6.

b. Interpret your result in part (a).

c. Why is the 95.44% confidence interval that you obtained in part (a) not necessarily exact?

EXTENDING THE CONCEPTS AND SKILLS

8.11 Refer to Examples 8.1 and 8.2. Use the data in Table 8.1 on page 444 to obtain a 99.74% confidence interval for the mean price of all new mobile homes. *(Hint:* Proceed as in Example 8.2 except use the "99.74 part" of the 68.26-95.44-99.74 rule instead of the "95.44 part."*)*

8.12 Refer to Examples 8.1 and 8.2. Use the data in Table 8.1 on page 444 to obtain a 68.26% confidence interval for the mean price of all new mobile homes.

8.2 CONFIDENCE INTERVALS FOR ONE POPULATION MEAN WHEN σ IS KNOWN

We discovered in Section 8.1 how to find a 95.44% confidence interval for a population mean, that is, a confidence interval at a confidence level of 95.44%. In this section we will generalize the arguments used there to obtain a confidence interval for a population mean at any prescribed confidence level.

To begin, it will be helpful to introduce some general notation that is used with confidence intervals. Frequently, we want to write the confidence level in the form $1 - \alpha$, where α is a number between 0 and 1; that is, if the confidence level is expressed as a decimal, then α is the number that must be subtracted from 1 to get the confidence level. To find α, simply subtract the confidence level from 1. If the confidence level is 95.44%, then $\alpha = 1 - 0.9544 = 0.0456$; if the confidence level is 90%, then $\alpha = 1 - 0.90 = 0.10$; and so forth.

Next recall that the symbol z_α is used to denote the z-score having area α to its right under the standard normal curve. So, for example, $z_{0.05}$ denotes the z-score having area 0.05 to its right, $z_{0.025}$ denotes the z-score having area 0.025 to its right, and $z_{\alpha/2}$ denotes the z-score having area $\alpha/2$ to its right.

Obtaining Confidence Intervals for a Population Mean When σ Is Known

We will now develop a simple step-by-step procedure for obtaining a confidence interval for a population mean when the population standard deviation is known. In doing so, we will assume the variable under consideration is normally distributed. But keep in mind that, because of the central limit theorem, the procedure applies as well for obtaining an approximately correct confidence interval for a population mean when the sample size is large, regardless of the distribution of the variable.

The basis of our confidence-interval procedure can be found in Key Fact 7.4: If x is a normally distributed variable with mean μ and standard deviation σ, then, for samples of size n, the variable $\bar{x}$ is also normally distributed and has mean μ and standard deviation $\sigma/\sqrt{n}$. In Section 8.1, we used this fact and the "95.44 part" of the 68.26-95.44-99.74 rule to conclude that 95.44% of all samples of size n have means within $2 \cdot \sigma/\sqrt{n}$ of μ, as seen in Fig. 8.2(a).

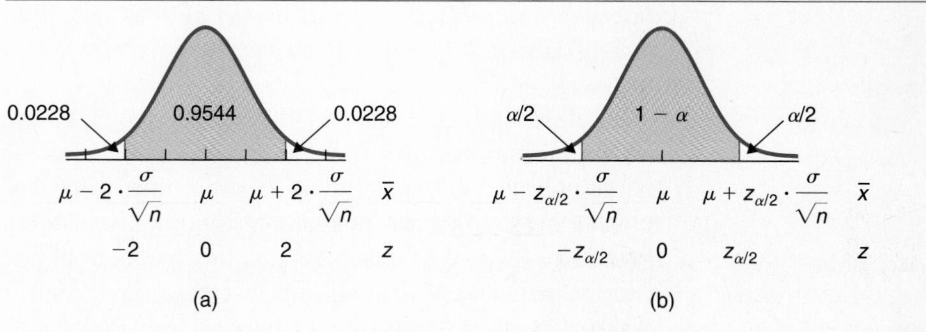

FIGURE 8.2 (a) 95.44% of all samples have means within 2 standard deviations of μ. (b) $100(1-\alpha)\%$ of all samples have means within $z_{\alpha/2}$ standard deviations of μ.

More generally, we can say that $100(1-\alpha)\%$ of all samples of size n have means within $z_{\alpha/2} \cdot \sigma/\sqrt{n}$ of μ, as can be seen in Fig. 8.2(b). Equivalently, we can say that $100(1-\alpha)\%$ of all samples of size n have the property that the interval from

$$\bar{x} - z_{\alpha/2} \cdot \frac{\sigma}{\sqrt{n}} \quad \text{to} \quad \bar{x} + z_{\alpha/2} \cdot \frac{\sigma}{\sqrt{n}}$$

contains μ. Consequently, we have the following procedure, sometimes referred to as the **one-sample *z*-interval procedure,** or more briefly as the ***z*-interval procedure.**

PROCEDURE 8.1 **THE ONE-SAMPLE z-INTERVAL PROCEDURE FOR A POPULATION MEAN**

ASSUMPTIONS

1. Normal population or large sample

2. σ known

Step 1 For a confidence level of $1 - \alpha$, use Table II to find $z_{\alpha/2}$.

Step 2 The confidence interval for μ is from

$$\bar{x} - z_{\alpha/2} \cdot \frac{\sigma}{\sqrt{n}} \quad \text{to} \quad \bar{x} + z_{\alpha/2} \cdot \frac{\sigma}{\sqrt{n}}$$

where $z_{\alpha/2}$ is found in Step 1, n is the sample size, and $\bar{x}$ is computed from the sample data.

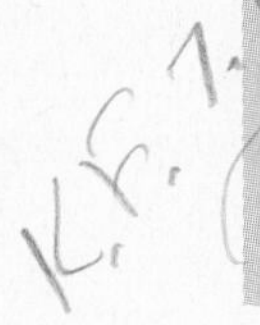

The confidence interval is exact for normal populations and is approximately correct for large samples from nonnormal populations.

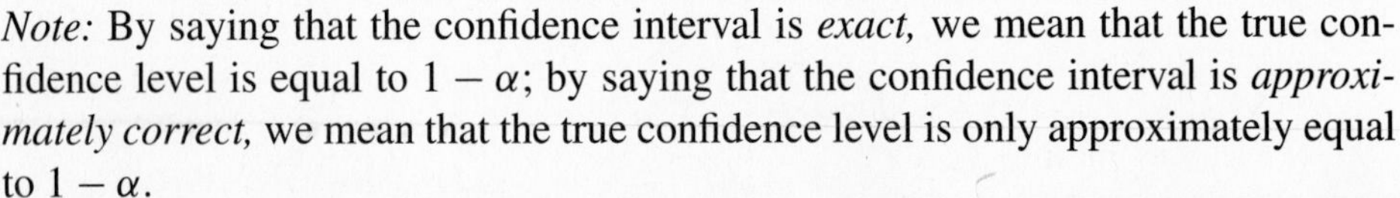

Note: By saying that the confidence interval is *exact,* we mean that the true confidence level is equal to $1 - \alpha$; by saying that the confidence interval is *approximately correct,* we mean that the true confidence level is only approximately equal to $1 - \alpha$.

Before applying Procedure 8.1, we need to make several comments. In Chapter 6 we mentioned that if a variable of a population is normally distributed and is the only variable under consideration, then it has become common statistical practice to say that the population is normally distributed. In Assumption 1 of Procedure 8.1, we have abbreviated this by "normal population." But remember, "normal population" really means that the variable under consideration is normally distributed on the population of interest. And, similarly, "nonnormal population" means that the variable under consideration is not normally distributed on the population of interest.

Assumption 1 of Procedure 8.1 is that the variable under consideration is normally distributed or the sample size is large. Actually, the procedure works reasonably well even when the variable is not normally distributed and the sample size is small or moderate, provided the variable is not too far from being normally distributed. Procedures that are insensitive to departures from the assumptions on which they are based are called **robust.** Thus the z-interval procedure is robust to moderate violations of the normality assumption.

When considering the z-interval procedure, it is also important to watch for outliers. The presence of outliers calls into question the normality assumption. And even for large samples, outliers can sometimes unduly affect a z-interval because the sample mean is not resistant to outliers. Key Fact 8.1 lists some general guidelines for when to use the z-interval procedure.

KEY FACT 8.1 **WHEN TO USE THE z-INTERVAL PROCEDURE (PROCEDURE 8.1)**

- For small samples, say, of size less than 15, the z-interval procedure should be used only when the variable under consideration is normally distributed or very close to being so.
- For moderate-size samples, say, between 15 and 30, the z-interval procedure can be used unless the data contain outliers or the variable under consideration is far from being normally distributed.
- For large samples, say, of size 30 or more, the z-interval procedure can be used essentially without restriction. However, if outliers are present and their removal is not justified, the effect of the outliers on the confidence interval should be examined; that is, you should compare the confidence intervals obtained with and without the outliers. If the effect is substantial, then it is probably best to use a different procedure or take another sample.
- If outliers are present but their removal is justified and results in a data set for which the z-interval procedure is appropriate (see above), then the procedure can be used.

Key Fact 8.1 makes it clear that you should conduct preliminary data analyses before applying the z-interval procedure. Normal probability plots, boxplots, stem-and-leaf diagrams, and histograms are often useful in this regard. More generally, we have the following fundamental principle of data analysis, which is relevant to all inferential procedures.

KEY FACT 8.2 **A FUNDAMENTAL PRINCIPLE OF DATA ANALYSIS**

Before performing a statistical-inference procedure, look at the sample data. If any of the conditions required for using the procedure appear to be violated, do not apply the procedure. Instead use a different, more appropriate procedure, or, if you are unsure of one, consult a statistician.

Even for small samples, where graphical displays must be interpreted carefully, it is still far better to look at the data than not to. Even for very small samples, graphical displays can sometimes detect violations of assumptions required for inferential procedures. Just remember to proceed cautiously when conducting graphical analyses of small samples, especially very small samples, say, of size 10 or less.

EXAMPLE 8.4 ILLUSTRATES PROCEDURE 8.1

The U.S. Bureau of Labor Statistics collects information on the ages of people in the civilian labor force and publishes the results in *Employment and Earnings.* Fifty people in the civilian labor force are randomly selected; their ages are displayed in Table 8.3. Find a 95% confidence interval for the mean age, μ, of all people in the civilian labor force. Assume $\sigma = 12.1$ years.

TABLE 8.3 Ages of 50 randomly selected people in the civilian labor force

22	58	40	42	43	32	34	45	38	19
33	16	49	29	30	43	37	19	21	62
60	41	28	35	37	51	37	65	57	26
27	31	33	24	34	28	39	43	26	38
42	40	31	34	38	35	29	33	32	33

SOLUTION We constructed (not shown) a normal probability plot, histogram, stem-and-leaf diagram, and boxplot for these data. The boxplot indicated potential outliers, but in view of the other three graphs, we concluded that in fact the data contain no outliers. Since the sample size is 50, which is large, and the population standard deviation, σ, is known, we can apply Procedure 8.1 to obtain the required confidence interval.

Step 1 *For a confidence level of* $1 - \alpha$, *use Table II to find* $z_{\alpha/2}$.

We want a 95% confidence interval, so $\alpha = 1 - 0.95 = 0.05$. Consulting Table II, we find that $z_{\alpha/2} = z_{0.05/2} = z_{0.025} = 1.96$.

Step 2 *The confidence interval for* μ *is from*

$$\bar{x} - z_{\alpha/2} \cdot \frac{\sigma}{\sqrt{n}} \quad \text{to} \quad \bar{x} + z_{\alpha/2} \cdot \frac{\sigma}{\sqrt{n}}.$$

We have $\sigma = 12.1$, $n = 50$, and, from Step 1, $z_{\alpha/2} = 1.96$. To compute $\bar{x}$ for the data in Table 8.3, we apply the usual formula:

$$\bar{x} = \frac{\Sigma x}{n} = \frac{1819}{50} = 36.4,$$

to one decimal place. Consequently, a 95% confidence interval for μ is from

$$36.4 - 1.96 \cdot \frac{12.1}{\sqrt{50}} \quad \text{to} \quad 36.4 + 1.96 \cdot \frac{12.1}{\sqrt{50}},$$

 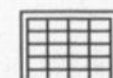 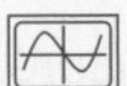

or 33.0 to 39.8. We can be 95% confident that the mean age, μ, of all people in the civilian labor force is somewhere between 33.0 years and 39.8 years. ■

Confidence and Precision

The confidence level of a confidence interval for a population mean, μ, signifies the confidence of the estimate; that is, it expresses the confidence we have that μ actually lies in the confidence interval. On the other hand, the length of the confidence interval indicates the precision of the estimate, that is, how well we have "pinned down" μ; long confidence intervals indicate poor precision, whereas short confidence intervals indicate good precision.

How does the confidence level affect the length of the confidence interval? To answer this question, let's return to Example 8.4, where we found a 95% confidence interval for the mean age, μ, of all people in the civilian labor force. The confidence level there is 0.95, and the confidence interval we computed is from 33.0 to 39.8 years. If we change the confidence level from 0.95 to, say, 0.90, then $z_{\alpha/2}$ changes from $z_{0.05/2} = z_{0.025} = 1.96$ to $z_{0.10/2} = z_{0.05} = 1.645$. The resulting confidence interval, using the same sample data (Table 8.3), is then from

$$36.4 - 1.645 \cdot \frac{12.1}{\sqrt{50}} \quad \text{to} \quad 36.4 + 1.645 \cdot \frac{12.1}{\sqrt{50}},$$

or from 33.6 to 39.2 years. We picture both the 90% and 95% confidence intervals in Fig. 8.3.

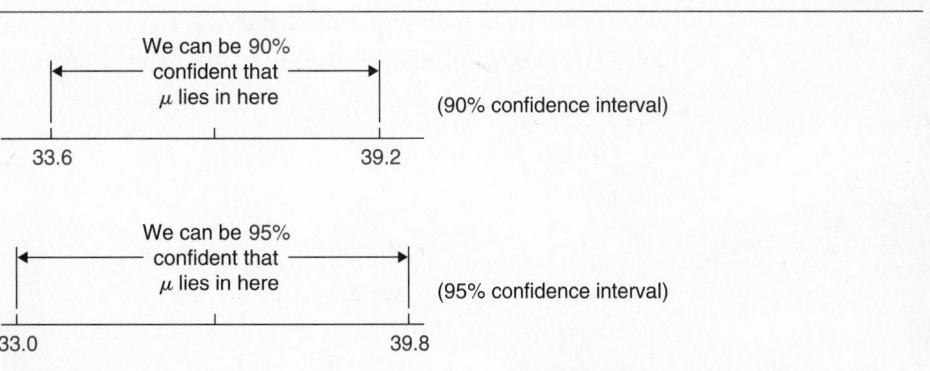

FIGURE 8.3 90% and 95% confidence intervals for μ using the data in Table 8.3

Thus, decreasing the confidence level decreases the length of the confidence interval, and vice-versa. This makes sense: If we are willing to settle for less confidence that μ lies in our confidence interval, then we can obtain a shorter interval; whereas, if we want to be more confident that μ lies in our confidence interval, then naturally we must settle for a more extensive interval. In other words:

KEY FACT 8.3 **CONFIDENCE AND PRECISION**

For a fixed sample size, decreasing the confidence level increases the precision, and vice-versa.

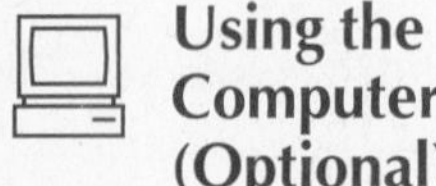

Using the Computer (Optional)

Procedure 8.1 on page 452 provides a step-by-step method for obtaining a confidence interval for a population mean, μ, when the population standard deviation is known. In Example 8.4 we applied that procedure to determine a confidence interval for the mean age of all people in the civilian labor force. Example 8.5 shows how Minitab can be employed to obtain that confidence interval.

EXAMPLE 8.5 USING MINITAB TO OBTAIN A ONE-SAMPLE *z*-INTERVAL

Table 8.3 on page 454 displays the ages of 50 randomly selected people in the civilian labor force. Use Minitab to determine a 95% confidence interval for the mean age, μ, of all people in the civilian labor force.

SOLUTION First we store the age data in Table 8.3 in a column named AGE. Then we proceed in the following manner.

1 Choose **Stat ➤ Basic Statistics ➤ 1-Sample Z...**
2 Specify AGE in the **Variables** text box
3 Select the **Confidence interval** option button
4 Click in the **Level** text box and type 95
5 Click in the **Sigma** text box and type 12.1
6 Click **OK**

Printout 8.1 shows the output that results.

PRINTOUT 8.1
Minitab output for the one-sample z-interval procedure

```
The assumed sigma = 12.1

Variable      N      Mean     StDev  SE Mean        95.0 % CI
AGE          50     36.38     11.07     1.71  (   33.03,    39.73)
```

The first line of Printout 8.1 displays the population standard deviation, σ: The assumed sigma = 12.1. Then we find the variable, sample size, sample mean, sample standard deviation, and standard error of the mean ($\sigma/\sqrt{n}$). The final item is the confidence interval. Hence a 95% confidence interval for μ is from 33.03 to 39.73. We can be 95% confident that the mean age, μ, of all people in the civilian labor force is somewhere between 33.03 years and 39.73 years. ◨

EXERCISES 8.2

STATISTICAL CONCEPTS AND SKILLS

8.13 Find the confidence level and α for
a. a 90% confidence interval.
b. a 99% confidence interval.

8.14 Find the confidence level and α for
a. an 85% confidence interval.
b. a 95% confidence interval.

8.15 What does it mean to say that a $1 - \alpha$ confidence interval is
a. exact? **b.** approximately correct?

8.16 In developing Procedure 8.1, we assumed that the variable under consideration is normally distributed.
a. Explain why we need that assumption.
b. Explain why the procedure yields an approximately correct confidence interval for large samples, regardless of the distribution of the variable under consideration.

8.17 What does the abbreviation "normal population" stand for?

8.18 Refer to Procedure 8.1.
a. Explain in detail the assumptions required for using the z-interval procedure.
b. How important is the normality assumption? Explain your answer.

8.19 What does it mean to say that a statistical procedure is robust?

8.20 In each part below, assume the population standard deviation is known. Decide whether it is reasonable to use the z-interval procedure to obtain a confidence interval for the population mean. Provide a reason for your answer.
a. The variable under consideration is very close to being normally distributed and the sample size is 10.
b. The variable under consideration is very close to being normally distributed and the sample size is 75.
c. The sample data contain outliers and the sample size is 20.
d. The sample data contain no outliers, the variable under consideration is roughly normally distributed, and the sample size is 20.
e. The variable under consideration is far from being normally distributed and the sample size is 20.
f. The sample data contain no outliers, the sample size is 250, and the variable under consideration is far from being normally distributed.

8.21 Suppose we have obtained data by taking a random sample from a population. Before we perform a statistical inference, what should we do?

8.22 Suppose we have obtained data by taking a random sample from a population and that we intend to find a confidence interval for the population mean, μ. We intend to use a confidence level of either 95% or 99%. Which confidence level will result in the confidence-interval giving a more precise estimate of μ?

Preliminary data analyses indicate that it is reasonable to apply the z-interval procedure (Procedure 8.1 on page 452) in Exercises 8.23–8.28.

8.23 The Gallup Organization conducts annual national surveys on home gardening. Results are published by the National Association for Gardening in *National Gardening Survey.* A random sample is taken of 25 households with vegetable gardens. The mean size of their gardens is 643 sq ft.
a. Determine a 90% confidence interval for the mean size, μ, of all household vegetable gardens in the United States. Assume $\sigma = 247$ sq ft.
b. Interpret your answer from part (a).

8.24 A quality-control engineer in a bakery goods plant needs to estimate the mean weight, μ, of bags of potato chips that are packed by a machine. He knows from experience that $\sigma = 0.1$ oz for this machine. A random sample of 12 bags has a mean weight of 16.01 oz.

a. Find a 99% confidence interval for μ.

b. Interpret your answer from part (a).

8.25 The U.S. National Center for Health Statistics estimates mean weights of Americans by age, height, and sex and publishes the results in *Vital and Health Statistics.* Forty U.S. women, 5 ft 4 in. tall and age 18–24, are randomly selected. Their weights, in pounds, are as follows.

140	136	147	138	143	122	115	125
136	152	130	134	150	153	148	132
116	159	128	136	134	126	120	146
131	167	145	132	138	137	115	145
154	139	139	147	123	154	127	116

a. Assuming the population standard deviation of all such weights is 12.0 lb, determine a 90% confidence interval for the mean weight, μ, of all U.S. women 5 ft 4 in. tall and in the age group 18–24 years. (*Note:* $\bar{x} = 136.88$ lb.)

b. Interpret your answer from part (a).

8.26 The Bureau of Labor Statistics collects data on employment and hourly earnings in private industry groups and publishes its findings in *Employment and Earnings.* Twenty people working in the manufacturing industry are selected at random; their hourly earnings, in dollars, are as follows.

16.70	7.44	13.78	16.49	7.49
17.92	17.21	5.51	10.40	10.75
15.27	19.72	10.68	13.10	14.70
15.55	16.67	15.07	14.02	15.99

a. Find a 95% confidence interval for the mean hourly earnings, μ, of all people employed in the manufacturing industry. Assume the population standard deviation of the hourly earnings is \$3.25. (*Note:* The sum of the data is \$274.46.)

b. Interpret your answer from part (a).

8.27 The U.S. National Center for Education Statistics surveys colleges and universities to obtain data on the costs of attending an institution of higher education. Results are published in *Digest of Education Statistics.* A random sample is taken of 150 four-year colleges and universities. The mean tuition and fees for the schools selected is \$16,107. Assuming $\sigma = \$4241$, obtain a 95% confidence interval for the mean tuition and fees of all four-year colleges and universities.

8.28 A telephone company in the Southwest undertook a study on various aspects of phone usage. One item of interest was how long calls last. According to the media relations manager, the company randomly selected 15,000 local calls involving Phoenix residential customers; their mean duration was 3.8 minutes. If the standard deviation of the durations of all local calls is 4.0 minutes, obtain a 90% confidence interval for the mean duration of all local phone calls made by Phoenix residential customers.

8.29 Refer to Exercise 8.25.

a. Find a 99% confidence interval for μ.

b. Why is the confidence interval you found in part (a) longer than the one in Exercise 8.25?

c. Draw a graph similar to Fig. 8.3 on page 455 that displays both confidence intervals.

d. Which confidence interval yields a more precise estimate of μ? Explain your answer.

8.30 Refer to Exercise 8.26.

a. Determine an 80% confidence interval for μ.

b. Why is the confidence interval you found in part (a) shorter than the one in Exercise 8.26?

c. Draw a graph similar to Fig. 8.3 on page 455 that displays both confidence intervals.

d. Which confidence interval yields a more precise estimate of μ? Explain your answer.

EXTENDING THE CONCEPTS AND SKILLS

8.31 The U.S. Bureau of the Census compiles data on family size and presents its findings in *Current Population Reports.* Suppose 500 U.S. families are randomly selected in order to estimate the mean size, μ, of all U.S. families. Further suppose the results are as shown in the following frequency distribution.

Size	2	3	4	5	6	7	8	9
Frequency	198	118	101	59	12	3	8	1

a. If the population standard deviation of family sizes is 1.3, determine a 95% confidence interval for the mean size, μ, of all U.S. families. *(Hint:* To find the sample mean, use the grouped-data formulas on page 157.*)*

b. Interpret your answer from part (a).

8.32 Key Fact 8.3 states that for a fixed sample size, decreasing the confidence level increases the precision of the confidence-interval estimate of μ, and vice-versa.

a. Suppose we want to increase the precision without reducing our level of confidence. What can we do?

b. Suppose we want to increase our level of confidence without reducing the precision. What can we do?

USING TECHNOLOGY  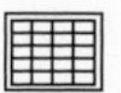 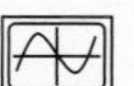

In Exercises 8.33 and 8.34, use Minitab or some other statistical software to

a. obtain a normal probability plot, boxplot, histogram, and stem-and-leaf diagram of the data, and

b. construct the required confidence interval.

c. Justify the use of your procedure in part (b).

8.33 The data set and confidence interval in part (a) of Exercise 8.25.

8.34 The data set and confidence interval in part (a) of Exercise 8.26.

8.35 A research physician wants to estimate the average age of people with diabetes. She takes a random sample of 35 diabetics and obtains the following ages.

48	41	57	83	41	55	59
61	38	48	79	75	77	7
54	23	47	56	79	68	61
64	45	53	82	68	38	70
10	60	83	76	21	65	47

Use Minitab or some other statistical software to

a. find a 95% confidence interval for the mean age, μ, of people with diabetes. Assume $\sigma = 21.2$ years.

b. obtain a normal probability plot, boxplot, histogram, and stem-and-leaf diagram of the data.

c. Remove the outliers (if any) from the data and then repeat part (a).

d. Comment on the advisability of using the z-interval procedure here.

8.36 A sociologist wants information on the number of children per farm family in her native state of Nebraska. Twenty-two randomly selected farm families have the following number of children.

1	2	1	2	0	2
1	5	4	1	0	2
1	3	1	0	1	
0	1	8	0	1	

Use Minitab or some other statistical software to

a. determine a 90% confidence interval for the mean number of children, μ, per farm family in Nebraska. Assume $\sigma = 1.95$.

b. obtain a normal probability plot, boxplot, histogram, and stem-and-leaf diagram of the data.

c. Remove the outliers (if any) from the data and then repeat part (a).

d. Comment on the advisability of using the z-interval procedure here.

8.37 We used Minitab's z-interval procedure to obtain the confidence interval required in Exercise 8.27 for the mean tuition and fees of all four-year colleges and universities. The Minitab output is shown in Printout 8.2 on the next page. Use the output to determine

a. the standard deviation of the tuitions and fees of all four-year colleges and universities.

b. the standard deviation of the tuitions and fees of all four-year colleges and universities in the sample.

c. the number of four-year colleges and universities in the sample.

d. the mean tuition and fees of all four-year colleges and universities in the sample.

e. a 95% confidence interval for the mean tuition and fees of all four-year colleges and universities.

8.38 We used Minitab's z-interval procedure to obtain the confidence interval required in Exercise 8.28 for the mean duration of all local phone calls made by Phoenix residential customers. The Minitab output is shown in Printout 8.3 on the next page. From the printout, determine

a. the standard deviation of the durations of all local phone calls made by Phoenix residential customers.
b. the standard deviation of the durations of the calls in the sample.
c. the number of calls in the sample.
d. the mean duration of the calls in the sample.
e. a 90% confidence interval for the mean duration of all local phone calls made by Phoenix residential customers.

8.39 This exercise can be done individually or, better yet, as a class project. Gestation periods of humans are normally distributed with a mean of 266 days and a standard deviation of 16 days.

a. Simulate 100 samples of nine human gestation periods each.
b. For each sample in part (a), obtain a 95% confidence interval for the population mean gestation period.
c. For the 100 confidence intervals that you obtained in part (b), roughly how many would you expect to contain the population mean gestation period of 266 days?
d. For the 100 confidence intervals that you obtained in part (b), determine the number that contain the population mean gestation period of 266 days.
e. Compare your answers from parts (c) and (d) and comment on any observed difference.

PRINTOUT 8.2 Minitab output for Exercise 8.37

```
The assumed sigma = 4241

Variable      N       Mean     StDev  SE Mean        95.0 % CI
TUITION     150      16107      4169      346  (   15428,    16786)
```

PRINTOUT 8.3 Minitab output for Exercise 8.38

```
The assumed sigma = 4.00

Variable      N       Mean     StDev  SE Mean        90.0 % CI
DURATION  15000     3.7992    3.9608   0.0327  (  3.7455,   3.8529)
```

8.3 MARGIN OF ERROR

In this section we will examine in detail how sample size affects the precision of estimating a population mean by a sample mean. We have already seen in Key Fact 7.1 that the larger the sample size, the smaller the sampling error tends to be. Now that we have studied confidence intervals, we can determine exactly how sample size affects the accuracy of the estimate. We begin by introducing *margin of error.*

EXAMPLE 8.6 MARGIN OF ERROR

In Example 8.4 we applied the one-sample z-interval procedure to the ages of a sample of 50 people in the civilian labor force to obtain a 95% confidence interval for the mean age, μ, of all people in the civilian labor force. Discuss the precision with which $\bar{x}$ estimates μ.

SOLUTION Recalling that $z_{\alpha/2} = z_{0.05/2} = z_{0.025} = 1.96$, $n = 50$, $\sigma = 12.1$, and $\bar{x} = 36.4$, we found that a 95% confidence interval for μ is from

$$\bar{x} - z_{\alpha/2} \cdot \frac{\sigma}{\sqrt{n}} \quad \text{to} \quad \bar{x} + z_{\alpha/2} \cdot \frac{\sigma}{\sqrt{n}}$$

or

$$36.4 - 1.96 \cdot \frac{12.1}{\sqrt{50}} \quad \text{to} \quad 36.4 + 1.96 \cdot \frac{12.1}{\sqrt{50}}$$

or

$$36.4 - 3.4 \quad \text{to} \quad 36.4 + 3.4$$

or

$$33.0 \quad \text{to} \quad 39.8.$$

We can be 95% confident that the mean age, μ, of all people in the civilian labor force is somewhere between 33.0 years and 39.8 years.

The confidence interval that we obtained is relatively long and hence provides a rather wide range for the possible values of μ. In other words, the precision of the estimate is poor. To improve the precision, we need to decrease the length of the confidence interval.

As we learned in Section 8.2, we can decrease the length of the confidence interval and thereby increase the precision of the estimate by decreasing the confidence level from 95% to some lower level. But suppose we want to retain the same level of confidence and still narrow the confidence interval. How can we accomplish this? To answer that question, we first look more closely at the confidence interval by displaying it graphically in Fig. 8.4.

FIGURE 8.4
95% confidence interval for the mean age, μ, of all people in the civilian labor force

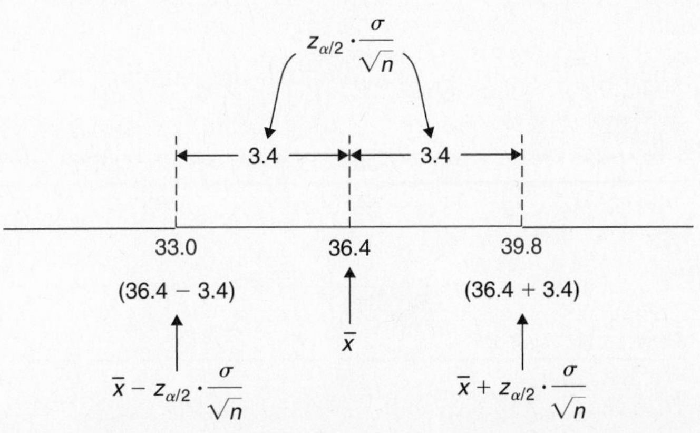

By studying Fig. 8.4 or referring to the computations done prior to it, we find that the length of the confidence interval is determined by the quantity

$$E = z_{\alpha/2} \cdot \frac{\sigma}{\sqrt{n}},$$

which is half the length of the confidence interval, in this case, 3.4. E is called the **margin of error,** also known as the **maximum error of the estimate.** We use this terminology because we can be 95% confident that μ lies in the confidence interval or, equivalently, that the margin of error in estimating μ by $\bar{x}$ is 3.4 years, as seen in Fig. 8.4. In newspapers and magazines, this is often expressed as "the poll has a margin of error of 3.4 years" or as "theoretically, in 95 out of 100 such polls the margin of error will be 3.4 years."

In any case we see that to narrow the confidence interval and thereby increase the precision of the estimate, we need only decrease the margin of error, E. Since the sample size, n, occurs in the denominator of the formula for E, we can decrease E by increasing the sample size. This makes sense, of course, because we expect to get more precise information from larger samples. □

In Example 8.6 we introduced some concepts and terminology important in confidence-interval analysis. We summarize that discussion in Definition 8.3 and Key Fact 8.4.

DEFINITION 8.3 MARGIN OF ERROR FOR THE ESTIMATE OF μ

The ***margin of error*** for the estimate of μ is

$$E = z_{\alpha/2} \cdot \frac{\sigma}{\sqrt{n}}.$$

The margin of error is equal to half the length of the confidence interval, as seen in Fig. 8.5.

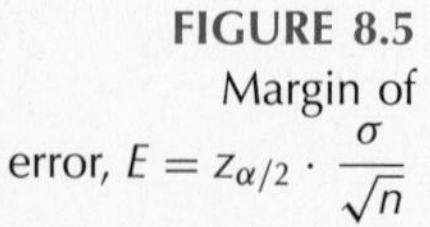

FIGURE 8.5 Margin of error, $E = z_{\alpha/2} \cdot \frac{\sigma}{\sqrt{n}}$

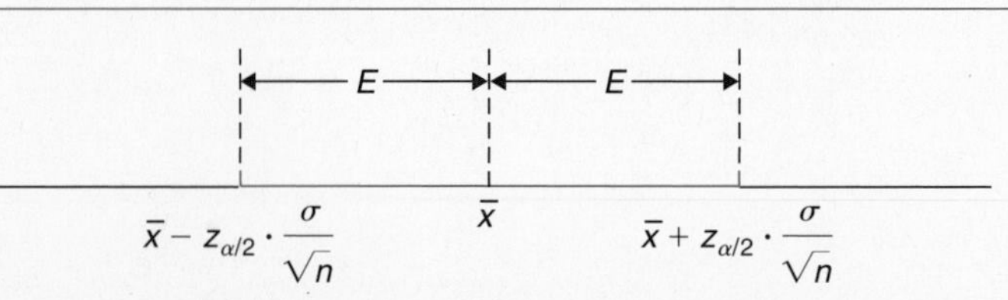

KEY FACT 8.4 MARGIN OF ERROR, PRECISION, AND SAMPLE SIZE

The length of a confidence interval for a population mean, μ, and hence the precision with which $\bar{x}$ estimates μ, is determined by the margin of error, E. For a fixed confidence level, increasing the sample size increases the precision, and vice-versa.

Determining the Required Sample Size

The margin of error and confidence level of a confidence interval are often specified in advance. We must then determine the sample size required to meet the specifications. The formula for the required sample size can be obtained by solving for n in the formula for the margin of error, $E = z_{\alpha/2} \cdot \sigma/\sqrt{n}$. The result is Formula 8.1.

FORMULA 8.1 SAMPLE SIZE FOR ESTIMATING μ

The sample size required for a $(1 - \alpha)$-level confidence interval for μ with a specified margin of error, E, is given by the formula

$$n = \left(\frac{z_{\alpha/2} \cdot \sigma}{E}\right)^2,$$

rounded up to the nearest whole number.

EXAMPLE 8.7 ILLUSTRATES FORMULA 8.1

Consider again the problem of estimating the mean age, μ, of all people in the civilian labor force.

a. Determine the sample size required to ensure that we can be 95% confident that μ is within 0.5 year of the estimate, $\bar{x}$. Recall that $\sigma = 12.1$ years.

b. Find a 95% confidence interval for μ if a sample of the size determined in part (a) has a mean age of 38.8 years.

SOLUTION **a.** To determine the required sample size, we apply Formula 8.1. In doing so, we must identify σ, E, and $z_{\alpha/2}$. We know that $\sigma = 12.1$ years; the margin of error, E, is specified at 0.5 year; and the confidence level is stipulated as 0.95, which means that $\alpha = 0.05$ and $z_{\alpha/2} = z_{0.05/2} = z_{0.025} = 1.96$. Thus the required sample size is

$$n = \left(\frac{z_{\alpha/2} \cdot \sigma}{E}\right)^2 = \left(\frac{1.96 \cdot 12.1}{0.5}\right)^2 = 2249.79.$$

Obviously, we cannot take a fractional sample size, so to be conservative we round up to 2250. Consequently, if 2250 people in the civilian labor force are randomly selected, then we can be 95% confident that the mean age, μ, of all people in the civilian labor force is within 0.5 year of the mean age, $\bar{x}$, of the people in the sample.

b. For this part we are to find a 95% confidence interval for the mean age of all people in the civilian labor force if a sample of the size determined in part (a) has a mean age of 38.8 years. Applying Procedure 8.1 with $\alpha = 0.05$, $\sigma = 12.1$, $\bar{x} = 38.8$, and $n = 2250$, we obtain the confidence interval:

$$\bar{x} - z_{\alpha/2} \cdot \frac{\sigma}{\sqrt{n}} \quad \text{to} \quad \bar{x} + z_{\alpha/2} \cdot \frac{\sigma}{\sqrt{n}}$$

or

$$38.8 - 1.96 \cdot \frac{12.1}{\sqrt{2250}} \quad \text{to} \quad 38.8 + 1.96 \cdot \frac{12.1}{\sqrt{2250}}$$

or

$$38.8 - 0.5 \quad \text{to} \quad 38.8 + 0.5$$

or

$$38.3 \quad \text{to} \quad 39.3.$$

We can be 95% confident that the mean age, μ, of all people in the civilian labor force is somewhere between 38.3 years and 39.3 years. ◨

Note: The sample size of 2250 was determined in part (a) of Example 8.7 to guarantee a margin of error of 0.5 year for a 95% confidence interval. Therefore, in view of Fig. 8.5 on page 462, the 95% confidence interval required in part (b) of Example 8.7 could be obtained simply by computing

$$\bar{x} \pm E = 38.8 \pm 0.5.$$

This gives the same confidence interval, 38.3 to 39.3, that we got in part (b) of Example 8.7 but with much less work. It is possible, however, that because the sample size is a rounded value, the simpler method we just used might have incorrectly yielded a slightly wider confidence interval. In practice, the simpler method is acceptable since, at worst, it provides a slightly conservative estimate.

The formula for finding the required sample size, Formula 8.1, involves the population standard deviation, σ. If σ is unknown, which is usually the case in practice, and we want to apply the formula, then we must first estimate σ. One way to do this is to take a preliminary large sample, say, of size 30 or more. The sample standard deviation, s, of the sample obtained provides an estimate of σ and can be used in place of σ in Formula 8.1.

EXERCISES 8.3

STATISTICAL CONCEPTS AND SKILLS

8.40 Discuss the relationship between the margin of error and the standard error of the mean.

8.41 Explain why the margin of error determines the precision with which a sample mean estimates a population mean.

8.42 In each of the following cases, explain the effect on the margin of error and, hence, on the precision of estimating a population mean by a sample mean.

a. Increasing the confidence level while keeping the same sample size.

b. Increasing the sample size while keeping the same confidence level.

8.43 A confidence interval for a population mean has a margin of error of 3.4.

a. Determine the length of the confidence interval.

b. If the sample mean is 52.8, obtain the confidence interval.

8.44 A confidence interval for a population mean has length 20.

a. Determine the margin of error.

b. If the sample mean is 60, obtain the confidence interval.

8.45 Answer true or false to each of the following statements concerning a confidence interval for a population mean. Give reasons for your answers.

a. The length of a confidence interval can be determined knowing only the margin of error.

b. The margin of error can be determined knowing only the length of the confidence interval.

c. The confidence interval can be obtained knowing only the margin of error.

d. The confidence interval can be obtained knowing only the margin of error and the sample mean.

e. The margin of error can be determined knowing only the confidence level.

f. The confidence level can be determined knowing only the margin of error.

g. The margin of error can be determined knowing only the confidence level, population standard deviation, and sample size.

h. The confidence level can be determined knowing only the margin of error, population standard deviation, and sample size.

8.46 Formula 8.1 provides a formula for computing the sample size required to obtain a confidence interval with a specified confidence level and margin of error. The number resulting from the formula should be rounded up to the nearest whole number.

a. Why do we want a whole number?

b. Why do we round up instead of down?

8.47 How do we apply Formula 8.1 if σ is unknown?

8.48 In estimating the mean monthly fuel expenditure, μ, per household vehicle, the U.S. Energy Information Administration takes a sample of size 6841. Assuming that $\sigma = \$20.65$, determine their margin of error in estimating μ at the 95% level of confidence.

8.49 A 90% confidence interval for the mean size of household vegetable gardens in the United States was found in Exercise 8.23 to be from 561.7 sq ft to 724.3 sq ft. Obtain the margin of error by

a. taking half the length of the confidence interval.

b. using the formula in Definition 8.3 on page 462. (Recall that $n = 25$ and $\sigma = 247$ sq ft.)

8.50 A 99% confidence interval for the mean weight of bags of potato chips packed by a machine was found in Exercise 8.24 to be from 15.94 oz to 16.08 oz. Obtain the margin of error by

a. taking half the length of the confidence interval.

b. using the formula in Definition 8.3 on page 462. (Recall that $n = 12$ and $\sigma = 0.1$ oz.)

8.51 In Exercise 8.25 you were given a sample of weights obtained from the random selection of 40 U.S. women 5 ft 4 in. tall and age 18–24. Based on that data, a 90% confidence interval for the mean weight, μ, of all such women is from 133.8 lb to 140.0 lb.

a. Determine the margin of error, E.

b. Explain the meaning of E in this context as far as the accuracy of the estimate is concerned.

c. Find the sample size required to have a margin of error of 2.0 lb and a 99% confidence level. (Recall that $\sigma = 12.0$ lb.)

d. Determine a 99% confidence interval for the mean weight, μ, of all U.S. women age 18–24 who are 5 ft 4 in. tall if a sample of the size determined in part (c) has a mean of 134.2 lb.

8.52 In Exercise 8.26 you were asked to determine a 95% confidence interval, based on a sample of size 20, for the mean hourly earnings, μ, of people employed in the manufacturing industry. The 95% confidence interval is from \$12.30 to \$15.15.

a. Determine the margin of error, E.

b. Explain the meaning of E in this context as far as the accuracy of the estimate is concerned.

c. Determine the sample size required to ensure that we can be 95% confident that μ is within \$0.50 of our estimate, $\bar{x}$. (Recall that $\sigma = \$3.25$.)

d. Find a 95% confidence interval for μ if a sample of the size determined in part (c) has a mean of \$13.87.

EXTENDING THE CONCEPTS AND SKILLS

8.53 Professor Thomas Stanley of Georgia State University has surveyed millionaires since 1973. Among other information, Professor Stanley obtains estimates for the mean age, μ, of all U.S. millionaires. Suppose one year's study involved a random sample of 36 U.S. millionaires whose mean age was 58.53 years with a sample standard deviation of 13.36 years.

a. If for next year's study, a confidence interval for μ is to have a margin of error of 2 years and a confidence level of 95%, determine the required sample size.

b. Why did you use the sample standard deviation, $s = 13.36$, in place of σ in your solution to part (a)? Why is it permissible to do so?

8.54 The U.S. Bureau of the Census estimates the mean value of the land and buildings per corporate farm. Those estimates are published in *Census of Agriculture.* Suppose an estimate, $\bar{x}$, is obtained and the margin of error is \$1000. Does this imply that the true mean, μ, is within \$1000 of the estimate? Explain your answer.

8.55 Suppose a random sample is taken from a normal population having standard deviation 10 in order to obtain a 95% confidence interval for the mean of the population.

a. If the sample size is 4, obtain the margin of error.

b. Repeat part (a) for a sample size of 16.

c. Can you guess the margin of error for a sample size of 64? Explain your reasoning.

8.56 For a fixed confidence level, show that it is necessary to (approximately) quadruple the sample size in order to halve the margin of error. *(Hint:* Use Formula 8.1.*)*

8.4 CONFIDENCE INTERVALS FOR ONE POPULATION MEAN WHEN σ IS UNKNOWN

In Section 8.2 we learned how to determine a confidence interval for a population mean, μ, when the population standard deviation, σ, is known. The basis of the procedure can be found in Key Fact 7.4: If x is a normally distributed variable with mean μ and standard deviation σ, then, for samples of size n, the variable $\bar{x}$ is also normally distributed and has mean μ and standard deviation $\sigma/\sqrt{n}$; equivalently, the **standardized version of $\bar{x}$,**

$$z = \frac{\bar{x} - \mu}{\sigma/\sqrt{n}}, \tag{1}$$

has the standard normal distribution.

But what if, as is usually the case in practice, the population standard deviation is unknown? Then we cannot base our confidence-interval procedure on the standardized version of $\bar{x}$. The best we can do is estimate the population standard deviation, σ, by the sample standard deviation, s; in other words, replace σ by s in Equation (1) and base our confidence-interval procedure on the resulting variable,

(2) $$t = \frac{\bar{x} - \mu}{s/\sqrt{n}},$$

called the **studentized version of $\bar{x}$.**

Unlike the standardized version, the studentized version of $\bar{x}$ does not have a normal distribution. To get an idea of how their distributions differ, we used Minitab to simulate each variable for samples of size four, assuming $\mu = 15$ and $\sigma = 0.8$. (Any sample size, population mean, and population standard deviation will do.) Here is what we did.

1. We simulated 5000 samples of size four each.
2. For each of the 5000 samples, we obtained the sample mean and sample standard deviation.
3. For each of the 5000 samples, we determined the observed values of both the standardized and studentized versions of $\bar{x}$, as given by Equations (1) and (2), respectively.
4. We obtained histograms of both the 5000 observed values of the standardized version of $\bar{x}$ and the 5000 observed values of the studentized version of $\bar{x}$, as shown in Printout 8.4.

PRINTOUT 8.4
Minitab histograms of z (standardized version of $\bar{x}$) and t (studentized version of $\bar{x}$) for 5000 samples of size four

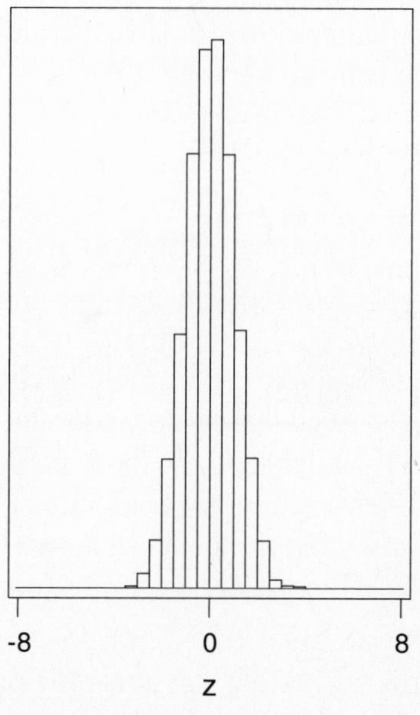

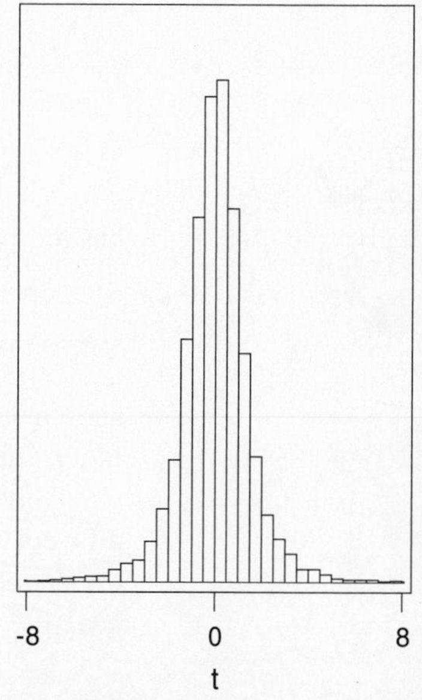

The two histograms in Printout 8.4 suggest that the distributions of the standardized version of $\overline{x}$—the variable z in Equation (1)—and the studentized version of $\overline{x}$—the variable t in Equation (2)—have things in common; both are bell-shaped and symmetric about 0. But there is an important difference, namely, that the distribution of the studentized version has more spread than the standardized version. This is not surprising since the variation in the possible values of the standardized version is due solely to the variation of sample means, whereas that of the studentized version is due to the variation of sample means and sample standard deviations.

As we know, the standardized version of $\overline{x}$ has the standard normal distribution. In 1908 William Gosset determined the distribution of the studentized version of $\overline{x}$, a distribution now called **Student's *t*-distribution** or, more briefly, just ***t*-distribution.**[†]

t-Distributions and *t*-Curves

Actually there is a different t-distribution for each sample size. We identify a particular t-distribution by giving its **degrees of freedom (df).** For the studentized version of $\overline{x}$, the degrees of freedom is one less than the sample size, which we indicate symbolically by $\text{df} = n - 1$. Later we will encounter t-statistics other than the studentized version of $\overline{x}$ whose degrees of freedom are different. But now, let's summarize the important fact we have learned about the studentized version of $\overline{x}$.

KEY FACT 8.5 **STUDENTIZED VERSION OF THE SAMPLE MEAN**

Suppose a variable x of a population is normally distributed with mean μ. Then, for samples of size n, the studentized version of $\overline{x}$,

$$t = \frac{\overline{x} - \mu}{s/\sqrt{n}},$$

has the t-distribution with $n - 1$ degrees of freedom.

Like normally distributed variables, a variable having a t-distribution has an associated curve, called a ***t*-curve.** In this book we don't need to know the equation of a t-curve, but we do need to understand the basic properties of a t-curve.

Although there is a different t-curve for each number of degrees of freedom, all t-curves are similar and resemble the standard normal curve. Figure 8.6 shows the standard normal curve and two t-curves. As illustrated by Fig. 8.6, t-curves have the properties delineated in Key Fact 8.6.

[†] See the biography on page 486 for more on Gosset and for the story of Student's t-distribution.

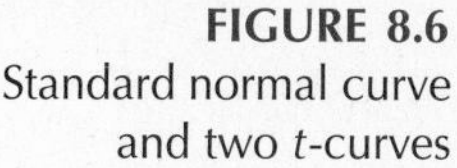

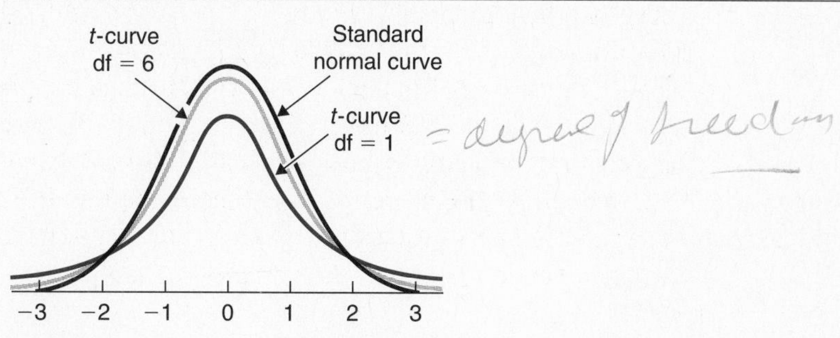

FIGURE 8.6
Standard normal curve and two t-curves

KEY FACT 8.6 **BASIC PROPERTIES OF t-CURVES**

Property 1: The total area under a t-curve is equal to 1.

Property 2: A t-curve extends indefinitely in both directions, approaching, but never touching, the horizontal axis as it does so.

Property 3: A t-curve is symmetric about 0.

Property 4: As the number of degrees of freedom becomes larger, t-curves look increasingly like the standard normal curve.

Using the t-Table

Percentages (and probabilities) for a variable having a t-distribution are equal to areas under its associated t-curve. For our purposes, one of which is obtaining confidence intervals for a population mean, we do not need a complete t-table for each t-curve; only certain areas will be important for us to know. Table IV, which appears in Appendix A and on the page facing the inside back cover, will be sufficient for our purposes.

The two outside columns of Table IV, labeled df, display the number of degrees of freedom. As expected, the symbol t_α denotes the t-value having area α to its right under a t-curve. Thus the column headed $t_{0.10}$ contains t-values having area 0.10 to their right; the column headed $t_{0.05}$ contains t-values having area 0.05 to their right; and so on. We illustrate a use of Table IV in Example 8.8.

EXAMPLE 8.8 FINDING THE t-VALUE HAVING A SPECIFIED AREA TO ITS RIGHT

For a t-curve with 13 degrees of freedom, determine $t_{0.05}$; that is, find the t-value having area 0.05 to its right, as shown in Fig. 8.7(a).

FIGURE 8.7 Finding the t-value having area 0.05 to its right

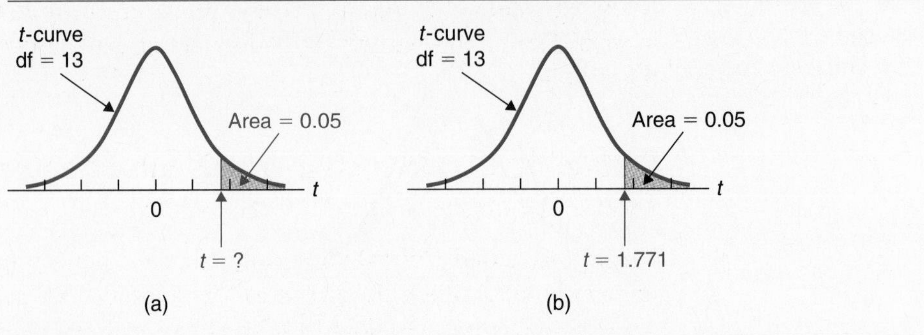

SOLUTION To find the t-value in question, we use Table IV. For ease of reference, we have repeated a portion of Table IV in Table 8.4.

TABLE 8.4 Values of t_α

df	$t_{0.10}$	$t_{0.05}$	$t_{0.025}$	$t_{0.01}$	$t_{0.005}$	df
.	.	.	.	.	.	.
.	.	.	.	.	.	.
.	.	.	.	.	.	.
12	1.356	1.782	2.179	2.681	3.055	12
13	1.350	1.771	2.160	2.650	3.012	13
14	1.345	1.761	2.145	2.624	2.977	14
15	1.341	1.753	2.131	2.602	2.947	15
.	.	.	.	.	.	.
.	.	.	.	.	.	.
.	.	.	.	.	.	.

Since the number of degrees of freedom is 13, we first go down the outside columns, labeled df, to "13." Then we go across that row until we are under the column headed $t_{0.05}$. The number in the body of the table there, 1.771, is the required t-value; that is, for a t-curve with df $= 13$, the t-value having area 0.05 to its right is $t_{0.05} = 1.771$, as seen in Fig. 8.7(b). □

  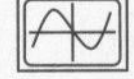

As we noted earlier, t-curves look increasingly like the standard normal curve as the number of degrees of freedom gets larger. For degrees of freedom greater than 1000, a t-curve and the standard normal curve are virtually indistinguishable. Consequently, we have stopped the t-table at df $= 1000$ and supplied the corresponding values of z_α beneath. These values can be used not only for the standard normal distribution but for any t-distribution having degrees of freedom greater than 1000.[†]

Obtaining Confidence Intervals for a Population Mean When σ Is Unknown

Having discussed t-distributions and t-curves, we can now develop a procedure for obtaining a confidence interval for a population mean when the population standard deviation is unknown. We proceed in essentially the same way as we did when the population standard deviation is known except now we invoke a t-distribution instead of the standard normal distribution.

The bottom line is that $t_{\alpha/2}$ instead of $z_{\alpha/2}$ is used in the formula for the confidence interval. Thus we have Procedure 8.2, which we refer to as the **one-sample *t*-interval procedure** or, more briefly, the ***t*-interval procedure.**

PROCEDURE 8.2

THE ONE-SAMPLE *t*-INTERVAL PROCEDURE FOR A POPULATION MEAN

ASSUMPTIONS

1. Normal population or large sample
2. σ unknown

Step 1 For a confidence level of $1 - \alpha$, use Table IV to find $t_{\alpha/2}$ with $\text{df} = n - 1$, where n is the sample size.

Step 2 The confidence interval for μ is from

$$\bar{x} - t_{\alpha/2} \cdot \frac{s}{\sqrt{n}} \quad \text{to} \quad \bar{x} + t_{\alpha/2} \cdot \frac{s}{\sqrt{n}}$$

where $t_{\alpha/2}$ is found in Step 1 and $\bar{x}$ and s are computed from the sample data.

The confidence interval is exact for normal populations and is approximately correct for large samples from nonnormal populations.

Before applying Procedure 8.2, we need to make several comments. Although the t-interval procedure was derived based on the assumption that the variable under consideration is normally distributed (normal population), it also applies approximately for large samples, regardless of the distribution of the variable under consideration, as noted at the bottom of Procedure 8.2.

† The values of z_α given at the bottom of Table IV are accurate to three decimal places and, because of that, some differ slightly from what we get by applying the method we learned using Table II.

Actually, like the z-interval procedure, the t-interval procedure works reasonably well even when the variable under consideration is not normally distributed and the sample size is small or moderate, provided the variable is not too far from being normally distributed. In other words, the t-interval procedure is robust to moderate violations of the normality assumption.

When considering the t-interval procedure, it is also important to watch for outliers. The presence of outliers calls into question the normality assumption. And even for large samples, outliers can sometimes unduly affect a t-interval because the sample mean and sample standard deviation are not resistant to outliers.

Guidelines for when to use the t-interval procedure are the same as those given for the z-interval procedure in Key Fact 8.1 on page 453. And remember, always look at the data before applying the t-interval procedure to ensure that it is reasonable to use it.

EXAMPLE 8.9 ILLUSTRATES PROCEDURE 8.2

The U.S. Federal Bureau of Investigation (FBI) compiles data on robbery and property crimes and publishes the information in *Population-at-Risk Rates and Selected Crime Indicators.* A sample of last year's pick-pocket offenses yields the values lost shown in Table 8.5. Use the data to obtain a 95% confidence interval for the mean value lost, μ, of all last year's pick-pocket offenses.

TABLE 8.5 Value lost ($) for a sample of 25 pick-pocket offenses

447	207	627	430	883
313	844	253	397	214
217	768	1064	26	587
833	277	805	653	549
649	554	570	223	443

SOLUTION Since the sample size, $n = 25$, is moderate, we first need to consider questions of normality and outliers. (See the second bulleted item in Key Fact 8.1 on page 453.) To do that, we constructed a normal probability plot for the data in Table 8.5, as shown in Fig. 8.8.

The normal probability plot in Fig. 8.8 shows no outliers and falls roughly in a straight line. Thus we can apply Procedure 8.2 to obtain the required confidence interval.

Step 1 *For a confidence level of* $1 - \alpha$*, use Table IV to find* $t_{\alpha/2}$ *with* $df = n - 1$*, where* n *is the sample size.*

We want a 95% confidence interval, so $\alpha = 1 - 0.95 = 0.05$. Since $n = 25$, $\text{df} = 25 - 1 = 24$. Consulting Table IV, we find that $t_{\alpha/2} = t_{0.05/2} = t_{0.025} = 2.064$.

FIGURE 8.8 Normal probability plot of the value-lost data in Table 8.5

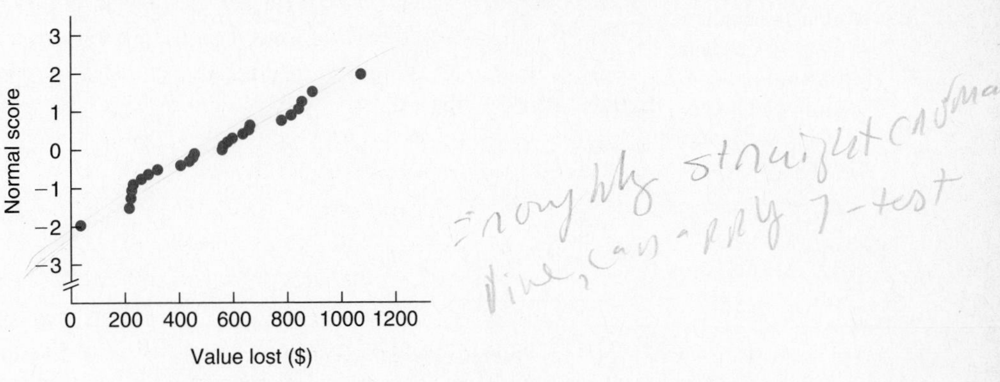

Step 2 *The confidence interval for* μ *is from*

$$\bar{x} - t_{\alpha/2} \cdot \frac{s}{\sqrt{n}} \quad \text{to} \quad \bar{x} + t_{\alpha/2} \cdot \frac{s}{\sqrt{n}}.$$

From Step 1, $t_{\alpha/2} = 2.064$. Applying the usual formulas for $\bar{x}$ and s to the data in Table 8.5, we find that $\bar{x} = 513.32$ and $s = 262.23$. Consequently, a 95% confidence interval for μ is from

$$513.32 - 2.064 \cdot \frac{262.23}{\sqrt{25}} \quad \text{to} \quad 513.32 + 2.064 \cdot \frac{262.23}{\sqrt{25}},$$

 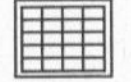 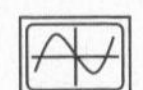

or 405.07 to 621.57. We can be 95% confident that the mean value lost, μ, of all last year's pick-pocket offenses is somewhere between \$405.07 and \$621.57. ◨

EXAMPLE 8.10 **ILLUSTRATES PROCEDURE 8.2**

The U.S. Department of Agriculture publishes data on chicken consumption in *Food Consumption, Prices, and Expenditures.* Last year's chicken consumptions, in pounds, for 17 randomly selected people are displayed in Table 8.6. Use the data to obtain a 90% confidence interval for last year's mean chicken consumption, μ.

TABLE 8.6 Sample of last year's chicken consumptions (lbs)

47	39	62	49	50	70
59	53	55	0	65	63
53	51	50	72	45	

SOLUTION A normal probability plot of the data in Table 8.6 is displayed in Fig. 8.9(a). The plot reveals an outlier—the observation of 0 lb. Since the sample size is only moderate, it is inappropriate to apply Procedure 8.2 to the data in Table 8.6.

FIGURE 8.9
Normal probability plots for chicken consumptions: (a) original data, (b) data with outlier removed

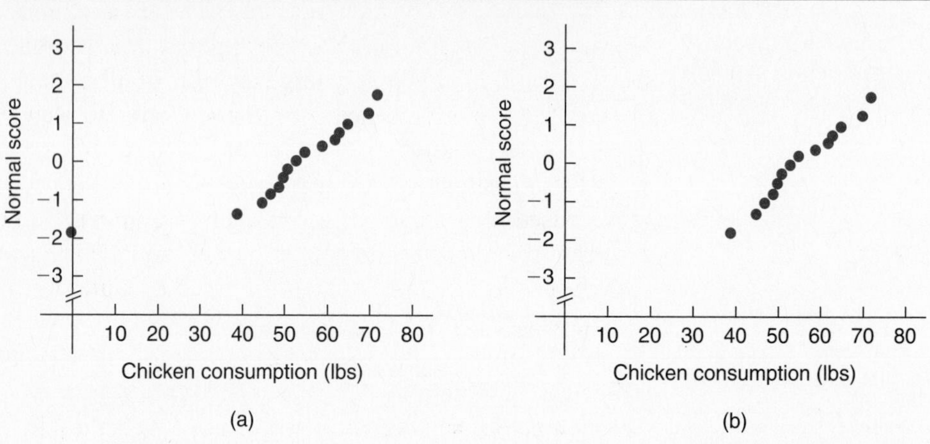

The outlier of 0 lb might be a recording error or it might reflect a person in the sample who does not eat chicken (e.g., a vegetarian). If we remove the outlier from the data, the normal probability plot for the abridged data shows no outliers and is roughly linear, as we see from Fig. 8.9(b).

This means that if we are willing to take as our population of interest only people who eat chicken, then we can use Procedure 8.2 to obtain a confidence interval. Applying that procedure to the sample data with the outlier removed, we find that a 90% confidence interval is from 51.2 to 59.2. We can be 90% confident that last year's mean chicken consumption, among people who eat chicken, is somewhere between 51.2 lbs and 59.2 lbs. ◨

By taking our population of interest in Example 8.10 to consist only of people who eat chicken, we were justified in removing the outlier of 0 lb. Generally, an outlier should not be removed unless careful consideration indicates that it is appropriate to do so. Simply removing an outlier because it is an outlier is unacceptable statistical practice.

If in Example 8.10 we had been careless in our analysis by blindly finding a confidence interval without first looking at the data, then our result would have been invalid and misleading. It is essential to perform preliminary data analyses to check assumptions before applying inferential procedures.

What If the Assumptions Are Not Satisfied?

We have now learned two methods for obtaining a confidence interval for a population mean. If the population standard deviation is known, we can use the z-interval procedure, Procedure 8.1 on page 452; if the population standard deviation is unknown, we can use the t-interval procedure, Procedure 8.2 on page 471.

But both of these procedures have another assumption for their use—the variable under consideration should be approximately normally distributed or the sample size should be relatively large and, for small samples, both procedures should be avoided in the presence of outliers. (Refer to Key Fact 8.1 on page 453 for general guidelines.)

So then, suppose we want to obtain a confidence interval for a population mean based on a small sample but that preliminary data analyses indicate either the presence of outliers or that the variable under consideration is far from normally distributed. Since neither the z-interval procedure nor the t-interval procedure is appropriate, what can we do?

Under certain conditions we can use a *nonparametric method.*[†] For example, if the variable under consideration has a symmetric distribution, then we can use a nonparametric method called the **one-sample Wilcoxon confidence-interval procedure** to obtain a confidence interval for the population mean.

Although most nonparametric methods have some assumptions for their use, they do not require even approximate normality, they are resistant to outliers and other extreme values, and they can be applied regardless of sample size. On the other hand, parametric methods, such as the z-interval and t-interval procedures, tend to give more accurate results than nonparametric methods when the normality assumption and other requirements for their use are met.

We will not cover the one-sample Wilcoxon confidence-interval procedure in this book. But we will discuss several other nonparametric procedures, beginning in Chapter 9 with the Wilcoxon signed-rank test.

EXAMPLE 8.11 CHOOSING A PROCEDURE

The Internal Revenue Service publishes information on federal individual income tax returns in *Statistics of Income, Individual Income Tax Returns.* A sample of 12 returns from last year revealed the adjusted gross incomes in thousands of dollars shown in Table 8.7. Which procedure should be used to obtain a confidence interval for the mean adjusted gross income, μ, of all last year's individual income tax returns?

TABLE 8.7 Adjusted gross incomes ($1000)

9.7	93.1	33.0	21.2
81.4	51.1	43.5	10.6
12.8	7.8	18.1	12.7

[†] Recall that descriptive measures for a population, such as μ and σ, are called parameters. Technically, inferential methods concerned with parameters are called **parametric methods;** those that are not are called **nonparametric methods.** However, it has become common practice to refer to most methods that can be applied without assuming normality (regardless of sample size) as nonparametric. Thus the term *nonparametric method* as used in contemporary statistics is somewhat of a misnomer.

SOLUTION Since the sample size is small ($n = 12$), we must first consider questions of normality and outliers. A normal probability plot of the sample data, as shown in Fig. 6.23 on page 387, suggests that adjusted gross incomes are far from being normally distributed. Consequently, neither the z-interval procedure nor the t-interval procedures should be used; instead some nonparametric confidence-interval procedure should be applied. ❑

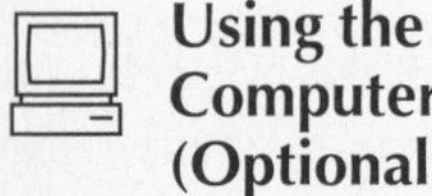

Using the Computer (Optional)

Procedure 8.2 on page 471 provides a step-by-step method for obtaining a confidence interval for a population mean, μ, when the population standard deviation is unknown. In Example 8.9 we applied Procedure 8.2 to determine a confidence interval for the mean value lost, μ, of all last year's pick-pocket offenses. Example 8.12 shows how Minitab can be employed to obtain that confidence interval.

EXAMPLE 8.12 USING MINITAB TO OBTAIN A ONE-SAMPLE t-INTERVAL

The values lost, in dollars, of 25 randomly selected pick-pocket offenses from last year are displayed in Table 8.5 on page 472. Use Minitab to obtain a 95% confidence interval for the mean value lost, μ, of all last year's pick-pocket offenses.

SOLUTION As we saw earlier, a normal probability plot of the data—which we could also get using Minitab—shows no outliers and is roughly linear. Thus we can apply Minitab's version of the one-sample t-interval procedure to obtain the required confidence interval.

First we store the value-lost data in Table 8.5 in a column named LOST. Then we proceed in the following manner.

1 Choose **Stat ➤ Basic Statistics ➤ 1-Sample t...**
2 Specify LOST in the **Variables** text box
3 Select the **Confidence interval** option button
4 Click in the **Level** text box and type 95
5 Click **OK**

The output obtained is displayed in Printout 8.5.

PRINTOUT 8.5
Minitab output for the one-sample t-interval procedure

```
Variable      N       Mean     StDev  SE Mean          95.0 % CI
LOST         25      513.3     262.2     52.4  (   405.1,     621.6)
```

The output in Printout 8.5 provides the variable, sample size, sample mean, sample standard deviation, estimated standard error of the mean ($s/\sqrt{n}$), and the confidence interval. This last item shows that we can be 95% confident that the mean value lost, μ, of all last year's pick-pocket offenses is somewhere between $405.1 and $621.6. □

EXERCISES 8.4

STATISTICAL CONCEPTS AND SKILLS

8.57 Explain the difference in the formulas for the standardized and studentized versions of $\bar{x}$.

8.58 Why do we need to consider the studentized version of $\bar{x}$ to develop a confidence-interval procedure for a population mean when the population standard deviation is unknown?

8.59 A variable has mean 100 and standard deviation 16. Four observations of this variable have a mean of 108 and a sample standard deviation of 12. Determine the observed value of the

a. standardized version of $\bar{x}$.

b. studentized version of $\bar{x}$.

8.60 A variable of a population has a normal distribution. Suppose you want to find a confidence interval for the population mean.

a. If you know the population standard deviation, which procedure would you use?

b. If you do not know the population standard deviation, which procedure would you use?

8.61 As reported by *Runner's World* magazine, the times of the finishers in the New York City 10-km run are normally distributed with a mean of 61 minutes and a standard deviation of 9 minutes. For samples of 12 finishing times, identify the distribution of each of the following variables.

a. $\dfrac{\bar{x} - 61}{9/\sqrt{12}}$ **b.** $\dfrac{\bar{x} - 61}{s/\sqrt{12}}$

8.62 In 1905 R. Pearl published the article "Biometrical Studies on Man. I. Variation and Correlation in Brain Weight" (*Biometrika,* Vol. 4, pp. 13–104). According to the study, brain weights of Swedish men are normally distributed with a mean of 1.40 kilograms (kg) and a standard deviation of 0.11 kg. For samples of 25 brain weights, identify the distribution of each of the following variables.

a. $\dfrac{\bar{x} - 1.40}{0.11/\sqrt{25}}$ **b.** $\dfrac{\bar{x} - 1.40}{s/\sqrt{25}}$

8.63 Explain why there is more variation in the possible values of the studentized version of $\bar{x}$ than in the possible values of the standardized version of $\bar{x}$.

8.64 Two t-curves have degrees of freedom 12 and 20, respectively. Which one more closely resembles the standard normal curve? Explain your answer.

8.65 For a t-curve with df = 6, use Table IV to find the following t-values.

a. $t_{0.10}$ **b.** $t_{0.025}$ **c.** $t_{0.01}$

8.66 For a t-curve with df = 17, use Table IV to find the following t-values.

a. $t_{0.05}$ **b.** $t_{0.025}$ **c.** $t_{0.005}$

8.67 For a t-curve with df = 21, find the following t-values and illustrate your results graphically.

a. the t-value having area 0.10 to its right

b. $t_{0.01}$

c. the t-value having area 0.025 to its left (*Hint:* A t-curve is symmetric about 0.)

d. the two t-values that divide the area under the curve into a middle 0.90 area and two outside areas of 0.05

8.68 For a t-curve with df = 8, find the following t-values and illustrate your results graphically.

a. the t-value having area 0.05 to its right

b. $t_{0.10}$

c. the t-value having area 0.01 to its left *(Hint:* A t-curve is symmetric about 0.*)*

d. the two t-values that divide the area under the curve into a middle 0.95 area and two outside 0.025 areas.

8.69 A random sample of size 100 is taken from a population with unknown standard deviation. A normal probability plot of the data displays significant curvature but no outliers. Is it reasonable to apply the t-interval procedure? Explain your answer.

8.70 A random sample of size 17 is taken from a population with unknown standard deviation. A normal probability plot of the data reveals an outlier but is otherwise roughly linear. Is it reasonable to apply the t-interval procedure? Explain your answer.

Preliminary data analyses indicate that it is reasonable to apply the t-interval procedure (Procedure 8.2 on page 471) in Exercises 8.71–8.76.

8.71 As reported by the Department of Agriculture in *Crop Production,* the mean yield of oats for U.S. farms is 58.4 bushels per acre. A farmer wants to estimate his mean yield using an organic method. He uses the method on a random sample of 1-acre plots and obtains the following yields, in bushels.

59	61	64	61	56	63	70	62
59	60	58	59	64	61	71	63
61	59	69	64	61	62	64	
61	59	67	61	58	62	59	
59	58	58	56	67	65	59	
65	57	67	62	69	59	58	
57	61	61	55	65	61	59	

a. Find a 99% confidence interval for the mean yield per acre, μ, that this farmer will get on his land with the organic method. *(Note:* $\bar{x} = 61.49$ bu, $s = 3.754$ bu.*)*

b. Does it appear that the farmer will get a mean yield different from the national average by using the organic method? Explain your answer.

8.72 As reported by the R. R. Bowker Company of New York in *Library Journal,* the mean annual subscription rate to law periodicals was \$97.33 in 1995. A sample of this year's law periodicals yields the following subscription rates to the nearest dollar.

106	122	120	123
118	114	138	131
128	124	119	130

a. Determine a 95% confidence interval for this year's mean annual subscription rate, μ, for all law periodicals. *(Note:* $\bar{x} = \$122.75$, $s = \$8.44$.*)*

b. Does your result from part (a) suggest a difference in the mean annual subscription rate from that in 1995? Justify your answer.

8.73 The *Physician's Handbook* provides statistics on heights and weights of children by age. The heights, in inches, of 20 randomly selected 6-year-old girls follow.

44	44	47	46	38
42	46	41	50	43
40	51	47	43	47
48	48	45	41	46

a. Obtain a 95% confidence interval for the mean height of all 6-year-old girls. *(Note:* $\bar{x} = 44.85$ inches, $s = 3.392$ inches.*)*

b. Interpret your answer from part (a).

8.74 To estimate the mean length, μ, of western rattlesnakes, 10 such snakes are randomly selected. Their lengths, in inches, are as follows.

40.2	43.1	45.5	44.5	39.5
40.2	41.0	41.6	43.1	44.9

a. Find a 90% confidence interval for the mean length of all western rattlesnakes. *(Note:* $\bar{x} = 42.36$ inches, $s = 2.158$ inches.*)*

b. Interpret your answer from part (a).

8.75 The U.S. Energy Information Administration surveys households to obtain data on monthly fuel expenditures for household vehicles. Results of those surveys are contained in *Residential Transportation Energy Consumption Survey, Consumption Patterns of Household Vehicles.* A random sample of 63 household vehicles yields the following monthly fuel expenditures to the nearest dollar.

61	51	59	50	99	58	40	42	62
79	73	56	47	45	58	48	59	57
61	66	62	44	66	45	77	76	63
67	83	55	49	77	51	60	39	71
71	56	59	47	55	60	51	55	48
66	54	48	49	72	39	52	56	89
45	82	65	38	69	64	51	54	60

a. Determine a 90% confidence interval for the mean monthly fuel expenditure of all household vehicles. *(Note:* $\bar{x} = \$58.90$, $s = \$12.75$.*)*

b. Interpret your answer from part (a).

8.76 In 1908 William S. Gosset published the article "The Probable Error of a Mean" (*Biometrika,* Vol. 6, pp. 1–25). It is in this pioneering paper, published under the pseudonym "Student," that he introduced what later became known as Student's t-distribution. As an example, Gosset used the following data set, which gives the additional sleep in hours obtained by 10 patients using laevohysocyamine hydrobromide.

1.9	0.8	1.1	0.1	−0.1
4.4	5.5	1.6	4.6	3.4

a. Find a 95% confidence interval for the additional sleep that would be obtained on the average for all people using laevohysocyamine hydrobromide. *(Note:* $\bar{x} = 2.33$ hr, $s = 2.002$ hr.*)*

b. Does it appear that the drug is effective in increasing sleep? Explain your answer.

EXTENDING THE CONCEPTS AND SKILLS

8.77 A city planner working on bikeways needs information about local bicycle commuters. She designs a questionnaire. One of the questions asks how many minutes it takes the rider to pedal from home to his or her destination. A sample of local bicycle commuters yields the following times.

22	19	24	31	29	29
21	15	27	23	37	31
30	26	16	26	12	
23	48	22	29	28	

a. Find a 90% confidence interval for the mean commuting time of all local bicycle commuters in the city. *(Note:* The sample mean and sample standard deviation of the data are 25.82 minutes and 7.71 minutes, respectively.*)*

b. Interpret your result in part (a).

c. Graphical analyses of the data indicate that the time of 48 minutes may be an outlier. Remove this potential outlier and repeat part (a). *(Note:* The sample mean and sample standard deviation of the abridged data are 24.76 and 6.05, respectively.*)*

d. Does it seem reasonable to perform the procedure you used in part (a)? Explain your answer.

8.78 Refer to Exercise 8.71.

a. Rework part (a) of Exercise 8.71 under the assumption that σ is known and equals 3.754 bu.

b. Compare the confidence interval from part (a) of this exercise to the one found in Exercise 8.71. Why is the margin of error smaller here?

8.79 The t-table, Table IV, contains degrees of freedom from 1 to 30 consecutively, but then contains only selected degrees of freedom.

a. Why couldn't we provide entries for every possible degrees of freedom?

b. Why did we construct the table so that consecutive entries appear for smaller degrees of freedom but only selected entries occur for larger degrees of freedom?

c. If you had only Table IV, what value would you use for $t_{0.05}$ with df = 85? with df = 52? with df = 78? Explain your answers.

8.80 As we mentioned earlier in this section, we stopped the t-table at df = 1000 and supplied the corresponding values of z_{α} beneath. Explain why it makes sense to do this.

8.81 A variable of a population has mean μ and standard deviation σ. For a sample of size n, under what conditions are the observed values of the studentized and standardized versions of $\bar{x}$ equal?

8.82 Let $0 < \alpha < 1$. For a t-curve, determine
a. the t-value having area α to its right in terms of t_α.
b. the t-value having area α to its left in terms of t_α.
c. the two t-values that divide the area under the curve into a middle $1 - \alpha$ area and two outside $\alpha/2$ areas.
d. Draw graphs to illustrate your results in parts (a)–(c).

USING TECHNOLOGY  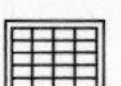 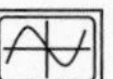

In Exercises 8.83 and 8.84, Use Minitab or some other statistical software to
a. obtain a normal probability plot, boxplot, histogram, and stem-and-leaf diagram of the data, and
b. construct the required confidence interval.
c. Justify the use of your procedure in part (b).

8.83 The data set and confidence interval in part (a) of Exercise 8.71.

8.84 The data set and confidence interval in part (a) of Exercise 8.72.

8.85 We used Minitab's t-interval procedure to obtain the confidence interval required in Exercise 8.73 for the mean height of all 6-year-old girls. The Minitab output is shown in Printout 8.6. Use the output to determine
a. the standard deviation of the heights of the girls in the sample.
b. the number of girls in the sample.
c. the mean height of the girls in the sample.
d. a 95% confidence interval for the mean height of all 6-year-old girls.

8.86 We used Minitab's t-interval procedure to obtain the confidence interval required in Exercise 8.74 for the mean length of all western rattlesnakes. The Minitab output is shown in Printout 8.7. From the printout, determine
a. the standard deviation of the lengths of the snakes in the sample.
b. the number of snakes in the sample.
c. the mean length of the snakes in the sample.
d. a 90% confidence interval for the mean length of all western rattlesnakes.

8.87 An issue of *Scientific American* reveals that the batting averages of major-league baseball players are approximately normally distributed with mean 0.270 and standard deviation 0.031.
a. Simulate 2000 samples of five batting averages each.
b. Determine the sample mean and sample standard deviation of each of the 2000 samples.
c. For each of the 2000 samples, determine the observed value of the standardized version of $\bar{x}$.
d. Obtain a histogram of the 2000 observations in part (c).
e. Theoretically, what is the distribution of the standardized version of $\bar{x}$?
f. Compare your results from parts (d) and (e).
g. For each of the 2000 samples, determine the observed value of the studentized version of $\bar{x}$.
h. Obtain a histogram of the 2000 observations in part (g).
i. Theoretically, what is the distribution of the studentized version of $\bar{x}$?
j. Compare your results from parts (h) and (i).
k. Compare your histograms from parts (d) and (h). How and why do they differ?

8.88 This exercise asks you to obtain graphs of several t-curves in order to illustrate the properties given in Key Fact 8.6 on page 469.
a. Using any technology available to you, obtain graphs of the standard normal curve and the t-curves having degrees of freedom 1, 2, 5, 10, and 20.
b. Examine the curves you obtained in part (a) in light of Key Fact 8.6.

8.89 This exercise uses a Minitab macro to obtain the graphs required in part (a) of Exercise 8.88. The macro file is `tcurves.mac` and can be found in the `Macro` folder of DataDisk CD. To invoke the macro, first enable the command language and then, in the Session window, type a percent sign (%) followed by the path for `tcurves.mac`.
a. Use the macro to obtain graphs of the standard normal curve and the t-curves having degrees of freedom 1, 2, 5, 10, and 20.
b. Interpret your result in part (a).

PRINTOUT 8.6 Minitab output for Exercise 8.85

```
Variable     N      Mean     StDev  SE Mean          95.0 % CI
HEIGHT      20    44.850     3.392    0.758   (  43.263,  46.437)
```

PRINTOUT 8.7 Minitab output for Exercise 8.86

```
Variable     N      Mean     StDev  SE Mean          90.0 % CI
LENGTH      10    42.360     2.158    0.683   (  41.109,  43.611)
```

CHAPTER REVIEW

You Should Be Able To

1. use and understand the formulas presented in this chapter.
2. obtain a point estimate for a population mean.
3. find and interpret a confidence interval for a population mean when the population standard deviation is known.
4. compute and interpret the margin of error for the estimate of μ.
5. understand the relationship between sample size, standard deviation, confidence level, and margin of error for a confidence interval for μ.
6. determine the sample size required for a specified confidence level and margin of error for the estimate of μ.
7. understand the difference between the standardized and studentized versions of $\bar{x}$.
8. state the basic properties of t-curves.
9. use Table IV to find $t_{\alpha/2}$ for df $= n - 1$ and selected values of α.
10. find and interpret a confidence interval for a population mean when the population standard deviation is unknown.
11. decide whether it is appropriate to use the z-interval procedure, t-interval procedure, or neither.
***12.** use the Minitab procedures covered in this chapter.
***13.** interpret the output obtained from the application of the Minitab procedures discussed in this chapter.

Key Terms

confidence-interval estimate, *445*
confidence level, *445*
degrees of freedom (df), *468*
margin of error, *462*
maximum error of the estimate, *462*
nonparametric methods, *475*
one-sample t-interval procedure, *471*
one-sample Wilcoxon confidence-interval procedure, *475*
one-sample z-interval procedure, *452*
parametric methods, *475*
point estimate, *445*
robust, *452*
standardized version of $\bar{x}$, *466*

REVIEW TEST

Statistical Concepts and Skills

1. Explain the difference between a point estimate of a parameter and a confidence-interval estimate of a parameter.

2. Answer true or false to the following statement and give a reason for your answer: If a 95% confidence interval for a population mean, μ, is from 33.8 to 39.0, then the mean of the population must lie somewhere between 33.8 and 39.0.

3. Is it necessary for the variable under consideration to be normally distributed in order to use the z-interval procedure or t-interval procedure? Explain your answer.

4. If you obtained one-thousand 95% confidence intervals for a population mean, μ, roughly how many of the intervals would actually contain μ?

5. Suppose you have obtained a sample with the intent of performing a particular statistical-inference procedure. What should you do before applying the procedure to the sample data? Why?

6. Suppose we intend to find a 95% confidence interval for a population mean by applying the one-sample z-interval procedure to a sample of size 100.

a. What would happen to the precision of the estimate if we used a sample of size 50 instead, keeping the same confidence level of 0.95?

b. What would happen to the precision of the estimate if we changed the confidence level to 0.90, keeping the same sample size of 100?

7. A confidence interval for a population mean has a margin of error of 10.7.

a. Obtain the length of the confidence interval.

b. If the mean of the sample is 75.2, determine the confidence interval.

8. Suppose that you plan to apply the one-sample z-interval procedure to obtain a 90% confidence interval for a population mean, μ. You know that $\sigma = 12$ and that you are going to use a sample of size nine.

a. What will your margin of error be?

b. What else do you need to know in order to obtain the confidence interval?

9. A variable of a population has mean 266 and standard deviation 16. Ten observations of this variable have a mean of 262.1 and a sample standard deviation of 20.4. Obtain the observed value of the

a. standardized version of $\bar{x}$.

b. studentized version of $\bar{x}$.

10. The monthly rents for studio apartments in a large city are normally distributed with a mean of \$585 and a standard deviation of \$45. For samples of three monthly rents, identify the distribution of each of the following variables.

a. $\dfrac{\bar{x} - 585}{45/\sqrt{3}}$ **b.** $\dfrac{\bar{x} - 585}{s/\sqrt{3}}$

11. The following figure shows the standard normal curve and two t-curves. Which of the two t-curves has the larger degrees of freedom? Explain your answer.

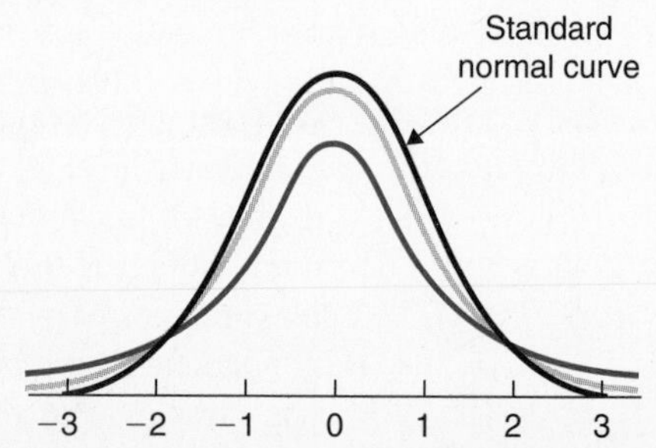

12. In each part of this problem, we have provided a scenario for a confidence interval. Decide whether the appropriate method for obtaining the confidence interval is the z-interval procedure, the t-interval procedure, or neither.

a. A random sample of size 17 is taken from a population. A normal probability plot of the sample data is found to be very close to linear (straight line). The population standard deviation is unknown.

b. A random sample of size 50 is taken from a population. A normal probability plot of the sample data is found to be roughly linear. The population standard deviation is known.

c. A random sample of size 25 is taken from a population. A normal probability plot of the sample data shows three outliers but is otherwise roughly linear. Furthermore, it is determined that the outliers are due to recording errors. The population standard deviation is known.

d. A random sample of size 20 is taken from a population. A normal probability plot of the sample data shows three outliers but is otherwise roughly linear. It is not clear whether it is reasonable to remove the outliers. The population standard deviation is unknown.

e. A random sample of size 128 is taken from a population. A normal probability plot of the data shows no outliers but significant curvature. The population standard deviation is known.

f. A random sample of size 13 is taken from a population. A normal probability plot of the sample data shows no outliers but significant curvature. The population standard deviation is unknown.

13. Dr. Thomas Stanley of Georgia State University has surveyed millionaires since 1973. Among other information, Stanley obtains estimates for the mean age, μ, of all U.S. millionaires. Suppose 36 U.S. millionaires are randomly selected and their ages are as follows.

31	45	79	64	48	38	39	68	52
59	68	79	42	79	53	74	66	66
71	61	52	47	39	54	67	55	71
77	64	60	75	42	69	48	57	48

Determine a 95% confidence interval for the mean age, μ, of all U.S. millionaires. Assume the standard deviation of ages of all U.S. millionaires is 13.0 years. (*Note:* The mean of the data is 58.53 years.)

14. From Problem 13 we know that "a 95% confidence interval for the mean age of all U.S. millionaires is from 54.3 years to 62.8 years." Decide which of the following provide a correct interpretation of the statement in quotes. Justify your answers.

a. 95% of all U.S. millionaires are between the ages of 54.3 years and 62.8 years.

b. There is a 95% chance that the mean age of all U.S. millionaires is between 54.3 years and 62.8 years.

c. We can be 95% confident that the mean age of all U.S. millionaires is between 54.3 years and 62.8 years.

d. The probability is 0.95 that the mean age of all U.S. millionaires is between 54.3 years and 62.8 years.

15. A certain brand of hand-held computer runs on four 1.5-volt AAA batteries. The mean battery life of a sample of 23 such computers is found to be 60.1 hours.

a. Assuming $\sigma = 4.3$ hours, obtain a 99% confidence interval for the mean battery life, μ, of this make of computer.

b. Interpret your answer from part (a).

c. What properties should a normal probability plot of the data have for it to be permissible to employ the procedure you used in part (a)?

16. Refer to Problem 15.

a. Find the margin of error, E.

b. Explain the meaning of E as far as the accuracy of the estimate is concerned.

c. Determine the sample size required to have a margin of error of 0.5 hours and a 99% confidence level.

d. Find a 99% confidence interval for μ if a sample of the size determined in part (c) yields a mean of 59.8 hours.

17. For a t-curve with df $= 18$, obtain the following t-values and illustrate your results graphically.

a. the t-value having area 0.025 to its right

b. $t_{0.05}$

c. the t-value having area 0.10 to its left

d. the two t-values that divide the area under the curve into a middle 0.99 area and two outside 0.005 areas

18. The A. C. Nielsen Company publishes information on television viewing by Americans in *Nielsen Report on Television.* A random sample of 20 U.S. households yields the following daily viewing times in hours.

7.5	3.5	6.8	5.0	7.9
5.3	8.9	8.8	10.3	8.8
9.5	9.5	6.1	9.4	8.4
8.2	6.5	9.0	6.4	8.4

a. A normal probability plot of this sample data shows no outliers and is roughly linear. Given that fact, find a 95% confidence interval for the mean daily viewing time, μ, of all U.S. households. *(Note:* $\bar{x} = 7.71$ *and* $s = 1.780$.*)*

b. Interpret your answer from part (a).

c. In 1992 the average U.S. household watched 7 hours and 4 minutes of TV per day. Does your confidence interval in part (a) provide evidence of a difference in average daily viewing time? Explain your answer.

Using Technology  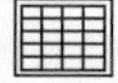 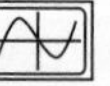

19. Use Minitab or some other statistical software to obtain the confidence interval required in Problem 13.

20. We used Minitab's one-sample z-interval procedure to obtain the confidence interval required in Problem 13 for the mean age of all U.S. millionaires. The Minitab output is shown in Printout 8.8. Use the output to determine

a. the (assumed) standard deviation of ages of all U.S. millionaires.

b. the sample standard deviation, s.

c. the number of millionaires sampled.

d. the mean age of the millionaires sampled.

e. a 95% confidence interval for the mean age of all U.S. millionaires.

21. Refer to Problem 18. Use Minitab or some other statistical software to

a. obtain and interpret a normal probability plot of the sample data.

b. find the required confidence interval.

22. We used Minitab's one-sample t-interval procedure to obtain the confidence interval required in Problem 18 for the mean daily viewing time of all U.S. households. The Minitab output is shown in Printout 8.9. Use the output to determine

a. the mean daily viewing time of the households in the sample.

b. the standard deviation of the daily viewing times of the households in the sample.

c. the estimated standard error of the mean.

d. the number of households in the sample.

e. a 95% confidence interval for the mean daily viewing time of all U.S. households.

23. Each year, thousands of college seniors take the Graduate Record Examination (GRE). The scores are transformed so they have a mean of 500 and a standard deviation of 100. Furthermore, the scores are known to be normally distributed.

a. Simulate 3000 samples of four GRE scores each.

b. Determine the sample mean and sample standard deviation of each of the 3000 samples.

c. For each of the 3000 samples, determine the observed value of the standardized version of $\bar{x}$.

d. Obtain a histogram of the 3000 observations in part (c).

e. Theoretically, what is the distribution of the standardized version of $\bar{x}$?

f. Compare your results from parts (d) and (e).

g. For each of the 3000 samples, determine the observed value of the studentized version of $\bar{x}$.

h. Obtain a histogram of the 3000 observations in part (g).

i. Theoretically, what is the distribution of the studentized version of $\bar{x}$?

j. Compare your results from parts (h) and (i).

k. Compare your histograms from parts (d) and (h). How and why do they differ?

PRINTOUT 8.8 Minitab output for Problem 20

```
The assumed sigma = 13.0

Variable      N       Mean     StDev  SE Mean        95.0 % CI
AGE          36      58.53     13.36     2.17  (   54.28,    62.77)
```

PRINTOUT 8.9 Minitab output for Problem 22

```
Variable      N       Mean     StDev  SE Mean        95.0 % CI
TIME         20      7.710     1.780    0.398  (   6.877,    8.543)
```

INTERNET PROJECT

Global Warming

The planet Earth is teeming with life, but that life exists in a complex ecosystem that depends in great part on a very narrow range of temperature. In recent years, concern has grown that the temperature on our planet is increasing.

In this project, you will explore data on global warming that has been collected from several sources. One of the most influential of these sources is the Intergovernmental Panel on Climate Change (IPCC).

Analysts at IPCC claim that the average temperature on Earth has already increased from approximately 0.5 to 1.1 degrees (Fahrenheit) since the late 1800s. They further predict that the temperature may rise another 1.8 to 6.3 degrees by the year 2100. If the predictions are correct, the planet may be in trouble. Now, in this Internet project, you will have an opportunity to explore the global-warming data for yourself.

URL for access to Internet Projects Page: `http://hepg.awl.com` *keyword:* Weiss

USING THE FOCUS DATABASE

In Chapter 1 we explained how to store the Focus database in a Minitab worksheet named `focus.mtw`. If you haven't already created that worksheet, follow the instructions on pages 54–55 to create it now.

The Focus database contains information on 500 randomly selected Arizona State University sophomores. Among the variables considered are high-school GPA, SAT math score, cumulative GPA, SAT verbal score, age, and total hours. Use Minitab or some other statistical software to solve the following problems.

a. Obtain a 99% confidence interval for the mean high-school GPA of all Arizona State University sophomores.
b. Repeat part (a) for the other five variables listed above.
c. Justify the use of the confidence-interval procedures you employed in parts (a) and (b).

CASE STUDY DISCUSSION

Zooplankton Nutrition in the Gulf of Mexico

At the beginning of this chapter, we discussed a study by two marine scientists investigating the nutritional state of zooplankton residing in three regions along the Gulf Coast. The scientists used lipid content as an index of nutritional quality. The data they obtained is displayed in the table on page 442. Referring to that table, answer the questions and conduct the analyses required in each part below.

a. Construct a normal probability plot for each of the three samples. *(Note:* The ordered normal scores for a sample of size four are −1.05, −0.30, 0.30, and 1.05.*)*
b. Identify outliers, if any, in each of the three samples.
c. For which samples might it be reasonable to use the t-interval procedure to obtain a confidence interval for the mean lipid content of zooplankton in the region? Explain your answer.
d. For each sample for which it might be reasonable to use the t-interval procedure, determine a 95% confidence interval for the mean lipid content of zooplankton in the region. Interpret your results.
e. Use Minitab or some other statistical software to solve parts (a), (b), and (d).

BIOGRAPHY WILLIAM GOSSET

William Sealy Gosset was born in Canterbury, England, on June 13, 1876, the eldest son of Colonel Frederic Gosset and Agnes Sealy. He studied mathematics and chemistry at Winchester College and New College, Oxford, receiving a first-class degree in natural sciences in 1899.

After graduation Gosset began work with Arthur Guinness and Sons, a brewery in Dublin, Ireland. He saw the need for accurate statistical analyses of various brewing processes ranging from barley production to yeast fermentation, and pressed the firm to solicit mathematical advice. So in 1906 the brewery sent him to work under Karl Pearson (see biographical sketch in Chapter 13) at University College in London.

In the next few years, Gosset developed what has come to be known as Student's t-distribution. This distribution has proved to be fundamental in statistical analyses involving normal distributions. In particular, Student's t-distribution is used in performing inferences for a population mean when the population being sampled is (approximately) normally distributed and the population standard deviation is unknown. Although the statistical theory for large samples had been completed in the early 1800s, no small-sample theory was available before Gosset.

Because Guinness's brewery prohibited its employees from publishing any of their research, Gosset published his contributions to statistical theory under the pseudonym "Student," thus the name "Student" in Student's t-distribution.

Gosset remained with Guinness his entire life. In 1935, he moved to London to take charge of a new brewery. But his tenure there was short-lived; he died in Beaconsfield, England, on October 16, 1937.

CASE STUDY

EFFECTS OF BREWERY EFFLUENT ON SOIL

Because many industrial wastes contain nutrients that enhance crop growth, efforts are being made, for environmental purposes, to use such wastes on agricultural soils. Two researchers, Mohammad Ajmal and Ahsan Ullah Khan, reported their findings on experiments with brewery wastes used for agricultural purposes in the article "Effects of Brewery Effluent on Agricultural Soil and Crop Plants" (*Environmental Pollution* (*Series A*), 33, pp. 341–351).

The researchers studied the physico-chemical properties of effluent from Mohan Meakin Breweries Ltd. (MMBL), Ghazibad, UP, India, and ". . . its effects on the physico-chemical characteristics of agricultural soil, seed germination pattern, and the growth of two common crop plants." They assessed the impact using different concentrations of the effluent: 25%, 50%, 75%, and 100%.

The following table shows the effects of the different dilutions of MMBL effluent on the available limestone, potassium, and phosphorus in the soil. Numbers not enclosed by and enclosed by parentheses are, respectively, the means and sample standard deviations of five observations. The data for limestone are in percentages and those for potassium and phosphorus are in milligrams per kilogram of soil.

		Treatment				
	Original	Control (water)	Effluent 100%	Effluent 75%	Effluent 50%	Effluent 25%
Limestone	2.30	2.80 (0.10)	2.50 (0.15)	2.60 (0.20)	3.00 (0.10)	2.95 (0.05)
Potassium	150	130 (1.5)	560 (2.0)	520 (1.0)	392 (1.5)	260 (3.0)
Phosphorus	530	560 (3.0)	810 (1.0)	690 (1.0)	630 (2.0)	590 (2.0)

After we study hypothesis testing in this chapter, we will be able to evaluate these findings statistically and gain insight into the effects of the application of brewery wastes on soil composition and plant growth.

Hypothesis Tests for One Population Mean

9

GENERAL OBJECTIVES

In Chapter 8 we examined methods for obtaining confidence intervals for one population mean. As we know, a confidence interval for a population mean, μ, is based on a sample mean, $\bar{x}$. Now we will learn how that statistic can be used to make decisions about hypothesized values of a population mean. For example, we might want to use the mean sentence, $\bar{x}$, of a sample of people imprisoned last year for drug offenses to decide whether last year's mean sentence, μ, for all such people exceeds the 1994 mean of 80.1 months. Statistical inferences of this kind are called *hypothesis tests.*

In this chapter we will study hypothesis tests for one population mean. Three different procedures will be considered. The first two are called the one-sample z-test and one-sample t-test; they are the hypothesis-testing analogues of the one-sample z-interval procedure and one-sample t-interval procedure discussed in Chapter 8. The third procedure we will consider is a nonparametric method called the Wilcoxon signed-rank test. It applies when the variable under consideration has a symmetric distribution.

We will also examine two different approaches to hypothesis testing, namely, the critical-value approach and the P-value approach.

CHAPTER OUTLINE

9.1 THE NATURE OF HYPOTHESIS TESTING

We often use inferential statistics to make decisions or judgments about the value of a parameter, such as a population mean. For example, we might need to decide whether the mean weight, μ, of all bags of pretzels packaged by a particular company differs from the advertised weight of 454 grams (g); or we might want to determine whether the mean age, μ, of all cars in use has increased from the 1995 mean of 8.5 years.

One of the most commonly used methods for making such decisions or judgments is to perform a **hypothesis test.** A **hypothesis** is a statement that something is true. For example, the statement "the mean weight of all bags of pretzels packaged differs from the advertised weight of 454 g" is a hypothesis.

Typically, a hypothesis test involves two hypotheses. One is called the **null hypothesis,** the other the **alternative hypothesis** (or **research hypothesis**).

DEFINITION 9.1 NULL AND ALTERNATIVE HYPOTHESES

Null hypothesis: A hypothesis to be tested. We use the symbol H_0 to represent the null hypothesis.

Alternative hypothesis: A hypothesis to be considered as an alternate to the null hypothesis. We use the symbol H_a to represent the alternative hypothesis.

For example, in the pretzel-packaging illustration, the null hypothesis might be "the mean weight of all bags of pretzels packaged equals the advertised weight of 454 g," and the alternative hypothesis might be "the mean weight of all bags of pretzels packaged differs from the advertised weight of 454 g."

Originally, the word *null* in *null hypothesis* stood for "no difference" or "the difference is null." Over the years, however, *null hypothesis* has come to mean simply a hypothesis to be tested. The problem in a hypothesis test is to decide whether or not the null hypothesis should be rejected in favor of the alternative hypothesis.

Choosing the Hypotheses

The first step in setting up a hypothesis test is to decide on the null hypothesis and the alternative hypothesis. Below we offer some guidelines for choosing these two hypotheses. Although the guidelines refer specifically to hypothesis tests for one population mean, μ, they apply to any hypothesis test concerning one parameter.

Null hypothesis: In this book the null hypothesis for a hypothesis test concerning a population mean, μ, should always specify a single value for that parameter. This means that the null hypothesis should always be of the form $\mu = \mu_0$, where μ_0 is some number. In other words, an equal sign (=) should appear in the null hypothesis. We can therefore express the null hypothesis concisely as

$$H_0\colon \mu = \mu_0.$$

Alternative hypothesis: The choice of the alternative hypothesis depends on and should reflect the purpose of the hypothesis test. Three choices are possible for the alternative hypothesis.

- If the primary concern is deciding whether a population mean, μ, is *different from* a specified value μ_0, then the alternative hypothesis should be $\mu \neq \mu_0$. In other words, a not-equal sign ($\neq$) should appear in the alternative hypothesis. We express such an alternative hypothesis as

  $$H_a\colon \mu \neq \mu_0.$$

  A hypothesis test whose alternative hypothesis is of this form is called a **two-tailed test.**

- If the primary concern is deciding whether a population mean, μ, is *less than* a specified value μ_0, then the alternative hypothesis should be $\mu < \mu_0$. In other words, a less-than sign (<) should appear in the alternative hypothesis. We express such an alternative hypothesis as

  $$H_a\colon \mu < \mu_0.$$

  A hypothesis test whose alternative hypothesis is of this form is called a **left-tailed test.**

- If the primary concern is deciding whether a population mean, μ, is *greater than* a specified value μ_0, then the alternative hypothesis should be $\mu > \mu_0$. In other words, a greater-than sign (>) should appear in the alternative hypothesis. We express such an alternative hypothesis as

  $$H_a\colon \mu > \mu_0.$$

  A hypothesis test whose alternative hypothesis is of this form is called a **right-tailed test.**

A hypothesis test is called a **one-tailed test** if it is either left-tailed or right-tailed, that is, if it is not two-tailed. In Section 9.2 we will explain the significance of the term *tailed.* But right now let's consider Examples 9.1–9.3, which illustrate the preceding discussion.

EXAMPLE 9.1 CHOOSING THE NULL AND ALTERNATIVE HYPOTHESES

A snack-food company produces a 454-gram bag of pretzels. Although the actual net weights deviate slightly from 454 g and vary from one bag to another, the company insists that the mean net weight of the bags be kept at 454 g. Indeed, if the mean net weight is less than 454 g, then the company will be short-changing its customers; and if the mean net weight exceeds 454 g, then the company will be unnecessarily overfilling the bags.

As part of its program, the quality-assurance department periodically performs a hypothesis test to decide whether the packaging machine is working properly, that is, to decide whether the mean net weight of all bags packaged is 454 g.

a. Determine the null hypothesis for the hypothesis test.

b. Determine the alternative hypothesis for the hypothesis test.

c. Classify the hypothesis test as two-tailed, left-tailed, or right-tailed.

SOLUTION Let μ denote the mean net weight of all bags packaged.

a. As we said, the null hypothesis for a hypothesis test concerning a population mean, μ, should always specify a single value for that parameter. Thus the null hypothesis for this hypothesis test is that the packaging machine is working properly, that is, that the mean net weight, μ, of all bags packaged *equals* 454 g. In symbols, H_0: $\mu = 454$ g.

b. The alternative hypothesis for this hypothesis test is that the packaging machine is not working properly, that is, that the mean net weight, μ, of all bags packaged is *different from* 454 g. In symbols, H_a: $\mu \neq 454$ g.

c. This hypothesis test is two-tailed since a not-equal sign ($\neq$) appears in the alternative hypothesis. □

EXAMPLE 9.2 CHOOSING THE NULL AND ALTERNATIVE HYPOTHESES

The R. R. Bowker Company of New York collects information on the retail prices of books and publishes the data in *Publishers Weekly.* In 1993 the mean retail price of history books was \$40.69. Suppose we want to perform a hypothesis test to decide whether this year's mean retail price of history books has increased over the 1993 mean.

a. Determine the null hypothesis for the hypothesis test.

b. Determine the alternative hypothesis for the hypothesis test.

c. Classify the hypothesis test as two-tailed, left-tailed, or right-tailed.

SOLUTION Let μ denote this year's mean retail price of history books.

a. Again, the null hypothesis for a hypothesis test about a population mean, μ, should always specify a single value for that parameter. Thus the null hypothesis for this hypothesis test is that this year's mean retail price of history books *equals* the 1993 mean of \$40.69; that is, H_0: $\mu = \$40.69$.

b. We want to decide whether this year's mean retail price of history books has increased over the 1993 mean. So the alternative hypothesis for this hypothesis test is that this year's mean retail price of history books is *greater than* \$40.69; that is, H_a: $\mu > \$40.69$.

c. This hypothesis test is right-tailed since a greater-than sign (>) appears in the alternative hypothesis. ❑

EXAMPLE 9.3 CHOOSING THE NULL AND ALTERNATIVE HYPOTHESES

Calcium is the most abundant mineral in the body and also one of the most important. It works with phosphorus to build and maintain bones and teeth. According to the Food and Nutrition Board of the National Academy of Sciences, the recommended daily allowance (RDA) of calcium for adults is 800 milligrams (mg). Suppose we want to perform a hypothesis test to decide whether the average person with an income below the poverty level gets less than the RDA of 800 mg.

a. Determine the null hypothesis for the hypothesis test.

b. Determine the alternative hypothesis for the hypothesis test.

c. Classify the hypothesis test as two-tailed, left-tailed, or right-tailed.

SOLUTION Let μ denote the mean calcium intake (per day) of all people with incomes below the poverty level.

a. The null hypothesis must specify a single value for the parameter μ. Hence the null hypothesis for this hypothesis test is that the mean calcium intake of all people with incomes below the poverty level *equals* 800 mg per day; that is, H_0: $\mu = 800$ mg.

b. We want to decide whether the mean calcium intake of all people with incomes below the poverty level is *less than* the RDA of 800 mg per day. Thus the alternative hypothesis for this hypothesis test is H_a: $\mu < 800$ mg.

c. This hypothesis test is left-tailed since a less-than sign (<) appears in the alternative hypothesis. ❑

The Logic of Hypothesis Testing

We have now seen how to choose appropriate null and alternative hypotheses for a hypothesis test. The next question is: How do we decide which of the two hypotheses is true; that is, how do we decide whether or not to reject the null hypothesis in favor of the alternative hypothesis?

Very roughly, the procedure for deciding goes like this: Take a random sample from the population. If the sample data are consistent with the null hypothesis, then do not reject the null hypothesis; if the sample data are inconsistent with the null hypothesis, then reject the null hypothesis and conclude that the alternative hypothesis is true.

Of course, in practice we must have a precise criterion for deciding whether or not to reject the null hypothesis. Example 9.4 illustrates how such a criterion can be devised for a two-tailed hypothesis test concerning a population mean, μ. The example also introduces the logic and some of the terminology of hypothesis testing. Several general procedures for performing hypothesis tests will be provided later in this chapter.

EXAMPLE 9.4 HYPOTHESIS TESTING

A company that produces snack foods uses a machine to package 454-gram bags of pretzels. We will assume that the net weights are normally distributed and that the population standard deviation of all such weights is 7.8 g.[†] A random sample of 25 bags of pretzels has the net weights, in grams, displayed in Table 9.1.

TABLE 9.1 Weights, in grams, of 25 randomly selected bags of pretzels

465	456	438	454	447
449	442	449	446	447
468	433	454	463	450
446	447	456	452	444
447	456	456	435	450

hypothesis test →

Do the data provide sufficient evidence to conclude that the packaging machine is not working properly? We will use the following steps to answer the question.

a. State the null and alternative hypotheses for the hypothesis test.

b. Discuss the logic behind carrying out the hypothesis test.

[†] We might know the population standard deviation from previous research or from a preliminary study of net weights. In Section 9.6, we will consider the more usual case of unknown σ.

c. Identify the distribution of the variable $\bar{x}$, that is, the sampling distribution of the mean for samples of size 25.

d. Obtain a precise criterion for deciding whether or not to reject the null hypothesis in favor of the alternative hypothesis.

e. Apply the criterion in part (d) to the sample data and state the conclusion.

SOLUTION Let μ denote the mean net weight of all bags packaged.

a. The null and alternative hypotheses for the hypothesis test were found in Example 9.1. They are

$$H_0\text{: } \mu = 454 \text{ g (the packaging machine is working properly)}$$
$$H_a\text{: } \mu \neq 454 \text{ g (the packaging machine is not working properly).}$$

b. Basically, the logic behind carrying out the hypothesis test is this: If the null hypothesis is true, that is, if $\mu = 454$ g, then the mean weight of the sample of 25 bags of pretzels should be approximately equal to 454 g. We say "approximately equal" because we cannot expect a sample mean to exactly equal the population mean; some sampling error is to be anticipated. However, if the sample mean weight differs "too much" from 454 g, then we would be inclined to reject the null hypothesis and conclude that the alternative hypothesis is true. As we will see in part (d), we can use our knowledge of the sampling distribution of the mean to decide how much difference is "too much."

c. Since $n = 25$, $\sigma = 7.8$, and the weights are normally distributed, Key Fact 7.4 on page 429 implies that

- $\mu_{\bar{x}} = \mu$ (which we don't know),
- $\sigma_{\bar{x}} = \sigma/\sqrt{n} = 7.8/\sqrt{25} = 1.56$, and
- $\bar{x}$ is normally distributed.

In other words, for samples of size 25, the variable $\bar{x}$ is normally distributed with mean μ and standard deviation 1.56 g.

d. The "95.44 part" of the 68.26-95.44-99.74 rule states that for a normally distributed variable, 95.44% of all possible observations lie within two standard deviations to either side of the mean. Applying this to the variable $\bar{x}$ and referring to part (c), we see that 95.44% of all samples of 25 bags of pretzels have mean weights within $2 \cdot 1.56 = 3.12$ g of μ. Or, equivalently, only 4.56% of all samples of 25 bags of pretzels have mean weights that are not within 3.12 g of μ. See Fig. 9.1.

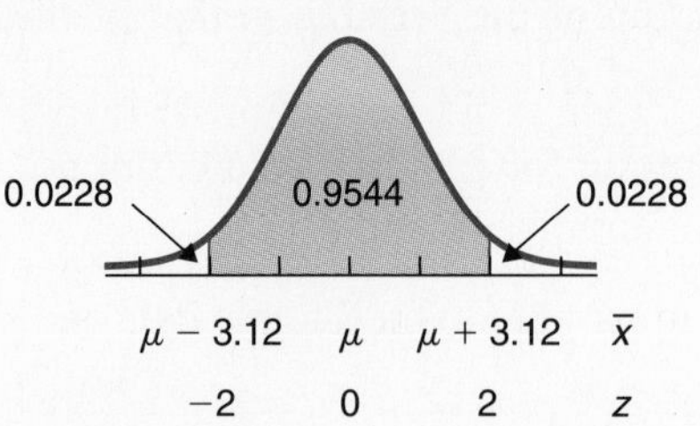

FIGURE 9.1 95.44% of all samples of 25 bags of pretzels have mean weights within two standard deviations (3.12 g) of μ

Consequently, if the mean weight of the 25 bags of pretzels sampled is not within two standard deviations (3.12 g) of 454 g, then we have evidence against the null hypothesis. Why? Because observing such a sample mean would occur by chance only 4.56% of the time if the null hypothesis, $\mu = 454$ g, were true.

In summary, then, we have obtained the following precise criterion for deciding whether or not to reject the null hypothesis:

> If the mean weight of the 25 bags of pretzels sampled is more than two standard deviations away from 454 g, then reject the null hypothesis, $\mu = 454$ g, and conclude that the alternative hypothesis, $\mu \neq 454$ g, is true. Otherwise, do not reject the null hypothesis.

This criterion is portrayed graphically in Fig. 9.2(a). If the null hypothesis is true, then the normal curve associated with $\bar{x}$ is the one with parameters 454 and 1.56; that normal curve is superimposed on Fig. 9.2(a) in Fig. 9.2(b).

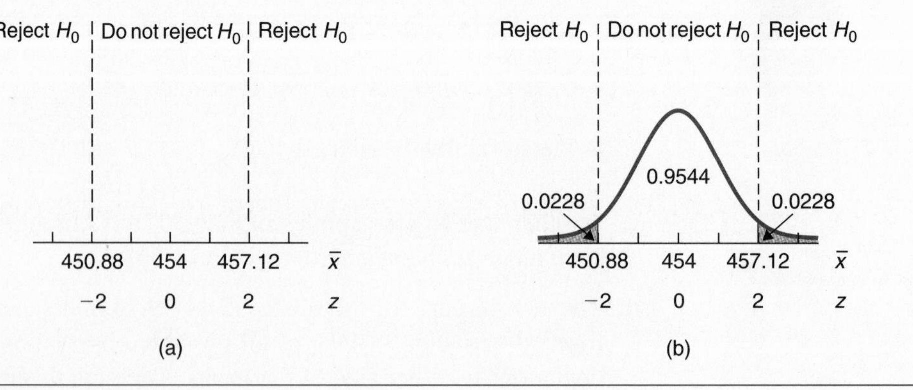

FIGURE 9.2 (a) Criterion for deciding whether or not to reject the null hypothesis (b) Normal curve associated with $\bar{x}$ if the null hypothesis is true, superimposed on the decision criterion

e. Here we are to apply the criterion obtained in part (d) to the sample data and state the conclusion. The mean weight of the sample of 25 bags of pretzels whose weights are given in Table 9.1 is 450 g. And, therefore,

$$z = \frac{\bar{x} - 454}{1.56} = \frac{450 - 454}{1.56} = -2.56.$$

Consequently, the sample mean of 450 g is 2.56 standard deviations below the null-hypothesis population mean of 454 g, as shown in Fig. 9.3.

FIGURE 9.3 Graph showing the number of standard deviations that the sample mean of 450 g is away from the null-hypothesis population mean of 454 g

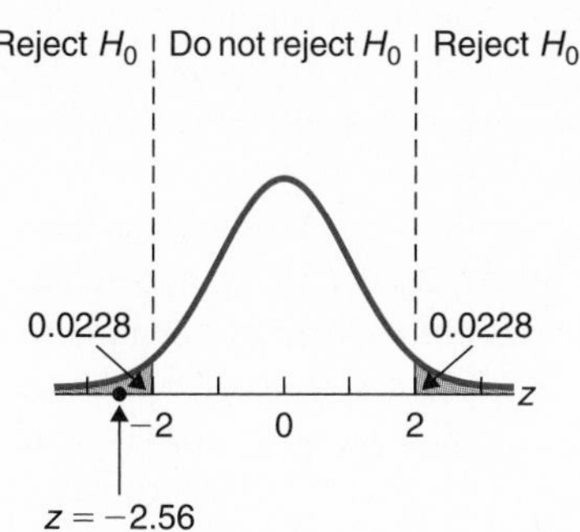

Since the mean weight of the 25 bags of pretzels sampled is more than two standard deviations away from 454 g, we reject the null hypothesis, $\mu = 454$ g, and conclude that the alternative hypothesis, $\mu \neq 454$ g, is true. In other words, the data provide sufficient evidence to conclude that the packaging machine is not working properly. ◨

Example 9.4 contains all the elements of a hypothesis test, including the necessary theory. But don't worry too much about the details at this point. What you should understand now is how to choose the null and alternative hypotheses for a hypothesis test and the logic behind performing a hypothesis test.

EXERCISES 9.1

STATISTICAL CONCEPTS AND SKILLS

9.1 Explain the meaning of the term *hypothesis* as used in inferential statistics.

9.2 What role does the decision criterion play in a hypothesis test?

9.3 Suppose we want to perform a hypothesis test for a population mean μ.

a. Express the null hypothesis both in words and in symbolic form.

b. Express each of the three possible alternative hypotheses in words and in symbolic form.

9.4 Suppose we are considering a hypothesis test for a population mean, μ. In each part below, express the alternative hypothesis symbolically and identify the hypothesis test as two-tailed, left-tailed, or right-tailed.

a. We want to decide whether the population mean is different from a specified value μ_0.
b. We want to decide whether the population mean is less than a specified value μ_0.
c. We want to decide whether the population mean is greater than a specified value μ_0.

In each of Exercises 9.5–9.10, a hypothesis test will be proposed. For each hypothesis test,
a. determine the null hypothesis.
b. determine the alternative hypothesis.
c. classify the hypothesis test as two-tailed, left-tailed, or right-tailed.

9.5 According to the U.S. Bureau of Labor Statistics publication *Consumer Expenditures,* the mean telephone expenditure per consumer unit was \$690 in 1994. We want to perform a hypothesis test to decide whether last year's mean expenditure has increased over the 1994 mean of \$690.

9.6 The Census Bureau publication *Census of Population and Housing* reports that the mean travel time to work in 1990 for all North Dakota residents was 13 minutes. A transportation official wants to use a sample of this year's travel times for North Dakota residents to decide whether the mean travel time to work for all North Dakota residents has changed from the 1990 mean of 13 minutes.

9.7 The Food and Nutrition Board of the National Academy of Sciences states that the recommended daily allowance (RDA) of iron for adult females under the age of 51 is 18 mg. A hypothesis test is to be performed to decide whether adult females under the age of 51 are, on the average, getting less than the RDA of 18 mg of iron.

9.8 Ten years ago, the mean age of all juveniles held in public custody was 16.0 years, as reported by the U.S. Office of Juvenile Justice and Delinquency Prevention in *Children in Custody.* The ages of a random sample of juveniles currently being held in public custody are to be used to decide whether this year's mean age of all juveniles being held in public custody is less than it was 10 years ago.

9.9 A study by researchers at the University of Maryland addressed the question of whether the mean body temperature of humans is 98.6°F. The results of the study by P. Mackowiak, S. Wasserman, and M. Levine appeared in the article "A Critical Appraisal of 98.6°F, the Upper Limit of the Normal Body Temperature, and Other Legacies of Carl Reinhold August Wunderlich" (*Journal of the American Medical Association,* 268, pp. 1578–1580). Among other data, the researchers obtained the body temperatures of roughly 100 healthy humans. Suppose we want to use that data to decide whether the mean body temperature of healthy humans differs from 98.6°F.

9.10 The American Hospital Association reports in *Hospital Stat* that the mean cost to community hospitals per patient per day in U.S. hospitals was \$931 in 1994. In that same year, a random sample was obtained of 30 daily costs in Massachusetts hospitals. A hypothesis test was then performed to decide whether the mean cost in Massachusetts hospitals exceeded the national mean.

EXTENDING THE CONCEPTS AND SKILLS

For Exercises 9.11 and 9.12, use the same method as that in Example 9.4 on pages 494–497 to perform the required hypothesis tests.

9.11 The Radio Advertising Bureau of New York reports in *Radio Facts* that in 1994 the mean number of radios per U.S. household was 5.6. A random sample of 45 U.S. households taken this year yields the following data on number of radios owned.

4	10	4	7	4	4	5	10	6
8	6	9	7	5	4	5	6	9
7	5	3	4	9	5	4	4	7
8	4	9	8	5	9	1	3	2
8	6	4	4	4	10	7	9	3

Do the data provide sufficient evidence to conclude that this year's mean number of radios per U.S. household has changed from the 1994 mean of 5.6? Use the following steps to answer the question.
a. State the null and alternative hypotheses.
b. Discuss the logic of conducting the hypothesis test.
c. Identify the distribution of the variable $\bar{x}$, that is, the sampling distribution of the mean for samples of size 45.

d. Obtain a precise criterion for deciding whether to reject the null hypothesis in favor of the alternative hypothesis.
e. Apply the criterion in part (d) to the sample data and state your conclusion. Assume the population standard deviation of this year's number of radios per U.S. household is 1.9.

9.12 The U.S. Energy Information Administration compiles data on energy consumption and publishes its findings in *Residential Energy Consumption Survey: Consumption and Expenditures.* In 1993 the mean energy consumed per U.S. household was 103.6 million BTU. For that same year, 20 randomly selected households in the West had the following energy consumptions, in millions of BTU.

104	84	72	95	69
80	78	74	76	81
82	61	94	65	100
70	65	83	76	84

Do the data provide sufficient evidence to conclude that in 1993 the mean energy consumed by western households differed from that of all U.S. households? Use the following steps to answer the question.
a. State the null and alternative hypotheses.
b. Discuss the logic of conducting the hypothesis test.
c. Identify the distribution of the variable $\bar{x}$, that is, the sampling distribution of the mean for samples of size 20.
d. Obtain a precise criterion for deciding whether to reject the null hypothesis in favor of the alternative hypothesis.
e. Apply the criterion in part (d) to the sample data and state your conclusion. Assume that in 1993 the standard deviation of energy consumptions of all western households was 15 million BTU.

9.13 Refer to Example 9.4 on pages 494–497. Suppose in the solution to part (d) we use the "68.26 part" of the 68.26-95.44-99.74 rule.
a. Determine the resulting decision criterion and portray it graphically using a graph similar to the one in Fig. 9.2(a) on page 496.
b. Construct a graph similar to the one in Fig. 9.2(b) that shows the implications of the decision criterion in part (a) if in fact the null hypothesis is true.
c. Apply the criterion in part (a) to the sample data in Table 9.1 on page 494 and state your conclusion.

9.14 Refer to Example 9.4 on pages 494–497. Suppose in the solution to part (d) we use the "99.74 part" of the 68.26-95.44-99.74 rule.
a. Determine the resulting decision criterion and portray it graphically using a graph similar to the one in Fig. 9.2(a) on page 496.
b. Construct a graph similar to the one in Fig. 9.2(b) that shows the implications of the decision criterion in part (a) if in fact the null hypothesis is true.
c. Apply the criterion in part (a) to the sample data in Table 9.1 on page 494 and state your conclusion.

9.15 Refer to Example 9.4 on pages 494–497. In that example we rejected the null hypothesis that the mean net weight of all bags packaged is 454 g in favor of the alternate hypothesis that the mean net weight of all bags packaged differs from 454 g. If in fact the null hypothesis is true, what is the chance of incorrectly rejecting it using a sample of 25 bags of pretzels? *(Hint:* Refer to Fig. 9.2(b).*)*

9.2 TERMS, ERRORS, AND HYPOTHESES

To fully understand the nature of hypothesis testing, we need to learn some additional terms and concepts. In this section we will define several more terms used in hypothesis testing, discuss the two types of errors that can occur in a hypothesis test, and interpret the possible conclusions for a hypothesis test.

Some Additional Terminology

To introduce some additional terminology used in hypothesis testing, we will refer to the pretzel-packaging hypothesis test of Example 9.4 on page 494. Recall that the null and alternative hypotheses for that hypothesis test are

$$H_0\text{: } \mu = 454 \text{ g (the packaging machine is working properly)}$$
$$H_a\text{: } \mu \neq 454 \text{ g (the packaging machine is not working properly),}$$

where μ is the mean net weight of all bags of pretzels packaged.

As a basis for deciding whether to reject the null hypothesis, we employed, in part (e) of Example 9.4, the variable

$$z = \frac{\bar{x} - \mu_0}{\sigma/\sqrt{n}} = \frac{\bar{x} - 454}{1.56},$$

which tells us how many standard deviations the sample mean is from the null hypothesis population mean of 454 g. That variable is called the **test statistic** for the hypothesis test.

Figure 9.3 includes a graph portraying the criterion used to decide whether or not the null hypothesis should be rejected. For ease of reference we repeat that graph in Fig. 9.4.

FIGURE 9.4
Criterion used to decide whether or not to reject the null hypothesis

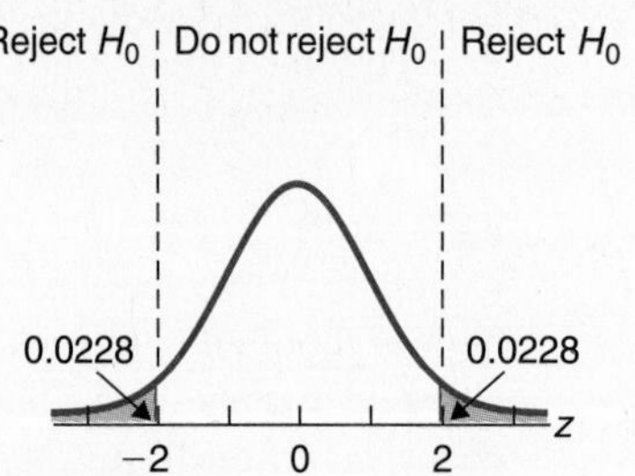

The set of values for the test statistic that leads us to reject the null hypothesis is called the **rejection region.** In this case the rejection region consists of all z-scores that lie either to the left of -2 or to the right of 2, that part of the horizontal axis under the shaded areas in Fig. 9.4.

The set of values for the test statistic that leads us not to reject the null hypothesis is called the **nonrejection region,** or **acceptance region.** In this case the nonrejection region consists of all z-scores that lie between -2 and 2, that part of the horizontal axis under the unshaded area in Fig. 9.4.

The values of the test statistic that separate the rejection and nonrejection regions are called the **critical values.** In this case the critical values are $z = \pm 2$, as we see from Fig. 9.4. We summarize the preceding discussion in Fig. 9.5.

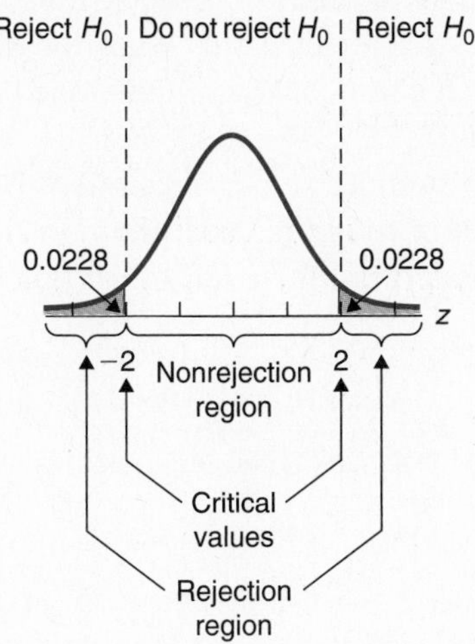

FIGURE 9.5 Rejection region, nonrejection region, and critical values for the pretzel-packaging illustration

The terminology introduced so far in this section applies to any hypothesis test, not just to hypothesis tests for a population mean.

DEFINITION 9.2 **TEST STATISTIC, REJECTION REGION, NONREJECTION REGION, CRITICAL VALUES**

Test statistic: The statistic used as a basis for deciding whether the null hypothesis should be rejected.

Rejection region: The set of values for the test statistic that leads to rejection of the null hypothesis.

Nonrejection region: The set of values for the test statistic that leads to nonrejection of the null hypothesis.

Critical values: The values of the test statistic that separate the rejection and nonrejection regions.

For a two-tailed test, as in the pretzel-packaging illustration, the null hypothesis is rejected when the test statistic is either too small or too large. So the rejection region for such a test consists of two parts, one on the left and one on the right, as shown in Fig. 9.5 and Fig. 9.6(a) on the following page.

For a left-tailed test, as in the calcium-intake illustration of Example 9.3 on page 493, the null hypothesis is rejected only when the test statistic is too small. So the rejection region for such a test consists of only one part, and that part is on the left, as shown in Fig. 9.6(b).

For a right-tailed test, as in the history-book illustration of Example 9.2 on page 492, the null hypothesis is rejected only when the test statistic is too large. So the rejection region for such a test consists of only one part, and that part is on the right, as shown in Fig. 9.6(c).

Table 9.2 and Fig. 9.6 summarize our discussion. By examining Fig. 9.6, we can understand why the term *tailed* is used: the rejection region is in both tails for a two-tailed test, in the left tail for a left-tailed test, and in the right tail for a right-tailed test.

TABLE 9.2
Rejection regions for two-tailed, left-tailed, and right-tailed tests

	Two-tailed test	Left-tailed test	Right-tailed test
Sign in H_a	$\neq$	$<$	$>$
Rejection region	Both sides	Left side	Right side

FIGURE 9.6
Graphical display of rejection regions for two-tailed, left-tailed, and right-tailed tests

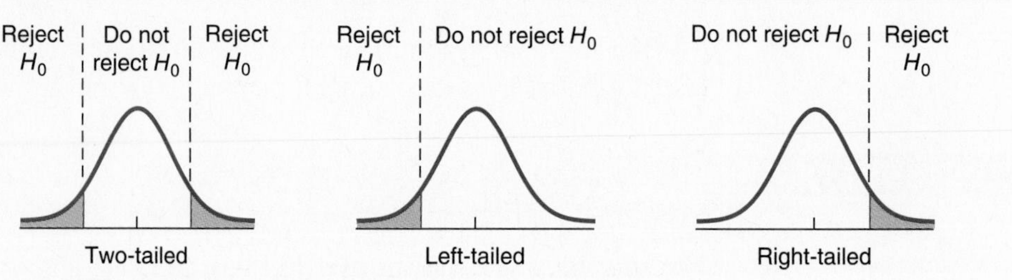

Type I and Type II Errors

Whenever we conduct a hypothesis test, it is possible that the decision we reach will be incorrect. This is because partial information, obtained from a sample, is used to draw conclusions about the entire population.

There are two types of incorrect decisions—rejection of a true null hypothesis and nonrejection of a false null hypothesis. To distinguish these two types of errors, we give them special names, as indicated in Table 9.3 and Definition 9.3.

TABLE 9.3
Correct and incorrect decisions for a hypothesis test

		H_0 is:	
		True	False
Decision:	Do not reject H_0	Correct decision	Type II error
	Reject H_0	Type I error	Correct decision

DEFINITION 9.3 TYPE I AND TYPE II ERRORS

Type I error: Rejecting the null hypothesis when it is in fact true.

Type II error: Not rejecting the null hypothesis when it is in fact false.

EXAMPLE 9.5 ILLUSTRATES DEFINITION 9.3

Consider once again the pretzel-packaging hypothesis test. The null and alternative hypotheses are

$$H_0\colon \mu = 454 \text{ g (the packaging machine is working properly)}$$
$$H_a\colon \mu \neq 454 \text{ g (the packaging machine is not working properly),}$$

where μ is the mean net weight of all bags of pretzels packaged. Explain what each of the following would mean.

a. Type I error

b. Type II error

c. Correct decision

Recall from Example 9.4 that the results of sampling 25 bags of pretzels led to rejection of the null hypothesis, $\mu = 454$ g, that is, to the conclusion that $\mu \neq 454$ g. Classify that conclusion by error type or as a correct decision if

d. the mean net weight, μ, is in fact 454 g.

e. the mean net weight, μ, is in fact not 454 g.

SOLUTION

a. A Type I error occurs when a true null hypothesis is rejected. In this case a Type I error would occur if in fact $\mu = 454$ g but the results of the sampling lead to the conclusion that $\mu \neq 454$ g—in words, if we conclude that the packaging machine is not working properly when in fact it is working properly.

b. A Type II error occurs when a false null hypothesis is not rejected. In this case a Type II error would occur if in fact $\mu \neq 454$ g but the results of the sampling fail to lead to that conclusion—in words, if we fail to conclude that the packaging machine is not working properly when in fact it isn't working properly.

c. A correct decision can occur in either of two ways:

- A true null hypothesis is not rejected. Here this would happen if in fact $\mu = 454$ g and the results of the sampling do not lead to the rejection of that fact—in words, if we fail to conclude that the packaging machine is not working properly when in fact it is working properly.

- A false null hypothesis is rejected. This would happen if in fact $\mu \neq 454$ g and the results of the sampling lead to that conclusion—in words, if we conclude that the packaging machine is not working properly when in fact it isn't working properly.

d. If in fact $\mu = 454$ g, then the null hypothesis is true. Consequently, by rejecting the null hypothesis, $\mu = 454$ g, a Type I error has been made—a true null hypothesis has been rejected.

e. If in fact $\mu \neq 454$ g, then the null hypothesis is false. Consequently, by rejecting the null hypothesis, $\mu = 454$ g, a correct decision has been made—a false null hypothesis has been rejected. ◨

Probabilities of Type I and Type II Errors

Part of evaluating the effectiveness of a hypothesis test involves an analysis of the chances of making an incorrect decision. As we have seen, two kinds of incorrect decisions are possible: rejection of a true null hypothesis (Type I error) and nonrejection of a false null hypothesis (Type II error).

A Type I error occurs if the test statistic falls in the rejection region when in fact the null hypothesis is true. The probability of this happening is called the **significance level** of the hypothesis test and is denoted by the Greek letter α (alpha).

Figure 9.4 on page 500 shows the rejection and nonrejection regions for the pretzel-packaging hypothesis test of Example 9.4. It also shows the normal curve for the test statistic

$$z = \frac{\bar{x} - 454}{1.56},$$

under the assumption that the null hypothesis, $\mu = 454$ g, is true. We see from Fig. 9.4 that if the null hypothesis is true, then the probability is $0.0228 + 0.0228$, or 0.0456, that the test statistic, z, will fall in the rejection region. Thus, for this hypothesis test, the significance level is 0.0456; in symbols, $\alpha = 0.0456$. There is only a 4.56% chance of concluding that $\mu \neq 454$ g when in fact $\mu = 454$ g.

DEFINITION 9.4 SIGNIFICANCE LEVEL

The probability of making a Type I error, that is, of rejecting a true null hypothesis, is called the ***significance level,*** α, of a hypothesis test.

A Type II error occurs if the test statistic falls in the nonrejection region when in fact the null hypothesis is false. The probability of this is denoted by the Greek letter β (beta) and depends on the true value of μ. Calculation of Type II error probabilities is considered briefly in Exercise 9.35 and in detail in Section 9.4.

Ideally, we would like both Type I and Type II errors to have small probabilities. Then the chance of making an incorrect decision would be small, regardless of whether the null hypothesis is true or false. As we will see in Section 9.3, it is always possible to design a hypothesis test to have any specified significance level. So, for instance, if it is important not to reject a true null hypothesis, then we should specify a small value for α. However, in making our choice for α, we should keep the following fact in mind.

KEY FACT 9.1 **RELATION BETWEEN TYPE I AND TYPE II ERROR PROBABILITIES**

For a fixed sample size, the smaller we specify the significance level, α, the larger will be the probability, β, of not rejecting a false null hypothesis.

Thus we must always assess the risks involved in committing both types of errors and use that assessment to balance the Type I and Type II error probabilities.

Possible Conclusions for a Hypothesis Test

As we know, the significance level, α, is the probability of making a Type I error, that is, of rejecting a true null hypothesis. So it is unlikely that a true null hypothesis will be rejected if the hypothesis test is conducted at a small significance level (e.g., $\alpha = 0.05$). Since in this text we will generally specify a small significance level, we can make the following statement concerning a hypothesis test: If we do reject the null hypothesis in a hypothesis test, then we can be reasonably confident that the null hypothesis is false and therefore that the alternative hypothesis is true.

On the other hand, we will usually not know the probability, β, of making a Type II error, that is, of not rejecting a false null hypothesis. Consequently, if we do not reject the null hypothesis in a hypothesis test, then we simply reserve judgment about which hypothesis is true. In other words, if we do not reject the null hypothesis, then we conclude only that the data did not provide sufficient evidence to support the alternative hypothesis; we do not conclude that the data provided sufficient evidence to support the null hypothesis. Key Fact 9.2 summarizes this discussion.

KEY FACT 9.2 **POSSIBLE CONCLUSIONS FOR A HYPOTHESIS TEST**

Suppose a hypothesis test is conducted at a small significance level.

- If the null hypothesis is rejected, we conclude that the alternative hypothesis is true.
- If the null hypothesis is not rejected, we conclude that the data do not provide sufficient evidence to support the alternative hypothesis.

When the null hypothesis is rejected in a hypothesis test performed at the significance level α, we frequently express that fact with the phrase "the test results are **statistically significant** at the α level." Similarly, when the null hypothesis is not rejected in a hypothesis test performed at the significance level α, we often express that fact with the phrase "the test results are **not statistically significant** at the α level."

EXERCISES 9.2

STATISTICAL CONCEPTS AND SKILLS

9.16 Explain in your own words the meaning of each of the following terms.

a. test statistic **b.** rejection region
c. nonrejection region **d.** critical values
e. significance level

9.17 Decide whether each of the following statements is true or false. Explain your answers.

a. If it is important not to reject a true null hypothesis, then the hypothesis test should be performed at a small significance level.
b. For a fixed sample size, decreasing the significance level of a hypothesis test results in an increase in the probability of making a Type II error.

9.18 Identify the two types of incorrect decisions in a hypothesis test. For each incorrect decision, what symbol is used to represent the probability of making that type of error?

Exercises 9.19–9.24 contain graphs portraying the decision criterion for a hypothesis test for a population mean, μ. The null hypothesis for each test is H_0: $\mu = \mu_0$; the test statistic is

$$z = \frac{\bar{x} - \mu_0}{\sigma/\sqrt{n}}.$$

The curve in each graph is the normal curve for the test statistic under the assumption that the null hypothesis is true. For each exercise, determine the

a. rejection region. b. nonrejection region.
c. critical value(s). d. significance level.
e. Construct a graph similar to Fig. 9.5 on page 501 that depicts your results from parts (a)–(d).
f. Identify the hypothesis test as two-tailed, left-tailed, or right-tailed.

9.19 A graphical display of the decision criterion is:

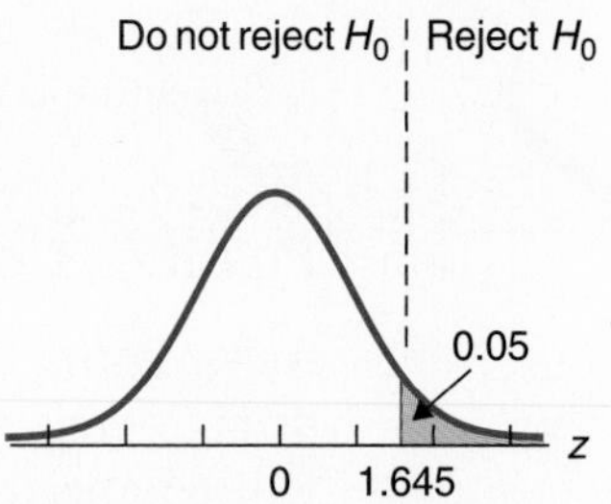

9.20 A graphical display of the decision criterion is:

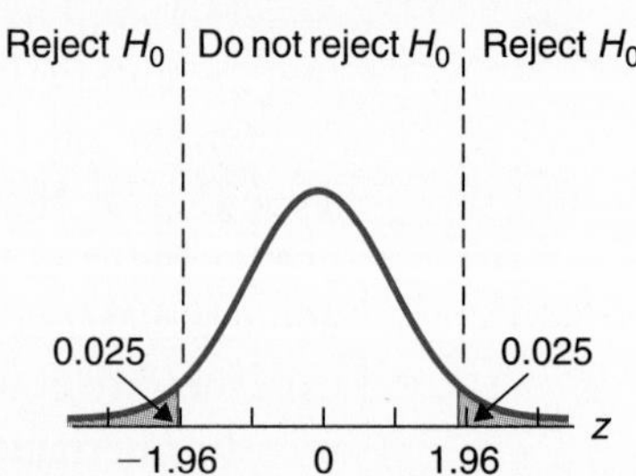

9.21 A graphical display of the decision criterion is:

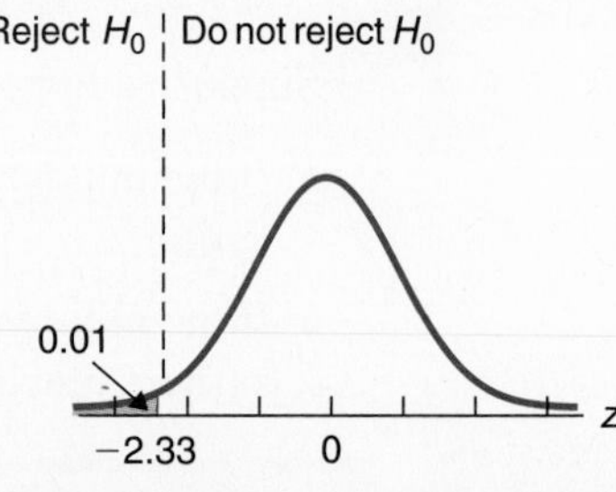

9.22 A graphical display of the decision criterion is:

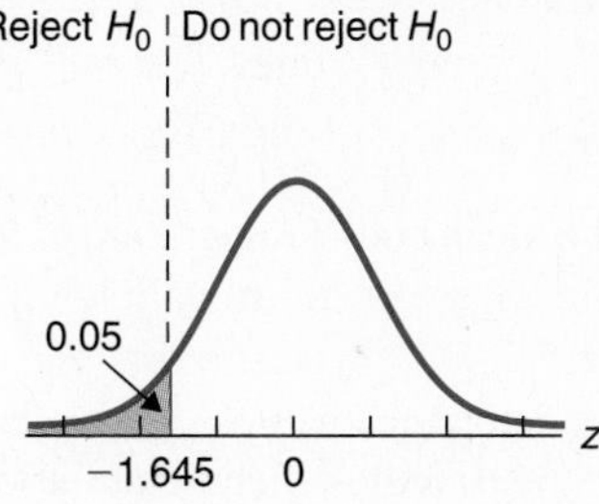

9.23 A graphical display of the decision criterion is:

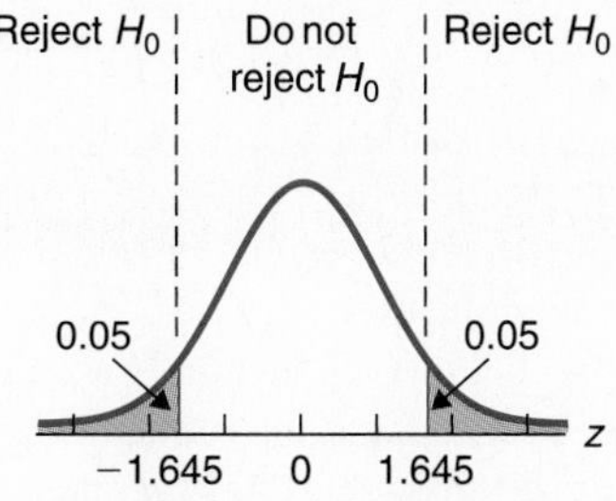

9.24 A graphical display of the decision criterion is:

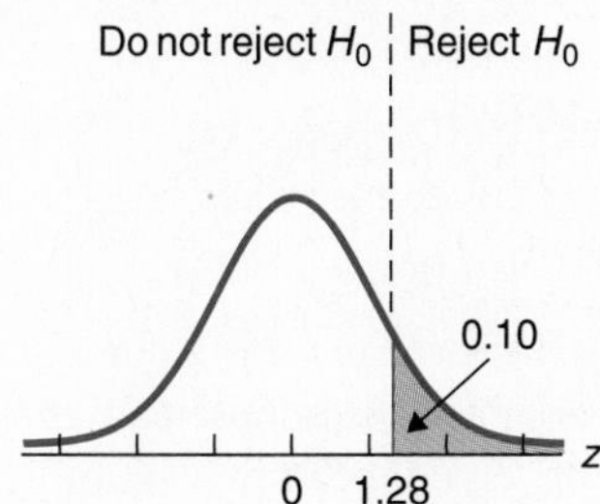

9.25 According to the Bureau of Labor Statistics publication *Consumer Expenditures,* the mean telephone expenditure per consumer unit was \$690 in 1994. We want to perform a hypothesis test to decide whether last year's mean expenditure has increased over the 1994 mean of \$690. The null and alternative hypotheses are

$$H_0: \mu = \$690$$
$$H_a: \mu > \$690,$$

where μ is last year's mean telephone expenditure per consumer unit. Explain what each of the following would mean.

a. Type I error **b.** Type II error
c. Correct decision

Now suppose the results of carrying out the hypothesis test lead to nonrejection of the null hypothesis. Classify that conclusion by error type or as a correct decision if in fact last year's mean telephone expenditure per consumer unit

d. is equal to the 1994 mean of \$690.
e. is greater than the 1994 mean of \$690.

9.26 The Census Bureau publication *Census of Population and Housing* reports that the mean travel time to work in 1990 for all North Dakota residents was 13 minutes. A transportation official wants to use a sample of this year's travel times for North Dakota residents to decide whether the mean travel time to work for all North Dakota residents has changed from the 1990 mean of 13 minutes. The null and alternative hypotheses for the hypothesis test are

$$H_0: \mu = 13 \text{ minutes}$$
$$H_a: \mu \neq 13 \text{ minutes},$$

where μ is this year's mean travel time to work for all North Dakota residents. Explain what each of the following would mean.

a. Type I error **b.** Type II error
c. Correct decision

Now suppose the results of the sampling lead to nonrejection of the null hypothesis. Classify that conclusion by error type or as a correct decision if this year's mean travel time to work, μ, for all North Dakota residents

d. has not changed from the 1990 mean of 13 minutes.
e. has changed from the 1990 mean of 13 minutes.

9.27 The Food and Nutrition Board of the National Academy of Sciences states that the RDA of iron for adult females under the age of 51 is 18 mg. A hypothesis test is to be performed to decide whether adult females under the age of 51 are, on the average, getting less than the RDA of 18 mg of iron. The null and alternative hypotheses for the hypothesis test are

$$H_0: \mu = 18 \text{ mg}$$
$$H_a: \mu < 18 \text{ mg},$$

where μ is the mean iron intake (per day) of all adult females under the age of 51. Explain what each of the following would mean.

a. Type I error **b.** Type II error
c. Correct decision

Now suppose the results of carrying out the hypothesis test lead to rejection of the null hypothesis, $\mu = 18$ mg, that is, to the conclusion that $\mu < 18$ mg. Classify that conclusion by error type or as a correct decision if in fact the mean iron intake, μ, of all adult females under the age of 51

d. is not less than the RDA of 18 mg per day.
e. is less than the RDA of 18 mg per day.

9.28 Ten years ago, the mean age of all juveniles held in public custody was 16.0 years, as reported by the U.S. Office of Juvenile Justice and Delinquency Prevention in *Children in Custody.* The ages of a random sample of juveniles currently being held in public custody are to be used to decide whether this year's mean age of all juveniles being held in public custody is less than it was 10 years ago. The null and alternative hypotheses for the hypothesis test are

$$H_0: \mu = 16.0 \text{ years}$$
$$H_a: \mu < 16.0 \text{ years},$$

where μ is this year's mean age of all juveniles being held in public custody. Explain what each of the following would mean.

a. Type I error **b.** Type II error
c. Correct decision

Now suppose that the results of carrying out the hypothesis test lead to rejection of the null hypothesis, $\mu = 16.0$ years, in other words, to the conclusion that $\mu < 16.0$ years. Classify that conclusion by error type or as a correct decision if in fact this year's mean age, μ, of all juveniles being held in public custody

d. is 16.0 years.
e. is less than 16.0 years.

9.29 A study by researchers at the University of Maryland addressed the question of whether the mean body temperature of humans is 98.6°F. The results of the study by P. Mackowiak, S. Wasserman, and M. Levine appeared in the article "A Critical Appraisal of 98.6°F, the Upper Limit of the Normal Body Temperature, and Other Legacies of Carl Reinhold August Wunderlich" (*Journal of the American Medical Association,* 268, pp. 1578–1580). Among other data, the researchers obtained the body temperatures of roughly 100 healthy humans. Suppose we want to use that data to decide whether the mean body temperature of healthy humans differs from 98.6°F. The null and alternative hypotheses for the hypothesis test are

$$H_0: \mu = 98.6°\text{F}$$
$$H_a: \mu \neq 98.6°\text{F},$$

where μ is the mean body temperature of all healthy humans. Explain what each of the following would mean.

a. Type I error **b.** Type II error
c. Correct decision

Now suppose the sample of temperatures leads to rejection of the null hypothesis. Classify that conclusion by error type or as a correct decision if in fact the mean body temperature, μ, of all healthy humans

d. is 98.6°F.
e. is not 98.6°F.

9.30 The American Hospital Association reports in *Hospital Stat* that the mean cost to community hospitals per patient per day in U.S. hospitals was $931 in 1994. In that same year, a random sample was obtained of 30 daily costs in Massachusetts hospitals. A hypothesis test was then performed to decide whether the mean cost in Massachusetts hospitals exceeded the national mean. The null and alternative hypotheses for the hypothesis test performed are

$$H_0: \mu = \$931$$
$$H_a: \mu > \$931,$$

where μ is the 1994 mean cost to community hospitals per patient per day in Massachusetts. Explain what each of the following would mean.

a. Type I error **b.** Type II error
c. Correct decision

Now suppose the results of the sampling led to rejection of the null hypothesis. Classify that conclusion by error type or as a correct decision if in fact the 1994 mean cost to community hospitals per patient per day in Massachusetts

d. did not exceed the national mean.
e. did exceed the national mean.

Extending the Concepts and Skills

9.31 Suppose we choose the significance level of a hypothesis test to be 0.

a. What is the probability of a Type I error?
b. What is the probability of a Type II error?

9.32 Identify an exercise in this section for which it is important to have
a. a small α probability.
b. a small β probability.
c. both α and β probabilities small.

9.33 Suppose you are performing a statistical test to decide whether a nuclear reactor should be approved for use. Further suppose that failing to reject the null hypothesis corresponds to approval. What property would you want the Type II error probability, β, to have?

9.34 In the U.S. court system, a defendant is assumed innocent until proven guilty. Suppose we regard a court trial as a hypothesis test with null and alternative hypotheses

H_0: Defendant is innocent
H_a: Defendant is guilty.

a. Explain the meaning of a Type I error.
b. Explain the meaning of a Type II error.
c. If you were the defendant, would you want α to be large or small? Explain your answer.
d. If you were the prosecuting attorney, would you want β to be large or small? Explain your answer.
e. What are the consequences to our court system if we make $\alpha = 0$? $\beta = 0$?

9.35 Type II error probabilities. In this exercise you are asked to compute Type II error probabilities for the pretzel-packaging hypothesis test. Recall that the null and alternative hypotheses are

H_0: $\mu = 454$ g (machine is working properly)
H_a: $\mu \neq 454$ g (machine is not working properly),

where μ is the mean net weight of all bags of pretzels packaged. Also recall that the net weights are normally distributed with a standard deviation of 7.8 g. Figure 9.2(a) on page 496 portrays the decision criterion for a hypothesis test at the 4.56% significance level ($\alpha = 0.0456$) using a sample size of 25.
a. Identify the probability of a Type I error.
b. Assuming that the mean net weight being packaged is in fact 447 g, identify the distribution of the variable $\bar{x}$, that is, the sampling distribution of the mean for samples of size 25.
c. Use part (b) to determine the probability, β, of a Type II error if in fact the mean net weight being packaged is 447 g. *(Hint:* Referring to Fig. 9.2(a), note that β equals the percentage of all samples of 25 bags of pretzels whose mean weights are between 450.88 g and 457.12 g.*)*
d. Repeat parts (b) and (c) if in fact the mean net weight being packaged is 448 g; 449 g; 450 g; 451 g; 452 g; 453 g; 455 g; 456 g; 457 g; 458 g; 459 g; 460 g; 461 g.
e. Use your answers from parts (b)–(d) to draw a graph of β versus the true value of μ. Interpret your graph.

USING TECHNOLOGY  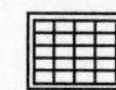 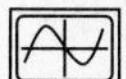

9.36 This exercise can be done individually or, better yet, as a class project. Refer to the pretzel-packaging hypothesis test. Recall that the null and alternative hypotheses are

H_0: $\mu = 454$ g (machine is working properly)
H_a: $\mu \neq 454$ g (machine is not working properly),

where μ is the mean net weight of all bags of pretzels packaged. Also recall that the net weights are normally distributed with a standard deviation of 7.8 g. Figure 9.4 on page 500 portrays the decision criterion for a test at the 4.56% significance level ($\alpha = 0.0456$) using a sample size of 25 and the test statistic

$$z = \frac{\bar{x} - \mu_0}{\sigma/\sqrt{n}} = \frac{\bar{x} - 454}{1.56}.$$

a. Assuming the null hypothesis, $\mu = 454$ g is true, simulate 100 samples of 25 net weights each. *(Note:* If you are using Minitab, you may find it helpful to refer to part (b) of Exercise 7.45 on page 424.*)*
b. Determine the mean of each sample in part (a). *(Note:* If you are using Minitab, you may find it helpful to refer to part (c) of Exercise 7.45.*)*
c. Use part (b) to determine the value of the test statistic, z, for each sample in part (a). If you are using Minitab, proceed as follows.

1 Choose **Calc ➤ Standardize...**
2 Specify XBAR in the **Input column(s)** text box
3 Click in the **Store in column(s)** text box and type <u>z</u>
4 Select the **Subtract** option button

5 Click in the **Subtract** text box and type 454

6 Click in the **and divide by** text box and type 1.56

7 Click **OK**

The values of the test statistic for the 100 samples will now be stored in a column named z.

d. For the 100 samples obtained in part (a), roughly how many would you expect to lead to rejection of the null hypothesis? Explain your answer.

e. For the 100 samples obtained in part (a), determine the number that lead to rejection of the null hypothesis. *(Hint:* Refer to part (c) and Fig. 9.4.*)*

f. Compare your answers from parts (d) and (e) and comment on any observed difference.

9.3 HYPOTHESIS TESTS FOR ONE POPULATION MEAN WHEN σ IS KNOWN

In this section we will discover how to conduct a hypothesis test for a population mean at any prescribed significance level. We have already learned most of what we need to know to do that. What remains is to discuss how to obtain the critical value(s) when the significance level is specified in advance.

Recall that the significance level of a hypothesis test is the probability of rejecting a true null hypothesis or, equivalently, the probability that the test statistic will fall in the rejection region when in fact the null hypothesis is true. With this in mind, we state Key Fact 9.3, which holds for any hypothesis test.

KEY FACT 9.3 OBTAINING CRITICAL VALUES

Suppose a hypothesis test is to be performed at a specified significance level, α. Then the critical value(s) must be chosen so that if the null hypothesis is true, the probability is equal to α that the test statistic will fall in the rejection region.

Hypothesis Tests for a Population Mean When σ Is Known

We will now develop a simple step-by-step procedure for performing a hypothesis test for a population mean when the population standard deviation is known. In doing so, we will assume the variable under consideration is normally distributed. But keep in mind that, because of the central limit theorem, the procedure will work reasonably well when the sample size is large, regardless of the distribution of the variable.

As we have seen, the null hypothesis for a hypothesis test concerning one population mean, μ, is of the form H_0: $\mu = \mu_0$, where μ_0 is some number. The test statistic for the hypothesis test is

$$z = \frac{\bar{x} - \mu_0}{\sigma/\sqrt{n}}.$$

That test statistic tells us how many standard deviations the observed sample mean, $\bar{x}$, is away from μ_0.

The basis of the hypothesis-testing procedure can be found in Key Fact 7.4: If x is a normally distributed variable with mean μ and standard deviation σ, then, for samples of size n, the variable $\bar{x}$ is also normally distributed and has mean μ and standard deviation $\sigma/\sqrt{n}$. This means that if the null hypothesis, $\mu = \mu_0$, is true, then the test statistic z has the standard normal distribution.

Consequently, in view of Key Fact 9.3, we see that for a specified significance level, α, we need to choose the critical value(s) so that the area under the standard normal curve that lies above the rejection region equals α. Here is an example.

EXAMPLE 9.6 OBTAINING THE CRITICAL VALUES

Determine the critical value(s) for a hypothesis test at the 5% significance level ($\alpha = 0.05$) if the test is

a. two-tailed. **b.** left-tailed. **c.** right-tailed.

SOLUTION Since $\alpha = 0.05$, we need to choose the critical value(s) so that the area under the standard normal curve that lies above the rejection region is equal to 0.05.

a. For a two-tailed test, the rejection region is on both the left and right. So in this case the critical values are the two z-scores that divide the area under the standard normal curve into a middle 0.95 area and two outside areas of 0.025. In other words, the critical values are $\pm z_{0.025}$. Consulting Table II, we find that $\pm z_{0.025} = \pm 1.96$, as shown in Fig. 9.7(a).

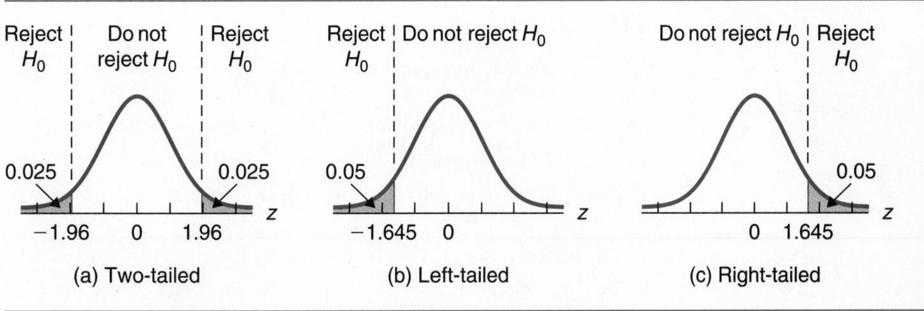

FIGURE 9.7 Critical value(s) for a hypothesis test at the 5% significance level if the test is (a) two-tailed, (b) left-tailed, (c) right-tailed

b. For a left-tailed test, the rejection region is on the left. So in this case the critical value is the z-score having area 0.05 to its left under the standard normal curve, which is $-z_{0.05}$. Consulting Table II, we find that $-z_{0.05} = -1.645$, as shown in Fig. 9.7(b).

c. For a right-tailed test, the rejection region is on the right. So in this case the critical value is the z-score having area 0.05 to its right under the standard normal curve, which is $z_{0.05}$. Consulting Table II, we find that $z_{0.05} = 1.645$, as shown in Fig. 9.7(c). □

By arguing as we did in Example 9.6, we can obtain the critical value(s) for any specified significance level, α. For a two-tailed test, the critical values are $\pm z_{\alpha/2}$; for a left-tailed test, the critical value is $-z_{\alpha}$; and for a right-tailed test, the critical value is z_{α}. See Fig. 9.8.

FIGURE 9.8 Critical value(s) for a hypothesis test at the significance level α if the test is (a) two-tailed, (b) left-tailed, (c) right-tailed

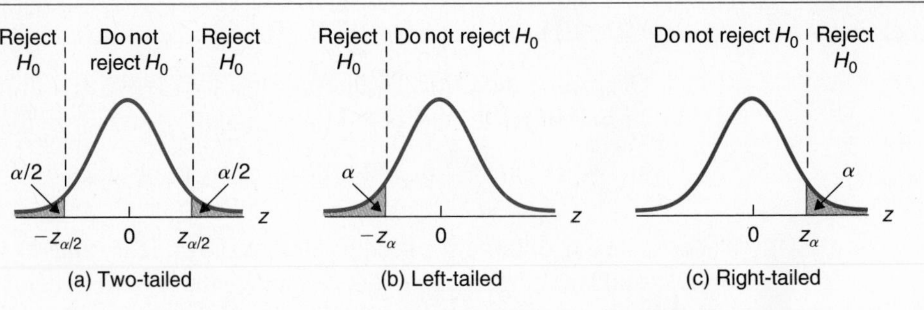

The most commonly used significance levels are 0.10, 0.05, and 0.01. If we consider both one-tailed and two-tailed tests, then these three significance levels give rise to five "tail areas." Using the standard-normal table, Table II, we obtained the value of z_{α} corresponding to each of those five tail areas. Table 9.4 provides a summary.

TABLE 9.4 Some important values of z_{α}

$z_{0.10}$	$z_{0.05}$	$z_{0.025}$	$z_{0.01}$	$z_{0.005}$
1.28	1.645	1.96	2.33	2.575

Alternatively, these five values of z_{α} can be found at the bottom of the t-table, Table IV, where they are displayed to three decimal places. Can you explain the slight discrepancy between the values given for $z_{0.005}$ in the two tables?

We now present Procedure 9.1, a simple method for performing a hypothesis test for a population mean when the population standard deviation is known. The procedure is obtained by carefully studying what we have done in this and the previous two sections. We often refer to Procedure 9.1 as the **one-sample z-test** or more briefly as the **z-test.**

PROCEDURE 9.1

THE ONE-SAMPLE z-TEST FOR A POPULATION MEAN

ASSUMPTIONS

1. Normal population or large sample
2. σ known

Step 1 The null hypothesis is H_0: $\mu = \mu_0$ and the alternative hypothesis is one of the following:

H_a: $\mu \neq \mu_0$ (Two-tailed)	or	H_a: $\mu < \mu_0$ (Left-tailed)	or	H_a: $\mu > \mu_0$ (Right-tailed)

Step 2 Decide on the significance level, α.

Step 3 The critical value(s) are

$\pm z_{\alpha/2}$ (Two-tailed)	or	$-z_\alpha$ (Left-tailed)	or	z_α (Right-tailed)

Use Table II to find the critical value(s).

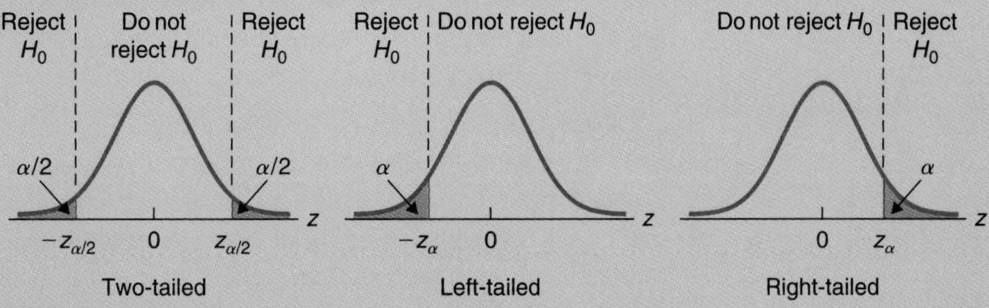

Step 4 Compute the value of the test statistic

$$z = \frac{\bar{x} - \mu_0}{\sigma/\sqrt{n}}.$$

Step 5 If the value of the test statistic falls in the rejection region, reject H_0; otherwise, do not reject H_0.

Step 6 State the conclusion in words.

The hypothesis test is exact for normal populations and is approximately correct for large samples from nonnormal populations.

Note: By saying that the hypothesis test is *exact,* we mean that the true significance level is equal to α; by saying that it is *approximately correct,* we mean that the true significance level is only approximately equal to α.

One of the assumptions for using the z-test is that either the variable under consideration is normally distributed or the sample size is large. But, as with the z-interval procedure, the z-test is robust to moderate violations of the normality assumption. Thus the z-test works reasonably well even when the sample size is small or moderate and the variable is not normally distributed, provided the variable is not too far from being normally distributed.

Again as with the z-interval procedure, when considering the z-test, it is important to watch for outliers. The presence of outliers calls into question the normality assumption. And even for large samples, outliers can sometimes unduly affect a z-test because the sample mean is not resistant to outliers.

KEY FACT 9.4 **WHEN TO USE THE z-TEST (PROCEDURE 9.1)**

- For small samples, say, of size less than 15, the z-test should be used only when the variable under consideration is normally distributed or very close to being so.
- For moderate-size samples, say, between 15 and 30, the z-test can be used unless the data contain outliers or the variable under consideration is far from being normally distributed.
- For large samples, say, of size 30 or more, the z-test can be used essentially without restriction. However, if outliers are present and their removal is not justified, the effect of the outliers on the hypothesis test should be examined; that is, the hypothesis test should be performed twice, once with the outliers retained and once with them removed. If the conclusion remains the same either way, we may be content to take that as our conclusion and close the investigation. But if the conclusion is affected, it is probably wise to make the more conservative conclusion, use a different procedure, or take another sample.
- If outliers are present but their removal is justified and results in a data set for which the z-test is appropriate (see above), then the procedure can be used.

Applying the z-Test

Examples 9.7–9.9 illustrate the use of the z-test, Procedure 9.1. These examples will be reexamined in Section 9.5 when we discuss P-values.

EXAMPLE 9.7 ILLUSTRATES PROCEDURE 9.1

The R. R. Bowker Company of New York collects information on the retail prices of books and publishes its findings in *Publishers Weekly.* In 1993 the mean retail price of all history books was \$40.69. This year's retail prices for 40 randomly selected history books are shown in Table 9.5. At the 1% significance level, do the data provide sufficient evidence to conclude that this year's mean retail price of all history books has increased over the 1993 mean of \$40.69? Assume the standard deviation of prices for this year's history books is \$7.61.

TABLE 9.5 This year's prices (\$) for 40 history books

45.23	35.48	36.57	43.22	42.94	37.11	44.05	44.96
42.99	40.23	50.93	36.26	51.91	37.03	40.12	41.59
40.18	61.40	40.51	40.17	49.93	61.61	36.93	45.39
41.56	40.93	50.49	43.03	40.13	52.97	42.10	30.31
54.16	46.67	43.32	31.88	64.60	45.71	58.27	31.94

SOLUTION We constructed (not shown) a normal probability plot, a histogram, a stem-and-leaf diagram, and a boxplot for these data. The boxplot indicated potential outliers, but in view of the other three graphs, we concluded that in fact the data contain no outliers. Since the sample size is 40, which is large, and the population standard deviation is known, we can apply Procedure 9.1 to perform the required hypothesis test.

Step 1 *State the null and alternative hypotheses.*

Let μ denote this year's mean retail price of all history books. The null and alternative hypotheses were stated in Example 9.2. They are

$$H_0\text{: } \mu = \$40.69 \text{ (mean price has not increased)}$$

$$H_a\text{: } \mu > \$40.69 \text{ (mean price has increased).}$$

Note that the hypothesis test is right-tailed since a greater-than sign (>) appears in the alternative hypothesis.

Step 2 *Decide on the significance level, α.*

We are to perform the test at the 1% significance level. Thus $\alpha = 0.01$.

Step 3 *The critical value for a right-tailed test is z_α.*

Since $\alpha = 0.01$, the critical value is $z_{0.01}$. From Table II (or Table 9.4 on page 512), we find that $z_{0.01} = 2.33$, as seen in Fig. 9.9.

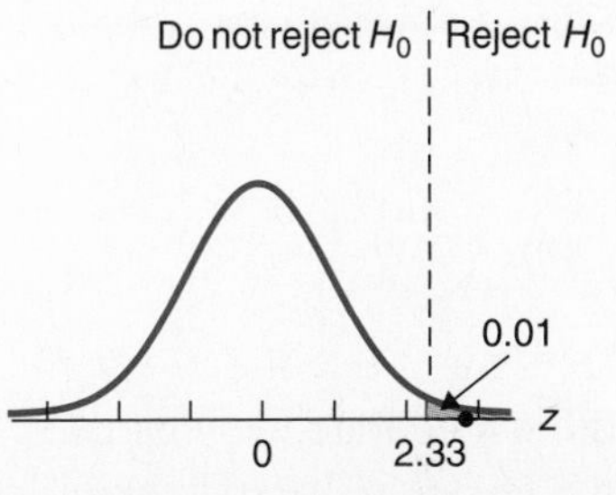

FIGURE 9.9 Criterion for deciding whether or not to reject the null hypothesis

Step 4 Compute the value of the test statistic

$$z = \frac{\bar{x} - \mu_0}{\sigma/\sqrt{n}}.$$

We have $\mu_0 = 40.69$, $\sigma = 7.61$, and $n = 40$. The mean of the sample data in Table 9.5 is $\bar{x} = 44.12$. Thus the value of the test statistic is

$$z = \frac{44.12 - 40.69}{7.61/\sqrt{40}} = 2.85.$$

This value of z is marked with a dot in Fig. 9.9.

Step 5 If the value of the test statistic falls in the rejection region, reject H_0; otherwise, do not reject H_0.

The value of the test statistic, found in Step 4, is $z = 2.85$. As we see from Fig. 9.9, this falls in the rejection region, and so we reject H_0.

Step 6 State the conclusion in words.

The test results are statistically significant at the 1% level; that is, at the 1% significance level, the data provide sufficient evidence to conclude that this year's mean retail price of all history books has increased over the 1993 mean of \$40.69. ◨

EXAMPLE 9.8 ILLUSTRATES PROCEDURE 9.1

Calcium is the most abundant mineral in the body and also one of the most important. It works with phosphorus to build and maintain bones and teeth. According to the Food and Nutrition Board of the National Academy of Sciences, the recommended daily allowance (RDA) of calcium for adults is 800 milligrams (mg).

A random sample of 18 people with incomes below the poverty level gives the daily calcium intakes shown in Table 9.6. At the 5% significance level, do the data provide sufficient evidence to conclude that the mean calcium intake of all people with incomes below the poverty level is less than the RDA of 800 mg? Assume that $\sigma = 188$ mg.

TABLE 9.6 Daily calcium intakes (mg)

686	433	743	647	734	641
993	620	574	634	850	858
992	775	1113	672	879	609

SOLUTION Since the sample size, $n = 18$, is moderate, we first need to consider questions of normality and outliers. (See the second bulleted item in Key Fact 9.4 on page 514.)

To that end we constructed a normal probability plot (not shown) for the data. The plot reveals no outliers and falls roughly in a straight line. Thus we can apply Procedure 9.1 to perform the required hypothesis test.

Step 1 *State the null and alternative hypotheses.*

Let μ denote the mean calcium intake (per day) of all people with incomes below the poverty level. The null and alternative hypotheses were obtained in Example 9.3. They are

$$H_0\colon \mu = 800 \text{ mg (mean calcium intake is not less than the RDA)}$$
$$H_a\colon \mu < 800 \text{ mg (mean calcium intake is less than the RDA).}$$

Note that the hypothesis test is left-tailed since a less-than sign (<) appears in the alternative hypothesis.

Step 2 *Decide on the significance level, α.*

We are to perform the test at the 5% significance level. Thus $\alpha = 0.05$.

Step 3 *The critical value for a left-tailed test is $-z_\alpha$.*

Since $\alpha = 0.05$, the critical value is $-z_{0.05}$. From Table II (or Table 9.4 or Table IV), we find that $z_{0.05} = 1.645$. Hence the critical value is $-z_{0.05} = -1.645$, as seen in Fig. 9.10.

FIGURE 9.10
Criterion for deciding whether or not to reject the null hypothesis

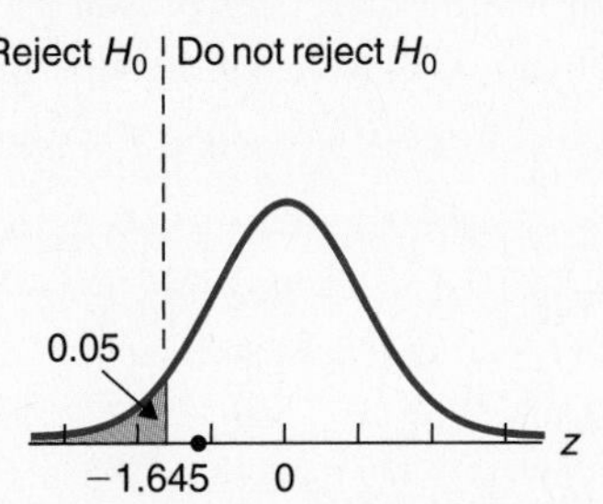

Step 4 *Compute the value of the test statistic*

$$z = \frac{\bar{x} - \mu_0}{\sigma/\sqrt{n}}.$$

We have $\mu_0 = 800$, $\sigma = 188$, and $n = 18$. From the data in Table 9.6, we find that $\bar{x} = 747.4$. Thus the value of the test statistic is

$$z = \frac{747.4 - 800}{188/\sqrt{18}} = -1.19.$$

This value of z is marked with a dot in Fig. 9.10.

Step 5 *If the value of the test statistic falls in the rejection region, reject H_0; otherwise, do not reject H_0.*

The value of the test statistic, found in Step 4, is $z = -1.19$. As we see from Fig. 9.10, this does not fall in the rejection region, and so we do not reject H_0.

Step 6 *State the conclusion in words.*

The test results are not statistically significant at the 5% level; that is, at the 5% significance level, the sample of 18 calcium intakes does not provide sufficient evidence to conclude that the mean calcium intake of all people with incomes below the poverty level is less than the RDA of 800 mg. □

EXAMPLE 9.9 ILLUSTRATES PROCEDURE 9.1

A 1998 issue of *Habitat World,* the publication of Habitat for Humanity International, contains an article on housing affordability. Included in the article is a table of 1997 fair market rents (FMR) for two bedroom units, by state, obtained from the National Low Income Housing Coalition. According to the table, the 1997 FMR for Maine is $590.

A sample of 32 randomly selected two-bedroom units in Maine yielded the data on monthly rents shown in Table 9.7. Do the data suggest that the mean monthly rent for two-bedroom units in Maine differs from the FMR of $590? Perform the appropriate hypothesis test at the 0.05 level of significance. Assume the standard deviation of monthly rents for two-bedroom units in Maine is $73.10.

TABLE 9.7 Monthly rents ($) for 32 two-bedroom units in Maine

289	597	648	669	745	577	626	661
657	595	604	739	598	545	696	450
521	669	656	565	610	503	589	472
675	586	663	609	560	507	643	749

SOLUTION A frequency histogram for the data in Table 9.7, displayed in Fig. 9.11, suggests that the first monthly rent, $289, is an outlier.

Thus, as suggested in the third bulleted item in Key Fact 9.4 (page 514), we will first apply Procedure 9.1 to the full data set in Table 9.7 and then examine the effect on the test results when the outlier, $289, is removed.

Step 1 *State the null and alternative hypotheses.*

The null and alternative hypotheses are

H_0: $\mu = \$590$ (mean monthly rent equals the FMR)

H_a: $\mu \neq \$590$ (mean monthly rent differs from the FMR),

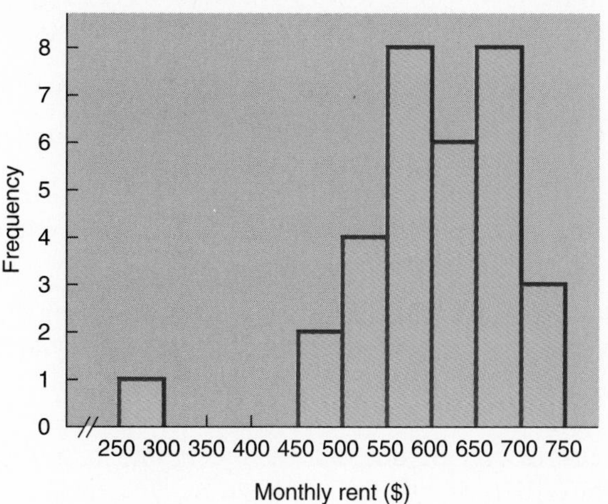

FIGURE 9.11 Frequency histogram for the monthly rents in Table 9.7

where μ denotes the mean monthly rent of all two-bedroom units in Maine. Note that the hypothesis test is two-tailed since a not-equal sign ($\neq$) appears in the alternative hypothesis.

Step 2 *Decide on the significance level,* α.

We are to perform the hypothesis test at the 0.05 level of significance; so $\alpha = 0.05$.

Step 3 *The critical values for a two-tailed test are* $\pm z_{\alpha/2}$.

Since $\alpha = 0.05$, we obtain from Table II (or Table 9.4 or Table IV) that the critical values are $\pm z_{0.05/2} = \pm z_{0.025} = \pm 1.96$, as seen in Fig. 9.12.

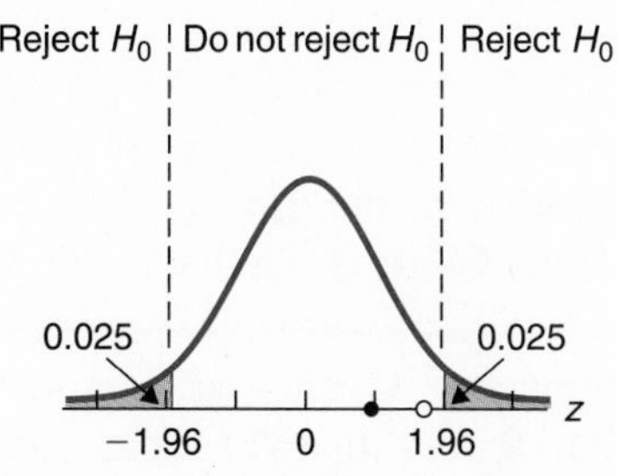

FIGURE 9.12 Criterion for deciding whether or not to reject the null hypothesis

Step 4 *Compute the value of the test statistic*

$$z = \frac{\bar{x} - \mu_0}{\sigma/\sqrt{n}}.$$

We have $\mu_0 = 590$, $\sigma = 73.10$, and $n = 32$. From the data in Table 9.7, we find that $\bar{x} = 602.28$. Thus the value of the test statistic is

$$z = \frac{602.28 - 590}{73.10/\sqrt{32}} = 0.95.$$

This value of z is marked with a solid dot in Fig. 9.12.

Step 5 *If the value of the test statistic falls in the rejection region, reject H_0; otherwise, do not reject H_0.*

From Step 4 the value of the test statistic is $z = 0.95$. This does not fall in the rejection region, as we see from Fig. 9.12. Hence we do not reject H_0.

Step 6 *State the conclusion in words.*

The test results are not statistically significant at the 5% level; that is, at the 5% significance level, the data do not provide sufficient evidence to conclude that the mean monthly rent for two-bedroom units in Maine differs from the FMR of $590. This completes the hypothesis test using all 32 monthly rents in Table 9.7.

Now recall that the first monthly rent, $289, is an outlier. Although for this problem we don't actually know whether removing this outlier is justified (a common situation), we can still remove it from the sample data and assess the effect on the hypothesis test. Doing this, we find that the value of the test statistic for the abridged data is $z = 1.71$, which we have marked with a hollow dot in Fig. 9.12. This value still lies in the nonrejection region, although it is much closer to the rejection region than the value of the test statistic for the unabridged data, $z = 0.95$.

Hence in this case, removing the outlier does not affect the conclusion of the hypothesis test. We can probably be content with accepting that the mean monthly rent for two-bedroom units in Maine is the same as the FMR of $590. ◨

Statistical Significance Versus Practical Significance

Recall that the results of a hypothesis test are *statistically significant* if the null hypothesis is rejected at the chosen level of α. This means that the data provide sufficient evidence to conclude that the truth is different from that stated in the null hypothesis. It does not necessarily mean that the difference is important in any practical sense.

For example, the manufacturer of a new car, the Orion, claims that a typical car gets 26 miles per gallon, that is, that the mean gas mileage of all Orions is $\mu = 26$ mpg. Suppose the mean gas mileage of a sample of 1000 Orions turns out to be 25.9 mpg. Assuming the standard deviation of gas mileages for all Orions is 1.4 mpg, the value of the test statistic for a z-test of

H_0: $\mu = 26$ mpg (mean gas mileage is 26 mpg)

H_a: $\mu < 26$ mpg (mean gas mileage is less than 26 mpg)

is $z = -2.26$. This is statistically significant at the 5% level (and even at the 1.19% level). We can easily reject the null hypothesis, that is, the manufacturer's claim that the mean gas mileage of all Orions is 26 mpg.

Because the sample size, 1000, is so large, the sample mean, $\bar{x} = 25.9$ mpg, is probably nearly the same as the population mean. This indicates that the manufacturer's claim was rejected because μ is about 25.9 mpg instead of 26 mpg. From a practical point of view, the difference between 25.9 mpg and 26 mpg is not important. Therefore in this case, the statistical significance is not practically significant. Remember: *Statistical significance does not necessarily imply practical significance!*

The Relation Between Hypothesis Tests and Confidence Intervals

Hypothesis tests and confidence intervals are closely related. Consider, for example, a two-tailed hypothesis test for a population mean at the significance level α. It can be shown that the null hypothesis will be rejected if and only if the value μ_0 given for the mean in the null hypothesis lies outside the $(1 - \alpha)$-level confidence interval for μ. Exercises 9.52 and 9.53 discuss this relation between hypothesis tests and confidence intervals in greater detail.

EXERCISES 9.3

STATISTICAL CONCEPTS AND SKILLS

In each of Exercises 9.37–9.42, suppose a hypothesis test is to be performed for a population mean, μ, with null hypothesis H_0: $\mu = \mu_0$. Further suppose the test statistic used will be

$$z = \frac{\bar{x} - \mu_0}{\sigma/\sqrt{n}}.$$

For each exercise, obtain the required critical value(s) and draw a graph that illustrates your answers.

9.37 A right-tailed test with $\alpha = 0.01$.

9.38 A left-tailed test with $\alpha = 0.10$.

9.39 A two-tailed test with $\alpha = 0.10$.

9.40 A right-tailed test with $\alpha = 0.05$.

9.41 A left-tailed test with $\alpha = 0.05$.

9.42 A two-tailed test with $\alpha = 0.01$.

Preliminary data analyses indicate that it is reasonable to apply the z-test (Procedure 9.1 on page 513) in Exercises 9.43–9.48. Comment on the practical significance of all hypothesis tests whose results are statistically significant.

9.43 According to the U.S. Bureau of Labor Statistics publication *Consumer Expenditures,* the mean telephone expenditure per consumer unit was \$690 in 1994. For last year a random sample of 40 consumer units revealed the following telephone expenditures, in dollars.

473	263	576	1244	1418	917	514	1741
1248	1015	292	905	1233	331	395	924
348	479	1155	586	178	169	783	1016
599	647	790	222	794	740	382	1013
502	687	406	174	565	1136	318	997

Do the data provide sufficient evidence to conclude that last year's mean telephone expenditure per consumer unit has increased over the 1994 mean of \$690? Assume that the standard deviation of last year's telephone expenditures per consumer unit is \$350. Perform the hypothesis test at the 5% significance level. *(Note:* The sum of the data is \$28,175.*)*

9.44 The Census Bureau publication *Census of Population and Housing* reports that the mean travel time to work in 1990 for all North Dakota residents was 13 minutes. A transportation official obtained this year's travel times, in minutes, for a random sample of 35 North Dakota residents. Here are the data.

29	40	0	12	10	6	41
25	21	5	4	19	2	7
10	8	3	6	52	4	12
0	33	6	2	17	21	8
38	2	13	8	14	11	2

At the 5% significance level, do the data provide sufficient evidence to conclude that the mean travel time to work for all North Dakota residents has changed from the 1990 mean of 13 minutes? Assume $\sigma = 11.6$ minutes. *(Note:* The sum of the data is 491 minutes.*)*

9.45 The Food and Nutrition Board of the National Academy of Sciences states that the RDA of iron for adult females under the age of 51 is 18 mg. The following iron intakes, in milligrams, during a 24-hour period were obtained for 45 randomly selected adult females under the age of 51.

15.0	18.1	14.4	14.6	10.9	18.1	18.2	18.3	15.0
16.0	12.6	16.6	20.7	19.8	11.6	12.8	15.6	11.0
15.3	9.4	19.5	18.3	14.5	16.6	11.5	16.4	12.5
14.6	11.9	12.5	18.6	13.1	12.1	10.7	17.3	12.4
17.0	6.3	16.8	12.5	16.3	14.7	12.7	16.3	11.5

At the 1% significance level, do the data suggest that adult females under the age of 51 are, on the average, getting less than the RDA of 18 mg of iron? Assume the population standard deviation is 4.2 mg. *(Note:* $\bar{x} = 14.68$ mg.*)*

9.46 Ten years ago, the mean age of all juveniles held in public custody was 16.0 years, as reported by the U.S. Office of Juvenile Justice and Delinquency Prevention in *Children in Custody.* The mean age of 250 randomly selected juveniles currently being held in public custody is 15.86 years. Assuming $\sigma = 1.01$ years, does it appear that the mean age, μ, of all juveniles being held in public custody this year is less than it was 10 years ago? Perform the appropriate hypothesis test using $\alpha = 0.10$.

9.47 A study by researchers at the University of Maryland addressed the question of whether the mean body temperature of humans is 98.6°F. The results of the study by P. Mackowiak, S. Wasserman, and M. Levine appeared in the article "A Critical Appraisal of 98.6°F, the Upper Limit of the Normal Body Temperature, and Other Legacies of Carl Reinhold August Wunderlich" (*Journal of the American Medical Association,* 268, pp. 1578–1580). The researchers obtained the following body temperatures of 93 healthy humans.

98.0	97.6	98.8	98.0	98.8	98.8	97.6	98.6	98.6
98.8	98.0	98.2	98.0	98.0	97.0	97.2	98.2	98.1
98.2	98.5	98.5	99.0	98.0	97.0	97.3	97.3	98.1
97.8	99.0	97.6	97.4	98.0	97.4	98.0	98.6	98.6
98.4	97.0	98.4	99.0	98.0	99.4	97.8	98.2	99.2
99.0	97.7	98.2	98.2	98.8	98.1	98.5	97.2	98.5
99.2	98.3	98.7	98.8	98.6	98.0	99.1	97.2	97.6
97.9	98.8	98.6	98.6	99.3	97.8	98.7	99.3	97.8
98.4	97.7	98.3	97.7	97.1	98.4	98.6	97.4	96.7
96.9	98.4	98.2	98.6	97.0	97.4	98.4	97.4	96.8
98.2	97.4	98.0						

At the 1% significance level, do the data provide sufficient evidence to conclude that the mean body temperature of healthy humans differs from 98.6°F? Assume $\sigma = 0.63$°F. *(Note:* The mean of the 93 temperatures is 98.12°F.*)*

9.48 The American Hospital Association reports in *Hospital Stat* that the mean cost to community hospitals per patient per day in U.S. hospitals was \$931 in 1994. In that same year, a random sample of 30 daily costs in Massachusetts hospitals yielded a mean of \$1131. Assuming a population standard deviation of \$333 for Massachusetts hospitals, do the data provide sufficient evidence to conclude that in 1994 the mean cost in Massachusetts hospitals exceeded the national mean of \$931? Perform the required hypothesis test at the 5% significance level.

EXTENDING THE CONCEPTS AND SKILLS

In each of Exercises 9.49 and 9.50,

a. *identify potential outliers, if any, for the given data.*
b. *perform the required hypothesis test using the given sample data.*
c. *remove observations that are potential outliers, if any, and perform the required hypothesis test using the abridged sample data.*
d. *comment on the effect that removing the potential outliers has on the hypothesis test.*
e. *indicate your conclusion regarding the hypothesis test and explain your answer.*

9.49 The manufacturer of a new car, the Orion, claims that a typical car gets 26 mpg. An independent consumer group is skeptical of this claim and thinks the mean gas mileage of all Orions may be less than 26 mpg. To try to justify its contention, the consumer group conducts mileage tests on 30 randomly selected Orions and obtains the following data.

25.3	25.1	29.6	24.6	26.0	26.0
26.3	23.6	26.0	25.4	26.1	23.8
25.1	24.1	25.8	26.4	23.4	24.8
22.6	26.6	25.1	26.6	28.0	23.3
23.8	25.4	26.2	25.1	25.3	21.5

At the 5% significance level, do the data support the consumer group's conjecture? Assume the standard deviation of gas mileages for all Orions is 1.4 mpg. *(Note:* $\bar{x} = 25.23$.*)*

9.50 A Louisiana cotton farmer has used a certain brand of fertilizer for the past 5 years. Based on experience, the farmer knows that the mean yield of cotton using this fertilizer is 623 lb/acre. Recently, a new brand of fertilizer appeared on the market that will supposedly increase cotton yield. The farmer uses the new fertilizer on 20 of his 1-acre plots. Here are the resulting cotton yields, in pounds.

633	673	630	619	648
637	641	611	588	642
590	631	591	604	607
633	661	635	618	653

At the 10% significance level, do the data provide sufficient evidence to conclude that the new fertilizer increases the mean yield of cotton on the farmer's land? Assume that, using the new fertilizer, the population standard deviation of cotton yields on the farmer's land is 20.6 lb/acre. *(Note:* $\bar{x} = 627.25$ lb/acre.*)* If the new fertilizer costs more than the one the farmer presently uses, would you buy the new fertilizer if you were the farmer? Why or why not?

9.51 Each part of this problem provides a scenario for a hypothesis test. Decide whether the z-test is an appropriate method for conducting the hypothesis test. Assume the population standard deviation is known in each case.

a. Preliminary analyses reveal that the sample data contain no outliers but that the distribution of the variable under consideration is probably highly skewed. The sample size is 20.

b. A normal probability plot reveals an outlier but is otherwise roughly linear. It is determined that the outlier is a legitimate observation and should not be removed. The sample size is 15.

c. Preliminary analyses reveal that the sample data contain no outliers but that the distribution of the variable under consideration is probably mildly skewed. The sample size is 70.

Exercises 9.52 and 9.53 examine the relationship between hypothesis tests and confidence intervals for a population mean.

9.52 In 1990 the average passenger vehicle was driven 10.3 thousand miles, as reported by the U.S. Federal Highway Administration in *Highway Statistics.* A random sample of 500 passenger vehicles had a mean of 10.1 thousand miles driven for last year. Let μ denote last year's mean distance driven for all passenger vehicles.

a. Use Procedure 9.1 on page 513 to perform the hypothesis test

$$H_0\text{: } \mu = 10.3 \text{ thousand miles}$$
$$H_a\text{: } \mu \neq 10.3 \text{ thousand miles}$$

at the 5% significance level. Assume last year's standard deviation of distances driven for all passenger vehicles is 6.0 thousand miles.

b. Use Procedure 8.1 on page 452 to find a 95% confidence interval for μ.

c. Does the value of 10.3 thousand miles, hypothesized for the mean, μ, in the null hypothesis of part (a), lie within your confidence interval from part (b)?

d. Repeat parts (a)–(c) if the 500 passenger vehicles sampled were driven a mean of 10.9 thousand miles last year.

e. Based on your observations in parts (a)–(d), complete the following statements concerning the relationship between a two-tailed hypothesis test,

$$H_0: \mu = \mu_0$$
$$H_a: \mu \neq \mu_0,$$

at the significance level α and a $(1-\alpha)$-level confidence interval for μ:

i. If μ_0 lies within the $(1-\alpha)$-level confidence interval for μ, then the null hypothesis (*will, will not*) be rejected.

ii. If μ_0 lies outside the $(1-\alpha)$-level confidence interval for μ, then the null hypothesis (*will, will not*) be rejected.

9.53 Hypothesis tests and confidence intervals. In this exercise we will examine the general relationship between a two-tailed hypothesis test for a population mean and a confidence-interval estimate for a population mean.

a. Show that the inequalities

$$\bar{x} - z_{\alpha/2} \cdot \frac{\sigma}{\sqrt{n}} < \mu_0 < \bar{x} + z_{\alpha/2} \cdot \frac{\sigma}{\sqrt{n}}$$

are equivalent to

$$-z_{\alpha/2} < \frac{\bar{x} - \mu_0}{\sigma/\sqrt{n}} < z_{\alpha/2}.$$

b. Deduce the following fact from part (a): For a two-tailed hypothesis test,

$$H_0: \mu = \mu_0$$
$$H_a: \mu \neq \mu_0,$$

at the significance level α, the null hypothesis will not be rejected if μ_0 lies within the $(1-\alpha)$-level confidence interval for μ, and conversely, the null hypothesis will be rejected if μ_0 does not lie within the $(1-\alpha)$-level confidence interval for μ.

9.4 TYPE II ERROR PROBABILITIES; POWER*

As we learned in Section 9.2, hypothesis tests do not always yield correct conclusions; they have built-in margins of error. An important part of planning a study is to take an advance look at the types of errors that can be made and the effects those errors might have.

Recall that two types of errors are possible with hypothesis tests. One is a Type I error: rejecting a true null hypothesis. The other is a Type II error: not rejecting a false null hypothesis. Also recall that the probability of making a Type I error is called the significance level of the hypothesis test and is denoted by the Greek letter α; the probability of making a Type II error is denoted by the Greek letter β.

In this section we will learn how to compute Type II error probabilities. We will also investigate the concept of the power of a hypothesis test. Although the discussion is limited to the one-sample z-test, the ideas apply to any hypothesis test.

* This section is optional and will not be needed in subsequent sections of the book.

Computing Type II Error Probabilities

The probability of making a Type II error depends on the sample size, the significance level, and the true value of the parameter under consideration. Example 9.10 explains how to compute the probability of making a Type II error for a one-sample z-test for a population mean.

EXAMPLE 9.10 COMPUTING TYPE II ERROR PROBABILITIES

The manufacturer of a new model car, the Orion, claims that a typical car gets 26 miles per gallon (mpg). A consumer advocacy group is skeptical of this claim and thinks the mean gas mileage, μ, of all Orions may be less than 26 mpg. The group plans to perform the hypothesis test

$$H_0: \mu = 26 \text{ mpg (manufacturer's claim)}$$
$$H_a: \mu < 26 \text{ mpg (consumer group's conjecture)},$$

at the 5% significance level using a sample of 30 Orions. Find the probability of making a Type II error if the true mean gas mileage of all Orions is

a. 25.8 mpg. **b.** 25.0 mpg.

Assume the gas mileages of Orions are normally distributed with a standard deviation of 1.4 mpg.

SOLUTION The inference under consideration is a left-tailed hypothesis test for a population mean at the 5% significance level. The test statistic is

$$z = \frac{\bar{x} - \mu_0}{\sigma/\sqrt{n}} = \frac{\bar{x} - 26}{1.4/\sqrt{30}},$$

and the critical value is $-z_\alpha = -z_{0.05} = -1.645$. Thus the decision criterion for the hypothesis test is this: If $z \leq -1.645$, reject H_0; if $z > -1.645$, do not reject H_0.

It is somewhat simpler to compute Type II error probabilities if the decision criterion is expressed in terms of $\bar{x}$ instead of z. To do that here, we must find the sample mean that is 1.645 standard deviations below the null-hypothesis population mean of 26:

$$\bar{x} = 26 - 1.645 \cdot \frac{1.4}{\sqrt{30}} = 25.6.$$

Consequently, the decision criterion can be expressed in terms of $\bar{x}$ as follows: If $\bar{x} \leq 25.6$ mpg, reject H_0; if $\bar{x} > 25.6$ mpg, do not reject H_0. This decision criterion is portrayed graphically in Fig. 9.13.

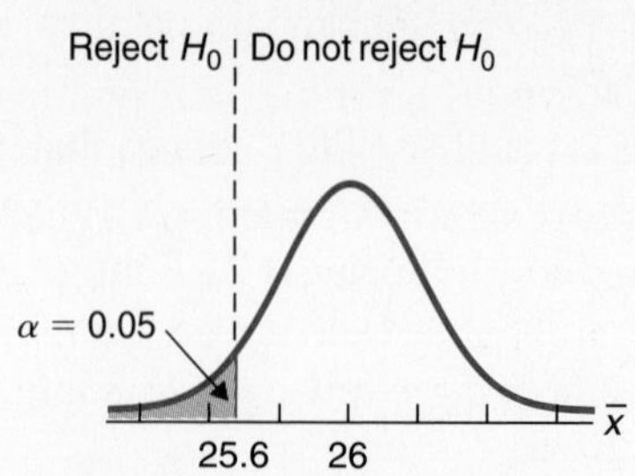

FIGURE 9.13 Graphical display of decision criterion for the gas-mileage illustration ($\alpha = 0.05$, $n = 30$)

a. For this part we want to determine the probability of making a Type II error if the true mean gas mileage of all Orions is 25.8 mpg. To that end, we first note that if $\mu = 25.8$ mpg, then

- $\mu_{\bar{x}} = \mu = 25.8$,
- $\sigma_{\bar{x}} = \sigma/\sqrt{n} = 1.4/\sqrt{30} = 0.26$, and
- $\bar{x}$ is normally distributed.

In other words, the variable $\bar{x}$ is normally distributed with mean 25.8 mpg and standard deviation 0.26 mpg. The normal curve for $\bar{x}$ is shown in Fig. 9.14.

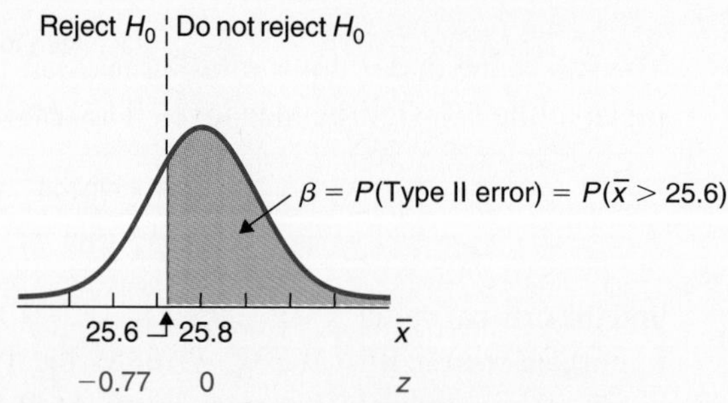

FIGURE 9.14 Determining the probability of a Type II error if $\mu = 25.8$ mpg

A Type II error occurs if we do not reject H_0, that is, if $\bar{x} > 25.6$ mpg. The probability of this happening equals the percentage of all samples whose means exceed 25.6 mpg, which we obtain in Fig. 9.14. Therefore, if the true mean gas mileage of all Orions is 25.8 mpg, then the probability of making a Type II error is 0.7794, that is, $\beta = 0.7794$.

In other words, there is roughly a 78% chance that the consumer group will fail to reject the manufacturer's claim that the mean gas mileage of all Orions is 26 mpg when, in fact, the true mean is 25.8 mpg. Although this is a rather high chance of error, we probably would not expect the hypothesis test to detect such a small difference in mean gas mileage (25.8 mpg as opposed to 26 mpg) with a sample size of only 30.

b. For this part we want to determine the probability of making a Type II error if the true mean gas mileage of all Orions is 25.0 mpg. We proceed as in part (a), but this time assuming $\mu = 25.0$ mpg. Figure 9.15 shows the required computations.

FIGURE 9.15
Determining the probability of a Type II error if $\mu = 25.0$ mpg

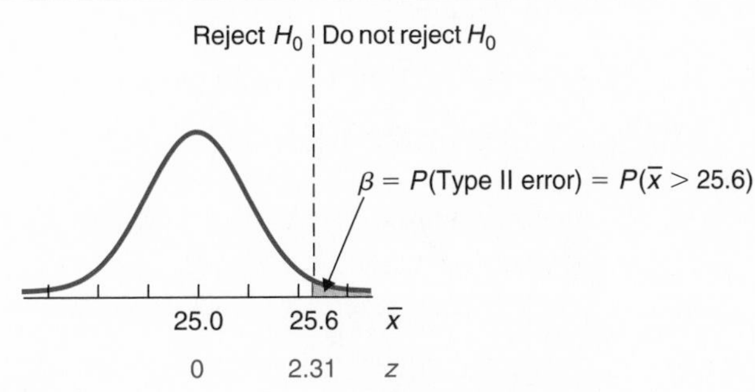

z-score computation:

$$\bar{x} = 25.6 \longrightarrow z = \frac{25.6 - 25.0}{0.26} = 2.31$$

Area to the left of z: 0.9896

Shaded area = 1 − 0.9896 = 0.0104

As we see from Fig. 9.15, if the true mean gas mileage of all Orions is 25.0 mpg, then the probability of making a Type II error is 0.0104, that is, $\beta = 0.0104$. In other words, there is only a 1% chance that the consumer group will fail to reject the manufacturer's claim that the mean gas mileage of all Orions is 26 mpg when, in fact, the true mean is 25.0 mpg. ◨

By combining figures such as Figs. 9.14 and 9.15, we can better understand Type II error probabilities. In Fig. 9.16 we combined those two figures with two others. The Type II error probabilities for the two additional values of μ were obtained using the same techniques as those used in Example 9.10.

FIGURE 9.16
Type II error probabilities for $\mu = 25.8, 25.6, 25.3$, and 25.0 $(\alpha = 0.05, n = 30)$

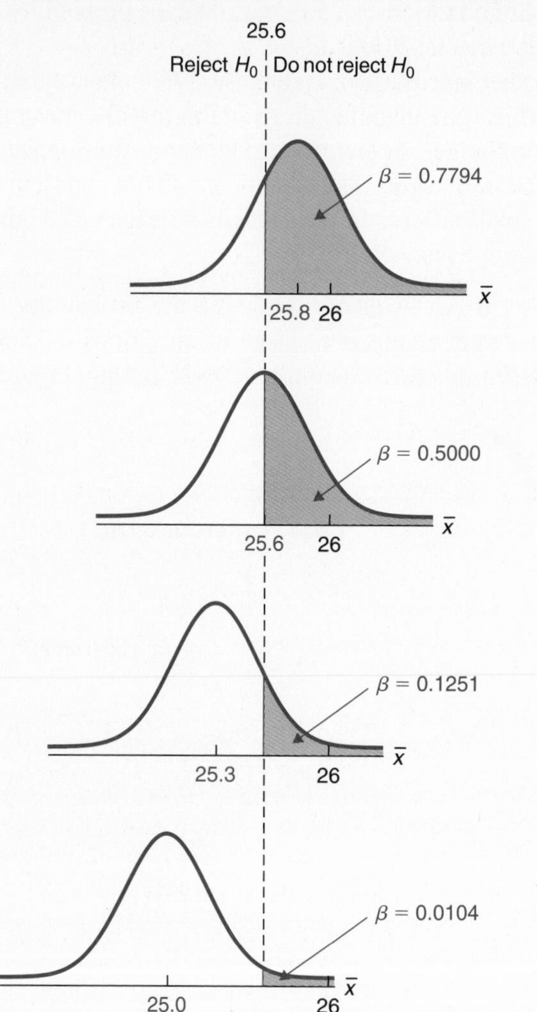

Figure 9.16 makes it clear that the farther the true mean is from the null-hypothesis mean of 26 mpg, the smaller the probability of making a Type II error. This is hardly surprising. We would expect it to be more likely for a false null hypothesis to be detected when the true mean is far from the null-hypothesis mean than when it is close to it.

Power and Power Curves

As we mentioned earlier, part of evaluating the effectiveness of a hypothesis test involves an analysis of the chances of making an incorrect decision. The probability of making a Type I error is specified by the significance level; the probability of making a Type II error depends on the true value of the parameter in question.

In modern statistical practice, analysts generally use the probability of not making a Type II error, called the **power,** to appraise the performance of a hypothesis test. Of course, once we know the Type II error probability, β, it is simple to obtain the power—just subtract β from 1.

DEFINITION 9.5 POWER

The ***power*** of a hypothesis test is the probability of not making a Type II error, that is, the probability of rejecting a false null hypothesis. We have

$$\text{Power} = 1 - P(\text{Type II error}) = 1 - \beta.$$

The power of a hypothesis test is between 0 and 1 and measures the ability of the hypothesis test to detect a false null hypothesis. If the power is near 0, the hypothesis test is not very good at detecting a false null hypothesis; if the power is near 1, the hypothesis test is extremely good at detecting a false null hypothesis.

Since in reality the true value of μ will be unknown, it is helpful to construct a table of powers for various values of μ in order to evaluate the effectiveness of the hypothesis test. For the gas-mileage illustration, we have already obtained the Type II error probability, β, when the true mean is 25.8 mpg, 25.6 mpg, 25.3 mpg, and 25.0 mpg, as depicted in Fig. 9.16. Similar calculations yield the other β probabilities shown in the second column of Table 9.8. The third column of Table 9.8 shows the power corresponding to each value of μ, obtained by subtracting β from 1.

TABLE 9.8 Selected Type II error probabilities and powers for the gas-mileage illustration ($\alpha = 0.05$, $n = 30$)

True mean μ	P(Type II error) β	Power $1 - \beta$
25.9	0.8749	0.1251
25.8	0.7794	0.2206
25.7	0.6480	0.3520
25.6	0.5000	0.5000
25.5	0.3520	0.6480
25.4	0.2206	0.7794
25.3	0.1251	0.8749
25.2	0.0618	0.9382
25.1	0.0274	0.9726
25.0	0.0104	0.9896
24.9	0.0036	0.9964
24.8	0.0010	0.9990

Table 9.8 can be used as an aid for evaluating the overall effectiveness of the hypothesis test. We can also obtain from Table 9.8 a visual display of that effectiveness by plotting points of power against μ and then connecting the points with a smooth curve. The resulting curve is called the **power curve** and is shown in Fig. 9.17. In general, the closer a power curve is to 1 (i.e., the horizontal line 1 unit above the horizontal axis), the better the hypothesis test is at detecting a false null hypothesis.

FIGURE 9.17
Power curve for the gas-mileage illustration ($\alpha = 0.05$, $n = 30$)

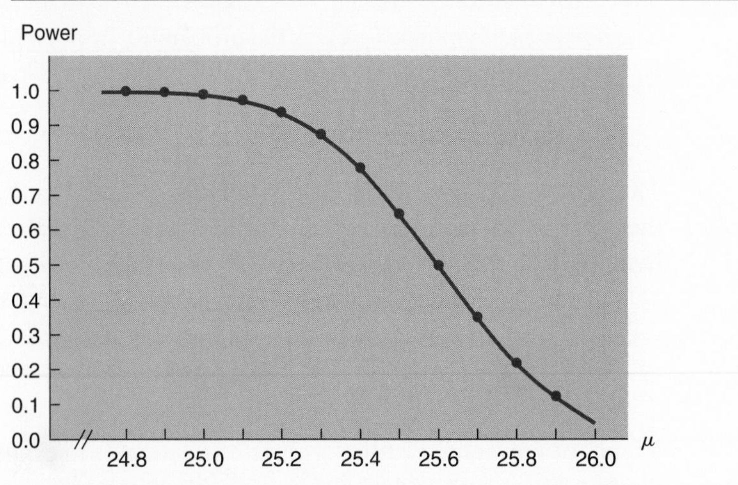

Sample Size and Power

Ideally, we would like both Type I and Type II errors to have small probabilities. In terms of significance level and power, this means that we want to specify a small significance level (close to 0) and yet have large power (close to 1).

As we noted in Section 9.2, for a fixed sample size, the smaller we specify the significance level, the larger will be the Type II error probability or, equivalently, the smaller will be the power. But there is a way that we can use a small significance level and have large power, namely, by taking a large sample. Example 9.11 provides an illustration.

EXAMPLE 9.11 THE EFFECT OF SAMPLE SIZE ON POWER

Consider again the hypothesis test for the gas-mileage illustration of Example 9.10:

$$H_0\text{: } \mu = 26 \text{ mpg (manufacturer's claim)}$$
$$H_a\text{: } \mu < 26 \text{ mpg (consumer group's conjecture),}$$

where μ is the mean gas mileage of all Orions. In Table 9.8, we presented selected powers when $\alpha = 0.05$ and $n = 30$. Now suppose the significance level is kept at 0.05 but the sample size is increased from 30 to 100.

a. Construct a table of powers similar to Table 9.8.

b. Use the table from part (a) to draw the power curve for $n = 100$ and compare it to the power curve drawn earlier for $n = 30$.

c. Interpret the results from parts (a) and (b).

SOLUTION The inference under consideration is a left-tailed hypothesis test for a population mean at the 5% significance level. The test statistic is

$$z = \frac{\bar{x} - \mu_0}{\sigma/\sqrt{n}} = \frac{\bar{x} - 26}{1.4/\sqrt{100}},$$

and the critical value is $-z_\alpha = -z_{0.05} = -1.645$. Thus the decision criterion for the hypothesis test is this: If $z \leq -1.645$, reject H_0; if $z > -1.645$, do not reject H_0.

As we noted earlier, it is somewhat simpler to compute Type II error probabilities if the decision criterion is expressed in terms of $\bar{x}$ instead of z. To do that here, we must find the sample mean that is 1.645 standard deviations below the null-hypothesis population mean of 26:

$$\bar{x} = 26 - 1.645 \cdot \frac{1.4}{\sqrt{100}} = 25.8.$$

Consequently, the decision criterion can be expressed in terms of $\bar{x}$ as follows: If $\bar{x} \leq 25.8$ mpg, reject H_0; if $\bar{x} > 25.8$ mpg, do not reject H_0. This decision criterion is portrayed graphically in Fig. 9.18.

FIGURE 9.18
Graphical display of decision criterion for the gas-mileage illustration ($\alpha = 0.05$, $n = 100$)

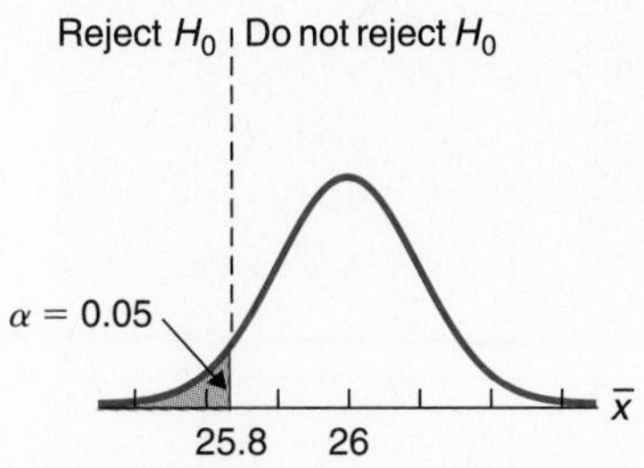

a. Now that the decision criterion has been expressed in terms of $\bar{x}$, Type II error probabilities can be obtained using the same techniques as in Example 9.10. We computed the Type II error probabilities corresponding to several values of μ, as shown in Table 9.9. The third column of Table 9.9 shows the powers.

TABLE 9.9
Selected Type II error probabilities and powers for the gas-mileage illustration ($\alpha = 0.05$, $n = 100$)

True mean μ	P(Type II error) β	Power $1 - \beta$
25.9	0.7611	0.2389
25.8	0.5000	0.5000
25.7	0.2389	0.7611
25.6	0.0764	0.9236
25.5	0.0162	0.9838
25.4	0.0021	0.9979
25.3	0.0002	0.9998
25.2	0.0000†	1.0000‡
25.1	0.0000	1.0000
25.0	0.0000	1.0000
24.9	0.0000	1.0000
24.8	0.0000	1.0000

† For $\mu \leq 25.2$, the β probabilities are zero to four decimal places.
‡ For $\mu \leq 25.2$, the powers are one to four decimal places.

b. Using Table 9.9, we can draw the power curve for the gas-mileage illustration when $n = 100$. This is shown in Fig. 9.19. For comparison purposes, we have also reproduced from Fig. 9.17 the power curve for $n = 30$.

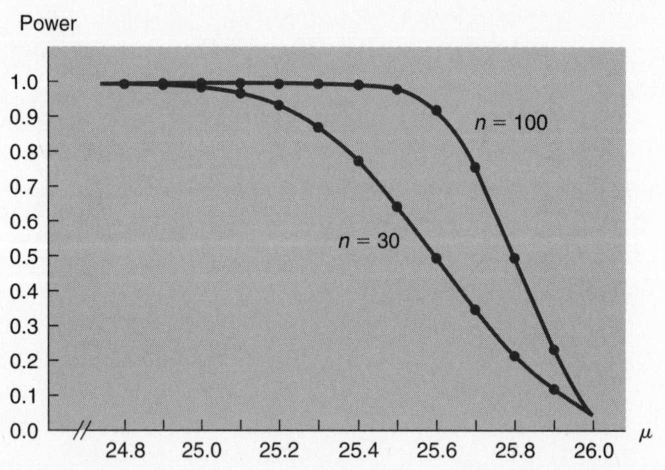

FIGURE 9.19
Power curves for the gas-mileage illustration when $n = 30$ and $n = 100$ ($\alpha = 0.05$)

c. Comparing Tables 9.8 and 9.9, we see that each power is greater when $n = 100$ than when $n = 30$. Figure 9.19 displays that fact visually. ■

In Example 9.11 we found that increasing the sample size without changing the significance level, increased the power. This is true in general.

KEY FACT 9.5 SAMPLE SIZE AND POWER

For a fixed significance level, increasing the sample size increases the power.

Key Fact 9.5 implies that by using a sufficiently large sample size, we can obtain a hypothesis test with as much power as we want. However, in practice we need to keep in mind that larger sample sizes tend to increase the cost of a study. Consequently, we must balance, among other things, the cost of a large sample against the cost of possible errors.

As we have discovered, power is useful for evaluating the overall effectiveness of a hypothesis-testing procedure. In addition, power can be used to compare different procedures. For example, a researcher might decide between two hypothesis-testing procedures on the basis of which test is more powerful for the situation under consideration.

EXERCISES 9.4

STATISTICAL CONCEPTS AND SKILLS

9.54 Why don't hypothesis tests always yield correct decisions?

9.55 Define each of the following terms.
a. Type I error **b.** Type II error
c. significance level

9.56 What does the power of a hypothesis test tell us? How is it related to the probability of making a Type II error?

9.57 Why is it useful to obtain the power curve for a hypothesis test?

9.58 What happens to the power of a hypothesis test if the sample size is increased without changing the significance level? Explain why your answer makes sense.

9.59 What happens to the power of a hypothesis test if the significance level is decreased without changing the sample size? Explain why your answer makes sense.

9.60 Suppose you must choose between two procedures for performing a hypothesis test, say Procedure A and Procedure B. Further suppose that for the same sample size and significance level, Procedure A has less power than Procedure B. Which procedure would you choose? Why?

In Exercises 9.61–9.66, we have given a hypothesis-testing situation and (i) the population standard deviation σ, (ii) a significance level, (iii) a sample size, and (iv) some values of μ. For each exercise,

a. express the decision criterion for the hypothesis test in terms of $\overline{x}$.
b. determine the probability of a Type I error.
c. construct a table similar to Table 9.8 on page 529 giving the probability of a Type II error and the power for each of the given values of μ.
d. use the table obtained in part (c) to draw the power curve.

9.61 According to the U.S. Bureau of Labor Statistics publication *Consumer Expenditures,* the mean telephone expenditure per consumer unit was \$690 in 1994. We want to perform a hypothesis test to decide whether last year's mean expenditure has increased over the 1994 mean of \$690. The null and alternative hypotheses are

$$H_0\text{: } \mu = \$690$$
$$H_a\text{: } \mu > \$690,$$

where μ is last year's mean telephone expenditure per consumer unit.

i. $\sigma = 350$ ii. $\alpha = 0.05$ iii. $n = 40$
iv. $\mu = 720, 750, 780, 810, 840, 870, 900, 930$

9.62 The Census Bureau publication *Census of Population and Housing* reports that the mean travel time to work in 1990 for all North Dakota residents was 13 minutes. A transportation official wants to use a sample of this year's travel times for North Dakota residents to decide whether the mean travel time to work for all North Dakota residents has changed from the 1990 mean of 13 minutes. The null and alternative hypotheses for the hypothesis test are

$$H_0: \mu = 13 \text{ minutes}$$
$$H_a: \mu \neq 13 \text{ minutes},$$

where μ is this year's mean travel time to work for all North Dakota residents.

i. $\sigma = 11.6$ ii. $\alpha = 0.05$ iii. $n = 35$
iv. $\mu = 3, 5, 7, 9, 11, 15, 17, 19, 21, 23$

9.63 The Food and Nutrition Board of the National Academy of Sciences states that the RDA of iron for adult females under the age of 51 is 18 mg. A hypothesis test is to be performed to decide whether adult females under the age of 51 are, on the average, getting less than the RDA of 18 mg of iron. The null and alternative hypotheses for the hypothesis test are

$$H_0: \mu = 18 \text{ mg}$$
$$H_a: \mu < 18 \text{ mg},$$

where μ is the mean iron intake (per day) of all adult females under the age of 51.

i. $\sigma = 4.2$ ii. $\alpha = 0.01$ iii. $n = 45$
iv. $\mu = 15.50, 15.75, 16.00, 16.25, 16.50, 16.75, 17.00, 17.25, 17.50, 17.75$

9.64 Ten years ago, the mean age of all juveniles held in public custody was 16.0 years, as reported by the U.S. Office of Juvenile Justice and Delinquency Prevention in *Children in Custody.* The ages of a random sample of juveniles currently being held in public custody are to be used to decide whether this year's mean age of all juveniles being held in public custody is less than it was 10 years ago. The null and alternative hypotheses for the hypothesis test are

$$H_0: \mu = 16.0 \text{ years}$$
$$H_a: \mu < 16.0 \text{ years},$$

where μ is this year's mean age of all juveniles being held in public custody.

i. $\sigma = 1$ ii. $\alpha = 0.10$ iii. $n = 250$
iv. $\mu = 15.70, 15.75, 15.80, 15.85, 15.90, 15.95$

9.65 A study by researchers at the University of Maryland addressed the question of whether the mean body temperature of humans is 98.6°F. The results of the study by P. Mackowiak, S. Wasserman, and M. Levine appeared in the article "A Critical Appraisal of 98.6°F, the Upper Limit of the Normal Body Temperature, and Other Legacies of Carl Reinhold August Wunderlich" (*Journal of the American Medical Association,* 268, pp. 1578–1580). Among other data, the researchers obtained the body temperatures of roughly 100 healthy humans. Suppose we want to use that data to decide whether the mean body temperature of healthy humans differs from 98.6°F. The null and alternative hypotheses for the hypothesis test are

$$H_0: \mu = 98.6°\text{F}$$
$$H_a: \mu \neq 98.6°\text{F},$$

where μ is the mean body temperature of all healthy humans.

i. $\sigma = 0.63$ ii. $\alpha = 0.01$ iii. $n = 93$
iv. $\mu = 98.30, 98.35, 98.40, 98.45, 98.50, 98.55, 98.65, 98.70, 98.75, 98.80, 98.85, 98.90$

9.66 The American Hospital Association reports in *Hospital Stat* that the mean cost to community hospitals per patient per day in U.S. hospitals was \$931 in 1994. In that same year, a random sample was obtained of 30 daily costs in Massachusetts hospitals. A hypothesis test was then performed to decide whether the mean cost in Massachusetts hospitals exceeded the national mean. The null and alternative hypotheses for the hypothesis test performed are

$$H_0: \mu = \$931$$
$$H_a: \mu > \$931,$$

where μ is the 1994 mean cost to community hospitals per patient per day in Massachusetts.

i. $\sigma = 333$ ii. $\alpha = 0.05$ iii. $n = 30$
iv. $\mu = 940, 970, 1000, 1030, 1060, 1090$ $1120, 1150, 1180$

9.67 Repeat parts (a)–(d) of Exercise 9.61 using a sample size of 80. Compare your power curves for the two sample sizes and explain the principle being illustrated.

9.68 Repeat parts (a)–(d) of Exercise 9.62 using a sample size of 100. Compare your power curves for the two sample sizes and explain the principle being illustrated.

EXTENDING THE CONCEPTS AND SKILLS

9.69 Consider a right-tailed hypothesis test for a population mean with null hypothesis H_0: $\mu = \mu_0$.

a. Draw the ideal power curve.

b. Explain what your curve in part (a) portrays.

9.70 Consider a left-tailed hypothesis test for a population mean with null hypothesis H_0: $\mu = \mu_0$.

a. Draw the ideal power curve.

b. Explain what your curve in part (a) portrays.

9.71 Consider a two-tailed hypothesis test for a population mean with null hypothesis H_0: $\mu = \mu_0$.

a. Draw the ideal power curve.

b. Explain what your curve in part (a) portrays.

USING TECHNOLOGY  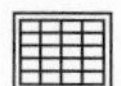 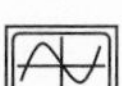

9.72 This exercise can be done individually or, better yet, as a class project. Refer to the gas-mileage hypothesis test of Example 9.10 on page 525. Recall that the null and alternative hypotheses are

H_0: $\mu = 26$ mpg (manufacturer's claim)

H_a: $\mu < 26$ mpg (consumer group's conjecture),

where μ is the mean gas mileage of all Orions. Also recall that the mileages are normally distributed with a standard deviation of 1.4 mpg. Figure 9.13 on page 526 portrays the decision criterion for a test at the 5% significance level using a sample size of 30. Suppose that in reality the mean gas mileage of all Orions is 25.4 mpg.

a. Determine the probability of making a Type II error.

b. Simulate 100 samples of 30 gas mileages each. *(Note:* If you are using Minitab, you may find it helpful to refer to part (b) of Exercise 7.45 on page 424.*)*

c. Determine the mean of each sample in part (b). *(Note:* If you are using Minitab, you may find it helpful to refer to part (c) of Exercise 7.45.*)*

d. For the 100 samples obtained in part (b), roughly how many would you expect to lead to nonrejection of the null hypothesis? Explain your answer.

e. For the 100 samples obtained in part (b), determine the number that lead to nonrejection of the null hypothesis.

f. Compare your answers from parts (d) and (e) and comment on any observed difference.

9.5 *P*-VALUES

In Section 9.3 we presented Procedure 9.1, a step-by-step method for performing a hypothesis test for a population mean when the population standard deviation is known. Step 3 of that procedure requires us to obtain critical values. And because of that, Procedure 9.1 is said to use the **critical-value approach** to hypothesis testing.

In this section we will discuss another approach to hypothesis testing, called the ***P*-value approach.** Very roughly speaking, the P-value indicates how likely (or unlikely) it would be to observe the value obtained for the test statistic if the null hypothesis were true. In particular, a small P-value (close to 0) indicates that it would be unlikely to observe the value obtained for the test statistic if the null hypothesis were true. Simply stated, small P-values provide evidence against the null hypothesis, others do not.

We can define the P-value as the percentage of samples that would yield a value of the test statistic as extreme or more extreme than that observed if the null hypothesis is true. But, more commonly, the P-value is defined using the language of probability as follows.

DEFINITION 9.6 *P*-VALUE

To obtain the ***P-value*** of a hypothesis test, we compute, assuming the null hypothesis is true, the probability of observing a value of the test statistic as extreme or more extreme than that observed. By "extreme" we mean "far from what we would expect to observe if the null hypothesis were true." We use the letter $\boldsymbol{P}$ to denote the P-value. The P-value is also referred to as the ***observed significance level*** or the ***probability value.***

In this section we will concentrate on P-values and the P-value approach for the one-sample z-test. But much of what we say here applies to P-values and the P-value approach for any hypothesis test.

Obtaining *P*-Values for a One-Sample *z*-Test

Recall that the test statistic for a one-sample z-test for a population mean with null hypothesis H_0: $\mu = \mu_0$ is

$$z = \frac{\overline{x} - \mu_0}{\sigma/\sqrt{n}}.$$

If the null hypothesis is true, this test statistic has the standard normal distribution and so its probabilities are equal to areas under the standard normal curve.

Let us denote by z_0 the observed value of the test statistic z. Then the P-value is obtained in the following way:

- *Two-tailed test:* The P-value is the probability of observing a value of the test statistic z at least as large in magnitude as the value actually observed, which is the area under the standard normal curve that lies outside the interval from $-|z_0|$ to $|z_0|$, as seen in Fig. 9.20(a).
- *Left-tailed test:* The P-value is the probability of observing a value of the test statistic z as small as or smaller than the value actually observed, which is the area under the standard normal curve that lies to the left of z_0, as seen in Fig. 9.20(b).
- *Right-tailed test:* The P-value is the probability of observing a value of the test statistic z as large as or larger than the value actually observed, which is the area under the standard normal curve that lies to the right of z_0, as seen in Fig. 9.20(c).

FIGURE 9.20
P-value for a z-test if the test is (a) two-tailed, (b) left-tailed, (c) right-tailed

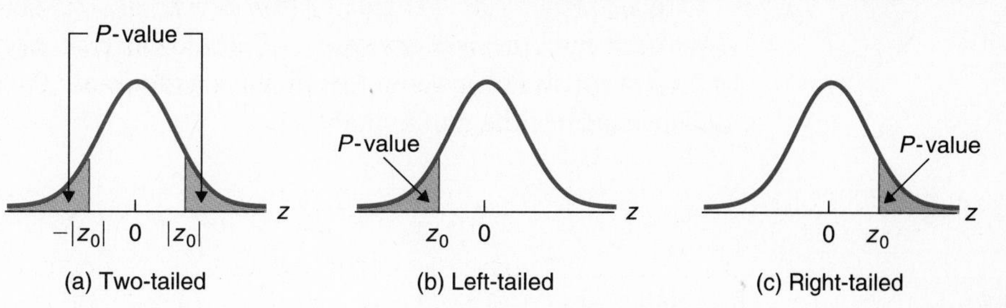

The best way to understand P-values is to look at several examples. In Examples 9.12 and 9.13, we will obtain the P-values for two of the hypothesis tests we conducted in Section 9.3.

EXAMPLE 9.12 ILLUSTRATES DEFINITION 9.6

Consider again the history-book hypothesis test of Example 9.7 where we wanted to decide whether this year's mean cost of all history books has increased over the 1993 mean of \$40.69. The null and alternative hypotheses are

$$H_0\text{: } \mu = \$40.69 \text{ (mean price has not increased)}$$
$$H_a\text{: } \mu > \$40.69 \text{ (mean price has increased)},$$

where μ is this year's mean retail price of all history books. Note that the test is right-tailed since a greater-than sign ($>$) appears in the alternative hypothesis. Table 9.5 on page 515 displays this year's prices for 40 randomly selected history books. Using that data and the fact that $\sigma = \$7.61$, we found the value of the test statistic to be 2.85. Obtain and interpret the P-value of the hypothesis test.

SOLUTION Since the hypothesis test is a right-tailed z-test, the P-value is the probability of observing a value of z of 2.85 or greater if the null hypothesis is true. That probability equals the area under the standard normal curve to the right of 2.85, the shaded area in Fig. 9.21. From Table II we find that area to be $1 - 0.9978 = 0.0022$.

FIGURE 9.21
P-value for the history-book hypothesis test

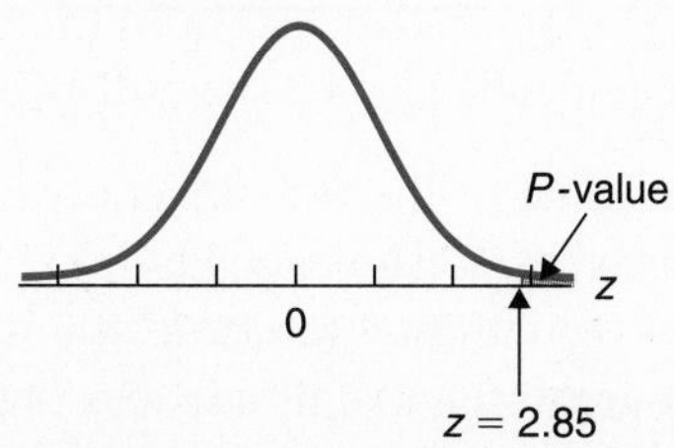

Consequently, the P-value of this hypothesis test is 0.0022. This means that, if the null hypothesis were true, we would observe a value of the test statistic of 2.85 or greater only about two times in a thousand; the data provide very strong evidence against the null hypothesis. ◨

EXAMPLE 9.13 ILLUSTRATES DEFINITION 9.6

In Example 9.9 we conducted a hypothesis test to decide whether the mean monthly rent for two-bedroom units in Maine differs from the fair market rent (FMR) of \$590. The null and alternative hypotheses are

$$H_0\colon \mu = \$590 \text{ (mean monthly rent equals the FMR)}$$
$$H_a\colon \mu \neq \$590 \text{ (mean monthly rent differs from the FMR),}$$

where μ denotes the mean monthly rent of all two-bedroom units in Maine. Note that the hypothesis test is two-tailed since a not-equal sign ($\neq$) appears in the alternative hypothesis.

Table 9.7 on page 518 shows the monthly rents of a random sample of 32 two-bedroom units in Maine. Recall that the first monthly rent, \$289, is an outlier.

a. Obtain and interpret the P-value of the hypothesis test using the unabridged data (i.e., including the outlier).

b. Obtain and interpret the P-value of the hypothesis test using the abridged data (i.e., with the outlier removed).

c. Comment on the effect that removing the outlier has on the evidence against the null hypothesis.

SOLUTION **a.** For the unabridged data, the value of the test statistic was found in Example 9.9 to be 0.95. Because the test is a two-tailed z-test, the P-value is the probability of observing a value of z of 0.95 or greater in magnitude if the null hypothesis is true. That probability is depicted in Fig. 9.22(a) and equals 0.3422. This means that, if the null hypothesis were true, we would observe a value of the test statistic of 0.95 or greater in magnitude roughly 34 times in 100; the unabridged data do not provide much evidence against the null hypothesis.

b. For the abridged data, the value of the test statistic was found in Example 9.9 to be 1.71. Thus in this case the P-value, shown in Fig. 9.22(b), equals 0.0872. This means that, if the null hypothesis were true, we would observe a value of the test statistic of 1.71 or greater in magnitude less than nine times in 100; the abridged data provide moderate evidence against the null hypothesis.

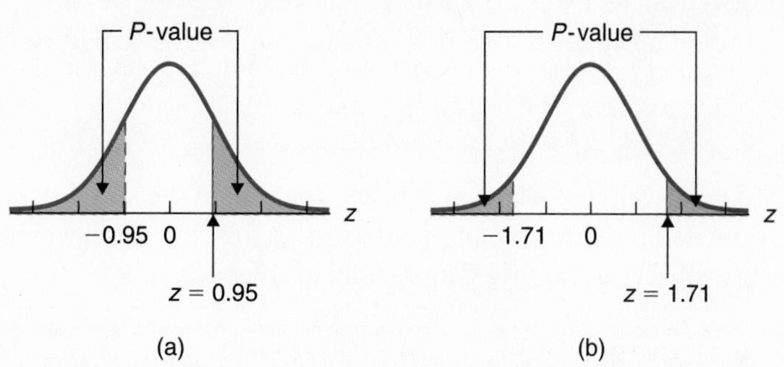

FIGURE 9.22 *P*-value for the monthly-rents hypothesis test (a) including outlier, (b) with outlier removed

c. From parts (a) and (b), we see that the strength of the evidence against the null hypothesis depends on whether the outlier is retained or removed. If the outlier is retained, there is virtually no evidence against the null hypothesis; if the outlier is removed, there is moderate evidence against the null hypothesis. ◨

The *P*-Value Approach to Hypothesis Testing

The P-value can be interpreted as the *observed significance level* of a hypothesis test. To illustrate, suppose the value of the test statistic for a right-tailed z-test turns out to be 1.88. Then the P-value of the hypothesis test is 0.03 (actually 0.0301), as depicted by the shaded area in Fig. 9.23.

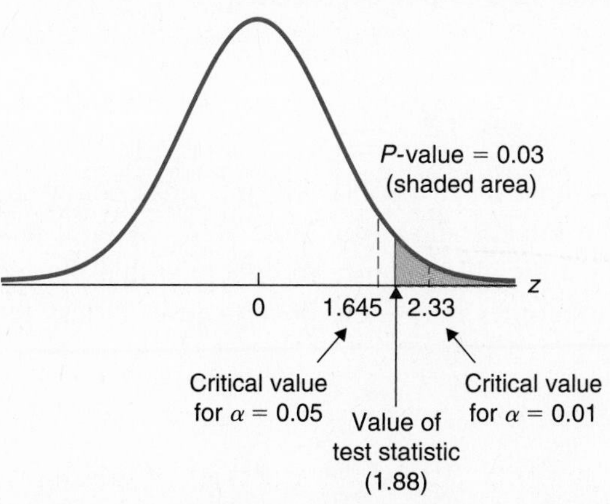

FIGURE 9.23 *P*-value as the observed significance level

From Fig. 9.23 we see that the null hypothesis would be rejected for a test at the 0.05 significance level but would not be rejected for a test at the 0.01 significance level. In fact, Fig. 9.23 makes it clear that the P-value is precisely the smallest significance level at which the null hypothesis would be rejected.

KEY FACT 9.6 **P-VALUE AS THE OBSERVED SIGNIFICANCE LEVEL**

The P-value of a hypothesis test is equal to the smallest significance level at which the null hypothesis can be rejected, that is, the smallest significance level for which the observed sample data results in rejection of H_0.

In view of Key Fact 9.6, we have the following criterion for deciding whether the null hypothesis should be rejected in favor of the alternative hypothesis.

KEY FACT 9.7 **DECISION CRITERION FOR A HYPOTHESIS TEST USING THE P-VALUE**

If the P-value is less than or equal to the specified significance level, then reject the null hypothesis; otherwise, do not reject the null hypothesis.

Key Fact 9.7 provides a foundation for the P-value approach to hypothesis testing. For the one-sample z-test, we have the procedure shown on the facing page.

In Example 9.8, we performed a one-sample z-test to decide whether the mean calcium intake of all people with incomes below the poverty level is less than the RDA of 800 mg. To carry out that hypothesis test, we used Procedure 9.1 on page 513, which employs the critical-value approach. Now we will perform that same test using the P-value approach.

EXAMPLE 9.14 **ILLUSTRATES PROCEDURE 9.2**

A random sample of 18 people with incomes below the poverty level gives the daily calcium intakes shown in Table 9.10. At the 5% significance level, do the data provide sufficient evidence to conclude that the mean calcium intake of all people with incomes below the poverty level is less than the RDA of 800 mg? Assume that $\sigma = 188$ mg.

TABLE 9.10 Daily calcium intakes (mg)

686	433	743	647	734	641
993	620	574	634	850	858
992	775	1113	672	879	609

SOLUTION We will apply Procedure 9.2 to perform the hypothesis test.

PROCEDURE 9.2 THE ONE-SAMPLE z-TEST FOR A POPULATION MEAN (P-VALUE APPROACH)

ASSUMPTIONS

1. Normal population or large sample
2. σ known

Step 1 The null hypothesis is H_0: $\mu = \mu_0$ and the alternative hypothesis is one of the following:

H_a: $\mu \neq \mu_0$ (Two-tailed)	or	H_a: $\mu < \mu_0$ (Left-tailed)	or	H_a: $\mu > \mu_0$ (Right-tailed)

Step 2 Decide on the significance level, α.

Step 3 Compute the value of the test statistic

$$z = \frac{\bar{x} - \mu_0}{\sigma/\sqrt{n}}$$

and denote that value by z_0.

Step 4 Use Table II to obtain the P-value.

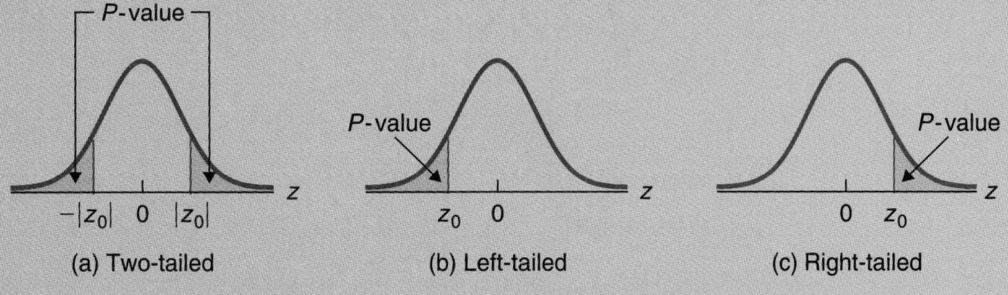

(a) Two-tailed (b) Left-tailed (c) Right-tailed

Step 5 If $P \leq \alpha$, reject H_0; otherwise, do not reject H_0.

Step 6 State the conclusion in words.

The hypothesis test is exact for normal populations and is approximately correct for large samples from nonnormal populations.

Step 1 *State the null and alternative hypotheses.*

Let μ denote the mean calcium intake (per day) of all people with incomes below the poverty level. The null and alternative hypotheses are

H_0: $\mu = 800$ mg (mean calcium intake is not less than the RDA)

H_a: $\mu < 800$ mg (mean calcium intake is less than the RDA).

Note that the hypothesis test is left-tailed since a less-than sign (<) appears in the alternative hypothesis.

Step 2 *Decide on the significance level, α.*

We are to perform the test at the 5% significance level. Thus $\alpha = 0.05$.

Step 3 *Compute the value of the test statistic*

$$z = \frac{\bar{x} - \mu_0}{\sigma/\sqrt{n}}.$$

We have $\mu_0 = 800$, $\sigma = 188$, and $n = 18$. From the data in Table 9.10, we find that $\bar{x} = 747.4$. Thus the value of the test statistic is

$$z = \frac{747.4 - 800}{188/\sqrt{18}} = -1.19.$$

This value is shown in Fig. 9.24.

FIGURE 9.24
Value of the test statistic and the *P*-value for the calcium-intake hypothesis test

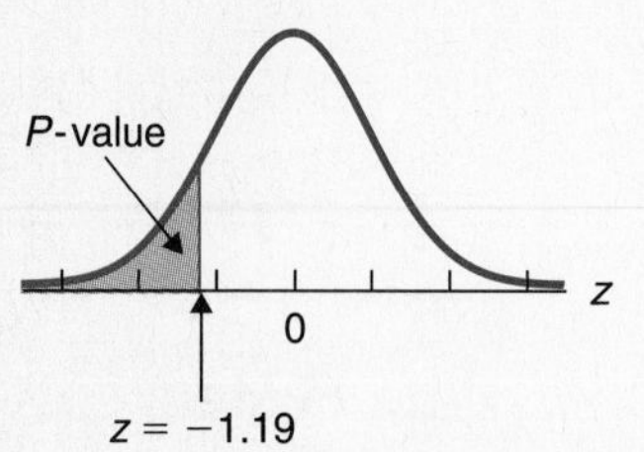

Step 4 *Use Table II to obtain the P-value.*

Since the test is left-tailed, the P-value is the probability of observing a value of z of -1.19 or less if the null hypothesis is true. That probability equals the shaded area in Fig. 9.24, which by Table II is 0.1170. Hence $P = 0.1170$.

Step 5 *If $P \leq \alpha$, reject H_0; otherwise, do not reject H_0.*

From Step 4, $P = 0.1170$. Since this exceeds the specified significance level of 0.05, we do not reject H_0.

Step 6 *State the conclusion in words.*

The test results are not statistically significant at the 5% level; that is, at the 5% significance level, the sample of 18 calcium intakes does not provide sufficient evidence to conclude that the mean calcium intake, μ, of all people with incomes below the poverty level is less than the RDA of 800 mg. □

Comparison of the Critical-Value and P-Value Approaches

We have now discussed both the critical-value and P-value approaches to hypothesis testing, but we have done so explicitly only in terms of the one-sample z-test. For future reference, we now provide the general elements of each approach.

TABLE 9.11 Comparison of critical-value and P-value approaches

CRITICAL-VALUE APPROACH		*P*-VALUE APPROACH	
Step 1	State the null and alternative hypotheses.	*Step 1*	State the null and alternative hypotheses.
Step 2	Decide on the significance level, α.	*Step 2*	Decide on the significance level, α.
Step 3	Determine the critical value(s).	*Step 3*	Compute the value of the test statistic.
Step 4	Compute the value of the test statistic.	*Step 4*	Determine the P-value.
Step 5	If the value of the test statistic falls in the rejection region, reject H_0; otherwise, do not reject H_0.	*Step 5*	If $P \leq \alpha$, reject H_0; otherwise, do not reject H_0.
Step 6	State the conclusion in words.	*Step 6*	State the conclusion in words.

Using the P-Value to Assess the Evidence Against H_0

One big advantage of the P-value is that it provides the actual significance of the hypothesis test—the smallest significance level at which the test results are statistically significant (i.e., at which the null hypothesis can be rejected). This allows us to assess significance at any level we desire. For example, if the P-value of a hypothesis test is 0.03, then we know that the test results are statistically significant at any level larger than 0.03 (e.g., $\alpha = 0.05$) and are not statistically significant at any level smaller than 0.03 (e.g., $\alpha = 0.01$).

Knowing the actual significance of the hypothesis test also allows us to evaluate the strength of the evidence against the null hypothesis—the smaller the P-value, the stronger the evidence against the null hypothesis. Table 9.12 presents guidelines for interpreting the P-value of a hypothesis test.

TABLE 9.12 Guidelines for using the P-value to assess the evidence against the null hypothesis

P-value	Evidence against H_0
$P > 0.10$	Weak or none
$0.05 < P \leq 0.10$	Moderate
$0.01 < P \leq 0.05$	Strong
$P \leq 0.01$	Very strong

Many researchers do not explicitly talk at all in terms of significance levels and critical values. Instead they simply obtain the P-value of the hypothesis test and use it to evaluate the strength of the evidence against the null hypothesis, as we did in Example 9.12 on page 537 and Example 9.13 on page 538.

Using the Computer (Optional)

We can use Minitab to perform a one-sample z-test for a population mean. To illustrate, we return once more to the hypothesis test concerning the mean calcium intake of all people with incomes below the poverty level.

EXAMPLE 9.15 USING MINITAB TO PERFORM A ONE-SAMPLE z-TEST

Use Minitab to perform the hypothesis test in Example 9.14 on page 540.

SOLUTION Let μ denote the mean calcium intake (per day) of all people with incomes below the poverty level. The problem is to perform the hypothesis test

$$H_0\colon \mu = 800 \text{ mg (mean calcium intake is not less than the RDA)}$$

$$H_a\colon \mu < 800 \text{ mg (mean calcium intake is less than the RDA)}$$

at the 5% significance level. Note that the hypothesis test is left-tailed since a less-than sign (<) appears in the alternative hypothesis. Recall that $\sigma = 188$ mg.

First we store the sample data from Table 9.10 on page 540 in a column named CALCIUM. Then we proceed in the following manner.

1 Choose **Stat ➤ Basic Statistics ➤ 1-Sample Z...**
2 Specify CALCIUM in the **Variables** text box
3 Select the **Test mean** option button
4 Click in the **Test mean** text box and type 800
5 Click the arrow button at the right of the **Alternative** drop-down list box and select **less than**
6 Click in the **Sigma** text box and type 188
7 Click **OK**

The resulting output is shown in Printout 9.1.

PRINTOUT 9.1 Minitab output for the one-sample z-test

```
Test of mu = 800.0 vs mu < 800.0
The assumed sigma = 188

Variable     N      Mean    StDev   SE Mean        Z          P
CALCIUM     18     747.4    172.0      44.3    -1.19       0.12
```

Printout 9.1 first displays a statement of the null and alternative hypotheses for the hypothesis test: `Test of mu = 800.0 vs mu < 800.0`. Next we find the value used for the population standard deviation, σ: `The assumed sigma = 188`. Then the output shows the variable, sample size, sample mean, sample standard deviation, and standard error of the mean. The next-to-last entry, `Z`, gives the value of the test statistic, z. So we see that $z = -1.19$.

The final entry shown in Printout 9.1 is the P-value (`P`). Since the P-value for the hypothesis test, 0.12, exceeds the specified significance level of 0.05, we do not reject H_0. The test results are not statistically significant at the 5% level; that is, at the 5% significance level, the data do not provide sufficient evidence to conclude that the mean calcium intake of all people with incomes below the poverty level is less than the RDA of 800 mg. □

EXERCISES 9.5

STATISTICAL CONCEPTS AND SKILLS

9.73 State two reasons why it is prudent to include the P-value when reporting the results of a hypothesis test.

9.74 What is the P-value of a hypothesis test? When does it provide evidence against the null hypothesis?

9.75 We have presented two different approaches to hypothesis testing. Identify and compare these two approaches.

9.76 Explain how the P-value is obtained for a one-sample z-test in case the hypothesis test is
a. left-tailed. **b.** right-tailed. **c.** two-tailed.

9.77 True or false: The P-value is the smallest significance level for which the observed sample data results in rejection of the null hypothesis.

9.78 In each part below we have given the significance level and P-value for a hypothesis test. For each case decide whether the null hypothesis should be rejected.
a. $\alpha = 0.05$, $P = 0.06$
b. $\alpha = 0.10$, $P = 0.06$
c. $\alpha = 0.06$, $P = 0.06$

9.79 Which provides stronger evidence against the null hypothesis, a P-value of 0.02 or a P-value of 0.03? Explain your answer.

9.80 In each part below we have given the P-value for a hypothesis test. For each case determine the strength of the evidence against the null hypothesis.
a. $P = 0.06$ **b.** $P = 0.35$
c. $P = 0.027$ **d.** $P = 0.004$

In Exercises 9.81–9.86, we have given the value obtained for the test statistic

$$z = \frac{\bar{x} - \mu_0}{\sigma/\sqrt{n}}$$

in a one-sample z-test for a population mean. We have also specified whether the test is two-tailed, left-tailed, or right-tailed. Determine the P-value in each case.

9.81 Right-tailed test:
a. $z = 2.03$ **b.** $z = -0.31$

9.82 Left-tailed test:
a. $z = -1.84$ **b.** $z = 1.25$

9.83 Left-tailed test:
a. $z = -0.74$ **b.** $z = 1.16$

9.84 Two-tailed test:
a. $z = 3.08$ **b.** $z = -2.42$

9.85 Two-tailed test:
a. $z = -1.66$ **b.** $z = 0.52$

9.86 Right-tailed test:
a. $z = 1.24$ **b.** $z = -0.69$

In each of Exercises 9.43–9.48 of Section 9.3, you were asked to perform a one-sample z-test for a population mean using Procedure 9.1, which employs the critical-value approach to hypothesis testing. Now in Exercises 9.87–9.92, you are asked to perform those same hypothesis tests using Procedure 9.2 on page 541, which employs the P-value approach to hypothesis testing. In addition, use Table 9.12 on page 543 to assess the strength of the evidence against the null hypotheses.

9.87 According to the U.S. Bureau of Labor Statistics publication *Consumer Expenditures,* the mean telephone expenditure per consumer unit was $690 in 1994. For last year a random sample of 40 consumer units revealed the following telephone expenditures, in dollars.

473	263	576	1244	1418	917	514	1741
1248	1015	292	905	1233	331	395	924
348	479	1155	586	178	169	783	1016
599	647	790	222	794	740	382	1013
502	687	406	174	565	1136	318	997

Do the data provide sufficient evidence to conclude that last year's mean telephone expenditure per consumer unit has increased over the 1994 mean of $690? Assume that the standard deviation of last year's telephone expenditures per consumer unit is $350. Perform the hypothesis test at the 5% significance level. (*Note:* The sum of the data is $28,175.)

9.88 The Census Bureau publication *Census of Population and Housing* reports that the mean travel time to work in 1990 for all North Dakota residents was 13 minutes. A transportation official obtained this year's travel times, in minutes, for a random sample of 35 North Dakota residents. Here are the data.

29	40	0	12	10	6	41
25	21	5	4	19	2	7
10	8	3	6	52	4	12
0	33	6	2	17	21	8
38	2	13	8	14	11	2

At the 5% significance level, do the data provide sufficient evidence to conclude that the mean travel time to work for all North Dakota residents has changed from the 1990 mean of 13 minutes? Assume $\sigma = 11.6$ minutes. (*Note:* The sum of the data is 491 minutes.)

9.89 The Food and Nutrition Board of the National Academy of Sciences states that the RDA of iron for adult females under the age of 51 is 18 mg. The following iron intakes, in milligrams, during a 24-hour period were obtained for 45 randomly selected adult females under the age of 51.

15.0	18.1	14.4	14.6	10.9	18.1	18.2	18.3	15.0
16.0	12.6	16.6	20.7	19.8	11.6	12.8	15.6	11.0
15.3	9.4	19.5	18.3	14.5	16.6	11.5	16.4	12.5
14.6	11.9	12.5	18.6	13.1	12.1	10.7	17.3	12.4
17.0	6.3	16.8	12.5	16.3	14.7	12.7	16.3	11.5

At the 1% significance level, do the data suggest that adult females under the age of 51 are, on the average, getting less than the RDA of 18 mg of iron? Assume the population standard deviation is 4.2 mg. (*Note:* $\bar{x} = 14.68$ mg.)

9.90 Ten years ago, the mean age of all juveniles held in public custody was 16.0 years, as reported by the U.S. Office of Juvenile Justice and Delinquency Prevention in *Children in Custody.* The mean age of 250 randomly selected juveniles currently being held in public custody is 15.86 years. Assuming $\sigma = 1.01$ years, does it appear that the mean age, μ, of all juveniles being held in public custody this year is less than it was 10 years ago? Perform the appropriate hypothesis test using $\alpha = 0.10$.

9.91 A study by researchers at the University of Maryland addressed the question of whether the mean body temperature of humans is 98.6°F. The results of the study by P. Mackowiak, S. Wasserman, and M. Levine appeared in the article "A Critical Appraisal of 98.6°F, the Upper Limit of the Normal Body Temperature, and Other Legacies of Carl Reinhold August Wunderlich" (*Journal of the American Medical Association,* 268, pp. 1578–1580). The researchers obtained the following body temperatures of 93 healthy humans.

98.0	97.6	98.8	98.0	98.8	98.8	97.6	98.6	98.6
98.8	98.0	98.2	98.0	98.0	97.0	97.2	98.2	98.1
98.2	98.5	98.5	99.0	98.0	97.0	97.3	97.3	98.1
97.8	99.0	97.6	97.4	98.0	97.4	98.0	98.6	98.6
98.4	97.0	98.4	99.0	98.0	99.4	97.8	98.2	99.2
99.0	97.7	98.2	98.2	98.8	98.1	98.5	97.2	98.5
99.2	98.3	98.7	98.8	98.6	98.0	99.1	97.2	97.6
97.9	98.8	98.6	98.6	99.3	97.8	98.7	99.3	97.8
98.4	97.7	98.3	97.7	97.1	98.4	98.6	97.4	96.7
96.9	98.4	98.2	98.6	97.0	97.4	98.4	97.4	96.8
98.2	97.4	98.0						

At the 1% significance level, do the data provide sufficient evidence to conclude that the mean body temperature of healthy humans differs from 98.6°F? Assume $\sigma = 0.63$°F. *(Note:* The mean of the 93 temperatures is 98.12°F.*)*

9.92 The American Hospital Association reports in *Hospital Stat* that the mean cost to community hospitals per patient per day in U.S. hospitals was $931 in 1994. In that same year, a random sample of 30 daily costs in Massachusetts hospitals yielded a mean of $1131. Assuming a population standard deviation of $333 for Massachusetts hospitals, do the data provide sufficient evidence to conclude that in 1994 the mean cost in Massachusetts hospitals exceeded the national mean of $931? Perform the required hypothesis test at the 5% significance level.

EXTENDING THE CONCEPTS AND SKILLS

In each of Exercises 9.93 and 9.94,

a. *identify potential outliers, if any.*
b. *perform the required hypothesis test by applying Procedure 9.2 to the unabridged sample data.*
c. *remove observations that are potential outliers, if any, and apply Procedure 9.2 to perform the required hypothesis test using the abridged sample data.*
d. *comment on the effect that removing the potential outliers has on the hypothesis test.*
e. *indicate your conclusion regarding the hypothesis test and explain your answer.*

9.93 The manufacturer of a new car, the Orion, claims that a typical car gets 26 mpg. An independent consumer group is skeptical of this claim and thinks the mean gas mileage of all Orions may be less than 26 mpg. To try to justify its contention, the consumer group conducts mileage tests on 30 randomly selected Orions and obtains the following data.

25.3	25.1	29.6	24.6	26.0	26.0
26.3	23.6	26.0	25.4	26.1	23.8
25.1	24.1	25.8	26.4	23.4	24.8
22.6	26.6	25.1	26.6	28.0	23.3
23.8	25.4	26.2	25.1	25.3	21.5

At the 5% significance level, do the data support the consumer group's conjecture? Assume the standard deviation of gas mileages for all Orions is 1.4 mpg. *(Note:* $\bar{x} = 25.23$.*)*

9.94 A Louisiana cotton farmer has used a certain brand of fertilizer for the past 5 years. Based on experience, the farmer knows that the mean yield of cotton using this fertilizer is 623 lb/acre. Recently, a new brand of fertilizer appeared on the market that will supposedly increase cotton yield. The farmer uses the new fertilizer on 20 of his 1-acre plots. Here are the resulting cotton yields, in pounds.

633	673	630	619	648
637	641	611	588	642
590	631	591	604	607
633	661	635	618	653

At the 10% significance level, do the data provide sufficient evidence to conclude that the new fertilizer increases the mean yield of cotton on the farmer's land? Assume that, using the new fertilizer, the population standard deviation of cotton yields on the farmer's land is 20.6 lb/acre. *(Note:* $\bar{x} = 627.25$ lb/acre.*)* If the new fertilizer costs more than the one the farmer presently uses, would you buy the new fertilizer if you were the farmer? Why or why not?

9.95 Consider a one-sample z-test for a population mean. Let us denote by z_0 the observed value of the test statistic z. If the test is right-tailed, then the P-value can be expressed as $P(z \geq z_0)$. Determine the corresponding expression for the P-value if the test is

a. left-tailed. **b.** two-tailed.

9.96 The symbol $\Phi(z)$ is often used to denote the area under the standard normal curve that lies to the left of a specified value of z. Consider a one-sample z-test for

a population mean. Let us denote by z_0 the observed value of the test statistic z. Express the P-value of the hypothesis test in terms of Φ if the test is
a. left-tailed. **b.** right-tailed. **c.** two-tailed.

9.97 Obtaining the *P*-value. Let x denote the test statistic for a hypothesis test and x_0 its observed value. Then the P-value of the hypothesis test equals
a. $P(x \geq x_0)$ for a right-tailed test,
b. $P(x \leq x_0)$ for a left-tailed test,
c. $2 \cdot \min\{P(x \leq x_0), P(x \geq x_0)\}$ for a two-tailed test,
where the probabilities are computed assuming the null hypothesis is true. Suppose we are considering a one-sample z-test for a population mean. Verify that the probability expressions in (a)–(c) are equivalent to those obtained in Exercise 9.95.

9.98 Discuss the relative advantages and disadvantages of using the P-value approach to hypothesis testing instead of the critical-value approach.

USING TECHNOLOGY  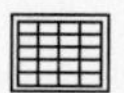 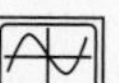

9.99 Use Minitab or some other statistical software to perform the hypothesis test in Exercise 9.87.

9.100 Use Minitab or some other statistical software to perform the hypothesis test in Exercise 9.88.

9.101 Use Minitab or some other statistical software to solve Exercise 9.93.

9.102 Use Minitab or some other statistical software to solve Exercise 9.94.

9.103 The American Hospital Association publishes information in *Hospital Stat* on the length of stay by patients in short-term hospitals. According to that publication, the mean hospital stay in 1994 was 6.7 days. We applied Minitab's one-sample z-test procedure to data on the lengths of stay by a sample of patients who were discharged from the hospital this year. The resulting output is shown in Printout 9.2. Use the printout to determine the
a. null and alternative hypotheses for the test.
b. (assumed) population standard deviation for this year's hospital stays.
c. sample standard deviation of the hospital stays.
d. number of patients sampled.
e. mean hospital stay of the patients sampled.
f. value obtained for the test statistic, z.
g. P-value of the hypothesis test.
h. smallest significance level at which the null hypothesis can be rejected.
i. conclusion if the test is performed using $\alpha = 0.05$.

9.104 A few years ago, the owner of a menswear store decided to advertise in local papers in an attempt to improve sales. His records showed that without advertising, average weekly sales had been \$1700. Since he did not wish to continue spending money on advertising if sales had not increased, he decided to perform a hypothesis test. The owner determined the weekly sales for a random sample of weeks in which advertising was used. Printout 9.3 displays the output obtained by applying Minitab's one-sample z-test procedure to the weekly-sales data. Use the printout to determine the
a. null and alternative hypotheses for the test.
b. (assumed) population standard deviation of weekly sales with advertising.
c. standard deviation of the weekly sales in the sample obtained by the store owner.
d. number of weeks sampled by the store owner.
e. mean sales of the weeks sampled.
f. value obtained for the test statistic, z.
g. P-value of the hypothesis test.
h. conclusion if the test is performed with $\alpha = 0.05$.

PRINTOUT 9.2 Minitab output for Exercise 9.103

```
Test of mu = 6.70 vs mu < 6.70
The assumed sigma = 7.70

Variable     N      Mean    StDev   SE Mean        Z        P
STAY        40      6.45     7.01      1.22    -0.21     0.42
```

PRINTOUT 9.3 Minitab output for Exercise 9.104

```
Test of mu = 1700.0 vs mu > 1700.0
The assumed sigma = 250

Variable      N      Mean    StDev   SE Mean        Z        P
SALES        32    1801.5    283.6      44.2     2.30    0.011
```

9.6 HYPOTHESIS TESTS FOR ONE POPULATION MEAN WHEN σ IS UNKNOWN

In Section 9.3 we learned how to perform a hypothesis test for a population mean when the population standard deviation, σ, is known. However, as we have mentioned, the population standard deviation is usually not known.

To develop a hypothesis-testing procedure for a population mean when σ is unknown, we begin by recalling Key Fact 8.5: Suppose a variable x of a population is normally distributed with mean μ. Then, for samples of size n, the studentized version of $\bar{x}$,

$$t = \frac{\bar{x} - \mu}{s/\sqrt{n}},$$

has the t-distribution with $n - 1$ degrees of freedom.

Because of Key Fact 8.5, we can perform a hypothesis test for a population mean when the population standard deviation is unknown by proceeding in essentially the same way as when it is known. The only difference is that we invoke a t-distribution instead of the standard normal distribution—we employ the variable

$$t = \frac{\bar{x} - \mu_0}{s/\sqrt{n}}$$

as our test statistic and use the t-table, Table IV, to obtain the critical value(s). Specifically, we have the procedure on the following page, which we often refer to as the **one-sample t-test** or more briefly as the **t-test.**

Before applying Procedure 9.3, we need to make several comments. Although the t-test was derived based on the assumption that the variable under consideration is normally distributed (normal population), it also applies approximately for large samples, regardless of the distribution of the variable under consideration, as noted at the bottom of Procedure 9.3.

Actually, like the z-test, the t-test works reasonably well even when the variable under consideration is not normally distributed and the sample size is small or moderate, provided the variable is not too far from being normally distributed. In other words, the t-test is robust to moderate violations of the normality assumption.

PROCEDURE 9.3 **THE ONE-SAMPLE t-TEST FOR A POPULATION MEAN**

ASSUMPTIONS

1. Normal population or large sample

2. σ unknown

Step 1 The null hypothesis is H_0: $\mu = \mu_0$ and the alternative hypothesis is one of the following:

H_a: $\mu \neq \mu_0$ (Two-tailed)	or	H_a: $\mu < \mu_0$ (Left-tailed)	or	H_a: $\mu > \mu_0$ (Right-tailed)

Step 2 Decide on the significance level, α.

Step 3 The critical value(s) are

$\pm t_{\alpha/2}$ (Two-tailed)	or	$-t_\alpha$ (Left-tailed)	or	t_α (Right-tailed)

with df $= n - 1$. Use Table IV to find the critical value(s).

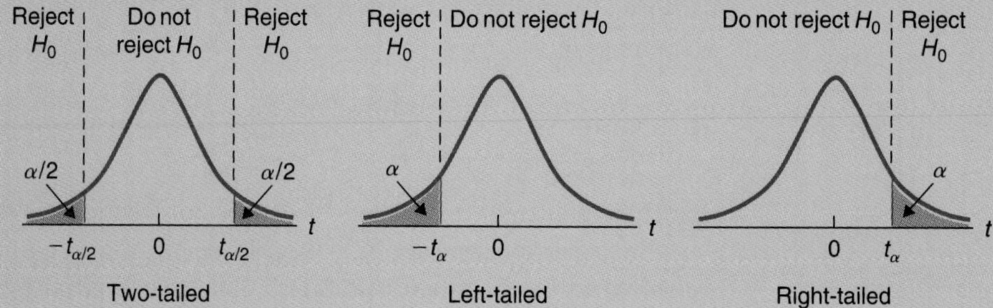

Step 4 Compute the value of the test statistic

$$t = \frac{\bar{x} - \mu_0}{s/\sqrt{n}}.$$

Step 5 If the value of the test statistic falls in the rejection region, reject H_0; otherwise, do not reject H_0.

Step 6 State the conclusion in words.

The hypothesis test is exact for normal populations and is approximately correct for large samples from nonnormal populations.

When considering the t-test, it is also important to watch for outliers. The presence of outliers calls into question the normality assumption. And even for large samples, outliers can sometimes unduly affect a t-test because the sample mean and sample standard deviation are not resistant to outliers.

Guidelines for when to use the t-test are the same as those given for the z-test in Key Fact 9.4 on page 514. And remember, always look at the data before applying the t-test to ensure that it is reasonable to use it.

EXAMPLE 9.16 ILLUSTRATES PROCEDURE 9.3

According to *Household Energy Consumption and Expenditures,* published by the U.S. Energy Information Administration, the mean energy expenditure of all U.S. households in 1993 was \$1282. That same year, 36 randomly selected households living in single-family detached homes reported the energy expenditures shown in Table 9.13. At the 5% significance level, do the data provide sufficient evidence to conclude that in 1993, households living in single-family detached homes spent more for energy, on the average, than the national average of \$1282?

TABLE 9.13 Energy expenditures (\$) for 36 households living in single-family detached homes

2016	1509	1658	1359	1564	1808	1155	1948	1162
1529	956	1284	2224	1549	1751	1408	1124	1083
1406	1370	1719	1647	1151	1735	1421	1403	1134
1877	1296	1734	1275	1341	932	1564	1012	1309

SOLUTION Graphical analyses of the data in Table 9.13 reveal no outliers. Since the sample size is 36, we can apply Procedure 9.3 to conduct the required hypothesis test.

Step 1 State the null and alternative hypotheses.

Let μ denote the mean energy expenditure in 1993 of all households living in single-family detached homes. Then the null and alternative hypotheses are

$$H_0\colon \mu = \$1282 \text{ (mean was not greater than the national mean)}$$
$$H_a\colon \mu > \$1282 \text{ (mean was greater than the national mean).}$$

Note that the hypothesis test is right-tailed since a greater-than sign (>) appears in the alternative hypothesis.

Step 2 Decide on the significance level, α.

We are to perform the test at the 5% significance level; so, $\alpha = 0.05$.

Step 3 The critical value for a right-tailed test is t_α, with df $= n - 1$.

We have $n = 36$ and $\alpha = 0.05$. Table IV shows that for df $= 36 - 1 = 35$, $t_{0.05} = 1.690$, as seen in Fig. 9.25.

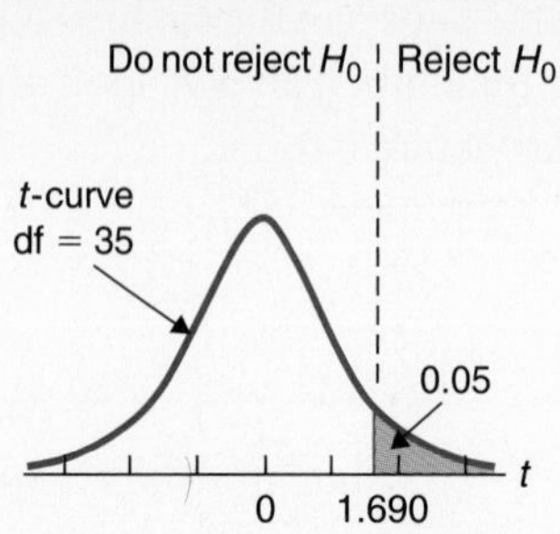

FIGURE 9.25
Criterion for deciding whether or not to reject the null hypothesis

Step 4 *Compute the value of the test statistic*

$$t = \frac{\bar{x} - \mu_0}{s/\sqrt{n}}.$$

We have $\mu_0 = \$1282$ and $n = 36$. Furthermore, the mean and standard deviation of the sample data in Table 9.13 are \$1455.92 and \$310.23, respectively. Consequently, the value of the test statistic is

$$t = \frac{1455.92 - 1282}{310.23/\sqrt{36}} = 3.364.$$

Step 5 *If the value of the test statistic falls in the rejection region, reject H_0; otherwise, do not reject H_0.*

The value of the test statistic, found in Step 4, is $t = 3.364$. As we see from Fig. 9.25, this falls in the rejection region. Hence we reject H_0.

Step 6 *State the conclusion in words.*

 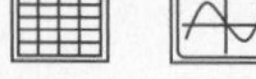

The test results are statistically significant at the 5% level; that is, at the 5% significance level, the data provide sufficient evidence to conclude that in 1993, households living in single-family detached homes spent more for energy, on the average, than the national average of \$1282. ◨

P-Values for a *t*-Test

We can also use the P-value approach to hypothesis testing to carry out a t-test. P-values for a t-test are obtained in a manner similar to that for a z-test.

Recall that the test statistic for a one-sample t-test for a population mean with null hypothesis H_0: $\mu = \mu_0$ is

$$t = \frac{\bar{x} - \mu_0}{s/\sqrt{n}}.$$

If the null hypothesis is true, this test statistic has the t-distribution with $n - 1$ degrees of freedom and so its probabilities are equal to areas under the t-curve with $\text{df} = n - 1$.

Let us denote by t_0 the observed value of the test statistic t. Then the P-value is obtained in the following way:

- *Two-tailed test:* The P-value is the probability of observing a value of the test statistic t at least as large in magnitude as the value actually observed, which is the area under the t-curve that lies outside the interval from $-|t_0|$ to $|t_0|$, as seen in Fig. 9.26(a).
- *Left-tailed test:* The P-value is the probability of observing a value of the test statistic t as small as or smaller than the value actually observed, which is the area under the t-curve that lies to the left of t_0, as seen in Fig. 9.26(b).
- *Right-tailed test:* The P-value is the probability of observing a value of the test statistic t as large as or larger than the value actually observed, which is the area under the t-curve that lies to the right of t_0, as seen in Fig. 9.26(c).

FIGURE 9.26
P-value for a *t*-test if the test is (a) two-tailed, (b) left-tailed, (c) right-tailed

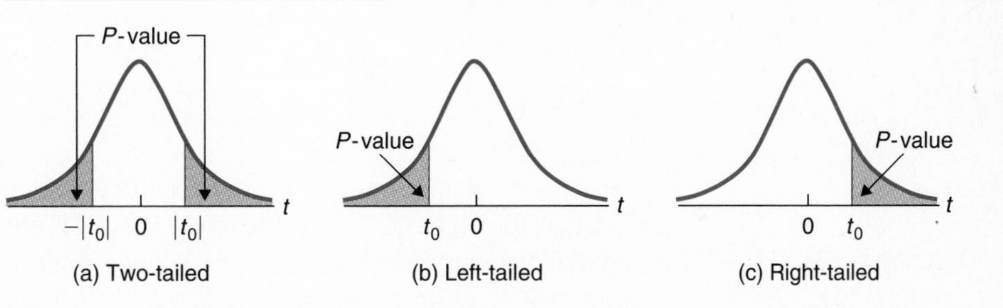

To obtain the exact P-value of a t-test, we need to use a computer with statistical software or a statistical calculator. However, we can use t-tables, such as Table IV, to estimate the P-value of a t-test; and an estimate of the P-value is usually sufficient for deciding whether to reject the null hypothesis.

For example, in the right-tailed t-test of Example 9.16, $\alpha = 0.05$, $\text{df} = 35$, and the value of the test statistic is $t = 3.364$. For $\text{df} = 35$, the t-value 3.364 is larger than any t-value in Table IV, the largest one being $t_{0.005} = 2.724$ (which means that the area under the t-curve that lies to the right of 2.724 equals 0.005). This, in turn, implies that the area to the right of 3.364 is less than 0.005; in other words, for the t-test in Example 9.16, we have $P < 0.005$. Because the P-value is less than the designated significance level of 0.05, we reject H_0.

Example 9.17 provides two more illustrations of how Table IV can be used to estimate the P-value of a t-test.

EXAMPLE 9.17 **USING TABLE IV TO ESTIMATE THE P-VALUE OF A t-TEST**

Use Table IV to estimate the P-value of each of the following t-tests.

a. Left-tailed test, $n = 12$, $t = -1.938$

b. Two-tailed test, $n = 25$, $t = -0.895$

SOLUTION **a.** Because the test is left-tailed, the P-value is the area under the t-curve with df $= 12 - 1 = 11$ that lies to the left of -1.938, as seen in Fig. 9.27(a).

FIGURE 9.27 Estimating the P-value of a left-tailed t-test with sample size 12 and test statistic $t = -1.938$

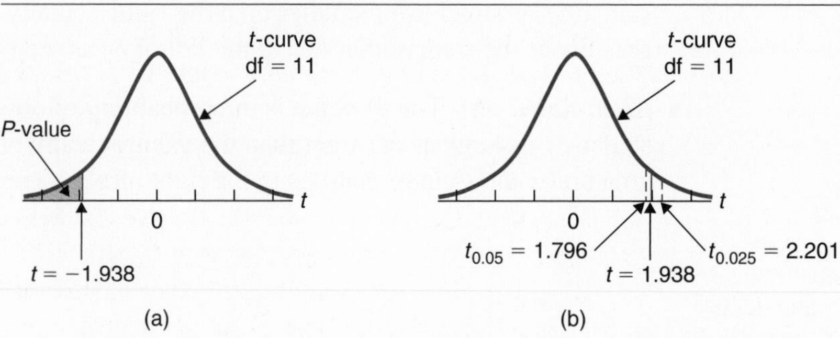

Since a t-curve is symmetric about 0, the area to the left of -1.938 equals the area to the right of 1.938; and we can use Table IV to estimate this latter area. Concentrating on the df $= 11$ row of Table IV, we search for the two t-values that straddle 1.938; they are $t_{0.05} = 1.796$ and $t_{0.025} = 2.201$. This implies that the area under the t-curve that lies to the right of 1.938 is somewhere between 0.025 and 0.05, as seen in Fig. 9.27(b).

Consequently, the area under the t-curve that lies to the left of -1.938 is also somewhere between 0.025 and 0.05, which means that $0.025 < P < 0.05$. So we can reject H_0 at any significance level of 0.05 or larger, and we cannot reject H_0 at any significance level of 0.025 or smaller. For significance levels between 0.025 and 0.05, Table IV is not sufficiently detailed for us to decide whether H_0 should be rejected.[†]

b. Because the test is two-tailed, the P-value is the area under the t-curve with df $= 25 - 1 = 24$ that lies either to the left of -0.895 or to the right of 0.895, as seen in Fig. 9.28(a).

[†] This latter case provides an example where an estimate of the P-value is not good enough. In cases such as this, use a computer with statistical software or a statistical calculator to obtain the exact P-value.

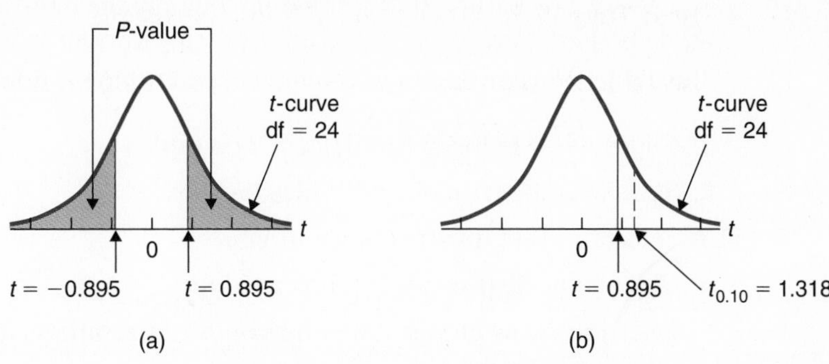

FIGURE 9.28 Estimating the P-value of a two-tailed t-test with sample size 25 and test statistic $t = -0.895$

Since a t-curve is symmetric about 0, the area to the left of -0.895 and the area to the right of 0.895 are equal. Concentrating on the df $= 24$ row of Table IV, we find that 0.895 is smaller than any t-value in Table IV, the smallest being $t_{0.10} = 1.318$. This implies that the area under the t-curve that lies to the right of 0.895 is greater than 0.10, as seen in Fig. 9.28(b).

Consequently, the area under the t-curve that lies either to the left of -0.895 or to the right of 0.895 is greater than 0.20, which means $P > 0.20$. So we cannot reject H_0 at any significance level of 0.20 or smaller. For significance levels larger than 0.20, Table IV is not sufficiently detailed for us to decide whether or not H_0 should be rejected. ◨

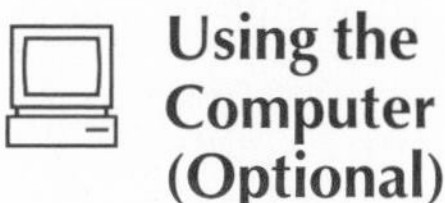

Using the Computer (Optional)

We can use Minitab to perform a one-sample t-test for a population mean. To illustrate, we return to the hypothesis test concerning the mean energy expenditure for households living in single-family detached homes.

EXAMPLE 9.18 USING MINITAB TO PERFORM A ONE-SAMPLE t-TEST

Use Minitab to perform the hypothesis test in Example 9.16 on page 551.

SOLUTION Let μ denote the mean energy expenditure in 1993 of all households living in single-family detached homes. The problem is to perform the hypothesis test

$$H_0\colon \mu = \$1282 \text{ (mean was not greater than the national mean)}$$
$$H_a\colon \mu > \$1282 \text{ (mean was greater than the national mean)}$$

at the 5% significance level ($\alpha = 0.05$). Note that the hypothesis test is right-tailed since a greater-than sign ($>$) appears in the alternative hypothesis.

As we saw earlier, it is reasonable to apply the one-sample t-test to carry out the hypothesis test. To do so using Minitab, we first store the sample data from Table 9.13 on page 551 in a column named ENERGY. Then we do the following.

1 Choose **Stat ➤ Basic Statistics ➤ 1-Sample t...**

2 Specify ENERGY in the **Variables** text box

3 Select the **Test mean** option button

4 Click in the **Test mean** text box and type 1282

5 Click the arrow button at the right of the **Alternative** drop-down list box and select **greater than**

6 Click **OK**

Printout 9.4 displays the output that results.

PRINTOUT 9.4
Minitab output for the one-sample t-test

```
Test of mu = 1282.0 vs mu > 1282.0

Variable      N      Mean     StDev    SE Mean         T          P
ENERGY       36    1455.9     310.2       51.7      3.36     0.0009
```

On the first line of Printout 9.4, we find a statement of the null and alternative hypotheses: `Test of mu = 1282.0 vs mu > 1282.0`. Then we find the variable, sample size, sample mean, sample standard deviation, and estimated standard error of the mean. The next-to-last entry, T, shows the value of the test statistic, t. So, we see that $t = 3.36$.

The final entry in Printout 9.4 displays the P-value of the hypothesis test, $P = 0.0009$. Since this is less than the specified significance level of 0.05, we reject H_0. At the 5% significance level, the data provide sufficient evidence to conclude that in 1993, households living in single-family detached homes spent more for energy, on the average, than the national average of $1282. □

EXERCISES 9.6

Statistical Concepts and Skills

9.105 What difference in assumptions is there between the one-sample t-test and the one-sample z-test?

9.106 Is there any restriction on sample size for using the one-sample t-test? Explain your answer.

Preliminary data analyses indicate that it is reasonable to use a t-test to conduct each of the hypothesis tests required in Exercises 9.107–9.112. Perform each t-test using either the critical-value approach or the P-value approach. Comment on the practical significance of those tests whose results are statistically significant.

9.107 According to the document *Consumer Expenditures,* a publication of the U.S. Bureau of Labor Statistics, the average consumer unit spent $1644 on apparel and services in 1994. That same year, 36 consumer units in the Midwest had the following annual expenditures, in dollars, on apparel and services.

2135	1561	1987	2087	2040	1951
1225	1455	2097	1757	1385	888
1539	1797	1334	1559	1811	1698
1769	1195	1312	1517	1514	1657
982	1612	2033	1191	1758	2212
1470	1606	1238	918	1398	1268

At the 5% significance level, do the data provide sufficient evidence to conclude that the 1994 mean annual expenditure on apparel and services for consumer units in the Midwest differed from the national mean of $1644? *(Note:* The sample mean and sample standard deviation of the data are $1582.11 and $351.69, respectively.*)*

9.108 A paint manufacturer claims that the average drying time for its new latex paint is 2 hours. To test that claim, the drying times are obtained for 20 randomly selected cans of paint. Here are the drying times, in minutes.

123	109	115	121	130
127	106	120	116	136
131	128	139	110	133
122	133	119	135	109

Do the data provide sufficient evidence to conclude that the mean drying time is greater than the manufacturer's claim of 120 minutes? Use $\alpha = 0.05$. *(Note:* The sample mean and sample standard deviation of the data are 123.1 minutes and 10.0 minutes, respectively.*)*

9.109 As reported by the College Entrance Examination Board in *National College-Bound Senior,* the mean (non-recentered) verbal score on the Scholastic Assessment Test (SAT) in 1995 was 428 points out of a possible 800. A random sample of 25 verbal scores for last year yielded the following data.

344	494	350	376	313
489	358	383	498	556
379	301	432	560	494
418	483	444	477	420
492	287	434	514	613

At the 10% significance level, does it appear that last year's mean for verbal SAT scores is greater than the 1995 mean of 428 points? *(Note:* $\bar{x} = 436.4$, $s = 85.5$.*)*

9.110 According to Nielsen Media Research, in 1997, during the time period from 8:00 P.M. to 11:00 P.M., the average person watched 7 hours and 37 minutes of television per week. A random sample of 40 women in the age group 18–24 years watched the following amount of TV during that same time period. The viewing times are rounded to the nearest 10 minutes.

0	110	790	450	0	750	550	510
130	160	510	70	740	550	350	120
710	310	580	260	240	130	190	240
120	240	240	540	430	350	50	240
680	400	450	340	10	440	330	420

Do the data provide sufficient evidence to conclude that, during the time period from 8:00 P.M. to 11:00 P.M., women in the age group 18–24 years watched less TV on the average than people in general? Perform the hypothesis test at the 1% significance level. *(Note:* $\bar{x} = 343.2$, $s = 222.2$.*)*

9.111 The average retail price for bananas in 1994 was 46.0 cents per pound, as reported by the U.S. Department of Agriculture in *Food Cost Review.* Recently, a random sample of 15 markets gave the following prices for bananas in cents per pound.

51	48	50	48	45
52	53	49	43	42
45	52	52	46	50

Can we conclude that the current mean retail price for bananas is different from the 1994 mean of 46.0 cents per pound? Use $\alpha = 0.05$. *(Note:* $\bar{x} = 48.4$, $s = 3.5$.*)*

9.112 Atlas Fishing Line, Inc., produces a 10-lb test line. Twelve randomly selected spools are subjected to tensile-strength tests. The results follow.

9.8	10.2	9.8	9.4
9.7	9.7	10.1	10.1
9.8	9.6	9.1	9.7

Use the data to decide whether Atlas Fishing Line's 10-lb test line is not up to specifications. Perform the required hypothesis test at the 5% significance level. *(Note:* $\bar{x} = 9.75$, $s = 0.31$.*)*

EXTENDING THE CONCEPTS AND SKILLS

9.113 According to *Food Consumption, Prices, and Expenditures,* published by the U.S. Department of Agriculture, the mean consumption of beef per person in 1990 was 64 lb (boneless, trimmed weight). A sample of 40 people taken this year yielded the following data, in pounds, on last year's beef consumptions.

77	65	57	54	68	79	56	0
50	49	51	56	56	78	63	72
0	62	74	61	61	60	56	37
76	77	67	67	62	89	56	75
69	73	75	62	8	71	20	47

a. Use the sample data to decide, at the 5% significance level, whether last year's mean beef consumption is less than the 1990 mean of 64 lb. *(Note:* The mean and standard deviation of the sample data are 58.40 lb and 20.42 lb, respectively.*)*

b. The sample data contain four potential outliers: 0, 0, 8, and 20. Remove those four observations and repeat the hypothesis test in part (a). *(Note:* The mean and standard deviation of the abridged sample data are 64.11 lb and 11.02 lb, respectively.*)*

c. Compare your results in parts (a) and (b).

d. Assuming the four potential outliers are not recording errors, comment on the advisability of removing them from the sample data prior to performing the hypothesis test.

e. What action would you take regarding this hypothesis test?

9.114 Suppose we want to perform a hypothesis test for a population mean based on a small sample but that preliminary data analyses indicate either the presence of outliers or that the variable under consideration is far from normally distributed.

a. Is either the z-test or t-test appropriate?

b. What type of procedure might be appropriate?

9.115 A manufacturer of light bulbs makes a 60-watt bulb having a mean life of 1000 hours. The research and development (R&D) department has developed a new bulb that it claims will, on the average, outlast the present bulb. To try to justify its claim, R&D tests 10 new bulbs. The results show that the 10 bulbs tested have a mean life of 1050.2 hours and a standard deviation of 65.8 hours.

a. At the 1% significance level, do the data obtained by R&D support its claim? Assume bulb life is normally distributed.

b. Suppose in part (a) you had mistakenly concluded that the test statistic

$$\frac{\bar{x} - 1000}{s/\sqrt{10}}$$

has the standard normal distribution.

i. What critical value would you have used?

ii. What critical value did you actually use?

iii. In general, does the mistaken use of a z critical value when a t critical value should have been used make it more or less likely that the null hypothesis will be rejected? Explain your answer.

9.116 Brown Swiss Dairy sells "half-gallon" cartons of milk. The contents of the cartons are known to be normally distributed with a standard deviation of 1.1 fluid oz. Suppose 15 randomly selected cartons have a mean content of 64.48 fluid oz.

a. Do the data provide sufficient evidence to infer that the cartons actually contain more milk, on the average, than 64 fluid oz? Perform the appropriate hypothesis test at the 0.05 level of significance.

b. Suppose in part (a) you had incorrectly concluded that the test statistic

$$\frac{\bar{x} - 64}{1.1/\sqrt{15}}$$

has the t-distribution with df $= 14$.

i. What critical value would you have used?

ii. What critical value did you actually use?

iii. In general, does the mistaken use of a t critical value when a z critical value should have been used make it more or less likely that the null hypothesis will be rejected? Explain your answer.

9.117 Refer to Exercise 9.116.

a. Suppose that you mistakenly use the t-table (with $df = 14$) instead of the standard-normal table to obtain the critical value for the hypothesis test in part (a) of Exercise 9.116. What will be the significance level of the resulting test? Compare this with the desired significance level of 0.05.

b. More generally, suppose you are performing a hypothesis test at the significance level α for the mean of a normally distributed variable. Further suppose the population standard deviation is known so that the appropriate hypothesis-testing procedure is the z-test. If you mistakenly use the t-table instead of the standard-normal table to obtain the critical value(s), will the actual significance level of the resulting test be higher or lower than α? Explain your answer.

9.118 Suppose you are performing a hypothesis test at the significance level α for the mean of a normally distributed variable. Also suppose the population standard deviation is unknown so that the appropriate hypothesis-testing procedure is the t-test. If you mistakenly use the standard-normal table instead of the t-table to obtain the critical value(s), will the actual significance level of the resulting test be higher or lower than α? Why?

USING TECHNOLOGY  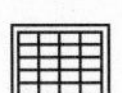 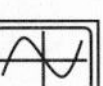

9.119 Refer to Exercise 9.107. Use Minitab or some other statistical software to

a. obtain a normal probability plot, a boxplot, a histogram, and a stem-and-leaf diagram of the data.

b. perform the required hypothesis test.

c. Is the use of your procedure in part (b) justified? Explain your answer.

9.120 Refer to Exercise 9.108. Use Minitab or some other statistical software to

a. obtain a normal probability plot, a boxplot, a histogram, and a stem-and-leaf diagram of the data.

b. perform the required hypothesis test.

c. Is the use of your procedure in part (b) justified? Explain your answer.

9.121 The *World Almanac, 1998,* reports that the average vehicle in the United States used 711 gallons of fuel in 1995. From a random sample of vehicles in Hawaii, the gallons used per vehicle in 1995 was obtained and Minitab's t-test procedure was applied to the resulting data. Printout 9.5 on the next page shows the output. Using the printout, determine the

a. null and alternative hypotheses.

b. sample standard deviation of the gallons used.

c. number of vehicles in the sample.

d. sample mean of the gallons used.

e. value obtained for the test statistic, t.

f. P-value of the hypothesis test.

g. smallest significance level at which the null hypothesis can be rejected.

h. conclusion if the hypothesis test is performed at the 5% significance level.

9.122 According to *Current Population Reports,* a publication of the U.S. Bureau of the Census, the mean annual earnings of year-round, full-time, male workers was \$41,118 in 1994. During that same year, a random sample was taken of such workers having an associate degree as their highest degree. The output in Printout 9.6 on the next page shows the results of applying Minitab's t-test procedure to their annual earnings. Use the printout to obtain the

a. null and alternative hypotheses.

b. standard deviation of the annual earnings of the people sampled.

c. number of workers sampled.

d. mean annual earnings of the workers sampled.

e. value obtained for the test statistic, t.

f. P-value of the hypothesis test.

g. smallest significance level at which the null hypothesis can be rejected.

h. conclusion if the hypothesis test is performed at the 5% significance level.

9.123 Refer to Exercise 9.113. Use Minitab or some other statistical software to

a. identify the potential outliers.

b. perform the hypothesis test considered in part (a) of Exercise 9.113.

c. perform the hypothesis test considered in part (b) of Exercise 9.113.

d. Compare the results obtained in parts (b) and (c).

PRINTOUT 9.5 Minitab output for Exercise 9.121

```
 Test of mu = 711.0 vs mu < 711.0

Variable     N       Mean     StDev   SE Mean         T          P
GALLONS     37      517.4     237.5      39.0     -4.96     0.0000
```

PRINTOUT 9.6 Minitab output for Exercise 9.122

```
 Test of mu = 41118 vs mu not = 41118

Variable     N       Mean     StDev   SE Mean         T          P
EARNINGS    50      38944     11237      1589     -1.37       0.18
```

9.7 THE WILCOXON SIGNED-RANK TEST*

Up to this point, we have learned two methods for performing a hypothesis test for a population mean. If the population standard deviation is known, we can use the z-test; if the population standard deviation is unknown, we can use the t-test.

But both of these procedures have another assumption for their use—the variable under consideration should be approximately normally distributed or the sample size should be relatively large and, for small samples, both procedures should be avoided in the presence of outliers.

In this section we will study a third method for performing a hypothesis test for a population mean—the **one-sample Wilcoxon signed-rank test** or, more briefly, the **Wilcoxon signed-rank test.** This test, which is sometimes more appropriate than either the z-test or the t-test, is an example of a nonparametric method.

What Is a Nonparametric Method?

Recall that descriptive measures for population data, such as μ and σ, are called parameters. Technically, inferential methods concerned with parameters are called **parametric methods;** those that are not are called **nonparametric methods.** However, it has become common statistical practice to refer to most methods that can be applied without assuming normality as nonparametric. Thus the term *nonparametric method* as used in contemporary statistics is somewhat of a misnomer.

Nonparametric methods have advantages beyond not requiring normality. They usually entail fewer and simpler computations than parametric methods and are resistant to outliers and other extreme values. On the other hand, parametric meth-

* This section begins the optional coverage of nonparametric statistics.

ods, such as the z-test and t-test, tend to give more accurate results (e.g., are more powerful) when the requirements for their use are met.

The Logic Behind the Wilcoxon Signed-Rank Test

The Wilcoxon signed-rank test assumes that the variable under consideration has a *symmetric distribution*—one that can be divided into two pieces that are mirror images of each other—but does not require that its distribution be normal or have any other specific shape. Thus, for example, the Wilcoxon signed-rank test applies to a variable having a normal, triangular, uniform, or symmetric bimodal distribution, but not to one whose distribution is right-skewed or left-skewed. Example 9.19 explains the reasoning behind the Wilcoxon signed-rank test.

EXAMPLE 9.19 INTRODUCES THE WILCOXON SIGNED-RANK TEST

In January 1994 the Department of Agriculture estimated that a typical U.S. family of four would spend \$117 per week for food. During that same year, a random sample of 10 Kansas families of four yielded the weekly food costs shown in Table 9.14. Do the data provide sufficient evidence to conclude that in 1994 the mean weekly food cost for Kansas families of four was less than the national mean of \$117?

TABLE 9.14 Sample of weekly food costs (\$)

103	129	109	95	121
98	112	110	101	119

SOLUTION Let μ denote the 1994 mean weekly food cost for all Kansas families of four. We want to perform the hypothesis test

$$H_0\text{: } \mu = \$117 \text{ (mean weekly food cost is not less than \$117)}$$

$$H_a\text{: } \mu < \$117 \text{ (mean weekly food cost is less than \$117).}$$

As we said, a condition for use of the Wilcoxon signed-rank test is that the variable under consideration has a symmetric distribution. If the weekly food costs for Kansas families of four have a symmetric distribution, then a graphical display of the sample data should be roughly symmetric.

Figure 9.29 shows a stem-and-leaf diagram of the sample data in Table 9.14. The diagram is roughly symmetric and so does not reveal any obvious violations of the symmetry condition.[†] We will therefore apply the Wilcoxon signed-rank test to carry out the hypothesis test.

[†] For ease in explaining the Wilcoxon signed-rank test, we have chosen an example in which the sample size is very small. This, however, makes it difficult to effectively check the symmetry condition. In general, we must proceed cautiously when dealing with very small samples.

FIGURE 9.29
Stem-and-leaf diagram of sample data in Table 9.14

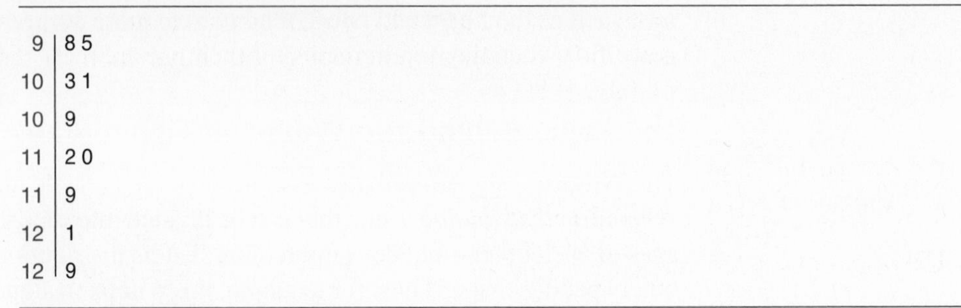

9 | 8 5
10 | 3 1
10 | 9
11 | 2 0
11 | 9
12 | 1
12 | 9

To begin, we rank the data in Table 9.14 according to distance and direction from the null hypothesis mean, $\mu_0 = \$117$. The steps for doing this are depicted in Table 9.15.

TABLE 9.15
Steps for ranking the data in Table 9.14 according to distance and direction from the null hypothesis mean

Cost (\$) x	Difference $D = x - 117$	$\lvert D \rvert$	Rank of $\lvert D \rvert$	Signed rank R
103	−14	14	7	−7
98	−19	19	9	−9
129	12	12	6	6
112	−5	5	3	−3
109	−8	8	5	−5
110	−7	7	4	−4
95	−22	22	10	−10
101	−16	16	8	−8
121	4	4	2	2
119	2	2	1	1

STEP 1 *Subtract μ_0 from x.*

STEP 2 *Make each difference positive by taking absolute values.*

STEP 3 *Rank the absolute differences in order from smallest (1) to largest (10).*

STEP 4 *Give each rank the same sign as the sign in Column 2.*

The absolute differences, $|D|$, displayed in the third column of Table 9.15 identify how far each observation is from 117; the ranks of those absolute differences, displayed in the fourth column, show which observations are closer to 117 and which are farther away; and the signed ranks, R, displayed in the last column, indicate additionally whether an observation is greater than 117 (+) or less than 117 (−). Figure 9.30 depicts this information for the second and third rows of Table 9.15.

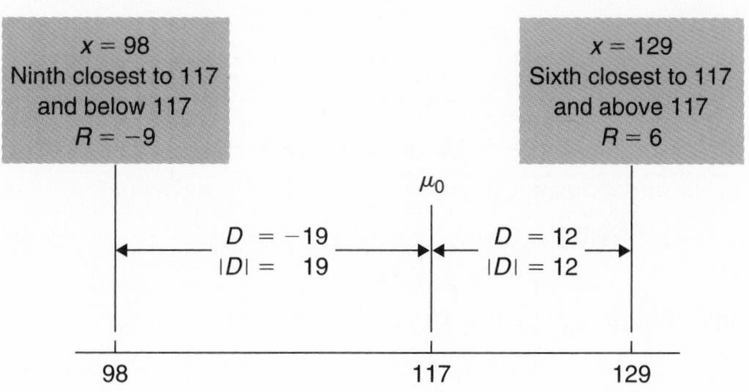

FIGURE 9.30 Meaning of signed ranks for the observations 98 and 129

The reasoning behind the Wilcoxon signed-rank test is this: If the null hypothesis, $\mu = \$117$, is true, then because the distribution of weekly food costs is symmetric, we expect the sum of the positive ranks and the sum of the negative ranks to be roughly the same in magnitude. Since the sample size here is 10, the sum of all the ranks must be $1 + 2 + \cdots + 10 = 55$; and half of 55 is 27.5.

So if the null hypothesis is true, we expect the sum of the positive ranks (and the sum of the negative ranks) to be roughly 27.5. To put it another way, if the sum of the positive ranks is too much smaller than 27.5, then we take this as evidence that the null hypothesis is false and conclude that the mean weekly food cost is less than \$117.

From the last column of Table 9.15, we see that the sum of the positive ranks, denoted by W, is equal to $6 + 2 + 1 = 9$. This is quite a bit smaller than 27.5, the value we would expect if the mean is in fact \$117.

Our question now is whether the difference between the observed and expected values of W can reasonably be attributed to sampling error or whether it indicates that the mean weekly food cost is actually less than \$117. To answer that question, we need a table of critical values for W, which we present as Table V in Appendix A. We will discuss that table and then return to complete the hypothesis test. ❑

Using the Wilcoxon Signed-Rank Table

Table V, located in Appendix A, provides critical values for a Wilcoxon signed-rank test. The first column of the table gives the sample size; the second and third columns give the significance levels for a one-tailed and two-tailed test, respectively; and the fourth and fifth columns give the lower and upper critical values, respectively.[†] Note that *a critical value from Table V is to be included as part of the rejection region.*

EXAMPLE 9.20 USING THE WILCOXON SIGNED-RANK TABLE, TABLE V

Refer to Example 9.19 on page 561. Determine the critical value, rejection region, and nonrejection region for the hypothesis test if it is to be performed at the 5% significance level.

SOLUTION The hypothesis test in Example 9.19 is left-tailed and the sample size is 10. To determine the critical value, we first go down the sample-size column of Table V to 10. Next we look for a significance level of 0.05 for a one-tailed test. Finally, we go across the row containing 0.05 to the W_ℓ column, where we find the corresponding critical value, 11, for a left-tailed test.

Thus, for a left-tailed test at the 5% significance level, the critical value is 11. This means that the rejection region consists of all W-values less than or equal to 11, and the nonrejection region consists of all W-values greater than 11; in other words, we reject H_0 if $W \leq 11$ and do not reject H_0 if $W > 11$, as seen in Fig. 9.31. □

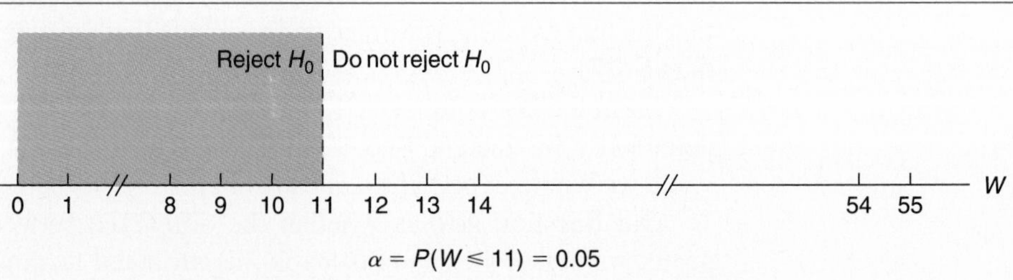

FIGURE 9.31 Rejection and nonrejection regions for a left-tailed Wilcoxon signed-rank test with $\alpha = 0.05$ and $n = 10$

Performing the Wilcoxon Signed-Rank Test

We now present a step-by-step procedure for performing a Wilcoxon signed-rank test. For brevity, we will sometimes use the phrase "symmetric population" to indicate that the variable under consideration has a symmetric distribution.

[†] Actually, the significance levels presented in Table V are only approximate, but they are considered close enough in practice.

PROCEDURE 9.4

THE WILCOXON SIGNED-RANK TEST FOR A POPULATION MEAN

ASSUMPTION

Symmetric population

Step 1 The null hypothesis is H_0: $\mu = \mu_0$ and the alternative hypothesis is one of the following:

H_a: $\mu \neq \mu_0$ (Two-tailed)	or	H_a: $\mu < \mu_0$ (Left-tailed)	or	H_a: $\mu > \mu_0$ (Right-tailed)

Step 2 Decide on the significance level, α.

Step 3 The critical value(s) are

W_ℓ and W_r (Two-tailed)	or	W_ℓ (Left-tailed)	or	W_r (Right-tailed)

Use Table V to find the critical value(s).

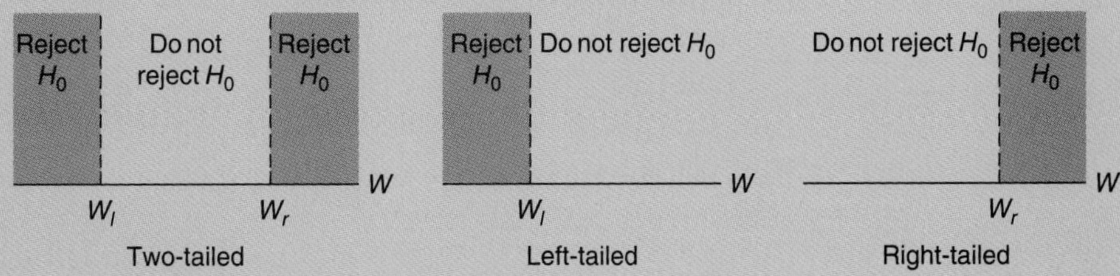

Step 4 Construct a work table of the following form:

Observation x	Difference $D = x - \mu_0$	$\|D\|$	Rank of $\|D\|$	Signed rank R
⋮	⋮	⋮	⋮	⋮

Step 5 Compute the value of the test statistic

$$W = \text{sum of the positive ranks.}$$

Step 6 If the value of the test statistic falls in the rejection region, reject H_0; otherwise, do not reject H_0.

Step 7 State the conclusion in words.

EXAMPLE 9.21 ILLUSTRATES PROCEDURE 9.4

As an application of Procedure 9.4, let's complete the hypothesis test of Example 9.19. A random sample of 10 Kansas families of four yielded the data on 1994 weekly food costs shown in Table 9.16. At the 5% significance level, do the data provide sufficient evidence to conclude that in 1994, the mean weekly food cost for Kansas families of four was less than the national mean of \$117?

TABLE 9.16 Sample of weekly food costs (\$)

103	129	109	95	121
98	112	110	101	119

SOLUTION We apply Procedure 9.4.

Step 1 *State the null and alternative hypotheses.*

Let μ denote the 1994 mean weekly food cost for all Kansas families of four. Then the null and alternative hypotheses are

$$H_0\colon \mu = \$117 \text{ (mean weekly food cost is not less than \$117)}$$
$$H_a\colon \mu < \$117 \text{ (mean weekly food cost is less than \$117).}$$

Note that the hypothesis test is left-tailed since a less-than sign (<) appears in the alternative hypothesis.

Step 2 *Decide on the significance level, α.*

The test is to be performed at the 5% significance level; so $\alpha = 0.05$.

Step 3 *The critical value for a left-tailed test is W_ℓ.*

We have already determined the critical value in Example 9.20. It is $W_\ell = 11$, as seen in Fig. 9.31 on page 564.

Step 4 *Construct a work table.*

We have already done this in Table 9.15 on page 562.

Step 5 *Compute the value of the test statistic*

$$W = \textit{sum of the positive ranks.}$$

The last column of Table 9.15 shows that the sum of the positive ranks equals

$$W = 6 + 2 + 1 = 9.$$

Step 6 *If the value of the test statistic falls in the rejection region, reject H_0; otherwise, do not reject H_0.*

The value of the test statistic is $W = 9$, as found in Step 5. This falls in the rejection region, as we observe from Fig. 9.31. Thus we reject H_0.

Step 7 *State the conclusion in words.*

The test results are statistically significant at the 5% level; that is, at the 5% significance level, the data provide sufficient evidence to conclude that in 1994 the mean weekly food cost for Kansas families of four was less than the national mean of \$117. ◨

As we mentioned earlier, one advantage of nonparametric methods is that they are resistant to outliers. We can illustrate that fact for the Wilcoxon signed-rank test by referring to Example 9.21. The stem-and-leaf diagram in Fig. 9.29 on page 562 shows that the sample data in Table 9.16 contain no outliers. The smallest observation, and also the farthest from the null hypothesis mean of 117, is 95. Replacing 95 by, say, 45, introduces an outlier but has absolutely no effect on the value of the test statistic and hence none on the hypothesis test itself. (Why is this so?)

The following points may be relevant when performing a Wilcoxon signed-rank test. Note them carefully.

- If an observation equals μ_0 (the value for the mean in the null hypothesis), then that observation should be removed and the sample size reduced by 1.
- If two or more absolute differences are tied, each should be assigned the mean of the ranks they would have had if there were no ties. For example, if two absolute differences are tied for second place, each should be assigned rank $(2 + 3)/2 = 2.5$, and rank 4 should be assigned to the next largest absolute difference, which really is fourth. If three absolute differences are tied for fifth place, each should be assigned rank $(5 + 6 + 7)/3 = 6$, and rank 8 should be assigned to the next largest absolute difference.

Because the mean and median of a symmetric distribution are identical, a Wilcoxon signed-rank test can be used to perform a hypothesis test for a population median, η, as well as for a population mean, μ. To employ Procedure 9.4 to carry out a hypothesis test for a population median, simply replace μ by η and μ_0 by η_0. In some of the exercises at the end of this section, you will be asked to use the Wilcoxon signed-rank test to perform hypothesis tests for a population median.

Comparison of the Wilcoxon Signed-Rank Test and the *t*-Test

As we learned in Section 9.6, a t-test can be used to conduct a hypothesis test for a population mean when the variable under consideration is normally distributed. Since a normally distributed variable necessarily has a symmetric distribution, we can also use the Wilcoxon signed-rank test to perform such a hypothesis test.

So now the question is this: If we want to perform a hypothesis test for a population mean and we know the variable under consideration is normally distributed, should we use the t-test or the Wilcoxon signed-rank test? As you might expect, we should use the t-test. For a normally distributed variable, the t-test is more powerful than the Wilcoxon signed-rank test because it is designed expressly for such variables; surprisingly, however, the t-test is not much more powerful than the Wilcoxon signed-rank test.

On the other hand, if the variable under consideration has a symmetric distribution but is not normally distributed, then the Wilcoxon signed-rank test is usually more powerful than the t-test and is often considerably more powerful. In summary, we have the following key fact.

KEY FACT 9.8 WILCOXON SIGNED-RANK TEST VERSUS THE t-TEST

Suppose a hypothesis test is to be performed for a population mean. When deciding between the t-test and the Wilcoxon signed-rank test, follow these guidelines.

- If you are reasonably sure that the variable under consideration is normally distributed, use the t-test.
- If you are not reasonably sure that the variable under consideration is normally distributed but are reasonably sure that it has a symmetric distribution, use the Wilcoxon signed-rank test.

Using the Computer (Optional)

Procedure 9.4 on page 565 provides a step-by-step method for performing a Wilcoxon signed-rank test for one population mean. Alternatively, we can use Minitab to carry out a Wilcoxon signed-rank test.

As we said earlier, a Wilcoxon signed-rank test can be used to perform a hypothesis test for a population median, η, as well as for a population mean, μ. Minitab presents the output of that procedure in terms of the median, but that output can be interpreted in terms of the mean: simply replace references to "median" by "mean." With this in mind, we now consider Example 9.22, which shows how Minitab is applied to perform a Wilcoxon signed-rank test.

EXAMPLE 9.22 USING MINITAB TO PERFORM A WILCOXON SIGNED-RANK TEST

Use Minitab to perform the hypothesis test in Example 9.21.

SOLUTION Let μ denote the 1994 mean weekly food cost for all Kansas families of four. The problem is to use the Wilcoxon signed-rank test to perform the hypothesis test

$$H_0\text{: } \mu = \$117 \text{ (mean weekly food cost is not less than \$117)}$$

$$H_a\text{: } \mu < \$117 \text{ (mean weekly food cost is less than \$117)}$$

at the 5% significance level ($\alpha = 0.05$). Note that the hypothesis test is left-tailed since a less-than sign ($<$) appears in the alternative hypothesis.

First we store the sample data from Table 9.16 on page 566 in a column named FOODCOST. Then we proceed as follows.

1 Choose **Stat ➤ Nonparametrics ➤ 1-Sample Wilcoxon...**
2 Specify FOODCOST in the **Variables** text box
3 Select the **Test median** option button
4 Click in the **Test median** text box and type 117
5 Click the arrow button at the right of the **Alternative** drop-down list box and select **less than**
6 Click **OK**

The output obtained as a result of this procedure is displayed in Printout 9.7.

PRINTOUT 9.7 Minitab output for the Wilcoxon signed-rank test

```
Test of median = 117.0 versus median < 117.0

                  N for   Wilcoxon              Estimated
              N    Test  Statistic        P        Median
FOODCOST     10      10        9.0    0.033         109.5
```

The first line of Printout 9.7 displays the null and alternative hypotheses for the hypothesis test: `Test of median = 117.0 versus median < 117.0`. (Remember, we can replace all references to "median" by "mean.") Below that we find several entries.

The entry headed `N` gives the sample size, which in this case is 10. The next entry, headed `N for Test`, shows the number of observations in the sample that are not equal to the null hypothesis mean of 117. Recall that for a Wilcoxon signed-rank test, if an observation is equal to the null hypothesis mean, then that observation should be removed and the sample size reduced by 1. Since here the sample size is 10 and none of the observations equal 117, the sample size for the test (`N for Test`) is also 10. The next entry, headed `Wilcoxon Statistic`, provides the value of the test statistic, W; so $W = 9$.

The final two entries in Printout 9.7, headed `P` and `Estimated Median`, give the P-value for the hypothesis test and a point estimate for the population median (or mean). Because the P-value of 0.033 is less than the designated significance level of 0.05, we reject H_0. At the 5% significance level, the data provide sufficient evidence to conclude that in 1994, the mean weekly food cost for Kansas families of four was less than the national mean of \$117. ❑

EXERCISES 9.7

STATISTICAL CONCEPTS AND SKILLS

9.124 Technically speaking, what is a *nonparametric method?* In modern statistical practice, how is that term used?

9.125 Discuss advantages and disadvantages of non-parametric methods relative to parametric methods.

9.126 What assumption must be met in order to use the Wilcoxon signed-rank test?

9.127 We mentioned that if in a Wilcoxon signed-rank test an observation equals μ_0 (the value given for the mean in the null hypothesis), then that observation should be removed and the sample size reduced by 1. Why do you think that is done?

9.128 Suppose you want to perform a hypothesis test for a population mean. Assume that the population standard deviation is unknown and the sample size is relatively small. In each part below we have given the distribution shape of the variable under consideration. Decide whether you would use the t-test, the Wilcoxon signed-rank test, or neither.
a. uniform **b.** normal **c.** reverse-J-shaped

9.129 Suppose you want to perform a hypothesis test for a population mean. Assume that the population standard deviation is unknown and the sample size is relatively small. In each part below we have given the distribution shape of the variable under consideration. Decide whether you would use the t-test, the Wilcoxon signed-rank test, or neither.
a. triangular **b.** symmetric bimodal
c. left-skewed

9.130 The Wilcoxon signed-rank test can be used to perform a hypothesis test for a population median, η, as well as for a population mean, μ. Why is that so?

In each of Exercises 9.131–9.134, use the Wilcoxon signed-rank test, Procedure 9.4 on page 565, to perform the required hypothesis test.

9.131 In 1995 the median age of U.S. residents was 34.3 years, as reported by the Census Bureau in *Current Population Reports.* A random sample taken this year of 10 U.S. residents yielded the following ages, in years.

40	60	12	55	34
43	47	37	9	24

At the 5% significance level, do the data provide sufficient evidence to conclude that the median age of today's U.S. residents has increased over the 1995 median age of 34.3 years?

9.132 The Bureau of Labor Statistics publishes information on average annual expenditures by consumers in *Consumer Expenditures.* In 1994 the mean amount spent per consumer unit on nonalcoholic beverages was $233. A random sample of 12 consumer units yielded the following data, in dollars, on last year's expenditures on nonalcoholic beverages.

378	193	201	282	276	298
257	290	276	266	211	275

At the 5% significance level, do the data provide sufficient evidence to conclude that last year's mean amount spent by consumers on nonalcoholic beverages has increased over the 1994 mean of $233?

9.133 A chemist working for a pharmaceutical company has developed a new antacid tablet that she feels will relieve pain more quickly than the company's present tablet. Experience indicates that the present tablet requires an average of 12 minutes to take effect. The chemist records the following times, in minutes, for relief with the new tablet.

10.9	11.4	12.0	8.8	4.4
15.0	7.1	10.1	9.8	14.8
14.2	9.2	9.2	6.6	8.0

Does it appear that the new antacid tablet works faster? Use $\alpha = 0.05$ for your hypothesis test.

9.134 The National Center for Health Statistics reports in *Vital Statistics of the United States* that the median birth weight of U.S. babies was 7.4 lb in 1993.

A random sample of this year's births provided the following weights, in pounds.

8.6	7.4	5.3	13.8	7.8	5.7	9.2
8.8	8.2	9.2	5.6	6.0	11.6	7.2

Can we conclude that this year's median birth weight differs from that in 1993? Use a significance level of 0.05.

9.135 The average retail price for bananas in 1994 was 46.0 cents per pound, as reported by the U.S. Department of Agriculture in *Food Cost Review.* Recently, a random sample of 15 markets gave the following prices for bananas in cents per pound.

51	48	50	48	45
52	53	49	43	42
45	52	52	46	50

a. Can we conclude that the current mean retail price for bananas is different from the 1994 mean of 46.0 cents per pound? Perform a Wilcoxon signed-rank test at the 5% significance level.

b. The hypothesis test considered in part (a) was done previously in Exercise 9.111 using a t-test. The assumption in that exercise is that retail prices for bananas are normally distributed. Assuming that to be the case, why is it permissible to perform a Wilcoxon signed-rank test for the mean retail price of bananas?

9.136 Atlas Fishing Line, Inc., produces a 10-lb test line. Twelve randomly selected spools are subjected to tensile-strength tests. The results follow.

9.8	10.2	9.8	9.4
9.7	9.7	10.1	10.1
9.8	9.6	9.1	9.7

a. Use the data to decide whether Atlas Fishing Line's 10-lb test line is not up to specifications. Perform a Wilcoxon signed-rank test for the mean tensile strength at a significance level of 0.05.

b. The hypothesis test of part (a) was done previously in Exercise 9.112 using a t-test. The assumption in that exercise is that tensile strengths are normally distributed. Assuming that to be the case, why is it permissible to perform a Wilcoxon signed-rank test for the mean tensile strength?

Extending the Concepts and Skills

9.137 A manufacturer of liquid soap produces a bottle with an advertised content of 310 mL. Sixteen bottles are randomly selected and found to have the following contents, in milliliters.

297	318	306	300
311	303	291	298
322	307	312	300
315	296	309	311

A normal probability plot of the data indicates that it is reasonable to assume the contents are normally distributed. Let μ denote the mean content of all bottles produced. To decide whether the mean content is less than advertised, perform the hypothesis test

$$H_0\colon \mu = 310 \text{ mL}$$
$$H_a\colon \mu < 310 \text{ mL}.$$

at the 5% significance level.

a. Use the t-test, Procedure 9.3 on page 550.

b. Use the Wilcoxon signed-rank test.

c. Assuming that the mean content is in fact less than 310 mL, how do you explain the discrepancy between the two tests?

9.138 Twenty years ago, the U.S. Bureau of Justice Statistics reported in *Profile of Jail Inmates* that the median educational attainment of jail inmates was 10.2 years. Ten current inmates are randomly selected and found to have the following educational attainments, in years.

14	10	5	6	8
10	10	8	9	9

Assume that educational attainments of current jail inmates have a symmetric, nonnormal distribution. At the 10% significance level, do the data provide sufficient evidence to conclude that this year's median educational attainment has changed from what it was 20 years ago?

a. Use the t-test, Procedure 9.3 on page 550.

b. Use the Wilcoxon signed-rank test.

c. Presuming this year's median educational attainment has in fact changed from what it was 20 years ago, how do you explain the discrepancy between the two tests?

9.139 A commuter train arrives punctually at a station every half hour. Each morning, a commuter, whom we shall call John, leaves his house and casually strolls to the train station. John thinks that he is unlucky and that he waits longer for the train on the average than he should.

a. Assuming John is not unlucky, how long should he expect to wait for the train on the average?

b. Assuming John is not unlucky, identify the distribution of the times he waits for the train.

c. Here is a sample of the times, in minutes, that John waited for the train.

24	20	3	19	28	22
26	4	11	5	16	24

Use the Wilcoxon signed-rank test to decide, at the 10% significance level, whether the data provide sufficient evidence to conclude that, on the average, John waits more than 15 minutes for the train.

d. Explain why the Wilcoxon signed-rank test is appropriate here.

e. Is the Wilcoxon signed-rank test more appropriate here than the t-test? Explain your answer.

Wilcoxon signed-rank test using a normal approximation. The table of critical values for the Wilcoxon signed-rank test, Table V, stops at $n = 20$. For larger samples a normal approximation can be used. In fact, the normal approximation works well even for sample sizes as small as 10.

NORMAL APPROXIMATION FOR W Suppose the variable under consideration has a symmetric distribution. Then, for samples of size n,

- $\mu_W = n(n+1)/4$.
- $\sigma_W = \sqrt{n(n+1)(2n+1)/24}$.
- W is approximately normally distributed for $n \geq 10$.

Thus, for samples of size 10 or more, the standardized variable

$$z = \frac{W - n(n+1)/4}{\sqrt{n(n+1)(2n+1)/24}}$$

has approximately the standard normal distribution.

In Exercises 9.140–9.142, we will develop and apply a "large-sample" procedure for a Wilcoxon signed-rank test based on the fact in the box above.

9.140 Formulate a hypothesis-testing procedure for a Wilcoxon signed-rank test that uses the test statistic z given in the box above.

9.141 Refer to Exercise 9.133.

a. Use the procedure you formulated in Exercise 9.140 to perform the hypothesis test in Exercise 9.133.

b. Compare your result in part (a) to the one you obtained in Exercise 9.133, where the normal approximation was not used.

9.142 Refer to Exercise 9.134.

a. Use the procedure you formulated in Exercise 9.140 to perform the hypothesis test in Exercise 9.134.

b. Compare your result in part (a) to the one you obtained in Exercise 9.134, where the normal approximation was not used.

9.143 In this exercise we will obtain the distribution of the variable W when $n = 3$. This will enable you to see how the critical values for the Wilcoxon signed-rank test are derived.

a. The rows of the following table give all possible signs for the signed ranks in a Wilcoxon signed-rank test with $n = 3$. For example, the first row covers the possibility that all three observations are greater than μ_0 and thus have positive sign ranks. There is an empty column for values of W. Fill it in. *(Hint:* The first entry is 6 and the last is 0.*)*

Rank			
1	2	3	W
+	+	+	
+	+	−	
+	−	+	
+	−	−	
−	+	+	
−	+	−	
−	−	+	
−	−	−	

b. If the null hypothesis, H_0: $\mu = \mu_0$, is true, what percentages of samples will match any particular row of the table? *(Hint:* The answer is the same for all rows.*)*

c. Use the answer from part (b) to obtain the distribution of W for samples of size three.
d. Draw a relative-frequency histogram of the distribution obtained in part (c).
e. Use part (d) to find the critical value for a left-tailed Wilcoxon test with $n = 3$ and $\alpha = 0.125$.

9.144 Repeat Exercise 9.143 with $n = 4$.

USING TECHNOLOGY  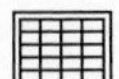 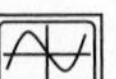

9.145 Refer to Exercise 9.131. Use Minitab or some other statistical software to perform the required hypothesis test using the Wilcoxon signed-rank test.

9.146 Refer to Exercise 9.132. Use Minitab or some other statistical software to perform the required hypothesis test using the Wilcoxon signed-rank test.

9.147 The National Center for Health Statistics publishes data on the duration of marriages in *Vital Statistics of the United States.* In 1995 the median duration of a marriage was 7.2 years. A random sample was taken of last year's divorce certificates and the marriage durations recorded. Minitab's Wilcoxon signed-rank procedure, applied to the resulting data, yielded the output in Printout 9.8. Use the printout to determine
a. the null and alternative hypotheses.
b. the number of divorce certificates sampled.
c. the number of certificates in the sample for which the marriage lasted exactly 7.2 years.
d. a point estimate for last year's median marriage duration.
e. the P-value of the hypothesis test.
f. the smallest significance level at which the null hypothesis can be rejected.
g. the conclusion if the hypothesis test is performed at the 5% significance level.
h. In this case, do you think it is appropriate to use the Wilcoxon signed-rank test? Why?

9.148 The Census Bureau estimates that the *U.S. Census Form* takes the average household 14 minutes to complete. The time it takes each household in a sample to complete the form is recorded. Printout 9.9 shows the output resulting from applying Minitab's Wilcoxon signed-rank procedure to the completion-time data. Use the printout to obtain
a. the null and alternative hypotheses.
b. the number of households sampled.
c. the number of households in the sample that took exactly 14 minutes to complete the form.
d. the estimated population median completion time.
e. the P-value of the hypothesis test.
f. the smallest significance level at which the null hypothesis can be rejected.
g. the conclusion if the hypothesis test is performed at the 10% significance level.
h. In this case, do you think it is appropriate to use the Wilcoxon signed-rank test? Why?

PRINTOUT 9.8 Minitab output for Exercise 9.147

```
Test of median = 7.200 versus median < 7.200

              N for   Wilcoxon              Estimated
          N    Test  Statistic        P        Median
YEARS    50      50      624.0    0.450         6.882
```

PRINTOUT 9.9 Minitab output for Exercise 9.148

```
Test of median = 14.00 versus median not = 14.00

              N for   Wilcoxon              Estimated
          N    Test  Statistic        P        Median
MINUTES  36      36      462.0    0.044         15.05
```

9.8 WHICH PROCEDURE SHOULD BE USED?*

In this chapter we have learned three procedures for performing a hypothesis test for one population mean: the z-test, the t-test, and the Wilcoxon signed-rank test.

The z-test and t-test are designed to be used when the variable under consideration has a normal distribution; in such cases, the z-test applies when the population standard deviation is known and the t-test applies when the population standard deviation is unknown. As we have seen, both tests are approximately correct when the sample size is large, regardless of the distribution of the variable under consideration. We have also discovered that these two tests should be used cautiously when outliers are present. Key Fact 9.4 on page 514 provides detailed guidelines for use of the z-test and t-test.

The Wilcoxon signed-rank test is designed to be used when the variable under consideration has a symmetric distribution. Unlike the z-test and t-test, the Wilcoxon signed-rank test is resistant to outliers.

We summarize the three procedures in Table 9.17. Each row of the table gives the type of test, the conditions required for using the test, the test statistic, and the procedure to use. Note that we have employed the abbreviations "normal population" for "the variable under consideration is normally distributed," "W-test" for "Wilcoxon signed-rank test," and "symmetric population" for "the variable under consideration has a symmetric distribution."

TABLE 9.17 Summary of hypothesis-testing procedures for one population mean, μ. The null hypothesis for all tests is H_0: $\mu = \mu_0$

Type	Assumptions	Test statistic	Procedure to use
z-test	1. Normal population or large sample 2. σ known	$z = \dfrac{\bar{x} - \mu_0}{\sigma/\sqrt{n}}$	9.1 (page 513)
t-test	1. Normal population or large sample 2. σ unknown	$t = \dfrac{\bar{x} - \mu_0}{s/\sqrt{n}}$ $(\text{df} = n - 1)$	9.3 (page 550)
W-test	Symmetric population	W = sum of positive ranks	9.4 (page 565)

In selecting the correct procedure, keep in mind that the best choice is the procedure expressly designed for the distribution-type under consideration, if such

* This section is optional and will not be needed in subsequent chapters of the book. All previous sections in this chapter, including the optional material on the Wilcoxon signed-rank test, are prerequisite to this section.

a procedure exists, and that the z-test and t-test are only approximately correct for large samples from nonnormal populations.

For instance, suppose that the variable under consideration is normally distributed and that the population standard deviation is known. Then both the z-test and Wilcoxon signed-rank test apply. The z-test applies because the variable under consideration is normally distributed and σ is known; the W-test applies because a normal distribution is symmetric. But the correct procedure is the z-test since that test is designed specifically for variables having a normal distribution.

The flowchart in Fig. 9.32 provides an organized strategy for choosing the correct hypothesis-testing procedure for a population mean. It has been constructed based on the above discussion.

FIGURE 9.32 Flowchart for choosing the correct hypothesis-testing procedure for one population mean

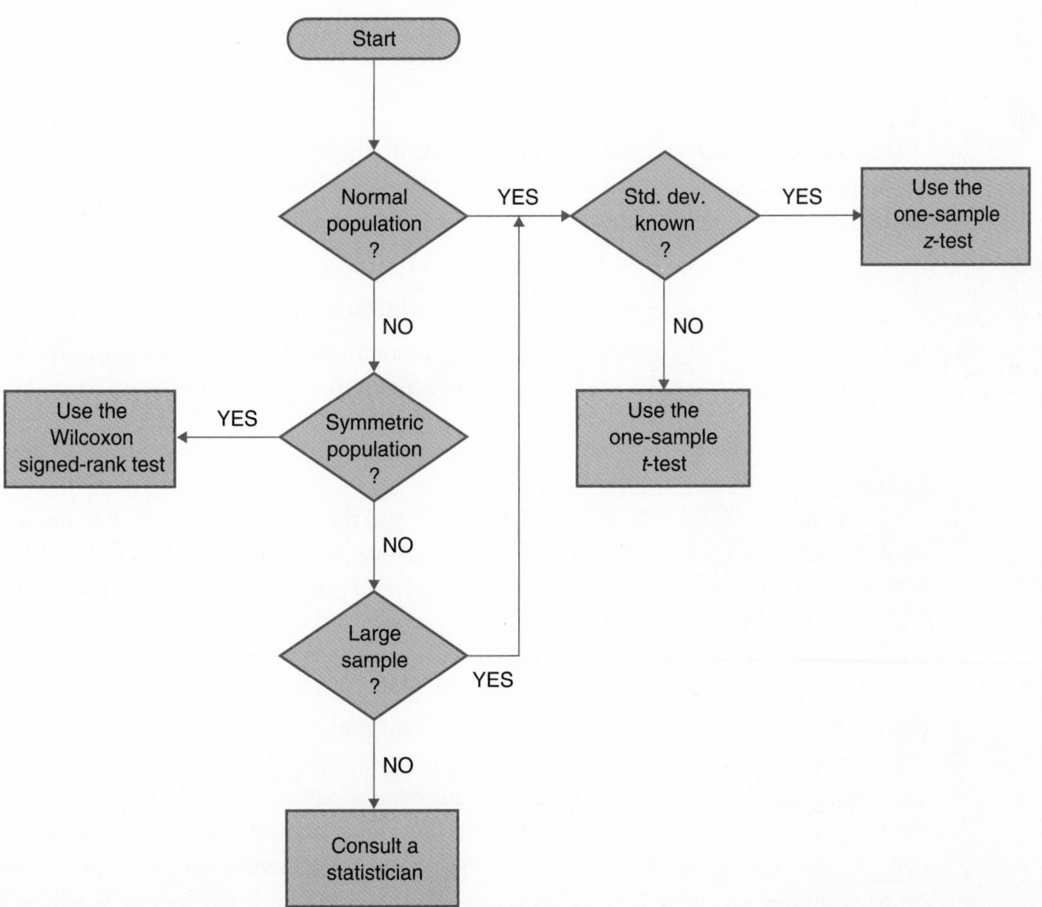

In practice we need to look at the sample data to ascertain distribution-type before we can select the appropriate procedure. We recommend using a normal probability plot and either a stem-and-leaf diagram (for small or moderate-size samples) or a histogram (for moderate-size or large samples).

EXAMPLE 9.23 CHOOSING THE CORRECT HYPOTHESIS-TESTING PROCEDURE

According to *Consumer Expenditure Survey,* published by the U.S. Bureau of Labor Statistics, the 1994 mean expenditure for entertainment was \$1567 per consumer unit. For that same year, a sample of 50 consumer units in the West yielded the annual expenditures for entertainment shown in Table 9.18.

TABLE 9.18 1994 entertainment expenditures (\$) for a sample of consumer units in the West

1568	2345	1834	2857	3023	2028	1626	2283	2238	1686
1732	489	1284	1057	1295	1924	1705	2061	2056	1853
2025	2371	3713	1797	1559	1289	455	1949	1816	2232
1581	1216	2054	2285	1819	2146	1850	617	2321	2428
1752	1682	1405	1488	1806	1901	1917	1797	1987	2104

Suppose we want to use the sample data in Table 9.18 to decide whether the 1994 mean entertainment expenditure by consumer units in the West exceeded that of the nation as a whole. Then we want to perform the hypothesis test

$$H_0\text{: } \mu = \$1567 \text{ (mean expenditure is not greater)}$$

$$H_a\text{: } \mu > \$1567 \text{ (mean expenditure is greater),}$$

where μ is the 1994 mean entertainment expenditure by consumer units in the West. Which procedure should be used to perform the hypothesis test?

SOLUTION We begin by drawing a normal probability plot and histogram of the sample data in Table 9.18, as seen in Fig. 9.33.

Next we consult the flowchart in Fig. 9.32 and the graphs in Fig. 9.33 to decide which procedure should be used. The first question we must answer in Fig. 9.32 is whether the variable under consideration is normally distributed. The normal probability plot in Fig. 9.33(a) shows some curvature and/or the presence of outliers; so the answer to the first question is "No."

This leads us to the next question in Fig. 9.32: Does the variable under consideration have a symmetric distribution? The histogram in Fig. 9.33(b) suggests that it is reasonable to presume the answer to that question is "Yes."

The "Yes" answer to the preceding question leads us to the box in Fig. 9.32 that states: Use the Wilcoxon signed-rank test. Thus an appropriate procedure for carrying out the hypothesis test is the Wilcoxon signed-rank test, Procedure 9.4 on page 565. ◨

FIGURE 9.33
(a) Normal probability plot and (b) histogram of expenditure data in Table 9.18

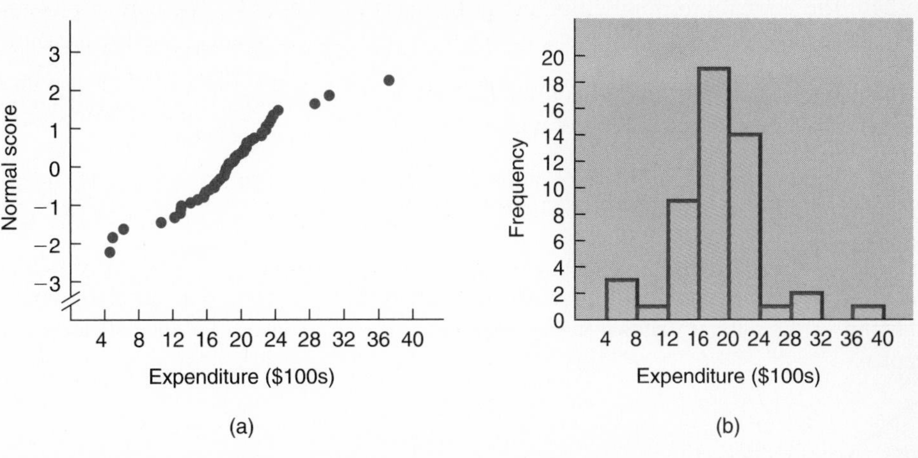

EXERCISES 9.8

Statistical Concepts and Skills

9.149 In this chapter, we have considered three procedures for conducting a hypothesis test for one population mean.

a. Identify the three procedures by name.
b. List the assumptions for using each procedure.
c. Identify the test statistic for each procedure.

9.150 Suppose we want to perform a hypothesis test for a population mean. Assume the variable under consideration is normally distributed and the population standard deviation is unknown.

a. Is it permissible to use the t-test to perform the hypothesis test? Explain your answer.
b. Is it permissible to use the Wilcoxon signed-rank test to perform the hypothesis test? Explain your answer.
c. Which procedure is preferable, the t-test or the Wilcoxon signed-rank test? Explain your answer.

9.151 Suppose we want to perform a hypothesis test for a population mean. Assume the variable under consideration has a symmetric nonnormal distribution and the population standard deviation is unknown. Further assume that the sample size is large and no outliers are present in the sample data.

a. Is it permissible to use the t-test to perform the hypothesis test? Explain your answer.
b. Is it permissible to use the Wilcoxon signed-rank test to perform the hypothesis test? Explain your answer.
c. Which procedure is preferable, the t-test or the Wilcoxon signed-rank test? Explain your answer.

9.152 Suppose we want to perform a hypothesis test for a population mean. Assume the variable under consideration has a highly-skewed distribution and the population standard deviation is known. Further assume that the sample size is large and no outliers are present in the sample data.

a. Is it permissible to use the z-test to perform the hypothesis test? Explain your answer.
b. Is it permissible to use the Wilcoxon signed-rank test to perform the hypothesis test? Explain your answer.

In each of Exercises 9.153–9.160, we have provided a normal probability plot and either a stem-and-leaf diagram or a frequency histogram for a set of sample data. The intent is to employ the sample data to perform a hypothesis test for the mean of the population from which the data were obtained. In each case consult the graphs provided and the flowchart in Fig. 9.32 to decide which procedure should be used.

9.153 The normal probability plot and stem-and-leaf diagram of the data are shown in Fig. 9.34 on page 578. σ is known.

9.154 The normal probability plot and histogram of the data are shown in Fig. 9.35. σ is known.

9.155 The normal probability plot and histogram of the data are shown in Fig. 9.36. σ is unknown.

9.156 The normal probability plot and stem-and-leaf diagram of the data are shown in Fig. 9.37. σ is unknown.

9.157 The normal probability plot and stem-and-leaf diagram of the data are shown in Fig. 9.38. σ is unknown.

9.158 The normal probability plot and stem-and-leaf diagram of the data are shown in Fig. 9.39 on page 580. σ is unknown. *(Note:* The decimal parts of the observations were removed before the stem-and-leaf diagram was constructed.*)*

9.159 The normal probability plot and stem-and-leaf diagram of the data are shown in Fig. 9.40 on page 580. σ is known.

9.160 The normal probability plot and stem-and-leaf diagram of the data are shown in Fig. 9.41 on page 580. σ is known.

FIGURE 9.34 Normal probability plot and stem-and-leaf diagram for Exercise 9.153

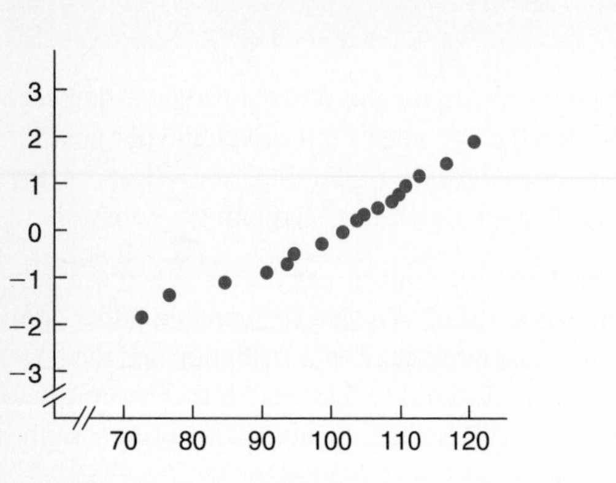

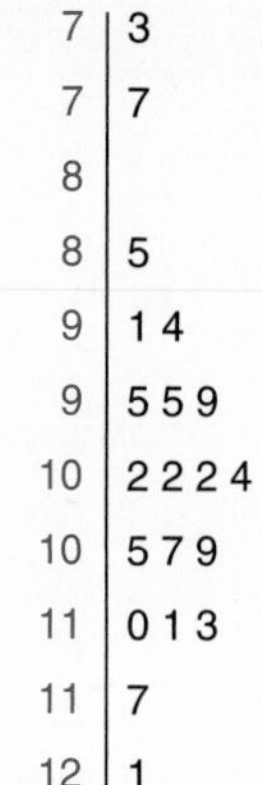

FIGURE 9.35 Normal probability plot and histogram for Exercise 9.154

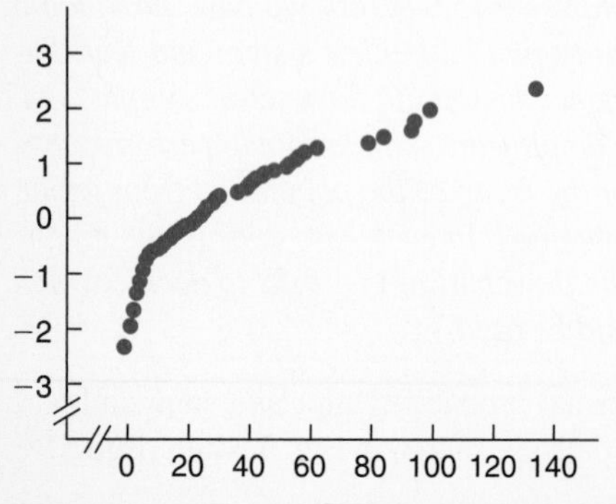

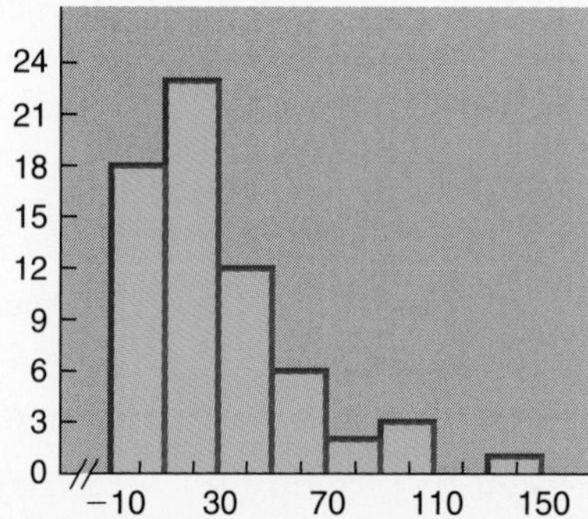

FIGURE 9.36 Normal probability plot and histogram for Exercise 9.155

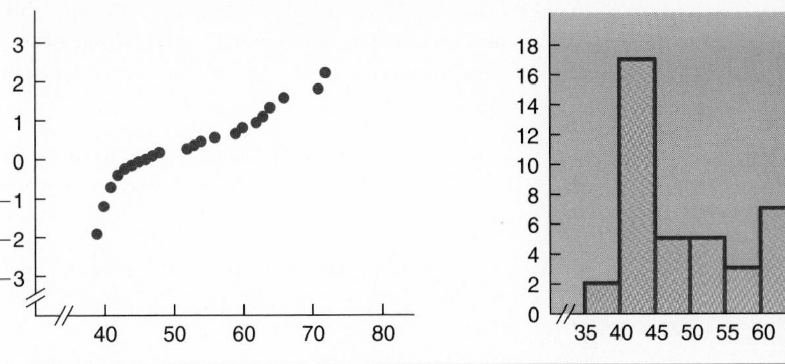

FIGURE 9.37 Normal probability plot and stem-and-leaf diagram for Exercise 9.156

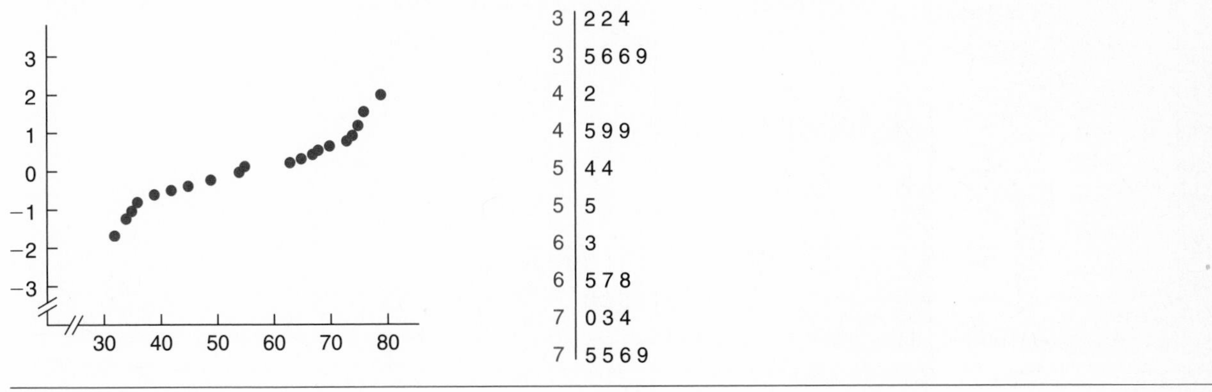

FIGURE 9.38 Normal probability plot and stem-and-leaf diagram for Exercise 9.157

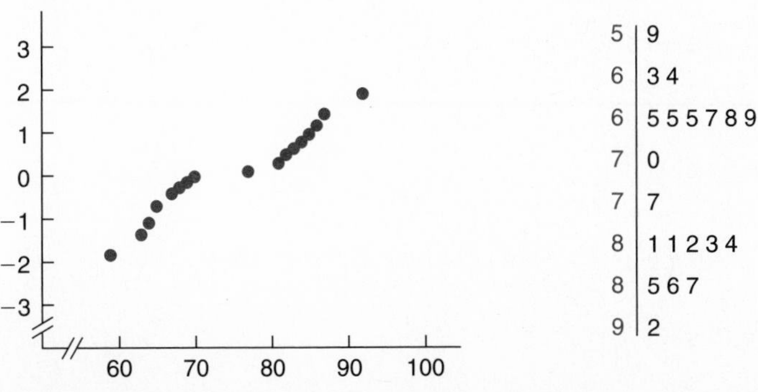


```
5 | 9
6 | 3 4
6 | 5 5 5 7 8 9
7 | 0
7 | 7
8 | 1 1 2 3 4
8 | 5 6 7
9 | 2
```

FIGURE 9.39 Normal probability plot and stem-and-leaf diagram for Exercise 9.158

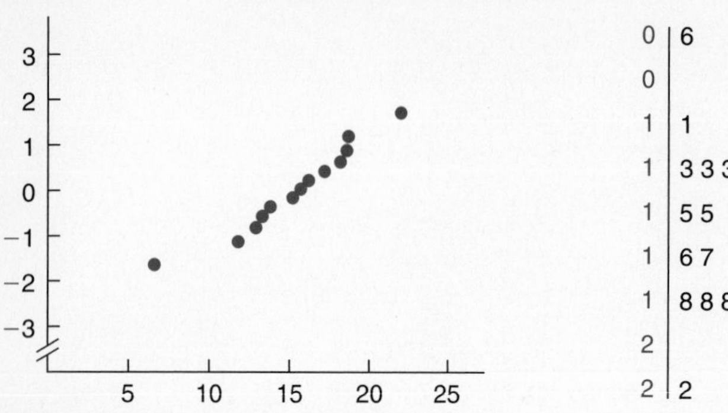

FIGURE 9.40 Normal probability plot and stem-and-leaf diagram for Exercise 9.159

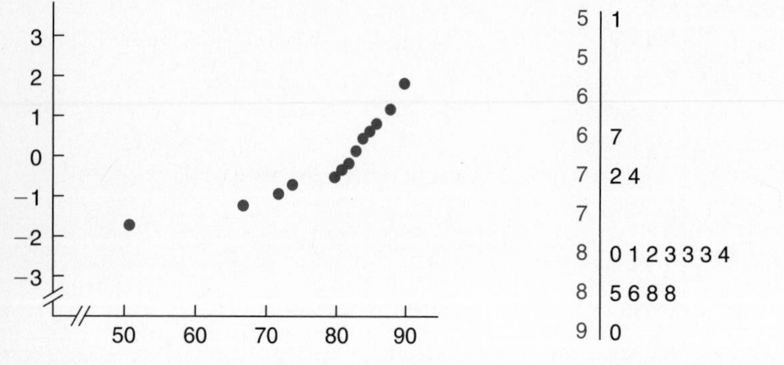

FIGURE 9.41 Normal probability plot and stem-and-leaf diagram for Exercise 9.160

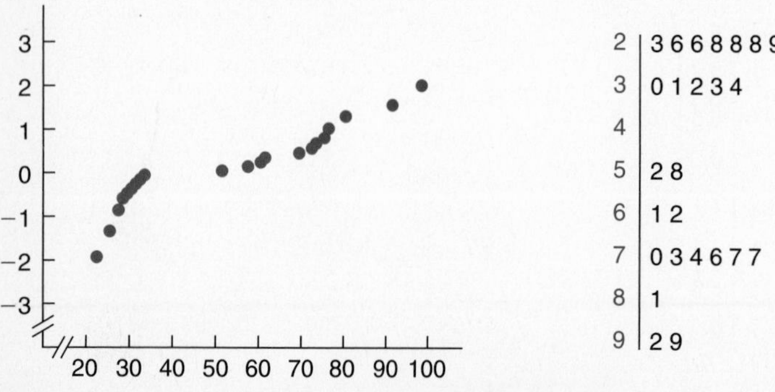

CHAPTER REVIEW

You Should Be Able To

1. use and understand the formulas presented in this chapter.
2. define the terms associated with hypothesis testing.
3. choose the null and alternative hypotheses for a hypothesis test.
4. explain the logic behind hypothesis testing.
5. identify the test statistic, rejection region, nonrejection region, and critical value(s) for a hypothesis test.
6. define and apply the concepts of Type I and Type II errors.
7. state and interpret the possible conclusions for a hypothesis test.
8. obtain the critical value(s) for a specified significance level.
9. perform a hypothesis test for a population mean when the population standard deviation is known.

*10. compute Type II error probabilities.
*11. calculate the power of a hypothesis test.
*12. draw a power curve.
*13. understand the relationship between sample size, significance level, and power.
14. obtain the P-value of a hypothesis test.
15. state and apply the steps for performing a hypothesis test using the critical-value approach to hypothesis testing.
16. state and apply the steps for performing a hypothesis test using the P-value approach to hypothesis testing.
17. perform a hypothesis test for a population mean when the population standard deviation is unknown.

*18. perform a hypothesis test for a population mean when the variable under consideration has a symmetric distribution.
*19. decide which procedure should be used to perform a hypothesis test for a population mean.
*20. use the Minitab procedures covered in this chapter.
*21. interpret the output obtained from the application of the Minitab procedures discussed in this chapter.

Key Terms

acceptance region, *500*
alternative hypothesis, *490*
critical-value approach to hypothesis testing, *535*
critical values, *501*
hypothesis, *490*
hypothesis test, *490*
left-tailed test, *491*
nonparametric methods,* *560*
nonrejection region, *501*
not statistically significant, *506*
null hypothesis, *490*
observed significance level, *536*
one-sample t-test, *550*
one-sample Wilcoxon signed-rank test,* *565*
one-sample z-test, *513, 541*
one-tailed test, *491*
P-value (P), *536*
P-value approach to hypothesis testing, *535*
parametric methods,* *560*
power,* *529*
power curve,* *530*

REVIEW TEST

Statistical Concepts and Skills

1. Explain the meaning of each of the following terms.
a. null hypothesis **b.** alternative hypothesis
c. test statistic **d.** rejection region
e. nonrejection region **f.** critical value(s)

2. The following statement appeared on a box of Tide® laundry detergent: "Individual packages of Tide may weigh slightly more or less than the marked weight due to normal variations incurred with high speed packaging machines, but each day's production of Tide will average slightly above the marked weight."
a. Explain in statistical terms what the statement means.
b. Describe in words a hypothesis test for checking the statement.
c. Suppose that the marked weight is 76 oz. State in words the null and alternative hypotheses for the hypothesis test. Then express those hypotheses using statistical terminology.

3. Regarding a hypothesis test:
a. Roughly speaking, what is the procedure for deciding whether the null hypothesis should be rejected?
b. How is this procedure made objective?

4. There are three possibilities for the alternative hypothesis in a hypothesis test for a population mean. Identify the three possibilities and explain when each is used.

5. There are two types of incorrect decisions that can be made in a hypothesis test, a Type I error and a Type II error.
a. Explain the meaning of each type of error.
b. Identify the letter used to represent the probability of each type of error.
c. If the null hypothesis is in fact true, then only one type of error is possible. Which type is that? Explain your answer.
d. If we fail to reject the null hypothesis, then only one type of error is possible. Which type is that? Explain your answer.

6. Suppose we want to conduct a right-tailed hypothesis test at the 5% significance level. How must the critical value be chosen?

7. In each part below, we have identified a hypothesis testing procedure for a population mean. State the assumptions required and the test statistic used in each case.
a. one-sample t-test
b. one-sample z-test
c. one-sample Wilcoxon signed-rank test

8. What is meant by saying that a hypothesis test is
a. exact? **b.** approximately correct?

9. Discuss the difference between statistical significance and practical significance.

10. For a fixed sample size, what happens to the probability of a Type II error if the significance level is decreased from 0.05 to 0.01?

***11.** Regarding the power of a hypothesis test:
a. What does it represent?
b. What happens to the power of a hypothesis test if the significance level is kept at 0.01 while the sample size is increased from 50 to 100?

12. Regarding the P-value of a hypothesis test:
a. What is the P-value of a hypothesis test?
b. Answer true or false: A P-value of 0.02 provides more evidence against the null hypothesis than a P-value of 0.03. Explain your answer.
c. Answer true or false: A P-value of 0.74 provides essentially no evidence against the null hypothesis. Explain your answer.
d. Explain why the P-value of a hypothesis test is also referred to as the observed significance level.

13. Discuss the differences between the critical-value and P-value approaches to hypothesis testing.

***14.** Identify two advantages of nonparametric methods over parametric methods. When is a parametric procedure preferred? Explain your answer.

15. The U.S. Department of Agriculture reports in *Food Consumption, Prices, and Expenditures* that the average American consumed 27.3 lb of cheese in 1995. Cheese consumption has increased steadily since 1960, when the average American ate only 8.3 lb of cheese annually. Suppose we want to decide whether last year's mean cheese consumption is greater than the 1995 mean.
a. Identify the null hypothesis.
b. Identify the alternative hypothesis.
c. Classify the hypothesis test as two-tailed, left-tailed, or right-tailed.

16. The graph at the top of the next column portrays the decision criterion for a hypothesis test concerning a population mean, μ. The null hypothesis for the test is H_0: $\mu = \mu_0$, and the test statistic is

$$z = \frac{\bar{x} - \mu_0}{\sigma/\sqrt{n}}.$$

The curve in the graph shows the implications of the decision criterion if in fact the null hypothesis is true.

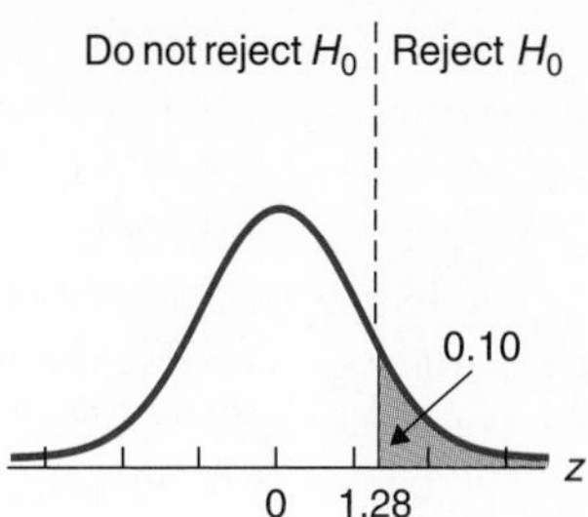

Determine the
a. rejection region.
b. nonrejection region.
c. critical value(s).
d. significance level.
e. Draw a graph that depicts your answers from parts (a)–(d).
f. Classify the hypothesis test as two-tailed, left-tailed, or right-tailed.

17. The null and alternative hypotheses for the hypothesis test in Problem 15 are

$$H_0\text{: } \mu = 27.3 \text{ lb (mean has not increased)}$$
$$H_a\text{: } \mu > 27.3 \text{ lb (mean has increased)},$$

where μ is last year's mean cheese consumption for all Americans. Explain what each of the following would mean.
a. Type I error **b.** Type II error
c. Correct decision

Now suppose the results of carrying out the hypothesis test lead to nonrejection of the null hypothesis. Classify that decision by error type or as a correct decision if in fact last year's mean cheese consumption
d. has not increased over the 1995 mean of 27.3 lb.
e. has increased over the 1995 mean of 27.3 lb.

***18.** Refer to Problem 15. Suppose we decide to use a significance level of 0.10 and a sample size of 35. Assume $\sigma = 6.9$ lb.
a. Determine the probability of a Type I error.
b. Assuming last year's mean cheese consumption is 27.5 lb, identify the distribution of the variable $\bar{x}$, that is, the sampling distribution of the mean for samples of size 35.

c. Use part (b) to determine the probability, β, of a Type II error if in fact last year's mean cheese consumption is 27.5 lb. *(Note:* Do not use the rounded value for $\sigma_{\bar{x}}$ obtained in part (b); rather, use $6.9/\sqrt{35}$ in your calculations.*)*

d. Repeat parts (b) and (c) if in fact last year's mean cheese consumption is 28.0 lb; 28.5 lb; 29.0 lb; 29.5 lb; 30.0 lb; 30.5 lb; 31.0 lb. 31.5 lb; 32.0 lb.

e. Use your answers from parts (c) and (d) to construct a table similar to Table 9.8 on page 529 of selected Type II error probabilities and powers.

f. Use your answer from part (e) to construct the power curve.

Using a sample size of 100 instead of 35,

g. repeat part (b). **h.** repeat part (c).

i. repeat part (d). **j.** repeat part (e).

k. repeat part (f).

l. Compare your power curves for the two sample sizes and explain the principle being illustrated.

19. Refer to Problem 15. The following table provides last year's cheese consumptions, in pounds, for 35 randomly selected Americans.

39	22	26	31	35	29	33
27	26	25	29	15	21	40
19	35	29	37	25	38	17
32	21	29	26	37	26	19
30	22	20	24	29	30	30

a. At the 10% significance level, do the data provide sufficient evidence to conclude that last year's mean cheese consumption for all Americans has increased over the 1995 mean? Assume $\sigma = 6.9$ lb. For your hypothesis test, use the critical-value approach. *(Note:* The sum of the data is 973 lb.*)*

b. Given the conclusion in part (a), if an error has been made, what type must it be? Explain your answer.

20. Refer to Problem 19.

a. Repeat the hypothesis test using the P-value approach to hypothesis testing.

b. Use Table 9.12 on page 543 to assess the strength of the evidence against the null hypothesis.

21. According to *Crime in the United States,* a publication of the FBI, the mean value lost due to purse snatching was \$279 in 1994. For this year, 41 randomly selected purse-snatching offenses have a mean value lost of \$260 with a standard deviation of \$84. Do the data provide sufficient evidence to conclude that the mean value lost due to purse snatching has decreased from the 1994 mean? Use $\alpha = 0.05$. *(Note:* Graphical analyses on the sample data reveal no outliers.*)*

22. Refer to Problem 21.

a. Determine the approximate P-value of the hypothesis test.

b. Perform the hypothesis test using the P-value approach to hypothesis testing.

c. Use Table 9.12 on page 543 to assess the strength of the evidence against the null hypothesis.

23. Each year, manufacturers perform mileage tests on new car models and submit the results to the Environmental Protection Agency (EPA). The EPA then tests the vehicles to determine whether the manufacturers' claims are correct. In 1998, one company reported that a particular model equipped with a four-speed manual transmission averaged 29 mpg on the highway. Suppose the EPA tested 15 of the cars and obtained the following gas mileages.

27.3	31.2	29.4	31.6	28.6
30.9	29.7	28.5	27.8	27.3
25.9	28.8	28.9	27.8	27.6

A normal probability plot of the data shows no outliers and is roughly linear. What decision would you make regarding the company's report on the gas mileage of the car? Perform the required hypothesis test at the 5% significance level.

a. Use the critical-value approach to hypothesis testing. *(Note:* $\bar{x} = 28.753$, $s = 1.595$.*)*

b. Use the P-value approach to hypothesis testing.

c. Use Table 9.12 on page 543 to assess the strength of the evidence against the null hypothesis.

***24.** Refer to Problem 23.

a. Perform the required hypothesis test using the Wilcoxon signed-rank test.

b. In performing the hypothesis test of part (a), what assumption are you making about the distribution of gas mileages for this model?

c. In Problem 23 we performed the hypothesis test in part (a) using the t-test. The assumption in that problem is that gas mileages for this particular model are normally distributed. Assuming that, in fact, the gas mileages are normally distributed, why is it permissible to perform a Wilcoxon signed-rank test for the mean gas mileage?

***25.** Refer to Problems 23 and 24. If in fact the gas mileages are normally distributed, which is the preferred procedure for performing the hypothesis test—the t-test or the Wilcoxon signed-rank test? Explain your answer.

Problems 26 and 27 each include a normal probability plot and either a frequency histogram or a stem-and-leaf diagram for a set of sample data. The intent is to use the sample data to perform a hypothesis test for the mean of the population from which the data were obtained. In each case consult the graphs provided to decide whether it would be best to use the z-test, the t-test, or neither. Explain your answer.

26. The normal probability plot and histogram of the data are depicted in Fig. 9.42 on the following page. σ is known.

27. The normal probability plot and stem-and-leaf diagram of the data are depicted in Fig. 9.43 on the next page. σ is unknown.

***28.** Refer to Problems 26 and 27.

a. In each case consult the appropriate graphs to decide whether it is reasonable to use the Wilcoxon signed-rank test to perform a hypothesis test for the mean of the population from which the data were obtained. Give reasons for your answers.

b. For each case where it is reasonable to use either the z-test or the t-test and where the Wilcoxon signed-rank test is also appropriate, decide which test is preferable. Give reasons for your answers.

USING TECHNOLOGY  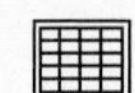 

29. Use Minitab or some other statistical software to carry out the hypothesis test considered in part (a) of Problem 20.

30. The output shown in Printout 9.10 on page 587 was obtained by applying Minitab's z-test procedure to the data in Problem 19. Using the printout, determine

a. the null and alternative hypotheses.

b. the (assumed) population standard deviation of last year's cheese consumptions.

c. the sample standard deviation of the cheese consumptions for the people in the sample.

d. the number of people in the sample.

e. last year's mean cheese consumption for the people in the sample.

f. the value of the test statistic, z.

g. the P-value of the hypothesis test.

h. the smallest significance level at which the null hypothesis can be rejected.

i. the conclusion if the hypothesis test is performed at the 10% significance level.

31. Refer to Problem 23. Use Minitab or some other statistical software to

a. obtain a normal probability plot of the data.

b. perform the required hypothesis test.

c. Justify the use of your procedure in part (b).

32. The output in Printout 9.11 on page 587 was obtained by applying Minitab's t-test procedure to the data in Problem 23. Use the printout to determine

a. the null and alternative hypotheses.

b. the standard deviation of the mileages for the cars tested.

c. the number of cars tested.

d. the mean mileage of the cars tested.

e. the value of the test statistic, t.

f. the P-value of the hypothesis test.

g. the smallest significance level at which the null hypothesis can be rejected.

h. the conclusion if the hypothesis test is performed at the 5% significance level.

***33.** Use Minitab or some other statistical software to carry out the hypothesis test considered in part (a) of Problem 24.

***34.** The output shown in Printout 9.12 was obtained by applying Minitab's Wilcoxon signed-rank test procedure to the data in Problem 23. Use the output to determine

a. the null and alternative hypotheses.

b. the number of cars tested.

c. the number of cars tested whose gas mileages exactly equal 29 mpg.

d. the value of the test statistic, W.

e. the P-value of the hypothesis test.

f. the smallest significance level at which the null hypothesis can be rejected.

g. the conclusion if the hypothesis test is performed at the 5% significance level.

h. a point estimate for the median gas mileage of all cars of this particular model.

FIGURE 9.42 Normal probability plot and histogram for Problem 26

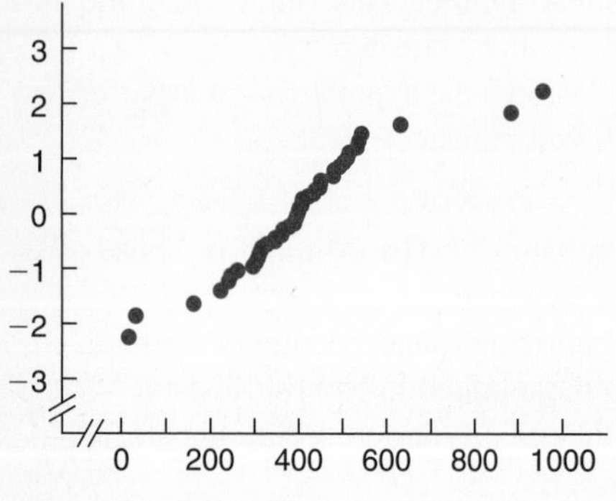

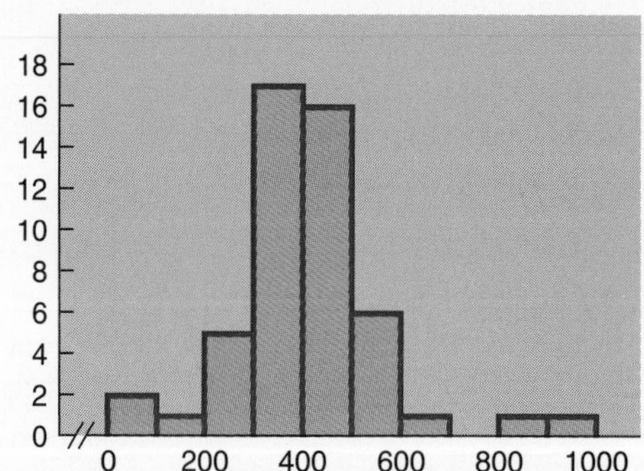

FIGURE 9.43 Normal probability plot and stem-and-leaf diagram for Problem 27

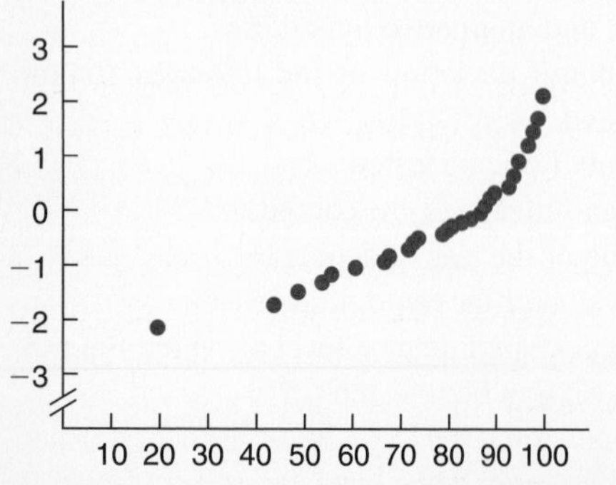

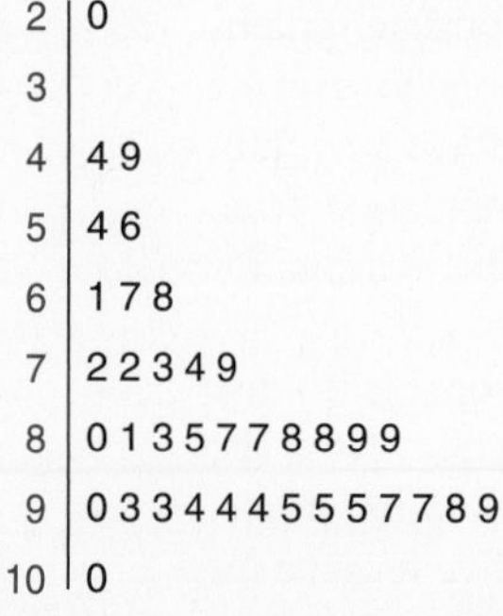

Stem	Leaves
2	0
3	
4	4 9
5	4 6
6	1 7 8
7	2 2 3 4 9
8	0 1 3 5 7 7 8 8 9 9
9	0 3 3 4 4 4 5 5 5 7 7 8 9
10	0

PRINTOUT 9.10 Minitab output for Problem 30

```
Test of mu = 27.30 vs mu > 27.30
The assumed sigma = 6.90

Variable      N       Mean     StDev   SE Mean          Z          P
CHEESE       35      27.80      6.48      1.17       0.43       0.33
```

PRINTOUT 9.11 Minitab output for Problem 32

```
Test of mu = 29.000 vs mu not = 29.000

Variable      N       Mean     StDev   SE Mean          T          P
MPG          15     28.753     1.595     0.412      -0.60       0.56
```

PRINTOUT 9.12 Minitab output for Problem 34

```
Test of median = 29.00 versus median not = 29.00

             N for   Wilcoxon              Estimated
         N    Test  Statistic        P        Median
MPG     15      15       48.5    0.532         28.65
```

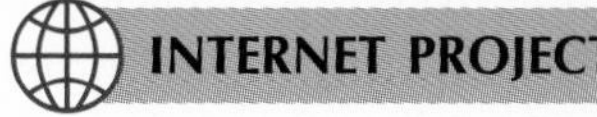

The Ozone Hole

Ozone is a molecule made up of three atoms of oxygen. It plays many roles in the ecosystem, but one of the most important for every life form on the Earth is that it protects the Earth from the harmful effects of the Sun's ultraviolet light.

For years, humans have used chemicals that only recently have been discovered to be harmful to the protective ozone layer. We are beginning to see what the possible effects of that damage may be. For example, the depletion of ozone in the stratosphere may increase the rate of skin cancer and cataracts, and it can harm crops and ocean life.

In this Internet project, you will explore data collected worldwide on the decrease of ozone in the Earth's atmosphere. You will also have an opportunity to investigate the possible consequences of ozone depletion: the amount of solar ultraviolet light reaching the Earth's surface and the effects it may have on humans.

URL for access to Internet Projects Page: `http://hepg.awl.com` *keyword:* Weiss

USING THE FOCUS DATABASE

In Chapter 1 we explained how to store the Focus database in a Minitab worksheet named `focus.mtw`. If you haven't already created that worksheet, follow the instructions on pages 54–55 to create it now.

The Focus database contains information on 500 randomly selected Arizona State University sophomores. Among the variables considered are high-school GPA, SAT math score, cumulative GPA, SAT verbal score, age, and total hours. Use Minitab or some other statistical software to solve the following problems.

a. In 1995 the (non-recentered) mean SAT math score was 482 nationally. Do the Focus data on SAT math scores provide sufficient evidence to conclude that the mean SAT math score of Arizona State University sophomores exceeds the 1995 national mean? Use $\alpha = 0.05$.

b. In 1995 the (non-recentered) mean SAT verbal score was 428 nationally. Do the Focus data on SAT verbal scores provide sufficient evidence to conclude that the mean SAT verbal score of Arizona State University sophomores exceeds the 1995 national mean? Use $\alpha = 0.05$.

c. Justify the use of the hypothesis-testing procedures that you employed in parts (a) and (b).

CASE STUDY DISCUSSION

Effects of Brewery Effluent on Soil

At the beginning of this chapter on page 488, we discussed research by Mohammad Ajmal and Ahsan Ullah Khan into the effects of the application of brewery wastes on soil composition and plant growth.

As we noted, the researchers studied the physico-chemical properties of effluent from Mohan Meakin Breweries Ltd. (MMBL), Ghazibad, UP, India, and "... its effects on the physico-chemical characteristics of agricultural soil, seed germination pattern, and the growth of two common crop plants." They assessed the impact using different concentrations of the effluent: 25%, 50%, 75%, and 100%.

Various chemical properties of the treated soil were measured, in particular, available nutrients. The table on page 488 shows the effects of the different dilutions of MMBL effluent on the available limestone, potassium, and phosphorus in the soil. A number not enclosed by parentheses is the mean of five observations; a number enclosed by parentheses is the sample standard deviation of the five observations. The data for limestone are in percentages and those for potassium and phosphorus are in milligrams per kilogram of soil.

a. Do the data provide sufficient evidence to conclude, at the 1% level of significance, that the mean available limestone in soil treated with 100% MMBL effluent exceeds that which is ordinarily found (original)?

b. Repeat part (a) for 50% MMBL effluent.

c. At the 1% significance level, do the data provide sufficient evidence to conclude that the mean available phosphorus in soil treated with 50% MMBL effluent exceeds that which is ordinarily found?

d. What assumptions are you making in solving parts (a)–(c)?

BIOGRAPHY JERZY NEYMAN

Jerzy Neyman was born on April 16, 1894, in Bendery, Russia. His father, Czeslaw, was a member of the Polish nobility, a lawyer, a judge, and an amateur archaeologist. Because the Russian authorities prohibited the family from living in Poland, Jerzy Neyman grew up in various cities in Russia. He entered the university in Kharkov in 1912. At Kharkov he was at first interested in physics, but, because of his clumsiness in the laboratory, he decided to pursue mathematics.

After World War I, when Russia was at war with Poland over borders, Neyman was jailed as an enemy alien. In 1921, as a result of a prisoner exchange, he went to Poland for the first time. In 1924, he received his doctorate from the University of Warsaw. Between 1924 and 1934, Neyman worked with Karl Pearson (see biography, Chapter 13) and his son Egon Pearson, and held a position at the University of Kraków. In 1934, Neyman took a position in Karl Pearson's statistical laboratory at University College in London. He stayed in England, where he worked with Egon Pearson until 1938, at which time he accepted an offer to join the faculty at the University of California at Berkeley.

When the United States entered World War II, Neyman set aside the development of a statistics program and did war work. After the war ended, Neyman organized a symposium to celebrate its end and "the return to theoretical research." That symposium, held in August 1945, and succeeding ones, held every 5 years until 1970, were instrumental in establishing Berkeley as a preeminent statistical center.

Neyman was a principal founder of the theory of modern statistics. His work on hypothesis testing, confidence intervals, and survey sampling transformed both the theory and the practice of statistics. His achievements were acknowledged by the receipt of many honors and awards, including election to the United States National Academy of Sciences, the Guy Medal in Gold of the Royal Statistical Society, and the United States National Medal of Science.

Neyman remained active until his death of heart failure on August 5, 1981, at the age of 87, in Oakland, California.

CASE STUDY

BREAST MILK AND IQ

Considerable controversy exists over whether long-term neurodevelopment is affected by nutritional factors in early life. Five researchers summarized their findings on that question for preterm babies in the paper "Breast Milk and Subsequent Intelligence Quotient in Children Born Preterm" (*The Lancet,* 339, pp. 261–264). Their study was a continuation of work begun in January 1982.

Previously these researchers showed that a mother's decision to provide breast milk for preterm infants is associated with higher developmental scores for the children at age 18 months. In the article referred to in the previous paragraph, they analyzed IQ data on the same children at age 7 ½–8 years. IQ was measured for 300 children using an abbreviated form of the Weschler Intelligence Scale for Children (revised Anglicized version: WISC-R UK).

The mothers of the children in the study had chosen whether to provide their infants with breast milk within 72 hours of delivery; 90 did not and 210 did. Of those 210 who chose to provide their infants with breast milk, 193 succeeded and 17 did not.

The children whose mothers declined to provide breast milk were designated by the researchers as Group I; those whose mothers had chosen but were unable to provide breast milk were designated as Group IIa; and those whose mothers had chosen and were able to provide breast milk were designated as Group IIb. The following table displays statistics for all three groups.

Group	Sample size	Mean IQ	St. Dev.
I	90	92.8	15.2
IIa	17	94.8	19.0
IIb	193	103.7	15.3

After studying the inferential methods discussed in this chapter, you will be able to conduct statistical analyses to see how breast-feeding affects subsequent IQ for children age 7 ½–8 years who were born preterm.

Inferences for Two Population Means

10

GENERAL OBJECTIVES

In Chapters 8 and 9, we learned how to obtain confidence intervals and perform hypothesis tests for one population mean. Frequently, however, inferential statistics is used to compare the means of two or more populations.

For example, we might want to perform a hypothesis test to decide whether the mean age of buyers of new domestic cars is greater than the mean age of buyers of new imported cars; or we might want to find a confidence interval for the difference between the two mean ages. In this chapter we will study methods for comparing the means of two populations.

CHAPTER OUTLINE

10.1 THE SAMPLING DISTRIBUTION OF THE DIFFERENCE BETWEEN TWO MEANS FOR INDEPENDENT SAMPLES

In this section we will lay the groundwork for making statistical inferences to compare the means of two populations. The methods we first consider require not only that the samples selected from the two populations be random but also that they be **independent samples,** meaning that the sample selected from one of the populations has no effect or bearing on the sample selected from the other population.

With independent random samples, each possible pair of samples, one from one of the populations and the other from the other, is equally likely to be the pair of samples selected. Example 10.1 provides an unrealistically simple illustration of independent samples but one that will help you understand the concept.

EXAMPLE 10.1 INDEPENDENT RANDOM SAMPLES

Let us consider two small populations, one consisting of three men and the other of four women, as shown in the following figure.

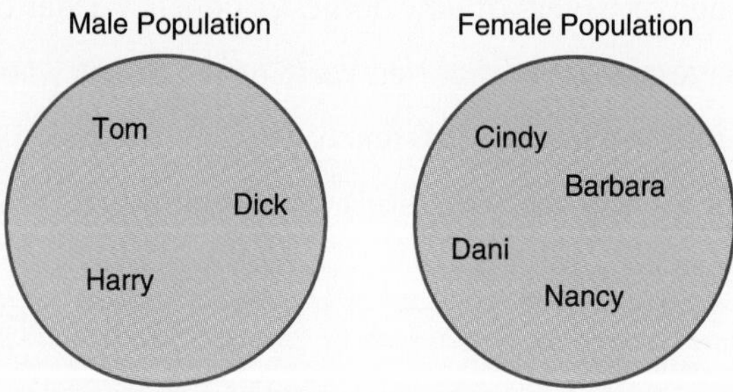

Suppose we take a sample of size two from the male population and a sample of size three from the female population.

a. List the possible pairs of independent samples.

b. If the samples are selected at random, determine the chance that any particular pair of samples will be the pair of independent samples obtained.

SOLUTION For convenience let us use the first letter of each name as an abbreviation for the actual name.

a. The possible samples of size two from the male population are listed in the first table of Table 10.1 and the possible samples of size three from the female population are listed in the second table of Table 10.1.

TABLE 10.1 Possible samples of size two from the male population and possible samples of size three from the female population

Male sample of size two
T, D
T, H
D, H

Female sample of size three
C, B, D
C, B, N
C, D, N
B, D, N

To obtain the possible pairs of independent samples, we list each possible male sample of size two with each possible female sample of size three. These are shown in Table 10.2. We see that there are 12 possible pairs of independent samples of two men and three women.

TABLE 10.2 Possible pairs of independent samples of two men and three women

Male sample of size two	Female sample of size three
T, D	C, B, D
T, D	C, B, N
T, D	C, D, N
T, D	B, D, N
T, H	C, B, D
T, H	C, B, N
T, H	C, D, N
T, H	B, D, N
D, H	C, B, D
D, H	C, B, N
D, H	C, D, N
D, H	B, D, N

b. For independent random samples, each of the 12 possible pairs of samples shown in Table 10.2 is equally likely to be the pair selected. Therefore the chances are $\frac{1}{12}$ (1 in 12) that any particular pair of samples will be the one obtained. ◨

The purpose of Example 10.1 is to provide a concrete illustration of independent samples and to emphasize that for independent random samples of any given sizes, each possible pair of independent samples is equally likely to be the one selected. In practice we neither obtain the number of possible pairs of independent samples nor explicitly compute the chance of selecting a particular pair of independent samples. But these concepts underlie the methods we do use.

Comparing Two Population Means Using Independent Samples

Now that we have discussed independent samples, we are ready to examine the process for comparing the means of two populations using independent samples. Example 10.2 introduces the pertinent ideas.

EXAMPLE 10.2 COMPARING TWO POPULATION MEANS USING INDEPENDENT SAMPLES

The American Association of University Professors (AAUP) conducts salary studies of college professors and publishes its findings in *AAUP Annual Report on the Economic Status of the Profession.* Suppose we want to decide whether the mean salaries of college faculty teaching in public and private institutions are different.

To formulate the problem statistically, we first note that we have one variable (salary) and two populations (faculty in public institutions and faculty in private institutions). Let the two populations in question be designated as Populations 1 and 2, respectively:

Population 1: Faculty in public institutions.
Population 2: Faculty in private institutions.

Next, we denote the means of the variable "salary" for the two populations by μ_1 and μ_2, respectively:

$$\mu_1 = \text{mean salary of faculty in public institutions.}$$
$$\mu_2 = \text{mean salary of faculty in private institutions.}$$

Then the hypothesis test we want to perform can be stated as

$$H_0\text{: } \mu_1 = \mu_2 \text{ (mean salaries are the same)}$$
$$H_a\text{: } \mu_1 \neq \mu_2 \text{ (mean salaries are different).}$$

Roughly speaking, the hypothesis test can be carried out by proceeding in the following manner.

1. Independently and randomly take a sample of faculty members from public institutions (Population 1) and a sample of faculty members from private institutions (Population 2).
2. Compute the mean salary, $\bar{x}_1$, of the sample of faculty members from public institutions and the mean salary, $\bar{x}_2$, of the sample of faculty members from private institutions.
3. Reject the null hypothesis if the sample means, $\bar{x}_1$ and $\bar{x}_2$, differ by too much; otherwise, do not reject the null hypothesis.

This process is pictured in Fig. 10.1.

FIGURE 10.1
Process for comparing two population means using independent samples

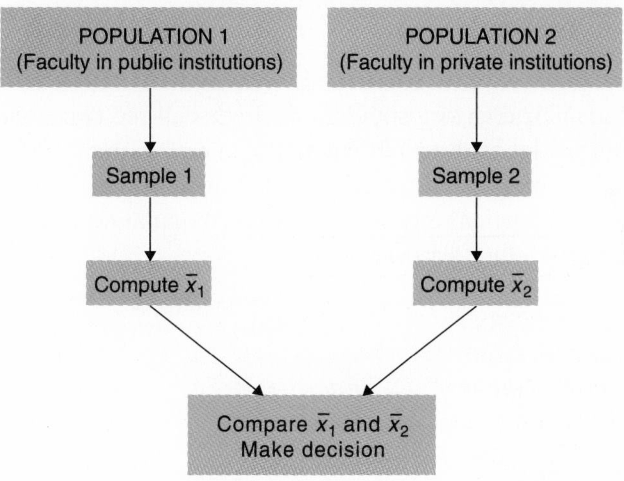

Suppose we randomly and independently sample 30 faculty members from public institutions (Population 1) and 35 faculty members from private institutions (Population 2) and that the salaries of these faculty members are as depicted in Table 10.3, where they are given in thousands of dollars rounded to the nearest hundred.

TABLE 10.3
Annual salaries ($1000s) for 30 faculty members in public institutions and 35 faculty members in private institutions

Sample 1 (public institutions)						Sample 2 (private institutions)						
34.2	63.6	24.4	79.4	33.8	88.2	92.9	102.2	51.5	77.6	71.1	59.3	71.0
90.0	56.8	56.0	42.2	40.2	44.6	52.0	62.9	46.4	61.6	73.5	97.5	97.3
100.4	41.4	58.2	81.8	51.2	64.4	63.1	53.8	45.2	78.3	67.6	27.2	92.6
24.6	35.0	76.8	29.2	41.2	74.0	118.5	101.0	76.0	66.3	52.4	81.2	56.0
107.4	54.2	84.2	15.8	60.2	71.0	37.7	68.6	56.1	31.1	47.2	24.8	62.3

The means of the two samples in Table 10.3 are

$$\bar{x}_1 = \frac{\Sigma x}{n_1} = \frac{1724.4}{30} = 57.48 \qquad \text{and} \qquad \bar{x}_2 = \frac{\Sigma x}{n_2} = \frac{2323.8}{35} = 66.39.$$

The question now is whether the difference of 8.91 ($8910) between these two sample means can reasonably be attributed to sampling error or whether the difference is large enough to indicate that the two populations have different means. To answer that question, we need to know the distribution of the difference between two sample means—the **sampling distribution of the difference between two means.** We will examine that sampling distribution in this section and complete the hypothesis test in Section 10.2. ◨

The Sampling Distribution of the Difference Between Two Means for Independent Samples

First we need to become familiar with the notation for parameters and statistics when analyzing two populations. Let's call the two populations under consideration Population 1 and Population 2. Then, as indicated in Example 10.2, we use a subscript 1 when referring to parameters or statistics for Population 1 and a subscript 2 when referring to them for Population 2. See Table 10.4.

TABLE 10.4
Notation for parameters and statistics when considering two populations

	Population 1	Population 2
Population mean	μ_1	μ_2
Population std. dev.	σ_1	σ_2
Sample mean	$\bar{x}_1$	$\bar{x}_2$
Sample std. dev.	s_1	s_2
Sample size	n_1	n_2

Armed with the notation in Table 10.4, we will now describe in Key Fact 10.1 the sampling distribution of the difference between two means. In doing so, we will assume the variable under consideration is normally distributed on each population. But keep in mind that, because of the central limit theorem, consequences of Key Fact 10.1 will still apply approximately for large samples regardless of distribution type.

In understanding Key Fact 10.1, it is also helpful to recall Key Fact 7.2: Suppose a variable x of a population is normally distributed and has mean μ and standard deviation σ. Then, for samples of size n,

- $\mu_{\bar{x}} = \mu$,
- $\sigma_{\bar{x}} = \sigma/\sqrt{n}$, and
- $\bar{x}$ is normally distributed.

Here now is Key Fact 10.1 which describes the sampling distribution of the difference between two means for independent samples.

KEY FACT 10.1 **THE SAMPLING DISTRIBUTION OF THE DIFFERENCE BETWEEN TWO MEANS FOR INDEPENDENT SAMPLES**

Suppose x is a normally distributed variable on each of two populations. Then, for independent samples of sizes n_1 and n_2 from the two populations,

- $\mu_{\bar{x}_1-\bar{x}_2} = \mu_1 - \mu_2$,
- $\sigma_{\bar{x}_1-\bar{x}_2} = \sqrt{(\sigma_1^2/n_1) + (\sigma_2^2/n_2)}$, and
- $\bar{x}_1 - \bar{x}_2$ is normally distributed.

In Key Fact 10.1, the first bulleted item says that the mean of all possible differences between the two sample means equals the difference between the two population means. The second bulleted item indicates that the standard deviation of all possible differences between the two sample means equals the square root of the sum of the population variances each divided by the corresponding sample size.

We should mention that the formulas for the mean and standard deviation of $\bar{x}_1 - \bar{x}_2$ hold regardless of the distributions of the variable on the two populations. The assumption that the variable is normally distributed on each of the two populations is needed only to conclude that $\bar{x}_1 - \bar{x}_2$ is normally distributed and, as we noted, that too holds true approximately for large samples, regardless of distribution type.

The Two-Sample z-Procedures

Under the conditions of Key Fact 10.1, the standardized version of $\bar{x}_1 - \bar{x}_2$,

$$z = \frac{(\bar{x}_1 - \bar{x}_2) - (\mu_1 - \mu_2)}{\sqrt{(\sigma_1^2/n_1) + (\sigma_2^2/n_2)}},$$

has the standard normal distribution. Based on knowing that, we can develop hypothesis-testing and confidence-interval procedures for comparing two population means when the population standard deviations are known. These procedures are often referred to as the **two-sample z-test** and the **two-sample z-interval procedure,** respectively, or, collectively, as the **two-sample z-procedures.**

Because population standard deviations are usually unknown, however, we will not discuss the two-sample z-procedures in detail. Instead we relegate them to the exercises and concentrate on the more useful two-sample t-procedures, which apply in the case of unknown population standard deviations. Sections 10.2 and 10.3 examine the two-sample t-procedures, specifically, the pooled t- and nonpooled t-procedures, respectively.

EXERCISES 10.1

STATISTICAL CONCEPTS AND SKILLS

10.1 In the introduction to this chapter, we considered the problem of comparing the mean age of buyers of new domestic cars to the mean age of buyers of new imported cars.

a. Identify the variable under consideration.
b. Identify the two populations.
c. Suppose we want to perform a hypothesis test to decide whether the mean age of buyers of new domestic cars is greater than the mean age of buyers of new imported cars. State the null and alternative hypotheses for the hypothesis test.

10.2 A 1997 issue of *USA TODAY* compared the mean amounts spent at the shopping mall by teens and adults.

a. Identify the variable under consideration.
b. Identify the two populations.
c. Suppose we want to perform a hypothesis test to decide whether the mean amount spent by teens is less than the mean amount spent by adults. State the null and alternative hypotheses for the hypothesis test.

10.3 Give an example of interest to you for comparing two population means. Identify the variable under consideration and the two populations.

10.4 Define the phrase *independent samples.*

10.5 Consider the quantities $\mu_1, \sigma_1, \bar{x}_1, s_1, \mu_2, \sigma_2, \bar{x}_2$, and s_2.

a. Which quantities represent parameters and which represent statistics?

b. Which quantities are fixed numbers and which are variables?

10.6 Discuss the basic strategy for performing a hypothesis test to compare the means of two populations based on independent samples.

10.7 To perform a hypothesis test to compare two population means, we need to know the sampling distribution of the difference between two means. Why?

10.8 Identify the assumption for using the two-sample z-procedures that renders those procedures generally impractical.

10.9 A variable of two populations has mean 40 and standard deviation 12 for one of the populations and mean 40 and standard deviation 6 for the other.

a. Find the mean and standard deviation of $\bar{x}_1 - \bar{x}_2$ for independent samples of sizes 9 and 4, respectively.

b. Must the variable under consideration be normally distributed on each of the two populations in order to answer part (a)? Explain your answer.

c. Can we conclude that the variable $\bar{x}_1 - \bar{x}_2$ is normally distributed? Explain your answer.

10.10 A variable of two populations has mean 40 and standard deviation 12 for one of the populations and mean 40 and standard deviation 6 for the other population. Moreover, the variable is normally distributed on each of the two populations.

a. For independent samples of sizes 9 and 4, respectively, determine the mean and standard deviation of $\bar{x}_1 - \bar{x}_2$.

b. Can we conclude that the variable $\bar{x}_1 - \bar{x}_2$ is normally distributed? Explain your answer.

c. Determine the percentage of all pairs of independent samples of sizes 9 and 4, respectively, from the two populations having the property that the difference between the sample means is between -10 and 10.

EXTENDING THE CONCEPTS AND SKILLS

The two-sample z-procedures. Using Key Fact 10.1, we can derive the two-sample z-procedures.

TWO-SAMPLE z-PROCEDURES

ASSUMPTIONS

1. Independent samples
2. Normal populations or large samples
3. Known population standard deviations

- **Two-sample z-test.** The test statistic for a hypothesis test with null hypothesis H_0: $\mu_1 = \mu_2$ (population means are equal) is

$$z = \frac{\bar{x}_1 - \bar{x}_2}{\sqrt{(\sigma_1^2/n_1) + (\sigma_2^2/n_2)}}.$$

- **Two-sample z-interval procedure.** The endpoints of a $(1 - \alpha)$-level confidence interval for the difference, $\mu_1 - \mu_2$, between the two population means are

$$(\bar{x}_1 - \bar{x}_2) \pm z_{\alpha/2} \cdot \sqrt{(\sigma_1^2/n_1) + (\sigma_2^2/n_2)}.$$

Exercises 10.11 and 10.12 illustrate the use of the two-sample z-procedures.

10.11 The Northwestern University Placement Center, Evanston, Illinois, conducts surveys on starting salaries for college graduates and publishes its observations in *The Northwestern Lindquist-Endicott Report.* The following table gives the starting annual salaries obtained from independent random samples of 35 liberal-arts graduates and 32 accounting graduates. Data are in thousands of dollars.

Liberal arts					Accounting			
33.0	29.8	34.3	33.6	34.0	35.9	31.4	34.3	31.9
31.7	36.8	31.3	30.7	31.0	34.4	36.3	37.8	33.6
32.1	34.1	32.6	32.0	31.7	35.0	33.0	32.7	34.0
33.8	33.4	30.1	32.5		37.2	36.9	35.8	36.3
32.9	32.2	32.1	30.2		36.5	31.4	33.4	35.5
31.3	35.7	33.0	32.2		35.9	36.4	36.8	35.2
33.9	32.1	33.8	33.5		33.2	33.1	32.0	35.7
34.4	29.3	29.3	33.5		32.5	33.5	36.4	37.6

a. At the 5% significance level, can we conclude that the mean starting salaries of liberal-arts and accounting graduates differ? Assume the population standard deviations of starting salaries are 1.82 ($1820) for liberal-arts graduates and 1.73 ($1730) for accounting graduates. (*Note:* The sum of the liberal-arts data is 1137.9; the sum of the accounting data is 1111.6.)

b. Determine a 95% confidence interval for the difference, $\mu_1 - \mu_2$, between the mean starting annual salaries of liberal-arts and accounting graduates.

10.12 The U.S. National Center for Health Statistics compiles data on the length of stay by patients in short-term hospitals and publishes its findings in *Vital and Health Statistics*. Independent samples of 39 male patients and 35 female patients gave the following data on length of stay, in days.

Male					Female				
4	4	12	18	9	14	7	15	1	12
6	12	10	3	6	1	3	7	21	4
15	7	3	13	1	1	5	4	4	3
2	10	13	5	7	5	18	12	5	1
1	23	9	2	1	7	7	2	15	4
17	2	24	11	14	9	10	7	3	6
6	2	1	8	1	5	9	6	2	14
3	19	3	1						

a. Do the data provide sufficient evidence to conclude that, on the average, the lengths of stay in short-term hospitals by males and females differ? Assume $\sigma_1 = 5.4$ days and $\sigma_2 = 4.6$ days. Perform the appropriate hypothesis test at the 5% significance level. (*Note:* The sum of the male data is 308 days, and the sum of the female data is 249 days.)

b. Determine a 95% confidence interval for the difference, $\mu_1 - \mu_2$, between the mean lengths of stay in short-term hospitals by males and females.

10.13 Hypothesis tests and confidence intervals. Use the results obtained in Exercises 10.11 and 10.12 to complete the following statement: A hypothesis test of H_0: $\mu_1 = \mu_2$ versus H_a: $\mu_1 \neq \mu_2$ at the significance level α will lead to rejection of the null hypothesis if and only if the number ______ does not lie in the $(1 - \alpha)$-level confidence interval for $\mu_1 - \mu_2$.

USING TECHNOLOGY  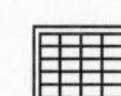 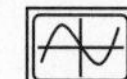

10.14 To determine the sampling distribution of the difference between two means for independent samples (Key Fact 10.1 on page 596), we need the fact that, for independent observations, the difference of two normally distributed variables is also a normally distributed variable. In this exercise you are asked to perform a computer simulation to make that fact plausible.

a. Simulate 2000 observations from a normally distributed variable having mean 100 and standard deviation 16.

b. Repeat part (a) for a normally distributed variable having mean 120 and standard deviation 12.

c. Determine the difference between each pair of observations in parts (a) and (b). If you are using Minitab, first give the two columns containing the 2000 observations obtained in parts (a) and (b) the names X1 and X2, respectively. Then proceed as follows.

1 Choose **Calc ➤ Calculator...**

2 Type DIFF in the **Store result in variable** text box

3 Type X1-X2 in the **Expression** text box

4 Click **OK**

The differences of the 2000 pairs of observations will now be stored in a column named DIFF.

d. Obtain a histogram of the 2000 differences found in part (c). Why is the histogram bell-shaped?

10.15 In this exercise you are to perform a computer simulation to illustrate the sampling distribution of the difference between two means for independent samples, Key Fact 10.1. You may want to refer to parts (b) and (c) of Exercise 7.45 on page 424.

a. Simulate 1000 samples of size 12 from a normally distributed variable having mean 640 and standard deviation 70. Then obtain the sample mean of each of the 1000 samples.

b. Simulate 1000 samples of size 15 from a normally distributed variable having mean 715 and standard deviation 150. Then obtain the sample mean of each of the 1000 samples.

c. Determine the difference, $\bar{x}_1 - \bar{x}_2$, for each of the 1000 pairs of sample means obtained in parts (a) and (b). *(Hint:* Refer to part (c) of Exercise 10.14.*)*

d. Obtain the mean, the standard deviation, and a histogram of the 1000 differences found in part (c).

e. Theoretically, what are the mean, standard deviation, and distribution of all possible differences, $\bar{x}_1 - \bar{x}_2$?

f. Compare your answers from parts (d) and (e).

10.2 INFERENCES FOR TWO POPULATION MEANS USING INDEPENDENT SAMPLES (STANDARD DEVIATIONS ASSUMED EQUAL)

In Section 10.1 we took the first steps required for developing inferential procedures to compare the means of two populations using independent samples. Armed with that information, we can derive two inferential methods. One requires that the two populations have equal standard deviations; the other does not impose this restriction. We will study the first method in this section and the second in Section 10.3.

Hypothesis Tests for the Means of Two Populations With Equal Standard Deviations Using Independent Samples

We will now develop a procedure for performing a hypothesis test to compare the means of two populations with equal, but unknown, standard deviations using independent samples. Our immediate goal is to find a test statistic for such a test. In doing so, we will assume the variable under consideration is normally distributed on each population. As we will see later, the resulting hypothesis-testing procedure is approximately correct for large samples, regardless of distribution-type.

Let's use σ to denote the common standard deviation of the two populations. We know from Key Fact 10.1 on page 596 that, for independent samples, the standardized version of $\bar{x}_1 - \bar{x}_2$,

$$z = \frac{(\bar{x}_1 - \bar{x}_2) - (\mu_1 - \mu_2)}{\sqrt{(\sigma_1^2/n_1) + (\sigma_2^2/n_2)}},$$

has the standard normal distribution. Replacing σ_1 and σ_2 in the above expression by their common value σ and using some algebra, we obtain the variable

$$z = \frac{(\bar{x}_1 - \bar{x}_2) - (\mu_1 - \mu_2)}{\sigma\sqrt{(1/n_1) + (1/n_2)}}. \quad (1)$$

This variable cannot be used as a basis for the required test statistic because σ is unknown.

Consequently, we need to use sample information to estimate the unknown population standard deviation, σ. We do this by first obtaining an estimate of the unknown population variance, σ^2. And the best way to do that is to regard the sample variances, s_1^2 and s_2^2, as two estimates of σ^2 and then **pool** those estimates by weighting them according to sample size (actually by degrees of freedom). Thus our estimate of σ^2 is

$$s_p^2 = \frac{(n_1 - 1)s_1^2 + (n_2 - 1)s_2^2}{n_1 + n_2 - 2}$$

and hence that of σ is

$$s_p = \sqrt{\frac{(n_1 - 1)s_1^2 + (n_2 - 1)s_2^2}{n_1 + n_2 - 2}}.$$

The subscript "p" stands for "pooled," and the quantity s_p is called the **pooled sample standard deviation.**

Replacing σ in Equation (1) by its estimate, s_p, we get the variable

$$\frac{(\bar{x}_1 - \bar{x}_2) - (\mu_1 - \mu_2)}{s_p\sqrt{(1/n_1) + (1/n_2)}},$$

which can be used as a basis for the required test statistic. However, unlike the variable in Equation (1), this one does not have the standard normal distribution. But its distribution is one with which we are familiar—a t-distribution.

KEY FACT 10.2 **DISTRIBUTION OF THE POOLED t-STATISTIC**

Suppose x is a normally distributed variable on each of two populations and that the population standard deviations are equal. Then, for independent samples of sizes n_1 and n_2 from the two populations, the variable

$$t = \frac{(\bar{x}_1 - \bar{x}_2) - (\mu_1 - \mu_2)}{s_p\sqrt{(1/n_1) + (1/n_2)}}$$

has the t-distribution with df $= n_1 + n_2 - 2$.

In view of Key Fact 10.2, we see that for a hypothesis test with null hypothesis H_0: $\mu_1 = \mu_2$ (population means are equal), we can use the variable

$$t = \frac{\bar{x}_1 - \bar{x}_2}{s_p\sqrt{(1/n_1) + (1/n_2)}}$$

as the test statistic and obtain the critical value(s) from the t-table. Specifically, we have the following procedure, which we often refer to as the **pooled t-test.**

PROCEDURE 10.1 **THE POOLED *t*-TEST FOR TWO POPULATION MEANS**

ASSUMPTIONS

1. Independent samples
2. Normal populations or large samples
3. Equal population standard deviations

Step 1 The null hypothesis is H_0: $\mu_1 = \mu_2$ and the alternative hypothesis is one of the following:

H_a: $\mu_1 \neq \mu_2$ (Two-tailed) or H_a: $\mu_1 < \mu_2$ (Left-tailed) or H_a: $\mu_1 > \mu_2$ (Right-tailed)

Step 2 Decide on the significance level, α.

Step 3 The critical value(s) are

$\pm t_{\alpha/2}$ (Two-tailed) or $-t_\alpha$ (Left-tailed) or t_α (Right-tailed)

with $\text{df} = n_1 + n_2 - 2$. Use Table IV to find the critical value(s).

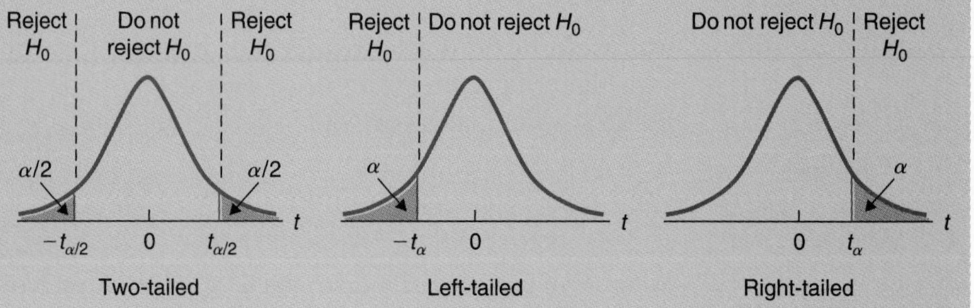

Step 4 Compute the value of the test statistic

$$t = \frac{\bar{x}_1 - \bar{x}_2}{s_p\sqrt{(1/n_1) + (1/n_2)}}$$

where

$$s_p = \sqrt{\frac{(n_1 - 1)s_1^2 + (n_2 - 1)s_2^2}{n_1 + n_2 - 2}}.$$

Step 5 If the value of the test statistic falls in the rejection region, reject H_0; otherwise, do not reject H_0.

Step 6 State the conclusion in words.

The hypothesis test is exact for normal populations and is approximately correct for large samples from nonnormal populations.

Before we apply the pooled t-test, several comments are in order. In Step 4 of Procedure 10.1, we need to calculate the pooled sample standard deviation, s_p. The pooled sample standard deviation always lies between the two sample standard deviations, s_1 and s_2. This fact is useful as a check when s_p is calculated by hand.

Next let's discuss the three assumptions for the pooled t-test. Assumption 1 (independent samples) is essential; the samples must be independent or the procedure does not apply.

Regarding Assumption 2, although the pooled t-test was derived under the condition that the variable under consideration is normally distributed on each of the two populations (normal populations), it also applies approximately for large samples regardless of distribution-type, as noted at the bottom of Procedure 10.1.

Actually, the pooled t-test works reasonably well even for small or moderate-size samples from nonnormal populations provided the populations are not too nonnormal. In other words, the pooled t-test is robust to moderate violations of the normality assumption. The pooled t-test is also robust to moderate violations of Assumption 3 (equal population standard deviations), provided the sample sizes are roughly equal. We will have more to say about the robustness of the pooled t-test at the end of Section 10.3.

As before, we can check normality using normal probability plots. The equal-standard-deviations assumption is more difficult to check, especially when the sample sizes are small. We recommend checking it by informally comparing the standard deviations of the two samples and by viewing together stem-and-leaf diagrams, histograms, or boxplots of the two samples. Use the same scales for each pair of graphs.

The equal-standard-deviations assumption is sometimes checked by performing a formal hypothesis test, called an F-test for the equality of two standard deviations. We do not recommend this procedure because although the pooled t-test is robust to moderate violations of normality, the F-test is extremely nonrobust to such violations. As the noted statistician George E. P. Box remarked: "To make a preliminary test on variances [standard deviations] is rather like putting to sea in a rowing boat to find out whether conditions are sufficiently calm for an ocean liner to leave port!"

When considering the pooled t-test, it is also important to watch for outliers. The presence of outliers calls into question the normality assumption. And even for large samples, outliers can sometimes unduly affect a pooled t-test because the sample mean and sample standard deviation are not resistant to them.

EXAMPLE 10.3 **ILLUSTRATES PROCEDURE 10.1**

We now return to the salary problem posed in Example 10.2. Recall that we want to perform a hypothesis test to decide whether there is a difference between the mean salaries of faculty teaching in public and private institutions.

Independent random samples of 30 faculty members teaching in public institutions and 35 faculty members teaching in private institutions yield the data in Table 10.5. At the 5% significance level, do the data provide sufficient evidence to conclude that mean salaries for faculty teaching in public and private institutions differ?

TABLE 10.5
Annual salaries ($1000s) for 30 faculty members in public institutions and 35 faculty members in private institutions

Sample 1 (public institutions)						Sample 2 (private institutions)						
34.2	63.6	24.4	79.4	33.8	88.2	92.9	102.2	51.5	77.6	71.1	59.3	71.0
90.0	56.8	56.0	42.2	40.2	44.6	52.0	62.9	46.4	61.6	73.5	97.5	97.3
100.4	41.4	58.2	81.8	51.2	64.4	63.1	53.8	45.2	78.3	67.6	27.2	92.6
24.6	35.0	76.8	29.2	41.2	74.0	118.5	101.0	76.0	66.3	52.4	81.2	56.0
107.4	54.2	84.2	15.8	60.2	71.0	37.7	68.6	56.1	31.1	47.2	24.8	62.3

SOLUTION First we present in Table 10.6 the required summary statistics for the two samples in Table 10.5. These statistics are obtained in the usual way.

TABLE 10.6
Summary statistics for the samples in Table 10.5

Public institutions	Private institutions
$\bar{x}_1 = 57.48$ $s_1 = 23.95$ $n_1 = 30$	$\bar{x}_2 = 66.39$ $s_2 = 22.26$ $n_2 = 35$

Next we check the three conditions required for using the pooled t-test. Since the samples are independent, Assumption 1 is satisfied. To check Assumption 2, we first note that the sample sizes are large. Thus we need only be concerned with the presence of outliers. Careful graphical analysis (not shown) suggest no outliers for either sample; so we can consider Assumption 2 satisfied. From Table 10.6 we see that the sample standard deviations are 23.95 and 22.26; these are close enough for us to consider Assumption 3 satisfied.

The preceding paragraph suggests that the pooled t-test can be used to carry out the hypothesis test. We apply Procedure 10.1.

Step 1 *State the null and alternative hypotheses.*

The null and alternative hypotheses are

$$H_0\colon \mu_1 = \mu_2 \text{ (mean salaries are the same)}$$
$$H_a\colon \mu_1 \neq \mu_2 \text{ (mean salaries are different),}$$

where μ_1 and μ_2 are, respectively, the mean salaries of all faculty in public and private institutions. Note that the hypothesis test is two-tailed since a not-equal sign ($\neq$) appears in the alternative hypothesis.

Step 2 Decide on the significance level, α.

The hypothesis test is to be performed at the 5% level, so $\alpha = 0.05$.

Step 3 The critical values for a two-tailed test are $\pm t_{\alpha/2}$ with $df = n_1 + n_2 - 2$.

From Step 2, $\alpha = 0.05$. Also, we see from Table 10.6 that $n_1 = 30$ and $n_2 = 35$. Hence $\text{df} = 30 + 35 - 2 = 63$. Consulting Table IV, we find that the critical value is $\pm t_{\alpha/2} = \pm t_{0.05/2} = \pm t_{0.025} = \pm 1.998$, as seen in Fig. 10.2.[†]

FIGURE 10.2
Criterion for deciding whether or not to reject the null hypothesis

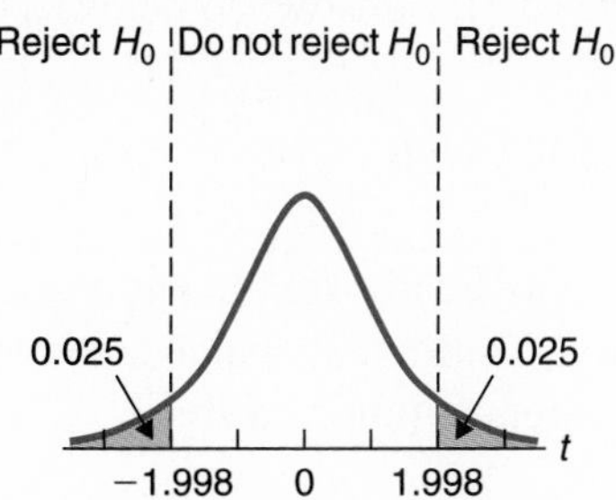

Step 4 Compute the value of the test statistic

$$t = \frac{\bar{x}_1 - \bar{x}_2}{s_p\sqrt{(1/n_1) + (1/n_2)}}$$

where

$$s_p = \sqrt{\frac{(n_1 - 1)s_1^2 + (n_2 - 1)s_2^2}{n_1 + n_2 - 2}}.$$

We first determine the pooled sample standard deviation, s_p. Referring to Table 10.6, we find that

$$s_p = \sqrt{\frac{(30 - 1) \cdot 23.95^2 + (35 - 1) \cdot 22.26^2}{30 + 35 - 2}} = 23.05.$$

[†] Table IV does not have an entry for df = 63 but skips from df = 60 to df = 70. So we used linear interpolation to approximate the required t-value: $t_{0.025} \approx 2.000 + \frac{3}{10} \cdot (1.994 - 2.000) = 1.998$.

Referring again to Table 10.6, we obtain the value of the test statistic:

$$t = \frac{\bar{x}_1 - \bar{x}_2}{s_p\sqrt{(1/n_1) + (1/n_2)}} = \frac{57.48 - 66.39}{23.05\sqrt{(1/30) + (1/35)}} = -1.554.$$

Step 5 *If the value of the test statistic falls in the rejection region, reject H_0; otherwise, do not reject H_0.*

From Step 4 the value of the test statistic is $t = -1.554$, which does not fall in the rejection region (see Fig. 10.2). Thus we do not reject H_0.

Step 6 *State the conclusion in words.*

The test results are not statistically significant at the 5% level; that is, at the 5% significance level, we have insufficient evidence to conclude that mean salaries for faculty in public and private institutions differ. □

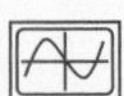

Using the *P*-Value Approach

The P-value approach to hypothesis testing can also be used to carry out the hypothesis test in Example 10.3 and to assess more precisely the evidence against the null hypothesis. From Step 4 we see that the value of the test statistic is $t = -1.554$. Recalling that df $= 63$ and that the test is two-tailed, we find from Table IV that $0.10 < P < 0.20$.

In particular, because the P-value exceeds the significance level of 0.05, we cannot reject H_0. Furthermore, by referring to Table 9.12 on page 543, we see that the data provide at most weak evidence against the null hypothesis of equal mean salaries.

Confidence Intervals for the Difference Between the Means of Two Populations With Equal Standard Deviations

We can also use Key Fact 10.2 on page 601 to derive the following confidence-interval procedure for the difference between two population means, which we often refer to as the **pooled *t*-interval procedure.**

PROCEDURE 10.2

THE POOLED t-INTERVAL PROCEDURE FOR TWO POPULATION MEANS

ASSUMPTIONS

1. Independent samples
2. Normal populations or large samples
3. Equal population standard deviations

Step 1 For a confidence level of $1 - \alpha$, use Table IV to find $t_{\alpha/2}$ with $\text{df} = n_1 + n_2 - 2$.

Step 2 The endpoints of the confidence interval for $\mu_1 - \mu_2$ are

$$(\bar{x}_1 - \bar{x}_2) \pm t_{\alpha/2} \cdot s_p\sqrt{(1/n_1) + (1/n_2)}.$$

The confidence interval is exact for normal populations and is approximately correct for large samples from nonnormal populations.

EXAMPLE 10.4 ILLUSTRATES PROCEDURE 10.2

Determine a 95% confidence interval for the difference, $\mu_1 - \mu_2$, between the mean salaries of faculty teaching in public and private institutions.

SOLUTION We apply Procedure 10.2.

Step 1 For a confidence level of $1 - \alpha$*, use Table IV to find* $t_{\alpha/2}$ *with* $df = n_1 + n_2 - 2$.

For a 95% confidence interval, $\alpha = 0.05$. From Table 10.6 on page 604, $n_1 = 30$ and $n_2 = 35$; so $\text{df} = n_1 + n_2 - 2 = 30 + 35 - 2 = 63$. Consulting Table IV, we find that for $\text{df} = 63$, $t_{\alpha/2} = t_{0.05/2} = t_{0.025} = 1.998$.

Step 2 The endpoints of the confidence interval for $\mu_1 - \mu_2$ *are*

$$(\bar{x}_1 - \bar{x}_2) \pm t_{\alpha/2} \cdot s_p\sqrt{(1/n_1) + (1/n_2)}.$$

From Step 1, $t_{\alpha/2} = 1.998$. Also, $n_1 = 30$, $n_2 = 35$, and from Example 10.3 we know that $\bar{x}_1 = 57.48$, $\bar{x}_2 = 66.39$, and $s_p = 23.05$. Hence the endpoints of the confidence interval for $\mu_1 - \mu_2$ are

$$(57.48 - 66.39) \pm 1.998 \cdot 23.05\sqrt{(1/30) + (1/35)},$$

or -8.91 ± 11.46. Thus the 95% confidence interval is from -20.37 to 2.55. We can be 95% confident that the difference, $\mu_1 - \mu_2$, between the mean salaries of faculty teaching in public and private institutions is somewhere between $-\$20,370$ and $\$2,550$. □

  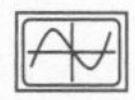

Using the Computer (Optional)

We can use Minitab to perform pooled t-procedures. To illustrate, we return once again to the hypothesis test and confidence interval comparing the mean salaries of college faculty teaching in public and private institutions.

EXAMPLE 10.5

USING MINITAB TO PERFORM POOLED t-PROCEDURES

Use Minitab to simultaneously perform the hypothesis test considered in Example 10.3 and obtain the confidence interval required in Example 10.4.

SOLUTION Let μ_1 and μ_2 denote, respectively, the mean salaries of all faculty in public and private institutions. The problem in Example 10.3 is to perform the hypothesis test

$$H_0\text{: } \mu_1 = \mu_2 \text{ (mean salaries are the same)}$$
$$H_a\text{: } \mu_1 \neq \mu_2 \text{ (mean salaries are different)}$$

at the 5% significance level; the problem in Example 10.4 is to obtain a 95% confidence interval for $\mu_1 - \mu_2$.

As we discovered in Example 10.3, it is reasonable to use pooled t-procedures here. To apply Minitab, we begin by entering the two sets of sample data from Table 10.5 on page 604 in columns named PUBLIC and PRIVATE. Then we proceed in the following way.

1 Choose **Stat ➤ Basic Statistics ➤ 2-Sample t...**
2 Select the **Samples in different columns** option button
3 Click in the **First** text box and specify PUBLIC
4 Click in the **Second** text box and specify PRIVATE
5 Click the arrow button at the right of the **Alternative** drop-down list box and select **not equal**
6 Click in the **Confidence level** text box and type 95
7 Select the **Assume equal variances** check box (to indicate that we are assuming the populations have equal standard deviations)
8 Click **OK**

The output that results is shown in Printout 10.1.

The first line of Printout 10.1 describes the test being performed: `Two sample T for PUBLIC vs PRIVATE`. The next three lines display a table that gives the sample size, sample mean, sample standard deviation, and estimated standard error of the mean for each of the two samples.

On the fifth line of Printout 10.1, we find the required 95% confidence interval for the difference between the two population means. Thus we can be 95% confident that the difference, $\mu_1 - \mu_2$, between the mean salaries of faculty teaching in public and private institutions is somewhere between −$20,400 and $2,500.

PRINTOUT 10.1
Minitab output for the pooled t-procedure

```
Two sample T for PUBLIC vs PRIVATE

           N      Mean      StDev   SE Mean
PUBLIC    30      57.5       24.0       4.4
PRIVATE   35      66.4       22.3       3.8

95% CI for mu PUBLIC - mu PRIVATE: ( -20.4,  2.5)
T-Test mu PUBLIC = mu PRIVATE (vs not =): T = -1.55  P = 0.13  DF = 63
Both use Pooled StDev = 23.1
```

The next-to-last line of Printout 10.1 provides a statement of the null and alternative hypotheses, followed by the value of the test statistic (`T = -1.55`), the P-value (`P = 0.13`), and the degrees of freedom (`DF = 63`). On the last line, we find the value of the pooled sample standard deviation, s_p.

Since the P-value of 0.13 exceeds the specified significance level of 0.05, we do not reject H_0. At the 5% level, the data do not provide sufficient evidence to conclude that mean salaries of faculty in public and private institutions differ. ❐

EXERCISES 10.2

STATISTICAL CONCEPTS AND SKILLS

10.16 Regarding the three conditions required for using the pooled t-procedures:
a. What are they?
b. How important are each of the conditions?

10.17 Why is s_p called the *pooled sample standard deviation?*

10.18 Independent samples are taken from two populations with the intent of performing a hypothesis test to compare their means. Prior data analyses have indicated that the variable under consideration is normally distributed on each of the two populations. The following table provides summary statistics for the two samples.

Sample 1	Sample 2
$\bar{x}_1 = 468.3$ $s_1 = 38.2$ $n_1 = 6$	$\bar{x}_2 = 394.6$ $s_2 = 84.7$ $n_2 = 14$

Is it reasonable to use the pooled t-test on these data? Explain your answer.

Preliminary data analyses indicate that it is reasonable to consider the assumptions for using pooled t-procedures satisfied in Exercises 10.19–10.24. For each exercise, perform the required hypothesis test using either the critical-value approach or the P-value approach.

10.19 The U.S. Bureau of Prisons publishes data in *Statistical Report* on the times served by prisoners released from federal institutions for the first time. Independent random samples of released prisoners in the fraud and firearms offense categories yielded the following information on time served, in months.

Fraud		Firearms	
3.6	17.9	25.5	23.8
5.3	5.9	10.4	17.9
10.7	7.0	18.4	21.9
8.5	13.9	19.6	13.3
11.8	16.6	20.9	16.1

At the 5% significance level, do the data provide sufficient evidence to conclude that the mean time served for fraud is less than that for firearms offenses? (*Note:* $\bar{x}_1 = 10.12$, $s_1 = 4.90$, $\bar{x}_2 = 18.78$, $s_2 = 4.64$.)

10.20 In a packing plant, a machine packs cartons with jars. Supposedly, a new machine will pack faster on the average than the machine currently used. To test that hypothesis, the times it takes each machine to pack 10 cartons are recorded. The results, in seconds, are shown in the following table.

New machine		Present machine	
42.0	41.0	42.7	43.6
41.3	41.8	43.8	43.3
42.4	42.8	42.5	43.5
43.2	42.3	43.1	41.7
41.8	42.7	44.0	44.1

Do the data provide sufficient evidence to conclude that, on the average, the new machine packs faster? Perform the required hypothesis test at the 5% level of significance. (*Note:* $\bar{x}_1 = 42.13$, $s_1 = 0.685$, $\bar{x}_2 = 43.23$, $s_2 = 0.750$.)

10.21 *Vital and Health Statistics,* published by the National Center for Health Statistics, provides information on heights and weights of Americans, by age and sex. Independent samples of 10 males age 25–34 years and 15 males age 45–54 years yield the following heights, in inches.

25–34		45–54		
73.3	70.4	73.2	69.5	64.7
64.8	66.8	68.5	74.5	73.0
72.1	70.7	62.4	70.6	66.7
68.9	74.4	65.5	69.3	68.1
68.7	71.8	71.3	67.1	64.3

At the 5% significance level, do the data provide sufficient evidence to conclude that males in the age group 25–34 years are, on the average, taller than those who were in that age group 20 years ago? (*Note:* $\bar{x}_1 = 70.19$, $s_1 = 2.951$, $\bar{x}_2 = 68.58$, $s_2 = 3.543$.)

10.22 The Federal Highway Administration compiles data on annual household vehicle miles of travel (VMT) and publishes its findings in *National Personal Transportation Survey, Summary of Travel Trends.* Independent samples of 15 midwestern households and 14 southern households provide the following data on last year's VMT, in thousands of miles.

Midwest			South		
16.2	12.9	17.3	22.2	19.2	9.3
14.6	18.6	10.8	24.6	20.2	15.8
11.2	16.6	16.6	18.0	12.2	20.1
24.4	20.3	20.9	16.0	17.5	18.2
9.6	15.1	18.3	22.8	11.5	

At the 5% significance level, does there appear to be a difference in last year's mean VMT for midwestern and southern households? (*Note:* $\bar{x}_1 = 16.23$, $s_1 = 4.06$, $\bar{x}_2 = 17.69$, $s_2 = 4.42$.)

10.23 The U.S. Energy Information Administration publishes data on residential energy consumption and expenditures in *Residential Energy Consumption Survey: Consumption and Expenditures.* Suppose you want to decide whether last year's mean annual fuel expenditure for households using natural gas is different from that for households using only electricity. At the 5% significance level, what conclusion would you draw given the data, in dollars, shown in the following table? (*Note:* $\bar{x}_1 = 1497.6$, $s_1 = 160.35$, $\bar{x}_2 = 1243.6$, $s_2 = 165.13$.)

Natural gas			Electricity			
2002	1456	1394	1376	1452	1235	1480
1541	1321	1338	1185	1327	1059	1400
1495	1526	1358	1227	1102	1168	1070
1801	1478	1376	1180	1221	1351	1014
1579	1375	1664	1461	1102	976	1394
1305	1458	1369	1379	987	1002	1532
1495	1507	1636	1450	1177	1150	
1698	1249	1377	1352	1266	1109	
1648	1557	1491	949	1351	1259	
1505	1355	1574	1179	1393	1456	

10.24 Agronomists study, among other things, conditions under which larger yields of crops might be obtained. Eighty 1-acre plots of corn are randomly divided into two groups of forty 1-acre plots. An insecticide is used on each 1-acre plot in the first group and sterilized male insects of an insect pest on each 1-acre plot in the second group. The yields, in bushels, are as shown in the following table.

Insecticide					Sterilized males				
109	101	97	89	100	105	109	110	118	109
98	98	94	99	104	113	111	111	99	112
103	88	108	102	106	106	117	99	107	119
97	105	102	104	101	110	111	103	110	108
101	100	105	110	96	104	102	111	114	114
102	95	100	95	109	122	117	101	109	109
91	98	113	91	95	102	109	103	109	106
106	98	101	99	96	107	107	111	128	109

a. Do the data provide sufficient evidence to conclude that the use of sterilized male insects is more effective than the insecticide in controlling the insect pest? Use $\alpha = 0.01$. *(Note:* $\bar{x}_1 = 100.15$, $s_1 = 5.73$, $\bar{x}_2 = 109.53$, $s_2 = 6.06$.*)*

b. Is this a designed experiment or an observational study? Explain your answer.

c. Interpret the results of the hypothesis test in view of your answer to part (b).

In each of Exercises 10.25–10.30, apply Procedure 10.2 on page 607 to obtain the required confidence interval.

10.25 Refer to Exercise 10.19.

a. Find a 98% confidence interval for the difference, $\mu_1 - \mu_2$, between the mean times served by prisoners in the fraud and firearms offense categories.

b. Interpret your result from part (a).

10.26 Refer to Exercise 10.20.

a. Determine a 90% confidence interval for the difference, $\mu_1 - \mu_2$, between the mean time it takes the new machine to pack 10 cartons and the mean time it takes the present machine to pack 10 cartons.

b. Interpret your result from part (a).

10.27 Refer to Exercise 10.21.

a. Obtain a 90% confidence interval for the difference between the mean height of males in the age group 25–34 years and the mean height of males in the age group 45–54 years.

b. Interpret your result from part (a).

10.28 Refer to Exercise 10.22.

a. Determine a 95% confidence interval for the difference between last year's mean VMTs by midwestern and southern households.

b. Interpret your result from part (a).

10.29 Refer to Exercise 10.23.

a. Obtain a 95% confidence interval for the difference between last year's mean fuel expenditures for households using natural gas and those using only electricity.

b. Interpret your answer in part (a).

10.30 Refer to Exercise 10.24.

a. Determine a 98% confidence interval for the difference between the mean yields of corn when the insecticide is used to control the insect pest and when sterilized males are used.

b. Interpret your answer in part (a).

EXTENDING THE CONCEPTS AND SKILLS

10.31 In this section we introduced the pooled t-test which provides a method for comparing two population means. In deriving the pooled t-test, we stated on page 600 that the variable

$$z = \frac{(\bar{x}_1 - \bar{x}_2) - (\mu_1 - \mu_2)}{\sigma\sqrt{(1/n_1) + (1/n_2)}}$$

cannot be used as a basis for the required test statistic because σ is unknown. Why do you think that is so?

10.32 The formula for the pooled variance, s_p^2, is given on page 601. Show that if the sample sizes, n_1 and n_2, are equal, then s_p^2 is just the mean of s_1^2 and s_2^2.

USING TECHNOLOGY  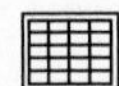 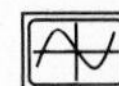

10.33 Refer to Exercises 10.19 and 10.25. Use Minitab or some other statistical software to

a. obtain normal probability plots, boxplots, and the standard deviations for the two samples.

b. perform the required hypothesis test and obtain the desired confidence interval.

c. Justify the use of your procedure in part (b).

10.34 Refer to Exercises 10.20 and 10.26. Use Minitab or some other statistical software to

a. obtain normal probability plots, boxplots, and the standard deviations for the two samples.

b. perform the required hypothesis test and obtain the desired confidence interval.

c. Justify the use of your procedure in part (b).

10.35 The U.S. Bureau of Labor Statistics conducts monthly surveys to estimate hourly earnings of nonsupervisory employees in various industry groups and publishes its findings in *Employment and Earnings.* Minitab's pooled t-procedure was applied to data obtained on hourly earnings of independent random samples of mine workers and construction workers. The resulting output is shown in Printout 10.2. Use the output to obtain

a. the null and alternative hypotheses.
b. the sample standard deviations of the hourly earnings for each group.
c. the number of mine workers sampled and the number of construction workers sampled.
d. the mean hourly earnings of the mine workers sampled and the mean hourly earnings of the construction workers sampled.
e. the value of the test statistic, t.
f. the P-value of the hypothesis test.
g. the smallest significance level at which the null hypothesis can be rejected.
h. the conclusion if the hypothesis test is performed at the 5% significance level.
i. a 90% confidence interval for the difference between the mean hourly earnings of mine and construction workers.
j. the pooled sample standard deviation.

10.36 Independent random samples were taken of 10 whites and 12 blacks, and their 24-hour protein intakes were recorded. Minitab's pooled t-procedure was applied to the data and yielded the output shown in Printout 10.3. From the output, determine

a. the null and alternative hypotheses.
b. the standard deviations of the two samples.
c. the mean protein intakes (per day) of the whites sampled and the blacks sampled.
d. the value of the test statistic, t.
e. the P-value of the hypothesis test.
f. the smallest significance level at which the null hypothesis can be rejected.
g. the conclusion if the hypothesis test is performed at the 5% significance level.
h. a 95% confidence interval for the difference between the mean protein intakes of whites and blacks.
i. the value of s_p.

PRINTOUT 10.2 Minitab output for Exercise 10.35

```
Two sample T for MINE vs CONSTRUC

          N      Mean      StDev   SE Mean
MINE      14    15.93       2.25      0.60
CONSTRUC  17    16.42       2.36      0.57

90% CI for mu MINE - mu CONSTRUC: ( -1.91,  0.93)
T-Test mu MINE = mu CONSTRUC (vs <): T = -0.59  P = 0.28  DF = 29
Both use Pooled StDev = 2.31
```

PRINTOUT 10.3 Minitab output for Exercise 10.36

```
Two sample T for WHITE vs BLACK

       N     Mean     StDev   SE Mean
WHITE  10    77.4      16.4      5.2
BLACK  12    68.8      15.2      4.4

95% CI for mu WHITE - mu BLACK: ( -5.5,  22.7)
T-Test mu WHITE = mu BLACK (vs not =): T = 1.27  P = 0.22  DF = 20
Both use Pooled StDev = 15.8
```

10.37 In this exercise you are to perform a simulation to illustrate the distribution of the pooled t-statistic, given in Key Fact 10.2 on page 601. You may want to refer to parts (b) and (c) of Exercise 7.45 on page 424.

a. Simulate 1000 samples of size four from a normally distributed variable with mean 100 and standard deviation 16. Then obtain the sample mean and sample standard deviation of each of the 1000 samples.

b. Simulate 1000 samples of size three from a normally distributed variable with mean 110 and standard deviation 16. Then obtain the sample mean and sample standard deviation of each of the 1000 samples.

c. Determine the value of the pooled t-statistic for each of the 1000 pairs of samples obtained in parts (a) and (b).

d. Obtain a histogram of the 1000 values found in part (c).

e. Theoretically, what is the distribution of all possible values of the pooled t-statistic?

f. Compare your results from parts (d) and (e).

10.3 INFERENCES FOR TWO POPULATION MEANS USING INDEPENDENT SAMPLES (STANDARD DEVIATIONS NOT ASSUMED EQUAL)

In Section 10.2 we examined methods of performing inferences to compare the means of two populations using independent samples. The methods discussed there, called pooled t-procedures, require that the standard deviations of the two populations be equal.

In this section we will develop inferential procedures (hypothesis tests and confidence intervals) to compare the means of two populations using independent samples that do not require the population standard deviations to be equal, even though they may be. As before, we suppose the population standard deviations are unknown, since that is usually the case in practice.

For our derivation we will assume the variable under consideration is normally distributed on each population. As we will see later, the resulting inferential procedures are approximately correct for large samples, regardless of distribution-type.

Hypothesis Tests for the Means of Two Populations Using Independent Samples

Let's begin by finding a test statistic. We know from Key Fact 10.1 on page 596 that, for independent samples, the standardized version of $\bar{x}_1 - \bar{x}_2$,

$$z = \frac{(\bar{x}_1 - \bar{x}_2) - (\mu_1 - \mu_2)}{\sqrt{(\sigma_1^2/n_1) + (\sigma_2^2/n_2)}},$$

has the standard normal distribution. Since we are assuming the population standard deviations, σ_1 and σ_2, are unknown, we cannot use this variable as a basis for the required test statistic. We therefore replace σ_1 and σ_2 by their sample estimates, s_1 and s_2, and obtain the variable

$$\frac{(\bar{x}_1 - \bar{x}_2) - (\mu_1 - \mu_2)}{\sqrt{(s_1^2/n_1) + (s_2^2/n_2)}},$$

which can be used as a basis for the required test statistic. This variable does not have the standard normal distribution, but it does have roughly a t-distribution, as indicated in the following key fact.

KEY FACT 10.3 DISTRIBUTION OF THE NONPOOLED t-STATISTIC

Suppose x is a normally distributed variable on each of two populations. Then, for independent samples of sizes n_1 and n_2 from the two populations, the variable

$$t = \frac{(\bar{x}_1 - \bar{x}_2) - (\mu_1 - \mu_2)}{\sqrt{(s_1^2/n_1) + (s_2^2/n_2)}}$$

has approximately a t-distribution. The degrees of freedom used is obtained from the sample data; it is

$$\Delta = \frac{\left[(s_1^2/n_1) + (s_2^2/n_2)\right]^2}{\dfrac{(s_1^2/n_1)^2}{n_1 - 1} + \dfrac{(s_2^2/n_2)^2}{n_2 - 1}},$$

rounded down to the nearest integer.

In view of Key Fact 10.3, we see that for a hypothesis test with null hypothesis H_0: $\mu_1 = \mu_2$, we can use the variable

$$t = \frac{\bar{x}_1 - \bar{x}_2}{\sqrt{(s_1^2/n_1) + (s_2^2/n_2)}}$$

as the test statistic and obtain the critical value(s) from the t-table, Table IV. Specifically, we have the procedure shown on the following page, which we often refer to as the **nonpooled t-test.**

Before we apply the nonpooled t-test, let's discuss the assumptions for its use. Assumption 1 (independent samples) is essential; the samples must be independent or the procedure does not apply.

Regarding Assumption 2, although the nonpooled t-test was derived under the condition that the variable under consideration is normally distributed on each of the two populations (normal populations), it also applies approximately for large samples regardless of distribution-type. Actually, the nonpooled t-test works reasonably well even for small or moderate-size samples from nonnormal populations provided the populations are not too nonnormal. In other words, the nonpooled t-test is robust to moderate violations of the normality assumption.

When considering the nonpooled t-test, it is also important to watch for outliers. The presence of outliers calls into question the normality assumption. And even for large samples, outliers can sometimes unduly affect a nonpooled t-test because the sample mean and sample standard deviation are not resistant to them.

PROCEDURE 10.3

THE NONPOOLED t-TEST FOR TWO POPULATION MEANS

ASSUMPTIONS

1. Independent samples
2. Normal populations or large samples

Step 1 The null hypothesis is H_0: $\mu_1 = \mu_2$ and the alternative hypothesis is one of the following:

H_a: $\mu_1 \neq \mu_2$ (Two-tailed)	or	H_a: $\mu_1 < \mu_2$ (Left-tailed)	or	H_a: $\mu_1 > \mu_2$ (Right-tailed)

Step 2 Decide on the significance level, α.

Step 3 The critical value(s) are

$\pm t_{\alpha/2}$ (Two-tailed)	or	$-t_\alpha$ (Left-tailed)	or	t_α (Right-tailed)

with df $= \Delta$, where

$$\Delta = \frac{\left[(s_1^2/n_1) + (s_2^2/n_2)\right]^2}{\dfrac{(s_1^2/n_1)^2}{n_1 - 1} + \dfrac{(s_2^2/n_2)^2}{n_2 - 1}},$$

rounded down to the nearest integer. Use Table IV to find the critical value(s).

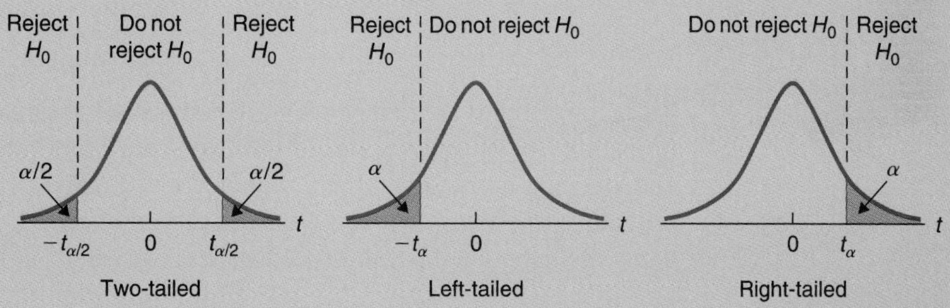

Step 4 Compute the value of the test statistic

$$t = \frac{\bar{x}_1 - \bar{x}_2}{\sqrt{(s_1^2/n_1) + (s_2^2/n_2)}}.$$

Step 5 If the value of the test statistic falls in the rejection region, reject H_0; otherwise, do not reject H_0.

Step 6 State the conclusion in words.

EXAMPLE 10.6 ILLUSTRATES PROCEDURE 10.3

Several neurosurgeons wanted to see whether a dynamic system (Z-plate) reduced the operative time relative to a static system (ALPS plate). R. Jacobowitz, Ph.D., an ASU professor, along with G. Vishteh, M.D., and other neurosurgeons, obtained the following data on operative times, in minutes, for the two systems. At the 1% significance level, do the data provide sufficient evidence to conclude that the mean operative time is less with the dynamic system than with the static system?

TABLE 10.7 Operative times, in minutes, for dynamic and static systems

Dynamic							Static		
370	360	510	445	295	315	490	430	445	455
345	450	505	335	280	325	500	455	490	535

SOLUTION First we present in Table 10.8 the required summary statistics for the two samples in Table 10.7. These statistics are obtained in the usual way.

TABLE 10.8 Summary statistics for the samples in Table 10.7

Dynamic	Static
$\bar{x}_1 = 394.6$	$\bar{x}_2 = 468.3$
$s_1 = 84.7$	$s_2 = 38.2$
$n_1 = 14$	$n_2 = 6$

Next we check the two conditions required for using the nonpooled t-test. Since the samples are independent, Assumption 1 is satisfied. Boxplots and normal probability plots (not shown) of the two samples in Table 10.7 reveal no outliers and, keeping in mind that the nonpooled t-test is robust to moderate violations of normality, show that we can consider Assumption 2 satisfied. We can therefore apply the nonpooled t-test, Procedure 10.3, to carry out the hypothesis test.

Step 1 State the null and alternative hypotheses.

Let μ_1 and μ_2 denote, respectively, the mean operative times for the dynamic and static systems. Then the null and alternative hypotheses are

H_0: $\mu_1 = \mu_2$ (mean dynamic time is not less than mean static time)

H_a: $\mu_1 < \mu_2$ (mean dynamic time is less than mean static time).

Note that the hypothesis test is left-tailed since a less-than sign (<) appears in the alternative hypothesis.

Step 2 Decide on the significance level, α.

The test is to be performed at the 1% significance level; thus $\alpha = 0.01$.

Step 3 *The critical value for a left-tailed test is $-t_\alpha$ with df $= \Delta$.*

From Step 2, $\alpha = 0.01$. Also, referring to Table 10.8, we find that

$$\text{df} = \Delta = \frac{\left[(84.7^2/14) + (38.2^2/6)\right]^2}{\dfrac{(84.7^2/14)^2}{14-1} + \dfrac{(38.2^2/6)^2}{6-1}} = 17 \text{ (rounded down).}$$

So the critical value is $-t_\alpha = -t_{0.01} = -2.567$, as seen in Fig. 10.3.

FIGURE 10.3
Criterion for deciding whether or not to reject the null hypothesis

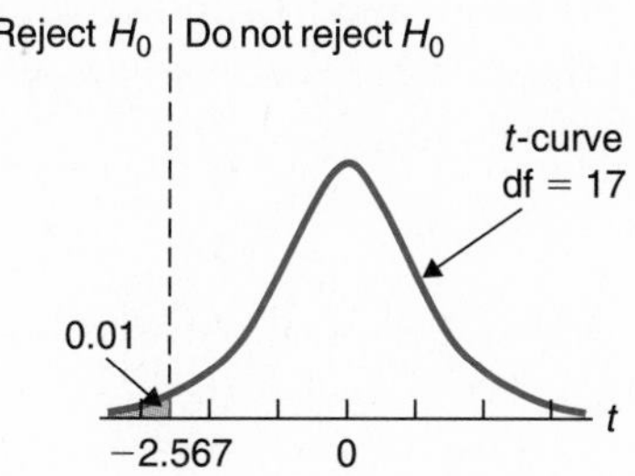

Step 4 *Compute the value of the test statistic*

$$t = \frac{\bar{x}_1 - \bar{x}_2}{\sqrt{(s_1^2/n_1) + (s_2^2/n_2)}}.$$

Referring again to Table 10.8, we obtain that

$$t = \frac{394.6 - 468.3}{\sqrt{(84.7^2/14) + (38.2^2/6)}} = -2.681.$$

Step 5 *If the value of the test statistic falls in the rejection region, reject H_0; otherwise, do not reject H_0.*

From Step 4 the value of the test statistic is $t = -2.681$, which, as we see from Fig. 10.3, falls in the rejection region. Thus we reject H_0.

Step 6 *State the conclusion in words.*

The test results are statistically significant at the 1% level; that is, at the 1% significance level, the data provide sufficient evidence to conclude that the mean operative time is less with the dynamic system than with the static system. □

Using the *P*-Value Approach

We can also use the P-value approach to conduct the hypothesis test in Example 10.6. From Step 4 we know that the value of the test statistic is $t = -2.681$.

Recalling that df = 17 and that the test is left-tailed, we find from Table IV that $0.005 < P < 0.01$.

In particular, because the P-value is less than the designated significance level of 0.01, we can reject H_0. And referring to Table 9.12 on page 543, we see that the data provide very strong evidence against the null hypothesis.

Confidence Intervals for the Difference Between the Means of Two Populations Using Independent Samples

We can also use Key Fact 10.3 on page 614 to derive the following confidence-interval procedure for the difference between two means, which we often refer to as the **nonpooled *t*-interval procedure.**

PROCEDURE 10.4

THE NONPOOLED *t*-INTERVAL PROCEDURE FOR TWO POPULATION MEANS

ASSUMPTIONS

1. Independent samples

2. Normal populations or large samples

Step 1 For a confidence level of $1-\alpha$, use Table IV to find $t_{\alpha/2}$ with df = Δ, where

$$\Delta = \frac{\left[(s_1^2/n_1) + (s_2^2/n_2)\right]^2}{\dfrac{(s_1^2/n_1)^2}{n_1 - 1} + \dfrac{(s_2^2/n_2)^2}{n_2 - 1}},$$

rounded down to the nearest integer.

Step 2 The endpoints of the confidence interval for $\mu_1 - \mu_2$ are

$$(\bar{x}_1 - \bar{x}_2) \pm t_{\alpha/2} \cdot \sqrt{(s_1^2/n_1) + (s_2^2/n_2)}.$$

EXAMPLE 10.7

ILLUSTRATES PROCEDURE 10.4

Use the sample data in Table 10.7 on page 616 to obtain a 98% confidence interval for the difference, $\mu_1 - \mu_2$, between the mean operative times of the dynamic and static systems.

SOLUTION We apply Procedure 10.4.

Step 1 *For a confidence level of* $1-\alpha$, *use Table IV to find* $t_{\alpha/2}$ *with df* $=\Delta$.

For a 98% confidence interval, $\alpha = 0.02$. As we saw in Example 10.6, df $= 17$. Consulting Table IV, we find that for df $= 17$, $t_{\alpha/2} = t_{0.02/2} = t_{0.01} = 2.567$.

Step 2 *The endpoints of the confidence interval for* $\mu_1 - \mu_2$ *are*

$$(\bar{x}_1 - \bar{x}_2) \pm t_{\alpha/2} \cdot \sqrt{(s_1^2/n_1) + (s_2^2/n_2)}.$$

From Step 1, $t_{\alpha/2} = 2.567$. Referring to Table 10.8 on page 616, we conclude that the endpoints of the confidence interval for $\mu_1 - \mu_2$ are

$$(394.6 - 468.3) \pm 2.567 \cdot \sqrt{(84.7^2/14) + (38.2^2/6)}$$

or -144.3 to -3.1. Consequently, we can be 98% confident that the difference between the mean operative times of the dynamic and static systems is somewhere between -144.3 minutes and -3.1 minutes. In other words, we can be 98% confident that compared to the static system, the dynamic system reduces the mean operative time by somewhere between 3.1 minutes and 144.3 minutes. ◨

  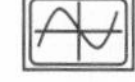

Pooled Versus Nonpooled

Suppose that we want to perform a hypothesis test to compare the means of two populations using independent samples. Further suppose either the variable under consideration is normally distributed for each of the two populations or the sample sizes are large. Then two tests are candidates for the job: the pooled t-test (Procedure 10.1 of Section 10.2) or the nonpooled t-test (Procedure 10.3 of this section).

In theory the pooled t-test requires that the population standard deviations be equal. What if the pooled t-test is used when in fact the population standard deviations are not equal? The answer to this question depends on several factors. If the population standard deviations are unequal, but not too unequal, and the sample sizes are nearly the same, then using the pooled t-test will not cause serious difficulties. If the population standard deviations are quite different, however, then using the pooled t-test can result in a significantly larger Type I error probability than the one specified (i.e., α).

On the other hand, the nonpooled t-test does not require that the population standard deviations be equal; it applies whether or not they are equal. Then why use the pooled t-test at all? The reason is that if the population standard deviations are equal or nearly so, then, on the average, the pooled t-test is slightly more powerful; that is, the probability of making a Type II error is somewhat smaller.

Thus we see that the pooled t-test should be used only when the two populations have nearly equal standard deviations; otherwise, the nonpooled t-test should be applied. Similar remarks apply to the pooled t-interval and nonpooled t-interval procedures.

KEY FACT 10.4 **CHOOSING BETWEEN A POOLED AND NONPOOLED PROCEDURE**

Suppose you want to compare the means of two populations using independent samples. When deciding between a pooled t-procedure and a nonpooled t-procedure, follow these guidelines: If you are reasonably sure the populations have nearly equal standard deviations, use a pooled t-procedure; otherwise, use a nonpooled t-procedure.

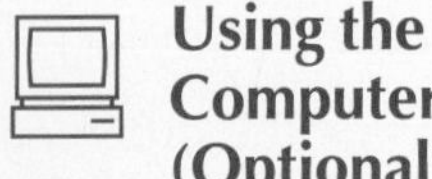

Using the Computer (Optional)

We can use Minitab to perform nonpooled t-procedures. To illustrate, we return once again to the hypothesis test and confidence interval comparing the mean operative times of two neurosurgery systems.

EXAMPLE 10.8 **USING MINITAB TO PERFORM NONPOOLED t-PROCEDURES**

Use Minitab to simultaneously perform the hypothesis test considered in Example 10.6 and obtain the confidence interval required in Example 10.7.

SOLUTION Let μ_1 and μ_2 denote, respectively, the mean operative times of the dynamic and static systems. The problem in Example 10.6 is to perform the hypothesis test

H_0: $\mu_1 = \mu_2$ (mean dynamic time is not less than mean static time)

H_a: $\mu_1 < \mu_2$ (mean dynamic time is less than mean static time)

at the 1% significance level; the problem in Example 10.7 is to obtain a 98% confidence interval for $\mu_1 - \mu_2$.

As we discovered earlier in Example 10.6, it is reasonable to use nonpooled t-procedures here. To apply Minitab, we begin by entering the two sets of sample data from Table 10.7 on page 616 in columns named DYNAMIC and STATIC. Then we proceed in the following way.

1 Choose **Stat ➤ Basic Statistics ➤ 2-Sample t...**
2 Select the **Samples in different columns** option button
3 Click in the **First** text box and specify DYNAMIC
4 Click in the **Second** text box and specify STATIC
5 Click the arrow button at the right of the **Alternative** drop-down list box and select **less than**
6 Click in the **Confidence level** text box and type 98
7 Deselect the **Assume equal variances** check box if it is selected (to indicate that we are not assuming the populations have equal standard deviations)
8 Click **OK**

The output that results is shown in Printout 10.4.

PRINTOUT 10.4
Minitab output for the nonpooled t-procedure

```
Two sample T for DYNAMIC vs STATIC

            N      Mean     StDev   SE Mean
DYNAMIC    14     394.6      84.7        23
STATIC      6     468.3      38.2        16

98% CI for mu DYNAMIC - mu STATIC: ( -144,  -3)
T-Test mu DYNAMIC = mu STATIC (vs <): T = -2.68  P = 0.0079  DF = 17
```

The first line of Printout 10.4 describes the test being performed: `Two sample T for DYNAMIC vs STATIC`. The next three lines display a table that gives the sample size, sample mean, sample standard deviation, and estimated standard error of the mean for each of the two samples.

On the fifth line of Printout 10.4, we find the required 98% confidence interval for the difference between the two population means. Thus we can be 98% confident that the difference, $\mu_1 - \mu_2$, between the mean operative times of the dynamic and static systems is somewhere between -144 minutes and -3 minutes.

The last line of Printout 10.4 provides a statement of the null and alternative hypotheses, followed by the value of the test statistic (`T = -2.68`), the P-value (`P = 0.0079`), and the degrees of freedom (`DF = 17`).

Since the P-value of 0.0079 is less than the specified significance level of 0.01, we reject H_0. At the 1% significance level, the data provide sufficient evidence to conclude that the mean operative time for the dynamic system is less than that for the static system. ❑

EXERCISES 10.3

STATISTICAL CONCEPTS AND SKILLS

10.38 Suppose you know that a variable is normally distributed on each of two populations. Further suppose that you want to perform a hypothesis test to compare the two population means using independent samples. In each case, decide whether you would use the pooled or nonpooled t-test and give a reason for your answer.

a. You know that the population standard deviations are equal.

b. You know that the population standard deviations are not equal.

c. The sample standard deviations are 23.6 and 25.2 and the sample sizes are each 25.

d. The sample standard deviations are 23.6 and 59.2.

10.39 What difference in assumptions is there for use of the pooled and nonpooled t-procedures?

10.40 Discuss the relative advantages and disadvantages of using pooled and nonpooled t-procedures.

Preliminary data analyses indicate that it is reasonable to use nonpooled t-procedures in Exercises 10.41–10.46. For each exercise, apply a nonpooled t-test to perform *the required hypothesis test using either the critical-value approach or the P-value approach.*

10.41 Independent random samples of 17 sophomores and 13 juniors attending a large state university yielded the following data on cumulative grade point average (GPA).

Sophomores			Juniors		
3.04	2.92	2.86	2.56	3.47	2.65
1.71	3.60	3.49	2.77	3.26	3.00
3.30	2.28	3.11	2.70	3.20	3.39
2.88	2.82	2.13	3.00	3.19	2.58
2.11	3.03	3.27	2.98		
2.60	3.13				

At the 5% significance level, do the data provide sufficient evidence to conclude that the mean GPAs of sophomores and juniors at the university differ? *(Note:* $\bar{x}_1 = 2.84$, $s_1 = 0.52$, $\bar{x}_2 = 2.98$, $s_2 = 0.31$.*)*

10.42 According to the publication, *High School Profile Report,* in past years college-bound males have outperformed college-bound females on the mathematics portion of tests given by the American College Testing (ACT) Program. Independent samples of this year's scores yield the following data.

Males			Females		
34	34	30	18	33	27
18	24	16	11	23	23
15	35	13	18	20	26
21	24	11	10	20	22
11	15	18	14	21	15

Does it appear that college-bound males are, on the average, still outperforming college-bound females on the mathematics portion of ACT tests? Use $\alpha = 0.05$. *(Note:* $\bar{x}_1 = 21.27$, $s_1 = 8.50$, $\bar{x}_2 = 20.07$, $s_2 = 6.13$.*)*

10.43 Refer to Example 10.6 on page 616. The researchers also obtained data on the number of acute postoperative days in the hospital using the dynamic and static systems. Here are the data.

Dynamic							Static		
7	5	8	8	6	7	7	6	18	9
9	10	7	7	7	7	8	7	14	9

At the 5% significance level, do the data provide sufficient evidence to conclude that the mean number of acute postoperative days in the hospital are fewer with the dynamic system than with the static system? *(Note:* $\bar{x}_1 = 7.36$, $s_1 = 1.22$, $\bar{x}_2 = 10.50$, $s_2 = 4.59$.*)*

10.44 Costs to community hospitals per patient per day are reported by the American Hospital Association in *Hospital Stat.* Independent random samples of 12 daily costs in Georgia hospitals and 15 daily costs in Illinois hospitals gave the following data, in dollars.

Georgia			Illinois			
524	753	673	1276	1200	1316	1167
887	590	522	766	552	1296	386
1230	1043	1088	633	1009	1178	1068
466	833	957	834	1283	855	

At the 10% significance level, do the data provide sufficient evidence to conclude that there is a difference in mean cost to community hospitals per patient per day in Georgia and Illinois? *(Note:* $\bar{x}_1 = 797.2$, $s_1 = 250.0$, $\bar{x}_2 = 987.9$, $s_2 = 300.2$.*)*

10.45 The U.S. Department of Agriculture compiles information on acreage, production, and value of potatoes and publishes its findings in *Agricultural Statistics.* Potato yield is measured in hundreds of pounds (cwt) per acre. Independent random samples of thirty-two 1-acre plots of potatoes from Nevada and forty 1-acre plots of potatoes from Idaho give the following yields.

Nevada				Idaho				
283	254	328	292	229	267	326	309	231
315	336	378	314	283	344	310	258	316
312	328	272	307	241	281	218	284	311
348	233	354	400	254	217	267	299	266
341	313	309	308	264	264	290	312	298
340	300	316	268	305	244	303	299	285
259	276	271	362	308	260	204	291	242
340	339	300	333	329	315	246	322	293

At the 5% significance level, do the data provide sufficient evidence to conclude that Nevada has a larger mean potato yield than Idaho? *(Note:* The sample mean and standard deviation of the Nevada data are 313.4 cwt and 37.2 cwt, respectively, and that for the Idaho data are 279.6 cwt and 34.6 cwt, respectively.*)*

10.46 The owner of a chain of car washes needed to decide between two brands of hot wax. One of the brands, Sureglow, costs less than the other brand, Mirror-sheen. So unless there was strong evidence that Mirror-sheen outlasts Sureglow, the owner would purchase Sureglow. With the cooperation of several local automobile dealers, 30 cars were selected to take part in a test. The cars were randomly divided into two groups of 15 cars each. One group was waxed with Sureglow and the other with Mirror-sheen. The cars were then exposed to the same environmental conditions. In the following table, you will find the data obtained on effectiveness times, in days.

Sureglow			Mirror-sheen		
93	85	86	90	95	96
96	88	93	97	88	91
87	91	91	91	92	97
91	82	87	94	94	92
88	91	88	100	89	92

a. At the 1% significance level, do the data provide sufficient evidence to conclude that Mirror-sheen has a longer effectiveness time, on the average, than Sureglow? *(Note:* $\bar{x}_1 = 89.13$, $s_1 = 3.60$, $\bar{x}_2 = 93.20$, $s_2 = 3.34$.*)*

b. Do the data provide strong evidence that Mirror-sheen outlasts Sureglow? Explain your answer.

c. Identify the study as a designed experiment or an observational study. Explain your answer.

In each of Exercises 10.47–10.52, apply Procedure 10.4 on page 618 to obtain the required confidence interval.

10.47 Refer to Exercise 10.41.

a. Find a 95% confidence interval for the difference, $\mu_1 - \mu_2$, between the mean GPAs of sophomores and juniors at the university.

b. Interpret your answer in words.

10.48 Refer to Exercise 10.42.

a. Find a 90% confidence interval for the difference, $\mu_1 - \mu_2$, between this year's mean mathematics ACT scores for males and females.

b. Interpret your answer in words.

10.49 Refer to Exercise 10.43.

a. Find a 90% confidence interval for the difference between the mean numbers of acute postoperative days in the hospital with the dynamic and static systems.

b. Interpret your answer in words.

10.50 Refer to Exercise 10.44.

a. Determine a 90% confidence interval for the difference between the mean costs to community hospitals per patient per day in Georgia and Illinois.

b. Interpret your answer in words.

10.51 Refer to Exercise 10.45.

a. Find a 90% confidence interval for the difference between the mean yields per acre of potatoes for Nevada and Idaho.

b. Interpret your answer in words.

10.52 Refer to Exercise 10.46.

a. Determine a 98% confidence interval for the difference between the mean effectiveness times of Sureglow and Mirror-sheen.

b. Interpret your answer in words.

EXTENDING THE CONCEPTS AND SKILLS

10.53 In Exercise 10.43 we conducted a nonpooled t-test to decide whether the mean number of acute postoperative days in the hospital are fewer with the dynamic system than with the static system.

a. Repeat that hypothesis test, but this time using a pooled t-test.

b. Compare your decisions with the pooled and nonpooled t-tests.

c. Which test do you think is more appropriate here, the pooled or nonpooled t-test? Explain your answer.

10.54 In Example 10.6 on page 616, we conducted a nonpooled t-test to decide whether the mean operative time is less with the dynamic system than with the static system.

a. Repeat that hypothesis test, but this time using a pooled t-test.

b. Compare your decisions with the pooled and nonpooled t-tests.

c. Which test do you think is more appropriate here, the pooled or nonpooled t-test? Explain your answer.

10.55 Each pair of graphs in Fig. 10.4 shows the distributions of a variable on two populations. Suppose that in each case, we want to perform a small-sample hypothesis test to compare the means of the two populations using independent samples. In each case decide whether the pooled t-test, nonpooled t-test, or neither should be used. Explain your answers.

10.56 Suppose a variable is normally distributed on each of two populations and that the population standard deviations are equal. Further suppose that we want to conduct a hypothesis test to compare the means of the two populations.

a. Are the assumptions met for use of the pooled t-test? Explain your answer.
b. Are the assumptions met for use of the nonpooled t-test? Explain your answer.
c. Which test would be best? Explain your answer.

USING TECHNOLOGY  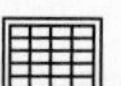 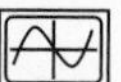

10.57 Refer to Exercises 10.41 and 10.47. Use Minitab or some other statistical software to

a. obtain boxplots and normal probability plots for the two samples.
b. perform the required hypothesis test and obtain the desired confidence interval.
c. Justify the use of your procedure in part (b).

10.58 Refer to Exercises 10.42 and 10.48. Use Minitab or some other statistical software to

a. obtain boxplots and normal probability plots for the two samples.
b. perform the required hypothesis test and obtain the desired confidence interval.
c. Justify the use of your procedure in part (b).

10.59 The National Center for Health Statistics compiles information on divorces in *Vital Statistics of the United States*. Independent random samples were taken of divorced males and females to decide whether the mean age at the time of first divorce for males is greater than that for females. Minitab's nonpooled t-procedure yielded the output shown in Printout 10.5. Use the printout to determine

a. the null and alternative hypotheses.
b. the standard deviation of the ages for each sample.
c. the number of males sampled and the number of females sampled.
d. the mean age at first divorce for both the males sampled and the females sampled.
e. the value of the test statistic, t.
f. the P-value of the hypothesis test.
g. the smallest significance level at which the null hypothesis can be rejected.
h. the conclusion if the hypothesis test is performed at the 5% significance level.
i. a 90% confidence interval for the difference between the mean ages at first divorce for males and females.
j. Do you think that it is reasonable to apply the nonpooled t-procedures here? Explain your answer.

10.60 The U.S. Federal Highway Administration publishes statistics on motor vehicle travel by type of vehicle in *Highway Statistics Summary*. To compare the mean numbers of miles driven last year by cars and trucks, Minitab's nonpooled t-procedure was applied to mileage data for independent random samples of cars and trucks. The resulting output is shown in Printout 10.6, where number of miles is given in thousands. Use the printout to determine

a. the null and alternative hypotheses.
b. the standard deviations of the number of miles driven last year for the cars sampled and for the trucks sampled.
c. the number of cars sampled and the number of trucks sampled.
d. the mean number of miles driven last year for both the cars sampled and the trucks sampled.
e. the value of the test statistic, t.
f. the P-value of the hypothesis test.
g. the smallest significance level at which the null hypothesis can be rejected.
h. the conclusion if the hypothesis test is performed at the 5% significance level.
i. a 95% confidence interval for the difference between the mean number of miles that cars were driven last year and the mean number of miles that trucks were driven last year.
j. Do you think that it is reasonable to apply the nonpooled t-procedures here? Explain your answer.

FIGURE 10.4 Figure for Exercise 10.55

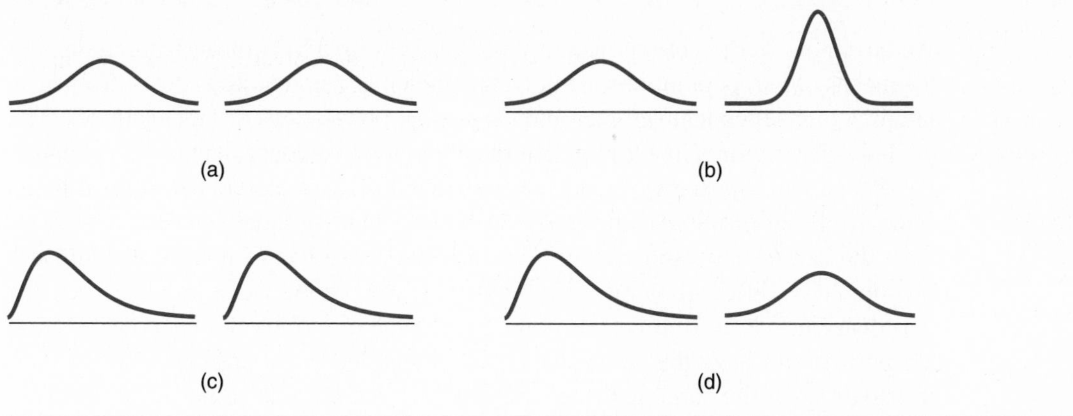

PRINTOUT 10.5 Minitab output for Exercise 10.59

```
Two sample T for MALES vs FEMALES

          N      Mean      StDev   SE Mean
MALES    11     35.69       4.62       1.4
FEMALES  12     32.81       7.25       2.1

90% CI for mu MALES - mu FEMALES: ( -1.5,  7.2)
T-Test mu MALES = mu FEMALES (vs >): T = 1.15  P = 0.13  DF = 18
```

PRINTOUT 10.6 Minitab output for Exercise 10.60

```
Two sample T for CARS vs TRUCKS

          N      Mean      StDev   SE Mean
CARS     15     11.81       4.90       1.3
TRUCKS   10      28.1       12.5       4.0

95% CI for mu CARS - mu TRUCKS: ( -25.5,  -7.0)
T-Test mu CARS = mu TRUCKS (vs not =): T = -3.91  P = 0.0029  DF = 10
```

10.4 THE MANN–WHITNEY TEST*

We have now studied two procedures for performing a hypothesis test to compare the means of two populations, namely, the pooled and nonpooled t-tests. Both tests require (1) independent samples and (2) normal populations or large samples. The pooled t-test requires in addition that the population standard deviations are equal.

Recall that the shape of a normal distribution is completely determined by its standard deviation. So two normal distributions having equal standard deviations have the same shape and two normal distributions having unequal standard deviations have different shapes. Consequently, the pooled t-test applies when the two distributions (one for each population) of the variable under consideration are normal and have the same shape; the nonpooled t-test applies when the two distributions of the variable under consideration are normal, same shape or not.

Another procedure for performing a hypothesis test to compare the means of two populations using independent samples is the **Mann–Whitney test.** This nonparametric test, introduced by Wilcoxon and further developed by Mann and Whitney, is also commonly referred to as the **Wilcoxon rank-sum test** or the **Mann–Whitney–Wilcoxon test.** The Mann–Whitney test applies when the two distributions of the variable under consideration have the same shape, but does not require that they be normal or have any other specific shape. Figure 10.5 summarizes our discussion graphically.

FIGURE 10.5 Appropriate procedure for comparing two population means using independent samples

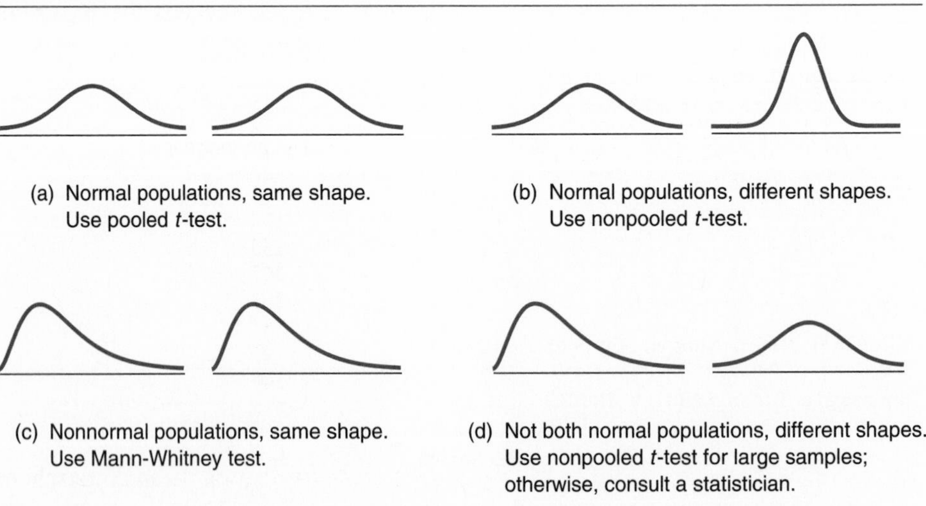

(a) Normal populations, same shape.
Use pooled t-test.
(b) Normal populations, different shapes.
Use nonpooled t-test.
(c) Nonnormal populations, same shape.
Use Mann-Whitney test.
(d) Not both normal populations, different shapes.
Use nonpooled t-test for large samples; otherwise, consult a statistician.

* This section continues the optional coverage of nonparametric statistics.

In Example 10.9 we introduce and explain the reasoning behind the Mann–Whitney test. Following that example, we examine how to obtain critical values for a Mann–Whitney test and then present a procedure for performing such a test.

EXAMPLE 10.9 INTRODUCES THE MANN–WHITNEY TEST

A nationwide shipping firm purchased a new computer system to track its current shipments, pickups, and deliveries. The system was linked to computer terminals in all regional offices, where office personnel could type in requests for information on the location of shipments and get answers immediately on display screens.

The company had to set up a training program to teach its staff how to use the computer terminals and decided to hire a technical writer to compose a short self-study manual for that purpose. The manual was designed so that a person could read it and be ready to use the computer terminal in 2 hours.

Some employees were able to apply the procedures outlined in the manual in very little time; other employees took considerably longer. Someone suggested that the reason for this difference in comprehension times might be that some employees had previous experience with computers whereas others did not. To test this suggestion, independent samples of employees with and without computer experience were randomly selected.

The times, in minutes, required for these employees to comprehend the manual are displayed in Table 10.9. At the 5% significance level, do the data provide sufficient evidence to conclude that the mean comprehension time for all employees without computer experience exceeds the mean comprehension time for all employees with computer experience?

TABLE 10.9
Times, in minutes, required to comprehend the self-study manual

Without experience	With experience
139	142
118	109
164	130
151	107
182	155
140	88
134	95
	104

SOLUTION Let μ_1 and μ_2 denote, respectively, the mean comprehension times for all employees without computer experience and with computer experience. Then the null and alternative hypotheses are

H_0: $\mu_1 = \mu_2$ (mean time for inexperienced employees is not greater)

H_a: $\mu_1 > \mu_2$ (mean time for inexperienced employees is greater).

As we said, use of the Mann–Whitney test requires that the two distributions of the variable under consideration have the same shape. If the comprehension-time distributions for employees without and with computer experience have the same shape, then the distributions of the two samples in Table 10.9 should have roughly the same shape.

To check this, we constructed in Fig. 10.6 a **back-to-back stem-and-leaf diagram** of the two samples in Table 10.9. In a back-to-back stem-and-leaf diagram, the leaves for the first sample are on the left, the stems are in the middle, and the leaves for the second sample are on the right.

FIGURE 10.6 Back-to-back stem-and-leaf diagram of the two comprehension-time samples in Table 10.9

Without experience		With experience
	8	8
	9	5
	10	9 7 4
8	11	
	12	
4 9	13	0
0	14	2
1	15	5
4	16	
	17	
2	18	

The stem-and-leaf diagrams in Fig. 10.6 have roughly the same shape and so do not reveal any obvious violations of the same-shape condition.† We will therefore apply the Mann–Whitney test to carry out the hypothesis test.

To apply the Mann–Whitney test, we first rank all the data from both samples combined. (It is very helpful to refer to Fig. 10.6 when ranking the data.) The results of ranking the data are depicted in Table 10.10, which shows, for instance, that the first employee without computer experience had the ninth shortest comprehension time among all 15 employees in the two samples combined.

The idea behind the Mann–Whitney test is a simple one: If the sum of the ranks for the sample of employees without experience is too large, then we take this as evidence that the mean comprehension time for all employees without experience exceeds that for all employees with experience (i.e., we reject the null hypothesis

† For ease in explaining the Mann–Whitney test, we have chosen an example in which the sample sizes are very small. This, however, makes it difficult to effectively check the same-shape condition. In general, we must proceed cautiously when dealing with very small samples.

TABLE 10.10
Results of ranking the combined data from Table 10.9

Without experience	Overall rank	With experience	Overall rank
139	9	142	11
118	6	109	5
164	14	130	7
151	12	107	4
182	15	155	13
140	10	88	1
134	8	95	2
		104	3

and conclude that the alternative hypothesis is true). From Table 10.10 we see that the sum of the ranks for the sample of employees without experience, denoted by M, is equal to

$$9 + 6 + 14 + 12 + 15 + 10 + 8 = 74.$$

To decide whether $M = 74$ is large enough to warrant rejection of the null hypothesis, we need a table of critical values for M, which can be found in Table VII in Appendix A. We will discuss that table and then return to complete the hypothesis test. □

Using the Mann–Whitney Table

The test statistic, M, for a Mann–Whitney test is the sum of the ranks associated with the smaller sample size. For instance, in Example 10.9, the smaller sample size is the one for the employees without computer experience, as seen in Table 10.9; so M is the sum of the ranks for the sample of employees without computer experience.

In general, it is convenient to designate the populations so that the sample size for Population 1 is less than or equal to the sample size for Population 2, that is, $n_1 \leq n_2$. Under those conditions, the test statistic, M, for a Mann–Whitney test is the sum of the ranks for the sample from Population 1.

Tables VI and VII in Appendix A provide critical values for a Mann–Whitney test. Table VI supplies critical values for a one-tailed test with $\alpha = 0.025$ or a two-tailed test with $\alpha = 0.05$; Table VII supplies critical values for a one-tailed test with $\alpha = 0.05$ or a two-tailed test with $\alpha = 0.10$.[†]

The sample size for the sample from Population 1 is given along the top of each table and the sample size for the sample from Population 2 along the left side. The numbers in the columns headed M_ℓ and M_r are, respectively, lower and

[†] Actually these are only approximate significance levels; but they are considered close enough in practice.

upper critical values. Note that *a critical value from Table VI or Table VII is to be included as part of the rejection region.* Example 10.10 illustrates the use of the Mann–Whitney tables.

EXAMPLE 10.10 USING THE MANN–WHITNEY TABLES

Determine the critical value, rejection region, and nonrejection region for the hypothesis test in Example 10.9.

SOLUTION Since the hypothesis test in Example 10.9 is right-tailed and is to be performed at the 5% significance level, we use Table VII. The sample sizes are $n_1 = 7$ and $n_2 = 8$. So to determine the critical value, we first locate the column labeled 7 along the top of the table and then go down that column until we are in the row labeled 8 along the left side of the table. There we find two numbers, 41 and 71, the second of which is the critical value for this right-tailed test.

Thus, for a right-tailed test at the 5% significance level, the critical value is 71. This means that the rejection region consists of all M-values greater than or equal to 71, and the nonrejection region consists of all M-values less than 71; in other words, we reject H_0 if $M \geq 71$ and do not reject H_0 if $M < 71$, as seen in Fig. 10.7. ◨

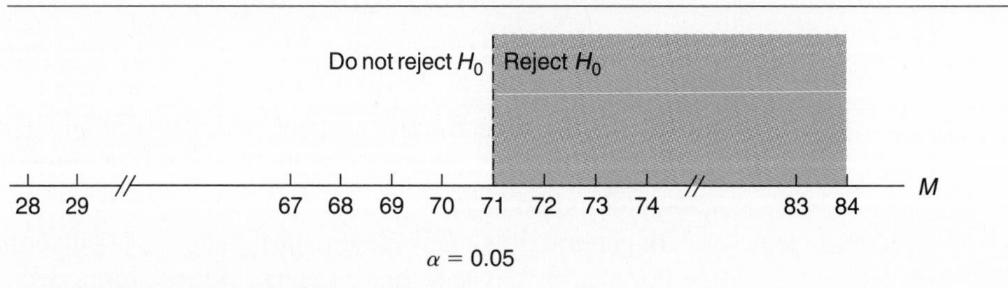

FIGURE 10.7 Rejection and nonrejection regions for a right-tailed Mann–Whitney test with $\alpha = 0.05$, $n_1 = 7$, and $n_2 = 8$

Performing the Mann–Whitney Test

We now present a step-by-step procedure for conducting a Mann–Whitney test. In stating the assumptions for the test, we will, for brevity, use the phrase "same shape populations" to indicate that the two distributions (one for each population) of the variable under consideration have the same shape.

The Mann–Whitney test can be used to compare two population medians as well as two population means. We will state the procedure in terms of population means. To employ the procedure for population medians, simply replace μ_1 by η_1 and μ_2 by η_2.

PROCEDURE 10.5

THE MANN–WHITNEY TEST FOR TWO POPULATION MEANS

ASSUMPTIONS

1. Independent samples
2. Same shape populations
3. $n_1 \leq n_2$

Step 1 The null hypothesis is H_0: $\mu_1 = \mu_2$ and the alternative hypothesis is one of the following:

H_a: $\mu_1 \neq \mu_2$ (Two-tailed)	or	H_a: $\mu_1 < \mu_2$ (Left-tailed)	or	H_a: $\mu_1 > \mu_2$ (Right-tailed)

Step 2 Decide on the significance level, α.

Step 3 The critical value(s) are

M_ℓ and M_r (Two-tailed)	or	M_ℓ (Left-tailed)	or	M_r (Right-tailed)

Use Table VI or Table VII to find the critical value(s).

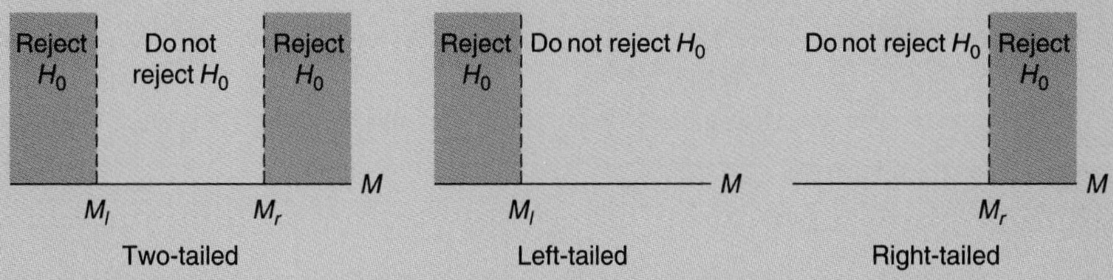

Step 4 Construct a work table of the following form:

Sample from Population 1	Overall rank	Sample from Population 2	Overall rank
⋮	⋮	⋮	⋮

Step 5 Compute the value of the test statistic

$$M = \text{sum of the ranks for sample data from Population 1.}$$

Step 6 If the value of the test statistic falls in the rejection region, reject H_0; otherwise, do not reject H_0.

Step 7 State the conclusion in words.

Note: When there are ties in the sample data, ranks are assigned in the same way as in the Wilcoxon signed-rank test. Namely, if two or more observations are tied, each is assigned the mean of the ranks they would have had if there were no ties.

EXAMPLE 10.11 ILLUSTRATES PROCEDURE 10.5

As an application of Procedure 10.5, let's complete the hypothesis test of Example 10.9. Independent samples of employees with and without computer experience were timed to see how long it would take them to comprehend a self-study manual that explained how to use a computer to track their company's products. The times, in minutes, are repeated in Table 10.11. At the 5% significance level, do the data provide sufficient evidence to conclude that the mean comprehension time for employees without computer experience exceeds that for employees with computer experience?

TABLE 10.11 Times, in minutes, required to comprehend the self-study manual

Without experience	With experience
139	142
118	109
164	130
151	107
182	155
140	88
134	95
	104

SOLUTION We apply Procedure 10.5.

Step 1 *State the null and alternative hypotheses.*

Let μ_1 and μ_2 denote the mean comprehension times for all employees without and with computer experience, respectively. Then the null and alternative hypotheses are

H_0: $\mu_1 = \mu_2$ (mean time for inexperienced employees is not greater)

H_a: $\mu_1 > \mu_2$ (mean time for inexperienced employees is greater).

Note that the hypothesis test is right-tailed since a greater-than sign (>) appears in the alternative hypothesis.

Step 2 *Decide on the significance level, α.*

We are to perform the hypothesis test at the 5% significance level; so $\alpha = 0.05$.

Step 3 *The critical value for a right-tailed test is M_r.*

We have already determined the critical value in Example 10.10. It is $M_r = 71$, as seen in Fig. 10.7 on page 630.

Step 4 *Construct a work table.*

We have already done this in Table 10.10 on page 629.

Step 5 *Compute the value of the test statistic*

$$M = \text{sum of the ranks for sample data from Population 1.}$$

Referring to the second column of Table 10.10, we find that

$$M = 9 + 6 + 14 + 12 + 15 + 10 + 8 = 74.$$

Step 6 *If the value of the test statistic falls in the rejection region, reject H_0; otherwise, do not reject H_0.*

From Step 5 the value of the test statistic is $M = 74$. Figure 10.7 shows that this falls in the rejection region. Thus we reject H_0.

Step 7 *State the conclusion in words.*

The test results are statistically significant at the 5% level; that is, at the 5% significance level, the data provide sufficient evidence to conclude that the mean comprehension time for employees without computer experience exceeds that for employees with computer experience. Evidently, those with computer experience can, on the average, comprehend the training manual more quickly than those

without. ◨

Comparison of the Mann–Whitney Test and the Pooled *t*-Test

In Section 10.2 we learned how to perform a pooled t-test to compare two population means using independent samples when the variable under consideration is normally distributed on each of the two populations and the population standard deviations are equal. Since two normal distributions having equal standard deviations have the same shape, we can also use the Mann–Whitney test to perform such a hypothesis test.

Which test is the better one to use under these circumstances? As you might expect, it is the pooled t-test because that test is designed expressly for normal populations; under conditions of normality, the pooled t-test is more powerful than the Mann–Whitney test. What is somewhat surprising is that the pooled t-test is not much more powerful than the Mann–Whitney test.

On the other hand, if the two distributions of the variable under consideration have the same shape but are not normal, then the Mann–Whitney test is usually more powerful than the pooled t-test, often considerably so. In summary, we have the following key fact.

KEY FACT 10.5 **THE MANN–WHITNEY TEST VERSUS THE POOLED t-TEST**

Suppose that the distributions of a variable of two populations have the same shape and that you want to compare the two population means using independent samples. When deciding between the pooled t-test and the Mann–Whitney test, follow these guidelines: If you are reasonably sure that the two distributions are normal, use the pooled t-test; otherwise, use the Mann–Whitney test.

Using the Computer (Optional)

Procedure 10.5 on page 631 provides a step-by-step method for performing a Mann–Whitney test. Alternatively, we can use Minitab to carry out a Mann–Whitney test.

As we said earlier, a Mann–Whitney test can be used to perform a hypothesis test to compare two population medians as well as two population means. Minitab presents the output of the Mann–Whitney test in terms of medians, but most of that output can be interpreted in terms of means. With this in mind, we now present Example 10.12.

EXAMPLE 10.12 USING MINITAB TO PERFORM A MANN-WHITNEY TEST

Use Minitab to perform the hypothesis test in Example 10.11.

SOLUTION Let μ_1 and μ_2 denote the mean times for comprehension of the self-study manual for all employees without and with computer experience, respectively. The problem is to use the Mann–Whitney procedure to perform the hypothesis test

H_0: $\mu_1 = \mu_2$ (mean time for inexperienced employees is not greater)

H_a: $\mu_1 > \mu_2$ (mean time for inexperienced employees is greater)

at the 5% significance level. Note that the hypothesis test is right-tailed since a greater-than sign (>) appears in the alternative hypothesis.

First we store the two samples from Table 10.11 on page 632 in columns named WITHOUT and WITH. Then we proceed in the following manner.

1 Choose **Stat ➤ Nonparametrics ➤ Mann-Whitney...**
2 Specify WITHOUT in the **First Sample** text box
3 Click in the **Second Sample** text box and specify WITH
4 Click the arrow button at the right of the **Alternative** drop-down list box and select **greater than**
5 Click **OK**

The output obtained is displayed in Printout 10.7.

PRINTOUT 10.7
Minitab output for the Mann-Whitney test

```
WITHOUT     N =   7     Median =        140.00
WITH        N =   8     Median =        108.00
Point estimate for ETA1-ETA2 is          31.50
95.7 Percent CI for ETA1-ETA2 is (3.99,56.00)
W = 74.0
Test of ETA1 = ETA2  vs  ETA1 > ETA2 is significant at 0.0214
```

All references to the median in Printout 10.7 can be interpreted as references to the mean except for the ones in the first and second lines. The first and second lines give the sample sizes and sample medians for the two samples. On the next two lines we find, respectively, a point estimate and a confidence interval for the difference between the population medians, $\eta_1 - \eta_2$ (ETA1-ETA2). By default the confidence level is chosen to be as close to 95% as possible; but other specifications are allowed.

The next-to-last line of Printout 10.7 displays the value of the test statistic,

$$M = \text{sum of the ranks for sample data from Population 1},$$

for a Mann–Whitney test (Minitab uses W instead of M); thus we see that the value of the test statistic is 74.0.

The final line of the output provides a statement of the null and alternative hypotheses (the ETAs can be replaced by MUs) and an approximate P-value; so $P = 0.0214$, approximately.[†] Since the P-value is less than the designated significance level of 0.05, we reject H_0. At the 5% significance level, the data provide sufficient evidence to conclude that employees with computer experience can, on the average, comprehend the training manual more quickly than those without computer experience. □

EXERCISES 10.4

STATISTICAL CONCEPTS AND SKILLS

10.61 Why do two normal distributions having equal standard deviations have the same shape?

10.62 State the conditions that are required for using the Mann–Whitney test.

10.63 Suppose you know that, for two populations, the distributions of the variable under consideration have the same shape. Further suppose that you want to perform a hypothesis test to compare the two population means using independent samples. In each case, decide whether you would use the pooled t-test or the Mann–

[†] Minitab employs a normal approximation with a continuity correction factor to obtain the approximate P-value. If there are ties in the data (there are none in this example), Minitab also prints an approximate P-value that adjusts for the ties. When there are ties, this latter approximation is usually closer to the actual P-value than the former approximation.

Whitney test and give a reason for your answer. You know the distributions of the variable are

a. normal. **b.** not normal.

10.64 Part of conducting a Mann–Whitney test involves ranking all the data from both samples combined. Explain how to deal with ties.

In Exercises 10.65–10.68, use the Mann–Whitney test, Procedure 10.5 on page 631, to carry out the required hypothesis tests.

10.65 A college chemistry instructor was concerned about the detrimental effects of poor mathematics background on her students. She randomly selected 15 students and divided them according to math background. Their semester averages turned out to be the following.

Fewer than two years of high-school algebra	Two or more years of high-school algebra
58 61	84 92 75
81 64	67 83 81
74 43	65 52 74

At the 5% significance level, do the data provide sufficient evidence to conclude that in this teacher's chemistry courses, students with fewer than two years of high-school algebra have a lower mean semester average than those with two or more years?

10.66 Independent random samples of two models of power lawn mower yielded the following data on number of years until major breakdown.

Brand 1	2.3 3.7 5.9 6.8 3.5
Brand 2	1.9 3.8 6.4 5.6 4.9

Do the data suggest that the mean times until major breakdown differ for the two models? Perform the required hypothesis test using a significance level of 0.05.

10.67 The National Center for Education Statistics surveys college libraries to obtain information on the number of volumes held. Results of the surveys are published in *Digest of Education Statistics* and *Academic Libraries*. Independent random samples of public and private colleges yield the following data on number of volumes held, in thousands.

Public	79 41 516 15 24 411 265
Private	139 603 113 27 67 500

At the 5% significance level, can we conclude that the median number of volumes held by public colleges is less than that held by private colleges? (*Note:* The sample size for the public colleges exceeds that for the private colleges.)

10.68 The U.S. Bureau of Labor Statistics publishes data on weekly earnings of full-time wage and salary workers in *Employment and Earnings*. Independent random samples of male and female workers give the following data on weekly earnings, in dollars.

Men	Women
826 2523	1994 2109
1790 288	510 291
477 317	426 274
307 718	290 1097
	1361 328

At the 5% significance level, do the data provide sufficient evidence to conclude that the median weekly earnings of male full-time wage and salary workers exceeds the median weekly earnings of female full-time wage and salary workers?

10.69 The U.S. Bureau of Prisons publishes data in *Statistical Report* on the times served by prisoners released from federal institutions for the first time. Independent random samples of released prisoners in the fraud and firearms offense categories yielded the following information on time served, in months.

Fraud	Firearms
3.6 17.9	25.5 23.8
5.3 5.9	10.4 17.9
10.7 7.0	18.4 21.9
8.5 13.9	19.6 13.3
11.8 16.6	20.9 16.1

a. Do the data provide sufficient evidence to conclude that the mean time served for fraud is less than that for firearms offenses? Perform a Mann–Whitney test using a significance level of 0.05.

b. The hypothesis test in part (a) was done in Exercise 10.19 using the pooled t-test. The assumption in that exercise is that times served for both offense categories are normally distributed and have equal standard deviations. Presuming that is in fact true, why is it permissible to perform a Mann–Whitney test to compare the means? Is it better in this case to use the pooled t-test or the Mann–Whitney test? Explain your answers.

10.70 In a packing plant, a machine packs cartons with jars. Supposedly, a new machine will pack faster on the average than the machine currently used. To test that hypothesis, the times it takes each machine to pack 10 cartons are recorded. The results, in seconds, are shown in the following table.

New machine		Present machine	
42.0	41.0	42.7	43.6
41.3	41.8	43.8	43.3
42.4	42.8	42.5	43.5
43.2	42.3	43.1	41.7
41.8	42.7	44.0	44.1

a. Do the data provide sufficient evidence to conclude that, on the average, the new machine packs faster? Perform a Mann–Whitney test at the 5% significance level.

b. The hypothesis test in part (a) was performed in Exercise 10.20 using the pooled t-test. The assumption in that exercise is that packing times for both machines are normally distributed and have equal standard deviations. Presuming that is in fact true, why is it permissible to perform a Mann–Whitney test to compare the means? Is it better in this case to use the pooled t-test or the Mann–Whitney test? Explain your answers.

Extending the Concepts and Skills

10.71 Suppose you want to perform a hypothesis test to compare the means of two populations using independent samples. For each part below, decide whether you would use the pooled t-test, the nonpooled t-test, the Mann–Whitney test, or none of these tests if preliminary data analyses of the samples suggest that the two distributions of the variable under consideration are

a. normal but do not have the same shape.
b. not normal but have the same shape.
c. not normal and do not have the same shape; both sample sizes are large.

10.72 Suppose you want to perform a hypothesis test to compare the means of two populations using independent samples. For each part below, decide whether you would use the pooled t-test, the nonpooled t-test, the Mann–Whitney test, or none of these tests if preliminary data analyses of the samples suggest that the two distributions of the variable under consideration are

a. normal and have the same shape.
b. not normal and do not have the same shape; one of the sample sizes is large and the other is small.
c. different, one being normal and the other not; both sample sizes are large.

10.73 Suppose you want to perform a hypothesis test to compare the means of two populations using independent samples. You know that the two distributions of the variable under consideration have the same shape and may be normal. You take the two samples and find that the data for one of the samples contain outliers. Which procedure would you use? Explain your answer.

Mann–Whitney test using a normal approximation. The tables of critical values for the Mann–Whitney test, Tables VI and VII, stop at $n_1 = 10$ and $n_2 = 10$. For larger samples, a normal approximation can be used.

NORMAL APPROXIMATION FOR M Suppose the two distributions of the variable under consideration have the same shape. Then, for samples of sizes n_1 and n_2,

- $\mu_M = n_1(n_1 + n_2 + 1)/2$.
- $\sigma_M = \sqrt{n_1 n_2(n_1 + n_2 + 1)/12}$.
- M is approximately normally distributed for $n_1 \geq 10$ and $n_2 \geq 10$.

Thus, for sample sizes of 10 or more, the standardized variable

$$z = \frac{M - n_1(n_1 + n_2 + 1)/2}{\sqrt{n_1 n_2(n_1 + n_2 + 1)/12}}$$

has approximately the standard normal distribution.

In Exercises 10.74–10.76, we will develop and apply a "large-sample" procedure for a Mann–Whitney test based on the fact in the box above.

10.74 Formulate a hypothesis-testing procedure for a Mann–Whitney test that uses the test statistic z given in the box above.

10.75 Refer to Exercise 10.69.

a. Use your procedure from Exercise 10.74 to perform the hypothesis test.

b. Compare your result in part (a) to the one you obtained in Exercise 10.69(a), where the normal approximation was not used.

10.76 Refer to Exercise 10.70.

a. Use your procedure from Exercise 10.74 to perform the hypothesis test.

b. Compare your result in part (a) to the one you obtained in Exercise 10.70(a), where the normal approximation was not used.

10.77 In this exercise we obtain the distribution of the variable M when the sample sizes are both three. This enables us to see how the critical values for the Mann–Whitney test are derived. We display all possible ranks for the data in the following table, where the letter A stands for a member from Population 1 and the letter B stands for a member from Population 2.

Rank						
1	2	3	4	5	6	M
A	A	A	B	B	B	6
A	A	B	A	B	B	7
A	A	B	B	A	B	8
·	·	·	·	·	·	·
·	·	·	·	·	·	·
·	·	·	·	·	·	·
B	B	A	B	A	A	14
B	B	B	A	A	A	15

a. Complete the table. (*Hint:* There are 20 rows.)

b. If the null hypothesis, H_0: $\mu_1 = \mu_2$, is true, what percentages of samples will match any given row of the table? (*Hint:* The answer is the same for all rows.)

c. Use the answer from part (b) to obtain the distribution of M when $n_1 = 3$ and $n_2 = 3$.

d. Draw a relative-frequency histogram of the distribution obtained in part (c).

e. Use your histogram from part (d) to obtain the entries in Table VII for M_ℓ and M_r when $n_1 = 3$ and $n_2 = 3$.

USING TECHNOLOGY  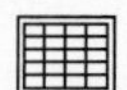 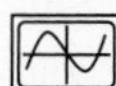

10.78 Transformations. Often the data we have do not satisfy the conditions for use of any of the standard hypothesis-testing procedures that we have discussed—the pooled t-test, nonpooled t-test, or Mann–Whitney test. However, by making a suitable *transformation,* we can often obtain data that do satisfy the assumptions of one or more of these standard tests.

In the paper "A Bayesian Analysis of a Multiplicative Treatment Effect in Weather Modification," Simpson, Alsen, and Eden presented the results of a study on cloud seeding with silver nitrate (*Technometrics,* 17, pp. 161–166). The following table gives the rainfall amounts, in acre-feet, for unseeded and seeded clouds.

Unseeded			Seeded		
1202.6	87.0	26.1	2745.6	274.7	115.3
830.1	81.2	24.4	1697.8	274.7	92.4
372.4	68.5	21.7	1656.0	255.0	40.6
345.5	47.3	17.3	978.0	242.5	32.7
321.2	41.1	11.5	703.4	200.7	31.4
244.3	36.6	4.9	489.1	198.6	17.5
163.0	29.0	4.9	430.0	129.6	7.7
147.8	28.6	1.0	334.1	119.0	4.1
95.0	26.3		302.8	118.3	

Suppose we want to perform a hypothesis test to decide whether cloud seeding with silver nitrate increases rainfall.

a. Obtain boxplots and normal probability plots for both samples.

b. Is it appropriate to use the pooled t-test?

c. Is it appropriate to use the nonpooled t-test?

d. Is it appropriate to use the Mann–Whitney test?

Now transform each sample by taking logarithms. If you are using Minitab, first store the two samples in columns named UNSEED and SEED, and then do the following:

1 Choose **Calc ➤ Calculator...**

2 Type LNUNSEED in the **Store result in variable** text box

3 Type LOGE(UNSEED) in the **Expression** text box

4 Click **OK**

Proceed in a similar way to logarithmically transform the data for the seeded clouds by typing LNSEED at the second step and LOGE(SEED) at the third step.

e. Obtain boxplots and normal probability plots for both transformed samples.

f. Is it appropriate to use the pooled t-test on the transformed data? Why or why not?

g. Is it appropriate to use the nonpooled t-test on the transformed data? Why or why not?

h. Is it appropriate to use the Mann–Whitney test on the transformed data? Why or why not?

i. Which of the three procedures would you use to conduct the hypothesis test for the transformed data? Explain your answer.

j. Use the test you designated in part (i) to conduct the hypothesis test for the transformed data.

k. What conclusions can you draw?

10.79 Refer to Exercise 10.67. Use Minitab or some other statistical software to perform the required hypothesis test using the Mann–Whitney procedure.

10.80 Refer to Exercise 10.68. Use Minitab or some other statistical software to perform the required hypothesis test using the Mann–Whitney procedure.

10.81 The U.S. Bureau of Labor Statistics conducts monthly surveys to estimate hourly earnings of nonsupervisory employees in various industry groups and publishes its findings in *Employment and Earnings.* Minitab's Mann–Whitney procedure was applied to data obtained on hourly earnings of independent samples of mine and construction workers. The resulting output is shown in Printout 10.8. Use the output to obtain

a. the null and alternative hypotheses.

b. the sample median hourly earnings for each group.

c. the number of mine workers sampled and the number of construction workers sampled.

d. the value of the Mann–Whitney test statistic.

e. the approximate P-value of the hypothesis test.

f. the smallest significance level at which the null hypothesis can be rejected.

g. the conclusion if the hypothesis test is performed at the 5% significance level.

h. a 90.1% confidence interval for the difference between the median hourly earnings of mine and construction workers.

10.82 Independent samples were taken of whites and blacks, and their 24-hour protein intakes were recorded. Minitab's Mann–Whitney procedure was applied to the data and yielded the output shown in Printout 10.9 on the next page. From the output, determine

a. the null and alternative hypotheses.

b. the median protein intakes (per day) of the whites sampled and the blacks sampled.

c. the number of whites sampled and the number of blacks sampled.

d. the value of the Mann–Whitney test statistic.

e. the approximate P-value of the hypothesis test.

f. the smallest significance level at which the null hypothesis can be rejected.

g. the conclusion if the hypothesis test is performed at the 5% significance level.

h. a 95.6% confidence interval for the difference between the median protein intakes per day of whites and blacks.

PRINTOUT 10.8 Minitab output for Exercise 10.81

```
MINE        N =  14     Median =      16.615
CONSTRUC    N =  17     Median =      16.880
Point estimate for ETA1-ETA2 is      -0.500
90.1 Percent CI for ETA1-ETA2 is (-1.790,1.080)
W = 208.0
Test of ETA1 = ETA2  vs  ETA1 < ETA2 is significant at 0.2692

Cannot reject at alpha = 0.05
```

PRINTOUT 10.9 Minitab output for Exercise 10.82

```
WHITE      N =  10     Median =          78.05
BLACK      N =  12     Median =          71.45
Point estimate for ETA1-ETA2 is           8.70
95.6 Percent CI for ETA1-ETA2 is (-5.20,25.00)
W = 134.0
Test of ETA1 = ETA2  vs  ETA1 not = ETA2 is significant at 0.2225

Cannot reject at alpha = 0.05
```

10.5 INFERENCES FOR TWO POPULATION MEANS USING PAIRED SAMPLES

Up to this point, the methods we have studied for comparing the means of two populations rely on independent samples. In this section and the next we will examine methods for comparing the means of two populations using **paired samples.** A paired sample may be appropriate when there is a natural pairing of the members of the two populations.

Each pair in a paired sample consists of a member of one population and that member's corresponding member in the other population. With a random paired sample, each possible paired sample is equally likely to be the one selected. Example 10.13 provides an unrealistically simple illustration of paired samples but one that will help you understand the concept.

EXAMPLE 10.13 RANDOM PAIRED SAMPLES

Let us consider two small populations, one consisting of five married women and the other of their five husbands, as shown in the following figure. The arrows in the figure indicate that Elizabeth and Karim are married, Carol and Harold are married, and so forth. The married couples constitute the pairs for these two populations.

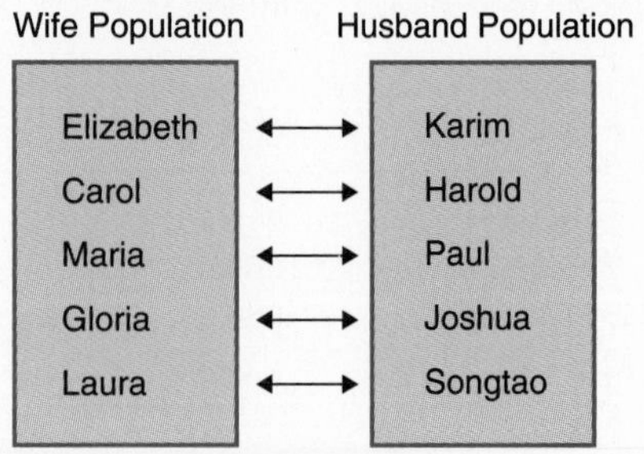

Suppose we take a paired sample of size three (i.e., a sample of three pairs) from these two populations.

a. List the possible paired samples.

b. If the paired sample is selected at random, determine the chance that any particular paired sample will be the one obtained.

SOLUTION For convenience let us use the first letter of each name as an abbreviation for the name and designate a wife-husband pair using parentheses. For example, (E, K) represents the couple Elizabeth and Karim.

a. There are 10 possible paired samples of size three, as seen in Table 10.12.

TABLE 10.12 Possible paired samples of size three from the wife and husband populations

Paired sample
(E, K), (C, H), (M, P)
(E, K), (C, H), (G, J)
(E, K), (C, H), (L, S)
(E, K), (M, P), (G, J)
(E, K), (M, P), (L, S)
(E, K), (G, J), (L, S)
(C, H), (M, P), (G, J)
(C, H), (M, P), (L, S)
(C, H), (G, J), (L, S)
(M, P), (G, J), (L, S)

b. For a random paired sample of size three, each of the 10 possible paired samples shown in Table 10.12 is equally likely to be the one selected. Therefore the chances are $\frac{1}{10}$ (1 in 10) that any particular paired sample of size three will be the one obtained. □

The purpose of Example 10.13 is to provide a concrete illustration of paired samples and to emphasize that for random paired samples of any given size, each possible paired sample is equally likely to be the one selected. In practice we neither obtain the number of possible paired samples nor explicitly compute the chance of selecting a particular paired sample. But these concepts underlie the methods we do use.

Comparing Two Population Means Using a Paired Sample

Now that we have discussed paired samples, we are ready to examine the process for comparing the means of two populations using a paired sample. Example 10.14 introduces the pertinent ideas.

EXAMPLE 10.14 COMPARING TWO POPULATION MEANS USING A PAIRED SAMPLE

Suppose we want to decide whether a newly developed gasoline additive increases gas mileage. We first note that we have one variable—gas mileage—and two populations:

Population 1: All cars when the additive is used.

Population 2: All cars when the additive is not used.

Let μ_1 and μ_2 denote the means of the variable "gas mileage" for Population 1 and Population 2, respectively:

μ_1 = mean gas mileage of all cars when the additive is used.

μ_2 = mean gas mileage of all cars when the additive is not used.

We want to perform the hypothesis test

H_0: $\mu_1 = \mu_2$ (mean gas mileage with additive is not greater)

H_a: $\mu_1 > \mu_2$ (mean gas mileage with additive is greater).

Independent samples could be used to carry out the hypothesis test: Take independent random samples of, say, 10 cars each; have one group driven with the additive (sample from Population 1) and the other group driven without the additive (sample from Population 2); and then apply a pooled or nonpooled t-test to the gas-mileage data obtained.

However, in this case, a paired sample is probably more appropriate. Here a pair consists of a car driven with the additive and the same car driven without the additive. The variable we analyze is the difference between the gas mileages of a car driven with and without the additive.

By using a paired sample, we can remove extraneous sources of variation, in this case the variation in the gas mileages of cars. As a consequence the sampling error made in estimating the difference between the population means will generally be smaller. This fact, in turn, makes it more likely that we will detect differences between the population means when such differences exist.

So suppose that 10 cars are selected at random and that the cars sampled are driven both with and without the additive, yielding a paired sample of size 10. Further suppose that the resulting gas mileages, in miles per gallon (mpg), are as displayed in the second and third columns of Table 10.13.

In the last column of Table 10.13, we have recorded the difference, d, between the gas mileages with and without the additive for each of the 10 cars sampled. Each difference is referred to as a **paired difference** since it is the difference of a pair of observations. For example, the first car got 25.7 mpg with the additive and 24.9 mpg without the additive, giving a paired difference of $d = 25.7 - 24.9 = 0.8$ mpg, an increase in gas mileage of 0.8 mpg with the additive.

TABLE 10.13
Gas mileages, with and without additive, for 10 randomly selected cars

Car	Gas mileage with additive	Gas mileage w/o additive	Paired difference d
1	25.7	24.9	0.8
2	20.0	18.8	1.2
3	28.4	27.7	0.7
4	13.7	13.0	0.7
5	18.8	17.8	1.0
6	12.5	11.3	1.2
7	28.4	27.8	0.6
8	8.1	8.2	−0.1
9	23.1	23.1	0.0
10	10.4	9.9	0.5
			6.6

If the null hypothesis is true, the paired differences of the gas mileages for the cars sampled should average out to about zero, that is, the sample mean $\bar{d}$ of the paired differences should be roughly zero. To put it another way, if $\bar{d}$ is too much greater than zero, we would take this as evidence that the null hypothesis is false.

From the last column of Table 10.13, we find that the sample mean of the paired differences is

$$\bar{d} = \frac{\Sigma d}{n} = \frac{6.6}{10} = 0.66 \text{ mpg},$$

a mean increase in gas mileage of 0.66 mpg when the additive is used. The question now is whether this mean increase in gas mileage can reasonably be attributed to sampling error or whether it is large enough to indicate that, on the average, the additive improves gas mileage (i.e., $\mu_1 > \mu_2$). To answer that question, we need to know the distribution of the variable $\bar{d}$. We will discuss that distribution and then return to solve the gas-mileage problem. ◨

The Paired t-Statistic

Suppose x is a variable on each of two populations whose members can be paired. To each pair we let d denote the difference between the values of the variable x on the members of the pair. We call d the **paired-difference variable.**

It can be shown that

$$\mu_d = \mu_1 - \mu_2,$$

that is, the mean of the paired differences equals the difference between the two population means. Furthermore, if d is normally distributed, then we can apply the previous equation and our knowledge of the studentized version of a sample mean (Key Fact 8.5 on page 468) to obtain the following fact.

KEY FACT 10.6 DISTRIBUTION OF THE PAIRED *t*-STATISTIC

Suppose x is a variable on each of two populations whose members can be paired. Further suppose that the paired-difference variable d is normally distributed. Then, for paired samples of size n, the variable

$$t = \frac{\bar{d} - (\mu_1 - \mu_2)}{s_d/\sqrt{n}}$$

has the t-distribution with $\text{df} = n - 1$.

Note: We will use the phrase **normal differences** as an abbreviation of "the paired-difference variable is normally distributed."

Hypothesis Tests for the Means of Two Populations Using a Paired Sample

We can now present a hypothesis-testing procedure for comparing the means of two populations using a paired sample when the paired-difference variable is normally distributed. In view of Key Fact 10.6, we see that for a hypothesis test with null hypothesis H_0: $\mu_1 = \mu_2$, we can use the variable

$$t = \frac{\bar{d}}{s_d/\sqrt{n}}$$

as the test statistic and obtain the critical value(s) from the t-table, Table IV. Specifically, we have the procedure shown on the next page, which we often refer to as the **paired *t*-test.**

Before applying Procedure 10.6, let's discuss the assumptions for its use. Assumption 1 (paired sample) is essential. The sample must be paired or the procedure does not apply.

Regarding Assumption 2, although the paired t-test was derived based on the assumption that the paired-difference variable is normally distributed, it also applies approximately for large samples regardless of distribution-type, as noted at the bottom of Procedure 10.6. And like the one-sample t-test, the paired t-test works reasonably well even for small or moderate-size samples when the paired-difference variable is not normally distributed, provided that variable is not too far from being normally distributed.

When considering the paired t-test, it is also important to watch for outliers in the sample of paired differences. The presence of outliers calls into question the normality assumption. And even for large samples, outliers can sometimes unduly affect a paired t-test because the sample mean and sample standard deviation are not resistant to them.

PROCEDURE 10.6 **THE PAIRED t-TEST FOR TWO POPULATION MEANS**

ASSUMPTIONS

1. Paired sample

2. Normal differences or large sample

Step 1 The null hypothesis is H_0: $\mu_1 = \mu_2$ and the alternative hypothesis is one of the following:

H_a: $\mu_1 \neq \mu_2$ (Two-tailed)	or	H_a: $\mu_1 < \mu_2$ (Left-tailed)	or	H_a: $\mu_1 > \mu_2$ (Right-tailed)

Step 2 Decide on the significance level, α.

Step 3 The critical value(s) are

$\pm t_{\alpha/2}$ (Two-tailed)	or	$-t_\alpha$ (Left-tailed)	or	t_α (Right-tailed)

with df $= n - 1$. Use Table IV to find the critical value(s).

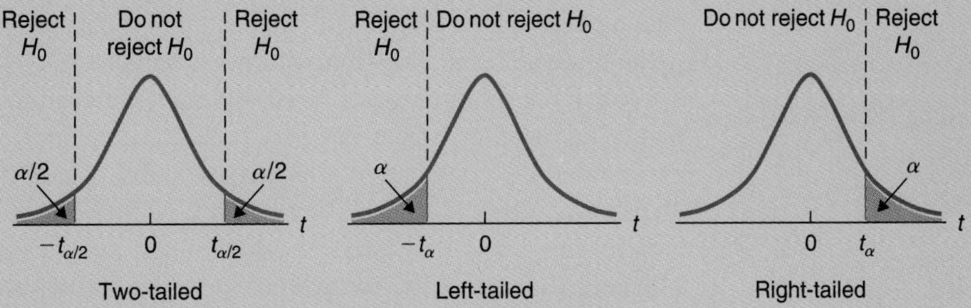

Step 4 Calculate the paired differences of the sample pairs.

Step 5 Compute the value of the test statistic

$$t = \frac{\bar{d}}{s_d/\sqrt{n}}.$$

Step 6 If the value of the test statistic falls in the rejection region, reject H_0; otherwise, do not reject H_0.

Step 7 State the conclusion in words.

The hypothesis test is exact when the paired-difference variable is normally distributed (normal differences) and is approximately correct for large samples when the paired-difference variable is not normally distributed (nonnormal differences).

Practical guidelines for when to use the paired t-test are the same as those given for the one-sample z-test in Key Fact 9.4 on page 514, when applied to paired differences. And remember, always look at the sample of paired differences before applying the paired t-test to ensure that it is reasonable to use it.

Finally, we emphasize that the normality assumption in Assumption 2 refers to the paired-difference variable. The two distributions of the variable under consideration need not be normally distributed.

EXAMPLE 10.15 ILLUSTRATES PROCEDURE 10.6

We now return to the gas-mileage problem posed in Example 10.14. The gas mileages of 10 randomly selected cars, both with and without a new gasoline additive, are displayed in the second and third columns of Table 10.13 on page 643. At the 5% significance level, do the data provide sufficient evidence to conclude that, on the average, the gasoline additive improves gas mileage?

SOLUTION To begin, we check the two conditions required for using the paired t-test. We are dealing here with a paired sample; each pair consists of a car driven both with and without the additive. So Assumption 1 is satisfied.

Since the sample size, $n = 10$, is small, we need to consider questions of normality and outliers. (See the first bulleted item in Key Fact 9.4 on page 514.) To that end we constructed a normal probability plot (not shown) for the sample of paired differences in the last column of Table 10.13. The normal probability plot reveals no outliers and is quite linear; so we can consider Assumption 2 satisfied. Consequently, the paired t-test can be applied to perform the required hypothesis test.

Step 1 State the null and alternative hypotheses.

Let μ_1 denote the mean gas mileage of all cars when the additive is used and μ_2 denote the mean gas mileage of all cars when the additive is not used. Then the null and alternative hypotheses are

$$H_0\colon \mu_1 = \mu_2 \text{ (mean gas mileage with additive is not greater)}$$

$$H_a\colon \mu_1 > \mu_2 \text{ (mean gas mileage with additive is greater).}$$

Note that the hypothesis test is right-tailed since a greater-than sign (>) appears in the alternative hypothesis.

Step 2 Decide on the significance level, α.

The test is to be performed at the 5% significance level. Thus $\alpha = 0.05$.

Step 3 The critical value for a right-tailed test is t_α with $df = n - 1$.

From Step 2, $\alpha = 0.05$. Also, since there are 10 pairs in the sample, we have $\text{df} = n - 1 = 10 - 1 = 9$. Consequently, the critical value is $t_{0.05} = 1.833$, as seen in Fig. 10.8.

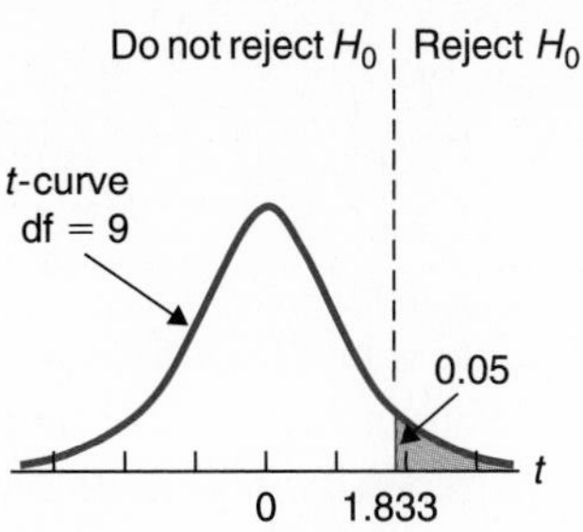

FIGURE 10.8 Criterion for deciding whether or not to reject the null hypothesis

Step 4 *Calculate the paired differences of the sample pairs.*

We have already done this in the last column of Table 10.13 on page 643.

Step 5 *Compute the value of the test statistic*

$$t = \frac{\bar{d}}{s_d/\sqrt{n}}.$$

We first need to determine the sample mean and sample standard deviation of the paired differences in the last column of Table 10.13. This is accomplished in the usual manner:

$$\bar{d} = \frac{\Sigma d}{n} = \frac{6.6}{10} = 0.66$$

and

$$s_d = \sqrt{\frac{\Sigma d^2 - (\Sigma d)^2/n}{n-1}} = \sqrt{\frac{6.12 - (6.6)^2/10}{10-1}} = 0.443.$$

Consequently, the value of the test statistic is

$$t = \frac{\bar{d}}{s_d/\sqrt{n}} = \frac{0.66}{0.443/\sqrt{10}} = 4.711.$$

Step 6 *If the value of the test statistic falls in the rejection region, reject H_0; otherwise, do not reject H_0.*

From Step 5 the value of the test statistic is $t = 4.711$, which falls in the rejection region. Hence we reject H_0.

Step 7 *State the conclusion in words.*

At the 5% significance level, the data provide sufficient evidence to conclude that the mean gas mileage of all cars when the additive is used is greater than the mean gas mileage of all cars when the additive is not used. Evidently, the additive is effective in increasing gas mileage. ❑

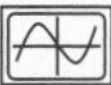

Using the *P*-Value Approach

The P-value approach to hypothesis testing can also be used to carry out the hypothesis test in Example 10.15 and to assess more precisely the evidence against the null hypothesis. From Step 5 the value of the test statistic is $t = 4.711$. Recalling that df $= 9$, we find from Table IV that $P < 0.005$.

In particular, because the P-value is less than the significance level of 0.05, we can reject H_0. Furthermore, by referring to Table 9.12 on page 543, we see that the data provide very strong evidence against the null hypothesis and hence in favor of the alternative hypothesis that, on the average, the additive increases gas mileage.

Confidence Intervals for the Difference Between the Means of Two Populations Using a Paired Sample

We can also use Key Fact 10.6 on page 644 to derive the following confidence-interval procedure for the difference between two population means, which we often refer to as the **paired *t*-interval procedure.**

PROCEDURE 10.7 THE PAIRED *t*-INTERVAL PROCEDURE FOR TWO POPULATION MEANS

ASSUMPTIONS

1. Paired sample
2. Normal differences or large sample

Step 1 For a confidence level of $1 - \alpha$, use Table IV to find $t_{\alpha/2}$ with df $= n - 1$.

Step 2 The endpoints of the confidence interval for $\mu_1 - \mu_2$ are

$$\overline{d} \pm t_{\alpha/2} \cdot \frac{s_d}{\sqrt{n}}.$$

The confidence interval is exact when the paired-difference variable is normally distributed (normal differences) and is approximately correct for large samples when the paired-difference variable is not normally distributed (nonnormal differences).

EXAMPLE 10.16 ILLUSTRATES PROCEDURE 10.7

Use the sample data in Table 10.13 on page 643 to obtain a 90% confidence interval for the difference, $\mu_1 - \mu_2$, between the mean gas mileage of all cars when the additive is used and the mean gas mileage of all cars when the additive is not used.

SOLUTION We apply Procedure 10.7.

Step 1 *For a confidence level of $1 - \alpha$, use Table IV to find $t_{\alpha/2}$ with $df = n - 1$.*

For a 90% confidence interval, we have $\alpha = 0.10$. From Table IV we find that for df $= n - 1 = 10 - 1 = 9$, $t_{\alpha/2} = t_{0.10/2} = t_{0.05} = 1.833$.

Step 2 *The endpoints of the confidence interval for $\mu_1 - \mu_2$ are*

$$\bar{d} \pm t_{\alpha/2} \cdot \frac{s_d}{\sqrt{n}}.$$

From Step 1, $t_{\alpha/2} = 1.833$. Also, $n = 10$, and from Example 10.15 we know that $\bar{d} = 0.66$ and $s_d = 0.44$. So the endpoints of the confidence interval for $\mu_1 - \mu_2$ are

$$0.66 \pm 1.833 \cdot \frac{0.44}{\sqrt{10}}$$

or 0.40 to 0.92. We can be 90% confident that the difference between the mean gas mileage of all cars when the additive is used and the mean gas mileage of all cars when the additive is not used is somewhere between 0.40 mpg and 0.92 mpg. In particular, we can be confident that, on the average, the additive increases gas mileage by at least 0.40 mpg. ◨

 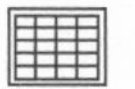 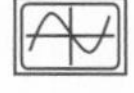

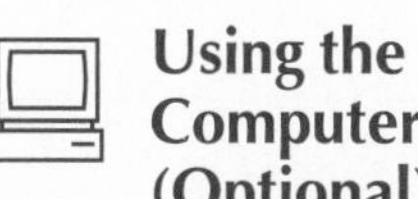

Using the Computer (Optional)

We can use Minitab to perform paired t-procedures. To illustrate, we return once again to the problem of deciding whether a newly-developed gasoline additive increases gas mileage on the average.

EXAMPLE 10.17 Using Minitab to Perform Paired t-Procedures

Use Minitab to simultaneously perform the hypothesis test considered in Example 10.15 and obtain the confidence interval required in Example 10.16.

SOLUTION Let μ_1 denote the mean gas mileage of all cars when the additive is used and μ_2 denote the mean gas mileage of all cars when the additive is not used. The problem in Example 10.15 is to perform the hypothesis test

$$H_0\colon \mu_1 = \mu_2 \text{ (mean gas mileage with additive is not greater)}$$
$$H_a\colon \mu_1 > \mu_2 \text{ (mean gas mileage with additive is greater)}$$

at the 5% significance level; the problem in Example 10.16 is to obtain a 90% confidence interval for $\mu_1 - \mu_2$.

As we discovered in Example 10.15, it is reasonable to use paired t-procedures here. To apply Minitab, we begin by entering the two sets of sample data from Table 10.13 on page 643 in columns named WITH and WITHOUT. Then we proceed in the following way.

1 Choose **Stat ➤ Basic Statistics ➤ Paired t...**
2 Click in the **First sample** text box and specify WITH
3 Click in the **Second sample** text box and specify WITHOUT
4 Click the **Options...** button
5 Click in the **Confidence level** text box and type 90
6 Click in the **Test mean** text box and type 0
7 Click the arrow button at the right of the **Alternative** drop-down list box and select **greater than**
8 Click **OK**
9 Click **OK**

The output that results is shown in Printout 10.10.

PRINTOUT 10.10
Minitab output for the paired t-procedure

```
Paired T for WITH - WITHOUT

                  N       Mean      StDev    SE Mean
WITH             10      18.91       7.47       2.36
WITHOUT          10      18.25       7.42       2.35
Difference       10      0.660      0.443      0.140

90% CI for mean difference: (0.403, 0.917)
T-Test of mean difference = 0 (vs > 0): T-Value = 4.71  P-Value = 0.001
```

The first line of Printout 10.10 describes the test being performed: `Paired T for WITH - WITHOUT`. The next four lines display a table that gives the sample size, sample mean, sample standard deviation, and estimated standard error of the mean for the two individual samples from the populations and for the paired differences.

On the sixth line of Printout 10.10, we find the required 90% confidence interval for the difference between the two population means. Thus we can be 90% confident that the difference between the mean gas mileages of all cars with and without the additive is somewhere between 0.403 mpg and 0.917 mpg.

The last line of Printout 10.10 provides a statement of the null and alternative hypotheses in terms of μ_d $(= \mu_1 - \mu_2)$, followed by the value of the test statistic (`T-Value = 4.71`) and the P-value (`P-Value = 0.001`).

Since the P-value of 0.001 is less than the specified significance level of 0.05, we reject H_0. At the 5% significance level, the data provide sufficient evidence to conclude that the mean gas mileage of all cars when the additive is used exceeds the mean gas mileage of all cars when the additive is not used. □

EXERCISES 10.5

Statistical Concepts and Skills

10.83 What does each pair in a paired sample consist of?

10.84 State one possible advantage of using paired samples instead of independent samples.

10.85 The U.S. Bureau of the Census publishes information on the ages of married people in *Current Population Reports.* Suppose we want to use a paired sample to compare the mean ages of married men and married women. Identify each of the following:

a. the variable under consideration
b. the two populations
c. the pairs
d. the paired-difference variable

10.86 The A. C. Nielsen Company collects data on the TV viewing habits of Americans and publishes the information in *Nielsen Report on Television.* Suppose we want to use a paired sample to compare the mean viewing times of married men and married women. Identify each of the following:

a. the variable under consideration
b. the two populations
c. the pairs
d. the paired-difference variable

10.87 State the two conditions required for performing a paired t-procedure. How important are those conditions?

10.88 Provide an example (different from those considered in this section) where a procedure using a paired sample would be more appropriate than one using independent samples.

Preliminary data analyses indicate that it is reasonable to employ a paired t-test in Exercises 10.89–10.94. Perform each hypothesis test using either the critical-value approach or the P-value approach.

10.89 Charles Darwin, author of *Origin of Species,* investigated the effect of cross-fertilization on the heights of plants. In one study he planted 15 pairs of Zea mays plants. Each pair consisted of one cross-fertilized plant and one self-fertilized plant grown in the same pot. The following table gives the height differences, in eighths of an inch, for the 15 pairs. Each difference is obtained by subtracting the height of the self-fertilized plant from that of the cross-fertilized plant.

49	−67	8	16	6
23	28	41	14	29
56	24	75	60	−48

a. Identify the variable under consideration.
b. Identify the two populations.
c. Identify the paired-difference variable.
d. Are the numbers in the above table paired differences? Why or why not?
e. At the 5% significance level, do the data provide sufficient evidence to conclude that the mean heights of cross-fertilized and self-fertilized Zea mays differ? *(Note: $\overline{d} = 20.93$ and $s_d = 37.74$.)*

10.90 In 1908 W. S. Gosset published "The Probable Error of a Mean" (*Biometrika,* Vol. 6, pp. 1–25). It is in this pioneering paper, published under the pseudonym "Student," that he introduced what later became known as Student's t-distribution. As an example, Gosset used the following data set which gives the additional sleep in hours obtained by 10 patients who used laevohyoscyamine hydrobromide.

1.9	0.8	1.1	0.1	−0.1
4.4	5.5	1.6	4.6	3.4

a. Identify the variable under consideration.
b. Identify the two populations.
c. Identify the paired-difference variable.
d. Are the numbers in the above table paired differences? Why or why not?
e. At the 1% significance level, do the data provide sufficient evidence to conclude that laevohysocyamine hydrobromide is effective in increasing sleep? *(Note:* $\bar{d} = 2.33$ and $s_d = 2.002$.*)*

10.91 Stichler, Richey, and Mandel compared two methods of measuring treadwear in their paper "Measurement of Treadwear of Commercial Tires" (*Rubber Age,* 73:2). Eleven tires were each measured for treadwear by two methods, one based on weight and the other on groove wear. Here are the data in thousands of miles.

Weight method	Groove method
30.5	28.7
30.9	25.9
31.9	23.3
30.4	23.1
27.3	23.7
20.4	20.9
24.5	16.1
20.9	19.9
18.9	15.2
13.7	11.5
11.4	11.2

At the 1% significance level, do the data provide sufficient evidence to conclude that, on the average, the two measurement methods give different results?

10.92 The A. C. Nielsen Company collects data on the TV viewing habits of Americans and publishes the information in *Nielsen Report on Television.* Twenty married couples are randomly selected. Their weekly viewing times, in hours, are as follows.

Husband	Wife	Husband	Wife	Husband	Wife
21	24	38	45	36	35
56	55	27	29	20	34
34	55	30	41	43	32
30	34	31	37	4	13
41	32	30	35	16	9
35	38	32	48	21	23
26	38	15	17		

At the 5% level of significance, does it appear that married men watch less TV, on the average, than married women?

10.93 *Current Population Reports,* published by the U.S. Bureau of the Census, presents data on the ages of married people. Ten married couples are randomly selected and have the ages shown here.

Husband	54	21	32	78	70	33	68	32	54	52
Wife	53	22	33	74	64	35	67	28	41	44

Do the data suggest that the mean age of married men is greater than the mean age of married women? Perform the appropriate hypothesis test at the 5% significance level.

10.94 A pediatrician measured the blood cholesterol levels of her young patients. She was surprised to find that many of them had levels over 200 mg per 100 mL, indicating increased risk of artery disease. Ten such patients were randomly selected to take part in a nutritional program designed to lower blood cholesterol. Two months after the program started, the pediatrician measured the blood cholesterol levels of the 10 patients again. Here are the data.

Patient	Before program	After program
1	210	212
2	217	210
3	208	210
4	215	213
5	202	200
6	209	208
7	207	203
8	210	199
9	221	218
10	218	214

Do the data suggest that the nutritional program is, on the average, effective in reducing cholesterol levels? Perform the appropriate hypothesis test at the 5% level of significance.

In each of Exercises 10.95–10.100, use Procedure 10.7 on page 648 to obtain the required confidence interval.

10.95 Refer to Exercise 10.89.

a. Determine a 95% confidence interval for the difference between the mean heights of cross-fertilized and self-fertilized Zea mays.

b. Interpret your answer in part (a).

10.96 Refer to Exercise 10.90.

a. Determine a 98% confidence interval for the additional sleep that would be obtained on the average by using laevohysocyamine hydrobromide.

b. Interpret your answer in part (a).

10.97 Refer to Exercise 10.91.

a. Determine a 99% confidence interval for the mean difference in measurement by the weight and groove methods.

b. Interpret your answer in part (a).

10.98 Refer to Exercise 10.92.

a. Determine a 90% confidence interval for the difference between the mean weekly TV viewing times of married men and married women.

b. Determine a 90% confidence interval for the mean difference in viewing times by spouses.

10.99 Refer to Exercise 10.93.

a. Find a 90% confidence interval for the difference between the mean ages of married men and married women.

b. Find a 90% confidence interval for the mean age difference of spouses.

10.100 Refer to Exercise 10.94.

a. Determine a 90% confidence interval for the difference, $\mu_1 - \mu_2$, between the mean cholesterol levels of high-level patients before and after the nutritional program.

b. Determine a 90% confidence interval for the mean drop in cholesterol level as a result of the nutritional program.

EXTENDING THE CONCEPTS AND SKILLS

10.101 Explain how a paired t-test can be formulated as a one-sample t-test. (*Hint:* Work solely with the paired-difference variable.)

10.102 In Example 10.3 (page 603), we performed a hypothesis test using independent samples to decide whether mean salaries differ for faculty teaching in public and private institutions. Now we will perform that same hypothesis test using a paired sample. Pairs are formed by matching faculty in public and private institutions by rank and specialty. A random sample of 30 pairs yields the following annual salaries, in thousands of dollars.

Public	Private	Public	Private	Public	Private
78.1	84.9	34.9	44.2	35.5	39.2
48.3	57.0	29.3	43.2	80.8	90.8
81.7	88.4	84.3	95.6	71.5	78.0
71.5	80.6	56.8	59.5	46.1	60.4
52.8	56.8	60.9	63.4	77.0	78.8
95.6	93.3	81.9	87.2	47.0	53.0
46.2	50.5	69.7	77.0	70.3	79.8
44.3	50.4	38.0	46.4	55.1	66.6
21.9	25.9	60.5	70.3	49.2	54.8
51.2	60.8	61.2	77.4	53.8	62.2

a. Do the data provide sufficient evidence to conclude that mean salaries differ for faculty teaching in public and private institutions? Perform the required hypothesis test at the 5% significance level. (*Note:* $\bar{d} = -7.367$ and $s_d = 3.992$.)

b. Compare your result in part (a) to the one obtained in Example 10.3.

c. Which test do you think is more appropriate? Explain your answer.

d. Find a 95% confidence interval for the difference between the mean salaries of faculty teaching in public and private institutions.

e. Compare your result in part (d) to the one obtained in Example 10.4 on page 607.

10.103 A hypothesis test is to be performed to compare the means of two populations using a paired sample. The sample of 15 paired differences contains an outlier but otherwise is roughly bell-shaped. Assuming it is not legitimate to remove the outlier, would it be better to use the paired t-test or a nonparametric test? Explain your answer.

10.104 This exercise shows what can happen when a hypothesis-testing procedure designed for use with independent samples is applied to perform a hypothesis test on a paired sample. In Example 10.15 on page 646, we applied the paired t-test to decide whether a gasoline additive is effective in increasing gas mileage. Specifically, if we let μ_1 and μ_2 denote, respectively, the mean gas mileages of all cars when the additive is and is not used, then the null and alternative hypotheses are

$$H_0: \mu_1 = \mu_2$$
$$H_a: \mu_1 > \mu_2.$$

a. Apply the nonpooled t-test, Procedure 10.3 given on page 615, to the sample data in the second and third columns of Table 10.13 on page 643 to perform the hypothesis test. Use $\alpha = 0.05$.
b. Why is it inappropriate to perform the hypothesis test the way you did in part (a)?
c. Compare your result in part (a) to the one obtained in Example 10.15. (See page 646.)

USING TECHNOLOGY  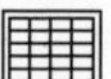 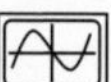

10.105 Refer to Exercises 10.91 and 10.97. Use Minitab or some other statistical software to
a. obtain a normal probability plot of the paired differences.
b. perform the required hypothesis test and obtain the desired confidence interval.
c. Justify the use of your procedure in part (b).

10.106 Refer to Exercises 10.92 and 10.98. Use Minitab or some other statistical software to
a. obtain a normal probability plot of the paired differences.
b. perform the required hypothesis test and obtain the desired confidence interval.
c. Justify the use of your procedure in part (b).

10.107 An algebra teacher at Arizona State University wanted to compare two methods of teaching college algebra—the lecture method and the personalized system of instruction (PSI) method. Students were paired by matching those with similar mathematics background and performance. A random sample of 11 pairs was selected. From each pair, one student was randomly chosen to take the lecture course; the other student took the PSI course. Both courses were taught by the algebra teacher. Minitab's paired t-procedure was applied to the final grades for the 11 pairs of students. The resulting output is shown in Printout 10.11. Use the output to determine
a. the null and alternative hypotheses.
b. the mean final grade for each group.
c. the mean of the paired differences.
d. the standard deviation of the paired differences.
e. the value of the test statistic, t.
f. the P-value of the hypothesis test.
g. the smallest significance level at which the null hypothesis can be rejected.
h. the conclusion if the hypothesis test is performed at the 5% significance level.
i. a 95% confidence interval for the difference between mean final grades of the two instructional methods.

10.108 It had been conjectured that a new running program would reduce heart rates. The heart rates of 15 randomly selected people were measured and then those people were placed on the running program. One year later the heart rates of the 15 people were measured again. Minitab's paired t-procedure was applied to the data and resulted in the output shown in Printout 10.12. Use the output to determine
a. the null and alternative hypotheses.
b. the mean heart rates of the 15 people before and after the running program.
c. the mean change in heart rate for the 15 people.
d. the sample standard deviation of the changes in heart rate for the 15 people.
e. the value of the test statistic, t.
f. the P-value of the hypothesis test.
g. the smallest significance level at which the null hypothesis can be rejected.
h. the conclusion if the hypothesis test is performed at the 5% significance level.
i. a 90% confidence interval for the mean decline in heart rate for all people using the running program.

PRINTOUT 10.11 Minitab output for Exercise 10.107

```
Paired T for LECTURE - PSI

                  N      Mean     StDev   SE Mean
LECTURE          11     69.73     16.22      4.89
PSI              11     67.82     16.47      4.97
Difference       11      1.91      3.73      1.12

95% CI for mean difference: (-0.59, 4.41)
T-Test of mean difference = 0 (vs not = 0): T-Value = 1.70  P-Value = 0.120
```

PRINTOUT 10.12 Minitab output for Exercise 10.108

```
Paired T for BEFORE - AFTER

                  N      Mean     StDev   SE Mean
BEFORE           15    74.667     3.478     0.898
AFTER            15    72.333     3.200     0.826
Difference       15     2.333     3.352     0.866

90% CI for mean difference: (0.809, 3.858)
T-Test of mean difference = 0 (vs > 0): T-Value = 2.70  P-Value = 0.009
```

10.6 THE PAIRED WILCOXON SIGNED-RANK TEST*

In Section 10.5 we discussed the paired t-procedures which provide methods for comparing two population means using paired samples. An assumption for use of those procedures is that the paired-difference variable is (approximately) normally distributed or the sample size is large. In cases where we have a paired sample but the sample size is small or moderate and the distribution of the paired-difference variable is far from normal, a paired t-procedure is inappropriate and a nonparametric procedure should be used instead.

For instance, if the distribution of the paired-difference variable is symmetric (but not necessarily normal), then we can perform a hypothesis test to compare the means of the two populations by applying the Wilcoxon signed-rank test (Procedure 9.4 on page 565) to the sample of paired differences. In this context the Wilcoxon signed-rank test is called the **paired Wilcoxon signed-rank test.**

Procedure 10.8 provides the steps for performing a paired Wilcoxon signed-rank test. We will use the phrase **symmetric differences** as an abbreviation of "the paired-difference variable has a symmetric distribution."

* This section continues the optional coverage of nonparametric statistics.

PROCEDURE 10.8

THE PAIRED WILCOXON SIGNED-RANK TEST FOR TWO POPULATION MEANS

ASSUMPTIONS

1. Paired sample
2. Symmetric differences

Step 1 The null hypothesis is H_0: $\mu_1 = \mu_2$ and the alternative hypothesis is one of the following:

H_a: $\mu_1 \neq \mu_2$ (Two-tailed)	or	H_a: $\mu_1 < \mu_2$ (Left-tailed)	or	H_a: $\mu_1 > \mu_2$ (Right-tailed)

Step 2 Decide on the significance level, α.

Step 3 Calculate the paired differences of the sample pairs.

Step 4 Discard all paired differences that equal 0 and reduce the sample size accordingly.

Step 5 The critical value(s) are

W_ℓ and W_r (Two-tailed)	or	W_ℓ (Left-tailed)	or	W_r (Right-tailed)

Use Table V to find the critical value(s).

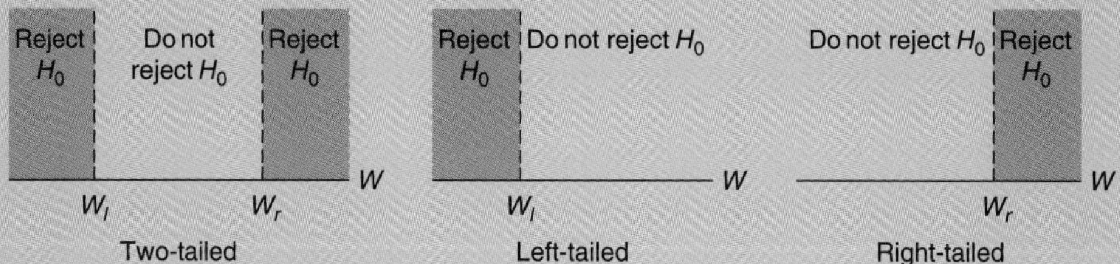

Step 6 Construct a work table of the following form:

Paired difference d	$\lvert d \rvert$	Rank of $\lvert d \rvert$	Signed rank R
⋮	⋮	⋮	⋮

Step 7 Compute the value of the test statistic

$$W = \text{sum of the positive ranks.}$$

Step 8 If the value of the test statistic falls in the rejection region, reject H_0; otherwise, do not reject H_0.

Step 9 State the conclusion in words.

In Example 10.15 we used a paired t-test to decide whether, on the average, a gasoline additive improves gas mileage. Now we will use the paired Wilcoxon signed-rank test to perform that hypothesis test.

EXAMPLE 10.18 ILLUSTRATES PROCEDURE 10.8

The gas mileages of 10 randomly selected cars, both with and without a new gasoline additive, are displayed in the second and third columns of Table 10.14. At the 5% significance level, do the data provide sufficient evidence to conclude that, on the average, the gasoline additive improves gas mileage? Use the paired Wilcoxon signed-rank test.

TABLE 10.14 Gas mileages, with and without additive, for 10 randomly selected cars

Car	Gas mileage with additive	Gas mileage w/o additive	Paired difference d
1	25.7	24.9	0.8
2	20.0	18.8	1.2
3	28.4	27.7	0.7
4	13.7	13.0	0.7
5	18.8	17.8	1.0
6	12.5	11.3	1.2
7	28.4	27.8	0.6
8	8.1	8.2	−0.1
9	23.1	23.1	0.0
10	10.4	9.9	0.5
			6.6

SOLUTION We apply Procedure 10.8.

Step 1 *State the null and alternative hypotheses.*

Let μ_1 denote the mean gas mileage of all cars when the additive is used and μ_2 denote the mean gas mileage of all cars when the additive is not used. Then the null and alternative hypotheses are

H_0: $\mu_1 = \mu_2$ (mean gas mileage with additive is not greater)

H_a: $\mu_1 > \mu_2$ (mean gas mileage with additive is greater).

Note that the hypothesis test is right-tailed since a greater-than sign (>) appears in the alternative hypothesis.

Step 2 *Decide on the significance level, α.*

The test is to be performed at the 5% significance level. So $\alpha = 0.05$.

Step 3 *Calculate the paired differences of the sample pairs.*

This is done in the last column of Table 10.14.

Step 4 *Discard all paired differences that equal 0 and reduce the sample size accordingly.*

As we observe from the last column of Table 10.14, there is one paired difference that equals 0. Discarding it, we now have a sample of size nine.

Step 5 *The critical value for a right-tailed test is W_r.*

Referring to Table V with $n = 9$ and $\alpha = 0.05$ and recalling that the test is right-tailed, we find that the critical value is $W_r = 37$, as seen in Fig. 10.9.

FIGURE 10.9
Criterion for deciding whether or not to reject the null hypothesis

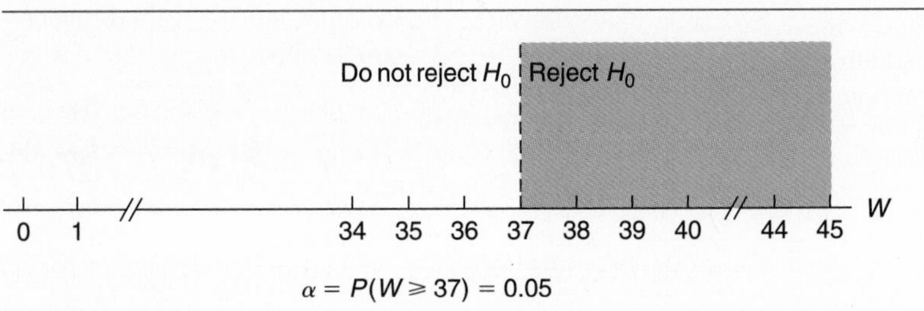

Step 6 *Construct a work table.*

Paired difference d	$\|d\|$	Rank of $\|d\|$	Signed rank R
0.8	0.8	6	6
1.2	1.2	8.5	8.5
0.7	0.7	4.5	4.5
0.7	0.7	4.5	4.5
1.0	1.0	7	7
1.2	1.2	8.5	8.5
0.6	0.6	3	3
−0.1	0.1	1	−1
0.5	0.5	2	2

Step 7 *Compute the value of the test statistic*

$$W = \textit{sum of the positive ranks.}$$

The last column of the work table shows that the sum of the positive ranks equals

$$W = 6 + 8.5 + 4.5 + 4.5 + 7 + 8.5 + 3 + 2 = 44.$$

Step 8 *If the value of the test statistic falls in the rejection region, reject H_0; otherwise, do not reject H_0.*

The value of the test statistic is $W = 44$, as found in Step 7. This falls in the rejection region, as we observe from Fig. 10.9. Thus we reject H_0.

Step 9 *State the conclusion in words.*

The test results are statistically significant at the 5% level; that is, at the 5% significance level, the data provide sufficient evidence to conclude that the mean gas mileage of all cars when the additive is used is greater than the mean gas mileage of all cars when the additive is not used. Evidently, the additive is effective in increasing gas mileage. ◨

Comparison of the Paired Wilcoxon Signed-Rank Test and the Paired t-Test

As we learned in Section 10.5, a paired t-test can be used to conduct a hypothesis test to compare two population means when we have a paired sample and the paired-difference variable is normally distributed. Since a normally distributed variable necessarily has a symmetric distribution, we can also use the paired Wilcoxon signed-rank test to perform such a hypothesis test.

So now the question is this: If we know the paired-difference variable is normally distributed, should we use the paired t-test or the paired Wilcoxon signed-rank test? As you might expect, we should use the paired t-test. For a normally distributed paired-difference variable, the paired t-test is more powerful than the paired Wilcoxon signed-rank test because it is designed expressly for such paired-difference variables; surprisingly, however, the paired t-test is not much more powerful than the paired Wilcoxon signed-rank test.

On the other hand, if the paired-difference variable has a symmetric distribution but is not normally distributed, then the paired Wilcoxon signed-rank test is usually more powerful than the paired t-test and is often considerably more powerful. In summary, we have the following key fact.

KEY FACT 10.7 **PAIRED WILCOXON SIGNED-RANK TEST VERSUS THE PAIRED t-TEST**

Suppose a hypothesis test is to be performed to compare the means of two populations using a paired sample. When deciding between the paired t-test and the paired Wilcoxon signed-rank test, follow these guidelines.

- If you are reasonably sure that the paired-difference variable is normally distributed, use the paired t-test.
- If you are not reasonably sure that the paired-difference variable is normally distributed but are reasonably sure that it has a symmetric distribution, use the paired Wilcoxon signed-rank test.

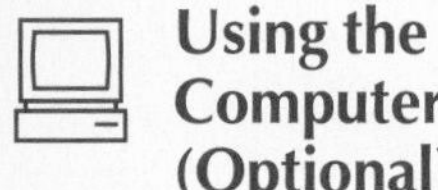

Using the Computer (Optional)

We can use Minitab to carry out a paired Wilcoxon signed-rank test because that procedure is simply a Wilcoxon signed-rank test on the sample of paired differences with null hypothesis H_0: $\mu_d = 0$.

Thus to employ Minitab, we first obtain the paired differences and then apply Minitab's Wilcoxon signed-rank procedure, as explained in Section 9.7. To illustrate, we return once more to the problem of determining whether a new gasoline additive improves gas mileage.

EXAMPLE 10.19 **USING MINITAB TO PERFORM A PAIRED WILCOXON SIGNED-RANK TEST**

Use Minitab to perform the hypothesis test in Example 10.18.

SOLUTION Let μ_1 denote the mean gas mileage of all cars when the additive is used and μ_2 denote the mean gas mileage of all cars when the additive is not used. Then the null and alternative hypotheses are

H_0: $\mu_1 = \mu_2$ (mean gas mileage with additive is not greater)

H_a: $\mu_1 > \mu_2$ (mean gas mileage with additive is greater).

Note that the hypothesis test is right-tailed since a greater-than sign (>) appears in the alternative hypothesis.

We want to perform a paired Wilcoxon signed-rank test. As we noted, this is equivalent to conducting a Wilcoxon signed-rank test on the paired differences. To obtain the paired differences using Minitab, we first store the sample data from the second and third columns of Table 10.14 on page 657 in Minitab columns named WITH and WITHOUT. Then we proceed as follows.

1 Choose **Calc ➤ Calculator...**

2 Type DIFF in the **Store results in variable** text box

3 Specify WITH - WITHOUT in the **Expression** text box

4 Click **OK**

The paired differences of the gas mileages with and without the additive are now stored in a column named DIFF. Next we apply Minitab's Wilcoxon signed-rank procedure to the data in DIFF:

1 Choose **Stat ➤ Nonparametrics ➤ 1-Sample Wilcoxon...**
2 Specify DIFF in the **Variables** text box
3 Select the **Test median** option button
4 Click in the **Test median** text box and type 0
5 Click the arrow button at the right of the **Alternative** drop-down list box and select **greater than**
6 Click **OK**

The output obtained as a result of this procedure is displayed in Printout 10.13.

PRINTOUT 10.13
Minitab output for the paired Wilcoxon signed-rank test

```
Test of median = 0.000000 versus median  >  0.000000

                N for   Wilcoxon              Estimated
            N    Test   Statistic        P      Median
DIFF       10       9        44.0    0.006      0.7000
```

All references to "median" in Printout 10.13 can be replaced by "$\mu_1 - \mu_2$." Thus the first line of Printout 10.13 displays the null and alternative hypotheses, respectively, $\mu_1 = \mu_2$ and $\mu_1 > \mu_2$.

The entry headed `N` gives the sample size, which in this case is 10. The next entry, headed `N for Test`, shows the number of paired differences in the sample that are not equal to 0—in this case the number of cars sampled whose gas mileages differ with and without the additive—which as we see is 9. The next entry, headed `Wilcoxon Statistic`, provides the value of the test statistic, W; so, we see that $W = 44.0$.

The final two entries in Printout 10.13, headed `P` and `Estimated Median`, give the P-value for the hypothesis test and a point estimate for the difference between the population means. Because the P-value of 0.006 is less than the designated significance level of 0.05, we reject H_0.

At the 5% significance level, the data provide sufficient evidence to conclude that the mean gas mileage of all cars when the additive is used is greater than the mean gas mileage of all cars when the additive is not used. And referring to Table 9.12 on page 543, we see that the data provide very strong evidence that the additive is effective in increasing gas mileage. □

EXERCISES 10.6

STATISTICAL CONCEPTS AND SKILLS

10.109 Suppose you want to perform a hypothesis test to compare the means of two populations using a paired sample and that you know that the paired-difference variable is normally distributed. Answer each of the following questions and explain your answers.

a. Is it acceptable to use the paired t-test?

b. Is it acceptable to use the paired Wilcoxon signed-rank test?

c. Which test is preferable, the paired t-test or the paired Wilcoxon signed-rank test?

10.110 Suppose you want to perform a hypothesis test to compare the means of two populations using a paired sample and you know that the paired-difference variable has a symmetric distribution that is far from normal.

a. Is it acceptable to use the paired t-test if the sample size is small or moderate? Why or why not?

b. Is it acceptable to use the paired t-test if the sample size is large? Why or why not?

c. Is it acceptable to use the paired Wilcoxon signed-rank test? Why or why not?

d. In case both the paired t-test and the paired Wilcoxon signed-rank test are acceptable, which is preferable?

Exercises 10.111–10.116 repeat the problems given in Exercises 10.89–10.94 of Section 10.5. In that section you were asked to apply the paired t-test to solve each problem. Now you are asked to solve each problem by applying the paired Wilcoxon signed-rank test.

10.111 Charles Darwin, author of *Origin of Species,* examined the effect of cross-fertilization on the heights of plants. In one study he planted 15 pairs of Zea mays plants. Each pair consisted of one cross-fertilized plant and one self-fertilized plant grown in the same pot. The following table gives the height differences, in eighths of an inch, for the 15 pairs. Each difference is obtained by subtracting the height of the self-fertilized plant from that of the cross-fertilized plant.

49	−67	8	16	6
23	28	41	14	29
56	24	75	60	−48

At the 5% significance level, do the data provide sufficient evidence to conclude that the mean heights of cross-fertilized and self-fertilized Zea mays differ?

10.112 In 1908 W. S. Gosset published "The Probable Error of a Mean" (*Biometrika,* Vol. 6, pp. 1–25). It is in this pioneering paper, published under the pseudonym "Student," that he introduced what later became known as Student's t-distribution. As an example, Gosset used the data set below, which gives the additional sleep in hours obtained by 10 patients using laevohysocyamine hydrobromide.

1.9	0.8	1.1	0.1	−0.1
4.4	5.5	1.6	4.6	3.4

At the 1% significance level, do the data provide sufficient evidence to conclude that laevohysocyamine hydrobromide is effective in increasing sleep?

10.113 Stichler, Richey, and Mandel compared two methods of measuring treadwear in their paper "Measurement of Treadwear of Commercial Tires" (*Rubber Age,* 73:2, 1953). Eleven tires were each measured for treadwear by two methods, one based on weight and the other on groove wear. Here are the data in thousands of miles.

Weight method	Groove method
30.5	28.7
30.9	25.9
31.9	23.3
30.4	23.1
27.3	23.7
20.4	20.9
24.5	16.1
20.9	19.9
18.9	15.2
13.7	11.5
11.4	11.2

At the 1% significance level, do the data provide sufficient evidence to conclude that, on the average, the two measurement methods give different results?

10.114 The A. C. Nielsen Company collects data on the TV viewing habits of Americans and publishes the information in *Nielsen Report on Television.* Twenty married couples are randomly selected. Their weekly viewing times, in hours, are as follows.

Husband	Wife	Husband	Wife	Husband	Wife
21	24	38	45	36	35
56	55	27	29	20	34
34	55	30	41	43	32
30	34	31	37	4	13
41	32	30	35	16	9
35	38	32	48	21	23
26	38	15	17		

At the 5% level of significance, does it appear that married men watch less TV, on the average, than married women?

10.115 *Current Population Reports,* published by the U.S. Bureau of the Census, presents data on the ages of married people. Ten married couples are randomly selected and have the ages shown here.

Husband	54	21	32	78	70	33	68	32	54	52
Wife	53	22	33	74	64	35	67	28	41	44

Do the data suggest that the mean age of married men is greater than the mean age of married women? Perform the appropriate hypothesis test at the 5% significance level.

10.116 A pediatrician measured the blood cholesterol levels of her young patients. She was surprised to find that many of them had levels over 200 mg per 100 mL, indicating increased risk of artery disease. Ten such patients were randomly selected to take part in a nutritional program designed to lower blood cholesterol. Two months after the program started, the pediatrician measured the blood cholesterol levels of the 10 patients again. The data are shown in the table at the top of the next column. Do the data suggest that the nutritional program is, on the average, effective in reducing cholesterol levels? Perform the appropriate hypothesis test at the 5% level of significance.

Patient	Before program	After program
1	210	212
2	217	210
3	208	210
4	215	213
5	202	200
6	209	208
7	207	203
8	210	199
9	221	218
10	218	214

EXTENDING THE CONCEPTS AND SKILLS

10.117 A hypothesis test is to be performed to compare the means of two populations using a paired sample. The sample of 15 paired differences contains an outlier but otherwise is roughly bell-shaped. Assuming it is not legitimate to remove the outlier, which test is better to use—the paired t-test or the paired Wilcoxon signed-rank test? Explain your answer.

10.118 Suppose you want to perform a hypothesis test to compare the means of two populations using a paired sample. For each part below, decide whether you would use the paired t-test, the paired Wilcoxon signed-rank test, or neither of these tests, if preliminary data analyses of the sample of paired differences suggest that the distribution of the paired-difference variable is

a. approximately normal.
b. highly skewed; the sample size is 20.
c. symmetric bimodal.

10.119 Suppose you want to perform a hypothesis test to compare the means of two populations using a paired sample. For each part below, decide whether you would use the paired t-test, the paired Wilcoxon signed-rank test, or neither of these tests, if preliminary data analyses of the sample of paired differences suggest that the distribution of the paired-difference variable is

a. uniform.
b. neither symmetric nor normal; the sample size is 132.
c. very moderately skewed but otherwise roughly bell-shaped.

10.120 Explain why the paired Wilcoxon signed-rank test is simply a Wilcoxon signed-rank test on the sample of paired differences with null hypothesis H_0: $\mu_d = 0$.

USING TECHNOLOGY  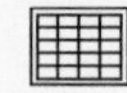 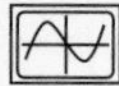

10.121 Refer to Exercise 10.113. Use Minitab or some other statistical software to

a. obtain a boxplot and stem-and-leaf diagram of the paired differences.
b. perform the required hypothesis test.
c. Justify the use of your procedure in part (b).

10.122 Refer to Exercise 10.114. Use Minitab or some other statistical software to

a. obtain a boxplot and stem-and-leaf diagram of the paired differences.
b. perform the required hypothesis test.
c. Justify the use of your procedure in part (b).

10.123 An algebra teacher at Arizona State University wanted to compare two methods of teaching college algebra—the lecture method and the personalized system of instruction (PSI) method. Students were paired by matching those with similar mathematics background and performance. A random sample of pairs was selected. From each pair, one student was randomly chosen to take the lecture course; the other student took the PSI course. Both courses were taught by the algebra teacher. Minitab's Wilcoxon signed-rank procedure was applied to the paired differences of the final grades for the pairs of students selected. The resulting output is shown in Printout 10.14. Use the output to determine

a. the null and alternative hypotheses.
b. the number of pairs of students selected.
c. the number of pairs in which both students had the same final grade.
d. the value of the test statistic, W.
e. the P-value of the hypothesis test.
f. the smallest significance level at which the null hypothesis can be rejected.
g. the conclusion if the hypothesis test is performed at the 5% significance level.
h. a point estimate for the difference between mean final grades of the two instructional methods.

10.124 It had been conjectured that a new running program would reduce heart rates. The heart rates of 15 randomly selected people were measured and then those people were placed on the running program. One year later the heart rates of the 15 people were measured again. Minitab's Wilcoxon signed-rank procedure was applied to the paired differences and resulted in the output shown in Printout 10.15. Use the output to determine

a. the null and alternative hypotheses.
b. the number of the 15 people whose heart rates remained unchanged.
c. the value of the test statistic, W.
d. the P-value of the hypothesis test.
e. the smallest significance level at which the null hypothesis can be rejected.
f. the conclusion if the hypothesis test is performed at the 5% significance level.
g. a point estimate for the decline in heart rate obtained, on the average, by people using the running program.

10.125 Refer to Exercise 10.115. Use Minitab or some other statistical software to

a. obtain a normal probability plot, stem-and-leaf diagram, histogram, and boxplot of the paired differences.
b. perform the required hypothesis test using the paired t-test.
c. perform the required hypothesis test using the paired Wilcoxon signed-rank test.
d. Compare your results from parts (b) and (c).
e. Based on your data analysis in part (a), which test do you think is more appropriate, the paired t-test or the paired Wilcoxon signed-rank test? Explain your answer.

10.126 Refer to Exercise 10.116. Use Minitab or some other statistical software to

a. obtain a normal probability plot, stem-and-leaf diagram, histogram, and boxplot of the paired differences.
b. perform the required hypothesis test using the paired t-test.
c. perform the required hypothesis test using the paired Wilcoxon signed-rank test.
d. Compare your results from parts (b) and (c).
e. Based on your data analysis in part (a), which test do you think is more appropriate, the paired t-test or the paired Wilcoxon signed-rank test? Explain your answer.

PRINTOUT 10.14 Minitab output for Exercise 10.123

```
Test of median = 0.000000 versus median not = 0.000000

                N for   Wilcoxon            Estimated
           N    Test   Statistic       P       Median
DIFF      11      10        42.5   0.139        1.500
```

PRINTOUT 10.15 Minitab output for Exercise 10.124

```
Test of median = 0.000000 versus median  >  0.000000

                N for   Wilcoxon            Estimated
           N    Test   Statistic       P       Median
DIFF      15      13        78.5   0.012        2.000
```

10.7 WHICH PROCEDURE SHOULD BE USED?*

In this chapter we have learned several inferential procedures for comparing the means of two populations. Table 10.15 on the next page summarizes the hypothesis-testing procedures; a similar table can be constructed for confidence-interval procedures.

Each row of Table 10.15 gives the type of test, the conditions required for using the test, the test statistic, and the procedure to use. For brevity we have written "paired W-test" instead of "paired Wilcoxon signed-rank test." And, as previously, we have used the following abbreviations:

- normal populations: the two distributions of the variable under consideration are normally distributed.
- same shape populations: the two distributions of the variable under consideration have the same shape.
- normal differences: the paired-difference variable is normally distributed.
- symmetric differences: the paired-difference variable has a symmetric distribution.

In selecting the correct procedure, keep in mind that the best choice is the procedure expressly designed for the distribution-types under consideration, if such a procedure exists, and that the three t-tests are only approximately correct for large samples.

* This section is optional and will not be needed in subsequent chapters of the book. All previous sections in this chapter, including the optional material on the Mann–Whitney test and paired Wilcoxon signed-rank test, are prerequisite to this section.

TABLE 10.15 Summary of hypothesis-testing procedures for comparing two population means. The null hypothesis for all tests is H_0: $\mu_1 = \mu_2$

Type	Assumptions	Test statistic	Procedure to use
Pooled t-test	1. Independent samples 2. Normal populations or large samples 3. Equal population standard deviations	$t = \dfrac{\bar{x}_1 - \bar{x}_2}{s_p\sqrt{(1/n_1) + (1/n_2)}}$ † ($\mathrm{df} = n_1 + n_2 - 2$)	10.1 (page 602)
Nonpooled t-test	1. Independent samples 2. Normal populations or large samples	$t = \dfrac{\bar{x}_1 - \bar{x}_2}{\sqrt{(s_1^2/n_1) + (s_2^2/n_2)}}$ ‡	10.3 (page 615)
Mann–Whitney test	1. Independent samples 2. Same shape populations 3. $n_1 \leq n_2$	M = sum of the ranks for sample data from Population 1	10.5 (page 631)
Paired t-test	1. Paired sample 2. Normal differences or large sample	$t = \dfrac{\bar{d}}{s_d/\sqrt{n}}$ ($\mathrm{df} = n - 1$)	10.6 (page 645)
Paired W-test	1. Paired sample 2. Symmetric differences	W = sum of positive ranks	10.8 (page 656)

† $s_p = \sqrt{\dfrac{(n_1 - 1)s_1^2 + (n_2 - 1)s_2^2}{n_1 + n_2 - 2}}$ ‡ $\mathrm{df} = \dfrac{[(s_1^2/n_1) + (s_2^2/n_2)]^2}{\dfrac{(s_1^2/n_1)^2}{n_1 - 1} + \dfrac{(s_2^2/n_2)^2}{n_2 - 1}}$

For instance, suppose that independent random samples are taken from two populations with equal standard deviations and that the two distributions (one for each population) of the variable under consideration are normally distributed. Then the pooled t-test, nonpooled t-test, and Mann–Whitney test are all applicable. But the correct procedure is the pooled t-test because that test is designed specifically for use with independent samples from two normally distributed populations having equal standard deviations.

The flowchart in Fig. 10.10 provides an organized strategy for choosing the correct hypothesis-testing procedure for comparing two population means. It has been constructed based on the above discussion.

In practice we need to look at the sample data to ascertain distribution-type before we can select the appropriate procedure. We recommend using normal probability plots and either stem-and-leaf diagrams (for small or moderate-size samples) or histograms (for moderate-size or large samples); boxplots can also be quite helpful, especially for moderate-size or large samples.

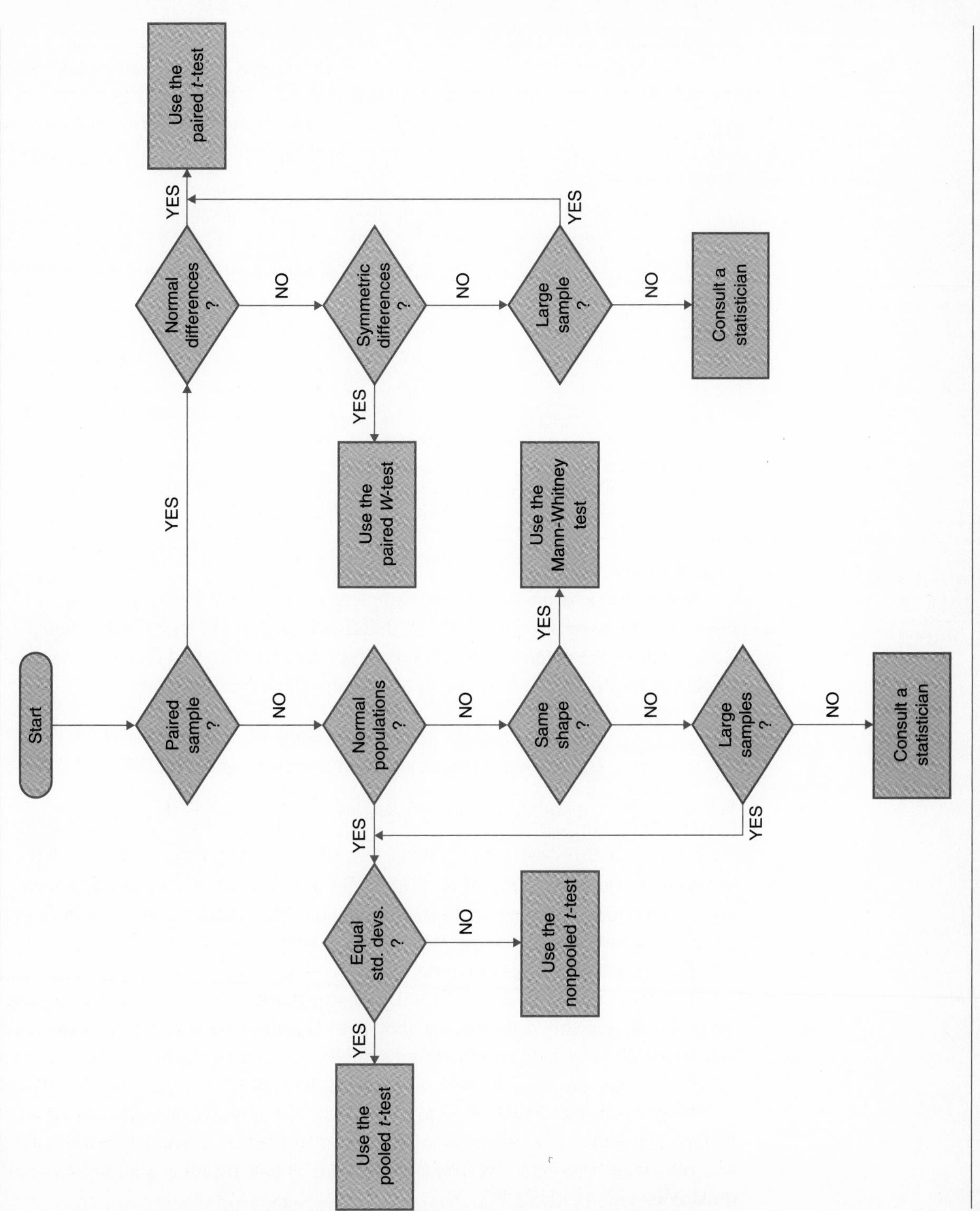

FIGURE 10.10 Flowchart for choosing the correct hypothesis-testing procedure for comparing two population means

EXAMPLE 10.20 CHOOSING THE CORRECT HYPOTHESIS-TESTING PROCEDURE

A study entitled "Body Composition of Elite Class Distance Runners" was conducted by M. L. Pollock and others to determine whether elite distance runners are truly thinner. Their results were published in *The Marathon: Physiological, Medical, Epidemiological, and Psychological Studies,* P. Milvey (ed.), New York: New York Academy of Sciences, 1977, p. 366.

The researchers measured the skinfold thickness, an indirect indicator of body fat, of runners and nonrunners in the same age group. The data in Table 10.16 are based on the skinfold thickness measurements on the thighs of the people sampled. Data are in millimeters.

TABLE 10.16 Skinfold thickness (mm) for independent samples of elite runners and others

Runners			Others			
7.3	6.7	8.7	24.0	19.9	7.5	18.4
3.0	5.1	8.8	28.0	29.4	20.3	19.0
7.8	3.8	6.2	9.3	18.1	22.8	24.2
5.4	6.4	6.3	9.6	19.4	16.3	16.3
3.7	7.5	4.6	12.4	5.2	12.2	15.6

Suppose we want to use the sample data in Table 10.16 to decide whether elite runners have smaller skinfold thickness, on the average, than other people. Let μ_1 denote the mean skinfold thickness of elite runners and μ_2 the mean skinfold thickness of others. Then we want to perform the hypothesis test

$$H_0: \mu_1 = \mu_2 \text{ (mean skinfold thickness is not smaller)}$$

$$H_a: \mu_1 < \mu_2 \text{ (mean skinfold thickness is smaller).}$$

Which procedure should be used to perform the hypothesis test?

SOLUTION We consult the flowchart in Fig. 10.10. The first question we must answer is: Do we have a paired sample? From the information provided, we see that the samples are not paired. Thus the answer to the first question is "No."

This leads us to the question: Are the populations normal? To answer this question, we constructed the normal probability plots in Fig. 10.11. The plots are quite linear and thereby indicate that it is certainly reasonable to presume that skinfold thickness is approximately normally distributed for both elite runners and others. So the answer to the second question is "Yes."

Next we must answer the question: Are the population standard deviations equal? The standard deviations of the two samples in Table 10.16 are 1.80 mm and 6.61 mm, respectively. These statistics suggest that the population standard deviations are not equal.

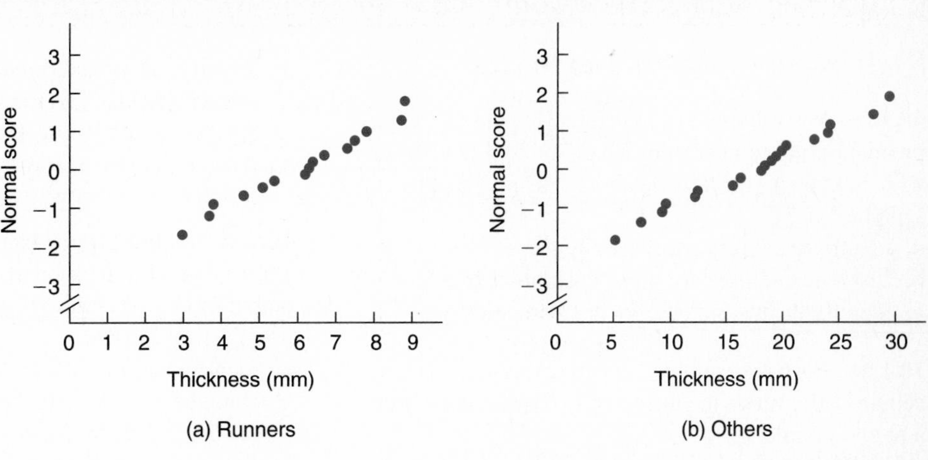

FIGURE 10.11 Normal probability plots of the sample data for (a) elite runners and (b) others

We can also see this by looking at a back-to-back stem-and-leaf diagram, histograms, or boxplots. Figure 10.12 displays boxplots for the two samples in Table 10.16. The vast difference in the spreads of the two plots again suggests that the standard deviations of skinfold thickness differ for elite runners and others. Thus the answer to the third question is "No."

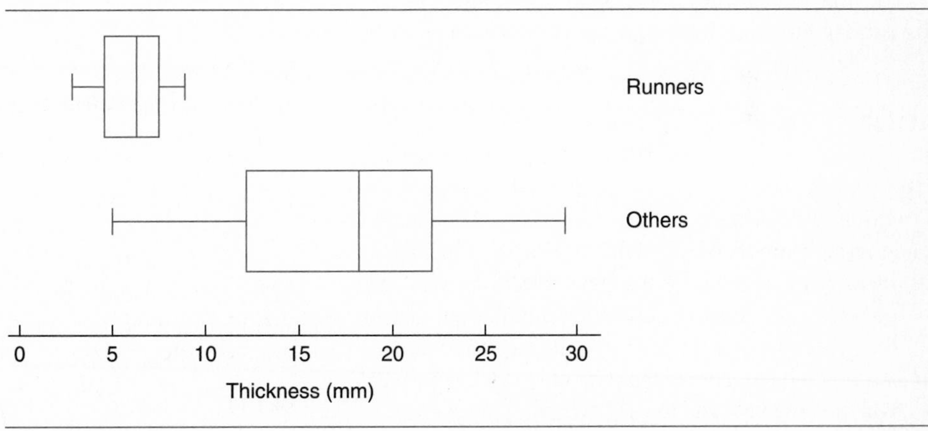

FIGURE 10.12 Boxplots of the sample data for elite runners and others

The "No" answer to the preceding question leads us to the box that states: Use the nonpooled t-test. In other words, we should use Procedure 10.3 on page 615 to conduct the hypothesis test. ◨

EXERCISES 10.7

STATISTICAL CONCEPTS AND SKILLS

10.127 We have considered three hypothesis-testing procedures to compare the means of two populations with unknown standard deviations using independent samples.

a. Identify the three procedures by name.
b. List the conditions for using each procedure.
c. Identify the test statistic for each procedure.

10.128 We have examined two hypothesis-testing procedures to compare the means of two populations using a paired sample.

a. Identify the two procedures by name.
b. List the conditions for using each procedure.
c. Identify the test statistic for each procedure.

10.129 Suppose you want to perform a hypothesis test to compare the means of two populations using independent samples. Assume the variable under consideration is normally distributed on each of the two populations and that the population standard deviations are equal.

a. Identify the procedures discussed in this chapter that could be used to carry out the hypothesis test, that is, the procedures whose assumptions are satisfied.
b. Among the procedures that you identified in part (a), which is the best one to use? Explain your answer.

10.130 Suppose you want to perform a hypothesis test to compare the means of two populations using independent samples. Assume the variable under consideration is normally distributed on each of the two populations and that the population standard deviations are unequal.

a. Identify the procedures discussed in this chapter that could be used to carry out the hypothesis test, that is, the procedures whose assumptions are satisfied.
b. Among the procedures that you identified in part (a), which is the best one to use? Explain your answer.

10.131 Suppose you want to perform a hypothesis test to compare the means of two populations using independent samples. Assume the two distributions of the variable under consideration have the same shape but are not normally distributed and that the sample sizes are both large.

a. Identify the procedures discussed in this chapter that could be used to carry out the hypothesis test, that is, the procedures whose assumptions are satisfied.
b. Among the procedures that you identified in part (a), which is the best one to use? Explain your answer.

10.132 Suppose you want to perform a hypothesis test to compare the means of two populations using a paired sample. Assume the paired-difference variable is normally distributed.

a. Identify the procedures discussed in this chapter that could be used to carry out the hypothesis test, that is, the procedures whose assumptions are satisfied.
b. Among the procedures that you identified in part (a), which is the best one to use? Explain your answer.

10.133 Suppose you want to perform a hypothesis test to compare the means of two populations using a paired sample. Assume the paired-difference variable has a nonnormal symmetric distribution and that the sample size is large.

a. Identify the procedures discussed in this chapter that could be used to carry out the hypothesis test, that is, the procedures whose assumptions are satisfied.
b. Among the procedures that you identified in part (a), which is the best one to use? Explain your answer.

Each of Exercises 10.134–10.139 provides a type of sampling (independent or paired), sample size(s), and a figure showing the results of preliminary data analyses on the sample(s). For independent samples the graphs are for the two samples; for a paired sample the graphs are for the paired differences. The intent is to employ the sample data to perform a hypothesis test to compare the means of the two populations. In each case, use the information provided and the flowchart in Fig. 10.10 on page 667 to decide which procedure should be applied.

10.134 Paired; $n = 75$; Fig. 10.13.

10.135 Independent; $n_1 = 25$, $n_2 = 20$; Fig. 10.14.

10.136 Independent; $n_1 = 17$, $n_2 = 17$; Fig. 10.15.

10.137 Independent; $n_1 = 40$, $n_2 = 45$; Fig. 10.16.

10.138 Independent; $n_1 = 20$, $n_2 = 15$; Fig. 10.17.

10.139 Paired; $n = 18$; Fig. 10.18.

FIGURE 10.13 Results of preliminary data analyses of the data in Exercise 10.134

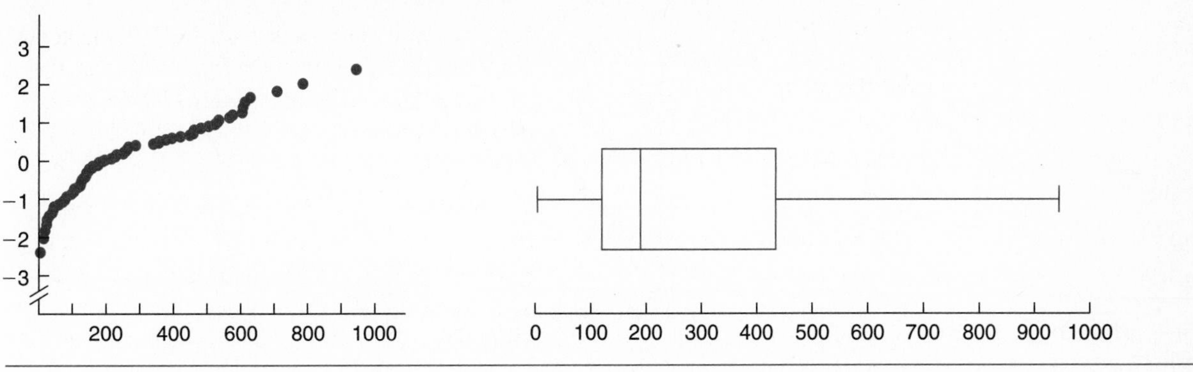

FIGURE 10.14 Results of preliminary data analyses of the data in Exercise 10.135

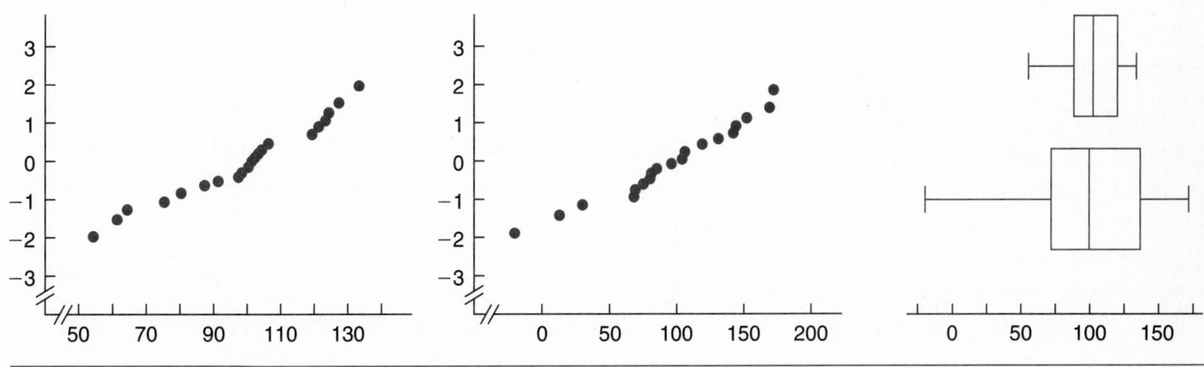

FIGURE 10.15 Results of preliminary data analyses of the data in Exercise 10.136

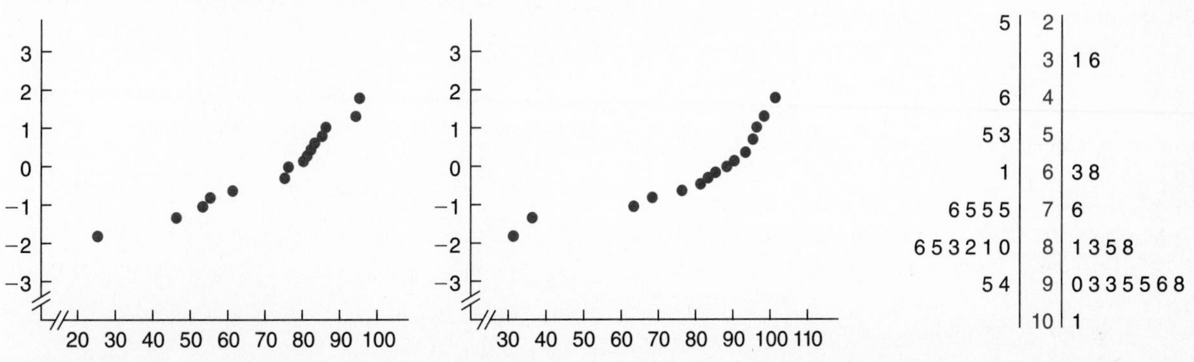

5	2	
	3	1 6
6	4	
5 3	5	
1	6	3 8
6 5 5 5	7	6
6 5 3 2 1 0	8	1 3 5 8
5 4	9	0 3 3 5 5 6 8
	10	1

FIGURE 10.16 Results of preliminary data analyses of the data in Exercise 10.137

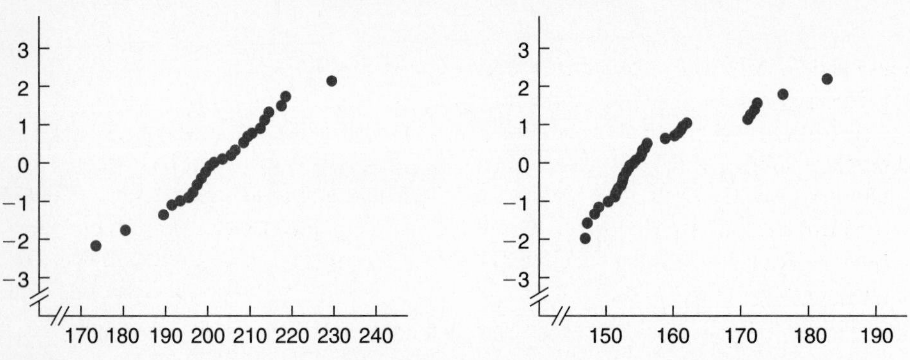

FIGURE 10.17 Results of preliminary data analyses of the data in Exercise 10.138

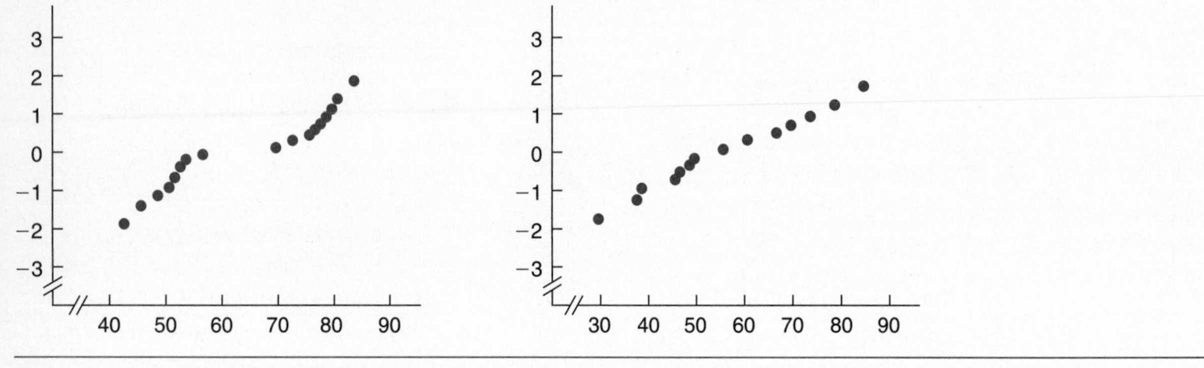

FIGURE 10.18 Results of preliminary data analyses of the data in Exercise 10.139

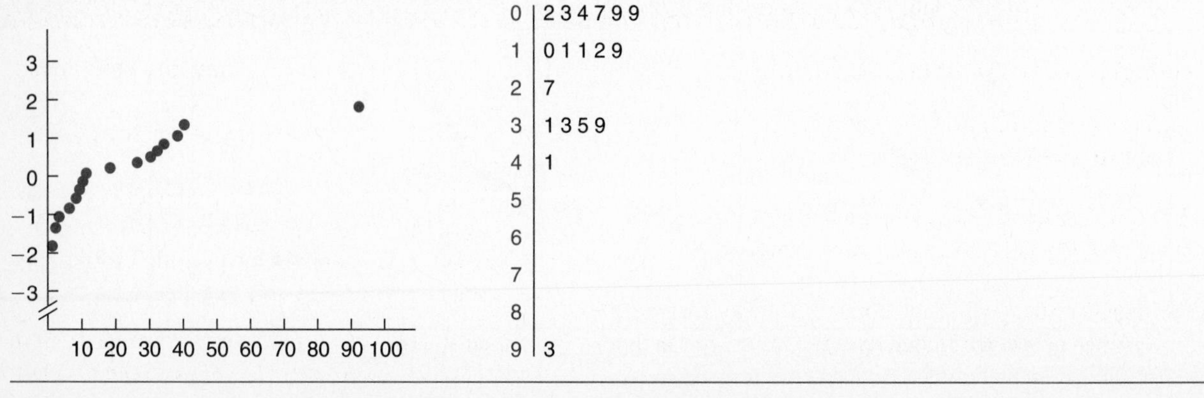

CHAPTER REVIEW

You Should Be Able To

1. use and understand the formulas presented in this chapter.
2. perform inferences to compare the means of two populations using independent samples when the population standard deviations are unknown but assumed equal.
3. perform inferences to compare the means of two populations using independent samples when the population standard deviations are unknown but not assumed equal.
*4. perform a hypothesis test to compare the means of two populations using independent samples when the distributions of the variable under consideration have the same shape.
5. perform inferences to compare the means of two populations using a paired sample.
*6. perform a hypothesis test to compare the means of two populations using a paired sample when the paired-difference variable has a symmetric distribution.
7. decide which procedure should be used to perform an inference to compare the means of two populations.
*8. use the Minitab procedures covered in this chapter.
*9. interpret the output obtained from the application of the Minitab procedures discussed in this chapter.

Key Terms

back-to-back stem-and-leaf diagram,* *628*
independent samples, *592*
Mann–Whitney test,* *631*
Mann–Whitney–Wilcoxon test,* *626*
nonpooled t-interval procedure, *618*
nonpooled t-test, *615*
normal differences, *644*
paired difference, *642*
paired-difference variable, *643*
paired samples, *640*
paired t-interval procedure, *648*
paired t-test, *645*
paired Wilcoxon signed-rank test,* *656*
pool, *601*
pooled sample standard deviation (s_p), *601*
pooled t-interval procedure, *607*
pooled t-test, *602*
sampling distribution of the difference between two means, *596*
symmetric differences,* *655*
Wilcoxon rank-sum test,* *626*

REVIEW TEST

STATISTICAL CONCEPTS AND SKILLS

1. Discuss the basic strategy for comparing the means of two populations using independent samples.

2. Discuss the basic strategy for comparing the means of two populations using a paired sample.

3. Regarding the pooled and nonpooled t-procedures:

a. What difference in assumptions is there between the two procedures?

b. How important is the assumption of independent samples for these procedures?

c. How important is the normality assumption for these procedures?

d. Fill in the following blank and explain your answer: Unless we are quite sure that the ____ are equal, the nonpooled t-procedures should be used instead of the pooled t-procedures.

***4.** Suppose that independent samples are taken from two populations in order to compare their means.

Further suppose that the two distributions of the variable under consideration have the same shape.

a. Would the nonpooled t-test ever be the procedure of choice in these circumstances? Explain your answer.

b. Under what conditions would it be preferable to use the pooled t-test instead of the Mann–Whitney test? Explain your answer.

5. Explain one possible advantage of using a paired sample instead of independent samples.

***6.** Suppose that a paired sample is taken from two populations in order to compare their means. Further suppose that the distribution of the paired-difference variable has a symmetric distribution. Under what conditions would it be preferable to use the paired t-test instead of the paired Wilcoxon signed-rank test? Explain your answer.

7. In the paper "Sex Differences in Static Strength and Fatigability in Three Different Muscle Groups" (*Research Quarterly for Exercise and Sport,* 61(3), pp. 238–242, 1990), J. Misner et al. published results of a study on grip and leg strength of males and females. The following data, in newtons, is based on their measurements of right-leg strength.

Male			Female		
2632	1796	2256	1344	1351	1369
2235	2298	1917	2479	1573	1665
1105	1926	2644	1791	1866	1544
1569	3129	2167	2359	1694	2799
1977			1868	2098	

a. Preliminary data analyses indicate that it is reasonable to presume that leg strength is normally distributed for both males and females and that the standard deviations of leg strength are approximately equal. At the 5% significance level, do the data provide sufficient evidence to conclude that mean right-leg strength of males exceeds that of females? Use the critical-value approach. (*Note:* $\bar{x}_1 = 2127$, $s_1 = 513$, $\bar{x}_2 = 1843$, $s_2 = 446$.)

b. Estimate the P-value of the hypothesis test.

c. Use part (b) and Table 9.12 on page 543 to assess the evidence against the null hypothesis.

8. Refer to Problem 7. Obtain a 90% confidence interval for the difference between the mean right-leg strengths of males and females.

9. Euromonitor Publications Limited, London, conducts surveys in various countries to obtain data on food consumption of major food commodities. Results of those surveys can be found in *European Marketing Data and Statistics.* Independent samples of 10 Germans and 15 Russians yielded the following fish consumptions, in kilograms, for last year.

Germans		Russians		
10	12	16	21	12
17	12	11	5	23
14	11	19	19	22
13	8	16	23	12
9	8	18	7	17

Preliminary data analyses indicate that it is reasonable to presume that last year's fish consumptions in both countries are approximately normally distributed but do not have equal standard deviations. Do the data provide sufficient evidence to conclude, at the 5% significance level, that last year the average German consumed less fish than the average Russian? (*Note:* $\bar{x}_1 = 11.40$, $s_1 = 2.84$, $\bar{x}_2 = 16.07$, $s_2 = 5.61$.)

10. Refer to Problem 9. Find a 90% confidence interval for the difference, $\mu_1 - \mu_2$, between last year's mean fish consumptions by Germans and Russians.

***11.** The National Association of REALTORS® compiles and publishes home prices of existing single-family homes in cities across the United States. Independent random samples of 10 homes each in New York City and Los Angeles yielded the following data on home prices in thousands of dollars.

New York		Los Angeles	
131.5	118.1	137.5	130.0
379.8	168.5	290.8	101.7
132.8	105.2	215.9	140.9
145.8	141.8	127.7	104.5
235.8	174.8	335.9	107.0

At the 5% significance level, can we conclude that the mean costs for existing single-family homes differ in New York City and Los Angeles? *(Note:* Preliminary data analyses suggest that it is reasonable to presume that the cost distributions for the two cities have roughly the same shape but that those distributions are right skewed.*)*

12. To compare the effectiveness of two speed-reading programs, 10 pairs of people were randomly selected. Each pair consisted of people whose then current reading speeds were essentially identical. From each pair, one person was randomly selected to take Program 1; the other person took Program 2. After the speed-reading programs were completed, the reading speeds, in words per minute, for the 10 pairs of people were found to be the following.

Pair	Program 1	Program 2
1	1114	1032
2	996	1148
3	979	1074
4	1125	1076
5	910	959
6	1056	1094
7	1091	1091
8	1053	1096
9	996	1032
10	894	1012

At the 10% significance level, can we conclude that there is a difference in effectiveness of the two speed-reading programs? *(Note:* A normal probability plot of the paired differences suggests that it is reasonable to presume that the paired-difference variable is approximately normally distributed.*)*

13. Refer to Problem 12. Find a 90% confidence interval for the difference in mean reading speeds for the two programs.

***14.** The U.S. Bureau of the Census conducts surveys to estimate average rents for housing units and publishes the statistics in *Current Housing Reports.* Renter-occupied housing units in the South and Midwest were paired according to square footage, quality of neighborhood, and so forth. The monthly rents of 15 randomly selected pairs yielded the following data, in dollars.

South	Midwest	South	Midwest	South	Midwest
216	192	240	256	406	358
750	723	330	304	677	681
258	254	1063	997	225	236
420	401	625	613	1374	1296
634	588	254	218	271	263

At the 1% significance level, do the data provide sufficient evidence to conclude that the mean monthly rent for renter-occupied housing units in the South exceeds that for those in the Midwest? Use the paired Wilcoxon signed-rank test.

USING TECHNOLOGY  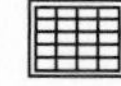 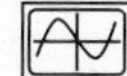

15. Refer to Problems 7 and 8. Use Minitab or some other statistical software to

a. obtain normal probability plots, stem-and-leaf diagrams, boxplots, and the standard deviations of the two samples.

b. perform the required hypothesis test and obtain the desired confidence interval.

c. Justify the use of your procedures in part (b).

16. Printout 10.16 on page 677 supplies the output obtained by applying Minitab's pooled t-procedure to the data in Problem 7. From the output, determine

a. the null and alternative hypotheses.

b. the standard deviation of each sample.

c. the mean right-leg strength of the males sampled and the mean right-leg strength of the females sampled.

d. the value of the test statistic, t.

e. the P-value of the hypothesis test.

f. the smallest significance level at which the null hypothesis can be rejected.

g. the conclusion if the hypothesis test is performed at the 5% significance level.

h. a 90% confidence interval for the difference between the mean right-leg strengths of males and females.

i. the value of s_p.

17. Refer to Problems 9 and 10. Use Minitab or some other statistical software to

a. obtain normal probability plots, stem-and-leaf diagrams, boxplots, and the standard deviations of the two samples.
b. perform the required hypothesis test and obtain the desired confidence interval.
c. Justify the use of your procedures in part (b).

18. Printout 10.17 shows the output obtained by applying Minitab's nonpooled t-procedure to the data in Problem 9. Employ the output to determine
a. the null and alternative hypotheses.
b. the standard deviation of fish consumption for each of the two samples.
c. the number of Germans sampled and the number of Russians sampled.
d. the mean fish consumptions for the Germans and Russians sampled.
e. the value of the test statistic, t.
f. the P-value of the hypothesis test.
g. the smallest significance level at which the null hypothesis can be rejected.
h. the conclusion if the hypothesis test is performed at the 5% significance level.
i. a 90% confidence interval for the difference between last year's mean fish consumptions by Germans and Russians.

***19.** Refer to Problem 11. Use Minitab or some other statistical software to
a. obtain normal probability plots, stem-and-leaf diagrams, boxplots, and the standard deviations of the two samples.
b. perform the required hypothesis test.
c. Justify the use of your procedure in part (b).

***20.** Printout 10.18 shows the output obtained by applying Minitab's Mann-Whitney procedure to the data in Problem 11. Employ the output to determine
a. the null and alternative hypotheses.
b. the median costs of the New York City homes sampled and the Los Angeles homes sampled.
c. the number of New York City homes sampled and the number of Los Angeles homes sampled.
d. the value of the Mann–Whitney test statistic.
e. the approximate P-value of the hypothesis test.
f. the smallest significance level at which the null hypothesis can be rejected.
g. the conclusion if the hypothesis test is performed at the 5% significance level.
h. a 95.5% confidence interval for the difference between the median (or mean) costs of existing single-family homes in New York City and Los Angeles.

21. Refer to Problems 12 and 13. Use Minitab or some other statistical software to
a. obtain a normal probability plot, stem-and-leaf diagram, and boxplot of the paired differences.
b. perform the required hypothesis test and obtain the desired confidence interval.
c. Justify the use of your procedures in part (b).

22. Printout 10.19 on page 678 shows the output obtained by applying Minitab's paired t-procedure to the data in Problem 12. Use the output to determine
a. the null and alternative hypotheses.
b. the sample mean reading speed for each group of 10 people.
c. the mean of the paired differences.
d. the standard deviation of the paired differences.
e. the value of the test statistic, t.
f. the P-value of the hypothesis test.
g. the smallest significance level at which the null hypothesis can be rejected.
h. the conclusion if the hypothesis test is performed at the 10% significance level.
i. a 90% confidence interval for the difference in mean reading speeds for the two programs.

***23.** Refer to Problem 14. Use Minitab or some other statistical software to
a. carry out the required hypothesis test.
b. obtain a normal probability plot, stem-and-leaf diagram, and boxplot of the paired differences.
c. From the graphs you obtained in part (b), does it appear that the conditions for using a paired Wilcoxon signed-rank test are satisfied? Explain your answer.
d. From the graphs you obtained in part (b), does it appear that the conditions for using a paired t-test are satisfied? Explain your answer.
e. Which procedure is preferable here, the paired Wilcoxon signed-rank test or the paired t-test? Explain your answer.

***24.** Minitab's Wilcoxon signed-rank procedure was applied to the paired differences of the data in Problem 14 and resulted in the output shown in Printout 10.20 on the next page. Use the output to obtain

a. the null and alternative hypotheses.

b. the number of pairs of housing units sampled that had the same monthly rent.

c. the value of the test statistic, W.

d. the P-value of the hypothesis test.

e. the smallest significance level at which the null hypothesis can be rejected.

f. the conclusion if the hypothesis test is performed at the 1% significance level.

g. a point estimate for the difference between the mean monthly rents of renter-occupied housing units in the South and Midwest.

PRINTOUT 10.16 Minitab output for Problem 16

```
Two sample T for MALE vs FEMALE

          N      Mean      StDev   SE Mean
MALE     13      2127        513       142
FEMALE   14      1843        446       119

90% CI for mu MALE - mu FEMALE: ( -31,  599)
T-Test mu MALE = mu FEMALE (vs >): T = 1.54  P = 0.068  DF = 25
Both use Pooled StDev =  479
```

PRINTOUT 10.17 Minitab output for Problem 18

```
Two sample T for GERMANS vs RUSSIANS

            N      Mean      StDev   SE Mean
GERMANS    10     11.40       2.84      0.90
RUSSIANS   15     16.07       5.61       1.4

90% CI for mu GERMANS - mu RUSSIANS: ( -7.60,  -1.7)
T-Test mu GERMANS = mu RUSSIANS (vs <): T = -2.74  P = 0.0062  DF = 21
```

PRINTOUT 10.18 Minitab output for Problem 20

```
NEW YORK   N =  10     Median =          143.8
LA         N =  10     Median =          133.8
Point estimate for ETA1-ETA2 is           11.4
95.5 Percent CI for ETA1-ETA2 is (-55.0,44.8)
W = 116.0
Test of ETA1 = ETA2  vs  ETA1 not = ETA2 is significant at 0.4274

Cannot reject at alpha = 0.05
```

PRINTOUT 10.19 Minitab output for Problem 22

```
Paired T for PROGRAM1 - PROGRAM2

                 N      Mean     StDev   SE Mean
PROGRAM1        10    1021.4      80.3      25.4
PROGRAM2        10    1061.4      53.4      16.9
Difference      10     -40.0      71.6      22.6

90% CI for mean difference: (-81.5, 1.5)
T-Test of mean difference = 0 (vs not = 0): T-Value = -1.77  P-Value = 0.111
```

PRINTOUT 10.20 Minitab output for Problem 24

```
Test of median = 0.000000 versus median  >  0.000000

              N for   Wilcoxon              Estimated
          N    Test  Statistic         P       Median
DIFF     15      15      108.5     0.003        23.50
```

INTERNET PROJECT

Women Workers and Equality in the United States

The rise of women in the U.S. labor force has been dramatic. According to the Women's Bureau of the U.S. Department of Labor, women workers made up 46% of the U.S. labor force in 1994. The Bureau predicts that this figure will increase to 48% by the year 2005.

Of course, these numbers represent progress for women in the United States, but problems remain. More women are working than in the past, but are they compensated the same as men? Some claim that most of the women in the U.S. work force have jobs that tend to pay less than male occupations. Additionally, it is suggested that such women have fewer opportunities for advancement. In this Internet project, you will explore relevant data and come to your own conclusion.

URL for access to Internet Projects Page: `http://hepg.awl.com` *keyword:* Weiss

USING THE FOCUS DATABASE

In Chapter 1 we explained how to store the Focus database in a Minitab worksheet named `focus.mtw`. If you haven't already created that worksheet, follow the instructions on pages 54–55 to create it now.

The Focus database contains information on 500 randomly selected Arizona State University sophomores for seven different variables: sex, high-school GPA, SAT math score, cumulative GPA, SAT verbal score, age, and total hours. Use Minitab or some other statistical software to solve the following problems.

a. Obtain normal probability plots, boxplots, and the sample standard deviations of the SAT math scores of the male sophomores in the sample and the female sophomores in the sample.

b. At the 5% significance level, do the data provide sufficient evidence to conclude that male sophomores at Arizona State University have a higher mean SAT math score than female sophomores? Justify the use of the procedure you chose to carry out the hypothesis test.

c. Obtain a 90% confidence interval for the difference between the mean SAT math scores of male and female sophomores at Arizona State University.

d. Obtain normal probability plots, boxplots, and the standard deviations of the SAT verbal scores of the male sophomores in the sample and the female sophomores in the sample.

e. At the 5% significance level, do the data provide sufficient evidence to conclude that male and female sophomores at Arizona State University have different mean SAT verbal scores? Justify the use of the procedure you selected to perform the hypothesis test.

f. Obtain a 95% confidence interval for the difference between the mean SAT verbal scores of male and female sophomores at Arizona State University.

g. Obtain a normal probability plot, a histogram, and a boxplot for the paired differences of the SAT verbal scores and SAT math scores of the sophomores in the sample.

h. Do the data provide sufficient evidence to conclude that the mean SAT verbal score is less than the mean SAT math score for Arizona State University sophomores? Perform the required hypothesis test at the 0.01 significance level. Justify the use of the procedure you employed to conduct the hypothesis test.

i. Find a 98% confidence interval for the difference between the mean SAT math and verbal scores of Arizona State University sophomores.

j. Obtain normal probability plots, boxplots, and the standard deviations of the cumulative GPAs of the sophomores in the sample who are under 21 years of age and those who are 21 years of age or over.

k. Do the data provide sufficient evidence to conclude that for Arizona State University sophomores, there is a difference between the mean cumulative GPAs of those under 21 years of age and those 21 years of age or over? Perform the required hypothesis test at the 5% significance level. Justify the use of the procedure you chose to carry out the hypothesis test.

CASE STUDY DISCUSSION

Breast Milk and IQ

On page 590 of this chapter, we presented data obtained by five researchers studying the effect of breast-feeding on subsequent IQ of preterm babies at age 7½–8 years. Three categories were considered: children whose mothers declined to provide breast milk (Group I);

those whose mothers had chosen but were unable to provide breast milk (Group IIa); and those whose mothers had chosen and were able to provide breast milk (Group IIb).

Presuming that IQs are normally distributed in all three categories, solve each of the following problems.

a. Do the data provide sufficient evidence to conclude that for children age 7 ½–8 years who are born preterm, a difference exists in mean IQ between those whose mothers decline to provide breast milk and those whose mothers choose but are unable to provide breast milk? Perform the required hypothesis test at the 5% significance level.
b. Do the data provide sufficient evidence to conclude that for children age 7 ½–8 years who are born preterm, the mean IQ of those whose mothers decline to provide breast milk is less than that of those whose mothers choose and are able to provide breast milk? Perform the required hypothesis test at the 5% significance level.
c. Do the data provide sufficient evidence to conclude that for children age 7 ½–8 years who are born preterm, the mean IQ of those whose mothers choose but are unable to provide breast milk is less than that of those whose mothers choose and are able to provide breast milk? Perform the required hypothesis test at the 5% significance level.
d. Is this study observational, or is it a designed experiment? Why?
e. Based on your answers in parts (a)–(d), what conclusions would you draw?

The researchers also adjusted the data for such factors as social class, mother's education, and infant's sex, and still reached the same conclusions: "... preterm babies whose mothers provided breast milk had a substantial advantage in subsequent IQ at 7 ½–8 years over those who did not ...." However, the researchers emphasized that they could not exclude the possibility that their findings could be explained by differences in parental behavior or genetic potential between the groups.

BIOGRAPHY GERTRUDE COX

Gertrude Mary Cox was born on January 13, 1900, in Dayton, Iowa, the daughter of John and Emmaline Cox. She graduated from Perry High School, Perry, Iowa, in 1918. Between 1918 and 1925, she prepared to become a deaconess in the Methodist Episcopal Church.

In 1929 and 1931, Cox received a B.S. and an M.S., respectively, from Iowa State College in Ames. Her work there was directed by George W. Snedecor, and her degree was the first master's degree in statistics given by the department of mathematics at Iowa State.

From 1931 to 1933, Cox studied psychological statistics at the University of California at Berkeley. Snedecor meanwhile had established a new Statistical Laboratory at Iowa State, and in 1933 he asked her to be his assistant. This began her internationally influential career in statistics. Cox worked in the lab until becoming an Iowa State assistant professor in 1939.

In 1940 the committee in charge of filling a newly created position as head of the department of experimental statistics at North Carolina State College in Raleigh asked

Snedecor for recommendations; he first named several male statisticians, then wrote, "... but if you would consider a woman for this position I would recommend Gertrude Cox of my staff." They did consider a woman and Cox accepted their offer.

In 1945 Cox organized and became director of the Institute of Statistics, which combined the teaching of statistics at the University of North Carolina and North Carolina State. Work conferences that Cox organized established the Institute as an international center for statistics. Cox also developed statistical programs throughout the South, referred to as "spreading the gospel according to St. Gertrude."

Cox's area of expertise was experimental design. She, with W. G. Cochran, wrote *Experimental Designs* (1950), recognized as the classic textbook on design and analysis of replicated experiments.

From 1960–1964, Cox was director of the Statistics Section of the Research Triangle Institute in Durham, North Carolina. She then retired, working only as a consultant. She died of leukemia on October 17, 1978, in Durham.

CASE STUDY — SPEAKER WOOFER DRIVER MANUFACTURING

Speaker driver manufacturing is an important industry in many countries. In Taiwan, for example, over 100 companies or factories produce and supply parts and driver units for speakers.

An essential component in driver units is the rubber edge, which impacts such aspects of sound quality as musical image and clarity. And an important characteristic of the rubber edge is weight. Generally, each process for manufacturing rubber edges calls for a *production weight specification* consisting of a lower specification limit (LSL), a target weight (T), and an upper specification limit (USL). The actual (population) mean and standard deviation of the weights of the rubber edges being produced are called, respectively, the process mean (μ) and process standard deviation (σ).

Several process capability indexes are used to measure process quality relative to the production weight specification and the process mean and standard deviation. A process that is on target ($\mu = T$) is called *super* if $(\text{USL} - \text{LSL})/6\sigma > 2$ or, equivalently, if $\sigma < (\text{USL} - \text{LSL})/12$.

W. L. Pearn and K. S. Chen investigated five rubber-edge manufacturing processes at Bopro—a company located in Taipei, Taiwan—for process capability. The following table, adapted from their paper "Multiprocess Performance Analysis: A Case Study" (*Quality Engineering,* 10(1), pp 1–8), provides weight data for one of the processes.

17.59	17.63	17.68	17.57	17.70	17.77	17.54	17.65	17.49	17.60
17.61	17.72	17.68	17.69	17.57	17.66	17.55	17.80	17.67	17.53
17.71	17.54	17.68	17.75	17.46	17.82	17.62	17.53	17.47	17.50
17.74	17.65	17.68	17.68	17.71	17.64	17.65	17.62	17.56	17.60
17.51	17.70	17.47	17.57	17.55	17.63	17.44	17.60	17.63	17.59
17.69	17.53	17.59	17.57	17.49	17.52	17.71	17.56	17.49	17.58

In this chapter you will study inferences for population standard deviations. After you have completed the chapter, you will be asked to use the data in the table to ascertain the process capability of the process.

Inferences for Population Standard Deviations*

11

GENERAL OBJECTIVES

Up to this point in our study of inferential statistics, we have been concentrating on inferences for population means. Another important class of inferences consists of those for population standard deviations (or variances).

For example, in Chapter 9 we discussed the problem of deciding whether the mean net weight of bags of pretzels being packaged by a machine equals the advertised weight of 454 g. This decision involves a hypothesis test for a population mean.

We should probably also be concerned with the variation in weights from bag to bag. If the variation is too large, then many bags will contain either considerably more or considerably less than they should. To investigate the variation, we can perform a hypothesis test or construct a confidence interval for the standard deviation of the weights. These are inferences for one population standard deviation.

Additionally, we might want to compare two different machines for packaging the pretzels to see whether one provides a smaller variation in weights than the other. This could be done using inferences for two population standard deviations.

In this chapter, we will discuss inferences for one and two population standard deviations.

* This chapter is optional. It is not needed for the remainder of the book.

CHAPTER OUTLINE

11.1 INFERENCES FOR ONE POPULATION STANDARD DEVIATION*

Recall that standard deviation is a measure of the variation (spread) of a data set. A data set with a great deal of variation will have a large standard deviation, whereas one with little variation will have a small standard deviation.

Also recall that for a variable x, the standard deviation of all possible observations for the entire population is called the *population standard deviation* or *standard deviation of the variable* x. It is denoted by σ_x or, when no confusion will arise, simply by σ.

Suppose we want to obtain information about a population standard deviation. If the population is small, then we can ordinarily determine σ exactly by first taking a census and then computing σ from the population data. However, if the population is large, which is usually the case, then a census is generally not feasible, and we must therefore use inferential methods to obtain the required information about σ.

In this section we will learn how to perform hypothesis tests and construct confidence intervals for the standard deviation of a normally distributed variable. Such inferences are based on a distribution called the chi-square distribution. We begin by studying that distribution.

The Chi-Square Distribution

A variable is said to have a **chi-square distribution** if its distribution has the shape of a special type of right-skewed curve, called a **chi-square (χ^2) curve.** Actually there are infinitely many chi-square distributions, and we identify the chi-square distribution (and χ^2-curve) in question by stating its number of degrees of freedom, just as we did for t-distributions. Figure 11.1 shows three χ^2-curves and illustrates some basic properties of χ^2-curves, which are presented in Key Fact 11.1.

FIGURE 11.1
χ^2-curves for df = 5, 10, and 19

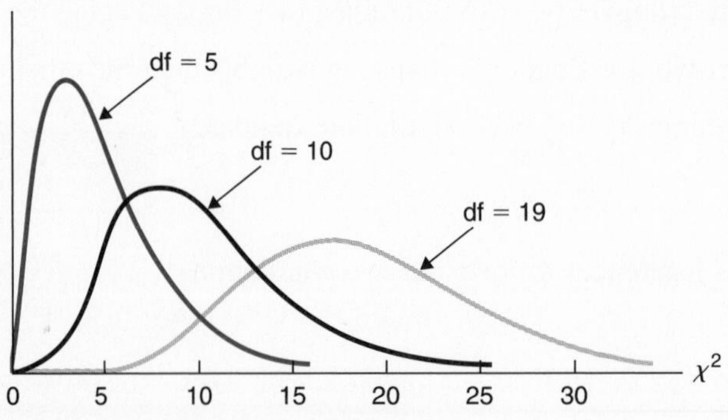

* See the note on the bottom of page 683.

KEY FACT 11.1 **BASIC PROPERTIES OF χ^2-CURVES**

Property 1: The total area under a χ^2-curve is equal to 1.

Property 2: A χ^2-curve starts at 0 on the horizontal axis and extends indefinitely to the right, approaching, but never touching, the horizontal axis as it does so.

Property 3: A χ^2-curve is right skewed.

Property 4: As the number of degrees of freedom becomes larger, χ^2-curves look increasingly like normal curves.

Percentages (and probabilities) for a variable having a chi-square distribution are equal to areas under its associated χ^2-curve. To perform a hypothesis test or construct a confidence interval for a population standard deviation, we need to know how to find the χ^2-value that corresponds to a specified area under a χ^2-curve. Table VIII in Appendix A provides χ^2-values corresponding to several areas for various degrees of freedom.

The χ^2-table (Table VIII) is similar to the t-table (Table IV). The two outside columns of Table VIII, labeled df, display the number of degrees of freedom. As expected, the symbol χ^2_α denotes the χ^2-value having area α to its right under a χ^2-curve. Thus the column headed $\chi^2_{0.995}$ contains χ^2-values having area 0.995 to their right; the column headed $\chi^2_{0.99}$ contains χ^2-values having area 0.99 to their right; and so on. Examples 11.1–11.3 explain how to use Table VIII.

EXAMPLE 11.1 **FINDING THE χ^2-VALUE HAVING A SPECIFIED AREA TO ITS RIGHT**

For a χ^2-curve with 12 degrees of freedom, find $\chi^2_{0.025}$; that is, find the χ^2-value having area 0.025 to its right, as shown in Fig. 11.2(a).

FIGURE 11.2 Finding the χ^2-value having area 0.025 to its right

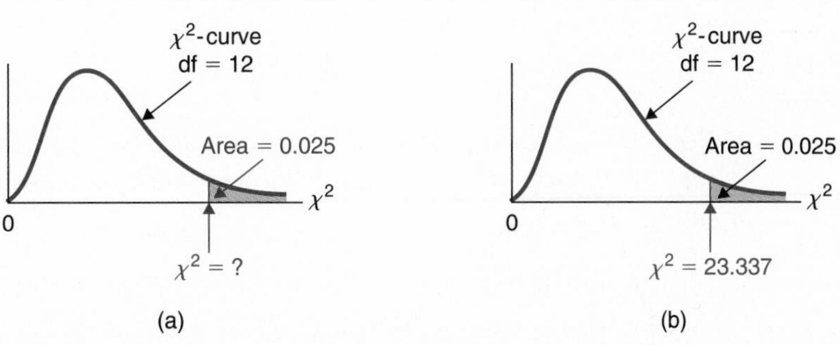

SOLUTION To find this χ^2-value, we use Table VIII. Since the number of degrees of freedom is 12, we first go down the outside columns, labeled df, to "12." Then we go across

that row until we are under the column headed $\chi^2_{0.025}$. The number in the body of the table there, 23.337, is the required χ^2-value; that is, for a χ^2-curve with df = 12, the χ^2-value having area 0.025 to its right is $\chi^2_{0.025} = 23.337$, as seen in Figure 11.2(b).

EXAMPLE 11.2 FINDING THE χ^2-VALUE HAVING A SPECIFIED AREA TO ITS LEFT

Determine the χ^2-value having area 0.05 to its left for a χ^2-curve with df = 7, as seen in Fig. 11.3(a).

FIGURE 11.3 Finding the χ^2-value having area 0.05 to its left

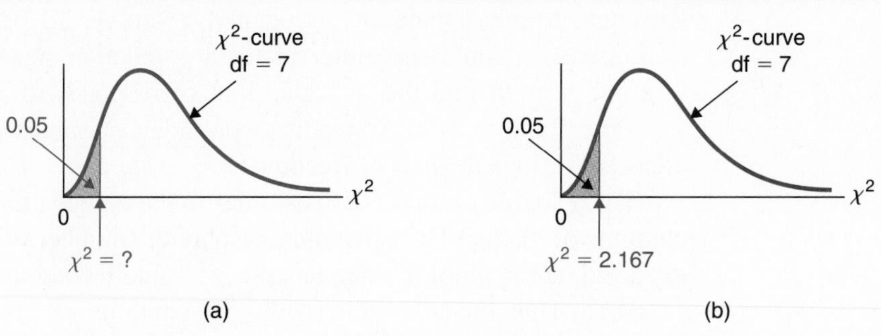

SOLUTION Since the total area under a χ^2-curve is equal to 1 (Property 1 of Key Fact 11.1), the unshaded area in Fig. 11.3(a) must equal 1 − 0.05 = 0.95. Thus the area to the right of the required χ^2-value is 0.95. This means that the required χ^2-value is $\chi^2_{0.95}$. From Table VIII we find that for df = 7, $\chi^2_{0.95} = 2.167$. Consequently, for a χ^2-curve with df = 7, the χ^2-value having area 0.05 to its left is 2.167, as shown in Fig. 11.3(b).

EXAMPLE 11.3 FINDING THE χ^2-VALUES FOR A SPECIFIED AREA

For a χ^2-curve with df = 20, determine the two χ^2-values that divide the area under the curve into a middle 0.95 area and two outside 0.025 areas, as shown in Fig. 11.4(a).

SOLUTION First we obtain the χ^2-value on the right in Fig. 11.4(a). From the figure, we see that the shaded area on the right is 0.025. This means that the χ^2-value on the right is $\chi^2_{0.025}$. Consulting Table VIII, we find that for df = 20, $\chi^2_{0.025} = 34.170$.

Next we obtain the χ^2-value on the left in Fig. 11.4(a). Because the area to the left of that χ^2-value is 0.025, the area to its right is 1 − 0.025 = 0.975. Hence the χ^2-value on the left is $\chi^2_{0.975}$, which, by Table VIII, equals 9.591 for df = 20.

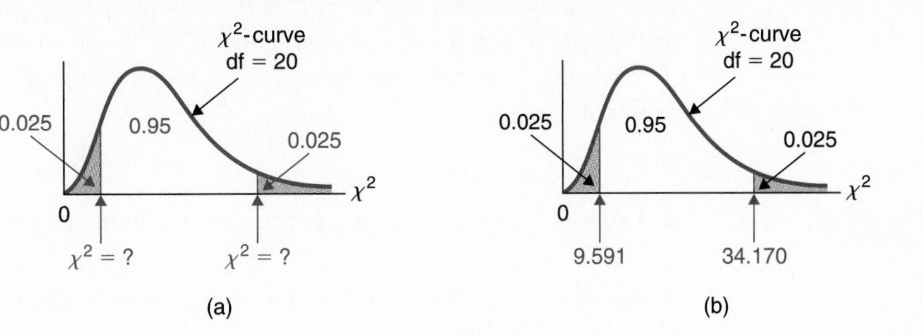

FIGURE 11.4 Finding the two χ^2-values dividing the area under the curve into a middle 0.95 area and two outside 0.025 areas

Consequently, for a χ^2-curve with df $= 20$, the two χ^2-values that divide the area under the curve into a middle 0.95 area and two outside 0.025 areas are 9.591 and 34.170, as shown in Fig. 11.4(b).

The Logic Behind Hypothesis Tests for One Population Standard Deviation

Next we need to discuss the logic behind hypothesis tests for one population standard deviation. We do this in the following example.

EXAMPLE 11.4 HYPOTHESIS TESTS FOR A POPULATION STANDARD DEVIATION: THE LOGIC

A hardware manufacturer produces 10-mm bolts. The manufacturer knows that the diameters of the bolts produced vary somewhat from 10 mm and also from each other. But even if he is willing to accept some variation in bolt diameters, he cannot tolerate too much variation. For if the variation is too large, then too many of the bolts produced will be unusable.

Consequently, the manufacturer must make sure that the standard deviation, σ, of the bolt diameters is not unduly large. Let's suppose it has been determined that an acceptable standard deviation for the bolt diameters is one that is less than 0.09 mm.[†] Knowing this, the manufacturer can decide whether or not there is too much variation in the diameters of the bolts being produced by performing the hypothesis test

$$H_0\text{: } \sigma = 0.09 \text{ mm (too much variation)}$$
$$H_a\text{: } \sigma < 0.09 \text{ mm (not too much variation).}$$

[†] See Exercise 11.30 for an explanation of how that information could be obtained.

If the null hypothesis can be rejected, then the manufacturer can be confident that the variation in bolt diameters is acceptable.[†]

Roughly speaking, the hypothesis test can be carried out by proceeding in the following manner.

1. Take a random sample of bolts.
2. Compute the standard deviation, s, of the diameters of the bolts sampled.
3. If s is too much smaller than 0.09 mm, reject the null hypothesis in favor of the alternative hypothesis; otherwise, do not reject the null hypothesis.

The manufacturer takes a random sample of 12 bolts and carefully measures their diameters. Table 11.1 displays the diameters, in millimeters.

TABLE 11.1 Diameters (in millimeters) of 12 randomly selected bolts

10.05	10.00	10.02	9.97
10.07	10.03	9.98	10.10
9.95	9.99	10.00	10.08

The sample standard deviation of the bolt diameters in Table 11.1 is

$$s = \sqrt{\frac{\Sigma x^2 - (\Sigma x)^2/n}{n-1}} = \sqrt{\frac{1204.8290 - (120.24)^2/12}{11}} = 0.047 \text{ mm.}$$

Is this value of s too much smaller than 0.09 mm, so that the null hypothesis should be rejected, or can the difference between $s = 0.047$ mm and the null hypothesis value of $\sigma = 0.09$ mm be attributed to sampling error? To answer that question, we need to know the distribution of the variable s, that is, the distribution of all possible sample standard deviations that could be obtained by sampling 12 bolts. We will examine that distribution and then return to complete the hypothesis test posed in this example. □

The Sampling Distribution of the Standard Deviation

Recall that to perform a hypothesis test for the mean, μ, of a normally distributed variable, we use the variable

$$t = \frac{\bar{x} - \mu_0}{s/\sqrt{n}}$$

[†] Instead, we could take the null hypothesis to be $H_0\colon \sigma = 0.09$ mm (not too much variation) and the alternative hypothesis to be $H_a\colon \sigma > 0.09$ mm (too much variation). Then rejection of the null hypothesis would indicate that the variation in bolt diameters is unacceptable.

as the test statistic, not simply the variable $\bar{x}$. Similarly, when performing a hypothesis test for the standard deviation, σ, of a normally distributed variable, we do not employ the variable s as the test statistic. Rather, we use the following modified version of that variable:

$$\chi^2 = \frac{n-1}{\sigma_0^2} s^2.$$

This variable has a chi-square distribution.

KEY FACT 11.2 **THE SAMPLING DISTRIBUTION OF THE STANDARD DEVIATION[†]**

Suppose a variable of a population is normally distributed with standard deviation σ. Then, for samples of size n, the variable

$$\chi^2 = \frac{n-1}{\sigma^2} s^2$$

has the chi-square distribution with $n - 1$ degrees of freedom.

EXAMPLE 11.5 **ILLUSTRATES KEY FACT 11.2**

Suppose, in reality, the bolt diameters are normally distributed with mean 10 mm and standard deviation 0.09 mm. Then, according to Key Fact 11.2, for samples of size 12, the variable χ^2 has a chi-square distribution with 11 degrees of freedom. Use simulation to make that fact plausible.

SOLUTION We first simulated 1000 samples of 12 bolt diameters each, that is, 1000 samples of 12 observations of a normally distributed variable with mean 10 and standard deviation 0.09. Then, for each of the 1000 samples, we determined the sample standard deviation, s, and obtained the value of the variable

$$\chi^2 = \frac{n-1}{\sigma^2} s^2 = \frac{11}{(0.09)^2} s^2.$$

A histogram of those 1000 values of χ^2 is shown at the top of the following page in Printout 11.1. As expected, the histogram is shaped like the superimposed χ^2-curve with df = 11. ◨

[†] Strictly speaking, the sampling distribution presented here is not the sampling distribution of the standard deviation but is the sampling distribution of a multiple of the variance (square of the standard deviation).

PRINTOUT 11.1
Minitab histogram of χ^2 for 1000 samples of 12 bolt diameters. The χ^2-curve for the sampling distribution is superimposed.

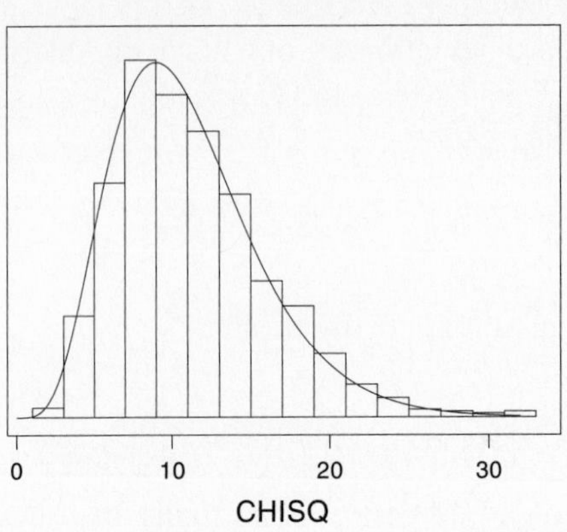

Hypothesis Tests for a Population Standard Deviation

Now that we know the sampling distribution of the standard deviation, we can state a step-by-step method for performing a hypothesis test for a population standard deviation. The method is presented in Procedure 11.1 and is sometimes referred to as the **χ^2-test for a population standard deviation.**

Unlike the z-tests and t-tests for one and two population means, the χ^2-test for one population standard deviation is not robust to moderate violations of the normality assumption. In fact, it is so nonrobust that many statisticians advise against using it unless there is considerable evidence that the variable under consideration is normally distributed or very nearly so.

Consequently, before applying Procedure 11.1, we should construct a normal probability plot. If the plot creates any doubt about the normality of the variable under consideration, then we should not use Procedure 11.1.

EXAMPLE 11.6 ILLUSTRATES PROCEDURE 11.1

We can now complete the hypothesis test proposed in Example 11.4. Recall that a hardware manufacturer needs to decide whether the standard deviation of bolt diameters is less than 0.09 mm. He randomly samples 12 bolts and measures their diameters. The results are shown in Table 11.1 on page 688. At the 5% significance level, do the data provide sufficient evidence to conclude that the standard deviation of the diameters of all 10-mm bolts produced by the manufacturer is less than 0.09 mm?

PROCEDURE 11.1

THE χ^2-TEST FOR A POPULATION STANDARD DEVIATION

ASSUMPTION

Normal population

Step 1 The null hypothesis is H_0: $\sigma = \sigma_0$ and the alternative hypothesis is one of the following:

H_a: $\sigma \neq \sigma_0$ (Two-tailed)	or	H_a: $\sigma < \sigma_0$ (Left-tailed)	or	H_a: $\sigma > \sigma_0$ (Right-tailed)

Step 2 Decide on the significance level, α.

Step 3 The critical value(s) are

$\chi^2_{1-\alpha/2}$ and $\chi^2_{\alpha/2}$ (Two-tailed)	or	$\chi^2_{1-\alpha}$ (Left-tailed)	or	χ^2_{α} (Right-tailed)

with df $= n - 1$. Use Table VIII to find the critical value(s).

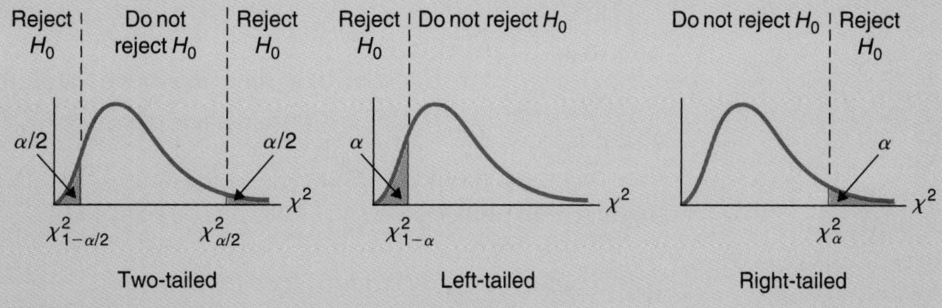

Step 4 Compute the value of the test statistic

$$\chi^2 = \frac{n-1}{\sigma_0^2} s^2.$$

Step 5 If the value of the test statistic falls in the rejection region, reject H_0; otherwise, do not reject H_0.

Step 6 State the conclusion in words.

SOLUTION To begin, we construct a normal probability plot for the data in Table 11.1, as shown in Fig. 11.5 at the top of the next page. As we see from Fig. 11.5, the normal probability plot is quite linear. This and other data on bolt diameters collected previously by the manufacturer make it reasonable to presume that the diameters of 10-mm bolts produced by this manufacturer are normally distributed. So, we can apply Procedure 11.1 to perform the required hypothesis test.

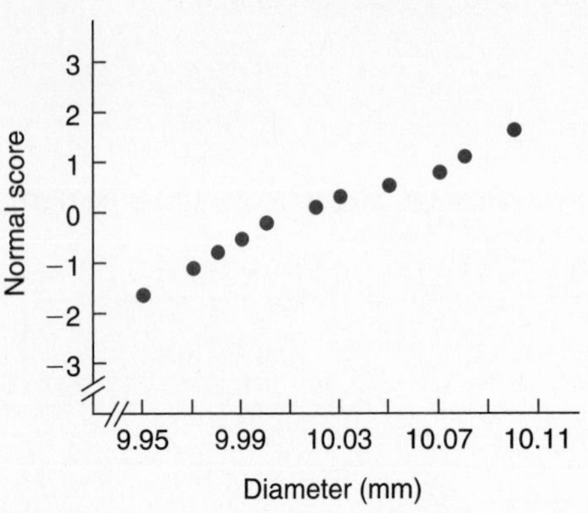

FIGURE 11.5 Normal probability plot for the sample of bolt diameters in Table 11.1

Step 1 *State the null and alternative hypotheses.*

Let σ denote the population standard deviation of bolt diameters. Then the null and alternative hypotheses are

$$H_0\colon \sigma = 0.09 \text{ mm (too much variation)}$$
$$H_a\colon \sigma < 0.09 \text{ mm (not too much variation).}$$

Note that the hypothesis test is left-tailed since a less-than sign ($<$) appears in the alternative hypothesis.

Step 2 *Decide on the significance level,* α.

The test is to be performed at the 5% level of significance; so, $\alpha = 0.05$.

Step 3 *The critical value for a left-tailed test is* $\chi^2_{1-\alpha}$, *with* $df = n - 1$.

We have $\alpha = 0.05$. Also, $n = 12$, and so $\text{df} = 12 - 1 = 11$. Consulting Table VIII, we find that the critical value is $\chi^2_{1-\alpha} = \chi^2_{1-0.05} = \chi^2_{0.95} = 4.575$, as seen in Fig. 11.6.

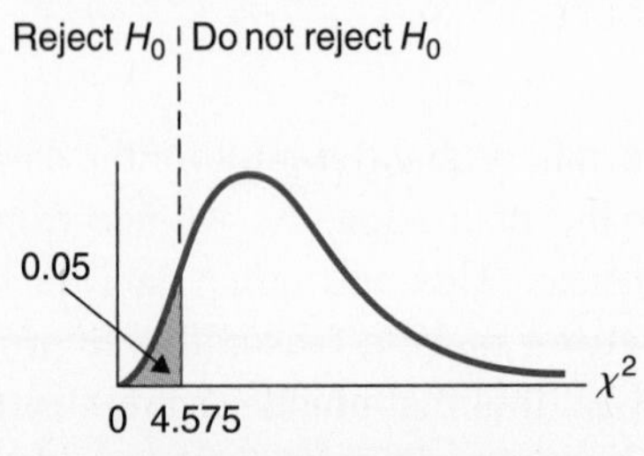

FIGURE 11.6 Criterion for deciding whether or not to reject the null hypothesis

Step 4 Compute the value of the test statistic

$$\chi^2 = \frac{n-1}{\sigma_0^2} s^2.$$

First we obtain the sample variance, s^2. From Table 11.1 we find that

$$s^2 = \frac{\Sigma x^2 - (\Sigma x)^2/n}{n-1} = \frac{1204.8290 - (120.24)^2/12}{11} = 0.0022.$$

Noting that $n = 12$ and $\sigma_0 = 0.09$, we see that the value of the test statistic is

$$\chi^2 = \frac{n-1}{\sigma_0^2} s^2 = \frac{12-1}{(0.09)^2} \cdot 0.0022 = 2.988.$$

Step 5 If the value of the test statistic falls in the rejection region, reject H_0; otherwise, do not reject H_0.

From Step 4 the value of the test statistic is $\chi^2 = 2.988$. This falls in the rejection region, as we see by referring to Fig. 11.6. Thus we reject H_0.

Step 6 State the conclusion in words.

The test results are statistically significant at the 5% level; that is, at the 5% significance level, the data provide sufficient evidence to conclude that the standard deviation, σ, of the diameters of all 10-mm bolts produced by the manufacturer is less than 0.09 mm. Evidently, the variation in bolt diameters is not too large. ◨

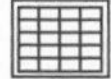

Using the *P*-Value Approach

The P-value approach to hypothesis testing can also be used to carry out the hypothesis test in Example 11.6 and to assess more precisely the evidence against the null hypothesis. From Step 4 we see that the value of the test statistic is $\chi^2 = 2.988$. Recalling that df = 11, we find from Table VIII that $0.005 < P < 0.01$.

In particular, because the P-value is less than the significance level of 0.05, we can reject H_0. Furthermore, by referring to Table 9.12 on page 543, we see that the data provide very strong evidence against the null hypothesis, that is, in support of the hypothesis that $\sigma < 0.09$ mm.

Confidence Intervals for a Population Standard Deviation

Using Key Fact 11.2 on page 689, we can also obtain the following confidence-interval procedure for a population standard deviation. This procedure is often referred to as the **χ^2-interval procedure for a population standard deviation.**

PROCEDURE 11.2 **THE χ^2-INTERVAL PROCEDURE FOR A POPULATION STANDARD DEVIATION**

ASSUMPTION

Normal population

Step 1 For a confidence level of $1 - \alpha$, use Table VIII to find $\chi^2_{1-\alpha/2}$ and $\chi^2_{\alpha/2}$ with df $= n - 1$.

Step 2 The confidence interval for σ is from

$$\sqrt{\frac{n-1}{\chi^2_{\alpha/2}}} \cdot s \quad \text{to} \quad \sqrt{\frac{n-1}{\chi^2_{1-\alpha/2}}} \cdot s$$

where $\chi^2_{1-\alpha/2}$ and $\chi^2_{\alpha/2}$ are found in Step 1, n is the sample size, and s is computed from the sample data obtained.

Like the χ^2-test for one population standard deviation, the χ^2-interval procedure is not at all robust to violations of the normality assumption; using it on data from a variable that is not normally distributed can result in misleading information. In other words, the χ^2-interval procedure should not be used unless there is considerable evidence that the variable under consideration is normally distributed or very nearly so.

EXAMPLE 11.7 ILLUSTRATES PROCEDURE 11.2

Use the sample data in Table 11.1 on page 688 to determine a 95% confidence interval for the standard deviation, σ, of the diameters of all 10-mm bolts produced by the manufacturer.

SOLUTION As we discovered in Example 11.6, it is reasonable to presume that the diameters of 10-mm bolts produced by the manufacturer are normally distributed. Thus we can apply Procedure 11.2 to obtain the required confidence interval.

Step 1 *For a confidence level of $1 - \alpha$, use Table VIII to find $\chi^2_{1-\alpha/2}$ and $\chi^2_{\alpha/2}$ with df $= n - 1$.*

We want a 95% confidence interval; so the confidence level is $0.95 = 1 - 0.05$. This means that $\alpha = 0.05$. Also, since $n = 12$, df $= 11$. Referring to Table VIII, we find that

$$\chi^2_{1-\alpha/2} = \chi^2_{1-0.05/2} = \chi^2_{0.975} = 3.816$$

and

$$\chi^2_{\alpha/2} = \chi^2_{0.05/2} = \chi^2_{0.025} = 21.920.$$

Step 2 *The confidence interval for σ is from*

$$\sqrt{\frac{n-1}{\chi^2_{\alpha/2}}} \cdot s \quad \textit{to} \quad \sqrt{\frac{n-1}{\chi^2_{1-\alpha/2}}} \cdot s.$$

We have $n = 12$, and from Step 1, $\chi^2_{1-\alpha/2} = 3.816$ and $\chi^2_{\alpha/2} = 21.920$. Also, we found in Example 11.4 that $s = 0.047$ mm. So a 95% confidence interval for σ is from

$$\sqrt{\frac{12-1}{21.920}} \cdot 0.047 \quad \text{to} \quad \sqrt{\frac{12-1}{3.816}} \cdot 0.047,$$

or 0.033 to 0.080. We can be 95% confident that the standard deviation, σ, of the diameters of all 10-mm bolts produced by the manufacturer is somewhere between 0.033 mm and 0.080 mm.

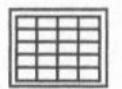

Minitab does not have a dedicated procedure for performing hypothesis tests and obtaining confidence intervals for one population standard deviation.[†] But we used Minitab's capabilities to write a macro for such inferences and called the macro `1stdev.mac`. It can be found in the `Macro` folder of DataDisk CD.

EXAMPLE 11.8 Using Minitab to Perform Inferences for One Population Standard Deviation

Use Minitab to perform the hypothesis test considered in Example 11.6 and to obtain the confidence interval required in Example 11.7.

SOLUTION Let σ denote the population standard deviation of bolt diameters. The problem in Example 11.6 is to perform the hypothesis test

$$H_0\colon \sigma = 0.09 \text{ mm (too much variation)}$$
$$H_a\colon \sigma < 0.09 \text{ mm (not too much variation)}$$

at the 5% significance level; the problem in Example 11.7 is to obtain a 95% confidence interval for σ.

[†] Actually, we can obtain a confidence interval for a population standard deviation directly from Minitab by choosing **Stat ➤ Basic Statistics ➤ Display Descriptive Statistics...**, clicking the **Graphs...** button, and selecting the **Graphical summary** check box. However, this procedure outputs many other descriptive and inferential results and does not provide for a hypothesis test for the population standard deviation.

As we discovered in Example 11.6, it is reasonable to use the χ^2-test and χ^2-interval procedure for one population standard deviation. To apply Minitab, we begin by entering the data from Table 11.1 on page 688 in a column named DIAMETER. Next we enable the command language as follows:

1 Click in the Session window
2 Choose **Editor ➤ Enable Command Language**

Now we can run `1stdev.mac` by typing, in the Session window, a percent sign (%) followed by the path for `1stdev.mac` and the location of the data enclosed by apostrophes. For instance, if DataDisk CD is in drive `E:`, we type

```
%E:\IS5\Macro\1stdev.mac 'DIAMETER'
```

and press the Enter key. Then we proceed as follows.

1 In response to `Do you want to perform a hypothesis test (Y/N)?`, type Y and press the Enter key.
2 In response to `Enter the null hypothesis population standard deviation.`, type 0.09 and press the Enter key.
3 In response to `Enter 0, 1, or -1, respectively, for a two-tailed, right-tailed, or left-tailed test.`, type -1 and press the Enter key.

Printout 11.2 shows the resulting output.

PRINTOUT 11.2
Hypothesis-test output for the 1stdev macro

```
Test of sigma =      0.09000 vs sigma <      0.09000


 Row  Variable     n  StDev  Chi-Sq       P

   1  DIAMETER    12  0.047   2.988  0.009
```

The output in Printout 11.2 provides the sample size (`n`), sample standard deviation (`StDev`), value of the chi-square test statistic (`Chi-Sq`), and P-value (`P`) for the hypothesis test. Since the P-value of 0.009 is less than the specified significance level of 0.05, we reject H_0. At the 5% significance level, the data provide sufficient evidence to conclude that the standard deviation of the diameters of all 10-mm bolts produced by the manufacturer is less than 0.09 mm.

The macro continues with the confidence-interval aspect. We proceed in the following manner.

1 In response to `Do you want a confidence interval (Y/N)?`, type Y and press the Enter key.
2 Since we want a 95% confidence interval, type 95 in response to `Enter the confidence level, as a percentage.`

Printout 11.3 shows the resulting output.

PRINTOUT 11.3
Confidence-interval output for the 1stdev macro

```
Row  Variable     n  StDev  Level     CI for sigma

  1  DIAMETER    12  0.047  95.0%  (0.033, 0.080)
```

The output in Printout 11.3 provides the sample size (n), sample standard deviation (StDev), confidence level (Level), and confidence interval for σ. We see that a 95% confidence interval for σ is from 0.033 to 0.080. We can be 95% confident that the standard deviation, σ, of the diameters of all 10-mm bolts produced by the manufacturer is somewhere between 0.033 mm and 0.080 mm. ◨

EXERCISES 11.1

Statistical Concepts and Skills

11.1 What does it mean to say that a variable has a chi-square distribution?

11.2 How do we identify different chi-square distributions?

11.3 Two χ^2-curves have, respectively, degrees of freedom 12 and 20. Which one more closely resembles a normal curve? Explain your answer.

11.4 The t-table has entries for areas of 0.10, 0.05, 0.025, 0.01, and 0.005. On the other hand, the χ^2-table has entries for those areas plus ones for 0.995, 0.99, 0.975, 0.95, and 0.90. Explain why we can obtain the t-values corresponding to these latter areas from the existing t-table, but that we must provide them explicitly in the χ^2-table.

In each of Exercises 11.5–11.12, use Table VIII to find the required χ^2-values. Illustrate your work graphically.

11.5 For a χ^2-curve with 19 degrees of freedom, find the χ^2-value having area
a. 0.025 to its right. **b.** 0.95 to its right.

11.6 For a χ^2-curve with 22 degrees of freedom, find the χ^2-value having area
a. 0.01 to its right. **b.** 0.995 to its right.

11.7 For a χ^2-curve with df = 10, determine
a. $\chi^2_{0.05}$. **b.** $\chi^2_{0.975}$.

11.8 For a χ^2-curve with df = 4, determine
a. $\chi^2_{0.005}$. **b.** $\chi^2_{0.99}$.

11.9 Consider a χ^2-curve with df = 8. Obtain the χ^2-value having area
a. 0.01 to its left. **b.** 0.95 to its left.

11.10 Consider a χ^2-curve with df = 16. Obtain the χ^2-value having area
a. 0.025 to its left. **b.** 0.975 to its left.

11.11 Determine the two χ^2-values that divide the area under the curve into a middle 0.95 area and two outside 0.025 areas for a χ^2-curve with
a. df = 5. **b.** df = 26.

11.12 Determine the two χ^2-values that divide the area under the curve into a middle 0.90 area and two outside 0.05 areas for a χ^2-curve with
a. df = 11. **b.** df = 28.

11.13 In using chi-square procedures to make inferences about a population standard deviation, why is it important that the variable under consideration be normally distributed or nearly so?

11.14 Give two situations in which it would be important to make an inference about a population standard deviation.

Preliminary data analyses and other information indicate that it is reasonable to assume the variables under consideration in Exercises 11.15–11.20 are normally distributed. For each exercise, perform the required hypothesis test using either the critical-value approach or the P-value approach.

11.15 Each year, thousands of high-school students bound for college take the Scholastic Assessment Test, or SAT. Scores are reported on a scale that ranges from a low of 200 to a high of 800; this scale was introduced in 1941. At that time the standard deviation of scores was 100 points. A random sample of 25 verbal scores for last year yields the following data.

344	494	350	376	313
489	358	383	498	556
379	301	432	560	494
418	483	444	477	420
492	287	434	514	613

Do the data provide sufficient evidence to conclude that the standard deviation, σ, of last year's verbal scores is different from the 1941 standard deviation of 100? Perform the required hypothesis test at the 5% significance level. (*Note:* The sample standard deviation of the 25 scores is 85.5.)

11.16 Gas-mileage estimates for cars and light-duty trucks are determined and published by the Environmental Protection Agency (EPA). According to the EPA, "... the mileages obtained by most drivers will be within plus or minus 15 percent of the [EPA] estimates...." The mileage estimate given for one 1998 model is 23 mpg on the highway. If the EPA claim is true, then the standard deviation of mileages should be about $0.15 \cdot 23/3 = 1.15$ mpg. A random sample of 12 cars of this model yields the following highway mileages.

24.1	23.3	22.5	23.2
22.3	21.1	21.4	23.4
23.5	22.8	24.5	24.3

At the 5% significance level, do the data suggest that the standard deviation of highway mileages for all 1998 cars of this model is different from 1.15 mpg? (*Note:* $s = 1.071$.)

11.17 R. Morris and E. Watson studied various aspects of process capability in the paper "Determining Process Capability in a Chemical Batch Process" (*Quality Engineering,* 10(2), pp. 389–396, 1997). In one part of the study, the researchers compared the variability in product of a particular piece of equipment to a known analytical capability in order to decide whether product consistency could be improved. The following data were obtained for 10 batches of product.

30.1	30.7	30.2	29.3	31.0
29.6	30.4	31.2	28.8	29.8

At the 1% significance level, do the data provide sufficient evidence to conclude that the process variation for this piece of equipment exceeds the analytical capability of 0.27? (*Note:* $s = 0.756$.)

11.18 Homestyle Pizza of Camp Verde, Arizona, provides baking instructions for its premade pizzas. According to the instructions, the average baking time is 12 to 18 minutes. Assuming the times are normally distributed, this means that the standard deviation of the times should be approximately 1 minute. A random sample of 15 pizzas yielded the following baking times to the nearest tenth of a minute.

15.4	15.1	14.0	15.8	16.0
13.7	15.6	11.6	14.8	12.8
17.6	15.1	16.4	13.1	15.3

At the 1% significance level, do the data provide sufficient evidence to conclude that the standard deviation of baking times exceeds 1 minute? *(Note:* The sample standard deviation of the 15 baking times is 1.54 minutes.*)*

11.19 A coffee machine is supposed to dispense 6 fl oz of coffee into a paper cup. In reality, the amounts dispensed vary from cup to cup. However, if the machine is working properly, then most of the cups will contain within 10% of the advertised 6 fl oz. This means that the standard deviation of the amounts dispensed should be less than 0.2 fl oz. A random sample of 15 cups provided the following data, in fluid ounces.

5.90	5.82	6.20	6.09	5.93
6.18	5.99	5.79	6.28	6.16
6.00	5.85	6.13	6.09	6.18

At the 5% significance level, do the data provide sufficient evidence to conclude that the standard deviation of the amounts being dispensed is less than 0.2 fl oz? *(Note:* $s = 0.154$.*)*

11.20 In Issue 10 of *STATS* from Iowa State University, data were published from an experiment studying the effects of machine adjustment on bolt production. An electronic counter records the number of bolts passing it on a conveyer belt and stops the run when the count reaches a preset number. The following data give the times, in seconds, that it took to count 20 bolts for eight different runs.

10.78	9.39	9.84	13.94
12.33	7.32	7.91	15.58

Do the data provide sufficient evidence to conclude that the standard deviation in the time it takes to count 20 bolts is less than 2 seconds? Use $\alpha = 0.05$. *(Note:* The sample standard deviation of the eight times is 2.8875 seconds.*)*

In each of Exercises 11.21–11.26, use Procedure 11.2 on page 694 to obtain the required confidence interval.

11.21 Refer to Exercise 11.15. Obtain a 95% confidence interval for the standard deviation, σ, of last year's verbal SAT scores.

11.22 Refer to Exercise 11.16. Find a 95% confidence interval for the standard deviation of highway gas mileages for all 1998 cars of the model in question.

11.23 Refer to Exercise 11.17. Determine a 98% confidence interval for the process variation of the piece of equipment under consideration.

11.24 Refer to Exercise 11.18. Obtain a 98% confidence interval for the standard deviation of baking times.

11.25 Refer to Exercise 11.19. Find a 90% confidence interval for the standard deviation of the amounts of coffee being dispensed.

11.26 Refer to Exercise 11.20. Find a 90% confidence interval for the standard deviation in the time it takes to count 20 bolts.

EXTENDING THE CONCEPTS AND SKILLS

11.27 Refer to Exercise 11.19. Why is it important that the standard deviation of the amounts of coffee being dispensed not be too large?

11.28 Refer to Exercises 11.16 and 11.22. Why is it useful to know the standard deviation of the gas mileages as well as the mean gas mileage?

11.29 Refer to Example 11.4 on page 687. In that example we chose the null hypothesis to be $H_0\colon \sigma = 0.09$ (too much variation) and the alternative hypothesis to be $H_a\colon \sigma < 0.09$ (not too much variation). Alternatively, we could take the null hypothesis to be $H_0\colon \sigma = 0.09$ (not too much variation) and the alternative hypothesis to be $H_a\colon \sigma > 0.09$ (too much variation). Explain the advantages and disadvantages for the two different choices of null and alternative hypotheses.

11.30 In the bolt-manufacturing problem of Example 11.4 on page 687, we assumed it had been determined that an acceptable standard deviation, σ, for the bolt diameters is one that is less than 0.09 mm. We will now see how such information might be obtained. Let's suppose the manufacturer has set the tolerance specifications for the 10-mm bolts at ± 0.3 mm; that is, a bolt's diameter is considered satisfactory if it is between 9.7 mm and 10.3 mm. Further suppose the manufacturer has decided that less than 0.1% (1 out of 1000) of the bolts produced should be defective.

a. Let X denote the diameter of a randomly selected bolt. Show that the manufacturer's production criteria can be expressed mathematically as

$$P(9.7 \leq X \leq 10.3) > 0.999.$$

b. Draw a normal-curve picture that illustrates the equation $P(9.7 \leq X \leq 10.3) = 0.999$. Include both an x-axis and a z-axis. Assume $\mu = 10$ mm.

c. Deduce from your picture in part (b) that the manufacturer's production criteria are equivalent to the condition that $0.3/\sigma > z_{0.0005}$.

d. Use part (c) to conclude that the manufacturer's production criteria are equivalent to requiring that the standard deviation of bolt diameters be less than 0.09 mm, that is, $\sigma < 0.09$ mm.

USING TECHNOLOGY  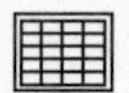 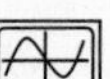

11.31 Refer to Exercises 11.15 and 11.21. Use Minitab or some other statistical software to

a. obtain both a normal probability plot and a boxplot of the data.

b. perform the required hypothesis test and obtain the desired confidence interval. (*Note:* Users of Minitab should employ `1stdev.mac` found in the `Macro` folder of DataDisk CD.)

c. Justify the use of your procedure in part (b).

11.32 Refer to Exercises 11.16 and 11.22. Use Minitab or some other statistical software to

a. obtain both a normal probability plot and a boxplot of the data.

b. perform the required hypothesis test and obtain the desired confidence interval. (*Note:* Users of Minitab should employ `1stdev.mac` found in the `Macro` folder of DataDisk CD.)

c. Justify the use of your procedure in part (b).

11.33 Intelligence quotients (IQs) measured on the Stanford Revision of the Binet-Simon Intelligence Scale are known to be normally distributed with a mean of 100 and a standard deviation of 16. Use Minitab or some other statistical software to do the following. (*Note:* If you are using Minitab, you may want to refer to Example 7.10 on page 431.)

a. Simulate 1000 samples of four IQs each.

b. Determine the sample standard deviation of each of the 1000 samples.

c. For each of the 1000 samples, obtain the quantity

$$\frac{n-1}{\sigma^2}s^2 = \frac{4-1}{16^2}s^2.$$

d. Obtain a histogram of the 1000 values found in part (c).

e. Theoretically, what is the distribution of the variable in part (c)?

f. Compare your answers from parts (d) and (e).

11.2 INFERENCES FOR TWO POPULATION STANDARD DEVIATIONS USING INDEPENDENT SAMPLES*

In Section 11.1 we studied hypothesis tests and confidence intervals for one population standard deviation. Now we will study those same inferences for comparing two population standard deviations. That is, we will learn how to perform hypothesis tests and obtain confidence intervals to compare the standard deviations of a

* See the note on the bottom of page 683.

single variable of two different populations. Such inferences are based on a distribution called the F-distribution, named in honor of Sir Ronald Fisher. (See the biography on page 988 for more on Fisher.)

The F-Distribution

A variable is said to have an ***F*-distribution** if its distribution has the shape of a special type of right-skewed curve, called an ***F*-curve.** Actually there are infinitely many F-distributions, and we identify the F-distribution (and F-curve) in question by stating its number of degrees of freedom, just as we did for t-distributions and chi-square distributions.

But an F-distribution has two numbers of degrees of freedom instead of one. Figure 11.7 depicts two different F-curves; one has df = (10, 2), and the other has df = (9, 50).

FIGURE 11.7
Two different F-curves

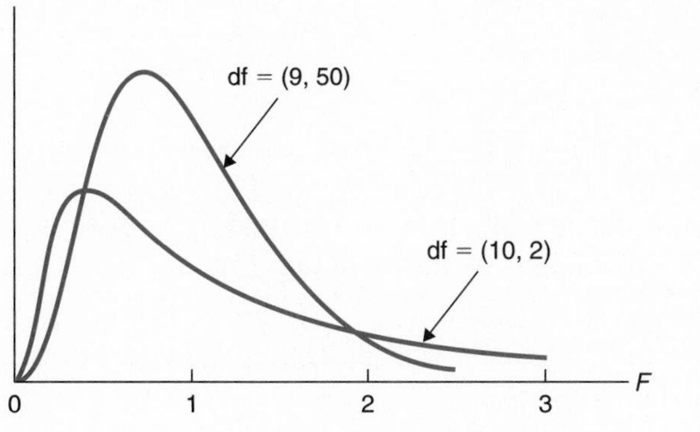

The first number of degrees of freedom for an F-curve is called the **degrees of freedom for the numerator** and the second the **degrees of freedom for the denominator.** (The reason for this terminology will become clear later in this section.) Thus for the F-curve in Fig. 11.7 with df = (10, 2), we have

$$\text{df} = (10, 2)$$

Degrees of freedom for the numerator ↗ ↖ Degrees of freedom for the denominator

Some basic properties of F-curves are presented in Key Fact 11.3.

KEY FACT 11.3 BASIC PROPERTIES OF *F*-CURVES

Property 1: The total area under an F-curve is equal to 1.

Property 2: An F-curve starts at 0 on the horizontal axis and extends indefinitely to the right, approaching, but never touching, the horizontal axis as it does so.

Property 3: An F-curve is right skewed.

Percentages (and probabilities) for a variable having an F-distribution are equal to areas under its associated F-curve. To perform a hypothesis test or construct a confidence interval for comparing two population standard deviations, we need to know how to find the F-value that corresponds to a specified area under an F-curve. The symbol $\boldsymbol{F_\alpha}$ is used to denote the F-value having area α to its right.

Table IX in Appendix A provides F-values having areas 0.005, 0.01, 0.025, 0.05, and 0.10 to their right for various degrees of freedom. The degrees of freedom for the denominator (dfd) are displayed in the outside columns of the table; the values of α in the next columns; and the degrees of freedom for the numerator (dfn) along the top. Examples 11.9–11.11 show how to use Table IX.

EXAMPLE 11.9 FINDING THE *F*-VALUE HAVING A SPECIFIED AREA TO ITS RIGHT

For an F-curve with df $= (4, 12)$, find $F_{0.05}$; that is, find the F-value having area 0.05 to its right, as shown in Fig. 11.8(a).

FIGURE 11.8 Finding the F-value having area 0.05 to its right

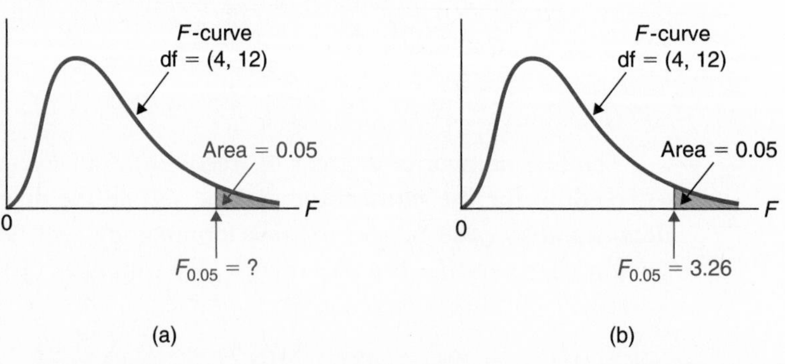

SOLUTION To obtain the F-value in question, we use Table IX. In this case $\alpha = 0.05$, the degrees of freedom for the numerator is 4, and the degrees of freedom for the denominator is 12.

So we first go down the outside columns to the row labeled "12." Next we concentrate on the row for α labeled 0.05. Then we go across that row until we are under the column headed "4." The number in the body of the table there, 3.26, is the required F-value; that is, for an F-curve with df = (4, 12), the F-value having area 0.05 to its right is 3.26: $F_{0.05} = 3.26$, as seen in Fig. 11.8(b).

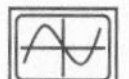

In many statistical analyses involving the F-distribution, we also need to determine F-values having areas 0.005, 0.01, 0.025, 0.05, and 0.10 to their left. Although such F-values cannot be found directly from Table IX, we can obtain them indirectly from that table by using the following fact.

KEY FACT 11.4 **RECIPROCAL PROPERTY OF F-CURVES**

For an F-curve with df = (ν_1, ν_2), the F-value having area α to its left is the reciprocal of the F-value having area α to its right for an F-curve with df = (ν_2, ν_1).

We illustrate the use of Key Fact 11.4 in Examples 11.10 and 11.11.

EXAMPLE 11.10 **FINDING THE F-VALUE HAVING A SPECIFIED AREA TO ITS LEFT**

For an F-curve with df = (60, 8), find the F-value having area 0.05 to its left.

SOLUTION We apply Key Fact 11.4. Accordingly, the required F-value is the reciprocal of the F-value having area 0.05 to its right for an F-curve with df = (8, 60). Consulting Table IX, we find that this latter F-value equals 2.10. Consequently, the required F-value is $\frac{1}{2.10}$, or 0.48. See Fig. 11.9.

FIGURE 11.9 Finding the F-value having area 0.05 to its left

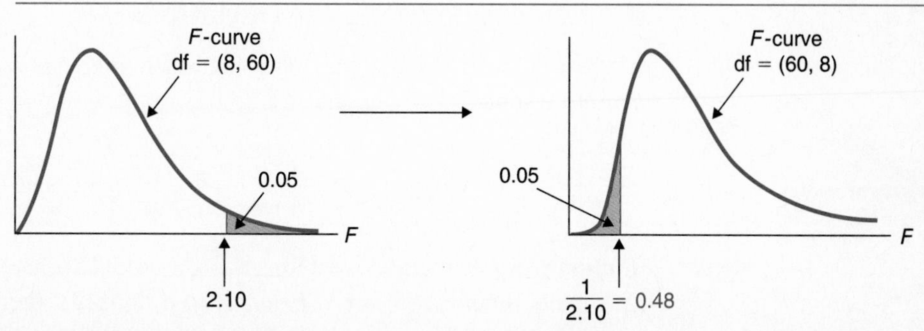

EXAMPLE 11.11 FINDING THE F-VALUES FOR A SPECIFIED AREA

For an F-curve with df = (9, 8), determine the two F-values that divide the area under the curve into a middle 0.95 area and two outside 0.025 areas, as shown in Fig. 11.10(a).

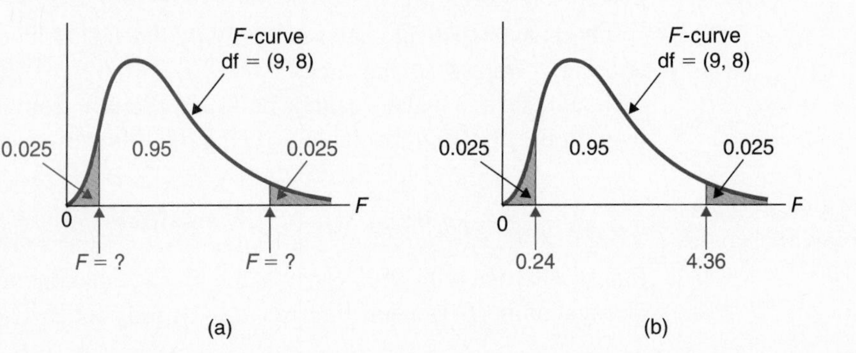

FIGURE 11.10 Finding the two F-values dividing the area under the curve into a middle 0.95 area and two outside 0.025 areas

SOLUTION First we obtain the F-value on the right in Fig. 11.10(a). From the figure, we see that the shaded area on the right is 0.025. This means that the F-value on the right is $F_{0.025}$. Consulting Table IX, we find that for df = (9, 8), $F_{0.025} = 4.36$.

Next we obtain the F-value on the left in Fig. 11.10(a). By Key Fact 11.4, that F-value is the reciprocal of the F-value having area 0.025 to its right for an F-curve with df = (8, 9). Consulting Table IX, we find that this latter F-value equals 4.10. So, the F-value on the left in Fig. 11.10(a) is $\frac{1}{4.10}$, or 0.24.

Consequently, for an F-curve with df = (9, 8), the two F-values that divide the area under the curve into a middle 0.95 area and two outside 0.025 areas are 0.24 and 4.36, as shown in Fig. 11.10(b). □

The Logic Behind Hypothesis Tests for Comparing Two Population Standard Deviations

Next we need to discuss the logic behind hypothesis tests for comparing two population standard deviations. We do this in the following example.

EXAMPLE 11.12 HYPOTHESIS TESTS FOR TWO POPULATION STANDARD DEVIATIONS: THE LOGIC

Variation within a method used for testing a product is an essential factor in deciding whether the method should be employed. Indeed, when the variation of such a test is high, it is difficult to ascertain the true quality of a product.

In the 1997 publication "Using Repeatability and Reproducibility Studies to Evaluate a Destructive Test Method" (*Quality Engineering,* 10(2), pp. 283–290), A. Phillips et al. studied the variability of the Elmendorf tear test. That test is used to evaluate material strength for fiberglass shingles, paper quality, and other manufactured products.

In one aspect of the study, the researchers independently and randomly obtained data on Elmendorf tear strength of three different vinyl floor coverings. Table 11.2 provides the data, in grams, for two of the three vinyl floor coverings.

TABLE 11.2
Results of Elmendorf tear test on two different vinyl floor coverings (data in grams)

Brand A		Brand B	
2288	2384	2592	2384
2368	2304	2512	2432
2528	2240	2576	2112
2144	2208	2176	2288
2160	2112	2304	2752

We want to decide whether the standard deviations of tear strength differ between the two vinyl floor coverings. Statistically, this means that we want to perform the hypothesis test

H_0: $\sigma_1 = \sigma_2$ (standard deviations of tear strength are the same)

H_a: $\sigma_1 \neq \sigma_2$ (standard deviations of tear strength are different)

where σ_1 and σ_2 denote, respectively, the population standard deviations of tear strength for Brand A and Brand B.

Roughly speaking, the hypothesis test can be carried out by comparing the sample standard deviations, s_1 and s_2, of the two samples in Table 11.2. There are several ways such a comparison could be made. Here we do it by looking at the square of the ratio of s_1 to s_2, or equivalently, the quotient of the sample variances. That statistic is called the ***F*-statistic.**

If the population standard deviations, σ_1 and σ_2, are equal, then the sample standard deviations, s_1 and s_2, should be roughly the same. This in turn implies that the value of the F-statistic should be close to 1. Looking at it another way, if the value of the F-statistic differs from 1 by too much, then this provides evidence against the null hypothesis of equal population standard deviations.

For the data in Table 11.2, we have $s_1 = 128.3$ g and $s_2 = 199.7$ g. Thus the value of the F-statistic is

$$F = \frac{s_1^2}{s_2^2} = \frac{128.3^2}{199.7^2} = 0.41.$$

Does this value of F differ from 1 by enough to conclude that the null hypothesis of equal population standard deviations is false? To answer that question, we need to know the distribution of the F-statistic. We will discuss that distribution and then return to complete the hypothesis test considered in this example. □

The Distribution of the F-Statistic

To perform hypothesis tests and obtain confidence intervals for two population standard deviations, we will need the following fact.

KEY FACT 11.5 **DISTRIBUTION OF THE F-STATISTIC FOR COMPARING TWO POPULATION STANDARD DEVIATIONS**

Suppose the variable under consideration is normally distributed on each of two populations. Then, for independent samples of sizes n_1 and n_2 from the two populations, the variable

$$F = \frac{s_1^2/\sigma_1^2}{s_2^2/\sigma_2^2}$$

has the F-distribution with df $= (n_1 - 1, n_2 - 1)$.

EXAMPLE 11.13 ILLUSTRATES KEY FACT 11.5

Suppose, in reality, that the Elmendorf tear strengths for Brand A vinyl floor covering are normally distributed with mean 2275 g and standard deviation 132 g, and that those for Brand B vinyl floor covering are normally distributed with mean 2405 g and standard deviation 194 g. Then, for independent random samples of sizes 10 and 10 from Brand A and Brand B, the variable F in Key Fact 11.5 has the F-distribution with df $= (9, 9)$. Use simulation to make that fact plausible.

SOLUTION We first simulated 1000 samples of 10 tear strengths each for Brand A vinyl floor covering, that is, 1000 samples of 10 observations of a normally distributed variable with mean 2275 and standard deviation 132. Next we simulated 1000 samples of 10 tear strengths each for Brand B vinyl floor covering, that is, 1000 samples of 10 observations of a normally distributed variable with mean 2405 and standard deviation 194. Then, for each of the 1000 pairs of samples from the two brands, we determined the sample standard deviations, s_1 and s_2, and obtained the value of the variable

$$F = \frac{s_1^2/\sigma_1^2}{s_2^2/\sigma_2^2} = \frac{s_1^2/132^2}{s_2^2/194^2}.$$

A histogram of those 1000 values of F is shown in Printout 11.4 which, as expected, is shaped like the superimposed F-curve with df $= (9, 9)$. □

PRINTOUT 11.4
Minitab histogram of F for 1000 independent samples. The corresponding F-curve is superimposed

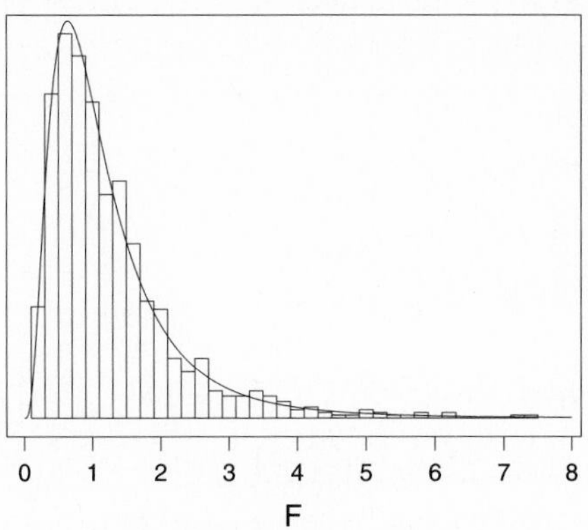

Hypothesis Tests for Two Population Standard Deviations

In view of Key Fact 11.5, we see that for a hypothesis test with null hypothesis H_0: $\sigma_1 = \sigma_2$ (population standard deviations are equal), we can use $F = s_1^2/s_2^2$ as the test statistic and obtain the critical value(s) from the F-table, Table IX. Specifically, we have Procedure 11.3 shown on the next page, which we will call the ***F*-test for two population standard deviations.** This procedure is also referred to as the F-test for two population variances.

Unlike the z-tests and t-tests for one and two population means, the F-test for two population standard deviations is not robust to moderate violations of the normality assumption. In fact, it is so nonrobust that many statisticians advise against using it unless there is considerable evidence that the variable under consideration is normally distributed on each population or very nearly so.

Consequently, before applying Procedure 11.3, we should construct a normal probability plot of each sample. If either plot creates any doubt about the normality of the variable under consideration, then we should not use Procedure 11.3.

EXAMPLE 11.14 ILLUSTRATES PROCEDURE 11.3

We can now complete the hypothesis test proposed in Example 11.12. Independent random samples of two vinyl floor coverings yield the data on Elmendorf tear strength shown in Table 11.2 on page 705. At the 5% significance level, do the data provide sufficient evidence to conclude that the population standard deviations of tear strength differ for the two vinyl floor coverings?

SOLUTION To begin, we construct normal probability plots (not shown) for the two samples in Table 11.2. The plots suggest that it is reasonable to presume that tear

PROCEDURE 11.3 **THE *F*-TEST FOR TWO POPULATION STANDARD DEVIATIONS**

ASSUMPTIONS

1. Independent samples
2. Normal populations

Step 1 The null hypothesis is H_0: $\sigma_1 = \sigma_2$ and the alternative hypothesis is one of the following:

H_a: $\sigma_1 \neq \sigma_2$ (Two-tailed)	or	H_a: $\sigma_1 < \sigma_2$ (Left-tailed)	or	H_a: $\sigma_1 > \sigma_2$ (Right-tailed)

Step 2 Decide on the significance level, α.

Step 3 The critical value(s) are

$F_{1-\alpha/2}$ and $F_{\alpha/2}$ (Two-tailed)	or	$F_{1-\alpha}$ (Left-tailed)	or	F_{α} (Right-tailed)

with df $= (n_1 - 1, n_2 - 1)$. Use Table IX to find the critical value(s).

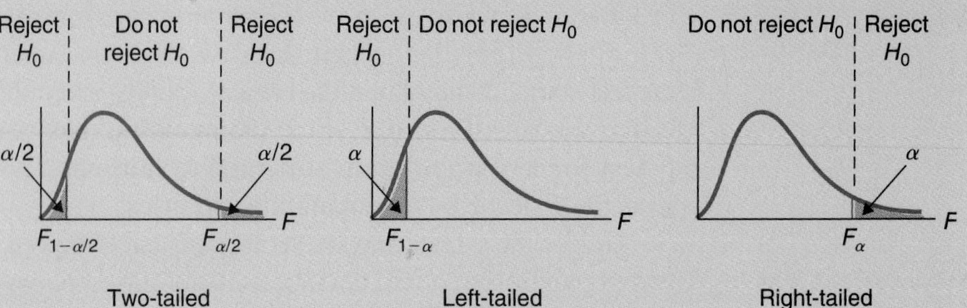

Step 4 Compute the value of the test statistic $F = s_1^2/s_2^2$.

Step 5 If the value of the test statistic falls in the rejection region, reject H_0; otherwise, do not reject H_0.

Step 6 State the conclusion in words.

strength is normally distributed for each brand of vinyl flooring. So, we can apply Procedure 11.3 to perform the required hypothesis test.

Step 1 *State the null and alternative hypotheses.*

Let σ_1 and σ_2 denote, respectively, the population standard deviations of tear strength for Brand A and Brand B. Then the null and alternative hypotheses are

H_0: $\sigma_1 = \sigma_2$ (standard deviations of tear strength are the same)

H_a: $\sigma_1 \neq \sigma_2$ (standard deviations of tear strength are different).

Note that the hypothesis test is two-tailed since a not-equal sign ($\neq$) appears in the alternative hypothesis.

Step 2 *Decide on the significance level, α.*

The test is to be performed at the 5% level of significance; so, $\alpha = 0.05$.

Step 3 *The critical values for a two-tailed test are $F_{1-\alpha/2}$ and $F_{\alpha/2}$ with $df = (n_1 - 1, n_2 - 1)$.*

We have $\alpha = 0.05$. Also, $n_1 = 10$ and $n_2 = 10$, so df $= (9, 9)$. Consequently, the critical values are $F_{1-\alpha/2} = F_{1-0.05/2} = F_{0.975}$ and $F_{\alpha/2} = F_{0.05/2} = F_{0.025}$. Consulting Table IX, we find that $F_{0.025} = 4.03$. To obtain $F_{0.975}$, we first note that it is the F-value having area 0.025 to its left. Applying the reciprocal property of F-curves (page 703), we conclude that $F_{0.975}$ equals the reciprocal of the F-value having area 0.025 to its right for an F-curve with df $= (9, 9)$. (We switched the degrees of freedom, but since they are the same, the difference is not apparent.) So, we see that $F_{0.975} = \frac{1}{4.03} = 0.25$. Figure 11.11 summarizes our results.

FIGURE 11.11
Criterion for deciding whether or not to reject the null hypothesis

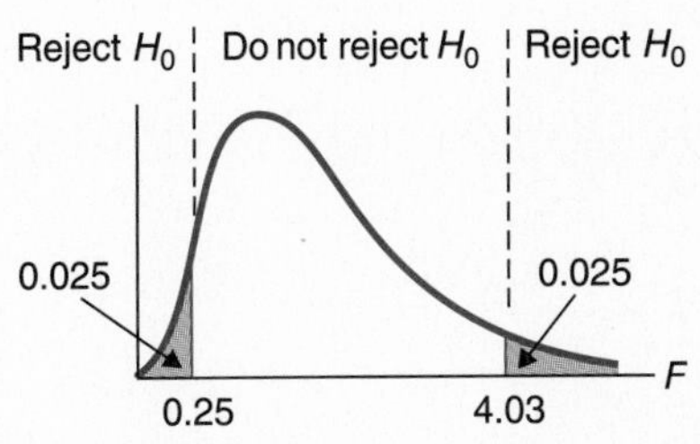

Step 4 *Compute the value of the test statistic $F = s_1^2/s_2^2$.*

We already computed the value of the test statistic at the end of Example 11.12, where we found that $F = 0.41$.

Step 5 *If the value of the test statistic falls in the rejection region, reject H_0; otherwise, do not reject H_0.*

From Step 4 the value of the test statistic is $F = 0.41$. This does not fall in the rejection region, as we see by referring to Fig. 11.11. Thus we do not reject H_0.

Step 6 *State the conclusion in words.*

The test results are not statistically significant at the 5% level; that is, at the 5% significance level, the data do not provide sufficient evidence to conclude that the (population) standard deviations of tear strength differ for the two vinyl floor coverings. ◨

  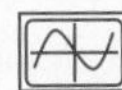

Using the *P*-Value Approach

The *P*-value approach to hypothesis testing can also be used to carry out the hypothesis test in Example 11.14 and to assess more precisely the evidence against the null hypothesis. From Step 4 we see that the value of the test statistic is $F = 0.41$. Recalling that df $= (9, 9)$, we find from Table IX that $P > 0.20$. (See Exercise 11.60.)

In particular, because the *P*-value exceeds the significance level of 0.05, we cannot reject H_0. Furthermore, by referring to Table 9.12 on page 543, we see that the data provide at most weak evidence against the null hypothesis of equal population standard deviations.

Confidence Intervals for Two Population Standard Deviations

Using Key Fact 11.5 on page 706, we can also obtain the following confidence-interval procedure for the ratio of two population standard deviations. This procedure is sometimes referred to as the ***F*-interval procedure for two population standard deviations.**

PROCEDURE 11.4

THE *F*-INTERVAL PROCEDURE FOR TWO POPULATION STANDARD DEVIATIONS

ASSUMPTIONS

1. Independent samples
2. Normal populations

Step 1 For a confidence level of $1 - \alpha$, use Table IX to find $F_{1-\alpha/2}$ and $F_{\alpha/2}$ for df $= (n_1 - 1, n_2 - 1)$.

Step 2 The confidence interval for σ_1/σ_2 is from

$$\frac{1}{\sqrt{F_{\alpha/2}}} \cdot \frac{s_1}{s_2} \quad \text{to} \quad \frac{1}{\sqrt{F_{1-\alpha/2}}} \cdot \frac{s_1}{s_2}$$

where $F_{1-\alpha/2}$ and $F_{\alpha/2}$ are found in Step 1, n_1 and n_2 are the sample sizes, and s_1 and s_2 are computed from the sample data obtained.

Like the *F*-test for two population standard deviations, the *F*-interval procedure is not at all robust to violations of the normality assumption; using it on data from a variable that is not normally distributed on each population can result in misleading information. In other words, the *F*-interval procedure should not be used unless there is considerable evidence that the variable under consideration is normally distributed on each population or very nearly so.

EXAMPLE 11.15 ILLUSTRATES PROCEDURE 11.4

Use the sample data in Table 11.2 on page 705 to determine a 95% confidence interval for the ratio, σ_1/σ_2, of the standard deviations of tear strength for Brand A and Brand B vinyl floor coverings.

SOLUTION As we discovered in Example 11.14, it is reasonable to presume that tear strengths are normally distributed for both brands of vinyl floor covering. Thus we can apply Procedure 11.4 to obtain the required confidence interval.

Step 1 *For a confidence level of $1-\alpha$, use Table IX to find $F_{1-\alpha/2}$ and $F_{\alpha/2}$ for df $= (n_1 - 1, n_2 - 1)$.*

We want a 95% confidence interval; so $\alpha = 0.05$. Consequently, we need to find $F_{0.975}$ and $F_{0.025}$ for df $= (n_1 - 1, n_2 - 1) = (9, 9)$. This was done earlier in Step 3 of Example 11.14 on page 707 where we found that $F_{0.975} = 0.25$ and $F_{0.025} = 4.03$.

Step 2 *The confidence interval for σ_1/σ_2 is from*

$$\frac{1}{\sqrt{F_{\alpha/2}}} \cdot \frac{s_1}{s_2} \quad \textit{to} \quad \frac{1}{\sqrt{F_{1-\alpha/2}}} \cdot \frac{s_1}{s_2}.$$

For the data in Table 11.2, we have $s_1 = 128.3$ g and $s_2 = 199.7$ g. And from Step 1, we have $F_{0.975} = 0.25$ and $F_{0.025} = 4.03$. Consequently, the required 95% confidence interval is from

$$\frac{1}{\sqrt{4.03}} \cdot \frac{128.3}{199.7} \quad \text{to} \quad \frac{1}{\sqrt{0.25}} \cdot \frac{128.3}{199.7}$$

or 0.32 to 1.28. We can be 95% confident that the ratio of the standard deviations of tear strength for Brand A and Brand B vinyl floor coverings is somewhere between 0.32 and 1.28. □

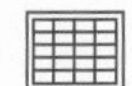

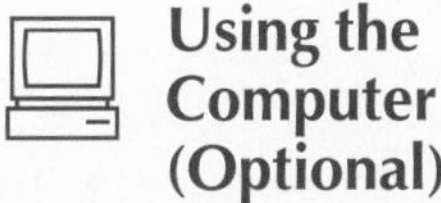

Using the Computer (Optional)

Minitab does not have a dedicated procedure for performing hypothesis tests and obtaining confidence intervals for two population standard deviations.† But we used Minitab's capabilities to write a macro for such inferences and called the macro `2stdev.mac`. It can be found in the `Macro` folder of DataDisk CD.

† Actually, we can perform a two-tailed test for two population standard deviations directly from Minitab by choosing **Stat ➤ ANOVA ➤ Homogeneity of Variance . . .** .

EXAMPLE 11.16 USING MINITAB TO PERFORM INFERENCES FOR TWO POPULATION STANDARD DEVIATIONS

Use Minitab to perform the hypothesis test considered in Example 11.14 and to obtain the confidence interval required in Example 11.15.

SOLUTION Let σ_1 and σ_2 denote, respectively, the population standard deviations of tear strength for Brand A and Brand B. The problem in Example 11.14 is to perform the hypothesis test

H_0: $\sigma_1 = \sigma_2$ (standard deviations of tear strength are the same)

H_a: $\sigma_1 \neq \sigma_2$ (standard deviations of tear strength are different)

at the 5% significance level; the problem in Example 11.15 is to obtain a 95% confidence interval for the ratio of σ_1 to σ_2.

As we discovered in Example 11.14, it is reasonable to use the F-test and F-interval procedure for two population standard deviations. To apply Minitab, we begin by entering the data from Table 11.2 on page 705 in columns named BRAND A and BRAND B, respectively. Next we enable the command language as follows:

1 Click in the Session window
2 Choose **Editor ➤ Enable Command Language**

Now we can run `2stdev.mac` by typing, in the Session window, a percent sign (%) followed by the path for `2stdev.mac` and the locations of the data enclosed by apostrophes. For instance, if DataDisk CD is in drive `E:`, we type

```
%E:\IS5\Macro\2stdev.mac 'BRAND A' 'BRAND B'
```

and press the [Enter] key. Then we proceed as follows.

1 In response to `Do you want to perform a hypothesis test (Y/N)?`, type Y and press the [Enter] key.
2 In response to `Enter 0, 1, or -1, respectively, for a two-tailed, right-tailed, or left-tailed test.`, type 0 and press the [Enter] key.

Printout 11.5 shows the resulting output.

The output in Printout 11.5 gives the sample sizes, sample standard deviations, value of the F-statistic, and P-value for the hypothesis test. Since the P-value of 0.204 exceeds the specified significance level of 0.05, we do not reject H_0. At the 5% significance level, the data do not provide sufficient evidence to conclude that the population standard deviations of tear strength differ for the two vinyl floor coverings.

PRINTOUT 11.5
Hypothesis-test output for the 2stdev macro

```
F-Test of sigma1 = sigma2 (vs not =)


Row  Variable    n    StDev       F      P

  1   BRAND A   10  128.322    0.41  0.204
  2   BRAND B   10  199.669
```

The macro continues with the confidence-interval aspect. We proceed in the following manner.

1 In response to `Do you want a confidence interval (Y/N)?`, type Y and press the Enter key.
2 Since we want a 95% confidence interval, type 95 in response to `Enter the confidence level, as a percentage.`, and then press the Enter key.

Printout 11.6 shows the resulting output.

PRINTOUT 11.6
Confidence-interval output for the 2stdev macro

```
Row  Variable    n    StDev  Level    CI for ratio

  1   BRAND A   10  128.322  95.0%  (0.320, 1.290)
  2   BRAND B   10  199.669
```

The output in Printout 11.6 provides the sample sizes, sample standard deviations, confidence level, and confidence interval for σ_1/σ_2. We see that a 95% confidence interval for σ_1/σ_2 is from 0.320 to 1.290. We can be 95% confident that the ratio of the standard deviations of tear strength for Brand A and Brand B vinyl floor coverings is somewhere between 0.320 and 1.290. ◨

EXERCISES 11.2

Statistical Concepts and Skills

11.34 What does it mean to say that a variable has an F-distribution?

11.35 How do we identify an F-distribution and its corresponding F-curve?

11.36 How many degrees of freedom does an F-curve have? What are those degrees of freedom called?

11.37 What symbol is used to denote the F-value with area 0.05 to its right? 0.025 to its right? α to its right?

11.38 Using the F_α-notation, identify the F-value having area 0.975 to its left.

11.39 An F-curve has df $= (12, 7)$. What is the number of degrees of freedom for the
a. numerator? **b.** denominator?

11.40 An F-curve has df $= (8, 19)$. What is the number of degrees of freedom for the
a. denominator? **b.** numerator?

In Exercises 11.41–11.48, use Table IX and the reciprocal property of F-curves, if necessary, to find the required F-values. Illustrate your work graphically.

11.41 An F-curve has df $= (24, 30)$. In each case find the F-value having the specified area to its right.
a. 0.05 **b.** 0.01 **c.** 0.025

11.42 An F-curve has df $= (12, 5)$. In each case find the F-value having the specified area to its right.
a. 0.01 **b.** 0.05 **c.** 0.005

11.43 For an F-curve with df $= (20, 21)$, find
a. $F_{0.01}$. **b.** $F_{0.05}$. **c.** $F_{0.10}$.

11.44 For an F-curve with df $= (6, 10)$, find
a. $F_{0.05}$. **b.** $F_{0.01}$. **c.** $F_{0.025}$

11.45 Consider an F-curve with df $= (6, 8)$. Obtain the F-value having area
a. 0.01 to its left. **b.** 0.95 to its left.

11.46 Consider an F-curve with df $= (15, 5)$. Obtain the F-value having area
a. 0.025 to its left. **b.** 0.975 to its left.

11.47 Determine the two F-values that divide the area under the curve into a middle 0.95 area and two outside 0.025 areas for an F-curve with
a. df $= (7, 4)$. **b.** df $= (12, 20)$.

11.48 Determine the two F-values that divide the area under the curve into a middle 0.90 area and two outside 0.05 areas for an F-curve with
a. df $= (10, 8)$. **b.** df $= (12, 12)$.

11.49 In using F procedures to make inferences for two population standard deviations, why is it important that the distributions (one for each population) of the variable under consideration be normally distributed or nearly so?

11.50 Give two situations in which it would be important to compare two population standard deviations.

Preliminary data analyses and other information indicate that it is reasonable to assume that in each of Exercises 11.51–11.54 the variable under consideration is normally distributed on both populations. For each exercise, perform the required hypothesis test using either the critical-value approach or the P-value approach.

11.51 One year at Arizona State University, the algebra course director decided to experiment with a new teaching method that might reduce variability in final-exam scores by eliminating lower ones. The director randomly divided the algebra students registered for class at 9:40 A.M. into two groups. One of the groups, called the control group, was taught the usual algebra course; the other group, called the experimental group, was taught by the new teaching method. Both classes covered the same material, took the same unit quizzes, and the same final exam at the same time. The final-exam scores (out of 40 possible) for the two groups are shown in the following table.

Control					Experimental		
36	35	35	33	32	36	35	35
32	31	29	29	28	31	30	29
28	28	27	27	27	27	27	26
26	26	25	24	24	23	21	21
24	23	20	20	19	35	32	
19	18	18	18	17	28	28	
17	16	15	15	15	25	23	
15	14	11			21	19	
10	9	4					

Do the data provide sufficient evidence to conclude that there is less variation among final-exam scores using the new teaching method? Perform an F-test at the 5% significance level. (*Note:* $s_1 = 7.813$ and $s_2 = 5.286$.)

11.52 In a paper appearing in *Obstetrics & Gynecology* (Vol. 9, No. 3, March 1998, pp. 336–341), researchers reported on a study of characteristics of neonates. In the study, infants treated for pulmonary hypertension, called the PH group, were compared with those not so treated, called the control group. One of the characteristics measured was head circumference. The following data, in centimeters (cm) are based on the results obtained by the researchers.

PH		Control				
33.9	35.1	35.2	35.6	36.7	35.1	36.0
33.4	34.5	33.4	31.3	33.5	35.8	36.3
37.9	31.3	34.3	33.1	32.4	35.1	33.6
32.5	32.9	31.8	34.1	35.2	34.8	34.5
36.3	34.2	31.6	31.9	31.9	32.8	34.0

Do the data provide sufficient evidence to conclude that variation in head circumference differs among neonates treated for pulmonary hypertension and those not so treated? Perform an F-test at the 5% significance level. *(Note:* $s_1 = 1.907$ and $s_2 = 1.594$.*)*

11.53 Patients who undergo chronic hemodialysis often experience severe anxiety. Videotapes of progressive relaxation exercises were shown to one group of patients and neutral videotapes to another group. Then both groups took the State-Trait Anxiety Inventory, a psychiatric questionnaire used to measure anxiety, where higher scores correspond to higher anxiety. In the paper "The Effectiveness of Progressive Relaxation in Chronic Hemodialysis Patients" (*Journal of Chronic Diseases,* 35(10)), R. Alarcon et al. presented the results of the study. The following data are based on those results.

Relaxation tapes				Neutral tapes			
30	41	28	14	36	44	47	45
40	36	38	24	50	54	54	45
61	36	24	45	50	46	28	35
38	43	32	28	42	35	32	43
37	34	20	23	41	33	35	36
34	47	25	31	32	17	45	
39	14	43	40	24	46		
29	21	40					

Do the data provide sufficient evidence to conclude that variation in anxiety-test scores differs between patients seeing videotapes showing progressive relaxation exercises and those seeing neutral videotapes? Perform an F-test at the 10% significance level. *(Note:* $s_1 = 10.154$ and $s_2 = 9.197$.*)*

11.54 The U.S. Federal Highway Administration publishes statistics on motor vehicle travel by type of vehicle in *Highway Statistics Summary.* Independent random samples of cars and trucks yielded the following data on number of miles driven last year, where number of miles is given in thousands.

Cars			Trucks	
3.9	11.3	2.0	5.2	26.6
11.0	17.9	15.8	42.9	38.1
11.9	12.7	8.7	13.1	28.7
7.2	14.9	9.7	43.1	31.9
14.9	18.1	17.1	32.4	18.8

Do the data provide sufficient evidence to conclude that variation in miles driven by cars is less than that for trucks? Perform an F-test at the 1% significance level. *(Note:* The sample standard deviations of the miles driven are 4.896 thousand and 12.541 thousand, respectively.*)*

11.55 Refer to Exercise 11.53. Find a 90% confidence interval for the ratio of the population standard deviations of scores for patients seeing videotapes showing progressive relaxation exercises and those seeing neutral videotapes.

11.56 Refer to Exercise 11.52. Find a 95% confidence interval for the ratio of the population standard deviations of head circumferences for neonates treated for pulmonary hypertension and those not so treated.

EXTENDING THE CONCEPTS AND SKILLS

11.57 Because of space restrictions, the numbers of degrees of freedom in Table IX are not consecutive. For instance, the degrees of freedom for the numerator skips from 24 to 30. If you had only Table IX to work with and you needed to find $F_{0.05}$ for df = (25, 20), how would you do it?

Estimating F-values from Table IX. One solution to Exercise 11.57 is to use linear interpolation as follows. For df = (24, 20), we have $F_{0.05} = 2.08$; and for df = (30, 20), we have $F_{0.05} = 2.04$. Since 25 is $\frac{1}{6}$th of the way between 24 and 30, we estimate that for an F-curve with df = (25, 20),

$$F_{0.05} = 2.08 + \frac{1}{6} \cdot (2.04 - 2.08) = 2.07.$$

Use linear interpolation to solve each of Exercises 11.58 and 11.59.

11.58 Refer to Exercise 11.54. Obtain a 98% confidence interval for the ratio of the population standard deviations of miles driven last year by cars and by trucks.

11.59 Refer to Exercise 11.51. Obtain a 90% confidence interval for the ratio of the population standard deviations of final-exam scores for students taught by the conventional method and by the new method.

11.60 Refer to Example 11.14 on page 707. Use Table IX to show that the P-value for the hypothesis test exceeds 0.20.

USING TECHNOLOGY  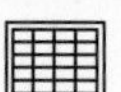 

11.61 Refer to Exercises 11.51 and 11.59. Use Minitab or some other statistical software to

a. obtain a normal probability plot and boxplot of each sample.
b. perform the required hypothesis test and obtain the desired confidence interval. (*Note:* Users of Minitab should employ `2stdev.mac`.)
c. Justify the use of your procedure in part (b).

11.62 Refer to Exercises 11.52 and 11.56. Use Minitab or some other statistical software to

a. obtain a normal probability plot and boxplot of each sample.
b. perform the required hypothesis test and obtain the desired confidence interval. (*Note:* Users of Minitab should employ `2stdev.mac`.)
c. Justify the use of your procedure in part (b).

11.63 Refer to Exercises 11.53 and 11.55. Use Minitab or some other statistical software to

a. obtain a normal probability plot and boxplot of each sample.
b. perform the required hypothesis test and obtain the desired confidence interval. (*Note:* Users of Minitab should employ `2stdev.mac`.)
c. Justify the use of your procedure in part (b).

11.64 Refer to Exercises 11.54 and 11.58. Use Minitab or some other statistical software to

a. obtain a normal probability plot and boxplot of each sample.
b. perform the required hypothesis test and obtain the desired confidence interval. (*Note:* Users of Minitab should employ `2stdev.mac`.)
c. Justify the use of your procedure in part (b).

11.65 Refer to Example 11.14 on page 707. Use Minitab or some other statistical software to determine the exact P-value for the hypothesis test.

11.66 The American Association of University Professors (AAUP) conducts salary studies of college professors and publishes its findings in *AAUP Annual Report on the Economic Status of the Profession.* Independent random samples of 30 faculty members from public institutions and 35 faculty members from private institutions yield the salaries shown in the following table, where they are given in thousands of dollars rounded to the nearest hundred.

Public			Private			
34.2	63.6	24.4	92.9	102.2	51.5	71.0
90.0	56.8	56.0	52.0	62.9	46.4	97.3
100.4	41.4	58.2	63.1	53.8	45.2	92.6
24.6	35.0	76.8	118.5	101.0	76.0	56.0
107.4	54.2	84.2	37.7	68.6	56.1	62.3
79.4	33.8	88.2	77.6	71.1	59.3	
42.2	40.2	44.6	61.6	73.5	97.5	
81.8	51.2	64.4	78.3	67.6	27.2	
29.2	41.2	74.0	66.3	52.4	81.2	
15.8	60.2	71.0	31.1	47.2	24.8	

Use Minitab or some other statistical software to solve parts (a) and (b).

a. Perform an F-test at the 5% significance level to decide whether the data provide sufficient evidence to conclude that there is a difference in variation of faculty salaries between public and private institutions.
b. Obtain a normal probability plot for each sample.
c. In view of your plots in part (b), does it seem reasonable to conduct the test you did in part (a)? Explain your answer.

11.67 A group of neurosurgeons wanted to decide whether a dynamic system (Z-plate) reduced the operative time relative to a static system (ALPS plate). R. Jacobowitz, Ph.D., an ASU professor, along with G. Vishteh, M.D., and other neurosurgeons, obtained the following data on operative times, in minutes, for the two systems.

Dynamic				Static	
370	360	505	335	430	445
345	450	295	315	455	490
510	445	280	325	455	535
490	500				

Use Minitab or some other statistical software to solve parts (a) and (b).

a. Perform an F-test at the 1% significance level to decide whether the data provide sufficient evidence to conclude that the variation in operative time differs for the dynamic system and the static system.
b. Obtain a normal probability plot for each sample.
c. In view of your plots in part (b), does it seem reasonable to conduct the test you did in part (a)? Explain your answer.

11.68 Use Minitab or some other statistical software to conduct the simulation discussed in Example 11.13 on page 706.

CHAPTER REVIEW

Note: In this chapter review, we have not used asterisks for optional material because the entire chapter is optional.

You Should Be Able To

1. use and understand the formulas presented in this chapter.
2. state the basic properties of χ^2-curves.
3. use the chi-square table, Table VIII.
4. perform a hypothesis test for a population standard deviation when the variable under consideration is normally distributed.
5. obtain a confidence interval for a population standard deviation when the variable under consideration is normally distributed.
6. state the basic properties of F-curves.
7. apply the reciprocal property of F-curves.
8. use the F-table, Table IX.
9. perform a hypothesis test to compare two population standard deviations when the variable under consideration is normally distributed on both populations.
10. obtain a confidence interval for the ratio of two population standard deviations when the variable under consideration is normally distributed on both populations.
11. use the Minitab procedures covered in this chapter.
12. interpret the output obtained from the application of the Minitab procedures discussed in this chapter.

Key Terms

χ^2_α, *685*
χ^2-interval procedure, *694*
χ^2-test, *691*
chi-square (χ^2) curve, *684*
chi-square distribution, *684*
degrees of freedom for the denominator, *701*
degrees of freedom for the numerator, *701*
F_α, *702*
F-curve, *701*
F-distribution, *701*
F-interval procedure, *710*
F-statistic, *705*
F-test, *708*

REVIEW TEST

Statistical Concepts and Skills

1. What distribution is used in this chapter to make inferences for one population standard deviation?

2. Fill in the following blanks:

a. A χ^2-curve is ____ skewed.

b. A χ^2-curve looks increasingly like a ____ curve as the number of degrees of freedom becomes larger.

3. When conducting inferences for one population standard deviation using the χ^2-test or χ^2-interval procedure, what assumption must be met by the variable under consideration? How important is it?

4. Consider a χ^2-curve with 17 degrees of freedom. Use Table VIII to determine

a. $\chi^2_{0.99}$. **b.** $\chi^2_{0.01}$.

c. the χ^2-value having area 0.05 to its right.

d. the χ^2-value having area 0.05 to its left.

e. the two χ^2-values that divide the area under the curve into a middle 0.95 area and two outside 0.025 areas.

5. What distribution is used in this chapter to make inferences for two population standard deviations?

6. Fill in the following blanks:

a. An F-curve is ____ skewed.

b. For an F-curve with df = (14, 5), the F-value having area 0.05 to its left equals the ____ of the F-value having area 0.05 to its right for an F-curve with df = (__, __).

c. The observed value of a variable having an F-distribution must be greater than or equal to ____.

7. When conducting inferences for two population standard deviations using the F-test or F-interval procedure, what assumption must be met by the variable under consideration? How important is it?

8. Consider an F-curve with df = (4, 8). Use Table IX to determine

a. $F_{0.01}$. **b.** $F_{0.99}$.

c. the F-value having area 0.05 to its right.

d. the F-value having area 0.05 to its left.

e. the two F-values that divide the area under the curve into a middle 0.95 area and two outside 0.025 areas.

9. IQs measured on the Stanford Revision of the Binet-Simon Intelligence Scale are supposed to have a standard deviation of 16 points. Twenty-five randomly selected people were given the IQ test; here are the data that were obtained.

91	96	106	116	97
102	96	124	115	121
95	111	105	101	86
88	129	112	82	98
104	118	127	66	102

Preliminary data analyses and other information indicate that it is reasonable to presume that IQs measured on the Stanford Revision of the Binet-Simon Intelligence Scale are normally distributed.

a. Do the data provide sufficient evidence to conclude that IQs measured on this scale have a standard deviation different from 16 points? Perform the required hypothesis test at the 10% significance level. *(Note:* $s = 15.006$.*)*

b. How crucial is the normality assumption for the hypothesis test you performed in part (a)?

10. Refer to Problem 9. Determine a 90% confidence interval for the standard deviation of IQs.

11. A study entitled "Body Composition of Elite Class Distance Runners" was conducted by M. L. Pollock et al. to decide whether elite distance runners are truly thinner. Their results were published in *The Marathon: Physiological, Medical, Epidemiological, and Psychological Studies,* P. Milvey (ed.), New York: New York Academy of Sciences, 1977, p. 366. The researchers measured the skinfold thickness, an indirect indicator of body fat, of runners and nonrunners in the same age group. The data, in millimeters, shown in the following table are based on the skinfold thickness measurements on the thighs of the people sampled.

Runners			Others			
7.3	6.7	8.7	24.0	19.9	7.5	18.4
3.0	5.1	8.8	28.0	29.4	20.3	19.0
7.8	3.8	6.2	9.3	18.1	22.8	24.2
5.4	6.4	6.3	9.6	19.4	16.3	16.3
3.7	7.5	4.6	12.4	5.2	12.2	15.6

a. For an F-test to compare the standard deviations of skinfold thickness of runners and others, identify the appropriate F-distribution.

b. For a left-tailed F-test at the 0.01 level, can we determine the critical value using Table IX and the reciprocal property of F-curves? Explain your answer.

c. For an F-curve with df $= (14, 19)$, we used Minitab to find that the F-value having area 0.01 to its left is 0.28. Given that information, decide at the 1% significance level whether the data provide sufficient evidence to conclude that runners have less variability in skinfold thickness than others. *(Note:* $s_1 = 1.798$ and $s_2 = 6.606$.*)*

d. What assumption about skinfold thickness are you making in carrying out the hypothesis test in part (c)? How would you check that assumption?

e. In addition to the assumption on skinfold thickness discussed in part (d), what other assumption is required for performing the F-test?

12. Refer to Problem 11. We used Minitab to determine that for an F-curve with df $= (14, 19)$, $F_{0.01} = 3.19$.

a. Find a 98% confidence interval for the ratio of the standard deviations of skinfold thickness for runners and for others.

b. Interpret your answer for part (a).

USING TECHNOLOGY  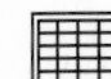 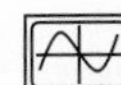

13. Use Minitab or some other statistical software to perform the required hypothesis test in Problem 9 and to obtain the desired confidence interval in Problem 10. *(Note:* Users of Minitab should employ `1stdev.mac`.*)*

14. Use Minitab or some other statistical software to perform the required hypothesis test in Problem 11 and to obtain the desired confidence interval in Problem 12. *(Note:* Users of Minitab should employ `2stdev.mac`.*)*

INTERNET PROJECT

Firearm Related Deaths for Men and Women

The Centers for Disease Control and Prevention (CDC) estimates that by the year 2003, firearms will become the number one cause of product-related deaths in the United States. (The leading cause now is motor vehicles.)

In this Internet project, you will see striking differences in firearm-related deaths for males and females. For instance, males are almost six times more likely to be killed by firearms than are females.

But the question posed in this project concerns the amount of variation in age at firearm-related death for males and females. You will estimate the standard deviation for each population and perform a hypothesis test to decide whether the two population standard deviations differ.

A more interesting question—which statistics alone cannot answer—is what do the standard deviations tell us in this context? Are there, for example, social or psychological reasons for one group to have a larger amount of variation than the other?

URL for access to Internet Projects Page: `http://hepg.awl.com` *keyword:* Weiss

USING THE FOCUS DATABASE

In Chapter 1 we explained how to store the Focus database in a Minitab worksheet named `focus.mtw`. If you haven't already created that worksheet, follow the instructions on pages 54–55 to create it now.

The Focus database contains information on 500 randomly selected Arizona State University sophomores for seven different variables: sex, high-school GPA, SAT math score, cumulative GPA, SAT verbal score, age, and total hours.

a. At the 5% significance level, do the data provide sufficient evidence to conclude that the standard deviation of SAT math scores of ASU sophomores differs from 100? *(Hint:* When the number of degrees of freedom is large, a chi-square distribution with df $= \nu$ is approximated well by a normal distribution with mean ν and standard deviation $\sqrt{2\nu}$.*)*

b. Obtain a 95% confidence interval for the standard deviation of SAT math scores of ASU sophomores.

c. At the 5% significance level, do the data provide sufficient evidence to conclude that the standard deviation of SAT verbal scores of ASU sophomores differs from 100?

d. Obtain a 95% confidence interval for the standard deviation of SAT verbal scores of ASU sophomores.

e. At the 5% significance level, do the data provide sufficient evidence to conclude that for ASU sophomores the standard deviations of SAT math scores and SAT verbal scores differ? *(Note:* For an F-curve with df $= (499, 499)$, $F_{0.025} = 1.19$.*)*

f. Obtain a 95% confidence interval for the ratio of the standard deviations of SAT math scores and SAT verbal scores of ASU sophomores.

g. Use Minitab or some other statistical software to solve parts (a)–(f). *(Note:* Users of Minitab should employ the macros `1stdev.mac` and `2stdev.mac` found in the `Macro` folder of DataDisk CD.*)*

CASE STUDY DISCUSSION

Speaker Woofer Driver Manufacturing

On page 682, at the beginning of this chapter, we discussed rubber-edge manufacturing of speaker woofer drivers and a criterion for classifying process capability.

We recall that each process for manufacturing rubber edges calls for a production weight specification consisting of a lower specification limit (LSL), a target weight (T), and an upper specification limit (USL). The actual mean and standard deviation of the weights of the rubber edges being produced are called, respectively, the process mean (μ) and process standard deviation (σ). A process that is on target ($\mu = T$) is called *super* if

$$\sigma < \frac{\text{USL} - \text{LSL}}{12}.$$

The table on page 682 provides data on rubber-edge weight for a sample of 60 observations. Use that data and the procedures discussed in this chapter to solve the following:

a. Obtain a 99% confidence interval for the process standard deviation.

b. The process under consideration is known to be on target and its production weight specification is LSL $= 16.72$, $T = 17.60$, and USL $= 18.48$. Do the data provide sufficient evidence to conclude that the process is super? Perform the required hypothesis test at the 1% significance level.

c. Obtain a normal probability plot of the data.

d. Referring to your plot in part (c), does it seem reasonable to conduct the inferences that you did in parts (a) and (b)? Explain your answer.

BIOGRAPHY W. EDWARDS DEMING

William Edwards Deming was born on October 14, 1900, in Sioux City, Iowa. Shortly after his birth, his father secured homestead land and moved the family first to Cody, Wyoming, then to Powell, Wyoming.

Deming obtained a B.S. in physics at the University of Wyoming in 1921, a Master's degree in physics and mathematics at the University of Colorado in 1924, and a doctorate in mathematical physics at Yale University in 1928.

Deming spent the next decade becoming an expert on sampling and quality control while working for various federal agencies. In 1939, he accepted the position of head mathematician and advisor in sampling at the United States Census Bureau. Deming began the use of sampling at the Census Bureau, and, expanding the work of Walter A. Shewhart (later known as the father of statistical quality control), also applied statistical methods of quality control to provide reliability and quality to the nonmanufacturing environment.

In 1946, Deming left the Census Bureau, joined the Graduate School of Business Administration at New York University, and offered his services to the private sector as a consultant in statistical studies. It was in this latter capacity that Deming transformed industry in Japan. Deming began his long association with Japanese businesses in 1947 when the U.S. War Department engaged him to instruct Japanese industrialists in statistical quality control methods. The reputation of Japan's goods changed from definitely shoddy to amazingly excellent over the next two decades as the businessmen of Japan implemented Deming's teachings.

It took more than 30 years for Deming's methods to gain widespread recognition by the business community in the United States. Finally, in 1980, as the result of the NBC white paper, *If Japan Can, Why Can't We?,* in which Deming's role was publicized, executives of major corporations (among them, Ford Motor Company) contracted with Deming to improve the quality of U.S. goods.

Deming maintained an intense work schedule throughout his eighties, giving four-day managerial seminars, teaching classes at NYU, sponsoring clinics for statisticians, and consulting with businesses internationally. His last book, *The New Economics,* was published in 1993. Dr. Deming died at his home in Washington, DC, on December 20, 1993.

CASE STUDY

CREDIT CARD BLUES

The Board of Governors of the Federal Reserve System reports that in 1995, almost two-thirds of Americans owned credit cards, holding a median of two cards and carrying a median balance of $1500. These cards are used to purchase goods and services in person, by phone, and over the Internet. According to projections in the *Nilson Report,* a publication of HSN Consultants, Inc., the total credit-card debt in the United States will be approximately $783.3 billion dollars at the beginning of the 21st century.

To get a picture of who uses credit cards, who keeps them paid up, and how much is owed by those who don't, ABCNEWS.com sponsored a national poll from October 1–5, 1997. Here are some of the highlights of the poll.

- About 37% of American adults do not own a credit card.
- A majority (approximately 52%) of those who own credit cards owe nothing on them.
- Roughly one-third of those who own credit cards carry a balance of more than $5000.
- Approximately 4% of those who own credit cards carry a balance of $10,000 or more.

The poll also revealed patterns of credit-card use by income, age, education, race, and sex. For example, approximately 57% of college graduates are paid in full, compared to 49% of those with a high-school education or less; no difference exists between average balances for men and women; and the age group 35–50 years is the one having the highest average credit-card balance.

Keep in mind that the statements made here give statistics, that is, they describe the sample. In this chapter we will learn inferential methods that permit us to generalize from the sample to all American adults.

Inferences for Population Proportions

12

GENERAL OBJECTIVES

In Chapters 8–10 we learned how to find confidence intervals and perform hypothesis tests for population means. Now we will discover how to conduct those inferences for population proportions. A *population proportion* is the proportion (percentage) of a population that has a specified attribute. For example, if the population under consideration consists of all Americans and the specified attribute is "retired," then the population proportion is the proportion of all Americans who are retired.

The first two sections of this chapter explain how to determine confidence intervals and perform hypothesis tests for one population proportion. The third section discusses how to perform hypothesis tests for comparing two population proportions and how to construct confidence intervals for the difference between two population proportions.

CHAPTER OUTLINE

12.1 CONFIDENCE INTERVALS FOR ONE POPULATION PROPORTION

Many statistical studies are concerned with obtaining the proportion (percentage) of a population that has a specified attribute. For example, we might be interested in

- the percentage of U.S. adults who have health insurance,
- the percentage of cars in the United States that are imports,
- the percentage of Americans who favor stricter clean air health standards, or
- the percentage of Canadian women in the labor force.

In the first case, the population consists of all U.S. adults and the specified attribute is "has health insurance." For the second case, the population consists of all cars in the United States and the specified attribute is "is an import." The population in the third case is all Americans and the specified attribute is "favors stricter clean air health standards." And, in the fourth case, the population consists of all Canadian women and the specified attribute is "is in the labor force."

Generally the population under consideration is large, and it is therefore usually impractical and often impossible to determine the population proportion by taking a census; for instance, imagine trying to interview every U.S. adult in order to ascertain the proportion who have health insurance. Thus, in practice, we mostly rely on sampling and use the sample data to make inferences about the population proportion. Example 12.1 introduces proportion notation and terminology.

EXAMPLE 12.1 PROPORTION NOTATION AND TERMINOLOGY

Many employers are concerned about employees who call in sick when in fact they are not ill. A survey commissioned by the Hilton Hotels Corporation investigated this issue. One question asked of the people taking part in the survey was whether they call in sick at least once a year when they simply need time to relax. For brevity we will use the phrase *play hooky* to refer to that practice. In the survey, 1010 randomly selected U.S. employees were polled. The proportion of the 1010 employees sampled who play hooky was used to estimate the proportion of all U.S. employees who play hooky.

We use the letter $\boldsymbol{p}$ to denote the proportion of all U.S. employees who play hooky; this is the **population proportion** and is the parameter whose value is to be estimated. The proportion of the 1010 U.S. employees sampled who play hooky is called a **sample proportion** and is designated by the symbol $\boldsymbol{\hat{p}}$ (read "p hat"); this is the statistic that will be used to estimate the unknown population proportion, p.

The population proportion, p, although unknown, is a fixed number. On the other hand, the sample proportion, $\hat{p}$, is a variable; its value varies from sample to sample. For instance, if in one sample, 202 of the 1010 employees play hooky, then

$$\hat{p} = \frac{202}{1010} = 0.2 \text{ (20\%)}.$$

And, if in another sample, 184 of the 1010 employees play hooky, then

$$\hat{p} = \frac{184}{1010} = 0.182 \text{ (18.2\%)}.$$

These two calculations also reveal how a sample proportion is computed: Divide the number of employees sampled who play hooky, denoted x, by the total number of employees sampled, n. In symbols, $\hat{p} = x/n$. ◨

Example 12.1 introduced some notation and terminology we use when making inferences about a population proportion. In general, we have the following definitions.

DEFINITION 12.1 **POPULATION PROPORTION AND SAMPLE PROPORTION**

Consider a population in which each member either has or does not have a specified attribute. Then we use the following notation and terminology.

Population proportion, p: The proportion (percentage) of the entire population that has the specified attribute.

Sample proportion, p̂: The proportion (percentage) of a sample from the population that has the specified attribute.

In Example 12.1 the population consists of all U.S. employees and the specified attribute is "plays hooky." The population proportion, p, is the proportion of all U.S. employees who play hooky; and a sample proportion, $\hat{p}$, is the proportion of employees sampled who play hooky. As we saw in Example 12.1, a sample proportion is computed using the following formula.

FORMULA 12.1 **SAMPLE PROPORTION**

A sample proportion, $\hat{p}$, is computed using the formula

$$\hat{p} = \frac{x}{n},$$

where x denotes the number of members in the sample that have the specified attribute and, as usual, n denotes the sample size.

Note: For convenience we will sometimes refer to x (the number of members in the sample that have the specified attribute) as the **number of successes** and to $n - x$ (the number of members in the sample that do not have the specified attribute) as the **number of failures.** But remember that in this context, the words *success* and *failure* need not conform to the ordinary connotations of the words.

Before proceeding, let's draw some parallels between proportions and means. Table 12.1 shows the correspondence between the notation for means and the notation for proportions.

TABLE 12.1
Correspondence between notations for means and proportions

	Parameter	Statistic
Means	μ	$\bar{x}$
Proportions	p	$\hat{p}$

As we know, a sample mean, $\bar{x}$, can be used to make inferences about a population mean, μ. Similarly, a sample proportion, $\hat{p}$, can be used to make inferences about a population proportion, p.

The Sampling Distribution of the Proportion

We have seen that to make inferences about a population mean, μ, we must know the sampling distribution of the mean, that is, the distribution of the variable $\bar{x}$. The same is true for proportions: To make inferences about a population proportion, p, we need to know the **sampling distribution of the proportion,** that is, the distribution of the variable $\hat{p}$.

The sampling distribution of the proportion can be derived from our knowledge of the sampling distribution of the mean, since a proportion can always be regarded as a mean. (See Exercise 12.33 for details.) Since, in practice, the sample size is large, we concentrate on that case.

KEY FACT 12.1 **THE SAMPLING DISTRIBUTION OF THE PROPORTION**

For samples of size n,

- The mean of $\hat{p}$ equals the population proportion: $\mu_{\hat{p}} = p$.
- The standard deviation of $\hat{p}$ equals the square root of the product of the population proportion and one minus the population proportion divided by the sample size: $\sigma_{\hat{p}} = \sqrt{p(1 - p)/n}$.
- $\hat{p}$ is approximately normally distributed for large n.

In particular, if n is large, then the possible sample proportions for samples of size n have approximately a normal distribution with mean p and standard deviation $\sqrt{p(1-p)/n}$.

The accuracy of the normal approximation depends on n and p. If p is close to 0.5, then the approximation is quite accurate even for moderate n. The farther p is from 0.5, the larger n needs to be for the approximation to be accurate. As a rule of thumb, we use the normal approximation when *np and $n(1-p)$ are both 5 or greater.*[†] This is what we will mean in this chapter when we say that n is large.

We can make Key Fact 12.1 plausible through simulation. To do so, let's return to the situation in Example 12.1.

EXAMPLE 12.2 ILLUSTRATES KEY FACT 12.1

Suppose, in reality, 19.1% of all U.S. employees play hooky, that is, the population proportion is $p = 0.191$. Then according to Key Fact 12.1, for samples of size 1010, $\mu_{\hat{p}} = p = 0.191$, $\sigma_{\hat{p}} = \sqrt{p(1-p)/n} = 0.012$, and $\hat{p}$ is approximately normally distributed. Use simulation to make these facts plausible.

SOLUTION We simulated 2000 samples of size 1010 each from the population of all U.S. employees, found the sample proportion for each of the 2000 samples, and obtained a histogram of those 2000 sample proportions. Printout 12.1 displays the histogram.

PRINTOUT 12.1 Minitab histogram of the sample proportions for 2000 samples of size 1010. The approximating normal curve for $\hat{p}$ is superimposed.

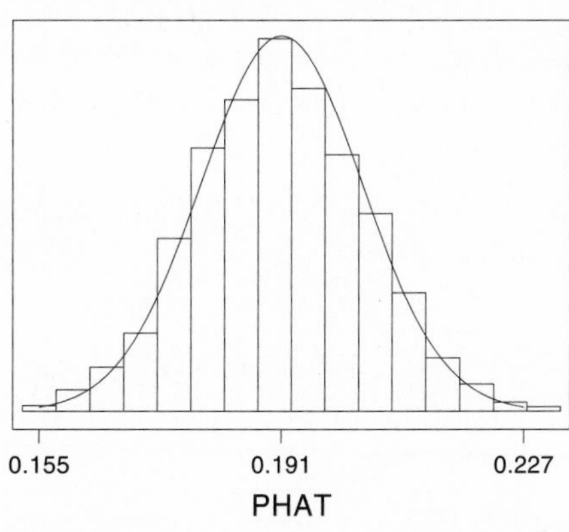

[†] Another commonly-used rule of thumb is that np and $n(1-p)$ are both 10 or greater; still another is that $np(1-p)$ is 25 or greater. Our rule of thumb, which is less conservative than either of these two, has been chosen so as to be consistent with the conditions required for performing a chi-square goodness-of-fit test (to be discussed in Chapter 13).

The histogram in Printout 12.1 is shaped roughly like a normal curve, specifically, like the normal curve with parameters 0.191 and 0.012, which we have superimposed on the histogram. This makes it plausible that $\hat{p}$ is approximately normally distributed with mean 0.191 and standard deviation 0.012, as guaranteed by Key Fact 12.1. ◨

Large-Sample Confidence Intervals for a Population Proportion

We now present Procedure 12.1, a step-by-step method for obtaining a confidence interval for a population proportion. We often refer to this procedure as the **one-sample z-interval procedure** for a population proportion or, more simply, as the **z-interval procedure** for a population proportion. The one-sample z-interval procedure for a population proportion is based on Key Fact 12.1 and is derived in a way similar to that of the one-sample z-interval procedure for a population mean, Procedure 8.1 on page 452.

PROCEDURE 12.1

THE ONE-SAMPLE z-INTERVAL PROCEDURE FOR A POPULATION PROPORTION

ASSUMPTION

The number of successes, x, and the number of failures, $n - x$, are both 5 or greater.

Step 1 For a confidence level of $1 - \alpha$, use Table II to find $z_{\alpha/2}$.

Step 2 The confidence interval for p is from

$$\hat{p} - z_{\alpha/2} \cdot \sqrt{\hat{p}(1 - \hat{p})/n} \quad \text{to} \quad \hat{p} + z_{\alpha/2} \cdot \sqrt{\hat{p}(1 - \hat{p})/n},$$

where $z_{\alpha/2}$ is found in Step 1, n is the sample size, and $\hat{p} = x/n$ is the sample proportion.

Note: As stated at the beginning of Procedure 12.1, the condition for using that procedure is "the number of successes, x, and the number of failures, $n - x$, are both 5 or greater." This can be restated as "$n\hat{p}$ and $n(1 - \hat{p})$ are both 5 or greater," which, for unknown p, corresponds to the rule of thumb given on page 727 for using the normal approximation.

As our first application of Procedure 12.1, we will obtain a confidence interval for the proportion of all U.S. employees who play hooky. We will use the information obtained in the poll commissioned by the Hilton Hotels Corporation.

EXAMPLE 12.3 ILLUSTRATES PROCEDURE 12.1

A poll was taken of 1010 U.S. employees. The employees sampled were asked whether they "play hooky," that is, call in sick at least once a year when they simply need time to relax; 202 responded "yes." Use these data to obtain a 95% confidence interval for the proportion, p, of all U.S. employees who play hooky.

SOLUTION We will apply Procedure 12.1, but first we need to check that the condition for its use is satisfied. The attribute in question is "plays hooky," the sample size is 1010, and the number of employees sampled who play hooky is 202. So $x = 202$ and $n - x = 1010 - 202 = 808$, both of which are greater than 5. So the condition for using Procedure 12.1 is met.

Step 1 *For a confidence level of $1 - \alpha$, use Table II to find $z_{\alpha/2}$.*

We want a 95% confidence interval, which means $\alpha = 0.05$. Consulting Table II or the bottom of Table IV, we find that $z_{\alpha/2} = z_{0.05/2} = z_{0.025} = 1.96$.

Step 2 *The confidence interval for p is from*

$$\hat{p} - z_{\alpha/2} \cdot \sqrt{\hat{p}(1-\hat{p})/n} \quad \text{to} \quad \hat{p} + z_{\alpha/2} \cdot \sqrt{\hat{p}(1-\hat{p})/n}.$$

We have $n = 1010$ and, from Step 1, $z_{\alpha/2} = 1.96$. Also, because 202 of the 1010 employees sampled play hooky, $\hat{p} = x/n = 202/1010 = 0.2$. Consequently, a 95% confidence interval for p is from

$$0.2 - 1.96 \cdot \sqrt{(0.2)(1-0.2)/1010} \quad \text{to} \quad 0.2 + 1.96 \cdot \sqrt{(0.2)(1-0.2)/1010}$$

or

$$0.2 - 0.025 \quad \text{to} \quad 0.2 + 0.025$$

or 0.175 to 0.225. We can be 95% confident that the percentage of all U.S. employees who play hooky is somewhere between 17.5% and 22.5%. ❑

Margin of Error

In Section 8.3 we discussed the margin of error in estimating a population mean by a sample mean. In general, the margin of error of an estimator represents the precision with which it estimates the parameter in question. Referring to the confidence-interval formula in Step 2 of Procedure 12.1, we see that the margin of error in estimating a population proportion by a sample proportion is $z_{\alpha/2} \cdot \sqrt{\hat{p}(1-\hat{p})/n}$.

DEFINITION 12.2 MARGIN OF ERROR FOR THE ESTIMATE OF p

The ***margin of error*** for the estimate of p is

$$E = z_{\alpha/2} \cdot \sqrt{\hat{p}(1-\hat{p})/n}.$$

The margin of error is equal to half the length of the confidence interval. It represents the precision with which a sample proportion, $\hat{p}$, estimates the population proportion, p, at the specified confidence level.

In Example 12.3 the margin of error is

$$E = z_{\alpha/2} \cdot \sqrt{\hat{p}(1 - \hat{p})/n} = 1.96 \cdot \sqrt{(0.2)(1 - 0.2)/1010} = 0.025,$$

which can also be obtained by taking one-half the length of the confidence interval: $(0.225 - 0.175)/2 = 0.025$. Consequently, we can be 95% confident that the error in estimating the proportion p of all U.S. employees who play hooky by the proportion 0.2 of those in the sample who play hooky is at most 0.025, that is, plus or minus 2.5 percentage points.

As we have seen, given a confidence interval, we can find the margin of error by taking half the length of the confidence interval. On the other hand, given the sample proportion and the margin of error, we can determine the confidence interval—its endpoints are $\hat{p} \pm E$.

Most newspaper and magazine polls provide the sample proportion and the margin of error associated with a 95% confidence interval. For example, a survey of U.S. women conducted by Gallup for the CNBC cable network stated, "... 36% of those polled believe their gender will hurt them; the margin of error for the poll is plus or minus 4 percentage points."

Translated into our terminology, $\hat{p} = 0.36$ and $E = 0.04$. Thus the confidence interval has endpoints $\hat{p} \pm E = 0.36 \pm 0.04$; we can be 95% confident that the percentage of all U.S. women who believe their gender will hurt them is somewhere between 32% and 40%.

Determining the Required Sample Size

The margin of error and confidence level of a confidence interval are often specified in advance. We must then determine the sample size required to meet the specifications. If we solve for n in the formula for the margin of error, we obtain

$$(1) \qquad n = \hat{p}(1 - \hat{p}) \left(\frac{z_{\alpha/2}}{E} \right)^2 .$$

This formula cannot be used to obtain the required sample size because the sample proportion, $\hat{p}$, is not known prior to sampling.

There are two ways around this problem. To begin, let's examine the graph of $\hat{p}(1 - \hat{p})$ versus $\hat{p}$ given in Fig. 12.1. As we see from the graph, the largest $\hat{p}(1 - \hat{p})$ can be is 0.25, which happens when $\hat{p} = 0.5$. The farther $\hat{p}$ is from 0.5, the smaller the value of $\hat{p}(1 - \hat{p})$.

FIGURE 12.1 Graph of $\hat{p}(1-\hat{p})$ versus $\hat{p}$

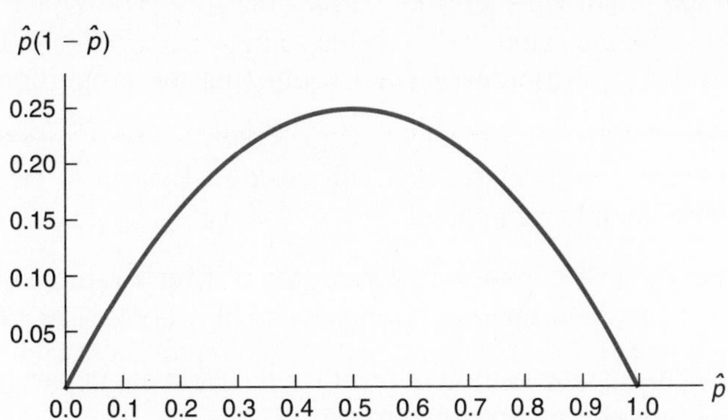

Since the largest possible value of $\hat{p}(1-\hat{p})$ equals 0.25, the most conservative approach for determining sample size is to use that value in Equation (1). The sample size thereby obtained will generally be larger than necessary and the margin of error less than required. Nonetheless, this approach guarantees that the specifications will be met or bettered.

On the other hand, because sampling tends to be time-consuming and expensive, it is usually best not to take a larger sample than necessary. If we can make an educated guess for the observed value of $\hat{p}$, say from a previous study or theoretical considerations, then we can use that guess to obtain a more realistic sample size.

In this same vein, if we have in mind a likely range for the observed value of $\hat{p}$, then in view of Fig. 12.1 we should take as our educated guess for $\hat{p}$ the value in the range closest to 0.5. But in either case we should be aware that if the observed value of $\hat{p}$ is closer to 0.5 than is our educated guess, then the margin of error will be larger than desired.

FORMULA 12.2 **SAMPLE SIZE FOR ESTIMATING p**

A $(1-\alpha)$-level confidence interval for a population proportion having a margin of error of at most E can be obtained by choosing

$$n = 0.25\left(\frac{z_{\alpha/2}}{E}\right)^2,$$

rounded up to the nearest whole number. If we can make an educated guess, $\hat{p}_g$ (g for guess), for the observed value of $\hat{p}$, then we should instead choose

$$n = \hat{p}_g(1-\hat{p}_g)\left(\frac{z_{\alpha/2}}{E}\right)^2,$$

rounded up to the nearest whole number.

EXAMPLE 12.4 ILLUSTRATES FORMULA 12.2

Consider again the problem of estimating the proportion of all U.S. employees who play hooky.

a. Obtain a sample size that will ensure a margin of error of at most 0.01 for a 95% confidence interval.

b. Find a 95% confidence interval for p if for a sample of the size determined in part (a), the proportion of those who play hooky is 0.194.

c. Determine the margin of error for the estimate in part (b) and compare it to the margin of error specified in part (a).

d. Repeat parts (a)–(c) if it is deemed reasonable to presume that the proportion of those sampled who play hooky will be somewhere between 0.1 and 0.3.

e. Compare the results obtained in parts (a)–(c) with those obtained in part (d).

SOLUTION **a.** Here we apply the first displayed equation in Formula 12.2. In doing so, we must identify $z_{\alpha/2}$ and the margin of error, E. The confidence level is stipulated to be 0.95, so $z_{\alpha/2} = z_{0.05/2} = z_{0.025} = 1.96$; the margin of error is specified at 0.01. Thus a sample size that will ensure a margin of error of at most 0.01 for a 95% confidence interval is

$$n = 0.25\left(\frac{z_{\alpha/2}}{E}\right)^2 = 0.25\left(\frac{1.96}{0.01}\right)^2 = 9604.$$

If we take a sample of 9604 U.S. employees, then the margin of error for our estimate of the proportion of all U.S. employees who play hooky will be 0.01 or less, that is, at most plus or minus 1 percentage point.

b. We find, by applying Procedure 12.1 on page 728 with $\alpha = 0.05$, $n = 9604$, and $\hat{p} = 0.194$, that a 95% confidence interval for p has endpoints

$$0.194 \pm 1.96 \cdot \sqrt{(0.194)(1 - 0.194)/9604},$$

or 0.194 ± 0.008, or 0.186 to 0.202. We can be 95% confident that the percentage of all U.S. employees who play hooky is somewhere between 18.6% and 20.2%.

c. The margin of error for the estimate in part (b) is 0.008. Not surprisingly, this is less than the margin of error of 0.01 specified in part (a).

d. If it is reasonable to presume that the proportion of those sampled who play hooky will be somewhere between 0.1 and 0.3, then we should obtain the sample size using the second displayed equation in Formula 12.2 with $\hat{p}_g = 0.3$:

$$n = \hat{p}_g(1 - \hat{p}_g)\left(\frac{z_{\alpha/2}}{E}\right)^2 = (0.3)(1 - 0.3)\left(\frac{1.96}{0.01}\right)^2 = 8068 \text{ (rounded up)}.$$

Applying Procedure 12.1 with $\alpha = 0.05$, $n = 8068$, and $\hat{p} = 0.194$, we find that a 95% confidence interval for p has endpoints

$$0.194 \pm 1.96 \cdot \sqrt{(0.194)(1 - 0.194)/8068},$$

or 0.194 ± 0.009, or 0.185 to 0.203. We can be 95% confident that the percentage of all U.S. employees who play hooky is somewhere between 18.5% and 20.3%. The margin of error for the estimate is 0.009.

e. By employing the guess for $\hat{p}$ in part (d), we reduced the required sample size by more than 1500 (from 9604 to 8068). Moreover, only $\frac{1}{10}$ of 1% of precision was lost—the margin of error rose from 0.008 to 0.009. The risk of using the guess 0.3 for $\hat{p}$ is that if the observed value of $\hat{p}$ had turned out to be larger than 0.3 (but smaller than 0.7), then the achieved margin of error would have exceeded the specified 0.01. ◨

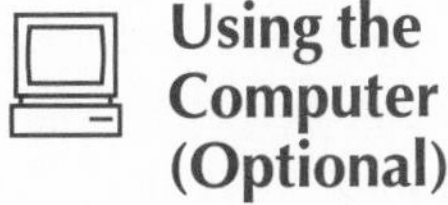

Using the Computer (Optional)

Procedure 12.1 on page 728 provides a step-by-step method for obtaining a confidence interval for a population proportion. Alternatively we can use Minitab. Actually, Minitab supplies two methods for obtaining a confidence interval for a population proportion, one using an exact procedure and the other using a normal approximation, as in Procedure 12.1. By default, Minitab uses the exact procedure.

Generally speaking, the exact method is impractical without access to statistical software and that is why we presented in Procedure 12.1 the method using the normal approximation. For large samples, there is little difference in the confidence intervals obtained by the two methods.

EXAMPLE 12.5 USING MINITAB TO OBTAIN A CONFIDENCE INTERVAL FOR A POPULATION PROPORTION

In Example 12.3 we applied Procedure 12.1 to determine a 95% confidence interval for the proportion of all U.S. employees who play hooky. Use Minitab to obtain that confidence interval.

SOLUTION We first recall that the sample size is 1010 and the number of employees sampled who play hooky is 202. Then we proceed as follows.

1 Choose **Stat ➤ Basic Statistics ➤ 1 Proportion...**

2 Select the **Summarized data** option button

3 Click in the **Number of trials** text box and type 1010
4 Click in the **Number of successes** text box and type 202
5 Click the **Options...** button
6 Click in the **Confidence level** text box and type 95
7 Click **OK**
8 Click **OK**

The resulting output is shown in Printout 12.2.

PRINTOUT 12.2
Minitab output for a confidence interval for one population proportion

```
Test of p = 0.5 vs p not = 0.5

                                                            Exact
Sample      X      N  Sample p         95.0 % CI          P-Value
1         202   1010  0.200000  (0.175739, 0.226023)        0.000
```

The output in Printout 12.2 provides both a hypothesis test and a confidence interval. We will address the hypothesis-test aspects of the output in Section 12.2.

In the columns headed `X`, `N`, and `Sample p`, we find, respectively, the number of successes, sample size, and sample proportion. Next we find the 95% confidence interval, which is from 0.175739 to 0.226023. We can be 95% confident that the proportion of all U.S. employees who play hooky is somewhere between 17.5739% and 22.6023%. Compare this confidence interval to the one we obtained in Example 12.3 using Procedure 12.1. Why the slight difference? □

EXERCISES 12.1

Statistical Concepts and Skills

12.1 In a newspaper or magazine of your choice, find a statistical study concerning the estimation of a population proportion.

12.2 Why do we generally resort to using statistical inference in order to obtain information about a population proportion?

12.3 Is a population proportion a parameter or a statistic? What about a sample proportion? Explain your answers.

12.4 This exercise concerns the basic notation and terminology for proportions.

a. What is a population proportion?
b. Identify the symbol used for a population proportion.
c. What is a sample proportion?
d. Identify the symbol used for a sample proportion.
e. For what is the phrase "number of successes" an abbreviation? Identify the symbol used for the number of successes.
f. For what is the phrase "number of failures" an abbreviation?
g. Explain the relationship between the sample proportion, the number of successes, and the sample size.

12.5 This exercise uses an unrealistically small population to provide a concrete illustration for the exact distribution of a sample proportion. A population consists of three men and two women. The first names of the men are Jose, Pete, and Carlo; the first names of the women are Gail and Frances. Suppose the specified attribute is "female."

a. Determine the population proportion, p.

b. The first column of the following table provides the possible samples of size two, where each person is represented by the first letter of his or her first name; the second column gives the number of successes—the number of females obtained—for each sample, and the third column shows the sample proportion. Complete the table.

Sample	Number of females x	Sample proportion $\hat{p}$
J, G	1	0.5
J, P	0	0.0
J, C	0	0.0
J, F	1	0.5
G, P		
G, C		
G, F		
P, C		
P, F		
C, F		

c. Construct a dotplot for the sampling distribution of the proportion for samples of size two. Mark the position of the population proportion on the dotplot.

d. Use the third column of the table to obtain the mean of the variable $\hat{p}$.

e. Compare your answers from parts (a) and (d). Why are they the same?

12.6 Repeat parts (b)–(e) of Exercise 12.5 for samples of size one.

12.7 Repeat parts (b)–(e) of Exercise 12.5 for samples of size three. (There are 10 possible samples.)

12.8 Repeat parts (b)–(e) of Exercise 12.5 for samples of size four. (There are five possible samples.)

12.9 Repeat parts (b)–(e) of Exercise 12.5 for samples of size five.

12.10 This exercise requires that you have done Exercises 12.5–12.9. What do your graphs in parts (c) of those exercises illustrate about the impact of increasing sample size on sampling error? Explain your answer.

12.11 According to *Directory of Governors of the American States, Commonwealths & Territories,* 64% of the 1997 U.S. governors were Republican.

a. Identify the population.

b. Identify the specified attribute.

c. Is the proportion 0.64 (64%) a population proportion or a sample proportion? Explain your answer.

12.12 *TELENATION/Market Facts Inc.,* reports that 83% of 18–24 year-olds go to the movies at least once a month.

a. Identify the population.

b. Identify the specified attribute.

c. Is the proportion 0.83 (83%) a population proportion or a sample proportion? Explain your answer.

12.13 In a 1998 poll conducted through the Web site excite.com, 452 of 779 respondents said that it makes sense to abolish the U.S. tax code and start all over.

a. Determine the margin of error for a 95% confidence interval.

b. Without doing any calculations, answer the following question and explain your answer: Will the margin of error be smaller or larger for a 90% confidence interval?

12.14 A *USA TODAY/CNN/Gallup Poll,* which was taken in January of 1998, reported that 623 of 1005 respondents approved of the way President Clinton was handling his job.

a. Determine the margin of error for a 90% confidence interval.

b. Without doing any calculations, answer the following question and explain your answer: Will the margin of error be smaller or larger for a 95% confidence interval?

12.15 In each of parts (a)–(f) of this exercise, we have given a likely range for the observed value of a sample proportion $\hat{p}$. Based on the given range, identify the educated guess that should be used for the observed value of $\hat{p}$ in calculating the required sample size for a prescribed confidence level and margin of error.

a. 0.2 to 0.4 **b.** 0.4 to 0.7 **c.** 0.7 or greater
d. 0.2 or less **e.** 0.4 or greater **f.** 0.7 or less
g. In each of parts (a)–(f), which observed values of the sample proportion will yield a larger margin of error than the one specified if the educated guess is used for the sample size computation?

In each of Exercises 12.16–12.19, apply Procedure 12.1 on page 728 to obtain the required confidence interval. Be sure to check the condition for using that procedure.

12.16 Studies are performed to estimate the percentage of the nation's 10 million asthmatics who are allergic to sulfites. In one survey, 38 of 500 randomly selected U.S. asthmatics were found to be allergic to sulfites.
a. Find a 95% confidence interval for the proportion of all U.S. asthmatics who are allergic to sulfites.
b. Interpret your result from part (a).

12.17 A *Reader's Digest/Gallup Survey* on the drinking habits of Americans estimated the percentage of adults across the country who drink beer, wine, or hard liquor, at least occasionally. Of the 1516 adults interviewed, 985 said they drank.
a. Determine a 95% confidence interval for the proportion of all Americans who drink beer, wine, or hard liquor, at least occasionally.
b. Interpret your result from part (a).

12.18 A *Gallup Poll* asked public-school teachers nationwide to grade their fellow educators' performance. The poll found that 634 of the 813 teachers surveyed gave their fellow educators an A or B grade.
a. Find a 90% confidence interval for the percentage of all public-school teachers who would give their fellow educators an A or B grade for performance.
b. Interpret your answer in part (a).

12.19 A *Gallup Poll* estimated the support among Americans for "right to die" laws. For the survey, 1528 adults were asked whether they favor voluntary withholding of life-support systems from the terminally ill. The results: 1238 said yes.
a. Determine a 99% confidence interval for the percentage of all adult Americans who are in favor of "right to die" laws.
b. Interpret your answer in part (a).

12.20 Suppose you have been hired to estimate the percentage of adults in your state who are literate. You take a random sample of 100 adults and find that 96 of them are literate. You then obtain a 95% confidence interval as follows:

$$0.96 \pm 1.96 \cdot \sqrt{(0.96)(0.04)/100},$$

or 0.922 to 0.998. From this you conclude that we can be 95% confident that the percentage of all adults in your state who are literate is somewhere between 92.2% and 99.8%. Is anything wrong with your reasoning?

12.21 Suppose I have been commissioned to estimate the infant mortality rate in Norway. From a random sample of 500 live births, I find that 0.8% of them resulted in infant deaths. I then construct a 90% confidence interval for the infant mortality rate in Norway:

$$0.008 \pm 1.645 \cdot \sqrt{(0.008)(0.992)/500},$$

or 0.001 to 0.015. Then I conclude that, "We can be 90% confident that the proportion of infant deaths in Norway is somewhere between 0.001 and 0.015." How did I do?

12.22 *Parade Magazine* conducted a nationwide survey of U.S. adults and reported that 82% of those polled said they have a "positive attitude" toward the police; the margin of error was plus or minus 1.5 percentage points (for a 0.95 confidence level). Use this information to obtain a 95% confidence interval for the percentage of all U.S. adults who would say they have a "positive attitude" toward the police.

12.23 A study of alternative-medicine use by Americans, directed by Dr. David M. Eisenberg of Boston's Beth Israel Hospital, appeared in a 1993 issue of the *New England Journal of Medicine*. The study revealed that 34% of the Americans surveyed used at least one unconventional therapy in 1990. The margin of error was plus or minus 2.4 percentage points (for a 0.95 confidence level). Use this information to obtain a 95% confidence interval for the percentage of all Americans who used at least one unconventional therapy in 1990.

12.24 Refer to Exercise 12.16, where you considered the problem of estimating the proportion of all U.S. asthmatics who are allergic to sulfites.
a. Determine the margin of error for the estimate of p.

b. Obtain a sample size that will ensure a margin of error of at most 0.01 for a 95% confidence interval without making a guess for the observed value of $\hat{p}$.
c. Find a 95% confidence interval for p if for a sample of the size determined in part (b), the proportion of asthmatics sampled who are allergic to sulfites is 0.071.
d. Determine the margin of error for the estimate in part (c) and compare it to the margin of error specified in part (b).
e. Repeat parts (b)–(d) if you deem it reasonable to presume that the proportion of asthmatics sampled who are allergic to sulfites will be at most 0.10.
f. Compare the results you obtained in parts (b)–(d) with those obtained in part (e).

12.25 Refer to Exercise 12.17, where a 95% confidence interval was determined for the proportion of U.S. adults who drink alcoholic beverages.
a. Determine the margin of error for the estimate of p.
b. Obtain a sample size that will ensure a margin of error of at most 0.02 for a 95% confidence interval without making a guess for the observed value of $\hat{p}$.
c. Find a 95% confidence interval for p if for a sample of the size determined in part (b), 63% of those sampled drink alcoholic beverages.
d. Determine the margin of error for the estimate in part (c) and compare it to the margin of error specified in part (b).
e. Repeat parts (b)–(d) if you deem it reasonable to presume that the percentage of adults sampled who drink alcoholic beverages will be at least 60%.
f. Compare the results you obtained in parts (b)–(d) with those obtained in part (e).

12.26 Refer to Exercise 12.18, where you obtained a 90% confidence interval for the percentage of all public-school teachers who would give their fellow educators an A or B grade for performance.
a. Find the margin of error for the estimate of the percentage.
b. Obtain a sample size that will ensure a margin of error of at most 1.5 percentage points for a 90% confidence interval without making a guess for the observed value of $\hat{p}$.
c. Find a 90% confidence interval for p if for a sample of the size determined in part (b), 79.2% of the public-school teachers sampled would give their fellow educators an A or B grade for performance.
d. Determine the margin of error for the estimate in part (c) and compare it to the margin of error specified in part (b).
e. Repeat parts (b)–(d) if you deem it reasonable to presume that the percentage of public-school teachers sampled who would give their fellow educators an A or B grade for performance will be between 70% and 85%.
f. Compare the results you obtained in parts (b)–(d) with those obtained in part (e).

12.27 Refer to Exercise 12.19, where you determined a 99% confidence interval for the percentage of adult Americans who are in favor of "right to die" laws.
a. Find the margin of error for the estimate of the percentage.
b. Obtain a sample size that will ensure a margin of error of at most 1 percentage point for a 99% confidence interval without making a guess for the observed value of $\hat{p}$.
c. Find a 99% confidence interval for p if for a sample of the size determined in part (b), 82.5% of the adult Americans sampled are in favor of "right to die" laws.
d. Determine the margin of error for the estimate in part (c) and compare it to the margin of error specified in part (b).
e. Repeat parts (b)–(d) if you deem it reasonable to presume that the percentage of adult Americans sampled who are in favor of "right to die" laws will be between 75% and 90%.
f. Compare the results you obtained in parts (b)–(d) with those obtained in part (e).

12.28 A company manufactures goods that are sold exclusively by mail order. The director of market research needed to test market a new product. She planned to send out brochures to a random sample of households and use the proportion of orders obtained as an estimate of the true proportion, known as the *product response rate.* The results of the market research were to be employed as a primary source for advance production planning and, consequently, the director wanted the figures she presented to be as accurate as possible. Specifically, she wanted to be 95% confident that the estimate of the product response rate would be accurate to within 1%.

a. Without making any assumptions, determine the sample size required.

b. Historically, product response rates for products sold by this company have ranged from 0.5% to 4.9%. If the director had been willing to assume that the sample product response rate for this product would also fall in that range, find the required sample size.

c. Compare the results from parts (a) and (b).

d. Discuss the possible consequences if the assumption made in part (b) turns out to be incorrect.

12.29 On Thursday, June 13, 1996, then-Arizona Governor Fife Symington was indicted on 23 counts of fraud and extortion. Just hours after the federal prosecutors announced the indictment, several polls were conducted of Arizonans asking whether they thought Symington should resign. One poll, conducted by Research Resources Inc., and appearing in *The Phoenix Gazette,* revealed that 58% of Arizonans felt that Symington should resign; it had a margin of error of plus or minus 4.9 percentage points. Another poll, conducted by Phoenix-based Behavior Research Center and appearing in the *Tempe Daily News,* reported that 54% of Arizonans felt that Symington should resign; it had a margin of error of plus or minus 4.4 percentage points. Is it possible that both of these polls were correct in their conclusions? Explain your answer.

EXTENDING THE CONCEPTS AND SKILLS

12.30 What important theorem in statistics implies that, for a large sample size, the possible sample proportions of that size have approximately a normal distribution?

12.31 In discussing the sample size required for obtaining a confidence interval with a prescribed confidence level and margin of error, we made the following statement: "If we have in mind a likely range for the observed value of $\hat{p}$, then in view of Fig. 12.1 we should take as our educated guess for $\hat{p}$ the value in the range closest to 0.5." Explain why this is so.

12.32 In discussing the sample size required for obtaining a confidence interval with a prescribed confidence level and margin of error, we made the following statement: "... we should be aware that if the observed value of $\hat{p}$ is closer to 0.5 than is our educated guess, then the margin of error will be larger than desired. Explain why this is so.

12.33 Consider a population in which the proportion of members having a specified attribute is equal to p. Let y be the variable whose value is 1 if a member has the specified attribute and 0 if a member does not.

a. If the size of the population is N, how many members of the population have the specified attribute?

b. Use part (a) and Definition 3.11 on page 175 to show that $\mu_y = p$.

c. Use part (b) and the computing formula in Definition 3.12 on page 178 to show that $\sigma_y = \sqrt{p(1-p)}$.

d. Explain why $\overline{y} = \hat{p}$.

e. Use parts (b)–(d) and Key Fact 7.4 on page 429, to justify Key Fact 12.1.

USING TECHNOLOGY

 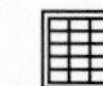 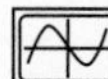

12.34 Refer to Exercise 12.16.

a. Use Minitab or some other statistical software to obtain the confidence interval required in part (a) of Exercise 12.16.

b. Compare your answer to the one you obtained in Exercise 12.16. Explain any discrepancy that you observe.

12.35 Refer to Exercise 12.17.

a. Use Minitab or some other statistical software to obtain the confidence interval required in part (a) of Exercise 12.17.

b. Compare your answer to the one you obtained in Exercise 12.17. Explain any discrepancy that you observe.

12.36 In 1998, Arizona lawmakers were discussing the imposition of a law banning smoking in virtually all indoor workspaces, bars and restaurants included. A poll was conducted to measure the pulse of Arizonans on this issue. Each person in the survey was asked whether he or she would favor such a law. Printout 12.3 shows the results of applying Minitab's confidence interval procedure for a population proportion to the survey data. Use the output to determine

a. the number of people polled.

b. the number polled who would favor the law.

c. the percentage polled who would favor the law.

d. a 90% confidence interval for the percentage of all Arizonans who would favor the law.

12.37 A 1996 *Associated Press* article entitled "Animal rights gain support, poll shows," reported on a survey taken by ICR Survey Research Group. In the survey, a sample of Americans were asked whether they agree with a basic tenet of the animal-rights movement: "An animal's right to live free of suffering should be just as important as a person's right to live free of suffering." We applied Minitab's confidence interval procedure for a population proportion to the survey data and obtained the output shown in Printout 12.4. Use the output to determine

a. the number of people polled.

b. the number of people polled who agreed with the basic tenet of the animal-rights movement.

c. the percentage of people polled who agreed with the basic tenet of the animal-rights movement.

d. a 95% confidence interval for the percentage of all Americans who agree with the basic tenet of the animal-rights movement.

12.38 Refer to Exercise 12.20.

a. Use Minitab or some other statistical software to obtain the required 95% confidence interval based on an exact method.

b. Compare the exact confidence interval found in part (a) to the one determined in Exercise 12.20 using the normal approximation (i.e., Procedure 12.1).

c. Explain the discrepancy between the two methods.

12.39 Refer to Exercise 12.21.

a. Use Minitab or some other statistical software to obtain the required 90% confidence interval based on an exact method.

b. Compare the exact confidence interval found in part (a) to the one determined in Exercise 12.21 using the normal approximation (i.e., Procedure 12.1).

c. Explain the discrepancy between the two methods.

PRINTOUT 12.3 Minitab output for Exercise 12.36

```
Test of p = 0.5 vs p not = 0.5

                                                      Exact
Sample       X       N  Sample p        90.0 % CI    P-Value
1           96     168  0.571429  (0.505075, 0.635871)   0.076
```

PRINTOUT 12.4 Minitab output for Exercise 12.37

```
Test of p = 0.5 vs p not = 0.5

                                                      Exact
Sample       X       N  Sample p        95.0 % CI    P-Value
1          670    1004  0.667331  (0.637230, 0.696447)   0.000
```

12.2 HYPOTHESIS TESTS FOR ONE POPULATION PROPORTION

In Section 12.1 we discovered how to obtain confidence intervals for a population proportion. Now we will learn how to perform hypothesis tests for a population proportion. The procedure we present is actually a special case of the one-sample z-test for a population mean, Procedure 9.1 on page 513.

From Key Fact 12.1 on page 726, we can deduce that for large n, the standardized version of $\hat{p}$,

$$z = \frac{\hat{p} - p}{\sqrt{p(1-p)/n}},$$

has approximately the standard normal distribution. Consequently, to perform a large-sample hypothesis test with null hypothesis H_0: $p = p_0$, we can use the variable

$$z = \frac{\hat{p} - p_0}{\sqrt{p_0(1-p_0)/n}}$$

as the test statistic and obtain the critical value(s) from the standard-normal table, Table II. Specifically, we have the following procedure, which we often refer to as the **one-sample z-test** for a population proportion or, more briefly, as the **z-test** for a population proportion.

PROCEDURE 12.2

THE ONE-SAMPLE z-TEST FOR A POPULATION PROPORTION

ASSUMPTION

Both np_0 and $n(1 - p_0)$ are 5 or greater.

Step 1 The null hypothesis is H_0: $p = p_0$ and the alternative hypothesis is one of the following:

H_a: $p \neq p_0$ (Two-tailed)	or	H_a: $p < p_0$ (Left-tailed)	or	H_a: $p > p_0$ (Right-tailed)

Step 2 Decide on the significance level, α.

Step 3 The critical value(s) are

$\pm z_{\alpha/2}$ (Two-tailed)	or	$-z_{\alpha}$ (Left-tailed)	or	z_{α} (Right-tailed)

Use Table II to find the critical value(s).

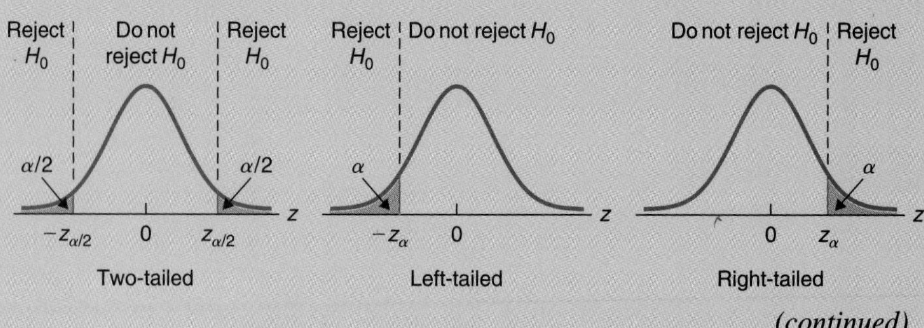

(continued)

Step 4 Compute the value of the test statistic

$$z = \frac{\hat{p} - p_0}{\sqrt{p_0(1 - p_0)/n}}.$$

Step 5 If the value of the test statistic falls in the rejection region, reject H_0; otherwise, do not reject H_0.

Step 6 State the conclusion in words.

EXAMPLE 12.6 ILLUSTRATES PROCEDURE 12.2

One of the more controversial issues in the United States is gun control; there are many avid proponents and opponents of banning handgun sales. In a survey conducted by Louis Harris of LH Research, 1250 U.S. adults were polled regarding their view on banning handgun sales. The results were that 650 of those sampled favored a ban. At the 5% significance level, do the data provide sufficient evidence to conclude that a majority of U.S. adults (i.e., more than 50%) favor banning handgun sales?

SOLUTION We will apply Procedure 12.2 to perform the required hypothesis test. But first we must check that the condition for its use is met. We have $n = 1250$ and $p_0 = 0.50$ (50%). Therefore,

$$np_0 = 1250 \cdot 0.50 = 625 \qquad \text{and} \qquad n(1 - p_0) = 1250 \cdot (1 - 0.50) = 625.$$

Since both np_0 and $n(1 - p_0)$ exceed 5, we can employ Procedure 12.2.

Step 1 *State the null and alternative hypotheses.*

Let p denote the proportion of all U.S. adults that favor banning handgun sales. Then the null and alternative hypotheses are

H_0: $p = 0.50$ (it is not true that a majority favor a ban)

H_a: $p > 0.50$ (a majority favor a ban).

Note that the hypothesis test is right-tailed since a greater-than sign (>) appears in the alternative hypothesis.

Step 2 *Decide on the significance level, α.*

We are to perform the test at the 5% significance level. So $\alpha = 0.05$.

Step 3 *The critical value for a right-tailed test is z_α.*

Since $\alpha = 0.05$, the critical value is $z_{0.05} = 1.645$, as seen in Fig. 12.2.

FIGURE 12.2
Criterion for deciding whether or not to reject the null hypothesis

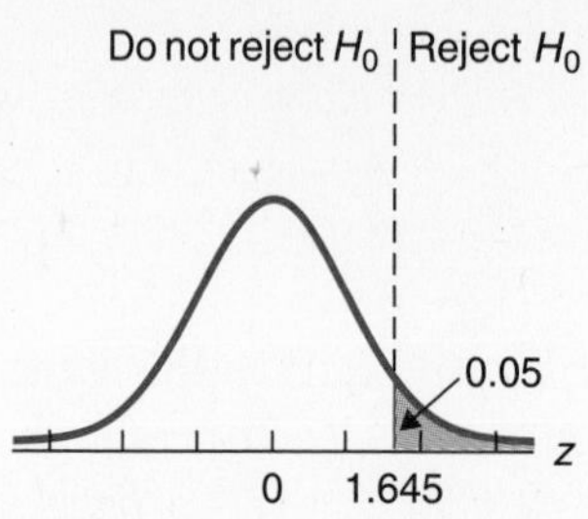

Step 4 *Compute the value of the test statistic*

$$z = \frac{\hat{p} - p_0}{\sqrt{p_0(1 - p_0)/n}}.$$

We have $n = 1250$ and $p_0 = 0.50$. The number of U.S. adults surveyed who favor banning handgun sales is 650. Therefore the proportion of those surveyed who favor a ban is $\hat{p} = x/n = 650/1250 = 0.520$ (52.0%). Consequently, the value of the test statistic is

$$z = \frac{0.520 - 0.50}{\sqrt{(0.50)(1 - 0.50)/1250}} = 1.41.$$

Step 5 *If the value of the test statistic falls in the rejection region, reject H_0; otherwise, do not reject H_0.*

From Step 4, the value of the test statistic is $z = 1.41$, which as we see from Fig. 12.2 does not fall in the rejection region. Thus we do not reject H_0.

Step 6 *State the conclusion in words.*

 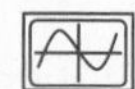

The test results are not statistically significant at the 5% level; that is, at the 5% significance level, the data do not provide sufficient evidence to conclude that a majority of U.S. adults favor banning handgun sales. ◨

Example 12.6 also provides a good illustration of how statistical results are sometimes misstated. The newspaper article we read that featured the survey had the headline "Fear prompts 52% in U.S. to back pistol-sale ban, poll says." In fact, the poll says no such thing. It says only that 52% of those *sampled* back a pistol-sale ban; and as we have seen, at the 5% significance level, those data do not provide sufficient evidence to conclude that even a majority of U.S. adults back a pistol-sale ban.

Using the P-Value Approach

By using the P-value approach to hypothesis testing, we can assess more precisely the evidence against the null hypothesis. From Step 4 of Example 12.6, we know that the value of the test statistic is $z = 1.41$. Consulting Table II, we find that $P = 0.0793$.

So we cannot reject H_0 at the 5% significance level, although we can reject it at the 8% significance level or for that matter at any level greater than or equal to 7.93%. Moreover, according to Table 9.12 on page 543, the data do provide moderate (but not strong) evidence against the null hypothesis and thereby for the alternative hypothesis that a majority of U.S. adults favor banning handgun sales.

Using the Computer (Optional) We can use Minitab to perform a hypothesis test for a population proportion. To illustrate, we return to the hypothesis test concerning the banning of handgun sales considered in the previous example.

EXAMPLE 12.7 USING MINITAB TO PERFORM A HYPOTHESIS TEST FOR A POPULATION PROPORTION

Use Minitab to perform the hypothesis test in Example 12.6 on page 741.

SOLUTION Let p denote the proportion of all U.S. adults that favor banning handgun sales. The problem is to perform the hypothesis test

H_0: $p = 0.50$ (it is not true that a majority favor a ban)

H_a: $p > 0.50$ (a majority favor a ban)

at the 5% significance level. Note that the hypothesis test is right-tailed since a greater-than sign (>) appears in the alternative hypothesis.

First we recall that 1250 U.S. adults were polled and that 650 of those adults favored banning handgun sales. Then we proceed as follows.

1 Choose **Stat ➤ Basic Statistics ➤ 1 Proportion...**
2 Select the **Summarized data** option button
3 Click in the **Number of trials** text box and type 1250
4 Click in the **Number of successes** text box and type 650
5 Click the **Options...** button
6 Click in the **Test proportion** text box and type 0.5
7 Click the arrow button at the right of the **Alternative** drop-down list box and select **greater than**
8 Click **OK**
9 Click **OK**

The resulting output is shown in Printout 12.5.

PRINTOUT 12.5
Minitab output for a hypothesis test for one population proportion

```
Test of p = 0.5 vs p > 0.5

                                                              Exact
Sample      X      N  Sample p         95.0 % CI            P-Value
1         650   1250  0.520000  (0.491885, 0.548021)         0.083
```

The output in Printout 12.5 provides both a hypothesis test and a confidence interval. We will concentrate on the hypothesis-test aspects.

The first line of the output presents the null and alternative hypotheses. Next, in the columns headed `X`, `N`, and `Sample p`, we find, respectively, the number of successes, sample size, and sample proportion. And the final item, headed `Exact P-Value`, gives the P-value for the hypothesis test.

Since the P-value of 0.083 exceeds the specified significance level of 0.05, we do not reject H_0. At the 5% significance level, the data do not provide sufficient evidence to conclude that a majority of U.S. adults favor banning handgun sales. □

EXERCISES 12.2

STATISTICAL CONCEPTS AND SKILLS

12.40 Of what procedure is Procedure 12.2 a special case? Why do you think that is so?

In Exercises 12.41–12.46, perform each hypothesis test using either the critical-value approach or the P-value approach. Comment on the practical significance of all tests whose results are statistically significant.

12.41 *The Arizona Republic* conducted a telephone poll of 758 Arizona adults who celebrate Christmas. The question asked was, "In your family, do you open presents on Christmas Eve or Christmas Day?" Of those surveyed, 394 said they wait until Christmas Day.

a. Determine the sample proportion.

b. At the 5% significance level, do the data provide sufficient evidence to conclude that a majority of Arizona families who celebrate Christmas wait until Christmas Day to open their presents?

12.42 In a 1998 poll conducted through the Web site excite.com, 4381 of 7553 respondents said that Independent Prosecutor Kenneth Starr should question all available witnesses, including President Clinton's aides.

a. Determine the sample proportion.

b. Do the data provide sufficient evidence to conclude that a majority of people believed that Starr should question all available witnesses, including President Clinton's aides? Perform the required hypothesis test at the 5% significance level.

12.43 The U.S. Substance Abuse and Mental Health Services Administration conducts surveys on drug use by type of drug and age group. Results are published in *National Household Survey on Drug Abuse.* According to that publication, 12.0% of 18–25 year olds were current users of marijuana or hashish in 1995. A recent poll of 1283 randomly selected 18–25 year olds revealed that 146 currently use marijuana or hashish. At the 10% significance level, do the data provide sufficient evidence to conclude that the percentage of 18–25 year olds who currently use marijuana or hashish has changed from the 1995 percentage of 12.0%?

12.44 In 1995, 10.8% of all U.S. families had incomes below the poverty level, as reported by the Census Bureau in *Current Population Reports.* During that same year, of 404 randomly selected families whose householder had a Bachelor's degree or more, 10 had incomes below the poverty level. At the 1% significance level, do the data provide sufficient evidence to conclude that in 1995, families whose householder had a Bachelor's degree or more had a lower percentage earning incomes below the poverty level than the national percentage of 10.8%?

12.45 The Chicago Title Insurance Company publishes statistics on recent home buyers in *The Guarantor.* According to that publication, 83.1% of home buyers in 1995 purchased single-family houses. Out of 2544 randomly selected home buyers for this year, 2081 purchased single-family houses. Do the data provide sufficient evidence to conclude that this year's percentage of home buyers purchasing single-family houses has decreased from the 1995 figure of 83.1%? Use $\alpha = 0.05$.

12.46 Of the 38 numbers on a roulette wheel, 18 are red, 18 are black, and 2 are green. If the wheel is balanced, the probability of the ball landing on red is $\frac{18}{38} = 0.474$. A gambler has been studying a roulette wheel. If the wheel is out of balance, then he can improve his odds of winning. The gambler observes 200 spins of the wheel and finds that the ball lands on red 93 times. At the 10% significance level, do the data provide sufficient evidence to conclude that the ball is not landing on red the correct percentage of the time for a balanced wheel?

USING TECHNOLOGY  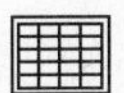 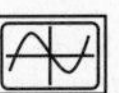

12.47 Use Minitab or some other statistical software to perform the hypothesis test in Exercise 12.43.

12.48 Use Minitab or some other statistical software to perform the hypothesis test in Exercise 12.44.

12.49 A 1998 poll taken through the World Wide Web site excite.com asked whether it makes sense to abolish the U.S. tax code and start all over. Printout 12.6 shown on the following page presents the results of applying Minitab's hypothesis test procedure for a population proportion to the survey data. Use the output to determine the

a. null and alternative hypotheses.
b. number of people in the poll.
c. number of people in the poll who thought it makes sense to abolish the U.S. tax code and start all over.
d. percentage of people in the poll who thought it makes sense to abolish the U.S. tax code and start all over.
e. P-value for the hypothesis test.
f. conclusion if the test is performed at the 1% significance level.

12.50 In 1995, the U.S. Bureau of the Census reported in *Health Insurance Coverage* that 15.4% of Americans are not covered by health insurance. That same year, a sample of South Carolina residents were asked whether they were covered by health insurance. We applied Minitab's hypothesis test procedure for a population proportion to the data and obtained Printout 12.7 shown on the following page. In the analysis, not having health insurance is the specified attribute. Use the output to determine the

a. null and alternative hypotheses.
b. number of people surveyed.
c. number of people surveyed who did not have health insurance.
d. percentage of people surveyed who did not have health insurance.
e. P-value for the hypothesis test.
f. conclusion if the test is performed at the 5% significance level.

12.51 Regarding the hypothesis test on banning handgun sales considered earlier in this section: Following Example 12.6, we stated that the P-value of the hypothesis test is 0.0793. But the Minitab output in Printout 12.5 shows a P-value of 0.083. Explain the discrepancy between the two P-values.

PRINTOUT 12.6 Minitab output for Exercise 12.49

```
Test of p = 0.5 vs p > 0.5

                                                      Exact
Sample     X      N  Sample p         95.0 % CI      P-Value
1        452    779  0.580231  (0.544683, 0.615169)   0.000
```

PRINTOUT 12.7 Minitab output for Exercise 12.50

```
Test of p = 0.154 vs p not = 0.154

                                                      Exact
Sample     X      N  Sample p         95.0 % CI      P-Value
1        544   3728  0.145923  (0.134737, 0.157666)   0.174
```

12.3 INFERENCES FOR TWO POPULATION PROPORTIONS USING INDEPENDENT SAMPLES

In Sections 12.1 and 12.2 we studied inferences for one population proportion. Now we will examine inferences for comparing two population proportions. In this setting, we have two populations and one specified attribute; the problem is to compare the proportion of one population having the specified attribute to the proportion of the other population having the specified attribute. We begin by discussing hypothesis testing.

EXAMPLE 12.8 INTRODUCES HYPOTHESIS TESTS FOR TWO POPULATION PROPORTIONS

The U.S. National Center for Health Statistics annually conducts the National Health Interview Survey (NHIS). Information on cigarette smoking is published in *Health, United States.* The following data are based on results obtained in the NHIS.

Independent random samples of 2235 U.S. women and 2065 U.S. men were obtained in order to compare the percentage of women who smoke cigarettes to the percentage of men who smoke cigarettes. Of the women sampled, 503 smoked; of the men sampled, 572 smoked. Do the data provide sufficient evidence to conclude that, in the United States, the percentage of women who smoke cigarettes is smaller than the percentage of men who smoke cigarettes?

SOLUTION We first note that the specified attribute is "smokes cigarettes" and the two populations are:

Population 1: All U.S. women.

Population 2: All U.S. men.

Let p_1 and p_2 denote the population proportions of cigarette smokers for the two populations:

p_1 = proportion of all U.S. women who smoke.

p_2 = proportion of all U.S. men who smoke.

We want to perform the hypothesis test

H_0: $p_1 = p_2$ (percentage of female smokers is not less)

H_a: $p_1 < p_2$ (percentage of female smokers is less).

Roughly speaking, the hypothesis test can be carried out as follows:

1. Compute the proportion of the women sampled who smoke cigarettes, $\hat{p}_1$, and the proportion of the men sampled who smoke cigarettes, $\hat{p}_2$.
2. If $\hat{p}_1$ is too much smaller than $\hat{p}_2$, reject H_0; otherwise, do not reject H_0.

The first step is easy. Since 503 of the 2235 women sampled and 572 of the 2065 men sampled were found to be smokers,

$$\hat{p}_1 = \frac{x_1}{n_1} = \frac{503}{2235} = 0.225\ (22.5\%)$$

and

$$\hat{p}_2 = \frac{x_2}{n_2} = \frac{572}{2065} = 0.277\ (27.7\%).$$

For the second step, we must decide whether the sample proportion $\hat{p}_1 = 0.225$ is less than the sample proportion $\hat{p}_2 = 0.277$ by a sufficient amount to warrant rejecting the null hypothesis in favor of the alternative hypothesis. In other words, we need to decide whether the difference between the two sample proportions can reasonably be attributed to sampling error or whether it indicates that the percentage of women who smoke is less than the percentage of men who smoke.

To make that decision, we need to know the distribution of the difference between two sample proportions—the **sampling distribution of the difference between two proportions.** We will discuss that sampling distribution and then complete the hypothesis test. ❐

The Sampling Distribution of the Difference Between Two Proportions for Large and Independent Samples

To begin our discussion of the sampling distribution of the difference between two proportions, we summarize the required notation in Table 12.2.

TABLE 12.2
Notation for parameters and statistics when considering two population proportions

	Population 1	Population 2
Population proportion	p_1	p_2
Sample size	n_1	n_2
Number of successes	x_1	x_2
Sample proportion	$\hat{p}_1$	$\hat{p}_2$

Recall that the *number of successes* refers to the number of members sampled that have the specified attribute. Consequently, the sample proportions are computed using the formulas

$$\hat{p}_1 = \frac{x_1}{n_1} \quad \text{and} \quad \hat{p}_2 = \frac{x_2}{n_2}.$$

Armed with the notation in Table 12.2, we will now describe in Key Fact 12.2 the sampling distribution of the difference between two proportions. In understanding Key Fact 12.2, it is helpful to recall Key Fact 12.1 on page 726, which gives the sampling distribution of one proportion.

KEY FACT 12.2 **THE SAMPLING DISTRIBUTION OF THE DIFFERENCE BETWEEN TWO PROPORTIONS FOR INDEPENDENT SAMPLES**

For independent samples of sizes n_1 and n_2 from the two populations,

- $\mu_{\hat{p}_1-\hat{p}_2} = p_1 - p_2$,
- $\sigma_{\hat{p}_1-\hat{p}_2} = \sqrt{p_1(1-p_1)/n_1 + p_2(1-p_2)/n_2}$, and
- $\hat{p}_1 - \hat{p}_2$ is approximately normally distributed for large n_1 and n_2.

In particular, for large samples, the possible differences between the two sample proportions have approximately a normal distribution with mean $p_1 - p_2$ and standard deviation $\sqrt{p_1(1-p_1)/n_1 + p_2(1-p_2)/n_2}$.

Key Fact 12.2 provides the necessary theory for deriving inferential procedures to compare two population proportions.

Large-Sample Hypothesis Tests for Two Population Proportions Using Independent Samples

We will now develop a hypothesis-testing procedure for comparing two population proportions. Our immediate goal is to identify a variable that can be used as the test statistic. From Key Fact 12.2, we know that for large, independent samples, the variable

$$z = \frac{(\hat{p}_1 - \hat{p}_2) - (p_1 - p_2)}{\sqrt{p_1(1 - p_1)/n_1 + p_2(1 - p_2)/n_2}} \tag{2}$$

has approximately the standard normal distribution.

The null hypothesis for a hypothesis test to compare two population proportions is

$$H_0\text{: } p_1 = p_2 \text{ (population proportions are equal).}$$

If the null hypothesis is true, then $p_1 - p_2 = 0$, and so the variable in Equation (2) becomes

$$z = \frac{\hat{p}_1 - \hat{p}_2}{\sqrt{p(1 - p)/n_1 + p(1 - p)/n_2}}, \tag{3}$$

where p denotes the common value of p_1 and p_2. Factoring $p(1 - p)$ out of the denominator of Equation (3) yields the variable

$$z = \frac{\hat{p}_1 - \hat{p}_2}{\sqrt{p(1 - p)}\sqrt{(1/n_1) + (1/n_2)}}. \tag{4}$$

We cannot use this variable as the test statistic since p is unknown.

Consequently, we must estimate p using sample information. The best estimate of p is obtained by pooling the data to get the proportion of successes in both samples combined; that is, we estimate p by

$$\hat{p}_p = \frac{x_1 + x_2}{n_1 + n_2}.$$

We call $\hat{p}_p$ the **pooled sample proportion.**

Replacing p in Equation (4) by its estimate $\hat{p}_p$ yields the variable

$$\frac{\hat{p}_1 - \hat{p}_2}{\sqrt{\hat{p}_p(1 - \hat{p}_p)}\sqrt{(1/n_1) + (1/n_2)}},$$

which can be used as the test statistic and, like the variable in Equation (4), has approximately the standard normal distribution for large samples if the null hypothesis is true. Therefore we have the following procedure, which we call the **two-sample z-test** for two population proportions.

PROCEDURE 12.3 **THE TWO-SAMPLE z-TEST FOR TWO POPULATION PROPORTIONS**

ASSUMPTIONS

1. Independent samples
2. x_1, $n_1 - x_1$, x_2, and $n_2 - x_2$ are all 5 or greater.

Step 1 The null hypothesis is H_0: $p_1 = p_2$ and the alternative hypothesis is one of the following:

H_a: $p_1 \neq p_2$ (Two-tailed)	or	H_a: $p_1 < p_2$ (Left-tailed)	or	H_a: $p_1 > p_2$ (Right-tailed)

Step 2 Decide on the significance level, α.

Step 3 The critical value(s) are

$\pm z_{\alpha/2}$ (Two-tailed)	or	$-z_{\alpha}$ (Left-tailed)	or	z_{α} (Right-tailed)

Use Table II to find the critical value(s).

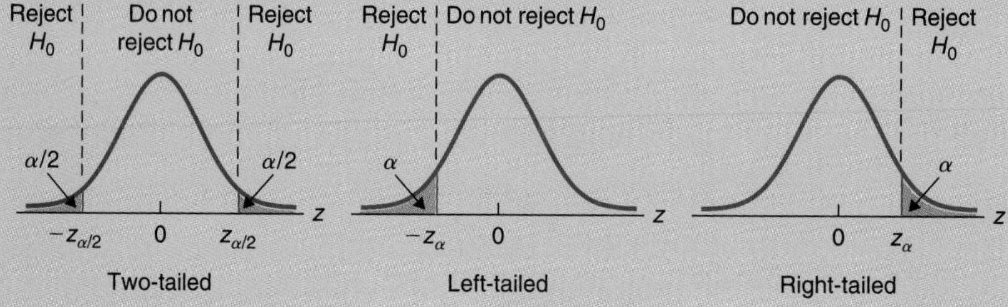

Step 4 Compute the value of the test statistic

$$z = \frac{\hat{p}_1 - \hat{p}_2}{\sqrt{\hat{p}_p(1 - \hat{p}_p)}\sqrt{(1/n_1) + (1/n_2)}},$$

where $\hat{p}_p = (x_1 + x_2)/(n_1 + n_2)$.

Step 5 If the value of the test statistic falls in the rejection region, reject H_0; otherwise, do not reject H_0.

Step 6 State the conclusion in words.

EXAMPLE 12.9 ILLUSTRATES PROCEDURE 12.3

We now return to the problem posed in Example 12.8. Independent random samples of 2235 U.S. women and 2065 U.S. men were obtained in order to compare the percentage of women who smoke cigarettes to the percentage of men who smoke cigarettes. Of the women sampled, 503 smoked; of the men sampled, 572 smoked. Do the data provide sufficient evidence to conclude that, in the United States, the percentage of women who smoke cigarettes is smaller than the percentage of men who smoke cigarettes? Conduct the hypothesis test at the 5% level of significance.

SOLUTION We apply Procedure 12.3, noting first that both assumptions for its use are satisfied.

Step 1 *State the null and alternative hypotheses.*

Let p_1 and p_2 denote, respectively, the proportions of all U.S. women and all U.S. men who smoke cigarettes. Then the null and alternative hypotheses are

$$H_0\colon\ p_1 = p_2 \text{ (percentage of female smokers is not less)}$$
$$H_a\colon\ p_1 < p_2 \text{ (percentage of female smokers is less).}$$

Note that the hypothesis test is left-tailed since a less-than sign (<) appears in the alternative hypothesis.

Step 2 *Decide on the significance level, α.*

The test is to be performed at the 5% significance level; so $\alpha = 0.05$.

Step 3 *The critical value for a left-tailed test is $-z_\alpha$.*

Since $\alpha = 0.05$, the critical value is $-z_{0.05} = -1.645$, as seen in Fig. 12.3.

FIGURE 12.3
Criterion for deciding whether or not to reject the null hypothesis

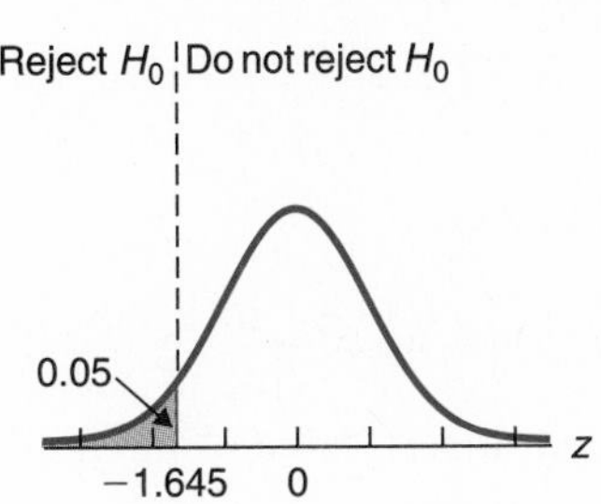

Step 4 *Compute the value of the test statistic*

$$z = \frac{\hat{p}_1 - \hat{p}_2}{\sqrt{\hat{p}_p(1-\hat{p}_p)}\sqrt{(1/n_1)+(1/n_2)}},$$

where $\hat{p}_p = (x_1 + x_2)/(n_1 + n_2)$.

We first obtain $\hat{p}_1$, $\hat{p}_2$, and $\hat{p}_p$. Since 503 of the 2235 women sampled and 572 of the 2065 men sampled were found to be cigarette smokers, we have $x_1 = 503$, $n_1 = 2235$, $x_2 = 572$, and $n_2 = 2065$. Therefore

$$\hat{p}_1 = \frac{x_1}{n_1} = \frac{503}{2235} = 0.225, \qquad \hat{p}_2 = \frac{x_2}{n_2} = \frac{572}{2065} = 0.277,$$

and

$$\hat{p}_p = \frac{x_1 + x_2}{n_1 + n_2} = \frac{503 + 572}{2235 + 2065} = \frac{1075}{4300} = 0.250.$$

Consequently, the value of the test statistic is

$$z = \frac{\hat{p}_1 - \hat{p}_2}{\sqrt{\hat{p}_p(1 - \hat{p}_p)}\sqrt{(1/n_1) + (1/n_2)}}$$

$$= \frac{0.225 - 0.277}{\sqrt{(0.250)(1 - 0.250)}\sqrt{(1/2235) + (1/2065)}} = -3.93.$$

Step 5 *If the value of the test statistic falls in the rejection region, reject H_0; otherwise, do not reject H_0.*

From Step 4 the value of the test statistic is $z = -3.93$, which, as we see from Fig. 12.3, falls in the rejection region. Thus we reject H_0.

Step 6 *State the conclusion in words.*

The test results are statistically significant at the 5% level; that is, at the 5% significance level, the data provide sufficient evidence to conclude that, in the United States, the percentage of women who smoke cigarettes is smaller than the percentage of men who smoke cigarettes. ◨

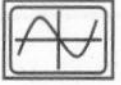

Using the *P*-Value Approach

We can also carry out the hypothesis test in Example 12.9 using the P-value approach. From Step 4 we see that the value of the test statistic is $z = -3.93$. Consulting Table II, we find that the P-value of the hypothesis test is zero to four decimal places.

Since the P-value is less than the specified significance level of 0.05, we can reject the null hypothesis. Moreover, by referring to Table 9.12 on page 543, we see that the data provide very strong evidence against the null hypothesis and hence in favor of the alternative hypothesis that a smaller percentage of women smoke than men.

Large-Sample Confidence Intervals for the Difference Between Two Population Proportions

Key Fact 12.2 on page 748 can also be used to derive the following confidence-interval procedure for the difference between two population proportions, a procedure we call the **two-sample z-interval procedure** for two population proportions.

PROCEDURE 12.4

THE TWO-SAMPLE z-INTERVAL PROCEDURE FOR TWO POPULATION PROPORTIONS

ASSUMPTIONS

1. Independent samples

2. $x_1, n_1 - x_1, x_2$, and $n_2 - x_2$ are all 5 or greater.

Step 1 For a confidence level of $1 - \alpha$, use Table II to find $z_{\alpha/2}$.

Step 2 The endpoints of the confidence interval for $p_1 - p_2$ are

$$(\hat{p}_1 - \hat{p}_2) \pm z_{\alpha/2} \cdot \sqrt{\hat{p}_1(1 - \hat{p}_1)/n_1 + \hat{p}_2(1 - \hat{p}_2)/n_2}.$$

EXAMPLE 12.10 ILLUSTRATES PROCEDURE 12.4

Use the smoking data from Example 12.9 to obtain a 90% confidence interval for the difference, $p_1 - p_2$, between the proportions of U.S. women and men who smoke cigarettes.

SOLUTION We apply Procedure 12.4, noting first that both conditions for its use are met.

Step 1 *For a confidence level of* $1 - \alpha$, *use Table II to find* $z_{\alpha/2}$.

For a 90% confidence interval, we have $\alpha = 0.10$. Consulting Table II, we find that $z_{\alpha/2} = z_{0.10/2} = z_{0.05} = 1.645$.

Step 2 *The endpoints of the confidence interval for* $p_1 - p_2$ *are*

$$(\hat{p}_1 - \hat{p}_2) \pm z_{\alpha/2} \cdot \sqrt{\hat{p}_1(1 - \hat{p}_1)/n_1 + \hat{p}_2(1 - \hat{p}_2)/n_2}.$$

From Step 1, we have $z_{\alpha/2} = 1.645$. Referring to Example 12.9, we find that $\hat{p}_1 = 0.225$, $n_1 = 2235$, $\hat{p}_2 = 0.277$, and $n_2 = 2065$. Therefore the endpoints of the 90% confidence interval for $p_1 - p_2$ are

$$(0.225 - 0.277) \pm 1.645 \cdot \sqrt{(0.225)(1 - 0.225)/2235 + (0.277)(1 - 0.277)/2065}$$

or -0.052 ± 0.022, or -0.074 to -0.030. We can be 90% confident that, in the United States, the percentage of men who smoke cigarettes exceeds the percentage of women who smoke cigarettes by somewhere between 3.0 and 7.4 percentage points.

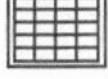
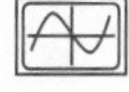

Margin of Error and Sample Size

The **margin of error** in estimating the difference between two population proportions can be obtained by referring to Step 2 of Procedure 12.4. And from the formula for the margin of error, we can determine the sample sizes required to obtain a confidence interval with a specified confidence level and margin of error. Formula 12.3 supplies the formulas.

FORMULA 12.3 **MARGIN OF ERROR AND SAMPLE SIZE FOR ESTIMATING $p_1 - p_2$**

The margin of error for the estimate of $p_1 - p_2$ is

$$E = z_{\alpha/2} \cdot \sqrt{\hat{p}_1(1-\hat{p}_1)/n_1 + \hat{p}_2(1-\hat{p}_2)/n_2}.$$

It equals half the length of the confidence interval and represents the precision with which the difference between the sample proportions, $\hat{p}_1 - \hat{p}_2$, estimates the difference between the population proportions, $p_1 - p_2$, at the specified confidence level.

A $(1-\alpha)$-level confidence interval for the difference between two population proportions having a margin of error of at most E can be obtained by choosing

$$n_1 = n_2 = 0.5\left(\frac{z_{\alpha/2}}{E}\right)^2,$$

rounded up to the nearest whole number. If we can make educated guesses, $\hat{p}_{1g}$ and $\hat{p}_{2g}$, for the observed values of $\hat{p}_1$ and $\hat{p}_2$, then we should instead choose

$$n_1 = n_2 = \left(\hat{p}_{1g}(1-\hat{p}_{1g}) + \hat{p}_{2g}(1-\hat{p}_{2g})\right)\left(\frac{z_{\alpha/2}}{E}\right)^2,$$

rounded up to the nearest whole number.

The second displayed formula in Formula 12.3 provides sample sizes that ensure we will obtain a $(1-\alpha)$-level confidence interval with a margin of error of at most E, but it may yield sample sizes that are unnecessarily large. The third displayed formula in Formula 12.3 yields smaller sample sizes, but should not be used unless the guesses for the sample proportions are considered reasonably accurate.

If likely ranges for the observed values of the two sample proportions are known, then the values in the ranges closest to 0.5 should be taken as the educated guesses. For further discussion of these ideas and for applications of Formula 12.3, see Exercise 12.70.

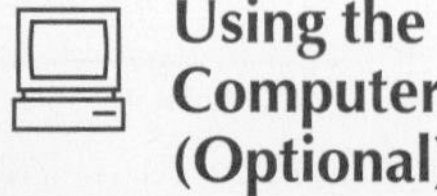

Using the Computer (Optional)

We can use Minitab to perform two-sample z-procedures for two population proportions. To illustrate, we return once again to the hypothesis test and confidence interval comparing the proportions of U.S. women and men who smoke cigarettes.

EXAMPLE 12.11 USING MINITAB TO PERFORM z-PROCEDURES FOR TWO POPULATION PROPORTIONS

Use Minitab to simultaneously perform the hypothesis test considered in Example 12.9 and obtain the confidence interval required in Example 12.10.

SOLUTION Let p_1 and p_2 denote, respectively, the proportions of all U.S. women and all U.S. men who smoke cigarettes. The problem in Example 12.9 is to perform the hypothesis test

$$H_0\colon p_1 = p_2 \text{ (percentage of female smokers is not less)}$$

$$H_a\colon p_1 < p_2 \text{ (percentage of female smokers is less)}$$

at the 5% significance level; the problem in Example 12.10 is to obtain a 90% confidence interval for $p_1 - p_2$.

We first recall that 2235 U.S. women and 2065 U.S. men were sampled and that of the women sampled, 503 smoked; of the men sampled, 572 smoked. Then we proceed in the following manner.

1 Choose **Stat ➤ Basic Statistics ➤ 2 Proportions...**
2 Select the **Summarized data** option button
3 Click in the **Trials** text box for **First sample** and type 2235
4 Click in the **Successes** text box for **First sample** and type 503
5 Click in the **Trials** text box for **Second sample** and type 2065
6 Click in the **Successes** text box for **Second sample** and type 572
7 Click the **Options...** button
8 Click in the **Confidence level** text box and type 90
9 Click in the **Test difference** text box and type 0
10 Click the arrow button at the right of the **Alternative** drop-down list box and select **less than**
11 Select the **Use pooled estimate of p for test** check box
12 Click **OK**
13 Click **OK**

The resulting output is shown in Printout 12.8 on the next page.

The first three lines of the output in Printout 12.8 provide a table giving the number of successes, sample size, and sample proportion for each sample. Next comes a point estimate for the difference, $p_1 - p_2$, between the two population proportions. On the following line is the required confidence interval: a 90% confidence interval for the difference between the proportions of U.S. women and men smokers is from −0.0737021 to −0.0301812.

PRINTOUT 12.8 Minitab output for the two-sample z-procedure for two population proportions

```
Sample        X       N  Sample p
1           503    2235  0.225056
2           572    2065  0.276998

Estimate for p(1) - p(2):  -0.0519417
90% CI for p(1) - p(2):  (-0.0737021, -0.0301812)
Test for p(1) - p(2) = 0 (vs < 0):  Z = -3.93  P-Value = 0.000
```

The last line of Printout 12.8 first gives the null and alternative hypotheses for the hypothesis test: `p(1) - p(2) = 0 (vs < 0)`, that is, $p_1 = p_2$ versus $p_1 < p_2$. Then we find the value of the test statistic z, and the P-value, which here is zero to three decimal places. Since the P-value is less than the specified significance level of 0.05, we reject the null hypothesis. At the 5% significance level, the data provide sufficient evidence to conclude that, in the United States, the percentage of women who smoke cigarettes is smaller than the percentage of men who smoke cigarettes. □

EXERCISES 12.3

STATISTICAL CONCEPTS AND SKILLS

12.52 Explain the basic idea for performing a hypothesis test to compare two population proportions using independent samples.

12.53 A Roper Starch Worldwide for A.B.C. Global Kids Study conducted surveys in various countries to estimate the percentage of children who attend church at least once a week. Two of the countries in the survey were the United States and Germany. Considering these two countries only:

a. Identify the specified attribute.
b. Identify the two populations.
c. What are the two population proportions under consideration here?

12.54 A recently conducted study by researchers at the Medical College of Wisconsin Eye Institute compared color blindness of American men and women.

a. Identify the specified attribute.
b. Identify the two populations.
c. What are the two population proportions under consideration here?

12.55 Industry Research polled teen-age girls and boys on sunscreen use. The survey revealed that 46% of teen-age girls and 30% of teen-age boys regularly use sunscreen before going out in the sun.

a. Identify the specified attribute.
b. Identify the two populations.
c. Are the proportions 0.46 (46%) and 0.30 (30%) sample proportions or population proportions? Why?

12.56 Consider a hypothesis test for two population proportions with null hypothesis H_0: $p_1 = p_2$. What parameter is being estimated by the

a. sample proportion $\hat{p}_1$?
b. sample proportion $\hat{p}_2$?
c. pooled sample proportion $\hat{p}_p$?

12.57 Of the quantities p_1, p_2, x_1, x_2, $\hat{p}_1$, $\hat{p}_2$, and $\hat{p}_p$,

a. which represent parameters and which represent statistics?
b. which are fixed numbers and which are variables?

For each of Exercises 12.58–12.63, perform the required hypothesis test using either the critical-value approach or the P-value approach.

12.58 Roughly 450,000 vasectomies are performed each year in the United States. In this surgical procedure for contraception, the tube carrying sperm from the testicles is cut. Several studies have been conducted to analyze the relationship between vasectomies and prostate cancer. One such study appeared in a February 1993 issue of the *Journal of the American Medical Association.* The following problem is based on data presented in that journal article. Of 21,300 men who have not had a vasectomy, 69 were found to have prostate cancer; and of 22,000 men who have had a vasectomy, 113 were found to have prostate cancer.

a. At the 1% significance level, do the data provide sufficient evidence to conclude that men who have had a vasectomy are at greater risk of having prostate cancer?

b. Is this a designed experiment or an observational study? Explain your answer.

c. In view of your answers to parts (a) and (b), would you say that it is reasonable to conclude that having a vasectomy causes an increased risk of prostate cancer? Why?

12.59 For several years, evidence has been mounting that folic acid reduces major birth defects. *The Arizona Republic* reported on a Hungarian study that provides the strongest evidence yet. The results of the study, directed by Dr. Andrew E. Czeizel and Dr. Istvan Dudas of the National Institute of Hygiene in Budapest, were published in the *New England Journal of Medicine.* For the study, the doctors enrolled 4753 women prior to conception. The women were divided randomly into two groups. One group, consisting of 2701 women, took daily multivitamins containing 0.8 mg of folic acid; the other group, consisting of 2052 women, received only trace elements. Major birth defects occurred in 35 cases where the women took folic acid and in 47 cases where the women did not.

a. At the 1% significance level, do the data provide sufficient evidence to conclude that women who take folic acid are at lesser risk of having children with major birth defects?

b. Is this a designed experiment or an observational study? Explain your answer.

c. In view of your answers to parts (a) and (b), would you say that it is reasonable to conclude that taking folic acid causes a reduction in major birth defects? Explain your answer.

12.60 The Organization for Economic Cooperation and Development, Paris, France, summarizes data on labor-force participation rates in *OECD in Figures, Statistics of the Member Countries.* Independent samples were taken of 300 U.S. women and 250 Canadian women. Of the U.S. women, 211 were found to be in the labor force; and of the Canadian women, 170 were found to be in the labor force. At the 5% significance level, do the data suggest that there is a difference between the labor-force participation rates of U.S. and Canadian women?

12.61 Overweight is defined for men as body mass index of 27.8 kilograms/meter squared or greater. This point is used because it represents the sex-specific 85th percentile for males 20–29 years of age in the 1976–1980 *National Health and Nutrition Examination Survey,* NHANES for short. In 1980, of 750 men 20–34 years old, 130 were found to be overweight, whereas, in 1990, of 700 men 20–34 years old, 160 were found to be overweight. At the 5% significance level, do the data provide sufficient evidence to conclude that for men 20–34 years old, a higher percentage were overweight in 1990 than 10 years earlier?

12.62 A 1996 *Wall Street Journal* article entitled "Hypertension Drug Linked to Cancer" reported on a study of several types of high-blood-pressure drugs and links to cancer. For one type, called calcium channel blockers, 27 of 202 elderly patients taking the drug developed cancer. For another type, called beta blockers, 28 of 424 other elderly patients developed cancer. At the 1% significance level, do the data provide sufficient evidence to conclude that elderly people taking beta blockers have a lower cancer rate than those taking calcium channel blockers? (*Note:* The results of this study were challenged and questioned by several sources, claiming, for example, that the study was flawed and that several other studies have suggested that calcium channel blockers are safe.)

12.63 Annual surveys are performed by the U.S. Bureau of the Census to obtain estimates of the percentage of the voting-age population that has registered to vote. The information from those surveys is published in *Current Population Reports.* Independent samples were taken of 400 employed people and 450 unemployed people. It was found that 262 of the employed people and 224 of the unemployed people had registered to vote. Can we conclude, at the 5% significance level, that a difference exists between the percentages of employed and unemployed workers who have registered to vote?

12.64 Refer to Exercise 12.58.

a. Determine a 98% confidence interval for the difference between the prostate cancer rates of men who have had a vasectomy and those who have not.

b. Interpret your answer from part (a).

12.65 Refer to Exercise 12.59.

a. Determine a 98% confidence interval for the difference between the rates of major birth defects for babies born to women who have taken folic acid and those born to women who have not.

b. Interpret your answer from part (a).

12.66 Refer to Exercise 12.60.

a. Find a 95% confidence interval for the difference, $p_1 - p_2$, between the labor-force participation rates of U.S. and Canadian women.

b. Interpret your result from part (a).

12.67 Refer to Exercise 12.61.

a. For men 20–34 years old, find a 90% confidence interval for the difference between the proportions that were overweight in the years 1990 and 1980.

b. Interpret your result from part (a).

12.68 Refer to Exercise 12.62.

a. Determine a 98% confidence interval for the difference between the cancer rates of elderly people taking calcium channel blockers and those taking beta blockers.

b. Interpret your result from part (a).

12.69 Refer to Exercise 12.63.

a. Find a 95% confidence interval for the difference between the proportions of employed and unemployed workers who have registered to vote.

b. Interpret your result from part (a).

EXTENDING THE CONCEPTS AND SKILLS

12.70 In this exercise we will apply Formula 12.3 on page 754 to the study on smoking considered in Examples 12.8–12.10.

a. Obtain the margin of error for the estimate of the difference between the proportions of women and men smokers by taking half the length of the confidence interval found in Example 12.10 on page 753. Interpret your answer in words.

b. Obtain the margin of error for the estimate of the difference between the proportions of women and men smokers by applying the first displayed formula in Formula 12.3.

c. Without making a guess for the observed values of the sample proportions, obtain the common sample size that will ensure a margin of error of at most 0.01 for a 90% confidence interval.

d. Find a 90% confidence interval for $p_1 - p_2$ if for samples of the size determined in part (c), 22.3% of the women and 27.2% of the men smoke cigarettes.

e. Determine the margin of error for the estimate in part (d) and compare it to the required margin of error specified in part (c).

f. Repeat parts (c)–(e) if it is deemed reasonable to presume that at most 25% of the women sampled will be smokers and at most 30% of the men sampled will be smokers.

g. Compare the results obtained in parts (c)–(e) with those obtained in part (f).

12.71 What formula shows that the difference between the two sample proportions is an unbiased estimator of the difference between the two population proportions? Explain your answer.

USING TECHNOLOGY   

12.72 Refer to Exercises 12.60 and 12.66. Use Minitab or some other statistical software to conduct the required hypothesis test and obtain the desired confidence interval.

12.73 Refer to Exercises 12.61 and 12.67. Use Minitab or some other statistical software to conduct the required hypothesis test and obtain the desired confidence interval.

12.74 Polls are taken regularly to determine the public's view on how the president is handling his job. Two *USA TODAY/CNN/Gallup* polls conducted two weeks apart in January, 1998, asked a sample of U.S. adults: "Do you approve of the way President Clinton is handling his job?" Minitab's two-sample z-procedure for two population proportions was applied to the data from those two polls. The output obtained is displayed in Printout 12.9. Use the output to determine,

a. the sample size for each poll.
b. the numbers of people in each poll that approved.
c. the percentage of people in each poll that approved.
d. a 90% confidence interval for the difference between the proportion of all U.S. adults who approved at the time of the first survey and the proportion of all U.S. adults who approved at the time of the second survey.
e. the null and alternative hypotheses for the test.
f. the P-value for the hypothesis test.
g. the conclusion if the hypothesis test is performed at the 5% significance level.

12.75 In 1998, before a signed agreement for diplomatic solutions with Iraq, the *Fort Worth Star-Telegram* conducted a poll asking residents whether they were in favor of a military strike against Iraq. Two weeks later, after the signed agreement, the paper again conducted the poll. Printout 12.10 shows the output obtained by applying Minitab's two-sample z-procedure for two population proportions to the data from the two polls. Use the output to determine,

a. the sample size for each poll.
b. the numbers of people in each poll that were in favor.
c. the percentage in each poll that were in favor.
d. a 95% confidence interval for the difference between the proportion of all residents who were in favor at the time of the survey before the signed agreement and the proportion of all residents who were in favor at the time of the survey after the signed agreement.
e. the null and alternative hypotheses for the test.
f. the P-value for the hypothesis test.
g. the conclusion if the hypothesis test is performed at the 5% significance level.

PRINTOUT 12.9 Minitab output for Exercise 12.74

```
Sample       X       N  Sample p
1          593    1022  0.580235
2          623    1005  0.619900

Estimate for p(1) - p(2):  -0.0396657
90% CI for p(1) - p(2):  (-0.0754302, -0.00390112)
Test for p(1) - p(2) = 0 (vs < 0):  Z = -1.82  P-Value = 0.034
```

PRINTOUT 12.10 Minitab output for Exercise 12.75

```
Sample       X       N  Sample p
1          193     272  0.709559
2          165     229  0.720524

Estimate for p(1) - p(2):  -0.0109652
95% CI for p(1) - p(2):  (-0.0902652, 0.0683348)
Test for p(1) - p(2) = 0 (vs not = 0):  Z = -0.27  P-Value = 0.787
```

CHAPTER REVIEW

You Should Be Able To

1. use and understand the formulas presented in this chapter.
2. find a large-sample confidence interval for a population proportion.
3. compute the margin of error for the estimate of a population proportion.
4. understand the relationship between the sample size, confidence level, and margin of error for a confidence interval for a population proportion.
5. determine the sample size required for a specified confidence level and margin of error for the estimate of a population proportion.
6. perform a large-sample hypothesis test for a population proportion.
7. perform large-sample inferences (hypothesis tests and confidence intervals) to compare two population proportions.
8. understand the relationship between the sample sizes, confidence level, and margin of error for a confidence interval for the difference between two population proportions.
9. determine the sample sizes required for a specified confidence level and margin of error for the estimate of the difference between two population proportions.

*10. use the Minitab procedures covered in this chapter.

*11. interpret the output obtained from the application of the Minitab procedures discussed in this chapter.

Key Terms

REVIEW TEST

STATISTICAL CONCEPTS AND SKILLS

1. In a recent survey, Accounttemps estimated the percentage of chief financial officers (CFOs) whose favorite recreation is golf. Identify the
   a. specified attribute. **b.** population.
   c. population proportion.
   d. sample proportion.

2. According to *USA TODAY*, about 72% of car owners wash their vehicle at least once a month. Identify in words the
   a. specified attribute. **b.** population.
   c. population proportion.
   d. sample proportion.
   e. Is the proportion 0.72 (72%) a population proportion or a sample proportion? Explain your answer.

3. Why do we generally use a sample proportion to estimate a population proportion instead of obtaining the population proportion directly?

4. Explain what each of the following phrases mean:
   a. number of successes
   b. number of failures

5. Fill in the following blanks.

a. The mean of all possible sample proportions equals the _____.

b. For large samples, the possible sample proportions have approximately a _____ distribution.

c. A rule of thumb for using a normal distribution to approximate the distribution of all possible sample proportions is that both _____ and _____ are _____ or greater.

6. What does the margin of error for the estimate of a population proportion tell us?

7. A poll was conducted by Opinion Research Corp. to estimate the proportions of men and women who get the "holiday blues." Identify the

a. specified attribute.

b. two populations.

c. two population proportions.

d. two sample proportions.

e. According to the poll, 34% of men and 44% of women get the "holiday blues." Are the proportions 0.34 and 0.44 sample proportions or population proportions? Explain your answer.

8. Suppose we are using independent samples to compare two population proportions. Fill in the following blanks.

a. The mean of all possible differences between the two sample proportions equals the _____.

b. For large samples, the possible differences between the two sample proportions have approximately a _____ distribution.

9. Suppose we want to obtain a 95% confidence interval for the difference between two population proportions using independent samples and that we want a margin of error of at most 0.01.

a. Without making an educated guess for the observed sample proportions, what common sample size is required?

b. Suppose that from past experience we are quite sure that the two sample proportions will be 0.75 or greater. What common sample size should we use?

10. Between June 11 and 15, 1993, the Times Mirror Center for People and the Press interviewed 1006 adults concerning their views on media treatment of then recently inaugurated President Clinton. According to an article in the *Los Angeles Times,* "... a notable 43 percent of Americans [surveyed (433 of those sampled)] said that they thought news organizations were criticizing the Clinton administration unfairly."

a. Find a 95% confidence interval for the proportion, p, of all Americans who thought news organizations were criticizing the Clinton administration unfairly.

b. Interpret your result from part (a) in words.

11. Refer to Problem 10.

a. Find the margin of error for the estimate of p.

b. Obtain a sample size that will ensure a margin of error of at most 0.02 for a 95% confidence interval without making a guess for the observed value of $\hat{p}$.

c. Find a 95% confidence interval for p if for a sample of the size determined in part (b), 41.6% of those surveyed thought news organizations were criticizing the Clinton administration unfairly.

d. Determine the margin of error for the estimate in part (c) and compare it to the required margin of error specified in part (b).

e. Repeat parts (b)–(d) if it is deemed reasonable to presume that the percentage of those surveyed who thought news organizations were criticizing the Clinton administration unfairly will be at most 45%.

f. Compare the results obtained in parts (b)–(d) with those obtained in part (e).

12. A poll of Americans conducted by Gallup for the CNBC cable network revealed that only 17% of those surveyed felt they had achieved the American dream. The poll had a margin of error of plus or minus 4 percentage points (for a 0.95 confidence level). Use this information to obtain a 95% confidence interval for the percentage of all Americans who feel they have achieved the American dream.

13. In an issue of *Parade Magazine,* the editors reported on a national survey on law and order. One question asked of the 2512 U.S. adults taking part was whether they believed that juries "almost always" convict the guilty and free the innocent. Only 578 said they did. Do the data provide sufficient evidence to conclude that less than one in four Americans believe that juries "almost always" convict the guilty and free the innocent. Perform the appropriate hypothesis test at the 5% significance level using

a. the critical-value approach to hypothesis testing.
b. the P-value approach to hypothesis testing.
c. Assess the strength of the evidence against the null hypothesis by referring to Table 9.12 on page 543.

14. An article published in an issue of the *Annals of Epidemiology* discussed the relationship between height and breast cancer. The study by the National Cancer Institute, which took 5 years and involved more than 1500 women having breast cancer and 2000 women not having breast cancer, revealed that there is a trend between height and breast cancer: "...taller women have a 50 to 80 percent greater risk of getting breast cancer than women who are closer to 5 feet tall." But Christine Swanson, a nutritionist who was involved with the study, added that "...height may be associated with the culprit, ...but no one really knows" the exact relationship between height and breast-cancer risk.

a. Classify this study as either an observational study or a designed experiment. Explain your answer.
b. Interpret the statement made by Christine Swanson in view of your answer to part (a).

15. State and local governments often poll their constituents about their views on the economy. In two polls, taken approximately 1 year apart, O'Neil Associates asked 600 Maricopa County (Arizona) residents whether they thought the state's economy would improve over the next 2 years. In the first poll, 48% said yes; in the second poll, 60% said yes. At the 1% significance level, do the data provide sufficient evidence to conclude that the percentage of Maricopa County residents who thought the state's economy would improve over the next 2 years was less during the time of the first poll than during the time of the second? Perform the required hypothesis test using

a. the critical-value approach to hypothesis testing.
b. the P-value approach to hypothesis testing.
c. Assess the strength of the evidence against the null hypothesis by referring to Table 9.12 on page 543.

16. Refer to Problem 15.

a. Determine a 98% confidence interval for the difference, $p_1 - p_2$, between the proportions of Maricopa County residents who thought that the state's economy would improve over the next 2 years during the time of the first poll and during the time of the second poll.
b. Interpret your answer from part (a).

17. Refer to Problems 15 and 16.

a. Take half the length of the confidence interval found in Problem 16(a) to obtain the margin of error for the estimate of the difference between the two population proportions. Interpret your result in words.
b. Solve part (a) by applying the first displayed formula in Formula 12.3 on page 754.
c. Obtain the common sample size that will ensure a margin of error of at most 0.03 for a 98% confidence interval without making a guess for the observed values of the sample proportions.
d. Find a 98% confidence interval for $p_1 - p_2$ if for samples of the size determined in part (c), the sample proportions are 0.475 and 0.603, respectively.
e. Determine the margin of error for the estimate in part (d) and compare it to the required margin of error specified in part (c).

USING TECHNOLOGY  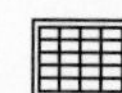 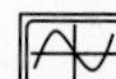

18. Use Minitab or some other statistical software to determine the confidence interval required in Problem 10.

19. Use Minitab or some other statistical software to conduct the hypothesis test of Problem 13.

20. Use Minitab or some other statistical software to perform the required hypothesis test in Problem 15 and to obtain the desired confidence interval in Problem 16.

21. Printout 12.11 on the following page shows the output obtained by applying Minitab's one-proportion procedure to the data in Problem 10. Use the output to determine

a. the number of people polled.

b. the number of people polled who thought news organizations were criticizing the Clinton administration unfairly.

c. the percentage of people polled who thought news organizations were criticizing the Clinton administration unfairly.

d. a 95% confidence interval for the proportion of all Americans who thought news organizations were criticizing the Clinton administration unfairly.

22. We applied Minitab's one-proportion procedure to the data in Problem 13 and obtained the output shown in Printout 12.12 on the next page. Using the output, determine the

a. number of people in the survey.

b. number of people in the survey who believe that juries "almost always" convict the guilty and free the innocent.

c. percentage of people in the survey who believe that juries "almost always" convict the guilty and free the innocent.

d. null and alternative hypotheses for the hypothesis test.

e. P-value for the hypothesis test.

f. conclusion if the hypothesis test is performed at the 5% significance level.

23. In Printout 12.13 on the following page we have displayed the output obtained by applying Minitab's two-proportion procedure to the data in Problem 15. From the output, determine

a. the sample size for each poll.

b. the number of people in each poll that thought the state's economy would improve over the next 2 years.

c. the percentage in each poll that thought the state's economy would improve over the next 2 years.

d. a 98% confidence interval for the difference between the proportions of Maricopa County residents who thought that the state's economy would improve over the next 2 years during the time of the first poll and during the time of the second poll.

e. the null and alternative hypotheses for the hypothesis test.

f. the P-value for the hypothesis test.

g. the conclusion if the hypothesis test is performed at the 1% significance level.

PRINTOUT 12.11 Minitab output for Problem 21

```
Test of p = 0.5 vs p not = 0.5

                                                           Exact
Sample        X       N  Sample p         95.0 % CI        P-Value
1           433    1006  0.430417  (0.399566, 0.461677)     0.000
```

PRINTOUT 12.12 Minitab output for Problem 22

```
Test of p = 0.25 vs p < 0.25

                                                           Exact
Sample        X       N  Sample p         95.0 % CI        P-Value
1           578    2512  0.230096  (0.213759, 0.247063)     0.011
```

PRINTOUT 12.13 Minitab output for Problem 23

```
Sample        X       N  Sample p
1           288     600  0.480000
2           360     600  0.600000

Estimate for p(1) - p(2):  -0.12
98% CI for p(1) - p(2):  (-0.186454, -0.0535462)
Test for p(1) - p(2) = 0 (vs < 0):  Z = -4.17  P-Value = 0.000
```

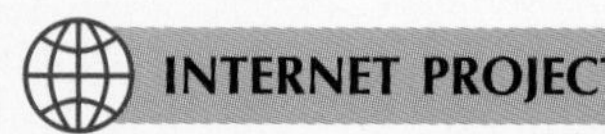

AIDS and Condom Use

Can condoms and condom promotion help prevent the spread of Acquired Immune Deficiency Syndrome, or AIDS? The virus responsible for AIDS is the human immunodeficiency virus (HIV). This Internet project provides you with an opportunity to investigate whether promotion of condom use can assist in the stemming of new HIV infections. The project includes an animated time-lapse map showing the spread of the virus throughout the United States.

Although controversy persists on the question of the effectiveness of condom promotion in averting HIV transmission, many believe that if condoms are easily available and effectively promoted, they can play an important role in HIV prevention. In this project, you can draw your own conclusion by exploring data from around the world on the effect of condom promotion and condom use on HIV transmission.

URL for access to Internet Projects Page: `http://hepg.awl.com` *keyword:* Weiss

USING THE FOCUS DATABASE

In Chapter 1 we explained how to store the Focus database in a Minitab worksheet named `focus.mtw`. If you haven't already created that worksheet, follow the instructions on pages 54–55 to create it now.

The Focus database contains information on 500 randomly selected Arizona State University sophomores for seven different variables: sex, high-school GPA, SAT math score, cumulative GPA, SAT verbal score, age, and total hours. Use Minitab or some other statistical software to solve the following problems.

a. Obtain a 95% confidence interval for the percentage of all Arizona State University sophomores whose cumulative GPAs are at least 3.00. Interpret your result.

b. Referring to part (a), determine the margin of error for the estimate. Interpret your answer.

c. At the 5% significance level, do the data provide sufficient evidence to conclude that more than 20% of Arizona State University sophomores score 600 or over on the SAT math?

d. Can we conclude that a difference exists between the proportions of male and female sophomores at Arizona State University who score 500 or over on the SAT verbal? Perform the required hypothesis test at the 5% significance level.

e. Determine a 95% confidence interval for the difference between the proportions of male and female sophomores at Arizona State University who score 500 or over on the SAT verbal.

CASE STUDY DISCUSSION

Credit Card Blues

At the beginning of this chapter, we discussed a poll sponsored by ABCNEWS.com on various aspects of credit-card usage. The poll, whose field work was done by International Communications Research of Media, PA, was conducted October 1–5, 1997. It used a sample of 1003 Americans, randomly selected nationally and, according to ABCNEWS.com, has a margin of error of plus or minus three percentage points.

The statements made on page 722, gleaned from the results of the poll, are based on information obtained from the Americans sampled and, as such, are statistics. In the following problems, you are asked to apply some of the inferential techniques discussed in this chapter to draw conclusions about the population of all American adults based on the data given on page 722.

a. For the American adults polled, identify the sample proportion of those who do not own a credit card.

b. Obtain a 95% confidence interval for the proportion of all American adults who do not own a credit card. Interpret your answer.

c. Find the margin of error for the estimate in part (b) and interpret your answer.

d. At the 5% significance level, do the data provide sufficient evidence to conclude that more than one-third of American adults do not own a credit card?

e. Referring to part (d), use Table 9.12 on page 543 to assess the strength of the evidence against the null hypothesis and hence in favor of the alternative hypothesis that more than one-third of American adults do not own a credit card.

In the remaining parts, we will investigate the proportion, p, of all American adults who carry a total credit-card balance of \$10,000 or more. Note that the population under consideration here consists of all American adults, not just those who own credit cards.

f. For a 95% confidence interval, find the margin of error for the estimate of p.

g. Obtain a sample size that will ensure a margin of error of at most 0.005 for a 95% confidence interval.

h. Suppose that, for a sample of the size determined in part (g), 2.7% of those sampled carry a balance of \$10,000 or more. Construct a 95% confidence interval for p.

i. Determine the margin of error for the estimate in part (h) and compare it to the margin of error specified in part (g).

j. Repeat parts (g)–(i) if it is deemed reasonable to presume that the proportion of those sampled who carry a balance of \$10,000 or more will be somewhere between 0.02 and 0.03.

k. Compare the results obtained in parts (g)–(i) with those obtained in part (j).

BIOGRAPHY ABRAHAM DE MOIVRE

Abraham de Moivre was born in Vitry-le-Francois, France, on May 26, 1667, the son of a country surgeon. He was educated in the Catholic school in his village and at the Protestant Academy at Sedan. In 1684 he went to Paris to study under Jacques Ozanam.

In late 1685 de Moivre, a French Huguenot (Protestant), was imprisoned in Paris because of his religion. (In October 1685 Louis XIV revoked an edict that had allowed Protestantism in addition to the Catholicism favored by the French Court.) The duration of his incarceration is unclear, but de Moivre was probably jailed for roughly 1 to 3 years. In any case, upon his release he fled to London, where he began tutoring students in mathematics.

In London, de Moivre mastered Sir Isaac Newton's *Principia* and became a close friend of Newton and of Edmond Halley, an English astronomer (in whose honor, incidentally, Halley's Comet is named). In Newton's later years, he would refuse to take new students, saying, "Go to Mr. de Moivre; he knows these things better than I do."

De Moivre's contributions to mathematics range from the definition of statistical independence to analytical trigonometric formulas to his major discovery—the normal approximation to the binomial distribution, of monumental importance in its own right and precursor to the central limit theorem. The definition of statistical independence appeared in *The Doctrine of Chances,* published in 1718 and dedicated to Newton; the normal approximation to the binomial distribution was contained in a Latin pamphlet published in 1733. Many of his other papers were published in *Philosophical Transactions of the Royal Society.*

De Moivre also did research on the analysis of mortality statistics and the theory of annuities. In 1725 the first edition of his *Annuities on Lives,* in which he derived annuity formulas and addressed other annuity problems, was published.

De Moivre was elected to the Royal Society in 1697, to the Berlin Academy of Sciences in 1735, and to the Paris Academy in 1754. But despite his obvious talents as a mathematician and his many champions, he was never able to obtain a position in any of England's universities. Instead he had to rely on his meager earnings as a tutor in mathematics and a consultant on gambling and insurance, supplemented by the sales of his books. De Moivre died in London on November 27, 1754.

CASE STUDY

TIME ON OR OFF THE JOB: WHICH DO AMERICANS ENJOY MORE?

A CNN/*USA TODAY* poll conducted by Gallup asked a sample of employed Americans the following question: "Which do you enjoy more, the hours when you are on your job, or the hours when you are not on your job?" The responses to this question were cross tabulated against several characteristics, among which were sex, age, type of community, amount of education, income, and type of employer. The following table summarizes the poll's results.

	On the job	Off the job	Don't know
Male	77	263	28
Female	77	215	27
18–29 years	33	136	5
30–49 years	77	274	25
50–64 years	35	56	19
65 and older	9	11	5
Urban	62	197	14
Suburban	47	171	25
Rural	42	109	15
Postgraduate	23	51	10
College graduate	41	126	18
Some college	20	108	5
No college	68	192	21
Under $20,000	40	80	11
$20,000–$29,999	32	116	7
$30,000–$49,999	41	131	8
$50,000 and over	34	138	22
Private	67	326	27
Government	23	82	11
Self	61	69	14

After you study the chi-square independence test in this chapter, you will be asked to analyze the survey results to decide whether preferences depend on sex, age, type of community, amount of education, income, or type of employer.

Chi-Square Procedures

13

GENERAL OBJECTIVES

The statistical-inference techniques presented so far have dealt exclusively with hypothesis tests and confidence intervals for population parameters, such as population means and population proportions. In this chapter we will consider two widely used inferential procedures that are not concerned with population parameters. These two procedures are often referred to as **chi-square procedures** because they rely on a distribution called the *chi-square distribution.*

First, in preparation for our examination of chi-square tests, we will study the chi-square distribution. Next we will present the chi-square goodness-of-fit test, a test that can be used to make inferences about the distribution of a variable. For instance, we could apply the chi-square goodness-of-fit test to a sample of university students to decide whether the political-preference distribution of all university students differs from that of the population as a whole.

In the third section of this chapter, as a preliminary to the study of our second chi-square procedure, we will discuss contingency tables—frequency distributions for bivariate data—and related topics. Then, in the final section, we will present the chi-square independence test, a test used to decide whether an association exists between two characteristics of a population. For example, we could apply the chi-square independence test to a sample of U.S. adults to decide whether an association exists between annual income and educational level for all U.S. adults.

CHAPTER OUTLINE

13.1 THE CHI-SQUARE DISTRIBUTION

The statistical-inference procedures discussed in this chapter rely on a distribution called the **chi-square distribution.** A variable is said to have a chi-square distribution if its distribution has the shape of a special type of right-skewed curve, called a **chi-square (χ^2) curve.**

Actually there are infinitely many chi-square distributions, and we identify the chi-square distribution (and χ^2-curve) in question by stating its number of degrees of freedom, just as we did for t-distributions. Figure 13.1 shows three χ^2-curves and illustrates some basic properties of χ^2-curves, which are presented in Key Fact 13.1.

FIGURE 13.1
χ^2-curves for df = 5, 10, and 19

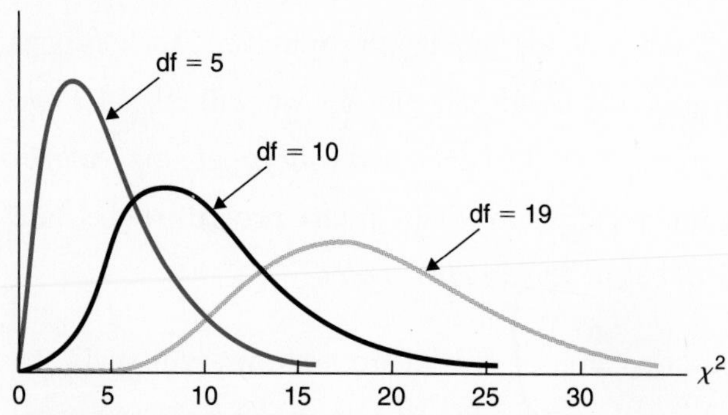

KEY FACT 13.1 BASIC PROPERTIES OF χ^2-CURVES

Property 1: The total area under a χ^2-curve is equal to 1.

Property 2: A χ^2-curve starts at 0 on the horizontal axis and extends indefinitely to the right, approaching, but never touching, the horizontal axis as it does so.

Property 3: A χ^2-curve is right skewed.

Property 4: As the number of degrees of freedom becomes larger, χ^2-curves look increasingly like normal curves.

Using the χ^2-Table

Percentages (and probabilities) for a variable having a chi-square distribution are equal to areas under its associated χ^2-curve. To perform a chi-square test, we need to know how to find the χ^2-value having a specified area to its right. Table VIII in Appendix A provides χ^2-values corresponding to several areas for various degrees of freedom.

The χ^2-table (Table VIII) is similar to the t-table (Table IV). The two outside columns of Table VIII, labeled df, display the number of degrees of freedom. As expected, the symbol χ^2_α denotes the χ^2-value having area α to its right under a χ^2-curve. Thus the column headed $\chi^2_{0.995}$ contains χ^2-values having area 0.995 to their right; the column headed $\chi^2_{0.99}$ contains χ^2-values having area 0.99 to their right; and so on. We illustrate a use of Table VIII in Example 13.1.

EXAMPLE 13.1 FINDING THE χ^2-VALUE HAVING A SPECIFIED AREA TO ITS RIGHT

For a χ^2-curve with 12 degrees of freedom, find $\chi^2_{0.025}$; that is, find the χ^2-value having area 0.025 to its right, as shown in Fig. 13.2(a).

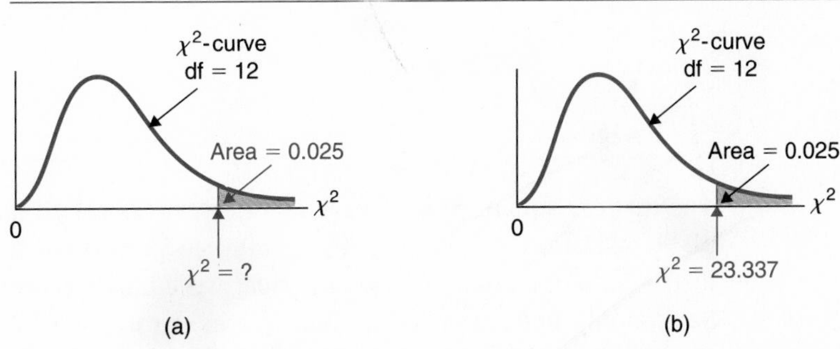

FIGURE 13.2 Finding the χ^2-value having area 0.025 to its right

SOLUTION To find this χ^2-value, we use Table VIII. Since the number of degrees of freedom is 12, we first go down the outside columns, labeled df, to "12." Then we go across that row until we are under the column headed $\chi^2_{0.025}$. The number in the body of the table there, 23.337, is the required χ^2-value; that is, for a χ^2-curve with df = 12, the χ^2-value having area 0.025 to its right is $\chi^2_{0.025} = 23.337$, as seen in Figure 13.2(b).

 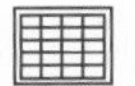 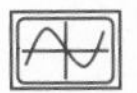

EXERCISES 13.1

STATISTICAL CONCEPTS AND SKILLS

13.1 What does it mean to say that a variable has a chi-square distribution?

13.2 How do we identify different chi-square distributions?

13.3 Consider two χ^2-curves with degrees of freedom 12 and 20, respectively. Which one more closely resembles a normal curve? Explain your answer.

13.4 The t-table has entries for areas of 0.10, 0.05, 0.025, 0.01, and 0.005. On the other hand, the χ^2-table has entries for those areas plus ones for 0.995, 0.99,

0.975, 0.95, and 0.90. Explain why we can obtain the t-values corresponding to these latter areas from the existing t-table, but that we must provide them explicitly in the χ^2-table.

In each of Exercises 13.5–13.8, use Table VIII to find the required χ^2-values. Illustrate your work graphically.

13.5 For a χ^2-curve with 19 degrees of freedom, find the χ^2-value having area
a. 0.025 to its right. **b.** 0.95 to its right.

13.6 For a χ^2-curve with 22 degrees of freedom, find the χ^2-value having area
a. 0.01 to its right. **b.** 0.995 to its right.

13.7 For a χ^2-curve with df = 10, determine
a. $\chi^2_{0.05}$. **b.** $\chi^2_{0.975}$.

13.8 For a χ^2-curve with df = 4, determine
a. $\chi^2_{0.005}$. **b.** $\chi^2_{0.99}$.

Extending the Concepts and Skills

13.9 Explain how you would use Table VIII to find the χ^2-value having area 0.05 to its left. Obtain this χ^2-value for a χ^2-curve having df = 26.

13.10 Explain how you would use Table VIII to find the two χ^2-values that divide the area under a χ^2-curve into a middle 0.95 area and two outside 0.025 areas. Find these two χ^2-values for a χ^2-curve having df = 14.

13.2 CHI-SQUARE GOODNESS-OF-FIT TEST

The first chi-square procedure we will discuss is called the **chi-square goodness-of-fit test.** This procedure can be used to perform a hypothesis test about the distribution of a qualitative (categorical) variable or a discrete quantitative variable having only finitely many possible values. Example 13.2 introduces and explains the reasoning behind the chi-square goodness-of-fit test.

EXAMPLE 13.2 Introduces the Chi-Square Goodness-of-Fit Test

The U.S. Federal Bureau of Investigation (FBI) compiles data on crimes and crime rates and publishes the information in *Crime in the United States.* A violent crime is classified by the FBI as murder, forcible rape, robbery, or aggravated assault. Table 13.1 provides a relative-frequency distribution for (reported) violent crimes in 1995. The table shows, for instance, that in 1995, 32.3% of violent crimes were robberies.

TABLE 13.1 Distribution of violent crimes in the United States, 1995

Type of violent crime	Relative frequency
Murder	0.012
Forcible rape	0.054
Robbery	0.323
Agg. assault	0.611
	1.000

hypothesis test

A random sample of 500 violent-crime reports from last year yielded the frequency distribution shown in Table 13.2. Do the data provide sufficient evidence to conclude that last year's distribution of violent crimes has changed from the 1995 distribution?

TABLE 13.2
Sample results for 500 randomly selected violent-crime reports from last year

Type of violent crime	Frequency
Murder	9
Forcible rape	26
Robbery	144
Agg. assault	321
	500

SOLUTION Here the population is last year's (reported) violent crimes. The variable under consideration is "type of violent crime" and its possible values are murder, forcible rape, robbery, and aggravated assault. We want to perform the hypothesis test

H_0: Last year's violent-crime distribution is the same as the 1995 distribution.

H_a: Last year's violent-crime distribution is different from the 1995 distribution.

The basic idea behind the chi-square goodness-of-fit test is to compare the observed frequencies in the second column of Table 13.2 to the frequencies that would be expected if last year's violent-crime distribution is the same as the 1995 distribution. If the observed and expected frequencies match up fairly well, then we do not reject the null hypothesis; otherwise, we reject the null hypothesis. To transform this idea into a precise procedure, we need to answer two questions:

1. What frequencies should we expect from a random sample of 500 violent-crime reports from last year if last year's violent-crime distribution is the same as the 1995 distribution?
2. How do we decide whether the observed frequencies match up reasonably well with those that we would expect?

The first question is easy to answer. If last year's violent-crime distribution is the same as the 1995 distribution, then, for instance, 32.3% of last year's violent crimes were robberies. (See Table 13.1.) Therefore, in a random sample of 500 violent-crime reports from last year, we would expect about 32.3% of the 500, or 161.5, to be robberies. In general, each expected frequency, denoted by the letter E, is computed using the formula

$$E = np,$$

where n is the sample size and p is the appropriate relative frequency from the second column of Table 13.1. For instance, the expected frequency for robberies is

$$E = np = 500 \cdot 0.323 = 161.5,$$

as we have already seen. The expected frequencies for all four types of violent crime are calculated in Table 13.3.

TABLE 13.3
Expected frequencies if last year's violent-crime distribution is the same as the 1995 distribution

Type of violent crime	Relative frequency p	Expected frequency $np = E$
Murder	0.012	$500 \cdot 0.012 = 6.0$
Forcible rape	0.054	$500 \cdot 0.054 = 27.0$
Robbery	0.323	$500 \cdot 0.323 = 161.5$
Agg. assault	0.611	$500 \cdot 0.611 = 305.5$

The third column of Table 13.3 provides the answer to our first question. It gives the frequencies we would expect if last year's violent-crime distribution is the same as the 1995 distribution.

The second question, whether the observed frequencies match up reasonably well with the expected frequencies, is harder to answer. We need to calculate a number that measures how good the fit is.

The second column of Table 13.4 repeats the **observed frequencies** from the second column of Table 13.2, and the third column of Table 13.4 lists the **expected frequencies** from the third column of Table 13.3.

TABLE 13.4
Calculating the goodness of fit

Type of violent crime x	Observed frequency O	Expected frequency E	Difference $O - E$	Square of difference $(O - E)^2$	Chi-square subtotal $(O - E)^2/E$
Murder	9	6.0	3.0	9.00	1.500
Forcible rape	26	27.0	−1.0	1.00	0.037
Robbery	144	161.5	−17.5	306.25	1.896
Agg. assault	321	305.5	15.5	240.25	0.786
	500	500.0	0		4.219

To measure how well the observed and expected frequencies match up, it is logical to look at the differences $O - E$, shown in the fourth column of Table 13.4. As we see, summing these differences to obtain a "total difference" is not very useful since the sum is 0. Instead, each difference is squared (fifth column of Table 13.4) and then divided by the corresponding expected frequency. This gives the values $(O - E)^2/E$, called **chi-square subtotals,** shown in the sixth column of Table 13.4. The sum of the chi-square subtotals,

$$\Sigma(O - E)^2/E = 4.219,$$

is the statistic used to measure how well (or poorly) the observed and expected frequencies match up.

If the null hypothesis is true, the observed and expected frequencies should be roughly equal, resulting in a small value of the test statistic, $\Sigma(O - E)^2/E$. In other words, large values of $\Sigma(O - E)^2/E$ provide evidence against the null hypothesis.

As we have seen, $\Sigma(O - E)^2/E = 4.219$. Can this value be reasonably attributed to sampling error, or is it large enough to suggest that the null hypothesis is false? To answer this question, we need to know the distribution of the test statistic $\Sigma(O - E)^2/E$. ❑

KEY FACT 13.2

DISTRIBUTION OF THE χ^2-STATISTIC FOR A CHI-SQUARE GOODNESS-OF-FIT TEST

For a chi-square goodness-of-fit test, the test statistic

$$\chi^2 = \Sigma(O - E)^2/E$$

has approximately a chi-square distribution if the null hypothesis is true. The number of degrees of freedom is one less than the number of possible values for the variable under consideration.

Procedure for the Chi-Square Goodness-of-Fit Test

We can now state a procedure for performing a chi-square goodness-of-fit test, specifically, Procedure 13.1 on the following page. Since the null hypothesis will be rejected only when the test statistic is too large, the rejection region is always on the right, that is, the hypothesis test is always right-tailed.

Regarding Assumptions 1 and 2 of Procedure 13.1, many texts give the rule that all expected frequencies should be 5 or greater. Research by the noted statistician W. G. Cochran shows that the "rule of 5" is too restrictive.

EXAMPLE 13.3 | **ILLUSTRATES PROCEDURE 13.1**

We can now complete the hypothesis test introduced in Example 13.2. Table 13.5 on page 777 gives the relative-frequency distribution for violent crimes in the United States in 1995.

PROCEDURE 13.1 **THE CHI-SQUARE GOODNESS-OF-FIT TEST**

ASSUMPTIONS

1. All expected frequencies are 1 or greater.
2. At most 20% of the expected frequencies are less than 5.

Step 1 The null and alternative hypotheses are

H_0: The variable under consideration has the specified distribution.

H_a: The variable under consideration does not have the specified distribution.

Step 2 Calculate the expected frequency for each possible value of the variable under consideration using the formula $E = np$, where n is the sample size and p is the relative frequency (or probability) given for the value in the null hypothesis.

Step 3 Check whether the expected frequencies satisfy Assumptions 1 and 2. If they do not, this procedure should not be used.

Step 4 Decide on the significance level, α.

Step 5 The critical value is χ^2_α with df $= k - 1$, where k is the number of possible values for the variable under consideration. Use Table VIII to find the critical value.

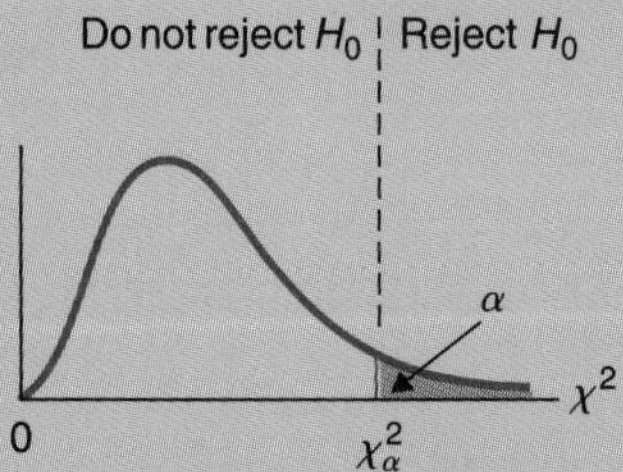

Step 6 Compute the value of the test statistic

$$\chi^2 = \Sigma(O - E)^2/E,$$

where O and E denote observed and expected frequencies, respectively.

Step 7 If the value of the test statistic falls in the rejection region, reject H_0; otherwise, do not reject H_0.

Step 8 State the conclusion in words.

TABLE 13.5 Distribution of violent crimes in the United States, 1995

Type of violent crime	Relative frequency
Murder	0.012
Forcible rape	0.054
Robbery	0.323
Agg. assault	0.611
	1.000

A random sample of 500 violent-crime reports from last year yielded the frequency distribution shown in Table 13.6. At the 5% significance level, do the data provide sufficient evidence to conclude that last year's violent-crime distribution is different from the 1995 distribution?

TABLE 13.6 Sample results for 500 randomly selected violent-crime reports from last year

Type of violent crime	Observed frequency
Murder	9
Forcible rape	26
Robbery	144
Agg. assault	321

SOLUTION We apply Procedure 13.1.

Step 1 *State the null and alternative hypotheses.*

The null and alternative hypotheses are

H_0: Last year's violent-crime distribution is the same as the 1995 distribution.

H_a: Last year's violent-crime distribution is different from the 1995 distribution.

Step 2 *Calculate the expected frequency for each possible value of the variable under consideration using the formula $E = np$, where n is the sample size and p is the relative frequency given for the value in the null hypothesis.*

We have $n = 500$ and the relative frequencies for the null hypothesis are shown in the second column of Table 13.5. The required calculations are summarized in Table 13.3 on page 774.

Step 3 *Check whether the expected frequencies satisfy Assumptions 1 and 2.*

1. All expected frequencies are 1 or greater? Yes, as we see from Table 13.3.
2. At most 20% of the expected frequencies are less than 5? Yes, in fact, none of the expected frequencies are less than 5, as we see from Table 13.3.

Step 4 *Decide on the significance level, α.*

We are to perform the test at the 5% significance level. Thus $\alpha = 0.05$.

Step 5 *The critical value is χ^2_α with df $= k - 1$, where k is the number of possible values for the variable under consideration.*

From Step 4, $\alpha = 0.05$. The variable under consideration is "type of violent crime." Since there are four types of violent crime, $k = 4$. Referring to Table VIII, we find that for df $= k - 1 = 4 - 1 = 3$, $\chi^2_{0.05} = 7.815$, as shown in Fig. 13.3.

FIGURE 13.3
Criterion for deciding whether or not to reject the null hypothesis

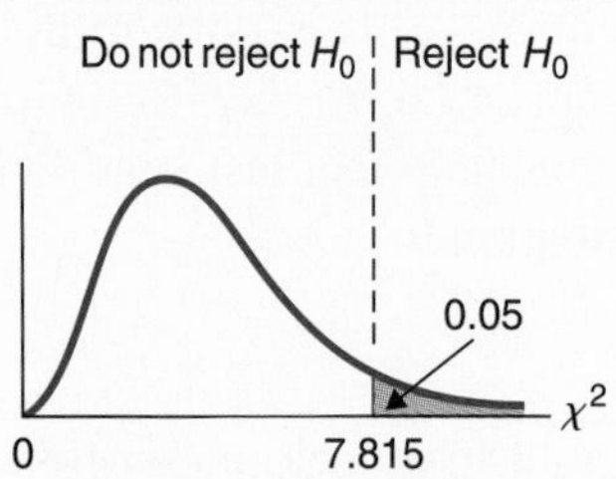

Step 6 *Compute the value of the test statistic*

$$\chi^2 = \Sigma(O - E)^2/E,$$

where O and E represent observed and expected frequencies, respectively.

The observed frequencies are displayed in the second column of Table 13.6. We calculated the value of the test statistic in Table 13.4 on page 774 where we found that

$$\chi^2 = \Sigma(O - E)^2/E = 4.219.$$

Step 7 *If the value of the test statistic falls in the rejection region, reject H_0; otherwise, do not reject H_0.*

From Step 6, the value of the test statistic is $\chi^2 = 4.219$. Since this does not fall in the rejection region, shown in Fig. 13.3, we do not reject H_0.

Step 8 *State the conclusion in words.*

The test results are not statistically significant at the 5% level; that is, at the 5% significance level, the data do not provide sufficient evidence to conclude that last year's violent-crime distribution differs from the 1995 distribution. ◨

Using the *P*-Value Approach

The *P*-value approach to hypothesis testing can also be used to carry out the hypothesis test in Example 13.3 and to assess more precisely the evidence against the null hypothesis. From Step 6 we see that the value of the test statistic is $\chi^2 = 4.219$. Recalling that df $= 3$, we find from Table VIII that $P > 0.10$.

In particular, because the P-value exceeds the significance level of 0.05, we cannot reject H_0. Furthermore, by referring to Table 9.12 on page 543, we see that the data provide at most weak evidence that last year's violent-crime distribution has changed from the 1995 distribution.

Using the Computer (Optional)

Minitab does not have a dedicated procedure for performing a chi-square goodness-of-fit test. But we used Minitab's capabilities to write a macro for such a test and called the macro `fittest.mac`. It can be found in the `Macro` folder of DataDisk CD.

EXAMPLE 13.4 USING MINITAB TO PERFORM A CHI-SQUARE GOODNESS-OF-FIT TEST

Use Minitab to conduct the chi-square goodness-of-fit test that was considered in Example 13.3.

SOLUTION First we enable the command language as follows:

1 Click in the Session window
2 Choose **Editor ➤ Enable Command Language**

Now we can run `fittest.mac` by typing, in the Session window, a percent sign (%) followed by the path for `fittest.mac`. For instance, if DataDisk CD is in drive E:, we type

`%E:\IS5\Macro\fittest.mac`

and press the [Enter] key. Then we proceed as follows.

1 In response to `Enter the number of possible values for the variable under consideration`, type 4 and press the [Enter] key. (See Table 13.5 on page 777.)
2 In response to `Enter the relative frequencies (or probabilities) for the null hypothesis`, type `0.012 0.054 0.323 0.611` and press the [Enter] key. (See Table 13.5 on page 777.)
3 In response to `Enter the observed frequencies`, type `9 26 144 321` and press the [Enter] key. (See Table 13.6 on page 777.)

Printout 13.1 shows the resulting output.

PRINTOUT 13.1
Minitab output for the fittest macro

```
Chi-square goodness-of-fit test


Row      n      k      ChiSq     P-value

  1    500      4    4.21974    0.238693
```

The output in Printout 13.1 provides the sample size, number of possible values for the variable under consideration, the value of the chi-square test statistic, and the P-value for the hypothesis test. Since the P-value of 0.238693 exceeds the specified significance level of 0.05, we do not reject H_0. At the 5% significance level, the data do not provide sufficient evidence to conclude that last year's violent-crime distribution differs from the 1995 distribution.

The `fittest` macro also offers to provide a graphical display of the results of the chi-square goodness-of-fit test. Type Y or N, respectively, and press the Enter key depending on whether or not you want to see a graphical display. Here we will reply affirmatively to the offer. Printout 13.2 depicts the output that results. As you can see, the graphical display includes a chi-square curve with the P-value shaded. ◨

PRINTOUT 13.2
Minitab graphical display of the chi-square statistic and P-value

Chi-square statistic and P-value

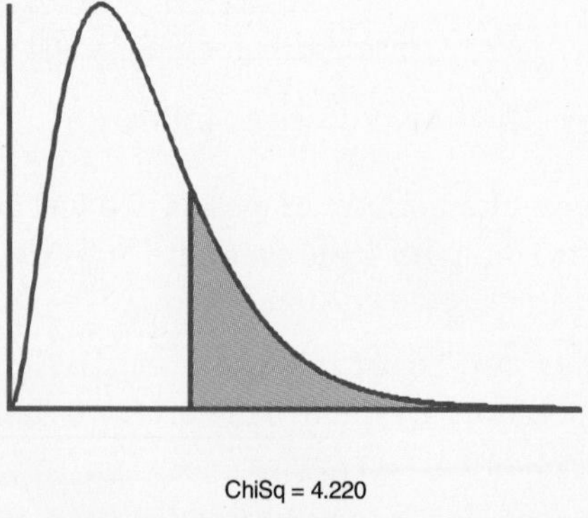

ChiSq = 4.220

P = 0.239

EXERCISES 13.2

Statistical Concepts and Skills

13.11 Why do you think the phrase "goodness of fit" is used to describe the type of hypothesis test considered in this section?

13.12 Explain in your own words the logic behind the chi-square goodness-of-fit test.

13.13 Are the observed frequencies variables? What about the expected frequencies? Explain your answers.

13.14 In each part of this exercise, we have given the relative frequencies for the null hypothesis of a chi-square goodness-of-fit test and the sample size. In each case, decide whether the two assumptions for using that test are satisfied.

a. Sample size: $n = 100$.
Relative frequencies: 0.65, 0.30, 0.05.

b. Sample size: $n = 50$.
Relative frequencies: 0.65, 0.30, 0.05.

c. Sample size: $n = 50$.
Relative frequencies: 0.20, 0.20, 0.25, 0.30, 0.05.

d. Sample size: $n = 50$.
Relative frequencies: 0.22, 0.21, 0.25, 0.30, 0.02.

e. Sample size: $n = 50$.
Relative frequencies: 0.22, 0.22, 0.25, 0.30, 0.01.

f. Sample size: $n = 100$.
Relative frequencies: 0.44, 0.25, 0.30, 0.01.

In each of Exercises 13.15–13.20, perform the required hypothesis test using either the critical-value approach or the P-value approach.

13.15 The American Automobile Manufacturers Association compiles data on U.S. car sales by type of buyer. Here is the 1995 distribution, as reported in the *World Almanac, 1998.*

Type of buyer	Consumer	Business	Government
Rel. frequency	0.497	0.485	0.018

A random sample of last year's U.S. car sales gave the following data.

Type of buyer	Consumer	Business	Government
Frequency	1422	1521	57

At the 5% significance level, do the data provide sufficient evidence to conclude that last year's type-of-buyer distribution for U.S. cars is different from the 1995 distribution?

13.16 The American Automobile Manufacturers Association compiles data on U.S. car sales by type of car. Here is the 1990 distribution, as reported in the *World Almanac, 1998.*

Type of car	Small	Midsize	Large	Luxury
Rel. frequency	0.328	0.448	0.094	0.130

A random sample of last year's U.S. car sales yielded the following data.

Type of car	Small	Midsize	Large	Luxury
Frequency	136	242	54	68

At the 5% significance level, do the data provide sufficient evidence to conclude that last year's type-of-car distribution for U.S. car sales differs from the 1990 distribution?

13.17 The U.S. Centers for Disease Control compiles figures on AIDS deaths by selected characteristics and publishes the information in *Surveillance Report.* According to that report, the percent distribution of AIDS deaths by race in 1992 is as follows.

Race	White	Black	Hispanic	Other
Percentage	51.5	33.1	14.4	1.0

A random sample of AIDS deaths occurring in 1995 gave the following data.

Race	White	Black	Hispanic	Other
Frequency	239	183	72	6

Do the data provide sufficient evidence to conclude that the distribution of AIDS deaths by race in 1995 is different than the 1992 distribution? Use a significance level of 10%.

13.18 According to *Current Housing Reports,* published by the Census Bureau, the primary-heating-fuel distribution of all occupied housing units is as follows.

Primary heating fuel	Percentage
Natural gas	56.7
Fuel oil, kerosene	14.3
Electricity	16.0
Liquid propane gas	4.5
Wood	6.7
Other	1.8

A random sample of 250 occupied housing units built after 1974 yielded the following frequency distribution.

Primary heating fuel	Frequency
Natural gas	91
Fuel oil, kerosene	16
Electricity	110
Liquid propane gas	14
Wood	17
Other	2

Do the data provide sufficient evidence to conclude that the primary-heating-fuel distribution of occupied housing units built after 1974 differs from that of all occupied housing units? Use $\alpha = 0.05$.

13.19 A gambler thought a die was loaded (i.e., the six numbers were not equally likely.) To test his suspicion, he rolled the die 150 times and obtained the data shown in the following table.

Number	1	2	3	4	5	6
Frequency	23	26	23	21	31	26

Do the data provide sufficient evidence to conclude that the die was loaded? Perform the hypothesis test at the 0.05 level of significance.

13.20 A roulette wheel contains 18 red numbers, 18 black numbers, and 2 green numbers. The table below shows the frequency with which the ball landed on each color in 200 trials.

Number	Red	Black	Green
Frequency	88	102	10

At the 5% significance level, do the data suggest that the wheel is out of balance?

EXTENDING THE CONCEPTS AND SKILLS

13.21 In Table 13.4 on page 774, we calculated the sum of the observed frequencies, the sum of the expected frequencies, and the sum of their differences. Strictly speaking, those sums are not needed. However, they serve as a check for computational errors.

a. In general, what common value should the sum of the observed frequencies and the sum of the expected frequencies equal? Explain your answer.

b. Fill in the following blank. The sum of the differences between each observed and expected frequency should equal _____.

c. Suppose you are conducting a chi-square goodness-of-fit test. If the sum of the expected frequencies does not equal the sample size, what can you conclude?

d. Suppose you are conducting a chi-square goodness-of-fit test. If the sum of the expected frequencies equals the sample size, can you conclude that you made no error in calculating the expected frequencies? Explain your answer.

13.22 The chi-square goodness-of-fit test provides a method for performing a hypothesis test about the distribution of a variable having k possible values. If the number of possible values is two, that is, $k = 2$, then the chi-square goodness-of-fit test is equivalent to a procedure that we studied earlier.

a. Which procedure do you think that is? Explain your answer.

b. Suppose we want to perform a hypothesis test to decide whether the proportion of a population having

a specified attribute is different from p_0. Discuss the method for performing such a test using the
i. one-sample z-test for a population proportion (Procedure 12.2 on page 740).
ii. chi-square goodness-of-fit test.

USING TECHNOLOGY  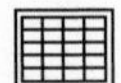 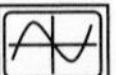

13.23 Use Minitab or some other statistical software to perform the hypothesis test in Exercise 13.15.

13.24 Use Minitab or some other statistical software to perform the hypothesis test in Exercise 13.16.

13.25 The American Medical Association compiles information on physicians and publishes its findings in *Physician Characteristics and Distribution in the U.S.* From that document we obtained the 1995 specialty distribution of U.S. physicians based on the categories general practice, medical, surgical, and other. A random sample was taken of U.S. physicians currently practicing medicine in order to decide whether the current specialty distribution has changed from the 1995 distribution. We applied the Minitab macro `fittest.mac` to the data and obtained the output shown in Printout 13.3. Use the output to determine the

a. number of physicians sampled.
b. value of the chi-square test statistic.
c. P-value for the hypothesis test.
d. smallest significance level at which the null hypothesis can be rejected.
e. conclusion if the test is performed at the 5% significance level.

13.26 *Current Population Reports,* released by the Census Bureau, contains the 1995 income-level distribution of U.S. households. A random sample of households was taken last year in order to decide whether last year's income-level distribution differs from the 1995 distribution. The incomes obtained were adjusted for inflation to 1995 levels. Printout 13.4 displays the output we obtained by applying the Minitab macro `fittest.mac` to the data. Use the output to determine the

a. number of households sampled.
b. value of the chi-square test statistic.
c. P-value for the hypothesis test.
d. smallest significance level at which the null hypothesis can be rejected.
e. conclusion if the test is performed at the 5% significance level.
f. number of categories in the income-level distribution.

PRINTOUT 13.3 Minitab output for Exercise 13.25

```
Chi-square goodness-of-fit test


 Row      n   k    ChiSq   P-value

   1    500   4  4.82575  0.185010
```

PRINTOUT 13.4 Minitab output for Exercise 13.26

```
Chi-square goodness-of-fit test


 Row      n   k    ChiSq   P-value

   1    825   8  6.03240  0.535860
```

13.3 CONTINGENCY TABLES; ASSOCIATION

The next chi-square procedure we will study is the **chi-square independence test.** Before we can do that, however, we need to discuss two prerequisite concepts: *contingency tables* and *association.*

Contingency Tables

In Section 2.2 we learned how to group data from one variable into a frequency distribution. Data obtained by observing values of one variable of a population are called **univariate data.**

Now we will learn how to simultaneously group data from two variables into a frequency distribution. Data obtained by observing values of two variables of a population are called **bivariate data,** and a frequency distribution for bivariate data is called a **contingency table** or **two-way table.** Example 13.5 introduces contingency tables.

EXAMPLE 13.5 INTRODUCES CONTINGENCY TABLES

In Example 2.8 on page 73, we considered data on political party affiliation for the students in Professor Weiss's introductory statistics course. These are univariate data obtained by observing values of the single variable "political party affiliation."

Now we will simultaneously consider data on political party affiliation and class level for the students in Professor Weiss's introductory statistics course, as shown in Table 13.7. These are bivariate data obtained by observing values of the two variables "political party affiliation" and "class level." Group these bivariate data into a contingency table.

SOLUTION A contingency table must provide for each possible pair of values for the two variables. So in this case, the contingency table has the form shown in Table 13.8. The small boxes inside the rectangle formed by the heavy lines of Table 13.8 are called **cells.** The cells will hold the frequencies.

To group the bivariate data in Table 13.7 into the contingency table, we need to determine how many students fall in each cell. We do this by going through the data in Table 13.7 and placing a tally mark in the appropriate cell of Table 13.8 for each student. For instance, the first student is both a Democrat and a freshman and, so, this calls for a tally mark in the upper left cell of Table 13.8. The results of the tallying procedure are shown in Table 13.8.

Now we count the tallies in each cell to determine the frequencies. Replacing the tallies in Table 13.8 by the frequencies, we obtain the contingency table for the bivariate data in Table 13.7, as shown in Table 13.9.

TABLE 13.7 Political party affiliation and class level for students in introductory statistics

Student	Political party	Class level	Student	Political party	Class level
1	Democratic	Freshman	21	Democratic	Junior
2	Other	Junior	22	Democratic	Senior
3	Democratic	Senior	23	Republican	Freshman
4	Other	Sophomore	24	Democratic	Sophomore
5	Democratic	Sophomore	25	Democratic	Senior
6	Republican	Sophomore	26	Republican	Sophomore
7	Republican	Junior	27	Republican	Junior
8	Other	Freshman	28	Other	Junior
9	Other	Sophomore	29	Other	Junior
10	Republican	Sophomore	30	Democratic	Sophomore
11	Republican	Sophomore	31	Republican	Sophomore
12	Republican	Junior	32	Democratic	Junior
13	Republican	Sophomore	33	Republican	Junior
14	Democratic	Junior	34	Other	Senior
15	Republican	Sophomore	35	Other	Sophomore
16	Republican	Senior	36	Republican	Freshman
17	Democratic	Sophomore	37	Republican	Freshman
18	Democratic	Junior	38	Republican	Freshman
19	Other	Senior	39	Democratic	Junior
20	Republican	Sophomore	40	Republican	Senior

TABLE 13.8 Form of contingency table for political party affiliation and class level

		Class level				
		Freshman	Sophomore	Junior	Senior	Total
Political party	Democratic	I	IIII	~~IIII~~	III	
	Republican	IIII	~~IIII~~ III	IIII	II	
	Other	I	III	III	II	
	Total					

TABLE 13.9 Contingency table for political party affiliation and class level

		Class level				
		Freshman	Sophomore	Junior	Senior	Total
Political party	Democratic	1	4	5	3	13
	Republican	4	8	4	2	18
	Other	1	3	3	2	9
	Total	6	15	12	7	40

The number 1 in the upper left cell of Table 13.9 indicates that one student in the course is both a Democrat and a freshman. The number 8, diagonally below and to the right of the 1, shows that eight students in the course are both Republicans and sophomores.

The row total in the first row of Table 13.9 indicates that 13 $(1 + 4 + 5 + 3)$ of the students are Democrats. Similarly, the column total in the third column shows that 12 of the students are juniors. The number 40 in the lower right corner of the table gives the total number of students in the course. That total can be found by summing either the row totals or the column totals; it can also be found by summing the frequencies in the 12 cells of the contingency table. ◨

It is useful for purposes of understanding to group bivariate data into a contingency table by hand, as we did in Example 13.5. However, in practice, computers are almost always used to accomplish such tasks.

Association

Next we need to discuss the concept of *association* for two variables. We do this for variables that are either categorical or quantitative with only finitely many possible values. Roughly speaking, there is an association between two variables of a population if knowing the value of one of the variables imparts information about the value of the other variable. Example 13.6 introduces the concept of association.

EXAMPLE 13.6 INTRODUCES ASSOCIATION

In Example 13.5 we presented data on political party affiliation and class level for the students in Professor Weiss's introductory statistics course. Considering those students a population of interest, decide whether there is an association between the variables "political party affiliation" and "class level."

SOLUTION To decide whether there is an association between the two variables, we first obtain the distribution of political party affiliation within each class level. This is done by dividing each column of the contingency table in Table 13.9 by its column total. The results are shown in Table 13.10.

The first column of Table 13.10 gives the distribution of political party affiliation for freshman: 16.7% are Democrats, 66.7% are Republicans, and 16.7% are Other. This distribution is called the **conditional distribution** of the variable "political party affiliation" corresponding to the value "freshman" of the variable "class level"; or, more simply, the conditional distribution of political party affiliation for freshmen.

TABLE 13.10
Conditional distributions of political party affiliation by class level

Political party	Class level: Freshman	Sophomore	Junior	Senior	Total
Democratic	0.167	0.267	0.417	0.429	0.325
Republican	0.667	0.533	0.333	0.286	0.450
Other	0.167	0.200	0.250	0.286	0.225
Total	1.000	1.000	1.000	1.000	1.000

Similarly, the second, third, and fourth columns of Table 13.10 give the conditional distributions of political party affiliation for sophomores, juniors, and seniors, respectively. The "Total" column provides the (unconditional) distribution of political party affiliation for the entire population which, in this context, is called the **marginal distribution** of the variable "political party affiliation." This is the same distribution that we found in Example 2.8, as seen in Table 2.11 on page 74.

Table 13.10 shows that there is an association between the variables "political party affiliation" and "class level" because knowing the value of the variable "class level" imparts information about the variable "political party affiliation." For instance, as we see from Table 13.10, given no information about the class level of a student in the course, there is a 32.5% chance that the student is a Democrat. On the other hand, if we know that the student is a junior, then there is a 41.7% chance that the student is a Democrat.

If there were no association between the variables "political party affiliation" and "class level," then the four conditional distributions of political party affiliation would be the same as each other and as the marginal distribution of political party affiliation; in other words, all five columns of Table 13.10 would be identical.

A **segmented bar graph** is helpful for understanding the concept of association. The first four bars in the segmented bar graph in Fig. 13.4 on the following page provide the conditional distributions of political party affiliation for freshman, sophomores, juniors, and seniors, respectively, and the fifth bar gives the marginal distribution of political party affiliation.

If there were no association between political party affiliation and class level, then the four bars displaying the conditional distributions of political party affiliation would be the same as each other and as the bar displaying the marginal distribution of political party affiliation; in other words, all five bars in Fig. 13.4 would be identical. The fact that there is an association between political party affiliation and class level is reflected in the segmented bar graph by nonidentical bars.

We note that, alternatively, we could decide whether there is an association between the two variables by obtaining the conditional distribution of class level

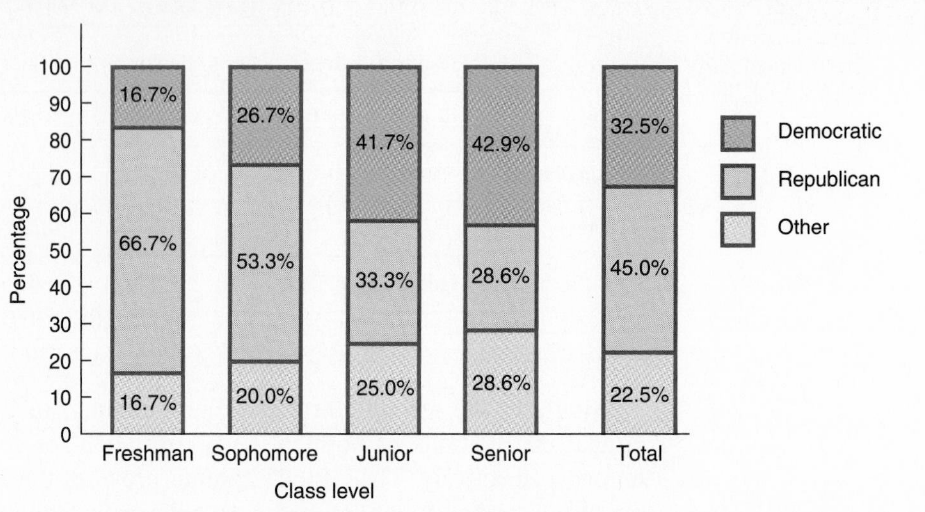

FIGURE 13.4 Segmented bar graph for the conditional distributions and marginal distribution of political party affiliation

within each political party affiliation. It does not matter for which variable we obtain the conditional distributions; the conclusion regarding association (or nonassociation) will be the same. ◨

Keeping in mind the terminology introduced in Example 13.6, we can now define the concept of *association* for two variables.

DEFINITION 13.1 ASSOCIATION

We say that there is an ***association*** between two variables of a population if the conditional distributions of one variable given the other are not identical.

The phrase **statistically dependent** is also used to express the fact that there is an association between two variables. And, likewise, the phrase **statistically independent** is often used to indicate that there is no association between two variables.

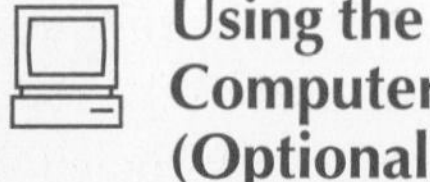

Using the Computer (Optional) We can use Minitab to group bivariate data into a contingency table and to obtain conditional and marginal distributions as well. To illustrate how this is done, we return once again to the data on political party affiliation and class level for the students in Professor Weiss's introductory statistics course.

EXAMPLE 13.7 **USING MINITAB TO OBTAIN CONTINGENCY TABLES AND DISTRIBUTIONS**

Refer to Table 13.7 on page 785. Use Minitab to

a. group the political-party-affiliation and class-level data into a contingency table.

b. determine the conditional distributions of political party affiliation within class levels and the marginal distribution of political party affiliation.

SOLUTION First we store the data in Table 13.7 in columns named PARTY and CLASS. In contingency tables and other output, Minitab's default procedure is to list the possible values of each variable in alphabetical order; so, we need to take some preliminary steps to obtain the orderings displayed in Table 13.9. To order the possible values of political party affiliation as Democratic, Republican, and Other, we do the following.

1 Click anywhere in the column named PARTY
2 Choose **Editor ➤ Set Column ➤ Value Order...**
3 Select the **User-specified order** option button
4 In the **Define an order (one value per line)** text box, edit the order of the possible political party affiliations so that it reads Democratic, Republican, and Other
5 Click **OK**

Proceed in a similar manner to order the possible values of class level as Freshman, Sophomore, Junior, and Senior. Now we are ready to accomplish the tasks at hand.

a. To obtain the contingency table, we proceed as follows.

1 Choose **Stat ➤ Tables ➤ Cross Tabulation...**
2 Specify PARTY and CLASS in the **Classification variables** text box
3 Select the **Counts** check box in the **Display** list
4 Click **OK**

The resulting output is shown in Printout 13.5 on the following page. Compare the contingency table in Printout 13.5 to the one we obtained by hand in Table 13.9 on page 785.

b. To determine the conditional distributions of political party affiliation within class levels and the marginal distribution of political party affiliation, we proceed as we did in part (a) except that at step 3, we select the **Column percents** check box instead of the **Counts** check box (and unselect the **Counts** check box). Printout 13.6 displays the output that results.

PRINTOUT 13.5
Minitab contingency table for political party affiliation and class level

```
Rows: PARTY      Columns: CLASS

             Freshman Sophomor    Junior    Senior       All

Democrat            1        4         5         3        13
Republic            4        8         4         2        18
Other               1        3         3         2         9
All                 6       15        12         7        40

  Cell Contents --
                   Count
```

PRINTOUT 13.6
Minitab conditional and marginal distributions for political party affiliation

```
Rows: PARTY      Columns: CLASS

             Freshman Sophomor    Junior    Senior       All

Democrat        16.67    26.67     41.67     42.86     32.50
Republic        66.67    53.33     33.33     28.57     45.00
Other           16.67    20.00     25.00     28.57     22.50
All            100.00   100.00    100.00    100.00    100.00

  Cell Contents --
                   % of Col
```

Note that Minitab displays percent distributions instead of relative-frequency distributions. Compare these distributions to the ones we obtained by hand in Table 13.10 on page 787. ◨

EXERCISES 13.3

Statistical Concepts and Skills

13.27 Provide an example of univariate data; of bivariate data.

13.28 Identify the type of table that is used to group bivariate data.

13.29 What are the small boxes inside the heavy lines of a contingency table called?

13.30 Suppose that bivariate data are to be grouped into a contingency table. Determine the number of cells that the contingency table will have if the number of possible values for the two variables are

a. two and three. **b.** four and three.
c. m and n.

13.31 Identify three ways in which the total number of observations of bivariate data can be obtained from the frequencies in a contingency table.

13.32 Congressional Quarterly, Inc., provides information about the popular vote cast for president, by political party and region, in *America Votes.* According to that publication, in the 1996 presidential election, 40.7% of voting Americans voted for the Republican candidate, whereas 45.2% of voting Americans living in the South did so. For that presidential election, is there an association between the variables "party of presidential candidate voted for" and "region of residence" for Americans who voted? Explain your answer.

13.33 The American Medical Association compiles information on U.S. physicians in *Physician Characteristics and Distribution in the U.S.* According to that document, in 1996, 15.9% of male physicians specialized in internal medicine and 19.1% of female physicians specialized in internal medicine. Is there an association between the variables "sex" and "specialty" for U.S. physicians practicing medicine in 1996? Explain your answer.

Table 13.11 provides data on sex, class level, and college for the students in one section of the course *Introduction to Computer Science* in the spring semester of 1998 at Arizona State University. In the table, we have used the abbreviations BUS for Business, ENG for Engineering and Applied Sciences, and LIB for Liberal Arts and Sciences.

TABLE 13.11 Sex, class level, and college for students in introduction to computer science

Sex	Class	College	Sex	Class	College
M	Junior	ENG	F	Soph	BUS
M	Soph	ENG	F	Junior	ENG
F	Senior	BUS	M	Junior	LIB
F	Junior	BUS	F	Junior	BUS
M	Junior	ENG	M	Soph	BUS
F	Junior	LIB	M	Junior	BUS
M	Senior	LIB	M	Soph	ENG
M	Soph	ENG	M	Junior	ENG
M	Junior	ENG	M	Junior	ENG
M	Soph	ENG	M	Soph	LIB
F	Soph	BUS	F	Senior	ENG
F	Junior	BUS	F	Senior	BUS
M	Junior	ENG			

Exercises 13.34–13.36 use the data in Table 13.11.

13.34 Refer to Table 13.11. Consider the variables "sex" and "class level."

a. Group the bivariate data for these two variables into a contingency table.

b. Determine the conditional distribution of sex within each class level and the marginal distribution of sex.

c. Determine the conditional distribution of class level within each sex and the marginal distribution of class level.

d. Is there an association between the variables "sex" and "class level" for this population? Explain your answer.

13.35 Refer to Table 13.11. Consider the variables "sex" and "college."

a. Group the bivariate data for these two variables into a contingency table.

b. Determine the conditional distribution of sex within each college and the marginal distribution of sex.

c. Determine the conditional distribution of college within each sex and the marginal distribution of college.

d. Is there an association between the variables "sex" and "college" for this population? Why?

13.36 Refer to Table 13.11. Consider the variables "class level" and "college."

a. Group the bivariate data for these two variables into a contingency table.

b. Determine the conditional distribution of class level within each college and the marginal distribution of class level.

c. Determine the conditional distribution of college within each class level and the marginal distribution of college.

d. Is there an association between the variables "class level" and "college" for this population? Explain your answer.

Table 13.12 provides theoretical data on political party affiliation and class level for the students in a night-school course.

TABLE 13.12 Political party affiliation and class level for the students in a night-school course

Party	Class	Party	Class	Party	Class
Rep	Jun	Rep	Soph	Rep	Jun
Dem	Soph	Other	Jun	Rep	Soph
Dem	Jun	Dem	Soph	Rep	Soph
Other	Jun	Rep	Soph	Rep	Fresh
Dem	Jun	Dem	Sen	Rep	Soph
Dem	Fresh	Rep	Jun	Rep	Jun
Dem	Soph	Dem	Jun	Rep	Sen
Dem	Sen	Dem	Jun	Rep	Jun
Other	Sen	Rep	Sen	Dem	Soph
Dem	Fresh	Rep	Fresh	Rep	Jun
Rep	Jun	Rep	Jun	Other	Jun
Rep	Jun	Dem	Jun	Dem	Jun
Dem	Sen	Rep	Sen	Other	Soph
Rep	Jun	Rep	Sen	Rep	Sen
Dem	Sen	Rep	Sen	Other	Soph
Rep	Jun	Dem	Soph	Rep	Soph
Rep	Soph	Other	Fresh	Other	Soph
Rep	Fresh	Rep	Soph	Other	Sen
Rep	Jun	Other	Jun	Rep	Soph
Dem	Soph	Dem	Jun	Dem	Jun

We will use the data in Table 13.12 in Exercises 13.37 and 13.38.

13.37 Refer to Table 13.12.

a. Group the bivariate data for the two variables into a contingency table.

b. Determine the conditional distribution of political party affiliation within each class level.

c. Is there an association between the variables "political party affiliation" and "class level" for this population of night-school students? Explain your answer.

d. Without doing further calculation, determine the marginal distribution of political party affiliation.

e. Without doing further calculation, respond true or false to the following statement and explain your answer. The conditional distributions of class level within political party affiliations are identical to each other and to the marginal distribution of class level.

13.38 Refer to Table 13.12.

a. If you have not done Exercise 13.37, group the bivariate data for the two variables into a contingency table.

b. Determine the conditional distribution of class level within each political party affiliation.

c. Is there an association between the variables "political party affiliation" and "class level" for this population of night-school students? Explain your answer.

d. Without doing further calculation, determine the marginal distribution of class level.

e. Without doing further calculation, respond true or false to the following statement and explain your answer. The conditional distributions of political party affiliation within class levels are identical to each other and to the marginal distribution of political party affiliation.

13.39 The U.S. National Center for Education Statistics compiles information on institutions of higher education and publishes its findings in *Digest of Education Statistics*. Following is a contingency table giving the number of institutions of higher education in the United States by region and type.

		Type		
		Public	Private	Total
Region	Northeast	266		821
	Midwest	359	504	863
	South	533	502	
	West	313	242	555
	Total		1803	

a. How many cells does this contingency table have?

b. Fill in the missing entries.

c. What is the total number of institutions of higher education in the United States?

d. How many institutions are in the Midwest?

e. How many are public?

f. How many are private schools in the South?

13.40 As reported by the Motor Vehicle Manufacturers Association of the United States in *Motor Vehicle Facts and Figures*, the number of cars and trucks in use by age are as shown in the following contingency table. Frequencies are in millions.

Age (yrs)	Type: Car	Truck	Total
Under 6	46.2	27.8	74.0
6–8	26.9		40.0
9–11	23.3	10.7	
12 & over	26.8	18.6	45.4
Total	123.2		

a. How many cells does this contingency table have?
b. Fill in the missing entries.
c. What is the total number of cars and trucks in use?
d. How many vehicles are trucks?
e. How many vehicles are between 6 and 8 years old?
f. How many vehicles are trucks that are between 9 and 11 years old?

13.41 The American Medical Association compiles information on U.S. physicians in *Physician Characteristics and Distribution in the U.S.* Following is a contingency table for U.S. surgeons cross classified by specialty and base of practice.

Specialty	Base of practice: Office	Hospital	Other	Total
General surgery	24,128	12,225	1,658	38,011
Obstetrics/ gynecology	24,150	6,734	1,140	32,024
Orthopedics	13,364	4,248	414	18,026
Ophthal- mology	12,328	2,694	518	15,540
Total	73,970	25,901	3,730	103,601

a. How many surgeons are office-based?
b. How many surgeons are ophthalmologists?
c. How many surgeons are office-based ophthalmologists?
d. How many surgeons are either office-based or ophthalmologists?
e. How many general surgeons are hospital-based?
f. How many hospital-based surgeons are OB/GYNs?
g. How many surgeons are not hospital based?

13.42 The American Hospital Association publishes information about U.S. hospitals and nursing homes in *Hospital Statistics.* Here is a contingency table providing a cross classification of U.S. hospitals and nursing homes by type of facility and number of beds.

Facility	Number of beds: 24−	25–74	75+	Total
General	260	1586	3557	5403
Psychiatric	24	242	471	737
Chronic	1	3	22	26
Tuberculosis	0	2	2	4
Other	25	177	208	410
Total	310	2010	4260	6580

In the following questions, we will for brevity use the term "hospital" to refer to either a hospital or nursing home.

a. How many hospitals have at least 75 beds?
b. How many psychiatric facilities are there?
c. How many hospitals are psychiatric facilities with at least 75 beds?
d. How many hospitals either are psychiatric facilities or have at least 75 beds?
e. How many general facilities have between 25 and 74 beds?
f. How many hospitals with between 25 and 74 beds are chronic facilities?
g. How many hospitals have more than 24 beds?

13.43 Refer to Exercise 13.39.

a. Determine the conditional distribution of type of educational institution within each region.

b. Determine the marginal distribution of type of educational institution.
c. Is there an association between the variables "type" and "region" for institutions of higher education in the United States? Explain your answer.
d. Obtain the percentage of institutions of higher education that are private.
e. Obtain the percentage of institutions of higher education in the Northeast that are private.
f. Without doing further calculations, respond true or false to the following statement and explain your answer. The conditional distributions of region within types of educational institutions are not identical.
g. Determine and interpret the marginal distribution of region and the conditional distributions of region within types of educational institutions.

13.44 Refer to Exercise 13.40.
Here we will use the term "vehicle" to mean either a U.S. car or truck currently in use.
a. Determine the conditional distribution of age group for each type of vehicle.
b. Determine the marginal distribution of age group for vehicles.
c. Is there an association between the variables "type" and "age group" for vehicles? Explain your answer.
d. Find the percentage of vehicles under 6 years old.
e. Find the percentage of cars under 6 years old.
f. Without doing any further calculations, respond true or false to the following statement and explain your answer. The conditional distributions of type of vehicle within age groups are not identical.
g. Determine and interpret the marginal distribution of type of vehicle and the conditional distributions of type of vehicle within age groups.

13.45 Refer to Exercise 13.41.
a. Find the conditional distribution of specialty within each base-of-practice category.
b. Is there an association between specialty and base of practice for U.S. surgeons? Explain your answer.
c. Determine the marginal distribution of specialty for U.S. surgeons.
d. Construct a segmented bar graph for the conditional distributions of specialty and marginal distribution of specialty that you obtained in parts (a) and (c), respectively. Interpret the graph in light of your answer to part (b).
e. Without doing any further calculations, respond true or false to the following statement and explain your answer. The conditional distributions of base of practice within specialties are identical.
f. Determine the marginal distribution of base of practice and the conditional distributions of base of practice within specialties.
g. Find the percentage of surgeons that are hospital-based.
h. Find the percentage of OB/GYNs that are hospital-based.
i. Find the percentage of hospital-based surgeons that are OB/GYNs.

13.46 Refer to Exercise 13.42.
a. Determine the conditional distribution of number of beds within each facility type.
b. Is there an association between facility type and number of beds for U.S. hospitals? Explain your answer.
c. Determine the marginal distribution of number of beds for U.S. hospitals.
d. Construct a segmented bar graph for the conditional distributions and marginal distribution of number of beds. Interpret the graph in light of your answer to part (b).
e. Without doing any further calculations, respond true or false to the following statement and explain your answer. The conditional distributions of facility type within number-of-beds categories are identical.
f. Obtain the marginal distribution of facility type and the conditional distributions of facility type within number-of-beds categories.
g. What percentage of hospitals are general facilities?
h. What percentage of hospitals having at least 75 beds are general facilities?
i. What percentage of general facilities have at least 75 beds?

EXTENDING THE CONCEPTS AND SKILLS

13.47 In this exercise, we will consider two variables, x and y, defined on a hypothetical population.

Following are the conditional distributions of the variable y corresponding to each value of the variable x.

		x: A	B	C	Total
y	0	0.316	0.316	0.316	
	1	0.422	0.422	0.422	
	2	0.211	0.211	0.211	
	3	0.047	0.047	0.047	
	4	0.004	0.004	0.004	
	Total	1.000	1.000	1.000	

a. Is there an association between the variables x and y? Explain your answer.
b. Determine the marginal distribution of y.
c. Can you determine the marginal distribution of x? Explain your answer.

13.48 The Bureau of the Census publishes census data on the resident population of the United States in *Current Population Reports*. According to that document, 6.9% of male residents are in the age group 20–24 years.

a. If there were no association between age group and sex, what percentage of the resident population would be in the age group 20–24 years? Explain your answer.
b. If there were no association between age group and sex, what percentage of female residents would be in the age group 20–24 years? Explain your answer.
c. There are (roughly) 135.5 million female residents of the United States. If there were no association between age group and sex, how many female residents would there be in the age group 20–24 years?
d. In fact there are (roughly) 8.6 million female residents in the age group 20–24 years. Given this and your answer to part (c), what can you conclude?

USING TECHNOLOGY  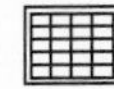 

13.49 Use Minitab or some other statistical software to carry out parts (a)–(c) of Exercise 13.35.

13.50 Use Minitab or some other statistical software to carry out parts (a)–(c) of Exercise 13.36.

13.51 Use Minitab or some other statistical software to carry out parts (a) and (b) of Exercise 13.37.

13.52 Use Minitab or some other statistical software to carry out parts (a) and (b) of Exercise 13.38.

13.4 CHI-SQUARE INDEPENDENCE TEST

In Section 13.3 we learned how to determine whether there is an association between two variables of a population, assuming that we have the bivariate data for the entire population. However, because in most cases, data for an entire population are not available, we must usually apply inferential methods to decide whether an association exists between two variables. One of the most commonly used procedures for making such decisions is the chi-square independence test. Example 13.8 introduces and explains the reasoning behind the chi-square independence test.

EXAMPLE 13.8 INTRODUCES THE CHI-SQUARE INDEPENDENCE TEST

A national survey was conducted to obtain information on the alcohol consumption patterns of U.S. adults by marital status. A random sample of 1772 residents,

18 years old and over, yielded the data displayed in Table 13.13.[†] The table shows, for instance, that of the 1772 adults sampled, 1173 are married, 590 abstain, and 411 are married and abstain.

TABLE 13.13 Contingency table of marital status and alcohol consumption for 1772 randomly selected U.S. adults

Marital status	Drinks per month: Abstain	1–60	Over 60	Total
Single	67	213	74	354
Married	411	633	129	1173
Widowed	85	51	7	143
Divorced	27	60	15	102
Total	590	957	225	1772

We want to use the sample data to decide whether there is an association between marital status and alcohol consumption; that is, we want to perform the hypothesis test

H_0: Marital status and alcohol consumption are not associated.

H_a: Marital status and alcohol consumption are associated.

The idea behind the chi-square independence test is to compare the observed frequencies in Table 13.13 with the frequencies that would be expected if the null hypothesis of nonassociation is true. The test statistic employed to make the comparison is the same one used for the goodness-of-fit test: $\chi^2 = \Sigma(O - E)^2/E$ where O represents observed frequency and E represents expected frequency.

We will now develop a formula for computing the expected frequencies. Consider, for instance, the cell of Table 13.13 corresponding to "Married *and* Abstain," the cell in the second row and first column of the table. To begin, we note that the population proportion of all adults who abstain can be estimated by the sample proportion of the 1772 adults sampled who abstain, that is, by

Number sampled who abstain ↘

$$\frac{590}{1772} = 0.333\ (33.3\%).$$

Total number sampled ↗

[†] Adapted from "Alcohol Use and Alcohol Problems among U.S. Adults: Results of the 1979 National Survey" by W. B. Clark and L. Midanik. In National Institute on Alcohol Abuse and Alcoholism, *Alcohol and Health Monograph No. 1, Alcohol Consumption and Related Problems.* DHHS Pub. No. (ADM) 82–1190, 1982.

If no association exists between marital status and alcohol consumption (i.e., if H_0 is true), then the proportion of married adults who abstain is the same as the proportion of all adults who abstain. Thus, if H_0 is true, the sample proportion, $\frac{590}{1772}$, or 33.3%, is also an estimate of the population proportion of married adults who abstain. This, in turn, implies that of the 1173 married adults sampled, we would expect roughly

$$\frac{590}{1772} \cdot 1173 = 390.6$$

to abstain from alcohol.

It is useful to rewrite the left side of this expected-frequency computation in a slightly different way. By using algebra and referring to Table 13.13, we obtain the following equalities:

$$\text{Expected frequency} = \frac{590}{1772} \cdot 1173 = \frac{1173 \cdot 590}{1772}$$

$$= \frac{(\text{Row total}) \cdot (\text{Column total})}{\text{Sample size}}$$

If we let R denote "Row total" and C denote "Column total," then we can express this equation compactly as

$$E = \frac{R \cdot C}{n},$$

where, as usual, E denotes expected frequency and n denotes sample size.

Using this simple formula, we can obtain the expected frequencies for all 12 cells in Table 13.13. We have already done that for the cell in the second row and first column. For the cell in the upper right corner of the table, we get

$$E = \frac{R \cdot C}{n} = \frac{354 \cdot 225}{1772} = 44.9.$$

Similar computations give the expected frequencies for the remaining cells.

In Table 13.14 we have modified Table 13.13 by placing the expected frequency for each cell beneath the corresponding observed frequency. Table 13.14 shows, for instance, that of the adults sampled, 74 were observed to be single and consume over 60 drinks per month, whereas if there is no association between marital status and alcohol consumption, the expected frequency is 44.9.

If the null hypothesis of nonassociation is true, then the observed and expected frequencies should be roughly equal, resulting in a relatively small value of the test statistic, $\chi^2 = \Sigma(O - E)^2/E$. Consequently, if χ^2 is too large, we will reject the null hypothesis and conclude that an association exists between marital status and

TABLE 13.14
Observed and expected frequencies for marital status and alcohol consumption; expected frequencies are printed below observed frequencies

Marital status	Drinks per month: Abstain	1–60	Over 60	Total
Single	67 117.9	213 191.2	74 44.9	354
Married	411 390.6	633 633.5	129 148.9	1173
Widowed	85 47.6	51 77.2	7 18.2	143
Divorced	27 34.0	60 55.1	15 13.0	102
Total	590	957	225	1772

alcohol consumption. From Table 13.14 we find that

$$\begin{aligned}
\chi^2 &= \Sigma(O-E)^2/E \\
&= (67-117.9)^2/117.9 + (213-191.2)^2/191.2 + (74-44.9)^2/44.9 \\
&\quad + (411-390.6)^2/390.6 + (633-633.5)^2/633.5 + (129-148.9)^2/148.9 \\
&\quad + (85-47.6)^2/47.6 + (51-77.2)^2/77.2 + (7-18.2)^2/18.2 \\
&\quad + (27-34.0)^2/34.0 + (60-55.1)^2/55.1 + (15-13.0)^2/13.0 \\
&= 21.952 + 2.489 + 18.776 + 1.070 + 0.000 + 2.670 \\
&\quad + 29.358 + 8.908 + 6.856 + 1.427 + 0.438 + 0.324 \\
&= 94.269.
\end{aligned}$$

Can this value be reasonably attributed to sampling error, or is it large enough to indicate that marital status and alcohol consumption are associated? Before we can answer that question, we must know the distribution of the χ^2-statistic. ◨

Note: In the calculation done at the end of Example 13.8, we displayed the expected frequencies to one decimal place and the chi-square subtotals to three decimal places. But the calculations were done using full calculator accuracy.

KEY FACT 13.3 **DISTRIBUTION OF THE χ^2-STATISTIC FOR A CHI-SQUARE INDEPENDENCE TEST**

For a chi-square independence test, the test statistic

$$\chi^2 = \Sigma(O-E)^2/E$$

has approximately a chi-square distribution if the null hypothesis of nonassociation is true. The number of degrees of freedom is $(r-1)(c-1)$, where r and c are the number of possible values for the two variables under consideration.

Procedure for the Chi-Square Independence Test

We can now state a procedure for performing a chi-square independence test, specifically, Procedure 13.2 on the following page. Since the null hypothesis will be rejected only when the test statistic is too large, the rejection region is always on the right, that is, the hypothesis test is always right-tailed.

EXAMPLE 13.9 ILLUSTRATES PROCEDURE 13.2

A random sample of 1772 U.S. adults yielded the data on marital status and alcohol consumption displayed in Table 13.13 on page 796. At the 5% significance level, do the data provide sufficient evidence to conclude that an association exists between marital status and alcohol consumption?

SOLUTION We employ Procedure 13.2.

Step 1 *State the null and alternative hypotheses.*

The null and alternative hypotheses are

H_0: Marital status and alcohol consumption are not associated.

H_a: Marital status and alcohol consumption are associated.

Step 2 *Calculate the expected frequencies using the formula*

$$E = \frac{R \cdot C}{n},$$

where R = row total, C = column total, and n = sample size. Place each expected frequency below its corresponding observed frequency in the contingency table.

We did this earlier in Table 13.14.

Step 3 *Check whether the expected frequencies satisfy Assumptions 1 and 2.*

1. All expected frequencies are 1 or greater? Yes, as we see from Table 13.14.
2. At most 20% of the expected frequencies are less than 5? Yes, in fact, none of the expected frequencies are less than 5, as we see from Table 13.14.

PROCEDURE 13.2 **THE CHI-SQUARE INDEPENDENCE TEST**

ASSUMPTIONS

1. All expected frequencies are 1 or greater.

2. At most 20% of the expected frequencies are less than 5.

Step 1 The null and alternative hypotheses are

H_0: The two variables under consideration are not associated.

H_a: The two variables under consideration are associated.

Step 2 Calculate the expected frequencies using the formula

$$E = \frac{R \cdot C}{n},$$

where R = row total, C = column total, and n = sample size. Place each expected frequency below its corresponding observed frequency in the contingency table.

Step 3 Check whether the expected frequencies satisfy Assumptions 1 and 2. If they do not, this procedure should not be used.

Step 4 Decide on the significance level, α.

Step 5 The critical value is χ^2_α with df $= (r - 1)(c - 1)$, where r and c are the number of possible values for the two variables under consideration. Use Table VIII to find the critical value.

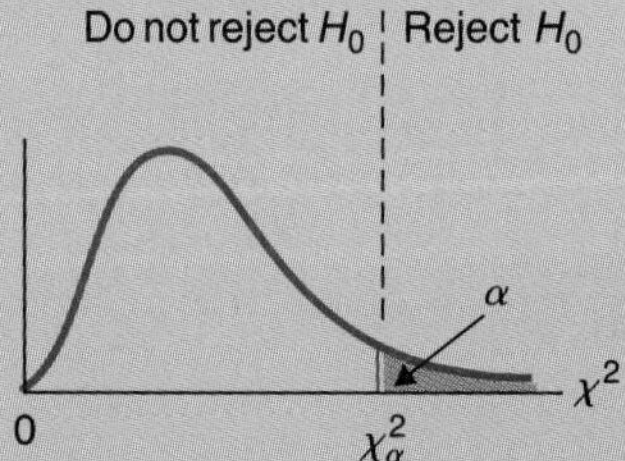

Step 6 Compute the value of the test statistic

$$\chi^2 = \Sigma(O - E)^2/E,$$

where O and E represent observed and expected frequencies, respectively.

Step 7 If the value of the test statistic falls in the rejection region, reject H_0; otherwise, do not reject H_0.

Step 8 State the conclusion in words.

Step 4 *Decide on the significance level, α.*

The test is to be performed at the 5% significance level. Hence $\alpha = 0.05$.

Step 5 *The critical value is χ^2_α with df $= (r-1)(c-1)$, where r and c are the number of possible values for the two variables under consideration.*

The number of marital-status categories is four and the number of drinks-per-month categories is three. So, $r = 4$, $c = 3$, and

$$\text{df} = (r-1)(c-1) = (4-1)(3-1) = 3 \cdot 2 = 6.$$

Since $\alpha = 0.05$, we find from Table VIII that the critical value is $\chi^2_{0.05} = 12.592$, as seen in Fig. 13.5.

FIGURE 13.5
Criterion for deciding whether or not to reject the null hypothesis

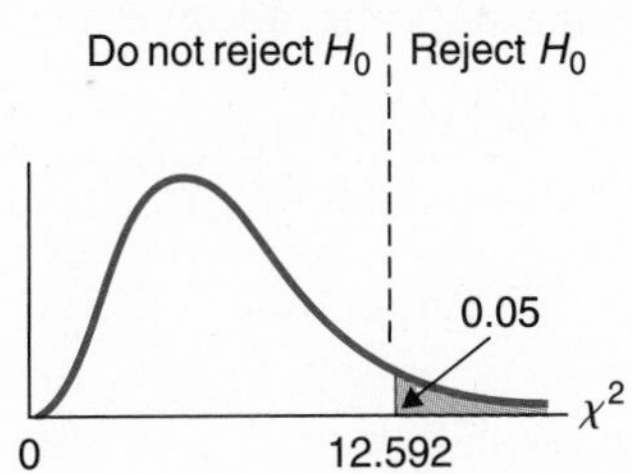

Step 6 *Compute the value of the test statistic*

$$\chi^2 = \Sigma(O-E)^2/E,$$

where O and E represent observed and expected frequencies, respectively.

The observed and expected frequencies are displayed in Table 13.14. Using these we compute the value of the test statistic:

$$\begin{aligned}\chi^2 &= (67-117.9)^2/117.9 + (213-191.2)^2/191.2 + \cdots + (15-13.0)^2/13.0 \\ &= 21.952 + 2.489 + \cdots + 0.324 = 94.269.\end{aligned}$$

Step 7 *If the value of the test statistic falls in the rejection region, reject H_0; otherwise, do not reject H_0.*

From Step 6, the value of the test statistic is $\chi^2 = 94.269$, which falls in the rejection region, as we observe from Fig. 13.5. Thus we reject H_0.

Step 8 *State the conclusion in words.*

 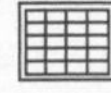 

The test results are statistically significant at the 5% level; that is, at the 5% significance level, the data provide sufficient evidence to conclude that there is an association between marital status and alcohol consumption. ❑

Using the *P*-Value Approach

The P-value approach to hypothesis testing can also be used to carry out the hypothesis test in Example 13.9 and to assess more precisely the evidence against the null hypothesis. From Step 6 we see that the value of the test statistic is $\chi^2 = 94.269$. Recalling that df $= 6$, we find from Table VIII that $P < 0.005$.

Because the P-value is less than the significance level of 0.05, we can reject H_0. Furthermore, by referring to Table 9.12 on page 543, we see that the data provide very strong evidence against the null hypothesis and hence in favor of the alternative hypothesis that marital status and alcohol consumption are associated.

Concerning the Assumptions

In Procedure 13.2 we made two assumptions about expected frequencies:

1. All expected frequencies are 1 or greater.

2. At most 20% of the expected frequencies are less than 5.

What can we do if one or both of these assumptions are violated? Three approaches are possible. We can combine rows or columns to increase the expected frequencies in those cells in which they are too small; we can eliminate certain rows or columns in which the small expected frequencies occur; or we can increase the sample size.

Association and Causation

The chi-square independence test is used to decide whether an association exists between two variables of a population—the null hypothesis is that the two variables are not associated and the alternative hypothesis is that they are associated. If the null hypothesis is rejected, we can conclude that the two variables are associated, but not that they are causally related.

For instance, in Example 13.9 we rejected the null hypothesis of nonassociation for the variables marital status and alcohol consumption. This means that knowing the marital status of a person imparts information about the alcohol consumption of that person, and vice versa. It does not necessarily mean, for instance, that being single causes a person to drink more.

We emphasize this very important fact: *Association does not imply causation!* On the other hand, if two variables are not associated, there is no point in looking for a causal relationship.

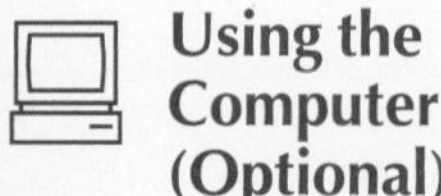

Procedure 13.2 gives a step-by-step method for performing a chi-square independence test. Alternatively, we can use Minitab's procedure for such a test. To illustrate, we return once again to the bivariate data on marital status and alcohol consumption.

EXAMPLE 13.10 USING MINITAB TO PERFORM A CHI-SQUARE INDEPENDENCE TEST

Use Minitab to carry out the hypothesis test considered in Example 13.9.

SOLUTION We want to perform the hypothesis test

H_0: Marital status and alcohol consumption are not associated.

H_a: Marital status and alcohol consumption are associated.

at the 5% significance level.

To employ Minitab, we first store the cell data appearing in the first three columns of Table 13.13 on page 796 in columns named ABSTAIN, 1-60, and OVER 60. Then we proceed in the following manner.

1 Choose **Stat ➤ Tables ➤ Chi-Square Test...**
2 Specify ABSTAIN, '1-60', and 'OVER 60' in the **Columns containing the table** text box
3 Click **OK**

The resulting output is shown in Printout 13.7.

PRINTOUT 13.7
Minitab output for the chi-square independence test

```
Expected counts are printed below observed counts

          ABSTAIN      1-60   OVER 60      Total
    1          67       213        74        354
           117.87    191.18     44.95

    2         411       633       129       1173
           390.56    633.50    148.94

    3          85        51         7        143
            47.61     77.23     18.16

    4          27        60        15        102
            33.96     55.09     12.95

Total         590       957       225       1772

Chi-Sq = 21.952 +  2.489 + 18.776 +
          1.070 +  0.000 +  2.670 +
         29.358 +  8.908 +  6.856 +
          1.427 +  0.438 +  0.324 = 94.269
DF = 6, P-Value = 0.000
```

The first part of Printout 13.7 provides a table of the observed and expected frequencies, which is Minitab's version of Table 13.14 on page 798. Below the table we find the value of the test statistic, $\chi^2 = \Sigma(O - E)^2/E$, including the chi-square subtotals. We see that $\chi^2 = 94.269$.

The last line of Printout 13.7 gives the number of degrees of freedom (`DF = 6`) and the P-value, which is zero to three decimal places. Since the P-value is less than the specified significance level of 0.05, we reject H_0. At the 5% significance level, the data provide sufficient evidence to conclude that there is an association between marital status and alcohol consumption. □

EXERCISES 13.4

STATISTICAL CONCEPTS AND SKILLS

13.53 In deciding whether two variables of a population are associated, we usually need to resort to inferential methods such as the chi-square independence test. Why is that so?

13.54 Step 1 of Procedure 13.2 on page 800 gives generic statements for the null and alternative hypotheses of a chi-square independence test. Restate those hypotheses using the terms *statistically dependent* and *statistically independent,* introduced on page 788.

13.55 In Example 13.8 we made the following statement: If no association exists between marital status and alcohol consumption, then the proportion of married adults who abstain is the same as the proportion of all adults who abstain. Explain why that statement is true.

13.56 Explain why a chi-square independence test is always right tailed.

13.57 A chi-square independence test is to be conducted to decide whether an association exists between two variables of a population. One variable has six possible values and the other variable has four. What is the degrees of freedom for the χ^2-statistic?

13.58 Studies have shown that a positive association exists between educational level and annual salary; in other words, people with more education tend to make more money.

a. Does this mean that more education *causes* a person to make more money? Explain your answer.

b. Do you think there is a causal relationship between educational level and annual salary? Explain your answer.

13.59 We stated earlier that if two variables are not associated, then there is no point in looking for a causal relationship. Why is that so?

13.60 Identify three techniques that can be tried as a remedy when one or more of the expected-frequency assumptions for a chi-square independence test are violated.

In Exercises 13.61–13.66, perform a chi-square independence test using either the critical-value approach or the P-value approach, provided the conditions for using the test are met.

13.61 The U.S. Bureau of the Census compiles data on money income of families by selected characteristics and publishes the information in *Current Population Reports.* A study to decide whether an association exists between family income and educational attainment of householder yielded the following sample data.

Educational attainment

Family income ($1000s)	Not HS grad	HS grad	College grad	Total
Under 25	65	63	14	142
25 < 50	35	84	38	157
50 < 75	11	44	43	98
75 or more	4	23	72	99
Total	115	214	167	496

At the 1% significance level, do the data provide sufficient evidence to conclude that an association exists between family income and educational attainment of householder?

13.62 In 1993 approximately 62.1 million Americans suffered injuries, as reported by the National Center for Health Statistics in *Vital and Health Statistics.* More males (33.4 million) were injured than females (28.7 million). Those statistics do not tell us whether males and females tend to be injured in similar circumstances. One set of categories commonly used for accident circumstance is "while at work," "home," "motor vehicle," and "other." A random sample of accident reports gave the data in the contingency table below.

Sex

Circumstance	Male	Female	Total
While at work	18	4	22
Home	26	28	54
Motor vehicle	4	6	10
Other	36	24	60
Total	84	62	146

At the 5% significance level, do the data provide sufficient evidence to conclude that an association exists between accident circumstance and sex?

13.63 The Gallup Organization conducts periodic surveys to gauge the support by U.S. adults for regional primary elections. The question asked is, "It has been proposed that four individual primaries be held in different weeks of June during presidential election years. Does this sound like a good idea or a poor idea?" Here is a contingency table for responses by political affiliation, adapted from the results of a Gallup Poll appearing in *The Arizona Republic.*

Response

Political affiliation	Good idea	Poor idea	No opinion	Total
Rep	266	266	186	718
Dem	308	250	176	734
Ind	28	27	21	76
Total	602	543	383	1528

At the 5% level of significance, do the data suggest that the feelings of U.S. adults on the issue of regional primaries are associated with political affiliation?

13.64 The U.S. Federal Bureau of Investigation compiles information on arrests for violent crimes by type of crime committed and age of person arrested. Results are published in *Crime in the United States.* To decide whether there is an association between the type of violent crime committed and the age of the person arrested, 750 arrest records were randomly selected. The data obtained are summarized in the following contingency table.

Age

Type of violent crime	18–24	25–44	45+	Total
Murder	11	16	4	31
Forcible rape	21	26	4	51
Robbery	128	92	6	226
Aggravated assault	162	234	46	442
Total	322	368	60	750

Is there evidence that an association exists between the type of violent crime committed and the age of the person arrested? Perform the required hypothesis test at the 1% significance level.

13.65 *Statistics of Income Bulletin,* a publication of the Internal Revenue Service, contains data on top wealthholders by marital status. A random sample of 487 top wealthholders yielded the following contingency table.

Net worth	Marital status: Married	Single/ Divorced	Widowed	Total
\$100,000–\$249,999	227	54	63	344
\$250,000–\$499,999	60	15	22	97
\$500,000–\$999,999	20	4	7	31
\$1,000,000 or more	10	2	3	15
Total	317	75	95	487

Do the data provide sufficient evidence to conclude that net worth and marital status are statistically dependent for top wealthholders? Perform the required hypothesis test at the 5% significance level.

13.66 The American Bar Foundation publishes information on the characteristics of lawyers in *The Lawyer Statistical Report.* The contingency table shown at the top of the next column cross-classifies 307 randomly selected U.S. lawyers by status in practice and size of city practicing in. Do the data provide sufficient evidence to conclude that size of city and status in practice are statistically dependent for U.S. lawyers? Perform the required hypothesis test at the 5% significance level.

Status in practice	Size of city: Less than 250,000	250,000–499,999	500,000 or more	Total
Government	12	4	14	30
Judicial	8	1	2	11
Private practice	122	31	69	222
Salaried	19	7	18	44
Total	161	43	103	307

EXTENDING THE CONCEPTS AND SKILLS

13.67 In Exercise 13.65 it was not possible to perform the chi-square independence test because the assumptions regarding expected frequencies were not met. As mentioned on page 802, we can try three approaches to remedy the situation: (1) combine rows or columns; (2) eliminate rows or columns; or (3) increase the sample size.

a. Combine the last two rows of the contingency table in Exercise 13.65 to form a new contingency table.

b. Use the table obtained in part (a) to perform the hypothesis test required in Exercise 13.65, if possible.

c. Eliminate the last row of the contingency table in Exercise 13.65 to form a new contingency table.

d. Use the table obtained in part (c) to perform the hypothesis test required in Exercise 13.65, if possible.

13.68 In Exercise 13.66 it was not possible to perform the chi-square independence test because the assumptions regarding expected frequencies were not met. As mentioned on page 802, we can try three approaches to remedy the situation: (1) combine rows or columns; (2) eliminate rows or columns; or (3) increase the sample size.

a. Combine the first two rows of the contingency table in Exercise 13.66 to form a new contingency table.

b. Use the table obtained in part (a) to perform the hypothesis test required in Exercise 13.66, if possible.
c. Eliminate the second row of the contingency table in Exercise 13.66 to form a new contingency table.
d. Use the table obtained in part (c) to perform the hypothesis test required in Exercise 13.66, if possible.

USING TECHNOLOGY  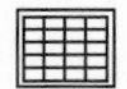 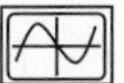

13.69 Use Minitab or some other statistical software to perform the hypothesis test in Exercise 13.63.

13.70 Use Minitab or some other statistical software to perform the hypothesis test in Exercise 13.64.

13.71 A study was conducted at Arizona State University to decide whether an association exists between grade and study time for intermediate algebra students who complete the course with a grade of C or better. Data were collected for a random sample of 422 students and compiled in a contingency table with categories for the number of hours studied per week across the top and categories for the grade earned along the side. For the number of hours studied, there were four categories: 0–3, 4–6, 7–9, and 10 or more. For the grade earned, there were three categories: A, B, and C. Printout 13.8 shown on the following page was obtained by applying Minitab's procedure for a chi-square independence test to the data. Determine the
a. number of students in the sample who studied at least 10 hours per week.
b. number of students in the sample who received a grade of B in the class.
c. observed number of students in the sample who studied at least 10 hours per week and received a grade of B in the class.
d. expected number of students in the sample who studied at least 10 hours per week and received a grade of B in the class if grade and study time are not associated.
e. chi-square subtotal for the "at least 10 hours per week *and* grade of B" cell.
f. number of degrees of freedom.
g. null and alternative hypotheses for a hypothesis test to decide whether there is an association between grade and study time for intermediate algebra students at Arizona State University.
h. value of the test statistic, χ^2.
i. P-value of the hypothesis test.
j. conclusion if the hypothesis test is performed at the 5% significance level.

13.72 Surveys are performed by the Book Industry Study Group to obtain information on characteristics of book readers. A *book reader* is defined as a person who read one or more books in the 6 months prior to the survey; a *non–book reader* is defined as a person who read newspapers or magazines but no books in the 6 months prior to the survey; and a *nonreader* is defined as a person who did not read a book, newspaper, or magazine in the 6 months prior to the survey. Printout 13.9 shown on the following page was obtained by applying Minitab's procedure for a chi-square independence test to data from a random sample of people 16 years old and over. Across the top of the contingency table are the reader-classification categories: book reader, non–book reader, and nonreader (labeled C1, C2, and C3). Along the side are household-income categories: less than \$15,000, \$15,000–\$24,999, \$25,000–\$39,999, and \$40,000 or over (labeled 1–4). Determine the
a. total number of people in the sample.
b. number of people in the sample that have a household income between \$25,000 and \$39,999.
c. number of people in the sample that are book readers.
d. observed number of people in the sample that are book readers who have a household income between \$25,000 and \$39,999.
e. expected number of people in the sample that are book readers who have a household income between \$25,000 and \$39,999 if household income and reader classification are statistically independent.
f. number of degrees of freedom.
g. null and alternative hypotheses for a hypothesis test to decide whether household income and reader classification are statistically dependent.
h. value of the test statistic, χ^2.
i. P-value of the hypothesis test.
j. conclusion if the hypothesis test is performed at the 1% significance level.

PRINTOUT 13.8 Minitab output for Exercise 13.71

```
Expected counts are printed below observed counts

            C1       C2       C3       C4    Total
    1       48       51       34       13      146
         37.02    59.51    37.71    11.76

    2       37       75       43        9      164
         41.58    66.84    42.36    13.21

    3       22       46       32       12      112
         28.40    45.65    28.93     9.02

Total      107      172      109       34      422

Chi-Sq =  3.257 +  1.216 +  0.365 +  0.130 +
          0.505 +  0.995 +  0.010 +  1.343 +
          1.441 +  0.003 +  0.326 +  0.982 = 10.574
DF = 6, P-Value = 0.104
```

PRINTOUT 13.9 Minitab output for Exercise 13.72

```
Expected counts are printed below observed counts

            C1       C2       C3    Total
    1      173      267       55      495
        247.33   217.88    29.79

    2      168      130       19      317
        158.39   139.53    19.08

    3      160      144        9      313
        156.39   137.77    18.84

    4      213       88        3      304
        151.89   133.81    18.30

Total      714      629       86     1429

Chi-Sq = 22.337 + 11.072 + 21.334 +
          0.583 +  0.651 +  0.000 +
          0.083 +  0.281 +  5.137 +
         24.583 + 15.684 + 12.787 = 114.534
DF = 6, P-Value = 0.000
```

CHAPTER REVIEW

Key Terms

association, *788*
bivariate data, *784*
cells, *784*
χ^2_α, *771*
chi-square (χ^2) curve, *770*
chi-square distribution, *770*
chi-square goodness-of-fit test, *776*
chi-square independence test, *800*
chi-square procedures, *769*
chi-square subtotals, *774*
conditional distribution, *786*
contingency table, *784*
expected frequencies, *774*
marginal distribution, *787*
observed frequencies, *774*
segmented bar graph, *787*
statistically dependent, *788*
statistically independent, *788*
two-way table, *784*
univariate data, *784*

You Should Be Able To

1. use and understand the formulas presented in this chapter.
2. identify the basic properties of χ^2-curves.
3. use the chi-square table, Table VIII.
4. explain the reasoning behind the chi-square goodness-of-fit test.
5. perform a chi-square goodness-of-fit test.
6. group bivariate data into a contingency table.
7. obtain and graph marginal and conditional distributions.
8. decide whether an association exists between two variables of a population, given bivariate data for the entire population.
9. explain the reasoning behind the chi-square independence test.
10. perform a chi-square independence test to decide whether an association exists between two variables of a population, given bivariate data for a sample of the population.
***11.** use the Minitab procedures covered in this chapter.
***12.** interpret the output obtained from the application of the Minitab procedures discussed in this chapter.

REVIEW TEST

STATISTICAL CONCEPTS AND SKILLS

1. We know there are infinitely many chi-square distributions and corresponding χ^2-curves. How do we distinguish among these distributions and curves?

2. Regarding a χ^2-curve:
a. At what point on the horizontal axis does the curve begin?
b. What shape does it have?
c. As the number of degrees of freedom increases, a χ^2-curve begins to look like another type of curve. What type of curve is that?

3. Recall that the number of degrees of freedom for the t-distribution used in a one-sample t-test for a population mean depends on the sample size.
a. Is that true for the chi-square distribution used in a chi-square goodness-of-fit test? Explain your answer.
b. Is that true for the chi-square distribution used in a chi-square independence test? Explain your answer.

4. Explain why a chi-square goodness-of-fit test or a chi-square independence test is always right-tailed.

5. If the observed and expected frequencies for a chi-square goodness-of-fit test or a chi-square independence test matched up perfectly, what would be the value of the test statistic?

6. Regarding the expected-frequency assumptions for a chi-square goodness-of-fit test or a chi-square independence test:
a. State them.
b. How important are they?

7. The U.S. Bureau of the Census publishes information on how Americans get to work in *Census of Population and Housing.* According to that document, 5.3% of Americans use public transportation to get to work.
a. If there were no association between means of transportation to work and region of residence, what percentage of Americans living in the West would use public transportation to get to work?
b. There are (roughly) 58.5 million Americans living in the West. If there were no association between means of transportation to work and region of residence, how many Americans living in the West would use public transportation to get to work?
c. In fact, there are (roughly) 2.4 million Americans living in the West who use public transportation to get to work. Given this and your answer to part (b), what can you conclude?

8. Suppose you have bivariate data for an entire population.
a. How would you decide whether there is an association between the two variables under consideration?
b. Assuming you make no calculation mistakes, is it possible that your conclusion is in error? Explain your answer.

9. Suppose you have bivariate data for a sample of a population.
a. How would you decide whether there is an association between the two variables under consideration?
b. Assuming you make no calculation mistakes, is it possible that your conclusion is in error? Explain your answer.

10. Consider a χ^2-curve with 17 degrees of freedom. Use Table VIII to determine
a. $\chi^2_{0.99}$. **b.** $\chi^2_{0.01}$.
c. the χ^2-value having area 0.05 to its right.
d. the χ^2-value having area 0.05 to its left.
e. the two χ^2-values that divide the area under the curve into a middle 0.95 area and two outside 0.025 areas.

11. The U.S. Bureau of the Census compiles census data on educational attainment of Americans. From the document *1990 Census of Population,* we obtained the 1990 percent distribution of educational attainment for adults 25 years old and over. Here is that distribution.

Highest level	Percentage
Not HS graduate	24.8
HS graduate	30.0
Some college	18.7
Associate's degree	6.2
Bachelor's degree	13.1
Advanced degree	7.2

A random sample of 500 adults (25 years old and over) taken this year gave the following frequency distribution.

Highest level	Frequency
Not HS graduate	92
HS graduate	168
Some college	86
Associate's degree	36
Bachelor's degree	79
Advanced degree	39

a. Use the critical-value approach to decide, at the 5% significance level, whether this year's distribution of educational attainment differs from the 1990 distribution.
b. Estimate the P-value of the hypothesis test and use that estimate to assess the evidence against the null hypothesis.
c. Can we conclude, at the 1% significance level, that this year's distribution of educational attainment differs from the 1990 distribution? Explain your answer.

12. The table that follows provides the region and governor's party affiliation for each state in the United States. [SOURCE: *Statistical Abstract of the United States, 1997,* and *World Almanac, 1998.*]

State	Region	Party	State	Region	Party
AL	SO	Rep	MT	WE	Rep
AK	WE	Dem	NE	MW	Dem
AZ	WE	Rep	NV	WE	Dem
AR	SO	Rep	NH	NE	Dem
CA	WE	Rep	NJ	NE	Rep
CO	WE	Dem	NM	WE	Rep
CT	NE	Rep	NY	NE	Rep
DE	SO	Dem	NC	SO	Dem
FL	SO	Dem	ND	MW	Rep
GA	SO	Dem	OH	MW	Rep
HI	WE	Dem	OK	SO	Rep
ID	WE	Rep	OR	WE	Dem
IL	MW	Rep	PA	NE	Rep
IN	MW	Dem	RI	NE	Rep
IA	MW	Rep	SC	SO	Rep
KS	MW	Rep	SD	MW	Rep
KY	SO	Dem	TN	SO	Rep
LA	SO	Rep	TX	SO	Rep
ME	NE	Ind	UT	WE	Rep
MD	SO	Dem	VT	NE	Dem
MA	NE	Rep	VA	SO	Rep
MI	MW	Rep	WA	WE	Dem
MN	MW	Rep	WV	SO	Rep
MS	SO	Rep	WI	MW	Rep
MO	MW	Dem	WY	WE	Rep

a. Group the data for the variables "region" and "party of governor" into a contingency table.

b. Find the conditional distributions of region by party and the marginal distribution of region.

c. Find the conditional distributions of party by region and the marginal distribution of party.

d. Is there an association between the variables "region" and "party of governor" for the states of the United States? Explain your answer.

e. What percentage of states have Republican governors?

f. If there were no association between region and party of governor, use part (e) to determine the percentage of Midwest states that would have Republican governors.

g. In reality, what percentage of Midwest states have Republican governors?

h. What percentage of states are in the Midwest?

i. If there were no association between region and party of governor, use part (h) to determine the percentage of states with Republican governors that would be in the Midwest.

j. In reality, what percentage of states with Republican governors are in the Midwest?

13. From data in *Hospital Statistics,* published by the American Hospital Association, we obtained the following contingency table for U.S. hospitals and nursing homes, by type of facility and type of control. We have used the abbreviations Gov for Government, Prop for Proprietary, and NP for nonprofit.

Facility	Control: Gov	Prop	NP	Total
General	1697	660	3046	5403
Psychiatric	266	358	113	737
Chronic	21	1	4	26
Tuberculosis	3	0	1	4
Other	59	148	203	410
Total	2046	1167	3367	6580

In the questions below, we will use the term "hospital" to refer to either a hospital or nursing home.

a. How many hospitals are government controlled?

b. How many psychiatric facilities are there?

c. How many hospitals are government-controlled psychiatric facilities?

d. How many general facilities are nonprofit?

e. How many hospitals are not under proprietary control?

f. How many hospitals are either general facilities or under proprietary control?

14. Refer to Problem 13.

a. Obtain the conditional distribution of control type within each facility type.

b. Is there an association between facility type and control type for U.S. hospitals? Why?

c. Determine the marginal distribution of control type for U.S. hospitals.

d. Construct a segmented bar graph for the conditional distributions and marginal distribution of control type. Interpret the graph in light of your answer to part (b).

e. Without doing any further calculations, respond true or false to the following statement and explain your answer. The conditional distributions of facility type within control types are identical.

f. Determine the marginal distribution of facility type and the conditional distributions of facility type within control types.

g. What percentage of hospitals are under proprietary control?

h. What percentage of psychiatric hospitals are under proprietary control?

i. What percentage of hospitals under proprietary control are psychiatric hospitals?

15. A Gallup Poll asked 1528 adults: "The New Jersey Supreme Court recently ruled that all life-sustaining medical treatment may be withheld or withdrawn from terminally ill patients, provided that is what the patients want or would want if they were able to express their wishes. Would you like to see such a ruling in the state in which you live, or not?" The following contingency table, which cross-classifies response by educational level, presents a modified version of the results.

Educational level	Response: Favor	Oppose	No opinion	Total
College graduate	264	17	6	287
Some college	205	26	7	238
HS graduate	461	81	34	576
Not HS graduate	290	81	56	427
Total	1220	205	103	1528

Can we conclude from the data that an association exists between opinion on this issue and educational level? Perform the required hypothesis test at the 1% significance level.

USING TECHNOLOGY  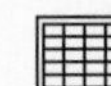 

16. Use Minitab or some other statistical software to conduct the required chi-square goodness-of-fit test in Problem 11.

17. We applied the Minitab macro `fittest.mac` to perform the chi-square goodness-of-fit test required in Problem 11 and obtained the output shown in Printout 13.10. Use the output to determine the

a. number of people sampled.

b. value of the chi-square test statistic.

c. P-value for the hypothesis test.

d. smallest significance level at which the null hypothesis can be rejected.

e. conclusion if the test is performed at the 5% significance level.

18. Use Minitab or some other statistical software to solve parts (a)–(c) of Problem 12.

19. Use Minitab or some other statistical software to perform the hypothesis test in Problem 15.

20. Printout 13.11 shows the output obtained by applying Minitab's chi-square independence test to the cell data in the columns of the contingency table in Problem 15. Use the printout to determine the

a. number of people sampled who would be opposed to the New Jersey Supreme Court ruling in their state.

b. number of people sampled who had some college. *(Note:* The row labeled "2" in the printout corresponds to the "Some college" category, as we can see from the contingency table in Problem 15.*)*

c. observed number of people sampled who had some college and would be opposed to such a ruling in their state.

d. expected number of people sampled who had some college and would be opposed to such a ruling in their state if opinion on the issue and educational level are not associated.

e. chi-square subtotal for the cell "Some college *and* Oppose."

f. number of degrees of freedom.

g. value of the test statistic, χ^2.

h. P-value of the hypothesis test.

i. conclusion if the hypothesis test is performed at the 1% significance level.

PRINTOUT 13.10 Minitab output for Problem 17

```
Chi-square goodness-of-fit test


 Row     n   k    ChiSq     P-value

   1   500   6  14.8586   0.0109841
```

PRINTOUT 13.11 Minitab output for Problem 20

```
Expected counts are printed below observed counts

          FAVOR   OPPOSE   NOOPIN    Total
     1      264       17        6      287
         229.15    38.50    19.35

     2      205       26        7      238
         190.03    31.93    16.04

     3      461       81       34      576
         459.90    77.28    38.83

     4      290       81       56      427
         340.93    57.29    28.78

Total      1220      205      103     1528

Chi-Sq =  5.300 + 12.010 +  9.207 +
          1.180 +  1.102 +  5.097 +
          0.003 +  0.179 +  0.600 +
          7.608 +  9.815 + 25.735 = 77.837
DF = 6, P-Value = 0.000
```

INTERNET PROJECT

Sex and the Death Penalty

In this Internet project, you will examine data for men and women on death row in order to determine whether sex differences exist in the death row population. At the time of this writing, there are 49 women on death row. This accounts for only 1.5% of the total death-row population.

In general, both the death-sentencing rate and the death-row population are very small for women compared to that for men. Additionally, actual execution of female offenders is rare, with only 533 documented instances beginning with the first in 1632. Now you can explore these differences yourself in an attempt to understand the many social aspects underlying the numbers.

URL for access to Internet Projects Page: `http://hepg.awl.com` *keyword:* Weiss

USING THE FOCUS DATABASE

In Chapter 1 we explained how to store the Focus database in a Minitab worksheet named `focus.mtw`. If you haven't already created that worksheet, follow the instructions on pages 54–55 to create it now.

The Focus database contains data on 500 randomly selected Arizona State University sophomores for seven different variables, among which are sex, high-school GPA, and cumulative GPA. Use Minitab or some other statistical software to perform the chi-square independence tests below. Employ the coding scheme 1, 2, 3, and 4, respectively, for grade point averages less than 1, at least 1 but less than 2, at least 2 but less than 3, and at least 3.

a. At the 5% significance level, do the data provide sufficient evidence to conclude that sex and high-school GPA are statistically dependent for ASU sophomores?

b. Repeat part (a) for sex and cumulative GPA.

CASE STUDY DISCUSSION

Time On or Off the Job: Which Do Americans Enjoy More?

On page 768, at the beginning of this chapter, we gave the results of a CNN/*USA TODAY* poll that asked a sample of Americans whether they enjoy their hours on the job more than their time spent off the job. Now that you have studied the chi-square independence test, you can analyze these data to determine whether preferences depend on sex, age, or amount of education.

Use the data in the table on page 768 to decide whether an association exists between each of the following pairs of variables. Conduct each hypothesis test at the 5% significance level.

a. sex and response (to the question "Which do you enjoy more, the hours when you are on your job, or the hours when you are not on your job?")
b. age and response
c. type of community and response
d. amount of education and response
e. income and response
f. type of employer and response
g. Use Minitab or some other statistical software to perform the hypothesis tests in parts (a)–(f).

BIOGRAPHY KARL PEARSON

Karl Pearson was born on March 27, 1857, in London, the second son of William Pearson, a prominent lawyer, and his wife, Fanny Smith. Karl Pearson's early education took place at home. At the age of 9, he was sent to University College School in London, where he remained for the next 7 years. Because of ill health, Pearson was then privately tutored for a year. He received a scholarship at King's College, Cambridge, in 1875. There he earned a B.A. (with honors) in mathematics in 1879 and an M.A. in law in 1882. He then studied physics and metaphysics in Heidelberg, Germany.

In addition to his expertise in mathematics, law, physics, and metaphysics, Pearson was competent in literature and knowledgeable about German history, folklore, and philosophy. He was also considered somewhat of a political radical because of his interest in the ideas of Karl Marx and the rights of women.

In 1884 Pearson was appointed Goldsmid professor of applied mathematics and mechanics at University College; from 1891–1894 he was also a lecturer in geometry at Gresham College, London. In 1911 he gave up the Goldsmid chair to become the first Galton professor of Eugenics at University College. Pearson was elected to the Royal Society, a prestigious association of scientists, in 1896 and awarded the society's Darwin Medal in 1898.

Pearson really began his pioneering work in statistics in 1893, mainly through an association with Walter Weldon (a zoology professor at University College), Francis Edgeworth (a professor of logic at University College), and Sir Francis Galton (see the Chapter 15 biography). An analysis of published data on roulette wheels at Monte Carlo led to Pearson's discovery of the chi-square goodness-of-fit test. He also coined the term "standard deviations," introduced his amazingly diverse skew curves, and developed the most widely used measure of correlation, the correlation coefficient.

Pearson, Weldon, and Galton co-founded the statistical journal *Biometrika,* of which Pearson was editor (1901–1936) and a major contributor. Pearson retired from University College in 1933. He died in London on April 27, 1936.

P A R T

Regression, Correlation, and ANOVA

CASE STUDY — FAT CONSUMPTION AND PROSTATE CANCER

Researchers have asked if there is a relationship between nutrition and cancer, and many studies have shown that there is. In fact, one of the conclusions of a study by B. Reddy et al. ("Nutrition and Its Relationship to Cancer," *Advances in Cancer Research, 32,* pp. 237–345) was that "... none of the risk factors for cancer is probably more significant than diet and nutrition."

Prostate cancer is one of the most virulent forms of cancer. One dietary factor that has been studied for its relationship with prostate cancer is fat consumption. The data in the following table were obtained from a graph—adapted from information in the aforementioned article—in John Robbins's classic book *Diet for a New America* (Walpole, NH: Stillpoint Publishing, 1987).

Country	Dietary fat (grams/day)	Death rate (per 100,000)	Country	Dietary fat (grams/day)	Death rate (per 100,000)
El Salvador	38	0.9	Spain	97	10.1
Philippines	29	1.3	Portugal	73	11.4
Japan	42	1.6	Finland	112	11.1
Mexico	57	4.5	Hungary	100	13.1
Greece	96	4.8	United Kingdom	143	12.4
Colombia	47	5.4	Germany	134	12.9
Bulgaria	67	5.5	Canada	142	13.4
Yugoslavia	72	5.6	Austria	119	13.9
Poland	93	6.4	France	137	14.4
Panama	58	7.8	Netherlands	152	14.4
Israel	95	8.4	Australia	129	15.1
Romania	67	8.8	Denmark	156	15.9
Venezuela	62	9.0	United States	147	16.3
Czechoslovakia	96	9.1	Norway	133	16.8
Italy	86	9.4	Sweden	132	18.4

After we study the techniques of regression and correlation, we will return to these data and investigate how fat consumption is related to prostate cancer among nations in the world.

Descriptive Methods in Regression and Correlation

14

GENERAL OBJECTIVES

We frequently want to know whether two or more variables are related, and if they are, how they are related. For instance, is there a relationship between SAT scores and college GPA? If these variables are related, how are they related? Or assume the president of a large corporation knows sales tend to increase as advertising expenditures increase but needs to know how strong that tendency is and how she can predict the approximate sales that will result from various advertising expenditures.

Some commonly used methods for examining the relationship between two or more variables and for making predictions are *linear regression* and *correlation.* Descriptive methods in linear regression and correlation will be discussed in this chapter. Inferential methods in linear regression and correlation will be discussed in Chapter 15.

CHAPTER OUTLINE

14.1 LINEAR EQUATIONS WITH ONE INDEPENDENT VARIABLE

To understand linear regression, we first need to review linear equations with one independent variable. The general form of a **linear equation** with one independent variable can be written as

$$y = b_0 + b_1x,$$

where b_0 and b_1 are constants (fixed numbers), x is the independent variable, and y is the dependent variable.[†]

The graph of a linear equation with one independent variable is a **straight line;** furthermore, any nonvertical straight line can be represented by such an equation. Three examples of linear equations with one independent variable are $y = 4 + 0.2x$, $y = -1.5 - 2x$, and $y = -3.4 + 1.8x$. In Fig. 14.1, we have drawn the straight-line graphs of these three linear equations.

FIGURE 14.1
Straight-line graphs of three linear equations

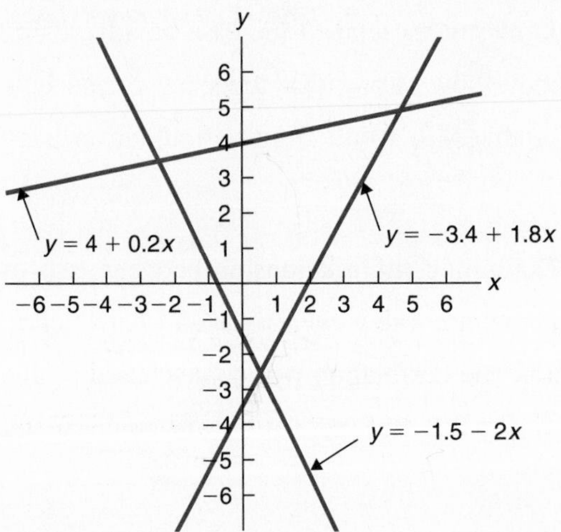

Linear equations with one independent variable occur frequently in applications of mathematics to many different fields, including the management, life, and social sciences, as well as the physical and mathematical sciences. In Examples 14.1 and 14.2, we illustrate the use of linear equations in a simple business application.

[†] You may be familiar with the form $y = mx + b$ instead of the form $y = b_0 + b_1x$. In statistics the latter form is preferred because it allows a smoother transition to multiple regression, in which there is more than one independent variable. We will not discuss multiple regression in this text but instead refer the interested reader to the modules *Multiple Regression Analysis* and *Model Building in Regression* by Dennis L. Young (Reading, MA: Addison-Wesley, 1999).

EXAMPLE 14.1 **LINEAR EQUATIONS**

CJ² Business Services does word processing as one of its basic functions. Its rate is \$20/hr plus a \$25 disk charge. The total cost to a customer depends, of course, on the number of hours it takes to complete the job. Find the equation that expresses the total cost in terms of the number of hours required to complete the job.

SOLUTION Let x denote the number of hours required to complete the job and y the total cost to the customer. Since the rate for word processing is \$20/hr, a job that takes x hours will cost \$$20x$ plus the \$25 disk charge. So the total cost, y, of a job that takes x hours is $y = 25 + 20x$. ◨

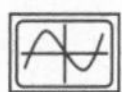

The equation, $y = 25 + 20x$, for the total cost of a word-processing job is a linear equation; here $b_0 = 25$ and $b_1 = 20$. Using the equation, we can find the exact cost for a job once we know the number of hours required. For instance, a job that takes 5 hours will cost $y = 25 + 20 \cdot 5 = \$125$; a job that takes 7.5 hours will cost $y = 25 + 20 \cdot 7.5 = \$175$. Table 14.1 displays these costs and a few others.

TABLE 14.1
Times and costs for five word-processing jobs

Time (hrs) x	5.0	7.5	15.0	20.0	22.5
Cost (\$) y	125	175	325	425	475

As we have already mentioned, a linear equation, such as $y = 25 + 20x$, has a straight-line graph. We can obtain the graph of $y = 25 + 20x$ by plotting the points in Table 14.1 and connecting them with a straight line, which is done in Fig. 14.2.

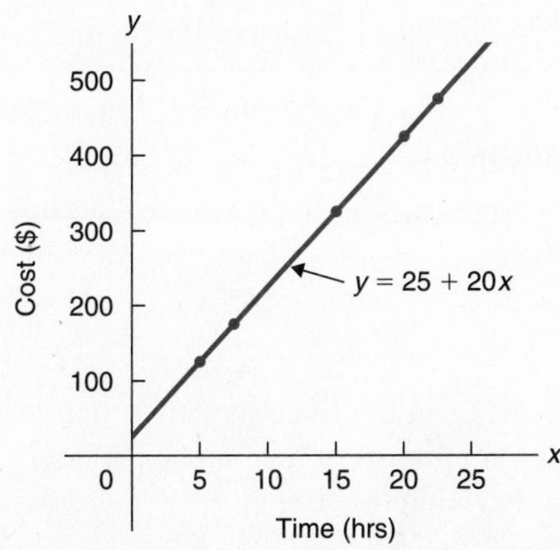

FIGURE 14.2
Graph of $y = 25 + 20x$, obtained from the points in Table 14.1

The graph in Fig. 14.2 is useful for quickly estimating cost. For example, a glance at the graph shows that a 10-hour job will cost somewhere between \$200 and \$300. The exact cost is $y = 25 + 20 \cdot 10 = \$225$.

Intercept and Slope

For a linear equation $y = b_0 + b_1x$, the numbers b_0 and b_1 have an important geometric interpretation. The number b_0 is the y-value at which the straight-line graph of the linear equation intersects the y-axis. The number b_1 measures the steepness of the straight line; more precisely, b_1 indicates how much the y-value on the straight line increases (or decreases) when the x-value increases by 1 unit. Figure 14.3 illustrates these relationships.

FIGURE 14.3
Graph of $y = b_0 + b_1x$

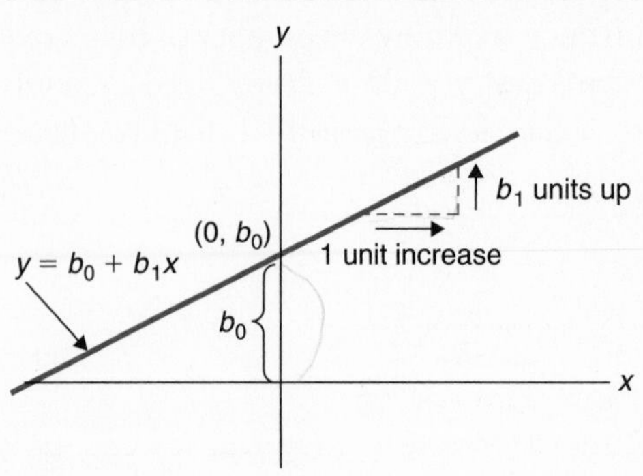

Because of the geometric interpretation, described above, of the numbers b_0 and b_1, they are given special names that reflect that interpretation.

DEFINITION 14.1 y-INTERCEPT AND SLOPE

For a linear equation $y = b_0 + b_1x$, the number b_0 is called the ***y*-intercept** and the number b_1 is called the ***slope.***

EXAMPLE 14.2 y-INTERCEPT AND SLOPE

In Example 14.1 we obtained the linear equation that expresses the total cost, y, of a word-processing job in terms of the number of hours, x, required to complete the job. The equation is $y = 25 + 20x$.

a. Find the y-intercept and slope of that linear equation.

b. Interpret the y-intercept and slope in terms of the graph of the equation.

c. Interpret the y-intercept and slope in terms of word-processing costs.

SOLUTION **a.** The y-intercept for the equation is $b_0 = 25$ and the slope is $b_1 = 20$.

b. The y-intercept $b_0 = 25$ is the y-value at which the straight line $y = 25 + 20x$ intersects the y-axis. The slope $b_1 = 20$ indicates that the y-value increases by 20 units for every increase in x of 1 unit. See Fig. 14.4.

FIGURE 14.4 Graph of $y = 25 + 20x$

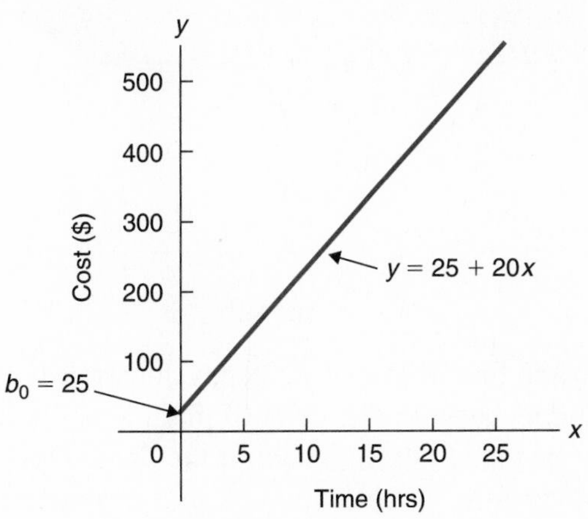

c. In terms of word-processing costs, the y-intercept $b_0 = 25$ represents the total cost of a job that takes 0 hours. In other words, the y-intercept of \$25 is a fixed cost that is always charged no matter how long the job takes. The slope $b_1 = 20$ represents the fact that the cost per hour is \$20; it is the amount that the total cost, y, goes up for every increase of 1 hour in the time, x, required to complete the job. ◨

A straight line is determined by any two distinct points that lie on the line. This means that the straight-line graph of a linear equation, $y = b_0 + b_1x$, can be obtained by first substituting two different x-values into the equation to get two distinct points and then connecting those two points with a straight line.

For example, to graph the linear equation $y = 5 - 3x$, we can use the x-values 1 and 3 (or any other two x-values). The y-values corresponding to those two x-values are $y = 5 - 3 \cdot 1 = 2$ and $y = 5 - 3 \cdot 3 = -4$, respectively. Therefore, the graph of the linear equation $y = 5 - 3x$ is the straight line that passes through the two points $(1, 2)$ and $(3, -4)$, as depicted in Fig. 14.5.

FIGURE 14.5 Graph of $y = 5 - 3x$

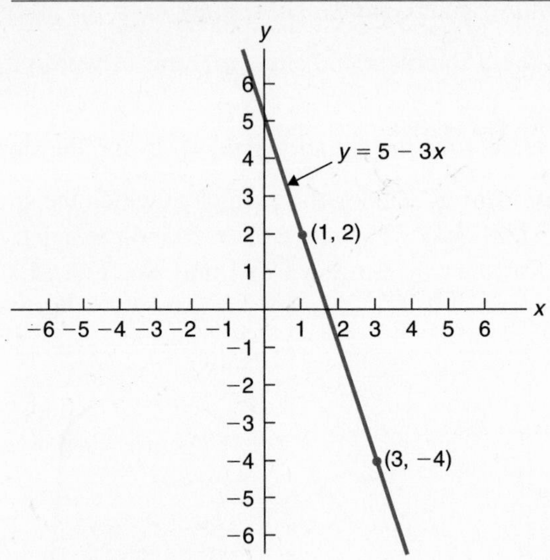

Notice that the line in Fig. 14.5 slopes downward—the y-values decrease as x increases. This is because the slope of the line is negative: $b_1 = -3 < 0$. Now look at the line in Fig. 14.4, the graph of the linear equation $y = 25 + 20x$. That line slopes upward—the y-values increase as x increases. This is because the slope of the line is positive: $b_1 = 20 > 0$. In general, we have the following fact.

KEY FACT 14.1 **GRAPHICAL INTERPRETATION OF SLOPE**

The straight-line graph of the linear equation $y = b_0 + b_1x$ slopes upward if $b_1 > 0$, slopes downward if $b_1 < 0$, and is horizontal if $b_1 = 0$, as seen in Fig. 14.6.

FIGURE 14.6 Graphical interpretation of slope

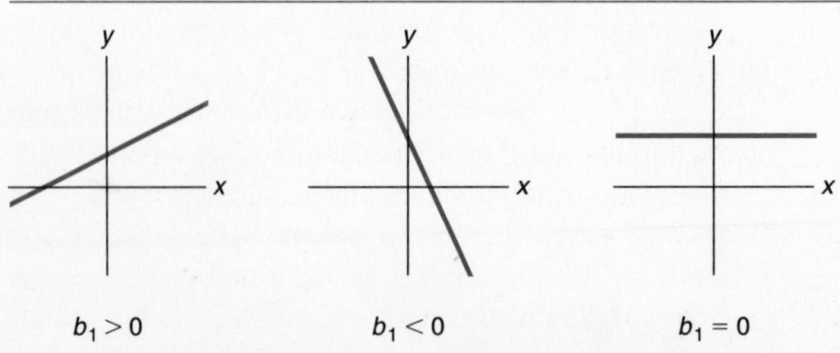

EXERCISES 14.1

STATISTICAL CONCEPTS AND SKILLS

14.1 Regarding linear equations with one independent variable.

a. Write the general form of such an equation.

b. In your expression in part (a), which letters represent constants and which represent variables?

c. In your expression in part (a), which letter represents the independent variable and which represents the dependent variable?

14.2 Fill in the blank: The graph of a linear equation with one independent variable is a ____.

14.3 Consider the linear equation $y = b_0 + b_1x$.

a. Identify and give the geometric interpretation of b_0.

b. Identify and give the geometric interpretation of b_1.

14.4 Answer true or false to each of the following statements and explain your answers.

a. The straight-line graph of a linear equation slopes upward unless the slope is 0.

b. The value of the y-intercept has no effect on which direction the straight-line graph of a linear equation slopes.

14.5 On June 24, 1998, the Avis Rent-A-Car rate for renting a fullsize car in Arkansas was \$45.90 per day plus \$0.25 per mile. For a 1-day rental, let x denote the number of miles driven and y total cost.

a. Obtain the equation that expresses y in terms of x.

b. Find b_0 and b_1.

c. Construct a table similar to Table 14.1 on page 821 for the x-values 50, 100, and 250 miles.

d. Draw the graph of the equation that you obtained in part (a) by plotting the points from part (c) and connecting them with a straight line.

e. Apply the graph from part (d) to visually estimate the cost of driving the car 150 miles. Then calculate that cost exactly using the equation from part (a).

14.6 Encore Air Conditioning charges \$36 per hour plus a \$30 service charge. Let x denote the number of hours it takes for a job and y the total cost to the customer.

a. Obtain the equation that expresses y in terms of x.

b. Find b_0 and b_1.

c. Construct a table similar to Table 14.1 on page 821 for the x-values 0.5, 1, and 2.25 hours.

d. Draw the graph of the equation that you obtained in part (a) by plotting the points from part (c) and connecting them with a straight line.

e. Apply the graph from part (d) to visually estimate the cost of a job that takes 1.75 hours. Then calculate that cost exactly using the equation from part (a).

14.7 The most commonly used scales for measuring temperature are the Fahrenheit and Celsius scales. If we let y denote Fahrenheit temperature and x Celsius temperature, then we can express the relationship between those two scales with the linear equation $y = 32 + 1.8x$.

a. Determine b_0 and b_1.

b. Find the Fahrenheit temperatures corresponding to the following Celsius temperatures: $-40°$, $0°$, $20°$, and $100°$.

c. Graph the linear equation $y = 32 + 1.8x$ using the four points found in part (b).

d. Apply the graph obtained in part (c) to visually estimate the Fahrenheit temperature corresponding to a Celsius temperature of $28°$. Then calculate that temperature exactly by employing the linear equation $y = 32 + 1.8x$.

14.8 A ball is thrown straight up in the air with an initial velocity of 64 ft per second. According to the laws of physics, if we let y denote the velocity of the ball after x seconds, then $y = 64 - 32x$.

a. Determine b_0 and b_1 for this linear equation.

b. Determine the velocity of the ball after 1, 2, 3, and 4 seconds.

c. Graph the linear equation $y = 64 - 32x$ using the four points obtained in part (b).

d. Use the graph from part (c) to visually estimate the velocity of the ball after 1.5 seconds. Then calculate that velocity exactly by employing the linear equation $y = 64 - 32x$.

In each of Exercises 14.9–14.12,

a. determine the y-intercept and slope of the specified linear equation.

b. explain what the y-intercept and slope represent in terms of the graph of the equation.

c. explain what the y-intercept and slope represent in terms relating to the application.

14.9 $y = 45.90 + 0.25x$ (from Exercise 14.5)

14.10 $y = 30 + 36x$ (from Exercise 14.6)

14.11 $y = 32 + 1.8x$ (from Exercise 14.7)

14.12 $y = 64 - 32x$ (from Exercise 14.8)

In each of Exercises 14.13–14.22, you are given a linear equation. For each exercise,

a. find the y-intercept and slope.

b. determine whether the line slopes upward, slopes downward, or is horizontal, without graphing the equation.

c. graph the equation using two points.

14.13 $y = 3 + 4x$ **14.14** $y = -1 + 2x$

14.15 $y = 6 - 7x$ **14.16** $y = -8 - 4x$

14.17 $y = 0.5x - 2$ **14.18** $y = -0.75x - 5$

14.19 $y = 2$ **14.20** $y = -3x$

14.21 $y = 1.5x$ **14.22** $y = -3$

In each of Exercises 14.23–14.30, we have identified the y-intercept and slope, respectively, of a straight line. For each of these exercises,

a. determine whether the line slopes upward, slopes downward, or is horizontal, without graphing the equation.

b. find the equation of the line.

c. graph the equation using two points.

14.23 5 and 2 **14.24** −3 and 4

14.25 −2 and −3 **14.26** 0.4 and 1

14.27 0 and −0.5 **14.28** −1.5 and 0

14.29 3 and 0 **14.30** 0 and 3

EXTENDING THE CONCEPTS AND SKILLS

14.31 On page 820 we stated that any nonvertical straight line can be described by an equation of the form $y = b_0 + b_1x$.

a. Why can't a vertical straight line be expressed in this form?

b. What is the form of the equation of a vertical straight line?

c. Does a vertical straight line have a slope? Explain your answer.

14.2 THE REGRESSION EQUATION

In Examples 14.1 and 14.2, we discussed the linear equation $y = 25 + 20x$, which expresses the total cost, y, of a word-processing job in terms of the time in hours, x, required to complete the job. Given the amount of time required, x, we can use the equation to determine the *exact* cost of the job, y.

Real-life applications are sometimes not so simple as the word-processing example, in which one variable (cost) can be predicted exactly in terms of another variable (time required). Rather, we must often be content with rough predictions. For instance, we cannot predict the exact price, y, of a particular make and model of car just by knowing its age, x. Indeed, even for a fixed age, say, 3 years old, the price varies from car to car. We must be satisfied with making a rough prediction for the price of a 3-year-old car of the particular make and model or with an estimate of the mean price of all such 3-year-old cars.

Table 14.2 displays data on age and price for a sample of cars of a particular make and model. We will refer to the car as the Orion, but the data, obtained from the *Asian Import* edition of the *Auto Trader* magazine, is for a real car. Ages are in years; prices are in hundreds of dollars, rounded to the nearest hundred dollars.

TABLE 14.2
Age and price data for a sample of 11 Orions

Car	Age (yrs) x	Price ($100s) y
1	5	85
2	4	103
3	6	70
4	5	82
5	5	89
6	5	98
7	6	66
8	6	95
9	2	169
10	7	70
11	7	48

It is useful to plot the data so we can visualize any apparent relationships between age and price. Such a plot is called a **scatter diagram** (or **scatterplot**). The scatter diagram for the data in Table 14.2 is depicted in Fig. 14.7.

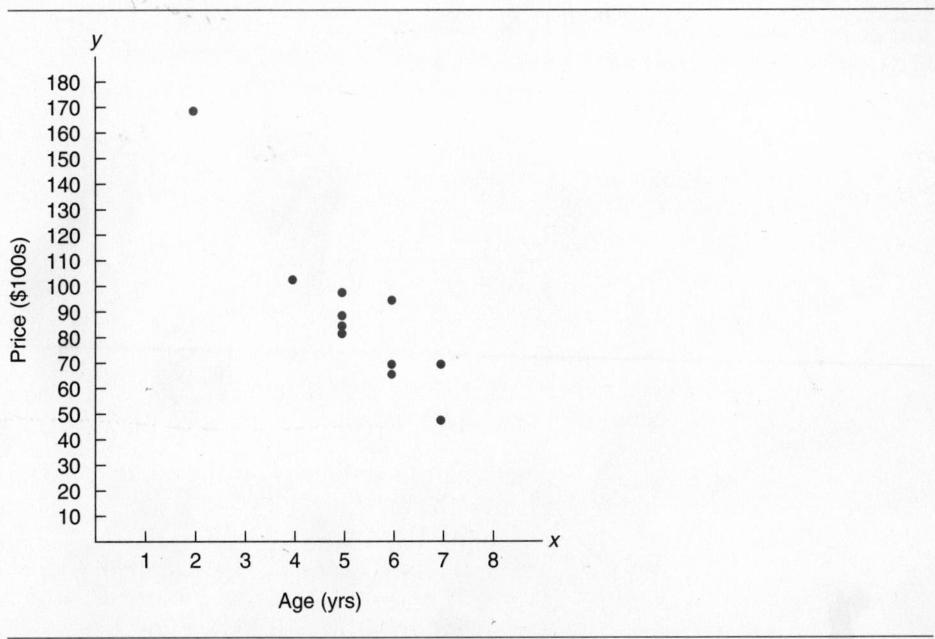

FIGURE 14.7
Scatter diagram for the age and price data of Orions from Table 14.2

Although it is clear from the scatter diagram that the data points do not lie on a straight line, it appears that they are clustered about a straight line. We would like to fit a straight line to the data points; then we could use that line to predict the price of an Orion given its age.

Since we could draw many different straight lines through the cluster of data points, we need a method to choose the "best" line. The method employed is called the **least-squares criterion.** It is based on an analysis of the errors made in using a straight line to fit the data points. To introduce the least-squares criterion, we will use a very simple data set. We will return to the Orion data shortly.

EXAMPLE 14.3 INTRODUCES THE LEAST-SQUARES CRITERION

Let's consider the problem of fitting a straight line to the four data points displayed in Table 14.3. A scatter diagram for those data is pictured in Fig. 14.8.

TABLE 14.3
Four data points

x	y
1	1
1	2
2	2
4	6

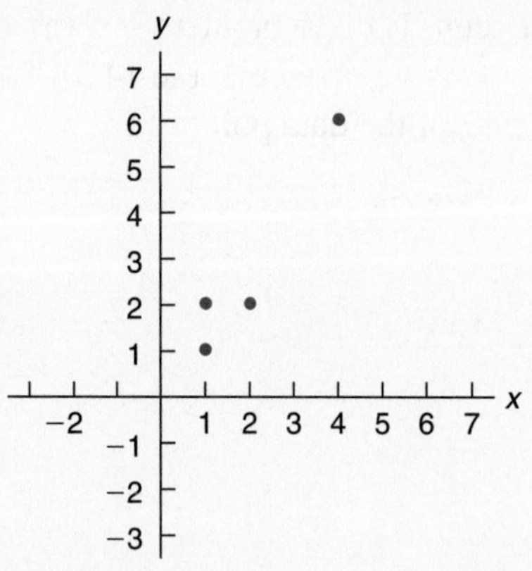

FIGURE 14.8
Scatter diagram for the data points in Table 14.3

It is possible to fit (infinitely) many straight lines to the data points in Table 14.3. Figures 14.9(a) and 14.9(b) show two possibilities.

To avoid confusion, we use $\hat{y}$ to denote the y-value predicted by a straight line for a value of x. For instance, the y-value predicted by Line A for $x = 2$ is

$$\hat{y} = 0.50 + 1.25 \cdot 2 = 3,$$

FIGURE 14.9 Two possible straight-line fits to the data points in Table 14.3

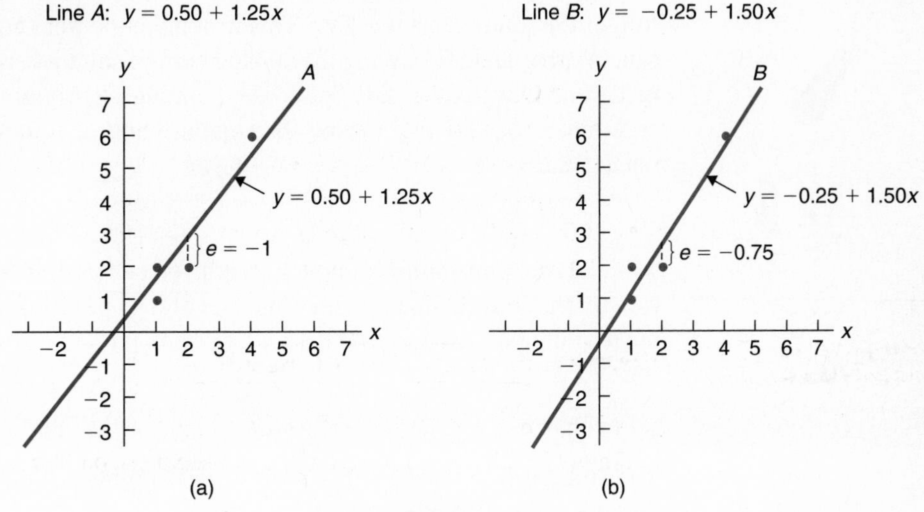

and the y-value predicted by Line B for $x = 2$ is

$$\hat{y} = -0.25 + 1.50 \cdot 2 = 2.75.$$

To measure quantitatively how well a line fits the data, we first look at the errors, e, made in using the line to predict the y-values of the data points. For instance, as we have just seen, Line A predicts a y-value of $\hat{y} = 3$ when $x = 2$. The actual y-value for $x = 2$ is $y = 2$ (see Table 14.3). So the error made in using Line A to predict the y-value of the data point (2, 2) is

$$e = y - \hat{y} = 2 - 3 = -1,$$

as seen in Fig. 14.9(a).

The fourth column of Table 14.4(a) shows the errors made by Line A for all four data points; the fourth column of Table 14.4(b) shows that for Line B.

TABLE 14.4 Determining how well the data points in Table 14.3 are fit by (a) Line A, (b) Line B

Line A: $y = 0.50 + 1.25x$

x	y	$\hat{y}$	e	e^2
1	1	1.75	−0.75	0.5625
1	2	1.75	0.25	0.0625
2	2	3.00	−1.00	1.0000
4	6	5.50	0.50	0.2500
				1.8750

(a)

Line B: $y = -0.25 + 1.50x$

x	y	$\hat{y}$	e	e^2
1	1	1.25	−0.25	0.0625
1	2	1.25	0.75	0.5625
2	2	2.75	−0.75	0.5625
4	6	5.75	0.25	0.0625
				1.2500

(b)

To decide which line, Line A or Line B, fits the data better, we first compute the sum of the squared errors, Σe^2. This is done in the final columns of Tables 14.4(a) and 14.4(b). The line having the smaller sum of squared errors, in this case Line B, is the one that fits the data better. And among all straight lines, the least-squares criterion is that the line having the smallest sum of squared errors is the one that fits the data best. ◨

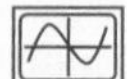

With the preceding example in mind, we can now state the least-squares criterion for the straight line that best fits a set of data points. Following that, we present the terminology used for the best-fitting line.

KEY FACT 14.2 LEAST-SQUARES CRITERION

The straight line that best fits a set of data points is the one having the smallest possible sum of squared errors.

DEFINITION 14.2 REGRESSION LINE AND REGRESSION EQUATION

Regression line: The straight line that best fits a set of data points according to the least-squares criterion.

Regression equation: The equation of the regression line.

Although the least-squares criterion tells us what property the regression line for a set of data points must have, it does not tell us how to find that line. This latter task is accomplished by Formula 14.1, which provides formulas for obtaining the regression line. In preparation, we introduce some notation that will be used throughout our study of regression and correlation.

DEFINITION 14.3 NOTATION USED IN REGRESSION AND CORRELATION

We define S_{xx}, S_{xy}, and S_{yy} by $S_{xx} = \Sigma(x - \bar{x})^2$, $S_{xy} = \Sigma(x - \bar{x})(y - \bar{y})$, and $S_{yy} = \Sigma(y - \bar{y})^2$. For hand computations, these three quantities are most easily obtained by using the following computing formulas:

$$S_{xx} = \Sigma x^2 - (\Sigma x)^2/n,$$

$$S_{xy} = \Sigma xy - (\Sigma x)(\Sigma y)/n,$$

$$S_{yy} = \Sigma y^2 - (\Sigma y)^2/n.$$

FORMULA 14.1 REGRESSION EQUATION

The regression equation for a set of n data points is $\hat{y} = b_0 + b_1 x$, where

$$b_1 = \frac{S_{xy}}{S_{xx}} \quad \text{and} \quad b_0 = \frac{1}{n}(\Sigma y - b_1 \Sigma x) = \overline{y} - b_1 \overline{x}.$$

EXAMPLE 14.4 ILLUSTRATES FORMULA 14.1

Table 14.2 displays data on age and price for a sample of 11 Orions. We repeat that data in the first two columns of Table 14.5.

a. Determine the regression equation for the data.

b. Graph the regression equation and the data points.

c. Describe the apparent relationship between age and price of Orions.

d. Interpret the slope of the regression line in terms of prices for Orions.

e. Use the regression equation to predict the price of a 3-year-old Orion and a 4-year-old Orion.

SOLUTION **a.** To determine the regression equation, we need to compute b_1 and b_0 using Formula 14.1. To that end it is convenient to construct a table of values for x (age), y (price), xy, x^2, and their sums. This is done in Table 14.5.

TABLE 14.5 Table for computing the regression equation for the Orion data

Age (yrs) x	Price ($100s) y	xy	x^2
5	85	425	25
4	103	412	16
6	70	420	36
5	82	410	25
5	89	445	25
5	98	490	25
6	66	396	36
6	95	570	36
2	169	338	4
7	70	490	49
7	48	336	49
58	975	4732	326

The slope of the regression line is, therefore,

$$b_1 = \frac{S_{xy}}{S_{xx}} = \frac{\Sigma xy - (\Sigma x)(\Sigma y)/n}{\Sigma x^2 - (\Sigma x)^2/n} = \frac{4732 - (58)(975)/11}{326 - (58)^2/11} = -20.26.$$

The y-intercept is

$$b_0 = \frac{1}{n}(\Sigma y - b_1 \Sigma x) = \frac{1}{11}\left[975 - (-20.26) \cdot 58\right] = 195.47.$$

So we see that the regression equation is $\hat{y} = 195.47 - 20.26x$. *Note:* The usual warnings about rounding apply. When computing the slope, b_1, of the regression line, do not round until the computation is finished. Moreover, when computing the y-intercept, b_0, do not use the rounded value of b_1; instead, keep full calculator accuracy.

b. To graph the regression equation, we need to substitute two different x-values into the regression equation to obtain two distinct points. Let's use the x-values 2 and 8. The corresponding y-values are

$$\hat{y} = 195.47 - 20.26 \cdot 2 = 154.95 \quad \text{and} \quad \hat{y} = 195.47 - 20.26 \cdot 8 = 33.39.$$

So, the regression line passes through the two points (2, 154.95) and (8, 33.39). In Fig. 14.10 we have plotted these two points using hollow dots. Drawing a straight line through the two hollow dots yields the regression line.

FIGURE 14.10
Regression line and data points for Orion data

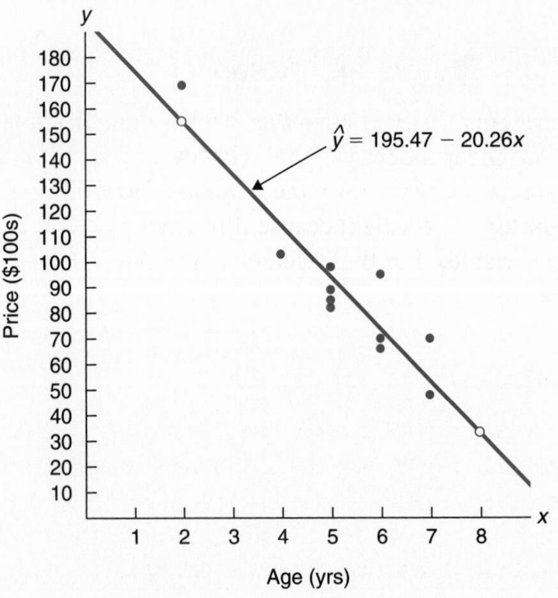

Also included in Fig. 14.10 are the data points from the first two columns of Table 14.5. As we know, the regression line in Fig. 14.10 is the straight line that best fits the data points according to the least-squares criterion; that is, it is the straight line having the smallest possible sum of squared errors.

c. Here we are to describe the apparent relationship between age and price of Orions. Since the slope of the regression line is negative, we see that price tends to decrease as age increases—no particular surprise.

d. For this part, we are to interpret the slope of the regression line in terms of prices for Orions. Recalling that x represents age, in years, and y represents price, in hundreds of dollars, we see that the slope of -20.26 indicates that Orions depreciate an estimated \$2026 per year, at least in the 2- to 7-year-old range.

e. Finally, we are to use the regression equation to predict the price of a 3-year-old Orion and a 4-year-old Orion. For a 3-year-old Orion, we have $x = 3$, and so the predicted price is

$$\hat{y} = 195.47 - 20.26 \cdot 3 = 134.69,$$

or \$13,469. Similarly, the price the regression equation predicts for a 4-year-old Orion is

$$\hat{y} = 195.47 - 20.26 \cdot 4 = 114.43,$$

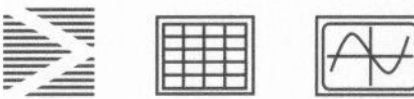

or \$11,443. Questions concerning the accuracy and reliability of such predictions will be discussed later in this chapter and also in Chapter 15. ◨

Predictor Variable and Response Variable

For a linear equation $y = b_0 + b_1x$, we have noted that y is the dependent variable and x is the independent variable. However, in the context of regression analysis, it is more customary to call y the **response variable** and x the **predictor variable** or **explanatory variable** (because it is used to predict or explain the values of the response variable). For the Orion example, age is the predictor variable and price is the response variable.

Extrapolation

If a scatter diagram indicates a linear relationship between two variables, then it is reasonable to use the regression equation to make predictions for values of the predictor variable within the range of the observed values of the predictor variable. However, it may not be reasonable to do so for values of the predictor variable outside that range, because the linear relationship between the variables may not hold there.

Using the regression equation to make predictions for values of the predictor variable outside the range of the observed values of the predictor variable is called **extrapolation.** Grossly incorrect predictions can result from extrapolation.

The Orion example provides an excellent illustration of where extrapolation can lead to grossly incorrect predictions. The regression equation for the sample of Orions is $\hat{y} = 195.47 - 20.26x$, and the observed ages (values of the predictor variable) range from 2 to 7 years old.

Suppose we extrapolate by using the regression equation to predict the price of an 11-year-old Orion. The predicted price is

$$\hat{y} = 195.47 - 20.26 \cdot 11 = -27.39,$$

or −$2739. Clearly, this is ridiculous—no one is going to pay us $2739 to take away their 11-year-old Orion.

Consequently, we see that although the relationship between age and price of Orions appears to be linear in the range from 2 to 7 years old, it is definitely not so in the range from 2 to 11 years old. Figure 14.11 summarizes the discussion on extrapolation as it applies to age and price of Orions.

FIGURE 14.11
Extrapolation in the Orion example

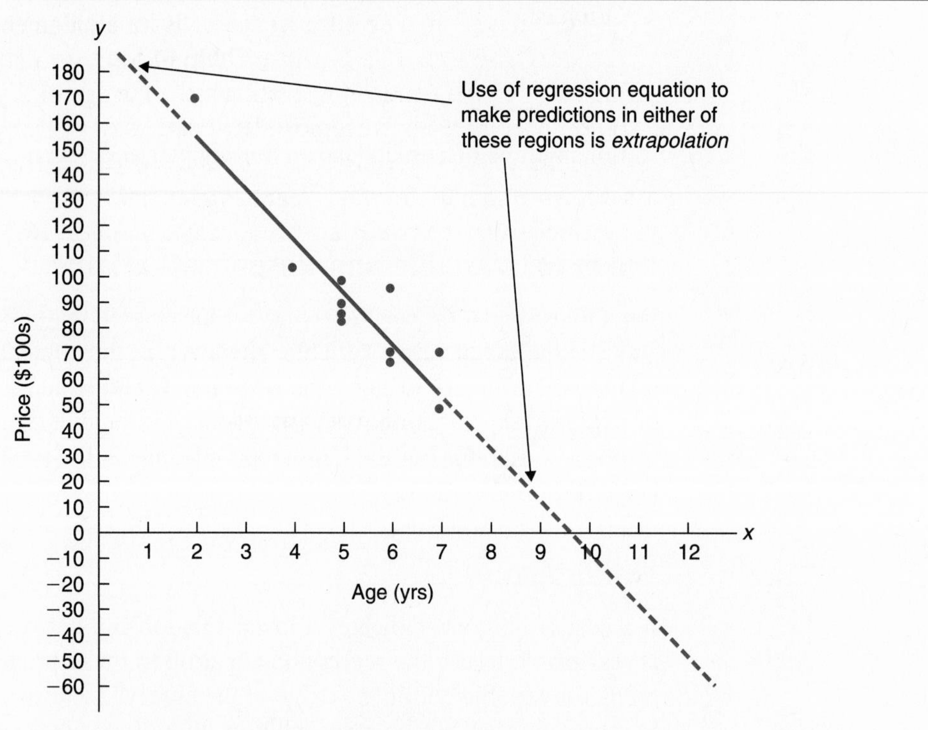

Outliers and Influential Observations

Recall that an outlier is an observation that lies outside the overall pattern of the data. In the context of regression, an **outlier** is a data point that lies far from the regression line, relative to the other data points. Figure 14.10 on page 832 shows that the Orion data has no outliers.

An outlier can sometimes have a significant effect on a regression analysis. So, as usual, it is important to identify outliers and to remove them from the analysis if appropriate, for example, if the outlier is found to be a measurement or recording error.

We must also watch for influential observations. In regression analysis, an **influential observation** is a data point whose removal causes the regression equation (and line) to change considerably. A data point that is separated in the x-direction from the other data points is often an influential observation because the regression line is "pulled" toward such a data point without counteraction by other data points.

As with an outlier, we should try to determine the reason for an influential observation. If it is discovered that an influential observation is due to a measurement or recording error or that for some other reason it clearly does not belong in the data set, then it can be removed without further ado. However, if no explanation for the influential observation is apparent, then the decision whether or not to retain it in the data set can often be difficult and calls for a judgment by the researcher.

For the Orion data, Fig. 14.10 (or Table 14.5) shows that the data point (2, 169) is potentially an influential observation since the age of 2 years is separated from the other observed ages. We removed that data point and recalculated the regression equation; the result is $\hat{y} = 160.33 - 14.24x$. As we see from Fig. 14.12, this equation differs markedly from the regression equation, $\hat{y} = 195.47 - 20.26x$, which we obtained using the full data set. So the data point (2, 169) is indeed an influential observation.

FIGURE 14.12 Regression lines with and without the influential observation removed

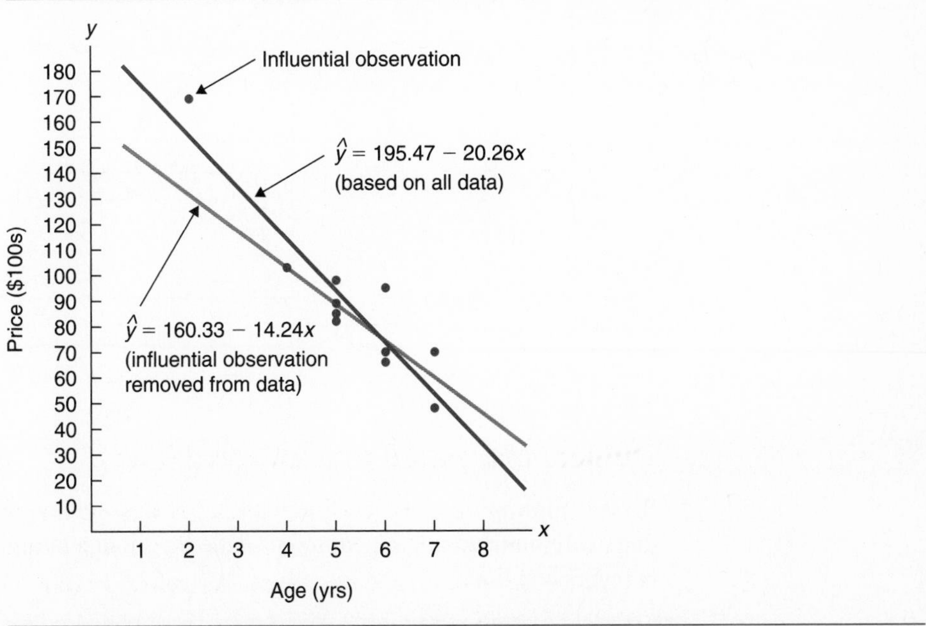

The influential observation (2, 169) is not a recording error, but a legitimate data point. Nonetheless it may be advisable either to remove it—thus limiting the analysis to Orions between 4 and 7 years old—or to obtain additional data on 2-year-old (and 3-year-old) Orions so that the regression analysis is not so dependent on one data point.

We added data for one 2-year-old and three 3-year-old Orions and obtained the regression equation $\hat{y} = 193.63 - 19.93x$. This regression equation differs very little from our original one, $\hat{y} = 195.47 - 20.26x$. So we could justify using the original regression equation to analyze the relationship between age and price of Orions between 2 and 7 years of age, even though the corresponding data set contains an influential observation.

An outlier may or may not be an influential observation; and an influential observation may or may not be an outlier. Many statistical software packages, including Minitab, identify potential outliers and influential observations.

A Warning on the Use of Linear Regression

The idea behind finding a regression line is based on the assumption that the data points are scattered about a straight line. In some cases, data points are scattered about a curve instead of a straight line, as in Fig. 14.13(a). The formulas for b_0 and b_1 will still work for this data set and fit an inappropriate straight line, as shown in Fig. 14.13(b), instead of a curve. This procedure would lead us to predict that y-values in Fig. 14.13(a) will keep increasing when they have actually begun to decrease. Key Fact 14.3 summarizes the criterion for finding a regression line.

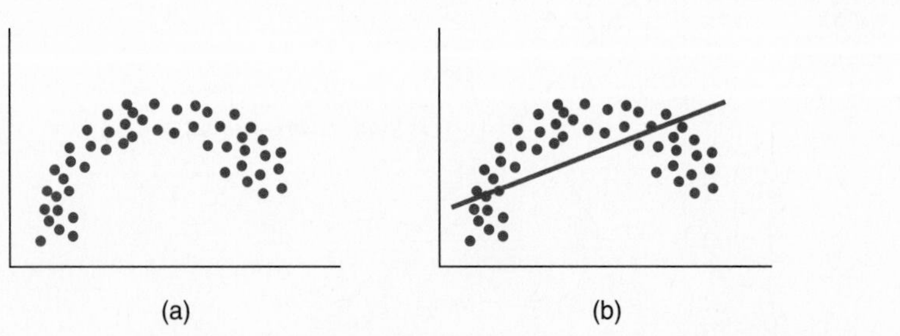

FIGURE 14.13
(a) Data points scattered about a curve
(b) Inappropriate straight line fit to the data points

KEY FACT 14.3 **CRITERION FOR FINDING A REGRESSION LINE**

Before finding a regression line for a set of data points, draw a scatter diagram. If the data points do not appear to be scattered about a straight line, do not determine a regression line.

Techniques are available for fitting curves to data points showing a curved pattern, like the data points in Fig. 14.13(a). Those techniques, referred to as **curvilinear regression,** are discussed in the module *Model Building in Regression* by Dennis L. Young (Reading, MA: Addison-Wesley, 1999).

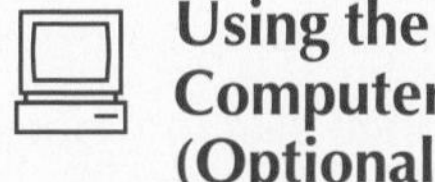

Using the Computer (Optional)

As we know, before determining a regression line, we should look at a scatter diagram of the data to check whether the data points appear to be scattered about a straight line. Minitab can be used to obtain such a diagram.

EXAMPLE 14.5 USING MINITAB TO OBTAIN A SCATTER DIAGRAM

Use Minitab to obtain a scatter diagram for the age and price data of Orions displayed in Table 14.2 on page 827.

SOLUTION We begin by storing the age and price data in columns named AGE and PRICE, respectively. Then we proceed in the following manner.

1 Choose **Graph ➤ Plot...**

2 Specify PRICE in the **Y** text box for **Graph 1**

3 Click in the **X** text box for **Graph 1** and specify AGE

4 Click **OK**

The resulting scatter diagram is displayed in Printout 14.1.

PRINTOUT 14.1
Minitab scatter diagram for the Orion data

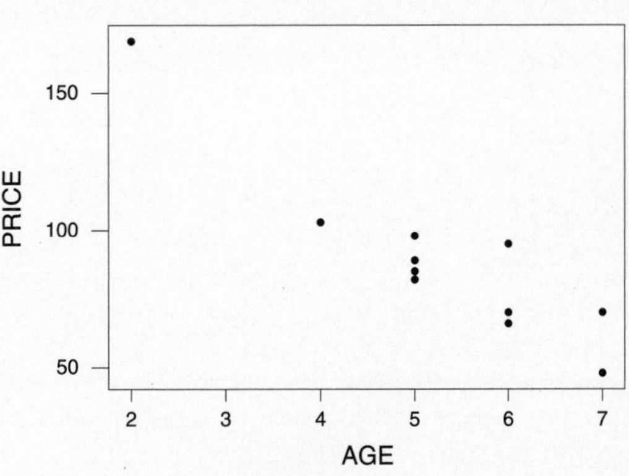

Printout 14.1 is Minitab's version of the scatter diagram we drew by hand in Fig. 14.7 on page 827. It shows that the data points are scattered about a straight line and hence that it is reasonable to find a regression line for these data. ❒

Formula 14.1 on page 831 provides the formulas to determine the regression equation for a set of data points. Alternatively, we can use Minitab's regression procedure to obtain a regression equation.

EXAMPLE 14.6 USING MINITAB TO OBTAIN A REGRESSION EQUATION

Use Minitab to obtain the regression equation for the age and price data of Orions displayed in Table 14.2 on page 827.

SOLUTION In Example 14.5 the sample data on age and price were stored in columns named AGE and PRICE, respectively. With that in mind, we proceed as follows.

1 Choose **Stat ➤ Regression ➤ Regression...**
2 Specify PRICE in the **Response** text box
3 Click in the **Predictors** text box and specify AGE
4 Click **OK**

The resulting output is depicted in Printout 14.2.

PRINTOUT 14.2
Minitab regression output

```
The regression equation is
PRICE = 195 - 20.3 AGE

Predictor        Coef       StDev          T        P
Constant       195.47       15.24      12.83    0.000
AGE           -20.261       2.800      -7.24    0.000

S = 12.58        R-Sq = 85.3%      R-Sq(adj) = 83.7%

Analysis of Variance

Source           DF          SS         MS          F        P
Regression        1      8285.0     8285.0      52.38    0.000
Residual Error    9      1423.5      158.2
Total            10      9708.5

Unusual Observations
Obs        AGE      PRICE          Fit    StDev Fit     Residual    St Resid
  9       2.00     169.00       154.95         9.92        14.05      1.82 X

X denotes an observation whose X value gives it large influence.
```

From the second line of Printout 14.2, we find that the required regression equation is PRICE = 195 - 20.3 AGE or, in our notation, $\hat{y} = 195 - 20.3x$. Next we find a table that provides information about the y-intercept, b_0, and slope, b_1, of the regression line. In particular, the entries `195.47` and `-20.261` under the column headed `Coef` are the values of b_0 and b_1, respectively.

Near the bottom of Printout 14.2 we find information on `Unusual Observations`, which may be either potential outliers or influential observations. Minitab labels potential outliers with an `R` and potential influential observations with an `X`. Here we see that there is one potential influential observation, namely, the data point (2, 169). As we discovered on page 835, this data point is in fact an influential observation since its removal causes a drastic change in the regression equation.

Potential outliers detected by Minitab are identified as data points having large standardized residuals, namely, standardized residuals with magnitude greater than 2.† A standardized residual with magnitude greater than 3 or relatively many standardized residuals with magnitudes greater than 2 is cause for concern because least-squares regression is not resistant to outliers. Evidently, there are no potential outliers in the Orion data.

We have discussed only a small portion of the information provided by the regression output in Printout 14.2. Later sections will address other aspects. ◨

Minitab offers four options for the amount of output from its regression procedure. The options are available in the **Regression - Results** dialog box, which is accessed from the **Regression** dialog box by clicking the **Results...** button. In future illustrations of Minitab's regression procedure, we will sometimes use an option other than the default.

EXERCISES 14.2

STATISTICAL CONCEPTS AND SKILLS

14.32 Regarding a scatter diagram:

a. Identify one of its uses.

b. What property should it have in order to proceed to obtain a regression line for the data?

14.33 Regarding the criterion used to decide on the line that best fits a set of data points:

a. What is that criterion called?

b. Specifically, what is the criterion?

14.34 Regarding the line that best fits a set of data points:

a. What is that line called?

b. What is the equation of that line called?

14.35 Regarding the two variables under consideration in a regression analysis:

a. What is the dependent variable called?

b. What is the independent variable called?

† The *residual* of a data point is the difference between the observed and predicted y-values, $y - \hat{y}$; the *standardized residual* is obtained by dividing the residual by the standard deviation of the residual.

14.36 Fill in the blank: Using the regression equation to make predictions for values of the predictor variable outside the range of the observed values of the predictor variable is called ____.

14.37 Fill in the blanks:

a. In the context of regression, an ____ is a data point that lies far from the regression line, relative to the other data points.

b. In regression analysis, an ____ is a data point whose removal causes the regression equation to change considerably.

In each of Exercises 14.38 and 14.39,

a. graph each linear equation and the data points.

b. construct tables for x, y, $\hat{y}$, e, and e^2 similar to Table 14.4 on page 829.

c. determine which line fits the set of data points better according to the least-squares criterion.

14.38 Line A: $y = 1.5 + 0.5x$
Line B: $y = 1.125 + 0.375x$

x	1	1	5	5
y	1	3	2	4

14.39 Line A: $y = 3 - 0.6x$
Line B: $y = 4 - x$

x	0	2	2	5	6
y	4	2	0	−2	1

For Exercises 14.40–14.47, be sure to save your work. You will need it in later sections.

14.40 Refer to Exercise 14.38.

a. Find the regression equation for the data points.

b. Graph the regression equation and the data points.

14.41 Refer to Exercise 14.39.

a. Find the regression equation for the data points.

b. Graph the regression equation and the data points.

14.42 The *Kelley Blue Book* provides information on wholesale and retail prices of cars. Following are age and price data for 10 randomly selected Corvettes between 1 and 6 years old. Here x denotes age, in years, and y denotes price, in hundreds of dollars.

x	6	6	6	2	2	5	4	5	1	4
y	175	165	180	310	269	200	240	213	310	210

a. Determine the regression equation for the data.

b. Graph the regression equation and the data points.

c. Describe the apparent relationship between age and price for Corvettes.

d. What does the slope of the regression line represent in terms of Corvette prices?

e. Use the regression equation that you obtained in part (a) to predict the price of a 2-year-old Corvette; a 3-year-old Corvette.

f. Identify the predictor and response variables.

g. Identify outliers and potential influential observations.

14.43 The National Center for Health Statistics publishes data on heights and weights in *Vital and Health Statistics.* A random sample of 11 males age 18–24 years gave the following data, where x denotes height, in inches, and y denotes weight, in pounds.

x	65	67	71	71	66	75	67	70	71	69	69
y	175	133	185	163	126	198	153	163	159	151	155

a. Determine the regression equation for the data.

b. Graph the regression equation and the data points.

c. Describe the apparent relationship between height and weight for 18–24-year-old males.

d. What does the slope of the regression line represent in terms of weights of 18–24-year-old males?

e. Use the regression equation determined in part (a) to predict the weight of an 18–24-year-old male who is 67 inches tall; 73 inches tall.

f. Identify the predictor and response variables.

g. Identify outliers and potential influential observations.

14.44 Hanna Properties specializes in custom-home resales in the Equestrian Estates, an exclusive subdivision in Phoenix, Arizona. A random sample of nine custom homes currently listed for sale provided the following information on size and price. Here x denotes size, in hundreds of square feet, rounded to the nearest hundred, and y denotes price, in thousands of dollars, rounded to the nearest thousand.

x	26	27	33	29	29	34	30	40	22
y	259	274	294	296	325	380	457	523	215

a. Determine the regression equation for the data.
b. Graph the regression equation and the data points.
c. Describe the apparent relationship between square footage and price for custom homes in the Equestrian Estates.
d. What does the slope of the regression line represent in terms of sizes and prices of custom homes in the Equestrian Estates?
e. Use the regression equation determined in part (a) to predict the price of a custom home in the Equestrian Estates that has 2600 sq ft.
f. Identify the predictor and response variables.
g. Identify outliers and potential influential observations.

14.45 Many studies have been done that indicate the maximum heart rate an individual can reach during intensive exercise decreases with age. See, for example, page 73 of the September 16, 1996 issue of *Newsweek*. A physician decided to do his own study and recorded the ages and peak heart rates of 10 randomly selected people. The results are shown in the following table, where x denotes age, in years, and y denotes peak heart rate.

x	30	38	41	38	29	39	46	41	42	24
y	186	183	171	177	191	177	175	176	171	196

a. Determine the regression equation for the data.
b. Graph the regression equation and the data points.
c. Describe the apparent relationship between age and peak heart rate.
d. What does the slope of the regression line represent in terms of age and peak heart rate?
e. Use the regression equation to predict the peak heart rate of a 28-year-old person.
f. Identify the predictor and response variables.
g. Identify outliers and potential influential observations.

14.46 An instructor at Arizona State University asked a random sample of eight students to record their study times in a beginning calculus course. She then made a table for total hours studied, x, over 2 weeks, and test score, y, at the end of the 2 weeks. Here are the results.

x	10	15	12	20	8	16	14	22
y	92	81	84	74	85	80	84	80

a. Determine the regression equation for the data.
b. Graph the regression equation and the data points.
c. Describe the apparent relationship between study time and test score. Does it surprise you?
d. What does the slope of the regression line represent in terms of study time and test score?
e. Use the regression equation to predict the test score of a student who studies for 15 hours.
f. Identify the predictor and response variables.
g. Identify outliers and potential influential observations.

14.47 An economist is interested in the relationship between the disposable income of a family and the amount of money spent annually on food. For a preliminary study, the economist takes a random sample of eight middle-income families of the same size (father, mother, two children). The results are as follows, where x denotes disposable income, in thousands of dollars, and y denotes food expenditure, in hundreds of dollars.

x	30	36	27	20	16	24	19	25
y	55	60	42	40	37	26	39	43

a. Determine the regression equation for the data.
b. Graph the regression equation and the data points.
c. Describe the apparent relationship between disposable income and annual food expenditure.

d. What does the slope of the regression line represent in terms of disposable income and annual food expenditure?
e. Use the regression equation to predict the annual food expenditure of a family with a disposable income of \$25,000.
f. Identify the predictor and response variables.
g. Identify outliers and potential influential observations.

14.48 For which of the following sets of data points is it reasonable to determine a regression line? Explain your answer.

14.49 For which of the following sets of data points is it reasonable to determine a regression line? Explain your answer.

14.50 In Exercise 14.42 you determined a regression equation that can be used to predict the price of a Corvette given its age.

a. Should that regression equation be used to predict the price of a 4-year-old Corvette? a 10-year-old Corvette? Explain your answers.
b. For which ages is it reasonable to use the regression equation to predict price?

14.51 In Exercise 14.43 you determined a regression equation that relates the variables height and weight for 18–24-year-old males.

a. Should that regression equation be used to predict the weight of an 18–24-year-old male who is 68 inches tall? 60 inches tall? Explain your answers.
b. For which heights is it reasonable to use the regression equation to predict weight?

EXTENDING THE CONCEPTS AND SKILLS

14.52 The negative relation between study time and grade, found in Exercise 14.46, has been discovered by many investigators. Can you think of a possible explanation for it?

Sample covariance: For n pairs of observations from two variables, x and y, the **sample covariance, s_{xy},** is defined by

$$s_{xy} = \frac{\Sigma(x - \bar{x})(y - \bar{y})}{n - 1}. \qquad (1)$$

14.53 Determine the sample covariance of the data points in Exercise 14.39.

14.54 Determine the sample covariance of the data points in Exercise 14.38.

The sample covariance, defined in Equation (1), can be used as an alternate method for obtaining the slope and y-intercept of the regression line for a set of data points. The formulas are

$$b_1 = s_{xy}/s_x^2 \quad \text{and} \quad b_0 = \bar{y} - b_1\bar{x}, \qquad (2)$$

where s_x denotes the sample standard deviation of the x-values.

14.55 Use the equations in (2) to find the regression equation for the data points in Exercise 14.39. Compare your answer to the one you obtained in part (a) of Exercise 14.41.

14.56 Use the equations in (2) to find the regression equation for the data points in Exercise 14.38. Compare your answer to the one you obtained in part (a) of Exercise 14.40.

USING TECHNOLOGY   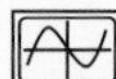

14.57 Refer to Exercise 14.47. Use Minitab or some other statistical software to

a. obtain a scatter diagram for the data.
b. determine the regression equation for the data.
c. identify potential outliers and influential observations.
d. Justify the use of your procedure in part (b).

14.58 Refer to Exercise 14.46. Use Minitab or some other statistical software to
a. obtain a scatter diagram for the data.
b. determine the regression equation for the data.
c. identify potential outliers and influential observations.
d. Justify the use of your procedure in part (b).

14.59 Greene and Touchstone conducted a study on the relationship between the estriol levels of pregnant women and the birth weights of their children. Their findings, "Urinary Tract Estriol: An Index of Placental Function," were published in the *American Journal of Obstetrics and Gynecology.* Printout 14.3 shows the output that results by applying Minitab's regression procedure to the data obtained by Greene and Touchstone. The estriol levels are in milligrams per 24 hours, and the birth weights are in grams.
a. Consult the printout to determine the regression equation for the data.
b. Use the regression equation to predict the birth weight of the child of a pregnant woman with an estriol level of 17 mg/24 hr.
c. Identify potential outliers and influential observations.

14.60 The Energy Information Administration publishes data on energy consumption by family income in *Residential Energy Consumption Survey: Consumption and Expenditures.* We applied Minitab's regression procedure to data on family income and last year's energy consumption from a random sample of 25 families. Printout 14.4 at the top of the next page shows the output. The income data are in thousands of dollars, and the energy-consumption data are in millions of BTU.
a. Consult the printout to obtain the regression equation for the data.
b. Predict last year's energy consumption for a family with an income of $42,000.
c. Identify potential outliers and influential observations.

PRINTOUT 14.3 Minitab output for Exercise 14.59 (and Exercises 14.77, 15.27, and 15.47)

```
The regression equation is
WEIGHT = 2152 + 60.8 ESTRIOL

Predictor        Coef       StDev         T        P
Constant       2152.3       262.0      8.21    0.000
ESTRIOL         60.82       14.68      4.14    0.000

S = 382.1       R-Sq = 37.2%     R-Sq(adj) = 35.0%

Analysis of Variance

Source              DF          SS          MS        F        P
Regression           1     2505745     2505745    17.16    0.000
Residual Error      29     4234255      146009
Total               30     6740000

Unusual Observations
Obs    ESTRIOL      WEIGHT         Fit   StDev Fit    Residual    St Resid
 14       24.0      2800.0      3612.0       120.8      -812.0      -2.24R

R denotes an observation with a large standardized residual
```

PRINTOUT 14.4 Minitab output for Exercise 14.60 (and Exercises 14.78, 15.26, and 15.46)

```
The regression equation is
CONSUMPT = 82.0 + 0.931 INCOME

Predictor        Coef       StDev          T        P
Constant       82.036       2.054      39.94    0.000
INCOME        0.93051     0.05727      16.25    0.000

S = 5.375      R-Sq = 92.0%      R-Sq(adj) = 91.6%

Analysis of Variance

Source            DF          SS          MS          F        P
Regression         1      7626.6      7626.6     264.02    0.000
Residual Error    23       664.4        28.9
Total             24      8291.0

Unusual Observations
Obs     INCOME    CONSUMPT        Fit    StDev Fit     Residual     St Resid
  3       15.0       81.69      95.99         1.40       -14.31       -2.76R
 19        5.0       97.07      86.69         1.82        10.38        2.05R

R denotes an observation with a large standardized residual
```

14.3 THE COEFFICIENT OF DETERMINATION

In Example 14.4 we determined the regression equation, $\hat{y} = 195.47 - 20.26x$, for data on age and price of a sample of 11 Orions. Here x represents age, in years, and $\hat{y}$ represents predicted price, in hundreds of dollars.

We can apply the regression equation to predict the price of an Orion of a particular age. For instance, we predict that a 4-year-old Orion will cost roughly

$$\hat{y} = 195.47 - 20.26 \cdot 4 = 114.43,$$

or \$11,443. But how valuable are such predictions? Is the regression equation useful for predicting price, or could we do just as well by ignoring age?

In general, we can evaluate the utility of a regression equation for making predictions in several ways. One way is to determine the percentage of variation in the observed values of the response variable that is explained by the regression (or predictor variable). To see how this is done, we return to the data on age and price of Orions.

EXAMPLE 14.7 INTRODUCES THE COEFFICIENT OF DETERMINATION

The scatter diagram for the age and price data of 11 Orions, first given in Fig. 14.7 on page 827, is included in Fig. 14.14. Also shown in Fig. 14.14 is the regression line for the data.

FIGURE 14.14 Scatter diagram and regression line for Orion data

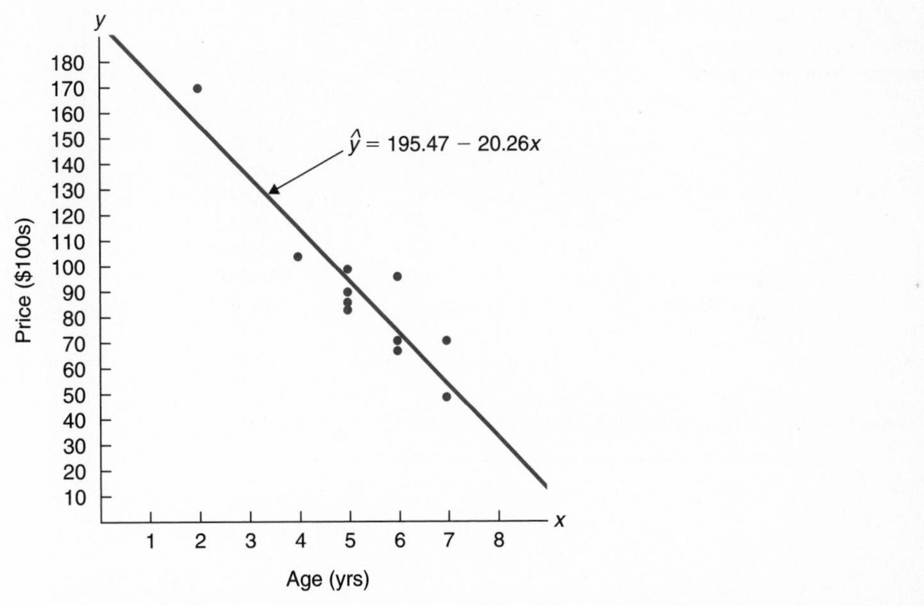

As we can see from the scatter diagram in Fig. 14.14, the prices of the 11 Orions sampled vary widely, ranging from a low of 48 ($4800) to a high of 169 ($16,900). But Fig. 14.14 also shows that much of the price variation is "explained" by the regression (or age); that is, the regression line, with age as the predictor variable, predicts a good portion of the type of variation found in the prices.

To describe quantitatively how much of the variation in the observed prices is explained by the regression, we need to define two measures of variation: (1) the total variation in the observed prices and (2) the amount of variation in the observed prices that is explained by the regression.

As a measure of total variation in the observed prices, we use the sum of squared deviations of the observed prices from the mean price. This is called the **total sum of squares, *SST*.** In symbols, we have $SST = \Sigma(y - \bar{y})^2$. If we divide SST by $n - 1$, we get the sample variance of the observed prices. So SST really is a measure of total variation.

To compute the total sum of squares, we must first find the sample mean price, which is done as usual:

$$\bar{y} = \frac{\Sigma y}{n} = \frac{975}{11} = 88.64.$$

Now we obtain the total sum of squares for the Orion price data using Table 14.6.[†]

TABLE 14.6
Table for computing SST for the Orion price data

Age (yrs) x	Price ($100s) y	$y - \bar{y}$	$(y - \bar{y})^2$
5	85	−3.64	13.2
4	103	14.36	206.3
6	70	−18.64	347.3
5	82	−6.64	44.0
5	89	0.36	0.1
5	98	9.36	87.7
6	66	−22.64	512.4
6	95	6.36	40.5
2	169	80.36	6458.3
7	70	−18.64	347.3
7	48	−40.64	1651.3
	975		9708.5

From the final column of Table 14.6, we see that

$$SST = \Sigma(y - \bar{y})^2 = 9708.5.$$

This is our measure of total variation in the observed prices.

To obtain the amount of variation in the observed prices that is explained by the regression, let's first look at a particular observed price, say, $y = 98$, corresponding to the data point (5, 98). In Fig. 14.15 we have magnified a portion of Fig. 14.14 showing that data point, the mean of the observed prices ($\bar{y} = 88.64$) and the predicted price for a 5-year-old Orion ($\hat{y} = 94.16$).

The total variation in the observed prices is based on the deviation of each observed price from the mean price, $y - \bar{y}$. As illustrated in Fig. 14.15, each such deviation can be decomposed into two parts: the deviation that is explained by the regression line, $\hat{y} - \bar{y}$, and the remaining unexplained deviation, $y - \hat{y}$. So the amount of variation (squared deviation) in the observed prices that is explained by the regression is $\Sigma(\hat{y} - \bar{y})^2$. This is called the **regression sum of squares, *SSR*.**

To compute SSR we need the predicted prices, $\hat{y}$, and the mean of the observed prices, $\bar{y}$. We have already computed the mean of the observed prices and, as we

[†] Values in Table 14.6 and all other tables in this section are displayed to various numbers of decimal places, but computations are done using full calculator accuracy.

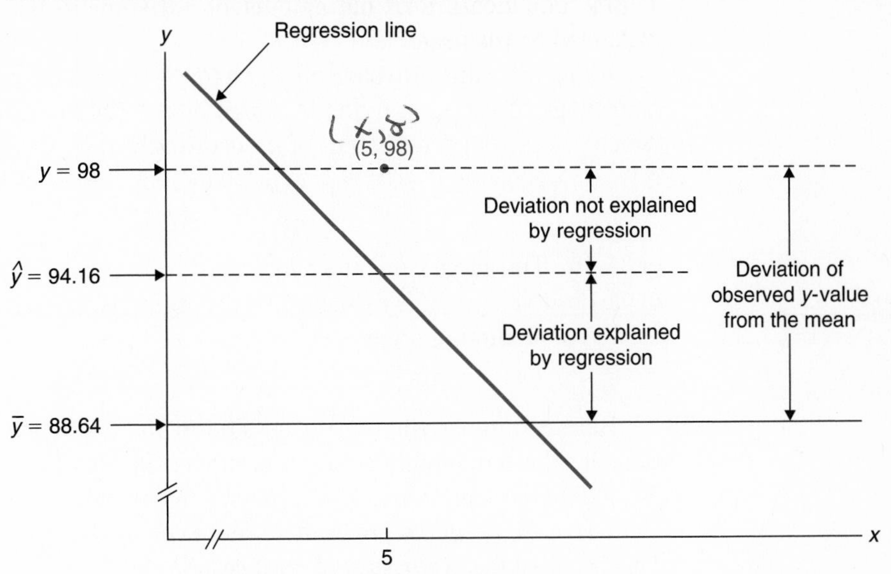

FIGURE 14.15 Magnification of a portion of Fig. 14.14 showing only the data point (5, 98)

have seen, each predicted price is obtained by substituting the age of the Orion in question into the regression equation $\hat{y} = 195.47 - 20.26x$. The third column of Table 14.7 shows the predicted prices for all 11 Orions.

TABLE 14.7 Table for computing *SSR* for the Orion data

Age (yrs) x	Price (\$100s) y	$\hat{y}$	$\hat{y} - \bar{y}$	$(\hat{y} - \bar{y})^2$
5	85	94.16	5.53	30.5
4	103	114.42	25.79	665.0
6	70	73.90	−14.74	217.1
5	82	94.16	5.53	30.5
5	89	94.16	5.53	30.5
5	98	94.16	5.53	30.5
6	66	73.90	−14.74	217.1
6	95	73.90	−14.74	217.1
2	169	154.95	66.31	4397.0
7	70	53.64	−35.00	1224.8
7	48	53.64	−35.00	1224.8
				8285.0

Recalling that $\bar{y} = 88.64$, we construct the fourth column of Table 14.7 and then obtain the regression sum of squares, *SSR,* from the fifth column:

$$SSR = \Sigma(\hat{y} - \bar{y})^2 = 8285.0.$$

This is our measure of the amount of variation in the observed prices that is explained by the regression.

Using the values we have obtained for *SST* and *SSR*, we can now determine the percentage of variation in the observed prices that is explained by the regression, which is called the **coefficient of determination, r^2.** We have

$$r^2 = \frac{SSR}{SST} = \frac{8285.0}{9708.5} = 0.853,$$

or 85.3%. This large percentage indicates that a substantial amount of the variation in the observed prices is explained by the regression and, therefore, that age is quite useful for predicting price. □

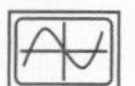

Before summarizing our discussion of the coefficient of determination, let's consider the remaining deviation portrayed in Fig. 14.15—the deviation not explained by the regression, $y - \hat{y}$. We see that the amount of variation (squared deviation) in the observed prices that is not explained by the regression is $\Sigma(y - \hat{y})^2$. This is called the **error sum of squares, *SSE*.**

To compute *SSE* we need the observed prices, y, and the predicted prices, $\hat{y}$. Both of these quantities can be found in Table 14.7 and are repeated in the second and third columns of Table 14.8.

TABLE 14.8
Table for computing *SSE* for the Orion data

Age (yrs) x	Price ($100s) y	$\hat{y}$	$y - \hat{y}$	$(y - \hat{y})^2$
5	85	94.16	−9.16	83.9
4	103	114.42	−11.42	130.5
6	70	73.90	−3.90	15.2
5	82	94.16	−12.16	147.9
5	89	94.16	−5.16	26.6
5	98	94.16	3.84	14.7
6	66	73.90	−7.90	62.4
6	95	73.90	21.10	445.2
2	169	154.95	14.05	197.5
7	70	53.64	16.36	267.7
7	48	53.64	−5.64	31.8
				1423.5

From the final column of Table 14.8, we obtain the error sum of squares:

$$SSE = \Sigma(y - \hat{y})^2 = 1423.5.$$

This is our measure of the amount of variation in the observed prices that is not explained by the regression. Since the regression line is the line that best fits the data according to the least squares criterion, we see that *SSE* is also the smallest possible sum of squared errors among all straight lines.

We now summarize our discussion of the three sums of squares and the coefficient of determination.

DEFINITION 14.4 SUMS OF SQUARES IN REGRESSION

Total sum of squares, SST: The variation in the observed values of the response variable. We have $SST = \Sigma(y - \overline{y})^2$.

Regression sum of squares, SSR: The variation in the observed values of the response variable that is explained by the regression. We have $SSR = \Sigma(\hat{y} - \overline{y})^2$.

Error sum of squares, SSE: The variation in the observed values of the response variable that is not explained by the regression. We have $SSE = \Sigma(y - \hat{y})^2$.

DEFINITION 14.5 COEFFICIENT OF DETERMINATION

The ***coefficient of determination, r^2,*** is the proportion of variation in the observed values of the response variable that is explained by the regression. We have

$$r^2 = \frac{SSR}{SST}.$$

The coefficient of determination always lies between 0 and 1 and is a descriptive measure of the utility of the regression equation for making predictions. Values of r^2 near 0 indicate that the regression equation is not very useful for making predictions, whereas values of r^2 near 1 indicate that the regression equation is extremely useful for making predictions.

The Regression Identity

For the Orion data, we have determined that $SST = 9708.5$, $SSR = 8285.0$, and $SSE = 1423.5$. Since $9708.5 = 8285.0 + 1423.5$, we see that $SST = SSR + SSE$. This equation is always true and is called the **regression identity.**

KEY FACT 14.4 REGRESSION IDENTITY

The total sum of squares equals the regression sum of squares plus the error sum of squares. In symbols, $SST = SSR + SSE$.

Because of the regression identity, we can also express the coefficient of determination in terms of the total sum of squares and error sum of squares:

$$r^2 = \frac{SSR}{SST} = \frac{SST - SSE}{SST} = 1 - \frac{SSE}{SST}.$$

This formula shows that the coefficient of determination can also be interpreted as the percentage reduction obtained in the total squared error by using the regression equation instead of the mean, $\bar{y}$, to predict the observed values of the response variable. We will examine this interpretation in the exercises.

Computing Formulas for the Sums of Squares

Calculating the three sums of squares, *SST, SSR,* and *SSE,* using the defining formulas is time-consuming and can lead to significant roundoff error unless full accuracy is retained. For those reasons we usually employ computing formulas or a computer to obtain the sums of squares. The computing formulas are presented in Formula 14.2.

FORMULA 14.2 **COMPUTING FORMULAS FOR THE SUMS OF SQUARES**

The three sums of squares, *SST, SSR,* and *SSE,* can be obtained using the following computing formulas:

Total sum of squares: $SST = S_{yy}$

Regression sum of squares: $SSR = S_{xy}^2/S_{xx}$

Error sum of squares: $SSE = S_{yy} - S_{xy}^2/S_{xx}$

The formulas for S_{yy}, S_{xy}, and S_{xx} are given in Definition 14.3 on page 830.

EXAMPLE 14.8 **ILLUSTRATES FORMULA 14.2**

The age and price data for a sample of 11 Orions are repeated in the first two columns of Table 14.9. Use the computing formulas in Formula 14.2 to determine the three sums of squares.

SOLUTION To apply the computing formulas, we will need a table of values for x (age), y (price), xy, x^2, y^2, and their sums. This is presented in Table 14.9.

Using the last row of Table 14.9 and Formula 14.2, we can now obtain the three sums of squares for the Orion data. The total sum of squares equals

$$SST = S_{yy} = \Sigma y^2 - (\Sigma y)^2/n = 96{,}129 - (975)^2/11 = 9708.5;$$

the regression sum of squares equals

$$SSR = \frac{S_{xy}^2}{S_{xx}} = \frac{\left[\Sigma xy - (\Sigma x)(\Sigma y)/n\right]^2}{\Sigma x^2 - (\Sigma x)^2/n} = \frac{\left[4732 - (58)(975)/11\right]^2}{326 - (58)^2/11} = 8285.0;$$

TABLE 14.9 Table for obtaining the three sums of squares for the Orion data using the computing formulas

Age (yrs) x	Price ($100s) y	xy	x^2	y^2
5	85	425	25	7,225
4	103	412	16	10,609
6	70	420	36	4,900
5	82	410	25	6,724
5	89	445	25	7,921
5	98	490	25	9,604
6	66	396	36	4,356
6	95	570	36	9,025
2	169	338	4	28,561
7	70	490	49	4,900
7	48	336	49	2,304
58	975	4732	326	96,129

and, using the two preceding results, we see that the error sum of squares equals

$$SSE = S_{yy} - \frac{S_{xy}^2}{S_{xx}} = 9708.5 - 8285.0 = 1423.5.$$

The values obtained here for the three sums of squares by applying the computing formulas are, of course, the same as the values we found earlier by using the defining formulas. However, when the computing formulas are employed, the computations are much simpler and less subject to roundoff error. ◨

  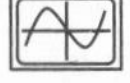

Using the Computer (Optional)

In Section 14.2 we learned that Minitab can be used to obtain the regression equation for a set of data points. The output that results from applying that procedure contains much more than just the regression equation. In particular, it also provides the coefficient of determination, r^2, and the three sums of squares, *SST, SSR,* and *SSE*.

EXAMPLE 14.9 USING MINITAB TO OBTAIN r^2 AND THE THREE SUMS OF SQUARES

Printout 14.2 on page 838 shows the output obtained by applying Minitab's regression procedure to the data on age and price for a sample of 11 Orions. Use the printout to find

a. the coefficient of determination, r^2.

b. the three sums of squares, *SST, SSR,* and *SSE*.

SOLUTION **a.** The coefficient of determination, r^2, is displayed as the second entry in the sixth line of Printout 14.2: `R-Sq = 85.3%`. In other words, $r^2 = 0.853$.

b. To obtain the three sums of squares, we use the table in Printout 14.2 entitled `Analysis of Variance`. The entries in the column headed `Source` identify the three sums of squares and their values are found in the column headed `SS`. Hence the first entry in the `SS` column is the regression sum of squares, *SSR*, the second is the error sum of squares, *SSE*, and the third is the total sum of squares, *SST*. So we see that $SSR = 8285.0$, $SSE = 1423.5$, and $SST = 9708.5$. □

EXERCISES 14.3

STATISTICAL CONCEPTS AND SKILLS

14.61 In this section we introduced a descriptive measure of the utility of the regression equation for making predictions.

a. Identify the term and symbol for that descriptive measure.

b. Provide an interpretation of that descriptive measure.

14.62 Fill in the blanks.

a. A measure of total variation in the observed values of the response variable is the ____. The mathematical abbreviation for this is ____.

b. A measure of the amount of variation in the observed values of the response variable that is explained by the regression is the ____. The mathematical abbreviation for this is ____.

c. A measure of the amount of variation in the observed values of the response variable that is not explained by the regression is the ____. The mathematical abbreviation for this is ____.

14.63 For a particular regression analysis, we found that $SST = 8291.0$ and $SSR = 7626.6$.

a. Obtain and interpret the coefficient of determination.

b. Determine *SSE*.

In Exercises 14.64 and 14.65, we have repeated the data from Exercises 14.38 and 14.39, respectively. We have also provided the regression equations for those data, which were found in Exercises 14.40 and 14.41, respectively. For each exercise,

a. compute the three sums of squares, SST, SSR, and SSE, using the defining formulas (page 849).

b. verify the regression identity, $SST = SSR + SSE$.

c. compute the coefficient of determination.

d. determine the percentage of variation in the observed values of the response variable that is explained by the regression.

e. state how useful the regression equation appears to be for making predictions. (Answers for this part may vary due to differing interpretation.)

14.64 The data from Exercise 14.38:

x	1	1	5	5
y	1	3	2	4

Regression equation is $\hat{y} = 1.75 + 0.25x$.

14.65 The data from Exercise 14.39:

x	0	2	2	5	6
y	4	2	0	−2	1

Regression equation is $\hat{y} = 2.875 - 0.625x$.

For each of Exercises 14.66–14.71,

a. compute SST, SSR, and SSE using Formula 14.2 on page 850.

b. compute the coefficient of determination, r^2.

c. determine the percentage of variation in the observed values of the response variable that is explained by the regression, and interpret your answer.

d. state how useful the regression equation appears to be for making predictions.

14.66 The age and price data for Corvettes from Exercise 14.42:

x	6	6	6	2	2	5	4	5	1	4
y	175	165	180	310	269	200	240	213	310	210

14.67 The height and weight data for males age 18–24 years from Exercise 14.43:

x	65	67	71	71	66	75	67	70	71	69	69
y	175	133	185	163	126	198	153	163	159	151	155

14.68 The size and price data for custom homes from Exercise 14.44:

x	26	27	33	29	29	34	30	40	22
y	259	274	294	296	325	380	457	523	215

14.69 The data on age and peak heart rate from Exercise 14.45:

x	30	38	41	38	29	39	46	41	42	24
y	186	183	171	177	191	177	175	176	171	196

14.70 The data on study time and test score from Exercise 14.46:

x	10	15	12	20	8	16	14	22
y	92	81	84	74	85	80	84	80

14.71 The data on disposable income and annual food expenditure from Exercise 14.47:

x	30	36	27	20	16	24	19	25
y	55	60	42	40	37	26	39	43

EXTENDING THE CONCEPTS AND SKILLS

14.72 Suppose that $r^2 = 1$ for a data set. What can you say about

a. SSE? **b.** SSR?

c. the utility of the regression equation for making predictions?

14.73 Suppose that $r^2 = 0$ for a data set. What can you say about

a. SSE? **b.** SSR?

c. the utility of the regression equation for making predictions?

14.74 On page 849, we noted that because of the regression identity, we can also express the coefficient of determination in terms of the total sum of squares and error sum of squares as follows:

$$r^2 = 1 - \frac{SSE}{SST}.$$

a. Explain why this formula shows that the coefficient of determination can also be interpreted as the percentage reduction obtained in the total squared error by using the regression equation instead of the mean, $\bar{y}$, to predict the observed values of the response variable.

b. Referring to Exercise 14.66, what percentage reduction is obtained in the total squared error by using the regression equation instead of the mean of the observed prices to predict the observed prices?

c. Referring to Exercise 14.67, what percentage reduction is obtained in the total squared error by using the regression equation instead of the mean of the observed weights to predict the observed weights?

USING TECHNOLOGY  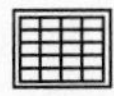 

14.75 Use Minitab or some other statistical software to obtain the coefficient of determination, r^2, and the three sums of squares, *SST, SSR,* and *SSE,* for the data in Exercise 14.71.

14.76 Use Minitab or some other statistical software to obtain the coefficient of determination, r^2, and the three sums of squares, *SST, SSR,* and *SSE,* for the data in Exercise 14.70.

14.77 Greene and Touchstone conducted a study on the relationship between the estriol levels of pregnant women and the birth weights of their children. Their findings, "Urinary Tract Estriol: An Index of Placental Function," were published in the *American Journal of Obstetrics and Gynecology.* Printout 14.3 on page 843 shows the computer output that results by applying Minitab's regression procedure to the data obtained by Greene and Touchstone. The estriol levels are in milligrams per 24 hours, and the birth weights are in grams. Using the printout, determine

a. the coefficient of determination for the data.

b. the regression sum of squares, the error sum of squares, and the total sum of squares.

c. the percentage of variation in the observed birth weights that is explained by estriol level.

14.78 The U.S. Energy Information Administration publishes data on energy consumption by family income in *Residential Energy Consumption Survey: Consumption and Expenditures.* We applied Minitab's regression procedure to data on family income and last year's energy consumption from a random sample of 25 families and obtained the output shown in Printout 14.4 on page 844. Use the printout to find

a. the coefficient of determination for the data.

b. the regression sum of squares, the error sum of squares, and the total sum of squares.

c. the percentage of variation in the observed energy consumptions that is explained by family income.

14.4 LINEAR CORRELATION

We often hear statements pertaining to the correlation or lack of correlation between two variables: "There is a positive correlation between advertising expenditures and sales" or "IQ and alcohol consumption are uncorrelated." In this section we will explain the meaning of such statements.

Several statistics can be employed to measure the correlation between two variables. The one most commonly used is the **linear correlation coefficient, *r*,** also referred to as the **Pearson product moment correlation coefficient.** The linear correlation coefficient is a descriptive measure of the strength of the linear (straight-line) relationship between two variables.

DEFINITION 14.6 LINEAR CORRELATION COEFFICIENT

The ***linear correlation coefficient, r,*** of n data points is defined by

$$r = \frac{\frac{1}{n-1}\Sigma(x-\bar{x})(y-\bar{y})}{s_x s_y}.$$

It can also be obtained from the computing formula

$$r = \frac{S_{xy}}{\sqrt{S_{xx}S_{yy}}},$$

where S_{xx}, S_{xy}, and S_{yy} are given in Definition 14.3 on page 830.

The computing formula in Definition 14.6 is almost always preferred for hand calculations, but it is the defining formula that reveals the meaning and basic properties of the linear correlation coefficient.

One important property of the linear correlation coefficient is that it is positive when the scatter diagram shows a positive slope and negative when the scatter diagram shows a negative slope. To see why this is true, we refer to the defining formula in Definition 14.6 and to Fig. 14.16, where we have drawn a coordinate system with a second set of axes centered at the point $(\bar{x}, \bar{y})$.

FIGURE 14.16 Coordinate system with a second set of axes centered at $(\bar{x}, \bar{y})$

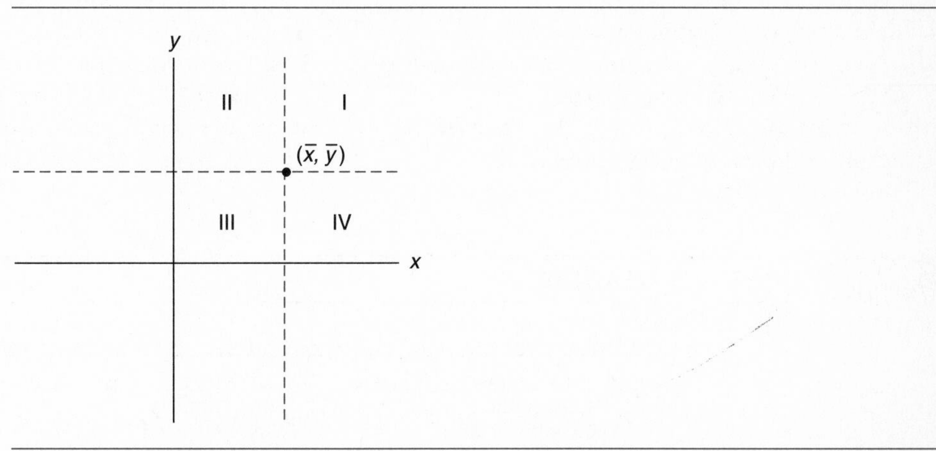

If the scatter diagram shows a positive slope, then, on the average, the data points will lie in either Region I or Region III. For such a data point, the deviations from the means, $x - \bar{x}$ and $y - \bar{y}$, will either both be positive or both be negative. This implies that, on the average, the product $(x - \bar{x})(y - \bar{y})$ will be positive and, consequently, the correlation coefficient will be positive.

If the scatter diagram shows a negative slope, then, on the average, the data points will lie in either Region II or Region IV. For such a data point, one of the deviations from the mean will be positive and the other negative. This implies that, on the average, the product $(x - \bar{x})(y - \bar{y})$ will be negative and, consequently, the correlation coefficient will be negative.

Another important factor in the defining formula for the linear correlation coefficient is the division by the sample standard deviations, s_x and s_y. This gives the linear correlation coefficient two other important properties: It is independent of the choice of units and always lies between -1 and 1.

Values of r close to -1 or 1 indicate a strong linear relationship between the variables and that the variable x is a good linear predictor of the variable y (i.e., the regression equation is extremely useful for making predictions). On the other hand, values of r near 0 indicate at most a weak linear relationship between the variables and that the variable x is a poor linear predictor of the variable y (i.e., the regression equation is either useless or not very useful for making predictions).

Positive values of r suggest that the variables are **positively linearly correlated,** meaning that y tends to increase linearly as x increases, with the tendency being greater the closer that r is to 1. Negative values of r suggest that the variables are **negatively linearly correlated,** meaning that y tends to decrease linearly as x increases, with the tendency being greater the closer that r is to -1. The sign of r is the same as the sign of the slope of the regression line.

To graphically portray the meaning of the linear correlation coefficient, we have presented various degrees of linear correlation in Fig. 14.17.

FIGURE 14.17 Various degrees of linear correlation

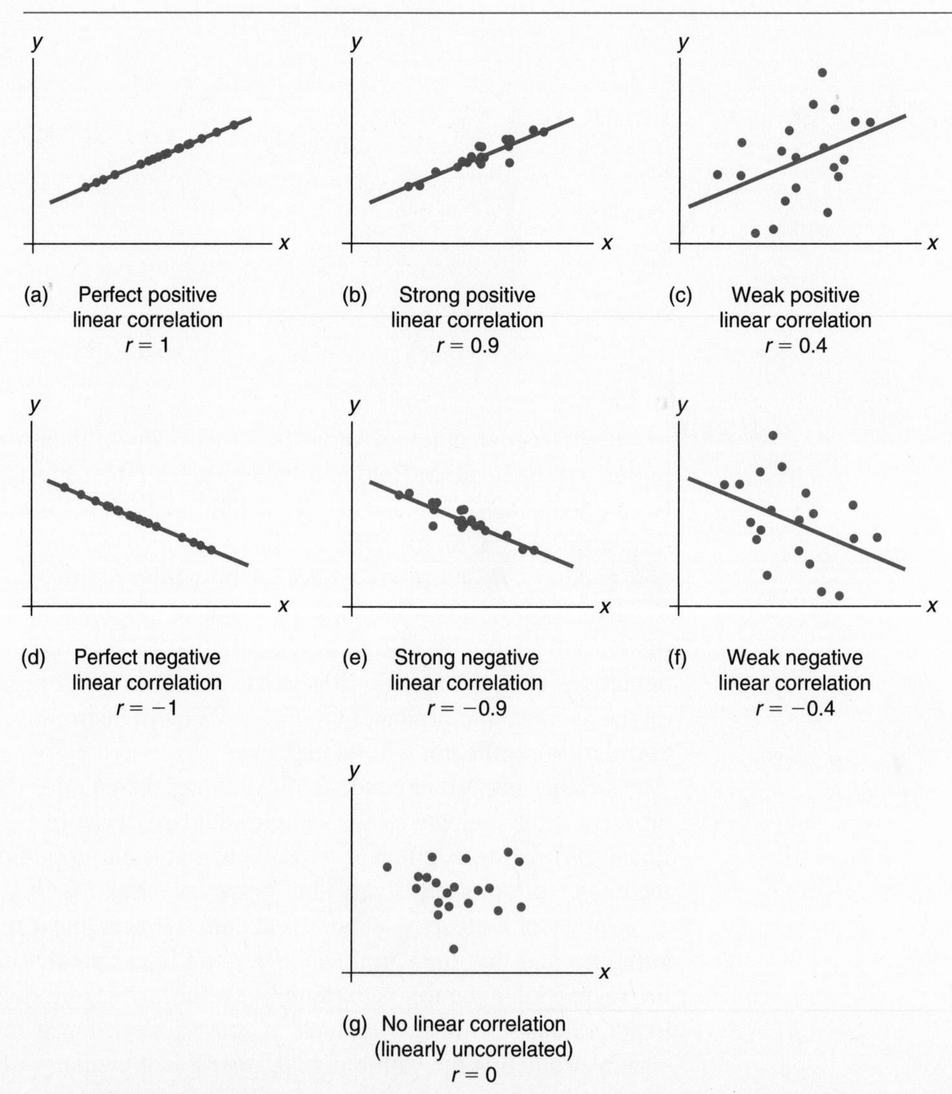

Referring to Fig. 14.17, we can also verbally summarize our discussion of the meaning and properties of the linear correlation coefficient as follows. If the linear correlation coefficient, r, is close to ± 1, then the data points are clustered closely about the regression line. If r is farther from ± 1, then the data points are more widely scattered about the regression line. And if r is near 0, then the slope of the regression line is also near 0, indicating that there is at most a weak linear relationship between the variables.

We will now illustrate how to compute and interpret the linear correlation coefficient of a set of data points. To do that, we return again to the data on age and price for a sample of Orions.

EXAMPLE 14.10 ILLUSTRATES DEFINITION 14.6

The age and price data for a sample of 11 Orions are repeated in the first two columns of Table 14.10.

a. Compute the linear correlation coefficient, r, of the data.

b. Interpret the value of r obtained in part (a) in terms of the linear relationship between the variables age and price of Orions.

c. Discuss the graphical implications of the value of r.

SOLUTION **a.** We will apply the computing formula in Definition 14.6 on page 854 to obtain the linear correlation coefficient. In doing so, we need a table of values for x, y, xy, x^2, y^2, and their sums, as shown in Table 14.10.

TABLE 14.10 Table for obtaining the linear correlation coefficient for the Orion data using the computing formula

Age (yrs) x	Price ($100s) y	xy	x^2	y^2
5	85	425	25	7,225
4	103	412	16	10,609
6	70	420	36	4,900
5	82	410	25	6,724
5	89	445	25	7,921
5	98	490	25	9,604
6	66	396	36	4,356
6	95	570	36	9,025
2	169	338	4	28,561
7	70	490	49	4,900
7	48	336	49	2,304
58	975	4732	326	96,129

Referring to the last row of Table 14.10, we get

$$r = \frac{S_{xy}}{\sqrt{S_{xx}S_{yy}}} = \frac{\Sigma xy - (\Sigma x)(\Sigma y)/n}{\sqrt{[\Sigma x^2 - (\Sigma x)^2/n][\Sigma y^2 - (\Sigma y)^2/n]}}$$

$$= \frac{4732 - (58)(975)/11}{\sqrt{[326 - (58)^2/11][96{,}129 - (975)^2/11]}} = -0.924.$$

b. The linear correlation coefficient, $r = -0.924$, suggests that there is a strong negative linear correlation between age and price of Orions. In particular, it indicates that as age increases there is a strong tendency for price to decrease, which is not surprising. It also implies that the regression equation, $\hat{y} = 195.47 - 20.26x$, is extremely useful for making predictions.

c. Since the correlation coefficient, $r = -0.924$, is quite close to -1, the data points should be clustered rather closely about the regression line. Figure 14.14 on page 845 shows that to be the case. □

Relationship Between the Correlation Coefficient and the Coefficient of Determination

In Section 14.3 we discussed the coefficient of determination, r^2, a descriptive measure of the utility of the regression equation for making predictions. Now we have introduced the linear correlation coefficient, r, as a descriptive measure of the strength of the linear relationship between two variables.

We expect the strength of the linear relationship to also indicate the usefulness of the regression equation for making predictions. In other words, there should be a relationship between the linear correlation coefficient and the coefficient of determination, and there is. The relationship is precisely the one suggested by the notation.

KEY FACT 14.5

RELATIONSHIP BETWEEN THE CORRELATION COEFFICIENT AND THE COEFFICIENT OF DETERMINATION

The coefficient of determination is the square of the linear correlation coefficient.

In Example 14.10 we found that the linear correlation coefficient for the data on age and price of a sample of 11 Orions is $r = -0.924$. From this and Key Fact 14.5, we can easily obtain the coefficient of determination: $r^2 = (-0.924)^2 = 0.854$. As expected, this is the same value (except for roundoff error) as the one we found for r^2 on page 848 by using the defining formula $r^2 = SSR/SST$.

In general, we can compute the coefficient of determination for a set of data points either by using the defining formula, $r^2 = SSR/SST$, or by first obtaining the linear correlation coefficient and then squaring the result.

A Warning on the Use of the Linear Correlation Coefficient

As we mentioned in Section 14.2, an assumption for finding the regression line for a set of data points is that the data points are scattered about a straight line. That same assumption applies to the linear correlation coefficient: The linear correlation coefficient is used to describe the strength of the *linear* relationship between two variables. It should be employed as a descriptive measure only when a scatter diagram indicates that the data points are scattered about a straight line.

Correlation Is Not Causation

Two variables may have a high correlation without being causally related. For example, Table 14.11 displays data on total pari-mutuel turnover (money wagered) at U.S. racetracks and college enrollment for five randomly selected years. [SOURCE: National Association of State Racing Commissioners and U.S. National Center for Education Statistics.]

TABLE 14.11
Pari-mutuel turnover and college enrollment for five randomly selected years

Pari-mutuel turnover ($millions) x	College enrollment (thousands) y
5,977	8,581
7,862	11,185
10,029	11,260
11,677	12,372
11,888	12,426

The linear correlation coefficient of the data points in Table 14.11 is $r = 0.931$, suggesting a strong positive linear correlation between pari-mutuel wagering and college enrollment. But this does not mean that a causal relationship exists between the two variables, such as that when people go to racetracks they are somehow inspired to go to college. On the contrary, we can only infer that the two variables have a strong tendency to increase (or decrease) simultaneously and that total pari-mutuel turnover is a good predictor of college enrollment.

Two variables may be strongly correlated because they are both associated with other variables, called **lurking variables,** that cause changes in the two variables under consideration. For example, a study showed that teachers' salaries and the dollar amount of liquor sales are positively linearly correlated. A possible explanation for this curious fact might be that both of the variables, teachers' salaries and liquor sales, are tied to other variables, such as the rate of inflation, that pull them along together.

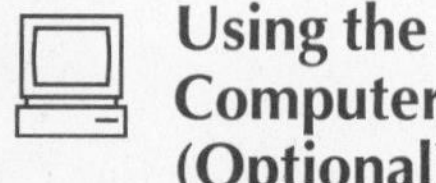

Using the Computer (Optional)

Minitab can be used to determine the linear correlation coefficient of a set of data points. Example 14.11 shows how this is done for the age and price data of a sample of Orions.

EXAMPLE 14.11 Using Minitab to Obtain a Linear Correlation Coefficient

The data on age and price for a sample of 11 Orions are displayed in the first two columns of Table 14.10 on page 857. Use Minitab to determine the linear correlation coefficient of the data.

SOLUTION To begin, we store the age and price data in columns named AGE and PRICE, respectively. Then we proceed as follows.

1 Choose **Stat ➤ Basic Statistics ➤ Correlation...**
2 Specify AGE and PRICE in the **Variables** text box
3 Click **OK**

Printout 14.5 shows the output that results.

PRINTOUT 14.5
Minitab output for the correlation coefficient

```
Correlation of AGE and PRICE = -0.924, P-Value = 0.000
```

From Printout 14.5 we see that the linear correlation coefficient for the age and price data is −0.924; that is, $r = -0.924$. □

EXERCISES 14.4

Statistical Concepts and Skills

14.79 What is one purpose of the linear correlation coefficient?

14.80 The linear correlation coefficient is also known by another name. What is that name?

14.81 Fill in the blanks.

a. The symbol used for the linear correlation coefficient is ____.

b. Values of r close to ± 1 indicate that there is a ____ linear relationship between the variables.

c. Values of r close to ____ indicate that there is either no linear relationship between the variables or a weak one.

14.82 Fill in the blanks.

a. Values of r close to ____ indicate that the regression equation is extremely useful for making predictions.

b. Values of r close to 0 indicate that the regression equation is either useless or ____ for making predictions.

14.83 Fill in the blanks.

a. If y tends to increase linearly as x increases, we say that the variables are ____ linearly correlated.

b. If y tends to decrease linearly as x increases, we say that the variables are ____ linearly correlated.

c. If there is no linear relationship between x and y, then we say that the variables are linearly ____.

14.84 Answer true or false to the following statement and provide a reason for your answer: If there is a very strong positive correlation between two variables, then one can infer that a causal relationship exists between the two variables.

14.85 The linear correlation coefficient of a set of data points is 0.846.

a. Is the slope of the regression line positive or negative? Explain your answer.

b. Determine the coefficient of determination.

In Exercises 14.86 and 14.87, we have repeated data from exercises in Section 14.2. For each exercise, obtain the linear correlation coefficient using the defining formula in Definition 14.6 on page 854.

14.86 The data from Exercise 14.38:

x	1	1	5	5
y	1	3	2	4

14.87 The data from Exercise 14.39:

x	0	2	2	5	6
y	4	2	0	−2	1

In Exercises 14.88–14.93, we have repeated data from exercises in Section 14.2. For each exercise,

a. obtain the linear correlation coefficient using the computing formula in Definition 14.6 on page 854.

b. interpret the value of r in terms of the linear relationship between the two variables in question.

c. discuss the graphical interpretation of the value of r and check that it is consistent with the graph you obtained in the corresponding exercise in Section 14.2.

d. square r and compare the result with the value of the coefficient of determination you obtained in the corresponding exercise in Section 14.3.

14.88 The *Kelley Blue Book* provides information on wholesale and retail prices of cars. Following are age and price data for 10 randomly selected Corvettes between 1 and 6 years old. Here x denotes age, in years, and y denotes price, in hundreds of dollars.

x	6	6	6	2	2	5	4	5	1	4
y	175	165	180	310	269	200	240	213	310	210

14.89 The National Center for Health Statistics publishes data on heights and weights in *Vital and Health Statistics*. A random sample of 11 males age 18–24 years gave the following data, where x denotes height, in inches, and y denotes weight, in pounds.

x	65	67	71	71	66	75	67	70	71	69	69
y	175	133	185	163	126	198	153	163	159	151	155

14.90 Hanna Properties specializes in custom-home resales in the Equestrian Estates, an exclusive subdivision in Phoenix, Arizona. A random sample of nine custom homes currently listed for sale provided the following information on size and price. Here x denotes size, in hundreds of square feet, rounded to the nearest hundred, and y denotes price, in thousands of dollars, rounded to the nearest thousand.

x	26	27	33	29	29	34	30	40	22
y	259	274	294	296	325	380	457	523	215

14.91 Many studies have been done that indicate the maximum heart rate an individual can reach during intensive exercise decreases with age. See, for example, page 73 of the September 16, 1996 issue of *Newsweek.* A physician decided to do his own study and recorded the ages and peak heart rates of 10 randomly selected people. The results are shown in the following table, where x denotes age, in years, and y denotes peak heart rate.

x	30	38	41	38	29	39	46	41	42	24
y	186	183	171	177	191	177	175	176	171	196

14.92 An instructor at Arizona State University asked a random sample of eight students to record their study times in a beginning calculus course. She then made a table for total hours studied, x, over 2 weeks, and test score, y, at the end of the 2 weeks. Here are the results.

x	10	15	12	20	8	16	14	22
y	92	81	84	74	85	80	84	80

14.93 An economist is interested in the relationship between the disposable income of a family and the amount of money spent annually on food. For a preliminary study, the economist takes a random sample of eight middle-income families of the same size (father, mother, two children). The results are as follows, where x denotes disposable income, in thousands of dollars, and y denotes food expenditure, in hundreds of dollars.

x	30	36	27	20	16	24	19	25
y	55	60	42	40	37	26	39	43

14.94 We took a sample of 10 students from an introductory statistics class and obtained the following data, where x denotes height, in inches, and y denotes score on the final exam.

x	71	68	71	65	66	68	68	64	62	65
y	87	96	66	71	71	55	83	67	86	60

a. What sort of value of r would you expect to find for these data? Explain your answer.
b. Compute r.

14.95 Consider the following set of data points.

x	−3	−2	−1	0	1	2	3
y	9	4	1	0	1	4	9

a. Compute the linear correlation coefficient, r.
b. Can you conclude from your answer in part (a) that the variables x and y are unrelated? Explain your answer.
c. Draw a scatter diagram for the data.
d. Is it appropriate to use the linear correlation coefficient as a descriptive measure for the data? Explain your answer.
e. Show that the data are related by the quadratic equation $y = x^2$. Graph that equation and the data points.

14.96 Determine whether r is positive, negative, or zero for each of the following data sets.

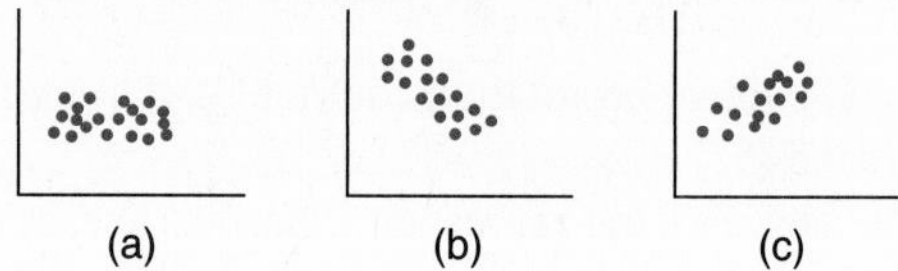

(a) (b) (c)

EXTENDING THE CONCEPTS AND SKILLS

14.97 The coefficient of determination of a set of data points is 0.716.
a. Can you determine the linear correlation coefficient? If yes, obtain it. If no, why not?
b. Can you determine whether the slope of the regression line is positive or negative? Why or why not?
c. If we tell you that the slope of the regression line is negative, can you determine the linear correlation coefficient? If yes, obtain it. If no, why not?
d. If we tell you that the slope of the regression line is positive, can you determine the linear correlation coefficient? If yes, obtain it. If no, why not?

14.98 A Knight-Ridder News Service article, appearing in the *Wichita Eagle,* discussed a study on the relationship between country music and suicide. The results of the study, coauthored by sociologist John Gundlach, appeared in an issue of *Social Forces.* According to the article, "... analysis of 49 metropolitan areas shows that the greater the airtime devoted to country music, the greater the white suicide rate." (Suicide rates in the black population were found to be uncorrelated with the amount of country-music airtime.)

a. Use the terminology introduced in this section to describe the statement quoted above.

b. One of the conclusions stated in the journal article was that country music "nurtures a suicidal mood" by dwelling on marital status and alienation from work. Do you think this conclusion is warranted solely on the basis of the positive correlation found between airtime devoted to country music and white suicide rate? Explain your answer.

USING TECHNOLOGY  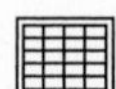 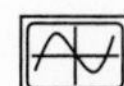

14.99 Use Minitab or some other statistical software to obtain the linear correlation coefficient of the data in Exercise 14.93.

14.100 Use Minitab or some other statistical software to obtain the linear correlation coefficient of the data in Exercise 14.92.

CHAPTER REVIEW

You Should Be Able To

1. use and understand the formulas presented in this chapter.
2. define and apply the concepts related to linear equations with one independent variable.
3. explain the least-squares criterion.
4. obtain and graph the regression equation for a set of data points, interpret the slope of the regression line, and use the regression equation to make predictions.
5. define and use the terminology *predictor variable* and *response variable.*
6. understand the concept of extrapolation.
7. identify outliers and influential observations.
8. understand when it is appropriate to obtain a regression line for a set of data points.
9. calculate and interpret the three sums of squares, *SST, SSE,* and *SSR,* and the coefficient of determination, r^2.
10. determine and interpret the linear correlation coefficient, r.
11. explain and apply the relationship between the linear correlation coefficient and the coefficient of determination.
***12.** use the Minitab procedures covered in this chapter.
***13.** interpret the output obtained from the application of the Minitab procedures discussed in this chapter.

Key Terms

coefficient of determination (r^2), *849*
curvilinear regression, *837*
error sum of squares (SSE), *849*
explanatory variable, *833*
extrapolation, *833*
influential observation, *835*
least-squares criterion, *830*
linear correlation coefficient (r), *854*

REVIEW TEST

STATISTICAL CONCEPTS AND SKILLS

1. For a linear equation $y = b_0 + b_1 x$, identify the
a. independent variable
b. dependent variable
c. slope
d. y-intercept

2. Consider the linear equation $y = 4 - 3x$.
a. At what y-value does its graph intersect the y-axis?
b. At what x-value does its graph intersect the y-axis?
c. What is its slope?
d. By how much does the y-value on the line change when the x-value increases by 1 unit?
e. By how much does the y-value on the line change when the x-value decreases by 2 units?

3. Answer true or false to each of the following statements and explain your answers.
a. The y-intercept of a straight line has no effect on the steepness of the line.
b. A horizontal line has no slope.
c. If a line has a positive slope, then y-values on the line decrease as the x-values decrease.

4. What kind of plot is useful for deciding whether it is reasonable to find a regression line for a set of data points?

5. Identify one use of a regression equation.

6. Regarding the variables in a regression analysis.
a. What is the independent variable called?
b. What is the dependent variable called?

7. Fill in the blanks.
a. Based on the least-squares criterion, the line that best fits a set of data points is the one having the _____ possible sum of squared errors.
b. The line that best fits a set of data points according to the least-squares criterion is called the _____ line.
c. When a regression equation is used to make predictions for values of the predictor variable outside the range of the observed values of the predictor variable, it is called _____.

8. In the context of regression, what is an
a. outlier? **b.** influential observation?

9. Identify a use of the coefficient of determination as a descriptive measure.

10. For each of the following sums of squares in regression, identify its name and what it measures.
a. *SST* **b.** *SSR* **c.** *SSE*

11. Fill in the blanks.
a. One use of the linear correlation coefficient is as a descriptive measure of the strength of the _____ relationship between two variables.

b. A positive linear relationship between two variables means that one variable tends to increase linearly as the other _____.
c. Values of r close to -1 suggest a strong _____ linear relationship between the variables.
d. Values of r close to _____ suggest at most a weak linear relationship between the variables.

12. Answer true or false to the following statement: A strong correlation between two variables does not necessarily mean that they are causally related.

13. A small company has purchased a microcomputer system for $7200 and plans to depreciate the value of the equipment by $1200 per year for 6 years. Let x denote the age of the equipment, in years, and y denote the value of the equipment, in hundreds of dollars.

a. Find the equation that expresses y in terms of x.
b. Find the y-intercept, b_0, and slope, b_1, of the linear equation in part (a).
c. Without graphing the equation in part (a), decide whether the line slopes upward, slopes downward, or is horizontal.
d. Find the value of the computer equipment after 2 years; after 5 years.
e. Obtain the graph of the equation in part (a) by plotting the points from part (d) and connecting them with a straight line.
f. Use the graph from part (e) to visually estimate the value of the equipment after 4 years. Then calculate that value exactly using the equation from part (a).

14. Graduation rates and what influences them have become a concern in U.S. colleges and universities. *U.S. News and World Report*'s "College Guide" provides data on graduation rates for colleges and universities as a function of the percentage of freshmen in the top 10% of their high-school class, total spending per student, and student-to-faculty ratio. (Here *graduation rate* refers to the percentage of entering freshmen, attending full time, that graduate within 5 years.) A random sample of 10 universities gave the following data on student-to-faculty ratio (S/F ratio) and graduation rate (grad rate).

S/F ratio x	Grad rate y
16	45
20	55
17	70
19	50
22	47
17	46
17	50
17	66
10	26
18	60

a. Draw a scatter diagram of the data.
b. Is it reasonable to find a regression line for the data? Explain your answer.
c. Determine the regression equation for the data and draw its graph on the scatter diagram you drew in part (a).
d. Describe the apparent relationship between the student-to-faculty ratio and graduation rate.
e. What does the slope of the regression line represent in terms of graduation rate?
f. Use the regression equation to predict the graduation rate of a university having a student-to-faculty ratio of 17.
g. Identify outliers and potential influential observations.

15. Refer to Problem 14.

a. Find *SST, SSR,* and *SSE* using the computing formulas.
b. Obtain the coefficient of determination.
c. Obtain the percentage of the total variation in the observed graduation rates that is explained by the student-to-faculty ratio (i.e., by the regression line).
d. State how useful the regression equation appears to be for making predictions.

16. Refer to Problem 14.

a. Compute the linear correlation coefficient, r.
b. Interpret your answer from part (a) in terms of the linear relationship between student-to-faculty ratio and graduation rate.

c. Discuss the graphical implications of the value of the linear correlation coefficient, r.

d. Use your answer from part (a) to obtain the coefficient of determination.

USING TECHNOLOGY  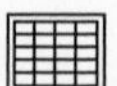 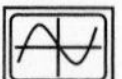

17. Refer to Problem 14. Use Minitab or some other statistical software to
    a. obtain a scatter diagram for the data.
    b. determine the regression equation for the data.
    c. find the coefficient of determination, r^2, and the three sums of squares, *SST, SSR,* and *SSE.*
    d. identify potential outliers and influential observations.
    e. obtain the regression equation with the potential influential observation removed.
    f. Is the potential influential observation actually an influential observation; that is, does its removal markedly change the regression equation?

18. Printout 14.6 shows the output obtained by applying Minitab's regression procedure to the data in Problem 14. Use the output to determine the
    a. regression equation.
    b. coefficient of determination.
    c. regression sum of squares, error sum of squares, and total sum of squares.
    d. potential outliers and influential observations, if any.

19. Use Minitab or some other statistical software to obtain the linear correlation coefficient of the data in Problem 14.

PRINTOUT 14.6 Minitab output for Problem 18

```
The regression equation is
GRADRATE = 16.4 + 2.03 SFRATIO

Predictor        Coef       StDev        T       P
Constant        16.45       21.15     0.78   0.459
SFRATIO         2.026       1.205     1.68   0.131

S = 11.31       R-Sq = 26.1%     R-Sq(adj) = 16.9%

Analysis of Variance

Source           DF         SS          MS        F       P
Regression        1      361.7       361.7     2.83   0.131
Residual Error    8     1022.8       127.9
Total             9     1384.5

Unusual Observations
Obs   SFRATIO   GRADRATE        Fit   StDev Fit    Residual    St Resid
  9      10.0      26.00      36.71        9.49      -10.71     -1.74 X

X denotes an observation whose X value gives it large influence.
```

INTERNET PROJECT

Assisted Reproductive Technology

Since 1981, assisted reproductive technology (ART) has been used in the United States to help women achieve pregnancy. Many American women have received some type of fertility service, such as in vitro fertilization (IVF) or egg transfer.

Several factors can influence the chance of having a child by using ART, one of the most important being the age of the prospective mother. In this Internet project, you will examine how a woman's age can influence the chance for success in assisted reproductive technology. Specifically, you will consider the relationship between the variables reproductive success rate and age.

URL for access to Internet Projects Page: `http://hepg.awl.com` *keyword:* Weiss

USING THE FOCUS DATABASE

In Chapter 1 we explained how to store the Focus database in a Minitab worksheet named `focus.mtw`. If you haven't already created that worksheet, follow the instructions on pages 54–55 to create it now.

The Focus database contains information on 500 randomly selected Arizona State University sophomores for seven different variables: sex, high-school GPA, SAT math score, cumulative GPA, SAT verbal score, age, and total hours. For these database exercises, you should eliminate all cases (students) in which one or more of the four variables, cumulative GPA, high-school GPA, SAT math score, and SAT verbal score, equal 0.

First we will perform a correlation analysis to choose the best predictor of cumulative GPA from the variables high-school GPA, SAT math score, and SAT verbal score.

a. Determine the correlation coefficient between the cumulative GPA data and each of the three data sets, high-school GPA, SAT math score, and SAT verbal score.

b. Among the variables high-school GPA, SAT math score, and SAT verbal score, identify the one that appears to be the best predictor of cumulative GPA for Arizona State University sophomores. Explain your answer.

Next we will perform a regression analysis on cumulative GPA using the predictor variable identified in part (b).

c. Obtain the regression equation for cumulative GPA using the predictor variable identified in part (b).

d. Find the coefficient of determination and interpret your answer.

e. Determine and interpret the three sums of squares, *SSR, SSE,* and *SST.*

CASE STUDY DISCUSSION

Fat Consumption and Prostate Cancer

At the beginning of this chapter, we presented data on fat consumption and prostate cancer death rate for nations of the world. Now that we have studied regression and correlation, we can analyze the relationship between those two variables. Referring to the table on page 818, solve each of the following problems.

a. Draw a scatter diagram for the data on dietary fat and prostate cancer death rate. What does the scatter diagram tell you?
b. Does it appear reasonable to obtain a regression equation for the data? Explain your answer.
c. Find the regression equation for the data using dietary fat as the predictor variable.
d. Interpret the slope of the regression line.
e. Compute the correlation coefficient of the data and interpret your result.
f. Identify outliers and potential influential observations, if any.
g. Use Minitab or some other statistical software to solve (a), (c), (e) and (f).

BIOGRAPHY ADRIEN LEGENDRE

Adrien-Marie Legendre was born in Paris, France, on September 18, 1752, the son of a moderately wealthy family. He studied at the Collège Mazarin and received degrees in mathematics and physics in 1770 at the age of 18.

Although Legendre's financial assets were sufficient to allow him to devote himself to research, he took a position teaching mathematics at the École Militaire in Paris from 1775 to 1780. In March 1783, he was elected to the Academie des Sciences in Paris, and in 1787, he was assigned to a project undertaken jointly by the observatories at Paris and at Greenwich, England. At this time he became a fellow of the Royal Society.

As a result of the French Revolution, which began in 1789, Legendre lost his "small fortune" and was forced to find work. He held various positions during the early 1790s, for example, commissioner of astronomical operations for the Academie des Sciences, professor of pure mathematics at the Institut de Marat, and head of the National Executive Commission of Public Instruction. During this same period, Legendre wrote a geometry book that became the major text used in elementary geometry courses for nearly a century.

Legendre's major contribution to statistics was the publication, in 1805, of the first statement and the first application of the most widely used, nontrivial technique of statistics: the method of least squares. Stigler writes in *The History of Statistics,* "[Legendre's] presentation . . . must be counted as one of the clearest and most elegant introductions of a new statistical method in the history of statistics." Because Gauss also claimed the method of least squares, there was strife between the two men. Although evidence shows that Gauss was not successful in any communication of the method prior to 1805, his development of the method was crucial to its usefulness.

In 1813 Legendre was appointed Chief of the Bureau des Longitudes, where he remained until his death, following a long illness, in Paris on January 10, 1833.

CASE STUDY

FAT CONSUMPTION AND PROSTATE CANCER

As we learned in the Chapter 14 case study, many investigations have shown that there is a relationship between nutrition and cancer. Prostate cancer is one of the most virulent forms of cancer. One dietary factor that has been studied for its relationship with prostate cancer is fat consumption.

In the case study of Chapter 14 we also examined data on fat consumption and prostate cancer death rate for various nations of the world. We repeat the data presented there in the following table, obtained from a graph in John Robbins's classic book *Diet for a New America* (Walpole, NH: Stillpoint Publishing, 1987).

Country	Dietary fat (grams/day)	Death rate (per 100,000)	Country	Dietary fat (grams/day)	Death rate (per 100,000)
El Salvador	38	0.9	Spain	97	10.1
Philippines	29	1.3	Portugal	73	11.4
Japan	42	1.6	Finland	112	11.1
Mexico	57	4.5	Hungary	100	13.1
Greece	96	4.8	United Kingdom	143	12.4
Colombia	47	5.4	Germany	134	12.9
Bulgaria	67	5.5	Canada	142	13.4
Yugoslavia	72	5.6	Austria	119	13.9
Poland	93	6.4	France	137	14.4
Panama	58	7.8	Netherlands	152	14.4
Israel	95	8.4	Australia	129	15.1
Romania	67	8.8	Denmark	156	15.9
Venezuela	62	9.0	United States	147	16.3
Czechoslovakia	96	9.1	Norway	133	16.8
Italy	86	9.4	Sweden	132	18.4

The regression and correlation analyses conducted on these data in Chapter 14 were descriptive. At the end of this chapter, you will be asked to return to the data to make regression and correlation inferences.

Inferential Methods in Regression and Correlation

15

GENERAL OBJECTIVES

In Chapter 14 we studied descriptive methods in regression and correlation. We discovered how to determine the regression equation for a set of data points and how to use that equation to make predictions. We also learned how to compute and interpret the coefficient of determination and the linear correlation coefficient for a set of data points.

In this chapter we will examine inferential methods in regression and correlation. As an illustration, we will again consider the age and price data for the Orion automobile introduced in Chapter 14. We will discover how the regression equation can be used to obtain a confidence interval for the mean price of all Orions of a particular age and to find a prediction interval for the price of an Orion of a particular age. We will also see how the linear correlation coefficient can be employed to decide whether there is a negative correlation between age and price of Orions.

Additionally, we will present an inferential procedure for testing whether a variable is normally distributed. This method is based on the linear correlation between the observations of the variable and their normal scores.

CHAPTER OUTLINE

15.1 THE REGRESSION MODEL; ANALYSIS OF RESIDUALS

To perform statistical inferences in regression and correlation, the variables under consideration must satisfy certain conditions. In this section we will discuss those conditions and examine methods for checking whether they hold.

The Regression Model

Let's return to the Orion illustration used throughout Chapter 14. In Table 15.1 we have reproduced the data on age and price for a sample of 11 Orions. Ages are in years, and prices are in hundreds of dollars, rounded to the nearest hundred dollars.

TABLE 15.1
Age and price data for a sample of 11 Orions

Car	Age (yrs) x	Price ($100s) y
1	5	85
2	4	103
3	6	70
4	5	82
5	5	89
6	5	98
7	6	66
8	6	95
9	2	169
10	7	70
11	7	48

On page 832 we found that the regression equation for this data, using age as the predictor variable and price as the response variable, is $\hat{y} = 195.47 - 20.26x$. As we know, the regression equation can be used to predict the price of an Orion given its age. However, we cannot expect such predictions to be completely accurate since prices vary even for Orions of the same age.

For instance, the sample data in Table 15.1 include four 5-year-old Orions. Their prices are $8500, $8200, $8900, and $9800. This variation in price for 5-year-old Orions should be expected because such cars generally have different mileages, interior conditions, paint quality, and so forth.

We will use the population of all 5-year-old Orions to introduce some important terminology employed in the context of regression. The distribution of prices for all 5-year-old Orions is called the **conditional distribution** of the response variable "price" corresponding to the value 5 of the predictor variable "age." Likewise, the mean price of all 5-year-old Orions is called the **conditional mean** of the response variable "price" corresponding to the value 5 of the predictor variable "age." Similar terminology applies to the standard deviation and other parameters.

In general, for each age, there is a population of Orions of that age. The distribution, mean, and standard deviation of prices for that population are called, respectively, the conditional distribution, conditional mean, and conditional standard deviation of the response variable "price" corresponding to the given value of the predictor variable "age."

With the preceding discussion in mind, we now state the conditions required for using inferential methods in regression analysis.

KEY FACT 15.1 ASSUMPTIONS FOR REGRESSION INFERENCES

1. ***Population regression line:*** There are constants β_0 and β_1 such that for each value x of the predictor variable, the conditional mean of the response variable is $\beta_0 + \beta_1 x$. We refer to the straight line $y = \beta_0 + \beta_1 x$ as the **population regression line** and to its equation as the **population regression equation.**
2. ***Equal standard deviations:*** The conditional standard deviations of the response variable are the same for all values of the predictor variable. We denote this common standard deviation by σ.
3. ***Normal populations:*** For each value of the predictor variable, the conditional distribution of the response variable is a normal distribution.
4. ***Independent observations:*** The observations of the response variable are independent of one another.

Assumptions 1–3 require that there are constants, β_0, β_1, and σ, such that for each value x of the predictor variable, the conditional distribution of the response variable, y, is a normal distribution having mean $\beta_0 + \beta_1 x$ and standard deviation σ. These assumptions are often referred to as the **regression model.**

The inferential procedures in regression are robust to moderate violations of Assumptions 1–3 for regression inferences. In other words, the inferential procedures will work reasonably well provided the variables under consideration don't violate any of those assumptions too badly.

EXAMPLE 15.1 ASSUMPTIONS FOR REGRESSION INFERENCES

Discuss what it would mean for the regression-inference Assumptions 1–3 to be satisfied for Orions with age as the predictor variable and price as the response variable. Display those assumptions graphically.

SOLUTION For the regression-inference Assumptions 1–3 to be satisfied, it would mean there are constants, β_0, β_1, and σ, such that for each age, x, the prices of all Orions of that age are normally distributed with mean $\beta_0 + \beta_1 x$ and standard deviation σ.

In other words, the prices of all 2-year-old Orions are normally distributed with mean $\beta_0 + \beta_1 \cdot 2$ and standard deviation σ; the prices of all 3-year-old Orions are normally distributed with mean $\beta_0 + \beta_1 \cdot 3$ and standard deviation σ; and so on.

To display the assumptions for regression inferences graphically, let's first consider Assumption 1. This assumption requires that for each age, the mean price of all Orions of that age lies on the straight line $y = \beta_0 + \beta_1 x$, as shown in Fig. 15.1.

FIGURE 15.1 Population regression line

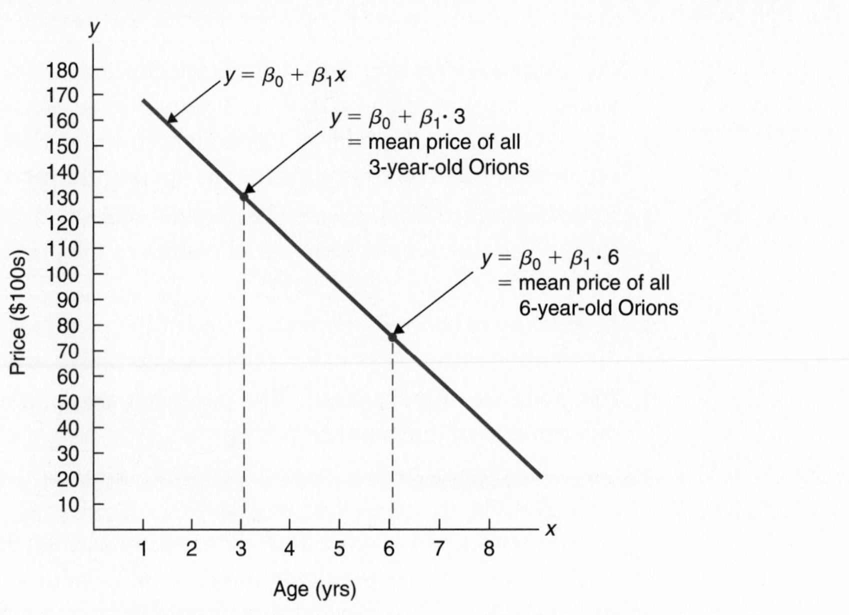

Next we present graphs that depict Assumptions 2 and 3. Those assumptions require that the price distributions for the various ages of Orions are all normally distributed with the same standard deviation, σ. Figure 15.2 shows this for the price distributions of 2-year-old, 5-year-old, and 7-year-old Orions.

FIGURE 15.2 Price distributions for 2-, 5-, and 7-year-old Orions under Assumptions 2 and 3. (The means shown reflect Assumption 1.)

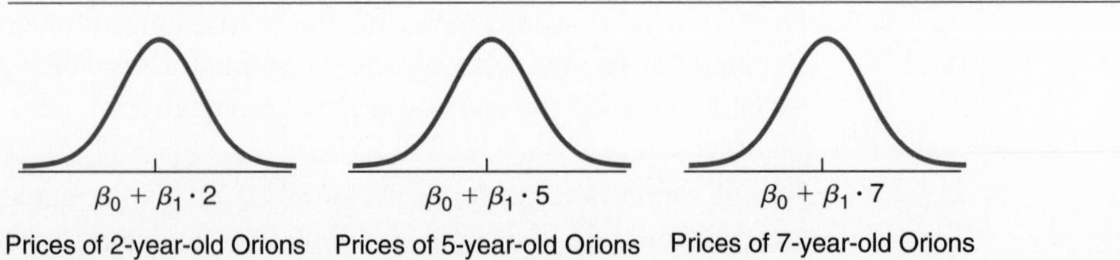

Notice that the shapes of the three normal curves in Fig. 15.2 are identical. This is because normal distributions having the same standard deviation have the same shape, and, under Assumptions 2 and 3, the price distributions for the various ages are all normally distributed and have equal standard deviations.

Assumptions 1–3 for regression inferences, as they pertain to the variables age and price of Orions, can be portrayed graphically by combining Figs. 15.1 and 15.2 into a three-dimensional graph, as seen in Fig. 15.3.

FIGURE 15.3 Graphical portrayal of Assumptions 1–3 for regression inferences as pertaining to age and price of Orions

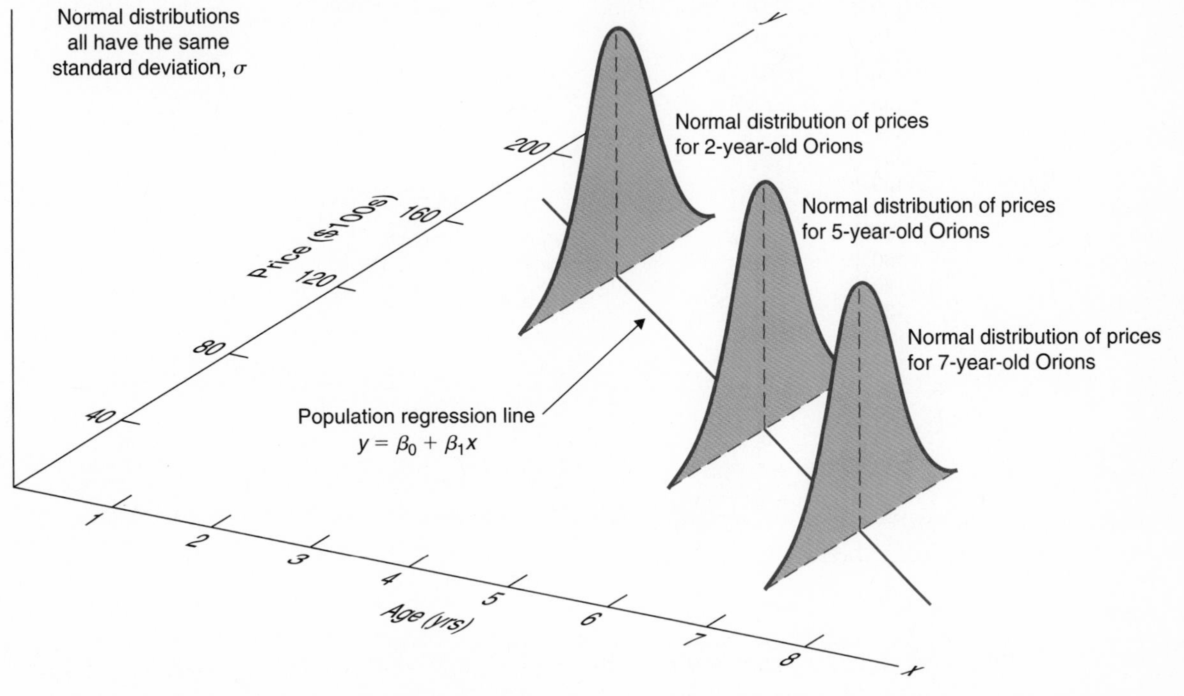

Figure 15.3 depicts the various price distributions for Orions under the condition that Assumptions 1–3 are true for the variables age and price of Orions. Whether this is actually the case remains to be seen. □

Estimating the Regression Parameters

Suppose we are considering two variables, x and y, for which the assumptions for regression inferences are met. Then there are constants, β_0, β_1, and σ, such that for each value x of the predictor variable, the conditional distribution of the response variable is a normal distribution having mean $\beta_0 + \beta_1 x$ and standard deviation σ.

The parameters β_0, β_1, and σ are usually unknown and must therefore be estimated from sample data. Point estimates for the y-intercept, β_0, and slope, β_1, of the population regression line are provided, respectively, by the y-intercept, b_0, and slope, b_1, of a sample regression line.

Another way of looking at this is that a sample regression line is used to estimate the population regression line. Of course, a sample regression line ordinarily will not be the same as the population regression line, just as a sample mean, $\bar{x}$, generally will not equal the population mean, μ. We picture this situation for the Orion example in Fig. 15.4.

FIGURE 15.4
Population regression line and sample regression line for age and price of Orions

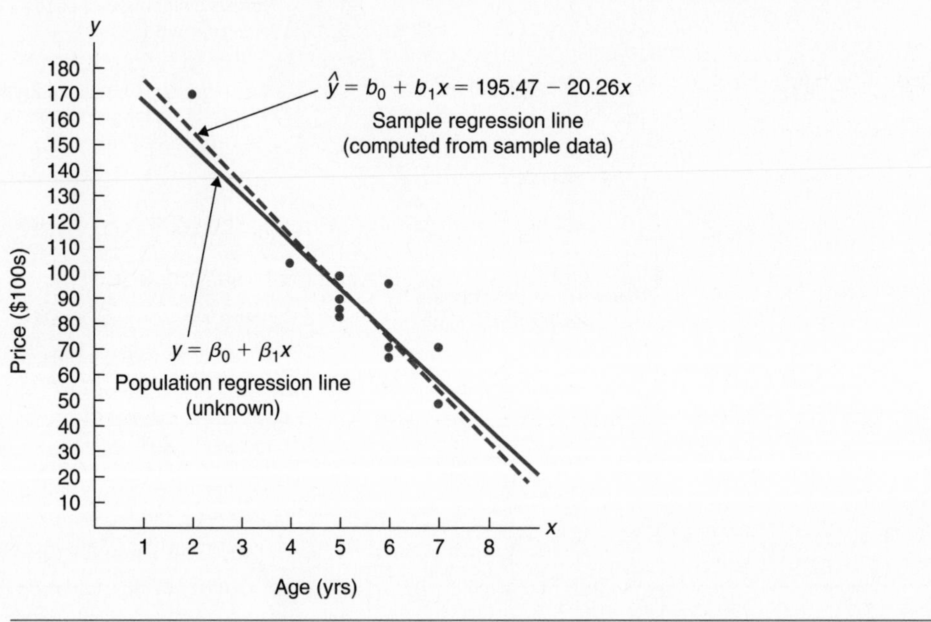

The solid line in Fig. 15.4 is the population regression line; the dashed line is a sample regression line. This sample regression line is the best approximation that can be made to the population regression line by using the sample data in Table 15.1 on page 872. A different sample of Orions would almost certainly yield a different sample regression line.

The statistic used to obtain a point estimate for the common conditional standard deviation σ is called the **standard error of the estimate** or the **residual standard deviation** and is defined as follows.

DEFINITION 15.1 STANDARD ERROR OF THE ESTIMATE

The ***standard error of the estimate, s_e,*** is defined by

$$s_e = \sqrt{\frac{SSE}{n-2}},$$

where $SSE = \Sigma(y - \hat{y})^2 = S_{yy} - S_{xy}^2/S_{xx}$.

Recall that SSE is the error sum of squares and equals the sum of squared errors when the regression equation is used to predict the observed values of the response variable. Thus, very roughly speaking, the standard error of the estimate indicates how much, on average, the predicted values of the response variable differ from the observed values of the response variable.

EXAMPLE 15.2 ILLUSTRATES DEFINITION 15.1

Refer to the age and price data for a sample of 11 Orions displayed in Table 15.1 on page 872.

a. Compute and interpret the standard error of the estimate.

b. Presuming that the variables age and price for Orions satisfy the assumptions for regression inferences, interpret the result from part (a).

SOLUTION **a.** The error sum of squares was computed on page 848, where we found that $SSE = 1423.5$. So the standard error of the estimate is

$$s_e = \sqrt{\frac{SSE}{n-2}} = \sqrt{\frac{1423.5}{11-2}} = 12.58.$$

As a rough estimate, we can say that, on the average, the predicted price of an Orion in the sample differs from the observed price by \$1258.

b. Presuming that the variables age and price for Orions satisfy the assumptions for regression inferences, the standard error of the estimate, $s_e = 12.58$, or \$1258, provides an estimate for the common population standard deviation, σ, of prices for all Orions of any particular age. ◨

 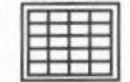 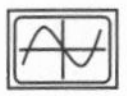

Analysis of Residuals

Now that we have examined the assumptions for regression inferences, we need to discuss how the sample data can be used to decide whether it is reasonable to presume that those assumptions are met. We will concentrate on Assumptions 1–3;

checking Assumption 4—the independence assumption—is somewhat more involved and is best left for a second course in statistics.

The method for checking Assumptions 1–3 relies on an analysis of the errors made in using the regression equation to predict the observed values of the response variable, that is, on the differences, $y - \hat{y}$, between the observed and predicted values of the response variable. Each such difference is called a **residual,** denoted generically by the letter ***e*.** Thus,

$$\text{Residual} = e = y - \hat{y}.$$

Figure 15.5 gives a graphical representation for the residual of a single data point.

FIGURE 15.5
Residual, e, of a data point

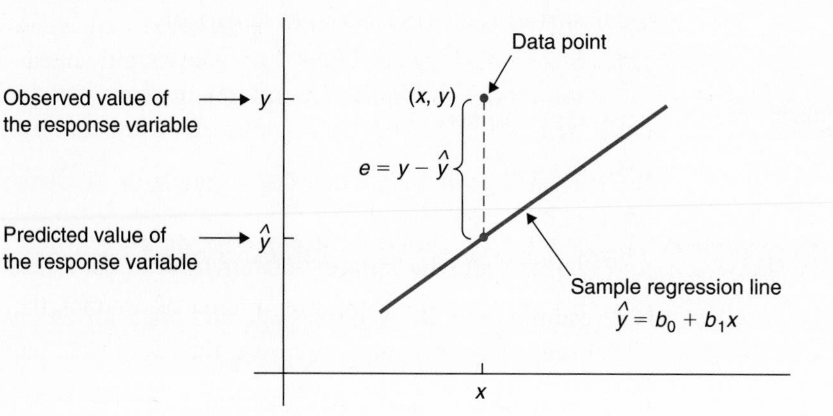

We can express the standard error of the estimate, s_e, in terms of the residuals. Referring to Definition 15.1, we find that the standard error of the estimate can be written as

$$s_e = \sqrt{\frac{SSE}{n-2}} = \sqrt{\frac{\Sigma(y-\hat{y})^2}{n-2}} = \sqrt{\frac{\Sigma e^2}{n-2}}.$$

It can be shown that the sum of the residuals is always 0 which, in turn, implies that $\bar{e} = 0$. Consequently, we see that the standard error of the estimate is essentially equal to the standard deviation of the residuals.† This explains why the standard error of the estimate is sometimes called the residual standard deviation.

The reason we can analyze the residuals to check whether Assumptions 1–3 for regression inferences are met is that those assumptions can be translated into con-

† The exact standard deviation of the residuals is obtained by dividing by $n - 1$ instead of $n - 2$.

ditions on the residuals. To see how this is done, consider a sample of data points obtained from two variables that satisfy the assumptions for regression inferences.

In view of Assumption 1, the data points should be scattered about the (sample) regression line, which means that the residuals should be scattered about the x-axis; in view of Assumption 2, the variation of the observed values of the response variable should remain approximately constant from one value of the predictor variable to the next, which means the residuals should fall roughly in a horizontal band; and in view of Assumption 3, for each value of the predictor variable the distribution of the corresponding observed values of the response variable should be approximately bell-shaped, which implies that the horizontal band should be centered and symmetric about the x-axis.

Furthermore, considering all four regression assumptions simultaneously, we see that the residuals can be regarded as independent observations of a variable having a normal distribution with mean 0 and standard deviation σ. Thus a normal probability plot of the residuals should be roughly linear.

In summary, we have the criteria in Key Fact 15.2 for deciding whether Assumptions 1–3 for regression inferences are met by the two variables under consideration.

KEY FACT 15.2 RESIDUAL ANALYSIS FOR THE REGRESSION MODEL

If the assumptions for regression inferences are met, then the following two conditions should hold.

- A plot of the residuals against the values of the predictor variable should fall roughly in a horizontal band centered and symmetric about the x-axis.
- A normal probability plot of the residuals should be roughly linear.

Failure of either of these two conditions casts doubt on the validity of one or more of the assumptions for regression inferences for the variables under consideration.

A plot of the residuals against the values of the predictor variable, called a **residual plot,** provides roughly the same information as does a scatter diagram of the data points. However, a residual plot makes it easier to spot patterns such as curvature and nonconstant standard deviation.

Figure 15.6(a) on the next page shows a residual plot in which the linearity assumption (Assumption 1) and the constant-standard-deviation assumption (Assumption 2) appear to be met; Fig. 15.6(b) shows a residual plot in which the relation between the variables appears to be curved instead of linear; and Fig. 15.6(c) shows a residual plot in which the conditional standard deviations appear to increase as x increases instead of remaining constant.

FIGURE 15.6
Residual plots indicating (a) no violation of linearity or constant standard deviation, (b) violation of linearity, and (c) violation of constant standard deviation

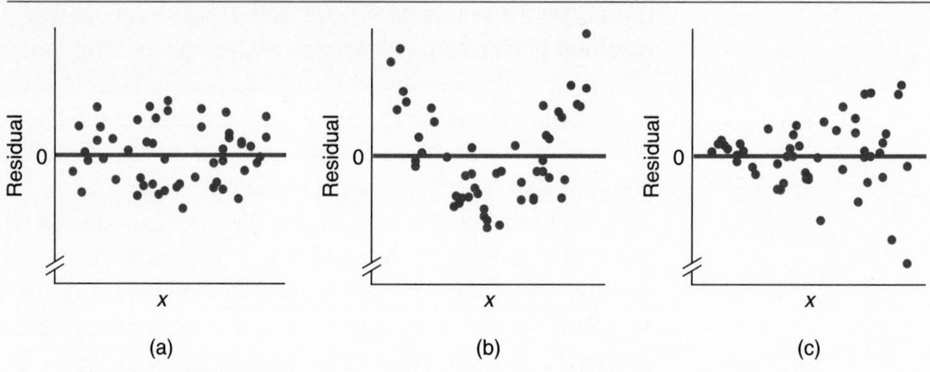

In our previous data analyses, we have seen that it is often difficult to decide on the appropriateness of a model when dealing with small samples, say, of size 20 or less. The same holds true in regression: For small samples we must be more liberal in allowing moderate departures from the idealized patterns when analyzing residual plots and normal probability plots to ascertain whether the assumptions for regression inferences are met. There are no definite rules—only judgment based on experience.

EXAMPLE 15.3 ANALYSIS OF RESIDUALS

Perform a residual analysis to decide whether it is reasonable to consider the assumptions for regression inferences met by the variables age and price of Orions.

SOLUTION We apply the criteria in Key Fact 15.2. The ages and residuals for the Orion data are displayed, respectively, in the first and fourth columns of Table 14.8 on page 848. We repeat that information here in Table 15.2.

TABLE 15.2
Age and residual data for Orions

Age x	Residual e
5	−9.16
4	−11.42
6	−3.90
5	−12.16
5	−5.16
5	3.84
6	−7.90
6	21.10
2	14.05
7	16.36
7	−5.64

Figure 15.7(a) shows a plot of the residuals against age, and Fig. 15.7(b) shows a normal probability plot for the residuals.

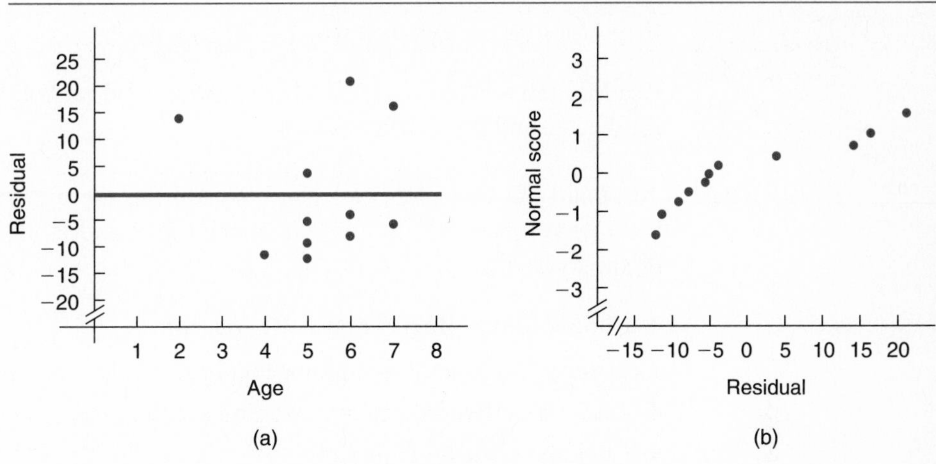

FIGURE 15.7
(a) Residual plot
(b) Normal probability plot for residuals

Taking into account the small sample size, we can say that the residuals fall roughly in a horizontal band centered and symmetric about the x-axis. We can also say that the normal probability plot for the residuals is (very) roughly linear, although the departure from linearity is sufficient for some concern. Therefore, based on the sample data, there are no obvious violations of the assumptions for regression inferences. ◨

  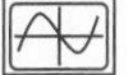

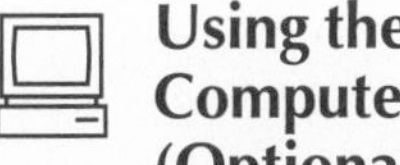

Using the Computer (Optional)

In Section 14.2 we learned that Minitab's regression procedure can be used to obtain the (sample) regression equation for a set of data points. The output resulting from the application of that procedure displays the standard error of the estimate as well as the regression equation and several other statistics.

EXAMPLE 15.4 Using Minitab to Obtain the Standard Error of the Estimate

Printout 14.2 on page 838 shows the output obtained by applying Minitab's regression procedure to the age and price data for a sample of 11 Orions. Use the output to determine the standard error of the estimate.

SOLUTION The standard error of the estimate is the first entry in the sixth line of Printout 14.2: `S = 12.58` (Minitab uses `S` instead of s_e to denote the standard error of the estimate). So we see that for the Orion data, $s_e = 12.58$. ◨

We can also use Minitab's regression procedure to obtain a residual plot and a normal probability plot of the residuals. Example 15.5 provides the details.

EXAMPLE 15.5 USING MINITAB TO OBTAIN PLOTS OF RESIDUALS

Use Minitab to obtain a residual plot and normal probability plot of the residuals for the age and price data of Orions.

SOLUTION To obtain the two plots, we begin by storing the age and price data from Table 15.1 on page 872 in columns named AGE and PRICE, respectively. Then we do the following.

1 Choose **Stat ➤ Regression ➤ Regression...**
2 Specify PRICE in the **Response** text box
3 Click in the **Predictors** text box and specify AGE
4 Click the **Graphs...** button
5 Select the **Regular** option button from the **Residuals for Plots** list
6 Select the **Normal plot of residuals** check box from the **Residual Plots** list
7 Click in the **Residuals versus the variables** text box and specify AGE
8 Click **OK**
9 Click **OK**

The plots obtained are shown in Printouts 15.1 and 15.2. Compare these plots to the ones we obtained by hand in Fig. 15.7 on page 881. ◨

PRINTOUT 15.1
Minitab residual plot for the Orion data

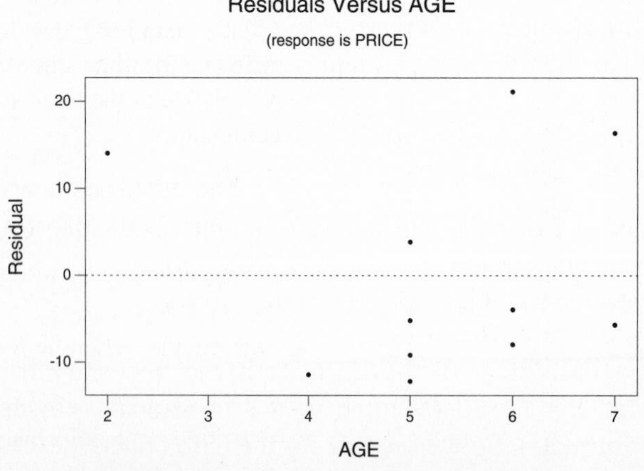

PRINTOUT 15.2
Minitab normal probability plot of the residuals for the Orion data

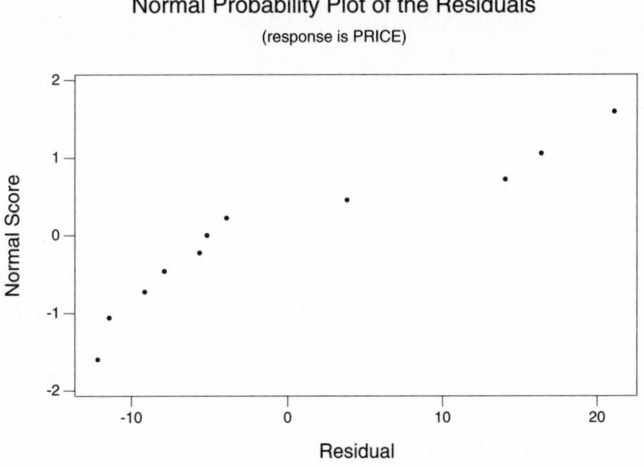

EXERCISES 15.1

STATISTICAL CONCEPTS AND SKILLS

15.1 Suppose that x and y are, respectively, predictor and response variables of a population. Consider the population consisting of all members of the original population having a specified value of the predictor variable. The distribution, mean, and standard deviation of the response variable for this population are called, respectively, the _____, _____, and _____ corresponding to the specified value of the predictor variable.

15.2 State the four conditions required for making regression inferences.

In Exercises 15.3–15.6, assume the variables under consideration satisfy the assumptions for regression inferences.

15.3 Fill in the blanks.

a. The straight line $y = \beta_0 + \beta_1 x$ is called the _____.

b. The common conditional standard deviation of the response variable is denoted by the letter _____.

c. For $x = 6$, the conditional distribution of the response variable is a _____ distribution having mean _____ and standard deviation _____.

15.4 What statistic is used to estimate

a. the y-intercept of the population regression line?

b. the slope of the population regression line?

c. the common conditional standard deviation, σ, of the response variable?

15.5 Based on a sample of data points, what is the best estimate of the population regression line?

15.6 Regarding the standard error of the estimate:

a. Give two interpretations of it.

b. Identify another name used for it, and explain the rationale for that name.

c. Which one of the three sums of squares figures in its computation?

15.7 The difference between the observed and predicted values of the response variable is called a _____.

15.8 Identify two graphs used in a residual analysis to check the Assumptions 1–3 for regression inferences and explain the reasoning behind their use.

15.9 Which graph used in a residual analysis provides roughly the same information as a scatter diagram? What advantages does it have over a scatter diagram?

In Exercises 15.10–15.15, we have repeated the information from Exercises 14.42–14.47. For each exercise, discuss what it would mean for Assumptions 1–3 for regression inferences to be satisfied by the variables under consideration.

15.10 The *Kelley Blue Book* provides information on wholesale and retail prices of cars. Following are age and price data for 10 randomly selected Corvettes between 1 and 6 years old. Here x denotes age, in years, and y denotes price, in hundreds of dollars.

x	6	6	6	2	2	5	4	5	1	4
y	175	165	180	310	269	200	240	213	310	210

15.11 The National Center for Health Statistics publishes data on heights and weights in *Vital and Health Statistics*. A random sample of 11 males age 18–24 years gave the following data, where x denotes height, in inches, and y denotes weight, in pounds.

x	65	67	71	71	66	75	67	70	71	69	69
y	175	133	185	163	126	198	153	163	159	151	155

15.12 Hanna Properties specializes in custom-home resales in the Equestrian Estates, an exclusive subdivision in Phoenix, Arizona. A random sample of nine custom homes currently listed for sale provided the following information on size and price. Here x denotes size, in hundreds of square feet, rounded to the nearest hundred, and y denotes price, in thousands of dollars, rounded to the nearest thousand.

x	26	27	33	29	29	34	30	40	22
y	259	274	294	296	325	380	457	523	215

15.13 Many studies have been done that indicate the maximum heart rate an individual can reach during intensive exercise decreases with age. See, for example, page 73 of the September 16, 1996 issue of *Newsweek*. A physician decided to do his own study and recorded the ages and peak heart rates of 10 randomly selected people. The results are shown in the following table, where x denotes age, in years, and y denotes peak heart rate.

x	30	38	41	38	29	39	46	41	42	24
y	186	183	171	177	191	177	175	176	171	196

15.14 An instructor at Arizona State University asked a random sample of eight students to record their study times in a beginning calculus course. She then made a table for total hours studied, x, over 2 weeks, and test score, y, at the end of the 2 weeks. Here are the results.

x	10	15	12	20	8	16	14	22
y	92	81	84	74	85	80	84	80

15.15 An economist is interested in the relationship between the disposable income of a family and the amount of money spent annually on food. For a preliminary study, the economist takes a random sample of eight middle-income families of the same size (father, mother, two children). The results are as follows, where x denotes disposable income, in thousands of dollars, and y denotes food expenditure, in hundreds of dollars.

x	30	36	27	20	16	24	19	25
y	55	60	42	40	37	26	39	43

For each of Exercises 15.16–15.21,

a. *compute the standard error of the estimate and interpret your answer.*

b. *interpret the result from part (a) if the assumptions for regression inferences hold.*

c. *obtain a residual plot and a normal probability plot of the residuals.*

d. *decide whether it appears reasonable to consider Assumptions 1–3 for regression inferences met by the variables under consideration. (The answer here is subjective, especially in view of the extremely small sample sizes.)*

15.16 The age and price data for 10 Corvettes from Exercise 15.10.

15.17 The height and weight data for males age 18–24 years from Exercise 15.11.

15.18 The size and price data for custom homes from Exercise 15.12.

15.19 The data on age and peak heart rate from Exercise 15.13.

15.20 The data on study time and test score from Exercise 15.14.

15.21 The data on disposable income and annual food expenditure from Exercise 15.15.

15.22 Figure 15.8 shows three residual plots and a normal probability plot of residuals. For each part, decide whether the graph suggests violation of one or more of the assumptions for regression inferences. Explain your answers.

15.23 Figure 15.9, shown at the top of the following page, displays three residual plots and one normal probability plot of residuals. For each part, decide whether the graph suggests violation of one or more of the assumptions for regression inferences. Explain your answers.

FIGURE 15.8 Plots for Exercise 15.22

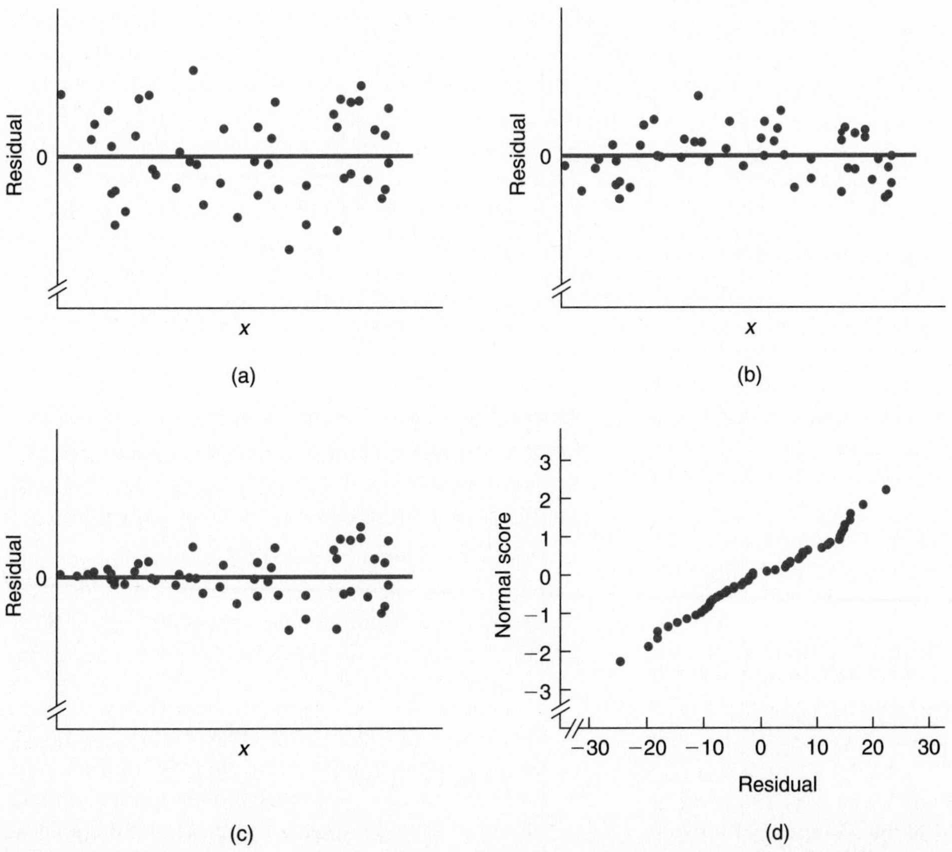

FIGURE 15.9 Plots for Exercise 15.23

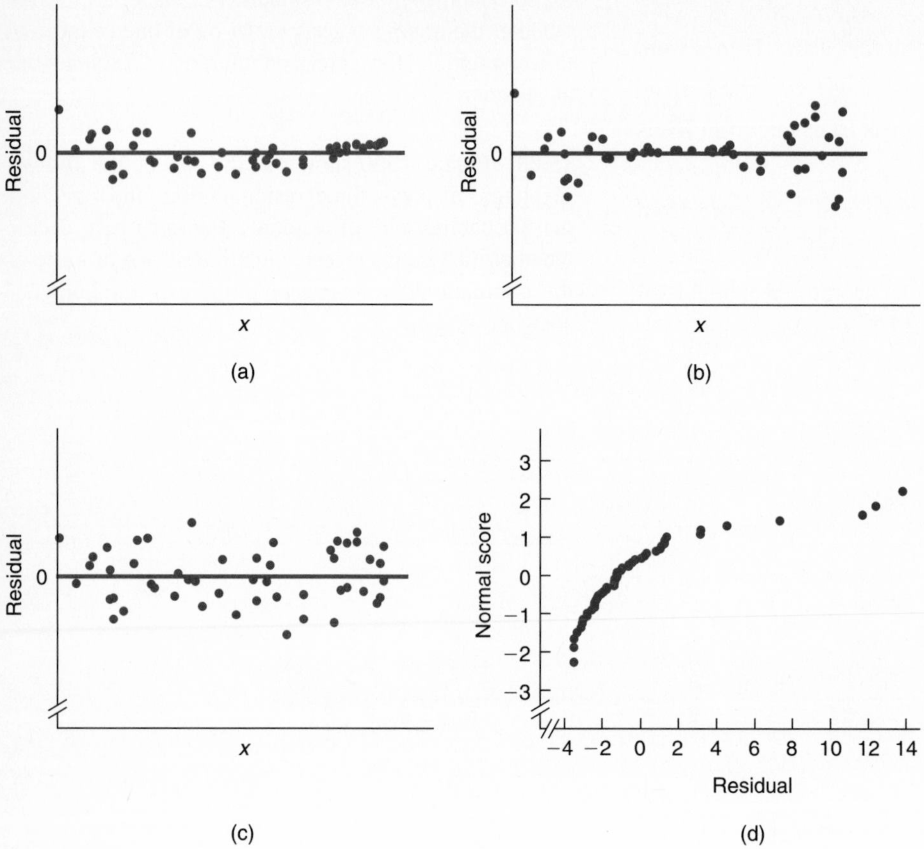

USING TECHNOLOGY

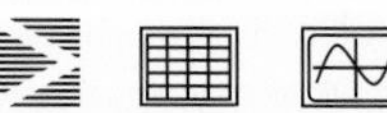

15.24 Refer to Exercise 15.14. Use Minitab or some other statistical software to

a. determine the standard error of the estimate.

b. obtain a residual plot and a normal probability plot of the residuals.

15.25 Refer to Exercise 15.15. Use Minitab or some other statistical software to

a. determine the standard error of the estimate.

b. obtain a residual plot and a normal probability plot of the residuals.

15.26 The Energy Information Administration publishes data on energy consumption by family income in *Residential Energy Consumption Survey: Consumption and Expenditures.* We applied Minitab's regression procedure to data on family income and last year's energy consumption from a random sample of 25 families.

a. Printout 14.4 on page 844 displays the computer output. The income data are in thousands of dollars and the energy-consumption data are in millions of BTU. Use the output to determine the standard error of the estimate.

b. We also used Minitab to obtain a residual plot (Printout 15.3) and a normal probability plot of the residuals (Printout 15.4). Do these plots suggest violations of any of the assumptions for regression inferences? Explain your answer.

PRINTOUT 15.3 Minitab output for Exercise 15.26(b)

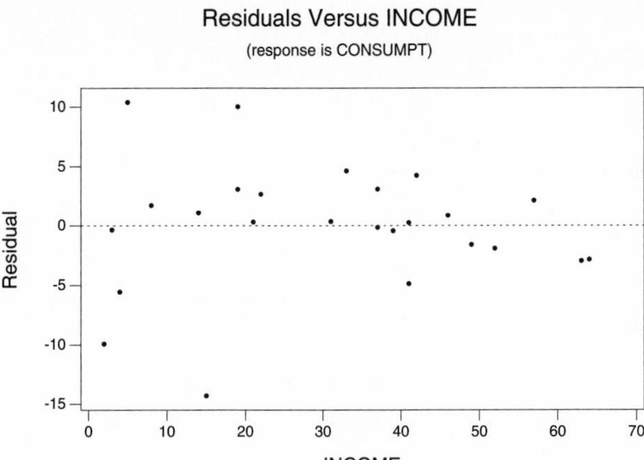

PRINTOUT 15.4 Minitab output for Exercise 15.26(b)

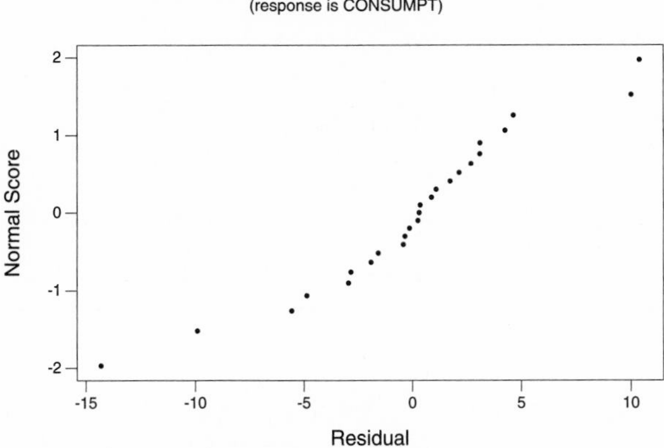

15.27 Greene and Touchstone conducted a study on the relationship between the estriol levels of pregnant women and the birth weights of their children. Their findings, "Urinary Tract Estriol: An Index of Placental Function," were published in the *American Journal of Obstetrics and Gynecology.*

a. Printout 14.3 on page 843 shows the computer output that results by applying Minitab's regression procedure to the data obtained by Greene and Touchstone. The estriol levels are in milligrams per 24 hours and the birth weights are in grams. Use the output to find the standard error of the estimate.

b. We also used Minitab to obtain a residual plot (Printout 15.5) and a normal probability plot of the residuals (Printout 15.6). Do these plots suggest violations of any of the assumptions for regression inferences? Explain your answer.

PRINTOUT 15.5 Minitab output for Exercise 15.27(b)

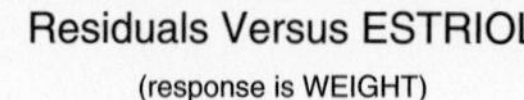

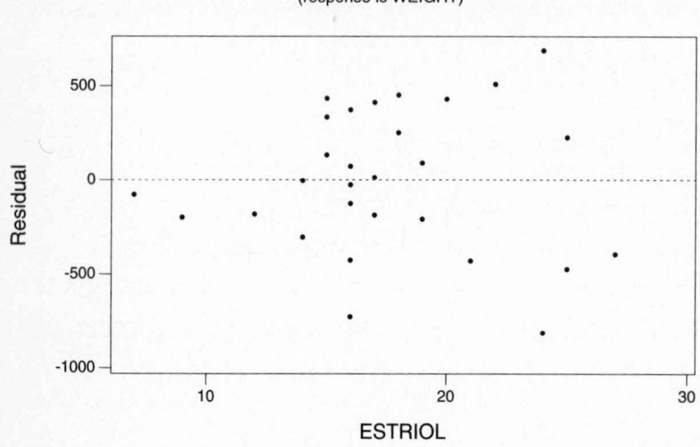

PRINTOUT 15.6 Minitab output for Exercise 15.27(b)

Normal Probability Plot of the Residuals

(response is WEIGHT)

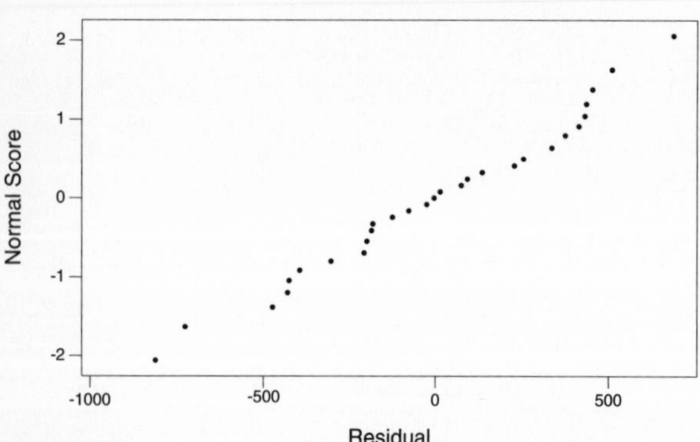

15.2 INFERENCES FOR THE SLOPE OF THE POPULATION REGRESSION LINE

In this section and in Section 15.3, we will examine several inferential procedures used in regression analysis. Strictly speaking, these inferential techniques require that the assumptions for regression inferences given in Key Fact 15.1 on page 873 are satisfied by the variables under consideration. However, as we noted earlier, these techniques are robust to moderate violations of those assumptions.

The first inferential methods we will study are those concerning the slope, β_1, of the population regression line. To begin, we consider hypothesis testing.

Hypothesis Tests for the Slope of the Population Regression Line

If the variables x and y satisfy the assumptions for regression inferences, then for each value x of the predictor variable, the conditional distribution of the response variable is a normal distribution with mean $\beta_0 + \beta_1 x$ and standard deviation σ.

Of particular interest is whether the slope, β_1, of the population regression line equals 0. Because if $\beta_1 = 0$, then for each value x of the predictor variable, the conditional distribution of the response variable is a normal distribution having mean β_0 $(= \beta_0 + 0 \cdot x)$ and standard deviation σ; and neither of those two parameters involves x. Consequently, in this case, x is useless as a predictor of y.†

So we can decide whether x is useful as a predictor of y (i.e., whether the regression has utility) by performing the hypothesis test

$$H_0\text{: } \beta_1 = 0 \text{ (}x\text{ is not useful for predicting }y\text{)}$$
$$H_a\text{: } \beta_1 \neq 0 \text{ (}x\text{ is useful for predicting }y\text{).}$$

Not surprisingly, we base hypothesis tests for β_1 (the slope of the population regression line) on the statistic b_1 (the slope of a sample regression line). To understand how this works, let's return to the Orion illustration. The data on age and price for a sample of 11 Orions are repeated in Table 15.3.

TABLE 15.3
Age and price data for a sample of 11 Orions

Car	Age (yrs) x	Price ($100s) y
1	5	85
2	4	103
3	6	70
4	5	82
5	5	89
6	5	98
7	6	66
8	6	95
9	2	169
10	7	70
11	7	48

On page 832 we found that the regression equation for this data, using age as the predictor variable and price as the response variable, is $\hat{y} = 195.47 - 20.26x$. In particular, the slope, b_1, of the sample regression line is -20.26.

† Although x alone may not be useful for predicting y, it may be useful in conjunction with another variable or variables. Thus, in this section, when we say that x is not useful for predicting y, we really mean that the regression equation with x as the only predictor variable is not useful for predicting y. Conversely, although x alone may be useful for predicting y, it may not be useful in conjunction with another variable or variables. Thus, in this section, when we say that x is useful for predicting y, we really mean that the regression equation with x as the only predictor variable is useful for predicting y.

Consider now all possible samples of 11 Orions whose ages are the same as those given in the second column of Table 15.3. For such samples, the slope, b_1, of the sample regression line varies from one sample to another and is therefore a variable. Its distribution is called the **sampling distribution of the slope of the regression line.** And, from the assumptions for regression inferences, it can be shown that this distribution is a normal distribution with mean equal to the slope, β_1, of the population regression line. More generally, we have the following fact.

KEY FACT 15.3 THE SAMPLING DISTRIBUTION OF THE SLOPE OF THE REGRESSION LINE

Suppose the variables x and y satisfy Assumptions 1–3 for regression inferences. Then, for samples of size n, each with the same values $x_1, x_2, \ldots, x_n$, for the predictor variable, the following properties hold for the slope, b_1, of the sample regression line.

- The mean of b_1 equals the slope of the population regression line: $\mu_{b_1} = \beta_1$.
- The standard deviation of b_1 is $\sigma_{b_1} = \sigma/\sqrt{S_{xx}}$.
- The variable b_1 is normally distributed.

In other words, the distribution of slopes of all possible sample regression lines is a normal distribution with mean β_1 and standard deviation $\sigma/\sqrt{S_{xx}}$.

As a consequence of Key Fact 15.3, the standardized variable

$$z = \frac{b_1 - \beta_1}{\sigma/\sqrt{S_{xx}}}$$

has the standard normal distribution. But this variable cannot be used as a basis for the required test statistic because the common conditional standard deviation, σ, is unknown. We therefore replace σ by its sample estimate s_e, the standard error of the estimate. As we might suspect, the resulting variable has a t-distribution.

KEY FACT 15.4 t-DISTRIBUTION FOR INFERENCES FOR β_1

Suppose the variables x and y satisfy Assumptions 1–3 for regression inferences. Then, for samples of size n, each with the same values $x_1, x_2, \ldots, x_n$, for the predictor variable, the variable

$$t = \frac{b_1 - \beta_1}{s_e/\sqrt{S_{xx}}}$$

has the t-distribution with df $= n - 2$.

In view of Key Fact 15.4, we see that for a hypothesis test with null hypothesis H_0: $\beta_1 = 0$, we can use the variable

$$t = \frac{b_1}{s_e/\sqrt{S_{xx}}}$$

as the test statistic and obtain the critical values from the t-table, Table IV. Specifically, we have the following procedure.

PROCEDURE 15.1

THE t-TEST FOR THE UTILITY OF A REGRESSION

ASSUMPTIONS

The four assumptions for regression inferences

Step 1 The null and alternative hypotheses are

H_0: $\beta_1 = 0$ (predictor variable is not useful for making predictions)

H_a: $\beta_1 \neq 0$ (predictor variable is useful for making predictions).

Step 2 Decide on the significance level, α.

Step 3 The critical values are $\pm t_{\alpha/2}$ with df $= n - 2$. Use Table IV to find the critical values.

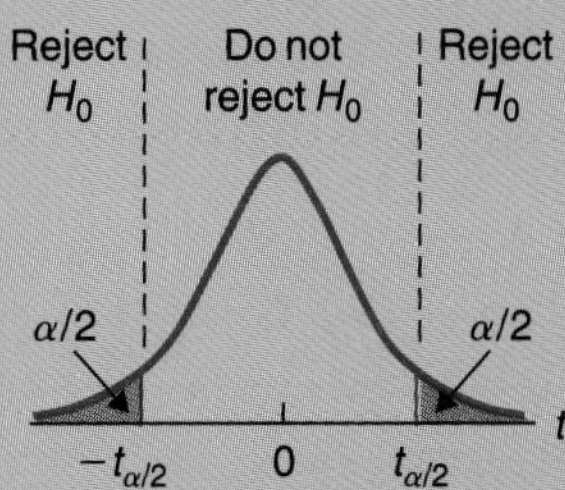

Step 4 Compute the value of the test statistic

$$t = \frac{b_1}{s_e/\sqrt{S_{xx}}}.$$

Step 5 If the value of the test statistic falls in the rejection region, reject H_0; otherwise, do not reject H_0.

Step 6 State the conclusion in words.

EXAMPLE 15.6 ILLUSTRATES PROCEDURE 15.1

The data on age and price for a sample of 11 Orions are displayed in Table 15.3 on page 889. At the 5% significance level, do the data provide sufficient evidence to conclude that age is useful as a predictor of price for Orions?

SOLUTION As we discovered in Example 15.3, it is reasonable to consider the assumptions for regression inferences satisfied by the variables age and price for Orions, at least for Orions between 2 and 7 years old. So we will apply Procedure 15.1 to carry out the required hypothesis test.

Step 1 State the null and alternative hypotheses.

Let β_1 denote the slope of the population regression line that relates price to age for Orions. Then the null and alternative hypotheses are

$$H_0\text{: } \beta_1 = 0 \text{ (age is not useful for predicting price)}$$
$$H_a\text{: } \beta_1 \neq 0 \text{ (age is useful for predicting price).}$$

Step 2 Decide on the significance level, α.

We are to perform the hypothesis test at the 5% significance level; so $\alpha = 0.05$.

Step 3 The critical values are $\pm t_{\alpha/2}$ with df $= n - 2$.

From Step 2, $\alpha = 0.05$. Also, $n = 11$; so df $= n - 2 = 11 - 2 = 9$. Using Table IV, we find that the critical values are $\pm t_{\alpha/2} = \pm t_{0.05/2} = \pm t_{0.025} = \pm 2.262$, as seen in Fig. 15.10.

FIGURE 15.10 Criterion for deciding whether or not to reject the null hypothesis

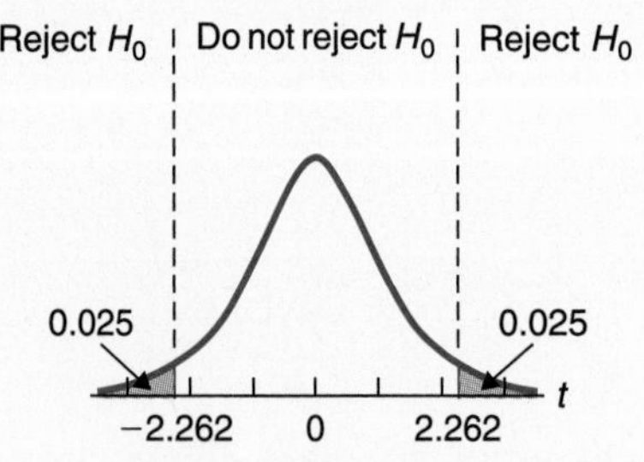

Step 4 Compute the value of the test statistic

$$t = \frac{b_1}{s_e/\sqrt{S_{xx}}}.$$

In Example 14.4 on page 831, we found that $b_1 = -20.26$, $\Sigma x^2 = 326$, and $\Sigma x = 58$. Also, in Example 15.2 on page 877, we determined that $s_e = 12.58$.

Therefore since $n = 11$, the value of the test statistic is

$$t = \frac{b_1}{s_e/\sqrt{S_{xx}}} = \frac{b_1}{s_e/\sqrt{\Sigma x^2 - (\Sigma x)^2/n}} = \frac{-20.26}{12.58/\sqrt{326 - (58)^2/11}} = -7.235.$$

Step 5 *If the value of the test statistic falls in the rejection region, reject H_0; otherwise, do not reject H_0.*

The value of the test statistic, found in Step 4, is $t = -7.235$. Since this falls in the rejection region, we reject H_0.

Step 6 *State the conclusion in words.*

The test results are statistically significant at the 5% level; that is, at the 5% significance level, the data provide sufficient evidence to conclude that the slope of the population regression line is not 0 and hence that age is useful as a predictor of price for Orions. ◨

Using the *P*-Value Approach

The *P*-value approach to hypothesis testing can also be used to carry out the hypothesis test in Example 15.6 and to assess more precisely the evidence against the null hypothesis. From Step 4 we see that the value of the test statistic is $t = -7.235$. Recalling that df $= 9$, we find from Table IV that $P < 0.01$.

In particular, because the *P*-value is less than the specified significance level of 0.05, we can reject H_0. Furthermore, by referring to Table 9.12 on page 543, we see that the data provide very strong evidence against the null hypothesis and hence in favor of the alternative hypothesis that age is useful as a predictor of price for Orions.

Other Procedures for Testing Utility of the Regression

Procedure 15.1, which is based on the statistic b_1, is used for performing a hypothesis test to decide whether the slope of the population regression line is not 0 or, equivalently, whether the regression equation is useful for making predictions. In Section 14.3 we introduced the coefficient of determination, r^2, as a descriptive measure of the utility of the regression equation for making predictions.

This suggests that we should also be able to employ the statistic r^2 as a basis for performing a hypothesis test to decide whether the regression equation is useful for making predictions, and indeed we can. However, we will not cover the hypothesis test based on r^2 since it is equivalent to the hypothesis test based on b_1.

We can also use the linear correlation coefficient, r, introduced in Section 14.4, as a basis for performing a hypothesis test to decide whether the regression equation is useful for making predictions. That test too is equivalent to the hypothesis test based on b_1 but, because it has other uses, we will discuss it in Section 15.4.

Confidence Intervals for the Slope of the Population Regression Line

Recall that the slope of a straight line represents the change in the dependent variable (y) resulting from an increase in the independent variable (x) by 1 unit. Also recall that the population regression line, whose slope is β_1, gives the conditional means of the response variable. Therefore β_1 represents the change in the conditional mean of the response variable for each increase in the value of the predictor variable by 1 unit.

For example, consider the variables age and price of Orions. In this case, β_1 is the amount that the mean price decreases for every increase in age by 1 year. In other words, β_1 is the mean yearly depreciation of Orions.

Consequently, we see that it is worthwhile to obtain an estimate for the slope of the population regression line. As we know, a point estimate for β_1 is provided by b_1. To determine a confidence-interval estimate for β_1, we apply Key Fact 15.4 on page 890 to obtain the following procedure.

PROCEDURE 15.2

THE t-INTERVAL PROCEDURE FOR THE SLOPE OF A POPULATION REGRESSION LINE

ASSUMPTIONS

The four assumptions for regression inferences

Step 1 For a confidence level of $1 - \alpha$, use Table IV to find $t_{\alpha/2}$ with $\text{df} = n - 2$.

Step 2 The endpoints of the confidence interval for β_1 are

$$b_1 \pm t_{\alpha/2} \cdot \frac{s_e}{\sqrt{S_{xx}}}.$$

EXAMPLE 15.7 **ILLUSTRATES PROCEDURE 15.2**

Use the data in Table 15.3 on page 889 to obtain a 95% confidence interval for the slope of the population regression line that relates price to age for Orions.

SOLUTION We apply Procedure 15.2.

Step 1 *For a confidence level of* $1 - \alpha$, *use Table IV to find* $t_{\alpha/2}$ *with* $df = n - 2$.

For a 95% confidence interval, $\alpha = 0.05$. Since $n = 11$, df $= 11 - 2 = 9$. Using Table IV, we find that $t_{\alpha/2} = t_{0.05/2} = t_{0.025} = 2.262$.

Step 2 *The endpoints of the confidence interval for* β_1 *are*

$$b_1 \pm t_{\alpha/2} \cdot \frac{s_e}{\sqrt{S_{xx}}}.$$

From Example 14.4, $b_1 = -20.26$, $\Sigma x^2 = 326$, and $\Sigma x = 58$. Also, from Example 15.2, $s_e = 12.58$. So the endpoints of the confidence interval for β_1 are

$$-20.26 \pm 2.262 \cdot \frac{12.58}{\sqrt{326 - (58)^2/11}}$$

or -20.26 ± 6.33 or -26.59 to -13.93. We can be 95% confident that the slope of the population regression line is somewhere between -26.59 and -13.93. In other words, we can be 95% confident that the yearly decrease in mean price for Orions is somewhere between \$1393 and \$2659.

Using the Computer (Optional)

Procedure 15.1 on page 891 provides a step-by-step method for performing a hypothesis test to decide whether the slope of a population regression line is not 0. Alternatively, we can use Minitab to carry out such a hypothesis test. In fact, the output obtained by applying Minitab's regression procedure contains all the information we need. Example 15.8 explains why.

EXAMPLE 15.8 USING MINITAB TO PERFORM A HYPOTHESIS TEST FOR β_1

In Example 14.6 we applied Minitab's regression procedure to the data on age and price for a sample of 11 Orions. The resulting output is shown in Printout 14.2 on page 838. Use that output to perform the hypothesis test considered in Example 15.6 on page 892.

SOLUTION Let β_1 denote the slope of the population regression line that relates price to age for Orions. The problem is to perform the hypothesis test

$$H_0\text{: } \beta_1 = 0 \text{ (age is not useful for predicting price)}$$
$$H_a\text{: } \beta_1 \neq 0 \text{ (age is useful for predicting price)}$$

at the 5% significance level.

Refer to the fifth line of Printout 14.2, the line labeled `AGE`. The second entry in that line, which is under the column headed `Coef`, displays the slope, b_1, of the sample regression line; so $b_1 = -20.261$. The third entry in that line, which is under the column headed StDev, shows the estimated standard deviation of b_1, $s_e/\sqrt{S_{xx}}$; so $s_e/\sqrt{S_{xx}} = 2.800$. The fourth entry in that line, which is under the column headed T, provides the value of the test statistic,

$$t = \frac{b_1}{s_e/\sqrt{S_{xx}}}.$$

So we see that $t = -7.24$.

The final entry in the line labeled AGE, under the column headed P, gives the P-value for the hypothesis test, which is 0.000 (i.e., zero to three decimal places). Since this is less than the specified significance level of 0.05, we reject H_0. At the 5% significance level, the data provide sufficient evidence to conclude that the slope of the population regression line is not 0 and hence that age is useful as a predictor of price for Orions. ◨

EXERCISES 15.2

STATISTICAL CONCEPTS AND SKILLS

15.28 Explain why the predictor variable is useless as a predictor of the response variable if the slope of the population regression line is zero.

15.29 For two variables satisfying Assumptions 1–3 for regression inferences, the population regression equation is $y = 20 - 3.5x$. For samples of size 10 and given values of the predictor variable, the distribution of slopes of all possible sample regression lines is a ____ distribution with mean ____.

15.30 Consider the standardized variable

$$z = \frac{b_1 - \beta_1}{\sigma/\sqrt{S_{xx}}}.$$

a. Identify its distribution.

b. Why can't it be used as the test statistic for a hypothesis test concerning β_1?

c. What statistic is used? What is the distribution of that statistic?

15.31 In this section we used the statistic b_1 as a basis for conducting a hypothesis test to decide whether a regression equation is useful for prediction. Identify two other statistics that can be used as a basis for such a test.

In Exercises 15.32–15.37, we have repeated the information from Exercises 15.10–15.15. Presuming that the assumptions for regression inferences are met, perform the required hypothesis tests using either the critical-value approach or the P-value approach. (Note: You previously obtained the sample regression equations in Exercises 14.42–14.47 and the standard errors of the estimate in Exercises 15.16–15.21.)

15.32 The *Kelley Blue Book* provides information on wholesale and retail prices of cars. Following are age and price data for 10 randomly selected Corvettes between 1 and 6 years old. Here x denotes age, in years, and y denotes price, in hundreds of dollars.

x	6	6	6	2	2	5	4	5	1	4
y	175	165	180	310	269	200	240	213	310	210

At the 10% significance level, do the data provide sufficient evidence to conclude that the slope of the population regression line is not 0 and hence that age is useful as a predictor of price for Corvettes?

15.33 The National Center for Health Statistics publishes data on heights and weights in *Vital and Health Statistics*. A random sample of 11 males age 18–24 years gave the following data, where x denotes height, in inches, and y denotes weight, in pounds.

x	65	67	71	71	66	75	67	70	71	69	69
y	175	133	185	163	126	198	153	163	159	151	155

Do the data provide sufficient evidence to conclude that the slope of the population regression line is not 0 and hence that height is useful as a predictor of weight for 18–24-year-old males? Use $\alpha = 0.10$.

15.34 Hanna Properties specializes in custom-home resales in the Equestrian Estates, an exclusive subdivision in Phoenix, Arizona. A random sample of nine custom homes currently listed for sale provided the following information on size and price. Here x denotes size, in hundreds of square feet, rounded to the nearest hundred, and y denotes price, in thousands of dollars, rounded to the nearest thousand.

x	26	27	33	29	29	34	30	40	22
y	259	274	294	296	325	380	457	523	215

Do the data suggest that size is useful as a predictor of price for custom homes in the Equestrian Estates? Perform the required hypothesis test at the 0.01 level of significance.

15.35 Many studies have been done that indicate the maximum heart rate an individual can reach during intensive exercise decreases with age. See, for example, page 73 of the September 16, 1996 issue of *Newsweek.* A physician decided to do his own study and recorded the ages and peak heart rates of 10 randomly selected people. The results are shown in the following table, where x denotes age, in years, and y denotes peak heart rate.

x	30	38	41	38	29	39	46	41	42	24
y	186	183	171	177	191	177	175	176	171	196

At the 5% significance level, do the data provide sufficient evidence to conclude that age is useful as a predictor of peak heart rate?

15.36 An instructor at Arizona State University asked a random sample of eight students to record their study times in a beginning calculus course. She then made a table for total hours studied, x, over 2 weeks, and test score, y, at the end of the 2 weeks. Here are the results.

x	10	15	12	20	8	16	14	22
y	92	81	84	74	85	80	84	80

Do the data provide sufficient evidence to conclude that study time is useful as a predictor of test score in beginning calculus courses? Use $\alpha = 0.05$.

15.37 An economist is interested in the relationship between the disposable income of a family and the amount of money spent annually on food. For a preliminary study, the economist takes a random sample of eight middle-income families of the same size (father, mother, two children). The results are as follows, where x denotes disposable income, in thousands of dollars, and y denotes food expenditure, in hundreds of dollars.

x	30	36	27	20	16	24	19	25
y	55	60	42	40	37	26	39	43

Do the data provide sufficient evidence to conclude that disposable income is useful as a predictor of annual food expenditure for middle-income families with a father, mother, and two children? Use $\alpha = 0.01$.

In each of Exercises 15.38–15.43, apply Procedure 15.2 on page 894 to obtain the required confidence interval.

15.38 Refer to Exercise 15.32.

a. Find a 90% confidence interval for the slope, β_1, of the population regression line that relates price to age for Corvettes.

b. Interpret your answer from part (a).

15.39 Refer to Exercise 15.33.

a. Obtain a 90% confidence interval for the slope, β_1, of the population regression line that relates weight to height for males age 18–24.

b. Interpret your answer from part (a).

15.40 Refer to Exercise 15.34.

a. Find a 99% confidence interval for the slope of the population regression line that relates price to size for custom homes in the Equestrian Estates.

b. Interpret your answer from part (a).

15.41 Refer to Exercise 15.35.

a. Find a 95% confidence interval for the slope of the population regression line that relates peak heart rate to age.

b. Interpret your answer from part (a).

15.42 Refer to Exercise 15.36.

a. Obtain a 95% confidence interval for the slope of the population regression line that relates test score to study time in beginning calculus courses.

b. Interpret your answer from part (a).

15.43 Refer to Exercise 15.37.

a. Obtain a 99% confidence interval for the slope of the population regression line that relates annual food expenditure to disposable income for middle-income families with a father, mother, and two children.

b. Interpret your answer from part (a).

USING TECHNOLOGY  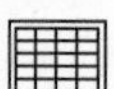 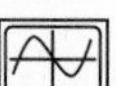

15.44 Use Minitab or some other statistical software to perform the hypothesis test in Exercise 15.36.

15.45 Use Minitab or some other statistical software to perform the hypothesis test in Exercise 15.37.

15.46 The Energy Information Administration publishes data on energy consumption by family income in *Residential Energy Consumption Survey: Consumption and Expenditures*. We applied Minitab's regression procedure to data on family income and last year's energy consumption from a random sample of 25 families. The income data are in thousands of dollars and the energy-consumption data are in millions of BTU. Printout 14.4 on page 844 shows the computer output. Using the printout,

a. find the slope of the sample regression line.

b. determine the estimated standard deviation of the variable b_1.

c. obtain the value of the test statistic, t, for a hypothesis test to decide whether the slope of the population regression line is not 0.

d. determine the P-value for the hypothesis test.

e. decide whether family income is useful for predicting energy consumption. Use $\alpha = 0.01$.

f. obtain a 99% confidence interval for the slope of the population regression line and interpret your result. (*Note:* You will need to use Table IV to determine $t_{\alpha/2}$, but everything else that is required to obtain the confidence interval can be found in the printout.)

g. In performing the inferences that you did in this exercise, what assumptions are you making? How would you check those assumptions?

15.47 Greene and Touchstone conducted a study on the relationship between the estriol levels of pregnant women and the birth weights of their children. Their findings, "Urinary Tract Estriol: An Index of Placental Function," were published in the *American Journal of Obstetrics and Gynecology.* Printout 14.3 on page 843 shows the computer output that results by applying Minitab's regression procedure to the 31 pairs of data obtained by Greene and Touchstone. The estriol levels are in milligrams per 24 hours and the birth weights are in grams. Using the printout,

a. find the slope of the sample regression line.

b. determine the estimated standard deviation of the variable b_1.

c. obtain the value of the test statistic, t, for a hypothesis test to decide whether the slope of the population regression line is not 0.

d. determine the P-value for the hypothesis test.

e. decide, at the 5% significance level, whether estriol level is useful for predicting birth weight.

f. obtain a 95% confidence interval for the slope of the population regression line and interpret your result. (*Note:* You will need to use Table IV to determine $t_{\alpha/2}$, but everything else that is required to obtain the confidence interval can be found in the printout.)

g. In performing the inferences that you did in this exercise, what assumptions are you making? How would you check those assumptions?

15.3 ESTIMATION AND PREDICTION

In this section we will learn how a sample regression equation can be used to make two important inferences:

- Estimating the conditional mean of the response variable corresponding to a particular value of the predictor variable.
- Predicting the value of the response variable for a particular value of the predictor variable.

We will use the Orion example to illustrate the pertinent ideas. In doing so, we will presume that the assumptions for regression inferences (Key Fact 15.1 on page 873) are satisfied by the variables age and price for Orions. Example 15.3 on page 880 shows it is not unreasonable to do that.

EXAMPLE 15.9 ESTIMATING CONDITIONAL MEANS IN REGRESSION

The data on age and price for a sample of 11 Orions are displayed in Table 15.3 on page 889. Use that data to estimate the mean price of all 3-year-old Orions.

SOLUTION By Assumption 1 of the assumptions for regression inferences, the population regression line gives the mean prices for the various ages of Orions. In particular, the mean price of all 3-year-old Orions is $\beta_0 + \beta_1 \cdot 3$. Since β_0 and β_1 are unknown, we estimate the mean price of all 3-year-old Orions, $\beta_0 + \beta_1 \cdot 3$, by the corresponding value, $b_0 + b_1 \cdot 3$, on the sample regression line.

Recalling that the sample regression equation for the age and price data in Table 15.3 is $\hat{y} = 195.47 - 20.26x$, our estimate for the mean price of all 3-year-old Orions is

$$\hat{y} = 195.47 - 20.26 \cdot 3 = 134.69,$$

or \$13,469. We note that the estimate for the mean price of all 3-year-old Orions is the same as the predicted price for a 3-year-old Orion. Both are obtained by substituting $x = 3$ into the sample regression equation. □

The estimate of \$13,469 for the mean price of all 3-year-old Orions is a point estimate. As we know, it would be more informative if we had some idea of how accurate that point estimate is; in other words, it would be better to provide a confidence-interval estimate for the mean price of all 3-year-old Orions. We will now see how to obtain such confidence-interval estimates.

Confidence Intervals for Conditional Means in Regression

To develop a confidence-interval procedure for conditional means in regression, we must first identify the distribution of the predicted value of the response variable for a particular value of the predictor variable. For motivation, let's return to the Orion illustration and consider the particular value 3 of the predictor variable, that is, 3-year-old Orions.

Data on age and price for a sample of 11 Orions are repeated in Table 15.4. As we just saw in Example 15.9, based on this sample data, the predicted price for a 3-year-old Orion is 134.69 ($13,469).

TABLE 15.4
Age and price data for a sample of 11 Orions

Car	Age (yrs) x	Price ($100s) y
1	5	85
2	4	103
3	6	70
4	5	82
5	5	89
6	5	98
7	6	66
8	6	95
9	2	169
10	7	70
11	7	48

Consider now all possible samples of 11 Orions whose ages are the same as those given in the second column of Table 15.4. For such samples, the predicted price of a 3-year-old Orion varies from one sample to another and is therefore a variable. Using the assumptions for regression inferences, it can be shown that its distribution is a normal distribution with mean equal to the mean price of all 3-year-old Orions. More generally, we have the following fact.

KEY FACT 15.5 **DISTRIBUTION OF THE PREDICTED VALUE OF A RESPONSE VARIABLE**

Suppose the variables x and y satisfy Assumptions 1–3 for regression inferences. Let x_p denote a particular value of the predictor variable and $\hat{y}_p$ the corresponding value predicted for the response variable by the sample regression equation, that is, $\hat{y}_p = b_0 + b_1 x_p$. Then, for samples of size n, each with the same values x_1, x_2, $\dots$, x_n, for the predictor variable, the following properties hold for $\hat{y}_p$.

- The mean of $\hat{y}_p$ equals the conditional mean of the response variable corresponding to the value x_p of the predictor variable: $\mu_{\hat{y}_p} = \beta_0 + \beta_1 x_p$.

- The standard deviation of $\hat{y}_p$ is

$$\sigma_{\hat{y}_p} = \sigma\sqrt{\frac{1}{n} + \frac{(x_p - \Sigma x/n)^2}{S_{xx}}}.$$

- The variable $\hat{y}_p$ is normally distributed.

In particular, the distribution of all possible predicted values of the response variable corresponding to x_p is a normal distribution with mean $\beta_0 + \beta_1 x_p$.

In view of Key Fact 15.5, if we standardize the variable $\hat{y}_p$, the resulting variable has the standard normal distribution. However, because the standardized variable contains the unknown parameter σ, it cannot be used as a basis for a confidence-interval formula. Consequently, we replace σ by its estimate s_e, the standard error of the estimate. The resulting variable has a t-distribution. More precisely, we have the following key fact.

KEY FACT 15.6 ***t*-DISTRIBUTION FOR CONFIDENCE INTERVALS FOR CONDITIONAL MEANS IN REGRESSION**

Suppose the variables x and y satisfy Assumptions 1–3 for regression inferences. Then, for samples of size n, each with the same values $x_1, x_2, \ldots, x_n$, for the predictor variable, the variable

$$t = \frac{\hat{y}_p - (\beta_0 + \beta_1 x_p)}{s_e\sqrt{\frac{1}{n} + \frac{(x_p - \Sigma x/n)^2}{S_{xx}}}}$$

has the t-distribution with $\text{df} = n - 2$.

Recalling that $\beta_0 + \beta_1 x_p$ is the conditional mean of the response variable corresponding to the value x_p of the predictor variable, we can use Key Fact 15.6 to derive a confidence-interval procedure for means in regression. Specifically, we have Procedure 15.3.

PROCEDURE 15.3 **THE *t*-INTERVAL PROCEDURE FOR A CONDITIONAL MEAN OF THE RESPONSE VARIABLE**

ASSUMPTIONS
The four assumptions for regression inferences

Step 1 For a confidence level of $1-\alpha$, use Table IV to find $t_{\alpha/2}$ with $\text{df} = n-2$.

Step 2 Compute the point estimate, $\hat{y}_p = b_0 + b_1 x_p$, for the conditional mean of the response variable corresponding to the particular value x_p of the predictor variable.

Step 3 The endpoints of the confidence interval for the conditional mean of the response variable are

$$\hat{y}_p \pm t_{\alpha/2} \cdot s_e \sqrt{\frac{1}{n} + \frac{(x_p - \Sigma x/n)^2}{S_{xx}}}.$$

EXAMPLE 15.10 **ILLUSTRATES PROCEDURE 15.3**

Use the sample data in Table 15.4 on page 900 to obtain a 95% confidence interval for the mean price of all 3-year-old Orions.

SOLUTION We apply Procedure 15.3.

Step 1 *For a confidence level of $1-\alpha$, use Table IV to find $t_{\alpha/2}$ with $df = n-2$.*

We want a 95% confidence interval, which means $\alpha = 0.05$. Since $n = 11$, we have $\text{df} = 9$. Consulting Table IV, we find that $t_{\alpha/2} = t_{0.05/2} = t_{0.025} = 2.262$.

Step 2 *Compute the point estimate, $\hat{y}_p = b_0 + b_1 x_p$, for the conditional mean of the response variable corresponding to the particular value x_p of the predictor variable.*

From Example 14.4, the sample regression equation for the data in Table 15.4 is $\hat{y} = 195.47 - 20.26x$. Here we want $x = x_p = 3$ (3-year-old Orions). So the point estimate for the mean price of all 3-year-old Orions is

$$\hat{y}_p = 195.47 - 20.26 \cdot 3 = 134.69.$$

Step 3 *The endpoints of the confidence interval for the conditional mean of the response variable are*

$$\hat{y}_p \pm t_{\alpha/2} \cdot s_e \sqrt{\frac{1}{n} + \frac{(x_p - \Sigma x/n)^2}{S_{xx}}}.$$

In Example 14.4, we found that $\Sigma x = 58$ and $\Sigma x^2 = 326$; and in Example 15.2 we determined that $s_e = 12.58$. Also, from Step 1, $t_{\alpha/2} = 2.262$, and from Step 2, $\hat{y}_p = 134.69$. Consequently, the endpoints of the confidence interval for the conditional mean are

$$134.69 \pm 2.262 \cdot 12.58\sqrt{\frac{1}{11} + \frac{(3 - 58/11)^2}{326 - (58)^2/11}}$$

or 134.69 ± 16.76 or 117.93 to 151.45. We can be 95% confident that the mean price of all 3-year-old Orions is somewhere between \$11,793 and \$15,145. ◨

Prediction Intervals

A primary use of a sample regression equation is for making predictions. The sample regression equation for the Orion data in Table 15.4 is $\hat{y} = 195.47 - 20.26x$. Substituting, for example, $x = 3$ into that equation, we see that the predicted price for a 3-year-old Orion is 134.69, or \$13,469. However, since the prices of such cars vary, it makes more sense to find a **prediction interval** for the price of a 3-year-old Orion than to give a single predicted value.

Prediction intervals are similar to confidence intervals. The term *confidence* is usually reserved for interval estimates of parameters, such as the mean price of all 3-year-old Orions. The term *prediction* is used for interval estimates of variables, such as the price of a 3-year-old Orion.

To develop a prediction-interval procedure, we must first identify the distribution of the difference between the observed and predicted values of the response variable for a particular value of the predictor variable. For motivation, let's return to the Orion illustration and consider the particular value 3 of the predictor variable, that is, 3-year-old Orions.

As we have seen, based on the sample data for 11 Orions shown in Table 15.4, the predicted price, in hundreds of dollars, for a 3-year-old Orion is 134.69. Suppose we observe the price of a 3-year-old Orion and find it to be 144.12. Then the difference between the observed price and predicted price is 144.12 – 134.69, or 9.43.

Consider now all possible samples of 11 Orions whose ages are the same as those given in the second column of Table 15.4. For such samples, the predicted price of a 3-year-old Orion varies from one sample to another and is therefore a variable. The observed price of a 3-year-old Orion is also a variable. Consequently, the difference between the observed price and predicted price is a variable as well. Using the assumptions for regression inferences, it can be shown that its distribution is a normal distribution with mean 0. More generally, we have the following fact.

KEY FACT 15.7 **DISTRIBUTION OF THE DIFFERENCE BETWEEN THE OBSERVED AND PREDICTED VALUES OF THE RESPONSE VARIABLE**

Suppose the variables x and y satisfy Assumptions 1–3 for regression inferences. Let x_p denote a particular value of the predictor variable and $\hat{y}_p$ the corresponding value predicted for the response variable by the sample regression equation. Furthermore, let y_p be an independently observed value of the response variable corresponding to the value x_p of the predictor variable. Then, for samples of size n, each with the same values $x_1, x_2, \ldots, x_n$, for the predictor variable, the following properties hold for $y_p - \hat{y}_p$, the difference between the observed and predicted values.

- The mean of $y_p - \hat{y}_p$ equals zero: $\mu_{y_p - \hat{y}_p} = 0$.
- The standard deviation of $y_p - \hat{y}_p$ is

$$\sigma_{y_p - \hat{y}_p} = \sigma\sqrt{1 + \frac{1}{n} + \frac{(x_p - \Sigma x/n)^2}{S_{xx}}}.$$

- The variable $y_p - \hat{y}_p$ is normally distributed.

In particular, the distribution of all possible differences between the observed and predicted values of the response variable corresponding to x_p is a normal distribution with mean 0.

In view of Key Fact 15.7, if we standardize the variable $y_p - \hat{y}_p$, the resulting variable has the standard normal distribution. However, because the standardized variable contains the unknown parameter σ, it cannot be used as a basis for a prediction-interval formula. So we replace σ by its estimate s_e, the standard error of the estimate. The resulting variable has a t-distribution.

KEY FACT 15.8 ***t*-DISTRIBUTION FOR PREDICTION INTERVALS IN REGRESSION**

Suppose the variables x and y satisfy Assumptions 1–3 for regression inferences. Then, for samples of size n, each with the same values $x_1, x_2, \ldots, x_n$, for the predictor variable, the variable

$$t = \frac{y_p - \hat{y}_p}{s_e\sqrt{1 + \frac{1}{n} + \frac{(x_p - \Sigma x/n)^2}{S_{xx}}}}$$

has the t-distribution with $\text{df} = n - 2$.

Using Key Fact 15.8, we can derive the following procedure for obtaining a prediction interval.

PROCEDURE 15.4

THE t-INTERVAL PROCEDURE FOR A PREDICTION OF THE RESPONSE VARIABLE

ASSUMPTIONS

The four assumptions for regression inferences

Step 1 For a prediction level of $1 - \alpha$, use Table IV to find $t_{\alpha/2}$ with $\mathrm{df} = n - 2$.

Step 2 Compute the predicted value, $\hat{y}_p = b_0 + b_1 x_p$, of the response variable corresponding to the particular value x_p of the predictor variable.

Step 3 The endpoints of the prediction interval for the observed value of the response variable are

$$\hat{y}_p \pm t_{\alpha/2} \cdot s_e \sqrt{1 + \frac{1}{n} + \frac{(x_p - \Sigma x/n)^2}{S_{xx}}}.$$

EXAMPLE 15.11 ILLUSTRATES PROCEDURE 15.4

Using the sample data in Table 15.4 on page 900, obtain a 95% prediction interval for the price of a 3-year-old Orion.

SOLUTION We apply Procedure 15.4.

Step 1 *For a prediction level of $1 - \alpha$, use Table IV to find $t_{\alpha/2}$ with $df = n - 2$.*

We want a 95% prediction interval; so $\alpha = 0.05$. Also, since $n = 11$, we have $\mathrm{df} = 9$. Consulting Table IV, we find that $t_{\alpha/2} = t_{0.05/2} = t_{0.025} = 2.262$.

Step 2 *Compute the predicted value, $\hat{y}_p = b_0 + b_1 x_p$, of the response variable corresponding to the particular value x_p of the predictor variable.*

As we have seen, the sample regression equation for the data in Table 15.4 is $\hat{y} = 195.47 - 20.26x$. So the predicted price for a 3-year-old Orion is

$$\hat{y}_p = 195.47 - 20.26 \cdot 3 = 134.69.$$

Step 3 *The endpoints of the prediction interval for the observed value of the response variable are*

$$\hat{y}_p \pm t_{\alpha/2} \cdot s_e \sqrt{1 + \frac{1}{n} + \frac{(x_p - \Sigma x/n)^2}{S_{xx}}}.$$

From Example 14.4, $\Sigma x = 58$ and $\Sigma x^2 = 326$; and from Example 15.2, we know that $s_e = 12.58$. Also, $n = 11$, $t_{\alpha/2} = 2.262$, $x_p = 3$, and $\hat{y}_p = 134.69$. Consequently, the endpoints of the prediction interval are

$$134.69 \pm 2.262 \cdot 12.58 \sqrt{1 + \frac{1}{11} + \frac{(3 - 58/11)^2}{326 - (58)^2/11}}$$

 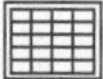

or 134.69 ± 33.02 or 101.67 to 167.71. We can be 95% certain that the observed price of a 3-year-old Orion will be somewhere between \$10,167 and \$16,771. ◨

We have just seen that a 95% prediction interval for the observed price of a 3-year-old Orion is from \$10,167 to \$16,771; and in Example 15.10 we found that a 95% confidence interval for the mean price of all 3-year-old Orions is from \$11,793 to \$15,145. We picture both intervals in Fig. 15.11.

FIGURE 15.11 Prediction and confidence intervals for 3-year-old Orions

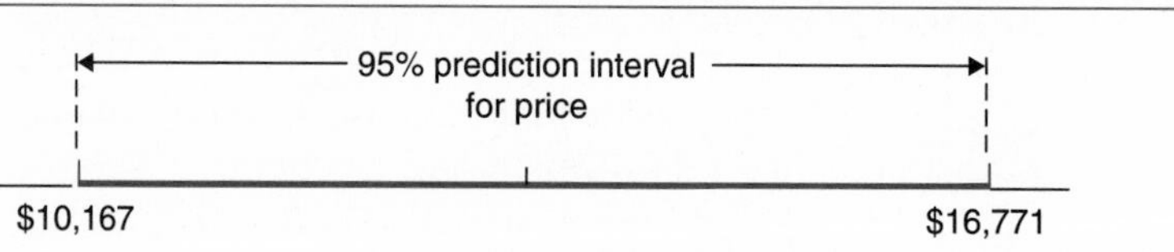

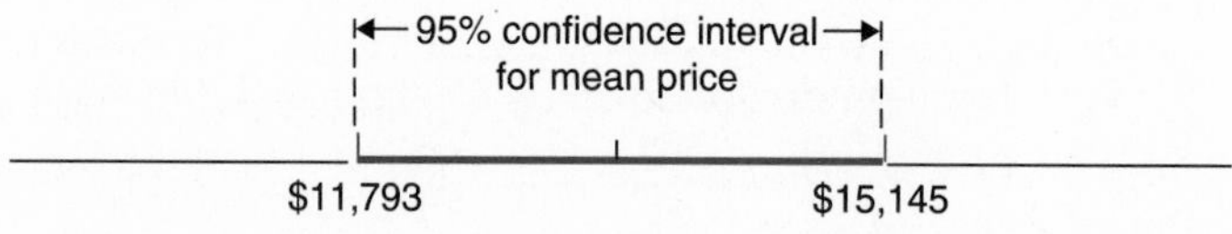

Notice that the prediction interval is wider than the confidence interval. This is to be expected for the following reason: The error in the estimate of the mean price of all 3-year-old Orions is due only to the fact that the population regression line is being estimated by a sample regression line. On the other hand, the error in the prediction of the observed price of a 3-year-old Orion is due to the previously mentioned error in estimating the mean price plus the variation in prices of 3-year-old Orions.

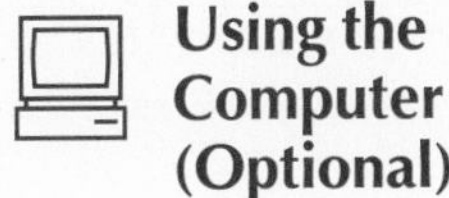

Using the Computer (Optional)

We can apply Minitab's regression procedure to obtain confidence intervals and prediction intervals in regression. The method for doing this is described in the following example.

EXAMPLE 15.12 USING MINITAB FOR ESTIMATION AND PREDICTION

Use Minitab to simultaneously obtain a 95% confidence interval for the mean price of all 3-year-old Orions and a 95% prediction interval for the price of a 3-year-old Orion.

SOLUTION We begin by storing the age and price data from Table 15.4 on page 900 in columns named AGE and PRICE, respectively. Then we do the following.

1 Choose **Stat ➤ Regression ➤ Regression...**
2 Specify PRICE in the **Response** text box
3 Click in the **Predictors** text box and specify AGE
4 Click the **Options...** button
5 Click in the **Prediction intervals for new observations** text box and type 3
6 Click in the **Confidence level** text box and type 95
7 Click **OK**
8 Click **OK**

The resulting output is depicted in Printout 15.7 on the next page.

The output in Printout 15.7 is the same as that in Printout 14.2 on page 838 except for the last two lines, the lines that provide the information about the confidence and prediction intervals. The first item in the last line gives a point estimate for the mean price of all 3-year-old Orions or for the predicted price of a 3-year-old Orion. The second item in the last line displays the estimated standard deviation of the predicted price for a 3-year-old Orion.

The third item in the last line of Printout 15.7 shows the required confidence interval. Hence a 95% confidence interval for the mean price of all 3-year-old Orions is from 117.93 to 151.44. We can be 95% confident that the mean price of all 3-year-old Orions is somewhere between \$11,793 and \$15,144.

The final item in the last line gives the required prediction interval. So a 95% prediction interval for the price of a 3-year-old Orion is from 101.67 to 167.70. We can be 95% certain that the observed price of a 3-year-old Orion will be somewhere between \$10,167 and \$16,770. ❏

PRINTOUT 15.7
Minitab output for the regression procedure with confidence and prediction intervals

```
The regression equation is
PRICE = 195 - 20.3 AGE

Predictor        Coef        StDev          T        P
Constant       195.47        15.24      12.83    0.000
AGE           -20.261        2.800      -7.24    0.000

S = 12.58        R-Sq = 85.3%       R-Sq(adj) = 83.7%

Analysis of Variance

Source           DF           SS           MS          F        P
Regression        1       8285.0       8285.0      52.38    0.000
Residual Error    9       1423.5        158.2
Total            10       9708.5

Unusual Observations
Obs        AGE       PRICE          Fit    StDev Fit     Residual    St Resid
  9       2.00      169.00       154.95         9.92        14.05      1.82 X

X denotes an observation whose X value gives it large influence.

Predicted Values

     Fit  StDev Fit          95.0% CI              95.0% PI
  134.68       7.41    ( 117.93,  151.44)  ( 101.67,  167.70)
```

EXERCISES 15.3

Statistical Concepts and Skills

15.48 What two regression inferences did we discuss in this section? What assumptions are required for such inferences?

15.49 Without doing any calculations, fill in the blank and explain your answer. Based on the sample data in Table 15.4, the predicted price for a 4-year-old Orion is \$11,443. A point estimate for the mean price of all 4-year-old Orions, based on the same sample data, is ______.

In Exercises 15.50–15.55, we have repeated the information from Exercises 15.10–15.15. Presuming that the assumptions for regression inferences are met, determine the required confidence and prediction intervals. (Note: You previously obtained the sample regression equations in Exercises 14.42–14.47 and the standard errors of the estimate in Exercises 15.16–15.21.)

15.50 The *Kelley Blue Book* provides information on wholesale and retail prices of cars. Following are age and price data for 10 randomly selected Corvettes between 1 and 6 years old. Here x denotes age, in years, and y denotes price, in hundreds of dollars.

x	6	6	6	2	2	5	4	5	1	4
y	175	165	180	310	269	200	240	213	310	210

a. Obtain a point estimate for the mean price of all 4-year-old Corvettes.
b. Determine a 90% confidence interval for the mean price of all 4-year-old Corvettes.
c. Find the predicted price of a 4-year-old Corvette.

d. Determine a 90% prediction interval for the price of a 4-year-old Corvette.

e. Draw graphs similar to those in Fig. 15.11 (page 906) showing both the 90% confidence and prediction intervals from parts (b) and (d), respectively.

f. Why is the prediction interval wider than the confidence interval?

15.51 The National Center for Health Statistics publishes data on heights and weights in *Vital and Health Statistics.* A random sample of 11 males age 18–24 years gave the following data, where x denotes height, in inches, and y denotes weight, in pounds.

x	65	67	71	71	66	75	67	70	71	69	69
y	175	133	185	163	126	198	153	163	159	151	155

a. Obtain a point estimate for the mean weight of all 18–24-year-old males who are 70 inches tall.

b. Find a 90% confidence interval for the mean weight of all 18–24-year-old males who are 70 inches tall.

c. Find the predicted weight of an 18–24-year-old male who is 70 inches tall.

d. Determine a 90% prediction interval for the weight of an 18–24-year-old male who is 70 inches tall.

e. Draw graphs similar to those in Fig. 15.11 (page 906) showing both the 90% confidence and prediction intervals from parts (b) and (d), respectively.

f. Why is the prediction interval wider than the confidence interval?

15.52 Hanna Properties specializes in custom-home resales in the Equestrian Estates, an exclusive subdivision in Phoenix, Arizona. A random sample of nine custom homes currently listed for sale provided the following information on size and price. Here x denotes size, in hundreds of square feet, rounded to the nearest hundred, and y denotes price, in thousands of dollars, rounded to the nearest thousand.

x	26	27	33	29	29	34	30	40	22
y	259	274	294	296	325	380	457	523	215

a. Obtain a point estimate for the mean price of all 2800-sq-ft Equestrian Estate homes.

b. Find a 99% confidence interval for the mean price of all 2800-sq-ft Equestrian Estate homes.

c. Find the predicted price of a 2800-sq-ft Equestrian Estate home.

d. Determine a 99% prediction interval for the price of a 2800-sq-ft Equestrian Estate home.

15.53 Many studies have been done that indicate the maximum heart rate an individual can reach during intensive exercise decreases with age. See, for example, page 73 of the September 16, 1996 issue of *Newsweek.* A physician decided to do his own study and recorded the ages and peak heart rates of 10 randomly selected people. The results are shown in the table below, where x denotes age, in years, and y denotes peak heart rate.

x	30	38	41	38	29	39	46	41	42	24
y	186	183	171	177	191	177	175	176	171	196

a. Obtain a point estimate for the mean peak heart rate of all 40-year-olds.

b. Find a 95% confidence interval for the mean peak heart rate of all 40-year-olds.

c. Find the predicted peak heart rate of a 40-year-old.

d. Find a 95% prediction interval for the peak heart rate of a 40-year-old.

15.54 An instructor at Arizona State University asked a random sample of eight students to record their study times in a beginning calculus course. She then made a table for total hours studied, x, over 2 weeks, and test score, y, at the end of the 2 weeks. Here are the results.

x	10	15	12	20	8	16	14	22
y	92	81	84	74	85	80	84	80

a. Obtain a 95% confidence interval for the mean test score of all beginning calculus students who study for 15 hours.

b. Obtain a 95% prediction interval for the test score of a beginning calculus student who studies for 15 hours.

15.55 An economist is interested in the relationship between the disposable income of a family and the amount of money spent annually on food. For a preliminary study, the economist takes a random sample of eight middle-income families of the same size (father, mother, two children). The results are as follows, where x denotes disposable income, in thousands of dollars, and y denotes food expenditure, in hundreds of dollars.

x	30	36	27	20	16	24	19	25
y	55	60	42	40	37	26	39	43

a. Determine a 99% confidence interval for the mean annual food expenditure of all middle-income families consisting of a father, mother, and two children that have a disposable income of $25,000.

b. Find a 99% prediction interval for the annual food expenditure of a middle-income family consisting of a father, mother, and two children that has a disposable income of $25,000.

USING TECHNOLOGY  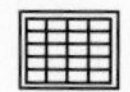 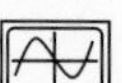

15.56 Use Minitab or some other statistical software to obtain the confidence and prediction intervals required in Exercise 15.54.

15.57 Use Minitab or some other statistical software to obtain the confidence and prediction intervals required in Exercise 15.55.

15.58 The Energy Information Administration publishes data on energy consumption by family income in *Residential Energy Consumption Survey: Consumption and Expenditures.* We applied Minitab's regression procedure to data on family income and last year's energy consumption from a random sample of 25 families. The income data are in thousands of dollars and the energy-consumption data are in millions of BTU. Printout 15.8 shows (abridged) computer output. The confidence and prediction intervals are for an income level of $40,000.

a. Determine a point estimate for last year's mean energy consumption of all families with an annual income of $40,000.

b. Find a 95% confidence interval for last year's mean energy consumption of all families with an annual income of $40,000.

c. Obtain a 95% prediction interval for last year's energy consumption by a family with an annual income of $40,000.

PRINTOUT 15.8 Minitab output for Exercise 15.58

```
The regression equation is
CONSUMPT = 82.0 + 0.931 INCOME

Predictor        Coef       StDev         T        P
Constant       82.036       2.054     39.94    0.000
INCOME        0.93051     0.05727     16.25    0.000

S = 5.375      R-Sq = 92.0%     R-Sq(adj) = 91.6%

Analysis of Variance

Source            DF         SS          MS         F        P
Regression         1     7626.6      7626.6    264.02    0.000
Residual Error    23      664.4        28.9
Total             24     8291.0

Predicted Values

     Fit  StDev Fit          95.0% CI              95.0% PI
  119.26       1.20   (  116.77,  121.75)  (  107.86,  130.65)
```

15.59 Greene and Touchstone conducted a study on the relationship between the estriol levels of pregnant women and the birth weights of their children. Their findings, "Urinary Tract Estriol: An Index of Placental Function," were published in the *American Journal of Obstetrics and Gynecology.* Printout 15.9 shows (abridged) output that results by applying Minitab's regression procedure to the 31 pairs of data obtained by Greene and Touchstone. Estriol levels are in milligrams per 24 hours and birth weights are in grams. The confidence and prediction intervals are for an estriol level of 15 mg/24 hours.

a. Obtain a point estimate for the mean birth weight of all babies whose mothers have an estriol level of 15 mg/24 hours.
b. Determine a 95% confidence interval for the mean birth weight of all babies whose mothers have an estriol level of 15 mg/24 hours.
c. Determine a 95% prediction interval for the birth weight of a baby whose mother has an estriol level of 15 mg/24 hours.

PRINTOUT 15.9 Minitab output for Exercise 15.59

```
The regression equation is
WEIGHT = 2152 + 60.8 ESTRIOL

Predictor        Coef        StDev          T        P
Constant       2152.3        262.0       8.21    0.000
ESTRIOL         60.82        14.68       4.14    0.000

S = 382.1       R-Sq = 37.2%      R-Sq(adj) = 35.0%

Analysis of Variance

Source            DF          SS          MS         F        P
Regression         1     2505745     2505745     17.16    0.000
Residual Error    29     4234255      146009
Total             30     6740000

Predicted Values

    Fit  StDev Fit         95.0% CI              95.0% PI
  3064.6       76.0   (  2909.2,  3220.1)   (  2267.8,  3861.4)
```

15.4 INFERENCES IN CORRELATION

Frequently, we want to decide whether two variables are linearly correlated, that is, whether there is a linear relationship between the two variables. In the context of regression, we can make that decision by performing a hypothesis test for the slope of the population regression line, as discussed in Section 15.2.

Alternatively, we can perform a hypothesis test for the **population linear correlation coefficient, ρ** (rho). The population linear correlation coefficient measures the linear correlation of all possible pairs of observations of two variables in the same way that a sample linear correlation coefficient, r, measures the linear correlation of a sample of pairs. So it is ρ that actually describes the strength of the linear relationship between two variables; r is only an estimate of ρ obtained from sample data.

The population linear correlation coefficient of two variables, x and y, always lies between -1 and 1. Values of ρ near -1 or 1 indicate a strong linear relationship between the variables, whereas values of ρ near 0 indicate a weak linear relationship between the variables.

If $\rho > 0$, the variables are **positively linearly correlated,** meaning that y tends to increase linearly as x increases (and vice versa), with the tendency being greater the closer ρ is to 1. If $\rho < 0$, the variables are **negatively linearly correlated,** meaning that y tends to decrease linearly as x increases (and vice versa), with the tendency being greater the closer ρ is to -1. If $\rho = 0$, the variables are **linearly uncorrelated,** meaning that there is no linear relationship between the variables.

Because a sample linear correlation coefficient, r, is an estimate of the population linear correlation coefficient, ρ, we can use r as a basis for performing a hypothesis test for ρ. For a test with null hypothesis H_0: $\rho = 0$ (i.e., the variables are linearly uncorrelated), we will employ the following fact.

KEY FACT 15.9 ***t*-DISTRIBUTION FOR A CORRELATION TEST**

Suppose the variables x and y satisfy Assumptions 1–3 for regression inferences. Then, for samples of size n, the variable

$$t = \frac{r}{\sqrt{\dfrac{1 - r^2}{n - 2}}}$$

has the t-distribution with df $= n - 2$ if the null hypothesis $\rho = 0$ is true.

In view of Key Fact 15.9, we see that for a hypothesis test with null hypothesis H_0: $\rho = 0$, we can use the variable

$$t = \frac{r}{\sqrt{\dfrac{1 - r^2}{n - 2}}}$$

as the test statistic and obtain the critical values from the t-table, Table IV. Specifically, we have the following procedure.

PROCEDURE 15.5

THE t-TEST FOR CORRELATION

ASSUMPTIONS

The four assumptions for regression inferences

Step 1 The null hypothesis is H_0: $\rho = 0$ and the alternative hypothesis is one of the following:

H_a: $\rho \neq 0$ (Two-tailed)	or	H_a: $\rho < 0$ (Left-tailed)	or	H_a: $\rho > 0$ (Right-tailed)

Step 2 Decide on the significance level, α.

Step 3 The critical value(s) are

$\pm t_{\alpha/2}$ (Two-tailed)	or	$-t_\alpha$ (Left-tailed)	or	t_α (Right-tailed)

with df $= n - 1$. Use Table IV to find the critical value(s).

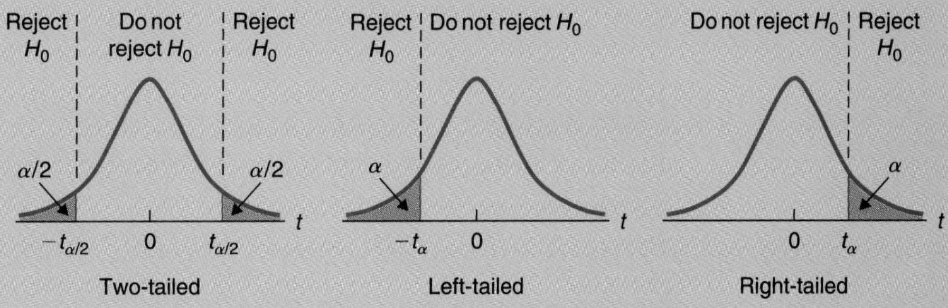

Step 4 Compute the value of the test statistic

$$t = \frac{r}{\sqrt{\dfrac{1 - r^2}{n - 2}}}.$$

Step 5 If the value of the test statistic falls in the rejection region, reject H_0; otherwise, do not reject H_0.

Step 6 State the conclusion in words.

EXAMPLE 15.13 ILLUSTRATES PROCEDURE 15.5

Refer to the age and price data for a sample of 11 Orions given in Table 15.4 on page 900. At the 5% significance level, do the data provide sufficient evidence to conclude that age and price of Orions are negatively linearly correlated?

SOLUTION As we discovered in Example 15.3 on page 880, it is not unreasonable to consider the assumptions for regression inferences met by the variables age and price for Orions, at least for Orions between 2 and 7 years old. Consequently, we will apply Procedure 15.5 to carry out the required hypothesis test.

Step 1 *State the null and alternative hypotheses.*

Let ρ denote the population linear correlation coefficient for the variables age and price of Orions. Then the null and alternative hypotheses are

H_0: $\rho = 0$ (age and price are linearly uncorrelated)

H_a: $\rho < 0$ (age and price are negatively linearly correlated).

Note that the hypothesis test is left-tailed since a less-than sign ($<$) appears in the alternative hypothesis.

Step 2 *Decide on the significance level, α.*

We are to use $\alpha = 0.05$.

Step 3 *The critical value for a left-tailed test is $-t_{\alpha}$, with df $= n - 2$.*

We have $n = 11$, so df $= 9$. Also, $\alpha = 0.05$. Consulting Table IV, we find that for df $= 9$, $t_{0.05} = 1.833$. Thus the critical value is $-t_{0.05} = -1.833$, as seen in Fig. 15.12.

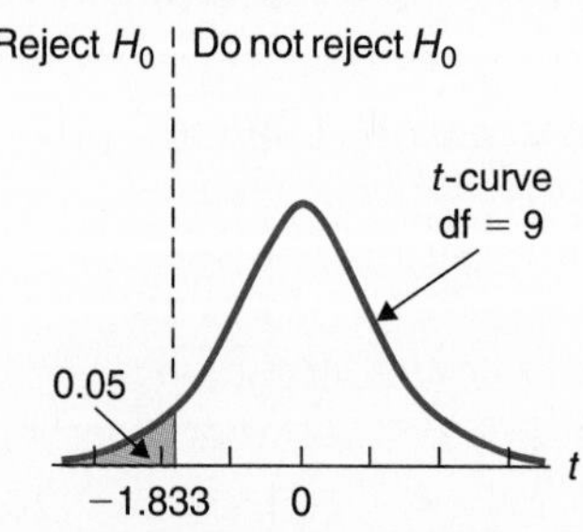

FIGURE 15.12
Criterion for deciding whether or not to reject the null hypothesis

Step 4 *Compute the value of the test statistic*

$$t = \frac{r}{\sqrt{\dfrac{1 - r^2}{n - 2}}}.$$

In Example 14.10 on page 857, we computed the sample linear correlation coefficient for the age and price data in Table 15.4. We found that $r = -0.924$. So the value of the test statistic is

$$t = \frac{-0.924}{\sqrt{\frac{1 - (-0.924)^2}{11 - 2}}} = -7.249.$$

Step 5 *If the value of the test statistic falls in the rejection region, reject H_0; otherwise, do not reject H_0.*

The value of the test statistic, found in Step 4, is $t = -7.249$. A glance at Fig. 15.12 shows that this falls in the rejection region. Consequently, we reject H_0.

Step 6 *State the conclusion in words.*

 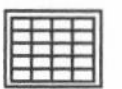 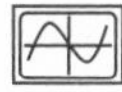

The test results are statistically significant at the 5% level; that is, at the 5% significance level, the data provide sufficient evidence to conclude that age and price of Orions are negatively linearly correlated. Prices for Orions tend to decrease linearly with increasing age, at least for Orions between 2 and 7 years old. ◨

Using the *P*-Value Approach

The P-value approach to hypothesis testing can also be used to carry out the hypothesis test in Example 15.13 and to assess more precisely the evidence against the null hypothesis. From Step 4 we see that the value of the test statistic is $t = -7.249$. Recalling that df $= 9$, we find from Table IV that $P < 0.005$.

In particular, because the P-value is less than the specified significance level of 0.05, we can reject H_0. Furthermore, by referring to Table 9.12 on page 543, we see that the data provide very strong evidence against the null hypothesis and hence in favor of the alternative hypothesis that age and price of Orions are negatively linearly correlated.

Using the Computer (Optional)

Procedure 15.5 on page 913 provides a step-by-step method for performing a hypothesis test for a population linear correlation coefficient, ρ. Alternatively, we can use Minitab to carry out such a hypothesis test. In fact, the Minitab output obtained for the sample linear correlation coefficient, r, also gives the P-value for a correlation test.

Note, however, that the P-value provided by Minitab is for a two-tailed test. Consequently, we must take care in interpreting the P-value for tests that are not two-tailed. Example 15.14 provides an illustration.

EXAMPLE 15.14 Using Minitab to Perform a Correlation Test

In Example 14.11 we applied Minitab's correlation procedure to obtain the sample linear correlation coefficient of data on age and price for a sample of 11 Orions.

The resulting output is shown in Printout 14.5 on page 860. Use that output to perform the hypothesis test considered in Example 15.13.

SOLUTION Let ρ denote the population linear correlation coefficient for the variables age and price of Orions. The problem is to perform the hypothesis test

$$H_0\colon \rho = 0 \text{ (age and price are linearly uncorrelated)}$$
$$H_a\colon \rho < 0 \text{ (age and price are negatively linearly correlated)}$$

at the 5% significance level.

Referring to Printout 14.5, we see that the P-value for a two-tailed correlation test is 0.000 (i.e., zero to three decimal places). We also observe from the printout that the sample linear correlation coefficient is negative. This and the fact that the hypothesis test under consideration is left-tailed implies that its P-value is one-half the P-value for the two-tailed test. So we see that the P-value for our test is also zero to three decimal places.

Since the P-value for our test is less than the specified significance level of 0.05, we reject H_0. At the 5% level of significance, the data provide sufficient evidence to conclude that age and price of Orions are negatively linearly correlated. □

EXERCISES 15.4

STATISTICAL CONCEPTS AND SKILLS

15.60 Is ρ a parameter or statistic? What about r?

15.61 Identify the statistic used to estimate the population linear correlation coefficient.

15.62 Suppose that for a sample of pairs of observations for two variables, the linear correlation coefficient, r, is positive. Does this necessarily imply that the variables are positively linearly correlated? Explain your answer.

15.63 Fill in the blanks.

a. If $\rho = 0$, then the two variables under consideration are linearly ____.

b. If two variables are positively linearly correlated, then one of the variables tends to increase as the other ____.

c. If two variables are ____ linearly correlated, then one of the variables tends to decrease as the other increases.

In Exercises 15.64–15.69, we have repeated the information from Exercises 15.10–15.15. Presuming that the assumptions for regression inferences are met, perform the required correlation test using either the critical-value approach or the P-value approach. (Note: You previously obtained the sample linear correlation coefficients in Exercises 14.88–14.93.)

15.64 The *Kelley Blue Book* provides information on wholesale and retail prices of cars. Following are age and price data for 10 randomly selected Corvettes between 1 and 6 years old. Here x denotes age, in years, and y denotes price, in hundreds of dollars.

x	6	6	6	2	2	5	4	5	1	4
y	175	165	180	310	269	200	240	213	310	210

At the 5% level of significance, do the data provide sufficient evidence to conclude that age and price of Corvettes are negatively linearly correlated?

15.65 The National Center for Health Statistics publishes data on heights and weights in *Vital and Health Statistics.* A random sample of 11 males age 18–24 years gave the following data, where x denotes height, in inches, and y denotes weight, in pounds.

x	65	67	71	71	66	75	67	70	71	69	69
y	175	133	185	163	126	198	153	163	159	151	155

At the 5% significance level, do the data provide sufficient evidence to conclude that the variables height and weight are positively linearly correlated for 18–24-year-old males?

15.66 Hanna Properties specializes in custom-home resales in the Equestrian Estates, an exclusive subdivision in Phoenix, Arizona. A random sample of nine custom homes currently listed for sale provided the following information on size and price. Here x denotes size, in hundreds of square feet, rounded to the nearest hundred, and y denotes price, in thousands of dollars, rounded to the nearest thousand.

x	26	27	33	29	29	34	30	40	22
y	259	274	294	296	325	380	457	523	215

At the 0.5% significance level, do the data provide sufficient evidence to conclude that for custom homes in the Equestrian Estates, size and price are positively linearly correlated?

15.67 Many studies have been done that indicate the maximum heart rate an individual can reach during intensive exercise decreases with age. See, for example, page 73 of the September 16, 1996 issue of *Newsweek.* A physician decided to do his own study and recorded the ages and peak heart rates of 10 randomly selected people. The results are shown in the following table, where x denotes age, in years, and y denotes peak heart rate.

x	30	38	41	38	29	39	46	41	42	24
y	186	183	171	177	191	177	175	176	171	196

At the 2.5% significance level, do the data provide sufficient evidence to conclude that age and peak heart rate are negatively linearly correlated?

15.68 An instructor at Arizona State University asked a random sample of eight students to record their study times in a beginning calculus course. She then made a table for total hours studied, x, over 2 weeks, and test score, y, at the end of the 2 weeks. Here are the results.

x	10	15	12	20	8	16	14	22
y	92	81	84	74	85	80	84	80

Do the data provide sufficient evidence to conclude that in beginning calculus courses, study time and test score are linearly correlated? Perform the hypothesis test using a significance level of 0.05.

15.69 An economist is interested in the relationship between the disposable income of a family and the amount of money spent annually on food. For a preliminary study, the economist takes a random sample of eight middle-income families of the same size (father, mother, two children). The results are as follows, where x denotes disposable income, in thousands of dollars, and y denotes food expenditure, in hundreds of dollars.

x	30	36	27	20	16	24	19	25
y	55	60	42	40	37	26	39	43

At the 1% significance level, do the data provide sufficient evidence to conclude that family disposable income and annual food expenditure are linearly correlated for middle-income families with a father, mother, and two children?

15.70 We took a sample of 10 students from an introductory statistics class and obtained the following data, where x denotes height, in inches, and y denotes score on the final exam.

x	71	68	71	65	66	68	68	64	62	65
y	87	96	66	71	71	55	83	67	86	60

At the 5% significance level, do the data provide sufficient evidence to conclude that for students in introductory statistics courses, height and final-exam score are linearly correlated?

USING TECHNOLOGY  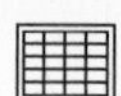 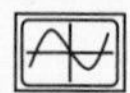

15.71 Use Minitab or some other statistical software to perform the correlation test from Exercise 15.69. *(Note:* Minitab provides the P-value only for a two-tailed test.*)*

15.72 Use Minitab or some other statistical software to perform the correlation test from Exercise 15.68. *(Note:* Minitab provides the P-value only for a two-tailed test.*)*

15.5 TESTING FOR NORMALITY*

As we know, several descriptive methods are available for assessing normality of a variable from sample data. One of the most commonly used methods is the normal probability plot, a plot of the normal scores against the sample data.

If the variable is normally distributed, then a normal probability plot of the sample data should be roughly linear. So we can assess normality as follows.

- If the plot is roughly linear, then accept as reasonable that the variable is normally distributed.
- If the plot shows systematic deviations from linearity (e.g., if it displays significant curvature), then conclude that the variable is probably not normally distributed.

This visual assessment of normality is subjective because what constitutes "roughly linear" is a matter of opinion. To overcome this difficulty, we can perform a hypothesis test for normality based on the linear correlation coefficient: If the variable under consideration is normally distributed, then the correlation between the sample data and their normal scores should be near 1 because the normal probability plot should be roughly linear.[†]

So to perform a hypothesis test for normality, we compute the linear correlation coefficient between the sample data and their normal scores. If the correlation is too much smaller than 1, then we reject the null hypothesis that the variable is normally distributed in favor of the alternative hypothesis that the variable is not normally distributed. Of course, we need a table of critical values to decide what is "too much smaller than 1." This is provided by Table X in Appendix A.

We will use the letter w to denote the normal score corresponding to an observed value of the variable x. And, for this special context, we will use $\boldsymbol{R_p}$ instead

* This more advanced section is optional. It is not needed for the remainder of the book.

† Since large normal scores are associated with large observations and vice versa, the correlation between the sample data and their normal scores cannot be negative.

of r to denote the linear correlation coefficient. So, in view of the computing formula for the linear correlation coefficient given in Definition 14.6 on page 854, the correlation between the sample data and their normal scores equals

$$R_p = \frac{S_{xw}}{\sqrt{S_{xx} S_{ww}}},$$

where (see Definition 14.3 on page 830) we have $S_{xw} = \Sigma xw - (\Sigma x)(\Sigma w)/n$, $S_{xx} = \Sigma x^2 - (\Sigma x)^2/n$, and $S_{ww} = \Sigma w^2 - (\Sigma w)^2/n$. However, because the sum of the normal scores for a data set always equals 0, we can simplify the preceding displayed equation to

$$R_p = \frac{\Sigma xw}{\sqrt{S_{xx}\, \Sigma w^2}}$$

and use this as our test statistic for a correlation test for normality.

Since the null hypothesis of normality will be rejected only when the test statistic is too small, the rejection region is always on the left, that is, the hypothesis test is always left-tailed. Consequently, we have the procedure on the next page.

Several other tests for normality exist in addition to the correlation test for normality. However, the correlation test for normality is one of the most powerful.

In Example 6.14 on page 386, we considered data on adjusted gross incomes for a sample of 12 federal individual income tax returns. We obtained the normal scores for the data (Table 6.4) and drew a normal probability plot (Fig. 6.23). Because the normal probability plot shows significant curvature, we concluded that adjusted gross incomes are probably not normally distributed. This is a subjective conclusion based on a graph. Now we will apply Procedure 15.6 so that we can make an objective conclusion.

EXAMPLE 15.15 ILLUSTRATES PROCEDURE 15.6

The Internal Revenue Service publishes data on federal individual income tax returns in *Statistics of Income, Individual Income Tax Returns.* A random sample of 12 returns from last year revealed the adjusted gross incomes, in thousands of dollars, shown in Table 15.5.

TABLE 15.5 Adjusted gross incomes ($1000s)

9.7	93.1	33.0	21.2
81.4	51.1	43.5	10.6
12.8	7.8	18.1	12.7

At the 5% significance level, do the data provide sufficient evidence to conclude that adjusted gross incomes are not normally distributed?

PROCEDURE 15.6 **THE CORRELATION TEST FOR NORMALITY**

Step 1 The null and alternative hypotheses are

H_0: The variable under consideration is normally distributed.

H_a: The variable under consideration is not normally distributed.

Step 2 Decide on the significance level, α.

Step 3 The critical value is R_p^*. Use Table X to find the critical value.

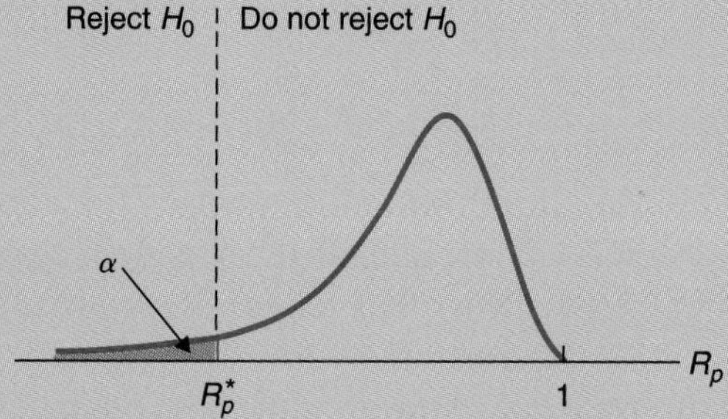

Step 4 Compute the value of the test statistic

$$R_p = \frac{\Sigma xw}{\sqrt{S_{xx}\,\Sigma w^2}},$$

where x and w denote, respectively, observations of the variable and the corresponding normal scores.

Step 5 If the value of the test statistic falls in the rejection region, reject H_0; otherwise, do not reject H_0.

Step 6 State the conclusion in words.

SOLUTION We apply Procedure 15.6.

Step 1 *State the null and alternative hypotheses.*

The null and alternative hypotheses are

H_0: Adjusted gross incomes are normally distributed.

H_a: Adjusted gross incomes are not normally distributed.

Step 2 *Decide on the significance level, α.*

We are to perform the hypothesis test at the 5% significance level; so $\alpha = 0.05$.

Step 3 *The critical value is R_p^*.*

We have $\alpha = 0.05$ and $n = 12$. Consulting Table X, we find that the critical value is $R_p^* = 0.927$, as seen in Fig. 15.13.

FIGURE 15.13
Criterion for deciding whether or not to reject the null hypothesis

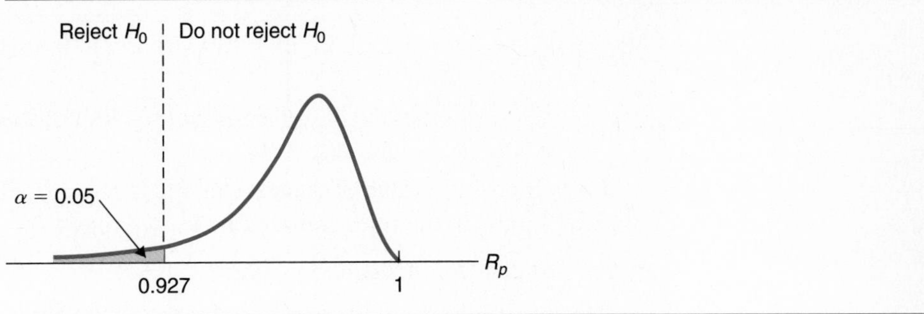

Step 4 *Compute the value of the test statistic*

$$R_p = \frac{\Sigma xw}{\sqrt{S_{xx}\,\Sigma w^2}}.$$

To compute the value of the test statistic, we need a table for x, w, xw, x^2, and w^2. The normal scores for the adjusted gross incomes in Table 15.5 are found by consulting Table III in Appendix A. From the data and normal scores, we obtain Table 15.6.

TABLE 15.6
Table for computing R_p

Adjusted gross income x	Normal score w	xw	x^2	w^2
7.8	−1.64	−12.792	60.84	2.6896
9.7	−1.11	−10.767	94.09	1.2321
10.6	−0.79	−8.374	112.36	0.6241
12.7	−0.53	−6.731	161.29	0.2809
12.8	−0.31	−3.968	163.84	0.0961
18.1	−0.10	−1.810	327.61	0.0100
21.2	0.10	2.120	449.44	0.0100
33.0	0.31	10.230	1,089.00	0.0961
43.5	0.53	23.055	1,892.25	0.2809
51.1	0.79	40.369	2,611.21	0.6241
81.4	1.11	90.354	6,625.96	1.2321
93.1	1.64	152.684	8,667.61	2.6896
395.0	0.00	274.370	22,255.50	9.8656

Referring to Table 15.6, we find that

$$R_p = \frac{\Sigma xw}{\sqrt{S_{xx}\,\Sigma w^2}} = \frac{\Sigma xw}{\sqrt{\left[\Sigma x^2 - (\Sigma x)^2/n\right]\left[\Sigma w^2\right]}}$$

$$= \frac{274.370}{\sqrt{\left[22{,}255.50 - (395.0)^2/12\right] \cdot 9.8656}} = 0.908.$$

Step 5 *If the value of the test statistic falls in the rejection region, reject H_0; otherwise, do not reject H_0.*

From Step 4 the value of the test statistic is $R_p = 0.908$, which, as we see from Fig. 15.13, falls in the rejection region. So, we reject H_0.

Step 6 *State the conclusion in words.*

The test results are statistically significant at the 5% level; that is, at the 5% significance level, the data provide sufficient evidence to conclude that adjusted gross incomes are not normally distributed. ◨

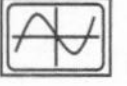

Using the *P*-Value Approach

The P-value approach to hypothesis testing can also be used to carry out the hypothesis test in Example 15.15 and to assess more precisely the evidence against the null hypothesis. From Step 4 we see that the value of the test statistic is $R_p = 0.908$. Recalling that $n = 12$, we find from Table X that $0.01 < P < 0.05$.

In particular, because the P-value is less than the specified significance level of 0.05, we can reject H_0. Furthermore, by referring to Table 9.12 on page 543, we see that the data provide strong evidence against the null hypothesis of normality.

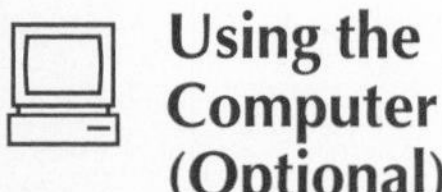

Using the Computer (Optional)

In Example 15.15 we went through the details of applying Procedure 15.6 so you could see exactly how a correlation test for normality works. But, generally, such tests are carried out by computer. We will explain two methods using Minitab.

One method is to first apply Minitab's normal scores procedure to obtain the normal scores for the data, next apply Minitab's linear correlation procedure to obtain the correlation between the data and their normal scores, and then compare that correlation to the appropriate critical value obtained from Table X.

A simpler method is to choose **Stat ➤ Basic Statistics ➤ Normality Test...**, specify the variable in the **Variable** text box, select the **Ryan-Joiner** option button from the **Tests for Normality** field, and then click **OK**. The resulting output provides, among other things, the value of the test statistic R_p (labeled R in the output) and an approximate P-value. We will explore correlation tests for normality by computer in the exercises.

EXERCISES 15.5

STATISTICAL CONCEPTS AND SKILLS

15.73 Regarding normal probability plots:

a. What are they?

b. Identify an important use of such plots.

c. How is a normal probability plot used for assessing the normality of a variable?

d. Why is the method described in part (c) subjective?

15.74 In a correlation test for normality, what correlation is computed?

15.75 If you examine Procedure 15.6, you will notice that a correlation test for normality is always left-tailed. Explain in words why this is so.

15.76 Suppose you perform a correlation test for normality at the 1% significance level. Further suppose you reject the null hypothesis that the population is normally distributed. Can you be confident in stating that the population from which the sample was drawn is not normally distributed? Explain your answer.

In Exercises 15.77–15.80, perform a correlation test for normality using either the critical-value approach or the P-value approach. (Note: You obtained the normal scores for these data sets in Exercises 6.93–6.96.)

15.77 A sample of the final-exam scores in a large introductory statistics course is as follows.

88	67	64	76	86
85	82	39	75	34
90	63	89	90	84
81	96	100	70	96

At the 5% significance level, do the data provide sufficient evidence to conclude that final-exam scores in this introductory statistics class are not normally distributed?

15.78 As reported by the R. R. Bowker Company of New York in *Library Journal,* the mean annual subscription rate to law periodicals was \$97.33 in 1995. A sample of this year's law periodicals yields the following subscription rates to the nearest dollar.

106	122	120	123
118	114	138	131
128	124	119	130

Do the data provide sufficient evidence to conclude that this year's subscription rates to law periodicals are not normally distributed? Use $\alpha = 0.05$.

15.79 The U.S. Federal Highway Administration conducts studies on motor vehicle travel by type of vehicle. Results are published annually in *Highway Statistics.* A sample of 15 cars yields the following data on number of miles driven, in thousands, for last year.

10.2	10.3	8.9	12.7	8.3
9.2	13.7	7.7	3.3	10.6
11.8	6.6	8.6	5.7	12.0

Can we conclude from the data that last year's distribution for the number of miles cars were driven is not normal? Use $\alpha = 0.10$.

15.80 The Bureau of Labor Statistics publishes information on average annual expenditures by consumers in *Consumer Expenditure Survey.* In 1995 the mean amount spent by consumers on nonalcoholic beverages was \$240. A random sample of 12 consumers yielded the following data, in dollars, on last year's expenditures on nonalcoholic beverages.

361	176	184	265
259	281	240	273
259	249	194	258

Can we conclude from the data that last year's distribution for amounts spent by consumers on nonalcoholic beverages is not normal? Perform the required hypothesis test at the 10% level of significance.

USING TECHNOLOGY  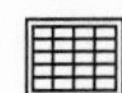 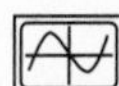

15.81 Use Minitab or some other statistical software to perform the hypothesis test in Exercise 15.77.

15.82 Use Minitab or some other statistical software to perform the hypothesis test in Exercise 15.78.

In each of Exercises 15.83–15.86, use any technology you have available to perform the required correlation test for normality.

15.83 The U.S. Energy Information Administration reports figures on residential energy consumption and expenditures in *Residential Energy Consumption Survey: Consumption and Expenditures.* A random sample of 18 households using electricity as their primary energy source yields the following data on one year's energy expenditures.

$1376	1452	1235	1480	1185	1327
1059	1400	1227	1102	1168	1070
949	1351	1259	1179	1393	1456

At the 1% significance level, do the data provide sufficient evidence to conclude that for the year in question, energy expenditures for households using electricity as their primary energy source are not normally distributed?

15.84 In January 1994 the Department of Agriculture estimated that a typical U.S. family of four would spend $117 per week for food. During that same year, a random sample of 10 Kansas families of four yielded the weekly food costs shown in the following table.

103	129	109	95	121
98	112	110	101	119

Based on these data, can we conclude that in 1994, weekly food costs for Kansas families of four were not normally distributed? Use $\alpha = 0.01$.

15.85 The U.S. National Center for Health Statistics compiles data on the length of stay by patients in short-term hospitals and publishes its findings in *Vital and Health Statistics.* A random sample of 21 patients yielded the following data on length of stay. The data are in days.

4	4	12	18	9	6	12
3	6	15	7	3	55	1
10	13	5	7	1	23	9

At the 5% significance level, do the data provide sufficient evidence to conclude that length of stay by patients in short-term hospitals is not normally distributed?

15.86 IQs measured on the Stanford Revision of the Binet-Simon Intelligence Scale have mean 100 points and standard deviation 16 points. Twenty-five randomly selected people are given the IQ test; the results are shown in the following table.

91	96	106	116	97
102	96	124	115	121
95	111	105	101	86
88	129	112	82	98
104	118	127	66	102

Can we conclude from these data, at the 10% significance level, that IQ measured on the Stanford Revision of the Binet-Simon Intelligence Scale is not normally distributed?

15.87 Gestation periods of humans are normally distributed with a mean of 266 days and a standard deviation of 16 days.

a. Simulate four random samples of 50 human gestation periods each. (*Note:* If you are using Minitab, refer to Example 6.2 on page 360 for an explanation of how to simulate a normally distributed variable.)

b. Perform a correlation test for normality on each sample in part (a). Use $\alpha = 0.05$.

c. Are the conclusions in part (b) what you expected? Explain your answer.

15.88 Desert Samaritan Hospital, located in Mesa, Arizona, keeps records of emergency-room traffic. From those records we find that the times between arriving patients have a special type of reverse J-shaped distribution called an *exponential* distribution. We also find that the mean time between arriving patients is approximately 8.7 minutes.

a. Simulate four random samples of 75 interarrival times each. (*Note:* If you are using Minitab, proceed in a similar way as for simulating a normal distribution, but begin with **Calc ➤ Random Data ➤ Exponential. . . .**)

b. Perform a correlation test for normality on each sample in part (a). Use $\alpha = 0.05$.

c. Are the conclusions in part (b) what you expected? Explain your answer.

CHAPTER REVIEW

You Should Be Able To

1. use and understand the formulas presented in this chapter.
2. state the assumptions for regression inferences.
3. understand the difference between the population regression line and a sample regression line.
4. estimate the regression parameters β_0, β_1, and σ.
5. determine the standard error of the estimate.
6. perform a residual analysis to check the assumptions for regression inferences.
7. perform a hypothesis test to decide whether the slope, β_1, of the population regression line is not 0 and hence whether x is useful for predicting y (i.e., whether the regression has utility).
8. obtain a confidence interval for β_1.
9. determine a point estimate and a confidence interval for the conditional mean of the response variable corresponding to a particular value of the predictor variable.
10. determine a predicted value and a prediction interval for the response variable corresponding to a particular value of the predictor variable.
11. understand the difference between the population correlation coefficient and a sample correlation coefficient.
12. perform a hypothesis test for a population linear correlation coefficient.

*13. perform a correlation test for normality.

*14. use the Minitab procedures covered in this chapter.

*15. interpret the output obtained from the application of the Minitab procedures discussed in this chapter.

Key Terms

conditional distribution, *872*
conditional mean, *872*
linearly uncorrelated, *912*
negatively linearly correlated, *912*
population linear correlation coefficient (ρ), *912*
population regression equation, *873,*
population regression line, *873*
positively linearly correlated, *912*
prediction interval, *903*
regression model, *873*
residual (e), *878*
residual plot, *879*
residual standard deviation, *876*
sampling distribution of the slope of the regression line, *890*
standard error of the estimate (s_e), *877*

REVIEW TEST

STATISTICAL CONCEPTS AND SKILLS

1. Suppose x and y are two variables of a population with x a predictor variable and y a response variable.

a. The distribution of all possible values of the response variable y corresponding to a particular value of the predictor variable x is called a _____ distribution of the response variable.

b. State the assumptions for regression inferences.

2. Suppose x and y are two variables of a population and that the assumptions for regression inferences are met with x as the predictor variable and y as the response variable.

a. What statistic is used to estimate the slope of the population regression line?

b. What statistic is used to estimate the y-intercept of the population regression line?

c. What statistic is used to estimate the common conditional standard deviation of the response variable corresponding to fixed values of the predictor variable?

3. What two plots did we use in this chapter to decide whether it is reasonable to presume that the assumptions for regression inferences are met by two variables of a population? What properties should those plots have?

4. Regarding analysis of residuals, decide in each case which assumption for regression inferences may be violated.

a. A residual plot—that is, a plot of the residuals against the values of the predictor variable—shows curvature.

b. A residual plot becomes wider with increasing values of the predictor variable.

c. A normal probability plot of the residuals shows extreme curvature.

d. A normal probability plot of the residuals shows outliers but is otherwise roughly linear.

5. Suppose that we perform a hypothesis test for the slope of the population regression line with null hypothesis H_0: $\beta_1 = 0$ and alternative hypothesis H_a: $\beta_1 \neq 0$. If we reject the null hypothesis, then what can we say about the utility of the regression equation for making predictions?

6. Identify three statistics each of which can be used as a basis for testing the utility of a regression.

7. For a particular value of a predictor variable, is there a difference between the predicted value of the response variable and the point estimate for the conditional mean of the response variable? Explain your answer.

8. Generally speaking, what is the difference between a confidence interval and a prediction interval?

9. Fill in the blank: $\bar{x}$ is to μ as r is to ____.

10. Identify the relationship between two variables and the terminology used to describe that relationship if **a.** $\rho > 0$. **b.** $\rho = 0$. **c.** $\rho < 0$.

11. Graduation rates and what influences them have become a concern in U.S. colleges and universities. *U.S. News and World Report*'s "College Guide" provides data on graduation rates for colleges and universities as a function of the percentage of freshmen in the top 10% of their high-school class, total spending per student, and student-to-faculty ratio. (Here *graduation rate* refers to the percentage of entering freshmen, attending full time, that graduate within 5 years.) A random sample of 10 universities gave the following data on student-to-faculty ratio (S/F ratio) and graduation rate (grad rate).

S/F ratio x	Grad rate y
16	45
20	55
17	70
19	50
22	47
17	46
17	50
17	66
10	26
18	60

Discuss what it would mean for the assumptions for regression inferences to be satisfied with student-to-faculty ratio as the predictor variable and graduation rate as the response variable.

12. Refer to Problem 11.

a. Determine the regression equation for the data.

b. Compute and interpret the standard error of the estimate. *(Note:* See Example 15.2, page 877.*)*

c. Presuming the assumptions for regression inferences are met, interpret your answer for part (b).

13. Refer to Problems 11 and 12. Perform a residual analysis to decide whether it is reasonable to consider the assumptions for regression inferences met by the variables student-to-faculty ratio and graduation rate.

In the remainder of this review test, we will presume that the variables student-to-faculty ratio and graduation rate satisfy the assumptions for regression inferences.

14. Refer to Problems 11 and 12.
  a. At the 5% significance level, do the data provide sufficient evidence to conclude that student-to-faculty ratio is useful as a predictor of graduation rate?
  b. Determine a 95% confidence interval for the slope, β_1, of the population regression line that relates graduation rate to student-to-faculty ratio. Interpret your answer.

15. Refer to Problems 11 and 12.
  a. Determine a point estimate for the mean graduation rate of all universities having a student-to-faculty ratio of 17.
  b. Determine a 95% confidence interval for the mean graduation rate of all universities having a student-to-faculty ratio of 17.
  c. Find the predicted graduation rate for a university having a student-to-faculty ratio of 17.
  d. Obtain a 95% prediction interval for the graduation rate of a university having a student-to-faculty ratio of 17.
  e. Explain why the prediction interval in part (d) is wider than the confidence interval in part (b).

16. Refer to Problem 11. At the 2.5% significance level, do the data provide sufficient evidence to conclude that the variables student-to-faculty ratio and graduation rate are positively linearly correlated?

***17.** In a correlation test for normality, the linear correlation coefficient is computed for the sample data and ______.

***18.** Each year, car makers perform mileage tests on their new car models and submit their results to the Environmental Protection Agency (EPA). The EPA then tests the vehicles to find whether the manufacturers' results are correct. In 1998, one company reported that a particular model averaged 29 miles per gallon (mpg) on the highway. Let's suppose the EPA tested 15 of the cars and obtained the following gas mileages.

27.3	31.2	29.4	31.6	28.6
30.9	29.7	28.5	27.8	27.3
25.9	28.8	28.9	27.8	27.6

At the 5% significance level, do the data provide sufficient evidence to conclude that gas mileages for this model are not normally distributed? Use a correlation test for normality.

USING TECHNOLOGY  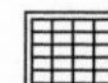 

19. Use Minitab or some other statistical software to carry out the residual analysis in Problem 13.

20. We used Minitab to obtain a residual plot and a normal probability plot of the residuals for the data in Problem 11, as seen in Printouts 15.10 and 15.11, respectively, on the next page. Do these plots suggest violations of any of the assumptions for regression inferences? Explain your answer.

21. Refer to Problem 11. Use Minitab or some other statistical software to
  a. obtain the sample regression equation.
  b. determine the standard error of the estimate.

22. Use Minitab or some other statistical software to
  a. carry out the hypothesis test in Problem 14(a).
  b. obtain the confidence interval in Problem 15(b) and the prediction interval in Problem 15(d).

23. Printout 15.12 on page 929 displays the computer output that results from applying Minitab's regression procedure to the data displayed in Problem 11. The confidence and prediction intervals are for a student-to-faculty ratio of 17. Use the output to
  a. determine the sample regression equation.
  b. obtain the standard error of the estimate, s_e.
  c. find the slope of the sample regression line.
  d. determine the estimated standard deviation of the slope of the sample regression line.
  e. obtain the value of the test statistic, t, for a hypothesis test to decide whether the slope of the population regression line is not 0.
  f. determine the P-value for the hypothesis test referred to in part (e).
  g. decide whether student-to-faculty ratio is useful for predicting graduation rate. Use $\alpha = 0.05$.

h. find and interpret a 95% confidence interval for the slope of the population regression line. *(Note:* Consult Table IV to find $t_{\alpha/2}$; all else required to obtain the confidence interval can be found in Printout 15.12.*)*

i. obtain a point estimate for the mean graduation rate of all universities having a student-to-faculty ratio of 17.

j. determine a 95% confidence interval for the mean graduation rate of all universities having a student-to-faculty ratio of 17.

k. find the predicted graduation rate for a university having a student-to-faculty ratio of 17.

l. obtain a 95% prediction interval for the graduation rate of a university having a student-to-faculty ratio of 17.

24. Use Minitab or some other statistical software to perform the test in Problem 16. *(Note:* Minitab provides the P-value only for a two-tailed test.*)*

***25.** Use Minitab or some other statistical software to carry out the hypothesis test in Problem 18.

PRINTOUT 15.10 Minitab residual plot for Problem 20

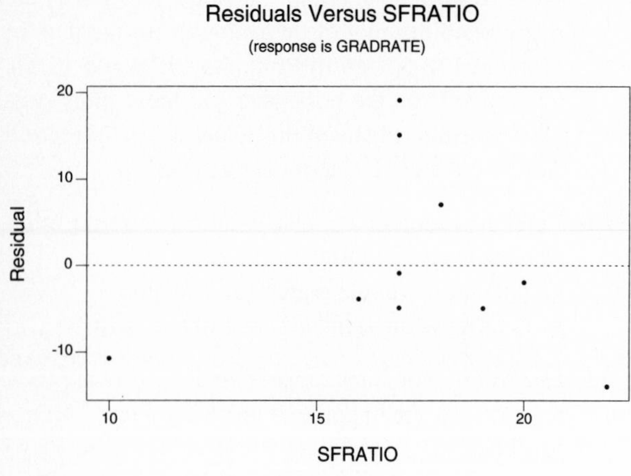

PRINTOUT 15.11 Minitab normal probability plot of the residuals for Problem 20

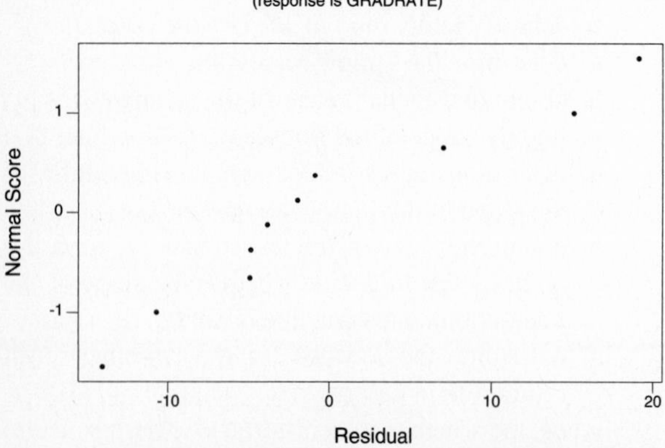

PRINTOUT 15.12 Minitab output for Problem 23

```
The regression equation is
GRADRATE = 16.4 + 2.03 SFRATIO

Predictor        Coef       StDev          T        P
Constant        16.45       21.15       0.78    0.459
SFRATIO         2.026       1.205       1.68    0.131

S = 11.31       R-Sq = 26.1%     R-Sq(adj) = 16.9%

Analysis of Variance

Source            DF          SS          MS         F        P
Regression         1       361.7       361.7      2.83    0.131
Residual Error     8      1022.8       127.9
Total              9      1384.5

Unusual Observations
Obs    SFRATIO   GRADRATE          Fit   StDev Fit     Residual    St Resid
  9       10.0      26.00        36.71        9.49       -10.71     -1.74 X

X denotes an observation whose X value gives it large influence.

Predicted Values

     Fit  StDev Fit          95.0% CI              95.0% PI
   50.89       3.59   (  42.60,   59.18)  (  23.53,   78.25)
```

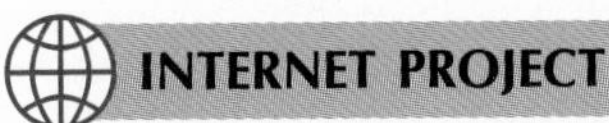

Assisted Reproductive Technology (Revisited)

In this Internet project, you will take a deeper look into the data on Assisted Reproductive Technology (ART). In your previous exploration of the data in Chapter 14, you saw that there is a linear relationship between the age of the prospective mother using ART and her chances for reproductive success. Now you will further analyze the results from that regression.

Another interesting relationship appears in the ART data. When eggs from a donor are implanted (i.e., when the implanted eggs are not the prospective mother's), the relationship between success rate and age changes in a surprising way. This part of the study will encourage you to think about the real situation that the analysis is attempting to explain.

URL for access to Internet Projects Page: `http://hepg.awl.com` *keyword:* Weiss

USING THE FOCUS DATABASE

In Chapter 1 we explained how to store the Focus database in a Minitab worksheet named `focus.mtw`. If you haven't already created that worksheet, follow the instructions on pages 54–55 to create it now.

The Focus database contains information on 500 randomly selected Arizona State University sophomores for seven different variables: sex, high-school GPA, SAT math score, cumulative GPA, SAT verbal score, age, and total hours. For these exercises, you should eliminate all cases (students) in which one or more of the four variables, cumulative GPA, high-school GPA, SAT math score, and SAT verbal score, equal 0. We will consider a regression analysis on cumulative GPA using high-school GPA as the predictor variable.

a. Determine the sample regression equation.

b. Perform a residual analysis to decide whether it appears reasonable to consider the assumptions for regression inferences satisfied.

Presuming now that the assumptions for regression inferences hold for the variables high-school GPA and cumulative GPA, solve the following problems.

c. Obtain and interpret the standard error of the estimate.

d. At the 5% significance level, do the data provide sufficient evidence to conclude that high-school GPA is useful for predicting cumulative GPA of sophomores at Arizona State University?

e. Determine a point estimate for the mean cumulative GPA of all sophomores at Arizona State University who had a high-school GPA of 3.0.

f. Find a 95% confidence interval for the mean cumulative GPA of all sophomores at Arizona State University who had a high-school GPA of 3.0.

g. Determine the predicted cumulative GPA of a sophomore at Arizona State University who had a high-school GPA of 3.0.

h. Obtain a 95% prediction interval for the cumulative GPA of a sophomore at Arizona State University who had a high-school GPA of 3.0.

i. Repeat parts (a)–(h) with SAT math score instead of high-school GPA as the predictor variable. For the estimation and prediction, use an SAT math score of 500.

j. Repeat parts (a)–(h) with SAT verbal score instead of high-school GPA as the predictor variable. For the estimation and prediction, use an SAT verbal score of 450.

CASE STUDY DISCUSSION

Fat Consumption and Prostate Cancer

At the beginning of this chapter (page 870), we presented data on fat consumption and prostate cancer death rate for nations of the world. In Chapter 14 you used those data to perform some descriptive regression and correlation analyses. Now you will employ those same data to carry out several inferential procedures in regression and correlation.

a. Obtain the sample regression equation using fat consumption as the predictor variable for prostate cancer death rate.

b. Perform a residual analysis to decide whether it appears reasonable to consider Assumptions 1–3 for regression inferences (page 873) satisfied by the variables fat consumption and prostate cancer death rate.

c. Obtain and interpret the standard error of the estimate.

d. At the 5% significance level, do the data provide sufficient evidence to conclude that fat consumption is useful for predicting prostate cancer death rate for nations of the world?

e. Find a point estimate for the mean prostate cancer death rate for nations with a fat consumption of 140 g per day.

f. Obtain a 95% confidence interval for the mean prostate cancer death rate for nations with a fat consumption of 140 g per day. Interpret your answer.

g. Determine the predicted prostate cancer death rate of a nation with a fat consumption of 140 g per day.

h. Find a 95% prediction interval for the prostate cancer death rate of a nation with a fat consumption of 140 g per day. Interpret your answer.

i. At the 5% significance level, do the data provide sufficient evidence to conclude that fat consumption and prostate cancer death rate are positively linearly correlated?

j. Use Minitab or some other statistical software to solve parts (a)–(i).

BIOGRAPHY SIR FRANCIS GALTON

Francis Galton was born on February 16, 1822, into a wealthy Quaker family of bankers and gunsmiths on his father's side and as a cousin of Charles Darwin on his mother's side. Although his IQ was estimated to be roughly 200, his formal education was unfinished.

He began training in medicine in Birmingham and London, but quit when, in his words, "A passion for travel seized me as if I had been a migratory bird." After a tour through Germany and southeastern Europe, he went to Trinity College in Cambridge to study mathematics. He left Cambridge in his third year, broken down from overwork. He recovered quickly, and resumed his medical studies in London. However, his father died before he had finished medical school and left to him, at 22 years old, "a sufficient fortune to make me independent of the medical profession."

Galton held no professional or academic positions; his experiments were conducted at his home or performed by friends. He was curious about almost everything, and carried out research in fields that included meteorology, biology, psychology, statistics, and genetics.

The origination of the concepts of regression and correlation, developed by Galton as tools for measuring the influence of heredity, are summed up in his work *Natural Inheritance.* He discovered regression during experiments with sweet-pea seeds to determine the law of inheritance of size. His other great discovery, correlation, was made while applying his techniques to the problem of measuring the degree of association between the sizes of two different body organs of an individual.

In his later years, Galton was associated with Karl Pearson, who became his champion and an extender of his ideas. Galton was knighted in 1909. He died in Haslemere, Surrey, England, in 1911.

CASE STUDY

HEAVY DRINKING AMONG COLLEGE STUDENTS

Excessive drinking causes many problems for society in general and for college students in particular. According to an article by T. M. O'Hare, published in the *Journal of Studies on Alcohol,* 81.5% of college students drink alcohol. Moreover, 37% of college students are moderate drinkers and 19% are heavy drinkers.

Professor Kate Carey of Syracuse University surveyed 78 college students from an introductory psychology class, all of whom were regular drinkers of alcohol. Her purpose was to identify interpersonal and intrapersonal situations associated with excessive drinking among college students and to detect those situations that differentiate heavy drinkers from light and moderate ones. She published her findings in the paper "Situational Determinants of Heavy Drinking Among College Students" (*Journal of Counseling Psychology,* 40, pp. 217–220).

To assess the frequency of excessive drinking in interpersonal and intrapersonal situations, Carey employed the short form of the Inventory of Drinking Situations (IDS). The table below provides the sample size and sample mean and standard deviation of IDS scores for each drinking category and situational context.

	Light drinkers ($n_1 = 16$)		Moderate drinkers ($n_2 = 47$)		Heavy drinkers ($n_3 = 15$)	
IDS subscale	$\bar{x}_1$	s_1	$\bar{x}_2$	s_2	$\bar{x}_3$	s_3
Interpersonal situations						
Conflict with others	1.23	0.27	1.53	0.49	1.79	0.49
Social pressure to drink	2.64	0.80	2.91	0.55	3.51	0.51
Pleasant times with others	2.21	0.67	2.53	0.51	3.03	0.38
Intrapersonal situations						
Unpleasant emotions	1.22	0.35	1.61	0.69	1.68	0.46
Physical discomfort	1.03	0.08	1.19	0.29	1.40	0.32
Pleasant emotions	2.09	0.73	2.61	0.58	3.03	0.30
Testing personal control	1.52	0.74	1.56	0.56	1.53	0.48
Urges and temptations	1.80	0.56	1.96	0.51	2.33	0.58

At the end of this chapter, you will analyze the data to decide for each IDS category whether a difference exists in mean IDS scores among the three drinker categories.

Analysis of Variance (ANOVA) 16

GENERAL OBJECTIVES

In Chapter 10 we studied inferential methods for comparing the means of two populations. Now we will study **analysis of variance,** or **ANOVA,** which provides methods for comparing the means of more than two populations. For instance, we could use ANOVA to compare the mean energy consumption by households among the four U.S. regions. Just as there are several different procedures for comparing the means of two populations, there are several different ANOVA procedures.

First, in preparation for our study of ANOVA, we will examine the F-distribution. Then we will present one-way analysis of variance, the simplest type of ANOVA. We will study the logic behind one-way ANOVA in Section 16.2 and the one-way ANOVA procedure in Section 16.3.

Next we will consider *multiple comparisons.* If we have conducted a one-way ANOVA and decided that the population means are not all equal, we may then want to know which means are different, which mean is largest, and, in general, the relation among all the means. We use multiple comparison methods for tackling these types of questions.

The last section presents the Kruskal–Wallis test. This hypothesis-testing procedure is a generalization of the Mann–Whitney test to more than two populations and provides a nonparametric alternative to one-way ANOVA.

CHAPTER OUTLINE

16.1 THE *F*-DISTRIBUTION

Analysis-of-variance procedures rely on a distribution called the ***F*-distribution,** named in honor of Sir Ronald Fisher.[†] A variable is said to have an F-distribution if its distribution has the shape of a special type of right-skewed curve, called an ***F*-curve.**

There are infinitely many F-distributions, and we identify the F-distribution (and F-curve) in question by stating its number of degrees of freedom, just as we did for t-distributions and chi-square distributions. But an F-distribution has two numbers of degrees of freedom instead of one. Figure 16.1 depicts two different F-curves; one has df = (10, 2), and the other has df = (9, 50).

FIGURE 16.1
Two different *F*-curves

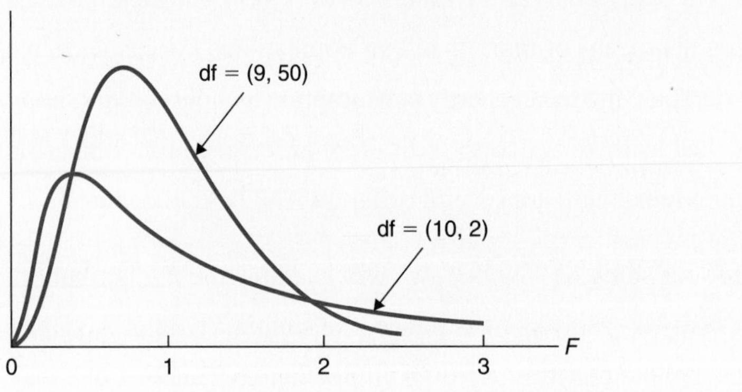

The first number of degrees of freedom for an F-curve is called the **degrees of freedom for the numerator** and the second the **degrees of freedom for the denominator.** (The reason for this terminology will become clear in Section 16.2.) Thus for the F-curve in Fig. 16.1 with df = (10, 2), we have

$$\text{df} = (10,\ 2)$$

Degrees of freedom for the numerator ↗ ↖ Degrees of freedom for the denominator

Some basic properties of F-curves are presented in Key Fact 16.1.

[†] See the biography at the end of this chapter for more on Fisher.

KEY FACT 16.1 BASIC PROPERTIES OF F-CURVES

Property 1: The total area under an F-curve is equal to 1.

Property 2: An F-curve starts at 0 on the horizontal axis and extends indefinitely to the right, approaching, but never touching, the horizontal axis as it does so.

Property 3: An F-curve is right skewed.

Using the F-Table

Percentages (and probabilities) for a variable having an F-distribution are equal to areas under its associated F-curve. To perform an ANOVA test, we need to know how to find the F-value having a specified area to its right. The symbol $\boldsymbol{F_\alpha}$ is used to denote the F-value having area α to its right.

Table IX in Appendix A provides F-values corresponding to several areas for various degrees of freedom. The degrees of freedom for the denominator (dfd) are displayed in the outside columns of the table; the values of α in the next columns; and the degrees of freedom for the numerator (dfn) along the top. Example 16.1 shows how to use Table IX.

EXAMPLE 16.1 FINDING THE F-VALUE HAVING A SPECIFIED AREA TO ITS RIGHT

For an F-curve with df $=$ (4, 12), find $F_{0.05}$; that is, find the F-value having area 0.05 to its right, as shown in Fig. 16.2(a).

FIGURE 16.2 Finding the F-value having area 0.05 to its right

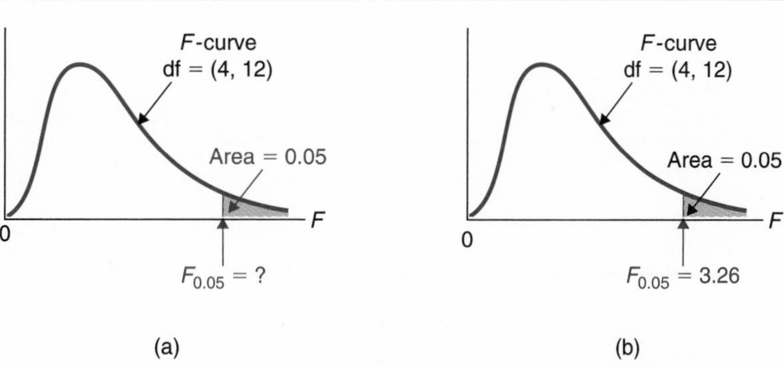

SOLUTION To obtain the F-value in question, we use Table IX. In this case $\alpha = 0.05$, the degrees of freedom for the numerator is 4, and the degrees of freedom for the denominator is 12. Thus we first go down the outside columns to the row labeled "12." Next we concentrate on the row for α labeled 0.05. Then we go

 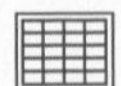 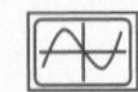

across that row until we are under the column headed "4." The number in the body of the table there, 3.26, is the required F-value; that is, for an F-curve with df = (4, 12), the F-value having area 0.05 to its right is 3.26: $F_{0.05} = 3.26$, as seen in Fig. 16.2(b).

EXERCISES 16.1

STATISTICAL CONCEPTS AND SKILLS

16.1 How do we identify an F-distribution and its corresponding F-curve?

16.2 How many degrees of freedom does an F-curve have? What are those degrees of freedom called?

16.3 What symbol is used to denote the F-value having area 0.05 to its right? 0.025 to its right? α to its right?

16.4 Using the F_α-notation, identify the F-value having area 0.975 to its left.

16.5 An F-curve has df = (12, 7). What is the number of degrees of freedom for the
a. numerator?
b. denominator?

16.6 An F-curve has df = (8, 19). What is the number of degrees of freedom for the
a. denominator?
b. numerator?

In Exercises 16.7–16.10, use Table IX in Appendix A to find the required F-values. Illustrate your work with graphs similar to Fig. 16.2 on page 935.

16.7 An F-curve has df = (24, 30). In each case find the F-value having the specified area to its right.
a. 0.05 **b.** 0.01 **c.** 0.025

16.8 An F-curve has df = (12, 5). In each case find the F-value having the specified area to its right.
a. 0.01 **b.** 0.05 **c.** 0.005

16.9 For an F-curve with df = (20, 21), find
a. $F_{0.01}$. **b.** $F_{0.05}$. **c.** $F_{0.10}$.

16.10 For an F-curve with df = (6, 10), find
a. $F_{0.05}$. **b.** $F_{0.01}$. **c.** $F_{0.025}$

EXTENDING THE CONCEPTS AND SKILLS

16.11 Refer to Table IX in Appendix A. Because of space restrictions, the numbers of degrees of freedom are not consecutive. For instance, the degrees of freedom for the numerator skips from 24 to 30. If you had only Table IX to work with and you needed to find $F_{0.05}$ for df = (25, 20), how would you do it?

16.2 ONE-WAY ANOVA: THE LOGIC

In Chapter 10 we learned how to compare two populations means, that is, the means of a single variable for two different populations. We examined a variety of methods for making such comparisons, one being the pooled t-procedure.

Analysis of variance (ANOVA) provides methods for comparing several population means, that is, the means of a single variable for several populations. In this section and the next, we will study the simplest kind of ANOVA, **one-way analysis of variance.** This type of ANOVA is called *one-way* analysis of variance because it compares the means of a variable for populations that result from a classification by *one* other variable, called the **factor.** The possible values of the factor are referred to as the **levels** of the factor.

For example, suppose we want to compare the mean energy consumption by households among the four regions of the United States. The variable under consideration here is "energy consumption" and there are four populations—households in the Northeast, Midwest, South, and West. The four populations result from classifying households in the United States by the factor "region," whose levels are Northeast, Midwest, South, and West.

One-way analysis of variance is the generalization to more than two populations of the pooled t-procedure. As in the pooled t-procedure, we make the following assumptions.

KEY FACT 16.2 ASSUMPTIONS FOR ONE-WAY ANOVA

1. ***Independent samples:*** The samples taken from the populations under consideration are independent of one another.
2. ***Normal populations:*** For each population, the variable under consideration is normally distributed.
3. ***Equal standard deviations:*** The standard deviations of the variable under consideration are the same for all the populations.

One-way ANOVA has the same robustness properties as those of the pooled t-procedure. The independent-samples assumption (Assumption 1) is essential; the samples must be independent or the procedure does not apply. One-way ANOVA is robust to moderate violations of the normality assumption (Assumption 2). It is also reasonably robust to moderate violations of the equal-standard-deviations assumption (Assumption 3), provided the sample sizes are roughly equal.

Generally, normal probability plots are effective in detecting gross violations of the normality assumption. The equal-standard-deviations assumption is usually more difficult to check. As a rule of thumb, we consider that assumption satisfied if *the ratio of the largest to the smallest sample standard deviation is less than 2.* For convenience, we'll call this rule of thumb the **rule of 2.**

Additionally, the normality and equal-standard-deviations assumptions can be assessed by performing a residual analysis, in a way similar to what is done in regression. See Section 15.1 for a discussion of the analysis of residuals.

In ANOVA the **residual** of an observation is the difference between the observation and the mean of the sample containing it. If the normality and equal-standard-deviations assumptions are met, then a normal probability plot of (all) the residuals should be roughly linear and a plot of the residuals against the sample means should fall roughly in a horizontal band centered and symmetric about the horizontal axis.

The Logic Behind One-Way ANOVA

The reason for the word *variance* in *analysis of variance* is that the procedure for comparing the means involves analyzing the variation in the sample data. To see how this works, suppose independent random samples are taken from two populations, Populations 1 and 2, having means μ_1 and μ_2. Further suppose the means of the two samples are $\bar{x}_1 = 20$ and $\bar{x}_2 = 25$. Can we reasonably conclude from these statistics that $\mu_1 \neq \mu_2$, that is, that the population means are different? To answer this question, we must consider the variation within the samples.

Suppose, for instance, that the sample data are as depicted in Table 16.1 and Fig. 16.3.

TABLE 16.1 Sample data from Populations 1 and 2

Sample from Population 1	21	37	11	20	8	23
Sample from Population 2	24	31	29	40	9	17

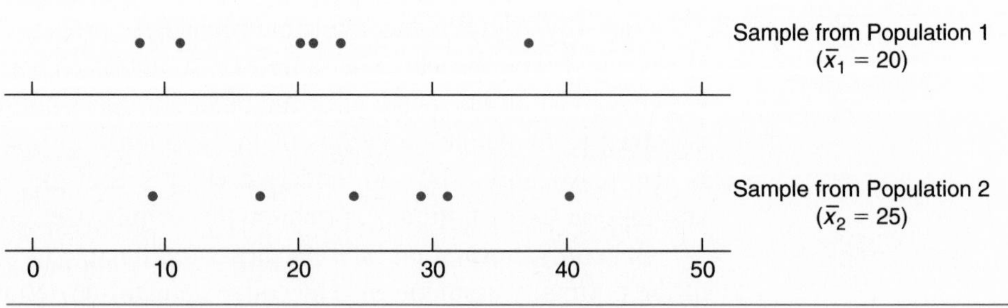

FIGURE 16.3 Dotplots for sample data in Table 16.1

For these two samples, $\bar{x}_1 = 20$ and $\bar{x}_2 = 25$. But we cannot infer that $\mu_1 \neq \mu_2$ because it is not clear whether the difference between the sample means is due to a difference between the population means or to the variation within the populations. In other words, because the variation between the sample means is not large relative to the variation within the samples, we cannot conclude that $\mu_1 \neq \mu_2$.

On the other hand, suppose that the sample data are as depicted in Table 16.2 and Fig. 16.4.

TABLE 16.2
Sample data from Populations 1 and 2

Sample from Population 1	21	21	20	18	20	20
Sample from Population 2	25	28	25	24	24	24

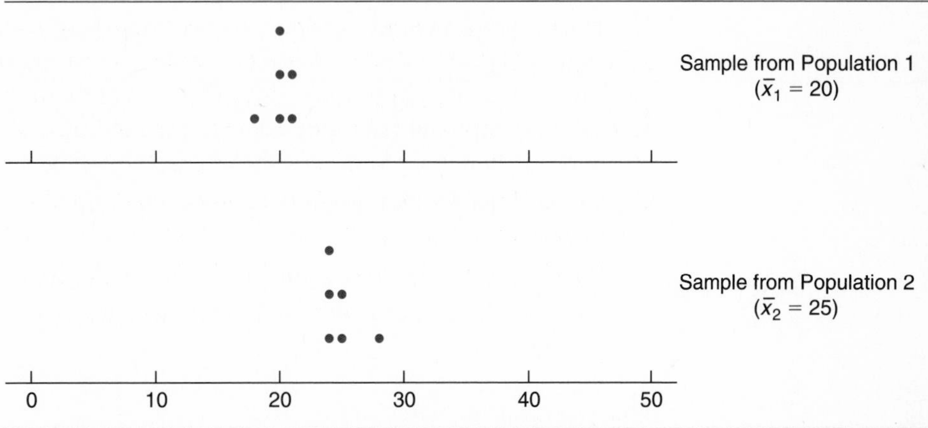

FIGURE 16.4
Dotplots for sample data in Table 16.2

Again, for these two samples, $\bar{x}_1 = 20$ and $\bar{x}_2 = 25$. But this time we *can* infer that $\mu_1 \neq \mu_2$ because it seems clear that the difference between the sample means is due to a difference between the population means and not to the variation within the populations. In other words, because the variation between the sample means is large relative to the variation within the samples, we can conclude that $\mu_1 \neq \mu_2$.

The preceding two illustrations reveal the basic idea for performing a one-way analysis of variance to compare the means of several populations: (1) take independent random samples from the populations; (2) compute the sample means; and (3) if the variation among the sample means is large relative to the variation within the samples, conclude that the means of the populations are not all equal.

To make this process precise, we need quantitative measures of the variation among the sample means and the variation within the samples; we also need an objective method for deciding whether the variation among the sample means is large relative to the variation within the samples. Example 16.2 addresses these two issues.

EXAMPLE 16.2 INTRODUCES ONE-WAY ANOVA

The U.S. Energy Information Administration gathers data on residential energy consumption and expenditures and publishes its findings in *Residential Energy Consumption Survey: Consumption and Expenditures.* Suppose we want to decide whether a difference exists in mean annual energy consumption by households among the four U.S. regions. Let μ_1, μ_2, μ_3, and μ_4 denote last year's mean energy consumptions by households in the Northeast, Midwest, South, and West, respectively. Then the hypotheses to be tested are

$$H_0\colon \mu_1 = \mu_2 = \mu_3 = \mu_4 \text{ (mean energy consumptions are all equal)}$$
$$H_a\colon \text{Not all the means are equal.}$$

The basic strategy for carrying out this hypothesis test follows the three steps just mentioned:

1. Independently and randomly take samples of households in the four U.S. regions.
2. Compute last year's mean energy consumptions, $\bar{x}_1$, $\bar{x}_2$, $\bar{x}_3$, and $\bar{x}_4$, of the four samples.
3. Reject the null hypothesis if the variation among the sample means is large relative to the variation within the samples; otherwise, do not reject the null hypothesis.

This process is depicted in Fig. 16.5.

FIGURE 16.5
Process for comparing four population means

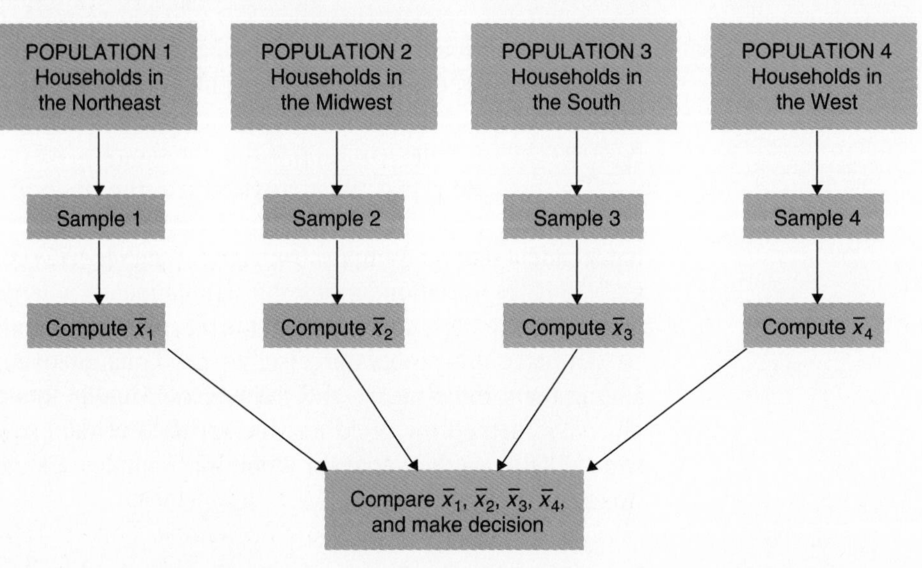

Steps 1 and 2 entail obtaining the sample data and computing the sample means. Suppose the results of those steps are as shown in Table 16.3, where the data are displayed to the nearest 10 million BTU.

TABLE 16.3
Samples and their means of last year's energy consumptions for households in the four U.S. regions

Northeast	Midwest	South	West
15	17	11	10
10	12	7	12
13	18	9	8
14	13	13	7
13	15		9
	12		
13.0	14.5	10.0	9.2

← *Means*

Step 3 involves comparing the variation among the four sample means, shown at the bottom of Table 16.3, to the variation within the samples. Let's first consider the variation among the sample means.

In hypothesis tests for two population means, we measure the variation between the two sample means by calculating their difference, $\bar{x}_1 - \bar{x}_2$. When more than two populations are involved, as in this problem, we cannot measure the variation among the sample means simply by taking a difference. However, we can measure that variation by computing the standard deviation or variance of the sample means or, for that matter, by computing any descriptive statistic that measures variation.

In one-way ANOVA, we measure the variation among the sample means by a weighted average of their squared deviations about the mean, $\bar{x}$, of all the sample data. That measure of variation is called the **treatment mean square, *MSTR*,** and is defined by

$$MSTR = \frac{SSTR}{k-1},$$

where k denotes the number of populations being sampled and

$$SSTR = n_1(\bar{x}_1 - \bar{x})^2 + n_2(\bar{x}_2 - \bar{x})^2 + \cdots + n_k(\bar{x}_k - \bar{x})^2.$$

The quantity ***SSTR*** is called the **treatment sum of squares.**

MSTR is similar to the sample variance of the sample means. In fact, if the sample sizes are all identical, then *MSTR* is equal to that common sample size times the sample variance of the sample means.

Let's determine *MSTR* for the sample data in Table 16.3. We have $k = 4$, $n_1 = 5$, $n_2 = 6$, $n_3 = 4$, $n_4 = 5$, $\bar{x}_1 = 13.0$, $\bar{x}_2 = 14.5$, $\bar{x}_3 = 10.0$, and $\bar{x}_4 = 9.2$. To obtain the overall mean, $\bar{x}$, we need to divide the sum of all the observations in Table 16.3 by the total number of observations:

$$\bar{x} = \frac{\Sigma x}{n} = \frac{15 + 10 + 13 + \cdots + 7 + 9}{20} = \frac{238}{20} = 11.9.$$

Therefore

$$\begin{aligned} SSTR &= n_1(\bar{x}_1 - \bar{x})^2 + n_2(\bar{x}_2 - \bar{x})^2 + n_3(\bar{x}_3 - \bar{x})^2 + n_4(\bar{x}_4 - \bar{x})^2 \\ &= 5(13.0 - 11.9)^2 + 6(14.5 - 11.9)^2 + 4(10.0 - 11.9)^2 + 5(9.2 - 11.9)^2 \\ &= 97.5, \end{aligned}$$

and so

$$MSTR = \frac{SSTR}{k-1} = \frac{97.5}{4-1} = 32.5.$$

This is our measure of variation among the four sample means shown at the bottom of Table 16.3.

Next we must obtain a measure of variation within the samples. This measure is the pooled estimate of the common population variance, σ^2. It is called the **error mean square, *MSE*,** and is defined by

$$MSE = \frac{SSE}{n-k},$$

where n denotes the total number of observations and

$$SSE = (n_1 - 1)s_1^2 + (n_2 - 1)s_2^2 + \cdots + (n_k - 1)s_k^2.$$

The quantity SSE is called the **error sum of squares.**[†‡]

For the sample data in Table 16.3, we have $k = 4$, $n_1 = 5$, $n_2 = 6$, $n_3 = 4$, $n_4 = 5$, and $n = 20$. Computing the variance of each sample in Table 16.3, we find that $s_1^2 = 3.5$, $s_2^2 = 6.7$, $s_3^2 = 6.\overline{6}$, and $s_4^2 = 3.7$. Consequently,

$$\begin{aligned} SSE &= (n_1 - 1)s_1^2 + (n_2 - 1)s_2^2 + (n_3 - 1)s_3^2 + (n_4 - 1)s_4^2 \\ &= (5-1)\cdot 3.5 + (6-1)\cdot 6.7 + (4-1)\cdot 6.\overline{6} + (5-1)\cdot 3.7 = 82.3, \end{aligned}$$

and so

$$MSE = \frac{SSE}{n-k} = \frac{82.3}{20-4} = 5.144.$$

This is our measure of variation within the samples.

Finally, we must compare the variation among the sample means, $MSTR$, to the variation within the samples, MSE. To accomplish that, we use the statistic $F = MSTR/MSE$, which we refer to as the ***F*-statistic.** Large values of F indicate that the variation among the sample means is large relative to the variation within the samples and hence that the null hypothesis of equal population means should be rejected.

---

† The terms **treatment** and **error** arose from the fact that many ANOVA techniques were first developed to analyze agricultural experiments. In any case the treatments refer to the different populations and the errors pertain to the variation within the populations.

‡ For two populations (i.e., $k = 2$), MSE is the pooled variance, s_p^2, defined in Section 10.2 on page 601.

For the energy-consumption data, we have just seen that $MSTR = 32.5$ and $MSE = 5.144$. Thus the value of the F-statistic is

$$F = \frac{MSTR}{MSE} = \frac{32.5}{5.144} = 6.32.$$

Is this value of F large enough to conclude that the null hypothesis of equal population means is false? To answer that question, we need to know the distribution of the F-statistic. We will discuss that in the next section and then return to complete the hypothesis test considered in this example. ◨

EXERCISES 16.2

STATISTICAL CONCEPTS AND SKILLS

16.12 State the three assumptions required for one-way ANOVA. How crucial are these assumptions?

16.13 One-way ANOVA is a procedure for comparing the means of several populations. It is the generalization of what procedure for comparing the means of two populations?

16.14 If we define $s = \sqrt{MSE}$, then of which parameter is s an estimate?

16.15 Explain the reason for the word *variance* in the phrase *analysis of variance.*

16.16 The null and alternative hypotheses for a one-way ANOVA test are

$$H_0\colon \mu_1 = \mu_2 = \cdots = \mu_k$$
$$H_a\colon \text{ Not all means are equal.}$$

Suppose in reality that the null hypothesis is false. Does this mean that no two of the populations have the same mean? If not, what does it mean?

16.17 In one-way ANOVA, identify the statistic used

a. as a measure of variation among the sample means.

b. as a measure of variation within the samples.

c. to compare the variation among the sample means to the variation within the samples.

16.18 Explain in your own words the logic behind one-way ANOVA.

16.19 What is the significance of the term *one-way* in *one-way ANOVA*?

16.20 Figure 16.6 shows side-by-side boxplots of independent samples from three normally distributed populations having equal standard deviations. Based on these boxplots, would you be inclined to reject the null hypothesis of equal population means? Explain your answer.

FIGURE 16.6 Side-by-side boxplots for Exercise 16.20

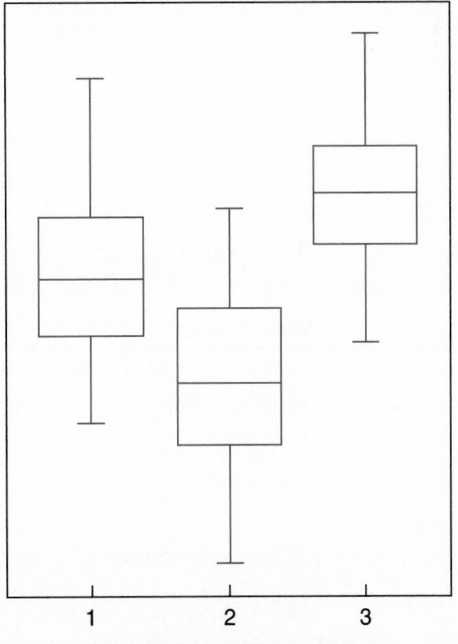

16.21 Figure 16.7 shows side-by-side boxplots of independent samples from three normally distributed populations having equal standard deviations. Based on these boxplots, would you be inclined to reject the null hypothesis of equal population means? Why?

FIGURE 16.7 Side-by-side boxplots for Exercise 16.21

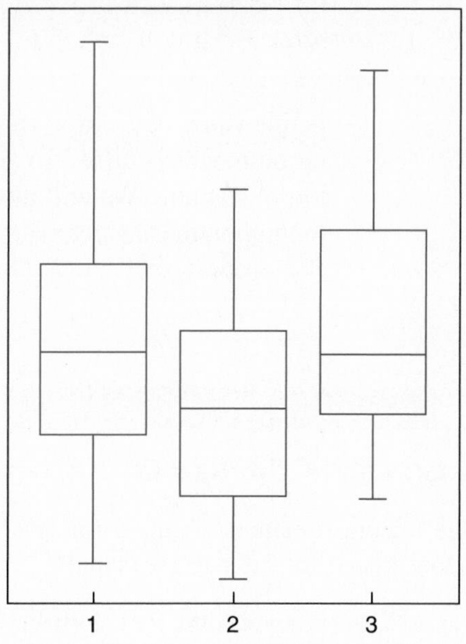

EXTENDING THE CONCEPTS AND SKILLS

16.22 Show that for two populations (i.e., $k = 2$), $MSE = s_p^2$, where s_p^2 is the pooled variance defined in Section 10.2 on page 601. Conclude that $\sqrt{MSE}$ is the pooled sample standard deviation, s_p.

16.23 Suppose that the variable under consideration is normally distributed on each of two populations and that the population standard deviations are equal. Suppose we want to perform a hypothesis test to decide whether the populations have different means, that is, whether $\mu_1 \neq \mu_2$. If independent samples are used, identify two hypothesis-testing procedures that can be employed to carry out the test.

16.3 ONE-WAY ANOVA: THE PROCEDURE

In this section we will present a step-by-step procedure for performing a one-way ANOVA to compare the means of several populations. To begin, we need to identify the distribution of the variable $F = MSTR/MSE$, introduced at the end of the preceding section.

KEY FACT 16.3 DISTRIBUTION OF THE *F*-STATISTIC FOR ONE-WAY ANOVA

Suppose the variable under consideration is normally distributed on each of k populations and that the population standard deviations are equal. Then, for independent samples from the k populations, the variable

$$F = \frac{MSTR}{MSE}$$

has the F-distribution with $\text{df} = (k - 1, n - k)$ if the null hypothesis of equal population means is true. Here n denotes the total number of observations.

We have now covered all the elements required to formulate a procedure for performing a one-way analysis of variance. Before presenting that procedure, however, it will be helpful to consider two additional concepts.

One-Way ANOVA Identity

First we define another sum of squares. That sum of squares provides a measure of total variation among all the sample data. It is called the **total sum of squares, *SST*,** and is defined by

$$SST = \Sigma(x - \overline{x})^2,$$

where the sum extends over all n observations. If we divide SST by $n - 1$, then we get the sample variance of all the observations. So SST really is a measure of total variation. For the energy-consumption data in Table 16.3 on page 941, $\overline{x} = 11.9$, and therefore

$$\begin{aligned} SST &= \Sigma(x - \overline{x})^2 = (15 - 11.9)^2 + (10 - 11.9)^2 + \cdots + (9 - 11.9)^2 \\ &= 9.61 + 3.61 + \cdots + 8.41 = 179.8. \end{aligned}$$

In Section 16.2 we found that for the energy-consumption data, $SSTR = 97.5$ and $SSE = 82.3$. Since $179.8 = 97.5 + 82.3$, we see that $SST = SSTR + SSE$. This equation is always true and is called the **one-way ANOVA identity.**

KEY FACT 16.4 ONE-WAY ANOVA IDENTITY

The total sum of squares equals the treatment sum of squares plus the error sum of squares; that is, $SST = SSTR + SSE$.

The one-way ANOVA identity shows that we can partition the total variation among all the sample data into a component representing variation among the sample means and a component representing variation within the samples. We can picture this partitioning as in Fig. 16.8.

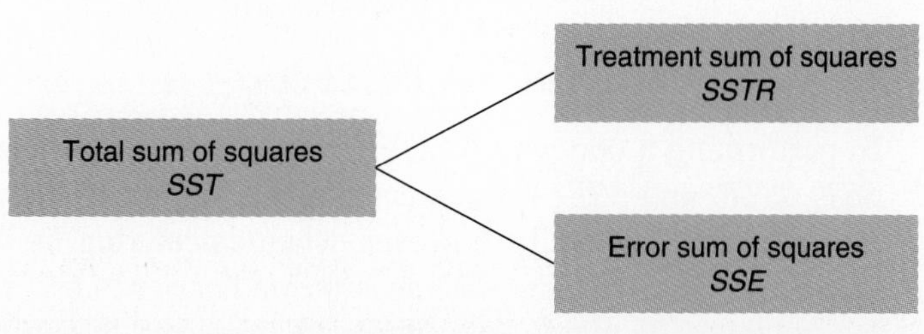

FIGURE 16.8 Partitioning of the total sum of squares into the treatment sum of squares and the error sum of squares

One-Way ANOVA Tables

Next we discuss **one-way ANOVA tables.** One-way ANOVA tables are useful for organizing and summarizing the quantities required for performing a one-way analysis of variance. The general format of a one-way ANOVA table is as depicted in Table 16.4.

TABLE 16.4 ANOVA table format for a one-way analysis of variance

Source	df	SS	$MS = SS/\text{df}$	F-statistic
Treatment	$k-1$	$SSTR$	$MSTR = \frac{SSTR}{k-1}$	$F = \frac{MSTR}{MSE}$
Error	$n-k$	SSE	$MSE = \frac{SSE}{n-k}$	
Total	$n-1$	SST		

For the energy-consumption data in Table 16.3, we have already computed all quantities appearing in the one-way ANOVA table. Table 16.5 displays the one-way ANOVA table for that data.

TABLE 16.5 One-way ANOVA table for the energy-consumption data

Source	df	SS	$MS = SS/\text{df}$	F-statistic
Treatment	3	97.5	32.500	6.32
Error	16	82.3	5.144	
Total	19	179.8		

The One-Way ANOVA Procedure

In performing a one-way ANOVA, we need to obtain the three sums of squares, SST, $SSTR$, and SSE. We can accomplish this using the defining formulas introduced earlier. Generally, however, when calculating by hand from the raw data, computing formulas are more accurate and easier to use. Both the defining formulas and their computing equivalents are presented in Formula 16.1.

FORMULA 16.1 SUMS OF SQUARES IN ONE-WAY ANOVA

For a one-way ANOVA of k population means, the defining and computing formulas for the three sums of squares are displayed in the table below.

Sum of square	Defining formula	Computing formula
Total, SST	$\Sigma(x - \bar{x})^2$	$\Sigma x^2 - (\Sigma x)^2/n$
Treatment, $SSTR$	$\Sigma n_j(\bar{x}_j - \bar{x})^2$	$\Sigma(T_j^2/n_j) - (\Sigma x)^2/n$
Error, SSE	$\Sigma(n_j - 1)s_j^2$	$SST - SSTR$

In the above table, we have employed the following notation:

n = total number of observations
$\bar{x}$ = mean of all n observations

and, for $j = 1, 2, \ldots, k$,

n_j = size of sample from Population j
$\bar{x}_j$ = mean of sample from Population j
s_j^2 = variance of sample from Population j
T_j = sum of sample data from Population j

Keep the following facts in mind when using Formula 16.1:

- Only two of the three sums of squares need ever be calculated; the remaining one can always be determined from the other two by employing the one-way ANOVA identity, Key Fact 16.4 on page 945.
- Summations involving no subscripted variables are over all n observations; those involving subscripts are over the k populations.
- When using the computing formulas, it is most efficient to calculate the sum of all n observations by employing the formula $\Sigma x = T_1 + T_2 + \cdots + T_k$.

We now present a step-by-step method that can be used to perform a one-way analysis of variance. Note that the hypothesis test is always right-tailed since the null hypothesis is rejected only when the test statistic, F, is too large.

PROCEDURE 16.1

THE ONE-WAY ANOVA TEST FOR k POPULATION MEANS

ASSUMPTIONS

1. Independent samples
2. Normal populations
3. Equal population standard deviations

Step 1 The null and alternative hypotheses are

$$H_0: \mu_1 = \mu_2 = \cdots = \mu_k$$
$$H_a: \text{Not all the means are equal.}$$

Step 2 Decide on the significance level, α.

Step 3 The critical value is F_α, with df $= (k-1, n-k)$, where n is the total number of observations. Use Table IX to find the critical value.

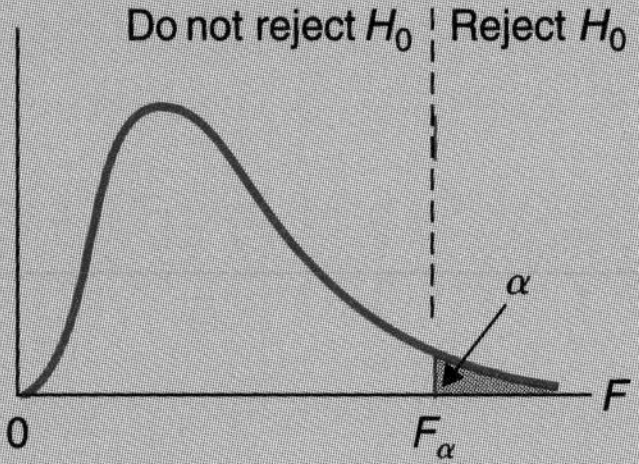

Step 4 Obtain the three sums of squares, *SST*, *SSTR*, and *SSE*.

Step 5 Construct the one-way ANOVA table:

Source	df	*SS*	*MS* = *SS*/df	*F*-statistic
Treatment	$k-1$	*SSTR*	$MSTR = \dfrac{SSTR}{k-1}$	$F = \dfrac{MSTR}{MSE}$
Error	$n-k$	*SSE*	$MSE = \dfrac{SSE}{n-k}$	
Total	$n-1$	*SST*		

Step 6 If the value of the F-statistic falls in the rejection region, reject H_0; otherwise, do not reject H_0.

Step 7 State the conclusion in words.

EXAMPLE 16.3 ILLUSTRATES PROCEDURE 16.1

Recall that independent random samples of households in the four U.S. regions yielded the data on last year's energy consumptions shown in Table 16.6. At the 5% significance level, do the data provide sufficient evidence to conclude that a difference exists in last year's mean energy consumption by households among the four U.S. regions?

TABLE 16.6 Last year's energy consumptions for samples of households in the four U.S. regions

Northeast	Midwest	South	West
15	17	11	10
10	12	7	12
13	18	9	8
14	13	13	7
13	15		9
	12		

SOLUTION First we check the three conditions required for performing a one-way ANOVA. Since the samples are independent, Assumption 1 is satisfied.

We next address the question of normality. Normal probability plots (not shown) of the four samples in Table 16.6 reveal no outliers and are roughly linear, thus indicating no gross violations of the normality assumption; so we can consider Assumption 2 satisfied.

The sample standard deviations of the four samples in Table 16.6 are, respectively, 1.87, 2.59, 2.58, and 1.92. We see that the ratio of the largest to the smallest standard deviation is $\frac{2.59}{1.87} = 1.39 < 2$. Therefore, by the rule of 2, we can consider Assumption 3 satisfied. A residual analysis further attests to it being reasonable to consider Assumptions 2 and 3 satisfied.

The preceding three paragraphs suggest that the one-way ANOVA procedure can be used to carry out the hypothesis test. We proceed as follows.

Step 1 *State the null and alternative hypotheses.*

Let μ_1, μ_2, μ_3, and μ_4 denote last year's mean energy consumptions for households in the Northeast, Midwest, South, and West, respectively. Then the null and alternative hypotheses are

H_0: $\mu_1 = \mu_2 = \mu_3 = \mu_4$ (mean energy consumptions are equal)

H_a: Not all the means are equal.

Step 2 *Decide on the significance level, α.*

We are to perform the test at the 5% significance level; thus $\alpha = 0.05$.

Step 3 *The critical value is F_α, with df $= (k-1, n-k)$, where n is the total number of observations.*

From Step 2, $\alpha = 0.05$. Also, as we see from Table 16.6, the number of populations under consideration is four ($k = 4$) and the total number of observations is 20 ($n = 20$). Hence df $= (k-1, n-k) = (4-1, 20-4) = (3, 16)$. Consulting Table IX, we find that the critical value is $F_\alpha = F_{0.05} = 3.24$, as seen in Fig. 16.9.

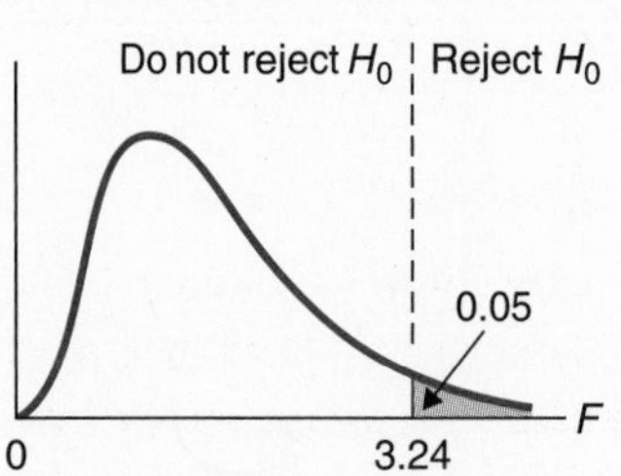

FIGURE 16.9 Criterion for deciding whether or not to reject the null hypothesis

Step 4 *Obtain the three sums of squares, SST, SSTR, and SSE.*

Although we obtained these earlier using the defining formulas, we will determine them again to illustrate the computing formulas. Referring to Formula 16.1 on page 947 and Table 16.6, we find that

$$k = 4$$

$$n_1 = 5 \qquad n_2 = 6 \qquad n_3 = 4 \qquad n_4 = 5$$
$$T_1 = 65 \qquad T_2 = 87 \qquad T_3 = 40 \qquad T_4 = 46$$

and

$$n = 5 + 6 + 4 + 5 = 20$$
$$\Sigma x = 65 + 87 + 40 + 46 = 238.$$

Summing the squares of all the data in Table 16.6 yields

$$\Sigma x^2 = 15^2 + 10^2 + 13^2 + \cdots + 7^2 + 9^2 = 3012.$$

Consequently,

$$SST = \Sigma x^2 - (\Sigma x)^2/n = 3012 - (238)^2/20 = 3012 - 2832.2 = 179.8,$$

$$\begin{aligned} SSTR &= \Sigma(T_j^2/n_j) - (\Sigma x)^2/n \\ &= 65^2/5 + 87^2/6 + 40^2/4 + 46^2/5 - (238)^2/20 \\ &= 2929.7 - 2832.2 = 97.5, \end{aligned}$$

and

$$SSE = SST - SSTR = 179.8 - 97.5 = 82.3.$$

Step 5 *Construct the one-way ANOVA table.*

The one-way ANOVA table for the energy-consumption data was constructed in Table 16.5 on page 946.

Step 6 *If the value of the F-statistic falls in the rejection region, reject H_0; otherwise, do not reject H_0.*

As we see from Table 16.5, $F = 6.32$. A glance at Fig. 16.9 shows that this value falls in the rejection region. Thus we reject H_0.

Step 7 *State the conclusion in words.*

The test results are statistically significant at the 5% level; that is, at the 5% significance level, the data provide sufficient evidence to conclude that a difference exists in last year's mean energy consumption by households among the four U.S. regions. It appears that at least two of the regions have different mean energy consumptions.

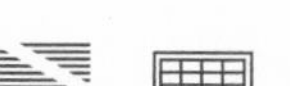

Using the *P*-Value Approach

The P-value approach to hypothesis testing can also be used to carry out the hypothesis test in Example 16.3 and to assess more precisely the evidence against the null hypothesis. From Step 6 we know that the value of the test statistic is $F = 6.32$. Consulting Table IX and recalling that df $= (3, 16)$, we find that $P < 0.005$.

In particular, because the P-value is less than the specified significance level of 0.05, we can reject H_0. Furthermore, by referring to Table 9.12 on page 543, we see that the data provide very strong evidence against the null hypothesis of equal mean energy consumptions.

Other Types of ANOVA

We can consider one-way ANOVA a method for comparing the means of populations classified according to one factor or, to put it another way, a method for analyzing the effect of one factor on the mean of the variable under consideration, called the **response variable.**

For instance, in Example 16.3, we compared last year's mean energy consumption by households among the four U.S. regions (Northeast, Midwest, South, and West). Here the factor is "region" and the response variable is "energy consumption." One-way ANOVA permits us to analyze the effect of region on mean energy consumption.

There are ANOVA procedures that provide methods for comparing the means of populations classified according to two or more factors or, more to the point, methods for simultaneously analyzing the effect of two or more factors on the mean of a response variable.

For example, suppose we want to consider the effect of "region" and "home-type" (the two factors) on energy consumption (the response variable). Two-way ANOVA permits us to determine simultaneously whether region affects mean energy consumption, whether home-type affects mean energy consumption, and whether region and home-type interact in their effect on mean energy consumption (e.g., whether the effect of home-type on mean energy consumption depends on region).

Time does not permit us to cover two-factor ANOVA and the many other ANOVA procedures. We refer those interested to the supplement *Design of Experiments and Analysis of Variance* by Dennis L. Young (Reading, MA: Addison-Wesley, 1999).

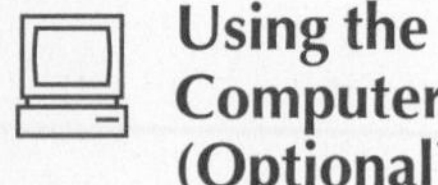

Using the Computer (Optional)

Procedure 16.1 provides a step-by-step method for performing a one-way ANOVA. Alternatively, we can apply Minitab to carry out such a hypothesis test. Example 16.4 shows how this is done.

EXAMPLE 16.4 USING MINITAB TO PERFORM A ONE-WAY ANOVA

Use Minitab to perform the hypothesis test in Example 16.3.

SOLUTION Let μ_1, μ_2, μ_3, and μ_4 denote last year's mean energy consumptions for households in the Northeast, Midwest, South, and West, respectively. We want to perform the hypothesis test

H_0: $\mu_1 = \mu_2 = \mu_3 = \mu_4$ (mean energy consumptions are equal)

H_a: Not all the means are equal.

at the 5% significance level.

To employ Minitab, we first store the four samples in Table 16.6 on page 949 in columns named NRTHEAST, MIDWEST, SOUTH, and WEST. Then we proceed in the following manner.

1 Choose **Stat ➤ ANOVA ➤ Oneway (Unstacked)...**
2 Specify NRTHEAST, MIDWEST, SOUTH, and WEST in the **Responses (in separate columns)** text box
3 Click **OK**

The output obtained is shown in Printout 16.1.

PRINTOUT 16.1
Minitab output for the one-way ANOVA procedure

```
Analysis of Variance
Source      DF        SS         MS          F        P
Factor       3     97.50      32.50       6.32    0.005
Error       16     82.30       5.14
Total       19    179.80
                                    Individual 95% CIs For Mean
                                    Based on Pooled StDev
Level        N      Mean      StDev  -------+---------+---------+---------
NRTHEAST     5    13.000      1.871                (------*-------)
MIDWEST      6    14.500      2.588                         (-----*------)
SOUTH        4    10.000      2.582     (-------*-------)
WEST         5     9.200      1.924  (-------*------)
                                     -------+---------+---------+---------
Pooled StDev =     2.268                   9.0      12.0      15.0
```

Printout 16.1 first displays a one-way ANOVA table. This table is Minitab's version of the one-way ANOVA table in Table 16.5 on page 946. We note from the printout that Minitab uses the term "Factor" instead of "Treatment."

To the right of the one-way ANOVA table in Printout 16.1, under the column headed P, is the P-value; so $P = 0.005$ (to three decimal places). Because the P-value of 0.005 is less than the specified significance level of 0.05, we reject H_0. The data provide sufficient evidence to conclude that last year's mean energy consumptions for households in the four U.S. regions are not all the same. □

Let's examine some of the other items shown in Printout 16.1. Below the ANOVA table is another table that provides the the sample sizes, sample means, and sample standard deviations of the four samples. Beneath that table we find the pooled estimate of the common standard deviation, σ, of the four populations.

Finally, the lower right side of the computer output depicts individual 95% confidence intervals for the means of the four populations under consideration. The formula used to obtain those confidence intervals is presented in the exercises.

EXERCISES 16.3

STATISTICAL CONCEPTS AND SKILLS

16.24 Suppose a one-way ANOVA is being performed to compare the means of three populations and that the sample sizes are 10, 12, and 15. Determine the degrees of freedom for the F-statistic.

16.25 We stated earlier that a one-way ANOVA test is always right-tailed since the null hypothesis is rejected only when the test statistic, F, is too large. Why is the null hypothesis rejected only when F is too large?

16.26 In each of parts (a)–(c), we have given the notation for one of the three sums of squares. For each sum of squares, state its name and the source of variation it represents.

a. SSE **b.** $SSTR$ **c.** SST

16.27 State the one-way ANOVA identity and interpret its meaning with regard to partitioning the total variation in the data.

16.28 True or false: If you know any two of the three sums of squares, *SST*, *SSTR*, and *SSE*, you can determine the remaining one. Explain your answer.

16.29 Fill in the missing entries in the following partially completed one-way ANOVA table.

Source	df	*SS*	$MS = SS/\text{df}$	*F*-statistic
Treatment	2		21.652	
Error		84.400		
Total	14			

16.30 Fill in the missing entries in the following partially completed one-way ANOVA table.

Source	df	*SS*	$MS = SS/\text{df}$	*F*-statistic
Treatment		2.124	0.708	0.75
Error	20			
Total				

16.31 Consider the following three samples.

A	*B*	*C*
1	5	2
9	2	8
	4	5
	3	
	1	

a. Compute *SST* using the defining formula.
b. Compute *SSTR* using the defining formula.
c. Compute *SSE* using the defining formula.
d. Verify that the one-way ANOVA identity holds for these samples.
e. Compute *SST*, *SSTR*, and *SSE* using the computing formulas and compare your answers with those obtained in parts (a)–(c).

16.32 Consider the following four samples.

A	*B*	*C*	*D*
6	9	4	8
3	5	4	4
3	7	2	6
	8	2	
	6	3	

a. Compute *SST* using the defining formula.
b. Compute *SSTR* using the defining formula.
c. Compute *SSE* using the defining formula.
d. Verify that the one-way ANOVA identity holds for these samples.
e. Compute *SST*, *SSTR*, and *SSE* using the computing formulas and compare your answers with those obtained in parts (a)–(c).

In each of Exercises 16.33–16.35, construct the one-way ANOVA table for the data. Compute SSTR and SSE using the defining formulas given in Formula 16.1 on page 947.

16.33 The times required by three workers to perform an assembly-line task were recorded on five randomly selected occasions. Here are the times, to the nearest minute.

Hank	Joseph	Susan
8	8	10
10	9	9
9	9	10
11	8	11
10	10	9

(Note: $\bar{x}_1 = 9.6$, $\bar{x}_2 = 8.8$, $\bar{x}_3 = 9.8$, $s_1^2 = 1.3$, $s_2^2 = 0.7$, $s_3^2 = 0.7$, and $\bar{x} = 9.4$.*)*

16.34 The U.S. Bureau of the Census collects data on monthly rents of newly completed apartments and publishes the results in *Current Housing Reports.* Independent random samples of monthly rents for newly completed apartments in the four U.S. regions yielded the following data, in dollars.

Northeast	Midwest	South	West
700	665	706	660
593	543	445	647
643	494	676	725
876	609	851	561
939	516		904
	401		

(Note: $\bar{x}_1 = 750.2$, $\bar{x}_2 = 538.0$, $\bar{x}_3 = 669.5$, $\bar{x}_4 = 699.4$, $s_1^2 = 22{,}548.7$, $s_2^2 = 8{,}476.8$, $s_3^2 = 28{,}239.0$, $s_4^2 = 16{,}492.3$, and $\bar{x} = 657.7$.*)*

16.35 Data on scores on the Scholastic Assessment Test (SAT) are published by the College Entrance Examination Board in *National College-Bound Senior.* SAT scores for randomly selected students from each of four high-school rank categories are displayed in the following table.

Top tenth	Second tenth	Second fifth	Third fifth
528	514	649	372
586	457	506	440
680	521	556	495
718	370	413	321
	532	470	424
			330

(Note: $\bar{x}_1 = 628.0$, $\bar{x}_2 = 478.8$, $\bar{x}_3 = 518.8$, $\bar{x}_4 = 397.0$, $s_1^2 = 7522.667$, $s_2^2 = 4540.700$, $s_3^2 = 8018.700$, $s_4^2 = 4614.400$, and $\bar{x} = 494.1$.*)*

Preliminary data analyses indicate that it is reasonable to consider the assumptions for one-way ANOVA satisfied in Exercises 16.36–16.40. For each exercise, perform the required hypothesis test using either the critical-value approach or the P-value approach.

16.36 In Section 16.2 we considered two hypothetical examples in order to explain the logic behind one-way ANOVA. We will examine those examples further in this exercise.

a. Refer to Table 16.1 on page 938. Perform a one-way ANOVA on the data and compare your conclusion to the informal one made in the text. Use $\alpha = 0.05$.

b. Repeat part (a) for the data displayed in Table 16.2 on page 939.

16.37 Four brands of flashlight batteries were compared by testing each brand in five flashlights. Twenty flashlights were randomly selected and divided randomly into four groups of five flashlights each. Then each group of flashlights used a different brand of battery. The lifetimes of the batteries to the nearest hour were as follows.

Brand *A*	Brand *B*	Brand *C*	Brand *D*
42	28	24	20
30	36	36	32
39	31	28	38
28	32	28	28
29	27	33	25

At the 5% significance level, does there appear to be a difference in mean lifetime among the four brands of batteries?

16.38 A chain of convenience stores wanted to test three different advertising policies:

- Policy 1: No advertising.
- Policy 2: Advertise in neighborhoods with circulars.
- Policy 3: Use circulars and advertise in newspapers.

Eighteen stores were randomly selected and divided randomly into three groups of six stores. Each group used one of the three policies. Following the implementation of the policies, sales figures were obtained for each of the stores during a 1-month period. The figures are displayed, in thousands of dollars, in the following table.

Policy 1	Policy 2	Policy 3
22	21	29
20	25	24
26	25	31
21	20	32
24	22	26
22	26	27

At the 1% significance level, do the data provide evidence of a difference in mean monthly sales among the three policies?

16.39 The U.S. Bureau of Labor Statistics publishes data on weekly earnings of nonsupervisory workers in *Employment and Earnings*. The following data, in dollars, were obtained from random samples of full and part-time nonsupervisory workers in five service-producing industries.

Transp. and Pub. util.	Wholesale trade	Retail trade	Finance, Insurance, Real estate	Services
543	524	260	482	408
583	469	188	436	420
544	449	170	518	427
588	518	298	404	343
635	502	185		380
566		279		317

Do the data provide sufficient evidence to conclude that a difference exists in mean weekly earnings of nonsupervisory workers among the five industries? Perform the required hypothesis test using $\alpha = 0.05$. *(Note:* $T_1 = 3459$, $T_2 = 2462$, $T_3 = 1380$, $T_4 = 1840$, $T_5 = 2295$, and $\Sigma x^2 = 5{,}290{,}870$.*)*

16.40 Manufacturers of golf balls always seem to be claiming that their ball goes the farthest. A writer for a sports magazine decided to conduct an impartial test. She randomly selected 20 golf professionals and then randomly assigned four golfers to each of five brands. Each golfer drove the assigned brand of ball. The driving distances, in yards, are displayed in the following table.

Brand 1	Brand 2	Brand 3	Brand 4	Brand 5
286	279	270	284	281
276	277	262	271	293
281	284	277	269	276
274	288	280	275	292

Do the data provide sufficient evidence to conclude that a difference exists in mean driving distances among the five brands of golf ball? Perform the required hypothesis test at the 5% significance level. *(Note:* $T_1 = 1117$, $T_2 = 1128$, $T_3 = 1089$, $T_4 = 1099$, $T_5 = 1142$, and $\Sigma x^2 = 1{,}555{,}185$.*)*

Extending the Concepts and Skills

Journal articles and other sources frequently provide only summary statistics (means, standard deviations, and sample sizes) when publishing ANOVA results. Exercises 16.41 and 16.42 give you practice in working with such data. Use the formula

$$\bar{x} = \frac{n_1\bar{x}_1 + n_2\bar{x}_2 + \cdots + n_k\bar{x}_k}{n_1 + n_2 + \cdots + n_k}$$

to obtain the mean of all the observations.

16.41 The U.S. Bureau of Prisons publishes data in *Statistical Report* on the times served by prisoners released from federal institutions for the first time. Independent random samples of released prisoners for five different offense categories yielded the following information on time served, in months.

Offense	n_j	$\bar{x}_j$	s_j
Counterfeiting	15	14.5	4.5
Drugs	17	18.4	3.8
Firearms	12	18.2	4.5
Forgery	10	15.6	3.6
Fraud	11	11.5	4.7

At the 1% significance level, do the data provide sufficient evidence to conclude that a difference exists in mean time served by prisoners among the five offense groups?

16.42 Data are collected by the Northwestern University Placement Center on starting salaries of college graduates, by major. Findings are reported in *The Northwestern Lindquist-Endicott Report.* Independent samples of college graduates earning Bachelor's degrees in marketing, mathematics, humanities, and computer science provided the information on annual starting salaries, in thousands of dollars, shown in the following table.

Major	n_j	$\bar{x}_j$	s_j
Marketing	35	26.7	2.9
Mathematics	25	31.8	2.7
Humanities	30	23.9	3.0
Computer science	34	35.4	3.5

Do the data imply that a difference exists in mean annual starting salaries among the four majors? Use $\alpha = 0.05$.

Confidence intervals in one-way ANOVA. Assume the conditions for one-way ANOVA are satisfied and let $s = \sqrt{MSE}$. Then we have the following confidence-interval formulas.

- A $(1-\alpha)$-level confidence interval for any particular population mean, say μ_i, has endpoints

$$\bar{x}_i \pm t_{\alpha/2} \cdot \frac{s}{\sqrt{n_i}}.$$

- A $(1-\alpha)$-level confidence interval for the difference between any two particular population means, say μ_i and μ_j, has endpoints

$$(\bar{x}_i - \bar{x}_j) \pm t_{\alpha/2} \cdot s\sqrt{(1/n_i) + (1/n_j)}.$$

In both formulas, $\text{df} = n - k$, where, as usual, k denotes the number of populations and n the total number of observations. We will apply these formulas in Exercises 16.43 and 16.44.

16.43 Refer to Exercise 16.35.
a. Determine a 90% confidence interval for the mean SAT score of all students ranked in the second fifth of their high-school class.
b. Find a 90% confidence interval for the difference between the mean SAT scores of students ranked in the top tenth and third fifth of their high-school class.
c. What assumptions are made in solving parts (a) and (b)?

16.44 Refer to Exercise 16.34.
a. Find a 99% confidence interval for the mean monthly rent of newly completed apartments in the Midwest.
b. Find a 99% confidence interval for the difference between the mean monthly rents of newly completed apartments in the Northeast and South.
c. What assumptions are made in solving parts (a) and (b)?

16.45 Refer to Exercise 16.43. Suppose you have obtained a 90% confidence interval for each of the two differences, $\mu_1 - \mu_2$ and $\mu_1 - \mu_3$. Can you be 90% confident of both results simultaneously, that is, that both differences are contained in their corresponding confidence intervals? Explain your answer.

USING TECHNOLOGY  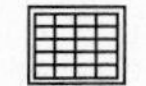 

16.46 Refer to Exercise 16.40. Use Minitab or some other statistical software to
a. obtain normal probability plots and the standard deviations of the samples.
b. perform a residual analysis.
c. perform the required hypothesis test.
d. Justify the use of your procedure in part (c).

16.47 Refer to Exercise 16.39. Use Minitab or some other statistical software to
a. obtain normal probability plots and the standard deviations of the samples.
b. perform a residual analysis.
c. perform the required hypothesis test.
d. Justify the use of your procedure in part (c).

16.48 The Census Bureau collects data on income by educational attainment, sex, and age. Results are published in *Current Population Reports.* Independent samples were taken of women from three categories of educational attainment: elementary school, high school, and college (4-year degree). We applied Minitab's one-way ANOVA procedure to the annual incomes of the women sampled and obtained Printout 16.2, shown on the next page, where the incomes are in thousands of dollars. Determine
a. the three sums of squares, *SSTR, SSE,* and *SST.*
b. the two mean squares, *MSTR* and *MSE.*
c. the value of the test statistic, F.
d. the null and alternative hypotheses.
e. the P-value for the hypothesis test.
f. the conclusion if the hypothesis test is performed at the 1% significance level.
g. the sample size, sample mean, and sample standard deviation for each of the three samples.
h. a 95% confidence interval for the mean annual income of all women whose educational attainment is at the high-school level.

16.49 The Motor Vehicle Manufacturers Association of the United States conducts surveys on the costs of owning and operating a motor vehicle. Data are published in *Motor Vehicle Facts and Figures.* Independent random samples of owners of large, intermediate, and compact cars were taken to obtain information on annual insurance premiums. Then Minitab's one-way ANOVA procedure was applied to the resulting data. Printout 16.3 displays the output. Determine

a. the three sums of squares, *SSTR, SSE,* and *SST*.
b. the two mean squares, *MSTR* and *MSE*.
c. the value of the test statistic, F.
d. the null and alternative hypotheses.
e. the P-value for the hypothesis test.
f. the conclusion if the hypothesis test is performed at the 5% significance level.
g. the sample size, sample mean, and sample standard deviation for each of the three samples.
h. a 95% confidence interval for the mean annual insurance premium of all compact-car owners.

PRINTOUT 16.2 Minitab output for Exercise 16.48

```
Analysis of Variance
Source     DF        SS        MS        F        P
Factor      2    9706.5    4853.2    83.55    0.000
Error     154    8946.0      58.1
Total     156   18652.5
                                   Individual 95% CIs For Mean
                                   Based on Pooled StDev
Level       N      Mean     StDev  --+---------+---------+---------+----
ELEMENT    40    15.598     6.703    (--*---)
HS         62    22.300     7.345              (--*--)
COLLEGE    55    35.198     8.502                              (--*--)
                                   --+---------+---------+---------+----
Pooled StDev =    7.622              14.0      21.0      28.0      35.0
```

PRINTOUT 16.3 Minitab output for Exercise 16.49

```
Analysis of Variance
Source     DF        SS        MS        F        P
Factor      2     36859     18430     0.55    0.580
Error      74   2482910     33553
Total      76   2519769
                                   Individual 95% CIs For Mean
                                   Based on Pooled StDev
Level       N      Mean     StDev  ----------+---------+---------+------
LARGE      20     902.4     152.5        (------------*-------------)
INTERMED   30     851.7     169.2  (----------*----------)
COMPACT    27     890.5     215.8        (----------*-----------)
                                   ----------+---------+---------+------
Pooled StDev =    183.2                     840       900       960
```

16.4 MULTIPLE COMPARISONS*

Suppose we perform a one-way ANOVA and reject the null hypothesis. Then we can conclude that the means of the populations under consideration are not all the same. Once we make that decision, we may also want to know which means are different, which mean is largest, or, more generally, the relation among all the means. Methods for dealing with these problems are called **multiple comparisons.**

Several multiple-comparison methods are available. In this book, we will discuss only the **Tukey multiple-comparison method.** Other commonly used multiple-comparison methods are the Bonferroni method, the Fisher method, and the Scheffé method.

One approach for implementing multiple comparisons is to obtain confidence intervals for the differences between all possible pairs of population means. Two means are declared different if the confidence interval for their difference does not contain 0. (If a confidence interval for the difference between two population means does not contain 0, then we can reject the null hypothesis that the two means are equal in favor of the alternative hypothesis that the two means are different; and vice versa. See Exercise 10.13 on page 599.)

In multiple comparisons it is important to distinguish between the individual confidence level and the family confidence level. The **individual confidence level** is the confidence we have that *any particular* confidence interval contains the difference between the corresponding population means; the **family confidence level** is the confidence we have that *all* the confidence intervals contain the differences between the corresponding population means. It is at the family confidence level that we can be confident in the truth of our conclusions when comparing all the population means simultaneously.

The Studentized Range Distribution

The distribution upon which the Tukey multiple-comparison method is based is called the **studentized range distribution,** which for brevity we refer to as the ***q*-distribution.** A variable is said to have a q-distribution if its distribution has the shape of a special type of right-skewed curve, called a ***q*-curve.** Actually, there are infinitely many q-distributions (and q-curves); a particular one is identified by two parameters, which we denote by κ (kappa) and ν (nu).

Percentages and probabilities for a variable having a q-distribution are equal to areas under its associated q-curve. To perform a Tukey multiple comparison, we need to know how to find the q-value having a specified area to its right. The symbol $\boldsymbol{q_\alpha}$ is used to denote the q-value having area α to its right.

---

* This section is optional and will not be needed in subsequent sections of the book. However, those using the supplement *Design of Experiments and Analysis of Variance* by Dennis L. Young (Reading, MA: Addison-Wesley, 1999) should cover this section.

Values of $q_{0.01}$ and $q_{0.05}$ are presented in Tables XI and XII in Appendix A, respectively. Example 16.5 explains how to use Table XII.

EXAMPLE 16.5 FINDING THE q-VALUE HAVING A SPECIFIED AREA TO ITS RIGHT

For the q-curve with parameters $\kappa = 4$ and $\nu = 16$, find $q_{0.05}$; that is, find the q-value having area 0.05 to its right, as shown in Fig. 16.10(a).

FIGURE 16.10 Finding the q-value having area 0.05 to its right

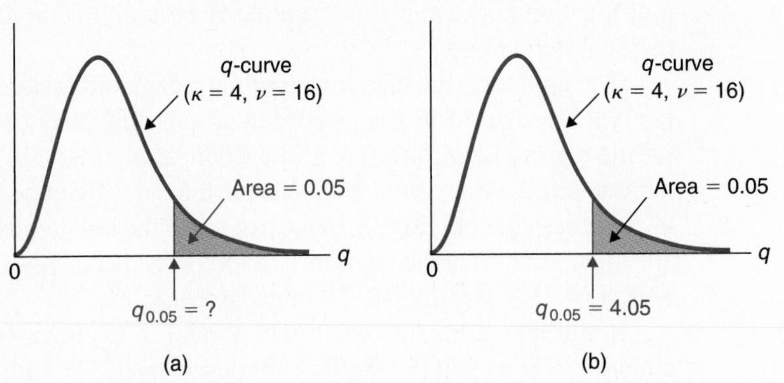

SOLUTION To obtain the q-value in question, we use Table XII. In this case $\kappa = 4$ and $\nu = 16$. Thus we first go down the outside columns to the row labeled "16." Then we go across that row until we are under the column headed "4." The number in the body of the table there, 4.05, is the required q-value; that is, for the q-curve with parameters $\kappa = 4$ and $\nu = 16$, the q-value having area 0.05 to its right is 4.05, as seen in Fig. 16.10(b). ◨

The Tukey Multiple-Comparison Method

The formulas used in the Tukey multiple-comparison method for obtaining confidence intervals for the differences between means are similar to the pooled t-interval formula (Procedure 10.2 on page 607). The essential difference is that in the Tukey multiple-comparison method we consult a q-table instead of a t-table. Specifically, we have the procedure shown on the facing page.

In Example 16.3 on page 949, we conducted a one-way ANOVA to decide whether a difference exists in last year's mean energy consumption by households among the four U.S. regions. Specifically, we performed the hypothesis test

H_0: $\mu_1 = \mu_2 = \mu_3 = \mu_4$ (mean energy consumptions are equal)

H_a: Not all the means are equal.

PROCEDURE 16.2

THE TUKEY MULTIPLE-COMPARISON METHOD FOR COMPARING k POPULATION MEANS

ASSUMPTIONS

1. Independent samples
2. Normal populations
3. Equal population standard deviations

Step 1 Decide on the family confidence level, $1 - \alpha$.

Step 2 Find q_α for the q-curve with parameters $\kappa = k$ and $\nu = n - k$, where k is the number of populations under consideration and n is the total number of observations.

Step 3 Obtain the endpoints of the confidence interval for $\mu_i - \mu_j$:

$$(\bar{x}_i - \bar{x}_j) \pm \frac{q_\alpha}{\sqrt{2}} \cdot s\sqrt{(1/n_i) + (1/n_j)},$$

where $s = \sqrt{MSE}$. Do this for all possible pairs of means with $i < j$.

Step 4 Declare two population means different if the confidence interval for their difference does not contain 0; otherwise, do not declare the two population means different.

Step 5 Summarize the results in Step 4 by ranking the sample means from smallest to largest and by connecting with lines those whose population means were not declared different.

at the 5% significance level, where μ_1, μ_2, μ_3, and μ_4 denote last year's mean energy consumptions for households in the Northeast, Midwest, South, and West, respectively.

The test results were statistically significant; that is, we rejected H_0. Thus we can conclude at the 5% significance level that for last year, at least two of the regions have different mean energy consumptions. The Tukey multiple-comparison method allows us to elaborate on this conclusion.

EXAMPLE 16.6 ILLUSTRATES PROCEDURE 16.2

Apply the Tukey multiple-comparison method to the energy-consumption data, repeated on the next page in Table 16.7.

SOLUTION We apply Procedure 16.2.

TABLE 16.7 Last year's energy consumptions for samples of households in the four U.S. regions

Northeast	Midwest	South	West
15	17	11	10
10	12	7	12
13	18	9	8
14	13	13	7
13	15		9
	12		

Step 1 *Decide on the family confidence level,* $1 - \alpha$.

As we have done previously in this illustration, we will use $\alpha = 0.05$; so the family confidence level is 0.95 (95%).

Step 2 *Find* q_α *for the* q*-curve with parameters* $\kappa = k$ *and* $\nu = n - k$*, where* n *is the total number of observations.*

As we see from Table 16.7, $\kappa = k = 4$ and $\nu = n - k = 20 - 4 = 16$. Consulting Table XII, we find that $q_\alpha = q_{0.05} = 4.05$.

Step 3 *Obtain the endpoints of the confidence interval for* $\mu_i - \mu_j$:

$$(\bar{x}_i - \bar{x}_j) \pm \frac{q_\alpha}{\sqrt{2}} \cdot s\sqrt{(1/n_i) + (1/n_j)},$$

where $s = \sqrt{MSE}$. *Do this for all possible pairs of means with* $i < j$.

To begin, it is helpful to construct a table giving the sample means and sample sizes. Referring to Table 16.7, we obtain Table 16.8.

TABLE 16.8 Sample means and sample sizes for the energy-consumption data

Region	Northeast	Midwest	South	West
j	1	2	3	4
$\bar{x}_j$	13.0	14.5	10.0	9.2
n_j	5	6	4	5

From Step 2, $q_\alpha = 4.05$. Also, on page 942 we found that $MSE = 5.144$ for the energy-consumption data. Now we are ready to obtain the required confidence intervals. The endpoints of the confidence interval for $\mu_1 - \mu_2$ are

$$(13.0 - 14.5) \pm \frac{4.05}{\sqrt{2}} \cdot \sqrt{5.144}\sqrt{(1/5) + (1/6)},$$

or -5.43 to 2.43.

Likewise, the endpoints of the confidence interval for $\mu_1 - \mu_3$ are

$$(13.0 - 10.0) \pm \frac{4.05}{\sqrt{2}} \cdot \sqrt{5.144}\sqrt{(1/5) + (1/4)},$$

or -1.36 to 7.36.

Proceeding in the same way, we obtain the remaining confidence intervals. All six confidence intervals are displayed in Table 16.9, where, for ease of reference, we have placed parenthetically the numbers used to represent the regions.

TABLE 16.9 Simultaneous 95% confidence intervals for the differences between the energy-consumption means

	Northeast (1)	Midwest (2)	South (3)
Midwest (2)	$(-5.43, 2.43)$		
South (3)	$(-1.36, 7.36)$	$(0.31, 8.69)$	
West (4)	$(-0.31, 7.91)$	$(1.37, 9.23)$	$(-3.56, 5.16)$

Each entry in Table 16.9 is the confidence interval for the difference between the mean labeled by the column and the mean labeled by the row. For instance, the entry in the column labeled "Midwest (2)" and the row labeled "West (4)" is $(1.37, 9.23)$. So the confidence interval for the difference, $\mu_2 - \mu_4$, between last year's mean energy consumptions for households in the Midwest and West is from 1.37 to 9.23.

Step 4 *Declare two population means different if the confidence interval for their difference does not contain 0; otherwise, do not declare the two population means different.*

Referring to Table 16.9, we declare means μ_2 and μ_3 different and means μ_2 and μ_4 different; all other pairs of means are not declared different.

Step 5 *Summarize the results in Step 4 by ranking the sample means from smallest to largest and by connecting with lines those whose population means were not declared different.*

In view of Table 16.8, Step 4, and the numbering used to represent the U.S. regions (shown parenthetically), we obtain the following diagram:

West (4)	South (3)	Northeast (1)	Midwest (2)
9.2	10.0	13.0	14.5

Interpreting this diagram, we conclude that last year's mean energy consumption in the Midwest exceeds that in the West and South, and that no other means can be declared different. *All of this* can be said with 95% confidence, the family confidence level.

As we see from Example 16.6, multiple comparisons require extensive computations, even when the number of populations and the sample sizes are quite small. This explains why multiple comparisons are almost always done by computer.

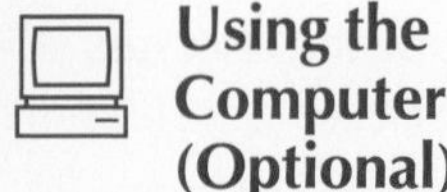

Using the Computer (Optional)

Procedure 16.2 on page 961 provides a step-by-step method for performing a Tukey multiple comparison. Alternatively, we can use Minitab. To do so, however, we need to apply a different version of Minitab's one-way ANOVA procedure than the one we discussed in Section 16.3 on pages 952–953. Recall that in that procedure, the samples are stored in different columns (*unstacked data*).

The version of Minitab's one-way ANOVA procedure that will also perform a Tukey multiple comparison requires that the samples be stored in a single column (*stacked data*), with another column indicating which observations belong to which samples. Example 16.7 provides the details.

EXAMPLE 16.7 USING MINITAB TO PERFORM A TUKEY MULTIPLE COMPARISON

Use Minitab to perform the Tukey multiple comparison done in Example 16.6.

SOLUTION First we store all 20 energy consumptions from Table 16.7 on page 962 in a column named ENERGY. Then, in a column named REGION, we store the regions corresponding to the energy consumptions in ENERGY. For instance, suppose we store the sample data for the Northeast in the first five rows of ENERGY, that for the Midwest in the next six rows of ENERGY, and so forth. Then we would store the word Nrtheast in the first five rows of REGION, the word Midwest in the next six rows of REGION, and so forth.

Also, if we want the output to order the regions as we have been doing (Northeast, Midwest, South, and West), we need to take some preliminary steps:

1 Click anywhere in the column named REGION
2 Choose **Editor ➤ Set Column ➤ Value Order...**
3 Select the **User-specified order** option button
4 In the **Define an order (one value per line)** text box, edit the order of the possible regions, if necessary, so that it reads Nrtheast, Midwest, South, and West.
5 Click **OK**

We are now ready to instruct Minitab to perform the Tukey multiple comparison. We proceed in the following manner.

1 Choose **Stat ➤ ANOVA ➤ One-way...**
2 Specify ENERGY in the **Response** text box

3 Click in the **Factor** text box and specify REGION

4 Click the **Comparisons...** button

5 Select the **Tukey's, family error rate** check box

6 Click in the **Tukey's, family error rate** text box and type 5 (to specify that the family confidence level is 95%)

7 Click **OK**

8 Click **OK**

The resulting output is displayed in Printout 16.4.

PRINTOUT 16.4 Minitab output for the Tukey multiple-comparison procedure

```
Analysis of Variance for ENERGY
Source     DF        SS        MS        F        P
REGION      3     97.50     32.50     6.32    0.005
Error      16     82.30      5.14
Total      19    179.80
                                    Individual 95% CIs For Mean
                                    Based on Pooled StDev
Level       N      Mean     StDev   -------+---------+---------+---------
Nrtheast    5    13.000     1.871                (------*-------)
Midwest     6    14.500     2.588                       (-----*------)
South       4    10.000     2.582     (-------*-------)
West        5     9.200     1.924   (-------*------)
                                    -------+---------+---------+---------
Pooled StDev =    2.268                   9.0      12.0      15.0

Tukey's pairwise comparisons

    Family error rate = 0.0500
Individual error rate = 0.0113

Critical value = 4.05

Intervals for (column level mean) - (row level mean)

            Nrtheast    Midwest     South

Midwest       -5.433
               2.433

South         -1.357      0.307
               7.357      8.693

West          -0.308      1.367    -3.557
               7.908      9.233     5.157
```

The portion of the output in Printout 16.4 from the one-way ANOVA table up to but not including `Tukey's pairwise comparisons` is essentially identical to that obtained using the "unstacked" one-way ANOVA procedure (Printout 16.1 on page 953). Then comes the multiple comparison.

The first and second items give the family error rate and individual error rate, which in our terminology are 1 minus the family confidence level and 1 minus the individual confidence level, respectively. So we see that the family confidence level is 0.95 (as specified) and the individual confidence level is 0.9887. The next item in the printout, `Critical value = 4.05`, is q_α; so $q_\alpha = q_{0.05} = 4.05$.

The final item in Printout 16.4 is a table providing the endpoints of the confidence intervals for the differences between the means. Compare this table with the one we obtained by hand in Table 16.9 on page 963. □

EXERCISES 16.4

STATISTICAL CONCEPTS AND SKILLS

16.50 What is the purpose of doing a multiple comparison?

16.51 If a confidence interval for the difference between two population means does not contain _____, then we can reject the null hypothesis that the two means are equal in favor of the alternative hypothesis that the two means are different; and vice versa.

16.52 What is the difference between the family confidence level and the individual confidence level?

16.53 Explain why the family confidence level is smaller than the individual confidence level for multiple comparisons involving three or more means.

16.54 What is the name of the distribution upon which the Tukey multiple-comparison method is based? What is the abbreviation used for that distribution?

16.55 The parameter ν for the q-curve in a Tukey multiple comparison equals one of the degrees of freedom for the F-curve in a one-way ANOVA. Which one?

16.56 Explain the essential difference between obtaining a confidence interval using the pooled t-interval procedure and obtaining a confidence interval using the Tukey multiple-comparison procedure.

16.57 Determine the following for a q-curve with parameters $\kappa = 6$ and $\nu = 13$:

a. the q-value having area 0.05 to its right.

b. $q_{0.01}$.

16.58 Determine the following for a q-curve with parameters $\kappa = 8$ and $\nu = 20$:

a. the q-value having area 0.01 to its right.

b. $q_{0.05}$.

In Exercises 16.59–16.62, we have repeated the information from Exercises 16.37–16.40. For each exercise, use Procedure 16.2 to perform a multiple comparison.

16.59 To compare four brands of flashlight batteries, 20 flashlights were randomly selected and divided randomly into four groups of five flashlights each. Then each group of flashlights used a different brand of battery. The lifetimes of the batteries to the nearest hour were as follows.

Brand A	Brand B	Brand C	Brand D
42	28	24	20
30	36	36	32
39	31	28	38
28	32	28	28
29	27	33	25

Perform a Tukey multiple comparison using a family confidence level of 0.95.

16.60 A chain of convenience stores wanted to test three different advertising policies:

- Policy 1: No advertising.
- Policy 2: Advertise in neighborhoods with circulars.
- Policy 3: Use circulars and advertise in newspapers.

Eighteen stores were randomly selected and divided randomly into three groups of six stores. Each group used one of the three policies. Following the implementation of the policies, sales figures were obtained for each of the stores during a 1-month period. The figures are displayed, in thousands of dollars, in the following table.

Policy 1	Policy 2	Policy 3
22	21	29
20	25	24
26	25	31
21	20	32
24	22	26
22	26	27

Perform a Tukey multiple comparison using a family confidence level of 0.99.

16.61 The Bureau of Labor Statistics publishes data on weekly earnings of nonsupervisory workers in *Employment and Earnings.* The following data, in dollars, were obtained from random samples of full and part-time nonsupervisory workers in five service-producing industries.

Transp. and Pub. util.	Wholesale trade	Retail trade	Finance, Insurance, Real estate	Services
543	524	260	482	408
583	469	188	436	420
544	449	170	518	427
588	518	298	404	343
635	502	185		380
566		279		317

Conduct a Tukey multiple comparison at the 95% family confidence level. *(Hint:* For the q-value, use the mean of the two closest q-values in Table XII.*)*

16.62 Manufacturers of golf balls always seem to be claiming that their ball goes the farthest. A writer for a sports magazine decided to conduct an impartial test. She randomly selected 20 golf professionals and then randomly assigned four golfers to each of five brands. Each golfer drove the assigned brand of ball. The driving distances, in yards, are displayed in the following table.

Brand 1	Brand 2	Brand 3	Brand 4	Brand 5
286	279	270	284	281
276	277	262	271	293
281	284	277	269	276
274	288	280	275	292

Conduct a Tukey multiple comparison at the 95% family confidence level.

Extending the Concepts and Skills

In Exercises 16.63 and 16.64, we have repeated the information from Exercises 16.41 and 16.42. For each exercise, use Procedure 16.2 on page 961 to perform a multiple comparison.

16.63 The U.S. Bureau of Prisons publishes data in *Statistical Report* on the times served by prisoners released from federal institutions for the first time. Independent random samples of released prisoners for five different offense categories yielded the following information on time served, in months.

Offense	n_j	$\bar{x}_j$	s_j
Counterfeiting	15	14.5	4.5
Drug laws	17	18.4	3.8
Firearms	12	18.2	4.5
Forgery	10	15.6	3.6
Fraud	11	11.5	4.7

Using a family confidence level of 99%, perform a Tukey multiple comparison.

16.64 Data are collected by the Northwestern University Placement Center on starting salaries of college graduates, by major. Findings are reported in *The Northwestern Lindquist-Endicott Report.* Independent samples of college graduates earning Bachelor's degrees in marketing, mathematics, humanities, and computer science provided the information on annual starting salaries, in thousands of dollars, shown in the following table.

Major	n_j	$\bar{x}_j$	s_j
Marketing	35	26.7	2.9
Mathematics	25	31.8	2.7
Humanities	30	23.9	3.0
Computer science	34	35.4	3.5

Using a family confidence level of 95%, perform a Tukey multiple comparison.

16.65 Explain precisely why the family confidence level and not the individual confidence level is the appropriate level for comparing all population means simultaneously.

16.66 In Step 3 of Procedure 16.2, we obtain confidence intervals only when $i < j$.

a. Explain how to determine the remaining confidence intervals from those.

b. Apply your answer from part (a) and Table 16.9 on page 963 to determine the remaining six confidence intervals for the differences between the energy-consumption means.

USING TECHNOLOGY  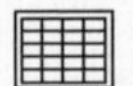 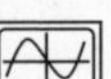

16.67 Use Minitab or some other statistical software to perform the multiple comparison required in Exercise 16.61.

16.68 Use Minitab or some other statistical software to perform the multiple comparison required in Exercise 16.62.

16.69 Explain the difference between stacked and unstacked data. Which type of data storage is required for Minitab's Tukey multiple-comparison procedure?

16.70 The U.S. Bureau of the Census collects data on income by educational attainment, sex, and age. Results are published in *Current Population Reports.* Independent random samples were taken of women from three categories of educational attainment: elementary school, high school, and college (4-year degree). Then Minitab's Tukey multiple-comparison procedure was applied to the annual incomes of the women sampled. Printout 16.5 displays the portion of the computer output needed for a Tukey multiple comparison. The incomes are in thousands of dollars. From the output, determine

a. the family confidence level and interpret the answer you obtain.

b. the individual confidence level and interpret the answer you obtain.

c. $q_{0.01}$.

d. the (simultaneous) confidence interval for the difference, $\mu_2 - \mu_3$, between the mean annual incomes of women whose educational attainments are at the high school and college levels.

e. the (simultaneous) confidence interval for the difference, $\mu_3 - \mu_1$, between the mean annual incomes of women whose educational attainments are at the college and elementary levels.

f. which pairs of means should be declared different.

g. Summarize the results of the Tukey multiple comparison in words.

16.71 The Motor Vehicle Manufacturers Association of the United States conducts surveys on the costs of owning and operating a motor vehicle. Data are published in *Motor Vehicle Facts and Figures* and include costs for gas and oil, tires, maintenance, insurance, license and registration, and depreciation. Independent random samples of owners of large, intermediate, and compact cars were taken to obtain information on annual insurance premiums. Then Minitab's Tukey multiple-comparison procedure was applied to the resulting data. Printout 16.6 displays the portion of the output required for a Tukey multiple comparison. From the output, determine

a. the family confidence level and interpret the answer you obtain.

b. the individual confidence level and interpret the answer you obtain.

c. $q_{0.05}$.

d. the (simultaneous) confidence interval for the difference, $\mu_1 - \mu_3$, between the mean annual insurance premiums of large and compact cars.

e. the (simultaneous) confidence interval for the difference, $\mu_3 - \mu_2$, between the mean annual insurance premiums of compact and intermediate cars.

f. which pairs of means should be declared different.

g. Summarize the results of the Tukey multiple comparison in words.

PRINTOUT 16.5 Minitab output for Exercise 16.70

```
Tukey's pairwise comparisons

   Family error rate = 0.0100
Individual error rate = 0.00361

Critical value = 4.18

Intervals for (column level mean) - (row level mean)

             Element           HS

 HS          -11.271
              -2.134

 College     -24.282      -17.071
             -14.919       -8.725
```

PRINTOUT 16.6 Minitab output for Exercise 16.71

```
Tukey's pairwise comparisons

   Family error rate = 0.0500
Individual error rate = 0.0194

Critical value = 3.38

Intervals for (column level mean) - (row level mean)

               Large    Intermed

Intermed         -76
                 177

Compact         -117        -155
                 141          77
```

16.5 THE KRUSKAL–WALLIS TEST*

In this section we will examine the **Kruskal–Wallis test,** a nonparametric alternative to the one-way ANOVA procedure discussed in Section 16.3. The Kruskal–Wallis test applies when the distributions (one for each population) of the variable

* This section continues the optional coverage of nonparametric statistics.

under consideration have the same shape, but does not require that they be normal or have any other specific shape.

Like the Mann–Whitney test, the Kruskal–Wallis test is based on ranks. When ties occur, ranks are assigned in the same way as in the Mann–Whitney test: *If two or more observations are tied, each is assigned the mean of the ranks they would have had if there were no ties.* The Kruskal–Wallis test is introduced in Example 16.8.

EXAMPLE 16.8 INTRODUCES THE KRUSKAL–WALLIS TEST

The U.S. Federal Highway Administration conducts annual surveys on motor vehicle travel by type of vehicle and publishes its findings in *Highway Statistics.* Independent random samples of cars, buses, and trucks provided the data on number of miles driven last year, in thousands, shown in Table 16.10. At the 5% significance level, do the data provide sufficient evidence to conclude that a difference exists in last year's mean number of miles driven among cars, buses, and trucks?

TABLE 16.10 Number of miles driven (1000s) last year for independent samples of cars, buses, and trucks

Cars	Buses	Trucks
19.9	1.8	24.6
15.3	7.2	37.0
2.2	7.2	21.2
6.8	6.5	23.6
34.2	13.3	23.0
8.3	25.4	15.3
12.0		57.1
7.0		14.5
9.5		26.0
1.1		

SOLUTION Let μ_1, μ_2, and μ_3 denote last year's mean number of miles driven for cars, buses, and trucks, respectively. Then the null and alternative hypotheses are

$$H_0\text{: } \mu_1 = \mu_2 = \mu_3 \text{ (mean miles driven are equal)}$$

$$H_a\text{: Not all the means are equal.}$$

Preliminary data analyses (not shown) suggest that the distributions of miles driven have roughly the same shape for cars, buses, and trucks but that those distributions are far from normal. Thus, although the one-way ANOVA test of Section 16.3 is probably inappropriate, the Kruskal–Wallis procedure appears suitable.

To apply the Kruskal–Wallis test, we first rank the data from all three samples combined. The results of this ranking appear in Table 16.11.

TABLE 16.11 Results of ranking the combined data from Table 16.10

Cars	Rank	Buses	Rank	Trucks	Rank
19.9	16	1.8	2	24.6	20
15.3	14.5	7.2	7.5	37.0	24
2.2	3	7.2	7.5	21.2	17
6.8	5	6.5	4	23.6	19
34.2	23	13.3	12	23.0	18
8.3	9	25.4	21	15.3	14.5
12.0	11			57.1	25
7.0	6			14.5	13
9.5	10			26.0	22
1.1	1				
	9.850		9.000		19.167

← *Mean ranks*

The idea behind the Kruskal–Wallis test is simple: If the null hypothesis of equal population means is true, then the means of the ranks for the three samples should be roughly equal. Put another way, if the variation among the mean ranks for the three samples is too large, then we have evidence against the null hypothesis.

To measure the variation among the mean ranks, we use the treatment sum of squares, *SSTR,* computed for the ranks. To decide whether that quantity is too large, we compare it to the variance of all the ranks, which can be expressed as $SST/(n-1)$, where *SST* is the total sum of squares for the ranks and n is the total number of observations.[†] More precisely, the test statistic for a Kruskal–Wallis test, denoted by H, is

$$H = \frac{SSTR}{SST/(n-1)}.$$

Large values of H indicate that the variation among the mean ranks is large (relative to the variance of all the ranks) and hence that the null hypothesis of equal population means should be rejected.

For the ranks in Table 16.11, we find that $SSTR = 537.475$, $SST = 1299$, and $n = 25$. Thus the value of the test statistic is

$$H = \frac{SSTR}{SST/(n-1)} = \frac{537.475}{1299/24} = 9.930.$$

Is this value of H large enough to conclude that the null hypothesis of equal population means is false? To answer this question, we need to know the distribution of the variable H. We will discuss that distribution and then complete the hypothesis test considered in this example. ◨

[†] Recall from Sections 16.2 and 16.3 that the treatment sum of squares, *SSTR*, is a measure of variation among means and that the total sum of squares, *SST*, is a measure of variation among all the data. The defining and computing formulas for *SSTR* and *SST* are given in Formula 16.1 on page 947. For the Kruskal–Wallis test, we apply those formulas to the ranks of the sample data, not to the sample data themselves.

KEY FACT 16.5 DISTRIBUTION OF THE *H*-STATISTIC FOR A KRUSKAL–WALLIS TEST

Suppose the k distributions (one for each population) of the variable under consideration have the same shape. Then, for independent samples from the k populations, the variable

$$H = \frac{SSTR}{SST/(n-1)}$$

has approximately a chi-square distribution with df $= k - 1$ if the null hypothesis of equal population means is true. Here n denotes the total number of observations.

The usual rule of thumb for using the chi-square distribution as an approximation to the true distribution of H is that all sample sizes should be 5 or greater. We will adopt this rule of thumb even though some statisticians consider it too restrictive and regard the chi-square approximation to be adequate unless $k = 3$ and none of the sample sizes exceed 5.

When computing the test statistic H by hand from the raw data, it is generally easier to use the computing formula

$$H = \frac{12}{n(n+1)} \sum \frac{R_j^2}{n_j} - 3(n+1),$$

where R_1 denotes the sum of the ranks for the sample data from Population 1, R_2 denotes the sum of the ranks for the sample data from Population 2, and so on. Strictly speaking, the computing formula for H is equivalent to the defining formula for H if no ties occur. In practice, however, the computing formula provides a sufficiently accurate approximation unless the number of ties is relatively large.

We now present, on the facing page, a step-by-step procedure for performing a Kruskal–Wallis test. The test can be used to compare several population medians as well as several population means. We state the procedure in terms of population means. To employ the procedure for population medians, simply replace μ_1 by η_1, μ_2 by η_2, and so on.

Regarding the second and third assumptions required for employing Procedure 16.3, we note the following facts:

- Assumption 2: For brevity, we will use the phrase "same shape populations" to indicate that the k distributions of the variable under consideration have the same shape.

- Assumption 3: This assumption is necessary only when using the chi-square distribution as an approximation to the distribution of H. Tables of critical values for H are available in cases where Assumption 3 fails.

PROCEDURE 16.3

THE KRUSKAL–WALLIS TEST FOR k POPULATION MEANS

ASSUMPTIONS

1. Independent samples
2. Same shape populations
3. All sample sizes are 5 or greater

Step 1 The null and alternative hypotheses are

$$H_0\colon \mu_1 = \mu_2 = \cdots = \mu_k$$
$$H_a\colon \text{Not all the means are equal.}$$

Step 2 Decide on the significance level, α.

Step 3 The critical value is χ^2_α with $\text{df} = k - 1$. Use Table VIII to determine the critical value.

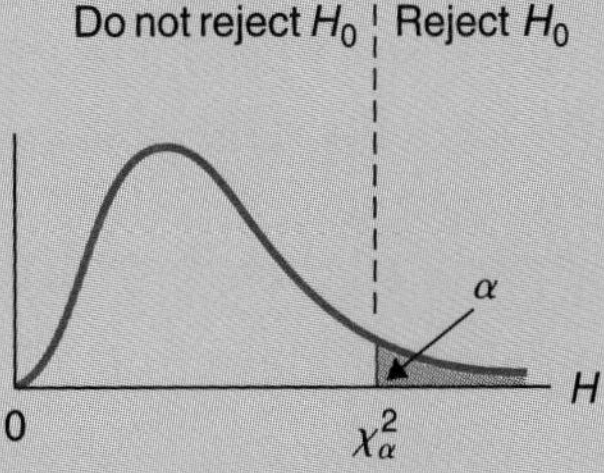

Step 4 Construct a work table of the following form:

Sample from Population 1	Overall rank	Sample from Population 2	Overall rank	...	Sample from Population k	Overall rank
⋮	⋮	⋮	⋮	⋮	⋮	⋮

Step 5 Compute the value of the test statistic

$$H = \frac{12}{n(n+1)} \sum \frac{R_j^2}{n_j} - 3(n+1),$$

where n denotes the total number of observations and $R_1, R_2, \ldots, R_k$ the sums of the ranks for the observations from Populations $1, 2, \ldots, k$.

Step 6 If the value of the test statistic falls in the rejection region, reject H_0; otherwise, do not reject H_0.

Step 7 State the conclusion in words.

EXAMPLE 16.9 ILLUSTRATES PROCEDURE 16.3

We now complete the hypothesis test introduced in Example 16.8. Independent random samples of cars, buses, and trucks provided the data on number of miles driven last year, in thousands, shown in Table 16.12. At the 5% significance level, do the data provide sufficient evidence to conclude that a difference exists in last year's mean number of miles driven among cars, buses, and trucks?

TABLE 16.12 Number of miles driven (1000s) last year for independent samples of cars, buses, and trucks

Cars	Buses	Trucks
19.9	1.8	24.6
15.3	7.2	37.0
2.2	7.2	21.2
6.8	6.5	23.6
34.2	13.3	23.0
8.3	25.4	15.3
12.0		57.1
7.0		14.5
9.5		26.0
1.1		

SOLUTION We apply Procedure 16.3.

Step 1 State the null and alternative hypotheses.

Let μ_1, μ_2, and μ_3 denote last year's mean number of miles driven for cars, buses, and trucks, respectively. Then the null and alternative hypotheses are

$$H_0\colon \mu_1 = \mu_2 = \mu_3 \text{ (mean miles driven are equal)}$$

$$H_a\colon \text{Not all the means are equal.}$$

Step 2 Decide on the significance level, α.

We are to perform the hypothesis test at the 5% significance level; so $\alpha = 0.05$.

Step 3 The critical value is χ^2_α with df $= k - 1$.

We have $k = 3$, the three types of vehicles; so df $= 3 - 1 = 2$. Consulting Table VIII, we find that the critical value is $\chi^2_{0.05} = 5.991$, as seen in Fig. 16.11.

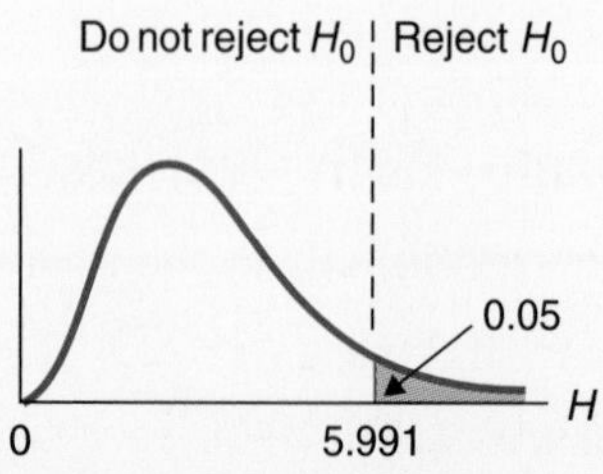

FIGURE 16.11 Criterion for deciding whether or not to reject the null hypothesis

Step 4 *Construct a work table.*

We have already done this in Table 16.11 on page 971. (Ignore the bottom row of that table now.)

Step 5 *Compute the value of the test statistic*

$$H = \frac{12}{n(n+1)} \sum \frac{R_j^2}{n_j} - 3(n+1).$$

We have $n = 10 + 6 + 9 = 25$. By summing the second, fourth, and sixth columns of Table 16.11, we find that $R_1 = 98.5$, $R_2 = 54.0$, and $R_3 = 172.5$. Thus the value of the test statistic is

$$H = \frac{12}{25(25+1)} \left(\frac{98.5^2}{10} + \frac{54.0^2}{6} + \frac{172.5^2}{9} \right) - 3(25+1) = 9.923.$$

Step 6 *If the value of the test statistic falls in the rejection region, reject H_0; otherwise, do not reject H_0.*

From Step 5 the value of the test statistic is $H = 9.923$. Figure 16.11 shows that this falls in the rejection region. Thus we reject H_0.

Step 7 *State the conclusion in words.*

The test results are statistically significant at the 5% level; that is, at the 5% significance level, the data provide sufficient evidence to conclude that a difference exists in last year's mean number of miles driven among cars, buses, and trucks. ◨

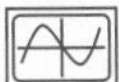

Using the *P*-Value Approach

The *P*-value approach to hypothesis testing can also be used to carry out the hypothesis test in Example 16.9. From Step 5 we see that the value of the test statistic is $H = 9.923$. Recalling that H has approximately a chi-square distribution with df $= 2$, we find from Table VIII that $0.005 < P < 0.01$.

In particular, because the *P*-value is less than the specified significance level of 0.05, we can reject H_0. Furthermore, by referring to Table 9.12 on page 543, we see that the data provide very strong evidence against the null hypothesis of equal mean number of miles driven.

Comparison of the Kruskal–Wallis Test and the One-Way ANOVA Test

In Section 16.3 we learned how to perform a one-way ANOVA to compare k population means using independent samples when the variable under consideration is normally distributed on each of the k populations and the population standard deviations are equal. Since normal distributions having equal standard deviations have the same shape, we can also use the Kruskal–Wallis test to perform such a hypothesis test.

Which test is the better one to use under these circumstances? As you might expect, it is the one-way ANOVA test because that test is designed expressly for normal populations; under conditions of normality, the one-way ANOVA test is more powerful than the Kruskal–Wallis test. What is somewhat surprising is that the one-way ANOVA test is not much more powerful than the Kruskal–Wallis test.

On the other hand, if the distributions of the variable under consideration have the same shape but are not normal, then the Kruskal–Wallis test is usually more powerful than the one-way ANOVA test, often considerably so. In summary, we have the following key fact.

KEY FACT 16.6 **THE KRUSKAL-WALLIS TEST VERSUS THE ONE-WAY ANOVA TEST**

Suppose that the distributions of a variable of several populations have the same shape and that you want to compare the population means using independent samples. When deciding between the one-way ANOVA test and the Kruskal–Wallis test, follow these guidelines: If you are reasonably sure that the distributions are normal, use the one-way ANOVA test; otherwise, use the Kruskal–Wallis test.

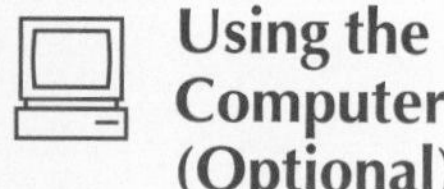

Using the Computer (Optional)

Procedure 16.3 provides a step-by-step method for performing a Kruskal–Wallis test. Alternatively, we can use Minitab to conduct such a test. Example 16.10 presents the details.

EXAMPLE 16.10 **USING MINITAB TO PERFORM A KRUSKAL-WALLIS TEST**

Use Minitab to perform the hypothesis test in Example 16.9.

SOLUTION Let μ_1, μ_2, and μ_3 denote last year's mean number of miles driven for cars, buses, and trucks, respectively. The problem is to use the Kruskal–Wallis procedure to perform the hypothesis test

$$H_0\colon \mu_1 = \mu_2 = \mu_3 \text{ (mean miles driven are equal)}$$

$$H_a\colon \text{Not all the means are equal.}$$

at the 5% significance level.

Minitab requires stacked data for its Kruskal–Wallis procedure. So we first store all 25 mileages from Table 16.12 on page 974 in a column named MILES. Then, in a column named VEHICLE, we store the vehicle types corresponding to the mileages in MILES.

For instance, suppose we store the mileage data for the cars in the first 10 rows of MILES, that for the buses in the next six rows of MILES, and that for the trucks in the next nine rows of MILES. Then we would store the word Cars in the first 10 rows of VEHICLE, the word Buses in the next six rows of VEHICLE, and the word Trucks in the next nine rows of VEHICLE.

Also, if we want the output to order the vehicle types as we have been doing (Cars, Buses, and Trucks), we need to take some preliminary steps, as follows.

1 Click anywhere in the column named VEHICLE
2 Choose **Editor ➤ Set Column ➤ Value Order...**
3 Select the **User-specified order** option button
4 In the **Define an order (one value per line)** text box, edit the order of the vehicle types, if necessary, so that it reads Cars, Buses, and Trucks.
5 Click **OK**

Now we are ready to instruct Minitab to perform the Kruskal–Wallis test. We proceed in the following manner.

1 Choose **Stat ➤ Nonparametrics ➤ Kruskal-Wallis...**
2 Specify MILES in the **Response** text box
3 Click in the **Factor** text box and specify VEHICLE
4 Click **OK**

The resulting output is displayed in Printout 16.7.

PRINTOUT 16.7
Minitab output for the Kruskal-Wallis test

```
Kruskal-Wallis Test on MILES

VEHICLE      N    Median    Ave Rank        Z
Cars        10     8.900         9.8    -1.75
Buses        6     7.200         9.0    -1.53
Trucks       9    23.600        19.2     3.14
Overall     25                  13.0

H = 9.92  DF = 2  P = 0.007
H = 9.93  DF = 2  P = 0.007 (adjusted for ties)
```

The first four columns of Printout 16.7 provide the sizes, medians, and mean ranks for the three samples. The fifth column, headed Z, displays the z-scores of the sample mean ranks. This provides a standardized measure of how much each sample mean rank differs from its expected mean rank if the null hypothesis of equal population means is true; each expected mean rank equals the mean rank of all the observations, which in this case is 13. Thus if the null hypothesis is true, each sample mean rank should not differ too much from 13, or equivalently, each z-score should not be too far from 0.

In the second to last line of Printout 16.7, we find the value of the Kruskal–Wallis test statistic (obtained using the calculating formula), the degrees of freedom, and the P-value for the hypothesis test. The last line provides the same information, but with an adjustment for ties; in this case, the adjusted value of the test statistic differs slightly from the unadjusted one, but the P-values are identical to three decimal places.

Since the P-value of 0.007 is less than the specified significance level of 0.05, we reject H_0. At the 5% significance level, the data provide sufficient evidence to conclude that a difference exists in last year's mean number of miles driven among cars, buses, and trucks. □

EXERCISES 16.5

STATISTICAL CONCEPTS AND SKILLS

16.72 Of what test is the Kruskal–Wallis test a nonparametric version?

16.73 State the conditions required for performing a Kruskal–Wallis test.

16.74 In the Kruskal–Wallis test, how do we deal with tied ranks?

16.75 Fill in the following blank: If the null hypothesis of equal population means is true, then the sample mean ranks should be roughly ____.

16.76 For a Kruskal–Wallis test, how do we

a. measure variation among sample mean ranks?

b. measure total variation of all the ranks?

c. decide whether the variation among sample mean ranks is large enough to warrant rejection of the null hypothesis of equal population means?

16.77 For a Kruskal–Wallis test to compare the means of five populations, what is the approximate distribution of the test statistic H?

In each of Exercises 16.78–16.81, perform a Kruskal–Wallis test using either the critical-value approach or the P-value approach.

16.78 The U.S. Bureau of Labor Statistics conducts surveys on consumer expenditures for various types of entertainment and publishes its findings in *Consumer Expenditure Survey*. Independent samples yielded the following data, in dollars, on last year's expenditures for three categories.

Fees and admissions	TV, radio, and sound equipment	Other equipment and services
173	100	0
112	1748	251
22	396	1293
495	0	31
111	470	75
1203	0	1024
609	562	1629
300		102
		1238

At the 5% significance level, do the data provide sufficient evidence to conclude that a difference exists in last year's mean expenditures among the three categories?

16.79 Indications are that Americans have become more aware of the dangers of excessive fat intake in their diets. The U.S. Department of Agriculture publishes data on annual consumption of selected beverages in *Food Consumption, Prices, and Expenditures.* Independent random samples of lowfat-milk consumptions for 1980, 1990, and 1995 revealed the following data, in gallons.

1980	1990	1995
11.1	12.8	14.9
10.7	13.3	18.6
8.6	13.4	14.2
9.4	12.0	16.7
9.2	8.8	18.9
15.1	14.4	15.4
11.6	13.5	16.1
8.3		12.7
		12.9

At the 1% significance level, do the data provide sufficient evidence to conclude that there is a difference in mean (per capita) consumption of lowfat milk for the years 1980, 1990, and 1995?

16.80 Information on characteristics of new-car buyers appeared in *Buyers of New Cars,* a publication of Newsweek, Inc. Independent random samples of new-car buyers yielded the following data on age of purchaser, in years, by origin of car purchased.

Domestic	Asian	European
41	78	72
42	42	42
51	51	58
47	45	39
33	21	67
83	24	39
35	21	45
69	39	27
50	45	33
60	30	55

Do the data provide sufficient evidence to conclude that a difference exists in the median ages of buyers of new domestic, Asian, and European cars? Use $\alpha = 0.05$.

16.81 The U.S. Bureau of the Census publishes information on the sizes of housing units in *Current Housing Reports.* Independent random samples of single-family detached homes (including mobile homes) in the four U.S. regions yielded the following data on square footage.

Northeast	Midwest	South	West
3182	2115	1591	1345
2130	2413	1354	694
1781	1639	722	2789
2989	1691	2135	1649
1581	1655	1982	2203
2149	1605	1639	2068
2286	3361	642	1565
1293	2058	1513	1655

At the 5% significance level, do the data provide sufficient evidence to conclude that a difference exists in median square footage of single-family detached homes among the four U.S. regions?

16.82 A chain of convenience stores wanted to test three different advertising policies:

- Policy 1: No advertising.
- Policy 2: Advertise in neighborhoods with circulars.
- Policy 3: Use circulars and advertise in newspapers.

Eighteen stores were randomly selected and divided randomly into three groups of six stores. Each group used one of the three policies. Following the implementation of the policies, sales figures were obtained for each of the stores during a 1-month period. The figures are displayed, in thousands of dollars, in the following table.

Policy 1	Policy 2	Policy 3
22	21	29
20	25	24
26	25	31
21	20	32
24	22	26
22	26	27

a. Do the data provide sufficient evidence to conclude that there is a difference in mean monthly sales among the three policies? Perform a Kruskal–Wallis test at the 1% significance level.

b. The hypothesis test in part (a) was done in Exercise 16.38 using the one-way ANOVA test. The assumption in that exercise is that for the three policies, monthly sales are normally distributed and have equal standard deviations. Presuming that is true, why is it permissible to perform a Kruskal–Wallis test to compare the means? Is it better in this case to use the one-way ANOVA test or the Kruskal–Wallis test? Explain your answers.

16.83 The Bureau of Labor Statistics publishes data on weekly earnings of nonsupervisory workers in *Employment and Earnings.* The following data, in dollars, were obtained from random samples of full and part-time nonsupervisory workers in five service-producing industries.

Transp. and Pub. util.	Wholesale trade	Retail trade	Finance, Insurance, Real estate	Services
543	524	260	482	408
583	469	188	436	420
544	449	170	518	427
588	518	298	404	343
635	502	185		380
566		279		317

a. Do the data provide sufficient evidence to conclude that a difference exists in mean weekly earnings among nonsupervisory workers in the five industries? Perform a Kruskal–Wallis test at the 5% significance level. *(Note:* Although $n_4 < 5$, most statisticians would consider it reasonable to conduct a Kruskal–Wallis test on these data.*)*

b. The hypothesis test in part (a) was done in Exercise 16.39 using the one-way ANOVA test. The assumption in that exercise is that weekly earnings in the five industries are normally distributed and have equal standard deviations. Presuming that is true, why is it permissible to perform a Kruskal–Wallis test to compare the means? Is it better in this case to use the one-way ANOVA test or the Kruskal–Wallis test? Explain your answers.

16.84 Suppose you want to perform a hypothesis test to compare four population means using independent samples. In each case, decide whether you would use the one-way ANOVA test, the Kruskal–Wallis test, or neither of these tests. Preliminary data analyses of the samples suggest that

a. the four distributions of the variable are not normal but have the same shape.

b. the four distributions of the variable are normal and have the same shape.

16.85 Suppose you want to perform a hypothesis test to compare six population means using independent samples. In each case, decide whether you would use the one-way ANOVA test, the Kruskal–Wallis test, or neither of these tests. Preliminary data analyses of the samples suggest that

a. the six distributions of the variable are not normal and have quite different shapes.

b. the six distributions of the variable are normal but have quite different shapes.

EXTENDING THE CONCEPTS AND SKILLS

16.86 In this section we illustrated the Kruskal–Wallis test using data on miles driven by samples of cars, buses, and trucks. The value of the test statistic H was computed on page 971 to be 9.930, whereas on page 975 we found its value to be 9.923.

a. Explain the discrepancy between the two values of H.

b. Does the difference in the values affect our conclusion in the hypothesis test we conducted? Explain.

c. Does the difference in the values affect our estimate of the P-value of the hypothesis test? Explain your answer.

USING TECHNOLOGY  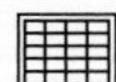 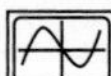

16.87 When using Minitab to perform a Kruskal–Wallis test, what type of data storage is required, stacked or unstacked?

16.88 Refer to Exercise 16.78. Use Minitab or some other statistical software to perform the required hypothesis test using the Kruskal–Wallis procedure.

16.89 Refer to Exercise 16.79. Use Minitab or some other statistical software to perform the required hypothesis test using the Kruskal–Wallis procedure.

16.90 The U.S. Bureau of the Census collects data on income by educational attainment, sex, and age. Results are published in *Current Population Reports.* Independent random samples were taken of women from three categories of educational attainment: elementary school, high school, and college (4-year degree). Then Minitab was applied to perform a Kruskal–Wallis test on the annual incomes of the women sampled. Printout 16.8 displays the resulting output, where the incomes are in thousands of dollars. Determine

a. the value of the test statistic, H.
b. the null and alternative hypotheses.
c. the P-value for the hypothesis test.
d. the conclusion if the hypothesis test is performed at the 1% significance level.
e. the sample size, sample median, and mean rank for each of the three samples.
f. Compare the result of the Kruskal–Wallis test to that of the one-way ANOVA test done on the same data in Exercise 16.48 on page 957.

16.91 The Motor Vehicle Manufacturers Association of the United States conducts surveys on the costs of owning and operating a motor vehicle. Data are published in *Motor Vehicle Facts and Figures* and include costs for gas and oil, tires, maintenance, insurance, license and registration, and depreciation. Independent random samples of owners of large, intermediate, and compact cars were taken to obtain information on annual insurance premiums. Then Minitab's Kruskal–Wallis procedure was applied to the resulting data. Printout 16.9 displays the output generated by Minitab. Determine

a. the value of the test statistic, H.
b. the null and alternative hypotheses.
c. the P-value for the hypothesis test.
d. the conclusion if the hypothesis test is performed at the 5% significance level.
e. the sample size, sample median, and mean rank for each of the three samples.
f. Compare the result of the Kruskal–Wallis test to that of the one-way ANOVA test done on the same data in Exercise 16.49 on page 957.

PRINTOUT 16.8 Minitab output for Exercise 16.90

```
Kruskal-Wallis Test on INCOME

EDUCATE     N   Median   Ave Rank       Z
Element    40    15.65       37.9   -6.63
HS         62    22.30       68.4   -2.37
College    55    36.70      120.9    8.48
Overall   157                79.0

H = 82.85  DF = 2  P = 0.000
H = 82.85  DF = 2  P = 0.000 (adjusted for ties)
```

PRINTOUT 16.9 Minitab output for Exercise 16.91

```
Kruskal-Wallis Test on PREMIUM

TYPE        N   Median   Ave Rank       Z
Large      20    926.2       43.0    0.93
Intermed   30    836.4       35.8   -1.01
Compact    27    859.0       39.6    0.18
Overall    77                39.0

H = 1.29  DF = 2  P = 0.525
```

CHAPTER REVIEW

You Should Be Able To

1. use and understand the formulas presented in this chapter.
2. use the F-table, Table IX in Appendix A.
3. explain the essential ideas behind a one-way analysis of variance.
4. state and check the assumptions required for a one-way ANOVA.
5. obtain the sums of squares for a one-way ANOVA using the defining formulas.
6. obtain the sums of squares for a one-way ANOVA using the computing formulas.
7. compute the mean squares and the F-statistic for a one-way ANOVA.
8. construct a one-way ANOVA table.
9. perform a one-way ANOVA test.

*10. use the q-tables, Tables XI and XII in Appendix A.
*11. perform a multiple comparison using the Tukey method.
*12. perform a Kruskal–Wallis test.
*13. use the Minitab procedures covered in this chapter.
*14. interpret the output obtained from the application of the Minitab procedures discussed in this chapter.

Key Terms

analysis of variance (ANOVA), *933*
degrees of freedom for the denominator, *934*
degrees of freedom for the numerator, *934*
error, *942*
error mean square (MSE), *942*
error sum of squares (SSE), *942*
F_α, *935*
F-curve, *934*
F-distribution, *934*
F-statistic, *942*
factor, *937*
family confidence level,* *959*
individual confidence level,* *959*
Kruskal–Wallis test,* *973*
level, *937*
multiple comparisons,* *959*
one-way analysis of variance, *937*
one-way ANOVA identity, *945*
one-way ANOVA tables, *946*
one-way ANOVA test, *948*
q_α,* *959*
q-curve,* *959*
q-distribution,* *959*
residual, *937*
response variable, *951*
rule of 2, *937*
studentized range distribution,* *959*
total sum of squares (SST), *945*
treatment, *942*
treatment mean square ($MSTR$), *941*
treatment sum of squares ($SSTR$), *941*
Tukey multiple-comparison method,* *961*

REVIEW TEST

Statistical Concepts and Skills

1. For what is one-way ANOVA used?
2. State the three assumptions for one-way ANOVA and explain how those assumptions can be checked.
3. On what distribution does one-way ANOVA rely?
4. Suppose we want to compare the means of three populations using one-way ANOVA. If the sample sizes are 5, 6, and 6, determine the degrees of freedom for the appropriate F-curve.

5. In one-way ANOVA, identify a statistic that measures each of the following:
a. the variation among the sample means.
b. the variation within the samples.

6. In one-way ANOVA,
a. list and interpret the three sums of squares.
b. state the one-way ANOVA identity and interpret its meaning with regard to partitioning the total variation among all the data.

7. For a one-way ANOVA:
a. Identify one purpose of one-way ANOVA tables.
b. Construct a generic one-way ANOVA table.

***8.** What is the purpose of conducting a multiple comparison?

***9.** Explain the difference between the individual confidence level and the family confidence level. Which confidence level is appropriate for multiple comparisons? Explain your answer.

***10.** Upon which distribution is the Tukey multiple-comparison procedure based?

***11.** Consider a Tukey multiple comparison of four population means with a family confidence level of 0.95. Is the individual confidence level smaller or larger than 0.95? Explain your answer.

***12.** Suppose that we want to compare the means of three populations using the Tukey multiple-comparison procedure. If the sample sizes are 5, 6, and 6, determine the parameters for the appropriate q-curve.

***13.** Identify a nonparametric alternative to the one-way ANOVA procedure.

***14.** Identify the distribution used as an approximation to the true distribution of the H statistic for a Kruskal–Wallis test.

***15.** Explain the logic behind a Kruskal–Wallis test.

***16.** Suppose you want to compare the means of several populations using independent samples. If given the choice between using the one-way ANOVA test and the Kruskal–Wallis test, which one would you choose if outliers occur in the sample data? Explain your answer.

17. Consider an F-curve with df = (2, 14).
a. Identify the number of degrees of freedom for the numerator.
b. Identify the number of degrees of freedom for the denominator.
c. Determine $F_{0.05}$.
d. Find the F-value having area 0.01 to its right.
e. Find the F-value having area 0.05 to its right.

18. Consider the following three hypothetical samples:

A	*B*	*C*
1	0	3
3	6	12
5	2	6
	5	3
	2	

a. Obtain the sample mean and sample standard deviation of each of the three samples.
b. Obtain *SST*, *SSTR*, and *SSE* using the defining formulas and verify that the one-way ANOVA identity holds.
c. Obtain *SST*, *SSTR*, and *SSE* using the computing formulas.
d. Construct the one-way ANOVA table.

19. The Federal Bureau of Investigation conducts surveys to obtain information on the value of losses due to various types of robberies. Results of the surveys are published in *Population-at-Risk Rates and Selected Crime Indicators.* Independent random samples of reports for three types of robberies—highway, gas station, and convenience store—gave the following data, in dollars, on value of losses.

Highway	Gas station	Convenience store
608	636	476
652	533	553
495	512	512
744	451	338
680	291	234
	618	246

a. What is *MSTR* measuring here?
b. What is *MSE* measuring here?

c. Suppose we want to perform a one-way ANOVA to compare the mean losses among the three types of robberies. What conditions are necessary? How crucial are those conditions?

20. Refer to Problem 19. At the 5% significance level, do the data provide sufficient evidence to conclude that a difference exists in mean losses among the three types of robberies? Use one-way ANOVA to perform the required hypothesis test. *(Note:* We have $T_1 = 3179$, $T_2 = 3041$, $T_3 = 2359$, and $\Sigma x^2 = 4{,}700{,}509$.*)*

***21.** Consider a q-curve with parameters 3 and 14.
a. Determine $q_{0.05}$.
b. Find the q-value having area 0.01 to its right.

***22.** Refer to Problem 19.
a. Apply the Tukey multiple-comparison method to the data. Use a family confidence level of 0.95.
b. Interpret your results from part (a) in words.

***23.** Refer to Problem 19.
a. At the 5% significance level, do the data provide sufficient evidence to conclude that a difference exists in mean losses among the three types of robberies? Use the Kruskal–Wallis procedure to perform the required hypothesis test.
b. The hypothesis test in part (a) was done in Problem 20 using the one-way ANOVA procedure. The assumption in that exercise is that for the three types of robberies, losses are normally distributed and have equal standard deviations. Presuming that is true, why is it permissible to perform a Kruskal–Wallis test to compare the means? Is it better in this case to use the one-way ANOVA test or the Kruskal–Wallis test? Explain your answers.

USING TECHNOLOGY  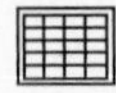 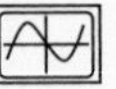

24. Refer to Problem 19. Use Minitab or some other statistical software to
a. Obtain normal probability plots for each of the three samples.
b. Perform a residual analysis.
c. Does it seem reasonable to consider the assumptions for one-way ANOVA met? Why?

25. Use Minitab or some other statistical software to carry out the one-way ANOVA test required in Problem 20.

26. In Problem 20 you performed a one-way ANOVA on the data in Problem 19 to decide whether a difference exists in mean losses among three types of robberies. Printout 16.10 shows the output obtained by applying Minitab's one-way ANOVA procedure to the data. Use the output to determine
a. the three sums of squares, *SSTR, SSE,* and *SST.*
b. the treatment mean square and the error mean square.
c. the value of the test statistic, F.
d. the P-value for the hypothesis test.
e. the conclusion, if the hypothesis test is performed at the 5% significance level.
f. the size, mean, and standard deviation for each of the three samples.
g. an individual 95% confidence interval for the mean loss for convenience-store robberies.

***27.** Use Minitab or some other statistical software to carry out the multiple comparison required in Problem 22.

***28.** We applied Minitab's Tukey multiple-comparison procedure to the data in Problem 19. Printout 16.11 displays the portion of the output required for a multiple comparison. Use the output to determine
a. the family confidence level.
b. the individual confidence level.
c. $q_{0.05}$.
d. the (simultaneous) confidence interval for the difference, $\mu_2 - \mu_3$, between the mean losses for gas-station and convenience-store robberies.
e. the (simultaneous) confidence interval for the difference, $\mu_2 - \mu_1$, between the mean losses for gas-station and highway robberies.
f. which means should be declared different.
g. Summarize the results of the Tukey multiple comparison in words.

***29.** Use Minitab or some other statistical software to carry out the Kruskal–Wallis test required in Problem 23.

***30.** In Problem 23 you performed a Kruskal–Wallis test on the data in Problem 19 to decide whether a difference exists in mean losses among three types of robberies. Printout 16.12 (next page) shows the output obtained by applying Minitab's Kruskal–Wallis procedure to the data. Use the output to find

a. the value of the test statistic, H.

b. the null and alternative hypotheses.

c. the P-value for the hypothesis test.

d. the conclusion, if the hypothesis test is performed at the 5% significance level.

e. the sample size, sample median, and mean rank for each of the three samples.

f. Compare the result of the Kruskal–Wallis test to that of the one-way ANOVA test done on the same data in Problem 26.

PRINTOUT 16.10 Minitab output for Problem 26

```
Analysis of Variance
Source     DF        SS          MS         F        P
Factor      2    160601       80301      5.34    0.019
Error      14    210540       15039
Total      16    371142
                                      Individual 95% CIs For Mean
                                      Based on Pooled StDev
Level       N      Mean       StDev   -+---------+---------+---------+-----
HIGHWAY     5     635.8        92.9                   (------*-------)
GAS         6     506.8       126.1           (------*------)
CONVEN      6     393.2       139.0   (------*------)
                                      -+---------+---------+---------+-----
Pooled StDev =    122.6               300       450       600       750
```

PRINTOUT 16.11 Minitab output for Problem 28

```
Tukey's pairwise comparisons

    Family error rate = 0.0500
Individual error rate = 0.0203

Critical value = 3.70

Intervals for (column level mean) - (row level mean)

             Highway          Gas

 Gas             -65
                 323

 Conven           48          -72
                 437          299
```

PRINTOUT 16.12 Minitab output for Problem 30

```
Kruskal-Wallis Test on LOSS

TYPE         N    Median    Ave Rank        Z
Highway      5     652.0        13.4     2.32
Gas          6     522.5         8.9    -0.05
Conven       6     407.0         5.4    -2.16
Overall     17                   9.0

H = 6.82  DF = 2  P = 0.033
H = 6.83  DF = 2  P = 0.033 (adjusted for ties)
```

INTERNET PROJECT

Brain Damage and the Courts

In this project, you will examine data collected within the Australian court system. The subjects in the study are the plaintiffs; they are people who suffered brain damage in automobile accidents and sued for monetary damages based on their injury.

You will be able to decide whether the (mean) amount of money plaintiffs request differ among age groups, in other words, whether there is a difference in requested compensation based on age. In a second analysis, you will determine whether there is a difference due to the plaintiffs' gender.

URL for access to Internet Projects Page: `http://hepg.awl.com` *keyword:* Weiss

USING THE FOCUS DATABASE

In Chapter 1 we explained how to store the Focus database in a Minitab worksheet named `focus.mtw`. If you haven't already created that worksheet, follow the instructions on pages 54–55 to create it now.

The Focus database contains information on 500 randomly selected Arizona State University sophomores for seven different variables: sex, high-school GPA, SAT math score, cumulative GPA, SAT verbal score, age, and total hours. For the following database exercises, you should first eliminate all cases (students) in which one or more of the four variables, cumulative GPA, high-school GPA, SAT math score, and SAT verbal score, equal 0. Then use Minitab or some other statistical software to solve each problem.

a. Conduct a one-way ANOVA at the 5% significance level to test for differences among mean cumulative GPAs of Arizona State University sophomores in the three high-school GPA categories: under 2, 2–2.99, and 3 or over. Also perform a Tukey multiple comparison using a family confidence level of 95%.

b. Conduct a one-way ANOVA at the 5% significance level to test for differences among mean cumulative GPAs of Arizona State University sophomores in the five SAT math score categories: under 400, 400–499, 500–599, 600–699, and 700 or over. Also perform a Tukey multiple comparison using a family confidence level of 95%.

c. Conduct a one-way ANOVA at the 5% significance level to test for differences among mean cumulative GPAs of Arizona State University sophomores in the four SAT verbal score categories: under 400, 400–499, 500–599, and 600–699. (None of the students under consideration have an SAT verbal score of 700 or above.) Also perform a Tukey multiple comparison using a family confidence level of 95%.

d. Conduct a one-way ANOVA at the 5% significance level to test for differences among mean cumulative GPAs of Arizona State University sophomores in the three age categories: under 20, 20, and over 20. Also perform a Tukey multiple comparison using a family confidence level of 95%.

CASE STUDY DISCUSSION

Heavy Drinking Among College Students

As we learned at the beginning of this chapter, Professor Kate Carey of Syracuse University surveyed 78 college students from an introductory psychology class, all of whom were regular drinkers of alcohol. Her purpose was to identify interpersonal and intrapersonal situations associated with excessive drinking among college students and to detect those situations that differentiate heavy drinkers from light and moderate ones.

To quantify drinking patterns, Carey used the time-line follow-back procedure (TLFB). The TLFB is a structured interview that is known to provide reliable estimates of daily drinking. In this case each student filled in each day of a blank calendar covering the previous month with the number of standard drink equivalents (SDEs) consumed on that day. One SDE is defined to be 1 fluid ounce of hard liquor, 12 fluid ounces of beer, or 4 fluid ounces of wine. Based on the results of the TLFB, the students were divided into three categories according to average quantity of alcohol consumed per drinking day: light drinkers (≤ 3 SDEs), moderate drinkers (4–6 SDEs), and heavy drinkers (> 6 SDEs).

To assess the frequency of excessive drinking in interpersonal and intrapersonal situations, Carey employed the short form of the Inventory of Drinking Situations (IDS). This form consists of 42 items, each of which is rated on a 4-point scale ranging from (1) never drink heavily in that type of situation to (4) almost always drink heavily in that type of situation. The 42 items are divided into eight subscales, three interpersonal and five intrapersonal, as displayed in the first column of the table on page 932. A subscale score represents the average rating for the items constituting the subscale.

a. For each of the eight IDS subscales, perform a (separate) one-way ANOVA to decide whether a difference exists in mean IDS scores among the three drinker categories. Use $\alpha = 0.05$. *(Note:* Use the formula given directly before Exercise 16.41 on page 956 to obtain the overall mean $\bar{x}$ for each ANOVA.*)*

***b.** For those one-way ANOVAs in part (a) that are statistically significant, carry out a Tukey multiple comparison using a family confidence level of 0.95.

c. Based on the data in the table on page 932, are there any of the eight ANOVAs that perhaps should not have been carried out? Explain your answer. *(Hint:* Rule of 2.*)*

BIOGRAPHY SIR RONALD FISHER

Ronald Fisher was born on February 17, 1890, in London, England; he was a surviving twin in a family of eight children; his father was a prominent auctioneer. Fisher graduated from Cambridge in 1912 with degrees in mathematics and physics.

From 1912 to 1919, Fisher worked at an investment house, did farm chores in Canada, and taught high school. In 1919, he took a position as a statistician at Rothamsted Experimental Station in Harpenden, West Hertford, England. His charge was to sort and reassess a 66-year accumulation of data on manurial field trials and weather records.

Fisher's work at Rothamsted during the next 15 years earned him the reputation as the leading statistician of his day and as a top-ranking geneticist. It was there, in 1925, that he published *Statistics for Research Workers,* a book that remained in print for 50 years. Fisher made important contributions to analysis of variance, or ANOVA, exact tests of significance for small samples, and maximum-likelihood solutions. He developed experimental designs to address issues in biological research, such as small samples, variable materials, and fluctuating environments.

Fisher has been described as "slight, bearded, eloquent, reactionary, and quirkish; genial to his disciples and hostile to his dissenters." He was also a prolific writer—over a span of 50 years, he wrote an average of one paper every two months!

In 1933, Fisher became Galton professor of Eugenics at University College in London, and, in 1943, Balfour professor of genetics at Cambridge. In 1952, he was knighted.

Fisher "retired" in 1959, moved to Australia, and spent the last three years of his life working at the Division of Mathematical Statistics of the Commonwealth Scientific and Industrial Research Organization. He died in 1962 in Adelaide, Australia.

Appendixes

Statistical Tables

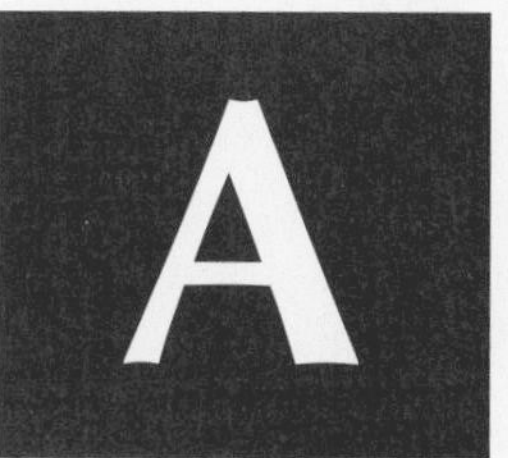

TABLE I
Random numbers

Line number	Column number									
	00–09		*10–19*		*20–29*		*30–39*		*40–49*	
00	15544	80712	97742	21500	97081	42451	50623	56071	28882	28739
01	01011	21285	04729	39986	73150	31548	30168	76189	56996	19210
02	47435	53308	40718	29050	74858	64517	93573	51058	68501	42723
03	91312	75137	86274	59834	69844	19853	06917	17413	44474	86530
04	12775	08768	80791	16298	22934	09630	98862	39746	64623	32768
05	31466	43761	94872	92230	52367	13205	38634	55882	77518	36252
06	09300	43847	40881	51243	97810	18903	53914	31688	06220	40422
07	73582	13810	57784	72454	68997	72229	30340	08844	53924	89630
08	11092	81392	58189	22697	41063	09451	09789	00637	06450	85990
09	93322	98567	00116	35605	66790	52965	62877	21740	56476	49296
10	80134	12484	67089	08674	70753	90959	45842	59844	45214	36505
11	97888	31797	95037	84400	76041	96668	75920	68482	56855	97417
12	92612	27082	59459	69380	98654	20407	88151	56263	27126	63797
13	72744	45586	43279	44218	83638	05422	00995	70217	78925	39097
14	96256	70653	45285	26293	78305	80252	03625	40159	68760	84716
15	07851	47452	66742	83331	54701	06573	98169	37499	67756	68301
16	25594	41552	96475	56151	02089	33748	65289	89956	89559	33687
17	65358	15155	59374	80940	03411	94656	69440	47156	77115	99463
18	09402	31008	53424	21928	02198	61201	02457	87214	59750	51330
19	97424	90765	01634	37328	41243	33564	17884	94747	93650	77668

TABLE II
Areas under the standard normal curve

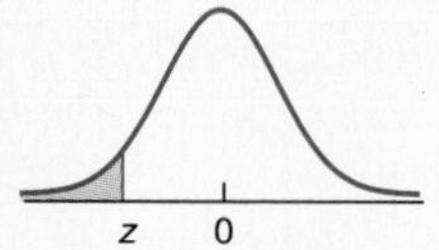

	Second decimal place in z									
z	0.09	0.08	0.07	0.06	0.05	0.04	0.03	0.02	0.01	0.00
−3.9										0.0000†
−3.8	0.0001	0.0001	0.0001	0.0001	0.0001	0.0001	0.0001	0.0001	0.0001	0.0001
−3.7	0.0001	0.0001	0.0001	0.0001	0.0001	0.0001	0.0001	0.0001	0.0001	0.0001
−3.6	0.0001	0.0001	0.0001	0.0001	0.0001	0.0001	0.0001	0.0001	0.0002	0.0002
−3.5	0.0002	0.0002	0.0002	0.0002	0.0002	0.0002	0.0002	0.0002	0.0002	0.0002
−3.4	0.0002	0.0003	0.0003	0.0003	0.0003	0.0003	0.0003	0.0003	0.0003	0.0003
−3.3	0.0003	0.0004	0.0004	0.0004	0.0004	0.0004	0.0004	0.0005	0.0005	0.0005
−3.2	0.0005	0.0005	0.0005	0.0006	0.0006	0.0006	0.0006	0.0006	0.0007	0.0007
−3.1	0.0007	0.0007	0.0008	0.0008	0.0008	0.0008	0.0009	0.0009	0.0009	0.0010
−3.0	0.0010	0.0010	0.0011	0.0011	0.0011	0.0012	0.0012	0.0013	0.0013	0.0013
−2.9	0.0014	0.0014	0.0015	0.0015	0.0016	0.0016	0.0017	0.0018	0.0018	0.0019
−2.8	0.0019	0.0020	0.0021	0.0021	0.0022	0.0023	0.0023	0.0024	0.0025	0.0026
−2.7	0.0026	0.0027	0.0028	0.0029	0.0030	0.0031	0.0032	0.0033	0.0034	0.0035
−2.6	0.0036	0.0037	0.0038	0.0039	0.0040	0.0041	0.0043	0.0044	0.0045	0.0047
−2.5	0.0048	0.0049	0.0051	0.0052	0.0054	0.0055	0.0057	0.0059	0.0060	0.0062
−2.4	0.0064	0.0066	0.0068	0.0069	0.0071	0.0073	0.0075	0.0078	0.0080	0.0082
−2.3	0.0084	0.0087	0.0089	0.0091	0.0094	0.0096	0.0099	0.0102	0.0104	0.0107
−2.2	0.0110	0.0113	0.0116	0.0119	0.0122	0.0125	0.0129	0.0132	0.0136	0.0139
−2.1	0.0143	0.0146	0.0150	0.0154	0.0158	0.0162	0.0166	0.0170	0.0174	0.0179
−2.0	0.0183	0.0188	0.0192	0.0197	0.0202	0.0207	0.0212	0.0217	0.0222	0.0228
−1.9	0.0233	0.0239	0.0244	0.0250	0.0256	0.0262	0.0268	0.0274	0.0281	0.0287
−1.8	0.0294	0.0301	0.0307	0.0314	0.0322	0.0329	0.0336	0.0344	0.0351	0.0359
−1.7	0.0367	0.0375	0.0384	0.0392	0.0401	0.0409	0.0418	0.0427	0.0436	0.0446
−1.6	0.0455	0.0465	0.0475	0.0485	0.0495	0.0505	0.0516	0.0526	0.0537	0.0548
−1.5	0.0559	0.0571	0.0582	0.0594	0.0606	0.0618	0.0630	0.0643	0.0655	0.0668
−1.4	0.0681	0.0694	0.0708	0.0721	0.0735	0.0749	0.0764	0.0778	0.0793	0.0808
−1.3	0.0823	0.0838	0.0853	0.0869	0.0885	0.0901	0.0918	0.0934	0.0951	0.0968
−1.2	0.0985	0.1003	0.1020	0.1038	0.1056	0.1075	0.1093	0.1112	0.1131	0.1151
−1.1	0.1170	0.1190	0.1210	0.1230	0.1251	0.1271	0.1292	0.1314	0.1335	0.1357
−1.0	0.1379	0.1401	0.1423	0.1446	0.1469	0.1492	0.1515	0.1539	0.1562	0.1587
−0.9	0.1611	0.1635	0.1660	0.1685	0.1711	0.1736	0.1762	0.1788	0.1814	0.1841
−0.8	0.1867	0.1894	0.1922	0.1949	0.1977	0.2005	0.2033	0.2061	0.2090	0.2119
−0.7	0.2148	0.2177	0.2206	0.2236	0.2266	0.2296	0.2327	0.2358	0.2389	0.2420
−0.6	0.2451	0.2483	0.2514	0.2546	0.2578	0.2611	0.2643	0.2676	0.2709	0.2743
−0.5	0.2776	0.2810	0.2843	0.2877	0.2912	0.2946	0.2981	0.3015	0.3050	0.3085
−0.4	0.3121	0.3156	0.3192	0.3228	0.3264	0.3300	0.3336	0.3372	0.3409	0.3446
−0.3	0.3483	0.3520	0.3557	0.3594	0.3632	0.3669	0.3707	0.3745	0.3783	0.3821
−0.2	0.3859	0.3897	0.3936	0.3974	0.4013	0.4052	0.4090	0.4129	0.4168	0.4207
−0.1	0.4247	0.4286	0.4325	0.4364	0.4404	0.4443	0.4483	0.4522	0.4562	0.4602
−0.0	0.4641	0.4681	0.4721	0.4761	0.4801	0.4840	0.4880	0.4920	0.4960	0.5000

† For $z \leq -3.90$, the areas are 0.0000 to four decimal places.

TABLE II (cont.)
Areas under the standard normal curve

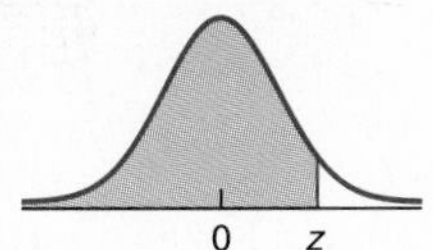

	Second decimal place in z									
z	0.00	0.01	0.02	0.03	0.04	0.05	0.06	0.07	0.08	0.09
0.0	0.5000	0.5040	0.5080	0.5120	0.5160	0.5199	0.5239	0.5279	0.5319	0.5359
0.1	0.5398	0.5438	0.5478	0.5517	0.5557	0.5596	0.5636	0.5675	0.5714	0.5753
0.2	0.5793	0.5832	0.5871	0.5910	0.5948	0.5987	0.6026	0.6064	0.6103	0.6141
0.3	0.6179	0.6217	0.6255	0.6293	0.6331	0.6368	0.6406	0.6443	0.6480	0.6517
0.4	0.6554	0.6591	0.6628	0.6664	0.6700	0.6736	0.6772	0.6808	0.6844	0.6879
0.5	0.6915	0.6950	0.6985	0.7019	0.7054	0.7088	0.7123	0.7157	0.7190	0.7224
0.6	0.7257	0.7291	0.7324	0.7357	0.7389	0.7422	0.7454	0.7486	0.7517	0.7549
0.7	0.7580	0.7611	0.7642	0.7673	0.7704	0.7734	0.7764	0.7794	0.7823	0.7852
0.8	0.7881	0.7910	0.7939	0.7967	0.7995	0.8023	0.8051	0.8078	0.8106	0.8133
0.9	0.8159	0.8186	0.8212	0.8238	0.8264	0.8289	0.8315	0.8340	0.8365	0.8389
1.0	0.8413	0.8438	0.8461	0.8485	0.8508	0.8531	0.8554	0.8577	0.8599	0.8621
1.1	0.8643	0.8665	0.8686	0.8708	0.8729	0.8749	0.8770	0.8790	0.8810	0.8830
1.2	0.8849	0.8869	0.8888	0.8907	0.8925	0.8944	0.8962	0.8980	0.8997	0.9015
1.3	0.9032	0.9049	0.9066	0.9082	0.9099	0.9115	0.9131	0.9147	0.9162	0.9177
1.4	0.9192	0.9207	0.9222	0.9236	0.9251	0.9265	0.9279	0.9292	0.9306	0.9319
1.5	0.9332	0.9345	0.9357	0.9370	0.9382	0.9394	0.9406	0.9418	0.9429	0.9441
1.6	0.9452	0.9463	0.9474	0.9484	0.9495	0.9505	0.9515	0.9525	0.9535	0.9545
1.7	0.9554	0.9564	0.9573	0.9582	0.9591	0.9599	0.9608	0.9616	0.9625	0.9633
1.8	0.9641	0.9649	0.9656	0.9664	0.9671	0.9678	0.9686	0.9693	0.9699	0.9706
1.9	0.9713	0.9719	0.9726	0.9732	0.9738	0.9744	0.9750	0.9756	0.9761	0.9767
2.0	0.9772	0.9778	0.9783	0.9788	0.9793	0.9798	0.9803	0.9808	0.9812	0.9817
2.1	0.9821	0.9826	0.9830	0.9834	0.9838	0.9842	0.9846	0.9850	0.9854	0.9857
2.2	0.9861	0.9864	0.9868	0.9871	0.9875	0.9878	0.9881	0.9884	0.9887	0.9890
2.3	0.9893	0.9896	0.9898	0.9901	0.9904	0.9906	0.9909	0.9911	0.9913	0.9916
2.4	0.9918	0.9920	0.9922	0.9925	0.9927	0.9929	0.9931	0.9932	0.9934	0.9936
2.5	0.9938	0.9940	0.9941	0.9943	0.9945	0.9946	0.9948	0.9949	0.9951	0.9952
2.6	0.9953	0.9955	0.9956	0.9957	0.9959	0.9960	0.9961	0.9962	0.9963	0.9964
2.7	0.9965	0.9966	0.9967	0.9968	0.9969	0.9970	0.9971	0.9972	0.9973	0.9974
2.8	0.9974	0.9975	0.9976	0.9977	0.9977	0.9978	0.9979	0.9979	0.9980	0.9981
2.9	0.9981	0.9982	0.9982	0.9983	0.9984	0.9984	0.9985	0.9985	0.9986	0.9986
3.0	0.9987	0.9987	0.9987	0.9988	0.9988	0.9989	0.9989	0.9989	0.9990	0.9990
3.1	0.9990	0.9991	0.9991	0.9991	0.9992	0.9992	0.9992	0.9992	0.9993	0.9993
3.2	0.9993	0.9993	0.9994	0.9994	0.9994	0.9994	0.9994	0.9995	0.9995	0.9995
3.3	0.9995	0.9995	0.9995	0.9996	0.9996	0.9996	0.9996	0.9996	0.9996	0.9997
3.4	0.9997	0.9997	0.9997	0.9997	0.9997	0.9997	0.9997	0.9997	0.9997	0.9998
3.5	0.9998	0.9998	0.9998	0.9998	0.9998	0.9998	0.9998	0.9998	0.9998	0.9998
3.6	0.9998	0.9998	0.9999	0.9999	0.9999	0.9999	0.9999	0.9999	0.9999	0.9999
3.7	0.9999	0.9999	0.9999	0.9999	0.9999	0.9999	0.9999	0.9999	0.9999	0.9999
3.8	0.9999	0.9999	0.9999	0.9999	0.9999	0.9999	0.9999	0.9999	0.9999	0.9999
3.9	1.0000†									

† For $z \geq 3.90$, the areas are 1.0000 to four decimal places.

TABLE III
Normal scores

Ordered position	n = 5	6	7	8	9	10	11	12	13
1	−1.18	−1.28	−1.36	−1.43	−1.50	−1.55	−1.59	−1.64	−1.68
2	−0.50	−0.64	−0.76	−0.85	−0.93	−1.00	−1.06	−1.11	−1.16
3	0.00	−0.20	−0.35	−0.47	−0.57	−0.65	−0.73	−0.79	−0.85
4	0.50	0.20	0.00	−0.15	−0.27	−0.37	−0.46	−0.53	−0.60
5	1.18	0.64	0.35	0.15	0.00	−0.12	−0.22	−0.31	−0.39
6		1.28	0.76	0.47	0.27	0.12	0.00	−0.10	−0.19
7			1.36	0.85	0.57	0.37	0.22	0.10	0.00
8				1.43	0.93	0.65	0.46	0.31	0.19
9					1.50	1.00	0.73	0.53	0.39
10						1.55	1.06	0.79	0.60
11							1.59	1.11	0.85
12								1.64	1.16
13									1.68

TABLE III (cont.)
Normal scores

Ordered position	n = 14	15	16	17	18	19	20	21	22
1	−1.71	−1.74	−1.77	−1.80	−1.82	−1.85	−1.87	−1.89	−1.91
2	−1.20	−1.24	−1.28	−1.32	−1.35	−1.38	−1.40	−1.43	−1.45
3	−0.90	−0.94	−0.99	−1.03	−1.06	−1.10	−1.13	−1.16	−1.18
4	−0.66	−0.71	−0.76	−0.80	−0.84	−0.88	−0.92	−0.95	−0.98
5	−0.45	−0.51	−0.57	−0.62	−0.66	−0.70	−0.74	−0.78	−0.81
6	−0.27	−0.33	−0.39	−0.45	−0.50	−0.54	−0.59	−0.63	−0.66
7	−0.09	−0.16	−0.23	−0.29	−0.35	−0.40	−0.45	−0.49	−0.53
8	0.09	0.00	−0.08	−0.15	−0.21	−0.26	−0.31	−0.36	−0.40
9	0.27	0.16	0.08	0.00	−0.07	−0.13	−0.19	−0.24	−0.28
10	0.45	0.33	0.23	0.15	0.07	0.00	−0.06	−0.12	−0.17
11	0.66	0.51	0.39	0.29	0.21	0.13	0.06	0.00	−0.06
12	0.90	0.71	0.57	0.45	0.35	0.26	0.19	0.12	0.06
13	1.20	0.94	0.76	0.62	0.50	0.40	0.31	0.24	0.17
14	1.71	1.24	0.99	0.80	0.66	0.54	0.45	0.36	0.28
15		1.74	1.28	1.03	0.84	0.70	0.59	0.49	0.40
16			1.77	1.32	1.06	0.88	0.74	0.63	0.53
17				1.80	1.35	1.10	0.92	0.78	0.66
18					1.82	1.38	1.13	0.95	0.81
19						1.85	1.40	1.16	0.98
20							1.87	1.43	1.18
21								1.89	1.45
22									1.91

TABLE III (cont.)
Normal scores

Ordered position	n							
	23	*24*	*25*	*26*	*27*	*28*	*29*	*30*
1	−1.93	−1.95	−1.97	−1.98	−2.00	−2.01	−2.03	−2.04
2	−1.48	−1.50	−1.52	−1.54	−1.56	−1.58	−1.59	−1.61
3	−1.21	−1.24	−1.26	−1.28	−1.30	−1.32	−1.34	−1.36
4	−1.01	−1.04	−1.06	−1.09	−1.11	−1.13	−1.15	−1.17
5	−0.84	−0.87	−0.90	−0.93	−0.95	−0.98	−1.00	−1.02
6	−0.70	−0.73	−0.76	−0.79	−0.82	−0.84	−0.87	−0.89
7	−0.57	−0.60	−0.63	−0.66	−0.69	−0.72	−0.75	−0.77
8	−0.44	−0.48	−0.52	−0.55	−0.58	−0.61	−0.64	−0.67
9	−0.33	−0.37	−0.41	−0.44	−0.48	−0.51	−0.54	−0.57
10	−0.22	−0.26	−0.30	−0.34	−0.38	−0.41	−0.44	−0.47
11	−0.11	−0.15	−0.20	−0.24	−0.28	−0.31	−0.35	−0.38
12	0.00	−0.05	−0.10	−0.14	−0.18	−0.22	−0.26	−0.29
13	0.11	0.05	0.00	−0.05	−0.09	−0.13	−0.17	−0.21
14	0.22	0.15	0.10	0.05	0.00	−0.04	−0.09	−0.12
15	0.33	0.26	0.20	0.14	0.09	0.04	0.00	−0.04
16	0.44	0.37	0.30	0.24	0.18	0.13	0.09	0.04
17	0.57	0.48	0.41	0.34	0.28	0.22	0.17	0.12
18	0.70	0.60	0.52	0.44	0.38	0.31	0.26	0.21
19	0.84	0.73	0.63	0.55	0.48	0.41	0.35	0.29
20	1.01	0.87	0.76	0.66	0.58	0.51	0.44	0.38
21	1.21	1.04	0.90	0.79	0.69	0.61	0.54	0.47
22	1.48	1.24	1.06	0.93	0.82	0.72	0.64	0.57
23	1.93	1.50	1.26	1.09	0.95	0.84	0.75	0.67
24		1.95	1.52	1.28	1.11	0.98	0.87	0.77
25			1.97	1.54	1.30	1.13	1.00	0.89
26				1.98	1.56	1.32	1.15	1.02
27					2.00	1.58	1.34	1.17
28						2.01	1.59	1.36
29							2.03	1.61
30								2.04

TABLE IV
Values of t_α

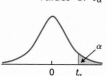

df	$t_{0.10}$	$t_{0.05}$	$t_{0.025}$	$t_{0.01}$	$t_{0.005}$	df
1	3.078	6.314	12.706	31.821	63.657	1
2	1.886	2.920	4.303	6.965	9.925	2
3	1.638	2.353	3.182	4.541	5.841	3
4	1.533	2.132	2.776	3.747	4.604	4
5	1.476	2.015	2.571	3.365	4.032	5
6	1.440	1.943	2.447	3.143	3.707	6
7	1.415	1.895	2.365	2.998	3.499	7
8	1.397	1.860	2.306	2.896	3.355	8
9	1.383	1.833	2.262	2.821	3.250	9
10	1.372	1.812	2.228	2.764	3.169	10
11	1.363	1.796	2.201	2.718	3.106	11
12	1.356	1.782	2.179	2.681	3.055	12
13	1.350	1.771	2.160	2.650	3.012	13
14	1.345	1.761	2.145	2.624	2.977	14
15	1.341	1.753	2.131	2.602	2.947	15
16	1.337	1.746	2.120	2.583	2.921	16
17	1.333	1.740	2.110	2.567	2.898	17
18	1.330	1.734	2.101	2.552	2.878	18
19	1.328	1.729	2.093	2.539	2.861	19
20	1.325	1.725	2.086	2.528	2.845	20
21	1.323	1.721	2.080	2.518	2.831	21
22	1.321	1.717	2.074	2.508	2.819	22
23	1.319	1.714	2.069	2.500	2.807	23
24	1.318	1.711	2.064	2.492	2.797	24
25	1.316	1.708	2.060	2.485	2.787	25
26	1.315	1.706	2.056	2.479	2.779	26
27	1.314	1.703	2.052	2.473	2.771	27
28	1.313	1.701	2.048	2.467	2.763	28
29	1.311	1.699	2.045	2.462	2.756	29
30	1.310	1.697	2.042	2.457	2.750	30
35	1.306	1.690	2.030	2.438	2.724	35
40	1.303	1.684	2.021	2.423	2.704	40
50	1.299	1.676	2.009	2.403	2.678	50
60	1.296	1.671	2.000	2.390	2.660	60
70	1.294	1.667	1.994	2.381	2.648	70
80	1.292	1.664	1.990	2.374	2.639	80
90	1.291	1.662	1.987	2.369	2.632	90
100	1.290	1.660	1.984	2.364	2.626	100
1000	1.282	1.646	1.962	2.330	2.581	1000

1.282	1.645	1.960	2.326	2.576
$z_{0.10}$	$z_{0.05}$	$z_{0.025}$	$z_{0.01}$	$z_{0.005}$

TABLE V
Critical values and approximate significance levels for a Wilcoxon signed-rank test

Sample size	Significance level, α		Critical value	
n	One-tailed	Two-tailed	W_l	W_r
7	0.01	0.02	0	28
	0.025	0.05	2	26
	0.05	0.10	4	24
	0.10	0.20	6	22
8	0.005	0.01	0	36
	0.01	0.02	2	34
	0.025	0.05	4	32
	0.05	0.10	6	30
	0.10	0.20	8	28
9	0.005	0.01	2	43
	0.01	0.02	3	42
	0.025	0.05	6	39
	0.05	0.10	8	37
	0.10	0.20	11	34
10	0.005	0.01	3	52
	0.01	0.02	5	50
	0.025	0.05	8	47
	0.05	0.10	11	44
	0.10	0.20	14	41
11	0.005	0.01	5	61
	0.01	0.02	7	59
	0.025	0.05	11	55
	0.05	0.10	14	52
	0.10	0.20	18	48
12	0.005	0.01	7	71
	0.01	0.02	10	68
	0.025	0.05	14	64
	0.05	0.10	17	61
	0.10	0.20	22	56
13	0.005	0.01	10	81
	0.01	0.02	13	78
	0.025	0.05	17	74
	0.05	0.10	21	70
	0.10	0.20	26	65

TABLE V (cont.)
Critical values and approximate significance levels for a Wilcoxon signed-rank test

Sample size n	Significance level, α One-tailed	Two-tailed	Critical value W_l	W_r
14	0.005	0.01	13	92
	0.01	0.02	16	89
	0.025	0.05	21	84
	0.05	0.10	26	79
	0.10	0.20	31	74
15	0.005	0.01	16	104
	0.01	0.02	20	100
	0.025	0.05	25	95
	0.05	0.10	30	90
	0.10	0.20	37	83
16	0.005	0.01	19	117
	0.01	0.02	24	112
	0.025	0.05	30	106
	0.05	0.10	36	100
	0.10	0.20	42	94
17	0.005	0.01	23	130
	0.01	0.02	28	125
	0.025	0.05	35	118
	0.05	0.10	41	112
	0.10	0.20	49	104
18	0.005	0.01	28	143
	0.01	0.02	33	138
	0.025	0.05	40	131
	0.05	0.10	47	124
	0.10	0.20	55	116
19	0.005	0.01	32	158
	0.01	0.02	38	152
	0.025	0.05	46	144
	0.05	0.10	54	136
	0.10	0.20	62	128
20	0.005	0.01	37	173
	0.01	0.02	43	167
	0.025	0.05	52	158
	0.05	0.10	60	150
	0.10	0.20	70	140

TABLE VI
Critical values for a one-tailed Mann–Whitney test with $\alpha = 0.025$ or a two-tailed Mann–Whitney test with $\alpha = 0.05$

n_2 \ n_1	3		4		5		6		7		8		9		10	
	M_l	M_r	M_l	M_r	M_l	M_r	M_l	M_r	M_l	M_r	M_l	M_r	M_l	M_r	M_l	M_r
3	–	–														
4	6	18	11	25												
5	6	21	12	28	18	37										
6	7	23	12	32	19	41	26	52								
7	7	26	13	35	20	45	28	56	37	68						
8	8	28	14	38	21	49	29	61	39	73	49	87				
9	8	31	15	41	22	53	31	65	41	78	51	93	63	108		
10	9	33	16	44	24	56	32	70	43	83	54	98	66	114	79	131

TABLE VII
Critical values for a one-tailed Mann–Whitney test with $\alpha = 0.05$ or a two-tailed Mann–Whitney test with $\alpha = 0.10$

n_2 \ n_1	3		4		5		6		7		8		9		10	
	M_l	M_r	M_l	M_r	M_l	M_r	M_l	M_r	M_l	M_r	M_l	M_r	M_l	M_r	M_l	M_r
3	6	15														
4	7	17	12	24												
5	7	20	13	27	19	36										
6	8	22	14	30	20	40	28	50								
7	9	24	15	33	22	43	30	54	39	66						
8	9	27	16	36	24	46	32	58	41	71	52	84				
9	10	29	17	39	25	50	33	63	43	76	54	90	66	105		
10	11	31	18	42	26	54	35	67	46	80	57	95	69	111	83	127

TABLE VIII
Values of χ^2_α

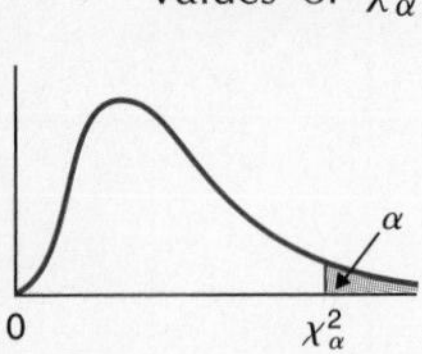

df	$\chi^2_{0.995}$	$\chi^2_{0.99}$	$\chi^2_{0.975}$	$\chi^2_{0.95}$	$\chi^2_{0.90}$
1	0.000	0.000	0.001	0.004	0.016
2	0.010	0.020	0.051	0.103	0.211
3	0.072	0.115	0.216	0.352	0.584
4	0.207	0.297	0.484	0.711	1.064
5	0.412	0.554	0.831	1.145	1.610
6	0.676	0.872	1.237	1.635	2.204
7	0.989	1.239	1.690	2.167	2.833
8	1.344	1.646	2.180	2.733	3.490
9	1.735	2.088	2.700	3.325	4.168
10	2.156	2.558	3.247	3.940	4.865
11	2.603	3.053	3.816	4.575	5.578
12	3.074	3.571	4.404	5.226	6.304
13	3.565	4.107	5.009	5.892	7.042
14	4.075	4.660	5.629	6.571	7.790
15	4.601	5.229	6.262	7.261	8.547
16	5.142	5.812	6.908	7.962	9.312
17	5.697	6.408	7.564	8.672	10.085
18	6.265	7.015	8.231	9.390	10.865
19	6.844	7.633	8.907	10.117	11.651
20	7.434	8.260	9.591	10.851	12.443
21	8.034	8.897	10.283	11.591	13.240
22	8.643	9.542	10.982	12.338	14.041
23	9.260	10.196	11.689	13.091	14.848
24	9.886	10.856	12.401	13.848	15.659
25	10.520	11.524	13.120	14.611	16.473
26	11.160	12.198	13.844	15.379	17.292
27	11.808	12.879	14.573	16.151	18.114
28	12.461	13.565	15.308	16.928	18.939
29	13.121	14.256	16.047	17.708	19.768
30	13.787	14.953	16.791	18.493	20.599
40	20.707	22.164	24.433	26.509	29.051
50	27.991	29.707	32.357	34.764	37.689
60	35.534	37.485	40.482	43.188	46.459
70	43.275	45.442	48.758	51.739	55.329
80	51.172	53.540	57.153	60.391	64.278
90	59.196	61.754	65.647	69.126	73.291
100	67.328	70.065	74.222	77.930	82.358

TABLE VIII (cont.)
Values of χ^2_α

$\chi^2_{0.10}$	$\chi^2_{0.05}$	$\chi^2_{0.025}$	$\chi^2_{0.01}$	$\chi^2_{0.005}$	df
2.706	3.841	5.024	6.635	7.879	1
4.605	5.991	7.378	9.210	10.597	2
6.251	7.815	9.348	11.345	12.838	3
7.779	9.488	11.143	13.277	14.860	4
9.236	11.070	12.833	15.086	16.750	5
10.645	12.592	14.449	16.812	18.548	6
12.017	14.067	16.013	18.475	20.278	7
13.362	15.507	17.535	20.090	21.955	8
14.684	16.919	19.023	21.666	23.589	9
15.987	18.307	20.483	23.209	25.188	10
17.275	19.675	21.920	24.725	26.757	11
18.549	21.026	23.337	26.217	28.300	12
19.812	22.362	24.736	27.688	29.819	13
21.064	23.685	26.119	29.141	31.319	14
22.307	24.996	27.488	30.578	32.801	15
23.542	26.296	28.845	32.000	34.267	16
24.769	27.587	30.191	33.409	35.718	17
25.989	28.869	31.526	34.805	37.156	18
27.204	30.143	32.852	36.191	38.582	19
28.412	31.410	34.170	37.566	39.997	20
29.615	32.671	35.479	38.932	41.401	21
30.813	33.924	36.781	40.290	42.796	22
32.007	35.172	38.076	41.638	44.181	23
33.196	36.415	39.364	42.980	45.559	24
34.382	37.653	40.647	44.314	46.928	25
35.563	38.885	41.923	45.642	48.290	26
36.741	40.113	43.195	46.963	49.645	27
37.916	41.337	44.461	48.278	50.994	28
39.087	42.557	45.722	49.588	52.336	29
40.256	43.773	46.979	50.892	53.672	30
51.805	55.759	59.342	63.691	66.767	40
63.167	67.505	71.420	76.154	79.490	50
74.397	79.082	83.298	88.381	91.955	60
85.527	90.531	95.023	100.424	104.213	70
96.578	101.879	106.628	112.328	116.320	80
107.565	113.145	118.135	124.115	128.296	90
118.499	124.343	129.563	135.811	140.177	100

TABLE IX
Values of F_α

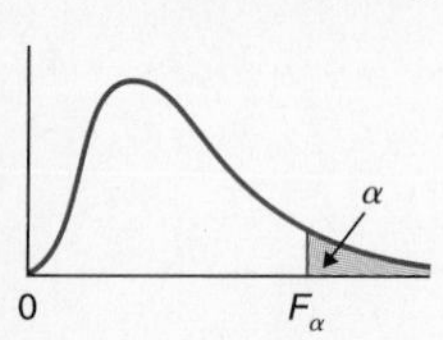

dfd	α	dfn 1	2	3	4	5	6	7	8	9
	0.10	39.86	49.50	53.59	55.83	57.24	58.20	58.91	59.44	59.86
	0.05	161.45	199.50	215.71	224.58	230.16	233.99	236.77	238.88	240.54
1	0.025	647.79	799.50	864.16	899.58	921.85	937.11	948.22	956.66	963.28
	0.01	4052.2	4999.5	5403.4	5624.6	5763.6	5859.0	5928.4	5981.1	6022.5
	0.005	16211	20000	21615	22500	23056	23437	23715	23925	24091
	0.10	8.53	9.00	9.16	9.24	9.29	9.33	9.35	9.37	9.38
	0.05	18.51	19.00	19.16	19.25	19.30	19.33	19.35	19.37	19.38
2	0.025	38.51	39.00	39.17	39.25	39.30	39.33	39.36	39.37	39.39
	0.01	98.50	99.00	99.17	99.25	99.30	99.33	99.36	99.37	99.39
	0.005	198.50	199.00	199.17	199.25	199.30	199.33	199.36	199.37	199.39
	0.10	5.54	5.46	5.39	5.34	5.31	5.28	5.27	5.25	5.24
	0.05	10.13	9.55	9.28	9.12	9.01	8.94	8.89	8.85	8.81
3	0.025	17.44	16.04	15.44	15.10	14.88	14.73	14.62	14.54	14.47
	0.01	34.12	30.82	29.46	28.71	28.24	27.91	27.67	27.49	27.35
	0.005	55.55	49.80	47.47	46.19	45.39	44.84	44.43	44.13	43.88
	0.10	4.54	4.32	4.19	4.11	4.05	4.01	3.98	3.95	3.94
	0.05	7.71	6.94	6.59	6.39	6.26	6.16	6.09	6.04	6.00
4	0.025	12.22	10.65	9.98	9.60	9.36	9.20	9.07	8.98	8.90
	0.01	21.20	18.00	16.69	15.98	15.52	15.21	14.98	14.80	14.66
	0.005	31.33	26.28	24.26	23.15	22.46	21.97	21.62	21.35	21.14
	0.10	4.06	3.78	3.62	3.52	3.45	3.40	3.37	3.34	3.32
	0.05	6.61	5.79	5.41	5.19	5.05	4.95	4.88	4.82	4.77
5	0.025	10.01	8.43	7.76	7.39	7.15	6.98	6.85	6.76	6.68
	0.01	16.26	13.27	12.06	11.39	10.97	10.67	10.46	10.29	10.16
	0.005	22.78	18.31	16.53	15.56	14.94	14.51	14.20	13.96	13.77
	0.10	3.78	3.46	3.29	3.18	3.11	3.05	3.01	2.98	2.96
	0.05	5.99	5.14	4.76	4.53	4.39	4.28	4.21	4.15	4.10
6	0.025	8.81	7.26	6.60	6.23	5.99	5.82	5.70	5.60	5.52
	0.01	13.75	10.92	9.78	9.15	8.75	8.47	8.26	8.10	7.98
	0.005	18.63	14.54	12.92	12.03	11.46	11.07	10.79	10.57	10.39
	0.10	3.59	3.26	3.07	2.96	2.88	2.83	2.78	2.75	2.72
	0.05	5.59	4.74	4.35	4.12	3.97	3.87	3.79	3.73	3.68
7	0.025	8.07	6.54	5.89	5.52	5.29	5.12	4.99	4.90	4.82
	0.01	12.25	9.55	8.45	7.85	7.46	7.19	6.99	6.84	6.72
	0.005	16.24	12.40	10.88	10.05	9.52	9.16	8.89	8.68	8.51
	0.10	3.46	3.11	2.92	2.81	2.73	2.67	2.62	2.59	2.56
	0.05	5.32	4.46	4.07	3.84	3.69	3.58	3.50	3.44	3.39
8	0.025	7.57	6.06	5.42	5.05	4.82	4.65	4.53	4.43	4.36
	0.01	11.26	8.65	7.59	7.01	6.63	6.37	6.18	6.03	5.91
	0.005	14.69	11.04	9.60	8.81	8.30	7.95	7.69	7.50	7.34

TABLE IX (cont.)
Values of F_α

				dfn						
10	*12*	*15*	*20*	*24*	*30*	*40*	*60*	*120*	α	dfd
60.19	60.71	61.22	61.74	62.00	62.26	62.53	62.79	63.06	*0.10*	
241.88	243.91	245.95	248.01	249.05	250.10	251.14	252.20	253.25	*0.05*	
968.63	976.71	984.87	993.10	997.25	1001.41	1005.60	1009.80	1014.02	*0.025*	*1*
6055.8	6106.3	6157.3	6208.7	6234.6	6260.6	6286.7	631.9	6339.4	*0.01*	
24224	24426	24630	24836	24940	25044	25148	25253	25359	*0.005*	
9.39	9.41	9.42	9.44	9.45	9.46	9.47	9.47	9.48	*0.10*	
19.40	19.41	19.43	19.45	19.45	19.46	19.47	19.48	19.49	*0.05*	
39.40	39.41	39.43	39.45	39.46	39.46	39.47	39.48	39.49	*0.025*	*2*
99.40	99.42	99.43	99.45	99.46	99.47	99.47	99.48	99.49	*0.01*	
199.40	199.42	199.43	199.45	199.46	199.47	199.47	199.48	199.49	*0.005*	
5.23	5.22	5.20	5.18	5.18	5.17	5.16	5.15	5.14	*0.10*	
8.79	8.74	8.70	8.66	8.64	8.62	8.59	8.57	8.55	*0.05*	
14.42	14.34	14.25	14.17	14.12	14.08	14.04	13.99	13.95	*0.025*	*3*
27.23	27.05	26.87	26.69	26.60	26.50	26.41	26.32	26.22	*0.01*	
43.69	43.39	43.08	42.78	42.62	42.47	42.31	42.15	41.99	*0.005*	
3.92	3.90	3.87	3.84	3.83	3.82	3.80	3.79	3.78	*0.10*	
5.96	5.91	5.86	5.80	5.77	5.75	5.72	5.69	5.66	*0.05*	
8.84	8.75	8.66	8.56	8.51	8.46	8.41	8.36	8.31	*0.025*	*4*
14.55	14.37	14.20	14.02	13.93	13.84	13.75	13.65	13.56	*0.01*	
20.97	20.70	20.44	20.17	20.03	19.89	19.75	19.61	19.47	*0.005*	
3.30	3.27	3.24	3.21	3.19	3.17	3.16	3.14	3.12	*0.10*	
4.74	4.68	4.62	4.56	4.53	4.50	4.46	4.43	4.40	*0.05*	
6.62	6.52	6.43	6.33	6.28	6.23	6.18	6.12	6.07	*0.025*	*5*
10.05	9.89	9.72	9.55	9.47	9.38	9.29	9.20	9.11	*0.01*	
13.62	13.38	13.15	12.90	12.78	12.66	12.53	12.40	12.27	*0.005*	
2.94	2.90	2.87	2.84	2.82	2.80	2.78	2.76	2.74	*0.10*	
4.06	4.00	3.94	3.87	3.84	3.81	3.77	3.74	3.70	*0.05*	
5.46	5.37	5.27	5.17	5.12	5.07	5.01	4.96	4.90	*0.025*	*6*
7.87	7.72	7.56	7.40	7.31	7.23	7.14	7.06	6.97	*0.01*	
10.25	10.03	9.81	9.59	9.47	9.36	9.24	9.12	9.00	*0.005*	
2.70	2.67	2.63	2.59	2.58	2.56	2.54	2.51	2.49	*0.10*	
3.64	3.57	3.51	3.44	3.41	3.38	3.34	3.30	3.27	*0.05*	
4.76	4.67	4.57	4.47	4.41	4.36	4.31	4.25	4.20	*0.025*	*7*
6.62	6.47	6.31	6.16	6.07	5.99	5.91	5.82	5.74	*0.01*	
8.38	8.18	7.97	7.75	7.64	7.53	7.42	7.31	7.19	*0.005*	
2.54	2.50	2.46	2.42	2.40	2.38	2.36	2.34	2.32	*0.10*	
3.35	3.28	3.22	3.15	3.12	3.08	3.04	3.01	2.97	*0.05*	
4.30	4.20	4.10	4.00	3.95	3.89	3.84	3.78	3.73	*0.025*	*8*
5.81	5.67	5.52	5.36	5.28	5.20	5.12	5.03	4.95	*0.01*	
7.21	7.01	6.81	6.61	6.50	6.40	6.29	6.18	6.06	*0.005*	

TABLE IX (cont.)
Values of F_α

dfd	α	1	2	3	4	5	6	7	8	9
		dfn								
	0.10	3.36	3.01	2.81	2.69	2.61	2.55	2.51	2.47	2.44
	0.05	5.12	4.26	3.86	3.63	3.48	3.37	3.29	3.23	3.18
9	0.025	7.21	5.71	5.08	4.72	4.48	4.32	4.20	4.10	4.03
	0.01	10.56	8.02	6.99	6.42	6.06	5.80	5.61	5.47	5.35
	0.005	13.61	10.11	8.72	7.96	7.47	7.13	6.88	6.69	6.54
	0.10	3.29	2.92	2.73	2.61	2.52	2.46	2.41	2.38	2.35
	0.05	4.96	4.10	3.71	3.48	3.33	3.22	3.14	3.07	3.02
10	0.025	6.94	5.46	4.83	4.47	4.24	4.07	3.95	3.85	3.78
	0.01	10.04	7.56	6.55	5.99	5.64	5.39	5.20	5.06	4.94
	0.005	12.83	9.43	8.08	7.34	6.87	6.54	6.30	6.12	5.97
	0.10	3.23	2.86	2.66	2.54	2.45	2.39	2.34	2.30	2.27
	0.05	4.84	3.98	3.59	3.36	3.20	3.09	3.01	2.95	2.90
11	0.025	6.72	5.26	4.63	4.28	4.04	3.88	3.76	3.66	3.59
	0.01	9.65	7.21	6.22	5.67	5.32	5.07	4.89	4.74	4.63
	0.005	12.23	8.91	7.60	6.88	6.42	6.10	5.86	5.68	5.54
	0.10	3.18	2.81	2.61	2.48	2.39	2.33	2.28	2.24	2.21
	0.05	4.75	3.89	3.49	3.26	3.11	3.00	2.91	2.85	2.80
12	0.025	6.55	5.10	4.47	4.12	3.89	3.73	3.61	3.51	3.44
	0.01	9.33	6.93	5.95	5.41	5.06	4.82	4.64	4.50	4.39
	0.005	11.75	8.51	7.23	6.52	6.07	5.76	5.52	5.35	5.20
	0.10	3.14	2.76	2.56	2.43	2.35	2.28	2.23	2.20	2.16
	0.05	4.67	3.81	3.41	3.18	3.03	2.92	2.83	2.77	2.71
13	0.025	6.41	4.97	4.35	4.00	3.77	3.60	3.48	3.39	3.31
	0.01	9.07	6.70	5.74	5.21	4.86	4.62	4.44	4.30	4.19
	0.005	11.37	8.19	6.93	6.23	5.79	5.48	5.25	5.08	4.94
	0.10	3.10	2.73	2.52	2.39	2.31	2.24	2.19	2.15	2.12
	0.05	4.60	3.74	3.34	3.11	2.96	2.85	2.76	2.70	2.65
14	0.025	6.30	4.86	4.24	3.89	3.66	3.50	3.38	3.29	3.21
	0.01	8.86	6.51	5.56	5.04	4.69	4.46	4.28	4.14	4.03
	0.005	11.06	7.92	6.68	6.00	5.56	5.26	5.03	4.86	4.72
	0.10	3.07	2.70	2.49	2.36	2.27	2.21	2.16	2.12	2.09
	0.05	4.54	3.68	3.29	3.06	2.90	2.79	2.71	2.64	2.59
15	0.025	6.20	4.77	4.15	3.80	3.58	3.41	3.29	3.20	3.12
	0.01	8.68	6.36	5.42	4.89	4.56	4.32	4.14	4.00	3.89
	0.005	10.80	7.70	6.48	5.80	5.37	5.07	4.85	4.67	4.54
	0.10	3.05	2.67	2.46	2.33	2.24	2.18	2.13	2.09	2.06
	0.05	4.49	3.63	3.24	3.01	2.85	2.74	2.66	2.59	2.54
16	0.025	6.12	4.69	4.08	3.73	3.50	3.34	3.22	3.12	3.05
	0.01	8.53	6.23	5.29	4.77	4.44	4.20	4.03	3.89	3.78
	0.005	10.58	7.51	6.30	5.64	5.21	4.91	4.69	4.52	4.38

TABLE IX (cont.)
Values of F_α

dfn										
10	*12*	*15*	*20*	*24*	*30*	*40*	*60*	*120*	α	dfd
2.42	2.38	2.34	2.30	2.28	2.25	2.23	2.21	2.18	*0.10*	
3.14	3.07	3.01	2.94	2.90	2.86	2.83	2.79	2.75	*0.05*	
3.96	3.87	3.77	3.67	3.61	3.56	3.51	3.45	3.39	*0.025*	*9*
5.26	5.11	4.96	4.81	4.73	4.65	4.57	4.48	4.40	*0.01*	
6.42	6.23	6.03	5.83	5.73	5.62	5.52	5.41	5.30	*0.005*	
2.32	2.28	2.24	2.20	2.18	2.16	2.13	2.11	2.08	*0.10*	
2.98	2.91	2.85	2.77	2.74	2.70	2.66	2.62	2.58	*0.05*	
3.72	3.62	3.52	3.42	3.37	3.31	3.26	3.20	3.14	*0.025*	*10*
4.85	4.71	4.56	4.41	4.33	4.25	4.17	4.08	4.00	*0.01*	
5.85	5.66	5.47	5.27	5.17	5.07	4.97	4.86	4.75	*0.005*	
2.25	2.21	2.17	2.12	2.10	2.08	2.05	2.03	2.00	*0.10*	
2.85	2.79	2.72	2.65	2.61	2.57	2.53	2.49	2.45	*0.05*	
3.53	3.43	3.33	3.23	3.17	3.12	3.06	3.00	2.94	*0.025*	*11*
4.54	4.40	4.25	4.10	4.02	3.94	3.86	3.78	3.69	*0.01*	
5.42	5.24	5.05	4.86	4.76	4.65	4.55	4.45	4.34	*0.005*	
2.19	2.15	2.10	2.06	2.04	2.01	1.99	1.96	1.93	*0.10*	
2.75	2.69	2.62	2.54	2.51	2.47	2.43	2.38	2.34	*0.05*	
3.37	3.28	3.18	3.07	3.02	2.96	2.91	2.85	2.79	*0.025*	*12*
4.30	4.16	4.01	3.86	3.78	3.70	3.62	3.54	3.45	*0.01*	
5.09	4.91	4.72	4.53	4.43	4.33	4.23	4.12	4.01	*0.005*	
2.14	2.10	2.05	2.01	1.98	1.96	1.93	1.90	1.88	*0.10*	
2.67	2.60	2.53	2.46	2.42	2.38	2.34	2.30	2.25	*0.05*	
3.25	3.15	3.05	2.95	2.89	2.84	2.78	2.72	2.66	*0.025*	*13*
4.10	3.96	3.82	3.66	3.59	3.51	3.43	3.34	3.25	*0.01*	
4.82	4.64	4.46	4.27	4.17	4.07	3.97	3.87	3.76	*0.005*	
2.10	2.05	2.01	1.96	1.94	1.91	1.89	1.86	1.83	*0.10*	
2.60	2.53	2.46	2.39	2.35	2.31	2.27	2.22	2.18	*0.05*	
3.15	3.05	2.95	2.84	2.79	2.73	2.67	2.61	2.55	*0.025*	*14*
3.94	3.80	3.66	3.51	3.43	3.35	3.27	3.18	3.09	*0.01*	
4.60	4.43	4.25	4.06	3.96	3.86	3.76	3.66	3.55	*0.005*	
2.06	2.02	1.97	1.92	1.90	1.87	1.85	1.82	1.79	*0.10*	
2.54	2.48	2.40	2.33	2.29	2.25	2.20	2.16	2.11	*0.05*	
3.06	2.96	2.86	2.76	2.70	2.64	2.59	2.52	2.46	*0.025*	*15*
3.80	3.67	3.52	3.37	3.29	3.21	3.13	3.05	2.96	*0.01*	
4.42	4.25	4.07	3.88	3.79	3.69	3.58	3.48	3.37	*0.005*	
2.03	1.99	1.94	1.89	1.87	1.84	1.81	1.78	1.75	*0.10*	
2.49	2.42	2.35	2.28	2.24	2.19	2.15	2.11	2.06	*0.05*	
2.99	2.89	2.79	2.68	2.63	2.57	2.51	2.45	2.38	*0.025*	*16*
3.69	3.55	3.41	3.26	3.18	3.10	3.02	2.93	2.84	*0.01*	
4.27	4.10	3.92	3.73	3.64	3.54	3.44	3.33	3.22	*0.005*	

TABLE IX (cont.)
Values of F_{α}

dfd	α	dfn 1	2	3	4	5	6	7	8	9
	0.10	3.03	2.64	2.44	2.31	2.22	2.15	2.10	2.06	2.03
	0.05	4.45	3.59	3.20	2.96	2.81	2.70	2.61	2.55	2.49
17	0.025	6.04	4.62	4.01	3.66	3.44	3.28	3.16	3.06	2.98
	0.01	8.40	6.11	5.18	4.67	4.34	4.10	3.93	3.79	3.68
	0.005	10.38	7.35	6.16	5.50	5.07	4.78	4.56	4.39	4.25
	0.10	3.01	2.62	2.42	2.29	2.20	2.13	2.08	2.04	2.00
	0.05	4.41	3.55	3.16	2.93	2.77	2.66	2.58	2.51	2.46
18	0.025	5.98	4.56	3.95	3.61	3.38	3.22	3.10	3.01	2.93
	0.01	8.29	6.01	5.09	4.58	4.25	4.01	3.84	3.71	3.60
	0.005	10.22	7.21	6.03	5.37	4.96	4.66	4.44	4.28	4.14
	0.10	2.99	2.61	2.40	2.27	2.18	2.11	2.06	2.02	1.98
	0.05	4.38	3.52	3.13	2.90	2.74	2.63	2.54	2.48	2.42
19	0.025	5.92	4.51	3.90	3.56	3.33	3.17	3.05	2.96	2.88
	0.01	8.18	5.93	5.01	4.50	4.17	3.94	3.77	3.63	3.52
	0.005	10.07	7.09	5.92	5.27	4.85	4.56	4.34	4.18	4.04
	0.10	2.97	2.59	2.38	2.25	2.16	2.09	2.04	2.00	1.96
	0.05	4.35	3.49	3.10	2.87	2.71	2.60	2.51	2.45	2.39
20	0.025	5.87	4.46	3.86	3.51	3.29	3.13	3.01	2.91	2.84
	0.01	8.10	5.85	4.94	4.43	4.10	3.87	3.70	3.56	3.46
	0.005	9.94	6.99	5.82	5.17	4.76	4.47	4.26	4.09	3.96
	0.10	2.96	2.57	2.36	2.23	2.14	2.08	2.02	1.98	1.95
	0.05	4.32	3.47	3.07	2.84	2.68	2.57	2.49	2.42	2.37
21	0.025	5.83	4.42	3.82	3.48	3.25	3.09	2.97	2.87	2.80
	0.01	8.02	5.78	4.87	4.37	4.04	3.81	3.64	3.51	3.40
	0.005	9.83	6.89	5.73	5.09	4.68	4.39	4.18	4.01	3.88
	0.10	2.95	2.56	2.35	2.22	2.13	2.06	2.01	1.97	1.93
	0.05	4.30	3.44	3.05	2.82	2.66	2.55	2.46	2.40	2.34
22	0.025	5.79	4.38	3.78	3.44	3.22	3.05	2.93	2.84	2.76
	0.01	7.95	5.72	4.82	4.31	3.99	3.76	3.59	3.45	3.35
	0.005	9.73	6.81	5.65	5.02	4.61	4.32	4.11	3.94	3.81
	0.10	2.94	2.55	2.34	2.21	2.11	2.05	1.99	1.95	1.92
	0.05	4.28	3.42	3.03	2.80	2.64	2.53	2.44	2.37	2.32
23	0.025	5.75	4.35	3.75	3.41	3.18	3.02	2.90	2.81	2.73
	0.01	7.88	5.66	4.76	4.26	3.94	3.71	3.54	3.41	3.30
	0.005	9.63	6.73	5.58	4.95	4.54	4.26	4.05	3.88	3.75
	0.10	2.93	2.54	2.33	2.19	2.10	2.04	1.98	1.94	1.91
	0.05	4.26	3.40	3.01	2.78	2.62	2.51	2.42	2.36	2.30
24	0.025	5.72	4.32	3.72	3.38	3.15	2.99	2.87	2.78	2.70
	0.01	7.82	5.61	4.72	4.22	3.90	3.67	3.50	3.36	3.26
	0.005	9.55	6.66	5.52	4.89	4.49	4.20	3.99	3.83	3.69

TABLE IX (cont.)
Values of F_α

dfn 10	12	15	20	24	30	40	60	120	α	dfd
2.00	1.96	1.91	1.86	1.84	1.81	1.78	1.75	1.72	0.10	
2.45	2.38	2.31	2.23	2.19	2.15	2.10	2.06	2.01	0.05	
2.92	2.82	2.72	2.62	2.56	2.50	2.44	2.38	2.32	0.025	17
3.59	3.46	3.31	3.16	3.08	3.00	2.92	2.83	2.75	0.01	
4.14	3.97	3.79	3.61	3.51	3.41	3.31	3.21	3.10	0.005	
1.98	1.93	1.89	1.84	1.81	1.78	1.75	1.72	1.69	0.10	
2.41	2.34	2.27	2.19	2.15	2.11	2.06	2.02	1.97	0.05	
2.87	2.77	2.67	2.56	2.50	2.44	2.38	2.32	2.26	0.025	18
3.51	3.37	3.23	3.08	3.00	2.92	2.84	2.75	2.66	0.01	
4.03	3.86	3.68	3.50	3.40	3.30	3.20	3.10	2.99	0.005	
1.96	1.91	1.86	1.81	1.79	1.76	1.73	1.70	1.67	0.10	
2.38	2.31	2.23	2.16	2.11	2.07	2.03	1.98	1.93	0.05	
2.82	2.72	2.62	2.51	2.45	2.39	2.33	2.27	2.20	0.025	19
3.43	3.30	3.15	3.00	2.92	2.84	2.76	2.67	2.58	0.01	
3.93	3.76	3.59	3.40	3.31	3.21	3.11	3.00	2.89	0.005	
1.94	1.89	1.84	1.79	1.77	1.74	1.71	1.68	1.64	0.10	
2.35	2.28	2.20	2.12	2.08	2.04	1.99	1.95	1.90	0.05	
2.77	2.68	2.57	2.46	2.41	2.35	2.29	2.22	2.16	0.025	20
3.37	3.23	3.09	2.94	2.86	2.78	2.69	2.61	2.52	0.01	
3.85	3.68	3.50	3.32	3.22	3.12	3.02	2.92	2.81	0.005	
1.92	1.87	1.83	1.78	1.75	1.72	1.69	1.66	1.62	0.10	
2.32	2.25	2.18	2.10	2.05	2.01	1.96	1.92	1.87	0.05	
2.73	2.64	2.53	2.42	2.37	2.31	2.25	2.18	2.11	0.025	21
3.31	3.17	3.03	2.88	2.80	2.72	2.64	2.55	2.46	0.01	
3.77	3.60	3.43	3.24	3.15	3.05	2.95	2.84	2.73	0.005	
1.90	1.86	1.81	1.76	1.73	1.70	1.67	1.64	1.60	0.10	
2.30	2.23	2.15	2.07	2.03	1.98	1.94	1.89	1.84	0.05	
2.70	2.60	2.50	2.39	2.33	2.27	2.21	2.14	2.08	0.025	22
3.26	3.12	2.98	2.83	2.75	2.67	2.58	2.50	2.40	0.01	
3.70	3.54	3.36	3.18	3.08	2.98	2.88	2.77	2.66	0.005	
1.89	1.84	1.80	1.74	1.72	1.69	1.66	1.62	1.59	0.10	
2.27	2.20	2.13	2.05	2.01	1.96	1.91	1.86	1.81	0.05	
2.67	2.57	2.47	2.36	2.30	2.24	2.18	2.11	2.04	0.025	23
3.21	3.07	2.93	2.78	2.70	2.62	2.54	2.45	2.35	0.01	
3.64	3.47	3.30	3.12	3.02	2.92	2.82	2.71	2.60	0.005	
1.88	1.83	1.78	1.73	1.70	1.67	1.64	1.61	1.57	0.10	
2.25	2.18	2.11	2.03	1.98	1.94	1.89	1.84	1.79	0.05	
2.64	2.54	2.44	2.33	2.27	2.21	2.15	2.08	2.01	0.025	24
3.17	3.03	2.89	2.74	2.66	2.58	2.49	2.40	2.31	0.01	
3.59	3.42	3.25	3.06	2.97	2.87	2.77	2.66	2.55	0.005	

TABLE IX (cont.)
Values of F_{α}

dfd	α	dfn 1	2	3	4	5	6	7	8	9
	0.10	2.92	2.53	2.32	2.18	2.09	2.02	1.97	1.93	1.89
	0.05	4.24	3.39	2.99	2.76	2.60	2.49	2.40	2.34	2.28
25	0.025	5.69	4.29	3.69	3.35	3.13	2.97	2.85	2.75	2.68
	0.01	7.77	5.57	4.68	4.18	3.85	3.63	3.46	3.32	3.22
	0.005	9.48	6.60	5.46	4.84	4.43	4.15	3.94	3.78	3.64
	0.10	2.91	2.52	2.31	2.17	2.08	2.01	1.96	1.92	1.88
	0.05	4.23	3.37	2.98	2.74	2.59	2.47	2.39	2.32	2.27
26	0.025	5.66	4.27	3.67	3.33	3.10	2.94	2.82	2.73	2.65
	0.01	7.72	5.53	4.64	4.14	3.82	3.59	3.42	3.29	3.18
	0.005	9.41	6.54	5.41	4.79	4.38	4.10	3.89	3.73	3.60
	0.10	2.90	2.51	2.30	2.17	2.07	2.00	1.95	1.91	1.87
	0.05	4.21	3.35	2.96	2.73	2.57	2.46	2.37	2.31	2.25
27	0.025	5.63	4.24	3.65	3.31	3.08	2.92	2.80	2.71	2.63
	0.01	7.68	5.49	4.60	4.11	3.78	3.56	3.39	3.26	3.15
	0.005	9.34	6.49	5.36	4.74	4.34	4.06	3.85	3.69	3.56
	0.10	2.89	2.50	2.29	2.16	2.06	2.00	1.94	1.90	1.87
	0.05	4.20	3.34	2.95	2.71	2.56	2.45	2.36	2.29	2.24
28	0.025	5.61	4.22	3.63	3.29	3.06	2.90	2.78	2.69	2.61
	0.01	7.64	5.45	4.57	4.07	3.75	3.53	3.36	3.23	3.12
	0.005	9.28	6.44	5.32	4.70	4.30	4.02	3.81	3.65	3.52
	0.10	2.89	2.50	2.28	2.15	2.06	1.99	1.93	1.89	1.86
	0.05	4.18	3.33	2.93	2.70	2.55	2.43	2.35	2.28	2.22
29	0.025	5.59	4.20	3.61	3.27	3.04	2.88	2.76	2.67	2.59
	0.01	7.60	5.42	4.54	4.04	3.73	3.50	3.33	3.20	3.09
	0.005	9.23	6.40	5.28	4.66	4.26	3.98	3.77	3.61	3.48
	0.10	2.88	2.49	2.28	2.14	2.05	1.98	1.93	1.88	1.85
	0.05	4.17	3.32	2.92	2.69	2.53	2.42	2.33	2.27	2.21
30	0.025	5.57	4.18	3.59	3.25	3.03	2.87	2.75	2.65	2.57
	0.01	7.56	5.39	4.51	4.02	3.70	3.47	3.30	3.17	3.07
	0.005	9.18	6.35	5.24	4.62	4.23	3.95	3.74	3.58	3.45
	0.10	2.79	2.39	2.18	2.04	1.95	1.87	1.82	1.77	1.74
	0.05	4.00	3.15	2.76	2.53	2.37	2.25	2.17	2.10	2.04
60	0.025	5.29	3.93	3.34	3.01	2.79	2.63	2.51	2.41	2.33
	0.01	7.08	4.98	4.13	3.65	3.34	3.12	2.95	2.82	2.72
	0.005	8.49	5.79	4.73	4.14	3.76	3.49	3.29	3.13	3.01
	0.10	2.75	2.35	2.13	1.99	1.90	1.82	1.77	1.72	1.68
	0.05	3.92	3.07	2.68	2.45	2.29	2.18	2.09	2.02	1.96
120	0.025	5.15	3.80	3.23	2.89	2.67	2.52	2.39	2.30	2.22
	0.01	6.85	4.79	3.95	3.48	3.17	2.96	2.79	2.66	2.56
	0.005	8.18	5.54	4.50	3.92	3.55	3.28	3.09	2.93	2.81

TABLE IX (cont.)
Values of F_α

dfn 10	12	15	20	24	30	40	60	120	α	dfd
1.87	1.82	1.77	1.72	1.69	1.66	1.63	1.59	1.56	0.10	
2.24	2.16	2.09	2.01	1.96	1.92	1.87	1.82	1.77	0.05	
2.61	2.51	2.41	2.30	2.24	2.18	2.12	2.05	1.98	0.025	25
3.13	2.99	2.85	2.70	2.62	2.54	2.45	2.36	2.27	0.01	
3.54	3.37	3.20	3.01	2.92	2.82	2.72	2.61	2.50	0.005	
1.86	1.81	1.76	1.71	1.68	1.65	1.61	1.58	1.54	0.10	
2.22	2.15	2.07	1.99	1.95	1.90	1.85	1.80	1.75	0.05	
2.59	2.49	2.39	2.28	2.22	2.16	2.09	2.03	1.95	0.025	26
3.09	2.96	2.81	2.66	2.58	2.50	2.42	2.33	2.23	0.01	
3.49	3.33	3.15	2.97	2.87	2.77	2.67	2.56	2.45	0.005	
1.85	1.80	1.75	1.70	1.67	1.64	1.60	1.57	1.53	0.10	
2.20	2.13	2.06	1.97	1.93	1.88	1.84	1.79	1.73	0.05	
2.57	2.47	2.36	2.25	2.19	2.13	2.07	2.00	1.93	0.025	27
3.06	2.93	2.78	2.63	2.55	2.47	2.38	2.29	2.20	0.01	
3.45	3.28	3.11	2.93	2.83	2.73	2.63	2.52	2.41	0.005	
1.84	1.79	1.74	1.69	1.66	1.63	1.59	1.56	1.52	0.10	
2.19	2.12	2.04	1.96	1.91	1.87	1.82	1.77	1.71	0.05	
2.55	2.45	2.34	2.23	2.17	2.11	2.05	1.98	1.91	0.025	28
3.03	2.90	2.75	2.60	2.52	2.44	2.35	2.26	2.17	0.01	
3.41	3.25	3.07	2.89	2.79	2.69	2.59	2.48	2.37	0.005	
1.83	1.78	1.73	1.68	1.65	1.62	1.58	1.55	1.51	0.10	
2.18	2.10	2.03	1.94	1.90	1.85	1.81	1.75	1.70	0.05	
2.53	2.43	2.32	2.21	2.15	2.09	2.03	1.96	1.89	0.025	29
3.00	2.87	2.73	2.57	2.49	2.41	2.33	2.23	2.14	0.01	
3.38	3.21	3.04	2.86	2.76	2.66	2.56	2.45	2.33	0.005	
1.82	1.77	1.72	1.67	1.64	1.61	1.57	1.54	1.50	0.10	
2.16	2.09	2.01	1.93	1.89	1.84	1.79	1.74	1.68	0.05	
2.51	2.41	2.31	2.20	2.14	2.07	2.01	1.94	1.87	0.025	30
2.98	2.84	2.70	2.55	2.47	2.39	2.30	2.21	2.11	0.01	
3.34	3.18	3.01	2.82	2.73	2.63	2.52	2.42	2.30	0.005	
1.71	1.66	1.60	1.54	1.51	1.48	1.44	1.40	1.35	0.10	
1.99	1.92	1.84	1.75	1.70	1.65	1.59	1.53	1.47	0.05	
2.27	2.17	2.06	1.94	1.88	1.82	1.74	1.67	1.58	0.025	60
2.63	2.50	2.35	2.20	2.12	2.03	1.94	1.84	1.73	0.01	
2.90	2.74	2.57	2.39	2.29	2.19	2.08	1.96	1.83	0.005	
1.65	1.60	1.55	1.48	1.45	1.41	1.37	1.32	1.26	0.10	
1.91	1.83	1.75	1.66	1.61	1.55	1.50	1.43	1.35	0.05	
2.16	2.05	1.94	1.82	1.76	1.69	1.61	1.53	1.43	0.025	120
2.47	2.34	2.19	2.03	1.95	1.86	1.76	1.66	1.53	0.01	
2.71	2.54	2.37	2.19	2.09	1.98	1.87	1.75	1.61	0.005	

TABLE X
Critical values for a correlation test for normality

	α		
n	0.10	0.05	0.01
5	0.903	0.880	0.832
6	0.911	0.889	0.841
7	0.918	0.897	0.852
8	0.924	0.905	0.862
9	0.930	0.911	0.871
10	0.935	0.917	0.879
11	0.938	0.923	0.887
12	0.942	0.927	0.894
13	0.945	0.931	0.900
14	0.948	0.935	0.905
15	0.951	0.938	0.910
16	0.953	0.941	0.914
17	0.955	0.944	0.918
18	0.957	0.946	0.922
19	0.959	0.949	0.925
20	0.960	0.951	0.928
21	0.962	0.952	0.931
22	0.963	0.954	0.933
23	0.964	0.956	0.936
24	0.966	0.957	0.938
25	0.967	0.958	0.940
26	0.968	0.960	0.942
27	0.969	0.961	0.944
28	0.969	0.962	0.945
29	0.970	0.963	0.947
30	0.971	0.964	0.949
40	0.977	0.972	0.958
50	0.981	0.976	0.966
60	0.983	0.980	0.971
70	0.985	0.982	0.975
80	0.987	0.984	0.978
90	0.988	0.986	0.980
100	0.989	0.987	0.982
200	0.994	0.993	0.990
300	0.996	0.995	0.993
400	0.997	0.996	0.995
500	0.998	0.997	0.996
1000	0.999	0.998	0.998

TABLE XI
Values of $q_{0.01}$

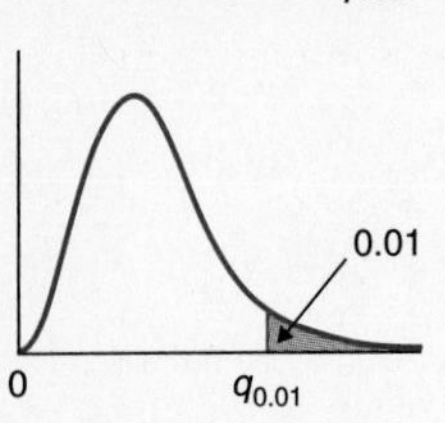

	κ									
ν	2	3	4	5	6	7	8	9	10	ν
1	90.0	135	164	186	202	216	227	237	246	1
2	14.0	19.0	22.3	24.7	26.6	28.2	29.5	30.7	31.7	2
3	8.26	10.6	12.2	13.3	14.2	15.0	15.6	16.2	16.7	3
4	6.51	8.12	9.17	9.96	10.6	11.1	11.5	11.9	12.3	4
5	5.70	6.97	7.80	8.42	8.91	9.32	9.67	9.97	10.2	5
6	5.24	6.33	7.03	7.56	7.97	8.32	8.61	8.87	9.10	6
7	4.95	5.92	6.54	7.01	7.37	7.68	7.94	8.17	8.37	7
8	4.74	5.63	6.20	6.63	6.96	7.24	7.47	7.68	7.87	8
9	4.60	5.43	5.96	6.35	6.66	6.91	7.13	7.32	7.49	9
10	4.48	5.27	5.77	6.14	6.43	6.67	6.87	7.05	7.21	10
11	4.39	5.14	5.62	5.97	6.25	6.48	6.67	6.84	6.99	11
12	4.32	5.04	5.50	5.84	6.10	6.32	6.51	6.67	6.81	12
13	4.26	4.96	5.40	5.73	5.98	6.19	6.37	6.53	6.67	13
14	4.21	4.89	5.32	5.63	5.88	6.08	6.26	6.41	6.54	14
15	4.17	4.83	5.25	5.56	5.80	5.99	6.16	6.31	6.44	15
16	4.13	4.78	5.19	5.49	5.72	5.92	6.08	6.22	6.35	16
17	4.10	4.74	5.14	5.43	5.66	5.85	6.01	6.15	6.27	17
18	4.07	4.70	5.09	5.38	5.60	5.79	5.94	6.08	6.20	18
19	4.05	4.67	5.05	5.33	5.55	5.73	5.89	6.02	6.14	19
20	4.02	4.64	5.02	5.29	5.51	5.69	5.84	5.97	6.09	20
24	3.96	4.54	4.91	5.17	5.37	5.54	5.69	5.81	5.92	24
30	3.89	4.45	4.80	5.05	5.24	5.40	5.54	5.65	5.76	30
40	3.82	4.37	4.70	4.93	5.11	5.27	5.39	5.50	5.60	40
60	3.76	4.28	4.60	4.82	4.99	5.13	5.25	5.36	5.45	60
120	3.70	4.20	4.50	4.71	4.87	5.01	5.12	5.21	5.30	120
∞	3.64	4.12	4.40	4.60	4.76	4.88	4.99	5.08	5.16	∞

TABLE XII
Values of $q_{0.05}$

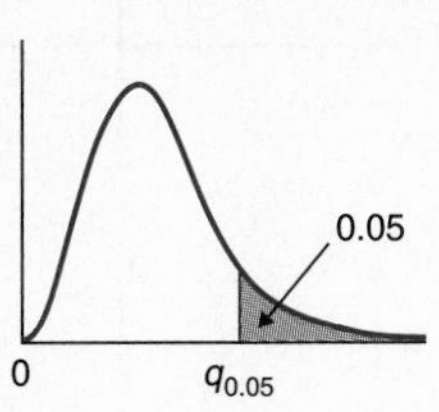

ν	κ = 2	3	4	5	6	7	8	9	10	ν
1	18.0	27.0	32.8	37.1	40.4	43.1	45.4	47.4	49.1	*1*
2	6.08	8.33	9.80	10.9	11.7	12.4	13.0	13.5	14.0	*2*
3	4.50	5.91	6.82	7.50	8.04	8.48	8.85	9.18	9.46	*3*
4	3.93	5.04	5.76	6.29	6.71	7.05	7.35	7.60	7.83	*4*
5	3.64	4.60	5.22	5.67	6.03	6.33	6.58	6.80	6.99	*5*
6	3.46	4.34	4.90	5.30	5.63	5.90	6.12	6.32	6.49	*6*
7	3.34	4.16	4.68	5.06	5.36	5.61	5.82	6.00	6.16	*7*
8	3.26	4.04	4.53	4.89	5.17	5.40	5.60	5.77	5.92	*8*
9	3.20	3.95	4.41	4.76	5.02	5.24	5.43	5.59	5.74	*9*
10	3.15	3.88	4.33	4.65	4.91	5.12	5.30	5.46	5.60	*10*
11	3.11	3.82	4.26	4.57	4.82	5.03	5.20	5.35	5.49	*11*
12	3.08	3.77	4.20	4.51	4.75	4.95	5.12	5.27	5.39	*12*
13	3.06	3.73	4.15	4.45	4.69	4.88	5.05	5.19	5.32	*13*
14	3.03	3.70	4.11	4.41	4.64	4.83	4.99	5.13	5.25	*14*
15	3.01	3.67	4.08	4.37	4.59	4.78	4.94	5.08	5.20	*15*
16	3.00	3.65	4.05	4.33	4.56	4.74	4.90	5.03	5.15	*16*
17	2.98	3.63	4.02	4.30	4.52	4.70	4.86	4.99	5.11	*17*
18	2.97	3.61	4.00	4.28	4.49	4.67	4.82	4.96	5.07	*18*
19	2.96	3.59	3.98	4.25	4.47	4.65	4.79	4.92	5.04	*19*
20	2.95	3.58	3.96	4.23	4.45	4.62	4.77	4.90	5.01	*20*
24	2.92	3.53	3.90	4.17	4.37	4.54	4.68	4.81	4.92	*24*
30	2.89	3.49	3.85	4.10	4.30	4.46	4.60	4.72	4.82	*30*
40	2.86	3.44	3.79	4.04	4.23	4.39	4.52	4.63	4.73	*40*
60	2.83	3.40	3.74	3.98	4.16	4.31	4.44	4.55	4.65	*60*
120	2.80	3.36	3.68	3.92	4.10	4.24	4.36	4.47	4.56	*120*
∞	2.77	3.31	3.63	3.86	4.03	4.17	4.29	4.39	4.47	∞

Answers to Selected Exercises

Note: Most of the numerical answers here were obtained using a computer. If you solve a problem by hand and do some intermediate rounding, your answer may differ somewhat.

CHAPTER 1

Exercises 1.1

1.1 See Definition 1.2 on page 5.

1.3 Descriptive statistics includes the construction of graphs, charts, and tables, and the calculation of various descriptive measures such as averages, measures of variation, and percentiles.

Exercises 1.2

1.5 inferential

1.7 inferential

1.9 descriptive

1.11 descriptive

1.13
a. inferential **b.** descriptive
c. descriptive **d.** inferential
e. inferential **f.** inferential
g. descriptive

Exercises 1.4

1.17 general-use spreadsheet software (e.g., Excel), graphing calculators (e.g, TI-83), and dedicated statistical software (e.g., Minitab)

1.19 See Table 1.2 on page 13.

1.21 First click **Calc** in the menu bar, next click **Random Data**, and then click **Normal**.

1.23 C5

Exercises 1.6

1.31 Conducting a census is generally time consuming and costly, frequently impractical, and sometimes impossible.

1.33 Since a sample will be used to draw conclusions about the entire population, it is essential that the sample be a representative sample—it should reflect as closely as possible the relevant characteristics of the population under consideration.

1.35 Dentists form a high-income group whose incomes are not representative of the incomes of Seattle residents in general.

1.37
a. Probability sampling uses a random device, such as tossing a coin or consulting a random-number table, to decide which members of the population will constitute the sample instead of leaving such decisions to human judgement.
b. False. Since probability sampling uses a random device, it is possible to obtain a nonrepresentative sample.
c. Probability sampling eliminates unintentional selection bias and permits the researcher to control the chance of obtaining a nonrepresentative sample. Also, use of probability sampling guarantees that the techniques of inferential statistics can be applied.

1.39 simple random sampling

1.41

a.

G, L, S	G, A, T
G, L, A	L, S, A
G, L, T	L, S, T
G, S, A	L, A, T
G, S, T	S, A, T

b. $\frac{1}{10}, \frac{1}{10}, \frac{1}{10}$

1.43

a.

MS, FS, LG, BI	MS, LG, JS, JC
MS, FS, LG, JS	MS, BI, JS, JC
MS, FS, LG, JC	FS, LG, BI, JS
MS, FS, BI, JS	FS, LG, BI, JC
MS, FS, BI, JC	FS, LG, JS, JC
MS, FS, JS, JC	FS, BI, JS, JC
MS, LG, BI, JS	LG, BI, JS, JC
MS, LG, BI, JC	

b. Write the initials of the six representatives on separate pieces of paper, place the six slips of paper into a box, and then, while blindfolded, pick four of the slips of paper.

c. $\frac{1}{15}, \frac{1}{15}$

1.45 Answers will vary.

1.47 Answers will vary.

Exercises 1.7

1.55

a. Answers will vary, but here is the procedure: (1) Divide the population size, 685, by the sample size, 25, and round down to the nearest whole number; this gives 27. (2) Use a table of random numbers (or a similar device) to obtain a number between 1 and 27; call it k. (3) List every 27th number, starting with k, until 25 numbers are obtained; thus the first number is k, the second is $k + 27$, the third is $k + 54$, and so forth. (For example, if $k = 6$, then the numbers on the list are 6, 33, 60, ...).

b. systematic random sampling

c. Yes, unless there is some kind of cyclical pattern in the listing of the employees.

1.57 Yes, since there is no cyclical pattern in the listing. In fact, because the listing is by sales, one could argue that, in this case, systematic random sampling is preferable to simple random sampling.

1.59

a. Number the suites from 1 to 48, use a table of random numbers to randomly select 3 of the 48 suites, and take as the sample the 24 dormitory residents living in the 3 suites obtained.

b. Probably not, since friends often have similar opinions.

c. Proportional allocation dictates that the number of freshmen, sophomores, juniors, and seniors selected be, respectively, 8, 7, 6, and 3. Thus a stratified sample of 24 dormitory residents can be obtained as follows: Number the freshman dormitory residents from 1 to 128 and use a table of random numbers to randomly select 8 of the 128 freshman dormitory residents; number the sophomore dormitory residents from 1 to 112 and use a table of random numbers to randomly select 7 of the 112 sophomore dormitory residents; and so forth.

Exercises 1.8

1.63 causation (cause and effect)

1.65 Here is one of several methods that could be used: Number the women from 1 to 4753; use a table of random numbers or a random-number generator to obtain 2376 different numbers between 1 and 4753; the 2376 women with those numbers are in one group, the remaining 2377 women are in the other group.

1.67 designed experiment

1.69 observational study

1.71

a. the individuals or items on which the experiment is performed

b. subject

1.73

a. the 20 flashlights

b. lifetime of a battery in a flashlight

c. battery brand

d. four brands of batteries

e. four brands of batteries

1.75

a. the cacti used in the study

b. total length of cuttings at the end of 16 months

c. hydrophilic polymer and irrigation regime

d. Hydrophilic polymer has two levels: with and without. Irrigation regime has five levels: none, light, medium, heavy, and very heavy.

e. Each treatment is a combination of a level of hydrophilic polymer and a level of irrigation regime. See Table 1.12 on page 43.

1.77

a. batches of the product being sold (Some might say that the stores are the experimental units.)

b. unit sales of the product

c. display type and pricing scheme

d. Display type has three levels: normal display space interior to an aisle, normal display space at the end of an aisle, and enlarged display space. Pricing scheme has three levels: regular price, reduced price, and cost.

e. Each treatment is a combination of a level of display type and a level of pricing scheme.

1.79 completely randomized design

Review Test for Chapter 1

1. Answers will vary.

2. It is almost always necessary to invoke techniques of descriptive statistics to organize and summarize the information obtained from a sample before carrying out an inferential analysis.

3. descriptive

4. inferential

5. inferential

6. descriptive

7. inferential

8. a literature search

9. a. A representative sample is a sample that reflects as closely as possible the relevant characteristics of the population under consideration.

b. In probability sampling, a random device, such as tossing a coin or consulting a table of random numbers, is used to decide which members of the population will constitute the sample instead of leaving such decisions to human judgement.

c. Simple random sampling is a sampling procedure for which each possible sample of a given size is equally likely to be the one obtained from the population.

10. No, because parents of students at Yale tend to have higher incomes than parents of college students in general.

11. only (b)

12. a.

WA, OR, CA	WA, AK, HI
WA, OR, AK	OR, CA, AK
WA, OR, HI	OR, CA, HI
WA, CA, AK	OR, AK, HI
WA, CA, HI	CA, AK, HI

b. $\frac{1}{10}, \frac{1}{10}, \frac{1}{10}$

13. a. Number the registered voters from 1 to 7246, use Table I to obtain 50 different numbers between 1 and 7246, and take as the sample the 50 registered voters who are numbered with the numbers obtained.

b.

6293	7075	4096	1946	5945
3331	6679	7132	3335	6562
6151	4106	0189	6417	5109
0940	6899	3094	0102	2930
1928	5236	0529	5669	0353
4124	2293	3909	4865	0538
0219	6984	1966	5203	3098
0341	1424	4204	2200	5306
0208	0315	5802	0788	1793
5470	4198	1065	6875	4830

14. See Section 1.7.

15. a. Answers will vary, but here is the procedure: (1) Divide the population size, 7246, by the sample size, 50, and round down to the nearest whole number; this gives 144. (2) Use a table of random numbers (or similar device) to select a number between 1 and 144; call it k. (3) List every 144th number, starting with k, until 50 numbers are obtained; thus the first number on the required list of 50 numbers is k, the second is $k + 144$, the third is $k + 288$, and so forth. (For example, if $k = 86$, then the numbers on the list are 86, 230, 374, ...)

b. Yes, unless for some reason there is a cyclical pattern in the listing of the registered voters.

16. a. Proportional allocation dictates that 10 full professors, 16 associate professors, 12 assistant professors, and 2 instructors be selected.

b. The procedure is as follows: Number the full professors from 1 to 205, and use Table I to randomly select 10 of the 205 full professors; number the associate professors from 1 to 328, and use Table I to randomly select 16 of the 328 associate professors; and so on.

17. a. In an observational study, researchers simply observe characteristics and take measurements. On the other hand, in a designed experiment, researchers impose treatments and controls and then observe characteristics and take measurements.

b. Observational studies can only reveal association, whereas designed experiments can help establish causation (cause and effect).

18. observational study

19. a. designed experiment

b. The treatment group consists of the 158 patients who took AVONEX™. The control group consists of the 143 patients who were given a placebo. The treatments are AVONEX™ and the placebo.

20. See Key Fact 1.1 on page 41.

21. a. the batches of doughnuts

b. amount of fat absorbed

c. fat type

d. four types of fat

e. four types of fat

22. a. the tomato plants in the study (Some might say the plots of land are the experimental units.)

b. yield of tomato plants

c. tomato variety and planting density

d. different tomato varieties and different planting densities

e. Each treatment is a combination of a level of tomato variety and a level of planting density.

23. a. completely randomized design

b. randomized block design; the six different car models

c. the randomized block design in part (b)

CHAPTER 2

Exercises 2.1

2.1 Answers will vary.

2.3 See Definition 2.2 on page 61.

2.5 qualitative variable

2.7 quantitative, continuous; annual U.S. tobacco production in millions of pounds

2.9 quantitative, discrete; number of employees in thousands

2.11

a. quantitative, continuous; land area in thousands of square miles

b. quantitative, discrete; land-area rank and population rank

c. quantitative, discrete; population in millions

d. qualitative; continent of birth

2.13

a. quantitative, discrete; rank by amount of deposits

b. quantitative, continuous (or discrete); amount of deposits in millions of dollars

Exercises 2.2

2.17 Grouping can help to make a large and complicated set of data more compact and easier to understand.

2.19 See 1–3 on pages 67–68.

2.21 Relative-frequency distributions are better than frequency distributions when comparing two data sets because relative frequencies are always between 0 and 1 and hence provide a standard for comparison.

2.25 Since each class is based on a single value, the midpoint of each class is the same as the class.

2.27

Consumption (mil. BTU)	Frequency	Relative frequency	Midpoint
40 < 50	1	0.02	45
50 < 60	7	0.14	55
60 < 70	7	0.14	65
70 < 80	3	0.06	75
80 < 90	6	0.12	85
90 < 100	10	0.20	95
100 < 110	5	0.10	105
110 < 120	4	0.08	115
120 < 130	2	0.04	125
130 < 140	3	0.06	135
140 < 150	0	0.00	145
150 < 160	2	0.04	155
	50	1.00	

2.29

Starting salary ($thousands)	Frequency	Relative frequency	Midpoint
25 < 26	3	0.086	25.5
26 < 27	3	0.086	26.5
27 < 28	5	0.143	27.5
28 < 29	9	0.257	28.5
29 < 30	9	0.257	29.5
30 < 31	4	0.114	30.5
31 < 32	1	0.029	31.5
32 < 33	1	0.029	32.5
	35	1.001	

2.31

Consumption (mil. BTU)	Frequency	Relative frequency	Midpoint
40–49	1	0.02	45
50–59	7	0.14	55
60–69	7	0.14	65
70–79	3	0.06	75
80–89	6	0.12	85
90–99	10	0.20	95
100–109	5	0.10	105
110–119	4	0.08	115
120–129	2	0.04	125
130–139	3	0.06	135
140–149	0	0.00	145
150–159	2	0.04	155
	50	1.00	

2.33

Starting salary ($thousands)	Frequency	Relative frequency	Midpoint
25–25.9	3	0.086	25.5
26–26.9	3	0.086	26.5
27–27.9	5	0.143	27.5
28–28.9	9	0.257	28.5
29–29.9	9	0.257	29.5
30–30.9	4	0.114	30.5
31–31.9	1	0.029	31.5
32–32.9	1	0.029	32.5
	35	1.001	

2.35

Number of cars sold	Frequency	Relative frequency
0	7	0.135
1	15	0.288
2	12	0.231
3	9	0.173
4	5	0.096
5	3	0.058
6	1	0.019
	52	1.000

2.37

Number of days missed	Frequency	Relative frequency
0	4	0.050
1	2	0.025
2	14	0.175
3	10	0.125
4	16	0.200
5	18	0.225
6	10	0.125
7	6	0.075
	80	1.000

2.39

Champion	Frequency	Relative frequency
Oklahoma State	5	0.167
Iowa State	6	0.200
Oklahoma	1	0.033
Iowa	17	0.567
Arizona State	1	0.033
	30	1.000

Exercises 2.3

2.53 A frequency histogram displays the class frequencies on the vertical axis, whereas a relative-frequency histogram displays the class relative frequencies on the vertical axis.

2.55 Since a bar graph is used for qualitative data, we separate the bars from each other to emphasize that there is no ordering for the classes; if the bars were to touch, some viewers might infer an ordering for the classes.

2.57

a. The heights of the bars in the histogram would be comparable to the heights of the columns of dots in the dotplot.

b. No. Each bar in the histogram would represent a range of possible values, whereas in the dotplot, each column of dots represents one possible value.

2.59

a.

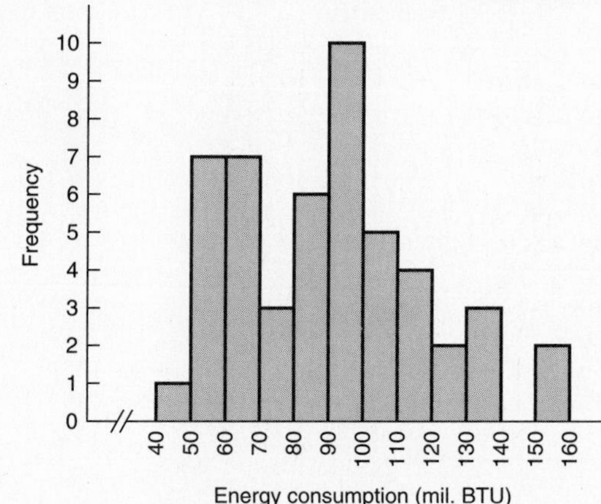

b.

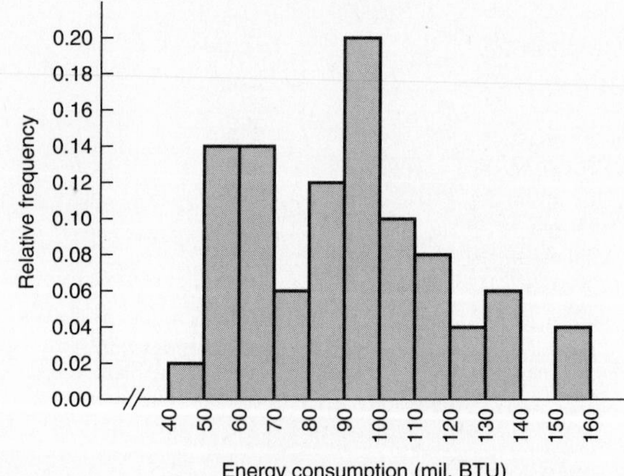

2.61

a.

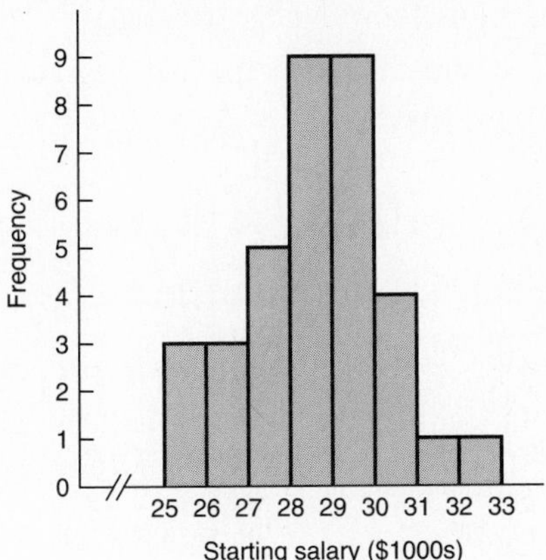

b.

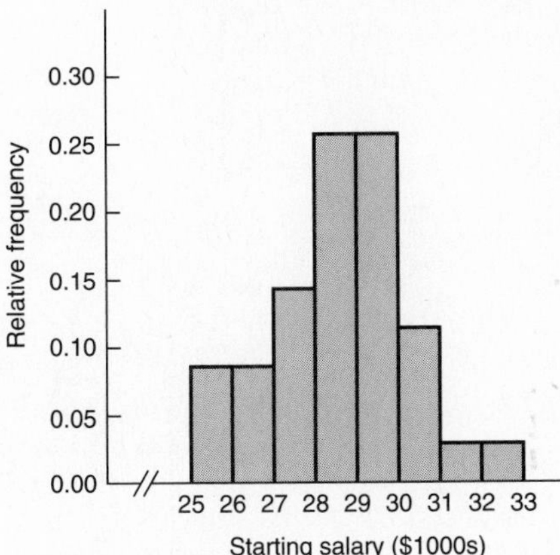

2.63

a.

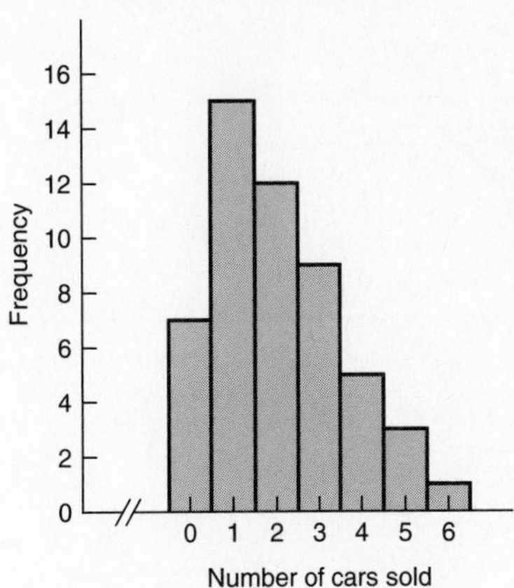

b.

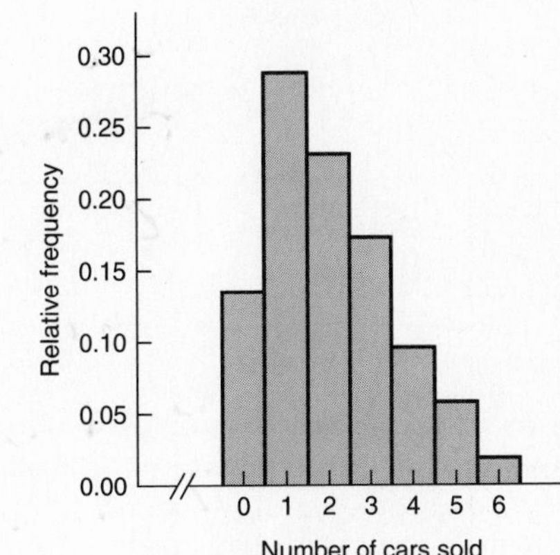

2.65

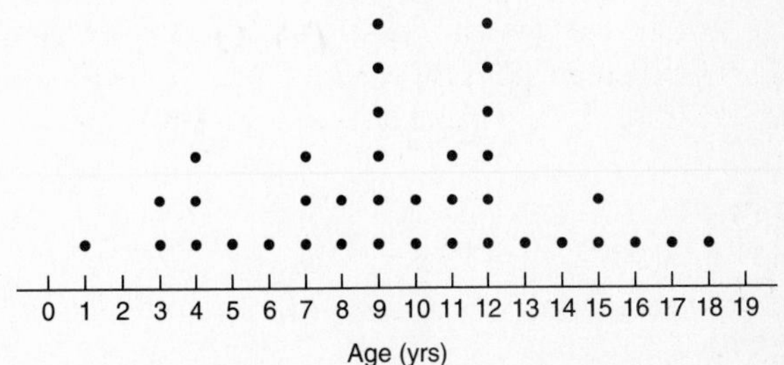

2.67

a.

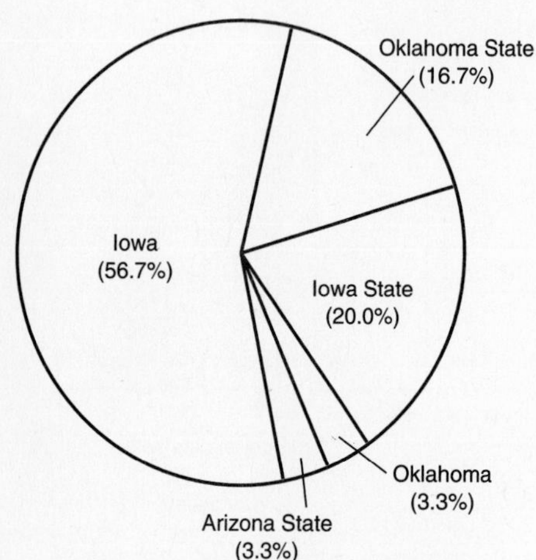

b.

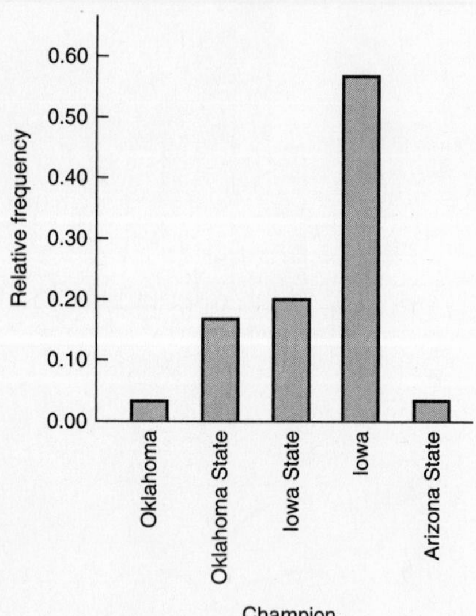

2.69

a. 20% **b.** 25% **c.** 7

2.79

a. 50 **b.** 95 **c.** 10 **d.** 5

2.83

a. 37 **b.** 7

Exercises 2.4

2.85 Histogram. Stem-and-leaf diagrams are generally not useful with large data sets.

2.87 Reconstruct the stem-and-leaf diagram using more lines per stem.

2.89

a.

91	4
92	
93	
94	6
95	9 7
96	4
97	7 5 4 7
98	6 9 4 8 7
99	0 6 1 9 5 7
100	1
101	4 8 0 7
102	5 8
103	0 1
104	
105	
106	0

b.

91	4
92	
93	
94	6
95	7 9
96	4
97	4 5 7 7
98	4 6 7 8 9
99	0 1 5 6 7 9
100	1
101	0 4 7 8
102	5 8
103	0 1
104	
105	
106	0

2.91

a.

4	5
5	8 4 5 1 0 5 5
6	4 7 9 6 0 6 2
7	7 5 8
8	6 7 1 0 3 3
9	7 6 3 4 9 7 6 0 1 7
10	1 0 9 4 2
11	1 3 1 3
12	9 5
13	0 9 6
14	
15	5 1

b.

4	5
5	0 1 4 5 5 5 8
6	0 2 4 6 6 7 9
7	5 7 8
8	0 1 3 3 6 7
9	0 1 3 4 6 6 7 7 7 9
10	0 1 2 4 9
11	1 1 3 3
12	5 9
13	0 6 9
14	
15	1 5

2.93

a.

2	7 7 5
3	7 5 3 3 1 3 9
4	9 8 5 1 1 6 9 4 3 8 4 7 5 1 0 3
5	7 3 6 4 3 0 1 6 6 1 9 3
6	2 0 7 7 1 7 2 3 0 0
7	9
8	3

b.

2	
2	7 7 5
3	3 3 1 3
3	7 5 9
4	1 1 4 3 4 1 0 3
4	9 8 5 6 9 8 7 5
5	3 4 3 0 1 1 3
5	7 6 6 6 9
6	2 0 1 2 3 0 0
6	7 7 7
7	
7	9
8	3
8	

2.95

a.

2	
2	9 9
3	3 1 3 2 3 4 3
3	6 9 5 9 5 7 9 9 5 7 8
4	0 0 1 2 0 4 4 0 3 2 1 2 4 1 1 4 4
4	8 8 5 9 5 6 5 5 7 9 8 5 9 5 9 9 7
5	1 1 3 1 2 1
5	8 7 5 7 9 5 7
6	
6	9
7	0 3
7	

b.

2	
2	
2	
2	
2	9 9
3	1
3	3 3 2 3 3
3	5 5 4 5
3	6 7 7
3	9 9 9 9 8
4	0 0 1 0 0 1 1 1
4	2 3 2 2
4	5 5 5 4 4 5 5 4 5 4 4
4	6 7 7
4	8 8 9 9 8 9 9 9
5	1 1 1 1
5	3 2
5	5 5
5	7 7 7
5	8 9
6	
6	
6	
6	
6	9
7	0
7	3
7	
7	
7	

2.101

a. 50 **b.** two **c.** 16 **d.** 8
e. 87 **f.** 64, 65, 66, 66, 67, 67, 68, 68

Exercises 2.5

2.103

a. The *distribution of a data set* is a table, graph, or formula that tells us the values of the observations and how often they occur.

b. *Sample data* is a data set obtained by observing the values of a variable for a sample of the population.

c. *Population data* is the data set obtained by observing the values of a variable for an entire population.

d. *Census data* is another name for population data.

e. A *sample distribution* is the distribution of sample data.

f. The *population distribution* is the distribution of population data.
g. *Distribution of a variable* is another name for population distribution.

2.105 a bell shape

2.107 Answers will vary.

2.109
a. right skewed **b.** right skewed

2.111
a. bell-shaped **b.** symmetric

2.113
a. left skewed **b.** left skewed
Note: The answers *bell-shaped* for part (a) and *symmetric* for part (b) are also acceptable.

2.115
a. bell-shaped **b.** symmetric

2.117
a. right skewed **b.** right skewed

Exercises 2.6

2.125
a. Part of the vertical axis of the graph has been cut off, or truncated.
b. It may be done to present a clearer picture of the ups and downs in a data pattern rather than to mislead the reader.
c. One should start the axis at 0 and put slashes in the axis to indicate that part of the axis is missing.

2.127
c. They give the misleading impression that the district average is much greater relative to the national average than it actually is.

2.129
a. It is a truncated graph.

b.

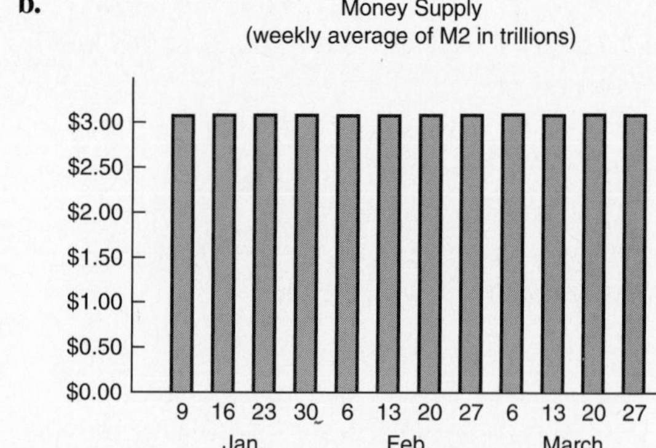

c.

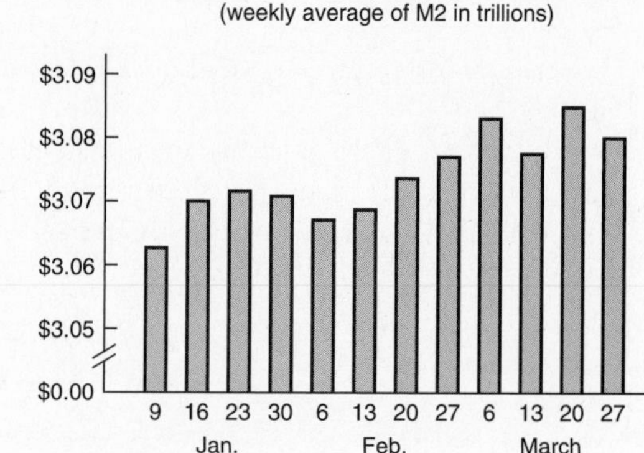

Review Test for Chapter 2

1. a. A variable is a characteristic that varies from one person or thing to another.
b. quantitative variables and qualitative (or categorical) variables
c. discrete variables and continuous variables
d. the information obtained by observing the values of a variable
e. by the type of variable being observed

2. It helps organize the data, making it much simpler to comprehend.

3. Qualitative data. It makes no sense to look for cutpoints or midpoints for nonnumerical data.

4. a. 6 and 14 **b.** 18 **c.** 22 and 30
d. the third class

5. a. 10 **b.** 20 **c.** 25 and 35

6. when grouping discrete data in which there are only relatively few distinct observations

7. a. The bar for each class extends from the lower cutpoint of the class to the upper cutpoint of the class.

b. The bar for each class is centered over the midpoint of the class.

8. pie charts and bar graphs

9. Histogram. Stem-and-leaf diagrams are generally not useful with large data sets.

10. See Figure 2.8 on page 105.

12. a. Left skewed. The distribution of a random sample taken from a population approximates the population distribution. The larger the sample, the better the approximation tends to be.

b. No. Sample distributions vary from sample to sample.

c. Yes. Left skewed. The overall shapes of the two sample distributions should be similar to that of the population distribution and hence to each other.

13. a. discrete quantitative

b. continuous quantitative

c. qualitative

14. a.

Age at inauguration	Frequency	Relative frequency	Midpoint
40–44	2	0.048	42.5
45–49	6	0.143	47.5
50–54	12	0.286	52.5
55–59	12	0.286	57.5
60–64	7	0.167	62.5
65–69	3	0.071	67.5
	42	1.001	

b. 40 and 45 **c.** 5

d.

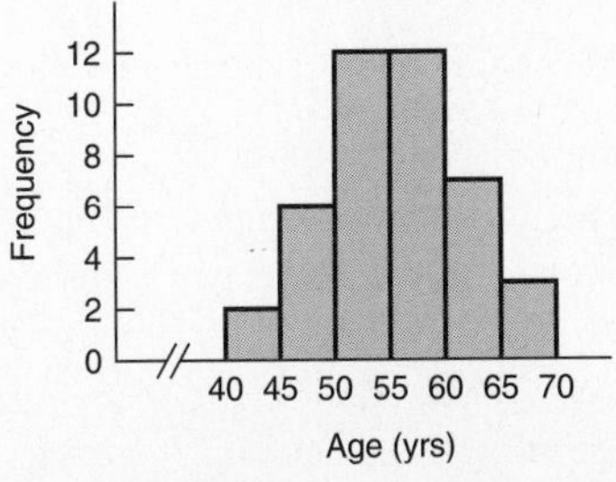

15.

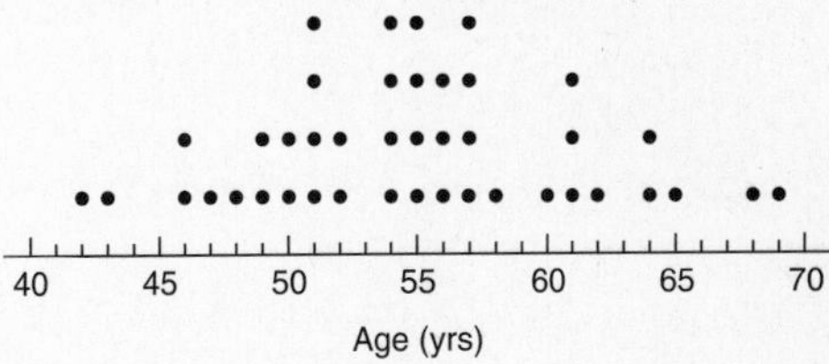

16. a.

4 | 2 3 6 6 7 8 9 9
5 | 0 0 1 1 1 1 2 2 4 4 4 4 5 5 5 5 6 6 6 7 7 7 7 8
6 | 0 1 1 1 2 4 4 5 8 9

b.

4 | 2 3
4 | 6 6 7 8 9 9
5 | 0 0 1 1 1 1 2 2 4 4 4 4
5 | 5 5 5 5 6 6 6 7 7 7 7 8
6 | 0 1 1 1 2 4 4
6 | 5 8 9

c. the one in part (b)

17. a.

Number busy	Frequency	Relative frequency
0	1	0.04
1	2	0.08
2	2	0.08
3	4	0.16
4	5	0.20
5	7	0.28
6	4	0.16
	25	1.00

b.

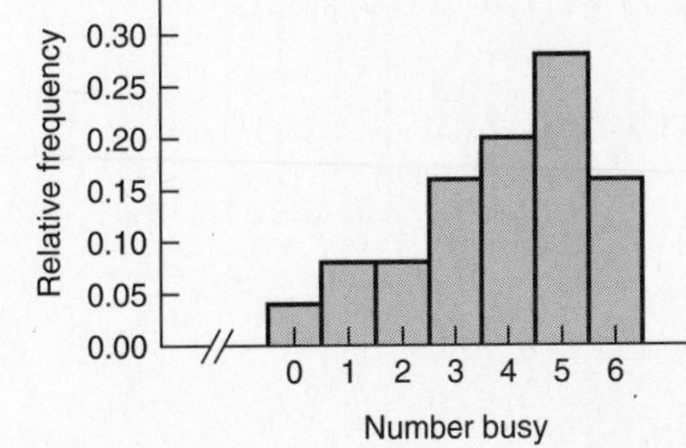

18. a.

Year	Frequency	Relative frequency
Freshman	6	0.150
Sophomore	15	0.375
Junior	12	0.300
Senior	7	0.175
	40	1.000

b.

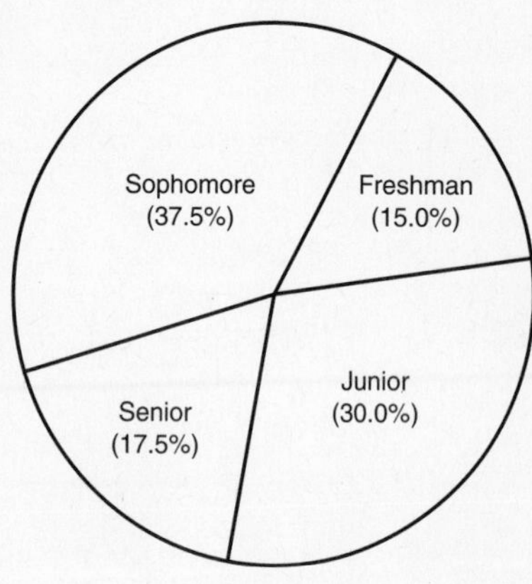

c.

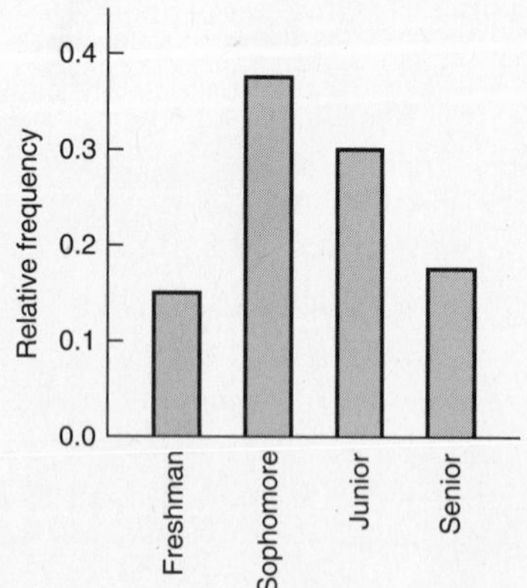

19. a.

High	Freq.	Relative frequency	Midpoint
400 < 1000	16	0.444	700
1000 < 1600	9	0.250	1300
1600 < 2200	2	0.056	1900
2200 < 2800	2	0.056	2500
2800 < 3400	2	0.056	3100
3400 < 4000	3	0.083	3700
4000 < 4600	0	0.000	4300
4600 < 5200	0	0.000	4900
5200 < 5800	1	0.028	5500
5800 < 6400	0	0.000	6100
6400 < 7000	1	0.028	6700
	36	1.001	

b.

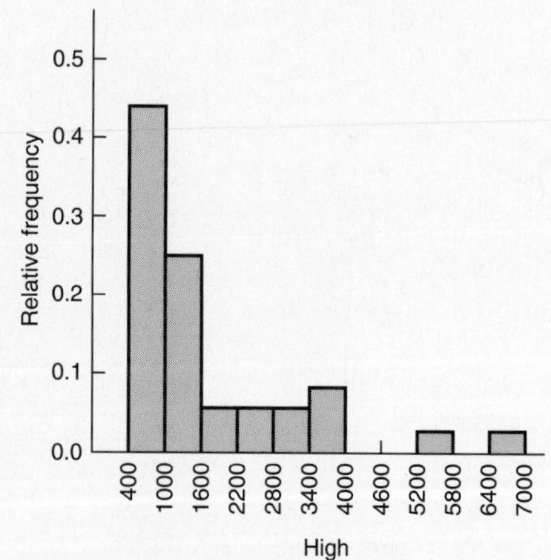

*Data from *The World Almanac, 1998*

20. a. bell-shaped **b.** left skewed

21. Answers will vary, but here is one possibility:

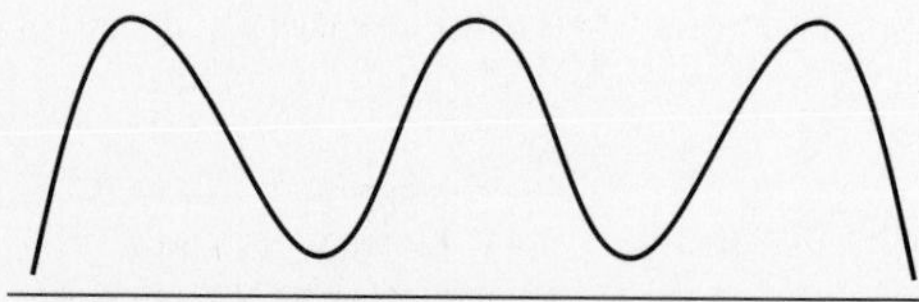

22. b. Having followed the directions in part (a), we might conclude that the percentage of women in the labor force for 2000 is about 3.5 times that for 1960.

c. Not covering up the vertical axis, we find that the percentage of women in the labor force for 2000 is about 1.8 times that for 1960.
d. The graph is potentially misleading because it is truncated. Notice that the vertical axis begins at 30 rather than at 0.
e. To make the graph less potentially misleading, we can start it at 0 instead of 30.

24. a. 22 **b.** 27.5 **c.** 3 **d.** TIME

25. a. two **b.** 2 **c.** 4 **d.** 48 minutes
e. 30, 31, 31, 37

CHAPTER 3

Exercises 3.1

3.1 to indicate where the center or most typical value of a data set lies

3.3 the mode

3.5
a. mean = 5, median = 5
b. mean = 15, median = 5. The median is a better measure of center because it is not influenced by the one unusually large value, 99.
c. resistance

3.7 Median. Unlike the mean, the median is not affected strongly by the relatively few homes that have extremely large or small floor spaces.

3.9 mean = 193.0 thousand; median = 79.0 thousand; no mode

3.11 mean = 33.5 years; median = 34.5 years; modes = 24, 28, 37 years

3.13
a. Iowa
b. No, neither the mean nor the median can be used as a measure of center for qualitative data.

Exercises 3.2

3.31 Mathematical notation allows us to express mathematical definitions and other mathematical relationships much more concisely.

3.33 No, the population mean is a constant. Yes, the sample mean is a variable because it varies from sample to sample.

3.35
a. 46 **b.** 4 **c.** 11.5

3.37
a. $12,971 **b.** 8 **c.** $1621.40

3.39
a. 23.3 hours **b.** 10 **c.** 2.33 hours

3.41
a. $98.00 **b.** 60,772; 0; 3148

Exercises 3.3

3.45 to indicate the amount of variation in a data set

3.47 the mean

3.49
a. 2.7 **b.** 31.6 **c.** resistance

3.51
a. 16.1 **b.** 16.1

3.53
a. 501 thousand **b.** 205.5 thousand
c. 205.5 thousand

3.55
a. 20 yrs **b.** 7.2 yrs **c.** 7.2 yrs

Exercises 3.4

3.75 The median and interquartile range are resistant measures, whereas the mean and standard deviation are not.

3.77 No. It may, for example, be an indication of skewness.

3.79
a. a measure of variation
b. the range of the middle 50% of the observations

3.81 If the data set has no potential outliers, then its boxplot and modified boxplot are identical. This is the case when the minimum and maximum observations both lie within the lower and upper limits. In other words, the minimum and maximum values are also the adjacent values.

3.83 $Q_1 = 67.75$, $Q_2 = 83.0$, $Q_3 = 89.75$

3.85 $Q_1 = 4$, $Q_2 = 7$, $Q_3 = 12.5$

3.87 $Q_1 = 7.7$, $Q_2 = 9.2$, $Q_3 = 11.8$

3.89

a. IQR = 22 **b.** 34, 67.75, 83.0, 89.75, 100
c. 34 is a potential outlier.
d. Fig. A.1(a) shows a boxplot; Fig. A.1(b) shows a modified boxplot.

3.91

a. IQR = 8.5 **b.** 1, 4, 7, 12.5, 55
c. 55 is a potential outlier.
d. Fig. A.2(a) shows a boxplot; Fig. A.2(b) shows a modified boxplot.

3.103

a. The least variation occurs in the second quarter of the data, the next least in the third quarter, the next least in the fourth quarter, and the greatest in the first quarter. Although answers may vary, the five-number summary is approximately 0, 54, 62, 73, and 89.
b. Yes. Potential outliers at approximately 0, 8, and 20. Assuming no recording errors, the potential outlier(s) at 0 may be a vegetarian or someone who does not eat beef for some other reason; the potential outliers at 8 and 20 may be people on low-beef diets.
c. 0, 54.5, 62, 72.75, and 89

Exercises 3.5

3.105 To describe the entire population.

3.107

a. $\bar{x}$ **b.** s

3.109 how many standard deviations the observation is from the mean and, by its sign, what direction it is from the mean

3.111 Parameter. A parameter is a descriptive measure for a population.

3.113

a. $\bar{x} = 75.0$ in. **b.** $s = 6.2$ in.
c. $\mu = 75.0$ in. **d.** $\sigma = 5.6$ in.
e. $\bar{x}$ and μ are computed in the same way: Sum the data and then divide by the total number of observations.
f. s and σ are computed differently: In the defining formula for s, we divide by one less than the total number of observations, whereas in the defining formula for σ, we divide by the total number of observations.

3.115

a. $\mu = \$97.1$ **b.** $\sigma = \$16.4$

3.117

a. $\mu = \$50.25$ million **b.** $\sigma = \$48.0$ million

3.119

a. $z = (y - 356)/1.63$
b. $\mu_z = 0, \sigma_z = 1$
c. $z = -2.45, z = 3.07$
d. A content of 352 mL is 2.45 standard deviations below the mean. A content of 361 mL is 3.07 standard deviations above the mean.
e. See Fig. A.3.

FIGURE A.1 (a) Boxplot and (b) modified boxplot for Exercise 3.89(d)

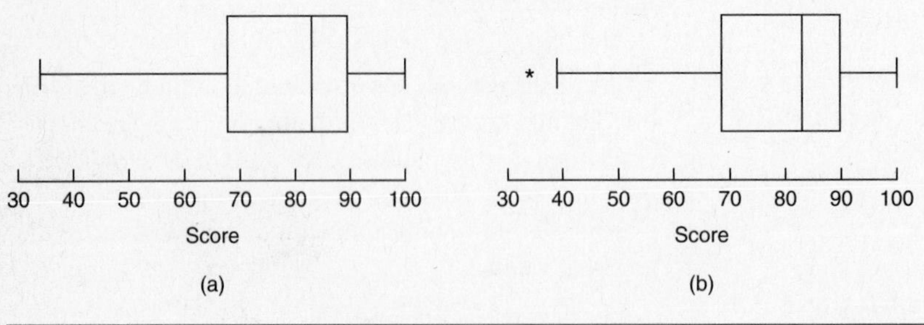

FIGURE A.2 (a) Boxplot and (b) modified boxplot for Exercise 3.91(d)

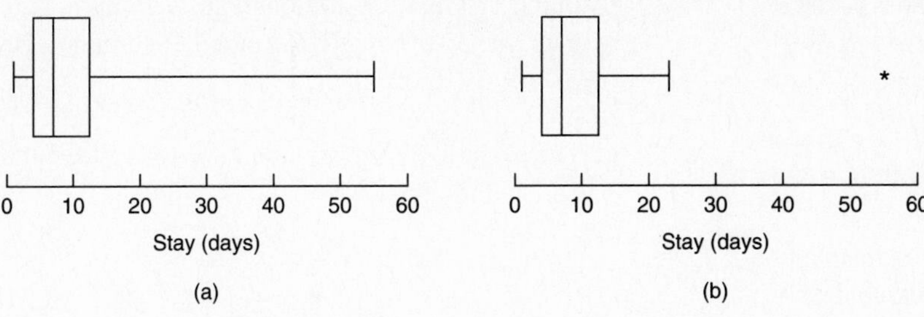

FIGURE A.3 Graph for Exercise 3.119(e)

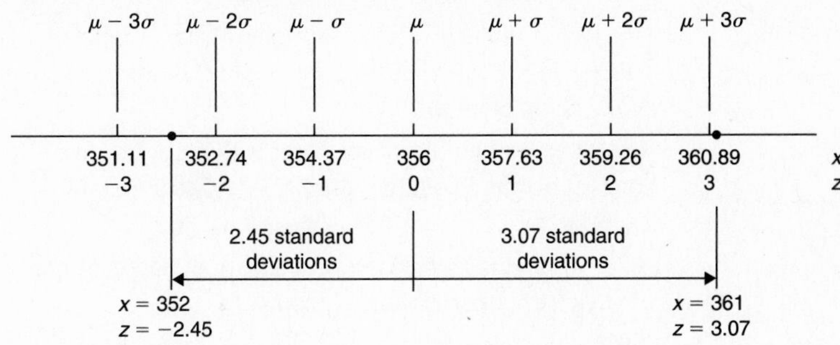

Review Test for Chapter 3

1. **a.** Numbers that are used to describe data sets are called descriptive measures.
   b. Descriptive measures that indicate where the center or most typical value of a data set lies are called measures of center.
   c. Descriptive measures that indicate the amount of variation or spread in a data set are called measures of variation.
2. Mean and median. The median is a resistant measure whereas the mean is not. The mean takes into account the actual numerical value of all observations whereas the median does not.
3. the mode
4. **a.** standard deviation **b.** interquartile range
5. **a.** $\bar{x}$ **b.** s **c.** μ **d.** σ
6. **a.** not necessarily true **b.** necessarily true
7. three
8. **a.** minimum, quartiles, and maximum, that is, Min, Q_1, Q_2, Q_3, Max
   b. Q_2 can be used to describe center. Max − Min, Q_1 − Min, Max − Q_3, $Q_2 - Q_1$, $Q_3 - Q_2$, and $Q_3 - Q_1$ are all measures of variation for different portions of the data.
   c. boxplot
9. **a.** An outlier is an observation that falls well outside the overall pattern of the data.
   b. First determine the lower and upper limits—the numbers that lie, respectively, 1.5 IQRs below the first quartile and 1.5 IQRs above the third quartile. Observations that lie outside the lower and upper limits are potential outliers.
10. **a.** Subtract from x its mean and then divide by its standard deviation.
    b. The z-score of an observation tells us how many standard deviations the observation is from the mean.

c. The observation is 2.9 standard deviations above the mean. It is larger than most of the other possible observations.

11. a. 11.4 kg; 16.1 kg **b.** 11.5 kg; 17.0 kg
c. 8, 12 kg; 12, 16, 19, 23 kg

12. The median, because it is resistant to outliers and other extreme values.

13. The mode; neither the mean nor the median can be used as a measure of center for qualitative data.

14. a. $\bar{x} = 4.0$ min **b.** Range = 14 min
c. $s = 4.1$ min

15. a.

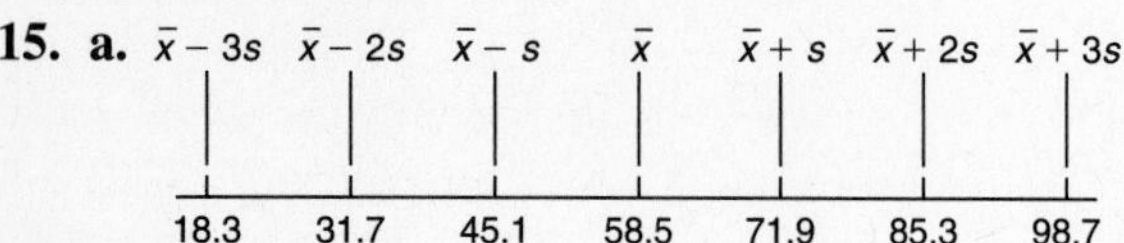

b. 18.3, 98.7

16. a. $Q_1 = 48$, $Q_2 = 59.5$, $Q_3 = 68.75$
b. 20.75
c. 31, 48, 59.5, 68.75, 79
d. Lower limit: 16.875. Upper limit: 99.875.
e. no potential outliers

f.

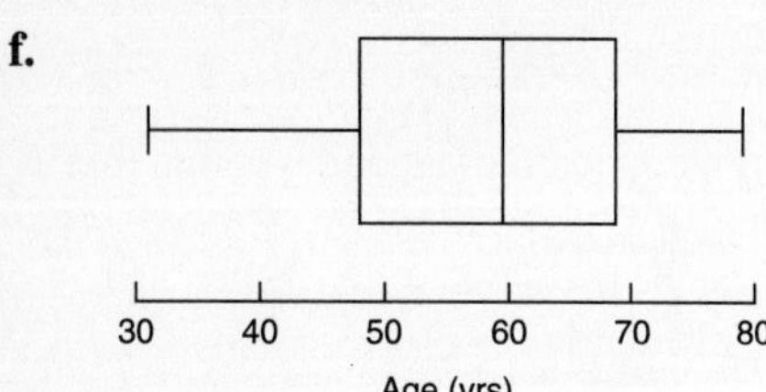

17. a. $\mu = 18.37$ **b.** $\sigma = 9.42$
c. $z = (x - 18.37)/9.42$
d. $\mu_z = 0, \sigma_z = 1$
f. 1.75, −0.03. The enrollment at UCLA is 1.75 standard deviations above the mean and the enrollment at UCSD is 0.03 standard deviations below the mean.

18. a. A sample mean. It is the mean price per gallon for the sample of 10,000 service stations.
b. $\bar{x}$
c. A statistic. It is a descriptive measure for a sample.

21. a. The least variation occurs in the third quarter, the next least in the fourth quarter, the next least in the second quarter, and the greatest in the first quarter. Although answers may vary, the approximate five-number summary is 31, 48, 59, 69, and 79.
b. no
c. The exact five-number summary is 31, 48, 59.5, 68.75, and 79.

CHAPTER 4

Exercises 4.1

4.1 An experiment is an action whose outcome cannot be predicted with certainty. An event is some specified result that may or may not occur when the experiment is performed.

4.3 There is no difference.

4.5 The probability of an event is the proportion of times it occurs in a large number of repetitions of the experiment.

4.7 (b) and (e), since the probability of an event must always be between 0 and 1, inclusive.

4.9
a. 0.190 **b.** 0.696 **c.** 0.021
d. 0 **e.** 1

4.11
a. 0.253 **b.** 0.323 **c.** 0.626 **d.** 0.173

4.13
a. 0.139 **b.** 0.500 **c.** 0.222 **d.** 0.111

4.15
a. 0.735 **b.** 0.932 **c.** 0.197

4.17 The event in part (e) is certain; the event in part (d) is impossible.

4.19 $\frac{1}{4}$

Exercises 4.2

4.25 Venn diagrams

4.27 Two events are mutually exclusive if they do not have outcomes in common. Three events are mutually exclusive if no two of them have outcomes in common.

4.29

$A =$ 

$B =$ 

$C =$

$D =$

4.31

$A = \{HHTT, HTHT, HTTH, THHT, THTH, TTHH\}$
$B = \{TTHH, TTHT, TTTH, TTTT\}$
$C = \{HHHH, HHHT, HHTH, HHTT, HTHH, HTHT, HTTH, HTTT\}$
$D = \{HHHH, TTTT\}$

4.33

a. (not A) =

The event the die comes up odd.

b. (A & B) =

The event the die comes up 4 or 6.

c. (B or C) =

The event the die does not come up 3.

4.35

a. (not B) = {HHHH, HHHT, HHTH, HHTT, HTHH, HTHT, HTTH, HTTT, THHH, THHT, THTH, THTT}
The event that at least one of the first two tosses is heads.

b. (A & B) = {TTHH}
The event that the first two tosses are tails and the last two are heads.

c. (C or D) = {HHHH, HHHT, HHTH, HHTT, HTHH, HTHT, HTTH, HTTT, TTTT}
The event that the first toss is a head or all four tosses are tails.

4.37

a. (not A) is the event the unit has at least five rooms. There are 76,217 thousand units that have at least five rooms.

b. (A & B) is the event the unit has two, three, or four rooms. There are 32,377 thousand units that have two, three, or four rooms.

c. (C or D) is the event the unit has at least five rooms. There are 76,217 thousand units that have at least five rooms. (*Note:* From part (a), we see that (not A) = (C or D).)

4.39

a. (not C) is the event that the person is 45 or older. There are 40,943 thousand such people.

b. (not B) is the event that the person is either under 20 or over 54. There are 21,929 thousand such people.

c. (B & C) is the event that the person is between 20 and 44, inclusive. There are 79,266 thousand such people.

d. (A or D) is the event that the person is either under 20 or over 54. There are 21,929 thousand such people. (*Note:* From part (b), we see that (not B) = (A or D).)

4.41

a. no **b.** yes **c.** no
d. Yes, events B, C, and D. No.

4.43

a. mutually exclusive **b.** not mutually exclusive
c. mutually exclusive **d.** not mutually exclusive
e. not mutually exclusive

Exercises 4.3

4.47 0.2; $P(E) = 0.2$

4.49

a. 0.56 **b.** $S = (A \text{ or } B \text{ or } C)$
c. 0.01, 0.14, 0.41 **d.** 0.56

4.51

a. 0.770 **b.** 0.077 **c.** 0.322
d. 77.0% of U.S. partnerships had receipts of under \$100,000; 7.7% had receipts of at least \$500,000; and 32.2% had receipts between \$25,000 and \$499,999, inclusive.

4.53

a. 0.99 **b.** 0.56

4.55

a. 0.959 **b.** 0.318

4.57

a. 0.167, 0.056, 0.028, 0.056, 0.028, 0.139, 0.167
b. 0.223 **c.** 0.112 **d.** 0.278 **e.** 0.278

4.59

a. 0.520, 0.095, 0.054

b. 0.561; 56.1% of U.S. adults are either female or divorced (or both).

c. 0.480

4.61 0.267

Exercises 4.4

4.67 summing the row totals, summing the column totals, or summing the frequencies in the cells

4.69

a. univariate **b.** bivariate

4.71

a. 8 **b.** 3274 **c.** 863 **d.** 1471 **e.** 502

4.73

a. 73,970 **b.** 15,540 **c.** 12,328
d. 77,182 **e.** 12,225 **f.** 6,734 **g.** 77,700

4.75

a. The missing entries in the second, third, and fourth rows are, respectively, 130, 183, and 27.

b. 15 **c.** 554 thousand **d.** 217 thousand
e. 111 thousand **f.** 813 thousand
g. 106 thousand

4.77

a. The institution selected is private; the institution selected is in the South; the institution selected is a public school in the West.

b. 0.551; 0.316; 0.096. 55.1% of institutions of higher education are private; 31.6% are in the South; and 9.6% are public schools in the West.

c.

Region	Type: Public T_1	Private T_2	$P(R_i)$
Northeast R_1	0.081	0.170	0.251
Midwest R_2	0.110	0.154	0.264
South R_3	0.163	0.153	0.316
West R_4	0.096	0.074	0.170
$P(T_j)$	0.449	0.551	1.000

4.79

a. The surgeon selected is office-based; the surgeon selected is an orthopedic surgeon; the surgeon selected is an office-based orthopedic surgeon.

b. 0.714, 0.174, 0.129

c. 0.759 **d.** 0.759

e.

Specialty	Base of practice: Office B_1	Hospital B_2	Other B_3	Total
General surgery S_1	0.233	0.118	0.016	0.367
Ob/Gyn S_2	0.233	0.065	0.011	0.309
Orthopedics S_3	0.129	0.041	0.004	0.174
Ophthalmology S_4	0.119	0.026	0.005	0.150
Total	0.714	0.250	0.036	1.000

4.81

a. (i) A_3 (ii) T_2 (iii) $(A_5 \;\&\; T_1)$

b. (i) 0.222 (ii) 0.310 (iii) 0.018

c.

		Tenure of operator			
		Full owner T_1	Part owner T_2	Tenant T_3	Total
Acreage	Under 50 A_1	23.1	3.0	2.7	28.8
	50 < 180 A_2	20.5	6.8	3.1	30.3
	180 < 500 A_3	9.9	9.5	2.9	22.2
	500 < 1000 A_4	2.5	5.8	1.4	9.7
	1000 & over A_5	1.8	5.9	1.2	9.0
	Total	57.8	31.0	11.3	100.0

Exercises 4.5

4.85

a. It is the probability that an event occurs under the assumption that another event has occurred.
b. the event that is assumed to have occurred

4.87

a. 0.077 **b.** 0.333 **c.** 0.077 **d.** 0
e. 0.231 **f.** 1 **g.** 0.231 **h.** 0.167

4.89

a. 0.190 **b.** 0.191 **c.** 0.298
d. 19.0% of U.S. housing units have exactly 4 rooms; 19.1% of U.S. housing units that have at least 2 rooms have exactly 4 rooms; 29.8% of U.S. housing units that have at least 2 rooms have at most 4 rooms.

4.91

a. 0.251 **b.** 0.308 **c.** 0.676
d. 25.1% of all institutions of higher education are in the Northeast; 30.8% of all private institutions of higher education are in the Northeast; 67.6% of all institutions of higher education in the Northeast are private schools.

4.93

a. 0.113 **b.** 0.029 **c.** 0.253 **d.** 0.257
e. 11.3% of farms are tenant-operated; 2.9% of farms are tenant-operated and have between 180 and 500 acres; 25.3% of tenant-operated farms have between 180 and 500 acres.

4.95

a. 0.187 **b.** 0.103 **c.** 0.551 **d.** 0.195
e. 18.7% of the members of the 105th Congress are senators; 10.3% of the members of the 105th Congress are Republican senators; 55.1% of the senators in the 105th Congress are Republicans; and 19.5% of the Republicans in the 105th Congress are senators.

4.97 52.4%

Exercises 4.6

4.101

a. See Formulas 4.5 and 4.4 on pages 245 and 240, respectively.
b. They are mathematical equivalents of each other.
c. When the joint and marginal probabilities are known or easily determined directly, we can use the conditional-probability rule to obtain conditional probabilities. On the other hand, when the marginal and conditional probabilities are known or easily determined, we can use the general multiplication rule to obtain joint probabilities.

4.103 0.229; 22.9% of all U.S. adults are women who suffer from holiday depression.

4.105

a. 0.1 **b.** 0.111 **c.** 0.011 **d.** 0.222

4.107

a. 0.222 **b.** 0.405
c. Let $R1$, $D1$, and $I1$ denote, respectively, the events that the first governor selected is Republican, Democrat, and Independent; and let $R2$, $D2$, and $I2$ denote, respectively, the events that the second governor selected is a Republican, Democrat, and Independent. See Fig. A.4 on the following page.
d. 0.516 **e.** 0.444

FIGURE A.4 Tree diagram for Exercise 4.107(c)

			Event	Probability
$D1$ ($\frac{17}{50}$)	$\frac{16}{49}$	$D2$	($D1$ & $D2$)	$\frac{17}{50} \cdot \frac{16}{49} = 0.111$
	$\frac{32}{49}$	$R2$	($D1$ & $R2$)	$\frac{17}{50} \cdot \frac{32}{49} = 0.222$
	$\frac{1}{49}$	$I2$	($D1$ & $I2$)	$\frac{17}{50} \cdot \frac{1}{49} = 0.007$
$R1$ ($\frac{32}{50}$)	$\frac{17}{49}$	$D2$	($R1$ & $D2$)	$\frac{32}{50} \cdot \frac{17}{49} = 0.222$
	$\frac{31}{49}$	$R2$	($R1$ & $R2$)	$\frac{32}{50} \cdot \frac{31}{49} = 0.405$
	$\frac{1}{49}$	$I2$	($R1$ & $I2$)	$\frac{32}{50} \cdot \frac{1}{49} = 0.013$
$I1$ ($\frac{1}{50}$)	$\frac{17}{49}$	$D2$	($I1$ & $D2$)	$\frac{1}{50} \cdot \frac{17}{49} = 0.007$
	$\frac{32}{49}$	$R2$	($I1$ & $R2$)	$\frac{1}{50} \cdot \frac{32}{49} = 0.013$
	$\frac{0}{49}$	$I2$	($I1$ & $I2$)	$\frac{1}{50} \cdot \frac{0}{49} = 0.000$

4.109

a. 0.151 **b.** 0.050

c. No, because $P(C_1 \mid S_2) \neq P(C_1)$.

d. No. $P(S_1) = 0.580$ and $P(S_1 \mid C_2) = 0.458$; so $P(S_1 \mid C_2) \neq P(S_1)$.

4.111

a. 0.5, 0.5, 0.375 **b.** 0.5

c. Yes, because $P(B \mid A) = P(B)$.

d. 0.25

e. No, because $P(C \mid A) \neq P(C)$.

4.113

a. 0.469, 0.187, 0.084

b. Events P_1 and C_2 are not independent because $P(P_1 \,\&\, C_2) \neq P(P_1) \cdot P(C2)$ $(0.084 \neq 0.469 \cdot 0.187)$.

4.115

a. 0.006 **b.** 0.005

4.117

a. 0.083 **b.** 0.5

4.119

a. 0.928 **b.** 0.072

c. There was a 7.2% chance that at least one "criticality 1" item would fail; in the long run, at least one "criticality 1" item will fail in 7.2 out of every 100 such missions.

4.121

a. 0.105 **b.** 0.105 **c.** 0.031 **d.** 0.420

Exercises 4.7

4.129 one or more of them must occur

4.131

a. $P(R_3)$ **b.** $P(S \mid R_3)$ **c.** $P(R_3 \mid S)$

4.133

a. 43.1% **b.** 33% **c.** 39.8%

4.135

a. 0.437 **b.** 0.256

c. 43.7% of adults are movie goers; 25.6% of movie-going adults are between 25 and 34 years old.

4.137

a. 34.0% **b.** 35.3%

4.139

a. 0.605 **b.** 0.189 **c.** 0.127

d. The method for 60.5% of suicides was firearms; 18.9% of suicides were committed by females; 12.7% of suicides by firearms were committed by females.

4.141 0.803

Exercises 4.8

4.147 Because we multiply the number of possibilities for each action to obtain the total number of possibilities.

4.149

a. See Fig. A.5.

b. 12 **c.** $3 \cdot 4 = 12$

4.151

a. 900 **b.** 9,000,000 **c.** 8,100,000,000

4.153 1,021,440

4.155

a. 24 **b.** 32,760 **c.** 30

d. 1 **e.** 40,320

4.157 657,720

4.159

a. 3,628,800 **b.** $\frac{1}{3,628,800} \approx 0.0000003$

4.161

a. 4 **b.** 1365 **c.** 15 **d.** 1 **e.** 1

4.163

a. 75,287,520 **b.** 67,800,320 **c.** 0.901

4.165 8,145,060

4.167

a. 0.243 **b.** 0.972 **c.** 0.271

4.169

a. 0.0000002 **b.** 0.002 **c.** 0.971

FIGURE A.5 Tree diagram for Exercise 4.149(a)

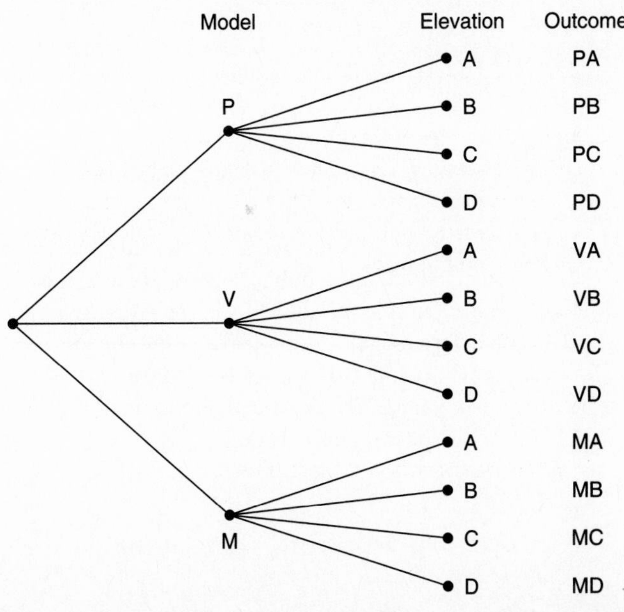

Review Test for Chapter 4

1. Probability theory enables us to evaluate and control the likelihood that a statistical inference is correct and, more generally, provides the mathematical basis for inferential statistics. It also allows us to extend concepts that apply to variables of finite populations—such as relative-frequency distribution, mean, and standard deviation—to other types of variables.

2. a. a probability model in which all possible outcomes of the experiment are equally likely to occur
b. Using the f/N-rule, see Definition 4.1 on page 199.

3. It is the proportion of times the event occurs in a large number of repetitions of the experiment.

4. (b) and (c), because the probability of an event must always be between 0 and 1, inclusive.

5. Venn diagrams

6. At most one of them can occur when the experiment is performed, that is, no two of them have outcomes in common.

7. a. $P(E)$ **b.** $P(E) = 0.436$

8. a. false **b.** true

9. It is sometimes easier to compute the probability that an event does not occur than the probability that it does occur.

10. a. univariate **b.** bivariate
c. contingency table or two-way table

11. marginal

12. a. $P(B \mid A)$ **b.** A

13. directly or using the conditional-probability rule

14. The joint probability equals the product of the marginal probabilities.

15. exhaustive

16. See Key Fact 4.2 on page 267.

17. a.

abc	*abd*	*acd*	*bcd*
acb	*adb*	*adc*	*bdc*
bac	*bad*	*cad*	*cbd*
bca	*bda*	*cda*	*cdb*
cba	*dab*	*dac*	*dbc*
cab	*dba*	*dca*	*dcb*

b. $\{a, b, c\}, \{a, b, d\}, \{a, c, d\}, \{b, c, d\}$
c. 24; 4

18. a. 0.120 **b.** 0.443
c. 0.120, 0.191, 0.195, 0.136, 0.103, 0.204, 0.051

19. a. (not J) is the event that the return shows an AGI of at least $100K. There are 4508 thousand such returns.
b. (H & I) is the event that the return shows an AGI of between $20K and $50K. There are 37,988 thousand such returns.
c. (H or K) is the event that the return shows an AGI of at least $20K. There are 60,374 thousand such returns.
d. (H & K) is the event that the return shows an AGI of between $50K and $100K. There are 17,878 thousand such returns.

20. a. not mutually exclusive
b. mutually exclusive
c. mutually exclusive
d. not mutually exclusive

21. a. 0.638, 0.745, 0.949, 0.255
b. $H = (C \text{ or } D \text{ or } E \text{ or } F)$
$I = (A \text{ or } B \text{ or } C \text{ or } D \text{ or } E)$
$J = (A \text{ or } B \text{ or } C \text{ or } D \text{ or } E \text{ or } F)$
$K = (F \text{ or } G)$
c. 0.638, 0.745, 0.949, 0.255

22. a. 0.051, 0.434, 0.689, 0.204
b. 0.949 **c.** 0.689 **d.** They are the same.

23. a. 6 **b.** 13,615 thousand
c. 48,778 thousand **d.** 2562 thousand

24. a. L_3 is the event that the student selected is in college; T_1 is the event that the student selected attends a public school; (T_1 & L_3) is the event that the student selected attends a public college.
b. 0.216; 0.866; 0.171. 21.6% of students attend college, 86.6% attend public schools; 17.1% attend public colleges.

c.

		Type		
		Public T_1	Private T_2	$P(L_i)$
Level	Elementary L_1	0.478	0.064	0.542
	High school L_2	0.217	0.025	0.242
	College L_3	0.171	0.045	0.216
	$P(T_j)$	0.866	0.134	1.000

d. 0.911 **e.** 0.911

25. **a.** 0.197; 19.7% of students attending public schools are in college.
b. 0.197

26. **a.** 0.134, 0.103
b. No, because $P(T_2 \mid L_2) \neq P(T_2)$. We see that 10.3% of high-school students attend private schools, whereas 13.4% of all students attend private schools.
c. No, because both events can occur if the student selected is any one of the 1400 thousand students who attend a private high school.
d. $P(L_1) = 0.542$, $P(L_1 \mid T_1) = 0.553$. Because $P(L_1 \mid T_1) \neq P(L_1)$, the event a student is in elementary school is not independent of the event a student attends public school.

27. **a.** 0.023 **b.** 0.309 **c.** 0.451

28. **a.** 0.686 **b.** 0.068 **c.** 0.271

29. **a.** No, since $P(A \,\&\, B) \neq 0$.
b. Yes, since $P(A \,\&\, B) = P(A) \cdot P(B)$.

30. **a.** 0.45 **b.** 0.507 **c.** 0.685 **d.** 0.608
e. 45% of the women surveyed answered "no" to the question; 50.7% of the people surveyed answered "no" to the question; 68.5% of the people surveyed were women; 60.8% of the people who answered "no" to the question were women.
f. The probabilities in parts (b) and (c) are prior; those in parts (a) and (d) are posterior.

31. **a.** 66 **b.** 1320 **c.** 28; 336

32. **a.** 635,013,559,600 **b.** 0.213
c. 0.00045 **d.** 0.032 **e.** 0.013

CHAPTER 5

Exercises 5.1

5.1
a. probability **b.** probability

5.3 $\{X = 3\}$ is the event that the student has three siblings; $P(X = 3)$ is the probability of the event that the student has three siblings.

5.5 the probability distribution of the random variable

5.7
a. 1, 2, 3, 4, 5, 6, 7 **b.** $\{X = 5\}$
c. 0.073. 7.3% of U.S. households consist of exactly five people.
d.

x	$P(X = x)$
1	0.232
2	0.317
3	0.175
4	0.154
5	0.073
6	0.030
7	0.019

e.

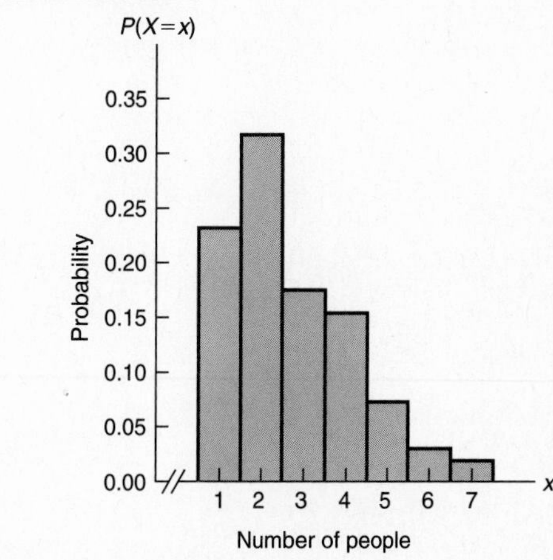

5.9
a. 2, 3, 4, 5, 6, 7, 8, 9, 10, 11, 12
b. $\{Y = 7\}$ **c.** $\frac{1}{6}$

d.

y	2	3	4	5	6	7	8	9	10	11	12
$P(Y = y)$	$\frac{1}{36}$	$\frac{1}{18}$	$\frac{1}{12}$	$\frac{1}{9}$	$\frac{5}{36}$	$\frac{1}{6}$	$\frac{5}{36}$	$\frac{1}{9}$	$\frac{1}{12}$	$\frac{1}{18}$	$\frac{1}{36}$

e.

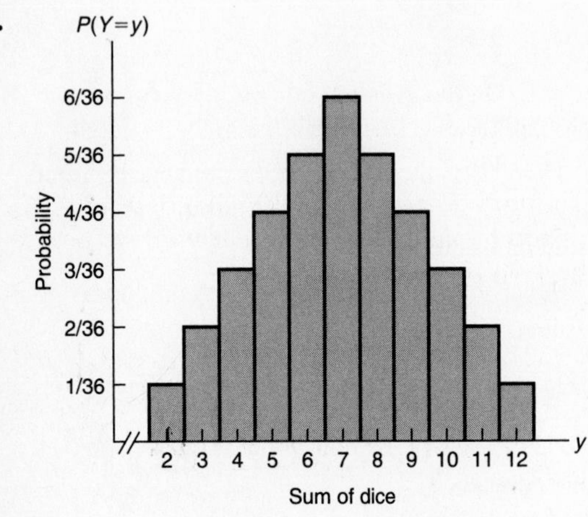

5.11

a. $\{X = 4\}$ **b.** $\{X \geq 2\}$ **c.** $\{X < 5\}$
d. $\{2 \leq X < 5\}$ **e.** 0.212 **f.** 0.922
g. 0.523 **h.** 0.445

Exercises 5.2

5.19

a. 2.7 **b.** 1.5

5.21

a. 7 **b.** 2.4

5.23

b. −0.052 **c.** 5.2¢ **d.** \$5.20, \$52

5.25

a. $\mu_W = 0.25, \sigma_W = 0.536$ **b.** 0.25 **c.** 62.5

Exercises 5.3

5.29 See Definition 5.8 on page 312.

5.31 two

5.33 5040; 40,320; 362,880

5.35

a. 10 **b.** 35 **c.** 120 **d.** 792

5.37

a. 10 **b.** 1 **c.** 1 **d.** 126

5.39

a. Each trial consists of observing whether a child with pinworm is cured by treatment with pyrantel and has two possible outcomes: cured or not cured. The trials are independent. The success probability is 0.9, that is, $p = 0.9$.

b.

Outcome	Probability
sss	$(0.9)(0.9)(0.9) = 0.729$
ssf	$(0.9)(0.9)(0.1) = 0.081$
sfs	$(0.9)(0.1)(0.9) = 0.081$
sff	$(0.9)(0.1)(0.1) = 0.009$
fss	$(0.1)(0.9)(0.9) = 0.081$
fsf	$(0.1)(0.9)(0.1) = 0.009$
ffs	$(0.1)(0.1)(0.9) = 0.009$
fff	$(0.1)(0.1)(0.1) = 0.001$

d. *ssf, sfs, fss*

e. 0.081. Because each probability is obtained by multiplying two success probabilities of 0.9 and one failure probability of 0.1.

f. 0.243

g.

x	$P(X = x)$
0	0.001
1	0.027
2	0.243
3	0.729

5.41

a. $p = 0.2$

b.

Outcome	Probability
ssss	$(0.2)(0.2)(0.2)(0.2) = 0.0016$
sssf	$(0.2)(0.2)(0.2)(0.8) = 0.0064$
ssfs	$(0.2)(0.2)(0.8)(0.2) = 0.0064$
ssff	$(0.2)(0.2)(0.8)(0.8) = 0.0256$
sfss	$(0.2)(0.8)(0.2)(0.2) = 0.0064$
sfsf	$(0.2)(0.8)(0.2)(0.8) = 0.0256$
sffs	$(0.2)(0.8)(0.8)(0.2) = 0.0256$
sfff	$(0.2)(0.8)(0.8)(0.8) = 0.1024$
fsss	$(0.8)(0.2)(0.2)(0.2) = 0.0064$
fssf	$(0.8)(0.2)(0.2)(0.8) = 0.0256$
fsfs	$(0.8)(0.2)(0.8)(0.2) = 0.0256$
fsff	$(0.8)(0.2)(0.8)(0.8) = 0.1024$
ffss	$(0.8)(0.8)(0.2)(0.2) = 0.0256$
ffsf	$(0.8)(0.8)(0.2)(0.8) = 0.1024$
fffs	$(0.8)(0.8)(0.8)(0.2) = 0.1024$
ffff	$(0.8)(0.8)(0.8)(0.8) = 0.4096$

d. *sssf, ssfs, sfss, fsss*

e. 0.0064. Because each probability is obtained by multiplying three success probabilities of 0.2 and one failure probability of 0.8.

f. 0.0256

g.

y	$P(Y = y)$
0	0.4096
1	0.4096
2	0.1536
3	0.0256
4	0.0016

5.47

a. $p = 0.5$ **b.** $p < 0.5$

5.49

a. 0.161 **b.** 0.332 **c.** 0.467 **d.** 0.821

e.

x	$P(X = x)$
0	0.004
1	0.040
2	0.161
3	0.328
4	0.332
5	0.135

f. left skewed

g.

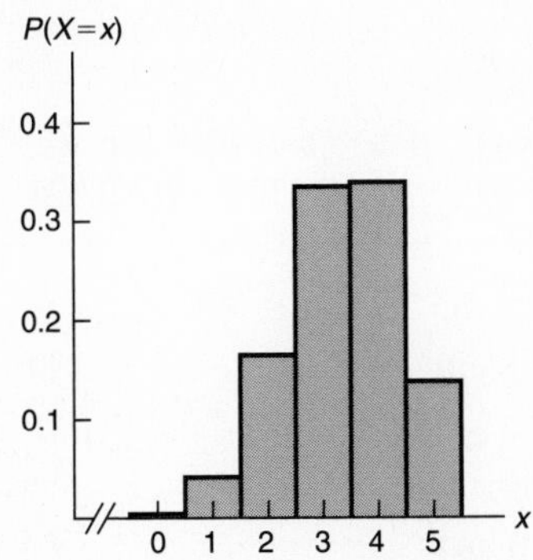

h. $\mu = 3.35, \sigma = 1.05$ **i.** $\mu = 3.35, \sigma = 1.05$

j. On the average, the favorite will finish in the money 3.35 races out of every five.

5.51

a. 0.211 **b.** 0.949 **c.** 0.680 **d.** 0.367

e.

y	$P(Y = y)$
0	0.316
1	0.422
2	0.211
3	0.047
4	0.004

f. $\mu = 1$. On the average, one out of every four U.S. children are not living with both parents.

g. $\sigma = 0.866$

h. The probability distribution is only approximately correct because sampling is without replacement. A hypergeometric distribution.

5.53

a. 0.230 **b.** 0.889 **c.** 0.341 **d.** 0.692

5.67

a. 6 **b.** 0.59 **c.** 0.4764

d. 0.8067 **e.** 0.2831

5.69

a. 0.8567 **b.** 0.3718 **c.** 0.2285

d. 0.6430 **e.** 0.3257

Exercises 5.4

5.71

a. 0.224 **b.** 0.647 **c.** 0.950

d. 3 **e.** 1.7

5.73

a. 0.195 **b.** 0.102 **c.** 0.704

5.75

a. 0.311 **b.** 0.757 **c.** 0.507

5.77

a. $\mu = 3.87$ particles. On the average, 3.87 particles will reach the screen during an 8-minute interval.

b. $\sigma = 1.97$ particles

5.79

a.

No. of calls x	Probability $P(X = x)$
0	0.183
1	0.311
2	0.264
3	0.150
4	0.064
5	0.022
6	0.006
7	0.001
8	0.000

5.81

a. 0.928 **b.** 0.072

5.83

a. 0.271 **b.** 0.677 **c.** 0.594

5.89

a. 2.1 **b.** 0.1890 **c.** 0.5999

Review Test for Chapter 5

1. **a.** random variable
b. finite; countably infinite

2. the possible values and corresponding probabilities of the discrete random variable

3. probability histogram

4. 1

5. **a.** $P(X = 2) = 0.386$ **b.** 38.6%
c. 19.3; 193

6. 3.6

7. X, because it has a smaller standard deviation, therefore less variation.

8. Each trial has the same two possible outcomes; the trials are independent; the probability of a success remains the same from trial to trial.

9. The binomial distribution is the probability distribution for the number of successes in a sequence of Bernoulli trials.

10. 120

11. Substitute the binomial (or Poisson) probability formula into the formulas for the mean and standard deviation of a discrete random variable and then simplify mathematically.

12. **a.** binomial distribution
b. hypergeometric distribution
c. When the sample size does not exceed 5% of the population size. Because, under this condition, there is little difference between sampling with and without replacement.

13. **a.** 1, 2, 3, 4 **b.** $\{X = 3\}$
c. 0.251; 25.1% of the undergraduates at ASU are juniors.

d.

x	$P(X = x)$
1	0.196
2	0.194
3	0.251
4	0.359

e.

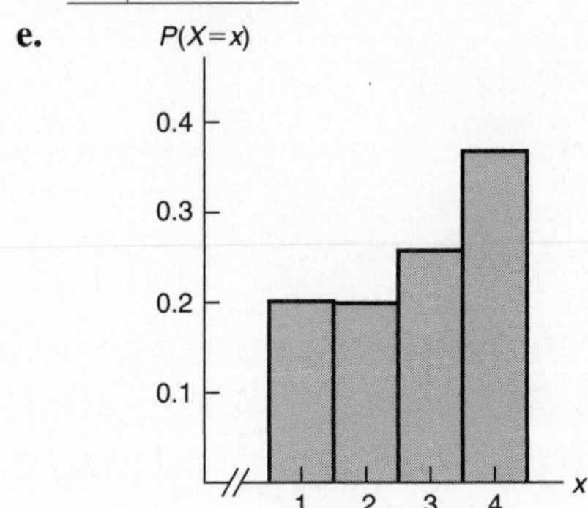

14. **a.** $\{Y = 4\}$ **b.** $\{Y \geq 4\}$ **c.** $\{2 \leq Y \leq 4\}$
d. $\{Y \geq 1\}$ **e.** 0.174 **f.** 0.322
g. 0.646 **h.** 0.948

15. **a.** 2.8 **b.** 2.8 **c.** 1.5

16. 1, 6, 24, 5040

17. **a.** 56 **b.** 56 **c.** 1 **d.** 45
e. 91,390 **f.** 1

18. **a.** $p = 0.493$

b.

Outcome	Probability
sss	$(0.493)(0.493)(0.493) = 0.120$
ssf	$(0.493)(0.493)(0.507) = 0.123$
sfs	$(0.493)(0.507)(0.493) = 0.123$
sff	$(0.493)(0.507)(0.507) = 0.127$
fss	$(0.507)(0.493)(0.493) = 0.123$
fsf	$(0.507)(0.493)(0.507) = 0.127$
ffs	$(0.507)(0.507)(0.493) = 0.127$
fff	$(0.507)(0.507)(0.507) = 0.130$

d. *ssf, sfs, fss*
e. 0.123. Each probability is obtained by multiplying two success probabilities of 0.493 and one failure probability of 0.507.
f. 0.369
g.

y	$P(Y = y)$
0	0.130
1	0.381
2	0.369
3	0.120

h. binomial with parameters $n = 3$ and $p = 0.493$

19. a. 0.410 **b.** 0.590 **c.** 0.819
d.

x	$P(X = x)$
0	0.002
1	0.026
2	0.154
3	0.410
4	0.410

e. left skewed
f.

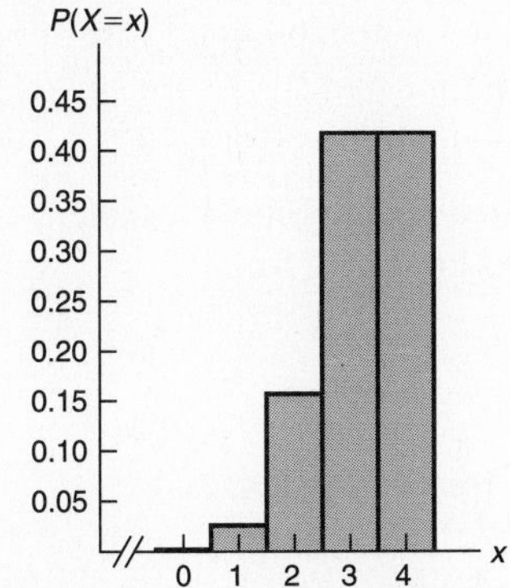

g. The probability distribution is only approximately correct because the sampling is without replacement. A hypergeometric distribution.
h. 3.2; on the average, 3.2 out of every four Surinamese are literate.
i. 0.8

20. a. $p = 0.5$ **b.** $p < 0.5$

21. a. 0.266 **b.** 0.099 **c.** 0.826
d.

x	$P(X = x)$
0	0.174
1	0.304
2	0.266
3	0.155
4	0.068
5	0.024
6	0.007
7	0.002
8	0.000

e.

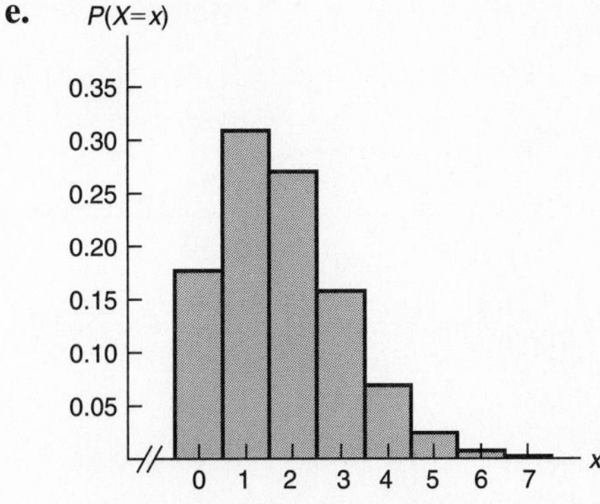

f. Right skewed. Yes, all Poisson distributions are right skewed.

22. a. $\mu = 1.75$; in a 1-minute period, there are, on the average, 1.75 calls to a wrong number.
b. $\sigma = 1.32$

23. a. 2.4 **b.** 0.2613 **c.** 0.6916

27. a. 5 **b.** 0.65 **c.** 0.0488 **d.** 0.5663
e. 0.0541 **f.** 0.9947

28. a. 8 **b.** 0.57 **c.** 0.4762
d. 0.7765 **e.** 0.2527

29. a. 0.0217 **b.** 0.9968
c. 0.0032 **d.** 0.6458

31. a. 2.4 **b.** 0.2613 **c.** 0.6916

CHAPTER 6

Exercises 6.1

6.1 The histogram will be roughly bell-shaped.

6.3 They are the same. A normal distribution is completely determined by the mean and standard deviation.

6.5 True. They have the same shape because both have the same standard deviation.

6.7 True, because the mean affects only where a normal distribution is centered. The shape is determined by the standard deviation.

6.9

a.

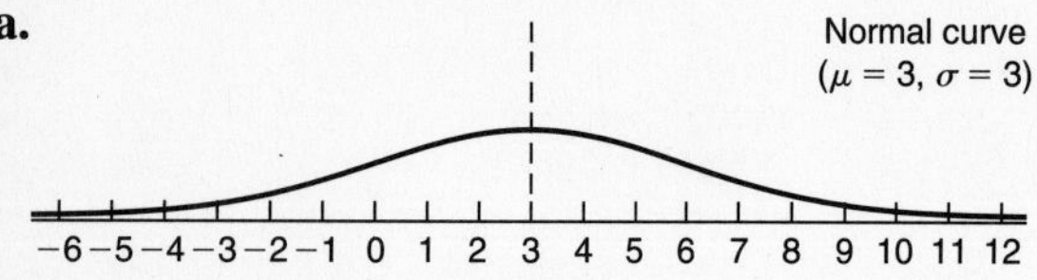

b.

c.

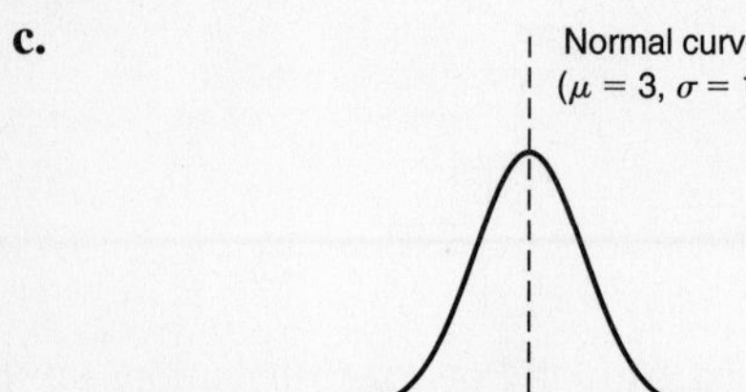

6.11 They are equal. They are approximately equal.

6.13

a. 55.70%

b. 0.5570; This is only an estimate because the distribution of heights is only approximately normally distributed.

6.15

a.

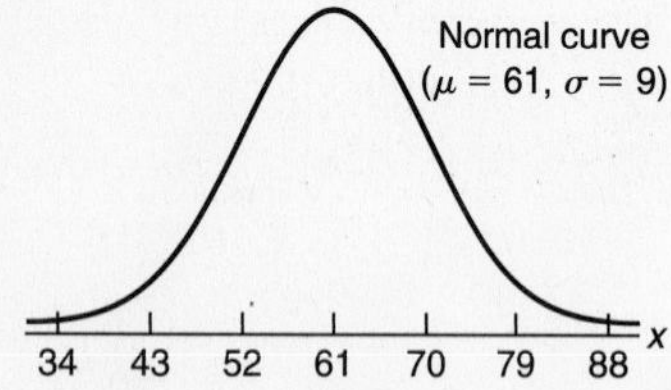

b. $z = (x - 61)/9$

c. standard normal distribution

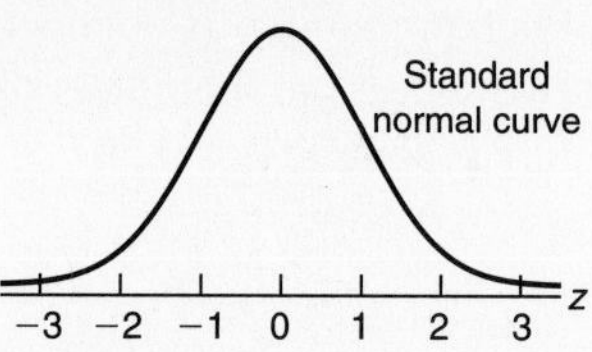

d. −1.22; 1 **e.** right; 1.56

Exercises 6.2

6.23 For a normally distributed variable, we can obtain the percentage of all possible observations that lie within any specified range by first converting to z-scores and then determining the corresponding area under the standard normal curve.

6.25 The total area under the standard normal curve equals 1 and the standard normal curve is symmetric about 0. Therefore, the area to the right of 0 is one-half of 1, or 0.5.

6.27 0.3336. The total area under the curve is 1, so the area to the right of 0.43 equals 1 minus the area to its left, which is $1 - 0.6664 = 0.3336$.

6.29 99.74%

6.31

a. Read the area directly from the table.

b. Subtract the table area from 1.

c. Subtract the smaller table area from the larger.

6.33

a. 0.9875 **b.** 0.0594 **c.** 0.5

d. 0.0000 (to four decimal places)

6.35

a. 0.8577 **b.** 0.2743 **c.** 0.5

d. 0.0000 (to four decimal places)

6.37

a. 0.9105 **b.** 0.0440 **c.** 0.2121 **d.** 0.1357

6.39

a. 0.0645 **b.** 0.7975

6.41

a. 0.7994 **b.** 0.8990 **c.** 0.0500 **d.** 0.0198

6.43

a. 0.6826

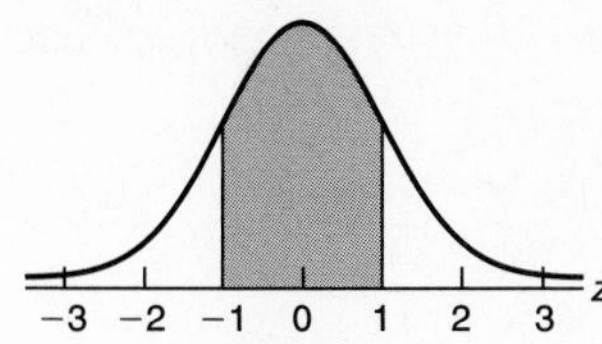

b. 0.9544

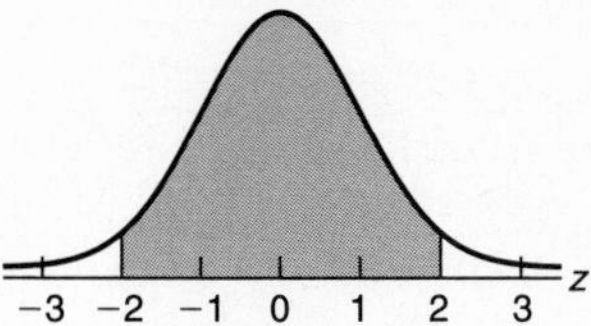

c. 0.9974

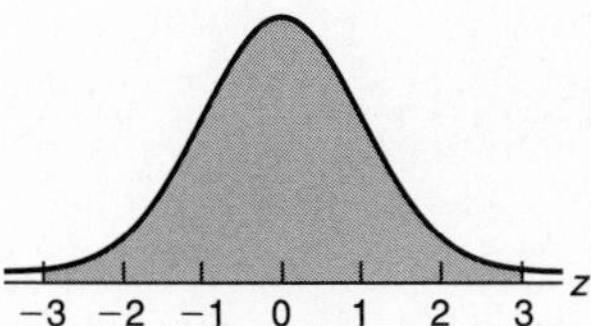

6.45 −1.96

6.47 0.67

6.49 −1.645

6.51 0.44

6.53

a. 1.88 **b.** 2.575

6.55 ±1.645

6.57 The four missing entries are 1.645, 1.96, 2.33, and 2.575.

Exercises 6.3

6.61 The z-scores corresponding to the x-values that lie two standard deviations below and above the mean are −2 and 2, respectively.

6.63

a. 69.43% **b.** 97.26%

6.65

a. 13.61% **b.** 3.67%

6.67

a. 0.0594 **b.** 0.2699

6.69

a. 68.26% **b.** 95.44% **c.** 99.74%

6.71

a. \$135.90, \$170.30 **b.** \$118.70, \$187.50

c. \$101.50, \$204.70

d. See Fig. A.6.

6.73

a. 160.44 lb **b.** 203.70 lb

c. $Q_1 = 165.62$ lb, $Q_2 = 175$ lb, $Q_3 = 184.38$ lb

6.75

a. $Q_1 = \$141.58$, $Q_2 = \$153.10$, $Q_3 = \$164.62$

b. $D_3 = \$144.16$ **c.** $P_{85} = \$170.99$

Exercises 6.4

6.89 Decisions about whether a variable is normally distributed often are important in subsequent analyses—from percentage or percentile calculations to statistical inferences.

FIGURE A.6 Graphs for Exercise 6.71(d)

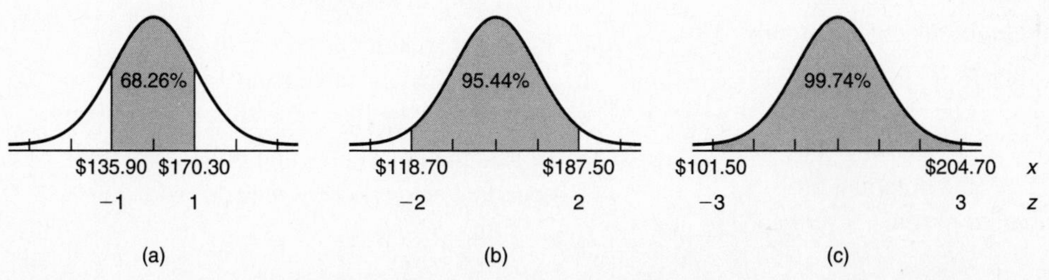

6.91 A normal probability plot is a plot of the observed values of the variable versus the normal scores—the observations we would expect to get for a variable having the standard normal distribution. If the variable is normally distributed, then the normal probability plot should be roughly linear, and vice versa.

6.93

a.

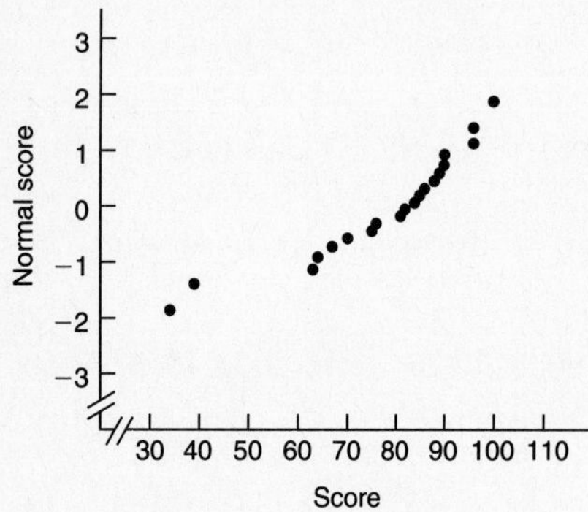

b. 34 and 39 are outliers.

c. Final-exam scores in this introductory statistics class appear not to be normally distributed.

6.95

a.

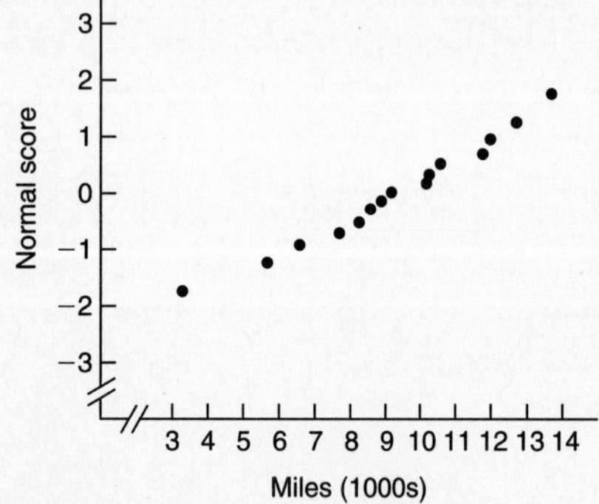

b. No outliers.

c. It appears plausible that the number of miles cars were driven last year is (approximately) normally distributed.

Exercises 6.5

6.105 It is not practical to use the binomial probability formula when the number of trials is very large.

6.107

a. 0.4512, 0.8907 **b.** 0.4544, 0.8858

6.109 the one with parameters $\mu = 15$ and $\sigma = 2.74$

6.111

a. 0.0398 **b.** 0.2835 **c.** 0.0099

6.113

a. 0.0263 **b.** 0.8242 **c.** 0.9991

Review Test for Chapter 6

1. It is often appropriate to use the normal distribution as the distribution of a variable; and the normal distribution is frequently employed in inferential statistics.

2. **a.** A variable is said to be normally distributed if its distribution has the shape of a normal curve.
b. If a variable of a population is normally distributed and is the only variable under consideration, then we say that we have a normally distributed population.
c. The parameters for a normal curve are the corresponding mean and standard deviation.

3. **a.** false
b. True. A normal distribution is completely determined by its mean and standard deviation.

4. They are the same when the areas are expressed as percentages.

5. standard normal distribution

6. **a.** true **b.** true

7. **a.** the second curve
b. the first and second curves
c. the first and third curves
d. the third curve **e.** the fourth curve

8. Key Fact 6.2, which states that the standardized version of a normally distributed variable has the standard normal distribution.

9. **a.** Read the area directly from the table.
b. Subtract the table area from 1.
c. Subtract the smaller table area from the larger.

10. **a.** Locate the table entry that is closest to the specified area and read the corresponding z-score.

b. Locate the table entry that is closest to 1 minus the specified area and read the corresponding z-score.

11. The z-score having area α to its right under the standard normal curve.

12. See Key Fact 6.4 on page 377.

13. They are the observations we would expect to get for a sample of the same size for a variable having the standard normal distribution.

14. linear

15. a.

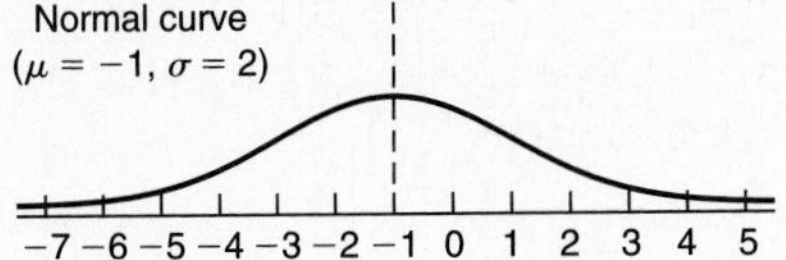

b.

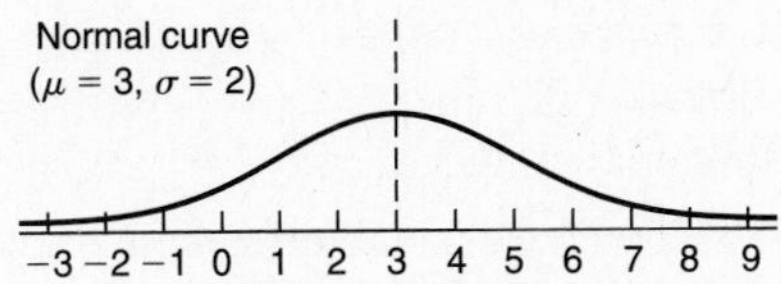

c.

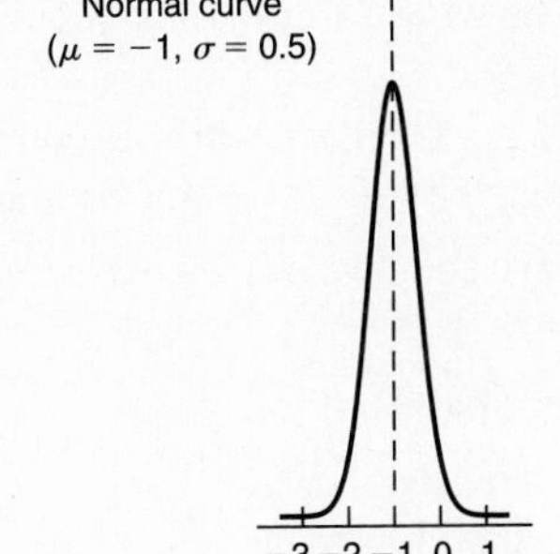

16. a.

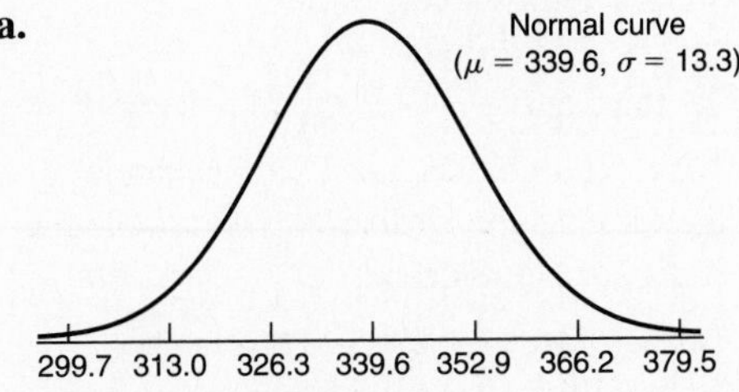

b. $z = (x - 339.6)/13.3$

c. standard normal distribution

d. 0.8672 **e.** right; 2.21

17. a. 0.1469 **b.** 0.1469 **c.** 0.7062

18. a. 0.0013 **b.** 0.2709 **c.** 0.1305

d. 0.9803 **e.** 0.0668 **f.** 0.8426

19. a. −0.52 **b.** 1.28

c. 1.96; 1.645; 2.33; 2.575 **d.** ±2.575

20. a. 82.76% **b.** 89.44% **c.** 99.38%

21. a. 400, 600 **b.** 300, 700 **c.** 200, 800

22. a. $Q_1 = 433$, $Q_2 = 500$, $Q_3 = 567$. Thus 25% of GRE scores are below 433, 25% are between 433 and 500, 25% are between 500 and 567, and 25% are above 567.

b. $P_{99} = 733$. Thus 99% of GRE scores are below 733 and 1% are above 733.

23. a.

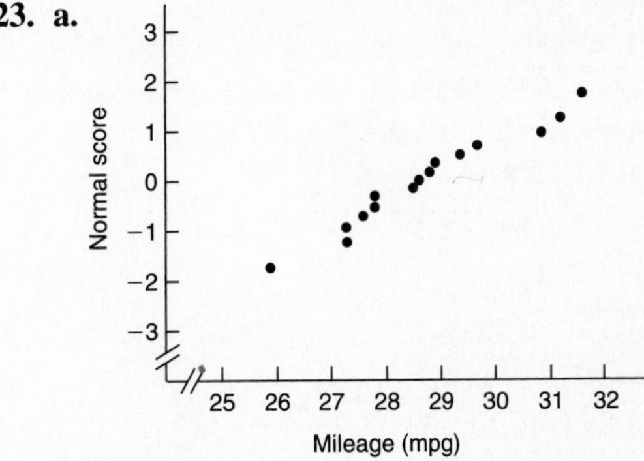

b. no outliers

c. It appears plausible that gas mileages for this particular model are (approximately) normally distributed.

24. a. 0.0076 **b.** 0.9505 **c.** 0.9988

CHAPTER 7

Exercises 7.1

7.1 Generally, sampling is less costly and can be done more quickly than a census.

7.3

a. $\mu = 64.4$ (i.e., \$64,400)

b.

Sample	Salaries ($1000s)	$\bar{x}$
G, L	70, 63	66.5
G, S	70, 44	57.0
G, A	70, 75	72.5
G, T	70, 70	70.0
L, S	63, 44	53.5
L, A	63, 75	69.0
L, T	63, 70	66.5
S, A	44, 75	59.5
S, T	44, 70	57.0
A, T	75, 70	72.5

c. See Fig. A.7. **d.** 0

e. 0.2. If we take a random sample of two salaries, there is a 20% chance that the mean of the sample obtained will be within 4 (i.e., $4000) of the population mean.

7.5

b.

Sample	Salaries ($1000s)	$\bar{x}$
G, L, S	70, 63, 44	59.0
G, L, A	70, 63, 75	69.3
G, L, T	70, 63, 70	67.7
G, S, A	70, 44, 75	63.0
G, S, T	70, 44, 70	61.3
G, A, T	70, 75, 70	71.7
L, S, A	63, 44, 75	60.7
L, S, T	63, 44, 70	59.0
L, A, T	63, 75, 70	69.3
S, A, T	44, 75, 70	63.0

c. See Fig. A.8. **d.** 0

e. 0.5. If we take a random sample of three salaries, there is a 50% chance that the mean of the sample obtained will be within 4 (i.e., $4000) of the population mean.

7.7

b.

Sample	Salaries ($1000s)	$\bar{x}$
G, L, S, A, T	70, 63, 4, 75, 70	64.4

c. See Fig. A.9. **d.** 1

e. 1. If we take a random sample of five salaries, there is a 100% chance that the mean of the sample obtained will be within 4 (i.e., $4000) of the population mean.

7.9

a. $\mu = 15$ cm

b.

Sample	Lengths	$\bar{x}$
A, B	19, 14	16.5
A, C	19, 15	17.0
A, D	19, 9	14.0
A, E	19, 16	17.5
A, F	19, 17	18.0
B, C	14, 15	14.5
B, D	14, 9	11.5
B, E	14, 16	15.0
B, F	14, 17	15.5
C, D	15, 9	12.0
C, E	15, 16	15.5
C, F	15, 17	16.0
D, E	9, 16	12.5
D, F	9, 17	13.0
E, F	16, 17	16.5

d. $\frac{1}{15} = 0.067$

e. $\frac{6}{15} = 0.4$. If we take a random sample of two bullfrogs, there is a 40% chance that their mean length will be within 1 cm of the population mean length.

7.11

b.

Sample	Lengths	$\bar{x}$
A, B, C	19, 14, 15	16.0
A, B, D	19, 14, 9	14.0
A, B, E	19, 14, 16	16.3
A, B, F	19, 14, 17	16.7
A, C, D	19, 15, 9	14.3
A, C, E	19, 15, 16	16.7
A, C, F	19, 15, 17	17.0
A, D, E	19, 9, 16	14.7
A, D, F	19, 9, 17	15.0
A, E, F	19, 16, 17	17.3
B, C, D	14, 15, 9	12.7
B, C, E	14, 15, 16	15.0
B, C, F	14, 15, 17	15.3
B, D, E	14, 9, 16	13.0
B, D, F	14, 9, 17	13.3
B, E, F	14, 16, 17	15.7
C, D, E	15, 9, 16	13.3
C, D, F	15, 9, 17	13.7
C, E, F	15, 16, 17	16.0
D, E, F	9, 16, 17	14.0

FIGURE A.7 Dotplot for Exercise 7.3(c)

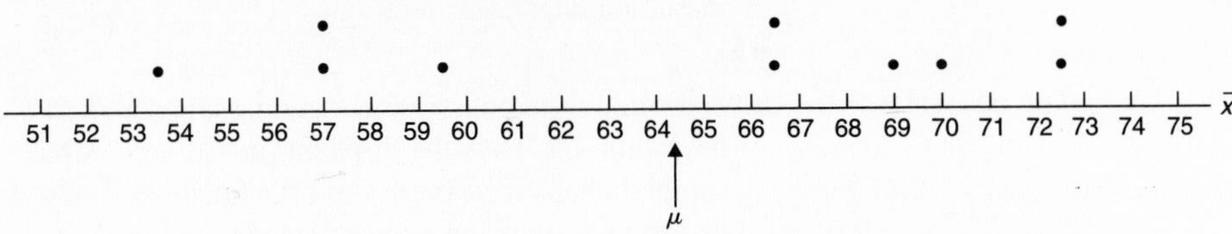

FIGURE A.8 Dotplot for Exercise 7.5(c)

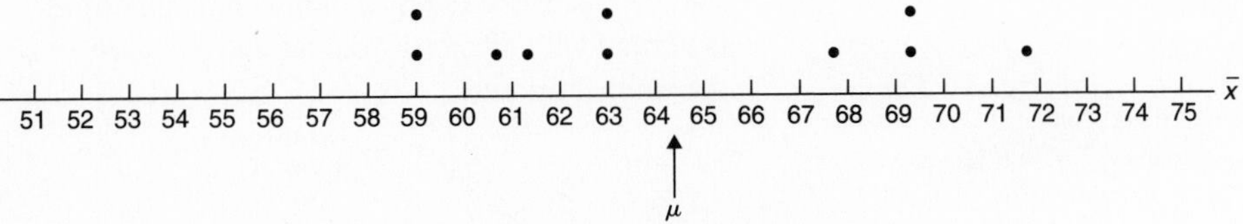

FIGURE A.9 Dotplot for Exercise 7.7(c)

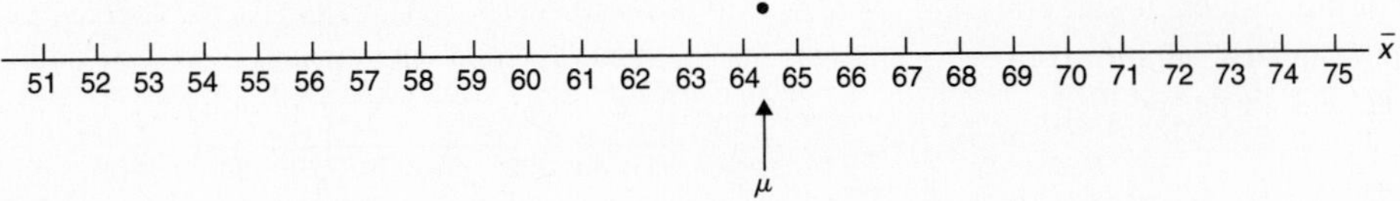

d. $\frac{2}{20} = 0.1$

e. $\frac{10}{20} = 0.5$. If we take a random sample of three bullfrogs, there is a 50% chance that their mean length will be within 1 cm of the population mean length.

7.13

b.

Sample	Lengths	$\bar{x}$
A, B, C, D, E	19, 14, 15, 9, 16	14.6
A, B, C, D, F	19, 14, 15, 9, 17	14.8
A, B, C, E, F	19, 14, 15, 16, 17	16.2
A, B, D, E, F	19, 14, 9, 16, 17	15.0
A, C, D, E, F	19, 15, 9, 16, 17	15.2
B, C, D, E, F	14, 15, 9, 16, 17	14.2

d. $\frac{1}{6} = 0.167$

e. $\frac{5}{6} = 0.833$. If we take a random sample of five bullfrogs, there is an 83.3% chance that their mean length will be within 1 cm of the population mean length.

7.15 The larger the sample size, the smaller the sampling error tends to be in estimating a population mean, μ, by a sample mean, $\bar{x}$.

Exercises 7.2

7.19 We know that a normal distribution is determined by the mean and standard deviation. Hence, a first step in learning how to approximate the sampling distribution of the mean by a normal distribution is to obtain the mean and standard deviation of the variable $\bar{x}$.

7.21 Yes. The standard deviation of $\bar{x}$ gets smaller as the sample size gets larger.

7.23 Standard error (SE) of the mean. Because the standard deviation of $\bar{x}$ determines the amount of sampling error to be expected when a population mean is estimated by a sample mean.

7.25

a. $\mu = 64.4$ **b.** $\mu = 64.4$ **c.** $\mu_{\bar{x}} = \mu = 64.4$

7.27

b. $\mu_{\bar{x}} = 64.4$ **c.** $\mu_{\bar{x}} = \mu = 64.4$

7.29

b. $\mu_{\bar{x}} = 64.4$ **c.** $\mu_{\bar{x}} = \mu = 64.4$

7.31

a. $\mu_{\bar{x}} = 5.8, \sigma_{\bar{x}} = 0.50$ **b.** $\mu_{\bar{x}} = 5.8, \sigma_{\bar{x}} = 0.19$

7.33

a. $\mu_{\bar{x}} = 8.5, \sigma_{\bar{x}} = 0.37$ **b.** $\mu_{\bar{x}} = 8.5, \sigma_{\bar{x}} = 0.18$

Exercises 7.3

7.47

a. approximately normally distributed with mean 100 and standard deviation 4

b. none

c. No, since the distribution of the variable under consideration is not specified, we need a sample size of at least 30 to apply Key Fact 7.4 on page 429.

7.49

a. normal with mean μ and standard deviation $\sigma/\sqrt{n}$

b. No, since the variable under consideration is normally distributed.

c. μ and $\sigma/\sqrt{n}$

d. no

7.51

a. All four graphs are centered at the same place because $\mu_{\bar{x}} = \mu$ and normal distributions are centered at their means.

b. Since $\sigma_{\bar{x}} = \sigma/\sqrt{n}$, we see that $\sigma_{\bar{x}}$ decreases as n increases. This results in a diminishing of the spread because the spread of a distribution is determined by its standard deviation. As a consequence, we see that the larger the sample size, the greater the likelihood for small sampling error.

c. If the variable under consideration is normally distributed, then so is the sampling distribution of the mean, regardless of sample size.

d. The central limit theorem indicates that if the sample size is relatively large, then the sampling distribution of the mean is approximately a normal distribution, regardless of the distribution of the variable under consideration.

7.53

a. A normal distribution with mean 61 and standard deviation 4.5. Thus the distribution of the possible sample means for samples of four finishing times is normal with mean 61 minutes and standard deviation 4.5 minutes.

b. A normal distribution with mean 61 and standard deviation 3. Thus the distribution of the possible sample means for samples of nine finishing times is normal with mean 61 minutes and standard deviation 3 minutes.

c.

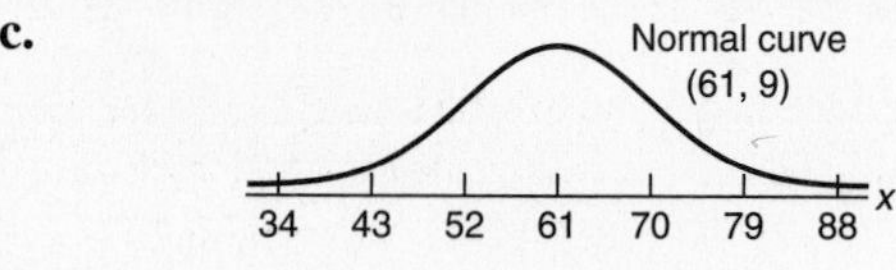

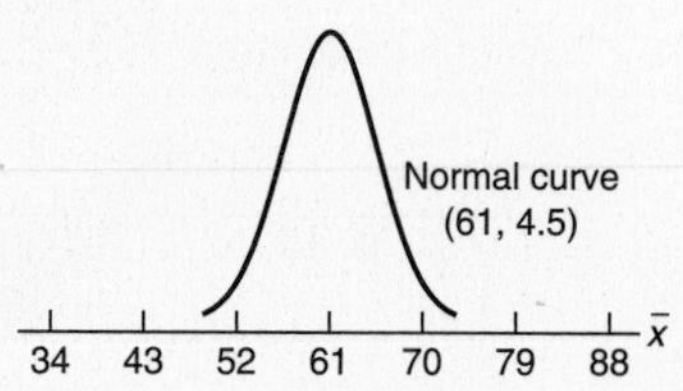

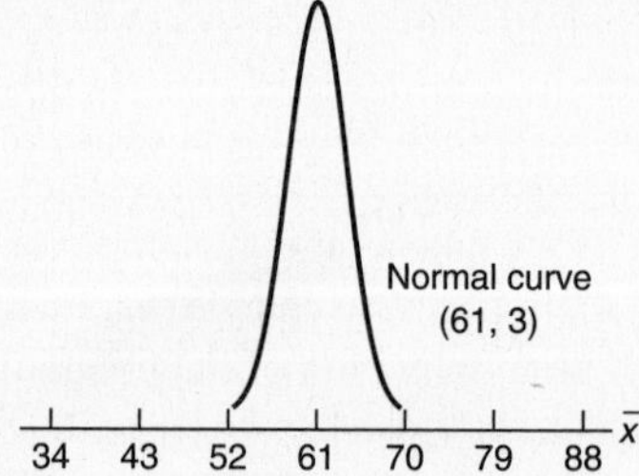

7.55

a. approximately a normal distribution with mean \$8657 and standard deviation \$750

b. approximately a normal distribution with mean \$8657 and standard deviation \$237.17

c. No, because the sample size is at least 30.

7.57

a. 73.30%. There is a 73.30% chance that the sampling error will be less than 5 minutes.

b. 90.50%. There is a 90.50% chance that the sampling error will be less than 5 minutes.

7.59
a. 0.4972 **b.** 0.9652

7.61 0.8064

Review Test for Chapter 7

1. Sampling error is the error resulting from using a sample to estimate a population characteristic.
2. The distribution of a statistic (i.e., of all possible observations of the statistic for samples of a given size) is called the sampling distribution of the statistic.
3. sampling distribution of the mean; distribution of the variable $\bar{x}$
4. The possible sample means cluster closer around the population mean as the sample size increases. Thus the larger the sample size, the smaller the sampling error tends to be in estimating a population mean, μ, by a sample mean, $\bar{x}$.
5. **a.** The error resulting from using the mean income tax, $\bar{x}$, of the 125,000 tax returns sampled as an estimate of the mean income tax, μ, of all 1994 tax returns.
   b. \$88
   c. No, not necessarily. However, increasing the sample size from 125,000 to 250,000 would increase the likelihood for small sampling error.
   d. Increase the sample size.
6. **a.** 18
   b. The completed table is shown here.

Sample	Salaries	$\bar{x}$
A, B, C, D	8, 12, 16, 20	14
A, B, C, E	8, 12, 16, 24	15
A, B, C, F	8, 12, 16, 28	16
A, B, D, E	8, 12, 20, 24	16
A, B, D, F	8, 12, 20, 28	17
A, B, E, F	8, 12, 24, 28	18
A, C, D, E	8, 16, 20, 24	17
A, C, D, F	8, 16, 20, 28	18
A, C, E, F	8, 16, 24, 28	19
A, D, E, F	8, 20, 24, 28	20
B, C, D, E	12, 16, 20, 24	18
B, C, D, F	12, 16, 20, 28	19
B, C, E, F	12, 16, 24, 28	20
B, D, E, F	12, 20, 24, 28	21
C, D, E, F	16, 20, 24, 28	22

   c.

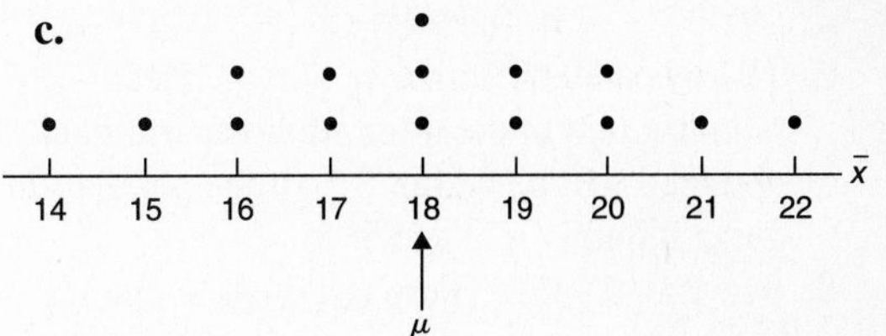

   d. $\frac{7}{15}$
   e. 18. The mean of all possible sample means equals the population mean.
   f. Yes, because $\mu_{\bar{x}} = \mu$ and, from part (a), $\mu = 18$.
7. **a.** 506 lb; 47.4 lb
   b. 506 lb; 16.8 lb
   c. Smaller, because $\sigma_{\bar{x}} = \sigma/\sqrt{n}$ and hence $\sigma_{\bar{x}}$ decreases with increasing sample size.
8. **a.** false **b.** not possible to tell **c.** true
9. **a.** false **b.** true **c.** true
10. **a.**

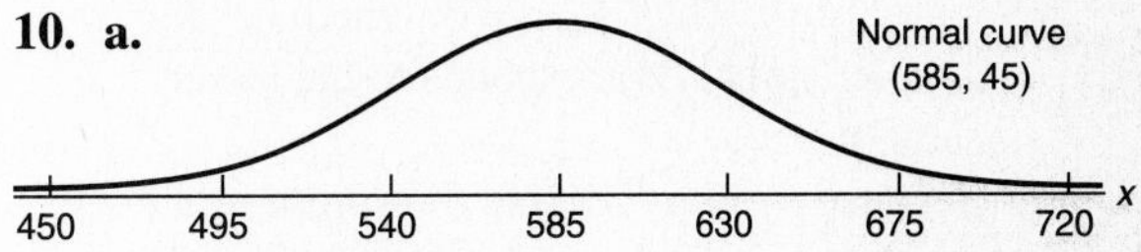

   b. Normal with mean \$585 and standard deviation \$25.98.

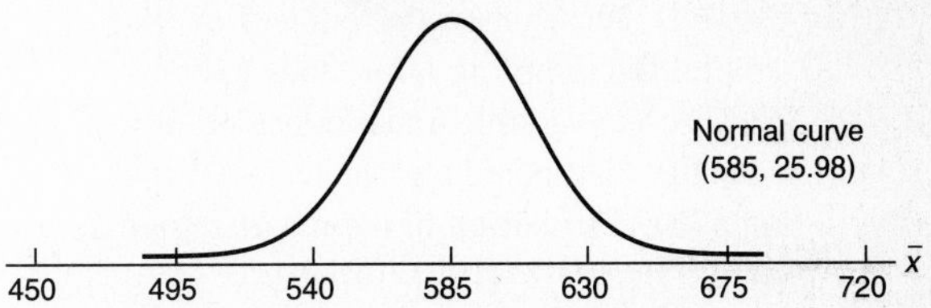

   c. Normal with mean \$585 and standard deviation \$15.

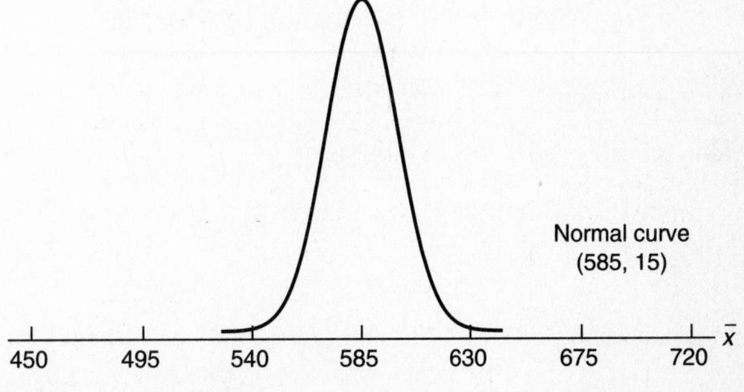

11. **a.** 29.60% **b.** 0.2960
c. There is a 29.60% chance that the mean monthly rent of the three studio apartments obtained will be within $10 of the population mean monthly rent of $585.
d. 94.52%; 0.9452. There is a 94.52% chance that the mean monthly rent of the 75 studio apartments obtained will be within $10 of the population mean monthly rent of $585.

12. **a.** For a normally distributed variable, the sampling distribution of the mean is a normal distribution, regardless of the sample size. Also, we know that $\mu_{\bar{x}} = \mu$. Consequently, since the normal curve for any normally distributed variable is centered at the mean, all three curves are centered at the same place.
b. Curve B. Since $\sigma_{\bar{x}} = \sigma/\sqrt{n}$, the larger the sample size, the smaller the value of $\sigma_{\bar{x}}$ and hence the smaller the spread of the normal curve for $\bar{x}$. Thus, Curve B, which has the smaller spread, corresponds to the larger sample size.
c. Because $\sigma_{\bar{x}} = \sigma/\sqrt{n}$ and the spread of a normal curve is determined by the standard deviation. Thus different sample sizes result in normal curves with different spreads.
d. Curve B. The smaller the value of $\sigma_{\bar{x}}$, the smaller the sampling error tends to be.
e. Because the variable under consideration is normally distributed and, hence, so is the sampling distribution of the mean, regardless of sample size.

13. **a.** 0.6212
b. No, because the sample size is large and therefore $\bar{x}$ is approximately normally distributed, regardless of the distribution of life-insurance amounts. Yes.
c. 0.9946

14. **a.** no **b.** yes **c.** no

CHAPTER 8

Exercises 8.1

8.1 point estimate

8.3
a. $82.00
b. No, because it is unlikely that a sample mean, $\bar{x}$, will be exactly equal to a population mean, μ; some sampling error is to be anticipated.

8.5 6.87 lb

8.7
a. $74.84 to $89.16
b. We can be 95.44% confident that the mean price, μ, of all science books is somewhere between $74.84 and $89.16.
c. It may or may not, but we can be 95.44% confident that it does.

8.9
a. 6.22 to 7.51 lb
b. We can be 95.44% confident that the mean weight, μ, of all newborns is somewhere between 6.22 and 7.51 lb.
c. Because birth weights may not be normally distributed.

Exercises 8.2

8.13
a. Confidence level = 0.90; $\alpha = 0.10$.
b. Confidence level = 0.99; $\alpha = 0.01$.

8.15
a. By saying that the CI is exact, we mean that the true confidence level is equal to $1 - \alpha$.
b. By saying that the CI is approximately correct, we mean that the true confidence level is only approximately equal to $1 - \alpha$.

8.17 The variable under consideration is normally distributed on the population of interest.

8.19 A statistical procedure is said to be *robust* if it is insensitive to departures from the assumptions on which it is based.

8.21 We should look at the data by obtaining one or more graphical displays of the data. If any of the conditions required for using the contemplated statistical-inference procedure appear to be violated, we should not apply the procedure. Instead we should use a different more appropriate procedure, or, if we are unsure of one, consult a statistician.

8.23
a. 561.7 to 724.3 sq ft
b. We can be 90% confident that the mean size, μ, of all household vegetable gardens in the United States is somewhere between 561.7 and 724.3 sq ft.

8.25
a. 133.8 to 140.0 lb
b. We can be 90% confident that the mean weight, μ, of all U.S. women 5 feet 4 inches tall and in the age group 18–24 years is somewhere between 133.8 and 140.0 lb.

8.27 \$15,428.30 to \$16,785.70

8.29
a. 132.0 to 141.8 lb
b. It is longer because the confidence level is greater.
c.

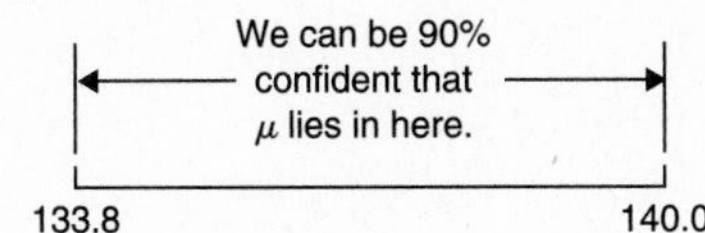

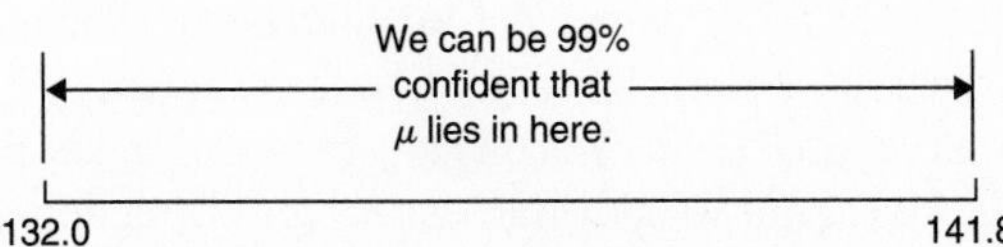

d. The 90% CI is a more precise estimate of μ because it is narrower than the 99% CI.

8.37
a. \$4241 **b.** \$4169 **c.** 150
d. \$16,107 **e.** \$15,428 to \$16,786

Exercises 8.3

8.41 Since the margin of error, E, is equal to half of the length of a CI, it determines the precision with which a sample mean estimates a population mean.

8.43
a. 6.8 **b.** 49.4 to 56.2

8.45
a. True. Since the margin of error is half the length of a CI, we can determine the length of a CI by doubling the margin of error, E.
b. True. By taking half the length of a CI, we can determine the margin of error.
c. False. We need to know the sample mean as well.
d. True. Since the CI is from $\bar{x} - E$ to $\bar{x} + E$, we can obtain a CI by knowing only the margin of error, E, and the sample mean, $\bar{x}$.
e. False. To determine the margin of error, E, we also need the population standard deviation and the sample size.
f. False. To determine the confidence level, we also need the population standard deviation and the sample size.
g. true **h.** true

8.47 If σ is unknown, which is usually the case in practice, and we want to apply Formula 8.1, then we must first estimate σ. One way to do this is to take a preliminary large sample, say, of size 30 or more. The sample standard deviation, s, of the sample obtained provides an estimate of σ and can be used in place of σ in Formula 8.1.

8.49
a. 81.3 sq ft **b.** 81.3 sq ft

8.51
a. $E = 3.1$ lb
b. We can be 90% confident that the error in estimating μ by $\bar{x}$ is at most 3.1 lb.
c. $n = 239$ **d.** 132.2 to 136.2 lb

Exercises 8.4

8.57 The difference in the formulas lies in their denominators. The denominator of the standardized version of $\bar{x}$ uses the population standard deviation, σ, whereas the denominator of the studentized version of $\bar{x}$ uses the sample standard deviation, s.

8.59
a. $z = 1$ **b.** $t = 1.333$

8.61
a. The standardized version of $\bar{x}$ has the standard normal distribution.
b. The studentized version of $\bar{x}$ has a t-distribution with $12 - 1 = 11$ degrees of freedom.

8.63 The variation in the possible values of the standardized version of $\bar{x}$ is due solely to the variation of sample means, whereas that of the studentized version of $\bar{x}$ is due to the variation of sample means and sample standard deviations.

8.65

a. 1.440 **b.** 2.447 **c.** 3.143

8.67

a. 1.323 **b.** 2.518 **c.** −2.080 **d.** ±1.721

8.69 Yes, since the sample size exceeds 30 and there are no outliers.

8.71

a. 60.1 to 62.9 bu

b. Yes, since the 99% CI in part (a) does not include the national average of 58.4 bu, it appears that the farmer will get a mean yield different from the national average.

8.73

a. 43.3 to 46.4 in.

b. We can be 95% confident that the mean height, μ, of all 6-year-old girls is somewhere between 43.3 and 46.4 in.

8.75

a. \$56.22 to \$61.58

b. We can be 90% confident that the mean monthly fuel expenditure, μ, of all household vehicles is somewhere between \$56.22 and \$61.58.

8.85

a. 3.392 in **b.** 20 **c.** 44.850 in

d. 43.263 to 46.437 in

Review Test for Chapter 8

1. A point estimate of a parameter is the value of a statistic that is used to estimate the parameter; it consists of a single number, or point. A confidence-interval estimate of a parameter consists of an interval of numbers obtained from a point estimate of the parameter together with a percentage that specifies how confident we are that the parameter lies in the interval.
2. False. The mean of the population may or may not lie somewhere between 33.8 and 39.0, but we can be 95% confident that it does.
3. No. See the guidelines in Key Fact 8.1 on page 453.
4. Roughly 950 intervals would actually contain μ.
5. Look at graphical displays of the data to ascertain whether the conditions required for using the procedure appear to be satisfied.
6. **a.** The precision of the estimate would decrease since the CI would be wider for a sample of size 50.
   b. The precision of the estimate would increase (narrower CI), however, the confidence level would be lower.
7. **a.** Since the length of a CI is twice the length of the margin of error, the length of the CI is 21.4.
   b. 64.5 to 85.9
8. **a.** 6.58
   b. You need to know the sample mean, $\bar{x}$.
9. **a.** $z = -0.77$ **b.** $t = -0.605$
10. **a.** standard normal distribution
    b. t-distribution with $3 - 1 = 2$ degrees of freedom
11. From Property 4 of Key Fact 8.6, we know that as the number of degrees of freedom becomes larger, t-curves look increasingly like the standard normal curve. So the curve that is closer to the standard normal curve has the larger degrees of freedom.
12. **a.** t-interval procedure
    b. z-interval procedure
    c. z-interval procedure **d.** neither procedure
    e. z-interval procedure **f.** neither procedure
13. 54.3 to 62.8 yrs
14. Part (c) provides the correct interpretation of the statement in quotes.
15. **a.** 57.8 to 62.4 hrs
    b. We can be 99% confident that the mean battery life, μ, of this make of computer is somewhere between 57.8 and 62.4 hrs.
    c. The normal probability plot of the data should fall roughly in a straight line.
16. **a.** Margin of error, $E = 2.3$ hrs
    b. We can be 99% confident that the error in estimating μ by $\bar{x}$ is at most 2.3 hrs.
    c. $n = 491$ **d.** 59.3 to 60.3 hrs
17. **a.** 2.101 **b.** 1.734
    c. −1.330 **d.** ±2.878
18. **a.** 6.88 to 8.54 hrs

b. We can be 95% confident that the mean daily viewing time, μ, of all U.S. households is somewhere between 6.88 and 8.54 hrs.
c. No, since the 1992 mean daily viewing time is contained in the CI in part (a).

20. **a.** 13.0 yrs **b.** 13.36 yrs **c.** 36
d. 58.53 yrs **e.** 54.28 to 62.77 yrs

22. **a.** 7.710 hrs **b.** 1.780 hrs **c.** 0.398 hrs
d. 20 **e.** 6.877 to 8.543 hrs

CHAPTER 9

Exercises 9.1

9.1 A hypothesis is a statement that something is true.

9.3 See the discussion on page 491.

9.5 Let μ denote last year's mean telephone expenditure per consumer unit.
a. H_0: $\mu = \$690$ **b.** H_a: $\mu > \$690$
c. right-tailed test

9.7 Let μ denote the mean daily iron intake of all adult females under the age of 51.
a. H_0: $\mu = 18$ mg **b.** H_a: $\mu < 18$ mg
c. left-tailed test

9.9 Let μ denote the mean body temperature of healthy humans.
a. H_0: $\mu = 98.6°$ F **b.** H_a: $\mu \neq 98.6°$ F
c. two-tailed test

Exercises 9.2

9.17
a. True. Since the significance level, α, is the probability of making a Type I error, it is unlikely that a true null hypothesis will be rejected if the hypothesis test is conducted at a small significance level.
b. True. By Key Fact 9.1, for a fixed sample size, the smaller we specify the significance level, α, the larger will be the probability, β, of not rejecting a false null hypothesis.

9.19
a. $z \geq 1.645$ **b.** $z < 1.645$ **c.** $z = 1.645$
d. $\alpha = 0.05$
e.

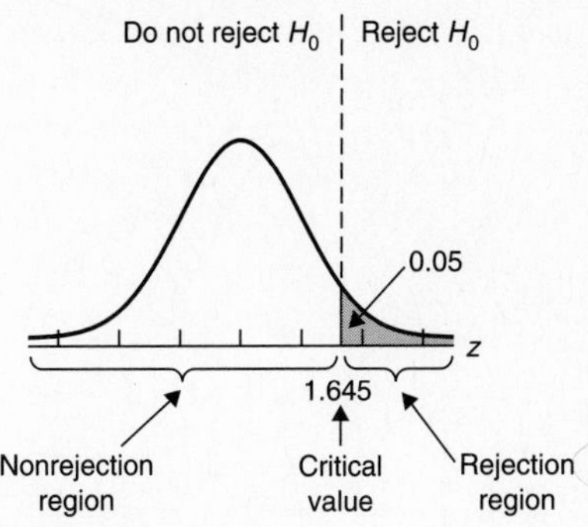

f. right-tailed test

9.21
a. $z \leq -2.33$ **b.** $z > -2.33$ **c.** $z = -2.33$
d. $\alpha = 0.01$
e.

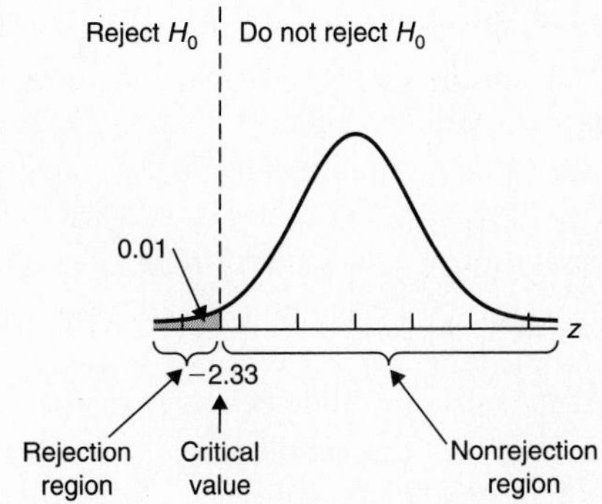

f. left-tailed test

9.23
a. $z \leq -1.645$ or $z \geq 1.645$
b. $-1.645 < z < 1.645$
c. $z = \pm 1.645$ **d.** $\alpha = 0.10$

e.

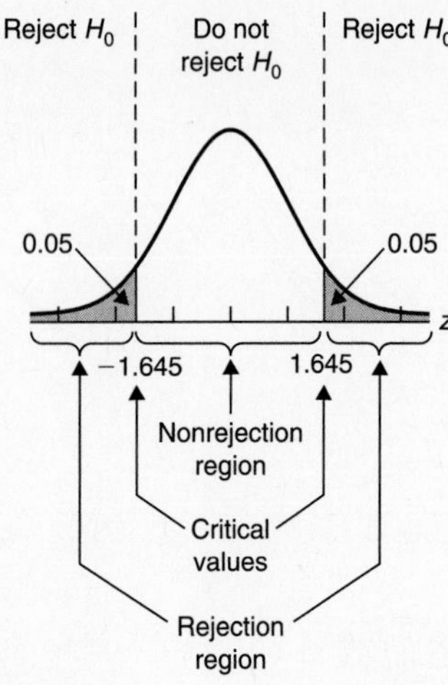

f. two-tailed test

9.25

a. A Type I error would occur if in fact $\mu = \$690$, but the results of the sampling lead to the conclusion that $\mu > \$690$.

b. A Type II error would occur if in fact $\mu > \$690$, but the results of the sampling fail to lead to that conclusion.

c. A correct decision would occur if in fact $\mu = \$690$ and the results of the sampling do not lead to the rejection of that fact; or if in fact $\mu > \$690$ and the results of the sampling lead to that conclusion.

d. correct decision **e.** Type II error

9.27

a. A Type I error would occur if in fact $\mu = 18$ mg, but the results of the sampling lead to the conclusion that $\mu < 18$ mg.

b. A Type II error would occur if in fact $\mu < 18$ mg, but the results of the sampling fail to lead to that conclusion.

c. A correct decision would occur if in fact $\mu = 18$ mg and the results of the sampling do not lead to the rejection of that fact; or if in fact $\mu < 18$ mg and the results of the sampling lead to that conclusion.

d. Type I error **e.** correct decision

9.29

a. A Type I error would occur if in fact $\mu = 98.6°$ F, but the results of the sampling lead to the conclusion that $\mu \neq 98.6°$ F.

b. A Type II error would occur if in fact $\mu \neq 98.6°$ F, but the results of the sampling fail to lead to that conclusion.

c. A correct decision would occur if in fact $\mu = 98.6°$ F, and the results of the sampling do not lead to the rejection of that fact; or if in fact $\mu \neq 98.6°$ F, and the results of the sampling lead to that conclusion.

d. Type I error **e.** correct decision

Exercises 9.3

9.37 Critical value: $z_{0.01} = 2.33$

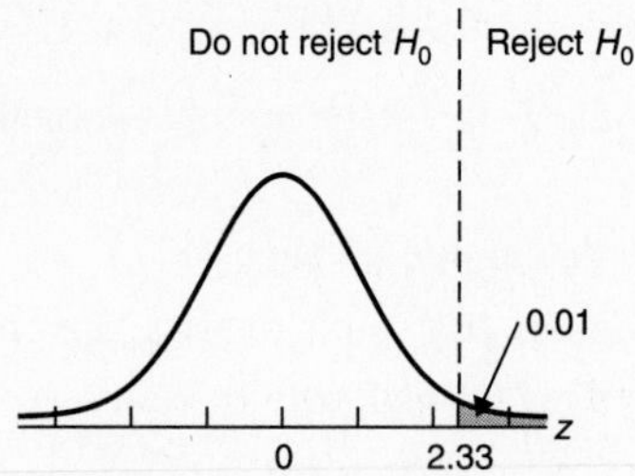

9.39 Critical values: $\pm z_{0.05} = \pm 1.645$

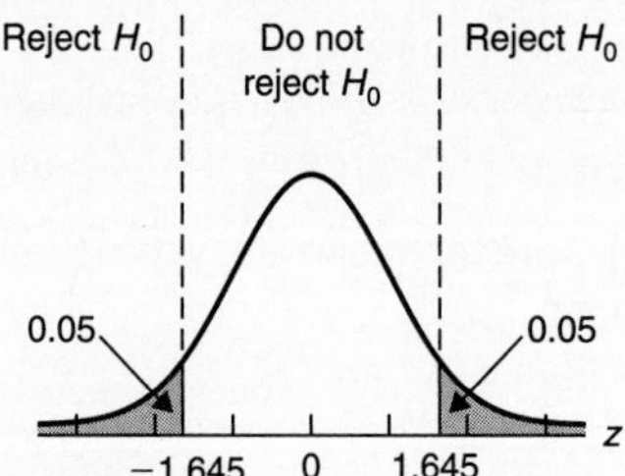

9.41 Critical value: $-z_{0.05} = -1.645$

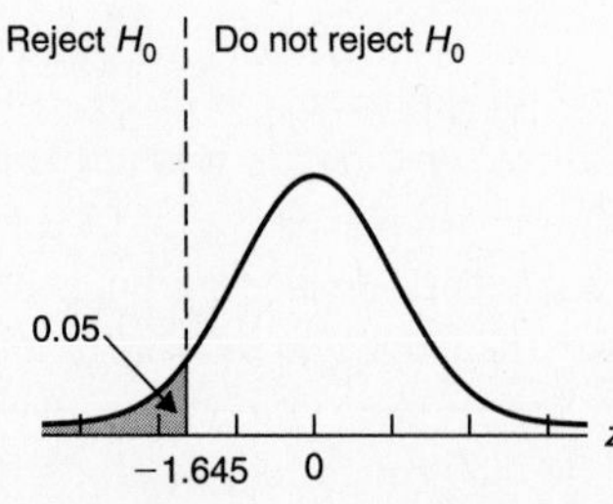

9.43 H_0: $\mu = \$690$, H_a: $\mu > \$690$; $\alpha = 0.05$; critical value $= 1.645$; $z = 0.26$; do not reject H_0; at the 5% significance level, the data do not provide sufficient evidence to conclude that last year's mean telephone expenditure per consumer has increased over the 1994 mean of $690.

9.45 H_0: $\mu = 18$ mg, H_a: $\mu < 18$ mg; $\alpha = 0.01$; critical value $= -2.33$; $z = -5.30$; reject H_0; at the 1% significance level, the data provide sufficient evidence to conclude that adult females under the age of 51 are, on the average, getting less than the RDA of 18 mg of iron.

9.47 H_0: $\mu = 98.6°$ F, H_a: $\mu \neq 98.6°$ F; $\alpha = 0.01$; critical values $= \pm 2.575$; $z = -7.35$; reject H_0; at the 1% significance level, the data provide sufficient evidence to conclude that the mean body temperature of healthy humans differs from 98.6° F.

Exercises 9.4

9.55

a. rejecting a true null hypothesis
b. not rejecting a false null hypothesis
c. the probability of rejecting a true null hypothesis

9.57 It provides a visual display of the overall effectiveness of the hypothesis test.

9.59 It decreases.

Note: The answers obtained to many of the parts of Exercises 9.61, 9.63, 9.65, and 9.67 may vary depending on when and how much intermediate rounding is done.

9.61

a. If $\bar{x} \geq \$781.0$, reject H_0; otherwise, do not reject H_0.
b. 0.05
c.

μ	β	Power	μ	β	Power
720	0.8643	0.1357	840	0.1423	0.8577
750	0.7123	0.2877	870	0.0537	0.9463
780	0.5080	0.4920	900	0.0158	0.9842
810	0.3015	0.6985	930	0.0036	0.9964

d.

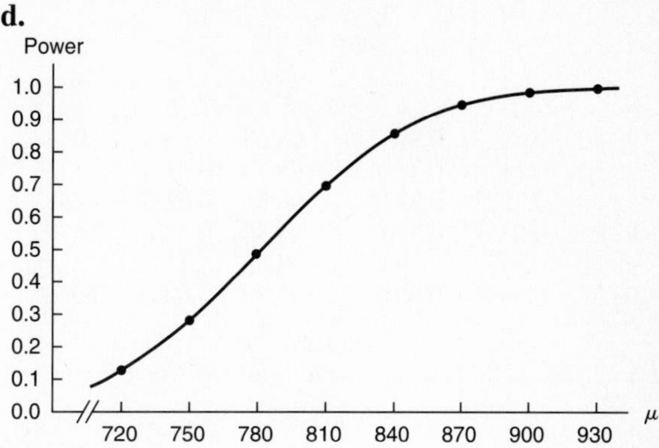

9.63

a. If $\bar{x} \leq 16.5$ mg, reject H_0; otherwise, do not reject H_0.
b. 0.01
c.

μ	β	Power	μ	β	Power
15.50	0.0548	0.9452	16.75	0.6554	0.3446
15.75	0.1151	0.8849	17.00	0.7881	0.2119
16.00	0.2119	0.7881	17.25	0.8849	0.1151
16.25	0.3446	0.6554	17.50	0.9452	0.0548
16.50	0.5000	0.5000	17.75	0.9772	0.0228

d.

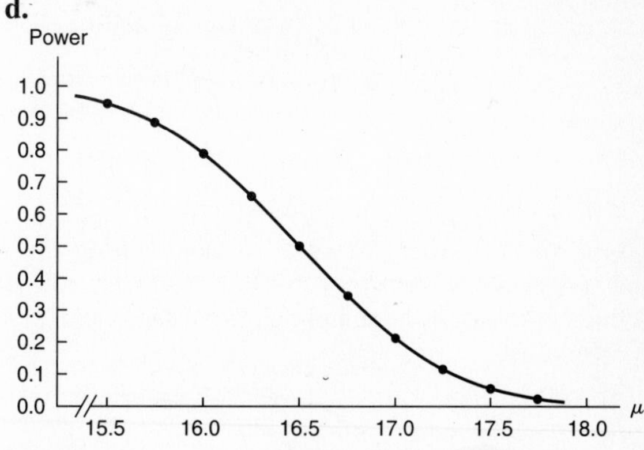

9.65

a. If $\bar{x} \leq 98.43°$F or $\bar{x} \geq 98.77°$F, reject H_0; otherwise, do not reject H_0.
b. 0.01

c.

μ	β	Power	μ	β	Power
98.30	0.0233	0.9767	98.65	0.9667	0.0333
98.35	0.1112	0.8888	98.70	0.8577	0.1423
98.40	0.3228	0.6772	98.75	0.6217	0.3783
98.45	0.6217	0.3783	98.80	0.3228	0.6772
98.50	0.8577	0.1423	98.85	0.1112	0.8888
98.55	0.9667	0.0333	98.90	0.0233	0.9767

d.

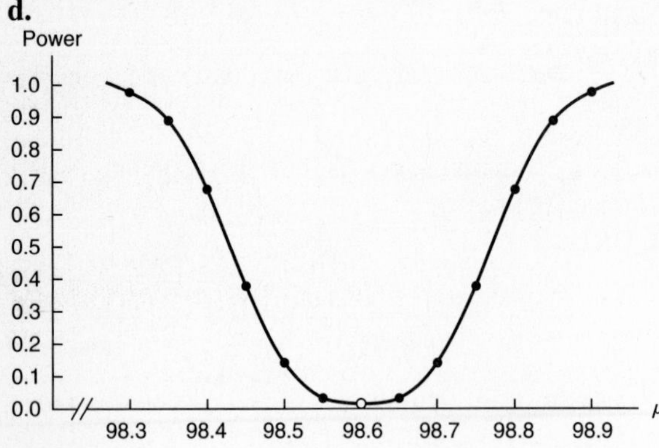

9.67

a. If $\overline{x} \geq \$754.4$, reject H_0; otherwise, do not reject H_0.

b. 0.05

c.

μ	β	Power	μ	β	Power
720	0.8106	0.1894	840	0.0143	0.9857
750	0.5438	0.4562	870	0.0016	0.9984
780	0.2578	0.7422	900	0.0001	0.9999
810	0.0778	0.9222	930	0.0000	1.0000

d.

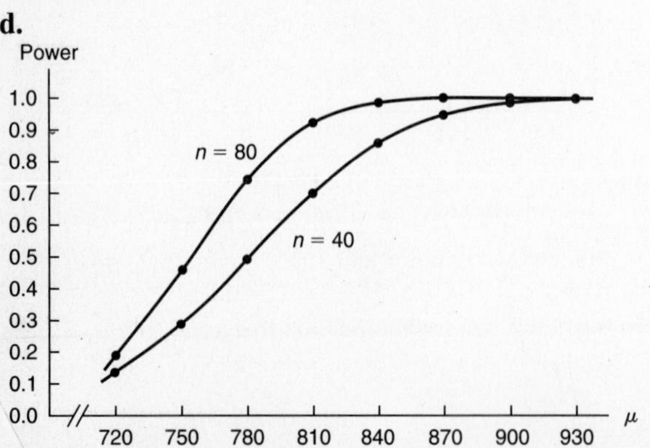

For a fixed significance level, increasing the sample size increases the power.

Exercises 9.5

9.75 The two different approaches to hypothesis testing are the critical-value approach and the P-value approach. For a comparison of the two approaches, see Table 9.11 on page 543.

9.77 true

9.79 A P-value of 0.02 provides stronger evidence against the null hypothesis because it reflects an observed value of the test statistic that is more inconsistent with the null hypothesis.

9.81

a. 0.0212 **b.** 0.6217

9.83

a. 0.2296 **b.** 0.8770

9.85

a. 0.0970 **b.** 0.6030

9.87 H_0: $\mu = \$690$, H_a: $\mu > \$690$; $\alpha = 0.05$; $z = 0.26$; $P = 0.3974$; do not reject H_0; at the 5% significance level, the data do not provide sufficient evidence to conclude that last year's mean telephone expenditure per consumer unit has increased over the 1994 mean of \$690. According to Table 9.12, the strength of the evidence against the null hypothesis is weak or none since $P > 0.10$.

9.89 H_0: $\mu = 18$ mg, H_a: $\mu < 18$ mg; $\alpha = 0.01$; $z = -5.30$; $P = 0.0000$ (to four decimal places); reject H_0; at the 1% significance level, the data provide sufficient evidence to conclude that adult females under the age of 51 are, on the average, getting less than the RDA of 18 mg of iron. According to Table 9.12, the strength of the evidence against the null hypothesis is very strong since $P < 0.01$.

9.91 H_0: $\mu = 98.6°$ F, H_a: $\mu \neq 98.6°$ F; $\alpha = 0.01$; $z = -7.35$; $P = 0.0000$ (to four decimal places); reject H_0; at the 1% significance level, the data provide sufficient evidence to conclude that the mean body temperature of healthy humans differs from 98.6° F. According to Table 9.12, the strength of the evidence against the null hypothesis is very strong since $P < 0.01$.

9.103

a. H_0: $\mu = 6.7$ days, H_a: $\mu < 6.7$ days
b. 7.70 days **c.** 7.01 days **d.** 40
e. 6.45 days **f.** −0.21 **g.** 0.42 **h.** 0.42
i. At the 5% significance level, the data do not provide sufficient evidence to conclude that the mean hospital stay, μ, for this year will be less than the 1994 mean hospital stay of 6.7 days.

Exercises 9.6

9.105 σ-unknown versus σ-known

9.107 H_0: $\mu = \$1644$, H_a: $\mu \neq \$1644$; $\alpha = 0.05$; critical values $= \pm 2.030$; $t = -1.056$; do not reject H_0; at the 5% significance level, the data do not provide sufficient evidence to conclude that the 1994 mean annual expenditure on apparel and services for consumer units in the Midwest differed from the national mean of \$1644. For the P-value approach, note that $P > 0.20$; do not reject H_0.

9.109 H_0: $\mu = 428$ points, H_a: $\mu > 428$ points; $\alpha = 0.10$; critical value $= 1.318$; $t = 0.491$; do not reject H_0; at the 10% significance level, the data do not provide sufficient evidence to conclude that last year's mean for verbal SAT scores is greater than the 1995 mean of 428 points. For the P-value approach, note that $P > 0.10$; do not reject H_0.

9.111 H_0: $\mu = 46.0$¢/lb, H_a: $\mu \neq 46.0$¢/lb; $\alpha = 0.05$; critical values $= \pm 2.145$; $t = 2.656$; reject H_0; at the 5% significance level, the data provide sufficient evidence to conclude that the current mean retail price for bananas is different from the 1994 mean of 46.0 cents per pound. For the P-value approach, note that $0.01 < P < 0.02$; reject H_0.

9.121

a. H_0: $\mu = 711.0$ gal, H_a: $\mu < 711.0$ gal
b. 237.5 gal **c.** 37 **d.** 517.4 gal
e. −4.96 **f.** 0.0000 **g.** 0.0000
h. At the 5% significance level, the data provide sufficient evidence to conclude that, in 1995, the mean fuel consumed per vehicle in Hawaii was less than that in the United States as a whole.

Exercises 9.7

9.125 Nonparametric methods do not require normality; they also usually entail fewer and simpler computations than parametric methods and are resistant to outliers and other extreme values. On the other hand, parametric methods tend to give more accurate results when the requirements for their use are met.

9.127 Because the D-value for such an observation equals 0 and so we cannot attach a sign to the rank of $|D|$.

9.129

a. Wilcoxon signed-rank test
b. Wilcoxon signed-rank test **c.** neither

9.131 H_0: $\eta = 34.3$ years, H_a: $\eta > 34.3$ years; $\alpha = 0.05$; critical value $= 44$; $W = 32$; do not reject H_0; at the 5% significance level, the data do not provide sufficient evidence to conclude that the median age of today's U.S. residents has increased over the 1995 median age of 34.3 years.

9.133 H_0: $\mu = 12$ min, H_a: $\mu < 12$ min; $\alpha = 0.05$; critical value $= 26$; $W = 20.5$; reject H_0; at the 5% significance level, the data provide sufficient evidence to conclude that the new antacid tablet works faster.

9.135

a. H_0: $\mu = 46.0$¢/lb, H_a: $\mu \neq 46.0$ ¢/lb; $\alpha = 0.05$; critical values $= 21, 84$; $W = 88.5$; reject H_0; at the 5% significance level, the data provide sufficient evidence to conclude that the mean retail price for bananas now is different from the 1994 mean of 46.0¢/lb.
b. Because a normal distribution is symmetric.

9.147

a. H_0: $\eta = 7.2$ yrs, H_a: $\eta < 7.2$ yrs **b.** 50
c. 0 **d.** 6.882 yrs **e.** 0.450 **f.** 0.450
g. Do not reject H_0; at the 5% significance level, the data do not provide sufficient evidence to conclude that last year's median marriage duration is less than the 1995 median of 7.2 years.
h. Probably not, since marriage durations most likely do not have a symmetric distribution.

Exercises 9.8

9.149 See Table 9.17 on page 574.

9.151

a. Yes, because the sample size is large and the population standard deviation is unknown.

b. Yes, because the variable under consideration has a symmetric distribution.

c. Since the variable under consideration has a nonnormal symmetric distribution, the preferred procedure is the Wilcoxon signed-rank test.

9.153 z-test

9.155 t-test

9.157 Wilcoxon signed-rank test

9.159 Consult a statistician.

Review Test for Chapter 9

1. a. a hypothesis to be tested

b. a hypothesis to be considered as an alternate to the null hypothesis

c. the statistic used as a basis for deciding whether the null hypothesis should be rejected

d. the set of values for the test statistic that leads to the rejection of the null hypothesis

e. the set of values for the test statistic that leads to nonrejection of the null hypothesis

f. the values of the test statistic that separate the rejection and nonrejection regions

2. a. The weight of a package of Tide® is a variable. A particular package may weigh slightly more or less than the marked weight. The mean weight of all packages produced on a given day (the population mean for that day) exceeds the marked weight.

b. The null hypothesis would be that a daily mean weight is equal to the marked weight; the alternative hypothesis would be that the daily mean weight is greater than the marked weight.

c. The null hypothesis would be that a daily mean weight equals the marked weight of 76 oz; the alternative hypothesis would be that the daily mean weight is greater than the marked weight of 76 oz. In statistical terminology, the hypothesis test would be:

$$H_0\colon \mu = 76 \text{ oz}$$
$$H_a\colon \mu > 76 \text{ oz},$$

where μ is the mean weight of all packages produced on the given day.

3. a. Take a random sample from the population. If the sample data are consistent with the null hypothesis, then do not reject the null hypothesis; if the sample data are inconsistent with the null hypothesis, then reject the null hypothesis and conclude that the alternative hypothesis is true.

b. We establish a precise criterion for deciding whether or not to reject the null hypothesis prior to taking the sample.

4. See the discussion on page 491.

5. a. A Type I error is the incorrect decision of rejecting a true null hypothesis. A Type II error is the incorrect decision of not rejecting a false null hypothesis.

b. α and β, respectively

c. a Type I error **d.** a Type II error

6. so that if the null hypothesis is true, the probability is equal to 0.05 that the test statistic will fall in the rejection region, in this case, to the right of the critical value

7. See Table 9.17 on page 574.

8. a. The true significance level is equal to α.

b. The true significance level is only approximately equal to α.

9. The results of a hypothesis test are statistically significant if the null hypothesis is rejected at the specified significance level. This means that the data provide sufficient evidence to conclude that the truth is different from that stated in the null hypothesis. It does not necessarily mean that the difference is important in any practical sense.

10. It increases.

11. a. the probability of rejecting a false null hypothesis

b. It increases.

12. a. See Definition 9.6 on page 536.

b. true **c.** true

d. Because it is the smallest significance level for which the observed sample data result in rejection of the null hypothesis.

13. See Table 9.11 on page 543 and the subsequent discussion.

15. Let μ denote last year's mean cheese consumption by Americans.
a. H_0: $\mu = 27.3$ lb **b.** H_a: $\mu > 27.3$ lb
c. right-tailed

16. a. $z \geq 1.28$ **b.** $z < 1.28$
c. $z = 1.28$ **d.** $\alpha = 0.10$
e.

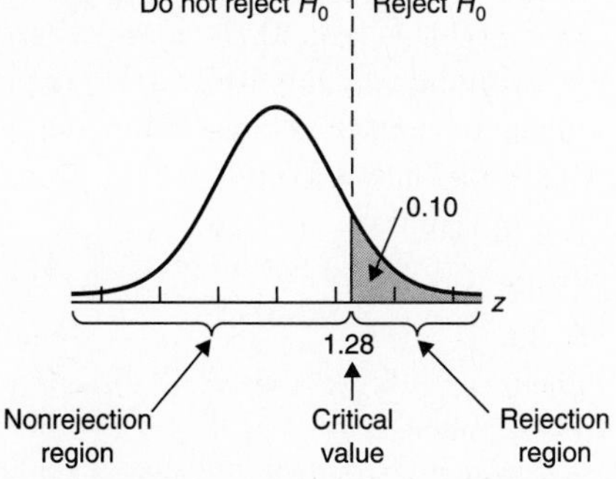

f. right-tailed

17. a. A Type I error would occur if in fact $\mu = 27.3$ lb, but the results of the sampling lead to the conclusion that $\mu > 27.3$ lb.
b. A Type II error would occur if in fact $\mu > 27.3$ lb, but the results of the sampling fail to lead to that conclusion.
c. A correct decision would occur if in fact $\mu = 27.3$ lb and the results of the sampling do not lead to the rejection of that fact; or if in fact $\mu > 27.3$ lb and the results of the sampling lead to that conclusion.
d. correct decision **e.** Type II error

18. Note that the answers obtained to many of the parts of this problem may vary depending on when and how much intermediate rounding is done.
a. 0.10
b. approximately normal with mean 27.5 and standard deviation 1.17
c. 0.8665
d. Approximately normal with the specified mean and standard deviation 1.17. The Type II error probabilities are shown in the table in part (e).
e.

μ	β	Power	μ	β	Power
27.5	0.8665	0.1335	30.0	0.1515	0.8485
28.0	0.7549	0.2451	30.5	0.0721	0.9279
28.5	0.6026	0.3974	31.0	0.0294	0.9706
29.0	0.4325	0.5675	31.5	0.0104	0.9896
29.5	0.2743	0.7257	32.0	0.0031	0.9969

f.

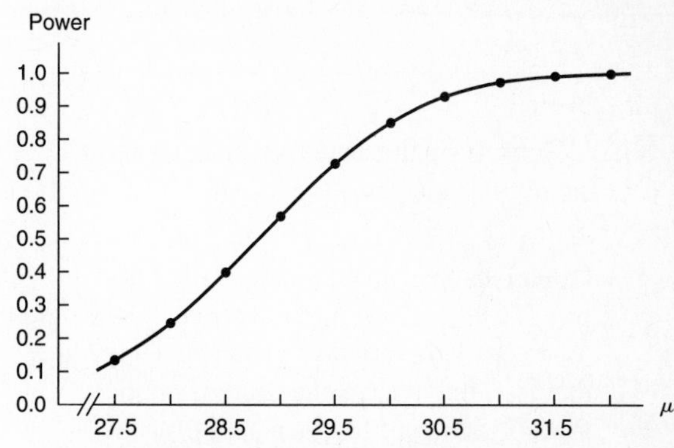

g. approximately normal with mean 27.5 and standard deviation 0.69
h. 0.8438
i. Approximately normal with the specified mean and standard deviation 0.69. The Type II error probabilities are shown in the table in part (j).
j.

μ	β	Power	μ	β	Power
27.5	0.8438	0.1562	30.0	0.0045	0.9955
28.0	0.6141	0.3859	30.5	0.0004	0.9996
28.5	0.3336	0.6664	31.0	0.0000	1.0000
29.0	0.1230	0.8770	31.5	0.0000	1.0000
29.5	0.0301	0.9699	32.0	0.0000	1.0000

k.

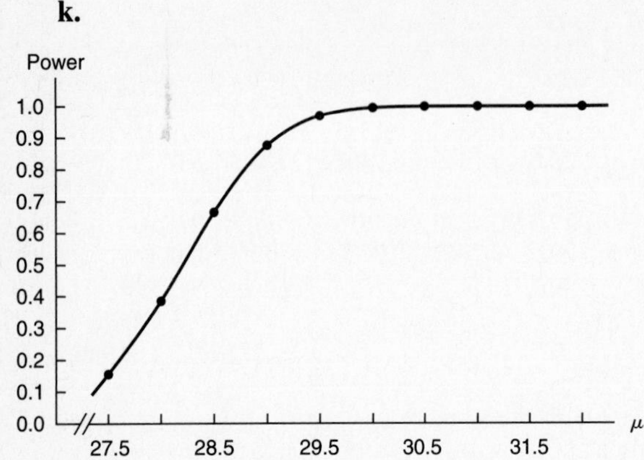

l. For a fixed significance level, increasing the sample size increases the power.

19. a. H_0: $\mu = 27.3$ lb, H_a: $\mu > 27.3$ lb; $\alpha = 0.10$; critical value $= 1.28$; $z = 0.43$; do not reject H_0; at the 10% significance level, the data do not provide sufficient evidence to conclude that last year's mean cheese consumption for all Americans has increased over the 1995 mean of 27.3 lb.

b. A Type II error, because given that the null hypothesis was not rejected, the only error that could be made is the error of not rejecting a false null hypothesis.

20. a. H_0: $\mu = 27.3$ lb, H_a: $\mu > 27.3$ lb; $\alpha = 0.10$; $z = 0.43$; $P = 0.3336$; do not reject H_0; at the 10% significance level, the data do not provide sufficient evidence to conclude that last year's mean cheese consumption for all Americans has increased over the 1995 mean of 27.3 lb.

b. The data provide at best only weak evidence against the null hypothesis.

21. H_0: $\mu = \$279$, H_a: $\mu < \$279$; $\alpha = 0.05$; critical value $= -1.684$; $t = -1.448$; do not reject H_0; at the 5% significance level, the data do not provide sufficient evidence to conclude that the mean value lost due to purse snatching has decreased from the 1994 mean of $279.

22. a. $0.05 < P < 0.10$

b. H_0: $\mu = \$279$, H_a: $\mu < \$279$; $\alpha = 0.05$; $t = -1.448$; $P > 0.05$; do not reject H_0; at the 5% significance level, the data do not provide sufficient evidence to conclude that the mean value lost due to purse snatching has decreased from the 1994 mean of $279.

c. The data provide moderate evidence against the null hypothesis.

23. a. H_0: $\mu = 29$ mpg, H_a: $\mu \neq 29$ mpg; $\alpha = 0.05$; critical values $= \pm 2.145$; $t = -0.600$; do not reject H_0; at the 5% significance level, the data do not provide sufficient evidence to conclude that the mean gas mileage is not equal to what is stated in the company's report.

b. H_0: $\mu = 29$ mpg, H_a: $\mu \neq 29$ mpg; $\alpha = 0.05$; $t = -0.600$; $P > 0.20$; do not reject H_0; at the 5% significance level, the data do not provide sufficient evidence to conclude that the mean gas mileage is not equal to what is stated in the company's report.

c. The data provide at best only weak evidence against the null hypothesis.

24. a. H_0: $\mu = 29$ mpg, H_a: $\mu \neq 29$ mpg; $\alpha = 0.05$; critical values $= 25, 95$; $W = 48.5$; do not reject H_0; at the 5% significance level, the data do not provide sufficient evidence to conclude that the mean gas mileage is not equal to what is stated in the company's report.

b. It is symmetric.

c. Because a normal distribution is symmetric.

25. The t-test is the preferred procedure because it is expressly designed for normally distributed variables.

26. It is probably okay to use the z-test because the sample size is large and σ is known. However, it does appear from the normal probability plot that there may be outliers so one should proceed cautiously in using the z-test. See the third bulleted item in Key Fact 9.4 on page 514.

27. It appears that the variable under consideration is far from being normally distributed and, in fact, has a left-skewed distribution. However, the sample size is large and the plots reveal no outliers. Keeping in mind that σ is unknown, it is probably reasonable to use the t-test.

28. a. In view of the graphs, it appears reasonable to assume that in Problem 26 the variable under consideration has (approximately) a symmetric distribution, but not so in Problem 27. Consequently, it would be reasonable to use the Wilcoxon signed-rank test in the first case, but not the second.

b. In Problem 26, it is a tough call between the Wilcoxon signed-rank test and the z-test, but considering the possible outliers, the Wilcoxon signed-rank test is probably the better one to use.

30. **a.** H_0: $\mu = 27.3$ lb, H_a: $\mu > 27.3$ lb
b. 6.90 lb **c.** 6.48 lb **d.** 35 **e.** 27.8 lb
f. 0.43 **g.** 0.33 **h.** 0.33
i. At the 10% significance level, the data do not provide sufficient evidence to conclude that last year's mean cheese consumption by Americans has increased over the 1995 mean of 27.3 lb.

32. **a.** H_0: $\mu = 29$ mpg, H_a: $\mu \neq 29$ mpg
b. 1.595 mpg **c.** 15 **d.** 28.753 mpg
e. −0.60 **f.** 0.56 **g.** 0.56
h. At the 5% significance level, the data do not provide sufficient evidence to conclude that the mean gas mileage is not equal to what is stated in the company's report.

34. **a.** H_0: $\eta = 29$ mpg, H_a: $\eta \neq 29$ mpg
b. 15 **c.** 0 **d.** 48.5
e. 0.532 **f.** 0.532
g. At the 5% significance level, the data do not provide sufficient evidence to conclude that the median (or mean) gas mileage is not equal to what is stated in the company's report.
h. 28.65 mpg

CHAPTER 10

Exercises 10.1

10.1

a. age
b. Population 1 consists of buyers of new domestic cars; Population 2 consists of buyers of new imported cars.
c. Let μ_1 and μ_2 denote, respectively, the mean ages of buyers of new domestic cars and new imported cars. Then the null and alternative hypotheses are, respectively, H_0: $\mu_1 = \mu_2$ and H_a: $\mu_1 > \mu_2$.

10.3 Answers will vary.

10.5

a. μ_1, σ_1, μ_2, and σ_2 are parameters; $\bar{x}_1, s_1, \bar{x}_2$, and s_2 are statistics.
b. μ_1, σ_1, μ_2, and σ_2 are fixed numbers; $\bar{x}_1, s_1, \bar{x}_2$, and s_2 are variables.

10.7 We need to know the sampling distribution of the difference between two means so that we can determine whether the observed difference between the two sample means can be reasonably attributed to sampling error.

10.9

a. 0 and 5 **b.** no
c. No, because we do not know the distributions of the variable on the two populations.

Exercises 10.2

10.17 Because it is obtained by taking the square root of the pooled variance, s_p^2. (For an explanation of the term "pooled," see page 601.)

10.19 H_0: $\mu_1 = \mu_2$, H_a: $\mu_1 < \mu_2$; $\alpha = 0.05$; critical value $= -1.734$; $t = -4.058$; reject H_0; at the 5% significance level, the data provide sufficient evidence to conclude that the mean time served for fraud is less than that for firearms offenses. For the P-value approach, note that $P < 0.005$.

10.21 H_0: $\mu_1 = \mu_2$, H_a: $\mu_1 > \mu_2$; $\alpha = 0.05$; critical value $= 1.714$; $t = 1.186$; do not reject H_0; at the 5% significance level, the data do not provide sufficient evidence to conclude that males in the age group 25–34 years are, on the average, taller than those who were in that age group 20 years ago. For the P-value approach, note that $P > 0.10$.

10.23 H_0: $\mu_1 = \mu_2$, H_a: $\mu_1 \neq \mu_2$; $\alpha = 0.05$; critical values $= \pm 1.998$; $t = 6.304$; reject H_0; at the 5% significance level, the data provide sufficient evidence to conclude that last year's mean annual fuel expenditure for households using natural gas is different from that for households using only electricity. For the P-value approach, note that $P < 0.01$.

10.25

a. −14.11 to −3.21 months

b. We can be 98% confident that the difference between the mean times served by prisoners in the fraud and firearms offense categories is somewhere between −14.11 and −3.21 months.

10.27

a. −0.72 to 3.94 inches

b. We can be 90% confident that the difference, $\mu_1 - \mu_2$, between the mean height of males in the age group 25–34 years and the mean height of males in the age group 45–54 years is somewhere between −0.72 and 3.94 inches.

10.29

a. \$173.50 to \$334.50

b. We can be 95% confident that the difference between last year's mean fuel expenditures for households using natural gas and those using only electricity is somewhere between \$173.50 and \$334.50.

10.35

a. H_0: $\mu_1 = \mu_2$, H_a: $\mu_1 < \mu_2$

b. $s_1 = \$2.25$, $s_2 = \$2.36$ **c.** $n_1 = 14$, $n_2 = 17$

d. $\bar{x}_1 = \$15.93$, $\bar{x}_2 = \$16.42$

e. −0.59 **f.** 0.28 **g.** 0.28

h. At the 5% significance level, the data do not provide sufficient evidence to conclude that the mean hourly earnings for mine workers is less than that for construction workers.

i. −\$1.91 to \$0.93 **j.** \$2.31

Exercises 10.3

10.39 The pooled t-test requires equal population standard deviations whereas the nonpooled t-test does not.

10.41 H_0: $\mu_1 = \mu_2$, H_a: $\mu_1 \neq \mu_2$; $\alpha = 0.05$; critical values $= \pm 2.056$; $t = -0.917$; do not reject H_0; at the 5% significance level, the data do not provide sufficient evidence to conclude that the mean GPAs of sophomores and juniors at the university differ. For the P-value approach, note that $P > 0.20$.

10.43 H_0: $\mu_1 = \mu_2$, H_a: $\mu_1 < \mu_2$; $\alpha = 0.05$; critical value $= -2.015$; $t = -1.651$; do not reject H_0; at the 5% significance level, the data do not provide sufficient evidence to conclude that the mean number of acute postoperative days in the hospital are fewer with the dynamic system than with the static system. For the P-value approach, note that $0.05 < P < 0.10$.

10.45 H_0: $\mu_1 = \mu_2$, H_a: $\mu_1 > \mu_2$; $\alpha = 0.05$; critical value $= 1.669$; $t = 3.951$; reject H_0; at the 5% significance level, the data provide sufficient evidence to conclude that Nevada has a larger mean potato yield than Idaho. For the P-value approach, note that $P < 0.005$.

10.47

a. −0.45 to 0.17

b. We can be 95% confident that the difference, $\mu_1 - \mu_2$, between the mean GPAs of sophomores and juniors at the university is somewhere between −0.45 and 0.17.

10.49

a. −6.97 to 0.69 days

b. We can be 90% confident that the difference between the mean numbers of acute postoperative days in the hospital with the dynamic and static systems is somewhere between −6.97 and 0.69 days.

10.51

a. 19.52 to 48.08 cwt

b. We can be 90% confident that the difference between the mean yields per acre of potatoes for Nevada and Idaho is somewhere between 19.52 and 48.08 cwt.

10.59

a. H_0: $\mu_1 = \mu_2$, H_a: $\mu_1 > \mu_2$

b. $s_1 = 4.62$ yrs, $s_2 = 7.25$ yrs

c. $n_1 = 11$, $n_2 = 12$

d. $\bar{x}_1 = 35.69$ yrs, $\bar{x}_2 = 32.81$ yrs

e. 1.15 **f.** 0.13 **g.** 0.13

h. At the 5% significance level, the data do not provide sufficient evidence to conclude that the mean age at the time of first divorce for males is greater than that for females.

i. −1.5 to 7.2 yrs

j. The age distributions under consideration here are probably not normal. But, remember, the nonpooled-t procedure is robust to moderate violations of the normality assumption. Therefore, whether it is reasonable to apply the nonpooled-t procedure here depends on how much the distributions deviate from normality.

Exercises 10.4

10.61 Because the shape of a normal distribution is determined by its standard deviation.

10.63

a. pooled t-test
b. Mann–Whitney test

10.65 H_0: $\mu_1 = \mu_2$, H_a: $\mu_1 < \mu_2$; $\alpha = 0.05$; critical value $= 33$; $M = 33$; reject H_0; at the 5% significance level, the data provide sufficient evidence to conclude that in this teacher's chemistry courses, students with fewer than two years of high-school algebra have a lower mean semester average than those with two or more years.

10.67 H_0: $\eta_1 = \eta_2$, H_a: $\eta_1 > \eta_2$; $\alpha = 0.05$; critical value $= 54$; $M = 47$; do not reject H_0; at the 5% significance level, the data do not provide sufficient evidence to conclude that the median number of volumes held by public colleges is less than that held by private colleges.

10.69

a. H_0: $\mu_1 = \mu_2$, H_a: $\mu_1 < \mu_2$; $\alpha = 0.05$; critical value $= 83$; $M = 65.5$; reject H_0; at the 5% significance level, the data provide sufficient evidence to conclude that the mean time served for fraud is less than that for firearms offenses.
b. Because two normal distributions with equal standard deviations have the same shape. The pooled t-test because, in the normal case, it is more powerful than the Mann–Whitney test.

10.81

a. H_0: $\eta_1 = \eta_2$, H_a: $\eta_1 < \eta_2$
b. \$16.615, \$16.880
c. $n_1 = 14$, $n_2 = 17$
d. 208.0 **e.** 0.2692 **f.** 0.2692
g. At the 5% significance level, the data do not provide sufficient evidence to conclude that the median hourly earnings for mine workers is less than that for construction workers.
h. −\$1.790 to \$1.080

Exercises 10.5

10.83 Each pair in a paired sample consists of a member of one population and that member's corresponding member in the other population.

10.85

a. age **b.** married men and married women
c. married couples
d. the difference between the ages of a married couple

10.87 Paired sample, and normal differences or large sample. See the second and third paragraphs from the bottom on page 644.

10.89

a. height
b. cross-fertilized Zea mays and self-fertilized Zea mays
c. the difference between the heights of a cross-fertilized Zea may and a self-fertilized Zea may grown in the same pot
d. Yes, because each number is the difference between the heights of a cross-fertilized Zea may and a self-fertilized Zea may grown in the same pot.
e. H_0: $\mu_1 = \mu_2$, H_a: $\mu_1 \neq \mu_2$; $\alpha = 0.05$; critical values $= \pm 2.145$; $t = 2.148$; reject H_0; at the 5% significance level, the data provide sufficient evidence to conclude that the mean heights of cross-fertilized and self-fertilized Zea mays differ. For the P-value approach, note that $0.02 < P < 0.05$.

10.91 H_0: $\mu_1 = \mu_2$, H_a: $\mu_1 \neq \mu_2$; $\alpha = 0.01$; critical values $= \pm 3.169$; $t = 3.866$; reject H_0; at the 1% significance level, the data provide sufficient evidence to conclude that, on the average, the two measurement methods give different results. For the P-value approach, note that $P < 0.01$.

10.93 H_0: $\mu_1 = \mu_2$, H_a: $\mu_1 > \mu_2$; $\alpha = 0.05$; critical value $= 1.833$; $t = 2.213$; reject H_0; at the 5% significance level, the data provide sufficient evidence to conclude that the mean age of married men is greater than the mean age of married women. For the P-value approach, note that $0.025 < P < 0.05$.

10.95

a. 0.03 to 41.83 eighths of an inch
b. We can be 95% confident that the difference between the mean heights of cross-fertilized and self-fertilized Zea mays is somewhere between 0.03 and 41.83 eighths of an inch.

10.97

a. 0.68 to 6.83 thousand miles

b. We can be 99% confident that the mean difference in measurement by the weight and groove methods is somewhere between 0.68 and 6.83 thousand miles.

10.99

a. 0.6 to 6.0 yrs **b.** 0.6 to 6.0 yrs

10.107

a. H_0: $\mu_1 = \mu_2$, H_a: $\mu_1 \neq \mu_2$

b. 69.73, 67.82 **c.** 1.91 **d.** 3.73

e. 1.70 **f.** 0.120 **g.** 0.120

h. At the 5% significance level, the data do not provide sufficient evidence to conclude that there is a difference between the mean final grades of the two instructional methods.

i. −0.59 to 4.41

Exercises 10.6

10.109

a. Yes, because both assumptions required for a paired t-test are satisfied.

b. Yes. Since the paired-difference variable has a normal distribution, its distribution is symmetric.

c. In this case, the paired t-test is preferable since it is more powerful than the paired Wilcoxon signed-rank test.

10.111 H_0: $\mu_1 = \mu_2$, H_a: $\mu_1 \neq \mu_2$; $\alpha = 0.05$; critical values = 25, 95; $W = 96$; reject H_0; at the 5% significance level, the data provide sufficient evidence to conclude that the mean heights of cross-fertilized and self-fertilized Zea mays differ. For the P-value approach, note that $0.02 < P < 0.05$.

10.113 H_0: $\mu_1 = \mu_2$, H_a: $\mu_1 \neq \mu_2$; $\alpha = 0.01$; critical values = 5, 61; $W = 64$; reject H_0; at the 1% significance level, the data provide sufficient evidence to conclude that, on the average, the two measurement methods give different results. For the P-value approach, note that $P < 0.01$.

10.115 H_0: $\mu_1 = \mu_2$, H_a: $\mu_1 > \mu_2$; $\alpha = 0.05$; critical value = 44; $W = 45$; reject H_0; at the 5% significance level, the data provide sufficient evidence to conclude that the mean age of married men is greater than the mean age of married women. For the P-value approach, note that $0.025 < P < 0.05$.

10.123

a. H_0: $\mu_1 = \mu_2$, H_a: $\mu_1 \neq \mu_2$

b. 11 **c.** 1 **d.** 42.5 **e.** 0.139 **f.** 0.139

g. At the 5% significance level, the data do not provide sufficient evidence to conclude that there is a difference between the mean final grades of the two instructional methods.

h. 1.500

Exercises 10.7

10.127 See the first three entries of Table 10.15 on page 666.

10.129

a. the pooled t-test, the nonpooled t-test, and the Mann–Whitney test

b. the pooled t-test

10.131

a. the pooled t-test, the nonpooled t-test, and the Mann–Whitney test

b. the Mann–Whitney test

10.133

a. the paired t-test and the paired Wilcoxon signed-rank test

b. the paired Wilcoxon signed-rank test

10.135 nonpooled t-test

10.137 nonpooled t-test

10.139 Consult a statistician.

Review Test for Chapter 10

1. Independently and randomly take samples from the two populations; compute the two sample means; compare the two sample means; and make the decision.
2. Randomly take a paired sample from the two populations; calculate the paired differences of the sample pairs; compute the mean of the sample of paired differences; compare the mean to 0; and make the decision.
3. **a.** The pooled t-test requires equal population standard deviations whereas the nonpooled t-test does not.

   b. It is essential that the assumption of independence be satisfied.

c. For very small sample sizes, the normality assumption is essential for both t-procedures. However, for larger samples, the normality assumption is less important.

d. population standard deviations

4. a. No. If the two distributions are normal and have the same shape, then they have equal population standard deviations; in this case, the pooled t-test is preferred. If the two distributions are nonnormal, but have the same shape, the Mann–Whitney test is preferred.

b. The two distributions are normal. Because, in this case, the pooled t-test is more powerful than the Mann–Whitney test.

5. By using a paired sample, we can remove extraneous sources of variation. As a consequence, the sampling error made in estimating the difference between the population means will generally be smaller. This fact, in turn, makes it more likely that we will detect differences between the population means when such differences exist.

6. If the paired-difference variable is normally distributed, it would be preferable to use the paired t-test because, in that case, it is more powerful than the paired Wilcoxon signed-rank test.

7. a. H_0: $\mu_1 = \mu_2$, H_a: $\mu_1 > \mu_2$; $\alpha = 0.05$; critical value $= 1.708$; $t = 1.538$; do not reject H_0; at the 5% significance level, the data do not provide sufficient evidence to conclude that the mean right-leg strength of males exceeds that of females.

b. For the P-value approach, note that $0.05 < P < 0.10$.

c. According to Table 9.12, the evidence against the null hypothesis is moderate.

8. -31.3 to 599.3 newtons (N)

9. H_0: $\mu_1 = \mu_2$, H_a: $\mu_1 < \mu_2$; $\alpha = 0.05$; critical value $= -1.721$; $t = -2.740$; reject H_0; at the 5% significance level, the data provide sufficient evidence to conclude that last year the average German consumed less fish than the average Russian. For the P-value approach, note that $0.005 < P < 0.01$.

10. -7.60 to -1.74 kilograms (kg)

11. H_0: $\mu_1 = \mu_2$, H_a: $\mu_1 \neq \mu_2$; $\alpha = 0.05$; critical values $= 79, 131$; $M = 116$; do not reject H_0; at the 5% significance level, the data do not provide sufficient evidence to conclude that the mean costs for existing single-family homes differ in New York City and Los Angeles.

12. H_0: $\mu_1 = \mu_2$, H_a: $\mu_1 \neq \mu_2$; $\alpha = 0.10$; critical values $= \pm 1.833$; $t = -1.766$; do not reject H_0; at the 10% significance level, the data do not provide sufficient evidence to conclude that there is a difference in effectiveness of the two speed-reading programs. For the P-value approach, note that $0.10 < P < 0.20$.

13. -81.5 to 1.5 words per minute (wpm)

14. H_0: $\mu_1 = \mu_2$, H_a: $\mu_1 > \mu_2$; $\alpha = 0.01$; critical value $= 100$; $W = 108.5$; reject H_0; at the 1% significance level, the data provide sufficient evidence to conclude that the mean monthly rent for renter-occupied housing units in the South exceeds that for those in the Midwest. For the P-value approach, note that $P < 0.005$.

16. a. H_0: $\mu_1 = \mu_2$, H_a: $\mu_1 > \mu_2$
b. 513 N, 446 N **c.** 2127 N, 1843 N
d. 1.54 **e.** 0.068 **f.** 0.068
g. At the 5% significance level, the data do not provide sufficient evidence to conclude that the mean right-leg strength of males exceeds that of females.
h. -31 to 599 N **i.** 479 N

18. a. H_0: $\mu_1 = \mu_2$, H_a: $\mu_1 < \mu_2$
b. 2.84 kg, 5.61 kg **c.** 10, 15
d. 11.40 kg, 16.07 kg **e.** -2.74
f. 0.0062 **g.** 0.0062
h. At the 5% significance level, the data provide sufficient evidence to conclude that last year the average German consumed less fish than the average Russian.
i. -7.60 to -1.7 kg

20. a. H_0: $\eta_1 = \eta_2$, H_a: $\eta_1 \neq \eta_2$
b. \$143.8 thousand, \$133.8 thousand
c. 10, 10 **d.** 116 **e.** 0.4274 **f.** 0.4274
g. At the 5% significance level, the data do not provide sufficient evidence to conclude that the

median costs for existing single-family homes differ in New York City and Los Angeles.
h. −\$55.0 thousand to \$44.8 thousand

22. **a.** H_0: $\mu_1 = \mu_2$, H_a: $\mu_1 \neq \mu_2$
b. 1021.4 wpm, 1061.4 wpm
c. −40.0 wpm **d.** 71.6 wpm
e. −1.77 **f.** 0.111 **g.** 0.111
h. At the 10% significance level, the data do not provide sufficient evidence to conclude that there is a difference in effectiveness of the two speed-reading programs.
i. −81.5 to 1.5 wpm

24. **a.** H_0: $\mu_1 = \mu_2$, H_a: $\mu_1 > \mu_2$
b. 0 **c.** 108.5 **d.** 0.003 **e.** 0.003
f. At the 1% significance level, the data provide sufficient evidence to conclude that the mean monthly rent for renter-occupied housing units in the South exceeds that for those in the Midwest.
g. \$23.50

CHAPTER 11

Exercises 11.1

11.1 A variable is said to have a chi-square distribution if its distribution has the shape of a special type of right-skewed curve, called a chi-square curve.

11.3 The χ^2-curve with 20 degrees of freedom more closely resembles a normal curve. As the number of degrees of freedom becomes larger, χ^2-curves look increasingly like normal curves.

11.5
a. 32.852 **b.** 10.117

11.7
a. 18.307 **b.** 3.247

11.9
a. 1.646 **b.** 15.507

11.11
a. 0.831, 12.833 **b.** 13.844, 41.923

11.13 Because the procedures are based on the assumption that the variable under consideration is normally distributed and are nonrobust to violations of that assumption.

11.15 H_0: $\sigma = 100$, H_a: $\sigma \neq 100$; $\alpha = 0.05$; critical values = 12.401, 39.364; $\chi^2 = 17.545$; do not reject H_0; at the 5% significance level, the data do not provide sufficient evidence to conclude that the standard deviation of last year's verbal scores is different from the 1941 standard deviation of 100. For the P-value approach, note that $P > 0.20$.

11.17 H_0: $\sigma = 0.27$, H_a: $\sigma > 0.27$; $\alpha = 0.01$; critical value = 21.666; $\chi^2 = 70.560$; reject H_0; at the 1% significance level, the data provide sufficient evidence to conclude that the process variation for this piece of equipment exceeds the analytical capability of 0.27. For the P-value approach, note that $P < 0.005$.

11.19 H_0: $\sigma = 0.2$ fl oz, H_a: $\sigma < 0.2$ fl oz; $\alpha = 0.05$; critical value = 6.571; $\chi^2 = 8.301$; do not reject H_0; at the 5% significance level, the data do not provide sufficient evidence to conclude that the standard deviation, σ, of the amounts of coffee being dispensed is less than 0.2 fl oz. For the P-value approach, note that $P > 0.10$.

11.21 66.8 to 118.9. We can be 95% confident that the standard deviation of last year's verbal SAT scores is somewhere between 66.8 and 118.9.

11.23 0.49 to 1.57. We can be 98% confident that the process variation for this piece of equipment is somewhere between 0.49 and 1.57.

11.25 0.118 to 0.225 fl oz. We can be 90% confident that the standard deviation, σ, of the amounts of coffee being dispensed is somewhere between 0.118 and 0.225 fl oz.

Exercises 11.2

11.35 by stating its numbers of degrees of freedom

11.37 $F_{0.05}$; $F_{0.025}$; F_α

11.39
a. 12 **b.** 7

11.41
a. 1.89 **b.** 2.47 **c.** 2.14

11.43
a. 2.88 **b.** 2.10 **c.** 1.78

11.45
a. 0.123 **b.** 3.58

11.47
a. 0.18, 9.07 **b.** 0.33, 2.68

11.49 Because the procedures are based in part on the assumption that the variable under consideration is normally distributed on each population and are nonrobust to violations of that assumption.

11.51 H_0: $\sigma_1 = \sigma_2$, H_a: $\sigma_1 > \sigma_2$; $\alpha = 0.05$; critical value $= 2.03$; $F = 2.18$; reject H_0; at the 5% significance level, the data provide sufficient evidence to conclude that there is less variation among final-exam scores using the new teaching method. For the P-value approach, note that $0.025 < P < 0.05$.

11.53 H_0: $\sigma_1 = \sigma_2$, H_a: $\sigma_1 \neq \sigma_2$; $\alpha = 0.10$; critical values $= 0.53, 1.94$; $F = 1.22$; do not reject H_0; at the 10% significance level, the data do not provide sufficient evidence to conclude that the variation in anxiety-test scores differs between patients seeing videotapes showing progressive relaxation exercises and those seeing neutral videotapes. For the P-value approach, note that $P > 0.10$.

11.55 0.793 to 1.518. We can be 90% confident that the ratio of the population standard deviations of scores for patients seeing videotapes showing progressive relaxation exercises and those seeing neutral videotapes is somewhere between 0.793 and 1.518.

Review Test for Chapter 11

1. chi-square distribution
2. **a.** right **b.** normal
3. The variable under consideration must be normally distributed or nearly so. It is very important because the procedures are nonrobust to violations of that assumption.
4. **a.** 6.408 **b.** 33.409 **c.** 27.587
   d. 8.672 **e.** 7.564, 30.191
5. F-distribution
6. **a.** right **b.** reciprocal; 5, 14
7. The distributions (one for each population) of the variable under consideration must be normally distributed or nearly so. It is very important because the procedures are nonrobust to violations of that assumption.
8. **a.** 7.01 **b.** 0.07 **c.** 3.84
   d. 0.17 **e.** 0.11, 5.05
9. **a.** H_0: $\sigma = 16$, H_a: $\sigma \neq 16$; $\alpha = 0.10$; critical values $= 13.848, 36.415$; $\chi^2 = 21.111$; do not reject H_0; at the 10% significance level, the data do not provide sufficient evidence to conclude that IQs measured on this scale have a standard deviation different from 16 points. For the P-value approach, note that $P > 0.20$.
   b. It is crucial because the χ^2-test is nonrobust to violations of that assumption.
10. 12.18 to 19.76 points. We can be 90% confident that the standard deviation of IQs measured on the Stanford Revision of the Binet-Simon Intelligence Scale is somewhere between 12.18 and 19.76 points.
11. **a.** F-distribution with df $= (14, 19)$
    b. No; there is no entry in the table for df $= (19, 14)$.
    c. H_0: $\sigma_1 = \sigma_2$, H_a: $\sigma_1 < \sigma_2$; $\alpha = 0.01$; critical value $= 0.28$; $F = 0.07$; reject H_0; at the 1% significance level, the data provide sufficient evidence to conclude that runners have less variability in skinfold thickness than others.
    d. Skinfold thickness is normally distributed for runners and for others. Construct normal probability plots of the two samples.
    e. The samples from the two populations must be independent.
12. **a.** 0.15 to 0.51
    b. We can be 98% confident that the ratio of the standard deviations of skinfold thickness for runners and for others is somewhere between 0.15 and 0.51.

CHAPTER 12

Exercises 12.1

12.1 Answers will vary.

12.3 A population proportion is a parameter since it is a descriptive measure for a population. A sample proportion is a statistic since it is a descriptive measure for a sample.

12.5

a. $p = 0.4$

b.

Sample	No. of females x	Sample proportion $\hat{p}$
J, G	1	0.5
J, P	0	0.0
J, C	0	0.0
J, F	1	0.5
G, P	1	0.5
G, C	1	0.5
G, F	2	1.0
P, C	0	0.0
P, F	1	0.5
C, F	1	0.5

c.

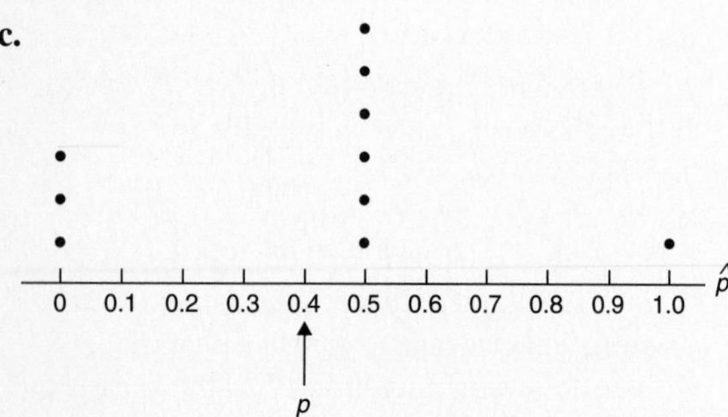

d. 0.4

e. They are the same because the mean of the variable $\hat{p}$ equals the population proportion; in symbols, $\mu_{\hat{p}} = p$.

12.7

b.

Sample	No. of females x	Sample proportion $\hat{p}$
J, P, C	0	0.00
J, P, G	1	0.33
J, P, F	1	0.33
J, C, G	1	0.33
J, C, F	1	0.33
J, G, F	2	0.67
P, C, G	1	0.33
P, C, F	1	0.33
P, G, F	2	0.67
C, G, F	2	0.67

c.

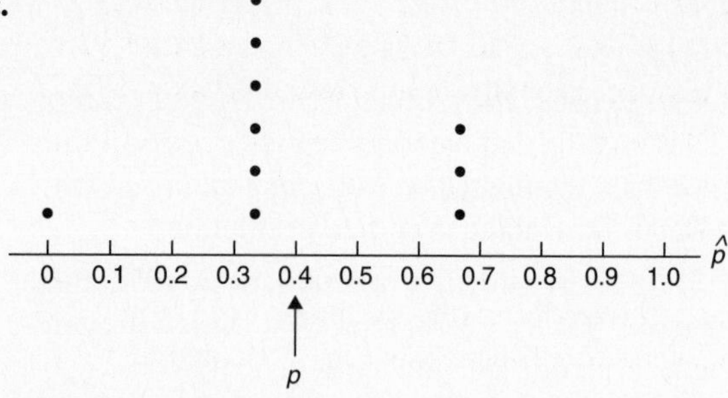

d. 0.4

e. They are the same because the mean of the variable $\hat{p}$ equals the population proportion; in symbols, $\mu_{\hat{p}} = p$.

12.9

b.

Sample	No. of females x	Sample proportion $\hat{p}$
J, P, C, G, F	2	0.4

c.

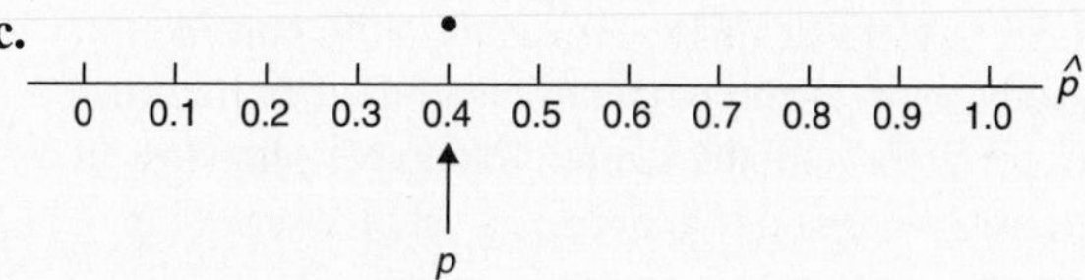

d. 0.4

e. They are the same because the mean of the variable $\hat{p}$ equals the population proportion; in symbols, $\mu_{\hat{p}} = p$.

12.11

a. all U.S. governors in 1997

b. Republican

c. Population proportion. It is the proportion of the population of U.S. governors (in 1997) that are Republican.

12.13

a. 0.035 **b.** smaller

12.15

a. 0.4 **b.** 0.5 **c.** 0.7

d. 0.2 **e.** 0.5 **f.** 0.5

g. (a) $0.4 < \hat{p} < 0.6$ (b) none (c) $0.3 < \hat{p} < 0.7$ (d) $0.2 < \hat{p} < 0.8$ (e) none (f) none

12.17

a. 0.626 to 0.674

b. We can be 95% confident that the percentage of Americans who drink beer, wine, or hard liquor, at least occasionally, is somewhere between 62.6% and 67.4%.

12.19

a. 78.4% to 83.6%

b. We can be 99% confident that the percentage of adult Americans who are in favor of "right to die" laws is somewhere between 78.4% and 83.6%.

12.21 Not very well! I applied Procedure 12.1 without checking the assumption for its use; namely, that the number of successes, x, and the number of failures, $n - x$, are both 5 or greater. Since the number of successes here is only $0.008 \cdot 500 = 4$, I should not have used Procedure 12.1.

12.23 31.6% to 36.4%

12.25

a. 0.024 **b.** 2401 **c.** 0.611 to 0.649

d. 0.019, which is less than 0.02

e. 2305; 0.610 to 0.650; 0.02

f. By employing the guess for $\hat{p}$ in part (e), we reduced the required sample size by 96. Moreover, only $\frac{1}{10}$ of 1% of precision was lost—the margin of error rose from 0.019 to 0.02.

12.27

a. plus or minus 2.6 percentage points

b. 16,577 **c.** 0.817 to 0.833 (81.7% to 83.3%)

d. 0.008 (0.8%), which is less than 0.01 (1%)

e. 12,433; 0.816 to 0.834; 0.009 (0.9%)

f. By employing the guess for $\hat{p}$ in part (e), we reduced the required sample size by 4144. Moreover, only $\frac{1}{10}$ of 1% of precision was lost—the margin of error rose from 0.008 (0.8%) to 0.009 (0.9%).

12.29 Yes, because the two confidence intervals, 53.1% to 62.9% and 49.6% to 58.4%, overlap.

12.37

a. 1004 **b.** 670

c. 66.7331% **d.** 63.7230% to 69.6447%

Exercises 12.2

12.41

a. 0.520

b. H_0: $p = 0.5$, H_a: $p > 0.5$; $\alpha = 0.05$; critical value $= 1.645$; $z = 1.09$; do not reject H_0; at the 5% significance level, the data do not provide sufficient evidence to conclude that a majority of Arizona families who celebrate Christmas wait until Christmas Day to open their presents. For the P-value approach, note that $P = 0.1379$.

12.43 H_0: $p = 0.12$, H_a: $p \neq 0.12$; $\alpha = 0.10$; critical values $= \pm 1.645$; $z = -0.68$; do not reject H_0; at the 10% significance level, the data do not provide sufficient evidence to conclude that the percentage of 18–25 year olds who currently use marijuana or hashish has changed from the 1995 percentage of 12%. For the P-value approach, note that $P = 0.4966$.

12.45 H_0: $p = 0.831$, H_a: $p < 0.831$; $\alpha = 0.05$; critical value $= -1.645$; $z = -1.75$; reject H_0; at the 5% significance level, the data provide sufficient evidence to conclude that this year's percentage of home buyers purchasing single-family houses has decreased from the 1995 figure of 83.1%. For the P-value approach, note that $P = 0.0401$.

12.49

a. H_0: $p = 0.5$, H_a: $p > 0.5$ **b.** 779

c. 452 **d.** 58.0231% **e.** 0.000

f. reject H_0

12.51 In Example 12.6, we used the z-test, which is based on the normal approximation to the binomial distribution; thus the P-value given there is only approximately correct. The Minitab output in Printout 12.5 is the result of an exact computation based directly on the binomial distribution; thus the P-value given there is exact.

Exercises 12.3

12.53

a. attend church at least once a week

b. children in the U.S. and children in Germany

c. proportion of children in the U.S. who attend church at least once a week and proportion of children in Germany who attend church at least once a week

12.55

a. uses sunscreen before going out in the sun

b. teen-age girls and teen-age boys

c. Sample proportions. Industry Research acquired those proportions by polling samples of the populations of all teen-age girls and all teen-age boys.

12.57

a. p_1 and p_2 are parameters and the other quantities are statistics.

b. p_1 and p_2 are fixed numbers and the other quantities are variables.

12.59

a. H_0: $p_1 = p_2$, H_a: $p_1 < p_2$; $\alpha = 0.01$; critical value $= -2.33$; $z = -2.61$; reject H_0; at the 1% significance level, the data provide sufficient evidence to conclude that women who take folic acid are at lesser risk of having children with major birth defects. For the P-value approach, note that $P = 0.0045$.

b. designed experiment

c. Yes, because for a designed experiment, it is reasonable to interpret statistical significance as a causal relationship.

12.61 H_0: $p_1 = p_2$, H_a: $p_1 < p_2$; $\alpha = 0.05$; critical value $= -1.645$; $z = -2.63$; reject H_0; at the 5% significance level, the data provide sufficient evidence to conclude that for men 20–34 years old, a higher percentage were overweight in 1990 than 10 years earlier. For the P-value approach, note that $P = 0.0043$.

12.63 H_0: $p_1 = p_2$, H_a: $p_1 \neq p_2$; $\alpha = 0.05$; critical values $= \pm 1.96$; $z = 4.62$; reject H_0; at the 5% significance level, we can conclude that a difference exists between the percentages of employed and unemployed workers who have registered to vote. For the P-value approach, note that $P = 0.0000$ (to four decimal places).

12.65

a. -0.019 to -0.001

b. We can be 98% confident that the rate of major birth defects for babies born to women who have not taken folic acid is somewhere between 1 per 1000 and 19 per 1000 higher than for babies born to women who have taken folic acid.

12.67

a. -0.090 to -0.021

b. We can be 90% confident that the difference, $p_1 - p_2$, between the proportions that were overweight in the years 1980 and 1990 is somewhere between -0.090 and -0.021.

12.69

a. 0.092 to 0.223

b. We can be 95% confident that the difference, $p_1 - p_2$, between the proportions of employed and unemployed workers who have registered to vote is somewhere between 0.092 and 0.223.

12.75

a. $n_1 = 272$; $n_2 = 229$

b. $x_1 = 193$; $x_2 = 165$

c. 70.9559%; 72.0524%

d. -0.0902652 to 0.0683348

e. H_0: $p_1 = p_2$, H_a: $p_1 \neq p_2$

f. 0.787

g. At the 5% significance level, the data do not provide sufficient evidence to conclude that a difference exists between the proportion of all residents who were in favor at the time of the survey before the signed agreement and the proportion of all residents who were in favor at the time of the survey after the signed agreement.

Review Test for Chapter 12

1. a. favorite recreation is golf
b. all chief financial officers (CFOs)
c. proportion of all CFOs whose favorite recreation is golf
d. proportion of all sampled CFOs whose favorite recreation is golf

2. a. wash vehicle at least once a month
b. all car owners
c. proportion of all car owners who wash their vehicle at least once a month
d. proportion of all sampled car owners who wash their vehicle at least once a month
e. Sample proportion. *USA TODAY* computed the proportion for a sample of car owners.

3. Generally, obtaining a sample proportion can be done more quickly and is less costly than obtaining the population proportion. Sampling is often the only practical way to proceed.

4. a. the number of members in the sample that have the specified attribute
b. the number of members in the sample that do not have the specified attribute

5. a. population proportion **b.** normal
c. np, $n(1-p)$, 5

6. the precision with which a sample proportion, $\hat{p}$, estimates the population proportion, p, at the specified confidence level

7. a. "holiday blues" **b.** all men, all women
c. proportion of all men who get the "holiday blues" and proportion of all women who get the "holiday blues"
d. proportion of all sampled men who get the "holiday blues" and proportion of all sampled women who get the "holiday blues"
e. Sample proportions. The poll used samples of men and women to obtain the proportions.

8. a. difference between the population proportions
b. normal

9. a. 19,208 **b.** 14,406

10. a. 0.400 to 0.461
b. We can be 95% confident that the percentage of all Americans who thought news organizations were criticizing the Clinton administration unfairly was somewhere between 40.0% and 46.1%.

11. a. 0.031 **b.** 2401 **c.** 0.396 to 0.436
d. 0.02, which is the same as that specified in part (b)
e. 2377; 0.396 to 0.436; 0.02
f. By employing the guess for $\hat{p}$ in part (e), we reduced the required sample size by 24 with (virtually) no sacrifice in precision.

12. 13% to 21%

13. a. H_0: $p = 0.25$, H_a: $p < 0.25$; $\alpha = 0.05$; critical value $= -1.645$; $z = -2.30$; reject H_0; at the 5% significance level, the data provide sufficient evidence to conclude that less than one in four Americans believe that juries "almost always" convict the guilty and free the innocent.
b. H_0: $p = 0.25$, H_a: $p < 0.25$; $\alpha = 0.05$; $z = -2.30$; $P = 0.0107$; reject H_0; at the 5% significance level, the data provide sufficient evidence to conclude that less than one in four Americans believe that juries "almost always" convict the guilty and free the innocent.
c. The data provide strong evidence against the null hypothesis.

14. a. observational study
b. Being observational, the study established only an association between height and breast cancer; no causal relationship can be inferred, although there may be one.

15. a. H_0: $p_1 = p_2$, H_a: $p_1 < p_2$; $\alpha = 0.01$; critical value $= -2.33$; $z = -4.17$; reject H_0; at the 1% significance level, the data provide sufficient evidence to conclude that the percentage of Maricopa County residents who thought Arizona's economy would improve over the next 2 years was less during the time of the first poll than during the time of the second poll.
b. H_0: $p_1 = p_2$, H_a: $p_1 < p_2$; $\alpha = 0.01$; $z = -4.17$; $P = 0.000$ (to four decimal places); reject H_0; at the 1% significance level, the data provide sufficient evidence to conclude that the percentage of Maricopa County residents who thought Arizona's economy would improve over the next 2 years was less during the time of the first poll than during the time of the second poll.
c. The data provide very strong evidence against the null hypothesis.

16. a. −0.187 to −0.053
b. We can be 98% confident that the difference between the percentages of Maricopa County residents who thought Arizona's economy would improve over the next 2 years during the time of the first poll and during the time of the second poll is somewhere between −18.7% and −5.3%.

17. a. 0.067; we can be 98% confident that the error in estimating the difference between the two population proportions, $p_1 - p_2$, by the difference between the two sample proportions, −0.12, is at most 0.067.

b. 0.067 **c.** 3017 **d.** −0.158 to −0.098
e. 0.03, which is the same as that specified in part (c)

21. a. 1006 **b.** 433 **c.** 43.0417%
d. 0.399566 to 0.461677

22. a. 2512 **b.** 578 **c.** 23.0096%
d. H_0: $p = 0.25$, H_a: $p < 0.25$ **e.** 0.011
f. Reject H_0; at the 5% significance level, the data provide sufficient evidence to conclude that less than one in four Americans believe that juries "almost always" convict the guilty and free the innocent.

23. a. $n_1 = 600$; $n_2 = 600$
b. $x_1 = 288$; $x_2 = 360$
c. 48%; 60%
d. −0.186454 to −0.0535462
e. H_0: $p_1 = p_2$, H_a: $p_1 < p_2$ **f.** 0.000
g. Reject H_0; at the 1% significance level, the data provide sufficient evidence to conclude that the percentage of Maricopa County residents who thought Arizona's economy would improve over the next 2 years was less during the time of the first poll than during the time of the second poll.

CHAPTER 13

Exercises 13.1

13.1 Its distribution has the shape of a special type of right-skewed curve, called a chi-square curve.

13.3 The one with 20 degrees of freedom, because as the number of degrees of freedom becomes larger, χ^2-curves look increasingly like normal curves.

13.5
a. 32.852 **b.** 10.117

13.7
a. 18.307 **b.** 3.247

Exercises 13.2

13.11 Because the hypothesis test is carried out by determining how well the observed frequencies fit the expected frequencies.

13.13 The observed frequencies are variables since their values vary from sample to sample. The expected frequencies are not variables, but are fixed numbers computed using the formula $E = np$.

13.15 H_0: Last year's type-of-buyer distribution for U.S. cars is the same as the 1995 distribution. H_a: Last year's type-of-buyer distribution for U.S. cars is different from the 1995 distribution. Assumptions 1 and 2 are satisfied since all expected frequencies are 5 or greater. $\alpha = 0.05$; critical value = 5.991; $\chi^2 = 6.354$; reject H_0; at the 5% significance level, the data provide sufficient evidence to conclude that last year's type-of-buyer distribution for U.S. cars is different from the 1995 distribution. For the P-value approach, note that $0.025 < P < 0.05$.

13.17 H_0: The 1995 distribution of AIDS deaths by race is the same as the 1992 distribution. H_a: The 1995 distribution of AIDS deaths by race is different from the 1992 distribution. Assumptions 1 and 2 are satisfied since all expected frequencies are 5 or greater. $\alpha = 0.10$; critical value = 6.251; $\chi^2 = 3.380$; do not reject H_0; at the 10% significance level, the data do not provide sufficient evidence to conclude that the 1995 distribution of AIDS deaths by race is different from the 1992 distribution. For the P-value approach, note that $P > 0.10$.

13.19 H_0: The die is not loaded. H_a: The die is loaded. Assumptions 1 and 2 are satisfied since all expected frequencies are 5 or greater. $\alpha = 0.05$; critical value = 11.070; $\chi^2 = 2.480$; do not reject H_0; at the 5% significance level, the data do not provide sufficient evidence to conclude that the die is loaded. For the P-value approach, note that $P > 0.10$.

13.25
a. 500 **b.** 4.82575
c. 0.185010 **d.** 0.185010
e. Do not reject H_0; at the 5% significance level, the data do not provide sufficient evidence to conclude that the current specialty distribution has changed from the 1995 distribution.

Exercises 13.3

13.27 Answers will vary.

13.29 cells

13.31 summing the row totals, summing the column totals, or summing the frequencies in the cells

13.33 Yes, because knowing the value of the variable "sex" imparts information about the variable "specialty."

13.35

a.

		College			
		Bus.	Engr.	Lib. Arts	Total
Sex	Male	2	10	3	15
	Female	7	2	1	10
	Total	9	12	4	25

b.

		College			
		Bus.	Engr.	Lib. Arts	Total
Sex	Male	0.222	0.833	0.750	0.600
	Female	0.778	0.167	0.250	0.400
	Total	1.000	1.000	1.000	1.000

c.

		College			
		Bus.	Engr.	Lib. Arts	Total
Sex	Male	0.133	0.667	0.200	1.000
	Female	0.700	0.200	0.100	1.000
	Total	0.360	0.480	0.160	1.000

d. Yes. The tables in parts (b) and (c) show that the conditional distributions of one variable given the other are not identical.

13.37

a.

		Class				
		Fresh.	Soph.	Jr.	Sr.	Total
Party	Republican	3	9	12	6	30
	Democrat	2	6	8	4	20
	Other	1	3	4	2	10
	Total	6	18	24	12	60

b.

		Class			
		Fresh.	Soph.	Jr.	Sr.
Party	Republican	0.500	0.500	0.500	0.500
	Democrat	0.333	0.333	0.333	0.333
	Other	0.167	0.167	0.167	0.167
	Total	1.000	1.000	1.000	1.000

c. No. The table in part (b) shows that the conditional distributions of political party affiliation within class levels are identical.

d. Republican 0.500, Democrat 0.333, Other 0.167, Total 1.000

e. True. From part (c), we know that political party affiliation and class level are not associated. Therefore, the conditional distributions of class level within political party affiliations are identical to each other and to the marginal distribution of class level.

13.39

a. 8

b. The missing entries, from top to bottom and left to right, are: 555, 1035, 1471, 3274

c. 3274 **d.** 863 **e.** 1471 **f.** 502

13.41

a. 73,970 **b.** 15,540 **c.** 12,328 **d.** 77,182

e. 12,225 **f.** 6734 **g.** 77,700

13.43

a.

Region	Public	Private	Total
Northeast	0.324	0.676	1.000
Midwest	0.416	0.584	1.000
South	0.515	0.485	1.000
West	0.564	0.436	1.000

b. Public 0.449, Private 0.551, Total 1.000

c. Yes, because the conditional distributions of type within regions are not identical.

d. 55.1% **e.** 67.6%

f. True, because by part (c), there is an association between the variables "type" and "region."

g.

Region	Public	Private	Total
Northeast	0.181	0.308	0.251
Midwest	0.244	0.280	0.264
South	0.362	0.278	0.316
West	0.213	0.134	0.170
Total	1.000	1.000	1.000

13.45

a.

Specialty	Office	Hospital	Other
General	0.326	0.472	0.445
Ob/Gyn	0.326	0.260	0.306
Orthopedics	0.181	0.164	0.111
Ophthalmology	0.167	0.104	0.139
Total	1.000	1.000	1.000

b. Yes, because the conditional distributions of specialty within base of practice categories are not identical.

c. General 0.367, Ob/Gyn 0.309, Orthopedics 0.174, Ophthalmology 0.150, Total 1.000

d.

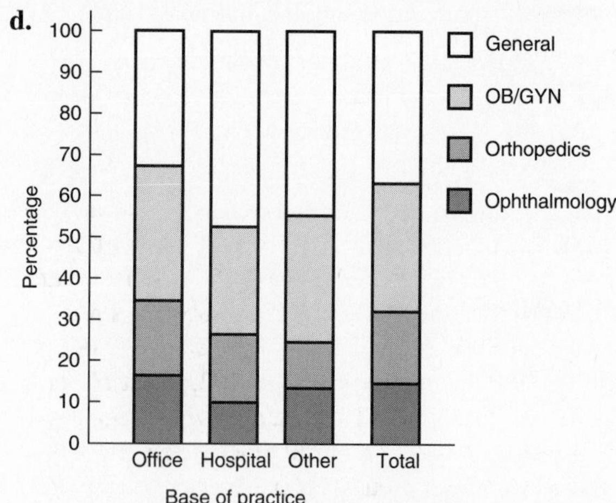

That the bars are not identical reflects the fact that there is an association between specialty and base of practice.

e. False, because by part (b), there is an association between specialty and base of practice.

f.

Specialty	Office	Hospital	Other	Total
General	0.635	0.322	0.044	1.000
Ob/Gyn	0.754	0.210	0.036	1.000
Orthopedics	0.741	0.236	0.023	1.000
Ophthalmology	0.793	0.173	0.033	1.000
Total	0.714	0.250	0.036	1.000

g. 25.0% **h.** 21.0% **i.** 26.0%

Exercises 13.4

13.53 In most cases, data for an entire population are not available. Therefore, we must usually apply inferential methods to decide whether an association exists between two variables.

13.55 If no association exists between marital status and alcohol consumption, then the (conditional) distribution of alcohol consumption for married adults is the same as the (marginal) distribution of alcohol consumption for all adults.

13.57 15

13.59 If there is a causal relationship between two variables, then they are necessarily associated. In other words, if there is no association between two variables, then they could not possibly be causally related.

13.61 H_0: Family income and educational attainment of householder are not associated. H_a: Family income and educational attainment of householder are associated. Assumptions 1 and 2 are satisfied since all expected frequencies are 5 or greater. $\alpha = 0.01$; critical value $= 16.812$; $\chi^2 = 142.102$; reject H_0; at the 1% significance level, the data provide sufficient evidence to conclude that family income and educational attainment of householder are associated. For the P-value approach, note that $P < 0.005$.

13.63 H_0: Response and political affiliation are not associated. H_a: Response and political affiliation are associated. Assumptions 1 and 2 are satisfied since all expected frequencies are 5 or greater. $\alpha = 0.05$; critical value $= 9.488$; $\chi^2 = 4.017$; do not reject H_0; at the 5% significance level, the data do not provide sufficient evidence to conclude that the feelings of U.S. adults on the issue of regional primaries are associated with political affiliation. For the P-value approach, note that $P > 0.10$.

13.65 H_0: Net worth and marital status for top wealthholders are statistically independent. H_a: Net worth and marital status for top wealthholders are statistically dependent. Assumption 2 is violated since 25% (3/12) of the expected frequencies are less than 5.

13.71

a. 34 **b.** 164 **c.** 9 **d.** 13.21
e. 1.343 **f.** 6
g. H_0: Grade and study time for intermediate algebra students at ASU are not associated. H_a: Grade and study time for intermediate algebra students at ASU are associated.
h. 10.574 **i.** 0.104
j. At the 5% level, the data do not provide sufficient evidence to conclude that grade and study time for intermediate algebra students at ASU are associated.

Review Test for Chapter 13

1. by their degrees of freedom

2. a. 0 **b.** right skewed **c.** normal curve

3. a. No. The degrees of freedom for the chi-square goodness-of-fit test depends on the number of possible values for the variable under consideration, not the sample size.
b. No. The degrees of freedom for the chi-square independence test depends on the number of possible values for the two variables under consideration, not the sample size.

4. For both tests, the null hypothesis is rejected only when the observed and expected frequencies match up poorly, which corresponds to large values of the chi-square test statistic. Thus both tests are always right-tailed.

5. 0

6. a. (1) All expected frequencies are 1 or greater. (2) At most 20% of the expected frequencies are less than 5.
b. They are very important. If the assumptions are not met, the results could be invalid.

7. a. 5.3% **b.** roughly 3.1 million
c. There is an association between means of transportation to work and region of residence.

8. a. Obtain the conditional distribution of one of the variables for each possible value of the other variable. If all of the conditional distributions are identical, there is no association between the variables; otherwise, there is an association.
b. No. Since the data are for an entire population, we are not making an inference from a sample to the population. The conclusion is a fact.

9. a. Perform a chi-square independence test.
b. Yes. As in any inference, it is always possible that the conclusion is in error.

10. a. 6.408 **b.** 33.409 **c.** 27.587
d. 8.672 **e.** 7.564, 30.191

11. **a.** H_0: This year's distribution of educational attainment is the same as the 1990 distribution. H_a: This year's distribution of educational attainment differs from the 1990 distribution. Assumptions 1 and 2 are satisfied since all expected frequencies are 5 or greater. $\alpha = 0.05$; critical value = 11.070; $\chi^2 = 14.859$; reject H_0; at the 5% significance level, the data provide sufficient evidence to conclude that this year's distribution of educational attainment differs from the 1990 distribution.
b. $0.01 < P < 0.025$; strong
c. No, since by part (b), $P > 0.01$.

12. **a.**

Party of governor

Region	Rep	Dem	Ind	Total
Northeast	6	2	1	9
Midwest	9	3	0	12
South	10	6	0	16
West	7	6	0	13
Total	32	17	1	50

b.

Party of governor

Region	Rep	Dem	Ind	Total
Northeast	0.188	0.118	1.000	0.180
Midwest	0.281	0.176	0.000	0.240
South	0.313	0.353	0.000	0.320
West	0.219	0.353	0.000	0.260
Total	1.000	1.000	1.000	1.000

c.

Party of governor

Region	Rep	Dem	Ind	Total
Northeast	0.667	0.222	0.111	1.000
Midwest	0.750	0.250	0.000	1.000
South	0.625	0.375	0.000	1.000
West	0.538	0.462	0.000	1.000
Total	0.640	0.340	0.020	1.000

d. Yes, because the conditional distributions of party of governor within regions are not identical.
e. 64.0% **f.** 64.0% **g.** 75.0%
h. 24.0% **i.** 24.0% **j.** 28.1%

13. **a.** 2046 **b.** 737 **c.** 266
d. 3046 **e.** 5413 **f.** 5910

14. **a.**

Control

Facility	Gov	Prop	NP	Total
General	0.314	0.122	0.564	1.000
Psychiatric	0.361	0.486	0.153	1.000
Chronic	0.808	0.038	0.154	1.000
Tuberculosis	0.750	0.000	0.250	1.000
Other	0.144	0.361	0.495	1.000

b. Yes, because the conditional distributions of control type within facility type are not identical.
c. Gov 0.311, Prop 0.177, NP 0.512, Total 1.000
d. See Fig. A.10. That the bars are not identical reflects the fact that there is an association between facility type and control type.
e. False, since by part (b) there is an association between facility type and control type.

FIGURE A.10 Segmented bar graph for Problem 14(d)

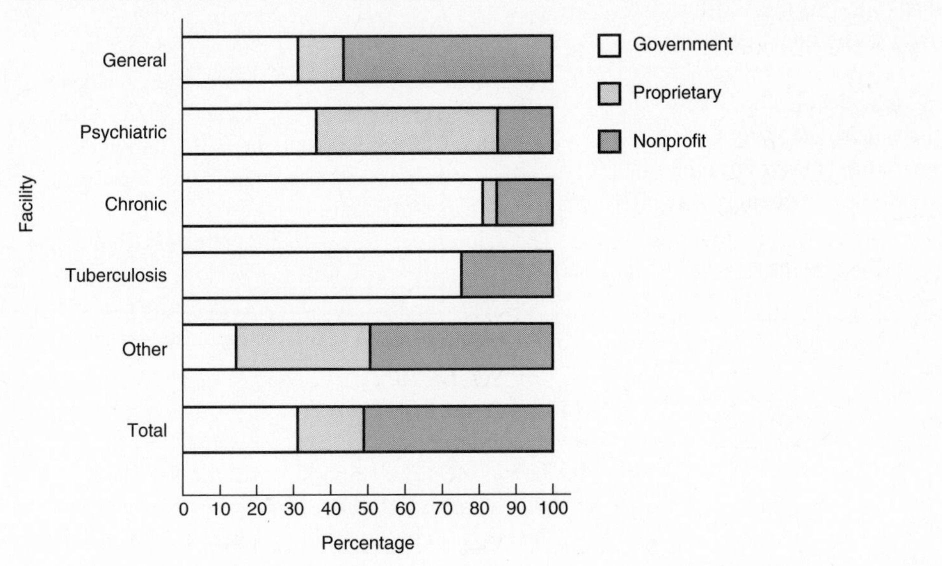

f.

Control (columns); Facility (rows)

Facility	Gov	Prop	NP	Total
General	0.829	0.566	0.905	0.821
Psychiatric	0.130	0.307	0.034	0.112
Chronic	0.010	0.001	0.001	0.004
Tuberculosis	0.001	0.000	0.000	0.001
Other	0.029	0.127	0.060	0.062
Total	1.000	1.000	1.000	1.000

g. 17.7% **g.** 48.6% **i.** 30.7%

15. H_0: Response and educational level are not associated. H_a: Response and educational level are associated. Assumptions 1 and 2 are satisfied since all expected frequencies are 5 or greater. $\alpha = 0.01$; critical value $= 16.812$; $\chi^2 = 77.837$; reject H_0; at the 1% significance level, the data provide sufficient evidence to conclude that response and educational level are associated. For the P-value approach, note that $P < 0.005$.

17. **a.** 500 **b.** 14.8586
c. 0.0109841 **d.** 0.0109841
e. Reject H_0; at the 5% significance level, the data provide sufficient evidence to conclude that this year's distribution of educational attainment differs from the 1990 distribution.

20. **a.** 205 **b.** 238 **c.** 26 **d.** 31.93
e. 1.102 **f.** 6 **g.** 77.837 **h.** 0.000
i. Reject H_0; at the 1% significance level, the data provide sufficient evidence to conclude that response and educational level are associated.

CHAPTER 14

Exercises 14.1

14.1

a. $y = b_0 + b_1x$
b. b_0 and b_1 are constants; x and y are variables
c. x is the independent variable; y is the dependent variable

14.3

a. The number b_0 is the y-intercept. It is the y-value at which the straight-line graph of the linear equation intersects the y-axis.

b. The number b_1 is the slope. It measures the steepness of the straight line; more precisely, b_1 indicates how much the y-value on the straight line changes (increases or decreases) when the x-value increases by 1 unit.

14.5

a. $y = 45.90 + 0.25x$

b. $b_0 = 45.90$, $b_1 = 0.25$

c.

x	50	100	250
y	58.4	70.9	108.4

d.

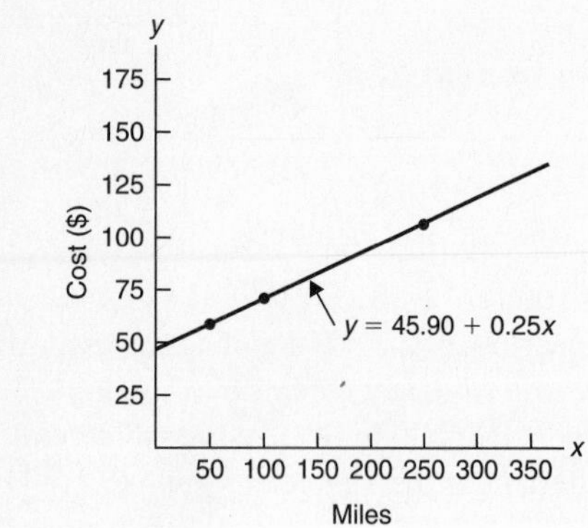

e. About \$85; exact cost is \$83.40.

14.7

a. $b_0 = 32$, $b_1 = 1.8$ **b.** $-40, 32, 68, 212$

c.

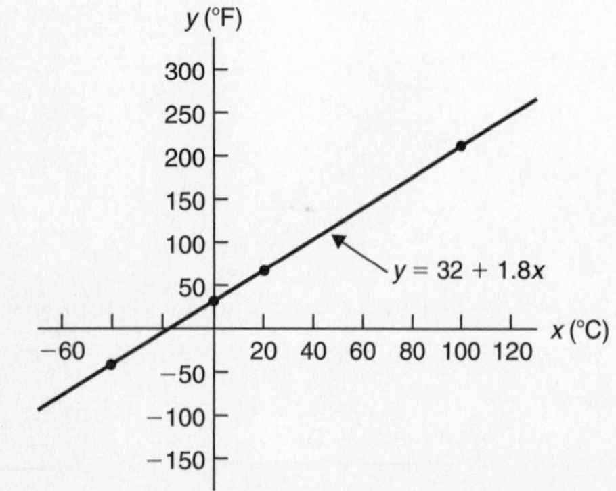

d. About 80° F; exact temperature is 82.4° F.

14.9

a. $b_0 = 45.90$, $b_1 = 0.25$

b. The y-intercept, $b_0 = 45.90$, gives the y-value at which the straight line, $y = 45.90 + 0.25x$, intersects the y-axis. The slope, $b_1 = 0.25$, indicates that the y-value increases by 0.25 unit for every increase in x of 1 unit.

c. The y-intercept, $b_0 = 45.90$, is the cost (in dollars) for driving the car 0 miles. The slope, $b_1 = 0.25$, represents the fact that the cost per mile is \$0.25; it is the amount the total cost increases for each additional mile driven.

14.11

a. $b_0 = 32$, $b_1 = 1.8$

b. The y-intercept, $b_0 = 32$, gives the y-value at which the straight line, $y = 32 + 1.8x$, intersects the y-axis. The slope, $b_1 = 1.8$, indicates that the y-value increases by 1.8 units for every increase in x of 1 unit.

c. The y-intercept, $b_0 = 32$, is the Fahrenheit temperature corresponding to 0° C. The slope, $b_1 = 1.8$, represents the fact that the Fahrenheit temperature increases by 1.8° for every increase of the Celsius temperature of 1°.

14.13

a. $b_0 = 3$, $b_1 = 4$ **b.** slopes upward

c.

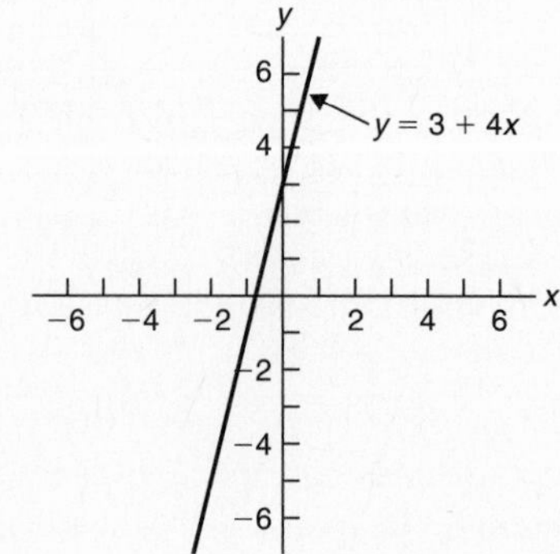

14.15

a. $b_0 = 6$, $b_1 = -7$ **b.** slopes downward

c.

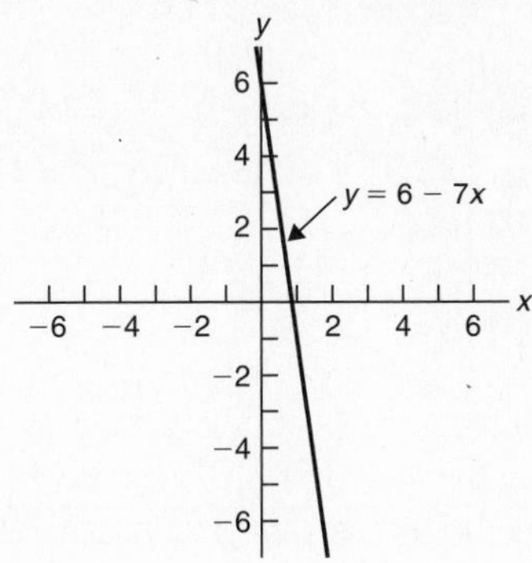

14.17

a. $b_0 = -2, b_1 = 0.5$ **b.** slopes upward

14.19

a. $b_0 = 2, b_1 = 0$ **b.** horizontal

14.21

a. $b_0 = 0, b_1 = 1.5$ **b.** slopes upward

14.23

a. slopes upward **b.** $y = 5 + 2x$

c.

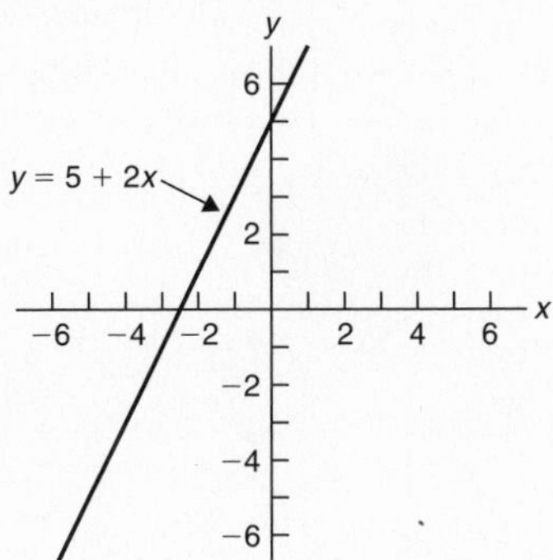

14.25

a. slopes downward **b.** $y = -2 - 3x$

c.

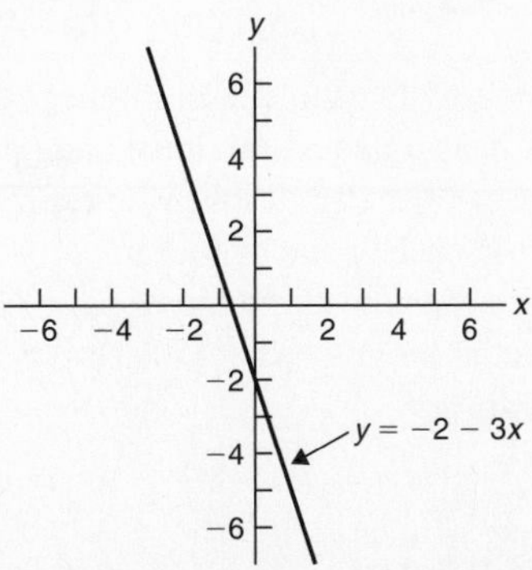

14.27

a. slopes downward **b.** $y = -0.5x$

14.29

a. horizontal **b.** $y = 3$

Exercises 14.2

14.33

a. least-squares criterion

b. The straight line that best fits a set of data points is the one having the smallest possible sum of squared errors.

14.35

a. response variable

b. predictor variable or explanatory variable

14.37

a. outlier

b. influential observation

14.39

a.

Line *A*: $y = 3 - 0.6x$

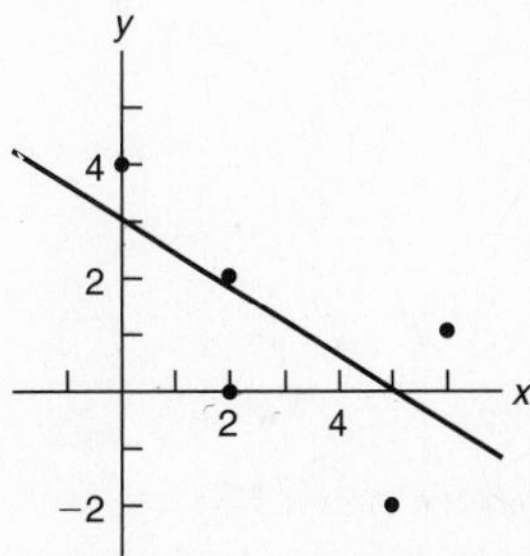

Line *B*: $y = 4 - x$

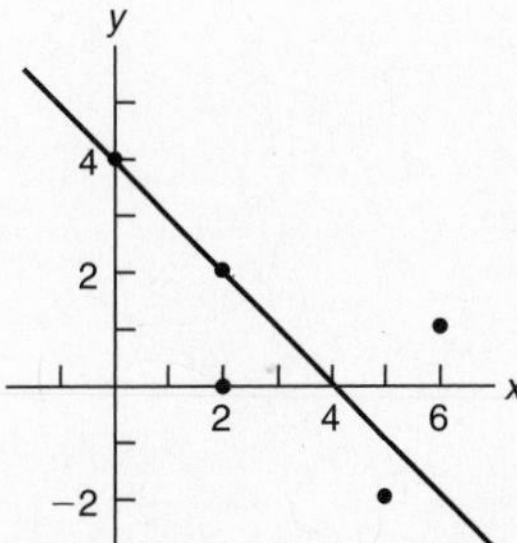

b. Line *A*: $y = 3 - 0.6x$

x	y	$\hat{y}$	e	e^2
0	4	3.0	1.0	1.00
2	2	1.8	0.2	0.04
2	0	1.8	−1.8	3.24
5	−2	0.0	−2.0	4.00
6	1	−0.6	1.6	2.56
				10.84

Line *B*: $y = 4 - x$

x	y	$\hat{y}$	e	e^2
0	4	4	0	0
2	2	2	0	0
2	0	2	−2	4
5	−2	−1	−1	1
6	1	−2	3	9
				14

c. Line *A*

14.41

a. $\hat{y} = 2.875 - 0.625x$

b.

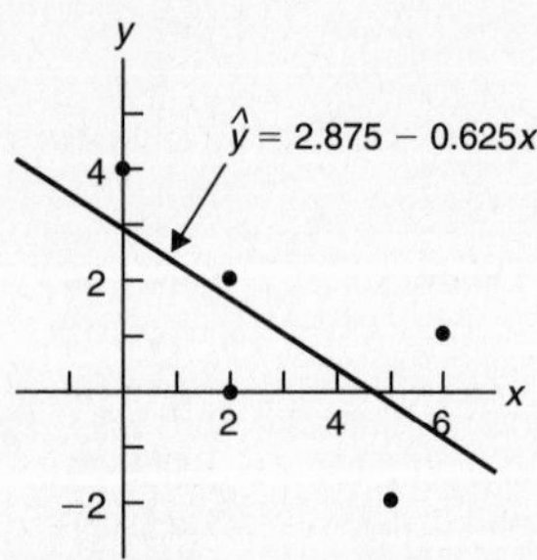

14.43

a. $\hat{y} = -174.49 + 4.84x$

b.

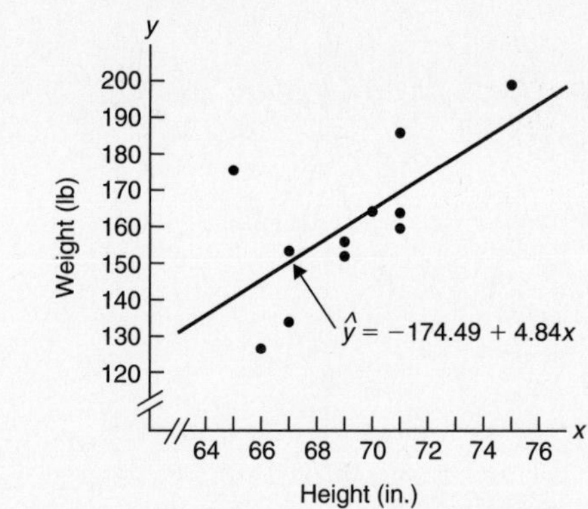

c. Weight tends to increase as height increases.

d. The weights of 18–24-year-old males increase an estimated 4.84 pounds for each increase in height of 1 inch.

e. 149.54 lb; 178.56 lb (see the note at the top of page A-31)

f. The predictor variable is height; the response variable is weight.

g. (65, 175) might be considered an outlier; (75, 198) is a potential influential observation.

14.45

a. $\hat{y} = 222.27 - 1.14x$

b.

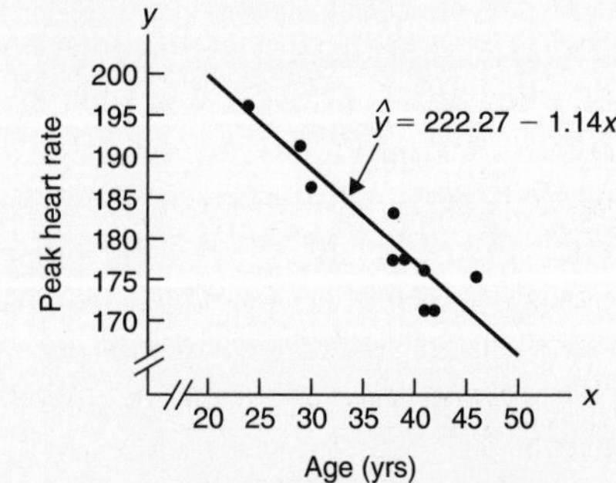

c. Peak heart rate tends to decrease as age increases.

d. The peak heart rate an individual can reach during intensive exercise decreases by an estimated 1.14 for each increase in age of 1 year.

e. 190.34 (see the note at the top of page A-31)

f. The predictor variable is age; the response variable is peak heart rate.

g. none

14.47

a. $\hat{y} = 12.86 + 1.21x$

b.

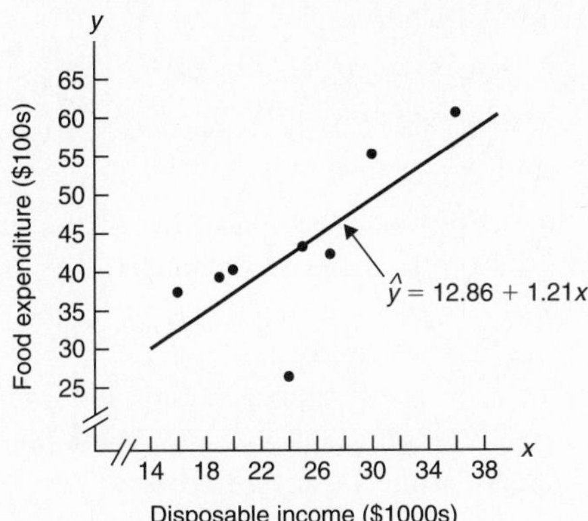

c. Annual food expenditure tends to increase as disposable income increases.
d. Annual food expenditure increases an estimated \$121 (1.21 hundred dollars) for each increase in disposable income of \$1000.
e. \$4321 (see the note at the top of page A-31)
f. The predictor variable is disposable income; the response variable is annual food expenditure.
g. (24, 26) appears to be an outlier; no potential influential observations.

14.49 Only the second one.

14.51
a. It is acceptable to use the regression equation to predict the weight of an 18–24-year-old male who is 68 inches tall since that height lies within the range of the heights in the sample data. It is not acceptable (and would be extrapolation) to use the regression equation to predict the weight of an 18–24-year-old male who is 60 inches tall since that height lies outside the range of the heights in the sample data.
b. Heights between 65 and 75 inches, inclusive.

14.59
a. $\hat{y} = 2152 + 60.8x$ **b.** 3185.6 g
c. (24, 2800) is a potential outlier; no potential influential observations.

Exercises 14.3

14.61
a. the coefficient of determination, r^2
b. the proportion of variation in the observed values of the response variable that is explained by the regression

14.63
a. $r^2 = 0.920$; 92.0% of the variation in the observed values of the response variable is explained by the regression. The fact that r^2 is near 1 indicates that the regression equation is extremely useful for making predictions.
b. 664.4

14.65
a. $SST = 20$, $SSR = 9.375$, $SSE = 10.625$
b. $20 = 9.375 + 10.625$ **c.** $r^2 = 0.469$
d. 46.9%
e. moderately useful

14.67
a. $SST = 4352.91$, $SSR = 1909.46$, $SSE = 2443.45$
b. 0.439
c. 43.9%; 43.9% of the variation in the weight data is explained by height.
d. moderately useful

14.69
a. $SST = 642.10$, $SSR = 553.60$, $SSE = 88.50$
b. 0.862
c. 86.2%; 86.2% of the variation in the peak-heart-rate data is explained by age.
d. extremely useful

14.71
a. $SST = 783.50$, $SSR = 429.95$, $SSE = 353.55$
b. 0.549
c. 54.9%; 54.9% of the variation in the data on food expenditure is explained by disposable income.
d. moderately useful

14.77
a. 0.372
b. $SSR = 2{,}505{,}745$, $SSE = 4{,}234{,}255$, $SST = 6{,}740{,}000$
c. 37.2%

Exercises 14.4

14.79 to provide a descriptive measure of the strength of the linear relationship between two variables

14.81

a. r **b.** strong **c.** 0

14.83

a. positively **b.** negatively **c.** uncorrelated

14.85

a. Positive. The sign of r is the same as the sign of the slope of the regression line.
b. $r^2 = 0.716$

14.87 $r = -0.685$

14.89

a. $r = 0.662$
b. Suggests a moderately strong positive linear correlation.
c. Data points are clustered moderately closely about the regression line.
d. $r^2 = 0.438$ (*Note:* In Exercise 14.67(b), we found that $r^2 = 0.439$. The discrepancy is due to the error resulting from rounding r to three decimal places before squaring.)

14.91

a. $r = -0.929$
b. Suggest an extremely strong negative linear correlation.
c. Data points are clustered extremely closely about the regression line.
d. $r^2 = 0.863$ (*Note:* In Exercise 14.69(b), we found that $r^2 = 0.862$. The discrepancy is due to the error resulting from rounding r to three decimal places before squaring.)

14.93

a. $r = 0.741$
b. Suggests a moderately strong positive linear correlation.
c. Data points are clustered moderately closely about the regression line.
d. $r^2 = 0.549$

14.95

a. $r = 0$
b. No, only that there is no *linear* relationship between the variables.
d. No, because the data points are not scattered about a straight line.
e. For each data point (x, y), we have $y = x^2$.

Review Test for Chapter 14

1. a. x **b.** y **c.** b_1 **d.** b_0

2. a. $y = 4$ **b.** $x = 0$ **c.** -3
d. -3 units **e.** 6 units

3. a. True. The y-intercept only indicates where the line crosses the y-axis, that is, the y-value when $x = 0$.
b. False. The slope is 0.
c. True. This is the same as saying: If a line has a positive slope, then y-values on the line increase as the x-values increase.

4. scatter diagram (or scatterplot)

5. A regression equation can be used to make predictions for values of the predictor variable within the range of the observed values of the predictor variable.

6. a. predictor variable or explanatory variable
b. response variable

7. a. smallest **b.** regression **c.** extrapolation

8. a. An outlier is a data point that lies far from the regression line, relative to the other data points.
b. An influential observation is a data point whose removal causes the regression equation (and line) to change considerably.

9. It is a descriptive measure of the utility of the regression equation for making predictions.

10. See Definition 14.4 on page 849.

11. a. linear **b.** increases
c. negative **d.** 0

12. true

13. a. $y = 72 - 12x$ **b.** $b_0 = 72, b_1 = -12$
c. The line slopes downward since $b_1 < 0$.
d. \$4800; \$1200

e.

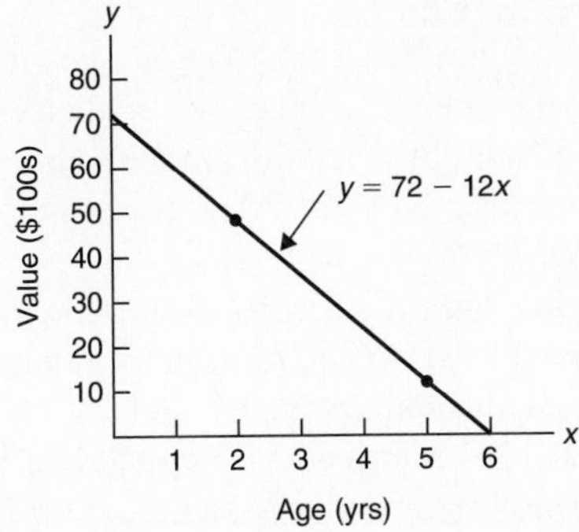

f. About $2500; exact value is $2400.

14. a.

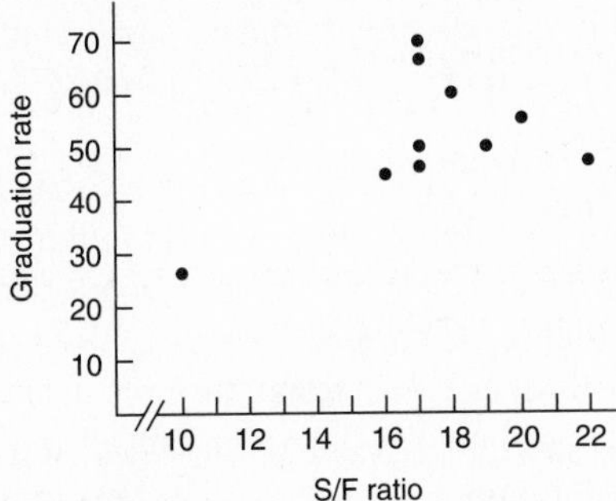

b. It is reasonable to find a regression line for the data because the data points appear to be scattered about a straight line.

c. $\hat{y} = 16.4 + 2.03x$

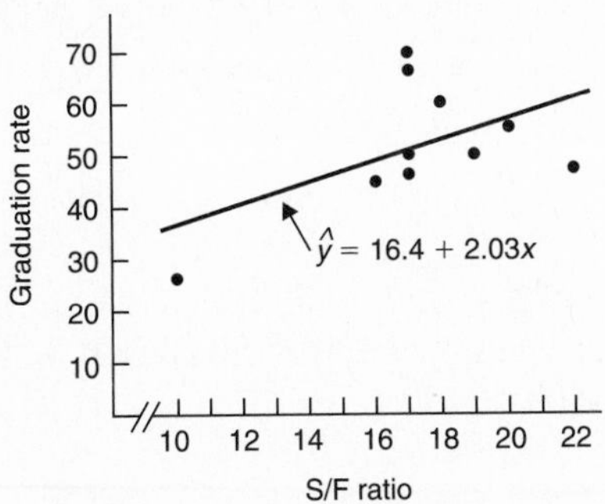

d. As the student-to-faculty ratio increases, the graduation rate tends to increase.

e. Graduation rate increases by an estimated 2.03 percentage points for each increase of 1 of the student-to-faculty ratio.

f. 50.9%

g. There are no outliers. The data point (10, 26) is a potential influential observation.

15. a. $SST = 1384.50$, $SSR = 361.66$, $SSE = 1022.84$
b. $r^2 = 0.261$ **c.** 26.1%
d. not very useful

16. a. $r = 0.511$
b. suggests a weak positive linear correlation
c. Data points are rather widely scattered about the regression line.
d. $r^2 = (0.511)^2 = 0.261$

18. a. $\hat{y} = 16.4 + 2.03x$ **b.** $r^2 = 0.261$
c. $SSR = 361.7$, $SSE = 1022.8$, $SST = 1384.5$
d. no outliers; potential influential observation at (10, 26)

CHAPTER 15

Exercises 15.1

15.1 conditional distribution, conditional mean, conditional standard deviation

15.3
a. population regression line **b.** σ
c. normal, $\beta_0 + 6\beta_1$, σ

15.5 the sample regression line, $\hat{y} = b_0 + b_1x$

15.7 residual

15.9 Residual plot. A residual plot makes it easier to spot patterns such as curvature and nonconstant standard deviation.

15.11 There are constants, β_0, β_1, and σ, such that for each height, x, the weights of all 18–24-year-old males of that height are normally distributed with mean $\beta_0 + \beta_1x$ and standard deviation σ.

15.13 There are constants, β_0, β_1, and σ, such that for each age, x, the peak heart rates that can be reached during intensive exercise by all individuals of that age are normally distributed with mean $\beta_0 + \beta_1x$ and standard deviation σ.

15.15 There are constants, β_0, β_1, and σ, such that for each disposable-income level, x, the annual food expenditures made by all middle-income families (father, mother, two children) at that level are normally distributed with mean $\beta_0 + \beta_1x$ and standard deviation σ.

15.17

a. $s_e = 16.48$ lb; very roughly speaking we can say that, on the average, the predicted weight of an 18–24-year-old male in the sample differs from the observed weight by about 16.48 lb.

b. Presuming that the variables height (x) and weight (y) for 18–24-year-old males satisfy the assumptions for regression inferences, the standard error of the estimate, $s_e = 16.48$ lb, provides an estimate for the common population standard deviation, σ, of weights for all 18–24-year-old males of any particular height.

c. See Fig. A.11. **d.** It appears reasonable.

15.19

a. $s_e = 3.33$; very roughly speaking we can say that, on the average, the predicted peak heart rate of a person in the sample differs from the observed peak heart rate by about 3.33.

b. Presuming that the variables age (x) and peak heart rate (y) for individuals satisfy the assumptions for regression inferences, the standard error of the estimate, $s_e = 3.33$, provides an estimate for the common population standard deviation, σ, of peak heart rates for all individuals of any particular age.

c. See Fig. A.12.

d. It appears reasonable, although the first residual plot in part (c) casts some doubt on the assumption of equal standard deviations (Assumption 2).

15.21

a. $s_e = \$7.68$ hundred; very roughly speaking we can say that, on the average, the predicted annual food expenditure of a family in the sample differs from the observed annual food expenditure by about $768.

b. Presuming that the variables disposable income (x) and annual food expenditure (y) for middle-income families with a father, mother, and two children satisfy the assumptions for regression inferences, the standard error of the estimate, $s_e = 7.68$ ($768), provides an estimate for the common population standard deviation, σ, of annual food expenditures for all middle-income families (father, mother, two children) with any particular disposable income.

c. See Fig. A.13.

d. If the outlier, (24, 26), is a legitimate data point, then the assumptions for regression inferences may be violated by the variables under consideration; if the outlier is a recording error or can be removed for some other valid reason, then the resulting data points reveal no obvious violations of the assumptions for regression inferences (as we can see by constructing a residual plot and normal probability plot of the residuals for the abridged data).

FIGURE A.11 (a) Residual plot and (b) normal probability plot of residuals for Exercise 15.17(c)

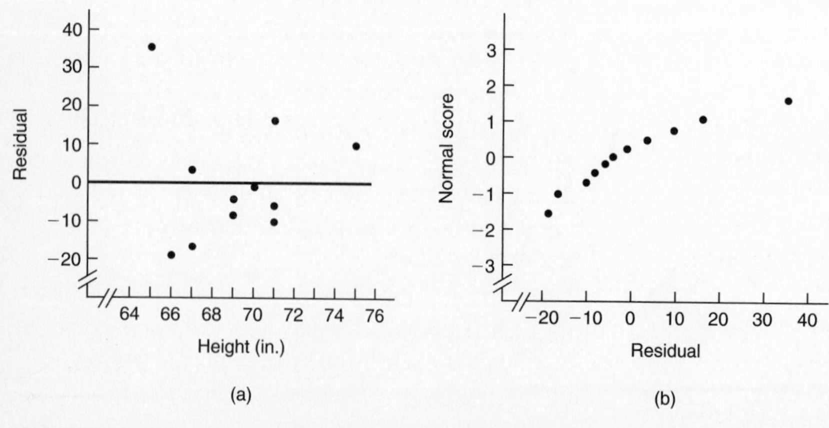

FIGURE A.12 (a) Residual plot and (b) normal probability plot of residuals for Exercise 15.19(c)

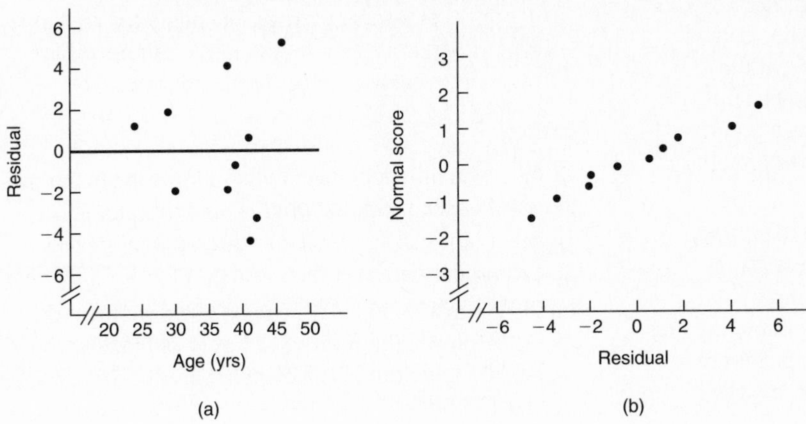

FIGURE A.13 (a) Residual plot and (b) normal probability plot of residuals for Exercise 15.21(c)

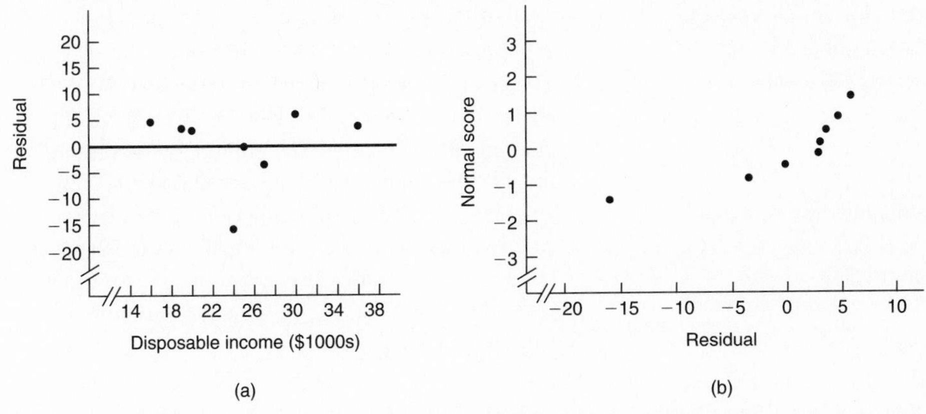

15.23 Part (a) is a tough call, but the assumption of linearity (Assumption 1) may be violated, as may the assumption of equal standard deviations (Assumption 2). In part (b) it appears that the assumption of equal standard deviations is violated. In part (d) it appears that the normality assumption (Assumption 3) is violated.

15.27

a. $s_e = 382.1$ g

b. No. The residual plot falls roughly in a horizontal band centered and symmetric about the x-axis, and the normal probability plot of the residuals is roughly linear.

Exercises 15.2

15.29 normal, −3.5

15.31 r^2, r

15.33 H_0: $\beta_1 = 0$, H_a: $\beta_1 \neq 0$; $\alpha = 0.10$; critical values $= \pm 1.833$; $t = 2.652$; reject H_0; at the 10% significance level, the data provide sufficient evidence to conclude that the slope of the population regression line is not 0 and hence that height is useful as a predictor of weight for 18–24-year-old males. For the P-value approach, note that $0.02 < P < 0.05$.

15.35 H_0: $\beta_1 = 0$, H_a: $\beta_1 \neq 0$; $\alpha = 0.05$; critical values $= \pm 2.306$; $t = -7.074$; reject H_0; at the 5% significance level, the data provide sufficient evidence to conclude that age is useful as a predictor of peak heart rate. For the P-value approach, note that $P < 0.01$.

15.37 H_0: $\beta_1 = 0$, H_a: $\beta_1 \neq 0$; $\alpha = 0.01$; critical values $= \pm 3.707$; $t = 2.701$; do not reject H_0; at the 1% significance level, the data do not provide sufficient evidence to conclude that disposable income is useful as a predictor of annual food expenditure for middle-income families with a father, mother, and two children. For the P-value approach, note that $0.02 < P < 0.05$.

15.39

a. 1.49 to 8.18

b. We can be 90% confident that for 18–24-year-old males, the increase in mean weight per 1-inch increase in height is somewhere between 1.49 and 8.18 lb.

15.41

a. −1.51 to −0.77

b. We can be 95% confident that the drop in mean peak heart rate per 1-year increase in age is somewhere between 0.77 and 1.51.

15.43

a. −0.45 to 2.88

b. We can be 99% confident that for middle-income families with a father, mother, and two children, the change in mean annual food expenditure per \$1000 increase in family disposable income is somewhere between −\$45 and \$288.

15.47

a. 60.82 **b.** 14.68 **c.** 4.14

d. 0.000 (to three decimal places)

e. At the 5% significance level, the data provide sufficient evidence to conclude that estriol level is useful for predicting birth weight.

f. 30.80 to 90.84. We can be 95% confident that the increase in mean birth weight per increase of estriol level by 1 mg/24 hr is somewhere between 30.80 and 90.84 g.

g. The assumptions for regression inferences, Assumptions 1–4 as indicated in Key Fact 15.1. By performing a residual analysis as described in Key Fact 15.2.

Exercises 15.3

15.49 \$11,443. A point estimate for the mean price is the same as the predicted price.

15.51

a. 164.05 lb

b. 154.54 to 173.56 lb. We can be 90% confident that the mean weight of all 18–24-year-old males who are 70 inches tall is somewhere between 154.54 and 173.56 lb.

c. 164.05 lb

d. 132.38 to 195.71 lb. We can be 90% certain that the observed weight of an 18–24-year-old male who is 70 inches tall will be somewhere between 132.38 and 195.71 lb.

e. See Fig. A.14.

f. The error in the estimate of the mean weight of all 18–24-year-old males who are 70 inches tall is due only to the fact that the population regression line is being estimated by a sample regression line, whereas the error in the prediction of the observed weight of an 18–24-year-old male who is 70 inches tall is due to the previously mentioned error in estimating the mean weight plus the variation in weights of such males.

15.53

a. 176.65

b. 173.95 to 179.35. We can be 95% confident that the mean peak heart rate of all 40-year-olds is somewhere between 173.95 and 179.35.

c. 176.65

d. 168.52 to 184.78. We can be 95% certain that the observed peak heart rate of a 40-year-old will be somewhere between 168.52 and 184.78.

15.55

a. 33.13 to 53.29. We can be 99% confident that the mean annual food expenditure of all middle-income families consisting of a father, mother, and two children that have a disposable income of \$25,000 is somewhere between \$3313 and \$5329.

FIGURE A.14 90% confidence and prediction intervals for Exercise 15.51(e)

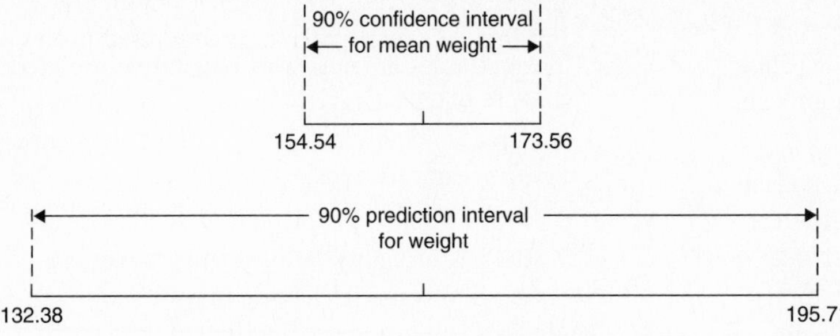

b. 13.02 to 73.39. We can be 99% certain that the observed annual food expenditure of a middle-income family consisting of a father, mother, and two children that has a disposable income of \$25,000 will be somewhere between \$1302 and \$7339.

15.59

a. 3064.6 g **b.** 2909.2 to 3220.1 g
c. 2267.8 to 3861.4 g

Exercises 15.4

15.61 r

15.63

a. uncorrelated **b.** increases **c.** negatively

15.65 H_0: $\rho = 0$, H_a: $\rho > 0$; $\alpha = 0.05$; critical value $= 1.833$; $t = 2.652$ (see the note at the top of page A-31); reject H_0; at the 5% significance level, the data provide sufficient evidence to conclude that the variables height and weight are positively linearly correlated for 18–24-year-old males. For the P-value approach, note that $0.01 < P < 0.025$.

15.67 H_0: $\rho = 0$, H_a: $\rho < 0$; $\alpha = 0.025$; critical value $= -2.306$; $t = -7.074$ (see the note at the top of page A-31); reject H_0; at the 2.5% significance level, the data provide sufficient evidence to conclude that age and peak heart rate are negatively linearly correlated. For the P-value approach, note that $P < 0.005$.

15.69 H_0: $\rho = 0$, H_a: $\rho \neq 0$; $\alpha = 0.01$; critical values $= \pm 3.707$; $t = 2.701$ (see the note at the top of page A-31); do not reject H_0; at the 1% significance level, the data do not provide sufficient evidence to conclude that family disposable income and annual food expenditure are linearly correlated for middle-income families with a father, mother, and two children. For the P-value approach, note that $0.02 < P < 0.05$.

Exercises 15.5

15.73

a. a plot of the normal scores against the sample data
b. assessing normality of a variable from sample data
c. If the plot is roughly linear, then accept as reasonable that the variable is normally distributed. If the plot shows systematic deviations from linearity, then conclude that the variable is probably not normally distributed.
d. What constitutes roughly linear is a matter of opinion.

15.75 If the variable under consideration is normally distributed, then the normal probability plot should be roughly linear, which means that the correlation between the sample data and its normal scores should be close to 1. Because the correlation can be at most 1, evidence against the null hypothesis of normality is provided if the correlation is "too much smaller than 1." Thus a correlation test for normality is always left-tailed.

15.77 H_0: Final-exam scores in the introductory statistics class are normally distributed. H_a: Final-exam scores in the introductory statistics class are not normally distributed. $\alpha = 0.05$; critical value $= 0.951$; $R_p = 0.939$; reject H_0; at the 5% significance level, the data provide sufficient evidence to conclude that the final-exam scores in the introductory statistics class are not normally distributed. For the P-value approach, note that $0.01 < P < 0.05$.

15.79 H_0: The number of miles cars were driven last year is normally distributed. H_a: The number of miles cars were driven last year is not normally distributed. $\alpha = 0.10$; critical value $= 0.951$; $R_p = 0.989$; do not reject H_0; at the 10% significance level, the data do not provide sufficient evidence to conclude that the number of miles cars were driven last year is not normally distributed. For the P-value approach, note that $P > 0.10$.

Review Test for Chapter 15

1. a. conditional
b. See Key Fact 15.1 on page 873.

2. a. b_1 **b.** b_0 **c.** s_e

3. Residual plot and normal probability plot. A plot of the residuals against the values of the predictor variable should fall roughly in a horizontal band centered and symmetric about the x-axis. A normal probability plot of the residuals should be roughly linear.

4. a. Assumption 1 **b.** Assumption 2
c. Assumption 3 **d.** Assumption 3

5. The regression equation is useful for making predictions.

6. b_1, r, r^2

7. no

8. The term *confidence* is usually reserved for interval estimates of parameters, whereas the term *prediction* is used for interval estimates of variables.

9. ρ

10. a. The variables are positively linearly correlated, meaning that y tends to increase linearly as x increases (and vice versa), with the tendency being greater the closer ρ is to 1.

b. The variables are linearly uncorrelated, meaning that there is no linear relationship between the variables.

c. The variables are negatively linearly correlated, meaning that y tends to decrease linearly as x increases (and vice versa), with the tendency being greater the closer ρ is to -1.

11. There are constants, β_0, β_1, and σ, such that for each student-to-faculty ratio, x, the graduation rates for all universities having that student-to-faculty ratio are normally distributed with mean $\beta_0 + \beta_1 x$ and standard deviation σ.

12. a. $\hat{y} = 16.4 + 2.03x$

b. $s_e = 11.31\%$; very roughly speaking we can say that, on the average, the predicted graduation rate for a university in the sample differs from the observed graduation rate by about 11.31 percentage points.

c. Presuming that the variables student-to-faculty ratio (x) and graduation rate (y) for universities satisfy the assumptions for regression inferences, the standard error of the estimate, $s_e = 11.31\%$, provides an estimate for the common population standard deviation, σ, of graduation rates for all universities having any particular student-to-faculty ratio.

13.

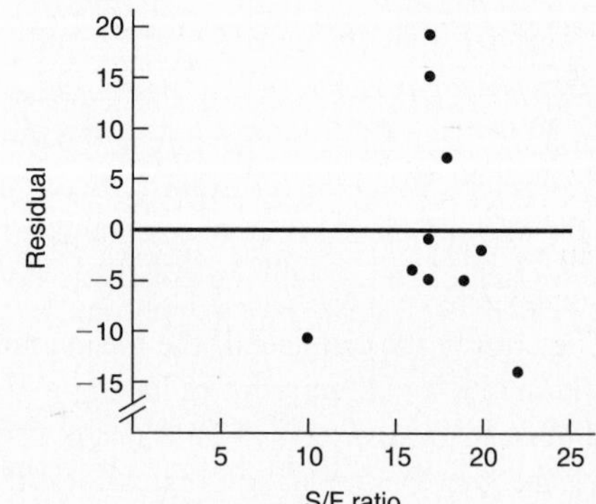

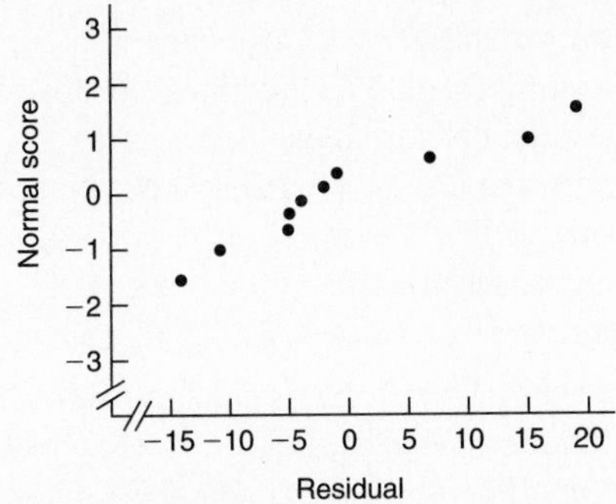

It appears reasonable.

14. **a.** H_0: $\beta_1 = 0$, H_a: $\beta_1 \neq 0$; $\alpha = 0.05$; critical values $= \pm 2.306$; $t = 1.682$; do not reject H_0; at the 5% significance level, the data do not provide sufficient evidence to conclude that, for universities, student-to-faculty ratio is useful as a predictor of graduation rate. For the P-value approach, note that $0.10 < P < 0.20$.

b. −0.75 to 4.80. We can be 95% confident that, for universities, the increase in mean graduation rate per increase in the student-to-faculty ratio by 1 is somewhere between −0.75 and 4.80 percentage points.

15. **a.** 50.9%

b. 42.6% to 59.2%. We can be 95% confident that the mean graduation rate for all universities having a student-to-faculty ratio of 17 is somewhere between 42.6% and 59.2%.

c. 50.9%

d. 23.5% to 78.3%. We can be 95% certain that the observed graduation rate for a university having a student-to-faculty ratio of 17 will be somewhere between 23.5% and 78.3%.

e. The error in the estimate of the mean graduation rate for all universities having a student-to-faculty ratio of 17 is due only to the fact that the population regression line is being estimated by a sample regression line, whereas the error in the prediction of the observed graduation rate for a university having a student-to-faculty ratio of 17 is due to the previously mentioned error in estimating the mean graduation rate plus the variation in graduation rates for such universities.

16. H_0: $\rho = 0$, H_a: $\rho > 0$; $\alpha = 0.025$; critical value $= 2.306$; $t = 1.682$ (see the note at the top of page A-31); do not reject H_0; at the 2.5% significance level, the data do not provide sufficient evidence to conclude that, for universities, the variables student-to-faculty ratio and graduation rate are positively linearly correlated. For the P-value approach, note that $0.05 < P < 0.10$.

17. their normal scores

18. H_0: Gas mileages for this model are normally distributed. H_a: Gas mileages for this model are not normally distributed. $\alpha = 0.05$; critical value $= 0.938$; $R_p = 0.981$; do not reject H_0; at the 5% significance level, the data do not provide sufficient evidence to conclude that the gas mileages for this model are not normally distributed. For the P-value approach, note that $P > 0.10$.

20. Although there is some cause for concern, it is probably reasonable to consider the assumptions for regression inferences met by the variables student-to-faculty ratio and graduation rate.

23. **a.** $\hat{y} = 16.4 + 2.03x$ **b.** 11.31%
c. 2.026 **d.** 1.205 **e.** 1.68 **f.** 0.131

g. At the 5% significance level, the data do not provide sufficient evidence to conclude that, for universities, student-to-faculty ratio is useful for predicting graduation rate.

h. −0.75 to 4.80. We can be 95% confident that, for universities, the increase in mean graduation rate per increase in the student-to-faculty ratio by 1 is somewhere between −0.75 and 4.80 percentage points.

i. 50.89% **j.** 42.60% to 59.18%
k. 50.89% **l.** 23.53% to 78.25%

CHAPTER 16

Exercises 16.1

16.1 by its numbers of degrees of freedom

16.3 $F_{0.05}$, $F_{0.025}$, F_α

16.5
a. 12 **b.** 7

16.7
a. 1.89 **b.** 2.47 **c.** 2.14

16.9
a. 2.88 **b.** 2.10 **c.** 1.78

Exercises 16.2

16.13 The pooled-t procedure of Section 10.2.

16.15 The procedure for comparing the means involves analyzing the variation in the data.

16.17
a. the treatment mean square, $MSTR$
b. the error mean square, MSE
c. the F-statistic, $F = MSTR/MSE$

16.19 It signifies that the ANOVA compares the means of a variable for populations that result from a classification by *one* other variable (called the factor).

16.21 No, because the variation among the sample means is not large relative to the variation within the samples.

Exercises 16.3

16.25 A small value of F results when $SSTR$ is small relative to SSE, that is, when the variation among sample means is small relative to the variation within samples. This describes what we expect to happen when the null hypothesis is true; thus it does not constitute evidence against the null hypothesis. Only when the variation among sample means is large relative to the variation within samples (i.e., only when F is large), do we have evidence that the null hypothesis is false.

16.27 $SST = SSTR + SSE$. The total variation among all the sample data can be partitioned into a component representing variation among the sample means and a component representing variation within the samples.

16.29 The missing entries in the first row are 43.304 and 3.08; in the second, 12 and 7.033; and in the third, 127.704.

16.31
a. 70 **b.** 10 **c.** 60
d. $70 = 10 + 60$
e. $SST = 70$, $SSTR = 10$, $SSE = 60$

16.33

Source	df	*SS*	*MS*	*F*
Treatment	2	2.8	1.4	1.56
Error	12	10.8	0.9	
Total	14	13.6		

16.35

Source	df	*SS*	*MS*	*F*
Treatment	3	132,508.2	44,169.40	7.37
Error	16	95,877.6	5,992.35	
Total	19	228,385.8		

16.37 H_0: $\mu_1 = \mu_2 = \mu_3 = \mu_4$, H_a: Not all the means are equal. $\alpha = 0.05$; critical value $= 3.24$; $SST = 560.2$, $SSTR = 68.2$, $SSE = 492.0$; $F = 0.74$; do not reject H_0; at the 5% significance level, the data do not provide sufficient evidence to conclude that there is a difference in mean lifetimes among the four brands of batteries. For the P-value approach, note that $P > 0.10$.

16.39 H_0: $\mu_1 = \mu_2 = \mu_3 = \mu_4 = \mu_5$, H_a: Not all the means are equal. $\alpha = 0.05$; critical value $= 2.82$; $SST = 447{,}088.667$, $SSTR = 404{,}258.467$, $SSE = 42{,}830.200$; $F = 51.91$; reject H_0; at the 5% significance level, the data provide sufficient evidence to conclude that a difference exists in mean weekly earnings among nonsupervisory workers in the five industries. For the P-value approach, note that $P < 0.005$.

16.49
a. $SSTR = 36{,}859$, $SSE = 2{,}482{,}910$, $SST = 2{,}519{,}769$
b. $MSTR = 18{,}430$, $MSE = 33{,}553$ **c.** 0.55
d. H_0: $\mu_1 = \mu_2 = \mu_3$, H_a: Not all the means are equal.
e. 0.580
f. Do not reject H_0; at the 5% significance level, the data do not provide sufficient evidence to conclude that there is a difference in mean annual insurance

premiums among owners of large, intermediate, and compact cars.

g.

Type	n	$\bar{x}$	s
Large	20	\$902.4	\$152.5
Intermediate	30	\$851.7	\$169.2
Compact	27	\$890.5	\$215.8

h. Approximately \$822 to \$960

Exercises 16.4

16.51 0

16.53 Because the family confidence level is the confidence we have that all the confidence intervals contain the differences between the corresponding population means, whereas the individual confidence level is the confidence we have that any particular confidence interval contains the difference between the corresponding population means.

16.55 degrees of freedom for the denominator

16.57
a. 4.69 **b.** 5.98

16.59 Family confidence level = 0.95; $q_{0.05} = 4.05$; simultaneous 95% CIs are as follows.

Means difference	Confidence interval
$\mu_1 - \mu_2$	−7.24 to 12.84
$\mu_1 - \mu_3$	−6.24 to 13.84
$\mu_1 - \mu_4$	−5.04 to 15.04
$\mu_2 - \mu_3$	−9.04 to 11.04
$\mu_2 - \mu_4$	−7.84 to 12.24
$\mu_3 - \mu_4$	−8.84 to 11.24

The above table shows that no two populations means can be declared different. This is summarized in the following diagram.

Brand *D* (4)	Brand *C* (3)	Brand *B* (2)	Brand *A* (1)
28.6	29.8	30.8	33.6

(A single line underlines all four means.)

16.61 Family confidence level = 0.95; $q_{0.05} = 4.20$; simultaneous 95% CIs are as follows.

Means difference	Confidence interval
$\mu_1 - \mu_2$	4.8 to 163.4
$\mu_1 - \mu_3$	270.8 to 422.2
$\mu_1 - \mu_4$	31.9 to 201.1
$\mu_1 - \mu_5$	118.3 to 269.7
$\mu_2 - \mu_3$	183.1 to 341.7
$\mu_2 - \mu_4$	−55.5 to 120.3
$\mu_2 - \mu_5$	30.6 to 189.2
$\mu_3 - \mu_4$	−314.6 to −145.4
$\mu_3 - \mu_5$	−228.2 to −76.8
$\mu_4 - \mu_5$	−7.1 to 162.1

The above table shows that we can declare the following pairs of means different: μ_1 and μ_2, μ_1 and μ_3, μ_1 and μ_4, μ_1 and μ_5, μ_2 and μ_3, μ_2 and μ_5, μ_3 and μ_4, μ_3 and μ_5; all other pairs of means are not declared different. This is summarized in the diagram below.

Retail trade (3)	Services (5)	Finance, Insurance, Real estate (4)	Wholesale trade (2)	Transp. and Pub. util. (1)
230.0	382.5	460.0	492.4	576.5

(One line underlines Services and Finance, Insurance, Real estate; a second line underlines Finance, Insurance, Real estate and Wholesale trade.)

Interpreting this diagram, we conclude with 95% confidence that the mean weekly earnings of transportation/public-utility workers exceeds those of the other four industries; the mean weekly earnings of retail-trade workers is less than those of the other four industries; the mean weekly earnings of service workers is less than those of wholesale trade workers; no other means can be declared different.

16.69 For stacked data, the observations for all the samples are stored in a single column with another column indicating which observations belong to which samples. For unstacked data, the samples are stored in different columns. Minitab requires stacked data storage for the Tukey multiple-comparison procedure.

16.71
a. 0.95; We can be 95% confidence that all the confidence intervals contain the differences between the corresponding population means.

b. 0.9806; We can be 98.06% confidence that any particular confidence interval contains the difference between the corresponding population means.
c. 3.38 **d.** −\$117 to \$141
e. −\$77 to \$155 **f.** none
g. No two population means can be declared different.

Exercises 16.5

16.73 Independent samples, same-shape population, and all sample sizes are 5 or greater.

16.75 equal

16.77 chi-square with df = 4

16.79 H_0: $\mu_1 = \mu_2 = \mu_3$, H_a: Not all the means are equal. $\alpha = 0.01$; critical value = 9.210; $H = 12.21$; reject H_0; at the 1% significance level, the data provide sufficient evidence to conclude that there is a difference in mean (per capita) consumption of low-fat milk for the years 1980, 1990, and 1995. For the P-value approach, note that $P < 0.005$.

16.81 H_0: $\eta_1 = \eta_2 = \eta_3 = \eta_4$, H_a: Not all the medians are equal. $\alpha = 0.05$; critical value = 7.815; $H = 6.37$; do not reject H_0; at the 5% significance level, the data do not provide sufficient evidence to conclude that a difference exists in median square footage of single-family detached homes among the four U.S. regions. For the P-value approach, note that $0.05 < P < 0.10$.

16.83

a. H_0: $\mu_1 = \mu_2 = \mu_3 = \mu_4 = \mu_5$, H_a: Not all the means are equal. $\alpha = 0.05$; critical value = 9.488; $H = 23.81$; reject H_0; at the 5% significance level, the data provide sufficient evidence to conclude that a difference exists in mean weekly earnings among nonsupervisory workers in the five industries. For the P-value approach, note that $P < 0.005$.
b. Because normal distributions with equal standard deviations have the same shape. It is better to use the one-way ANOVA test because when the assumptions for that test are met it is more powerful than the Kruskal–Wallis test.

16.85

a. neither **b.** neither

16.91

a. 1.29
b. H_0: $\mu_1 = \mu_2 = \mu_3$,
H_a: Not all the means are equal.
c. 0.525
d. Do not reject H_0; at the 5% significance level, the data do not provide sufficient evidence to conclude that there is a difference in mean annual insurance premiums among owners of large, intermediate, and compact cars.
e.

Type	n	Median	Mean rank
Large	20	\$926.2	43.0
Intermediate	30	\$836.4	35.8
Compact	27	\$859.0	39.6

f. The results are essentially the same. Neither test rejects the null hypothesis at the 5% significance level. Moreover, the P-values for the two tests are quite close: 0.580 and 0.525 for the one-way ANOVA and Kruskal–Wallis tests, respectively.

Review Test for Chapter 16

1. to compare the means of a variable for populations that result from a classification by one other variable (called the factor)
2. *Independent samples:* Check by carefully studying the way the sampling was done.

   Normal populations: Check by constructing normal probability plots.

   Equal standard deviations: As a rule of thumb, this assumption is considered satisfied if the ratio of the largest to the smallest sample standard deviation is less than 2.

   Additionally, the normality and equal-standard-deviations assumptions can be assessed by performing a residual analysis.
3. the F-distribution
4. df = (2, 14)
5. **a.** *MSTR* (or *SSTR*) **b.** *MSE* (or *SSE*)
6. **a.** The total sum of squares, *SST,* represents the total variation among all the sample data; the treatment sum of squares, *SSTR,* represents the variation among the sample means; and

the error sum of squares, *SSE*, represents the variation within the samples.

b. $SST = SSTR + SSE$; the one-way ANOVA identity shows that the total variation among all the sample data can be partitioned into a component representing variation among the sample means and a component representing variation within the samples.

7. a. These tables are useful for organizing and summarizing the quantities required for performing a one-way analysis of variance.

b.

Source	df	SS	MS = SS/df	F-statistic
Treatment	$k-1$	$SSTR$	$MSTR = \dfrac{SSTR}{k-1}$	$F = \dfrac{MSTR}{MSE}$
Error	$n-k$	SSE	$MSE = \dfrac{SSE}{n-k}$	
Total	$n-1$	SST		

8. If the null hypothesis of a one-way ANOVA is rejected, then we can conclude that the means of the populations under consideration are not all the same. Once we make that decision, we may also want to know which means are different, which mean is largest, or, more generally, the relation among all the means. Methods for dealing with these problems are called multiple comparisons.

9. The individual confidence level is the confidence we have that any particular confidence interval contains the difference between the corresponding population means; the family confidence level is the confidence we have that all the confidence intervals contain the differences between the corresponding population means. It is at the family confidence level that we can be confident in the truth of our conclusions when comparing all the population means simultaneously; thus the family confidence level is the appropriate one for multiple comparisons.

10. studentized range distribution (or q-distribution)

11. Larger. We can be more confident about the truth of one of several statements than about that of all statements simultaneously.

12. $\kappa = 3$, $\nu = 14$

13. Kruskal–Wallis test

14. chi-square distribution

15. The Kruskal–Wallis test is based on ranks. If the null hypothesis of equal population means is true, then the means of the ranks for the samples should be roughly equal. Put another way, if the variation among the mean ranks is too large, then we have evidence against the null hypothesis.

16. The Kruskal–Wallis test because, unlike the one-way ANOVA test, it is resistant to outliers and other extreme values.

17. a. 2 **b.** 14 **c.** 3.74
d. 6.51 **e.** 3.74

18. a. The sample means are, respectively, 3, 3, and 6; the sample standard deviations are, respectively, 2, 2.449, and 4.243.

b. $SST = 110$, $SSTR = 24$, $SSE = 86$; $110 = 24 + 86$

c. $SST = 110$, $SSTR = 24$, $SSE = 86$

d.

Source	df	SS	MS = SS/df	F-statistic
Treatment	2	24	12.000	1.26
Error	9	86	9.556	
Total	11	110		

19. a. The variation among the sample means.

b. The variation within the samples.

c. Independent samples, normal populations, and equal standard deviations. Independent samples are essential to the one-way ANOVA procedure. Normality is not too critical because one-way ANOVA is robust to moderate violations of the normality assumption. Equal standard deviations are also not that important provided the sample sizes are roughly equal.

20. H_0: $\mu_1 = \mu_2 = \mu_3$, H_a: Not all the means are equal. $\alpha = 0.05$; critical value $= 3.74$; $SST = 371{,}141.882$, $SSTR = 160{,}601.416$, $SSE = 210{,}540.467$; $F = 5.34$; reject H_0; at the 5% significance level, the data provide sufficient evidence to conclude that a difference in mean losses exists among the three types of robberies. For the P-value approach, note that $0.01 < P < 0.025$.

21. **a.** 3.70 **b.** 4.89

22. **a.** Family confidence level $= 0.95$; $q_{0.05} = 3.70$; simultaneous 95% CIs are as follows.

Means difference	Confidence interval
$\mu_1 - \mu_2$	−65.3 to 323.2
$\mu_1 - \mu_3$	48.4 to 436.9
$\mu_2 - \mu_3$	−71.6 to 298.9

b. The table in part (a) shows that we can declare as different only μ_1 and μ_3. This is summarized in the following diagram.

Convenience store (3)	Gas station (2)	Highway (1)
393.17	506.83	635.80

Interpreting this diagram, we conclude with 95% confidence that the mean loss due to convenience-store robberies is less than that due to highway robberies; no other means can be declared different.

23. **a.** H_0: $\mu_1 = \mu_2 = \mu_3$, H_a: Not all the means are equal. $\alpha = 0.05$; critical value $= 5.991$; $H = 6.819$; reject H_0; at the 5% significance level, the data provide sufficient evidence to conclude that a difference in mean losses exists among the three types of robberies. For the P-value approach, note that $0.025 < P < 0.05$.

b. Because normal distributions with equal standard deviations have the same shape. It is better to use the one-way ANOVA test because when the assumptions for that test are met, it is more powerful than the Kruskal–Wallis test.

26. **a.** 160,601, 210,540, 371,142
b. 80,301, 15,039
c. 5.34 **d.** 0.019
e. Reject H_0; at the 5% significance level, the data provide sufficient evidence to conclude that a difference in mean losses exists among the three types of robberies.
f. $n_1 = 5$, $\bar{x}_1 = 635.8$, $s_1 = 92.9$
$n_2 = 6$, $\bar{x}_2 = 506.8$, $s_2 = 126.1$
$n_3 = 6$, $\bar{x}_3 = 393.2$, $s_3 = 139.0$
g. approximately \$285 to \$495

28. **a.** 0.95 **b.** 0.9797 **c.** 3.70
d. −\$72 to \$299 **e.** −\$323 to \$65
f. only μ_1 and μ_3
g. With 95% confidence we can state that the mean loss due to convenience-store robberies is less than that due to highway robberies; no other means can be declared different.

30. **a.** 6.82 (6.83 when adjusted for ties)
b. H_0: $\mu_1 = \mu_2 = \mu_3$,
H_a: Not all the means are equal.
c. 0.033
d. At the 5% significance level, the data provide sufficient evidence to conclude that a difference in mean losses exists among the three types of robberies.
e. $n_1 = 5$, $M_1 = \$652.0$, $\bar{R}_1 = 13.4$
$n_2 = 6$, $M_2 = \$522.5$, $\bar{R}_2 = 8.9$
$n_3 = 6$, $M_3 = \$407.0$, $\bar{R}_3 = 5.4$
where M_j and $\bar{R}_j$ denote, respectively, the median and the mean rank of the jth sample.
f. Both tests reject the null hypothesis, but the one-way ANOVA test provides stronger evidence against the null hypothesis than the Kruskal–Wallis test; the P-values are 0.019 and 0.033, respectively.

Index

Biographical Sketches

Following is an index that provides page-number references for the biographical sketches appearing in the book.

Internet Projects

Following is an index that provides page-number references for the Internet projects appearing in the book.

Case Studies

Following is an index that provides page-number references for the case studies appearing in the book.

Procedure Index

Following is an index that provides page-number references for the various statistical procedures discussed in the book.

TABLE IV
Values of t_α

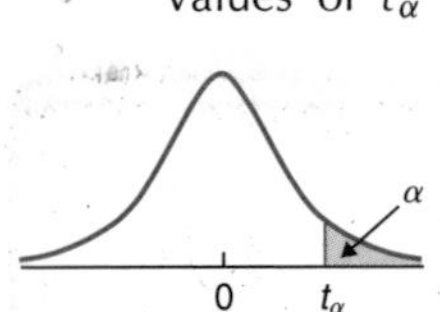

df	$t_{0.10}$	$t_{0.05}$	$t_{0.025}$	$t_{0.01}$	$t_{0.005}$	df
1	3.078	6.314	12.706	31.821	63.657	1
2	1.886	2.920	4.303	6.965	9.925	2
3	1.638	2.353	3.182	4.541	5.841	3
4	1.533	2.132	2.776	3.747	4.604	4
5	1.476	2.015	2.571	3.365	4.032	5
6	1.440	1.943	2.447	3.143	3.707	6
7	1.415	1.895	2.365	2.998	3.499	7
8	1.397	1.860	2.306	2.896	3.355	8
9	1.383	1.833	2.262	2.821	3.250	9
10	1.372	1.812	2.228	2.764	3.169	10
11	1.363	1.796	2.201	2.718	3.106	11
12	1.356	1.782	2.179	2.681	3.055	12
13	1.350	1.771	2.160	2.650	3.012	13
14	1.345	1.761	2.145	2.624	2.977	14
15	1.341	1.753	2.131	2.602	2.947	15
16	1.337	1.746	2.120	2.583	2.921	16
17	1.333	1.740	2.110	2.567	2.898	17
18	1.330	1.734	2.101	2.552	2.878	18
19	1.328	1.729	2.093	2.539	2.861	19
20	1.325	1.725	2.086	2.528	2.845	20
21	1.323	1.721	2.080	2.518	2.831	21
22	1.321	1.717	2.074	2.508	2.819	22
23	1.319	1.714	2.069	2.500	2.807	23
24	1.318	1.711	2.064	2.492	2.797	24
25	1.316	1.708	2.060	2.485	2.787	25
26	1.315	1.706	2.056	2.479	2.779	26
27	1.314	1.703	2.052	2.473	2.771	27
28	1.313	1.701	2.048	2.467	2.763	28
29	1.311	1.699	2.045	2.462	2.756	29
30	1.310	1.697	2.042	2.457	2.750	30
35	1.306	1.690	2.030	2.438	2.724	35
40	1.303	1.684	2.021	2.423	2.704	40
50	1.299	1.676	2.009	2.403	2.678	50
60	1.296	1.671	2.000	2.390	2.660	60
70	1.294	1.667	1.994	2.381	2.648	70
80	1.292	1.664	1.990	2.374	2.639	80
90	1.291	1.662	1.987	2.369	2.632	90
100	1.290	1.660	1.984	2.364	2.626	100
1000	1.282	1.646	1.962	2.330	2.581	1000

1.282	1.645	1.960	2.326	2.576
$z_{0.10}$	$z_{0.05}$	$z_{0.025}$	$z_{0.01}$	$z_{0.005}$